ECE/TRANS/257 (Vol.I)

Economic Commission for Europe
Inland Transport Committee

ADR

applicable as from 1 January 2017

European Agreement

Concerning the International Carriage
of Dangerous Goods by Road

Volume I

UNITED NATIONS
New York and Geneva, 2016

NOTE

The designations employed and the presentation of the material in this publication do not imply the expression of any opinion whatsoever on the part of the Secretariat of the United Nations concerning the legal status of any country, territory, city or area, or of its authorities, or concerning the delimitation of its frontiers or boundaries.

ECE/TRANS/257 (Vol.I)

Copyright © United Nations, 2016

UNITED NATIONS PUBLICATION
Sales No.: E.16.VIII.1
ISBN 978-92-1-139156-5
(complete set of 2 volumes)
e-ISBN 978-92-1-058130-1

Volumes I and II not to be sold separately.

United Nations Economic Commission for Europe (UNECE)

The United Nations Economic Commission for Europe (UNECE) is one of the five United Nations regional commissions, administered by the Economic and Social Council (ECOSOC). It was established in 1947 with the mandate to help rebuild post-war Europe, develop economic activity and strengthen economic relations among European countries, and between Europe and the rest of the world. During the Cold War, UNECE served as a unique forum for economic dialogue and cooperation between East and West. Despite the complexity of this period, significant achievements were made, with consensus reached on numerous harmonization and standardization agreements.

In the post-Cold War era, UNECE acquired not only many new member States, but also new functions. Since the early 1990s the organization has focused on analyses of the transition process, using its harmonization experience to facilitate the integration of central and eastern European countries into global markets.

UNECE is the forum where the countries of western, central and eastern Europe, Central Asia and North America – 56 countries in all – come together to forge the tools of their cooperation. That cooperation concerns economic cooperation and integration, statistics, environment, transport, trade, sustainable energy, forestry and timber, housing and land management and population. The Commission offers a regional framework for the elaboration and harmonization of conventions, norms and standards. The Commission's experts provide technical assistance to the countries of South-East Europe and the Commonwealth of Independent States. This assistance takes the form of advisory services, training seminars and workshops where countries can share their experiences and best practices.

Transport in UNECE

The UNECE Sustainable Transport Division is the secretariat of the Inland Transport Committee (ITC) and the ECOSOC Committee of Experts on the Transport of Dangerous Goods and on the Globally Harmonized System of Classification and Labelling of Chemicals. The ITC and its 17 working parties, as well as the ECOSOC Committee and its sub-committees are intergovernmental decision-making bodies that work to improve the daily lives of people and businesses around the world, in measurable ways and with concrete actions, to enhance traffic safety, environmental performance, energy efficiency and the competitiveness of the transport sector.

The ECOSOC Committee was set up in 1953 by the Secretary-General of the United Nations at the request of the Economic and Social Council to elaborate recommendations on the transport of dangerous goods. Its mandate was extended to the global (multi-sectoral) harmonization of systems of classification and labelling of chemicals in 1999. It is composed of experts from countries which possess the relevant expertise and experience in the international trade and transport of dangerous goods and chemicals. Its membership is restricted in order to reflect a proper geographical balance between all regions of the world and to ensure adequate participation of developing countries. Although the Committee is a subsidiary body of ECOSOC, the Secretary-General decided in 1963 that the secretariat services would be provided by the UNECE Transport Division.

ITC is a unique intergovernmental forum that was set up in 1947 to support the reconstruction of transport connections in post-war Europe. Over the years, it has specialized in facilitating the harmonized and sustainable development of inland modes of transport. The main results of this persevering and ongoing work are reflected, among other things, (i) in 58 United Nations conventions and many more technical regulations, which are updated on a regular basis and provide an international legal framework for the sustainable development of national and international road, rail, inland water and intermodal transport, including the transport of dangerous goods, as well as the construction and inspection of road motor vehicles; (ii) in the Trans-European North-south Motorway, Trans-European Railway and the Euro-Asia Transport Links projects, that facilitate multi-country coordination of transport infrastructure investment programmes; (iii) in the TIR system, which is a global customs transit facilitation solution; (iv) in the tool called For Future Inland Transport Systems (ForFITS), which can assist national and local governments to monitor carbon dioxide (CO_2) emissions coming from inland transport modes and to select and design climate change mitigation policies, based on their impact and adapted to local conditions; (v) in transport statistics – methods and data – that are internationally agreed on; (vi) in studies and reports that help transport policy development by addressing timely issues, based on cutting-edge research and analysis. ITC also devotes special attention to Intelligent Transport Services (ITS), sustainable urban mobility and city logistics, as well as to increasing the resilience of transport networks and services in response to climate change adaptation and security challenges.

In addition, the UNECE Sustainable Transport and Environment Divisions, together with the World Health Organization (WHO) – Europe, co-service the Transport Health and Environment Pan-European Programme (THE PEP).

Finally, as of 2015, the UNECE Sustainable Transport Division is providing the secretariat services for the Secretary General's Special Envoy for Road Safety, Mr. Jean Todt.

INTRODUCTION

General

The European Agreement concerning the International Carriage of Dangerous Goods by Road (ADR) was done at Geneva on 30 September 1957 under the auspices of the United Nations Economic Commission for Europe, and it entered into force on 29 January 1968. The Agreement itself was amended by the Protocol amending article 14 (3) done at New York on 21 August 1975, which entered into force on 19 April 1985.

According to article 2 of the Agreement, dangerous goods barred from carriage by Annex A shall not be accepted for international transport, while international transport of other dangerous goods shall be authorized subject to compliance with:

- the conditions laid down in Annex A for the goods in question, in particular as regards their packaging and labelling; and

- the conditions laid down in Annex B, in particular as regards the construction, equipment and operation of the vehicle carrying the goods in question.

Nevertheless, according to article 4, each Contracting Party shall retain the right to regulate or prohibit, for reasons other than safety during carriage, the entry of dangerous goods into its territory. Contracting Parties also retain the right to arrange, by bilateral or multilateral agreements, that certain dangerous goods which are prohibited from carriage by Annex A be internationally carried, subject to certain conditions, on their territories, or that dangerous goods authorized to be carried internationally according to Annex A be carried on their territories under conditions less stringent than those specified in Annexes A and B.

Annexes A and B have been regularly amended and updated since the entry into force of ADR.

Structure of Annexes A and B

The Working Party on the Transport of Dangerous Goods (WP.15) of the Economic Commission for Europe's Committee on Inland Transport decided, at its fifty-first session (26-30 October 1992), to restructure Annexes A and B, on the basis of a proposal by the International Road Transport Union (TRANS/WP.15/124, paras. 100-108). The main objectives were to make the requirements more accessible and more user-friendly so that they could be applied more easily not only to international road transport operations under ADR, but also to domestic traffic in all European States through national or European Community legislation, and ultimately to ensure a consistent regulatory framework at European level. It was also considered necessary to identify more clearly the duties of the various participants in the transport chain, to group more systematically the requirements concerning these various participants, and to differentiate the legal requirements of ADR from the European or international standards that could be applied to meet such requirements.

The structure is consistent with that of the United Nations *Recommendations on the Transport of Dangerous Goods, Model Regulations*, the *International Maritime Dangerous Goods Code (IMDG Code)* and the *Regulations concerning the International Carriage of Dangerous Goods by Rail (RID)*.

It has been split into nine parts, but still grouped under two annexes to align with the wording of article 2 of the Agreement itself. The layout is as follows:

Annex A: General provisions and provisions concerning dangerous articles and substances

Part 1   General provisions

Part 2   Classification

Part 3   Dangerous goods list, special provisions and exemptions related to limited and excepted quantities

Part 4   Packing and tank provisions

Part 5   Consignment procedures

Part 6   Requirements for the construction and testing of packagings, intermediate bulk containers (IBCs), large packagings, tanks and bulk containers

Part 7   Provisions concerning the conditions of carriage, loading, unloading and handling

Annex B: Provisions concerning transport equipment and transport operations

Part 8   Requirements for vehicle crews, equipment, operation and documentation

Part 9   Requirements concerning the construction and approval of vehicles

Part 1, which contains general provisions and definitions, is an essential part, since it contains all definitions for terms used throughout the other parts, and it defines precisely the scope and applicability of ADR, including the possibility of exemptions, as well as the applicability of other regulations. It also contains provisions concerning training, derogations and transitional measures, the respective safety obligations of the various participants in a chain of transport of dangerous goods, control measures, safety advisers, restrictions for the passage of vehicles carrying dangerous goods through road tunnels and transport of dangerous goods security.

Central to the use of the restructured ADR is table A of Chapter 3.2 which contains the dangerous goods list in the numerical order of UN numbers. Once the UN number of a specific dangerous substance or article has been determined, the table provides cross-references to specific requirements to be applied for the carriage of that substance or article, and to the chapters or sections where these specific requirements may be found. Nevertheless, it should be borne in mind that the general requirements or class specific requirements of the various Parts have to be applied in addition to specific requirements, as relevant.

An alphabetical index which indicates the UN number assigned to specific dangerous goods has been prepared by the secretariat and added as table B of Chapter 3.2 to facilitate the access to table A when the UN number is unknown. This table B is not an official part of ADR and has been added in the publication for easy reference only.

When goods which are known or suspected to be dangerous cannot be found by name in any of tables A or B, they have to be classified in accordance with Part 2, which contains all relevant procedures and criteria to determine whether such goods are deemed to be dangerous or not and which UN number should be assigned.

Applicable texts

This version ("2017 ADR") takes into account all new amendments adopted by WP.15 in 2014, 2015 and 2016, circulated under the symbols ECE/TRANS/WP.15/231 and -/Corr.1 and ECE/TRANS/WP.15/231/Add.1, which, subject to acceptance by the Contracting Parties in accordance with article 14(3) of the Agreement, should enter into force on 1 January 2017.

Nevertheless, due to the transitional measures provided for in 1.6.1.1 of Annex A, the previous version ("2015 ADR") as amended in accordance with Depositary Notifications C.N.354.2015.TREATIES-XI.B.14 and C.N.710.2015.TREATIES-XI.B.14 may continue to be used until 30 June 2017.

Territorial applicability

ADR is an Agreement between States, and there is no overall enforcing authority. In practice, highway checks are carried out by Contracting Parties, and non-compliance may then result in legal action by national authorities against offenders in accordance with their domestic legislation. ADR itself does not prescribe any penalties. At the time of publishing, the Contracting Parties are Albania, Andorra, Austria, Azerbaijan, Belarus, Belgium, Bosnia and Herzegovina, Bulgaria, Croatia, Cyprus, Czech Republic, Denmark, Estonia, Finland, France, Germany, Greece, Hungary, Iceland, Ireland, Italy, Kazakhstan, Latvia, Liechtenstein, Lithuania, Luxembourg, Malta, Montenegro, Morocco, Netherlands, Norway, Poland, Portugal, the Republic of Moldova, Romania, Russian Federation, Serbia, Slovakia, Slovenia, Spain, Sweden, Switzerland, Tajikistan, the former Yugoslav Republic of Macedonia, Tunisia, Turkey, Ukraine and United Kingdom.

ADR applies to transport operations performed on the territory of at least two of the above-mentioned Contracting Parties. In addition, it should be noted that, in the interest of uniformity and free trading across the European Union (EU), Annexes A and B of ADR have also been adopted by EU Member States as the basis for regulation of the carriage of dangerous goods by road within and between their territories (Directive 2008/68/EC of the European Parliament and of the Council of 24 September 2008 on the inland transport of dangerous goods, as amended). A number of non-EU countries have also adopted Annexes A and B of ADR as the basis for their national legislation.

Additional practical information

Any query concerning the application of ADR should be directed to the relevant competent authority. Additional information may be found on the UNECE Transport Division website on the following page:

> http://www.unece.org/trans/danger/danger.htm

This website is regularly updated and contains the following information:

- General information on ADR

- Agreement (without annexes)

- Protocol of signature

- Present status of ADR

- Depositary notifications

- Country information (Competent Authorities, notifications)

- Linguistic versions (ADR, instructions in writing)

- Multilateral agreements

- ADR 2017 (files)

- ADR 2015 (files)

- ADR 2015 (amendments)

- Previous versions (files and amendments)

- Publication details and Corrigenda

TABLE OF CONTENTS
VOLUME I

Page

Table of contents (cont'd)

Table of contents (cont'd)

TABLE OF CONTENTS
VOLUME II

Page

Table of contents (cont'd)

Table of contents (cont'd)

EUROPEAN AGREEMENT CONCERNING THE INTERNATIONAL CARRIAGE OF DANGEROUS GOODS BY ROAD (ADR)

THE CONTRACTING PARTIES,

DESIRING to increase the safety of international transport by road,

HAVE AGREED as follows:

Article 1

For the purpose of this Agreement,

(a)   the term "vehicle" shall mean motor vehicles, articulated vehicles, trailers and semi-trailers, as defined in article 4 of the Convention on Road Traffic of 19 September 1949, other than vehicles belonging to or under the orders of the armed forces of a Contracting Party;

(b)   the term "dangerous goods" shall mean those substances and articles the international carriage by road of which is prohibited by, or authorized only on certain conditions by, Annexes A and B;

(c)   the term "international transport" shall mean any transport operation performed on the territory of at least two Contracting Parties by vehicles defined in (a) above.

Article 2

1.   Subject to the provisions of article 4, paragraph 3, dangerous goods barred from carriage by Annex A shall not be accepted for international transport.

2.   International transport of other dangerous goods shall be authorized subject to compliance with:

(a)   the conditions laid down in Annex A for the goods in question, in particular as regards their packaging and labelling, and

(b)   the conditions laid down in Annex B, in particular as regards the construction, equipment and operation of the vehicle carrying the goods in question, subject to the provisions of article 4, paragraph 2.

Article 3

The Annexes to this Agreement shall form an integral part thereof.

Article 4

1.   Each Contracting Party shall retain the right to regulate or prohibit, for reasons other than safety during carriage, the entry of dangerous goods into its territory.

2.   Vehicles in service on the territory of a Contracting Party at the time of entry into force of this Agreement or brought into service on such territory within two months after its entry into force shall be allowed, for a period of three years from such entry into force, to perform the international transport of dangerous goods even if their construction and equipment do not entirely conform to the requirements laid down in Annex B for the transport operation in question. Under special clauses of Annex B, however, this period may be reduced.

3.   The Contracting Parties shall retain the right to arrange, by special bilateral or multilateral agreements, that certain of the dangerous goods which under this Agreement are barred from all international transport may, subject to certain conditions, be accepted for international transport on their territories, or that dangerous goods which under this Agreement are acceptable for international transport only on specified conditions may be accepted for international transport on their territories under conditions less stringent than those laid down in the Annexes to this Agreement. The special bilateral or multilateral agreements referred to in this paragraph shall be communicated to the Secretary-General of the United Nations, who shall communicate them to the Contracting Parties which are not signatories to the said agreements.

Article 5

The transport operations to which this Agreement applies shall remain subject to national or international regulations applicable in general to road traffic, international road transport and international trade.

Article 6

1.   Countries members of the Economic Commission for Europe and countries admitted to the Commission in a consultative capacity under paragraph 8 of the Commission's terms of reference may become Contracting Parties to this Agreement:

(a)   by signing it;

(b)   by ratifying it after signing it subject to ratification;

(c)   by acceding to it.

2.   Such countries as may participate in certain activities of the Economic Commission for Europe in accordance with paragraph 11 of the Commission's terms of reference may become Contracting Parties to this Agreement by acceding to it after its entry into force.

3.   The Agreement shall be open for signature until 15 December 1957. Thereafter, it shall be open for accession.

4.   Ratification or accession shall be effected by the depositing of an instrument with the Secretary-General of the United Nations.

Article 7

1.   This agreement shall enter into force one month after the date on which the number of countries mentioned in article 6, paragraph 1, which have signed it without reservation of ratification or have deposited their instruments of ratification or accession has reached a total of five. However, the Annexes thereto shall not apply until six months after the entry into force of the Agreement itself.

2.   For any country ratifying or acceding to this Agreement after five of the countries referred to in article 6, paragraph 1, have signed it without reservation of ratification or have deposited their instruments of ratification or accession, this Agreement shall enter into force one month after the said country has deposited its instrument of ratification or accession and the Annexes thereto shall apply for the said country either on the same date, if they are already in force by that date, or, if they are not in force by that date, on the date on which they apply under the provisions of paragraph 1 of this article.

Article 8

1.   Any contracting Party may denounce this Agreement by so notifying the Secretary-General of the United Nations.

2.   Denunciation shall take effect twelve months after the date of receipt by the Secretary-General of the notification of denunciation.

Article 9

1.   This Agreement shall cease to have effect if, after its entry into force, the number of Contracting Parties is less than five during twelve consecutive months.

2.   In the event of the conclusion of a worldwide agreement for the regulation of the transport of dangerous goods, any provision of this Agreement which is contrary to any provision of the said worldwide agreement shall, from the date on which the latter enters into force, automatically cease to apply to relations between the Parties to this Agreement which become parties to the worldwide agreement, and shall automatically be replaced by the relevant provision of the said worldwide agreement.

Article 10

1.   Any country may, at the time of signing this Agreement without reservation of ratification or of depositing its instrument of ratification or accession or at any time thereafter, declare by notification addressed to the Secretary-General of the United Nations that this Agreement shall extend to all or any of the territories for the international relations of which it is responsible. The Agreement and the annexes thereto shall extend to the territory or territories named in the notification one month after it is received by the Secretary-General.

2.   Any country which has made a declaration under paragraph 1 of this article extending this Agreement to any territory for whose international relations it is responsible may denounce the Agreement separately in respect of the said territory in accordance with the provisions of article 8.

Article 11

1.   Any dispute between two or more Contracting Parties concerning the interpretation or application of this Agreement shall so far as possible be settled by negotiation between them.

2.   Any dispute which is not settled by negotiation shall be submitted to arbitration if any one of the Contracting Parties in dispute so requests and shall be referred accordingly to one or more arbitrators selected by agreement between the Parties in dispute. If within three months from the date of the request for arbitration the Parties in dispute are unable to agree on the selection of an arbitrator or arbitrators, any of those Parties may request the Secretary-General of the United Nations to nominate a single arbitrator to whom the dispute shall be referred for decision.

3.   The decision of the arbitrator or arbitrators appointed under paragraph 2 of this article shall be binding on the Contracting Parties in dispute.

Article 12

1.   Each Contracting Party may, at the time of signing, ratifying, or acceding to, this Agreement, declare that it does not consider itself bound by article 11. Other Contracting Parties shall not be bound by article 11 in respect of any Contracting Party which has entered such a reservation.

2.   Any Contracting Party having entered a reservation as provided for in paragraph 1 of this article may at any time withdraw such reservation by notifying the Secretary-General of the United Nations.

Article 13

1.   After this Agreement has been in force for three years, any Contracting Party may, by notification to the Secretary-General of the United Nations, request that a conference be convened for the purpose of reviewing the text of the Agreement. The Secretary-General shall notify all Contracting Parties of the request and a review conference shall be convened by the Secretary-General if, within a period of four months following the date of notification by the Secretary-General, not less than one-fourth of the Contracting Parties notify him of their concurrence with the request.

2.   If a conference is convened in accordance with paragraph 1 of this article, the Secretary-General shall notify all the Contracting Parties and invite them to submit within a period of three months such proposals as they may wish the Conference to consider. The Secretary-General shall circulate to all Contracting Parties the provisional agenda for the conference, together with the texts of such proposals, at least three months before the date on which the conference is to meet.

3.   The Secretary-General shall invite to any conference convened in accordance with this article all countries referred to in article 6, paragraph 1, and countries which have become Contracting Parties under article 6, paragraph 2.

Article 14[1]

1.   Independently of the revision procedure provided for in article 13, any Contracting Party may propose one or more amendments to the Annexes to this Agreement. To that end it shall transmit the text thereof to the Secretary-General of the United Nations. The Secretary-General may also propose amendments to the Annexes to this Agreement for the purpose of ensuring concordance between those Annexes and other international agreements concerning the carriage of dangerous goods.

2.   The Secretary-General shall transmit any proposal made under paragraph 1 of this article to all Contracting Parties and inform thereof the other countries referred to in article 6, paragraph 1.

3.   Any proposed amendment to the Annexes shall be deemed to be accepted unless, within three months from the date on which the Secretary-General circulates it, at least one-third of the Contracting Parties, or five of them if one-third exceeds that figure, have given the Secretary-General written notification of their objection to the proposed amendment. If

---

[1]   ***Note by the Secretariat:*** *The text of Article 14, paragraph 3 incorporates a modification which entered into force on 19 April 1985 in accordance with a Protocol transmitted to Contracting Parties under cover of Depositary Notification C.N.229.1975.TREATIES-8 of 18 September 1975.*

the amendment is deemed to be accepted, it shall enter into force for all the Contracting Parties, on the expiry of a further period of three months, except in the following cases:

(a)   In cases where similar amendments have been or are likely to be made to the other international agreements referred to in paragraph 1 of this article, the amendment shall enter into force on the expiry of a period the duration of which shall be determined by the Secretary-General in such a way as to allow, wherever possible, the simultaneous entry into force of the amendment and those that have been made or are likely to be made to such other agreements; such period shall not, however, be of less than one month's duration;

(b)   The Contracting Party submitting the proposed amendment may specify in its proposal, for the purpose of entry into force of the amendment, should it be accepted, a period of more than three months' duration.

4.   The Secretary-General shall, as soon as possible, notify all Contracting Parties and all the countries referred to in article 6, paragraph 1, of any objection which may be received from the Contracting Parties to a proposed amendment.

5.   If the proposed amendment to the Annexes is not deemed to be accepted, but if at least one Contracting Party other than the Contracting Party which proposed the amendment has given the Secretary-General written notification of its agreement to the proposal, a meeting of all the Contracting Parties and all the countries referred to in article 6, paragraph 1, shall be convened by the Secretary-General within three months after the expiry of the period of three months within which, under paragraph 3 of this article, notification must be given of objection to the amendment. The Secretary-General may also invite to such meeting representatives of:

(a)   intergovernmental organizations which are concerned with transport matters;

(b)   international non-governmental organizations whose activities are directly related to the transport of dangerous goods in the territories of the Contracting Parties.

6.   Any amendment adopted by more than half the total number of Contracting Parties at a meeting convened in accordance with paragraph 5 of this article shall enter into force for all Contracting Parties in accordance with the procedure agreed at such meeting by the majority of the Contracting Parties attending it.

Article 15

In addition to the notifications provided for in articles 13 and 14, the Secretary-General of the United Nations shall notify the countries referred to in article 6, paragraph 1, and the countries which have become Contracting Parties under article 6, paragraph 2, of

(a)   signatures, ratifications and accessions in accordance with article 6;

(b)   the dates on which this Agreement and the Annexes thereto enter into force in accordance with article 7;

(c)   denunciations in accordance with article 8;

(d)   the termination of the Agreement in accordance with article 9;

(e)   notifications and denunciations received in accordance with article 10;

(f)   declarations and notifications received in accordance with article 12, paragraphs 1 and 2;

(g)   the acceptance and date of entry into force of amendments in accordance with article 14, paragraphs 3 and 6.

Article 16

1.   The Protocol of Signature of this Agreement shall have the same force, effect and duration as the Agreement itself, of which it shall be considered to be an integral part.

2.   No reservation to this Agreement, other than those entered in the Protocol of Signature and those made in accordance with article 12, shall be permitted.

Article 17

After 15 December 1957, the original of this Agreement shall be deposited with the Secretary-General of the United Nations, who shall transmit certified true copies thereof to each of the countries referred to in article 6, paragraph 1.

IN WITNESS WHEREOF the undersigned, being duly authorized thereto, have signed this Agreement.

DONE at Geneva, this thirtieth day of September one thousand nine hundred and fifty-seven, in a single copy, in the English and French languages for the text of the Agreement proper, and in the French language for the Annexes, each text being equally authentic for the Agreement proper.

The Secretary-General of the United Nations is requested to prepare an authoritative translation of the Annexes in the English language and attach it to the certified true copies referred to in article 17.

PROTOCOL OF SIGNATURE

PROTOCOL OF SIGNATURE

TO THE EUROPEAN AGREEMENT ON THE INTERNATIONAL
CARRIAGE OF DANGEROUS GOODS BY ROAD (ADR)

On proceeding to sign the European Agreement on the International Carriage of Dangerous Goods by Road (ADR) the undersigned, duly authorized,

1.   **CONSIDERING** that the conditions governing the carriage of dangerous goods by sea to or from the United Kingdom differ basically from those set forth in Annex A to ADR and that it is impossible to modify them so as to conform to the latter in the near future;

HAVING REGARD to the undertaking given by the United Kingdom to submit as an amendment to the said Annex A a special appendix containing special provisions for road-sea carriage of dangerous goods between the Continent and the United Kingdom;

HAVE AGREED that, until the entry into force of such special appendix, dangerous goods carried under ADR to or from the United Kingdom shall comply with the provisions of Annex A to ADR and also with the United Kingdom conditions for the carriage of dangerous goods by sea;

2.   **TAKE NOTE OF** a declaration by the representative of France to the effect that the Government of the French Republic reserves the right, notwithstanding the provisions of article 4, paragraph 2, to refuse to allow vehicles in service on the territory of another Contracting Party, whatever the date on which they were put into service, to be used for the carriage of dangerous goods on French territory unless such vehicles comply either with the conditions laid down for such carriage in Annex B or with the conditions laid down for the carriage of the goods in question in the French regulations governing the carriage of dangerous goods by road;

3.   **RECOMMEND** that, before submission in accordance with article 14, paragraph 1, or article 13, paragraph 2, proposed amendments to this Agreement or its Annexes shall as far as possible first be discussed at meetings of experts of the Contracting Parties and, if necessary, of the other countries mentioned in article 6, paragraph 1, of the Agreement and of the international organizations mentioned in article 14, paragraph 5, of the Agreement.

ANNEX A

GENERAL PROVISIONS AND PROVISIONS CONCERNING DANGEROUS SUBSTANCES AND ARTICLES

PART 1

General provisions

CHAPTER 1.1

SCOPE AND APPLICABILITY

1.1.1   **Structure**

Annexes A and B of ADR are grouped into nine parts. Annex A consists of Parts 1 to 7, and Annex B of Parts 8 and 9. Each part is subdivided into chapters and each chapter into sections and sub-sections. Within each part the number of the part is included with the numbers of the chapters, sections and sub-sections, for example Part 4, Chapter 2, Section 1 is numbered "4.2.1".

1.1.2   **Scope**

1.1.2.1   For the purposes of Article 2 of ADR, Annex A specifies:

(a)   Dangerous goods which are barred from international carriage;

(b)   Dangerous goods which are authorized for international carriage and the conditions attaching to them (including exemptions) particularly with regard to:

- classification of goods, including classification criteria and relevant test methods;

- use of packagings (including mixed packing);

- use of tanks (including filling);

- consignment procedures (including marking and labelling of packages and placarding and marking of means of transport as well as documentation and information required);

- provisions concerning the construction, testing and approval of packagings and tanks;

- use of means of transport (including loading, mixed loading and unloading).

1.1.2.2   Annex A contains certain provisions which, according to Article 2 of ADR, pertain to Annex B or to both Annexes A and B, as follows:

1.1.1	Structure
1.1.2.3	(Scope of Annex B)
1.1.2.4	
1.1.3.1	Exemptions related to the nature of the transport operation
1.1.3.6	Exemptions related to quantities carried per transport unit
1.1.4	Applicability of other regulations
1.1.4.5	Carriage other than by road
Chapter 1.2	Definitions and units of measurements
Chapter 1.3	Training of persons involved in the carriage of dangerous goods
Chapter 1.4	Safety obligations of the participants
Chapter 1.5	Derogations
Chapter 1.6	Transitional measures
Chapter 1.8	Checks and other support measures to ensure compliance with safety requirements
Chapter 1.9	Transport restrictions by the competent authorities
Chapter 1.10	Security provisions
Chapter 3.1	General
Chapter 3.2	Columns (1), (2), (14), (15) and (19) (application of provisions of Parts 8 and 9 to individual substances or articles).

1.1.2.3   For the purposes of Article 2 of ADR, Annex B specifies the conditions regarding the construction, equipment and operation of vehicles carrying dangerous goods authorized for carriage:

- requirements for vehicle crews, equipment, operation and documentation;

- requirements concerning the construction and approval of vehicles.

1.1.2.4   In Article 1(c) of ADR, the word "vehicles" need not refer to one and the same vehicle. An international transport operation may be performed by several different vehicles provided that the operation takes place on the territory of at least two Contracting Parties to ADR between the consignor and the consignee indicated in the transport document.

1.1.3   Exemptions

1.1.3.1   *Exemptions related to the nature of the transport operation*

The provisions laid down in ADR do not apply to:

(a)   The carriage of dangerous goods by private individuals where the goods in question are packaged for retail sale and are intended for their personal or domestic use or for their leisure or sporting activities provided that measures have been taken to prevent any leakage of contents in normal conditions of carriage. When these goods are flammable liquids carried in refillable receptacles filled by, or for, a private individual, the total quantity shall not exceed 60 litres per receptacle and 240 litres per transport unit. Dangerous goods in IBCs, large packagings or tanks are not considered to be packaged for retail sale;

(b)   The carriage of machinery or equipment not specified in this Annex and which happen to contain dangerous goods in their internal or operational equipment, provided that measures have been taken to prevent any leakage of contents in normal conditions of carriage;

(c)   The carriage undertaken by enterprises which is ancillary to their main activity, such as deliveries to or returns from building or civil engineering sites, or in relation to surveying, repairs and maintenance, in quantities of not more than 450 litres per packaging, including intermediate bulk containers (IBCs) and large packagings, and within the maximum quantities specified in 1.1.3.6. Measures shall be taken to prevent any leakage of contents in normal conditions of carriage. These exemptions do not apply to Class 7.

Carriage undertaken by such enterprises for their supply or external or internal distribution does not fall within the scope of this exemption;

(d)   The carriage undertaken by the competent authorities for the emergency response or under their supervision, insofar as such carriage is necessary in relation to the emergency response, in particular carriage undertaken:

-    by breakdown vehicles carrying vehicles which have been involved in accidents or have broken down and contain dangerous goods; or

-    to contain and recover the dangerous goods involved in an incident or accident and move them to the nearest appropriate safe place;

(e)   Emergency transport intended to save human lives or protect the environment provided that all measures are taken to ensure that such transport is carried out in complete safety;

(f)   The carriage of uncleaned empty static storage vessels which have contained gases of Class 2, groups A, O or F, substances of Class 3 or Class 9 belonging to packing group II or III or pesticides of Class 6.1 belonging to packing group II or III, subject to the following conditions:

-    All openings with the exception of pressure relief devices (when fitted) are hermetically closed;

-    Measures have been taken to prevent any leakage of contents in normal conditions of carriage; and

-    The load is fixed in cradles or crates or other handling devices or to the vehicle or container in such a way that they will not become loose or shift during normal conditions of carriage.

This exemption does not apply to static storage vessels which have contained desensitized explosives or substances the carriage of which is prohibited by ADR.

NOTE: *For radioactive material, see also 1.7.1.4.*

1.1.3.2   *Exemptions related to the carriage of gases*

The provisions laid down in ADR do not apply to the carriage of:

(a)   Gases contained in the fuel tanks or cylinders of a vehicle performing a transport operation and destined for its propulsion or for the operation of any of its equipment used or intended for use during carriage (e.g. refrigerating equipment).

The gases may be carried in fixed fuel tanks or cylinders, directly connected to the vehicle's engine and/or auxiliary equipment or transportable pressure receptacles, which comply with the pertinent legal provisions.

The total capacity of the fuel tanks or cylinders for a transport unit, including those allowed in accordance with 1.1.3.3 (a), shall not exceed the amount of energy (MJ) or mass (kg) corresponding to 54 000 MJ energy-equivalent.

NOTE 1:   The value of 54 000 MJ energy-equivalent corresponds to the fuel limit of 1.1.3.3 (a) (1500 litres). For the energy content of fuels see the following Table:

Fuel	*Energy content*
Diesel	*36 MJ/litre*
Petrol	*32 MJ/litre*
Natural Gas/Biogas	*35 MJ/Nm^3*
Liquefied Petroleum Gas (LPG)	*24 MJ/litre*
Ethanol	*21 MJ/litre*
Biodiesel	*33 MJ/litre*
Emulsion fuel	*32 MJ/litre*
Hydrogen	*11 MJ/Nm^3*

The total capacity shall not exceed:
- 1 080 kg for LNG and CNG;
- 2 250 litres for LPG;

NOTE 2:   A container fitted with equipment for use during carriage, secured on a vehicle, is considered as an integral part of the vehicle and benefits from the same exemptions as regards the fuel necessary to operate the equipment.

(b)   *(Deleted)*

(c)   Gases of Groups A and O (according to 2.2.2.1), if the pressure of the gas in the receptacle or tank at a temperature of 20 °C does not exceed 200 kPa (2 bar) and if the gas is not a liquefied or a refrigerated liquefied gas. This includes every kind of receptacle or tank, e.g. also parts of machinery and apparatus;

NOTE: This exemption does not apply to lamps. For lamps see 1.1.3.10.

(d)   Gases contained in the equipment used for the operation of the vehicle (e.g. fire extinguishers), including in spare parts (e.g. inflated pneumatic tyres); this exemption also applies to inflated pneumatic tyres carried as a load;

(e)   Gases contained in the special equipment of vehicles and necessary for the operation of this special equipment during transport (cooling systems, fish-tanks, heaters, etc.) as well as spare receptacles for such equipment or uncleaned empty exchange receptacles, transported in the same transport unit;

(f)   Gases contained in foodstuffs (except UN 1950), including carbonated beverages; and

(g)   Gases contained in balls intended for use in sports.

(h)   *(Deleted)*

1.1.3.3   *Exemptions related to the carriage of liquid fuels*

The provisions laid down in ADR do not apply to the carriage of:

(a)   Fuel contained in the tanks of a vehicle performing a transport operation and destined for its propulsion or for the operation of any of its equipment used or intended for use during carriage.

The fuel may be carried in fixed fuel tanks, directly connected to the vehicle's engine and/or auxiliary equipment, which comply with the pertinent legal provisions, or may be carried in portable fuel containers (such as jerricans).

The total capacity of the fixed tanks shall not exceed 1500 litres per transport unit and the capacity of a tank fitted to a trailer shall not exceed 500 litres. A maximum of 60 litres per transport unit may be carried in portable fuel containers. These restrictions shall not apply to vehicles operated by the emergency services.

NOTE 1:   A container fitted with equipment for use during carriage, secured on a vehicle, is considered as an integral part of the vehicle and benefits from the same exemptions as regards the fuel necessary to operate the equipment.

NOTE 2:   The total capacity of the tanks or cylinders, including those containing gaseous fuels, shall not exceed 54 000 MJ energy-equivalent (see NOTE 1 in 1.1.3.2 (a)).

(b) and (c)   *(Deleted)*

1.1.3.4   *Exemptions related to special provisions or to dangerous goods packed in limited or excepted quantities*

NOTE: For radioactive material, see also 1.7.1.4.

1.1.3.4.1   Certain special provisions of Chapter 3.3 exempt partially or totally the carriage of specific dangerous goods from the requirements of ADR. The exemption applies when the special provision is referred to in Column (6) of Table A of Chapter 3.2 against the dangerous goods entry concerned.

1.1.3.4.2   Certain dangerous goods may be subject to exemptions provided that the conditions of Chapter 3.4 are met.

1.1.3.4.3   Certain dangerous goods may be subject to exemptions provided that the conditions of Chapter 3.5 are met.

1.1.3.5   *Exemptions related to empty uncleaned packagings*

Empty uncleaned packagings (including IBCs and large packagings) which have contained substances of Classes 2, 3, 4.1, 5.1, 6.1, 8 and 9 are not subject to the conditions of ADR if adequate measures have been taken to nullify any hazard. Hazards are nullified if adequate measures have been taken to nullify all hazards of Classes 1 to 9.

1.1.3.6   *Exemptions related to quantities carried per transport unit*

1.1.3.6.1   For the purposes of this sub-section, dangerous goods are assigned to transport categories 0, 1, 2, 3, or 4, as indicated in Column (15) of Table A of Chapter 3.2. Empty uncleaned packagings having contained substances assigned to transport category "0" are also assigned to transport category "0". Empty uncleaned packagings having contained substances assigned to a transport category other than "0" are assigned to transport category "4".

1.1.3.6.2 Where the quantity of dangerous goods carried on a transport unit does not exceed the values indicated in column (3) of the table in 1.1.3.6.3 for a given transport category (when the dangerous goods carried in the transport unit belong to the same category) or the value calculated in accordance with 1.1.3.6.4 (when the dangerous goods carried in the transport unit belong to different transport categories), they may be carried in packages in one transport unit without application of the following provisions:

- Chapter 1.10 except for Class 1 explosives of UN Nos. 0029, 0030, 0059, 0065, 0073, 0104, 0237, 0255, 0267, 0288, 0289, 0290, 0360, 0361, 0364, 0365, 0366, 0439, 0440, 0441, 0455, 0456 and 0500 and except for Class 7 excepted packages of UN Nos. 2910 and 2911 if the activity level exceeds the A_2 value;
- Chapter 5.3;
- Section 5.4.3;
- Chapter 7.2, except for V5 and V8 of 7.2.4;
- CV1 of 7.5.11;
- Part 8 except for   8.1.2.1 (a),
  8.1.4.2 to 8.1.4.5,
  8.2.3,
  8.3.3,
  8.3.4,
  8.3.5,
  Chapter 8.4,
  S1(3) and (6),
  S2(1),
  S4; S5,
  S14 to S21 and
  S24 of Chapter 8.5;
- Part 9.

1.1.3.6.3 Where the dangerous goods carried in the transport unit belong to the same category, the maximum total quantity per transport unit is indicated in column (3) of the table below.

Transport category (1)	Substances or articles packing group or classification code/group or UN No. (2)	Maximum total quantity per transport unit (3)
0	Class 1: 1.1A/1.1L/1.2L/1.3L and UN No. 0190 Class 3: UN No. 3343 Class 4.2: Substances belonging to packing group I Class 4.3: UN Nos. 1183, 1242, 1295, 1340, 1390, 1403, 1928, 2813, 2965, 2968, 2988, 3129, 3130, 3131, 3134, 3148, 3396, 3398 and 3399 Class 5.1: UN No. 2426 Class 6.1: UN Nos. 1051, 1600, 1613, 1614, 2312, 3250 and 3294 Class 6.2: UN Nos. 2814 and 2900 Class 7: UN Nos. 2912 to 2919, 2977, 2978 and 3321 to 3333 Class 8: UN No. 2215 (MALEIC ANHYDRIDE, MOLTEN) Class 9: UN Nos. 2315, 3151, 3152 and 3432 and articles containing such substances or mixtures and empty uncleaned packagings, except those classified under UN No. 2908, having contained substances classified in this transport category.	0
1	Substances and articles belonging to packing group I and not classified in transport category 0 and substances and articles of the following classes: Class 1: 1.1B to 1.1J [a] /1.2B to 1.2J/1.3C/1.3G/1.3H/1.3J/1.5D [a] Class 2: groups T, TC [a], TO, TF, TOC [a] and TFC aerosols: groups C, CO, FC, T, TF, TC, TO, TFC and TOC chemicals under pressure: UN Nos. 3502, 3503, 3504 and 3505 Class 4.1: UN Nos. 3221 to 3224, 3231 to 3240, 3533 and 3534 Class 5.2: UN Nos. 3101 to 3104 and 3111 to 3120	20
2	Substances belonging to packing group II and not classified in transport categories 0, 1 or 4 and substances and articles of the following classes: Class 1: 1.4B to 1.4G and 1.6N Class 2: group F aerosols: group F chemicals under pressure: UN No. 3501 Class 4.1: UN Nos. 3225 to 3230, 3531 and 3532 Class 4.3: UN Nos. 3292 Class 5.1: UN Nos. 3356 Class 5.2: UN Nos. 3105 to 3110 Class 6.1: UN Nos. 1700, 2016 and 2017 and substances belonging to packing group III Class 9: UN No. 3090, 3091, 3245, 3480 and 3481	333
3	Substances belonging to packing group III and not classified in transport categories 0, 2 or 4 and substances and articles of the following classes: Class 2: groups A and O aerosols: groups A and O chemicals under pressure: UN No. 3500 Class 3: UN No. 3473 Class 4.3: UN No. 3476 Class 8: UN Nos. 2794, 2795, 2800, 3028, 3477 and 3506 Class 9: UN Nos. 2990 and 3072	1 000
4	Class 1: 1.4S Class 4.1: UN Nos. 1331, 1345, 1944, 1945, 2254 and 2623 Class 4.2: UN Nos. 1361 and 1362 packing group III Class 7: UN Nos. 2908 to 2911 Class 9: UN Nos. 3268, 3499, 3508 and 3509 and empty, uncleaned packagings having contained dangerous goods, except for those classified in transport category 0	unlimited

[a] *For UN Nos. 0081, 0082, 0084, 0241, 0331, 0332, 0482, 1005 and 1017, the total maximum quantity per transport unit shall be 50 kg.*

In the above table, "maximum total quantity per transport unit" means:

- For articles, gross mass in kilograms (for articles of Class 1, net mass in kilograms of the explosive substance; for dangerous goods in machinery and equipment specified in this Annex, the total quantity of dangerous goods contained therein in kilograms or litres as appropriate);

- For solids, liquefied gases, refrigerated liquefied gases and dissolved gases, net mass in kilograms;

- For liquids, the total quantity of dangerous goods contained in litres;

- For compressed gases, adsorbed gases and chemicals under pressure, the water capacity of the receptacle in litres.

1.1.3.6.4 Where dangerous goods of different transport categories are carried in the same transport unit, the sum of:

- The quantity of substances and articles of transport category 1 multiplied by "50";

- The quantity of substances and articles of transport category 1 referred to in Note a to the table in 1.1.3.6.3 multiplied by "20";

- The quantity of substances and articles of transport category 2 multiplied by "3"; and

- The quantity of substances and articles of transport category 3;

shall not exceed "1 000".

1.1.3.6.5 For the purposes of this sub-section, dangerous goods exempted in accordance with 1.1.3.1 (a), (b) and (d) to (f), 1.1.3.2 to 1.1.3.5, 1.1.3.7, 1.1.3.9 and 1.1.3.10 shall not be taken into account.

1.1.3.7 *Exemptions related to the carriage of electric energy storage and production systems*

The provisions laid down in ADR do not apply to electric energy storage and production systems (e.g., lithium batteries, electric capacitors, asymmetric capacitors, metal hydride storage systems and fuel cells):

(a) installed in a vehicle, performing a transport operation and destined for its propulsion or for the operation of any of its equipment;

(b) contained in equipment for the operation of this equipment used or intended for use during carriage (e.g. a laptop).

1.1.3.8 *(Reserved)*

1.1.3.9 *Exemptions related to dangerous goods used as a coolant or conditioner during carriage*

When used in vehicles or containers for cooling or conditioning purposes, dangerous goods that are only asphyxiant (which dilute or replace the oxygen normally in the atmosphere) are only subject to the provisions of section 5.5.3.

1.1.3.10 *Exemptions related to the carriage of lamps containing dangerous goods*

The following lamps are not subject to ADR provided that they do not contain radioactive material and do not contain mercury in quantities above those specified in special provision 366 of Chapter 3.3:

(a) Lamps that are collected directly from individuals and households when carried to a collection or recycling facility;

    NOTE: *This also includes lamps brought by individuals to a first collection point, and then carried to another collection point, intermediate processing or recycling facility.*

(b) Lamps each containing not more than 1 g of dangerous goods and packaged so that there is not more than 30 g of dangerous goods per package, provided that:

(i)   the lamps are manufactured according to a certified quality management system;

   NOTE: ISO 9001 may be used for this purpose.

and

(ii)   each lamp is either individually packed in inner packagings, separated by dividers, or surrounded with cushioning material to protect the lamps and packed into strong outer packagings meeting the general provisions of 4.1.1.1 and capable of passing a 1.2 m drop test;

(c)   Used, damaged or defective lamps each containing not more than 1 g of dangerous goods with not more than 30 g of dangerous goods per package when carried from a collection or recycling facility. The lamps shall be packed in strong outer packagings sufficient for preventing release of the contents under normal conditions of carriage meeting the general provisions of 4.1.1.1 and that are capable of passing a drop test of not less than 1.2 m;

(d)   Lamps containing only gases of Groups A and O (according to 2.2.2.1) provided they are packaged so that the projectile effects of any rupture of the lamp will be contained within the package.

   NOTE: Lamps containing radioactive material are addressed in 2.2.7.2.2.2 (b).

1.1.4   **Applicability of other regulations**

1.1.4.1   *(Reserved)*

1.1.4.2   ***Carriage in a transport chain including maritime or air carriage***

1.1.4.2.1   Packages, containers, portable tanks, tank-containers and MEGCs, which do not entirely meet the requirements for packing, mixed packing, marking, labelling of packages or placarding and orange plate marking, of ADR, but are in conformity with the requirements of the IMDG Code or the ICAO Technical Instructions shall be accepted for carriage in a transport chain including maritime or air carriage subject to the following conditions:

(a)   If the packages are not marked and labelled in accordance with ADR, they shall bear marks and danger labels in accordance with the requirements of the IMDG Code or the ICAO Technical Instructions;

(b)   The requirements of the IMDG Code or the ICAO Technical Instructions shall be applicable to mixed packing within a package;

(c)   For carriage in a transport chain including maritime carriage, if the containers, portable tanks, tank-containers or MEGCs are not marked and placarded in accordance with Chapter 5.3 of this Annex, they shall be marked and placarded in accordance with Chapter 5.3 of the IMDG Code. In such case, only 5.3.2.1.1 of this Annex is applicable to the marking of the vehicle itself. For empty, uncleaned portable tanks, tank-containers and MEGCs, this requirement shall apply up to and including the subsequent transfer to a cleaning station.

This derogation does not apply in the case of goods classified as dangerous goods in classes 1 to 9 of ADR and considered as non-dangerous goods according to the applicable requirements of the IMDG Code or the ICAO Technical Instructions.

1.1.4.2.2   Transport units composed of a vehicle or vehicles other than those carrying containers, portable tanks, tank-containers or MEGCs as provided for in 1.1.4.2.1 (c), which are not placarded in accordance with the provisions of 5.3.1 of ADR but which are marked and placarded in accordance with Chapter 5.3 of the IMDG Code, shall be accepted for carriage in a transport chain including maritime transport provided that the orange-coloured plate marking provisions of 5.3.2 of ADR are complied with.

1.1.4.2.3	For carriage in a transport chain including maritime or air carriage, the information required under 5.4.1 and 5.4.2 and under any special provision of Chapter 3.3 may be substituted by the transport document and information required by the IMDG Code or the ICAO Technical Instructions respectively provided that any additional information required by ADR is also included.

NOTE: *For carriage in accordance with 1.1.4.2.1, see also 5.4.1.1.7. For carriage in containers, see also 5.4.2.*

1.1.4.3 *Use of IMO type portable tanks approved for maritime transport*

IMO type portable tanks (types 1, 2, 5 and 7) which do not meet the requirements of Chapters 6.7 or 6.8, but which were built and approved before 1 January 2003 in accordance with the provisions of the IMDG Code (Amdt. 29-98) may continue to be used provided that they meet the applicable periodic inspection and test provisions of the IMDG Code[1]. In addition, they shall meet the provisions corresponding to the instructions set out in columns (10) and (11) of Table A in Chapter 3.2 and the provisions of Chapter 4.2 of ADR. See also 4.2.0.1 of the IMDG Code.

1.1.4.4 *(Reserved)*

1.1.4.5 *Carriage other than by road*

1.1.4.5.1 If the vehicle carrying out a transport operation subject to the requirements of ADR is conveyed over a section of the journey otherwise than by road haulage, then any national or international regulations which, on the said section, govern the carriage of dangerous goods by the mode of transport used for conveying the road vehicle shall alone be applicable to the said section of the journey.

1.1.4.5.2 In the cases referred to in 1.1.4.5.1 above, the involved ADR Contracting Parties may agree to apply the requirements of ADR to the section of a journey where a vehicle is conveyed otherwise than by road haulage, supplemented, if they consider it necessary, by additional requirements, unless such agreements between the involved ADR Contracting Parties would contravene clauses of the international conventions governing the carriage of dangerous goods by the mode of transport used for conveying the road vehicle on the said section of the journey, e.g. the International Convention for the Safety of Life at Sea (SOLAS), to which these ADR Contacting Parties would also be contracting parties.

These agreements shall be notified by the Contracting Party which has taken the initiative thereof to the Secretariat of the United Nations Economic Commission for Europe which shall bring them to the attention of the Contracting Parties.

1.1.4.5.3 In cases where a transport operation subject to the provisions of ADR is likewise subject over the whole or a part of the road journey to the provisions of an international convention which regulates the carriage of dangerous goods by a mode of transport other than road carriage by virtue of clauses extending the applicability of that convention to certain motor-vehicle services, then the provisions of that international convention shall apply over the journey in question concurrently with those of ADR which are not incompatible with them; the other clauses of ADR shall not apply over the journey in question.

1.1.5 Application of standards

Where the application of a standard is required and there is any conflict between the standard and the provisions of ADR, the provisions of ADR take precedence. The requirements of the standard that do not conflict with ADR shall be applied as specified, including the requirements of any other standard, or part of a standard, referenced within that standard as normative.

---

[1] *The International Maritime Organization (IMO) has issued "Guidance on the Continued Use of Existing IMO Type Portable Tanks and Road Tank Vehicles for the Transport of Dangerous Goods" as circular DSC.1/Circ.12 and Corrigenda. The text of this guidance can be found on the IMO website at: www.imo.org.*

CHAPTER 1.2

DEFINITIONS AND UNITS OF MEASUREMENT

1.2.1   **Definitions**

NOTE: This section contains all general or specific definitions.

For the purposes of ADR:

A

"ADN" means the European Agreement concerning the International Carriage of Dangerous Goods by Inland Waterways;

"Aerosol or aerosol dispenser" means an article consisting of any non-refillable receptacle meeting the requirements of 6.2.6, made of metal, glass or plastics and containing a gas, compressed, liquefied or dissolved under pressure, with or without a liquid, paste or powder, and fitted with a release device allowing the contents to be ejected as solid or liquid particles in suspension in a gas, as a foam, paste or powder or in a liquid state or in a gaseous state;

"Animal material" means animal carcasses, animal body parts, or animal foodstuffs;

"Applicant" means, in the case of conformity assessment, the manufacturer or its authorised representative in a country Contracting Party. In the case of periodic inspections, intermediate inspections and exceptional checks, *applicant* means the testing facility, the operator or their authorised representative in a country Contracting Party;

NOTE: Exceptionally a third party (for instance a tank-container operator in accordance with the definition of 1.2.1) may apply for the conformity assessment.

"Approval"

Multilateral approval, for the carriage of radioactive material, means approval by the relevant competent authority of the country of origin of the design or shipment, as applicable, and by the competent authority of each country through or into which the consignment is to be carried;

Unilateral approval, for the carriage of radioactive material, means an approval of a design which is required to be given by the competent authority of the country of origin of the design only. If the country of origin is not a Contracting Party to ADR, the approval shall require validation by the competent authority of a Contracting Party to ADR (see 6.4.22.8);

"ASTM" means the American Society for Testing and Materials (ASTM International, 100 Barr Harbor Drive, PO Box C700, West Conshohocken, PA, 19428-2959, United States of America);

B

"Bag" means a flexible packaging made of paper, plastics film, textiles, woven material or other suitable material;

"Battery-vehicle" means a vehicle containing elements which are linked to each other by a manifold and permanently fixed to this vehicle. The following elements are considered to be elements of a battery-vehicle: cylinders, tubes, bundles of cylinders (also known as frames), pressure drums as well as tanks destined for the carriage of gases as defined in 2.2.2.1.1 with a capacity of more than 450 litres;

"Body" (for all categories of IBC other than composite IBCs) means the receptacle proper, including openings and closures, but does not include service equipment;

"Box" means a packaging with complete rectangular or polygonal faces, made of metal, wood, plywood, reconstituted wood, fibreboard, plastics or other suitable material. Small holes for purposes of ease of handling or opening or to meet classification requirements, are permitted as long as they do not compromise the integrity of the packaging during carriage;

"Bulk container" means a containment system (including any liner or coating) intended for the carriage of solid substances which are in direct contact with the containment system. Packagings, intermediate bulk containers (IBCs), large packagings and tanks are not included.

A bulk container is:

-        of a permanent character and accordingly strong enough to be suitable for repeated use;

-        specially designed to facilitate the carriage of goods by one or more means of transport without intermediate reloading;

-        fitted with devices permitting its ready handling;

-        of a capacity of not less than 1.0 m^3;

Examples of bulk containers are containers, offshore bulk containers, skips, bulk bins, swap bodies, trough-shaped containers, roller containers, load compartments of vehicles;

***NOTE:** This definition only applies to bulk containers meeting the requirements of Chapter 6.11.*

"Closed bulk container" means a totally closed bulk container having a rigid roof, sidewalls, end walls and floor (including hopper-type bottoms). The term includes bulk containers with an opening roof, side or end wall that can be closed during carriage. Closed bulk containers may be equipped with openings to allow for the exchange of vapours and gases with air and which prevent under normal conditions of carriage the release of solid contents as well as the penetration of rain and splash water;

"Flexible bulk container" means a flexible container with a capacity not exceeding 15 m^3 and includes liners and attached handling devices and service equipment;

"Sheeted bulk container" means an open top bulk container with rigid bottom (including hopper-type bottom), side and end walls and a non-rigid covering;

"Bundle of cylinders" means an assembly of cylinders that are fastened together and which are interconnected by a manifold and carried as a unit. The total water capacity shall not exceed 3 000 litres except that bundles intended for the carriage of toxic gases of Class 2 (groups starting with letter T according to 2.2.2.1.3) shall be limited to 1 000 litres water capacity;

C

"Calculation pressure" means a theoretical pressure at least equal to the test pressure which, according to the degree of danger exhibited by the substance being carried, may to a greater or lesser degree exceed the working pressure. It is used solely to determine the thickness of the walls of the shell, independently of any external or internal reinforcing device (see also *"Discharge pressure"*, *"Filling pressure"*, *"Maximum working pressure (gauge pressure)"* and *"Test pressure"*);

***NOTE:** For portable tanks, see Chapter 6.7.*

"Capacity of shell or shell compartment" for tanks, means the total inner volume of the shell or shell compartment expressed in litres or cubic metres. When it is impossible to completely fill the shell or the shell compartment because of its shape or construction, this reduced capacity shall be used for the determination of the degree of filling and for the marking of the tank;

"Cargo transport unit" means a vehicle, a wagon, a container, a tank-container, a portable tank or an MEGC;

"Carriage" means the change of place of dangerous goods, including stops made necessary by transport conditions and including any period spent by the dangerous goods in vehicles, tanks and containers made necessary by traffic conditions before, during and after the change of place.

This definition also covers the intermediate temporary storage of dangerous goods in order to change the mode or means of transport (trans-shipment). This shall apply provided that transport documents showing the place of dispatch and the place of reception are presented on request and provided that packages and tanks are not opened during intermediate storage, except to be checked by the competent authorities;

"Carriage in bulk" means the carriage of unpackaged solids or articles in vehicles, containers or bulk containers. The term does not apply to packaged goods nor to substances carried in tanks;

"Carrier" means the enterprise which carries out the transport operation with or without a transport contract;

"CGA" means the Compressed Gas Association (CGA, 14501 George Carter Way, Suite 103, Chantilly, VA 20151, United States of America);

"CIM" means the Uniform Rules Concerning the Contract of International Carriage of Goods by Rail (Appendix B to the Convention concerning International Carriage by Rail (COTIF)), as amended;

"Closed bulk container", see *"Bulk container"*;

"Closed container", see *"Container"*;

"Closed vehicle" means a vehicle having a body capable of being closed;

"Closure" means a device which closes an opening in a receptacle;

"CMR" means the Convention on the Contract for the International Carriage of Goods by Road (Geneva, 19 May 1956), as amended;

"Collective entry" means an entry for a defined group of substances or articles (see 2.1.1.2, B, C and D);

"Combination packaging" means a combination of packagings for carriage purposes, consisting of one or more inner packagings secured in an outer packaging in accordance with 4.1.1.5;

NOTE: The term "inner packaging" used for combination packagings shall not be confused with the term "inner receptacle" used for composite packagings.

"Combustion heater" means a device directly using liquid or gaseous fuel and not using the waste heat from the engine used for propulsion of the vehicle;

"Competent authority" means the authority or authorities or any other body or bodies designated as such in each State and in each specific case in accordance with domestic law;

"Compliance assurance" (radioactive material) means a systematic programme of measures applied by a competent authority which is aimed at ensuring that the requirements of ADR are met in practice;

"Composite IBC with plastics inner receptacle" means an IBC comprising structural equipment in the form of a rigid outer casing encasing a plastics inner receptacle together with any service or other structural equipment. It is so constructed that the inner receptacle and outer casing once assembled form, and are used as, an integrated single unit to be filled, stored, transported or emptied as such;

NOTE: "Plastics material", when used in connection with inner receptacles for composite IBCs, is taken to include other polymeric materials such as rubber.

"Composite packaging" means a packaging consisting of an outer packaging and an inner receptacle so constructed that the inner receptacle and the outer packaging form an integral packaging. Once assembled it remains thereafter an integrated single unit; it is filled, stored, carried and emptied as such;

NOTE: The term "inner receptacle" used for composite packagings shall not be confused with the term "inner packaging" used for combination packagings. For example, the inner of a 6HA1 composite packaging (plastics material) is such an inner receptacle since it is normally not designed to perform a containment function without its outer packaging and is not therefore an inner packaging.

Where a material is mentioned in brackets after the term "composite packaging", it refers to the inner receptacle.

"Compressed Natural Gas (CNG)" means a compressed gas composed of natural gas with a high methane content assigned to UN No. 1971;

"Confinement system", for the carriage of radioactive material, means the assembly of fissile material and packaging components specified by the designer and agreed to by the competent authority as intended to preserve criticality safety;

"Conformity assessment" means the process of verifying the conformity of a product according to the provisions of sections 1.8.6 and 1.8.7 related to type approval, supervision of manufacture and initial inspection and testing;

"Consignee" means the consignee according to the contract for carriage. If the consignee designates a third party in accordance with the provisions applicable to the contract for carriage, this person shall be deemed to be the consignee within the meaning of ADR. If the transport operation takes place without a contract for carriage, the enterprise which takes charge of the dangerous goods on arrival shall be deemed to be the consignee;

"Consignment" means any package or packages, or load of dangerous goods, presented by a consignor for carriage;

"Consignor" means the enterprise which consigns dangerous goods either on its own behalf or for a third party. If the transport operation is carried out under a contract for carriage, consignor means the consignor according to the contract for carriage;

"Container" means an article of transport equipment (lift van or other similar structure):

-   of a permanent character and accordingly strong enough to be suitable for repeated use;

-   specially designed to facilitate the carriage of goods, by one or more means of transport, without breakage of load;

-   fitted with devices permitting its ready stowage and handling, particularly when being transloaded from one means of transport to another;

-   so designed as to be easy to fill and empty;

-   having an internal volume of not less than 1 m^3, except for containers for the carriage of radioactive material.

In addition:

   "Small container" means a container which has an internal volume of not more than 3 m^3;

   "Large container" means

   (a)   A container which does not meet the definition of a small container;

   (b)   In the meaning of the CSC, a container of a size such that the area enclosed by the four outer bottom corners is either

      (i)   at least 14 m^2 (150 square feet); or

      (ii)   at least 7 m^2 (75 square feet) if fitted with top corner fittings;

   "Closed container" means a totally enclosed container having a rigid roof, rigid side walls, rigid end walls and a floor. The term includes containers with an opening roof where the roof can be closed during transport;

   "Open container" means an open top container or a platform based container;

   "Sheeted container" means an open container equipped with a sheet to protect the goods loaded;

A *"swap body"* is a container which, in accordance with EN 283:1991 has the following characteristics:

-        from the point of view of mechanical strength, it is only built for carriage on a wagon or a vehicle on land or by roll-on roll-of ship;

-        it cannot be stacked;

-        it can be removed from vehicles by means of equipment on board the vehicle and on its own supports, and can be reloaded;

NOTE: The term "container" does not cover conventional packagings, IBCs, tank-containers or vehicles. Nevertheless, a container may be used as a packaging for the carriage of radioactive material.

"Containment system", for the carriage of radioactive material, means the assembly of components of the packaging specified by the designer as intended to retain the radioactive material during carriage;

"Control temperature" means the maximum temperature at which the organic peroxide or the self-reactive substance can be safely carried;

"Conveyance" means, for carriage by road or by rail, a vehicle or a wagon;

"Criticality safety index (CSI) assigned to a package, overpack or container containing fissile material", for the carriage of radioactive material, means a number which is used to provide control over the accumulation of packages, overpacks or containers containing fissile material;

"CSC" means the International Convention for Safe Containers (Geneva, 1972) as amended and published by the International Maritime Organization (IMO), London;

"Crate" means an outer packaging with incomplete surfaces;

"Critical temperature" means the temperature above which the substance cannot exist in the liquid state;

"Cryogenic receptacle" means a transportable thermally insulated pressure receptacle for refrigerated liquefied gases of a water capacity of not more than 1 000 litres (see also *"Open cryogenic receptacle"*);

"Cylinder" means a transportable pressure receptacle of a water capacity not exceeding 150 litres (see also *"Bundle of cylinders"*);

D

"Dangerous goods" means those substances and articles the carriage of which is prohibited by ADR, or authorized only under the conditions prescribed therein;

"Dangerous reaction" means:

(a)   Combustion or evolution of considerable heat;

(b)   Evolution of flammable, asphyxiant, oxidizing or toxic gases;

(c)   The formation of corrosive substances;

(d)   The formation of unstable substances; or

(e)   Dangerous rise in pressure (for tanks only);

"Demountable tank" means a tank, other than a fixed tank, a portable tank, a tank-container or an element of a battery-vehicle or a MEGC which has a capacity of more than 450 litres, is not designed for the carriage of goods without breakage of load, and normally can only be handled when it is empty;

"Design", for the carriage of radioactive material, means the description of fissile material excepted under 2.2.7.2.3.5 (f), special form radioactive material, low dispersible radioactive material, package

or packaging which enables such an item to be fully identified. The description may include specifications, engineering drawings, reports demonstrating compliance with regulatory requirements, and other relevant documentation;

"Design life", for composite cylinders and tubes, means the maximum life (in number of years) for which the cylinder or tube is designed and approved in accordance with the applicable standard;

"Discharge pressure" means the maximum pressure actually built up in the tank when it is being discharged under pressure (see also *"Calculation pressure"*, *"Filling pressure"*, *"Maximum working pressure (gauge pressure)"* and *"Test pressure"*);

"Drum" means a flat-ended or convex-ended cylindrical packaging made out of metal, fibreboard, plastics, plywood or other suitable materials. This definition also includes packagings of other shapes, e.g. round, taper-necked packagings or pail-shaped packagings. Wooden barrels and jerricans are not covered by this definition;

E

"EC Directive" means provisions decided by the competent institutions of the European Community and which are binding, as to the result to be achieved, upon each Member State to which it is addressed, but shall leave to the national authorities the choice of form and methods;

"ECE Regulation" means a regulation annexed to the Agreement concerning the adoption of uniform technical prescriptions for wheeled vehicles equipment and parts which can be fitted and or used on wheeled vehicles and the conditions for reciprocal recognition of approvals granted on the basis of these prescriptions (1958 Agreement, as amended);

"Emergency temperature" means the temperature at which emergency procedures shall be implemented in the event of loss of temperature control;

"EN" (standard) means a European standard published by the European Committee for Standardization (CEN) (CEN, Avenue Marnix 17, B-1000 Brussels);

"Enterprise" means any natural person, any legal person, whether profit-making or not, any association or group of persons without legal personality, whether profit-making or not, or any official body, whether it has legal personality itself or is dependent upon an authority that has such personality;

"Exclusive use", for the carriage of radioactive material, means the sole use, by a single consignor, of a vehicle or of a large container, in respect of which all initial, intermediate and final loading and unloading and shipment are carried out in accordance with the directions of the consignor or consignee, where so required by ADR;

F

"Fibreboard IBC" means a fibreboard body with or without separate top and bottom caps, if necessary an inner liner (but no inner packagings), and appropriate service and structural equipment;

"Filler" means any enterprise which fills dangerous goods into a tank (tank-vehicle, demountable tank, portable tank or tank-container) and/or into a vehicle, large container or small container for carriage in bulk, or into a battery-vehicle or MEGC;

"Filling pressure" means the maximum pressure actually built up in the tank when it is being filled under pressure (see also *"Calculation pressure"*, *"Discharge pressure"*, *"Maximum working pressure (gauge pressure)"* and *"Test pressure"*);

"Filling ratio" means the ratio of the mass of gas to the mass of water at 15 °C that would fill completely a pressure receptacle fitted ready for use;

"Fixed tank" means a tank having a capacity of more than 1 000 litres which is permanently attached to a vehicle (which then becomes a tank-vehicle) or is an integral part of the frame of such vehicle;

"Flammable component" (for aerosols) means flammable liquids, flammable solids or flammable gases and gas mixtures as defined in Notes 1 to 3 of sub-section 31.1.3 of Part III of the Manual of Tests and Criteria. This designation does not cover pyrophoric, self-heating or water-reactive

substances. The chemical heat of combustion shall be determined by one of the following methods ASTM D 240, ISO/FDIS 13943:1999 (E/F) 86.1 to 86.3 or NFPA 30B;

"Flash-point" means the lowest temperature of a liquid at which its vapours form a flammable mixture with air;

"Flexible bulk container", see *"Bulk container"*;

"Flexible IBC" means a body constituted of film, woven fabric or any other flexible material or combinations thereof, and if necessary, an inner coating or liner, together with any appropriate service equipment and handling devices;

"Fuel cell" means an electrochemical device that converts the chemical energy of a fuel to electrical energy, heat and reaction products;

"Fuel cell engine" means a device used to power equipment and which consists of a fuel cell and its fuel supply, whether integrated with or separate from the fuel cell, and includes all appurtenances necessary to fulfil its function;

"Full load" means any load originating from one consignor for which the use of a vehicle or of a large container is exclusively reserved and all operations for the loading and unloading of which are carried out in conformity with the instructions of the consignor or of the consignee;

NOTE: The corresponding term for radioactive material is "exclusive use".

G

"Gas" means a substance which:

(a)   At 50 °C has a vapour pressure greater than 300 kPa (3 bar); or

(b)   Is completely gaseous at 20 °C under standard pressure of 101.3 kPa;

"Gas cartridge", see *"Small receptacle containing gas"*;

"GHS" means the sixth revised edition of the Globally Harmonized System of Classification and Labelling of Chemicals, published by the United Nations as document ST/SG/AC.10/30/Rev.6;

H

"Handling device" (for flexible IBCs) means any sling, loop, eye or frame attached to the body of the IBC or formed from the continuation of the IBC body material;

"Hermetically closed tank" means a tank intended for the carriage of liquid substances with a calculation pressure of at least 4 bar or intended for the carriage of solid substances (powdery or granular) regardless of its calculation pressure, the openings of which are hermetically closed and which:

-     is not equipped with safety valves, bursting discs, other similar safety devices or vacuum valves; or

-     is not equipped with safety valves, bursting discs or other similar safety devices, but is equipped with vacuum valves, in accordance with the requirements of 6.8.2.2.3; or

-     is equipped with safety valves preceded by a bursting disc according to 6.8.2.2.10, but is not equipped with vacuum valves; or

-     is equipped with safety valves preceded by a bursting disc according to 6.8.2.2.10 and vacuum valves, in accordance with the requirements of 6.8.2.2.3;

"Holding time" means the time that will elapse from the establishment of the initial filling condition until the pressure has risen due to heat influx to the lowest set pressure of the pressure limiting devices (s) of tanks intended for the carriage of refrigerated liquefied gases;

NOTE: For portable tanks, see 6.7.4.1.

I

"IAEA" means the International Atomic Energy Agency (IAEA), (IAEA, P.O. Box 100 – A -1400 Vienna);

"IBC", see *"Intermediate bulk container"*;

"ICAO" means the International Civil Aviation Organization (ICAO, 999 University Street, Montreal, Quebec H3C 5H7, Canada);

"ICAO Technical Instructions" means the Technical Instructions for the Safe Transport of Dangerous Goods by Air, which complement Annex 18 to the Chicago Convention on International Civil Aviation (Chicago 1944), published by the International Civil Aviation Organization (ICAO) in Montreal;

"IMDG Code" means the International Maritime Dangerous Goods Code, for the implementation of Chapter VII, Part A, of the International Convention for the Safety of Life at Sea, 1974 (SOLAS Convention), published by the International Maritime Organization (IMO), London;

"IMO" means the International Maritime Organization (IMO, 4 Albert Embankment, London SE1 7SR, United Kingdom);

"Inner packaging" means a packaging for which an outer packaging is required for carriage;

"Inner receptacle" means a receptacle which requires an outer packaging in order to perform its containment function;

"Inspection body" means an independent inspection and testing body approved by the competent authority;

"Intermediate bulk container" *(IBC)* means a rigid, or flexible portable packaging, other than those specified in Chapter 6.1, that:

(a)   Has a capacity of:

    (i)   not more than 3 m^3 for solids and liquids of packing groups II and III;

    (ii)   not more than 1.5 m^3 for solids of packing group I when packed in flexible, rigid plastics, composite, fibreboard and wooden IBCs;

    (iii)   not more than 3 m^3 for solids of packing group I when packed in metal IBCs;

    (iv)   not more than 3 m^3 for radioactive material of Class 7;

(b)   Is designed for mechanical handling;

(c)   Is resistant to the stresses produced in handling and transport as determined by the tests specified in Chapter 6.5;

(see also *"Composite IBC with plastics inner receptacle"*, *"Fibreboard IBC"*, *"Flexible IBC"*, *"Metal IBC"*, *"Rigid plastics IBC"* and *"Wooden IBC"*).

NOTE 1: *Portable tanks or tank-containers that meet the requirements of Chapter 6.7 or 6.8 respectively are not considered to be intermediate bulk containers (IBCs).*

NOTE 2: *Intermediate bulk containers (IBCs) which meet the requirements of Chapter 6.5 are not considered to be containers for the purposes of ADR.*

"Remanufactured IBC" means a metal, rigid plastics or composite IBC that:

(a)   Is produced as a UN type from a non-UN type; or

(b)   Is converted from one UN design type to another UN design type.

Remanufactured IBCs are subject to the same requirements of ADR that apply to new IBCs of the same type (see also design type definition in 6.5.6.1.1);

"Repaired IBC" means a metal, rigid plastics or composite IBC that, as a result of impact or for any other cause (e.g. corrosion, embrittlement or other evidence of reduced strength as compared to the design type) is restored so as to conform to the design type and to be able to withstand the design type tests. For the purposes of ADR, the replacement of the rigid inner receptacle of a composite IBC with a receptacle conforming to the original design type from the same manufacturer is considered repair. However, routine maintenance of rigid IBCs is not considered repair. The bodies of rigid plastics IBCs and the inner receptacles of composite IBCs are not repairable. Flexible IBCs are not repairable unless approved by the competent authority;

"Routine maintenance of flexible IBCs" means the routine performance on plastics or textile flexible IBCs of operations, such as:

(a)   Cleaning; or

(b)   Replacement of non-integral components, such as non-integral liners and closure ties, with components conforming to the original manufacturer's specification;

provided that these operations do not adversely affect the containment function of the flexible IBC or alter the design type.

"Routine maintenance of rigid IBCs" means the routine performance on metal, rigid plastics or composite IBCs of operations such as:

(a)   Cleaning;

(b)   Removal and reinstallation or replacement of body closures (including associated gaskets), or of service equipment, conforming to the original manufacturer's specifications, provided that the leaktightness of the IBC is verified; or

(c)   Restoration of structural equipment not directly performing a dangerous goods containment or discharge pressure retention function so as to conform to the design type (e.g. the straightening of legs or lifting attachments) provided that the containment function of the IBC is not affected;

"Intermediate packaging" means a packaging placed between inner packagings or articles, and an outer packaging;

"ISO" (standard) means an international standard published by the International Organization for Standardization (ISO) (ISO - 1, rue de Varembé. CH-1204 Geneva 20);

J

"Jerrican" means a metal or plastics packaging of rectangular or polygonal cross-section with one or more orifices;

L

"Large container ", see *"Container"*;

"Large packaging" means a packaging consisting of an outer packaging which contains articles or inner packagings and which

(a)   Is designed for mechanical handling;

(b)   Exceeds 400 kg net mass or 450 litres capacity but has a volume of not more than 3 m^3;

"Large salvage packaging" means a special packaging which

(a)   is designed for mechanical handling; and

(b)   exceeds 400 kg net mass or 450 litres capacity but has a volume of not more than 3 m^3;

into which damaged, defective, leaking or non-conforming dangerous goods packages, or dangerous goods that have spilled or leaked are placed for purposes of carriage for recovery or disposal;

"Leakproofness test" means a test to determine the leakproofness of a tank, a packaging or an IBC and of the equipment and closure devices;

NOTE: *For portable tanks, see Chapter 6.7.*

"Light-gauge metal packaging" means a packaging of circular, elliptical, rectangular or polygonal cross-section (also conical) and taper-necked and pail-shaped packaging made of metal, having a wall thickness of less than 0.5 mm (e.g. tinplate), flat or convex bottomed and with one or more orifices, which is not covered by the definitions for drums or jerricans;

"Liner" means a tube or bag inserted into a packaging, including large packagings or IBCs, but not forming an integral part of it, including the closures of its openings;

"Liquid" means a substance which at 50 °C has a vapour pressure of not more than 300 kPa (3 bar), which is not completely gaseous at 20 °C and 101.3 kPa, and which

(a)   Has a melting point or initial melting point of 20 °C or less at a pressure of 101.3 kPa; or

(b)   Is liquid according to the ASTM D 4359-90 test method; or

(c)   Is not pasty according to the criteria applicable to the test for determining fluidity (penetrometer test) described in 2.3.4;

NOTE: *"Carriage in the liquid state", for the purpose of tank requirements, means:*

-       *Carriage of liquids according to the above definition; or*

-       *Solids handed over for carriage in the molten state.*

"Liquefied Natural Gas (LNG)" means a refrigerated liquefied gas composed of natural gas with a high methane content assigned to UN No. 1972;

"Liquefied Petroleum Gas (LPG)" means a low pressure liquefied gas composed of one or more light hydrocarbons which are assigned to UN Nos. 1011, 1075, 1965, 1969 or 1978 only and which consists mainly of propane, propene, butane, butane isomers, butene with traces of other hydrocarbon gases;

NOTE 1: *Flammable gases assigned to other UN numbers shall not be regarded as LPG.*

NOTE 2: *For UN No. 1075 see NOTE 2 under 2F, UN No. 1965, in the table for Liquefied gases in 2.2.2.3.*

"Loader" means any enterprise which:

(a)   Loads packaged dangerous goods, small containers or portable tanks into or onto a vehicle or a container; or

(b)   Loads a container, bulk-container, MEGC, tank-container or portable tank onto a vehicle.

"Loading" means all actions carried out by the loader, in accordance with the definition of loader;

M

"Management system", for the carriage of radioactive material, means a set of interrelated or interacting elements (system) for establishing policies and objectives and enabling the objectives to be achieved in an efficient and effective manner;

"Manual of Tests and Criteria" means the sixth revised edition of the Recommendations on the Transport of Dangerous Goods, Manual of Tests and Criteria, published by the United Nations (ST/SG/AC.10/11/Rev.6);

"Mass of package" means gross mass of the package unless otherwise stated. The mass of containers and tanks used for the carriage of goods is not included in the gross mass;

"Maximum capacity" means the maximum inner volume of receptacles or packagings including intermediate bulk containers (IBCs) and large packagings expressed in cubic metres or litres;

"Maximum net mass" means the maximum net mass of contents in a single packaging or maximum combined mass of inner packagings and the contents thereof expressed in kilograms;

"Maximum normal operating pressure", for the carriage of radioactive material, means the maximum pressure above atmospheric pressure at mean sea-level that would develop in the containment system in a period of one year under the conditions of temperature and solar radiation corresponding to environmental conditions in the absence of venting, external cooling by an ancillary system, or operational controls during carriage;

"Maximum permissible gross mass"

(a)   (for IBCs) means the mass of the IBC and any service or structural equipment together with the maximum net mass;

(b)   (for tanks) means the tare of the tank and the heaviest load authorized for carriage;

NOTE: *For portable tanks, see Chapter 6.7.*

"Maximum working pressure (gauge pressure)" means the highest of the following three pressures that may occur at the top of the tank in the operating position:

(a)   The highest effective pressure allowed in the tank during filling (maximum filling pressure allowed);

(b)   The highest effective pressure allowed in the tank during discharge (maximum discharge pressure allowed); and

(c)   The effective gauge pressure to which the tank is subjected by its contents (including such extraneous gases as it may contain) at the maximum working temperature.

Unless the special requirements prescribed in Chapter 4.3 provide otherwise, the numerical value of this working pressure (gauge pressure) shall not be lower than the vapour pressure (absolute pressure) of the filling substance at 50 °C.

For tanks equipped with safety valves (with or without bursting disc) other than tanks for the carriage of compressed, liquefied or dissolved gases of Class 2, the maximum working pressure (gauge pressure) shall however be equal to the prescribed opening pressure of such safety valves.

(See also *"Calculation pressure"*, *"Discharge pressure"*, *"Filling pressure"* and *"Test pressure"*);

NOTE 1:   *Maximum working pressure is not applicable to gravity-discharge tanks according to 6.8.2.1.14 (a).*

NOTE 2:   *For portable tanks, see Chapter 6.7.*

NOTE 3:   *For closed cryogenic receptacles, see NOTE to 6.2.1.3.6.5.*

"MEGC", see *"Multiple-element gas container"*;

"Member of a vehicle crew" means a driver or any other person accompanying the driver for safety, security, training or operational reasons;

"MEMU", see *"Mobile explosives manufacturing unit"*;

"Metal hydride storage system" means a single complete hydrogen storage system, including a receptacle, metal hydride, pressure relief device, shut-off valve, service equipment and internal components used for the carriage of hydrogen only;

"Metal IBC" means a metal body together with appropriate service and structural equipment;

"Mild steel" means a steel having a minimum tensile strength between 360 N/mm^2 and 440 N/mm^2;

NOTE: *For portable tanks, see Chapter 6.7.*

"Mobile explosives manufacturing unit" (MEMU) means a unit, or a vehicle mounted with a unit, for manufacturing and charging explosives from dangerous goods that are not explosives. The unit

consists of various tanks and bulk containers and process equipment as well as pumps and related equipment. The MEMU may have special compartments for packaged explosives;

NOTE: Even though the definition of MEMU includes the expression "manufacturing and charging explosives" the requirements for MEMUs apply only to carriage and not to manufacturing and charging of explosives.

"Multiple-element gas container" (MEGC) means a unit containing elements which are linked to each other by a manifold and mounted on a frame. The following elements are considered to be elements of a multiple-element gas container: cylinders, tubes, pressure drums or bundles of cylinders as well as tanks for the carriage of gases as defined in 2.2.2.1.1 having a capacity of more than 450 litres;

NOTE: For UN MEGCs, see Chapter 6.7.

N

"Net explosive mass (NEM)" means the total mass of the explosive substances, without the packagings, casings, etc. *(Net explosive quantity (NEQ), net explosive contents (NEC), net explosive weight (NEW)* or *net mass of explosive contents* are often used to convey the same meaning;

"Neutron radiation detector" means a device that detects neutron radiation. In such a device, a gas may be contained in a hermetically sealed electron tube transducer that converts neutron radiation into a measureable electric signal;

"N.O.S. entry (not otherwise specified entry)" means a collective entry to which substances, mixtures, solutions or articles may be assigned if they:

(a)   Are not mentioned by name in Table A of Chapter 3.2; and

(b)   Exhibit chemical, physical and/or dangerous properties corresponding to the Class, classification code, packing group and the name and description of the n.o.s. entry;

O

"Offshore bulk container" means a bulk container specially designed for repeated use for carriage to, from and between offshore facilities. An offshore bulk container is designed and constructed in accordance with the guidelines for the approval of offshore containers handled in open seas specified by the International Maritime Organization (IMO) in document MSC/Circ.860;

"Open container", see *"Container"*;

"Open cryogenic receptacle" means a transportable thermally insulated receptacle for refrigerated liquefied gases maintained at atmospheric pressure by continuous venting of the refrigerated liquefied gas;

"Open vehicle" means a vehicle the platform of which has no superstructure or is merely provided with side boards and a tailboard;

"Outer packaging" means the outer protection of the composite or combination packaging together with any absorbent materials, cushioning and any other components necessary to contain and protect inner receptacles or inner packagings;

"Overpack" means an enclosure used (by a single consignor in the case of radioactive material) to contain one or more packages, consolidated into a single unit easier to handle and stow during carriage;

Examples of overpacks:

(a)   A loading tray such as a pallet, on which several packages are placed or stacked and secured by a plastics strip, shrink or stretch wrapping or other appropriate means; or

(b)   An outer protective packaging such as a box or a crate;

P

"Package" means the complete product of the packing operation, consisting of the packaging or large packaging or IBC and its contents prepared for dispatch. The term includes receptacles for gases as defined in this section as well as articles which, because of their size, mass or configuration may be carried unpackaged or carried in cradles, crates or handling devices. Except for the carriage of radioactive material, the term does not apply to goods which are carried in bulk, nor to substances carried in tanks;

NOTE: *For radioactive material, see 2.2.7.2, 4.1.9.1.1 and Chapter 6.4.*

"Packaging" means one or more receptacles and any other components or materials necessary for the receptacles to perform their containment and other safety functions (see also *"Combination packaging"*, *"Composite packaging"*, *"Inner packaging"*, *"Intermediate bulk container (IBC)"*, *"Intermediate packaging"*, *"Large packaging"*, *"Light-gauge metal packaging"*, *"Outer packaging"*, *"Reconditioned packaging"*, *"Remanufactured packaging"*, *"Reused packaging"*, *"Salvage packaging"* and *"Sift-proof packaging"*);

"Packer" means any enterprise which puts dangerous goods into packagings, including large packagings and intermediate bulk containers (IBCs) and, where necessary, prepares packages for carriage;

"Packing group" means a group to which, for packing purposes, certain substances may be assigned in accordance with their degree of danger. The packing groups have the following meanings which are explained more fully in Part 2:

Packing group I: Substances presenting high danger;

Packing group II: Substances presenting medium danger; and

Packing group III: Substances presenting low danger;

NOTE: *Certain articles containing dangerous goods are assigned to a packing group.*

"Portable tank" means a multimodal tank having, when used for the carriage of gases as defined in 2.2.2.1.1, a capacity of more than 450 litres in accordance with the definitions in Chapter 6.7 or the IMDG Code and indicated by a portable tank instruction (T-Code) in Column (10) of Table A of Chapter 3.2;

"Portable tank operator", see *"Tank-container/portable tank operator"*;

"Pressure drum" means a welded transportable pressure receptacle of a water capacity exceeding 150 litres and of not more than1 000 litres, (e.g. cylindrical receptacles equipped with rolling hoops, spheres on skids);"

"Pressure receptacle" means a collective term that includes cylinders, tubes, pressure drums, closed cryogenic receptacles, metal hydride storage systems, bundles of cylinders and salvage pressure receptacles;

"Pressurized gas cartridge", see *"Aerosol or aerosol dispenser"*;

"Protected IBC" (for metal IBCs) means an IBC provided with additional protection against impact, the protection taking the form of, for example, a multi-layer (sandwich) or double-wall construction, or a frame with a metal lattice-work casing;

Q

"Quality assurance" means a systematic programme of controls and inspections applied by any organization or body which is aimed at providing confidence that the safety prescriptions in ADR are met in practice;

R

"Radiation detection system" means an apparatus that contains radiation detectors as components;

"Radiation level", for the carriage of radioactive material, means the corresponding dose rate expressed in millisieverts per hour or microsieverts per hour;

"Radioactive contents", for the carriage of radioactive material, mean the radioactive material together with any contaminated or activated solids, liquids, and gases within the packaging;

"Receptacle" (Class 1) includes boxes, bottles, cans, drums, jars and tubes, including any means of closure used in the inner or intermediate packaging;

"Receptacle" means a containment vessel for receiving and holding substances or articles, including any means of closing. This definition does not apply to shells (see also *"Cryogenic receptacle"*, *"Inner receptacle"*, *"Pressure receptacle"*, *"Rigid inner receptacle"* and *"Gas cartridge"*);

"Reconditioned packaging" means in particular

(a)   Metal drums that are:

    (i)   cleaned to original materials of construction, with all former contents, internal and external corrosion, and external coatings and labels removed;

    (ii)   restored to original shape and contour, with chimes (if any) straightened and sealed and all non-integral gaskets replaced; and

    (iii)   inspected after cleaning but before painting, with rejection of packagings with visible pitting, significant reduction in the material thickness, metal fatigue, damaged threads or closures or other significant defects;

(b)   Plastics drums and jerricans that:

    (i)   are cleaned to original materials of construction, with all former contents, external coatings and labels removed;

    (ii)   have all non-integral gaskets replaced; and

    (iii)   are inspected after cleaning with rejection of packagings with visible damage such as tears, creases or cracks, or damaged threads or closures or other significant defects;

"Recycled plastics material" means material recovered from used industrial packagings that has been cleaned and prepared for processing into new packagings;

"Reel" (Class 1) means a device made of plastics, wood, fibreboard, metal or other suitable material comprising a central spindle with, or without, side walls at each end of the spindle. Articles and substances can be wound onto the spindle and may be retained by side walls;

"Reference steel" means a steel with a tensile strength of 370 N/mm^2 and an elongation at fracture of 27%;

"Remanufactured IBC", see *"Intermediate Bulk Container (IBC)"*;

"Remanufactured large packaging" means a metal or rigid plastics large packaging that:

(a)   Is produced as a UN type from a non-UN type; or

(b)   Is converted from one UN design type to another UN design type.

Remanufactured large packagings are subject to the same requirements of ADR that apply to new large packagings of the same type (see also design type definition in 6.6.5.1.2);

"Remanufactured packaging" means in particular

(a)   Metal drums that:

    (i)   are produced as a UN type complying with the requirements of Chapter 6.1 from a non-UN type;

(ii) are converted from one UN type complying with the requirements of Chapter 6.1 to another UN type; or

(iii) undergo the replacement of integral structural components (such as non-removable heads);

(b) Plastics drums that:

(i) are converted from one UN type to another UN type (e.g. 1H1 to 1H2); or

(ii) undergo the replacement of integral structural components.

Remanufactured drums are subject to the requirements of Chapter 6.1 which apply to new drums of the same type;

"Repaired IBC", see *"Intermediate Bulk Container (IBC)"*;

"Reused large packaging" means a large packaging to be refilled which has been examined and found free of defects affecting the ability to withstand the performance tests; the term includes those which are refilled with the same or similar compatible contents and are carried within distribution chains controlled by the consignor of the product;

"Reused packaging" means a packaging which has been examined and found free of defects affecting the ability to withstand the performance tests. The term includes those which are refilled with the same or similar compatible contents and are carried within distribution chains controlled by the consignor of the product;

"RID" means Regulations concerning the International Carriage of Dangerous Goods by Rail (Appendix C of COTIF (Convention concerning international carriage by rail));

"Rigid inner receptacle" (for composite IBCs) means a receptacle which retains its general shape when empty without its closures in place and without benefit of the outer casing. Any inner receptacle that is not "rigid" is considered to be "flexible";

"Rigid plastics IBC" means a rigid plastics body, which may have structural equipment together with appropriate service equipment;

"Routine maintenance of flexible IBCs", see *"Intermediate Bulk Container (IBC)"*;

"Routine maintenance of rigid IBCs", see *"Intermediate Bulk Container (IBC)"*;

S

"Safety valve" means a spring-loaded device which is activated automatically by pressure the purpose of which is to protect the tank against unacceptable excess internal pressure;

"SADT" see *"Self-accelerating decomposition temperature"*;

"Salvage packaging" means a special packaging into which damaged, defective, leaking or non-conforming dangerous goods packages, or dangerous goods that have spilled or leaked are placed for purposes of carriage for recovery or disposal;

"Salvage pressure receptacle" means a pressure receptacle with a water capacity not exceeding 3 000 litres into which are placed damaged, defective, leaking or non-conforming pressure receptacle(s) for the purpose of carriage e.g. for recovery or disposal;

"SAPT", see *"Self-accelerating polymerization temperature"*;

"Self-accelerating decomposition temperature" (SADT), means the lowest temperature at which self-accelerating decomposition may occur with substance in the packaging as used during carriage. Provisions for determining the SADT and the effects of heating under confinement are contained in Part II of the Manual of Tests and Criteria;

"Self-accelerating polymerization temperature (SAPT)" means the lowest temperature at which polymerization may occur with a substance in the packaging, IBC or tank as offered for carriage. The SAPT shall be determined in accordance with the test procedures established for the self-accelerating decomposition temperature for self-reactive substances in accordance with Part II, section 28 of the Manual of Tests and Criteria;

"Service equipment"

(a)   Of the tank means filling and discharge, breather, safety, heating, heat insulating and additive devices and measuring instruments;

(b)   Of the elements of a battery-vehicle or of a MEGC means filling and discharge devices, including the manifold, safety devices and measuring instruments;

(c)   Of an IBC means the filling and discharge devices and any pressure-relief or venting, safety, heating and heat insulating devices and measuring instruments;

NOTE: For portable tanks, see Chapter 6.7.

"Service life", for composite cylinders and tubes, means the number of years the cylinder or tube is permitted to be in service;

"Settled pressure" means the pressure of the contents of a pressure receptacle in thermal and diffusive equilibrium;

"Sheeted bulk container", see *"Bulk container"*;

"Sheeted container", see *"Container"*;

"Sheeted vehicle" means an open vehicle provided with a sheet to protect the load;

"Shell" (for tanks), means the part of the tank which retains the substance intended for carriage, including openings and their closures, but does not include service equipment or external structural equipment;

NOTE: For portable tanks, see Chapter 6.7.

"Sift-proof packaging" means a packaging impermeable to dry contents, including fine solid material produced during carriage;

"Small container", see *"Container"*;

"Small receptacle containing gas (gas cartridge)" means a non-refillable receptacle having a water capacity not exceeding 1000 ml for receptacles made of metal and not exceeding 500 ml for receptacles made of synthetic material or glass, containing, under pressure, a gas or a mixture of gases. It may be fitted with a valve;

"Solid" means:

(a)   A substance with a melting point or initial melting point of more than 20 °C at a pressure of 101.3 kPa; or

(b)   A substance which is not liquid according to the ASTM D 4359-90 test method or which is pasty according to the criteria applicable to the test for determining fluidity (penetrometer test) described in 2.3.4;

"Structural equipment"

(a)   For tanks of a tank-vehicle or demountable tank, means the external or internal reinforcing, fastening, protective or stabilizing members of the shell;

(b)   For tanks of a tank-container, means the external or internal reinforcing, fastening, protective or stabilizing members of the shell;

(c)   For elements of a battery-vehicle or an MEGC means the external or internal reinforcing, fastening, protective or stabilizing members of the shell or receptacle;

(d)   For IBCs other than flexible IBCs means the reinforcing, fastening, handling, protective or stabilizing members of the body (including the base pallet for composite IBCs with plastics inner receptacle);

NOTE: For portable tanks, see Chapter 6.7.

"Swap body", see *"Container"*;

T

"Tank" means a shell, including its service and structural equipment. When used alone, the term tank means a tank-container, portable tank, demountable tank or fixed tank as defined in this Section, including tanks forming elements of battery-vehicles or MEGCs (see also *"Demountable tank"*, *"Fixed tank"*, *"Portable tank"* and *"Multiple-element gas container"*);

NOTE: For portable tanks, see 6.7.4.1.

"Tank-container" means an article of transport equipment meeting the definition of a container, and comprising a shell and items of equipment, including the equipment to facilitate movement of the tank-container without significant change of attitude, used for the carriage of gases, liquid, powdery or granular substances and, when used for the carriage of gases as defined in 2.2.2.1.1, having a capacity of more than 0.45 m^3 (450 litres);

NOTE: IBCs which meet the requirements of Chapter 6.5 are not considered to be tank-containers.

"Tank-container/portable tank operator" means any enterprise in whose name the tank-container/portable tank is registered;

"Tank record" means a file containing all the important technical information concerning a tank, a battery-vehicle or a MEGC, such as certificates referred to in 6.8.2.3, 6.8.2.4 and 6.8.3.4;

"Tank swap body" is considered to be a tank-container;

"Tank-vehicle" means a vehicle built to carry liquids, gases or powdery or granular substances and comprising one or more fixed tanks. In addition to the vehicle proper, or the units of running gear used in its stead, a tank-vehicle comprises one or more shells, their items of equipment and the fittings for attaching them to the vehicle or to the running-gear units;

"Technical name" means a recognized chemical name, if relevant a biological name, or other name currently used in scientific and technical handbooks, journals and texts (see 3.1.2.8.1.1);

"Test pressure" means the required pressure applied during a pressure test for initial or periodic inspection (see also *"Calculation pressure"*, *"Discharge pressure"*, *"Filling pressure"* and *"Maximum working pressure (gauge pressure)"*);

NOTE: For portable tanks, see Chapter 6.7.

"Through or into", for the carriage of radioactive material, means through or into the countries in which a consignment is carried but specifically excludes countries "over" which a consignment is carried by air provided that there are no scheduled stops in those countries;

"Transport index (TI) assigned to a package, overpack or container, or to unpackaged LSA-I or SCO-I", for the carriage of radioactive material, means a number which is used to provide control over radiation exposure;

"Transport unit" means a motor vehicle without an attached trailer, or a combination consisting of a motor vehicle and an attached trailer;

"Tray" (Class 1) means a sheet of metal, plastics, fibreboard or other suitable material which is placed in the inner, intermediate or outer packaging and achieves a close-fit in such packaging. The surface of the tray may be shaped so that packagings or articles can be inserted, held secure and separated from each other;

"Tube" (Class 2) means a transportable pressure receptacle of seamless or composite construction having a water capacity exceeding 150 litres and of not more than 3 000 litres;

U

"UIC" means the International Union of Railways (UIC, 16 rue Jean Rey, F-75015 Paris, France);

"UNECE" means the United Nations Economic Commission for Europe (UNECE, Palais des Nations, 8-14 avenue de la Paix, CH-1211 Geneva 10, Switzerland);

"Undertaking", see *"Enterprise"*;

"Unloader" means any enterprise which:

(a)   Removes a container, bulk-container, MEGC, tank-container or portable tank from a vehicle; or

(b)   Unloads packaged dangerous goods, small containers or portable tanks out of or from a vehicle or a container; or

(c)   Discharges dangerous goods from a tank (tank-vehicle, demountable tank, portable tank or tank-container) or from a battery-vehicle, MEMU or MEGC or from a vehicle, large container or small container for carriage in bulk or a bulk-container;

"Unloading" means all actions carried out by the unloader, in accordance with the definition of unloader;

"UN Model Regulations" means the Model Regulations annexed to the nineteenth revised edition of the Recommendations on the Transport of Dangerous Goods published by the United Nations (ST/SG/AC.10/1/Rev.19);

"UN number" means the four-figure identification number of the substance or article taken from the UN Model Regulations;

V

"Vacuum-operated waste tank" means a fixed tank, demountable tank, tank-container or tank swap body primarily used for the carriage of dangerous wastes, with special constructional features and/or equipment to facilitate the filling and discharging of wastes as specified in Chapter 6.10. A tank which fully complies with the requirements of Chapter 6.7 or 6.8 is not considered to be a vacuum-operated waste tank;

"Vacuum valve" means a spring-loaded device which is activated automatically by pressure the purpose of which is to protect the tank against unacceptable negative internal pressure;

"Vehicle" see *"Battery-vehicle"*, *"Closed vehicle"*, *"Open vehicle"*, *"Sheeted vehicle"* and *"Tank-vehicle"*;

W

"Wastes" means substances, solutions, mixtures or articles for which no direct use is envisaged but which are transported for reprocessing, dumping, elimination by incineration or other methods of disposal;

"Wooden barrel" means a packaging made of natural wood, of round cross-section, having convex walls, consisting of staves and heads and fitted with hoops;

"Wooden IBC" means a rigid or collapsible wooden body, together with an inner liner (but no inner packaging) and appropriate service and structural equipment;

"Working pressure" means the settled pressure of a compressed gas at a reference temperature of 15 °C in a full pressure receptacle;

NOTE: For tanks, see "Maximum working pressure".

"Woven plastics" (for flexible IBCs) means a material made from stretch tapes or monofilaments of suitable plastics material.

1.2.2 Units of measurement

1.2.2.1 The following units of measurement [a] are applicable in ADR:

Measurement of	SI Unit [b]	Acceptable alternative unit	Relationship between units
Length	m (metre)	-	-
Area	m^2 (square metre)	-	-
Volume	m^3 (cubic metre)	l [c] (litre)	1 l = 10^{-3} m^3
Time	s (second)	min (minute)	1 min = 60 s
		h (hour)	1 h = 3 600 s
		d (day)	1 d = 86 400 s
Mass	kg (kilogram)	g (gram)	1g = 10^{-3} kg
		t (ton)	1 t = 10^3 kg
Mass density	kg/m^3	kg/l	1 kg/l = 10^3 kg/m^3
Temperature	K (kelvin)	°C (degree Celsius)	0 °C = 273.15 K
Temperature difference	K (kelvin)	°C (degree Celsius)	1 °C = 1 K
Force	N (newton)	-	1 N = 1 kg.m/s^2
Pressure	Pa (pascal)		1 Pa = 1 N/m^2
		bar (bar)	1 bar = 10^5 Pa
Stress	N/m^2	N/mm^2	1 N/mm^2 = 1 MPa
Work		kWh (kilowatt hours)	1 kWh = 3.6 MJ
Energy	J (joule)		1 J = 1 N.m = 1 W.s
Quantity of heat		eV (electronvolt)	1 eV = 0.1602 H 10^{-18}J
Power	W (watt)	-	1 W = 1 J/s = 1 N.m/s
Kinematic viscosity	m^2/s	mm^2/s	1 mm^2/s = 10^{-6} m^2/s
Dynamic viscosity	Pa.s	mPa.s	1 mPa.s = 10^{-3} Pa.s
Activity	Bq (becquerel)		
Dose equivalent	Sv (sievert)		

[a]   *The following round figures are applicable for the conversion of the units hitherto used into SI Units.*

Force
1 kg	=	9.807 N
1 N	=	0.102 kg

Stress
1 kg/mm^2	=	9.807 N/mm^2
1 N/mm^2	=	0.102 kg/mm^2

Pressure

1 Pa	=	1 N/m^2	=	10^{-5} bar	=	1.02 × 10^{-5} kg/cm^2	=	0.75 × 10^{-2} torr
1 bar	=	10^5 Pa	=	1.02 kg/cm^2	=	750 torr		
1 kg/cm^2	=	9.807 × 10^4 Pa	=	0.9807 bar	=	736 torr		
1 torr	=	1.33 × 10^2 Pa	=	1.33 × 10^{-3} bar	=	1.36 × 10^{-3} kg/cm^2		

Energy, Work, Quantity of heat

1 J	=	1 N.m	=	0.278 × 10^{-6} kWh	=	0.102 kgm	=	0.239 × 10^{-3} kcal
1 kWh	=	3.6 × 10^6 J	=	367 × 10^3 kgm	=	860 kcal		
1 kgm	=	9.807 J	=	2.72 × 10^{-6} kWh	=	2.34 × 10^{-3} kcal		
1 kcal	=	4.19 × 10^3 J	=	1.16 × 10^{-3} kWh	=	427 kgm		

Power
1 W	=	0.102 kgm/s	=	0.86 kcal/h
1 kgm/s	=	9.807 W	=	8.43 kcal/h
1 kcal/h	=	1.16 W	=	0.119 kgm/s

Kinematic viscosity
1 m^2/s	=	10^4 St (Stokes)
1 St	=	10^{-4} m^2/s

Dynamic viscosity

1 Pa.s	=	1 N.s/m^2	=	10 P (poise)	=	0.102 kg.s/m^2
1 P	=	0.1 Pa.s	=	0.1 N.s/m^2	=	1.02 × 10^{-2} kg.s/m^2
1 kg.s/m^2	=	9.807 Pa.s	=	9.807 N.s/m^2	=	98.07 P

[b]   *The International System of Units (SI) is the result of decisions taken at the General Conference on Weights and Measures (Address: Pavillon de Breteuil, Parc de St-Cloud, F-92 310 Sèvres).*

[c]   *The abbreviation "L" for litre may also be used in place of the abbreviation "l" when a typewriter cannot distinguish between figure "1" and letter "l".*

The decimal multiples and sub-multiples of a unit may be formed by prefixes or symbols, having the following meanings, placed before the name or symbol of the unit:

Factor		Prefix	Symbol
1 000 000 000 000 000 000	$= 10^{18}$ quintillion	exa	E
1 000 000 000 000 000	$= 10^{15}$ quadrillion	peta	P
1 000 000 000 000	$= 10^{12}$ trillion	tera	T
1 000 000 000	$= 10^{9}$ billion	giga	G
1 000 000	$= 10^{6}$ million	mega	M
1 000	$= 10^{3}$ thousand	kilo	k
100	$= 10^{2}$ hundred	hecto	h
10	$= 10^{1}$ ten	deca	da
0.1	$= 10^{-1}$ tenth	deci	d
0.01	$= 10^{-2}$ hundredth	centi	c
0.001	$= 10^{-3}$ thousandth	milli	m
0.000 001	$= 10^{-6}$ millionth	micro	μ
0.000 000 001	$= 10^{-9}$ billionth	nano	n
0.000 000 000 001	$= 10^{-12}$ trillionth	pico	p
0.000 000 000 000 001	$= 10^{-15}$ quadrillionth	femto	f
0.000 000 000 000 000 001	$= 10^{-18}$ quintillionth	atto	a

NOTE: *10^{9} billion is United Nations usage in English. By analogy, so is $10^{-9} = 1$ billionth.*

1.2.2.2 Unless expressly stated otherwise, the sign "%" in ADR represents:

(a) In the case of mixtures of solids or of liquids, and also in the case of solutions and of solids wetted by a liquid, a percentage mass based on the total mass of the mixture, the solution or the wetted solid;

(b) In the case of mixtures of compressed gases, when filled by pressure, the proportion of the volume indicated as a percentage of the total volume of the gaseous mixture, or, when filled by mass, the proportion of the mass indicated as a percentage of the total mass of the mixture;

(c) In the case of mixtures of liquefied gases and dissolved gases, the proportion of the mass indicated as a percentage of the total mass of the mixture.

1.2.2.3 Pressures of all kinds relating to receptacles (such as test pressure, internal pressure, safety valve opening pressure) are always indicated in gauge pressure (pressure in excess of atmospheric pressure); however, the vapour pressure of substances is always expressed in absolute pressure.

1.2.2.4 Where ADR specifies a degree of filling for receptacles, this is always related to a reference temperature of the substances of 15 °C, unless some other temperature is indicated.

CHAPTER 1.3

TRAINING OF PERSONS INVOLVED IN THE CARRIAGE OF DANGEROUS GOODS

1.3.1   **Scope and applicability**

Persons employed by the participants referred to in Chapter 1.4, whose duties concern the carriage of dangerous goods, shall be trained in the requirements governing the carriage of such goods appropriate to their responsibilities and duties. Employees shall be trained in accordance with 1.3.2 before assuming responsibilities and shall only perform functions, for which required training has not yet been provided, under the direct supervision of a trained person. Training requirements specific to security of dangerous goods in Chapter 1.10 shall also be addressed.

NOTE 1:   With regard to the training for the safety adviser, see 1.8.3 instead of this section.

NOTE 2:   With regard to the training of the vehicle crew, see Chapter 8.2 instead of this section.

NOTE 3:   For training with regard to Class 7, see also 1.7.2.5.

1.3.2   **Nature of the training**

The training shall take the following form, appropriate to the responsibility and duties of the individual concerned.

1.3.2.1   *General awareness training*

Personnel shall be familiar with the general requirements of the provisions for the carriage of dangerous goods.

1.3.2.2   *Function-specific training*

Personnel shall be trained, commensurate directly with their duties and responsibilities in the requirements of the regulations concerning the carriage of dangerous goods.

Where the carriage of dangerous goods involves a multimodal transport operation, the personnel shall be aware of the requirements concerning other transport modes.

1.3.2.3   *Safety training*

Commensurate with the degree of risk of injury or exposure arising from an incident involving the carriage of dangerous goods, including loading and unloading, personnel shall be trained in the hazards and dangers presented by dangerous goods.

The training provided shall aim to make personnel aware of the safe handling and emergency response procedures.

1.3.2.4   The training shall be periodically supplemented with refresher training to take account of changes in regulations.

1.3.3   **Documentation**

Records of training received according to this Chapter shall be kept by the employer and made available to the employee or competent authority, upon request. Records shall be kept by the employer for a period of time established by the competent authority. Records of training shall be verified upon commencing a new employment.

CHAPTER 1.4

SAFETY OBLIGATIONS OF THE PARTICIPANTS

1.4.1   **General safety measures**

1.4.1.1   The participants in the carriage of dangerous goods shall take appropriate measures according to the nature and the extent of foreseeable dangers, so as to avoid damage or injury and, if necessary, to minimize their effects. They shall, in all events, comply with the requirements of ADR in their respective fields.

1.4.1.2   When there is an immediate risk that public safety may be jeopardized, the participants shall immediately notify the emergency services and shall make available to them the information they require to take action.

1.4.1.3   ADR may specify certain of the obligations falling to the various participants.

If a Contracting Party considers that no lessening of safety is involved, it may in its domestic legislation transfer the obligations falling to a specific participant to one or several other participants, provided that the obligations of 1.4.2 and 1.4.3 are met. These derogations shall be communicated by the Contracting Party to the Secretariat of the United Nations Economic Commission for Europe which will bring them to the attention of the Contracting Parties.

The requirements of 1.2.1, 1.4.2 and 1.4.3 concerning the definitions of participants and their respective obligations shall not affect the provisions of domestic law concerning the legal consequences (criminal nature, liability, etc.) stemming from the fact that the participant in question is e.g. a legal entity, a self-employed worker, an employer or an employee.

1.4.2   **Obligations of the main participants**

NOTE 1: Several participants to which safety obligations are assigned in this section may be one and the same enterprise. Also, the activities and the corresponding safety obligations of a participant can be assumed by several enterprises.

NOTE 2: For radioactive material, see also 1.7.6.

1.4.2.1   *Consignor*

1.4.2.1.1   The consignor of dangerous goods is required to hand over for carriage only consignments which conform to the requirements of ADR. In the context of 1.4.1, he shall in particular:

   (a)   Ascertain that the dangerous goods are classified and authorized for carriage in accordance with ADR;

   (b)   Furnish the carrier with information and data in a traceable form and, if necessary, the required transport documents and accompanying documents (authorizations, approvals, notifications, certificates, etc.), taking into account in particular the requirements of Chapter 5.4 and of the tables in Part 3;

   (c)   Use only packagings, large packagings, intermediate bulk containers (IBCs) and tanks (tank-vehicles, demountable tanks, battery-vehicles, MEGCs, portable tanks and tank-containers) approved for and suited to the carriage of the substances concerned and bearing the marks prescribed by ADR;

   (d)   Comply with the requirements on the means of dispatch and on forwarding restrictions;

   (e)   Ensure that even empty uncleaned and not degassed tanks (tank-vehicles, demountable tanks, battery-vehicles, MEGCs, portable tanks and tank-containers) or empty uncleaned vehicles and bulk containers are placarded, marked and labelled in accordance with Chapter 5.3 and that empty uncleaned tanks are closed and present the same degree of leakproofness as if they were full.

1.4.2.1.2   If the consignor uses the services of other participants (packer, loader, filler, etc.), he shall take appropriate measures to ensure that the consignment meets the requirements of ADR. He may, however, in the case of 1.4.2.1.1 (a), (b), (c) and (e), rely on the information and data made available to him by other participants.

1.4.2.1.3   When the consignor acts on behalf of a third party, the latter shall inform the consignor in writing that dangerous goods are involved and make available to him all the information and documents he needs to perform his obligations.

1.4.2.2   *Carrier*

1.4.2.2.1   In the context of 1.4.1, where appropriate, the carrier shall in particular:

(a)   Ascertain that the dangerous goods to be carried are authorized for carriage in accordance with ADR;

(b)   Ascertain that all information prescribed in ADR related to the dangerous goods to be carried has been provided by the consignor before carriage, that the prescribed documentation is on board the transport unit or if electronic data processing (EDP) or if electronic data interchange (EDI) techniques are used instead of paper documentation, that data is available during transport in a manner at least equivalent to that of paper documentation;

(c)   Ascertain visually that the vehicles and loads have no obvious defects, leakages or cracks, missing equipment, etc.;

(d)   Ascertain that the deadline for the next test for tank-vehicles, battery-vehicles, demountable tanks, portable tanks, tank-containers and MEGCs has not expired;

NOTE: Tanks, battery-vehicles and MEGCs may however be carried after the expiry of this deadline under the conditions of 4.1.6.10 (in the case of battery-vehicles and MEGCs containing pressure receptacles as elements), 4.2.4.4, 4.3.2.3.7, 4.3.2.4.4, 6.7.2.19.6, 6.7.3.15.6 or 6.7.4.14.6.

(e)   verify that the vehicles are not overloaded;

(f)   ascertain that the placards, marks and orange-coloured plates prescribed for the vehicles in Chapter 5.3 have been affixed;

(g)   ascertain that the equipment prescribed in ADR for the transport unit, vehicle crew and certain classes is on board the transport unit.

Where appropriate, this shall be done on the basis of the transport documents and accompanying documents, by a visual inspection of the vehicle or the containers and, where appropriate, the load.

1.4.2.2.2   The carrier may, however, in the case of 1.4.2.2.1 (a), (b), (e) and (f), rely on information and data made available to him by other participants.

1.4.2.2.3   If the carrier observes an infringement of the requirements of ADR, in accordance with 1.4.2.2.1, he shall not forward the consignment until the matter has been rectified.

1.4.2.2.4   If, during the journey, an infringement which could jeopardize the safety of the operation is observed, the consignment shall be halted as soon as possible bearing in mind the requirements of traffic safety, of the safe immobilisation of the consignment, and of public safety. The transport operation may only be continued once the consignment complies with applicable regulations. The competent authority(ies) concerned by the rest of the journey may grant an authorization to pursue the transport operation.

In case the required compliance cannot be achieved and no authorization is granted for the rest of the journey, the competent authority(ies) shall provide the carrier with the necessary administrative assistance. The same shall apply in case the carrier informs this/these competent authority(ies) that the dangerous nature of the goods carried was not communicated to him by the consignor and that he wishes, by virtue of the law applicable in particular to the contract of carriage, to unload, destroy or render the goods harmless.

1.4.2.2.5   *(Reserved)*

1.4.2.2.6   The carrier shall provide the vehicle crew with the instructions in writing as prescribed in ADR.

1.4.2.3   *Consignee*

1.4.2.3.1   The consignee has the obligation not to defer acceptance of the goods without compelling reasons and to verify, after unloading, that the requirements of ADR concerning him have been complied with.

1.4.2.3.2   If, in the case of a container, this verification brings to light an infringement of the requirements of ADR, the consignee shall return the container to the carrier only after the infringement has been remedied.

1.4.2.3.3   If the consignee makes use of the services of other participants (unloader, cleaner, decontamination facility, etc.) he shall take appropriate measures to ensure that the requirements of 1.4.2.3.1 and 1.4.2.3.2 of ADR have been complied with.

1.4.3   **Obligations of the other participants**

A non-exhaustive list of the other participants and their respective obligations is given below. The obligations of the other participants flow from section 1.4.1 above insofar as they know or should have known that their duties are performed as part of a transport operation subject to ADR.

1.4.3.1   *Loader*

1.4.3.1.1   In the context of 1.4.1, the loader has the following obligations in particular:

(a)   He shall hand the dangerous goods over to the carrier only if they are authorized for carriage in accordance with ADR;

(b)   He shall, when handing over for carriage packed dangerous goods or uncleaned empty packagings, check whether the packaging is damaged. He shall not hand over a package the packaging of which is damaged, especially if it is not leakproof, and there are leakages or the possibility of leakages of the dangerous substance, until the damage has been repaired; this obligation also applies to empty uncleaned packagings;

(c)   He shall comply with the special requirements concerning loading and handling;

(d)   He shall, after loading dangerous goods into a container comply with the requirements concerning placarding, marking and orange-coloured plates conforming to Chapter 5.3;

(e)   He shall, when loading packages, comply with the prohibitions on mixed loading taking into account dangerous goods already in the vehicle or large container and requirements concerning the separation of foodstuffs, other articles of consumption or animal feedstuffs.

1.4.3.1.2   The loader may, however, in the case of 1.4.3.1.1 (a), (d) and (e), rely on information and data made available to him by other participants.

1.4.3.2   *Packer*

In the context of 1.4.1, the packer shall comply with in particular:

(a)   The requirements concerning packing conditions, or mixed packing conditions; and

(b)   When he prepares packages for carriage, the requirements concerning marking and labelling of the packages.

1.4.3.3   *Filler*

In the context of 1.4.1, the filler has the following obligations in particular:

(a)   He shall ascertain prior to the filling of tanks that both they and their equipment are technically in a satisfactory condition;

(b)   He shall ascertain that the date of the next test for tank-vehicles, battery-vehicles, demountable tanks, portable tanks, tank-containers and MEGCs has not expired;

(c)   He shall only fill tanks with the dangerous goods authorized for carriage in those tanks;

(d)   He shall, in filling the tank, comply with the requirements concerning dangerous goods in adjoining compartments;

(e)   He shall, during the filling of the tank, observe the maximum permissible degree of filling or the maximum permissible mass of contents per litre of capacity for the substance being filled;

(f)   He shall, after filling the tank, ensure that all closures are in a closed position and that there is no leakage;

(g)   He shall ensure that no dangerous residue of the filling substance adheres to the outside of the tanks filled by him;

(h)   He shall, in preparing the dangerous goods for carriage, ensure that the placards, marks, orange-coloured plates and labels are affixed on the tanks, on the vehicles and on the containers for carriage in bulk in accordance with Chapter 5.3;

(i)   *(Reserved)*;

(j)   He shall, when filling vehicles or containers with dangerous goods in bulk, ascertain that the relevant provisions of Chapter 7.3 are complied with.

1.4.3.4   ***Tank-container/portable tank operator***

In the context of 1.4.1, the tank-container/portable tank operator shall in particular:

(a)   Ensure compliance with the requirements for construction, equipment, tests and marking;

(b)   Ensure that the maintenance of shells and their equipment is carried out in such a way as to ensure that, under normal operating conditions, the tank-container/portable tank satisfies the requirements of ADR until the next inspection;

(c)   Have an exceptional check made when the safety of the shell or its equipment is liable to be impaired by a repair, an alteration or an accident.

1.4.3.5 and 1.4.3.6   *(Reserved)*

1.4.3.7   ***Unloader***

1.4.3.7.1   In the context of 1.4.1, the unloader shall in particular:

(a)   Ascertain that the correct goods are unloaded by comparing the relevant information on the transport document with the information on the package, container, tank, MEMU, MEGC or vehicle;

(b)   Before and during unloading, check whether the packagings, the tank, the vehicle or container have been damaged to an extent which would endanger the unloading operation. If this is the case, ascertain that unloading is not carried out until appropriate measures have been taken;

(c)   Comply with all relevant requirements concerning unloading and handling;

(d)   Immediately following the unloading of the tank, vehicle or container:

    (i)   Remove any dangerous residues which have adhered to the outside of the tank, vehicle or container during the process of unloading; and

    (ii)   Ensure the closure of valves and inspection openings;

(e)   Ensure that the prescribed cleaning and decontamination of the vehicles or containers is carried out; and

(f)   Ensure that the containers once completely unloaded, cleaned and decontaminated, no longer display the placards, marks and orange-coloured plates that had been displayed in accordance with Chapter 5.3.

1.4.3.7.2   If the unloader makes use of the services of other participants (cleaner, decontamination facility, etc.) he shall take appropriate measures to ensure that the requirements of ADR have been complied with.

CHAPTER 1.5

DEROGATIONS

1.5.1   **Temporary derogations**

1.5.1.1   In accordance with Article 4, paragraph 3 of ADR, the competent authorities of the Contracting Parties may agree directly among themselves to authorize certain transport operations in their territories by temporary derogation from the requirements of ADR, provided that safety is not compromised thereby. The authority which has taken the initiative with respect to the temporary derogation shall notify such derogations to the Secretariat of the United Nations Economic Commission for Europe which shall bring them to the attention of the Contracting Parties[1].

NOTE: "Special arrangement" in accordance with 1.7.4 is not considered to be a temporary derogation in accordance with this section.

1.5.1.2   The period of validity of the temporary derogation shall not be more than five years from the date of its entry into force. The temporary derogation shall automatically cease as from the date of the entry into force of a relevant amendment to ADR.

1.5.1.3   Transport operations on the basis of temporary derogations shall constitute transport operations in the sense of ADR.

1.5.2   **(Reserved)**

---

[1]   ***Note by the Secretariat:*** *The special agreements concluded under this Chapter may be consulted on the web site of the Secretariat of the United Nations Economic Commission for Europe (http://www.unece.org/trans/danger/danger.htm).*

CHAPTER 1.6

TRANSITIONAL MEASURES

1.6.1   **General**

1.6.1.1   Unless otherwise provided, the substances and articles of ADR may be carried until 30 June 2017 in accordance with the requirements of ADR applicable up to 31 December 2016.

1.6.1.2   *(Deleted)*

1.6.1.3   Substances and articles of Class 1, belonging to the armed forces of a Contracting Party, that were packaged prior to 1 January 1990 in accordance with the requirements of ADR in effect at that time may be carried after 31 December 1989 provided the packagings maintain their integrity and are declared in the transport document as military goods packaged prior to 1 January 1990. The other requirements applicable as from 1 January 1990 for this class shall be complied with.

1.6.1.4   Substances and articles of Class 1 that were packaged between 1 January 1990 and 31 December 1996 in accordance with the requirements of ADR in effect at that time may be carried after 31 December 1996, provided the packagings maintain their integrity and are declared in the transport document as goods of Class 1 packaged between 1 January 1990 and 31 December 1996.

1.6.1.5   *(Reserved)*

1.6.1.6   Intermediate bulk containers (IBCs) manufactured before 1 January 2003 in accordance with the requirements of marginal 3612 (1) applicable up to 30 June 2001 and which do not conform to the requirements of 6.5.2.1.1 regarding the height of letters, numerals and symbols applicable as from 1 July 2001 may continue to be used.

1.6.1.7   Type approvals for drums, jerricans and composite packagings made of high or medium molecular mass polyethylene issued before 1 July 2005 in accordance with the requirements of 6.1.5.2.6 in force up to 31 December 2004, but which are not in accordance with the requirements of 4.1.1.21, continue to be valid until 31 December 2009. Any such packagings manufactured and marked on the basis of these type approvals may be used until the end of their period of use determined in 4.1.1.15.

1.6.1.8   Existing orange-coloured plates which meet the requirements of sub-section 5.3.2.2 applicable up to 31 December 2004 may continue to be used provided that the requirements of 5.3.2.2.1 and 5.3.2.2.2 that the plate, numbers and letters shall remain affixed irrespective of the orientation of the vehicle are met.

1.6.1.9 and 1.6.1.10   *(Deleted)*

1.6.1.11   Type approvals for drums, jerricans and composite packagings made of high or medium molecular mass polyethylene, and for high molecular mass polyethylene IBCs, issued before 1 July 2007 in accordance with the requirements of 6.1.6.1 (a) in force up to 31 December 2006, but which are not in accordance with the requirements of 6.1.6.1 (a) applicable as from 1 January 2007, continue to be valid.

1.6.1.12 and 1.6.1.13   *(Deleted)*

1.6.1.14   IBCs manufactured before 1 January 2011 and conforming to a design type which has not passed the vibration test of 6.5.6.13 or which was not required to meet the criteria of 6.5.6.9.5 (d) at the time it was subjected to the drop test, may still be used.

1.6.1.15   IBCs manufactured, remanufactured or repaired before 1 January 2011 need not be marked with the maximum permitted stacking load in accordance with 6.5.2.2.2. Such IBCs, not marked in accordance with 6.5.2.2.2, may still be used after 31 December 2010 but must be marked in accordance with 6.5.2.2.2 if they are remanufactured or repaired after that date. IBCs manufactured, remanufactured or repaired between 1 January 2011 and 31 December 2016 and marked with the maximum permitted stacking load in accordance with 6.5.2.2.2 in force up to 31 December 2014 may continue to be used.

1.6.1.16 to 1.6.1.20   *(Deleted)*

1.6.1.21	Training certificates for drivers conforming to the model applicable until 31 December 2010 issued by Contracting Parties until 31 December 2012, instead of those conforming to the requirements of 8.2.2.8.5, may continue in use to the end of their five year validity.
1.6.1.22	Inner receptacles of composite IBCs manufactured before 1 July 2011 and marked in accordance with the requirements of 6.5.2.2.4 in force up to 31 December 2010 may still be used.
1.6.1.23	Fire extinguishers constructed before 1 July 2011 in accordance with the requirements of 8.1.4.3 applicable until 31 December 2010 may continue to be used.
1.6.1.24	*(Deleted)*
1.6.1.25	Cylinders of 60 litres water capacity or less marked with a UN number in accordance with the provisions of ADR applicable up to 31 December 2012 and which do not conform to the requirements of 5.2.1.1 regarding the size of the UN number and of the letters "UN" applicable as from 1 January 2013 may continue to be used until the next periodic inspection but no later than 30 June 2018.
1.6.1.26	Large packagings manufactured or remanufactured before 1 January 2014 and which do not conform to the requirements of 6.6.3.1 regarding the height of letters, numerals and symbols applicable as from 1 January 2013 may continue to be used. Those manufactured or remanufactured before 1 January 2015 need not be marked with the maximum permitted stacking load in accordance with 6.6.3.3. Such large packagings not marked in accordance with 6.6.3.3 may still be used after 31 December 2014 but must be marked in accordance with 6.6.3.3 if they are remanufactured after that date. Large packagings manufactured or remanufactured between 1 January 2011 and 31 December 2016 and marked with the maximum permitted stacking load in accordance with 6.6.3.3 in force up to 31 December 2014 may continue to be used.
1.6.1.27	Means of containment integral to equipment or machinery containing liquid fuels of UN Nos. 1202, 1203, 1223, 1268, 1863 and 3475 constructed before 1 July 2013, which do not conform to the requirements of paragraph (a) of special provision 363 of Chapter 3.3 applicable as from 1 January 2013, may still be used.
1.6.1.28	*(Deleted)*
1.6.1.29	Lithium cells and batteries manufactured according to a type meeting the requirements of sub-section 38.3 of the Manual of Tests and Criteria, Revision 3, Amendment 1 or any subsequent revision and amendment applicable at the date of the type testing may continue to be carried, unless otherwise provided in ADR. Lithium cells and batteries manufactured before 1 July 2003 meeting the requirements of the Manual of Tests and Criteria, Revision 3, may continue to be carried if all other applicable requirements are fulfilled.
1.6.1.30	Labels which meet the requirements of 5.2.2.2.1.1 applicable up to 31 December 2014, may continue to be used until 30 June 2019.
1.6.1.31 and 1.6.1.32	*(Deleted)*
1.6.1.33	Electric double layer capacitors of UN No. 3499, manufactured before 1 January 2014, need not be marked with the energy storage capacity in Wh as required by sub-paragraph (e) of special provision 361 of Chapter 3.3.
1.6.1.34	Asymmetric capacitors of UN No. 3508, manufactured before 1 January 2016, need not be marked with the energy storage capacity in Wh as required by sub-paragraph (c) of special provision 372 of Chapter 3.3.
1.6.1.35	The instructions in writing in accordance with the requirements of ADR applicable up to 31 December 2014, but which do not however conform to the requirements of 5.4.3 applicable as from 1 January 2015, may continue to be used until 30 June 2017.
1.6.1.36	Driver training certificates issued before 1 January 2014 that do not comply with the requirements under 8.2.2.8.5 applicable from 1 January 2013 in respect of the sequence used for the presentation of dates under numbers 4. and 8., the colour (white with black lettering), and the use of the numbers 9. and 10. on the back of the certificate to introduce the corresponding lists of classes for which the certificate is valid, may continue to be used until their date of expiry.

1.6.1.37	*(Reserved)*
1.6.1.38	Contracting Parties may continue to issue training certificates for dangerous goods safety advisers conforming to the model applicable until 31 December 2016, instead of those conforming to the requirements of 1.8.3.18 applicable from 1 January 2017, until 31 December 2018. Such certificates may continue in use to the end of their five year validity.
1.6.1.39	Notwithstanding the requirements of special provision 188 of Chapter 3.3 applicable as from 1 January 2017, packages containing lithium cells or batteries may continue to be marked until 31 December 2018 in accordance with the requirements of special provision 188 of Chapter 3.3 in force up to 31 December 2016
1.6.1.40	Notwithstanding the requirements of ADR applicable as from 1 January 2017, articles of UN Nos. 0015, 0016 and 0303 containing smoke-producing substance(s) toxic by inhalation according to the criteria for Class 6.1 manufactured before 31 December 2016 may be carried until 31 December 2018 without a "TOXIC" subsidiary risk label (model No. 6.1, see 5.2.2.2.2).
1.6.1.41	Notwithstanding the requirements of ADR applicable as from 1 January 2017, large packagings conforming to the packing group III performance level in accordance with special packing provision L2 of packing instruction LP02 of 4.1.4.3 applicable until 31 December 2016 may continue to be used until 31 December 2022 for UN No. 1950.
1.6.1.42	Notwithstanding the requirements of column (5) of Table A of Chapter 3.2 applicable as from 1 January 2017 to UN Nos. 3090, 3091, 3480 and 3481, the Class 9 label (model No 9, see 5.2.2.2.2) may continue to be used for these UN numbers until 31 December 2018.
1.6.1.43	Vehicles registered or brought into service before 1 July 2017, as defined in special provisions 240, 385 and 669 of Chapter 3.3, and their equipment intended for use during carriage, which conform to the requirements of ADR applicable until 31 December 2016 but containing lithium cells and batteries which do not conform to the requirement of 2.2.9.1.7 may continue to be carried as a load in accordance with the requirements of special provision 666 of Chapter 3.3.

1.6.2   Pressure receptacles and receptacles for Class 2

1.6.2.1	Receptacles built before 1 January 1997 and which do not conform to the requirements of ADR applicable as from 1 January 1997, but the carriage of which was permitted under the requirements of ADR applicable up to 31 December 1996, may continue to be transported after that date if the periodic test requirements in packing instructions P200 and P203 are complied with.
1.6.2.2	*(Deleted)*
1.6.2.3	Receptacles intended for the carriage of Class 2 substances constructed before 1 January 2003, may continue to bear, after 1 January 2003, the marks conforming to the requirements applicable until 31 December 2002.
1.6.2.4	Pressure receptacles designed and constructed in accordance with technical codes no longer recognized according to 6.2.5 may still be used.
1.6.2.5	Pressure receptacles and their closures designed and constructed in accordance with standards applicable at the time of their construction (see 6.2.4) according to the provisions of ADR which were applicable at that time may still be used unless restricted by a specific transitional measure.
1.6.2.6	Pressure receptacles for substances other than those of Class 2, built before 1 July 2009 in accordance with the requirements of 4.1.4.4 in force up to 31 December 2008, but which do not conform to the requirements of 4.1.3.6 applicable as from 1 January 2009, may continue to be used provided that the requirements of 4.1.4.4 in force up to 31 December 2008 are complied with.
1.6.2.7 and 1.6.2.8	*(Deleted)*
1.6.2.9	The provisions of packing instruction P200 (10), special packing provision v of 4.1.4.1 applicable until 31 December 2010 may be applied by Contracting Parties to ADR to cylinders constructed before 1 January 2015.

1.6.2.10 Refillable welded steel cylinders for the carriage of gases of UN Nos. 1011, 1075, 1965, 1969 or 1978, granted 15 year intervals for periodic inspection in accordance with packing instruction P200 (10), special packing provision v of 4.1.4.1 as applicable until 31 December 2010 by the competent authority of the country (countries) of carriage, may continue to be periodically inspected according to those provisions.

1.6.2.11 Gas cartridges constructed and prepared for carriage before 1 January 2013 for which the requirements of 1.8.6, 1.8.7 or 1.8.8 for the conformity assessment of gas cartridges have not been applied may still be carried after this date, provided all the other applicable provisions of ADR are met.

1.6.2.12 Salvage pressure receptacles may continue to be constructed and approved according to national regulations up to 31 December 2013. Salvage pressure receptacles constructed and approved in accordance with national regulations before 1 January 2014 may continue to be used with the approval of the competent authorities of the countries of use.

1.6.2.13 Bundles of cylinders manufactured before 1 July 2013 which are not marked in accordance with 6.2.3.9.7.2 and 6.2.3.9.7.3 applicable from 1 January 2013 or 6.2.3.9.7.2 applicable from 1 January 2015 may be used until the next periodic inspection after 1 July 2015.

1.6.2.14 Cylinders constructed before 1 January 2016 in accordance with 6.2.3 and a specification approved by the competent authorities of the countries of transport and use, but not in accordance with ISO 11513:2011 or ISO 9809-1:2010 as required in 4.1.4.1, packing instruction P208 (1), may be used for the carriage of adsorbed gases provided the general packing requirements of 4.1.6.1 are met.

1.6.2.15 Bundles of cylinders periodically inspected before 1 July 2015 which are not marked in accordance with 6.2.3.9.7.3 applicable from 1 January 2015 may be used until the next periodic inspection after 1 July 2015.

1.6.3 Fixed tanks (tank-vehicles), demountable tanks and battery-vehicles

1.6.3.1 Fixed tanks (tank-vehicles), demountable tanks and battery-vehicles built before the entry into force of the requirements applicable as from 1 October 1978 may be kept in service if the equipment of the shell meets the requirements of Chapter 6.8. The thickness of the shell wall, except in the case of shells intended for the carriage of refrigerated liquefied gases of Class 2, shall be appropriate to a calculation pressure of not less than 0.4 MPa (4 bar) (gauge pressure) in the case of mild steel or of not less than 200 kPa (2 bar) (gauge pressure) in the case of aluminium and aluminium alloys. For other than circular cross-sections of tanks, the diameter to be used as a basis for calculation shall be that of a circle whose area is equal to that of the actual cross-section of the tank.

1.6.3.2 The periodic tests for fixed tanks (tank-vehicles), demountable tanks and battery-vehicles kept in service under these transitional requirements shall be conducted in accordance with the requirements of 6.8.2.4 and 6.8.3.4 and with the pertinent special requirements for the various classes. Unless the earlier requirements prescribed a higher test pressure, a test pressure of 200 kPa (2 bar) (gauge pressure) shall suffice for aluminium shells and aluminium alloy shells.

1.6.3.3 Fixed tanks (tank-vehicles), demountable tanks and battery-vehicles which meet the transitional requirements in 1.6.3.1 and 1.6.3.2 may be used until 30 September 1993 for the carriage of the dangerous goods for which they have been approved. This transitional period shall not apply to fixed tanks (tank-vehicles), demountable tanks and battery-vehicles intended for the carriage of substances of Class 2, or to fixed tanks (tank-vehicles), demountable tanks and battery-vehicles whose wall thickness and items of equipment meet the requirements of Chapter 6.8.

1.6.3.4 (a) Fixed tanks (tank-vehicles), demountable tanks and battery-vehicles constructed before 1 May 1985 in accordance with the requirements of ADR in force between 1 October 1978 and 30 April 1985 but not conforming to the requirements applicable as from 1 May 1985 may continue to be used after that date;

(b) Fixed tanks (tank-vehicles), demountable tanks and battery-vehicles, constructed between 1 May 1985 and the entry into force of the requirements applicable as from 1 January 1988 which do not conform to those requirements but were constructed according to the requirements of ADR in force until that date, may continue to be used after that date.

1.6.3.5	Fixed tanks (tank-vehicles), demountable tanks and battery-vehicles, constructed before 1 January 1993 in accordance with the requirements in force up to 31 December 1992 but which do not conform to the requirements applicable as from 1 January 1993 may still be used.

1.6.3.6   (a)   Fixed tanks (tank-vehicles), demountable tanks and battery-vehicles constructed between 1 January 1978 and 31 December 1984, if used after 31 December 2004, shall conform to the requirements of marginal 211 127 (5), applicable as from 1 January 1990, concerning shell thickness and protection against damage;

    (b)   Fixed tanks (tank-vehicles), demountable tanks and battery-vehicles constructed between 1 January 1985 and 31 December 1989, if used after 31 December 2010, shall conform to the requirements of marginal 211 127 (5), applicable as from 1 January 1990, concerning shell thickness and protection against damage.

1.6.3.7   Fixed tanks (tank-vehicles), demountable tanks and battery-vehicles constructed before 1 January 1999 in accordance with the requirements in force up to 31 December 1998 but which do not, however, conform to the requirements applicable as from 1 January 1999 may still be used.

1.6.3.8   When, because of amendments to ADR, some proper shipping names of gases have been modified, it is not necessary to modify the names on the plate or on the shell itself (see 6.8.3.5.2 or 6.8.3.5.3), provided that the names of the gases on the fixed tanks (tank-vehicles), demountable tanks and battery-vehicles or on the plates (see 6.8.3.5.6 (b) or (c)) are adapted at the first periodic test thereafter.

1.6.3.9 and 1.6.3.10   *(Reserved)*

1.6.3.11   Fixed tanks (tank-vehicles) and demountable tanks constructed before 1 January 1997 in accordance with the requirements in force up to 31 December 1996 but which do not, however, conform to the requirements of marginals 211 332 and 211 333 applicable as from 1 January 1997, may still be used.

1.6.3.12   *(Reserved)*

1.6.3.13   *(Deleted)*

1.6.3.14   *(Reserved)*

1.6.3.15   *(Deleted)*

1.6.3.16   For fixed tanks (tank-vehicles), demountable tanks and battery-vehicles constructed before 1 January 2007 which do not conform to the requirements of 4.3.2, 6.8.2.3, 6.8.2.4 and 6.8.3.4 concerning the tank record, the retention of files for the tank record shall start at the latest at the next periodic inspection.

1.6.3.17   Fixed tanks (tank-vehicles) and demountable tanks intended for the carriage of substances of Class 3, packing group I having a vapour pressure of not more than 175 kPa (1.75 bar) (absolute) at 50 °C, constructed before 1 July 2007 in accordance with the requirements applicable up to 31 December 2006, to which tank code L1.5BN had been assigned in accordance with the requirements applicable up to 31 December 2006, may continue to be used for the carriage of the substances mentioned above, until 31 December 2018.

1.6.3.18   Fixed tanks (tank-vehicles), demountable tanks and battery-vehicles constructed before 1 January 2003 in accordance with the requirements in force up to 30 June 2001, but which do not, however, conform to the requirements applicable as from 1 July 2001, may still be used provided that the assignment to the relevant tank code has been carried out.

1.6.3.19   Fixed tanks (tank-vehicles) and demountable tanks constructed before 1 January 2003 in accordance with the requirements of 6.8.2.1.21 in force up to 31 December 2002 but which do not, however, conform to the requirements applicable as from 1 January 2003 may still be used.

1.6.3.20   Fixed tanks (tank-vehicles) and demountable tanks constructed before 1 July 2003 in accordance with the requirements in force up to 31 December 2002 but which do not, however, conform to the requirements of 6.8.2.1.7 applicable as from 1 January 2003 and special provision TE15 of 6.8.4 (b) applicable from 1 January 2003 to 31 December 2006 may still be used.

1.6.3.21   *(Deleted)*

1.6.3.22 to 1.6.3.24   *(Reserved)*

1.6.3.25   *(Deleted)*

1.6.3.26   Fixed tanks (tank-vehicles) and demountable tanks constructed before 1 January 2007 in accordance with the requirements in force up to 31 December 2006 but which do not, however, conform to the requirements applicable as from 1 January 2007 regarding the marking of the external design pressure in accordance with 6.8.2.5.1, may still be used.

1.6.3.27 to 1.6.3.29   *(Reserved)*

1.6.3.30   Vacuum-operated waste fixed tanks (tank-vehicles) and demountable tanks constructed before 1 July 2005 in accordance with the requirements applicable up to 31 December 2004 but which do not conform to the requirements of 6.10.3.9 applicable as from 1 January 2005, may still be used.

1.6.3.31   Fixed tanks (tank-vehicles), demountable tanks and tanks forming elements of battery-vehicles designed and constructed in accordance with a technical code which was recognized at the time of their construction according to the provisions of 6.8.2.7 which were applicable at that time may still be used.

1.6.3.32   Fixed tanks (tank vehicles) and demountable tanks constructed before 1 July 2007 in accordance with the requirements in force up to 31 December 2006, equipped with manhole cover assemblies in accordance with the provisions of standard EN 13317:2002 referred to in the table of paragraph 6.8.2.6, applicable until 31 December 2006, including those of the figure and table B.2 of annex B of the said standard which are no longer accepted as from 1 January 2007, or the material of which does not meet the requirements of EN 13094:2004, paragraph 5.2, may still be used.

1.6.3.33   When the shell of a fixed tank (tank-vehicle) or demountable tank was already divided by partitions or surge plates into sections of not more than 7 500 litres capacity before 1 January 2009, the capacity of the shell need not be supplemented with the symbol "S" in the particulars required by 6.8.2.5.1 until the next periodic inspection according to 6.8.2.4.2 is performed.

1.6.3.34   Notwithstanding the provisions of 4.3.2.2.4, fixed tanks (tank-vehicles) and demountable tanks intended for the carriage of liquefied gases or refrigerated liquefied gases, which meet the applicable construction requirements of ADR but which were divided, before 1 July 2009, by partitions or surge plates into sections of more than 7 500 litres capacity may still be filled to more than 20% and less than 80% of their capacity.

1.6.3.35   *(Deleted)*

1.6.3.36   Fixed tanks (tank-vehicles) intended for the carriage of liquefied non-toxic flammable gases constructed before 1 July 2011 and which are equipped with non-return valves instead of internal stop-valves and which do not conform to the requirements of 6.8.3.2.3, may still be used.

1.6.3.37   *(Deleted)*

1.6.3.38   Fixed tanks (tank-vehicles), demountable tanks and battery-vehicles designed and constructed in accordance with standards applicable at the time of their construction (see 6.8.2.6 and 6.8.3.6) according to the provisions of ADR which were applicable at that time may still be used unless restricted by a specific transitional measure.

1.6.3.39   Fixed tanks (tank-vehicles) and demountable tanks constructed before 1 July 2011 in accordance with the requirements of 6.8.2.2.3 in force up to 31 December 2010 but which do not, however, conform to the requirements of 6.8.2.2.3, third paragraph, concerning the position of the flame trap or flame arrester may still be used.

1.6.3.40   *(Deleted)*

1.6.3.41	Fixed tanks (tank-vehicles) and demountable tanks constructed before 1 July 2013 in accordance with the requirements in force up to 31 December 2012, but which do not, however, meet the marking provisions of 6.8.2.5.2 or 6.8.3.5.6 applicable as from 1 January 2013, may continue to be marked in accordance with the requirements applicable up to 31 December 2012 until the next periodic inspection after 1 July 2013.
1.6.3.42	For UN No. 2381, the tank code specified in column (12) of Table A of Chapter 3.2 applicable up to 31 December 2012 may continue to be applied until 31 December 2018 for fixed tanks (tank-vehicles) and demountable tanks constructed before 1 July 2013.
1.6.3.43	Fixed tanks (tank-vehicles) and demountable tanks constructed before 1 January 2012 in accordance with the requirements in force up to 31 December 2012, but which do not however conform to the requirements of 6.8.2.6 relating to standards EN 14432:2006 and EN 14433:2006 applicable as from 1 January 2011, may still be used.
1.6.3.44	Fixed tanks (tank-vehicles) and demountable tanks intended for the carriage of UN Nos. 1202, 1203, 1223, 3475 and aviation fuel classified under UN Nos. 1268 or 1863, equipped with additive devices designed and constructed before 1 July 2015 in accordance with national provisions, but which do not, however, conform to the construction, approval and testing requirements of special provision 664 of Chapter 3.3 applicable as from 1 January 2015, may continue to be used until their first intermediate or periodic inspection after 31 December 2015. After this date, they shall only be used with the agreement of the competent authorities in the countries of use.
1.6.3.45	*(Reserved)*
1.6.3.46	Fixed tanks (tank-vehicles) and demountable tanks constructed before 1 July 2017 in accordance with the requirements in force up to 31 December 2016 but which do not however conform to the requirements of 6.8.2.1.23 applicable as from 1 January 2017 may still be used.
1.6.3.47 to 1.6.3.49	*(Reserved)*

1.6.3.50   ***Fibre-reinforced plastics (FRP) tanks***

FRP tanks which have been constructed before 1 July 2002 in conformity with a design type approved before 1 July 2001 in accordance with the requirements of Appendix B.1c which were in force until 30 June 2001 may continue to be used until the end of their lifetime provided that all the requirements in force up to 30 June 2001 have been and continue to be complied with.

However, as from 1 July 2001, no new design type may be approved in accordance with the requirements in force until 30 June 2001.

1.6.4   **Tank-containers, portable tanks and MEGCs**

1.6.4.1	Tank-containers constructed before 1 January 1988 in accordance with the requirements in force up to 31 December 1987 but which do not, however, conform to the requirements applicable as from 1 January 1988, may still be used.
1.6.4.2	Tank-containers constructed before 1 January 1993 in accordance with the requirements in force up to 31 December 1992 but which do not, however, conform to the requirements applicable as from 1 January 1993, may still be used.
1.6.4.3	Tank-containers constructed before 1 January 1999 in accordance with the requirements in force up to 31 December 1998 but which do not, however, conform to the requirements applicable as from 1 January 1999, may still be used.
1.6.4.4	*(Reserved)*
1.6.4.5	When, because of amendments to ADR, some proper shipping names of gases have been modified, it is not necessary to modify the names on the plate or on the shell itself (see 6.8.3.5.2 or 6.8.3.5.3), provided that the names of the gases on the tank-containers and MEGCs or on the plates [see 6.8.3.5.6 (b) or (c)] are adapted at the first periodic test thereafter.
1.6.4.6	Tank-containers constructed before 1 January 2007 in accordance with the requirements in force up to 31 December 2006 but which do not, however, conform to the requirements applicable as from 1

January 2007 regarding the marking of the external design pressure in accordance with 6.8.2.5.1, may still be used.

1.6.4.7 Tank-containers constructed before 1 January 1997 in accordance with the requirements in force up to 31 December 1996 but which do not, however, conform to the requirements of marginals 212 332 and 212 333 applicable as from 1 January 1997, may still be used.

1.6.4.8 *(Reserved)*

1.6.4.9 Tank-containers and MEGCs designed and constructed in accordance with a technical code which was recognized at the time of their construction according to the provisions of 6.8.2.7 which were applicable at that time may still be used.

1.6.4.10 *(Deleted)*

1.6.4.11 *(Reserved)*

1.6.4.12 Tank-containers and MEGCs constructed before 1 January 2003 in accordance with the requirements applicable up to 30 June 2001, but which do not, however, conform to the requirements applicable as from 1 July 2001, may still be used.

However, they shall be marked with the relevant tank code and if applicable the relevant alphanumeric codes of special provisions TC and TE in accordance with 6.8.4.

1.6.4.13 Tank-containers constructed before 1 July 2003 in accordance with the requirements in force up to 31 December 2002 but which do not, however, conform to the requirements of 6.8.2.1.7 applicable as from 1 January 2003 and special provision TE15 of 6.8.4 (b) applicable from 1 January 2003 to 31 December 2006 may still be used.

1.6.4.14 *(Reserved)*

1.6.4.15 The type of the test ("P" or "L") required by 6.8.2.5.1 need not be added to the tank plate until the first test after 1 January 2007 is performed.

1.6.4.16 and 1.6.4.17 *(Deleted)*

1.6.4.18 For tank-containers and MEGCs constructed before 1 January 2007 which do not conform to the requirements of 4.3.2, 6.8.2.3, 6.8.2.4 and 6.8.3.4 concerning the tank record, the retention of files for the tank record shall start at the latest at the next periodic inspection.

1.6.4.19 *(Deleted)*

1.6.4.20 Vacuum-operated waste tank-containers constructed before 1 July 2005 in accordance with the requirements applicable up to 31 December 2004 but which do not conform to the requirements of 6.10.3.9 applicable as from 1 January 2005, may still be used.

1.6.4.21 to 1.6.4.29 *(Reserved)*

1.6.4.30 Portable tanks and UN MEGCs which do not meet the design requirements applicable as from 1 January 2007 but which have been constructed according to a design approval certificate which has been issued before 1 January 2008 may continue to be used.

1.6.4.31 *(Deleted)*

1.6.4.32 When the shell of a tank-container was already divided by partitions or surge plates into sections of not more than 7 500 litres capacity before 1 January 2009, the capacity of the shell need not be supplemented with the symbol "S" in the particulars required by 6.8.2.5.1 until the next periodic inspection according to 6.8.2.4.2 is performed.

1.6.4.33 Notwithstanding the provisions of 4.3.2.2.4, tank-containers intended for the carriage of liquefied gases or refrigerated liquefied gases, which meet the applicable construction requirements of ADR but which were divided, before 1 July 2009, by partitions or surge plates into sections of more than 7 500 litres capacity may still be filled to more than 20% and less than 80% of their capacity.

1.6.4.34 to 1.6.4.36 *(Deleted)*

1.6.4.37   Portable tanks and MEGCs manufactured before 1 January 2012, that conform to the marking requirements of 6.7.2.20.1, 6.7.3.16.1, 6.7.4.15.1 or 6.7.5.13.1 applicable up to 31 December 2010, as relevant, may continue to be used if they comply with all other relevant requirements of ADR applicable as from 1 January 2011 including, when applicable, the requirement of 6.7.2.20.1 (g) for marking the symbol "S" on the plate when the shell or the compartment is divided by surge plates into sections of not more than 7 500 litres capacity.

1.6.4.38   Portable tanks manufactured before 1 January 2014 need not be marked with the portable tank instruction as required in 6.7.2.20.2, 6.7.3.16.2 and 6.7.4.15.2 until the next periodic inspection and test.

1.6.4.39   Tank-containers and MEGCs designed and constructed in accordance with standards applicable at the time of their construction (see 6.8.2.6 and 6.8.3.6) according to the provisions of ADR which were applicable at that time may still be used unless restricted by a specific transitional measure.

1.6.4.40   Tank-containers constructed before 1 July 2011 in accordance with the requirements of 6.8.2.2.3 in force up to 31 December 2010 but which do not, however, conform to the requirements of 6.8.2.2.3, third paragraph, concerning the position of the flame trap or flame arrester may still be used.

1.6.4.41   *(Deleted)*

1.6.4.42   Tank-containers constructed before 1 July 2013 in accordance with the requirements in force up to 31 December 2012, but which do not, however, meet the marking provisions of 6.8.2.5.2 or 6.8.3.5.6 applicable as from 1 January 2013, may continue to be marked in accordance with the requirements applicable up to 31 December 2012 until the next periodic inspection after 1 July 2013.

1.6.4.43   Portable tanks and MECGs manufactured before 1 January 2014 need not comply with the requirements of 6.7.2.13.1 (f), 6.7.3.9.1 (e), 6.7.4.8.1 (e) and 6.7.5.6.1 (d) concerning the marking of the pressure relief devices.

1.6.4.44   For substances where TP38 or TP39 is assigned in column (11) of Table A of Chapter 3.2, the portable tank instruction prescribed in ADR applicable up to 31 December 2012 may continue to be applied until 31 December 2018.

1.6.4.45   For UN No. 2381, the tank code specified in column (12) of Table A of Chapter 3.2 applicable up to 31 December 2012 may continue to be applied until 31 December 2018 for tank-containers constructed before 1 July 2013.

1.6.4.46   Tank-containers constructed before 1 January 2012 in accordance with the requirements in force up to 31 December 2012, but which do not however conform to the requirements of 6.8.2.6 relating to standards EN 14432:2006 and EN 14433:2006 applicable as from 1 January 2011, may still be used.

1.6.4.47   Tank containers for refrigerated liquefied gases constructed before 1 July 2017 in accordance with the requirements in force up to 31 December 2016 but which do not conform to the requirements of 6.8.3.4.10, 6.8.3.4.11 and 6.8.3.5.4 applicable from 1 January 2017 may continue to be used until the next inspection after 1 July 2017. Until this time, to meet the requirements of 4.3.3.5 and 5.4.1.2.2(d), the actual holding times may be estimated without recourse to the reference holding time.

1.6.4.48   Tank-containers constructed before 1 July 2017 in accordance with the requirements in force up to 31 December 2016 but which do not however conform to the requirements of 6.8.2.1.23 applicable as from 1 January 2017 may still be used.

1.6.5   Vehicles

1.6.5.1 and 1.6.5.2   *(Reserved)*

1.6.5.3   *(Deleted)*

1.6.5.4   As regards the construction of EX/II, EX/III, FL, OX and AT vehicles, the requirements of Part 9 in force up to 31 December 2016 may be applied until 31 March 2018.

1.6.5.5   Vehicles registered or entering into service before 1 January 2003 the electric equipment of which does not comply with the requirements of 9.2.2, 9.3.7 or 9.7.8 but complies with the requirements applicable until 30 June 2001 may still be used.

1.6.5.6	(Deleted)

1.6.5.7   Complete or completed vehicles which have been type-approved before 31 December 2002 according to ECE Regulation No. 105[1] as amended by the 01 series of amendments or the corresponding provisions of Directive 98/91/EC[2] and which do not comply with the requirements of Chapter 9.2 but comply with the requirements applicable to the construction of base vehicles (marginals 220 100 to 220 540 of Appendix B.2) applicable until 30 June 2001 may continue to be approved and used provided they are first registered or they entered into service before 1 July 2003.

1.6.5.8   EX/II and EX/III vehicles which have been first approved before 1 July 2005 and which comply with the requirements of Part 9 in force up to 31 December 2004 but which do not however conform to the requirements applicable as from 1 January 2005 may still be used.

1.6.5.9   Tank-vehicles with fixed tanks with a capacity of more than 3 m^3 intended for the carriage of dangerous goods in the liquid or molten state tested with a pressure of less than 4 bar, which do not comply with the requirements of 9.7.5.2, first registered (or which entered into service if the registration is not mandatory) before 1 July 2004, may still be used.

1.6.5.10   Certificates of approval which conform to the model shown in 9.1.3.5 applicable up to 31 December 2006 and those which conform to the model shown in 9.1.3.5 applicable from 1 January 2007 to 31 December 2008 may continue to be used. Certificates of approval which conform to the model shown in 9.1.3.5 applicable from 1 January 2009 up to 31 December 2014 may continue to be used.

1.6.5.11   MEMUs which have been constructed and approved before 1 July 2009 in accordance with the provisions of national law but which do not, however, conform to the construction and approval requirements applicable as from 1 January 2009 may be used with the approval of the competent authorities in the countries of use.

1.6.5.12   EX/III and FL vehicles registered or entering into service before 1 April 2012, the electrical connections of which do not comply with the requirements of 9.2.2.6.3, but comply with the requirements applicable until 31 December 2010, may still be used.

1.6.5.13   Trailers first registered (or which entered into service if registration was not mandatory) before 1 July 1995 equipped with anti-lock braking system in conformity with ECE Regulation No. 13, 06 series of amendments but which do not comply with the technical requirements for category A anti-lock braking system may still be used.

1.6.5.14   MEMUs which have been approved before 1 July 2013 in accordance with the provisions of ADR in force up to 31 December 2012, but which do not conform to the requirements of 6.12.3.1.2 or 6.12.3.2.2 applicable as from 1 January 2013, may still be used.

1.6.5.15   As regards the application of the provisions of Part 9, vehicles first registered or entered into service before 1 November 2014 and which have been approved according to the provisions of the directives repealed by the Regulation (EC) No. 661/2009[3], may continue to be used.

1.6.5.16   EX/II, EX/III, FL and OX vehicles registered before 1 April 2018, fitted with fuel tanks not approved according to ECE Regulation No. 34 may still be used.

1.6.5.17   Vehicles first registered or entering into service before 1 April 2018 that do not comply with subsection 9.2.2.8.5 or standards ISO 6722-1:2011 + Cor 01:2012 or ISO 6722-2:2013 for cables of subsection 9.2.2.2.1, but comply with the requirements applicable until 31 December 2016, may continue to be used.

---

[1]   *ECE Regulation No. 105 (Uniform provisions concerning the approval of vehicles intended for the carriage of dangerous goods with regard to their specific constructional features).*

[2]   *Directive 98/91/EC of the European Parliament and of the Council of 14 December 1998 relating to motor vehicles and their trailers intended for the transport of dangerous goods by road and amending Directive 70/156/EEC relating to the type approval of motor vehicles and their trailers (Official Journal of the European Communities No. L 011 of 16 January 1999, pp. 0025-0036).*

[3]   *Regulation (EC) 661/2009 of 13 July 2009 concerning type-approval requirements for the general safety of motor vehicles, their trailers and systems, components and separate technical units intended therefor (Official Journal L 200 of 31.7.2009, p. 1).*

1.6.5.18	Vehicles first registered or entering into service before 1 April 2018 approved specifically as OX vehicle may continue to be used for the carriage of substances of UN No. 2015.

1.6.5.19   As regards the annual technical inspection of the vehicles first registered or entering into service before 1 April 2018 approved specifically as OX vehicle, the requirements of Part 9 in force up to 31December 2016 may still be applied.

1.6.5.20   Certificates of approval for OX vehicles which conform to the model shown in 9.1.3.5 applicable up to 31 December 2016 may continue to be used.

1.6.6   **Class 7**

1.6.6.1   *Packages not requiring competent authority approval of design under the 1985 and 1985 (as amended 1990) editions of IAEA Safety Series No. 6*

Packages not requiring competent authority approval of design (excepted packages, Type IP-1, Type IP-2, Type IP-3 and Type A packages) shall meet the requirements of ADR in full, except that packages that meet the requirements of the 1985 or 1985 (as amended 1990) Editions of IAEA Regulations for the Safe Transport of Radioactive Material (IAEA Safety Series No.6):

(a)   May continue in carriage provided that they were prepared for carriage prior to 31 December 2003, and subject to the requirements of 1.6.6.3, if applicable;

(b)   May continue to be used provided that:

     (i)   They were not designed to contain uranium hexafluoride;

     (ii)   The applicable requirements of 1.7.3 are applied;

     (iii)   The activity limits and classification in 2.2.7 are applied;

     (iv)   The requirements and controls for carriage in Parts 1, 3, 4, 5 and 7 are applied;

     (v)   The packaging was not manufactured or modified after 31 December 2003.

1.6.6.2   *Packages approved under the 1973, 1973 (as amended), 1985 and 1985 (as amended 1990) editions of IAEA Safety Series No. 6*

1.6.6.2.1   Packages requiring competent authority approval of the design shall meet the requirements of ADR in full unless the following conditions are met:

(a)   The packagings were manufactured to a package design approved by the competent authority under the provisions of the 1973 or 1973 (as amended) or the 1985 or 1985 (as amended 1990) Editions of IAEA Safety Series No.6;

(b)   The package design is subject to multilateral approval;

(c)   The applicable requirements of 1.7.3 are applied;

(d)   The activity limits and classification in 2.2.7 are applied;

(e)   The requirements and controls for carriage in Parts 1, 3, 4, 5 and 7 are applied;

(f)   *(Reserved)*

(g)   For packages that meet the requirements of the 1973 or 1973 (as amended) Editions of IAEA Safety Series No. 6:

     (i)   The packages retain sufficient shielding to ensure that the radiation level at 1 m from the surface of the package would not exceed 10 mSv/h in the accident conditions of carriage defined in the 1973 Revised or 1973 Revised (as amended) Editions of IAEA Safety Series No.6 with the maximum radioactive contents which the package is authorized to contain;

     (ii)   The packages do not utilize continuous venting;

| | (iii) | A serial number in accordance with the provision of 5.2.1.7.5 is assigned to and marked on the outside of each packaging. |

1.6.6.2.2 No new manufacture of packagings to a package design meeting the provisions of the 1973, 1973 (as amended), 1985, and 1985 (as amended 1990) Editions of IAEA Safety Series No.6 shall be permitted to commence.

1.6.6.3 ***Packages excepted from the requirements for fissile materials under the 2011 and 2013 editions of ADR (2009 Edition of IAEA Safety Standard Series No.TS-R-1)***

Packages containing fissile material that is excepted from classification as "FISSILE" according to 2.2.7.2.3.5 (a) (i) or (iii) of the 2011 and 2013 editions of ADR (paras. 417 (a) (i) or (iii) of the 2009 Edition of IAEA Regulations for the Safe Transport of Radioactive Material) prepared for carriage before 31 December 2014 may continue in carriage and may continue to be classified as non-fissile or fissile-excepted except that the consignment limits in Table 2.2.7.2.3.5 of these editions shall apply to the vehicle. The consignment shall be carried under exclusive use.

1.6.6.4 ***Special form radioactive material approved under the 1973, 1973 (as amended), 1985 and 1985 (as amended 1990) Editions of IAEA Safety Series No. 6***

Special form radioactive material manufactured to a design which had received unilateral approval by the competent authority under the 1973, 1973 (as amended), 1985 or 1985 (as amended 1990) Editions of IAEA Safety Series No. 6 may continue to be used when in compliance with the mandatory management system in accordance with the applicable requirements of 1.7.3. No new manufacture of such special form radioactive material shall be permitted to commence.

CHAPTER 1.7

GENERAL PROVISIONS CONCERNING RADIOACTIVE MATERIAL

1.7.1   **Scope and application**

> *NOTE 1: In the event of accidents or incidents during the carriage of radioactive material, emergency provisions, as established by relevant national and/or international organizations, shall be observed to protect persons, property and the environment. Appropriate guidelines for such provisions are contained in "Planning and Preparing for Emergency Response to Transport Accidents Involving Radioactive Material", Safety Standard Series No. TS-G-1.2 (ST-3), IAEA, Vienna (2002).*

> *NOTE 2: Emergency procedures shall take into account the formation of other dangerous substances that may result from the reaction between the contents of a consignment and the environment in the event of an accident.*

1.7.1.1   ADR establishes standards of safety which provide an acceptable level of control of the radiation, criticality and thermal hazards to persons, property and the environment that are associated with the carriage of radioactive material. These standards are based on the IAEA Regulations for the Safe Transport of Radioactive material, 2012 Edition, IAEA Safety Standards Series No. SSR–6, IAEA, Vienna (2012). Explanatory material can be found in "Advisory Material for the IAEA Regulations for the Safe Transport of Radioactive Material (2012 Edition)", IAEA Safety Standards Series No. SSG-26, IAEA, Vienna (2014).

1.7.1.2   The objective of ADR is to establish requirements that shall be satisfied to ensure safety and to protect persons, property and the environment from the effects of radiation in the carriage of radioactive material. This protection is achieved by requiring:

     (a)   Containment of the radioactive contents;

     (b)   Control of external radiation levels;

     (c)   Prevention of criticality; and

     (d)   Prevention of damage caused by heat.

These requirements are satisfied firstly by applying a graded approach to contents limits for packages and vehicles and to performance standards applied to package designs depending upon the hazard of the radioactive contents. Secondly, they are satisfied by imposing conditions on the design and operation of packages and on the maintenance of packagings, including a consideration of the nature of the radioactive contents. Finally, they are satisfied by requiring administrative controls including, where appropriate, approval by competent authorities.

1.7.1.3   ADR applies to the carriage of radioactive material by road including carriage which is incidental to the use of the radioactive material. Carriage comprises all operations and conditions associated with and involved in the movement of radioactive material; these include the design, manufacture, maintenance and repair of packaging, and the preparation, consigning, loading, carriage including in-transit storage, unloading and receipt at the final destination of loads of radioactive material and packages. A graded approach is applied to the performance standards in ADR that are characterized by three general severity levels:

     (a)   Routine conditions of carriage (incident free);

     (b)   Normal conditions of carriage (minor mishaps);

     (c)   Accident conditions of carriage.

1.7.1.4   The provisions laid down in ADR do not apply to any of the following:

     (a)   Radioactive material that is an integral part of the means of transport;

(b)   Radioactive material moved within an establishment which is subject to appropriate safety regulations in force in the establishment and where the movement does not involve public roads or railways;

(c)   Radioactive material implanted or incorporated into a person or live animal for diagnosis or treatment;

(d)   Radioactive material in or on a person who is to be transported for medical treatment because the person has been subject to accidental or deliberate intake of radioactive material or to contamination;

(e)   Radioactive material in consumer products which have received regulatory approval, following their sale to the end user;

(f)   Natural material and ores containing naturally occurring radionuclides (which may have been processed), provided the activity concentration of the material does not exceed 10 times the values specified in Table 2.2.7.2.2.1, or calculated in accordance with 2.2.7.2.2.2 (a) and 2.2.7.2.2.3 to 2.2.7.2.2.6. For natural materials and ores containing naturally occurring radionuclides that are not in secular equilibrium the calculation of the activity concentration shall be performed in accordance with 2.2.7.2.2.4;

(g)   Non-radioactive solid objects with radioactive substances present on any surfaces in quantities not in excess of the limit set out in the definition for "contamination" in 2.2.7.1.2.

1.7.1.5   *Specific provisions for the carriage of excepted packages*

1.7.1.5.1   Excepted packages which may contain radioactive material in limited quantities, instruments, manufactured articles or empty packagings as specified in 2.2.7.2.4.1 shall be subject only to the following provisions of Parts 5 to 7:

(a)   The applicable provisions specified in 5.1.2.1, 5.1.3.2, 5.1.5.2.2, 5.1.5.2.3, 5.1.5.4, 5.2.1.10, 7.5.11 CV33 (3.1), (5.1) to (5.4) and (6); and

(b)   The requirements for excepted packages specified in 6.4.4.

except when the radioactive material possesses other hazardous properties and has to be classified in a class other than Class 7 in accordance with special provision 290 or 369 of Chapter 3.3, where the provisions listed in (a) and (b) above apply only as relevant and in addition to those relating to the main class.

1.7.1.5.2   Excepted packages are subject to the relevant provisions of all other parts of ADR. If the excepted package contains fissile material, one of the fissile exceptions provided by 2.2.7.2.3.5 shall apply and the requirements of 7.5.11 CV33 (4.3) shall be met.

1.7.2   **Radiation protection programme**

1.7.2.1   The carriage of radioactive material shall be subject to a Radiation protection programme which shall consist of systematic arrangements aimed at providing adequate consideration of radiation protection measures.

1.7.2.2   Doses to persons shall be below the relevant dose limits. Protection and safety shall be optimized in order that the magnitude of individual doses, the number of persons exposed and the likelihood of incurring exposure shall be kept as low as reasonably achievable, economic and social factors being taken into account within the restriction that the doses to individuals be subject to dose constraints. A structured and systematic approach shall be adopted and shall include consideration of the interfaces between carriage and other activities.

1.7.2.3   The nature and extent of the measures to be employed in the programme shall be related to the magnitude and likelihood of radiation exposures. The programme shall incorporate the requirements in 1.7.2.2, 1.7.2.4, 1.7.2.5 and 7.5.11 CV33 (1.1). Programme documents shall be available, on request, for inspection by the relevant competent authority.

1.7.2.4   For occupational exposures arising from transport activities, where it is assessed that the effective dose either:

(a)   Is likely to be between 1 mSv and 6 mSv in a year, a dose assessment programme via work place monitoring or individual monitoring shall be conducted; or

(b)   Is likely to exceed 6 mSv in a year, individual monitoring shall be conducted.

When individual monitoring or work place monitoring is conducted, appropriate records shall be kept.

NOTE: For occupational exposures arising from transport activities, where it is assessed that the effective dose is most unlikely to exceed 1 mSv in a year, no special work patterns, detailed monitoring, dose assessment programmes or individual record keeping need be required.

1.7.2.5   Workers (see 7.5.11, CV33 Note 3) shall be appropriately trained in radiation protection including the precautions to be observed in order to restrict their occupational exposure and the exposure of other persons who might be affected by their actions.

1.7.3   Management system

A management system based on international, national or other standards acceptable to the competent authority shall be established and implemented for all activities within the scope of ADR, as identified in 1.7.1.3, to ensure compliance with the relevant provisions of ADR. Certification that the design specification has been fully implemented shall be available to the competent authority. The manufacturer, consignor or user shall be prepared:

(a)   To provide facilities for inspection during manufacture and use; and

(b)   To demonstrate compliance with ADR to the competent authority.

Where competent authority approval is required, such approval shall take into account and be contingent upon the adequacy of the management system.

1.7.4   Special arrangement

1.7.4.1   Special arrangement shall mean those provisions, approved by the competent authority, under which consignments which do not satisfy all the requirements of ADR applicable to radioactive material may be transported.

NOTE: Special arrangement is not considered to be a temporary derogation in accordance with 1.5.1.

1.7.4.2   Consignments for which conformity with any provision applicable to radioactive material is impracticable shall not be transported except under special arrangement. Provided the competent authority is satisfied that conformity with the radioactive material provisions of ADR is impracticable and that the requisite standards of safety established by ADR have been demonstrated through alternative means the competent authority may approve special arrangement transport operations for single or a planned series of multiple consignments. The overall level of safety in carriage shall be at least equivalent to that which would be provided if all the applicable requirements had been met. For international consignments of this type, multilateral approval shall be required.

1.7.5   Radioactive material possessing other dangerous properties

In addition to the radioactive and fissile properties, any subsidiary risk of the contents of the package, such as explosiveness, flammability, pyrophoricity, chemical toxicity and corrosiveness, shall also be taken into account in the documentation, packing, labelling, marking, placarding, stowage, segregation and carriage, in order to be in compliance with all relevant provisions for dangerous goods of ADR.

1.7.6   **Non-compliance**

1.7.6.1   In the event of non-compliance with any limit in ADR applicable to radiation level or contamination,

      (a)   The consignor, consignee, carrier and any organization involved during carriage who may be affected, as appropriate, shall be informed of the non-compliance by:

            (i)   the carrier if the non-compliance is identified during carriage; or

            (ii)   the consignee if the non-compliance is identified at receipt;

      (b)   The carrier, consignor or consignee, as appropriate shall:

            (i)   take immediate steps to mitigate the consequences of the non-compliance;

            (ii)   investigate the non-compliance and its causes, circumstances and consequences;

            (iii)   take appropriate action to remedy the causes and circumstances that led to the non-compliance and to prevent a recurrence of similar circumstances that led to the non-compliance; and

            (iv)   communicate to the competent authority(ies) on the causes of the non-compliance and on corrective or preventive actions taken or to be taken;

      (c)   The communication of the non-compliance to the consignor and competent authority(ies), respectively, shall be made as soon as practicable and it shall be immediate whenever an emergency exposure situation has developed or is developing.

CHAPTER 1.8

CHECKS AND OTHER SUPPORT MEASURES TO ENSURE COMPLIANCE WITH SAFETY REQUIREMENTS

1.8.1   **Administrative controls of dangerous goods**

1.8.1.1   The competent authorities of the Contracting Parties may, on their national territory, at any time, conduct spot checks to verify whether the requirements concerning the carriage of dangerous goods have been met including, in accordance with 1.10.1.5, those concerning security measures.

These checks shall, however, be made without endangering persons, property or the environment and without major disruption of road services.

1.8.1.2   Participants in the carriage of dangerous goods (Chapter 1.4) shall, without delay, in the context of their respective obligations, provide the competent authorities and their agents with the necessary information for carrying out the checks.

1.8.1.3   The competent authorities may also, for the purposes of carrying out checks on the premises of the enterprises participating in the carriage of dangerous goods (Chapter 1.4), make inspections, consult the necessary documents and remove samples of dangerous goods or packagings for examination, provided that safety is not jeopardized thereby. The participants in the carriage of dangerous goods (Chapter 1.4) shall also make the vehicles or parts of vehicles and the equipment and installations accessible for the purpose of checking where this is possible and reasonable. They may, if they deem necessary, designate a person from the enterprise to accompany the representative of the competent authority.

1.8.1.4   If the competent authorities observe that the requirements of ADR have not been met, they may prohibit a consignment or interrupt a transport operation until the defects observed are rectified, or they may prescribe other appropriate measures. Immobilization may take place on the spot or at another place selected by the authorities for safety reasons. These measures shall not cause a major disruption in road services.

1.8.2   **Mutual administrative support**

1.8.2.1   The Contracting Parties shall agree on mutual administrative support for the implementation of ADR.

1.8.2.2   When a Contracting Party has reasons to observe that the safety of the carriage of dangerous goods on its territory is compromised as a result of very serious or repeated infringements by an enterprise which has its headquarters on the territory of another Contracting Party, it shall notify the competent authorities of this Contracting Party of such infringements. The competent authorities of the Contracting Party on the territory of which the very serious or repeated infringements were observed may request the competent authorities of the Contracting Party on the territory of which the enterprise has its headquarters to take appropriate measures against the offender(s). The transmission of data referring to persons shall not be permitted unless it is necessary for the prosecution of very serious or repeated infringements.

1.8.2.3   The authorities notified shall communicate to the competent authorities of the Contracting Party on the territory of which the infringements were observed, the measures which have, if necessary, been taken with respect to the enterprise.

1.8.3   **Safety adviser**

1.8.3.1   Each undertaking, the activities of which include the carriage, or the related packing, loading, filling or unloading, of dangerous goods by road shall appoint one or more safety advisers for the carriage of dangerous goods, responsible for helping to prevent the risks inherent in such activities with regard to persons, property and the environment.

1.8.3.2   The competent authorities of the Contracting Parties may provide that these requirements shall not apply to undertakings:

(a)   The activities of which concern quantities in each transport unit smaller than those referred to in 1.1.3.6, 1.7.1.4 and in Chapters 3.3, 3.4 and 3.5; or

(b)   The main or secondary activities of which are not the carriage or the related packing, filling, loading or unloading of dangerous goods but which occasionally engage in the national carriage or the related packing, filling, loading or unloading of dangerous goods posing little danger or risk of pollution.

1.8.3.3   The main task of the adviser shall be, under the responsibility of the head of the undertaking, to seek by all appropriate means and by all appropriate action, within the limits of the relevant activities of that undertaking, to facilitate the conduct of those activities in accordance with the requirements applicable and in the safest possible way.

With regard to the undertaking's activities, the adviser has the following duties in particular:

-     monitoring compliance with the requirements governing the carriage of dangerous goods;

-     advising his undertaking on the carriage of dangerous goods;

-     preparing an annual report to the management of his undertaking or a local public authority, as appropriate, on the undertaking's activities in the carriage of dangerous goods. Such annual reports shall be preserved for five years and made available to the national authorities at their request.

The adviser's duties also include monitoring the following practices and procedures relating to the relevant activities of the undertaking:

-     the procedures for compliance with the requirements governing the identification of dangerous goods being transported;

-     the undertaking's practice in taking account, when purchasing means of transport, of any special requirements in connection with the dangerous goods being transported;

-     the procedures for checking the equipment used in connection with the carriage, packing, filling, loading or unloading of dangerous goods;

-     the proper training of the undertaking's employees, including on the changes to the regulations, and the maintenance of records of such training;

-     the implementation of proper emergency procedures in the event of any accident or incident that may affect safety during the carriage, packing, filling, loading or unloading of dangerous goods;

-     investigating and, where appropriate, preparing reports on serious accidents, incidents or serious infringements recorded during the carriage, packing, filling, loading or unloading of dangerous goods;

-     the implementation of appropriate measures to avoid the recurrence of accidents, incidents or serious infringements;

-     the account taken of the legal prescriptions and special requirements associated with the carriage of dangerous goods in the choice and use of sub-contractors or third parties;

-     verification that employees involved in the carriage, packing, filling, loading or unloading of dangerous goods have detailed operational procedures and instructions;

-     the introduction of measures to increase awareness of the risks inherent in the carriage, packing, filling, loading and unloading of dangerous goods;

-     the implementation of verification procedures to ensure the presence on board the means of transport of the documents and safety equipment which must accompany transport and the compliance of such documents and equipment with the regulations;

-     the implementation of verification procedures to ensure compliance with the requirements governing packing, filling, loading and unloading;

-     the existence of the security plan indicated in 1.10.3.2.

1.8.3.4	The adviser may also be the head of the undertaking, a person with other duties in the undertaking, or a person not directly employed by that undertaking, provided that that person is capable of performing the duties of adviser.

1.8.3.5 Each undertaking concerned shall, on request, inform the competent authority or the body designated for that purpose by each Contracting Party of the identity of its adviser.

1.8.3.6 Whenever an accident affects persons, property or the environment or results in damage to property or the environment during carriage, packing, filling, loading or unloading carried out by the undertaking concerned, the adviser shall, after collecting all the relevant information, prepare an accident report to the management of the undertaking or to a local public authority, as appropriate. That report shall not replace any report by the management of the undertaking which might be required under any other international or national legislation.

1.8.3.7 An adviser shall hold a vocational training certificate, valid for transport by road. That certificate shall be issued by the competent authority or the body designated for that purpose by each Contracting Party.

1.8.3.8 To obtain a certificate, a candidate shall undergo training and pass an examination approved by the competent authority of the Contracting Party.

1.8.3.9 The main aims of the training shall be to provide candidates with sufficient knowledge of the risks inherent in the carriage, packing, filling, loading or unloading of dangerous goods, of the applicable laws, regulations and administrative provisions and of the duties listed in 1.8.3.3.

1.8.3.10 The examination shall be organized by the competent authority or by an examining body designated by the competent authority. The examining body shall not be a training provider.

The examining body shall be designated in writing. This approval may be of limited duration and shall be based on the following criteria:

- competence of the examining body;

- specifications of the form of the examinations the examining body is proposing, including, if necessary, the infrastructure and organisation of electronic examinations in accordance with 1.8.3.12.5, if these are to be carried out;

- measures intended to ensure that examinations are impartial;

- independence of the body from all natural or legal persons employing safety advisers.

1.8.3.11 The aim of the examination is to ascertain whether candidates possess the necessary level of knowledge to carry out the duties incumbent upon a safety adviser as listed in 1.8.3.3, for the purpose of obtaining the certificate prescribed in sub-section 1.8.3.7, and it shall cover at least the following subjects:

(a) Knowledge of the types of consequences which may be caused by an accident involving dangerous goods and knowledge of the main causes of accidents;

(b) Requirements under national law, international conventions and agreements, with regard to the following in particular:

- classification of dangerous goods (procedure for classifying solutions and mixtures, structure of the list of substances, classes of dangerous goods and principles for their classification, nature of dangerous goods transported, physical, chemical and toxicological properties of dangerous goods);

- general packing provisions, provisions for tanks and tank-containers (types, code, marking, construction, initial and periodic inspection and testing);

- marking and labelling, placarding and orange-coloured plate marking (marking and labelling of packages, placing and removal of placards and orange-coloured plates);

- particulars in transport documents (information required);

- method of consignment and restrictions on dispatch (full load, carriage in bulk, carriage in intermediate bulk containers, carriage in containers, carriage in fixed or demountable tanks);

- transport of passengers;

- prohibitions and precautions relating to mixed loading;

- segregation of goods;

- limitation of the quantities carried and quantities exemptions;

- handling and stowage (packing, filling, loading and unloading - filling ratios -, stowage and segregation);

- cleaning and/or degassing before packing, filling, loading and after unloading;

- crews, vocational training;

- vehicle documents (transport documents, instructions in writing, vehicle approval certificate, driver training certificate, copies of any derogations, other documents);

- instructions in writing (implementation of the instructions and crew protection equipment);

- supervision requirements (parking);

- traffic regulations and restrictions;

- operational discharges or accidental leaks of pollutants;

- requirements relating to transport equipment.

1.8.3.12 *Examinations*

1.8.3.12.1   The examination shall consist of a written test which may be supplemented by an oral examination.

1.8.3.12.2   The competent authority or an examining body designated by the competent authority shall invigilate every examination. Any manipulation and deception shall be ruled out as far as possible. Authentication of the candidate shall be ensured. The use in the written test of documentation other than international or national regulations is not permitted. All examination documents shall be recorded and kept as a print-out or electronically as a file.

1.8.3.12.3   Electronic media may be used only if provided by the examining body. There shall be no means of a candidate introducing further data to the electronic media provided; the candidate may only answer the questions posed.

1.8.3.12.4   The written test shall consist of two parts:

(a)   Candidates shall receive a questionnaire. It shall include at least 20 open questions covering at least the subjects mentioned in the list in 1.8.3.11. However, multiple choice questions may be used. In this case, two multiple choice questions count as one open question. Amongst these subjects particular attention shall be paid to the following subjects:

- general preventive and safety measures;

- classification of dangerous goods;

- general packing provisions, including tanks, tank-containers, tank-vehicles, etc.;

- danger marking, labelling and placardings;

- information in transport document;

- handling and stowage;

- crew, vocational training;

- vehicle documents and transport certificates;

- instructions in writing;

- requirements concerning transport equipment;

(b)   Candidates shall undertake a case study in keeping with the duties of the adviser referred to in 1.8.3.3, in order to demonstrate that they have the necessary qualifications to fulfil the task of adviser.

1.8.3.12.5   Written examinations may be performed, in whole or in part, as electronic examinations, where the answers are recorded and evaluated using electronic data processing (EDP) processes, provided the following conditions are met:

(a)   The hardware and software shall be checked and accepted by the competent authority or by an examining body designated by the competent authority;

(b)   Proper technical functioning shall be ensured. Arrangements as to whether and how the examination can be continued shall be made for a failure of the devices and applications. No aids shall be available on the input devices (e.g. electronic search function), the equipment provided according to 1.8.3.12.3 shall not allow the candidates to communicate with any other device during the examination;

(c)   Final inputs of each candidate shall be logged. The determination of the results shall be transparent.

1.8.3.13   The Contracting Parties may decide that candidates who intend working for undertakings specializing in the carriage of certain types of dangerous goods need only be questioned on the substances relating to their activities. These types of goods are:

- Class 1;

- Class 2;

- Class 7;

- Classes 3, 4.1, 4.2, 4.3, 5.1, 5.2, 6.1, 6.2, 8 and 9;

- UN Nos. 1202, 1203, 1223, 3475, and aviation fuel classified under UN Nos. 1268 or 1863.

The certificate prescribed in 1.8.3.7 shall clearly indicate that it is only valid for one type of the dangerous goods referred to in this sub-section and on which the adviser has been questioned under the conditions defined in 1.8.3.12.

1.8.3.14   The competent authority or the examining body shall keep a running list of the questions that have been included in the examination.

1.8.3.15   The certificate prescribed in 1.8.3.7 shall take the form laid down in 1.8.3.18 and shall be recognized by all Contracting Parties.

1.8.3.16   *Validity and renewal of certificates*

1.8.3.16.1   The certificate shall be valid for five years. The period of the validity of a certificate shall be extended from the date of its expiry for five years at a time where, during the year before its expiry, its holder has passed an examination. The examination shall be approved by the competent authority.

1.8.3.16.2   The aim of the examination is to ascertain that the holder has the necessary knowledge to carry out the duties set out in 1.8.3.3. The knowledge required is set out in 1.8.3.11 (b) and shall include the amendments to the regulations introduced since the award of the last certificate. The examination shall be held and supervised on the same basis as in 1.8.3.10 and 1.8.3.12 to 1.8.3.14. However, holders need not undertake the case study specified in 1.8.3.12.4 (b).

1.8.3.17   *(Deleted)*

1.8.3.18 *Form of certificate*

Certificate of training as safety adviser for the transport of dangerous goods

Certificate No: ....

Distinguishing sign of the State issuing the certificate: ....

Surname: ....

Forename(s): ....

Date and place of birth: ....

Nationality: ....

Signature of holder: ....

Valid until .... for undertakings which transport dangerous goods and for undertakings which carry out related packing, filling, loading or unloading:

☐ by road   ☐ by rail   ☐ by inland waterway

Issued by: ....

Date: ....   Signature: ....

1.8.4 **List of competent authorities and bodies designated by them**

The Contracting Parties shall communicate to the Secretariat of the United Nations Economic Commission for Europe the addresses of the authorities and bodies designated by them which are competent in accordance with national law to implement ADR, referring in each case to the relevant requirement of ADR and giving the addresses to which the relevant applications should be made.

The Secretariat of the United Nations Economic Commission for Europe shall establish a list on the basis of the information received and shall keep it up-to-date. It shall communicate this list and the amendments thereto to the Contracting Parties.

1.8.5 **Notifications of occurrences involving dangerous goods**

1.8.5.1   If a serious accident or incident takes place during loading, filling, carriage or unloading of dangerous goods on the territory of a Contracting Party, the loader, filler, carrier or consignee, respectively, shall ascertain that a report conforming to the model prescribed in 1.8.5.4 is made to the competent authority of the Contracting Party concerned at the latest one month after the occurrence.

1.8.5.2   The Contracting Party shall in turn, if necessary, make a report to the Secretariat of the United Nations Economic Commission for Europe with a view to informing the other Contracting Parties.

1.8.5.3   An occurrence subject to report in accordance with 1.8.5.1 has occurred if dangerous goods were released or if there was an imminent risk of loss of product, if personal injury, material or environmental damage occurred, or if the authorities were involved and one or more of the following criteria has/have been met:

Personal injury means an occurrence in which death or injury directly relating to the dangerous goods carried has occurred, and where the injury

(a)   Requires intensive medical treatment;

(b)   Requires a stay in hospital of at least one day; or

(c)   Results in the inability to work for at least three consecutive days.

Loss of product means the release of dangerous goods

(a)   Of transport category 0 or 1 in quantities of 50 kg / 50 *l* or more;

(b)   Of transport category 2 in quantities of 333 kg / 333 *l* or more; or

(c)   Of transport category 3 or 4 in quantities of 1 000 kg / 1 000 *l* or more.

The loss of product criterion also applies if there was an imminent risk of loss of product in the above-mentioned quantities. As a rule, this has to be assumed if, owing to structural damage, the means of containment is no longer suitable for further carriage or if, for any other reason, a sufficient level of safety is no longer ensured (e.g. owing to distortion of tanks or containers, overturning of a tank or fire in the immediate vicinity).

If dangerous goods of Class 6.2 are involved, the obligation to report applies without quantity limitation.

In occurrences involving radioactive material, the criteria for loss of product are:

(a)   Any release of radioactive material from the packages;

(b)   Exposure leading to a breach of the limits set out in the regulations for protection of workers and members of the public against ionizing radiation (Schedule II of IAEA Safety Series No. 115 – "International Basic Safety Standards for Protection Against Ionizing Radiation and for Safety of Radiation Sources"); or

(c)   Where there is reason to believe that there has been a significant degradation in any package safety function (containment, shielding, thermal protection or criticality) that may have rendered the package unsuitable for continued carriage without additional safety measures.

NOTE: See the requirements of 7.5.11 CV33 (6) for undeliverable consignments.

Material damage or environmental damage means the release of dangerous goods, irrespective of the quantity, where the estimated amount of damage exceeds 50,000 Euros. Damage to any directly involved means of carriage containing dangerous goods and to the modal infrastructure shall not be taken into account for this purpose.

Involvement of authorities means the direct involvement of the authorities or emergency services during the occurrence involving dangerous goods and the evacuation of persons or closure of public traffic routes (roads/railways) for at least three hours owing to the danger posed by the dangerous goods.

If necessary, the competent authority may request further relevant information.

1.8.5.4   *Model for report on occurrences during the carriage of dangerous goods*

**Report on occurrences during the carriage of dangerous goods
in accordance with RID/ADR section 1.8.5**

- 67 -

Carrier/Railway infrastructure operator:
....
Address:
....
Contact name: ....………… Telephone: .... Fax: ....…………

(The competent authority shall remove this cover sheet before forwarding the report)

1. Mode	
□ Rail Wagon number (optional)	□ Road Vehicle registration (optional) ....

2. Date and location of occurrence	
Year: .... Month: .... Day: .... Time:	
Rail □ Station □ Shunting/marshalling yard □ Loading/unloading/transhipment site Location / Country: .... or □ Open line: Description of line: .... Kilometres:	Road □ Built-up area □ Loading/unloading/transhipment site □ Open road Location / Country: ....

3. Topography
□ Gradient/incline □ Tunnel □ Bridge/Underpass □ Crossing

4. Particular weather conditions
□ Rain □ Snow □ Ice □ Fog □ Thunderstorm □ Storm Temperature: ..... °C

5. Description of occurrence
□ Derailment/Leaving the road □ Collision □ Overturning/Rolling over □ Fire □ Explosion □ Loss □ Technical fault Additional description of occurrence: ....

6. Dangerous goods involved

UN Number [1]	Class	Packing Group	Estimated quantity of loss of products (kg or *l*) [2]	Means of containment [3]	Means of containment material	Type of failure of means of containment [4]

[1] For dangerous goods assigned to collective entries to which special provision 274 applies, also the technical name shall be indicated.

[2] For Class 7, indicate values according to the criteria in 1.8.5.3.

[3]   Indicate the appropriate number
1 Packaging
2 IBC
3 Large packaging
4 Small container
5 Wagon
6 Vehicle
7 Tank-wagon
8 Tank-vehicle
9 Battery-wagon
10 Battery-vehicle
11 Wagon with demountable tanks
12 Demountable tank
13 Large container
14 Tank-container
15 MEGC
16 Portable tank

[4]   Indicate the appropriate number
1 Loss
2 Fire
3 Explosion
4 Structural failure

7. Cause of occurrence (if clearly known)

☐ Technical fault
☐ Faulty load securing
☐ Operational cause (rail operation)
☐ Other:

....
....
....

8. Consequences of occurrence

Personal injury in connection with the dangerous goods involved:
☐ Deaths (number: ......)
☐ Injured (number: ......)

Loss of product:
☐ Yes
☐ No
☐ Imminent risk of loss of product

Material/Environmental damage:
☐ Estimated level of damage ≤ 50,000 Euros
☐ Estimated level of damage > 50,000 Euros

Involvement of authorities:
☐ Yes   ☐ Evacuation of persons for a duration of at least three hours caused by the dangerous goods involved
          ☐ Closure of public traffic routes for a duration of at least three hours caused by the dangerous goods involved
☐ No

If necessary, the competent authority may request further relevant information.

1.8.6 **Administrative controls for application of the conformity assessments, periodic inspections, intermediate inspections and exceptional checks described in 1.8.7**

1.8.6.1 *Approval of inspection bodies*

The competent authority may approve inspection bodies for conformity assessments, periodic inspections, intermediate inspections, exceptional checks and surveillance of the in-house inspection service as specified in 1.8.7.

1.8.6.2 *Operational obligations for the competent authority, its delegate or inspection body*

1.8.6.2.1 The competent authority, its delegate or inspection body shall carry out conformity assessments, periodic inspections, intermediate inspections and exceptional checks in a proportionate manner, avoiding unnecessary burdens. The competent authority, its delegate or inspection body shall perform its activities taking into consideration the size, the sector and the structure of the undertakings involved, the relative complexity of the technology and the serial character of production.

1.8.6.2.2 Nevertheless the competent authority, its delegate or inspection body shall respect the degree of rigour and the level of protection required for the compliance of the transportable pressure equipment by the provisions of parts 4 and 6 as applicable.

1.8.6.2.3 Where a competent authority, its delegate or inspection body finds out that requirements laid down in parts 4 or 6 have not been met by the manufacturer, it shall require the manufacturer to take appropriate corrective measures and it shall not issue any type approval certificate or certificate of conformity.

1.8.6.3 *Information obligation*

Contracting Parties to ADR shall publish their national procedures for the assessment, appointment and monitoring of inspection bodies and of any changes to that information.

1.8.6.4 *Delegation of inspection tasks*

NOTE: In-house inspection services according to 1.8.7.6 are not covered by 1.8.6.4.

1.8.6.4.1 Where an inspection body uses the services of any other entity (e.g. subcontractor, subsidiary), to carry out specific tasks connected with the conformity assessment, periodic inspection, intermediate inspection or exceptional checks, this entity shall be included in the accreditation of the inspection body, or it shall be accredited separately. In the case of separate accreditation, this entity shall be duly accredited according to standard EN ISO/IEC 17025:2005 and shall be recognised by the inspection body as an independent and impartial testing laboratory in order to perform testing tasks in accordance with its accreditation, or it shall be accredited according to standard EN ISO/IEC 17020:2012 (except clause 8.1.3). The inspection body shall ensure that this entity meets the requirements set out for the tasks given to it with the same level of competence and safety as laid down for inspection bodies (see 1.8.6.8) and the inspection body shall monitor it. The inspection body shall inform the competent authority about the above mentioned arrangements.

1.8.6.4.2 The inspection body shall take full responsibility for the tasks performed by such entities wherever the tasks are performed by them.

1.8.6.4.3 The inspection body shall not delegate the whole task of conformity assessment, periodic inspection, intermediate inspection or exceptional checks. In any case, the assessment and the issue of certificates shall be carried out by the inspection body itself.

1.8.6.4.4 Activities shall not be delegated without the agreement of the applicant.

1.8.6.4.5 The inspection body shall keep at the disposal of the competent authority the relevant documents concerning the assessment of the qualifications and the work carried out by the above mentioned entities.

1.8.6.5   *Information obligations for inspection bodies*

Any inspection body shall inform the competent authority, which had approved it, of the following:

(a)   Except when the provisions of 1.8.7.2.4 apply, any refusal, restriction, suspension or withdrawal of type approval certificates;

(b)   Any circumstance(s) affecting the scope of and conditions for the approval as granted by the competent authority;

(c)   Any request for information on conformity assessment activities performed which they have received from competent authorities monitoring compliance according to 1.8.1 or 1.8.6.6;

(d)   On request, conformity assessment activities performed within the scope of their approval and any other activity performed, including delegation of tasks.

1.8.6.6   The competent authority shall ensure the monitoring of the inspection bodies and shall revoke or restrict the approval given, if it notes that an approved body is no longer in compliance with the approval and the requirements of 1.8.6.8 or does not follow the procedures specified in the provisions of ADR.

1.8.6.7   If the approval of the inspection body is revoked or restricted or if the inspection body ceased activity, the competent authority shall take the appropriate steps to ensure that the files are either processed by another inspection body or kept available.

1.8.6.8   The inspection body shall:

(a)   Have a staff with an organizational structure, capable, trained, competent and skilled, to satisfactorily perform its technical functions;

(b)   Have access to suitable and adequate facilities and equipment;

(c)   Operate in an impartial manner and be free from any influence which could prevent it from doing so;

(d)   Ensure commercial confidentiality of the commercial and proprietary activities of the manufacturer and other bodies;

(e)   Maintain clear demarcation between actual inspection body functions and unrelated functions;

(f)   Have a documented quality system;

(g)   Ensure that the tests and inspections specified in the relevant standard and in ADR are performed; and

(h)   Maintain an effective and appropriate report and record system in accordance with 1.8.7 and 1.8.8.

The inspection body shall additionally be accredited according to the standard EN ISO/IEC 17020:2012 (except clause 8.1.3), as specified in 6.2.2.11, 6.2.3.6 and TA4 and TT9 of 6.8.4.

An inspection body starting a new activity may be approved temporarily. Before temporary designation, the competent authority shall ensure that the inspection body meets the requirements of the standard EN ISO/IEC 17020:2012 (except clause 8.1.3). The inspection body shall be accredited in its first year of activity to be able to continue this new activity.

1.8.7   **Procedures for conformity assessment and periodic inspection**

NOTE: In this section, "relevant body" means a body assigned in 6.2.2.11 when certifying UN pressure receptacles, in 6.2.3.6 when approving non-UN pressure receptacles and in special provisions TA4 and TT9 of 6.8.4.

1.8.7.1   *General provisions*

1.8.7.1.1   The procedures in section 1.8.7 shall be applied according to 6.2.3.6 when approving non-UN pressure receptacles and according to TA4 and TT9 of 6.8.4 when approving tanks, battery-vehicles and MEGCs.

The procedures in section 1.8.7 may be applied according to the table in 6.2.2.11 when certifying UN pressure receptacles.

1.8.7.1.2   Each application for

(a)   The type approval in accordance with 1.8.7.2 or;

(b)   The supervision of manufacture in accordance with 1.8.7.3 and the initial inspection and test in accordance with 1.8.7.4; or

(c)   The periodic inspection, intermediate inspection and exceptional checks in accordance with 1.8.7.5

shall be lodged by the applicant with a single competent authority, its delegate or an approved inspection body of his choice.

1.8.7.1.3   The application shall include:

(a)   The name and address of the applicant;

(b)   For conformity assessment where the applicant is not the manufacturer, the name and address of the manufacturer;

(c)   A written declaration that the same application has not been lodged with any other competent authority, its delegate or inspection body;

(d)   The relevant technical documentation specified in 1.8.7.7;

(e)   A statement allowing the competent authority, its delegate or inspection body access for inspection purposes to the locations of manufacture, inspection, testing and storage and providing it with all necessary information.

1.8.7.1.4   Where the applicant can demonstrate to the satisfaction of the competent authority or its delegated inspection body conformity with 1.8.7.6 the applicant may establish an in-house inspection service which may perform part or all of the inspections and tests when specified in 6.2.2.11 or 6.2.3.6.

1.8.7.1.5   Design type approval certificates and certificates of conformity - including the technical documentation - shall be retained by the manufacturer or by the applicant for the type approval, if he is not the manufacturer, and by the inspection body, who issued the certificate, for a period of at least 20 years starting from the last date of production of products of the same type.

1.8.7.1.6   When a manufacturer or owner intends to cease operation, he shall send the documentation to the competent authority. The competent authority shall then retain the documentation for the rest of the period specified in 1.8.7.1.5.

1.8.7.2   *Type approval*

Type approvals authorise the manufacture of pressure receptacles, tanks, battery-vehicles or MEGCs within the period of validity of that approval.

1.8.7.2.1   The applicant shall:

(a)   In the case of pressure receptacles, place at the disposal of the relevant body representative samples of the production envisaged. The relevant body may request further samples if required by the test programme;

(b)   In the case of tanks, battery-vehicles or MEGCs, give access to the prototype for type testing.

1.8.7.2.2   The relevant body shall:

(a)   Examine the technical documentation specified in 1.8.7.7.1 to verify that the design is in accordance with the relevant provisions of ADR, and the prototype or the prototype lot has been manufactured in conformity with the technical documentation and is representative of the design;

(b)   Perform the examinations and witness the tests specified in ADR, to determine that the provisions have been applied and fulfilled, and the procedures adopted by the manufacturer meet the requirements;

(c)   Check the certificate(s) issued by the materials manufacturer(s) against the relevant provisions of ADR;

(d)   As applicable, approve the procedures for the permanent joining of parts or check that they have been previously approved, and verify that the staff undertaking the permanent joining of parts and the non-destructive tests are qualified or approved;

(e)   Agree with the applicant the location and testing facilities where the examinations and necessary tests are to be carried out.

The relevant body shall issue a type-examination report to the applicant.

1.8.7.2.3   Where the type satisfies all applicable provisions, the competent authority, its delegate or the inspection body, shall issue a type approval certificate to the applicant.

This certificate shall contain:

(a)   The name and address of the issuer;

(b)   The name and address of the manufacturer and of the applicant when the applicant is not the manufacturer;

(c)   A reference to the version of ADR and standards used for the type examination;

(d)   Any requirements resulting from the examination;

(e)   The necessary data for identification of the type and variation, as defined by the relevant standard;

(f)   The reference to the type examination report(s); and

(g)   The maximum period of validity of the type approval.

A list of the relevant parts of the technical documentation shall be annexed to the certificate (see 1.8.7.7.1).

1.8.7.2.4   The type approval shall be valid for a maximum of ten years. If within that period the relevant technical requirements of ADR (including referenced standards) have changed so that the approved type is no longer in conformity with them, the relevant body which issued the type approval shall withdraw it and inform the holder of the type approval.

NOTE: *For the ultimate dates for withdrawal of existing type approvals, see column (5) of the tables in 6.2.4 and 6.8.2.6 or 6.8.3.6 as appropriate.*

If a type approval has expired or has been withdrawn, the manufacture of the pressure receptacles, tanks, battery-vehicles or MEGCs according to that type approval is no longer authorised.

In such a case, the relevant provisions concerning the use, periodic inspection and intermediate inspection of pressure receptacles, tanks, battery-vehicles or MEGCs contained in the type approval which has expired or has been withdrawn shall continue to apply to these pressure receptacles, tanks, battery-vehicles or MEGCs constructed before the expiry or the withdrawal if they may continue to be used.

They may continue to be used as long as they remain in conformity with the requirements of ADR. If they are no longer in conformity with the requirements of ADR they may continue to be used only if such use is permitted by relevant transitional measures in Chapter 1.6.

Type approvals may be renewed by a complete review and assessment for conformity with the provisions of ADR applicable at the date of renewal. Renewal is not permitted after a type approval has been withdrawn. Interim amendments of an existing type approval (e.g. for pressure receptacles minor amendments such as the addition of further sizes or volumes not affecting conformity, or for tanks see 6.8.2.3.2) do not extend or modify the original validity of the certificate.

NOTE: The review and assessment of conformity can be done by a body other than the one which issued the original type approval.

The issuing body shall keep all documents for the type approval (see 1.8.7.7.1) for the whole period of validity including its renewals if granted.

1.8.7.2.5   In the case of a modification of a pressure receptacle, tank, battery-vehicle or MEGC with a valid, expired or withdrawn type approval, the testing, inspection and approval are limited to the parts of the pressure receptacle, tank, battery-vehicle or MEGC that have been modified. The modification shall meet the provisions of ADR applicable at the time of the modification. For all parts of the pressure receptacle, tank, battery-vehicle or MEGC not affected by the modification, the documentation of the initial type approval remains valid.

A modification may apply to one or more pressure receptacles, tanks, battery-vehicles or MEGCs covered by a type approval.

A certificate approving the modification shall be issued to the applicant by the competent authority of any Contracting Party to ADR or by a body designated by this authority. For tanks, battery-vehicles or MEGCs, a copy shall be kept as part of the tank record.

Each application for an approval certificate for a modification shall be lodged by the applicant with a single competent authority or body designated by this authority.

1.8.7.3   *Supervision of manufacture*

1.8.7.3.1   The manufacturing process shall be subject to a survey by the relevant body to ensure the product is produced in conformity with the provisions of the type approval.

1.8.7.3.2   The applicant shall take all the necessary measures to ensure that the manufacturing process complies with the applicable provisions of ADR and of the type approval certificate and its annexes.

1.8.7.3.3   The relevant body shall:

(a)   Verify the conformity with the technical documentation specified in 1.8.7.7.2;

(b)   Verify that the manufacturing process produces products in conformity with the requirements and the documentation which apply to it;

(c)   Verify the traceability of materials and check the material certificate(s) against the specifications;

(d)   As applicable, verify that the personnel undertaking the permanent joining of parts and the non-destructive tests are qualified or approved;

(e)   Agree with the applicant on the location where the examinations and necessary tests are to be carried out; and

(f)   Record the results of its survey.

1.8.7.4   *Initial inspection and tests*

1.8.7.4.1   The applicant shall:

(a)   Affix the marks specified in ADR; and

(b)   Supply to the relevant body the technical documentation specified in 1.8.7.7.

1.8.7.4.2   The relevant body shall:

(a)   Perform the necessary examinations and tests in order to verify that the product is manufactured in accordance with the type approval and the relevant provisions;

(b)   Check the certificates supplied by the manufacturers of service equipment against the service equipment;

(c)   Issue an initial inspection and test report to the applicant relating to the detailed tests and verifications carried out and the verified technical documentation;

(d)   Draw up a written certificate of conformity of the manufacture and affix its registered mark when the manufacture satisfies the provisions; and

(e)   Check if the type approval remains valid after provisions of ADR (including referenced standards) relevant to the type approval have changed.

The certificate in (d) and report in (c) may cover a number of items of the same type (group certificate or report).

1.8.7.4.3   The certificate shall contain as a minimum:

(a)   The name and address of the relevant body;

(b)   The name and address of the manufacturer and the name and address of the applicant, if not the manufacturer;

(c)   A reference to the version of the ADR and standards used for the initial inspections and tests;

(d)   The results of the inspections and tests;

(e)   The data for identification of the inspected product(s), at least the serial number or for non refillable cylinders the batch number; and

(f)   The type approval number.

1.8.7.5   *Periodic inspection, intermediate inspection and exceptional checks*

1.8.7.5.1   The relevant body shall:

(a)   Perform the identification and verify the conformity with the documentation;

(b)   Carry out the inspections and witness the tests in order to check that the requirements are met;

(c)   Issue reports of the results of the inspections and tests, which may cover a number of items; and

(d)   Ensure that the required marks are applied.

1.8.7.5.2   Reports of periodic inspections and tests of pressure receptacles shall be retained by the applicant at least until the next periodic inspection.

NOTE: For tanks, see provisions for tank records in 4.3.2.1.7.

1.8.7.6   *Surveillance of the applicant's in-house inspection service*

1.8.7.6.1   The applicant shall:

(a)   Implement an in-house inspection service with a quality system for inspections and tests documented in 1.8.7.7.5 and subject to surveillance;

(b)   Fulfil the obligations arising out of the quality system as approved and to ensure that it remains satisfactory and efficient;

(c)   Appoint trained and competent personnel for the in-house inspection service; and

(d)   Affix the registered mark of the inspection body where appropriate.

1.8.7.6.2   The inspection body shall carry out an initial audit. If satisfactory the inspection body shall issue an authorisation for a period not exceeding three years. The following provisions shall be met:

(a)   This audit shall confirm that the inspections and tests performed on the product are in compliance with the requirements of ADR;

(b)   The inspection body may authorise the in-house inspection service of the applicant to affix the registered mark of the inspection body to each approved product;

(c)   The authorisation may be renewed after a satisfactory audit in the last year prior to the expiry. The new period of validity shall begin with the date of expiry of the authorisation; and

(d)   The auditors of the inspection body shall be competent to carry out the assessment of conformity of the product covered by the quality system.

1.8.7.6.3   The inspection body shall carry out periodic audits within the duration of the authorisation to make sure that the applicant maintains and applies the quality system. The following provisions shall be met:

(a)   A minimum of two audits shall be carried out in a 12 month period;

(b)   The inspection body may require additional visits, training, technical changes, modifications of the quality system, restrict or prohibit the inspections and tests to be done by the applicant;

(c)   The inspection body shall assess any changes in the quality system and decide whether the modified quality system will still satisfy the requirements of the initial audit or whether a full reassessment is required;

(d)   The auditors of the inspection body shall be competent to carry out the assessment of conformity of the product covered by the quality system; and

(e)   The inspection body shall provide the applicant with a visit or audit report and, if a test has taken place, with a test report.

1.8.7.6.4   In cases of non conformity with the relevant requirements the inspection body shall ensure that corrective measures are taken. If corrective measures are not taken in due time, the inspection body shall suspend or withdraw the permission for the in-house inspection service to carry out its activities. The notice of suspension or withdrawal shall be transmitted to the competent authority. A report shall be provided to the applicant giving detailed reasons for the decisions taken by the inspection body.

1.8.7.7   *Documents*

The technical documentation shall enable an assessment to be made of conformity with the relevant requirements.

1.8.7.7.1   *Documents for type approval*

The applicant shall provide as appropriate:

(a)   The list of standards used for the design and manufacture;

(b)   A description of the type including all variations;

(c) The instructions according to the relevant column of table A of Chapter 3.2 or a list of dangerous goods to be transported for dedicated products;

(d) A general assembly drawing or drawings;

(e) The detailed drawings, including the dimensions used for the calculations, of the product, the service equipment, the structural equipment, the marking and/or the labelling necessary to verify the conformity;

(f) The calculation notes, results and conclusions;

(g) The list of the service equipment with the relevant technical data and information on the safety devices including the calculation of the relief capacity if relevant;

(h) The list of material requested in the standard for manufacture used for every part, sub-part, lining, service and structural equipment and the corresponding material specifications or the corresponding declaration of conformity to ADR;

(i) The approved qualification of permanent joining process;

(j) The description of the heat treatment process(es); and

(k) The procedures, descriptions and records of all relevant tests listed in the standards or ADR for the type approval and for the manufacture.

1.8.7.7.2   *Documents for the supervision of manufacture*

The applicant shall make available as appropriate:

(a) The documents listed in 1.8.7.7.1;

(b) A copy of the type approval certificate;

(c) The manufacturing procedures including test procedures;

(d) The manufacturing records;

(e) The approved qualifications of permanent joining operators;

(f) The approved qualifications of the non destructive test operators;

(g) The reports of the destructive and non destructive tests;

(h) The heat treatment records; and

(i) The calibration records.

1.8.7.7.3   *Documents for initial inspection and tests*

The applicant shall make available as appropriate:

(a) The documents listed in 1.8.7.7.1 and 1.8.7.7.2;

(b) The material certificates of the product and any sub-parts;

(c) The declarations of conformity and material certificates of the service equipment; and

(d) A declaration of conformity including the description of the product and all the variations adopted from the type approval.

1.8.7.7.4   *Documents for periodic inspections, intermediate inspections and exceptional checks*

The applicant shall make available as appropriate:

(a) For pressure receptacles, the documents specifying special requirements when the manufacturing and periodic inspections and tests standards so require;

(b) For tanks:

    (i) the tank record; and

    (ii) one or more of the documents mentioned in 1.8.7.7.1 to 1.8.7.7.3.

1.8.7.7.5   *Documents for the assessment of in-house inspection service*

The applicant for in-house inspection service shall make available the quality system documentation as appropriate:

(a) The organizational structure and responsibilities;

(b) The relevant inspection and test, quality control, quality assurance and process operation instructions, and systematic actions that will be used;

(c) The quality records, such as inspection reports, test data, calibration data and certificates;

(d) The management reviews to ensure the effective operation of the quality system arising from the audits in accordance with 1.8.7.6;

(e) The process describing how customer and regulation requirements are met;

(f) The process for control of documents and their revision;

(g) The procedures for dealing with non-conforming products; and

(h) The training programmes and qualification procedures for relevant personnel.

1.8.7.8   ***Products manufactured, approved, inspected and tested according to standards***

The requirements of 1.8.7.7 are considered to have been complied with if the following standards, as relevant, are applied:

Applicable subsection and paragraph	References	Title of the document
1.8.7.7.1 to 1.8.7.7.4	EN 12972:2007	Tanks for transport of dangerous goods - Testing, inspection and marking of metallic tanks

1.8.8   **Procedures for conformity assessment of gas cartridges**

When assessing the conformity of gas cartridges, one of the following procedures shall be applied:

(a) The procedure in section 1.8.7 for non-UN pressure receptacles, with the exception of 1.8.7.5; or

(b) The procedure in sub-sections 1.8.8.1 to 1.8.8.7.

1.8.8.1   *General provisions*

1.8.8.1.1   The supervision of manufacture shall be carried out by an Xa body and the tests as required in 6.2.6 shall be carried out either by that Xa body or by an IS-body approved by that Xa body; for definition of Xa and IS bodies see definitions in 6.2.3.6.1. Conformity assessment shall be carried out by the competent authority, its delegate or its approved inspection body of a Contracting Party to ADR.

1.8.8.1.2   By the application of 1.8.8, the applicant shall demonstrate, ensure and declare on his sole responsibility the conformity of gas cartridges with the provisions of 6.2.6 and all further applicable provisions of ADR.

1.8.8.1.3   The applicant shall

(a) Carry out a design type examination of each type of gas cartridges (including materials to be used and variations of that type, e.g. volumes, pressures, drawings and closing and release devices) according to 1.8.8.2;

<table>
<tr><td>(b)</td><td>Operate an approved quality system for design, manufacture, inspection and testing according to 1.8.8.3;</td></tr>
<tr><td>(c)</td><td>Operate an approved testing regime according to 1.8.8.4 for the tests required in 6.2.6;</td></tr>
<tr><td>(d)</td><td>Apply for the approval of his quality system for supervision of manufacture and for testing to one Xa body of his choice of the Contracting Party; if the applicant is not established in a Contracting Party he shall apply to one Xa body of a Contracting Party prior to first transport into a Contracting Party;</td></tr>
<tr><td>(e)</td><td>If the gas cartridge is finally assembled from parts manufactured by the applicant by one or more other enterprise(s), provide written instructions how to assemble and fill the gas cartridges to meet the provisions of his type examination certificate.</td></tr>
</table>

1.8.8.1.4   Where the applicant and enterprises assembling or filling gas cartridges according to the instructions of the applicant, can demonstrate to the satisfaction of the Xa body conformity with the provisions of 1.8.7.6 excluding 1.8.7.6.1 (d) and 1.8.7.6.2 (b), they may establish an in-house inspection service which may perform part or all of the inspections and tests specified in 6.2.6.

1.8.8.2   *Design type examination*

1.8.8.2.1   The applicant shall establish a technical documentation for each type of gas cartridges including the technical standard(s) applied. If he chooses to apply a standard not referenced in 6.2.6, he shall add the standard applied to the documentation.

1.8.8.2.2   The applicant shall retain the technical documentation together with samples of that type at the disposal of the Xa body during production and afterwards for a period of minimum five years starting from the last date of production of gas cartridges according to that type examination certificate.

1.8.8.2.3   The applicant shall after careful examination issue a design type certificate which shall be valid for a maximum period of ten years; he shall add this certificate to the documentation. This certificate authorises him to produce gas cartridges of that type for that period.

1.8.8.2.4   If within that period the relevant technical requirements of ADR (including referenced standards) have changed so that the design type is no longer in conformity with them, the applicant shall withdraw his type examination certificate and inform the Xa body.

1.8.8.2.5   The applicant may after careful and complete review reissue the certificate for another period of maximum ten years.

1.8.8.3   *Supervision of manufacture*

1.8.8.3.1   The procedure of design type examination as well as the manufacturing process shall be subject to a survey by the Xa body to ensure the type certified by the applicant and the product as produced are in conformity with the provisions of the design type certificate and the applicable provisions of ADR. If 1.8.8.1.3 (e) applies, the assembling and filling enterprises shall be included in that procedure.

1.8.8.3.2   The applicant shall take all the necessary measures to ensure that the manufacturing process complies with the applicable provisions of ADR and of his design type certificate and its annexes. If 1.8.8.1.3 (e) applies, the assembling and filling enterprises shall be included in that procedure.

1.8.8.3.3   The Xa body shall:

<table>
<tr><td>(a)</td><td>Verify the conformity of the design type examination of the applicant and conformity of the type of gas cartridges with the technical documentation specified in 1.8.8.2;</td></tr>
<tr><td>(b)</td><td>Verify that the manufacturing process produces products in conformity with the requirements and the documentation which apply to it; if the gas cartridge is finally assembled from parts manufactured by the applicant by one or more enterprise(s), the Xa body shall also verify that the gas cartridges are in full conformity with all applicable provisions after final assembly and filling and that the instructions of the applicant are correctly applied;</td></tr>
<tr><td>(c)</td><td>Verify that the personnel undertaking the permanent joining of parts and the tests are qualified or approved;</td></tr>
<tr><td>(d)</td><td>Record the results of its surveys.</td></tr>
</table>

1.8.8.3.4   If the findings of the Xa body show non-conformity of the design type certificate of the applicant or the manufacturing process, he shall require appropriate corrective measures or withdrawal of the certificate from the applicant.

1.8.8.4   *Leakproofness test*

1.8.8.4.1   The applicant and enterprises finally assembling and filling gas cartridges according to the instructions of the applicant shall:

(a)   Carry out the tests required in 6.2.6;

(b)   Record the test results;

(c)   Issue a certificate of conformity only for gas cartridges, which are in full compliance with the provisions of his design type examination and the applicable provisions of ADR and have successfully passed the tests as required in 6.2.6;

(d)   Retain the documentation as specified in 1.8.8.7 during production and afterwards for a period of minimum five years from the last date of production of gas cartridges belonging to one type approval for inspection by the Xa body at random intervals;

(e)   Affix a durable and legible mark identifying the type of gas cartridge, the applicant and the date of production or batch number; where due to limited available space the mark cannot be fully applied to the body of the gas cartridge, he shall affix a durable tag with this information to the gas cartridge or place it together with a gas cartridge in an inner packaging.

1.8.8.4.2   The Xa body shall:

(a)   Perform the necessary examinations and tests at random intervals, but at least shortly after starting of manufacture of a type of gas cartridges and thereafter at least once every three years, in order to verify that the procedure for design type examination of the applicant as well as that the manufacture and testing of the product are carried out in accordance with the design type certificate and the relevant provisions;

(b)   Check the certificates supplied by the applicant;

(c)   Carry out the tests as required in 6.2.6 or approve the program of testing and the in-house inspection service to carry out the tests.

1.8.8.4.3   The certificate shall contain as a minimum:

(a)   The name and address of the applicant and, when the final assembly is not carried out by the applicant but by an enterprise or enterprises in accordance with the written instructions of the applicant, the name(s) and address(es) of these enterprises;

(b)   A reference to the version of ADR and the standard(s) used for manufacture and tests;

(c)   The result of inspections and tests;

(d)   The data for marking as required in 1.8.8.4.1 (e).

1.8.8.5   *(Reserved)*

1.8.8.6   *Surveillance of the in-house inspection service*

When the applicant or enterprise assembling or filling gas cartridges has established an in-house inspection service, the provisions of 1.8.7.6 excluding 1.8.7.6.1 (d) and 1.8.7.6.2 (b) shall be applied. The enterprise assembling or filling gas cartridges shall comply with the provisions relevant to the applicant.

1.8.8.7   *Documents*

The provisions of 1.8.7.7.1, 1.8.7.7.2, 1.8.7.7.3 and 1.8.7.7.5 shall be applied.

CHAPTER 1.9

TRANSPORT RESTRICTIONS BY THE COMPETENT AUTHORITIES

1.9.1   In accordance with Article 4, paragraph 1 of ADR, the entry of dangerous goods into the territory of Contracting Parties may be subject to regulations or prohibitions imposed for reasons other than safety during carriage. Such regulations or prohibitions shall be published in an appropriate form.

1.9.2   Subject to the provisions of 1.9.3, a Contracting Party may apply to vehicles engaged in the international carriage of dangerous goods by road on its territory certain additional provisions not included in ADR, provided that those provisions do not conflict with Article 2, paragraph 2 of the Agreement, and are contained in its domestic legislation applying equally to vehicles engaged in the domestic carriage of dangerous goods by road on the territory of that Contracting Party.

1.9.3   Additional provisions falling within the scope of 1.9.2 are as follows:

     (a)   Additional safety requirements or restrictions concerning vehicles using certain structures such as bridges, vehicles using combined transport modes such as ferries or trains, or vehicles entering or leaving ports or other transport terminals;

     (b)   Requirements for vehicles to follow prescribed routes to avoid commercial or residential areas, environmentally sensitive areas, industrial zones containing hazardous installations or roads presenting severe physical hazards;

     (c)   Emergency requirements regarding routeing or parking of vehicles carrying dangerous goods resulting from extreme weather conditions, earthquake, accident, industrial action, civil disorder or military hostilities;

     (d)   Restrictions on movement of dangerous goods traffic on certain days of the week or year.

1.9.4   The competent authority of the Contracting Party applying on its territory any additional provisions within the scope of 1.9.3 (a) and (d) above shall notify the secretariat of the United Nations Economic Commission for Europe of the additional provisions, which secretariat shall bring them to the attention of the Contracting Parties[1].

1.9.5   Tunnel restrictions

NOTE: Provisions concerning restrictions for the passage of vehicles through road tunnels are also included in Chapter 8.6.

1.9.5.1   *General provisions*

When applying restrictions to the passage of vehicles carrying dangerous goods through tunnels, the competent authority shall assign the road tunnel to one of the tunnel categories defined in 1.9.5.2.2. Account should be taken of the tunnel characteristics, risk assessment including availability and suitability of alternative routes and modes and traffic management considerations. The same tunnel may be assigned to more than one tunnel category, e.g. depending on the hours of the day, or the day of the week etc.

1.9.5.2   *Categorization*

1.9.5.2.1   The categorization shall be based on the assumption that in tunnels there are three major dangers which may cause numerous victims or serious damage to the tunnel structure:

     (a)   Explosions;

     (b)   Release of toxic gas or volatile toxic liquid;

     (c)   Fires.

---

[1]   *A General Guideline for the Calculation of Risks in the Transport of Dangerous Goods by Road may be consulted on the website of the secretariat of the United Nations Economic Commission for Europe (http://www.unece.org/trans/danger/danger.htm).*

1.9.5.2.2   The five tunnel categories are the following:

Tunnel category A:

No restrictions for the carriage of dangerous goods;

Tunnel category B:

Restriction for the carriage of dangerous goods which may lead to a very large explosion;

The following dangerous goods are considered to fulfil this criterion[2]:

Class 1:	Compatibility groups A and L;
Class 3:	Classification code D (UN Nos. 1204, 2059, 3064, 3343, 3357 and 3379);
Class 4.1:	Classification codes D and DT; and
	Self-reactive substances, type B (UN Nos. 3221, 3222, 3231 and 3232);
Class 5.2:	Organic peroxides, type B (UN Nos. 3101, 3102, 3111 and 3112).
When the total net explosive mass per transport unit is greater than 1000 kg:	
Class 1:	Divisions 1.1, 1.2 and 1.5 (except compatibility groups A and L).
When carried in tanks:	
Class 2:	Classification codes F, TF and TFC;
Class 4.2:	Packing group I;
Class 4.3:	Packing group I;
Class 5.1:	Packing group I.
Class 6.1:	UN No. 1510

Tunnel category C:

Restriction for the carriage of dangerous goods which may lead to a very large explosion, a large explosion or a large toxic release;

The following dangerous goods are considered to fulfil this criterion[2]:

-    the dangerous goods restricted in tunnel category B, and

-    the following dangerous goods:

Class 1:	Divisions 1.1, 1.2 and 1.5 (except compatibility groups A and L); and
	Division 1.3 (compatibility groups H and J);
Class 7:	UN Nos. 2977 and 2978.
When the net explosive mass per transport unit is greater than 5000 kg:	
Class 1:	Division 1.3 (compatibility groups C and G).
When carried in tanks:	
Class 2:	Classification codes 2A, 2O, 3A and 3O, and classification codes containing the letter T only or letter groups TC, TO and TOC
Class 3:	Packing group I for classification codes FC, FT1, FT2 and FTC;
Class 6.1:	Packing group I, except UN No. 1510
Class 8:	Packing group I for classification codes CT1, CFT and COT.

---

[2]   *The assessment is based on the intrinsic dangerous properties of the goods, the type of containment and the quantity carried.*

Tunnel category D:

Restriction for the carriage of dangerous goods which may lead to a very large explosion, to a large explosion, to a large toxic release or to a large fire;

The following dangerous goods are considered to fulfil this criterion[2]:

- the dangerous goods restricted in tunnel category C, and

- the following dangerous goods:

Class 1:	Division 1.3 (compatibility groups C and G);
Class 2:	Classification codes F, FC, T, TF, TC, TO, TFC and TOC;
Class 4.1:	Self-reactive substances, types C, D, E and F; and
	UN Nos. 2956, 3241, 3242, 3251, 3531, 3532, 3533 and 3534;
Class 5.2:	Organic peroxides, types C, D, E and F;
Class 6.1:	Packing group I for classification codes TF1, TFC and TFW and UN No.3507; and
	Toxic by inhalation entries for which special provision 354 is assigned in column (6) of Table A of Chapter 3.2 and toxic by inhalation entries of UN Nos. 3381 to 3390;
Class 8:	Packing group I for classification codes CT1, CFT and COT;
Class 9:	Classification codes M9 and M10.
When carried in bulk or in tanks:	
Class 3	
Class 4.2:	Packing group II;
Class 4.3:	Packing group II;
Class 6.1:	Packing group II; and
	Packing group III for classification code TF2;
Class 8:	Packing group I for classification codes CF1, CFT and CW1; and
	Packing group II for classification codes CF1 and CFT
Class 9:	Classification codes M2 and M3.

Tunnel category E:

Restriction for the carriage of all dangerous goods other than those for which '(-)' is marked in Column (15) of Table A of Chapter 3.2 and for all dangerous goods in accordance with the provisions of Chapter 3.4 if the quantities carried exceed 8 tonnes total gross mass per transport unit..

NOTE: *For the dangerous goods assigned to UN Nos. 2919 and 3331, restrictions to the passage through tunnels may, however, be part of the special arrangement approved by the competent authority(ies) on the basis of 1.7.4.2.*

1.9.5.3   *Provisions for road signs and notification of restrictions*

1.9.5.3.1   Contracting Parties shall indicate tunnel prohibitions and alternative routes by means of signs and signals.

1.9.5.3.2   For this purpose, they may use signs C, 3h and D, 10a, 10b and 10c and signals according to the Vienna Convention on Road Signs and Signals (Vienna, 1968) and the European Agreement supplementing the Convention on Road Signs and Signals (Geneva, 1971) as interpreted by the Resolution on Road Signs and Signals (R.E.2) of the UNECE Inland Transport Committee Principal Working Party on Road Transport, as amended.

1.9.5.3.3   In order to facilitate international understanding of signs, the system of signs and signals prescribed in the Vienna Convention is based on the use of shapes, and colours characteristic of each class of signs and wherever possible, on the use of graphic symbols rather than inscriptions. Where Contracting Parties consider it necessary to modify the signs and symbols prescribed, the modifications made shall not alter their essential characteristics. Where Contracting Parties do not apply the Vienna

---

[2]   *The assessment is based on the intrinsic dangerous properties of the goods, the type of containment and the quantity carried.*

Convention, the prescribed signs and symbols may be modified, provided that the modifications made shall not alter their essential intent.

1.9.5.3.4   Traffic signs and signals intended to prohibit access of vehicles carrying dangerous goods to road tunnels shall be affixed at a place where the choice of alternative routes is possible.

1.9.5.3.5   When access to tunnels is restricted or alternative routes are prescribed, the signs shall be displayed with additional panels as follows:

No sign: no restriction

Sign with additional panel bearing the letter B: applies to vehicles carrying dangerous goods not allowed in tunnels of category B;

Sign with additional panel bearing the letter C: applies to vehicles carrying dangerous goods not allowed in tunnels of category C;

Sign with additional panel bearing the letter D: applies to vehicles carrying dangerous goods not allowed in tunnels of category D;

Sign with additional panel bearing the letter E: applies to vehicles carrying dangerous goods not allowed in tunnels of category E.

1.9.5.3.6   Tunnel restrictions shall apply to transport units for which an orange-coloured plate marking in accordance with 5.3.2 is required, except for the carriage of dangerous goods for which '(-)' is marked in Column (15) of Table A of Chapter 3.2. For the dangerous goods assigned to UN Nos. 2919 and 3331, restrictions to the passage through tunnels may, however, be part of the special arrangement approved by the competent authority(ies) on the basis of 1.7.4.2. For tunnels of category E, they shall apply also to transport units for which a marking in accordance with 3.4.13 is required or carrying containers for which a marking in accordance with 3.4.13 is required.

Tunnel restrictions shall not apply when dangerous goods are carried in accordance with 1.1.3, except when transport units carrying such goods are marked in accordance with 3.4.13 subject to 3.4.14.

1.9.5.3.7   Restrictions shall be published officially and made publicly available. Contracting Parties shall notify the secretariat of UNECE of such restrictions and the secretariat shall make this information publicly available on its website.

1.9.5.3.8   When Contracting Parties apply specific operating measures designed to reduce the risks and related to some or all vehicles using tunnels, such as declaration before entering or passage in convoys escorted by accompanying vehicles, such operating measures shall be published officially and made publicly available.

CHAPTER 1.10

SECURITY PROVISIONS

NOTE: *For the purposes of this Chapter, security means measures or precautions to be taken to minimise theft or misuse of dangerous goods that may endanger persons, property or the environment.*

1.10.1   **General provisions**

1.10.1.1   All persons engaged in the carriage of dangerous goods shall consider the security requirements set out in this Chapter commensurate with their responsibilities.

1.10.1.2   Dangerous goods shall only be offered for carriage to carriers that have been appropriately identified.

1.10.1.3   Areas within temporary storage terminals, temporary storage sites, vehicle depots, berthing areas and marshalling yards used for the temporary storage during carriage of dangerous goods shall be properly secured, well lit and, where possible and appropriate, not accessible to the general public.

1.10.1.4   Each member of a vehicle crew shall carry with them means of identification, which includes their photograph, during carriage of dangerous goods.

1.10.1.5   Safety inspections in accordance with 1.8.1 and 7.5.1.1 shall cover appropriate security measures.

1.10.1.6   The competent authority shall maintain up-to-date registers of all valid training certificates for drivers stipulated in 8.2.1 issued by it or by any recognized organization.

1.10.2   **Security training**

1.10.2.1   The training and the refresher training specified in Chapter 1.3 shall also include elements of security awareness. The security refresher training need not be linked to regulatory changes only.

1.10.2.2   Security awareness training shall address the nature of security risks, recognising security risks, methods to address and reduce such risks and actions to be taken in the event of a security breach. It shall include awareness of security plans (if appropriate) commensurate with the responsibilities and duties of individuals and their part in implementing security plans.

1.10.2.3   Such training shall be provided or verified upon employment in a position involving dangerous goods transport and shall be periodically supplemented with refresher training.

1.10.2.4   Records of all security training received shall be kept by the employer and made available to the employee or competent authority, upon request. Records shall be kept by the employer for a period of time established by the competent authority.

1.10.3   **Provisions for high consequence dangerous goods**

1.10.3.1   *Definition of high consequence dangerous goods*

1.10.3.1.1   High consequence dangerous goods are those which have the potential for misuse in a terrorist event and which may, as a result, produce serious consequences such as mass casualties, mass destruction or, particularly for Class 7, mass socio-economic disruption.

1.10.3.1.2   High consequence dangerous goods in classes other than Class 7 are those listed in Table 1.10.3.1.2 below and carried in quantities greater than those indicated therein.

Table 1.10.3.1.2: List of high consequence dangerous goods

Class	Division	Substance or article	Quantity		
			Tank (*l*) [c]	Bulk (kg) [d]	Packages (kg)
1	1.1	Explosives	a	a	0
	1.2	Explosives	a	a	0
	1.3	Compatibility group C explosives	a	a	0
	1.4	Explosives of UN Nos. 0104, 0237, 0255, 0267, 0289, 0361, 0365, 0366, 0440, 0441, 0455, 0456 and 0500	a	a	0
	1.5	Explosives	0	a	0
2		Flammable gases (classification codes including only the letter F)	3000	a	b
		Toxic gases (classification codes including letters T, TF, TC, TO, TFC or TOC) excluding aerosols	0	a	0
3		Flammable liquids of packing groups I and II	3000	a	b
		Desensitized explosives	0	a	0
4.1		Desensitized explosives	a	a	0
4.2		Packing group I substances	3000	a	b
4.3		Packing group I substances	3000	a	b
5.1		Oxidizing liquids of packing group I	3000	a	b
		Perchlorates, ammonium nitrate, ammonium nitrate fertilisers and ammonium nitrate emulsions or suspensions or gels	3000	3000	b
6.1		Toxic substances of packing group I	0	a	0
6.2		Infectious substances of Category A (UN Nos. 2814 and 2900, except for animal material)	a	0	0
8		Corrosive substances of packing group I	3000	a	b

[a] *Not relevant.*

[b] *The provisions of 1.10.3 do not apply, whatever the quantity is.*

[c] *A value indicated in this column is applicable only if carriage in tanks is authorized, in accordance with Chapter 3.2, Table A, column (10) or (12). For substances that are not authorized for carriage in tanks, the instruction in this column is not relevant.*

[d] *A value indicated in this column is applicable only if carriage in bulk is authorized, in accordance with Chapter 3.2, Table A, column (10) or (17). For substances that are not authorized for carriage in bulk, the instruction in this column is not relevant.*

1.10.3.1.3   For dangerous goods of Class 7, high consequence radioactive material is that with an activity equal to or greater than a transport security threshold of 3 000 A_2 per single package (see also 2.2.7.2.2.1) except for the following radionuclides where the transport security threshold is given in Table 1.10.3.1.3 below.

Table 1.10.3.1.3: Transport security thresholds for specific radionuclides

Element	Radionuclide	Transport security threshold (TBq)
Americium	Am-241	0.6
Gold	Au-198	2
Cadmium	Cd-109	200
Californium	Cf-252	0.2
Curium	Cm-244	0.5
Cobalt	Co-57	7
Cobalt	Co-60	0.3
Caesium	Cs-137	1
Iron	Fe-55	8000
Germanium	Ge-68	7
Gadolinium	Gd-153	10
Iridium	Ir-192	0.8
Nickel	Ni-63	600
Palladium	Pd-103	900
Promethium	Pm-147	400
Polonium	Po-210	0.6
Plutonium	Pu-238	0.6
Plutonium	Pu-239	0.6
Radium	Ra-226	0.4
Ruthenium	Ru-106	3
Selenium	Se-75	2
Strontium	Sr-90	10
Thallium	Tl-204	200
Thulium	Tm-170	200
Ytterbium	Yb-169	3

1.10.3.1.4   For mixtures of radionuclides, determination of whether or not the transport security threshold has been met or exceeded can be calculated by summing the ratios of activity present for each radionuclide divided by the transport security threshold for that radionuclide. If the sum of the fractions is less than 1, then the radioactivity threshold for the mixture has not been met nor exceeded.

This calculation can be made with the formula:

$$\sum_i \frac{A_i}{T_i} < 1$$

Where:

Ai = activity of radionuclide i that is present in a package (TBq)

Ti = transport security threshold for radionuclide i (TBq).

1.10.3.1.5   When radioactive material possesses subsidiary risks of other classes, the criteria of table 1.10.3.1.2 shall also be taken into account (see also 1.7.5).

1.10.3.2   *Security plans*

1.10.3.2.1   Carriers, consignors and other participants specified in 1.4.2 and 1.4.3 engaged in the carriage of high consequence dangerous goods (see Table 1.10.3.1.2) or high consequence radioactive material (see 1.10.3.1.3) shall adopt, implement and comply with a security plan that addresses at least the elements specified in 1.10.3.2.2.

1.10.3.2.2   The security plan shall comprise at least the following elements:

(a)   Specific allocation of responsibilities for security to competent and qualified persons with appropriate authority to carry out their responsibilities;

(b)   Records of dangerous goods or types of dangerous goods concerned;

(c)   Review of current operations and assessment of security risks, including any stops necessary to the transport operation, the keeping of dangerous goods in the vehicle, tank or container before, during and after the journey and the intermediate temporary storage of dangerous goods during the course of intermodal transfer or transhipment between units as appropriate;

(d)   Clear statement of measures that are to be taken to reduce security risks, commensurate with the responsibilities and duties of the participant, including:

-   training;

-   security policies (e.g. response to higher threat conditions, new employee/employment verification, etc.);

-   operating practices (e.g. choice/use of routes where known, access to dangerous goods in intermediate temporary storage (as defined in (c)), proximity to vulnerable infrastructure etc.);

-   equipment and resources that are to be used to reduce security risks;

(e)   Effective and up to date procedures for reporting and dealing with security threats, breaches of security or security incidents;

(f)   Procedures for the evaluation and testing of security plans and procedures for periodic review and update of the plans;

(g)   Measures to ensure the physical security of transport information contained in the security plan; and

(h)   Measures to ensure that the distribution of information relating to the transport operation contained in the security plan is limited to those who need to have it. Such measures shall not preclude the provision of information required elsewhere in ADR.

NOTE: Carriers, consignors and consignees should co-operate with each other and with competent authorities to exchange threat information, apply appropriate security measures and respond to security incidents.

1.10.3.3   Devices, equipment or arrangements to prevent the theft of the vehicle carrying high consequence dangerous goods (see Table 1.10.3.1.2) or high consequence radioactive material (see 1.10.3.1.3) and its cargo, shall be applied and measures taken to ensure that these are operational and effective at all times. The application of these protective measures shall not jeopardize emergency response.

NOTE: When appropriate and already fitted, the use of transport telemetry or other tracking methods or devices should be used to monitor the movement of high consequence dangerous goods (see Table 1.10.3.1.2) or high consequence radioactive material (see 1.10.3.1.3).

1.10.4   In accordance with the provisions of 1.1.3.6, the requirements of 1.10.1, 1.10.2, 1.10.3 and 8.1.2.1 (d) do not apply when the quantities carried in packages on a transport unit do not exceed those referred to in 1.1.3.6.3, except for UN Nos. 0029, 0030, 0059, 0065, 0073, 0104, 0237, 0255, 0267, 0288, 0289, 0290, 0360, 0361, 0364, 0365, 0366, 0439, 0440, 0441, 0455, 0456 and 0500 and except for UN Nos. 2910 and 2911 if the activity level exceeds the A_2 value (see first indent of 1.1.3.6.2). In addition, the requirements of 1.10.1, 1.10.2, 1.10.3 and 8.1.2.1 (d) do not apply when the quantities carried in tanks or in bulk on a transport unit do not exceed those referred to in 1.1.3.6.3. In addition the provisions of this Chapter do not apply to the carriage of UN No. 2912 RADIOACTIVE MATERIAL, LOW SPECIFIC ACTIVITY (LSA-I) and UN No. 2913 RADIOACTIVE MATERIAL, SURFACE CONTAMINATED OBJECTS (SCO-I).

1.10.5   For radioactive material, the provisions of this Chapter are deemed to be complied with when the provisions of the Convention on Physical Protection of Nuclear Material[1] and the IAEA circular on "The Physical Protection of Nuclear Material and Nuclear Facilities"[2] are applied.

---

[1]   *INFCIRC/274/Rev.1, IAEA, Vienna (1980).*
[2]   *INFCIRC/225/Rev.4 (Corrected), IAEA, Vienna (1999).*

PART 2

Classification

CHAPTER 2.1

GENERAL PROVISIONS

2.1.1   **Introduction**

2.1.1.1   The classes of dangerous goods according to ADR are the following:

Class 1	Explosive substances and articles
Class 2	Gases
Class 3	Flammable liquids
Class 4.1	Flammable solids, self-reactive substances, polymerizing substances and solid desensitized explosives
Class 4.2	Substances liable to spontaneous combustion
Class 4.3	Substances which, in contact with water, emit flammable gases
Class 5.1	Oxidizing substances
Class 5.2	Organic peroxides
Class 6.1	Toxic substances
Class 6.2	Infectious substances
Class 7	Radioactive material
Class 8	Corrosive substances
Class 9	Miscellaneous dangerous substances and articles

2.1.1.2   Each entry in the different classes has been assigned a UN number. The following types of entries are used:

A.   Single entries for well defined substances or articles including entries for substances covering several isomers, e.g.:

      UN No. 1090 ACETONE
      UN No. 1104 AMYL ACETATES
      UN No. 1194 ETHYL NITRITE SOLUTION

B.   Generic entries for a well defined group of substances or articles, which are not n.o.s. entries, e.g.:

      UN No. 1133 ADHESIVES
      UN No. 1266 PERFUMERY PRODUCTS
      UN No. 2757 CARBAMATE PESTICIDE, SOLID, TOXIC
      UN No. 3101 ORGANIC PEROXIDE TYPE B, LIQUID

C.   Specific n.o.s. entries covering a group of substances or articles of a particular chemical or technical nature, not otherwise specified, e.g.:

      UN No. 1477 NITRATES, INORGANIC, N.O.S.
      UN No. 1987 ALCOHOLS, N.O.S.

D.   General n.o.s. entries covering a group of substances or articles having one or more dangerous properties, not otherwise specified, e.g.:

      UN No. 1325 FLAMMABLE SOLID, ORGANIC, N.O.S.
      UN No. 1993 FLAMMABLE LIQUID, N.O.S.

The entries defined under B., C. and D. are defined as collective entries.

2.1.1.3   For packing purposes, substances other than those of Classes 1, 2, 5.2, 6.2 and 7, and other than self-reactive substances of Class 4.1 are assigned to packing groups in accordance with the degree of danger they present:

Packing group I: Substances presenting high danger;
Packing group II: Substances presenting medium danger;
Packing group III: Substances presenting low danger.

The packing group(s) to which a substance is assigned is (are) indicated in Table A of Chapter 3.2.

Articles are not assigned to packing groups. For packing purposes any requirement for a specific packaging performance level is set out in the applicable packing instruction.

2.1.2   **Principles of classification**

2.1.2.1   The dangerous goods covered by the heading of a class are defined on the basis of their properties according to sub-section 2.2.x.1 of the relevant class. Assignment of dangerous goods to a class and a packing group is made according to the criteria mentioned in the same sub-section 2.2.x.1. Assignment of one or several subsidiary risk(s) to a dangerous substance or article is made according to the criteria of the class or classes corresponding to those risks, as mentioned in the appropriate sub-section(s) 2.2.x.1.

2.1.2.2   All dangerous goods entries are listed in Table A of Chapter 3.2 in the numerical order of their UN Number. This table contains relevant information on the goods listed, such as name, class, packing group(s), label(s) to be affixed, packing and carriage provisions[1]. The substances listed by name in column (2) of Table A of Chapter 3.2 shall be carried according to their classification in Table A or under the conditions specified in 2.1.2.8.

2.1.2.3   A substance may contain technical impurities (for example those deriving from the production process) or additives for stability or other purposes that do not affect their classification. However, a substance mentioned by name, i.e. listed as a single entry in Table A of Chapter 3.2, containing technical impurities or additives for stability or other purposes affecting its classification shall be considered a solution or mixture (see 2.1.3.3).

2.1.2.4   Dangerous goods which are listed or defined in sub-section 2.2.x.2 of each class are not to be accepted for carriage.

2.1.2.5   Goods not mentioned by name, i.e. goods not listed as single entries in Table A of Chapter 3.2 and not listed or defined in one of the above-mentioned sub-sections 2.2.x.2 shall be assigned to the relevant class in accordance with the procedure of section 2.1.3. In addition, the subsidiary risk (if any) and the packing group (if any) shall be determined. Once the class, subsidiary risk (if any) and packing group (if any) have been established the relevant UN number shall be determined. The decision trees in sub-sections 2.2.x.3 (list of collective entries) at the end of each class indicate the relevant parameters for selecting the relevant collective entry (UN number). In all cases the most specific collective entry covering the properties of the substance or article shall be selected, according to the hierarchy indicated in 2.1.1.2 by the letters B, C and D respectively. If the substance or article cannot be classified under entries of type B or C according to 2.1.1.2, then, and only then shall it be classified under an entry of type D.

2.1.2.6   On the basis of the test procedures of Chapter 2.3 and the criteria set out in sub-sections 2.2.x.1 of classes when it is so specified, it may be determined that a substance, solution or mixture of a certain class, mentioned by name in Table A of Chapter 3.2, does not meet the criteria of that class. In such a case, the substance, solution or mixture is deemed not to belong to that class.

2.1.2.7   For the purposes of classification, substances with a melting point or initial melting point of 20 °C or lower at a pressure of 101.3 kPa shall be considered to be liquids. A viscous substance for which a specific melting point cannot be determined shall be subjected to the ASTM D 4359-90 test or to the test for determining fluidity (penetrometer test) prescribed in 2.3.4.

2.1.2.8   A consignor who has identified, on the basis of test data, that a substance listed by name in column 2 of Table A of Chapter 3.2 meets classification criteria for a class that is not identified in column 3a or 5 of Table A of Chapter 3.2, may, with the approval of the competent authority, consign the substance:

–   Under the most appropriate collective entry listed in sub-sections 2.2.x.3 reflecting all hazards; or

---

[1]   ***Note by the Secretariat:*** *An alphabetic list of these entries has been prepared by the secretariat and is reproduced in Table B of Chapter 3.2. This table is not an official part of the ADR.*

–   Under the same UN number and name but with additional hazard communication information as appropriate to reflect the additional subsidiary risk(s) (documentation, label, placard) provided that the class remains unchanged and that any other carriage conditions (e.g. limited quantity, packaging and tank provisions) that would normally apply to substances possessing such a combination of hazards are the same as those applicable to the substance listed.

NOTE 1:   The competent authority granting the approval may be the competent authority of any ADR Contracting Party who may also recognize an approval granted by the competent authority of a country which is not an ADR Contracting Party provided that this approval has been granted in accordance with the procedures applicable according to RID, ADR, ADN, the IMDG Code or the ICAO Technical Instructions.

NOTE 2:   When a competent authority grants such approvals, it should inform the United Nations Sub-Committee of Experts on the Transport of Dangerous Goods accordingly and submit a relevant proposal of amendment to the Dangerous Goods List of the UN Model Regulations. Should the proposed amendment be rejected, the competent authority should withdraw its approval.

NOTE 3:   For carriage in accordance with 2.1.2.8, see also 5.4.1.1.20.

2.1.3   Classification of substances, including solutions and mixtures (such as preparations and wastes), not mentioned by name

2.1.3.1   Substances including solutions and mixtures not mentioned by name shall be classified according to their degree of danger on the basis of the criteria mentioned in sub-section 2.2.x.1 of the various classes. The danger(s) presented by a substance shall be determined on the basis of its physical and chemical characteristics and physiological properties. Such characteristics and properties shall also be taken into account when such experience leads to a more stringent assignment.

2.1.3.2   A substance not mentioned by name in Table A of Chapter 3.2 presenting a single hazard shall be classified in the relevant class under a collective entry listed in sub-section 2.2.x.3 of that class.

2.1.3.3   A solution or mixture meeting the classification criteria of ADR composed of a single predominant substance mentioned by name in Table A of Chapter 3.2 and one or more substances not subject to ADR or traces of one or more substances mentioned by name in Table A of Chapter 3.2, shall be assigned the UN number and proper shipping name of the predominant substance mentioned by name in Table A of Chapter 3.2 unless:

(a)   The solution or mixture is mentioned by name in Table A of Chapter 3.2;

(b)   The name and description of the substance mentioned by name in Table A of Chapter 3.2 specifically indicate that they apply only to the pure substance;

(c)   The class, classification code, packing group, or physical state of the solution or mixture is different from that of the substance mentioned by name in Table A of Chapter 3.2; or

(d)   The hazard characteristics and properties of the solution or mixture necessitate emergency response measures that are different from those required for the substance mentioned by name in Table A of Chapter 3.2.

In those other cases, except the one described in (a), the solution or mixture shall be classified as a substance not mentioned by name in the relevant class under a collective entry listed in sub-section 2.2.x.3 of that class taking account of the subsidiary risks presented by that solution or mixture, if any, unless the solution or mixture does not meet the criteria of any class, in which case it is not subject to ADR.

2.1.3.4   Solutions and mixtures containing substances belonging to one of the entries mentioned in 2.1.3.4.1 or 2.1.3.4.2 shall be classified in accordance with the provisions of these paragraphs.

2.1.3.4.1   Solutions and mixtures containing one of the following substances mentioned by name shall always be classified under the same entry as the substance they contain, provided they do not have the hazard characteristics as indicated in 2.1.3.5.3:

- Class 3

  UN No. 1921 PROPYLENEIMINE, STABILIZED; UN No. 3064 NITROGLYCERIN SOLUTION IN ALCOHOL with more than 1% but not more than 5% nitroglycerin;

- Class 6.1

  UN No. 1051 HYDROGEN CYANIDE, STABILIZED, containing less than 3% water; UN No. 1185 ETHYLENEIMINE, STABILIZED; UN No. 1259 NICKEL CARBONYL; UN No. 1613 HYDROCYANIC ACID, AQUEOUS SOLUTION (HYDROGEN CYANIDE, AQUEOUS SOLUTION), with not more than 20% hydrogen cyanide; UN No. 1614 HYDROGEN CYANIDE, STABILIZED, containing not more than 3% water and absorbed in a porous inert material; UN No. 1994 IRON PENTACARBONYL; UN No. 2480 METHYL ISOCYANATE; UN No. 2481 ETHYL ISOCYANATE; UN No. 3294 HYDROGEN CYANIDE, SOLUTION IN ALCOHOL, with not more than 45% hydrogen cyanide;

- Class 8

  UN No. 1052 HYDROGEN FLUORIDE, ANHYDROUS; UN No. 1744 BROMINE or UN No. 1744 BROMINE SOLUTION; UN No. 1790 HYDROFLUORIC ACID with more than 85% hydrogen fluoride; UN No. 2576 PHOSPHORUS OXYBROMIDE, MOLTEN;

2.1.3.4.2   Solutions and mixtures containing a substance belonging to one of the following entries of Class 9:

UN No. 2315 POLYCHLORINATED BIPHENYLS, LIQUID;
UN No. 3151 POLYHALOGENATED BIPHENYLS, LIQUID;
UN No. 3151 HALOGENATED MONOMETHYLDIPHENYLMETHANES, LIQUID;
UN No. 3151 POLYHALOGENATED TERPHENYLS, LIQUID;
UN No. 3152 POLYHALOGENATED BIPHENYLS, SOLID;
UN No. 3152 HALOGENATED MONOMETHYLDIPHENYLMETHANES, SOLID;
UN No. 3152 POLYHALOGENATED TERPHENYLS, SOLID; or
UN No. 3432 POLYCHLORINATED BIPHENYLS, SOLID

shall always be classified under the same entry of Class 9 provided that:

- they do not contain any additional dangerous component other than components of packing group III of classes 3, 4.1, 4.2, 4.3, 5.1, 6.1 or 8; and

- they do not have the hazard characteristics as indicated in 2.1.3.5.3.

2.1.3.5   Substances not mentioned by name in Table A of Chapter 3.2, having more than one hazard characteristic and solutions or mixtures meeting the classification criteria of ADR containing several dangerous substances shall be classified under a collective entry (see 2.1.2.5) and packing group of the appropriate class in accordance with their hazard characteristics. Such classification according to the hazard characteristics shall be carried out as follows:

2.1.3.5.1   The physical and chemical characteristics and physiological properties shall be determined by measurement or calculation and the substance, solution or mixture shall be classified according to the criteria mentioned in sub-section 2.2.x.1 of the various classes.

2.1.3.5.2   If this determination is not possible without disproportionate cost or effort (as for some kinds of wastes), the substance, solution or mixture shall be classified in the class of the component presenting the major hazard.

2.1.3.5.3   If the hazard characteristics of the substance, solution or mixture fall within more than one class or group of substances listed below then the substance, solution or mixture shall be classified in the class or group of substances corresponding to the major hazard on the basis of the following order of precedence:

(a)   Material of Class 7 (apart from radioactive material in excepted packages for which, except for UN 3507 URANIUM HEXAFLUORIDE, RADIOACTIVE MATERIAL, EXCEPTED PACKAGE, special provision 290 of Chapter 3.3 applies , where the other hazardous properties take precedence);

(b)   Substances of Class 1;

(c)   Substances of Class 2;

(d)   Liquid desensitized explosives of Class 3;

(e)   Self-reactive substances and solid desensitized explosives of Class 4.1;

(f)   Pyrophoric substances of Class 4.2;

(g)   Substances of Class 5.2;

(h)   Substances of Class 6.1 meeting the inhalation toxicity criteria of packing group I (Substances meeting the classification criteria of Class 8 and having an inhalation toxicity of dust and mist (LC_{50}) in the range of Packing group I and a toxicity through oral ingestion or dermal contact only in the range of Packing group III or less, shall be allocated to Class 8);

(i)   Infectious substances of Class 6.2.

2.1.3.5.4   If the hazard characteristics of the substance fall within more than one class or group of substances not listed in 2.1.3.5.3 above, the substance shall be classified in accordance with the same procedure but the relevant class shall be selected according to the precedence of hazards table in 2.1.3.10.

2.1.3.5.5   If the substance to be carried is a waste, with a composition that is not precisely known, its assignment to a UN number and packing group in accordance with 2.1.3.5.2 may be based on the consignor's knowledge of the waste, including all available technical and safety data as requested by safety and environmental legislation in force[2].

In case of doubt, the highest danger level shall be taken.

If however, on the basis of the knowledge of the composition of the waste and the physical and chemical properties of the identified components, it is possible to demonstrate that the properties of the waste do not correspond to the properties of the packing group I level, the waste may be classified by default in the most appropriate n.o.s. entry of packing group II. However, if it is known that the waste possesses only environmentally hazardous properties, it may be assigned to packing group III under UN Nos. 3077 or 3082.

This procedure may not be used for wastes containing substances mentioned in 2.1.3.5.3, substances of Class 4.3, substances of the case mentioned in 2.1.3.7 or substances which are not accepted for carriage in accordance with 2.2.x.2.

2.1.3.6   The most specific applicable collective entry (see 2.1.2.5) shall always be used, i.e. a general n.o.s. entry shall only be used if a generic entry or a specific n.o.s. entry cannot be used.

2.1.3.7   Solutions and mixtures of oxidizing substances or substances with an oxidizing subsidiary risk may have explosive properties. In such a case they are not to be accepted for carriage unless they meet the requirements for Class 1.

2.1.3.8   Substances of classes 1 to 6.2, 8 and 9, other than those assigned to UN Nos. 3077 and 3082, meeting the criteria of 2.2.9.1.10 are additionally to their hazards of classes 1 to 6.2, 8 and 9 considered to be environmentally hazardous substances. Other substances meeting the criteria of no other class, but those of 2.2.9.1.10 are to be assigned to UN Nos. 3077 and 3082 as appropriate.

2.1.3.9   Wastes that do not meet the criteria for classification in classes 1 to 9 but are covered by the *Basel Convention on the Control of Transboundary Movements of Hazardous Wastes and their Disposal* may be carried under UN Nos. 3077 or 3082.

---

[2]   *Such legislation is for instance the Commission Decision 2000/532/EC of 3 May 2000 replacing Decision 94/3/EC establishing a list of wastes pursuant to Article 1(a) of Council Directive 75/442/EEC on waste and Council Decision 94/904/EC establishing a list of hazardous wastes pursuant to Article 1(4) of Council Directive 91/689/EEC on hazardous wastes (Official Journal of the European Communities No. L 226 of 6 September 2000, page 3); and Directive 2008/98/EC of the European Parliament and of the Council of 19 November 2008 on waste and repealing certain Directives (Official Journal of the European Union No. L312 of 22 November 2008, pages 3-30).*

2.1.3.10   Table of precedence of hazards

Class and packing group	4.1, II	4.1, III	4.2, II	4.2, III	4.3, I	4.3, II	4.3, III	5.1, I	5.1, II	5.1, III	6.1, I DERMAL	6.1, I ORAL	6.1, II	6.1, III	8, I	8, II	8, III	9
3, I	SOL 4.1 / LIQ 3,1	SOL 4.1 / LIQ 3,1	SOL 4.2 / LIQ 3,1	SOL 4.2 / LIQ 3,1	4.3,1	4.3,1	4.3,1	SOL 5.1,1 / LIQ 3,1	SOL 5.1,1 / LIQ 3,1	SOL 5.1,1 / LIQ 3,1	3,1	3,1	3,1	3,1	3,1	3,1	3,1	3,1
3, II	SOL 4.1 / LIQ 3,II	SOL 4.1 / LIQ 3,II	SOL 4.2 / LIQ 3,II	SOL 4.2 / LIQ 3,II	4.3,1	4.3,II	4.3,II	SOL 5.1,1 / LIQ 3,1	SOL 5.1,1 / LIQ 3,II	SOL 5.1,1 / LIQ 3,II	3,1	3,1	3,II	3,II	8,1	3,II	3,II	3,II
3, III	SOL 4.1 / LIQ 3,III	SOL 4.1 / LIQ 3,III	SOL 4.2 / LIQ 3,III	SOL 4.2 / LIQ 3,III	4.3,1	4.3,II	4.3,III	SOL 5.1,1 / LIQ 3,1	SOL 5.1,II / LIQ 3,II	SOL 5.1,III / LIQ 3,III	6.1,1	6.1,1	6.1,II	3,III^a	8,1	8,II	3,III	3,III
4.1, II			4.2,II	4.2,II	4.3,1	4.3,II	4.3,II	5.1,1	4.1,II	4.1,II	6.1,1	6.1,1	SOL 4.1,II / LIQ 6.1,II	SOL 4.1,II / LIQ 6.1,II	8,1	SOL 4.1,II / LIQ 8,II	SOL 4.1,II / LIQ 8,II	4.1,II
4.1, III			4.2,II	4.2,III	4.3,1	4.3,II	4.3,III	5.1,1	4.1,II	4.1,III	6.1,1	6.1,1	6.1,II	SOL 4.1,III / LIQ 6.1,III	8,1	8,II	SOL 4.1,III / LIQ 8,III	4.1,III
4.2, II					4.3,1	4.3,II	4.3,II	5.1,1	4.2,II	4.2,II	6.1,1	6.1,1	4.2,II	4.2,II	8,1	4.2,II	4.2,II	4.2,II
4.2, III					4.3,1	4.3,II	4.3,III	5.1,1	5.1,II	4.2,III	6.1,1	6.1,1	4.2,III	4.2,III	8,1	8,II	4.2,III	4.2,III
4.3, I								5.1,1	4.3,1	4.3,1	6.1,1	4.3,1	4.3,1	4.3,1	4.3,1	4.3,1	4.3,1	4.3,1
4.3, II								5.1,1	4.3,II	4.3,II	6.1,1	4.3,1	4.3,II	4.3,II	8,1	4.3,II	4.3,II	4.3,II
4.3, III								5.1,1	5.1,II	4.3,III	6.1,1	4.3,1	4.3,III	4.3,III	8,1	8,II	4.3,III	4.3,III
5.1, I											5.1,1	5.1,1	5.1,1	5.1,1	5.1,1	5.1,1	5.1,1	5.1,1
5.1, II											6.1,1	5.1,1	5.1,II	5.1,II	8,1	5.1,II	5.1,II	5.1,II
5.1, III											6.1,1	6.1,1	6.1,II	5.1,III	8,1	8,II	5.1,III	5.1,III
6.1, I DERMAL															SOL 6.1,1 / LIQ 8,1			6.1,1
6.1, I ORAL															SOL 6.1,1 / LIQ 8,1			6.1,1
6.1, II INHAL															SOL 6.1,1 / LIQ 8,1			6.1,II
6.1, II DERMAL															SOL 6.1,1 / LIQ 8,1	SOL 6.1,II / LIQ 8,II		6.1,II
6.1, II ORAL															8,1	SOL 6.1,II / LIQ 8,II		6.1,II
6.1, III															8,1	8,II		6.1,III
8, I																		8,1
8, II																		8,II
8, III																		8,III

SOL = Solid substances and mixtures
LIQ = Liquid substances, mixtures and solutions
DERMAL = Dermal toxicity
ORAL = Oral toxicity
INHAL = Inhalation toxicity
^a Class 6.1 for pesticides

NOTE 1: Examples to explain the use of the table

Classification of a single substance

Description of the substance to be classified:

An amine not mentioned by name meeting the criteria for Class 3, packing group II as well as those for Class 8, packing group I.

Procedure:

The intersection of line 3 II with column 8 I gives 8 I.

This amine has therefore to be classified in Class 8 under:

UN No. 2734 AMINES LIQUID, CORROSIVE, FLAMMABLE, N.O.S. or UN No. 2734 POLYAMINES, LIQUID, CORROSIVE, FLAMMABLE, N.O.S.

packing group I

Classification of a mixture

Description of the mixture to be classified:

Mixture consisting of a flammable liquid classified in Class 3, packing group III, a toxic substance in Class 6.1, packing group II and a corrosive substance in Class 8, packing group I.

Procedure:

The intersection of line 3 III with column 6.1 II gives 6.1 II.

The intersection of line 6.1 II with column 8 I gives 8 I LIQ.

This mixture not further defined has therefore to be classified in Class 8 under:

UN No. 2922 CORROSIVE LIQUID, TOXIC, N.O.S. packing group I.

NOTE 2: Examples for the classification of mixtures and solutions under a class and a packing group:

A phenol solution of Class 6.1, (II), in benzene of Class 3, (II) is to be classified in Class 3, (II); this solution is to be classified under UN No. 1992 FLAMMABLE LIQUID, TOXIC, N.O.S., Class 3, (II), by virtue of the toxicity of the phenol.

A solid mixture of sodium arsenate of Class 6.1, (II) and sodium hydroxide of Class 8, (II) is to be classified under UN No. 3290 TOXIC SOLID, CORROSIVE, INORGANIC, N.O.S., in Class 6.1 (II).

A solution of crude or refined naphthalene of Class 4.1, (III) in petrol of Class 3, (II), is to be classified under UN No. 3295 HYDROCARBONS, LIQUID, N.O.S. in Class 3, (II).

A mixture of hydrocarbons of Class 3, (III), and of polychlorinated biphenyls (PCB) of Class 9, (II), is to be classified under UN No. 2315 POLYCHLORINATED BIPHENYLS LIQUID or UN No. 3432 POLYCHLORINATED BIPHENYLS SOLID in Class 9, (II).

A mixture of propyleneimine of Class 3, and polychlorinated biphenyls (PCB) of Class 9, (II), is to be classified under UN No. 1921 PROPYLENEIMINE, INHIBITED in Class 3.

2.1.4   **Classification of samples**

2.1.4.1   When the class of a substance is uncertain and it is being carried for further testing, a tentative class, proper shipping name and UN number shall be assigned on the basis of the consignor's knowledge of the substance and application of:

(a)   the classification criteria of Chapter 2.2; and

(b) the requirements of this Chapter.

The most severe packing group possible for the proper shipping name chosen shall be used.

Where this provision is used the proper shipping name shall be supplemented with the word "SAMPLE" (e.g., "FLAMMABLE LIQUID, N.O.S., SAMPLE"). In certain instances, where a specific proper shipping name is provided for a sample of a substance considered to meet certain classification criteria (e.g., GAS SAMPLE, NON-PRESSURIZED, FLAMMABLE, UN No. 3167) that proper shipping name shall be used. When an N.O.S. entry is used to carry the sample, the proper shipping name need not be supplemented with the technical name as required by special provision 274 of Chapter 3.3.

2.1.4.2   Samples of the substance shall be carried in accordance with the requirements applicable to the tentative assigned proper shipping name provided:

(a) The substance is not considered to be a substance not accepted for carriage by sub-sections 2.2.x.2 of Chapter 2.2 or by Chapter 3.2;

(b) The substance is not considered to meet the criteria for Class 1 or considered to be an infectious substance or a radioactive material;

(c) The substance is in compliance with 2.2.41.1.15 or 2.2.52.1.9 if it is a self-reactive substance or an organic peroxide, respectively;

(d) The sample is carried in a combination packaging with a net mass per package not exceeding 2.5 kg; and

(e) The sample is not packed together with other goods.

2.1.5   **Classification of packagings, discarded, empty, uncleaned**

Empty uncleaned packagings, large packagings or IBCs, or parts thereof, carried for disposal, recycling or recovery of their material, other than reconditioning, repair, routine maintenance, remanufacturing or reuse, may be assigned to UN 3509 if they meet the requirements for this entry.

CHAPTER 2.2

CLASS SPECIFIC PROVISIONS

2.2.1   **Class 1**   **Explosive substances and articles**

2.2.1.1   *Criteria*

2.2.1.1.1   The heading of Class 1 covers:

(a)   Explosive substances: solid or liquid substances (or mixtures of substances) capable by chemical reaction of producing gases at such a temperature and pressure and at such a speed as to cause damage to the surroundings.

Pyrotechnic substances: substances or mixtures of substances designed to produce an effect by heat, light, sound, gas or smoke or a combination of these as the result of non-detonating self-sustaining exothermic chemical reactions;

NOTE 1: Substances which are not themselves explosive but which may form an explosive mixture of gas, vapour or dust are not substances of Class 1.

NOTE 2: Also excluded from Class 1 are: water- or alcohol-wetted explosives of which the water or alcohol content exceeds the limits specified and those containing plasticizers - these explosives are assigned to Class 3 or Class 4.1 - and those explosives which, on the basis of their predominant hazard, are assigned to Class 5.2.

(b)   Explosive articles: articles containing one or more explosive or pyrotechnic substances;

NOTE: Devices containing explosive or pyrotechnic substances in such small quantity or of such a character that their inadvertent or accidental ignition or initiation during carriage would not cause any manifestation external to the device by projection, fire, smoke, heat or loud noise are not subject to the requirements of Class 1.

(c)   Substances and articles not mentioned above which are manufactured with a view to producing a practical effect by explosion or a pyrotechnic effect.

For the purposes of Class 1, the following definition applies:

Phlegmatized means that a substance (or "phlegmatizer") has been added to an explosive to enhance its safety in handling and carriage. The phlegmatizer renders the explosive insensitive, or less sensitive, to the following actions: heat, shock, impact, percussion or friction. Typical phlegmatizing agents include, but are not limited to: wax, paper, water, polymers (such as chlorofluoropolymers), alcohol and oils (such as petroleum jelly and paraffin).

2.2.1.1.2   Any substance or article having or suspected of having explosive properties shall be considered for assignment to Class 1 in accordance with the tests, procedures and criteria prescribed in Part I, Manual of Tests and Criteria.

A substance or article assigned to Class 1 can only be accepted for carriage when it has been assigned to a name or n.o.s. entry listed in Table A of Chapter 3.2 and meets the criteria of the Manual of Tests and Criteria.

2.2.1.1.3   The substances and articles of Class 1 shall be assigned to a UN Number and a name or n.o.s. entry listed in Table A of Chapter 3.2. Interpretation of the names of substances and articles in Table A of Chapter 3.2 shall be based upon the glossary in 2.2.1.4.

Samples of new or existing explosive substances or articles carried for purposes including: testing, classification, research and development quality control, or as a commercial sample, other than initiating explosive, may be assigned to UN No. 0190 SAMPLES, EXPLOSIVE.

The assignment of explosive substances and articles not mentioned by name as such in Table A of Chapter 3.2 to an n.o.s entry of Class 1 or UN No. 0190 SAMPLES, EXPLOSIVE as well as the assignment of certain substances the carriage of which is subject to a specific authorization by the competent authority according to the special provisions referred to in Column (6) of Table A of Chapter 3.2 shall be made by the competent authority of the country of origin. This competent authority shall also approve in writing the conditions of carriage of these substances and articles. If the country of origin is not a Contracting Party to ADR, the classification and the conditions of carriage shall be recognized by the competent authority of the first country Contracting Party to ADR reached by the consignment.

2.2.1.1.4   Substances and articles of Class 1 shall have been assigned to a division in accordance with 2.2.1.1.5 and to a compatibility group in accordance with 2.2.1.1.6. The division shall be based on the results of the tests described in 2.3.0 and 2.3.1 applying the definitions in 2.2.1.1.5. The compatibility group shall be determined in accordance with the definitions in 2.2.1.1.6. The classification code shall consist of the division number and the compatibility group letter.

2.2.1.1.5   *Definition of divisions*

Division 1.1   Substances and articles which have a mass explosion hazard (a mass explosion is an explosion which affects almost the entire load virtually instantaneously).

Division 1.2   Substances and articles which have a projection hazard but not a mass explosion hazard.

Division 1.3   Substances and articles which have a fire hazard and either a minor blast hazard or a minor projection hazard or both, but not a mass explosion hazard:

(a)   combustion of which gives rise to considerable radiant heat; or

(b)   which burn one after another, producing minor blast or projection effects or both.

Division 1.4   Substances and articles which present only a slight risk of explosion in the event of ignition or initiation during carriage. The effects are largely confined to the package and no projection of fragments of appreciable size or range is to be expected. An external fire shall not cause virtually instantaneous explosion of almost the entire contents of the package.

Division 1.5   Very insensitive substances having a mass explosion hazard which are so insensitive that there is very little probability of initiation or of transition from burning to detonation under normal conditions of carriage. As a minimum requirement they must not explode in the external fire test.

Division 1.6   Extremely insensitive articles which do not have a mass explosion hazard. The articles predominantly contain extremely insensitive substances and demonstrate a negligible probability of accidental initiation or propagation.

NOTE: *The risk from articles of Division 1.6 is limited to the explosion of a single article.*

2.2.1.1.6   *Definition of compatibility groups of substances and articles*

A   Primary explosive substance.

B   Article containing a primary explosive substance and not having two or more effective protective features. Some articles, such as detonators for blasting, detonator assemblies for blasting and primers, cap-type, are included, even though they do not contain primary explosives.

C   Propellant explosive substance or other deflagrating explosive substance or article containing such explosive substance.

D   Secondary detonating explosive substance or black powder or article containing a secondary detonating explosive substance, in each case without means of initiation and without a propelling charge, or article containing a primary explosive substance and having two or more effective protective features.

E   Article containing a secondary detonating explosive substance, without means of initiation, with a propelling charge (other than one containing a flammable liquid or gel or hypergolic liquids).

F   Article containing a secondary detonating explosive substance with its own means of initiation, with a propelling charge (other than one containing a flammable liquid or gel or hypergolic liquids) or without a propelling charge.

G   Pyrotechnic substance, or article containing a pyrotechnic substance, or article containing both an explosive substance and an illuminating, incendiary, tear- or smoke-producing substance (other than a water-activated article or one which contains white phosphorus, phosphides, a pyrophoric substance, a flammable liquid or gel or hypergolic liquids).

H   Article containing both an explosive substance and white phosphorus.

J   Article containing both an explosive substance and a flammable liquid or gel.

K   Article containing both an explosive substance and a toxic chemical agent.

L   Explosive substance or article containing an explosive substance and presenting a special risk (e.g. due to water activation or the presence of hypergolic liquids, phosphides or a pyrophoric substance) necessitating isolation of each type.

N   Articles predominantly containing extremely insensitive substances.

S   Substance or article so packed or designed that any hazardous effects arising from accidental functioning are confined within the package unless the package has been degraded by fire, in which case all blast or projection effects are limited to the extent that they do not significantly hinder or prevent fire-fighting or other emergency response efforts in the immediate vicinity of the package.

NOTE 1: Each substance or article, packed in a specified packaging, may be assigned to one compatibility group only. Since the criterion of compatibility group S is empirical, assignment to this group is necessarily linked to the tests for assignment of a classification code.

NOTE 2: Articles of compatibility groups D and E may be fitted or packed together with their own means of initiation provided that such means have at least two effective protective features designed to prevent an explosion in the event of accidental functioning of the means of initiation. Such articles and packages shall be assigned to compatibility groups D or E.

NOTE 3: Articles of compatibility groups D and E may be packed together with their own means of initiation, which do not have two effective protective features (i.e. means of initiation assigned to compatibility group B), provided that they comply with mixed packing provision MP21 of Section 4.1.10. Such packages shall be assigned to compatibility groups D or E.

NOTE 4: Articles may be fitted or packed together with their own means of ignition provided that the means of ignition cannot function during normal conditions of carriage.

NOTE 5: Articles of compatibility groups C, D and E may be packed together. Such packages shall be assigned to compatibility group E.

2.2.1.1.7   *Assignment of fireworks to divisions*

2.2.1.1.7.1   Fireworks shall normally be assigned to divisions 1.1, 1.2, 1.3, and 1.4 on the basis of test data derived from Test Series 6 of the Manual of Tests and Criteria.

However:

(a)   waterfalls giving a positive result when tested in the HSL Flash composition test in Appendix 7 of the Manual of Tests and Criteria shall be classified as 1.1G regardless of the results of Test Series 6;

(b)   since the range of such articles is very extensive and the availability of test facilities may be limited, assignment to divisions may also be made in accordance with the procedure in 2.2.1.1.7.2.

2.2.1.1.7.2   Assignment of fireworks to UN Nos. 0333, 0334, 0335 and 0336 may be made on the basis of analogy, without the need for Test Series 6 testing, in accordance with the default fireworks classification table in 2.2.1.1.7.5. Such assignment shall be made with the agreement of the competent authority. Items not specified in the table shall be classified on the basis of test data derived from Test Series 6.

NOTE 1:   The addition of other types of fireworks to column 1 of the table in 2.2.1.1.7.5 shall only be made on the basis of full test data submitted to the UN Sub-Committee of Experts on the Transport of Dangerous Goods for consideration.

NOTE 2:   Test data derived by competent authorities which validates, or contradicts the assignment of fireworks specified in column 4 of the table in 2.2.1.1.7.5 to divisions in column 5 should be submitted to the UN Sub-Committee of Experts on the Transport of Dangerous Goods for information.

2.2.1.1.7.3   Where fireworks of more than one division are packed in the same package, they shall be classified on the basis of the most dangerous division unless test data derived from Test Series 6 indicate otherwise.

2.2.1.1.7.4   The classification shown in the table in 2.2.1.1.7.5 applies only for articles packed in fibreboard boxes (4G).

2.2.1.1.7.5   Default fireworks classification table[1]

NOTE 1:   References to percentages in the table, unless otherwise stated, are to the mass of all pyrotechnic substances (e.g. rocket motors, lifting charge, bursting charge and effect charge).

NOTE 2: "Flash composition" in this table refers to pyrotechnic substances in powder form or as pyrotechnic units as presented in the firework that are used to produce an aural effect or used as a bursting charge, or propellant charge unless the time taken for the pressure rise is demonstrated to be more than 6 ms for 0.5 g of pyrotechnic substance in the HSL Flash Composition Test in Appendix 7 of the Manual of Tests and Criteria.

NOTE 3:   Dimensions in mm refer to:

-       *for spherical and peanut shells the diameter of the sphere of the shell;*

-       *for cylinder shells the length of the shell;*

-       *for a shell in mortar, Roman candle, shot tube firework or mine the inside diameter of the tube comprising or containing the firework;*

-       *for a bag mine or cylinder mine, the inside diameter of the mortar intended to contain the mine.*

---

[1]   *This table contains a list of firework classifications which may be used in the absence of Test Series 6 data (see 2.2.1.1.7.2).*

Type	Includes: / Synonym:	Definition	Specification	Classification
Shell, spherical or cylindrical	Spherical display shell: aerial shell, colour shell, dye shell, multi-break shell, multi-effect shell, nautical shell, parachute shell, smoke shell, star shell; report shell: maroon, salute, sound shell, thunderclap, aerial shell kit	Device with or without propellant charge, with delay fuse and bursting charge, pyrotechnic unit(s) or loose pyrotechnic substance and designed to be projected from a mortar	All report shells	1.1G
			Colour shell: ≥ 180 mm	1.1G
			Colour shell: < 180 mm with > 25% flash composition, as loose powder and/or report effects	1.1G
			Colour shell: < 180 mm with ≤ 25% flash composition, as loose powder and/or report effects	1.3G
			Colour shell: ≤ 50 mm, or ≤ 60 g pyrotechnic substance, with ≤ 2% flash composition as loose powder and/or report effects	1.4G
	Peanut shell	Device with two or more spherical aerial shells in a common wrapper propelled by the same propellant charge with separate external delay fuses	The most hazardous spherical aerial shell determines the classification	
	Preloaded mortar, shell in mortar	Assembly comprising a spherical or cylindrical shell inside a mortar from which the shell is designed to be projected	All report shells	1.1G
			Colour shell: ≥ 180 mm	1.1G
			Colour shell: > 25% flash composition as loose powder and/or report effects	1.1G
			Colour shell: > 50 mm and < 180 mm	1.2G
			Colour shell: ≤ 50 mm, or ≤ 60 g pyrotechnic substance, with ≤ 25% flash composition as loose powder and/or report effects	1.3G

Type	Includes: / Synonym:	Definition	Specification	Classification
Shell, spherical or cylindrical (cont'd)	Shell of shells (spherical) (*Reference to percentages for shell of shells are to the gross mass of the fireworks article*)	Device without propellant charge, with delay fuse and bursting charge, containing report shells and inert materials and designed to be projected from a mortar	> 120 mm	1.1G
		Device without propellant charge, with delay fuse and bursting charge, containing report shells ≤ 25g flash composition per report unit, with ≤ 33% flash composition and ≥ 60% inert materials and designed to be projected from a mortar	≤ 120 mm	1.3G
		Device without propellant charge, with delay fuse and bursting charge, containing colour shells and/or pyrotechnic units and designed to be projected from a mortar	> 300 mm	1.1G
		Device without propellant charge, with delay fuse and bursting charge, containing colour shells ≤ 70mm and/or pyrotechnic units, with ≤ 25% flash composition and ≤ 60% pyrotechnic substance and designed to be projected from a mortar	> 200 mm and ≤ 300 mm	1.3G
		Device with propellant charge, with delay fuse and bursting charge, containing colour shells ≤ 70 mm and/or pyrotechnic units, with ≤ 25% flash composition and ≤ 60% pyrotechnic substance and designed to be projected from a mortar	≤ 200 mm	1.3G
Battery/ combination	Barrage, bombardos, cakes, finale box, flowerbed, hybrid, multiple tubes, shell cakes, banger batteries, flash banger batteries	Assembly including several elements either containing the same type or several types each corresponding to one of the types of fireworks listed in this table, with one or two points of ignition		The most hazardous firework type determines the classification

Type	Includes: / Synonym:	Definition	Specification	Classification
Roman candle	Exhibition candle, candle, bombettes	Tube containing a series of pyrotechnic units consisting of alternate pyrotechnic substance, propellant charge, and transmitting fuse	≥ 50 mm inner diameter, containing flash composition, or < 50 mm with $> 25\%$ flash composition	1.1G
			≥ 50 mm inner diameter, containing no flash composition	1.2G
			< 50 mm inner diameter and $\leq 25\%$ flash composition	1.3G
			≤ 30 mm inner diameter, each pyrotechnic unit ≤ 25 g and $\leq 5\%$ flash composition	1.4G
Shot tube	Single shot Roman candle, small preloaded mortar	Tube containing a pyrotechnic unit consisting of pyrotechnic substance, propellant charge with or without transmitting fuse	≤ 30 mm inner diameter and pyrotechnic unit > 25 g, or $> 5\%$ and $\leq 25\%$ flash composition	1.3G
			≤ 30 mm inner diameter, pyrotechnic unit ≤ 25 g and $\leq 5\%$ flash composition	1.4G
Rocket	Avalanche rocket, signal rocket, whistling rocket, bottle rocket, sky rocket, missile type rocket, table rocket	Tube containing pyrotechnic substance and/or pyrotechnic units, equipped with stick(s) or other means for stabilization of flight, and designed to be propelled into the air	Flash composition effects only	1.1G
			Flash composition $> 25\%$ of the pyrotechnic substance	1.1G
			> 20 g pyrotechnic substance and flash composition $\leq 25\%$	1.3G
			≤ 20 g pyrotechnic substance, black powder bursting charge and ≤ 0.13 g flash composition per report and ≤ 1 g in total	1.4G

Type	Includes: / Synonym:	Definition	Specification	Classification
Mine	Pot-a-feu, ground mine, bag mine, cylinder mine	Tube containing propellant charge and pyrotechnic units and designed to be placed on the ground or to be fixed in the ground. The principal effect is ejection of all the pyrotechnic units in a single burst producing a widely dispersed visual and/or aural effect in the air or:	> 25% flash composition, as loose powder and/ or report effects	1.1G
			≥ 180 mm and ≤ 25% flash composition, as loose powder and/ or report effects	1.1G
		Cloth or paper bag or cloth or paper cylinder containing propellant charge and pyrotechnic units, designed to be placed in a mortar and to function as a mine	< 180 mm and ≤ 25% flash composition, as loose powder and/ or report effects	1.3G
			≤ 150 g pyrotechnic substance, containing ≤ 5% flash composition as loose powder and/ or report effects. Each pyrotechnic unit ≤ 25 g, each report effect < 2g; each whistle, if any, ≤ 3 g	1.4G
Fountain	Volcanos, gerbs, lances, Bengal fire, flitter sparkle, cylindrical fountains, cone fountains, illuminating torch	Non-metallic case containing pressed or consolidated pyrotechnic substance producing sparks and flame	≥ 1 kg pyrotechnic substance	1.3G
			< 1 kg pyrotechnic substance	1.4G
		NOTE: Fountains intended to produce a vertical cascade or curtain of sparks are considered to be waterfalls (see row below).		

Type	Includes: / Synonym:	Definition	Specification	Classification
Waterfall	Cascades, showers	Pyrotechnic fountain intended to produce a vertical cascade or curtain of sparks	Containing a pyrotechnic substance which gives a positive result when tested in the HSL Flash composition test in Appendix 7 of the Manual of Tests and Criteria regardless of the results of Test Series 6 (see 2.2.1.1.7.1 (a))	1.1G
			Containing a pyrotechnic substance which gives a negative result when tested in the HSL Flash composition test in Appendix 7 of the Manual of Tests and Criteria	1.3G
Sparkler	Handheld sparklers, non-handheld sparklers, wire sparklers	Rigid wire partially coated (along one end) with slow burning pyrotechnic substance with or without an ignition tip	Perchlorate based sparklers: > 5 g per item or > 10 items per pack	1.3G
			Perchlorate based sparklers: ≤ 5 g per item and ≤ 10 items per pack; Nitrate based sparklers: ≤ 30 g per item	1.4G
Bengal stick	Dipped stick	Non-metallic stick partially coated (along one end) with slow-burning pyrotechnic substance and designed to be held in the hand	Perchlorate based items: > 5 g per item or > 10 items per pack	1.3 G
			Perchlorate based items: ≤ 5 g per item and ≤ 10 items per pack; nitrate based items: ≤ 30 g per item	1.4G

Type	Includes: / Synonym:	Definition	Specification	Classification
Low hazard fireworks and novelties	Table bombs, throwdowns, crackling granules, smokes, fog, snakes, glow worm, serpents, snaps, party poppers	Device designed to produce very limited visible and/ or audible effect which contains small amounts of pyrotechnic and/or explosive composition.	Throwdowns and snaps may contain up to 1.6 mg of silver fulminate; snaps and party poppers may contain up to 16 mg of potassium chlorate/red phosphorous mixture; other articles may contain up to 5 g of pyrotechnic substance, but no flash composition	1.4G
Spinner	Aerial spinner, helicopter, chaser, ground spinner	Non-metallic tube or tubes containing gas- or spark-producing pyrotechnic substance, with or without noise producing composition, with or without aerofoils attached	Pyrotechnic substance per item > 20 g, containing ≤ 3% flash composition as report effects, or whistle composition ≤ 5 g	1.3G
			Pyrotechnic substance per item ≤ 20 g, containing ≤ 3% flash composition as report effects, or whistle composition ≤ 5 g	1.4G
Wheels	Catherine wheels, Saxon	Assembly including drivers containing pyrotechnic substance and provided with a means of attaching it to a support so that it can rotate	≥ 1 kg total pyrotechnic substance, no report effect, each whistle (if any) ≤ 25 g and ≤ 50 g whistle composition per wheel	1.3G
			< 1 kg total pyrotechnic substance, no report effect, each whistle (if any) ≤ 5 g and ≤ 10 g whistle composition per wheel	1.4G

Type	Includes: / Synonym:	Definition	Specification	Classification
Aerial wheel	Flying Saxon, UFO's, rising crown	Tubes containing propellant charges and sparks-flame- and/or noise producing pyrotechnic substances, the tubes being fixed to a supporting ring	> 200 g total pyrotechnic substance or > 60 g pyrotechnic substance per driver, ≤ 3% flash composition as report effects, each whistle (if any) ≤ 25 g and ≤ 50 g whistle composition per wheel	1.3G
			≤ 200 g total pyrotechnic substance and ≤ 60 g pyrotechnic substance per driver, ≤ 3% flash composition as report effects, each whistle (if any) ≤ 5 g and ≤ 10 g whistle composition per wheel	1.4G
Selection pack	Display selection box, display selection pack, garden selection box, indoor selection box; assortment	A pack of more than one type each corresponding to one of the types of fireworks listed in this table	The most hazardous firework type determines the classification	
Firecracker	Celebration cracker, celebration roll, string cracker	Assembly of tubes (paper or cardboard) linked by a pyrotechnic fuse, each tube intended to produce an aural effect	Each tube ≤ 140 mg of flash composition or ≤ 1 g black powder	1.4G
Banger	Salute, flash banger, lady cracker	Non-metallic tube containing report composition intended to produce an aural effect	> 2 g flash composition per item	1.1G
			≤ 2 g flash composition per item and ≤ 10 g per inner packaging	1.3G
			≤ 1 g flash composition per item and ≤ 10 g per inner packaging or ≤ 10 g black powder per item	1.4G

2.2.1.1.8   *Exclusion from Class 1*

2.2.1.1.8.1   An article or a substance may be excluded from Class 1 by virtue of test results and the Class 1 definition with the approval of the competent authority of any ADR Contracting Party who may also recognize an approval granted by the competent authority of a country which is not an ADR Contracting Party provided that this approval has been granted in accordance with the procedures applicable according to RID, ADR, ADN, the IMDG Code or the ICAO Technical Instructions.

2.2.1.1.8.2   With the approval of the competent authority in accordance with 2.2.1.1.8.1, an article may be excluded from Class 1 when three unpackaged articles, each individually activated by its own means of initiation or ignition or external means to function in the designed mode, meet the following test criteria:

(a)   No external surface shall have a temperature of more than 65 °C. A momentary spike in temperature up to 200 °C is acceptable;

(b)   No rupture or fragmentation of the external casing or movement of the article or detached parts thereof of more than one metre in any direction;

   NOTE: *Where the integrity of the article may be affected in the event of an external fire these criteria shall be examined by a fire test, such as described in ISO 12097-3.*

(c)   No audible report exceeding 135 dB(C) peak at a distance of one metre;

(d)   No flash or flame capable of igniting a material such as a sheet of 80 ± 10 g/m² paper in contact with the article; and

(e)   No production of smoke, fumes or dust in such quantities that the visibility in a one cubic metre chamber equipped with appropriately sized blow out panels is reduced more than 50% as measured by a calibrated light (lux) meter or radiometer located one metre from a constant light source located at the midpoint on opposite walls. The general guidance on Optical Density Testing in ISO 5659-1 and the general guidance on the Photometric System described in Section 7.5 in ISO 5659-2 may be used or similar optical density measurement methods designed to accomplish the same purpose may also be employed. A suitable hood cover surrounding the back and sides of the light meter shall be used to minimize effects of scattered or leaking light not emitted directly from the source.

   NOTE 1: *If during the tests addressing criteria (a), (b), (c) and (d) no or very little smoke is observed the test described in (e) may be waived.*

   NOTE 2: *The competent authority referred to in 2.2.1.1.8.1 may require testing in packaged form if it is determined that, as packaged for carriage, the article may pose a greater risk.*

2.2.1.1.9   *Classification documentation*

2.2.1.1.9.1   A competent authority assigning an article or substance to Class 1 shall confirm that classification with the applicant in writing.

2.2.1.1.9.2   A competent authority classification document may be in any form and may consist of more than one page, provided pages are numbered consecutively. The document shall have a unique reference.

2.2.1.1.9.3   The information provided shall be easy to identify, legible and durable.

2.2.1.1.9.4   Examples of the information that may be provided in the classification documents are as follows:

(a)   The name of the competent authority and the provisions in national legislation under which it is granted its authority;

(b)   The modal or national regulations for which the classification document is applicable;

(c)   Confirmation that the classification has been approved, made or agreed in accordance with the UN Model Regulations or the relevant modal regulations;

(d)   The name and address of the person in law to which the classification has been assigned and any company registration which uniquely identifies a company or other body corporate under national legislation;

(e)   The name under which the explosives will be placed onto the market or otherwise supplied for carriage;

(f)   The proper shipping name, UN number, class, division and corresponding compatibility group of the explosives;

(g)   Where appropriate, the maximum net explosive mass of the package or article;

(h)   The name, signature, stamp, seal or other identification of the person authorised by the competent authority to issue the classification document is clearly visible;

(i)   Where safety in carriage or the division is assessed as being dependent upon the packaging, the packaging mark or a description of the permitted:

-       Inner packagings

-       Intermediate packagings

-       Outer packagings

(j)   The classification document states the part number, stock number or other identifying reference under which the explosives will be placed onto the market or otherwise supplied for carriage;

(k)   The name and address of the person in law who manufactured the explosives and any company registration which uniquely identifies a company or other body corporate under national legislation;

(l)   Any additional information regarding the applicable packing instruction and special packing provisions where appropriate;

(m)   The basis for assigning the classification, i.e. whether on the basis of test results, default for fireworks, analogy with classified explosive, by definition from Table A of Chapter 3.2 etc.;

(n)   Any special conditions or limitations that the competent authority has identified as relevant to the safety for carriage of the explosives, the communication of the hazard and international carriage;

(o)   The expiry date of the classification document is given where the competent authority considers one to be appropriate.

2.2.1.2   *Substances and articles not accepted for carriage*

2.2.1.2.1   Explosive substances which are unduly sensitive according to the criteria of the Manual of Tests and Criteria, Part I, or are liable to spontaneous reaction, as well as explosive substances and articles which cannot be assigned to a name or n.o.s. entry listed in Table A of Chapter 3.2, shall not be accepted for carriage.

2.2.1.2.2   Articles of compatibility group K shall not be accepted for carriage (1.2K, UN No. 0020 and 1.3K, UN No. 0021).

Classification code (see 2.2.1.1.4)	UN No.	Name of the substance or article
1.1A	0473	SUBSTANCES, EXPLOSIVE, N.O.S.
1.1B	0461	COMPONENTS, EXPLOSIVE TRAIN, N.O.S.
1.1C	0474	SUBSTANCES, EXPLOSIVE, N.O.S.
	0497	PROPELLANT, LIQUID
	0498	PROPELLANT, SOLID
	0462	ARTICLES, EXPLOSIVE, N.O.S.
1.1D	0475	SUBSTANCES, EXPLOSIVE, N.O.S.
	0463	ARTICLES, EXPLOSIVE, N.O.S.
1.1E	0464	ARTICLES, EXPLOSIVE, N.O.S.
1.1F	0465	ARTICLES, EXPLOSIVE, N.O.S.
1.1G	0476	SUBSTANCES, EXPLOSIVE, N.O.S.
1.1L	0357	SUBSTANCES, EXPLOSIVE, N.O.S.
	0354	ARTICLES, EXPLOSIVE, N.O.S.
1.2B	0382	COMPONENTS, EXPLOSIVE TRAIN, N.O.S.
1.2C	0466	ARTICLES, EXPLOSIVE, N.O.S.
1.2D	0467	ARTICLES, EXPLOSIVE, N.O.S.
1.2E	0468	ARTICLES, EXPLOSIVE, N.O.S.
1.2F	0469	ARTICLES, EXPLOSIVE, N.O.S.
1.2L	0358	SUBSTANCES, EXPLOSIVE, N.O.S.
	0248	CONTRIVANCES, WATER-ACTIVATED with burster, expelling charge or propelling charge
	0355	ARTICLES, EXPLOSIVE, N.O.S.
1.3C	0132	DEFLAGRATING METAL SALTS OF AROMATIC NITRO-DERIVATIVES, N.O.S.
	0477	SUBSTANCES, EXPLOSIVE, N.O.S.
	0495	PROPELLANT, LIQUID
	0499	PROPELLANT, SOLID
	0470	ARTICLES, EXPLOSIVE, N.O.S.
1.3G	0478	SUBSTANCES, EXPLOSIVE, N.O.S.
1.3L	0359	SUBSTANCES, EXPLOSIVE, N.O.S.
	0249	CONTRIVANCES, WATER-ACTIVATED with burster, expelling charge or propelling charge
	0356	ARTICLES, EXPLOSIVE, N.O.S.
1.4B	0350	ARTICLES, EXPLOSIVE, N.O.S.
	0383	COMPONENTS, EXPLOSIVE TRAIN, N.O.S.
1.4C	0479	SUBSTANCES, EXPLOSIVE, N.O.S.
	0501	PROPELLANT, SOLID
	0351	ARTICLES, EXPLOSIVE, N.O.S.
1.4D	0480	SUBSTANCES, EXPLOSIVE, N.O.S.
	0352	ARTICLES, EXPLOSIVE, N.O.S.
1.4E	0471	ARTICLES, EXPLOSIVE, N.O.S.
1.4F	0472	ARTICLES, EXPLOSIVE, N.O.S.

Classification code (see 2.2.1.1.4)	UN No.	Name of the substance or article
1.4G	0485	SUBSTANCES, EXPLOSIVE, N.O.S.
	0353	ARTICLES, EXPLOSIVE, N.O.S.
1.4S	0481	SUBSTANCES, EXPLOSIVE, N.O.S.
	0349	ARTICLES, EXPLOSIVE, N.O.S.
	0384	COMPONENTS, EXPLOSIVE TRAIN, N.O.S.
1.5D	0482	SUBSTANCES, EXPLOSIVE, VERY INSENSITIVE (SUBSTANCES, EVI) N.O.S.
1.6N	0486	ARTICLES, EXPLOSIVE, EXTREMELY INSENSITIVE (ARTICLES, EEI)
	0190	SAMPLES, EXPLOSIVE other than initiating explosive
		NOTE: Division and Compatibility Group shall be defined as directed by the competent authority and according to the principles in 2.2.1.1.4.

NOTE 1: The descriptions in the glossary are not intended to replace the test procedures, nor to determine the hazard classification of a substance or article of Class 1. Assignment to the correct division and a decision on whether Compatibility Group S is appropriate shall be based on testing of the product in accordance with the Manual of Tests and Criteria, Part 1 or by analogy with similar products which have already been tested and assigned in accordance with the procedures of the Manual of Tests and Criteria.

NOTE 2: The figures given after the names refer to the relevant UN numbers (Column 1 of Table A of Chapter 3.2). For the classification code, see 2.2.1.1.4.

AMMUNITION, ILLUMINATING, with or without burster, expelling charge or propelling charge: UN Nos. 0171, 0254, 0297

Ammunition designed to produce a single source of intense light for lighting up an area. The term includes illuminating cartridges, grenades and projectiles; and illuminating and target identification bombs.

NOTE: The following articles: CARTRIDGES, SIGNAL; SIGNAL DEVICES HAND; SIGNALS, DISTRESS; FLARES, AERIAL; FLARES, SURFACE are not included in this definition. They are listed separately.

AMMUNITION, INCENDIARY, liquid or gel, with burster, expelling charge or propelling charge: UN No. 0247

Ammunition containing liquid or gelatinous incendiary substance. Except when the incendiary substance is an explosive per se, it also contains one or more of the following: a propelling charge with primer and igniter charge; a fuze with burster or expelling charge.

AMMUNITION, INCENDIARY, WHITE PHOSPHORUS with burster, expelling charge or propelling charge: UN Nos. 0243, 0244

Ammunition containing white phosphorus as incendiary substance. It also contains one or more of the following: a propelling charge with primer and igniter charge; a fuze with burster or expelling charge.

AMMUNITION, INCENDIARY with or without burster, expelling charge or propelling charge: UN Nos. 0009, 0010, 0300

Ammunition containing incendiary composition. Except when the composition is an explosive per se, it also contains one or more of the following: a propelling charge with primer and igniter charge; a fuze with burster or expelling charge.

AMMUNITION, PRACTICE: UN Nos. 0362, 0488

Ammunition without a main bursting charge, containing a burster or expelling charge. Normally it also contains a fuze and a propelling charge.

NOTE: GRENADES, PRACTICE are not included in this definition. They are listed separately.

AMMUNITION, PROOF: UN No. 0363

Ammunition containing pyrotechnic substances, used to test the performance or strength of new ammunition, weapon components or assemblies.

AMMUNITION, SMOKE, WHITE PHOSPHORUS, with burster, expelling charge or propelling charge: UN Nos. 0245, 0246

Ammunition containing white phosphorus as a smoke-producing substance. It also contains one or more of the following: a propelling charge with primer and igniter charge; a fuze with burster or expelling charge. The term includes grenades, smoke.

AMMUNITION, SMOKE with or without burster, expelling charge or propelling charge: UN Nos. 0015, 0016, 0303

Ammunition containing a smoke-producing substance such as chlorosulphonic acid mixture or titanium tetrachloride; or a smoke-producing pyrotechnic composition based on hexachloroethane or red phosphorus. Except when the substance is an explosive per se, the ammunition also contains one or more of the following: a propelling charge with primer and igniter charge; a fuze with burster or expelling charge. The term includes grenades, smoke.

NOTE: SIGNALS, SMOKE are not included in this definition. They are listed separately.

AMMUNITION, TEAR-PRODUCING, with burster, expelling charge or propelling charge: UN Nos. 0018, 0019, 0301

Ammunition containing a tear-producing substance. It also contains one or more of the following: a pyrotechnic substance; a propelling charge with primer and igniter charge; a fuze with burster or expelling charge.

ARTICLES, EXPLOSIVE, EXTREMELY INSENSITIVE (ARTICLES EEI): UN No. 0486

Articles containing only extremely insensitive substances which demonstrate a negligible probability of accidental initiation or propagation under normal conditions of transport, and which have passed Test Series 7.

ARTICLES, FYROPHORIC: UN No. 0380

Articles which contain a pyrophoric substance (capable of spontaneous ignition when exposed to air) and an explosive substance or component. The term excludes articles containing white phosphorus.

ARTICLES, PYROTECHNIC, for technical purposes: UN Nos. 0428, 0429, 0430, 0431, 0432

Articles which contain pyrotechnic substances and are used for technical purposes such as heat generation, gas generation, theatrical effects, etc.

NOTE: The following articles: all ammunition; CARTRIDGES, SIGNAL; CUTTERS, CABLE, EXPLOSIVE; FIREWORKS; FLARES, AERIAL; FLARES, SURFACE; RELEASE DEVICES, EXPLOSIVE; RIVETS, EXPLOSIVE; SIGNAL DEVICES, HAND; SIGNALS, DISTRESS; SIGNALS, RAILWAY TRACK, EXPLOSIVES; SIGNALS, SMOKE are not included in this definition. They are listed separately.

BLACK POWDER (GUNPOWDER), COMPRESSED or BLACK POWDER (GUNPOWDER), IN PELLETS: UN No. 0028

Substance consisting of a pelletized form of black powder.

BLACK POWDER (GUNPOWDER), granular or as meal: UN No. 0027

Substance consisting of an intimate mixture of charcoal or other carbon and either potassium nitrate or sodium nitrate, with or without sulphur.

BOMBS, WITH FLAMMABLE LIQUID, with bursting charge: UN Nos. 0399, 0400

Articles which are dropped from aircraft, consisting of a tank filled with inflammable liquid and bursting charge.

BOMBS, PHOTO-FLASH: UN No. 0038

Explosive articles which are dropped from aircraft to provide brief, intense illumination for photography. They contain a charge of detonating explosive without means of initiation or with means of initiation containing two or more effective protective features.

BOMBS, PHOTO-FLASH: UN No. 0037

Explosive articles which are dropped from aircraft to provide brief, intense illumination for photography. They contain a charge of detonating explosive with means of initiation not containing two or more effective protective features.

BOMBS, PHOTO-FLASH: UN Nos. 0039, 0299

Explosive articles which are dropped from aircraft to provide brief, intense illumination for photography. They contain a photo-flash composition.

BOMBS with bursting charge: UN Nos. 0034; 0035

Explosive articles which are dropped from aircraft, without means of initiation or with means of initiation containing two or more effective protective features.

BOMBS with bursting charge: UN Nos. 0033, 0291

Explosive articles which are dropped from aircraft, with means of initiation not containing two or more effective protective features.

BOOSTERS WITH DETONATOR: UN Nos. 0225, 0268

Articles consisting of a charge of detonating explosive with means of initiation. They are used to increase the initiating power of detonators or detonating cord.

BOOSTERS without detonator: UN Nos. 0042, 0283

Articles consisting of a charge of detonating explosive without means of initiation. They are used to increase the initiating power of detonators or detonating cord.

BURSTERS, explosive: UN No. 0043

Articles consisting of a small charge of explosive used to open projectiles or other ammunition in order to disperse their contents.

CARTRIDGES, FLASH: UN Nos. 0049, 0050

Articles consisting of a casing, a primer and flash powder, all assembled in one piece ready for firing.

CARTRIDGES FOR TOOLS, BLANK: UN No. 0014

Article, used in tools, consisting of a closed cartridge case with a centre or rim fire primer with or without a charge of smokeless or black powder but with no projectile.

CARTRIDGES FOR WEAPONS, BLANK: UN Nos. 0326, 0413, 0327, 0338, 0014

Ammunition consisting of a closed cartridge case with a centre or rim fire primer and a charge of smokeless or black powder but no projectile. It produces a loud noise and is used for training, saluting, propelling charge, starter pistols, etc. The term includes ammunition, blank.

CARTRIDGES FOR WEAPONS, INERT PROJECTILE: UN Nos. 0328, 0417, 0339, 0012

Ammunition consisting of a projectile without bursting charge but with a propelling charge with or without a primer. The articles may include a tracer, provided that the predominant hazard is that of the propelling charge.

CARTRIDGES FOR WEAPONS with bursting charge: UN Nos. 0006, 0321, 0412

Ammunition consisting of a projectile with a bursting charge without means of initiation or with means of initiation containing two or more effective protective features; and a propelling charge with or without a primer. The term includes fixed (assembled) ammunition, semi-fixed (partially assembled) ammunition and separate loading ammunition when the components are packed together.

CARTRIDGES FOR WEAPONS with bursting charge: UN Nos. 0005, 0007, 0348

Ammunition consisting of a projectile with a bursting charge with means of initiation not containing two or more effective protective features; and a propelling charge with or without a primer. The term includes fixed (assembled) ammunition, semi-fixed (partially assembled) ammunition and separate loading ammunition when the components are packed together.

CARTRIDGES, OIL WELL: UN Nos. 0277, 0278

Articles consisting of a thin casing of fibreboard, metal or other material containing only propellant powder which projects a hardened projectile to perforate an oil well casing.

NOTE: CHARGES, SHAPED are not included in this definition. They are listed separately.

CARTRIDGES, POWER DEVICE: UN Nos. 0275, 0276, 0323, 0381

Articles designed to accomplish mechanical actions. They consist of a casing with a charge of deflagrating explosive and a means of ignition. The gaseous products of the deflagration produce inflation, linear or rotary motion or activate diaphragms, valves or switches or project fastening devices or extinguishing agents.

CARTRIDGES, SIGNAL: UN Nos. 0054, 0312, 0405

Articles designed to fire coloured flares or other signals from signal pistols, etc.

CARTRIDGES, SMALL ARMS: UN Nos. 0417, 0339, 0012

Ammunition consisting of a cartridge case fitted with a centre or rim fire primer and containing both a propelling charge and solid projectile. They are designed to be fired in weapons of calibre not larger than 19.1 mm. Shot-gun cartridges of any calibre are included in this description.

NOTE: CARTRIDGES, SMALL ARMS, BLANK, are not included in this definition. They are listed separately. Some military small arms cartridges are not included in this definition. They are listed under CARTRIDGES FOR WEAPONS, INERT PROJECTILE.

CARTRIDGES, SMALL ARMS, BLANK: UN Nos. 0014, 0327, 0338

Ammunition consisting of a closed cartridge case with a centre or rim fire primer and a charge of smokeless or black powder. The cartridge cases contain no projectiles. The cartridges are designed to be fired from weapons with a calibre of at most 19.1 mm and serve to produce a loud noise and are used for training, saluting, propelling charge, starter pistols, etc.

CASES, CARTRIDGE, EMPTY, WITH PRIMER: UN Nos. 0379; 0055

Articles consisting of a cartridge case made from metal, plastics or other non-inflammable material, in which the only explosive component is the primer.

CASES, COMBUSTIBLE, EMPTY, WITHOUT PRIMER: UN Nos. 0447, 0446

Articles consisting of a cartridge case made partly or entirely from nitrocellulose.

CHARGES, BURSTING, PLASTICS BONDED: UN Nos. 0457, 0458, 0459, 0460

Articles consisting of a charge of detonating explosive, plastics bonded, manufactured in a specific form without a casing and without means of initiation. They are designed as components of ammunition such as warheads.

CHARGES, DEMOLITION: UN No. 0048

Articles containing a charge of a detonating explosive in a casing of fibreboard, plastics, metal or other material. The articles are without means of initiation or with means of initiation containing two or more effective protective features.

NOTE: The following articles: BOMBS; MINES; PROJECTILES are not included in this definition. They are listed separately.

CHARGES, DEPTH: UN No. 0056

Articles consisting of a charge of detonating explosive contained in a drum or projectile without means of initiation or with means of initiation containing two or more effective protective features. They are designed to detonate under water.

CHARGES, EXPLOSIVE, COMMERCIAL without detonator: UN Nos. 0442, 0443, 0444, 0445

Articles consisting of a charge of detonating explosive without means of initiation, used for explosive welding, jointing, forming and other metallurgical processes.

CHARGES, PROPELLING, FOR CANNON: UN Nos. 0242, 0279, 0414

Charges of propellant in any physical form for separate-loading ammunition for cannon.

CHARGES, PROPELLING: UN Nos. 0271, 0272, 0415, 0491

Articles consisting of a charge of a propellant charge in any physical form, with or without a casing, as a component of rocket motors or for reducing the drag of projectiles.

CHARGES, SHAPED, without detonator: UN Nos. 0059, 0439, 0440, 0441

Articles consisting of a casing containing a charge of detonating explosive with a cavity lined with rigid material, without means of initiation. They are designed to produce a powerful, penetrating jet effect.

CHARGES, SHAPED, FLEXIBLE, LINEAR: UN Nos. 0237, 0288

Articles consisting of a V-shaped core of a detonating explosive clad by a flexible sheath.

CHARGES, SUPPLEMENTARY, EXPLOSIVE: UN No. 0060

Articles consisting of a small removable booster placed in the cavity of a projectile between the fuze and the bursting charge.

COMPONENTS, EXPLOSIVE TRAIN, N.O.S.: UN Nos. 0382, 0383, 0384, 0461

Articles containing an explosive designed to transmit detonation or deflagration within an explosive train.

CONTRIVANCES, WATER-ACTIVATED with burster, expelling charge or propelling charge: UN Nos. 0248, 0249

Articles whose functioning depends upon physico-chemical reaction of their contents with water.

CORD, DETONATING, flexible: UN Nos. 0065, 0289

Article consisting of a core of detonating explosive enclosed in spun fabric and a plastics or other covering. The covering is not necessary if the spun fabric is sift-proof.

CORD (FUSE) DETONATING, metal clad: UN Nos. 0102, 0290

Article consisting of a core of detonating explosive clad by a soft metal tube with or without protective covering.

CORD (FUSE) DETONATING, MILD EFFECT, metal clad: UN No. 0104

Article consisting of a core of detonating explosive clad by a soft metal tube with or without a protective covering. The quantity of explosive substance is so small that only a mild effect is manifested outside the cord.

CORD, IGNITER: UN No. 0066

Article consisting of textile yarns covered with black powder or another fast burning pyrotechnic composition and of a flexible protective covering; or it consists of a core of black powder surrounded by a flexible woven fabric. It burns progressively along its length with an external flame and is used to transmit ignition from a device to a charge or primer.

CUTTERS, CABLE, EXPLOSIVE: UN No. 0070

Articles consisting of a knife-edged device which is driven by a small charge of deflagrating explosive into an anvil.

DETONATOR ASSEMBLIES, NON-ELECTRIC for blasting: UN Nos. 0360, 0361, 0500

Non-electric detonators assembled with and activated by such means as safety fuse, shock tube, flash tube or detonating cord. They may be of instantaneous design or incorporate delay elements. Detonating relays incorporating detonating cord are included.

DETONATORS, ELECTRIC for blasting: UN Nos. 0030, 0255, 0456

Articles specially designed for the initiation of blasting explosives. These detonators may be constructed to detonate instantaneously or may contain a delay element. Electric detonators are activated by an electric current.

DETONATORS FOR AMMUNITION: UN Nos. 0073, 0364, 0365, 0366

Articles consisting of a small metal or plastics tube containing explosives such as lead azide, PETN or combinations of explosives. They are designed to start a detonation train.

DETONATORS, NON-ELECTRIC for blasting: UN Nos. 0029, 0267, 0455

Articles specially designed for the initiation of blasting explosives. These detonators may be constructed to detonate instantaneously or may contain a delay element. Non-electric detonators are activated by such means as shock tube, flash tube, safety fuse, other igniferous device or flexible detonating cord. Detonating relays without detonating cord are included.

EXPLOSIVE, BLASTING, TYPE A: UN No. 0081

Substances consisting of liquid organic nitrates such as nitroglycerine or a mixture of such ingredients with one or more of the following: nitrocellulose; ammonium nitrate or other inorganic nitrates; aromatic nitro-derivatives, or combustible materials, such as wood-meal and aluminium powder. They may contain inert components such as kieselguhr, and additives such as colouring agents and stabilizers. Such explosives shall be in powdery, gelatinous or elastic form. The term includes dynamite; gelatine, blasting and gelatine dynamites.

EXPLOSIVE, BLASTING, TYPE B: UN Nos. 0082, 0331

Substances consisting of

(a)   a mixture of ammonium nitrate or other inorganic nitrates with an explosive such as trinitrotoluene, with or without other substances such as wood-meal and aluminium powder; or

(b)   a mixture of ammonium nitrate or other inorganic nitrates with other combustible substances which are not explosive ingredients. In both cases they may contain inert components such as kieselguhr, and additives such as colouring agents and stabilizers. Such explosives must not contain nitroglycerine, similar liquid organic nitrates or chlorates.

EXPLOSIVE, BLASTING, TYPE C: UN No. 0083

Substances consisting of a mixture of either potassium or sodium chlorate or potassium, sodium or ammonium perchlorate with organic nitro-derivatives or combustible materials such as wood-meal or aluminium powder or a hydrocarbon. They may contain inert components such as kieselguhr and additives such as colouring agents and stabilizers. Such explosives must not contain nitroglycerine or similar liquid organic nitrates.

EXPLOSIVE, BLASTING, TYPE D: UN No. 0084

Substances consisting of a mixture of organic nitrated compounds and combustible materials such as hydrocarbons and aluminium powder. They may contain inert components such as kieselguhr and additives such as colouring agents and stabilizers. Such explosives must not contain nitroglycerine, similar liquid organic nitrates, chlorates and ammonium nitrate. The term generally includes plastic explosives.

EXPLOSIVES, BLASTING, TYPE E: UN Nos. 0241, 0332

Substances consisting of water as an essential ingredient and high proportions of ammonium nitrate or other oxidizers, some or all of which are in solution. The other constituents may include nitro-derivatives such as trinitrotoluene, hydrocarbons or aluminium powder. They may contain inert components such as kieselguhr and additives such as colouring agents and stabilizers. The term includes explosives, emulsion, explosives, slurry and explosives, watergel.

FIREWORKS: UN Nos. 0333, 0334, 0335, 0336, 0337

Pyrotechnic articles designed for entertainment.

FLARES, AERIAL: UN Nos. 0093, 0403, 0404, 0420, 0421

Articles containing pyrotechnic substances which are designed to be dropped from an aircraft to illuminate, identify, signal or warn.

FLARES, SURFACE: UN Nos. 0092, 0418, 0419

Articles containing pyrotechnic substances which are designed for use on the surface to illuminate, identify, signal or warn.

FLASH POWDER: UN Nos. 0094, 0305

Pyrotechnic substance which, when ignited, produces an intense light.

FRACTURING DEVICES, EXPLOSIVE without detonator, for oil wells: UN No. 0099

Articles consisting of a charge of detonating explosive contained in a casing without means of initiation. They are used to fracture the rock around a drill shaft to assist the flow of crude oil from the rock.

FUSE, IGNITER, tubular, metal clad: UN No. 0103

Article consisting of a metal tube with a core of deflagrating explosive.

FUSE, NON-DETONATING: UN No. 0101

Article consisting of cotton yarns impregnated with fine black powder (quickmatch). It burns with an external flame and is used in ignition trains for fireworks, etc.

FUSE, SAFETY: UN No. 0105

Article consisting of a core of fine grained black powder surrounded by a flexible woven fabric with one or more protective outer coverings. When ignited, it burns at a predetermined rate without any external explosive effect.

FUZES, DETONATING: UN Nos. 0106, 0107, 0257, 0367

Articles with explosive components designed to produce a detonation in ammunition. They incorporate mechanical, electrical, chemical or hydrostatic components to initiate the detonation. They generally incorporate protective features.

FUZES, DETONATING with protective features: UN Nos. 0408, 0409, 0410

Articles with explosive components designed to produce a detonation in ammunition. They incorporate mechanical, electrical, chemical or hydrostatic components to initiate the detonation. The detonating fuze must incorporate two or more effective protective features.

FUZES, IGNITING: UN Nos. 0316, 0317, 0368

Articles with primary explosive components designed to produce a deflagration in ammunition. They incorporate mechanical, electrical, chemical or hydrostatic components to start the deflagration. They generally incorporate protective features.

GRENADES, hand or rifle, with bursting charge: UN Nos. 0284, 0285

Articles which are designed to be thrown by hand or to be projected by a rifle. They are without means of initiation or with means of initiation containing two or more effective protective features.

GRENADES, hand or rifle, with bursting charge: UN Nos. 0292, 0293

Articles which are designed to be thrown by hand or to be projected by a rifle. They are with means of initiation not containing two or more effective protective features.

GRENADES, PRACTICE, hand or rifle: UN Nos. 0110, 0372, 0318, 0452

Articles without a main bursting charge which are designed to be thrown by hand or to be projected by a rifle. They contain the priming device and may contain a spotting charge.

HEXOTONAL: UN No. 0393

Substance consisting of an intimate mixture of cyclotrimethylene-trinitramine (RDX), trinitrotoluene (TNT) and aluminium.

HEXOLITE (HEXOTOL), dry or wetted with less than 15% water, by mass: UN No. 0118

Substance consisting of an intimate mixture of cyclotrimethylene-trinitramine (RDX) and trinitrotoluene (TNT). The term includes "Composition B".

IGNITERS: UN Nos. 0121, 0314, 0315, 0325, 0454

Articles containing one or more explosive substances designed to produce a deflagration in an explosive train. They may be actuated chemically, electrically or mechanically.

NOTE: The following articles: CORD, IGNITER; FUSE, IGNITER; FUSE, NON-DETONATING; FUZES, IGNITING; LIGHTERS, FUSE; PRIMERS, CAP TYPE; PRIMERS, TUBULAR are not included in this definition. They are listed separately.

JET PERFORATING GUNS, CHARGED, oil well, without detonator: UN Nos. 0124, 0494

Articles consisting of a steel tube or metallic strip, into which are inserted shaped charges connected by detonating cord, without means of initiation.

LIGHTERS, FUSE: UN No. 0131

Articles of various design actuated by friction, percussion or electricity and used to ignite a safety fuse.

MINES with bursting charge: UN Nos. 0137, 0138

Articles consisting normally of metal or composition receptacles filled with a detonating explosive, without means of initiation or with means of initiation containing two or more effective protective features. They are designed to be operated by the passage of ships, vehicles or personnel. The term includes "Bangalore torpedoes".

MINES with bursting charge: UN Nos. 0136, 0294

Articles consisting normally of metal or composition receptacles filled with a detonating explosive, with means of initiation not containing two or more effective protective features. They are designed to be operated by the passage of ships, vehicles or personnel. The term includes "Bangalore torpedoes".

OCTOLITE (OCTOL), dry or wetted with less than 15% water, by mass: UN No. 0266

Substance consisting of an intimate mixture of cyclotetramethylene-tetranitramine (HMX) and trinitrotoluene (TNT).

OCTONAL: UN No. 0496

Substance consisting of an intimate mixture of cyclotetramethylenetetranitramine (HMX), trinitrotoluene (TNT) and aluminium.

PENTOLITE, dry or wetted with less than 15% water, by mass: UN No. 0151

Substance consisting of an intimate mixture of pentaerythrite tetranitrate (PETN) and trinitrotoluene (TNT).

POWDER CAKE (POWDER PASTE), WETTED with not less than 17% alcohol, by mass; POWDER CAKE (POWDER PASTE), WETTED with not less than 25% water, by mass: UN Nos. 0433, 0159

Substance consisting of nitrocellulose impregnated with not more than 60% of nitroglycerine or other liquid organic nitrates or a mixture of these.

POWDER, SMOKELESS: UN Nos. 0160, 0161, 0509

Substance based on nitrocellulose used as propellant. The term includes propellants with a single base (nitrocellulose (NC) alone), those with a double base (such as NC and nitroglycerine/(NG)) and those with a triple base (such as NC/NG/nitroguanidine).

NOTE: Cast, pressed or bag-charges of smokeless powder are listed under CHARGES, PROPELLING or CHARGES, PROPELLING, FOR CANON.

PRIMERS, CAP TYPE: UN Nos. 0044, 0377, 0378

Articles consisting of a metal or plastics cap containing a small amount of primary explosive mixture that is readily ignited by impact. They serve as igniting elements in small arms cartridges and in percussion primers for propelling charges.

PRIMERS, TUBULAR: UN Nos. 0319, 0320, 0376

Articles consisting of a primer for ignition and an auxiliary charge of deflagrating explosive such as black powder used to ignite the propelling charge in a cartridge case for cannon, etc.

PROJECTILES, inert with tracer: UN Nos. 0345, 0424, 0425

Articles such as a shell or bullet, which are projected from a cannon or other gun, rifle or other small arm.

PROJECTILES with burster or expelling charge: UN Nos. 0346, 0347

Articles such as a shell or bullet, which are projected from a cannon or other gun. They are without means of initiation or with means of initiation containing two or more effective protective features. They are used to scatter dyes for spotting or other inert materials.

PROJECTILES with burster or expelling charge: UN Nos. 0426, 0427

Articles such as a shell or bullet, which are projected from a cannon or other gun. They are with means of initiation not containing two or more effective protective features. They are used to scatter dyes for spotting or other inert materials.

PROJECTILES with burster or expelling charge: UN Nos. 0434, 0435

Articles such as a shell or bullet, which are projected from a cannon or other gun, rifle or other small arm. They are used to scatter dyes for spotting or other inert materials.

PROJECTILES with bursting charge: UN Nos. 0168, 0169, 0344

Articles such as a shell or bullet, which are projected from a cannon or other gun. They are without means of initiation or with means of initiation containing two or more effective protective features.

PROJECTILES with bursting charge: UN Nos. 0167, 0324

Articles such as a shell or bullet, which are projected from a cannon or other gun. They are with means of initiation not containing two or more effective protective features.

PROPELLANT, LIQUID: UN Nos. 0495, 0497

Substance consisting of a deflagrating liquid explosive, used for propulsion.

PROPELLANT, SOLID: UN Nos. 0498, 0499, 0501

Substance consisting of a deflagrating solid explosive, used for propulsion.

RELEASE DEVICES, EXPLOSIVE: UN No. 0173

Articles consisting of a small charge of explosive with means of initiation and rods or links. They sever the rods or links to release equipment quickly.

RIVETS, EXPLOSIVE: UN No. 0174

Articles consisting of a small charge of explosive inside a metallic rivet.

ROCKET MOTORS: UN Nos. 0186, 0280, 0281, 0510

Articles consisting of a charge of explosive, generally a solid propellant, contained in a cylinder fitted with one or more nozzles. They are designed to propel a rocket or a guided missile.

ROCKET MOTORS, LIQUID FUELLED: UN Nos. 0395, 0396

Articles consisting of a liquid fuel within a cylinder fitted with one or more nozzles. They are designed to propel a rocket or a guided missile.

ROCKET MOTORS WITH HYPERGOLIC LIQUIDS with or without expelling charge: UN Nos. 0322, 0250

Articles consisting of a hypergolic fuel contained in a cylinder fitted with one or more nozzles. They are designed to propel a rocket or a guided missile.

ROCKETS, LINE THROWING: UN Nos. 0238, 0240, 0453

Articles consisting of a rocket motor which is designed to extend a line.

ROCKETS, LIQUID FUELLED with bursting charge: UN Nos. 0397, 0398

Articles consisting of a liquid fuel within a cylinder fitted with one or more nozzles and fitted with a warhead. The term includes guided missiles.

ROCKETS with bursting charge: UN Nos. 0181, 0182

Articles consisting of a rocket motor and a warhead without means of initiation or with means of initiation containing two or more effective protective features. The term includes guided missiles.

ROCKETS with bursting charge: UN Nos. 0180, 0295

Articles consisting of a rocket motor and a warhead with means of initiation not containing two or more effective protective features. The term includes guided missiles.

ROCKETS with expelling charge: UN Nos. 0436, 0437, 0438

Articles consisting of a rocket motor and a charge to expel the payload from a rocket head. The term includes guided missiles.

ROCKETS with inert head: UN Nos. 0183, 0502

Articles consisting of a rocket motor and an inert head. The term includes guided missiles.

SAFETY DEVICES, PYROTECHNIC: UN No. 0503

Articles which contain pyrotechnic substances or dangerous goods of other classes and are used in vehicles, vessels or aircraft to enhance safety to persons. Examples are: air bag inflators, air bag modules, seat-belt pretensioners and pyromechanical devices. These pyromechanical devices are assembled components for tasks such as but not limited to separation, locking, or occupant restraint.

SAMPLES, EXPLOSIVE, other than initiating explosive UN No. 0190

New or existing explosive substances or articles, not yet assigned to a name in Table A of Chapter 3.2 and carried in conformity with the instructions of the competent authority and generally in small quantities, inter alia, for the purposes of testing, classification, research and development, or quality control, or as commercial samples.

NOTE: Explosive substances or articles already assigned to another name in Table A of Chapter 3.2 are not included in this definition.

SIGNAL DEVICES, HAND: UN Nos. 0191, 0373

Portable articles containing pyrotechnic substances which produce visual signals or warnings. The term includes small surface flares such as highway or railway flares and small distress flares.

SIGNALS, DISTRESS, ship: UN Nos. 0194, 0195, 0505, 0506

Articles containing pyrotechnic substances designed to produce signals by means of sound, flame or smoke or any combination thereof.

SIGNALS, RAILWAY TRACK, EXPLOSIVE: UN Nos. 0192, 0193, 0492, 0493

Articles containing a pyrotechnic substance which explodes with a loud report when the article is crushed. They are designed to be placed on a rail.

SIGNALS, SMOKE: UN Nos. 0196, 0197, 0313, 0487, 0507

Articles containing pyrotechnic substances which emit smoke. In addition they may contain devices for emitting audible signals.

SOUNDING DEVICES, EXPLOSIVE: UN Nos. 0374, 0375

Articles consisting of a charge of detonating explosive, without means of initiation or with means of initiation containing two or more effective protective features. They are dropped from ships and function when they reach a predetermined depth or the sea bed.

SOUNDING DEVICES, EXPLOSIVE: UN Nos. 0204, 0296

Articles consisting of a charge of detonating explosive with means of initiation not containing two or more effective protective features. They are dropped from ships and function when they reach a predetermined depth or the sea bed.

SUBSTANCES, EXPLOSIVE, VERY INSENSITIVE (Substances, EVI), N.O.S.: UN No. 0482

Substances presenting a mass explosion hazard but which are so insensitive that there is very little probability of initiation or of transition from burning to detonation under normal conditions of transport, and which have passed Test Series 5.

TORPEDOES, LIQUID FUELLED with inert head: UN No. 0450

Articles consisting of a liquid explosive system to propel the torpedo through the water, with an inert head.

TORPEDOES, LIQUID FUELLED with or without bursting charge: UN No. 0449

Articles consisting of either a liquid explosive system to propel the torpedo through the water, with or without a warhead; or a liquid non-explosive system to propel the torpedo through the water, with a warhead.

TORPEDOES with bursting charge: UN No. 0451

Articles consisting of a non-explosive system to propel the torpedo through the water, and a warhead without means of initiation or with means of initiation containing two or more effective protective features.

TORPEDOES with bursting charge: UN No. 0329

Articles consisting of an explosive system to propel the torpedo through the water, and a warhead without means of initiation or with means of initiation containing two or more effective protective features.

TORPEDOES with bursting charge: UN No. 0330

Articles consisting of an explosive or non-explosive system to propel the torpedo through the water, and a warhead with means of initiation not containing two or more effective protective features.

TRACERS FOR AMMUNITION: UN Nos. 0212, 0306

Sealed articles containing pyrotechnic substances, designed to reveal the trajectory of a projectile.

TRITONAL: UN No. 0390

Substance consisting of trinitrotoluene (TNT) mixed with aluminium.

WARHEADS, ROCKET with burster or expelling charge: UN No. 0370

Articles consisting of an inert payload and a small charge of detonating or deflagrating explosive, without means of initiation or with means of initiation containing two or more effective protective features. They are designed to be fitted to a rocket motor to scatter inert material. The term includes warheads for guided missiles.

WARHEADS, ROCKET with burster or expelling charge: UN No. 0371

Articles consisting of an inert payload and a small charge of detonating or deflagrating explosive, with means of initiation not containing two or more effective protective features. They are designed to be fitted to a rocket motor to scatter inert material. The term includes warheads for guided missiles.

WARHEADS, ROCKET with bursting charge: UN Nos. 0286, 0287

Articles consisting of a detonating explosive, without means of initiation or with means of initiation containing two or more effective protective features. They are designed to be fitted to a rocket. The term includes warheads for guided missiles.

WARHEADS, ROCKET with bursting charge: UN No. 0369

Articles consisting of a detonating explosive, with means of initiation not containing two or more effective protective features. They are designed to be fitted to a rocket. The term includes warheads for guided missiles.

WARHEADS, TORPEDO with bursting charge: UN No. 0221

Articles consisting of a detonating explosive, without means of initiation or with means of initiation containing two or more effective protective features. They are designed to be fitted to a torpedo.

2.2.2   **Class 2   Gases**

2.2.2.1   *Criteria*

2.2.2.1.1   The heading of Class 2 covers pure gases, mixtures of gases, mixtures of one or more gases with one or more other substances and articles containing such substances.

A gas is a substance which:

(a)   at 50 °C has a vapour pressure greater than 300 kPa (3 bar); or

(b)   is completely gaseous at 20 °C at the standard pressure of 101.3 kPa.

NOTE 1: UN No. 1052 HYDROGEN FLUORIDE, ANHYDROUS is nevertheless classified in Class 8.

NOTE 2:   A pure gas may contain other components deriving from its production process or added to preserve the stability of the product, provided that the level of these components does not change its classification or its conditions of carriage, such as filling ratio, filling pressure, test pressure.

NOTE 3:   N.O.S. entries in 2.2.2.3 may cover pure gases as well as mixtures.

2.2.2.1.2   The substances and articles of Class 2 are subdivided as follows:

1.   *Compressed gas:* a gas which when packaged under pressure for carriage is entirely gaseous at -50 °C; this category includes all gases with a critical temperature less than or equal to -50 °C;

2.   *Liquefied gas:* a gas which when packaged under pressure for carriage is partially liquid at temperatures above -50 °C. A distinction is made between:

   High pressure liquefied gas: a gas with a critical temperature above -50 °C and equal to or below +65 °C; and

   Low pressure liquefied gas: a gas with a critical temperature above +65 °C;

3.   *Refrigerated liquefied gas*: a gas which when packaged for carriage is made partially liquid because of its low temperature;

4.   *Dissolved gas:* a gas which when packaged under pressure for carriage is dissolved in a liquid phase solvent;

5.   Aerosol dispensers and receptacles, small, containing gas (gas cartridges);

6.   Other articles containing gas under pressure;

7.   Non-pressurized gases subject to special requirements (gas samples);

8.   Chemicals under pressure: liquids, pastes or powders, pressurized with a propellant that meets the definition of a compressed or liquefied gas and mixtures thereof.

9.   *Adsorbed gas:* a gas which when packaged for carriage is adsorbed onto a solid porous material resulting in an internal receptacle pressure of less than 101.3 kPa at 20 °C and less than 300 kPa at 50 °C.

2.2.2.1.3   Substances and articles (except aerosols and chemicals under pressure) of Class 2 are assigned to one of the following groups according to their hazardous properties, as follows:

A   asphyxiant;

O   oxidizing;

F   flammable;

T   toxic;

TF   toxic, flammable;

TC   toxic, corrosive;

TO   toxic, oxidizing;

TFC   toxic, flammable, corrosive;

TOC toxic, oxidizing, corrosive.

For gases and gas mixtures presenting hazardous properties associated with more than one group according to the criteria, the groups designated by letter T take precedence over all other groups. The groups designated by letter F take precedence over the groups designated by letters A or O.

NOTE 1: *In the UN Model Regulations, the IMDG Code and the ICAO Technical Instructions, gases are assigned to one of the following three divisions, based on the primary hazard:*

Division 2.1:   *flammable gases (corresponding to the groups designated by the capital letter F);*

Division 2.2:   *non-flammable, non-toxic gases (corresponding to the groups designated by the capital letters A or O);*

Division 2.3:   *toxic gases (corresponding to the groups designated by the capital letter T i.e. T, TF, TC, TO, TFC and TOC).*

NOTE 2: *Receptacles, small containing gas (UN No. 2037) shall be assigned to the groups A to TOC according to the hazard of the contents. For aerosols (UN No. 1950), see 2.2.2.1.6. For chemicals under pressure (UN Nos. 3500 to 3505), see 2.2.2.1.7.*

NOTE 3: *Corrosive gases are considered to be toxic, and are therefore assigned to the group TC, TFC or TOC.*

2.2.2.1.4   If a mixture of Class 2 mentioned by name in Table A of Chapter 3.2 meets different criteria as mentioned in 2.2.2.1.2 and 2.2.2.1.5, this mixture shall be classified according to the criteria and assigned to an appropriate N.O.S. entry.

2.2.2.1.5   Substances and articles (except aerosols and chemicals under pressure) of Class 2 which are not mentioned by name in Table A of Chapter 3.2 shall be classified under a collective entry listed in 2.2.2.3 in accordance with 2.2.2.1.2 and 2.2.2.1.3. The following criteria shall apply:

Asphyxiant gases

Gases which are non-oxidizing, non-flammable and non-toxic and which dilute or replace oxygen normally in the atmosphere.

Flammable gases

Gases which at 20 °C and a standard pressure of 101.3 kPa:

(a)   are ignitable when in a mixture of 13% or less by volume with air; or

(b)   have a flammable range with air of at least 12 percentage points regardless of the lower flammable limit.

Flammability shall be determined by tests or by calculation, in accordance with methods adopted by ISO (see ISO 10156:2010).

Where insufficient data are available to use these methods, tests by a comparable method recognized by the competent authority of the country of origin may be used.

If the country of origin is not a Contracting Party to ADR these methods shall be recognized by the competent authority of the first country Contracting Party to ADR reached by the consignment.

Oxidizing gases

Gases, which may, generally by providing oxygen, cause or contribute to the combustion of other material more than air does. These are pure gases or gas mixtures with an oxidizing power greater than 23.5% as determined by a method specified in ISO 10156:2010.

Toxic gases

NOTE: *Gases meeting the criteria for toxicity in part or completely owing to their corrosivity are to be classified as toxic. See also the criteria under the heading "Corrosive gases" for a possible subsidiary corrosivity risk.*

Gases which:

(a)   are known to be so toxic or corrosive to humans as to pose a hazard to health; or

(b)   are presumed to be toxic or corrosive to humans because they have a LC_{50} value for acute toxicity equal to or less than 5 000 ml/m³ (ppm) when tested in accordance with 2.2.61.1.

In the case of gas mixtures (including vapours of substances from other classes) the following formula may be used:

$$LC_{50} \text{ Toxic (mixture)} = \frac{1}{\sum_{i=1}^{n} \frac{f_i}{T_i}}$$

where f_i =   mole fraction of the ith component substance of the mixture;

          T_i =   toxicity index of the ith component substance of the mixture.

                          The T_i equals the LC_{50} value as found in packing instruction P200 of 4.1.4.1.

                          When no LC_{50} value is listed in packing instruction P200 of 4.1.4.1, a LC_{50} value available in scientific literature shall be used.

                          When the LC_{50} value is unknown, the toxicity index is determined by using the lowest LC_{50} value of substances of similar physiological and chemical effects, or through testing if this is the only practical possibility.

Corrosive gases

Gases or gas mixtures meeting the criteria for toxicity completely owing to their corrosivity are to be classified as toxic with a subsidiary corrosivity risk.

A gas mixture that is considered to be toxic due to the combined effects of corrosivity and toxicity has a subsidiary risk of corrosivity when the mixture is known by human experience to be destructive to the skin, eyes or mucous membranes or when the LC_{50} value of the corrosive components of the mixture is equal to or less than 5 000 ml/m³ (ppm) when the LC_{50} is calculated by the formula:

$$LC_{50} \text{ Corrosive (mixture)} = \frac{1}{\sum_{i=1}^{n} \frac{f_{ci}}{T_{ci}}}$$

where f_{ci} =   mole fraction of the ith corrosive component substance of the mixture;

          T_{ci} =   toxicity index of the ith corrosive component substance of the mixture.

                          The T_{ci} equals the LC_{50} value as found in packing instruction P200 of 4.1.4.1.

When no LC_{50} value is listed in packing instruction P200 of 4.1.4.1, a LC_{50} value available in scientific literature shall be used.

When the LC_{50} value is unknown the toxicity index is determined by using the lowest LC_{50} value of substances of similar physiological and chemical effects, or through testing if this is the only practical possibility.

2.2.2.1.6   *Aerosols*

Aerosols (UN No. 1950) are assigned to one of the following groups according to their hazardous properties, as follows:

A   asphyxiant;

O   oxidizing;

F   flammable;

T   toxic;

C   corrosive;

CO   corrosive, oxidizing;

FC   flammable, corrosive;

TF   toxic, flammable;

TC   toxic, corrosive;

TO   toxic, oxidizing;

TFC   toxic, flammable, corrosive;

TOC   toxic, oxidizing, corrosive.

The classification depends on the nature of the contents of the aerosol dispenser.

NOTE: Gases, which meet the definition of toxic gases according to 2.2.2.1.5 and gases identified as "Considered as pyrophoric" by table note c of Table 2 of packing instruction P200 in 4.1.4.1, shall not be used as a propellant in an aerosol dispenser. Aerosols with contents meeting the criteria for packing group I for toxicity or corrosivity shall not be accepted for carriage (see also 2.2.2.2.2).

The following criteria shall apply:

(a)   Assignment to group A shall apply when the contents do not meet the criteria for any other group according to sub-paragraphs (b) to (f) below;

(b)   Assignment to group O shall apply when the aerosol contains an oxidizing gas according to 2.2.2.1.5;

(c)   Assignment to group F shall apply if the contents include 85% by mass or more flammable components and the chemical heat of combustion is 30 kJ/g or more.

It shall not apply if the contents contain 1% by mass or less flammable components and the heat of combustion is less than 20 kJ/g.

Otherwise the aerosol shall be tested for flammability in accordance with the tests described in the *Manual of Tests and Criteria*, Part III, section 31. Extremely flammable and flammable aerosols shall be assigned to group F;

NOTE: Flammable components are flammable liquids, flammable solids or flammable gases and gas mixtures as defined in Notes 1 to 3 of sub-section 31.1.3 of Part III of the Manual of Tests and Criteria. This designation does not cover pyrophoric, self-heating or water-reactive substances. The chemical heat of combustion shall be determined by one of the following methods ASTM D 240, ISO/FDIS 13943:1999 (E/F) 86.1 to 86.3 or NFPA 30B.

(d) Assignment to group T shall apply when the contents, other than the propellant of aerosol dispensers to be ejected, are classified as Class 6.1, packing groups II or III;

(e) Assignment to group C shall apply when the contents, other than the propellant of aerosol dispensers to be ejected, meet the criteria for Class 8, packing groups II or III;

(f) When the criteria for more than one group amongst groups O, F, T, and C are met, assignment to groups CO, FC, TF, TC TO, TFC or TOC shall apply, as relevant.

2.2.2.1.7   *Chemicals under pressure*

Chemicals under pressure (UN Nos. 3500 to 3505) are assigned to one of the following groups according to their hazardous properties, as follows:

A   asphyxiant;

F   flammable;

T   toxic;

C   corrosive;

FC   flammable, corrosive;

TF   toxic, flammable.

The classification depends on the hazard characteristics of the components in the different states:

The propellant;

The liquid; or

The solid.

NOTE 1: *Gases, which meet the definition of toxic gases or of oxidizing gases according to 2.2.2.1.5 or gases identified as "Considered as pyrophoric" by table note c of Table 2 of packing instruction P200 in 4.1.4.1, shall not be used as a propellant in chemicals under pressure.*

NOTE 2: *Chemicals under pressure with contents meeting the criteria for packing group I for toxicity or corrosivity or with contents meeting both the criteria for packing group II or III for toxicity and for packing group II or III for corrosivity shall not be accepted for carriage under these UN numbers.*

NOTE 3: *Chemicals under pressure with components meeting the properties of Class 1; liquid desensitized explosives of Class 3; self-reactive substances and solid desensitized explosives of Class 4.1; Class 4.2; Class 4.3; Class 5.1; Class 5.2; Class 6.2; or Class 7, shall not be used for carriage under these UN numbers.*

NOTE 4: *A chemical under pressure in an aerosol dispenser shall be carried under UN No. 1950.*

The following criteria shall apply:

(a)   Assignment to group A shall apply when the contents do not meet the criteria for any other group according to sub-paragraphs (b) to (e) below;

(b)   Assignment to group F shall apply if one of the components, which can be a pure substance or a mixture, needs to be classified as flammable. Flammable components are flammable liquids and liquid mixtures, flammable solids and solid mixtures or flammable gases and gas mixtures meeting the following criteria:

(i)   A flammable liquid is a liquid having a flashpoint of not more than 93 °C;

(ii)   A flammable solid is a solid which meets the criteria in 2.2.41.1;

(iii)   A flammable gas is a gas which meets the criteria in 2.2.2.1.5;

(c)   Assignment to group T shall apply when the contents, other than the propellant, are classified as dangerous goods of Class 6.1, packing groups II or III;

(d)   Assignment to group C shall apply when the contents, other than the propellant, are classified as dangerous goods of Class 8, packing groups II or III;

(e)   When the criteria for two groups amongst groups F, T, and C are met, assignment to groups FC or TF shall apply, as relevant.

2.2.2.2   *Gases not accepted for carriage*

2.2.2.2.1   Chemically unstable gases of Class 2 shall not be accepted for carriage unless the necessary precautions have been taken to prevent the possibility of a dangerous decomposition or polymerization under normal conditions of carriage or unless carried in accordance with special packing provision (r) of packing instruction P200 (10) of 4.1.4.1, as applicable. For the precautions necessary to prevent polymerization, see special provision 386 of Chapter 3.3. To this end particular care shall be taken to ensure that receptacles and tanks do not contain any substances liable to promote these reactions.

2.2.2.2.2   The following substances and mixtures shall not be accepted for carriage:

-    UN No. 2186 HYDROGEN CHLORIDE, REFRIGERATED LIQUID;

-    UN No. 2421 NITROGEN TRIOXIDE;

-    UN No. 2455 METHYL NITRITE;

-    Refrigerated liquefied gases which cannot be assigned to classification codes 3A, 3O or 3F;

-    Dissolved gases which cannot be classified under UN Nos. 1001, 2073 or 3318;

-    Aerosols where gases which are toxic according to 2.2.2.1.5 or pyrophoric according to packing instruction P200 in 4.1.4.1 are used as propellants;

-    Aerosols with contents meeting the criteria for packing group I for toxicity or corrosivity (see 2.2.61 and 2.2.8);

-    Receptacles, small, containing gases which are very toxic (LC_{50} lower than 200 ppm) or pyrophoric according to packing instruction P200 in 4.1.4.1.

Compressed gases		
Classification code	**UN No.**	**Name of the substance or article**
1A	1956	COMPRESSED GAS, N.O.S.
1O	3156	COMPRESSED GAS, OXIDIZING, N.O.S.
1F	1964	HYDROCARBON GAS MIXTURE, COMPRESSED, N.O.S.
	1954	COMPRESSED GAS, FLAMMABLE, N.O.S.
1T	1955	COMPRESSED GAS, TOXIC, N.O.S.
1TF	1953	COMPRESSED GAS, TOXIC, FLAMMABLE, N.O.S.
1TC	3304	COMPRESSED GAS, TOXIC, CORROSIVE, N.O.S.
1TO	3303	COMPRESSED GAS, TOXIC, OXIDIZING, N.O.S.
1TFC	3305	COMPRESSED GAS, TOXIC, FLAMMABLE, CORROSIVE, N.O.S.
1TOC	3306	COMPRESSED GAS, TOXIC, OXIDIZING, CORROSIVE, N.O.S.

Liquefied gases		
Classification code	**UN No.**	**Name of the substance or article**
2A	1058	LIQUEFIED GASES, non-flammable, charged with nitrogen, carbon dioxide or air
	1078	REFRIGERANT GAS, N.O.S.
		such as mixtures of gases, indicated by the letter R..., which as:
		Mixture F1, have a vapour pressure at 70 °C not exceeding 1.3 MPa (13 bar) and a density at 50 °C not lower than that of dichlorofluoromethane (1.30 kg/l);
		Mixture F2, have a vapour pressure at 70 °C not exceeding 1.9 MPa (19 bar) and a density at 50 °C not lower than that of dichlorodifluoromethane (1.21 kg/l);
		Mixture F3, have a vapour pressure at 70 °C not exceeding 3 MPa (30 bar) and a density at 50 °C not lower than that of chlorodifluoromethane (1.09 kg/l).
		NOTE: *Trichlorofluoromethane (Refrigerant R 11), 1,1,2-trichloro-1,2,2-trifluoroethane (Refrigerant R 113), 1,1,1-trichloro-2,2,2-trifluoroethane (Refrigerant R 113a), 1-chloro-1,2,2-trifluoroethane (Refrigerant R 133) and 1-chloro-1,1,2-trifluoroethane (Refrigerant R 133b) are not substances of Class 2. They may, however, enter into the composition of mixtures F1 to F3.*
	1968	INSECTICIDE GAS, N.O.S.
	3163	LIQUEFIED GAS, N.O.S.
2O	3157	LIQUEFIED GAS, OXIDIZING, N.O.S.
2F	1010	BUTADIENES AND HYDROCARBON MIXTURE, STABILIZED, having a vapour pressure at 70 °C not exceeding 1.1 MPa (11 bar) and a density at 50 °C not lower than 0.525 kg/l.
		NOTE: *Butadienes, stabilized are also classified under UN No. 1010, see Table A of Chapter 3.2.*
	1060	METHYLACETYLENE AND PROPADIENE MIXTURE, STABILIZED
		such as mixtures of methylacetylene and propadiene with hydrocarbons, which as:
		Mixture P1, contain not more than 63% methylacetylene and propadiene by volume and not more than 24% propane and propylene by volume, the percentage of C_4- saturated hydrocarbons being not less than 14% by volume; and as
		Mixture P2, contain not more than 48% methylacetylene and propadiene by volume and not more than 50% propane and propylene by volume, the percentage of C_4- saturated hydrocarbons being not less than 5% by volume,
		as well as mixtures of propadiene with 1 to 4% methylacetylene.

Liquefied gases *(cont'd)*			
Classification code	**UN No.**	**Name of the substance or article**	
2F *(cont'd)*	1965	HYDROCARBON GAS MIXTURE, LIQUEFIED, N.O.S	
		such as mixtures, which as:	
		Mixture A, have a vapour pressure at 70 °C not exceeding 1.1 MPa (11 bar) and a density at 50 °C not lower than 0.525 kg/l;	
		Mixture A01, have a vapour pressure at 70 °C not exceeding 1.6 MPa (16 bar) and a relative density at 50 °C not lower than 0.516 kg/l;	
		Mixture A02, have a vapour pressure at 70 °C not exceeding 1.6 MPa (16 bar) and a relative density at 50 °C not lower than 0.505 kg/l;	
		Mixture A0, have a vapour pressure at 70 °C not exceeding 1.6 MPa (16 bar) and a density at 50 °C not lower than 0.495 kg/l;	
		Mixture A1, have a vapour pressure at 70 °C not exceeding 2.1 MPa (21 bar) and a density at 50 °C not lower than 0.485 kg/l;	
		Mixture B1 have a vapour pressure at 70 °C not exceeding 2.6 MPa (26 bar) and a relative density at 50 °C not lower than 0.474 kg/l;	
		Mixture B2 have a vapour pressure at 70 °C not exceeding 2.6 MPa (26 bar) and a relative density at 50 °C not lower than 0.463 kg/l;	
		Mixture B, have a vapour pressure at 70 °C not exceeding 2.6 MPa (26 bar) and a density at 50 °C not lower than 0.450 kg/l;	
		Mixture C, have a vapour pressure at 70 °C not exceeding 3.1 MPa (31 bar) and a relative density at 50 °C not lower than 0.440 kg/l;	
		NOTE 1: In the case of the foregoing mixtures, the use of the following names customary in the trade is permitted for describing these substances: for mixtures A, A01, A02 and A0: BUTANE; for mixture C: PROPANE.	
		NOTE 2: UN No. 1075 PETROLEUM GASES, LIQUEFIED may be used as an alternative entry for UN No. 1965 HYDROCARBON GAS MIXTURE LIQUEFIED, N.O.S. for carriage prior to or following maritime or air carriage.	
	3354	INSECTICIDE GAS, FLAMMABLE, N.O.S.	
	3161	LIQUEFIED GAS, FLAMMABLE, N.O.S.	
2T	1967	INSECTICIDE GAS, TOXIC, N.O.S.	
	3162	LIQUEFIED GAS, TOXIC, N.O.S.	
2TF	3355	INSECTICIDE GAS, TOXIC, FLAMMABLE, N.O.S.	
	3160	LIQUEFIED GAS, TOXIC, FLAMMABLE, N.O.S.	
2TC	3308	LIQUEFIED GAS, TOXIC, CORROSIVE, N.O.S.	
2TO	3307	LIQUEFIED GAS, TOXIC, OXIDIZING, N.O.S.	
2TFC	3309	LIQUEFIED GAS, TOXIC, FLAMMABLE, CORROSIVE, N.O.S.	
2TOC	3310	LIQUEFIED GAS, TOXIC, OXIDIZING, CORROSIVE, N.O.S.	

Refrigerated liquefied gases			
Classification code	**UN No.**	**Name of the substance or article**	
3A	3158	GAS, REFRIGERATED LIQUID, N.O.S.	
3O	3311	GAS, REFRIGERATED LIQUID, OXIDIZING, N.O.S.	
3F	3312	GAS, REFRIGERATED LIQUID, FLAMMABLE, N.O.S.	

Dissolved gases			
Classification code	**UN No.**	**Name of the substance or article**	
4		Only substances listed in Table A of Chapter 3.2 are to be accepted for carriage.	

Aerosols and receptacles, small, containing gas		
Classification code	**UN No.**	**Name of the substance or article**
5	1950	AEROSOLS
	2037	RECEPTACLES, SMALL CONTAINING GAS (GAS CARTRIDGES) without a release device, non-refillable

Other articles containing gas under pressure		
Classification code	**UN No.**	**Name of the substance or article**
6A	2857	REFRIGERATING MACHINES containing non-flammable, non-toxic gases or ammonia solutions (UN 2672)
	3164	ARTICLES, PRESSURIZED, PNEUMATIC (containing non-flammable gas) or
	3164	ARTICLES, PRESSURIZED, HYDRAULIC (containing non-flammable gas)
6F	3150	DEVICES, SMALL, HYDROCARBON GAS POWERED or
	3150	HYDROCARBON GAS REFILLS FOR SMALL DEVICES, with release device
	3478	FUEL CELL CARTRIDGES, containing liquefied flammable gas or
	3478	FUEL CELL CARTRIDGES CONTAINED IN EQUIPMENT, containing liquefied flammable gas or
	3478	FUEL CELL CARTRIDGES PACKED WITH EQUIPMENT, containing liquefied flammable gas
	3479	FUEL CELL CARTRIDGES, containing hydrogen in metal hydride or
	3479	FUEL CELL CARTRIDGES CONTAINED IN EQUIPMENT, containing hydrogen in metal hydride or
	3479	FUEL CELL CARTRIDGES PACKED WITH EQUIPMENT, containing hydrogen in metal hydride
	3529	ENGINE, INTERNAL COMBUSTION, FLAMMABLE GAS POWERED or
	3529	ENGINE, FUEL CELL, FLAMMABLE GAS POWERED or
	3529	MACHINERY, INTERNAL COMBUSTION, FLAMMABLE GAS POWERED or
	3529	MACHINERY, FUEL CELL, FLAMMABLE GAS POWERED

Gas samples		
Classification code	**UN No.**	**Name of the substance or article**
7F	3167	GAS SAMPLE, NON-PRESSURIZED, FLAMMABLE, N.O.S., not refrigerated liquid
7T	3169	GAS SAMPLE, NON-PRESSURIZED, TOXIC, N.O.S., not refrigerated liquid
7TF	3168	GAS SAMPLE, NON-PRESSURIZED, TOXIC, FLAMMABLE, N.O.S., not refrigerated liquid

Chemicals under pressure		
Classification code	**UN No.**	**Name of the substance or article**
8A	3500	CHEMICAL UNDER PRESSURE, N.O.S.
8F	3501	CHEMICAL UNDER PRESSURE, FLAMMABLE, N.O.S.
8T	3502	CHEMICAL UNDER PRESSURE, TOXIC, N.O.S.
8C	3503	CHEMICAL UNDER PRESSURE, CORROSIVE, N.O.S.
8TF	3504	CHEMICAL UNDER PRESSURE, FLAMMABLE, TOXIC, N.O.S.
8FC	3505	CHEMICAL UNDER PRESSURE, FLAMMABLE, CORROSIVE, N.O.S.

Adsorbed gases		
Classification code	**UN No.**	**Name of the substance or article**
9A	3511	ADSORBED GAS, N.O.S.
9O	3513	ADSORBED GAS, OXIDIZING, N.O.S.
9F	3510	ADSORBED GAS, FLAMMABLE, N.O.S.
9T	3512	ADSORBED GAS, TOXIC, N.O.S.
9TF	3514	ADSORBED GAS, TOXIC, FLAMMABLE, N.O.S.
9TC	3516	ADSORBED GAS, TOXIC, CORROSIVE, N.O.S.
9TO	3515	ADSORBED GAS, TOXIC, OXIDIZING, N.O.S.
9TFC	3517	ADSORBED GAS, TOXIC, FLAMMABLE, CORROSIVE, N.O.S.
9TOC	3518	ADSORBED GAS, TOXIC, OXIDIZING, CORROSIVE, N.O.S.

2.2.3   **Class 3   Flammable liquids**

2.2.3.1   *Criteria*

2.2.3.1.1   The heading of Class 3 covers substances and articles containing substances of this Class which:

- are liquids according to subparagraph (a) of the definition for "liquid" in 1.2.1;

- have at 50 °C a vapour pressure of not more than 300 kPa (3 bar) and are not completely gaseous at 20 °C and at standard pressure of 101.3 kPa; and

- have a flash-point of not more than 60 °C (see 2.3.3.1 for the relevant test).

The heading of Class 3 also covers liquid substances and molten solid substances with a flash-point of more than 60°C and which are carried or handed over for carriage whilst heated at temperatures equal to or higher than their flash-point. These substances are assigned to UN No. 3256.

The heading of Class 3 also covers liquid desensitized explosives. Liquid desensitized explosives are explosive substances which are dissolved or suspended in water or other liquid substances, to form an homogeneous liquid mixture to suppress their explosive properties. Such entries in Table A of Chapter 3.2 are UN Nos. 1204, 2059, 3064, 3343, 3357 and 3379.

NOTE 1: Substances having a flash-point above 35 °C, which do not sustain combustion according to the criteria of sub-section 32.2.5 of Part III of the Manual of Tests and Criteria, are not substances of Class 3; if, however, these substances are handed over for carriage and carried whilst heated at temperatures equal to or higher than their flash-point, they are substances of Class 3.

NOTE 2: By derogation from paragraph 2.2.3.1.1 above, diesel fuel, gasoil, heating oil (light) including synthetically manufactured products having a flash-point above 60 °C and not more than 100 °C shall be deemed substances of Class 3, UN No. 1202.

NOTE 3: Flammable liquids which are highly toxic by inhalation, as defined in 2.2.61.1.4 to 2.2.61.1.9, and toxic substances having a flash-point of 23 °C or above are substances of Class 6.1 (see 2.2.61.1). Liquids which are highly toxic by inhalation are indicated as "toxic by inhalation" in their proper shipping name in Column (2) or by special provision 354 in Column (6) of Table A of Chapter 3.2.

NOTE 4: Flammable liquid substances and preparations used as pesticides, which are highly toxic, toxic or slightly toxic and have a flash-point of 23 °C or above are substances of Class 6.1 (see 2.2.61.1).

2.2.3.1.2   The substances and articles of Class 3 are subdivided as follows:

F   Flammable liquids, without subsidiary risk and articles containing such substances:

F1   Flammable liquids having a flash-point of or below 60 °C;

F2   Flammable liquids having a flash-point above 60 °C which are carried or handed over for carriage at or above their flash-point (elevated temperature substances);

F3   Articles containing flammable liquids;

FT   Flammable liquids, toxic:

FT1   Flammable liquids, toxic;

FT2   Pesticides;

FC   Flammable liquids, corrosive;

FTC   Flammable liquids, toxic, corrosive;

D   Liquid desensitized explosives.

2.2.3.1.3 Substances and articles classified in Class 3 are listed in Table A of Chapter 3.2. Substances not mentioned by name in Table A of Chapter 3.2 shall be assigned to the relevant entry of 2.2.3.3 and the relevant packing group in accordance with the provisions of this section. Flammable liquids shall be assigned to one of the following packing groups according to the degree of danger they present for carriage:

Packing group	Flash point (closed cup)	Initial boiling point
I	--	≤ 35°C
II [a]	< 23°C	> 35°C
III [a]	≥ 23°C ≤ 60°C	> 35°C

[a] *See also 2.2.3.1.4.*

For a liquid with (a) subsidiary risk(s), the packing group determined in accordance with the table above and the packing group based on the severity of the subsidiary risk(s) shall be considered; the classification and packing group shall then be determined in accordance with the table of precedence of hazards in 2.1.3.10.

2.2.3.1.4 Viscous flammable liquids such as paints, enamels, lacquers, varnishes, adhesives and polishes having a flash-point of less than 23 °C may be assigned to packing group III in conformity with the procedures prescribed in the Manual of Tests and Criteria, Part III, sub-section 32.3, provided that:

(a) the viscosity[2] and flash-point are in accordance with the following table:

Kinematic viscosity (extrapolated) v (at near-zero shear rate) mm²/s at 23°C	Flow-time t in seconds	Jet diameter (mm)	Flash-point, closed-cup (°C)
20 < v ≤ 80	20 < t ≤ 60	4	above 17
80 < v ≤ 135	60 < t ≤ 100	4	above 10
135 < v ≤ 220	20 < t ≤ 32	6	above 5
220 < v ≤ 300	32 < t ≤ 44	6	above -1
300 < v ≤ 700	44 < t ≤ 100	6	above -5
700 < v	100 < t	6	no limit

(b) Less than 3% of the clear solvent layer separates in the solvent separation test;

(c) The mixture or any separated solvent does not meet the criteria for Class 6.1 or Class 8;

(d) The substances are packed in receptacles of not more than 450 litre capacity.

NOTE: These provisions also apply to mixtures containing no more than 20% nitrocellulose with a nitrogen content not exceeding 12.6% by dry mass. Mixtures containing more than 20% but not more than 55% nitrocellulose with a nitrogen content not exceeding 12.6% by dry mass are substances assigned to UN No. 2059.

Mixtures having a flash-point below 23 °C and containing:

— more than 55% nitrocellulose, whatever their nitrogen content; or

— not more than 55% nitrocellulose with a nitrogen content above 12.6% by dry mass,

are substances of Class 1 (UN Nos. 0340 or 0342) or of Class 4.1 (UN Nos. 2555, 2556 or 2557).

---

[2] *Viscosity determination: Where the substance concerned is non-Newtonian, or where a flow cup method of viscosity determination is otherwise unsuitable, a variable shear-rate viscometer shall be used to determine the dynamic viscosity coefficient of the substance, at 23 °C, at a number of shear rates. The values obtained are plotted against shear rate and then extrapolated to zero shear rate. The dynamic viscosity thus obtained, divided by the density, gives the apparent kinematic viscosity at near-zero shear rate.*

2.2.3.1.5	*Viscous liquids*

2.2.3.1.5.1 Except as provided for in 2.2.3.1.5.2, viscous liquids which:

- have a flash-point of 23 °C or above and less than or equal to 60 °C;

- are not toxic, corrosive or environmentally hazardous;

- contain not more than 20% nitrocellulose provided the nitrocellulose contains not more than 12.6% nitrogen by dry mass; and

- are packed in receptacles of not more than 450 litre capacity;

are not subject to ADR, if:

(a) in the solvent separation test (see *Manual of Tests and Criteria*, Part III, sub-section 32.5.1), the height of the separated layer of solvent is less than 3% of the total height; and

(b) the flowtime in the viscosity test (see *Manual of Tests and Criteria*, Part III, sub-section 32.4.3), with a jet diameter of 6 mm is equal to or greater than:

(i) 60 seconds; or

(ii) 40 seconds if the viscous liquid contains not more than 60% of Class 3 substances.

2.2.3.1.5.2 Viscous liquids which are also environmentally hazardous, but meet all other criteria in 2.2.3.1.5.1, are not subject to any other provisions of ADR when they are carried in single or combination packagings containing a net quantity per single or inner packaging of 5 litres or less, provided the packagings meet the general provisions of 4.1.1.1, 4.1.1.2 and 4.1.1.4 to 4.1.1.8.

2.2.3.1.6 If substances of Class 3, as a result of admixtures, come into categories of risk different from those to which the substances mentioned by name in Table A of Chapter 3.2 belong, these mixtures or solutions shall be assigned to the entries to which they belong on the basis of their actual degree of danger.

NOTE: For the classification of solutions and mixtures (such as preparations and wastes) see also 2.1.3.

2.2.3.1.7 On the basis of the test procedures in accordance with 2.3.3.1 and 2.3.4, and the criteria set out in 2.2.3.1.1, it may also be determined whether the nature of a solution or a mixture mentioned by name or containing a substance mentioned by name is such that the solution or mixture is not subject to the provisions for this Class (see also 2.1.3).

2.2.3.2 **Substances not accepted for carriage**

2.2.3.2.1 Substances of Class 3 which are liable to form peroxides easily (as happens with ethers or with certain heterocyclic oxygenated substances) shall not be accepted for carriage if their peroxide content, calculated as hydrogen peroxide (H_2O_2), exceeds 0.3%. The peroxide content shall be determined as indicated in 2.3.3.3.

2.2.3.2.2 Chemically unstable substances of Class 3 shall not be accepted for carriage unless the necessary precautions have been taken to prevent the possibility of a dangerous decomposition or polymerization under normal conditions of carriage. For the precautions necessary to prevent polymerization, see special provision 386 of Chapter 3.3. To this end particular care shall be taken to ensure that receptacles and tanks do not contain any substances liable to promote these reactions.

2.2.3.2.3 Liquid desensitized explosives other than those listed in Table A of Chapter 3.2 shall not be accepted for carriage as substances of Class 3.

2.2.3.3   *List of collective entries*

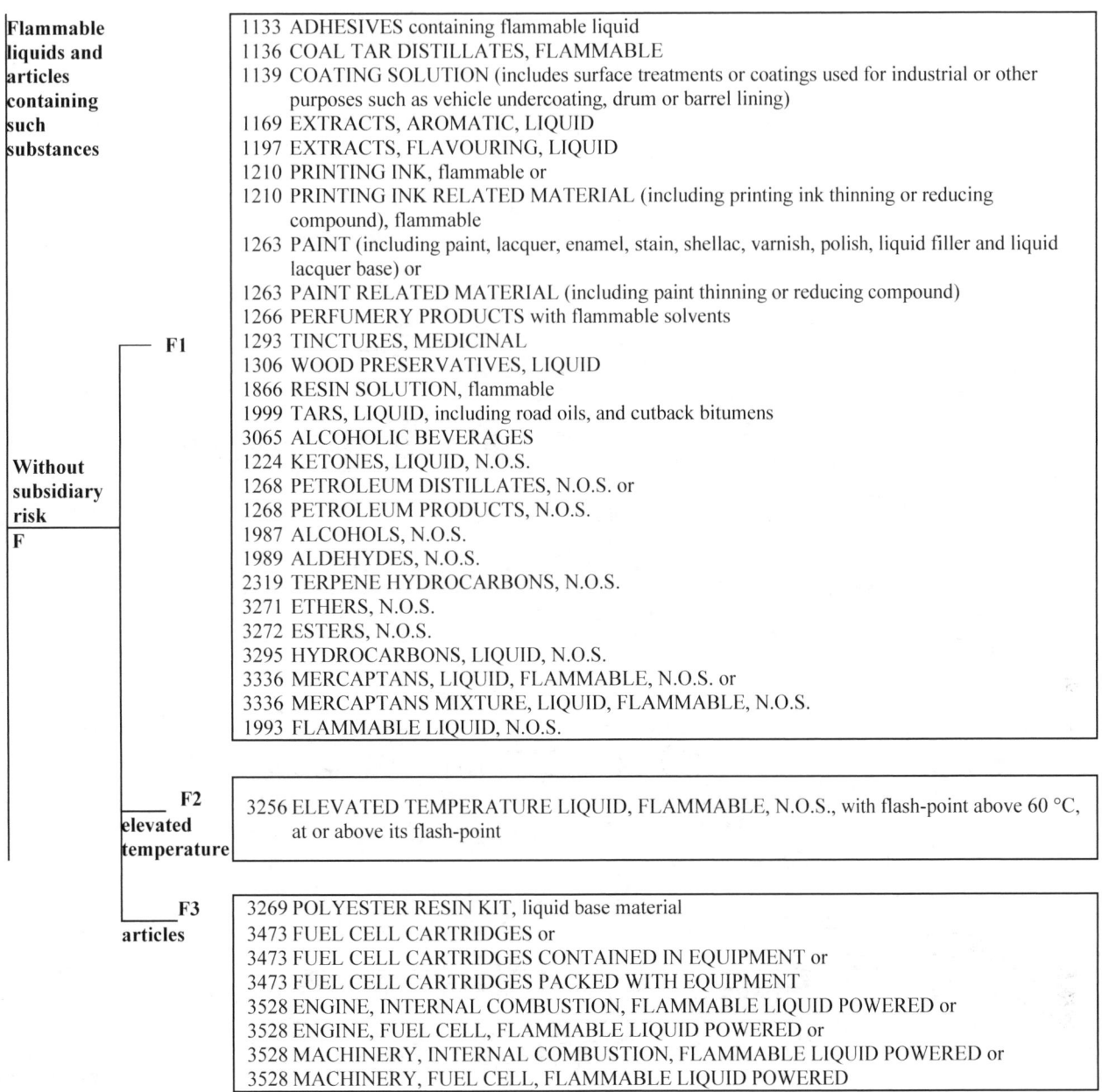

Flammable liquids and articles containing such substances		
Without subsidiary risk **F**	**F1**	1133 ADHESIVES containing flammable liquid 1136 COAL TAR DISTILLATES, FLAMMABLE 1139 COATING SOLUTION (includes surface treatments or coatings used for industrial or other purposes such as vehicle undercoating, drum or barrel lining) 1169 EXTRACTS, AROMATIC, LIQUID 1197 EXTRACTS, FLAVOURING, LIQUID 1210 PRINTING INK, flammable or 1210 PRINTING INK RELATED MATERIAL (including printing ink thinning or reducing compound), flammable 1263 PAINT (including paint, lacquer, enamel, stain, shellac, varnish, polish, liquid filler and liquid lacquer base) or 1263 PAINT RELATED MATERIAL (including paint thinning or reducing compound) 1266 PERFUMERY PRODUCTS with flammable solvents 1293 TINCTURES, MEDICINAL 1306 WOOD PRESERVATIVES, LIQUID 1866 RESIN SOLUTION, flammable 1999 TARS, LIQUID, including road oils, and cutback bitumens 3065 ALCOHOLIC BEVERAGES 1224 KETONES, LIQUID, N.O.S. 1268 PETROLEUM DISTILLATES, N.O.S. or 1268 PETROLEUM PRODUCTS, N.O.S. 1987 ALCOHOLS, N.O.S. 1989 ALDEHYDES, N.O.S. 2319 TERPENE HYDROCARBONS, N.O.S. 3271 ETHERS, N.O.S. 3272 ESTERS, N.O.S. 3295 HYDROCARBONS, LIQUID, N.O.S. 3336 MERCAPTANS, LIQUID, FLAMMABLE, N.O.S. or 3336 MERCAPTANS MIXTURE, LIQUID, FLAMMABLE, N.O.S. 1993 FLAMMABLE LIQUID, N.O.S.
	F2 **elevated temperature**	3256 ELEVATED TEMPERATURE LIQUID, FLAMMABLE, N.O.S., with flash-point above 60 °C, at or above its flash-point
	F3 **articles**	3269 POLYESTER RESIN KIT, liquid base material 3473 FUEL CELL CARTRIDGES or 3473 FUEL CELL CARTRIDGES CONTAINED IN EQUIPMENT or 3473 FUEL CELL CARTRIDGES PACKED WITH EQUIPMENT 3528 ENGINE, INTERNAL COMBUSTION, FLAMMABLE LIQUID POWERED or 3528 ENGINE, FUEL CELL, FLAMMABLE LIQUID POWERED or 3528 MACHINERY, INTERNAL COMBUSTION, FLAMMABLE LIQUID POWERED or 3528 MACHINERY, FUEL CELL, FLAMMABLE LIQUID POWERED

(cont'd on next page)

Toxic FT

FT1

1228 MERCAPTANS, LIQUID, FLAMMABLE, TOXIC, N.O.S. or
1228 MERCAPTAN MIXTURE, LIQUID, FLAMMABLE, TOXIC, N.O.S.
1986 ALCOHOLS, FLAMMABLE, TOXIC, N.O.S.
1988 ALDEHYDES, FLAMMABLE, TOXIC, N.O.S.
2478 ISOCYANATES, FLAMMABLE, TOXIC, N.O.S. or
2478 ISOCYANATE SOLUTION, FLAMMABLE, TOXIC, N.O.S.
3248 MEDICINE, LIQUID, FLAMMABLE, TOXIC, N.O.S.
3273 NITRILES, FLAMMABLE, TOXIC, N.O.S.
1992 FLAMMABLE LIQUID, TOXIC, N.O.S.

FT2
pesticide (f.p.<23 °C)

2758 CARBAMATE PESTICIDE, LIQUID, FLAMMABLE, TOXIC
2760 ARSENICAL PESTICIDE, LIQUID, FLAMMABLE, TOXIC
2762 ORGANOCHLORINE PESTICIDE, LIQUID, FLAMMABLE, TOXIC
2764 TRIAZINE PESTICIDE, LIQUID, FLAMMABLE, TOXIC
2772 THIOCARBAMATE PESTICIDE, LIQUID, FLAMMABLE, TOXIC
2776 COPPER BASED PESTICIDE, LIQUID, FLAMMABLE, TOXIC
2778 MERCURY BASED PESTICIDE, LIQUID, FLAMMABLE, TOXIC
2780 SUBSTITUTED NITROPHENOL PESTICIDE, LIQUID, FLAMMABLE, TOXIC
2782 BIPYRIDILIUM PESTICIDE, LIQUID, FLAMMABLE, TOXIC
2784 ORGANOPHOSPHORUS PESTICIDE, LIQUID, FLAMMABLE, TOXIC
2787 ORGANOTIN PESTICIDE, LIQUID, FLAMMABLE, TOXIC
3024 COUMARIN DERIVATIVE PESTICIDE, LIQUID, FLAMMABLE, TOXIC
3346 PHENOXYACETIC ACID DERIVATIVE PESTICIDE, LIQUID, FLAMMABLE, TOXIC
3350 PYRETHROID PESTICIDE, LIQUID, FLAMMABLE TOXIC
3021 PESTICIDE, LIQUID, FLAMMABLE, TOXIC, N.O.S.

NOTE: The classification of a pesticide under an entry shall be effected on the basis of the active ingredient, of the physical state of the pesticide and any subsidiary risks it may exhibit.

Corrosive FC

3469 PAINT, FLAMMABLE, CORROSIVE (including paint, lacquer, enamel, stain, shellac, varnish, polish, liquid filler and liquid lacquer base) or
3469 PAINT RELATED MATERIAL, FLAMMABLE, CORROSIVE (including paint thinning or reducing compound)
2733 AMINES, FLAMMABLE, CORROSIVE, N.O.S. or
2733 POLYAMINES, FLAMMABLE, CORROSIVE, N.O.S.
2985 CHLOROSILANES, FLAMMABLE, CORROSIVE, N.O.S.
3274 ALCOHOLATES SOLUTION, N.O.S., in alcohol
2924 FLAMMABLE LIQUID, CORROSIVE, N.O.S.

Toxic, corrosive FTC

3286 FLAMMABLE LIQUID, TOXIC, CORROSIVE, N.O.S.

Liquid desensitised explosive D

3343 NITROGLYCERIN MIXTURE, DESENSITIZED, LIQUID, FLAMMABLE, N.O.S. with not more than 30% nitroglycerin by mass
3357 NITROGLYCERIN MIXTURE, DESENSITIZED, LIQUID, N.O.S. with not more than 30% nitroglycerin by mass
3379 DESENSITIZED EXPLOSIVE, LIQUID, N.O.S.

2.2.41   **Class 4.1 Flammable solids, self-reactive substances, polymerizing substances and solid desensitized explosives**

2.2.41.1   *Criteria*

2.2.41.1.1   The heading of Class 4.1 covers flammable substances and articles, desensitized explosives which are solids according to subparagraph (a) of the definition "solid" in 1.2.1, self-reactive liquids or solids and polymerizing substances..

The following are assigned to Class 4.1:

-      readily flammable solid substances and articles (see paragraphs 2.2.41.1.3 to 2.2.41.1.8);

-      self-reactive solids or liquids (see paragraphs 2.2.41.1.9 to 2.2.41.1.17);

-      solid desensitized explosives (see 2.2.41.1.18);

-      substances related to self-reactive substances (see 2.2.41.1.19);

-      polymerizing substances (see 2.2.41.1.20 and 2.2.41.1.21).

2.2.41.1.2   The substances and articles of Class 4.1 are subdivided as follows:

F   Flammable solids, without subsidiary risk:

   F1   Organic;

   F2   Organic, molten;

   F3   Inorganic;

   F4   Articles;

FO   Flammable solids, oxidizing;

FT   Flammable solids, toxic:

   FT1   Organic, toxic;

   FT2   Inorganic, toxic;

FC   Flammable solids, corrosive:

   FC1   Organic, corrosive;

   FC2   Inorganic, corrosive;

D   Solid desensitized explosives without subsidiary risk;

DT   Solid desensitized explosives, toxic;

SR   Self-reactive substances:

   SR1   Not requiring temperature control;

   SR2   Requiring temperature control.

PM   Polymerizing substances

   PM1   Not requiring temperature control;

   PM2   Requiring temperature control.

Flammable solids

Definition and properties

2.2.41.1.3   *Flammable solids* are readily combustible solids and solids which may cause fire through friction.

Readily combustible solids are powdered, granular, or pasty substances which are dangerous if they can be easily ignited by brief contact with an ignition source, such as a burning match, and if the flame spreads rapidly. The danger may come not only from the fire but also from toxic combustion products. Metal powders are especially dangerous because of the difficulty of extinguishing a fire since normal extinguishing agents such as carbon dioxide or water can increase the hazard.

Classification

2.2.41.1.4   Substances and articles classified as flammable solids of Class 4.1 are listed in Table A of Chapter 3.2. The assignment of organic substances and articles not mentioned by name in Table A of Chapter 3.2 to the relevant entry of sub-section 2.2.41.3 in accordance with the provisions of Chapter 2.1 can be based on experience or on the results of the test procedures in accordance with Part III, sub-section 33.2.1 of the Manual of Tests and Criteria. The assignment of inorganic substances not mentioned by name shall be based on the results of the test procedures in accordance with Part III, sub-section 33.2.1 of the Manual of Tests and Criteria; experience shall also be taken into account when it leads to a more stringent assignment.

2.2.41.1.5   When substances not mentioned by name are assigned to one of the entries listed in 2.2.41.3 on the basis of the test procedures in accordance with the Manual of Tests and Criteria, Part III, sub-section 33.2.1, the following criteria apply:

(a)   With the exception of metal powders or powders of metal alloys, powdery, granular or pasty substances shall be classified as readily flammable substances of Class 4.1 if they can be easily ignited by brief contact with an ignition source (e.g. a burning match), or if, in the event of ignition, the flame spreads rapidly, the burning time is less than 45 seconds for a measured distance of 100 mm or the rate of burning is greater than 2.2 mm/s;

(b)   Metal powders or powders of metal alloys shall be assigned to Class 4.1 if they can be ignited by a flame and the reaction spreads over the whole length of the sample in 10 minutes or less.

Solids which may cause fire through friction shall be classified in Class 4.1 by analogy with existing entries (e.g. matches) or in accordance with any appropriate special provision.

2.2.41.1.6   On the basis of the test procedure in accordance with the Manual of Tests and Criteria, Part III, Section 33.2.1 and the criteria set out in 2.2.41.1.4 and 2.2.41.1.5, it may also be determined whether the nature of a substance mentioned by name is such that the substance is not subject to the provisions for this Class.

2.2.41.1.7   If substances of Class 4.1, as a result of admixtures, come into different categories of risk from those to which the substances mentioned by name in Table A of Chapter 3.2 belong, these mixtures shall be assigned to the entries to which they belong on the basis of their actual degree of danger.

NOTE: *For the classification of solutions and mixtures (such as preparations and wastes), see also 2.1.3.*

Assignment of packing groups

2.2.41.1.8   Flammable solids classified under the various entries in Table A of Chapter 3.2 shall be assigned to packing groups II or III on the basis of test procedures of the Manual of Tests and Criteria, Part III, sub-section 33.2.1, in accordance with the following criteria:

(a)   Readily flammable solids which, when tested, have a burning time of less than 45 seconds over a measured distance of 100 mm shall be assigned to:

Packing group II: if the flame passes the wetted zone;

Packing group III: if the wetted zone stops the flame for at least four minutes;

(b)   Metal powders or powders of metal alloys shall be assigned to:

Packing group II: if, when tested, the reaction spreads over the whole length of the sample in five minutes or less;

Packing group III: if, when tested, the reaction spreads over the whole length of the sample in more than five minutes.

For solids which may cause fire through friction, the packing group shall be assigned by analogy with existing entries or in accordance with any special provision.

Self-reactive substances

Definitions

2.2.41.1.9   For the purposes of ADR, *self-reactive substances* are thermally unstable substances liable to undergo a strongly exothermic decomposition even without participation of oxygen (air). Substances are not considered to be self-reactive substances of Class 4.1, if:

(a)   they are explosives according to the criteria of Class 1;

(b)   they are oxidizing substances according to the classification procedure for Class 5.1 (see 2.2.51.1) except that mixtures of oxidizing substances which contain 5% or more of combustible organic substances shall be subjected to the classification procedure defined in Note 2;

(c)   they are organic peroxides according to the criteria of Class 5.2 (see 2.2.52.1);

(d)   their heat of decomposition is less than 300 J/g; or

(e)   their self-accelerating decomposition temperature (SADT) (see Note 3 below) is greater than 75 °C for a 50 kg package.

NOTE 1: *The heat of decomposition can be determined using any internationally recognised method e.g. differential scanning calorimetry and adiabatic calorimetry.*

NOTE 2: *Mixtures of oxidizing substances meeting the criteria of Class 5.1 which contain 5% or more of combustible organic substances, which do not meet the criteria mentioned in (a), (c), (d) or (e) above, shall be subjected to the self-reactive substance classification procedure.*

A mixture showing the properties of a self-reactive substance, type B to F, shall be classified as a self-reactive substance of Class 4.1.

A mixture showing the properties of a self-reactive substance, type G, according to the principle given in section 20.4.3 (g) of Part II of the Manual of Tests and Criteria shall be considered for classification as a substance of Class 5.1 (see 2.2.51.1).

NOTE 3: *The self-accelerating decomposition temperature (SADT) is the lowest temperature at which self-accelerating decomposition may occur with a substance in the packaging as used during carriage. Requirements for the determination of the SADT are given in the Manual of Tests and Criteria, Part II, Chapter 20 and section 28.4.*

NOTE 4: *Any substance which shows the properties of a self-reactive substance shall be classified as such, even if this substance gives a positive test result according to 2.2.42.1.5 for inclusion in Class 4.2.*

Properties

2.2.41.1.10   The decomposition of self-reactive substances can be initiated by heat, contact with catalytic impurities (e.g. acids, heavy-metal compounds, bases), friction or impact. The rate of decomposition increases with temperature and varies with the substance. Decomposition, particularly if no ignition occurs, may result in the evolution of toxic gases or vapours. For certain self-reactive substances, the temperature shall be controlled. Some self-reactive substances may decompose explosively, particularly if confined. This characteristic may be modified by the addition of diluents or by the use of appropriate packagings. Certain self-reactive substances burn vigorously. Self-reactive substances are, for example, some compounds of the types listed below:

aliphatic azo compounds (-C-N=N-C-);

organic azides (-C-N₃);

diazonium salts (-CN₂⁺ Z⁻);

N-nitroso compounds (-N-N=O); and

aromatic sulphohydrazides (-SO₂-NH-NH₂).

This list is not exhaustive and substances with other reactive groups and some mixtures of substances may have similar properties.

Classification

2.2.41.1.11   Self-reactive substances are classified into seven types according to the degree of danger they present. The types of self-reactive substances range from type A, which is not accepted for carriage in the packaging in which it is tested, to type G, which is not subject to the provisions for self-reactive substances of Class 4.1. The classification of types B to F is directly related to the maximum quantity allowed in one packaging. The principles to be applied for classification as well as the applicable classification procedures, test methods and criteria and an example of a suitable test report are given in Part II of the Manual of Tests and Criteria.

2.2.41.1.12   Self-reactive substances which have already been classified and are already permitted for carriage in packagings are listed in 2.2.41.4, those already permitted for carriage in IBCs are listed in 4.1.4.2, packing instruction IBC520 and those already permitted for carriage in tanks according to Chapter 4.2 are listed in 4.2.5.2, portable tank instruction T23. Each permitted substance listed is assigned to a generic entry of Table A of Chapter 3.2 (UN Nos. 3221 to 3240), and appropriate subsidiary risks and remarks providing relevant transport information are given.

The collective entries specify:

-        self-reactive substances types B to F, see 2.2.41.1.11 above;

-        physical state (liquid/solid); and

-        temperature control (when required), see 2.2.41.1.17 below.

The classification of the self-reactive substances listed in 2.2.41.4 is based on the technically pure substance (except where a concentration of less than 100% is specified).

2.2.41.1.13   Classification of self-reactive substances not listed in 2.2.41.4, 4.1.4.2, packing instruction IBC520 or 4.2.5.2, portable tank instruction T23 and assignment to a collective entry shall be made by the competent authority of the country of origin on the basis of a test report. The statement of approval shall contain the classification and the relevant conditions of carriage. If the country of origin is not a Contracting Party to ADR, the classification and the conditions of carriage shall be recognized by the competent authority of the first country Contracting Party to ADR reached by the consignment.

2.2.41.1.14   Activators, such as zinc compounds, may be added to some self-reactive substances to change their reactivity. Depending on both the type and the concentration of the activator, this may result in a decrease in thermal stability and a change in explosive properties. If either of these properties is altered, the new formulation shall be assessed in accordance with the classification procedure.

2.2.41.1.15   Samples of self-reactive substances or formulations of self-reactive substances not listed in 2.2.41.4, for which a complete set of test results is not available and which are to be carried for further testing or evaluation, shall be assigned to one of the appropriate entries for self-reactive substances type C provided the following conditions are met:

-        the available data indicate that the sample would be no more dangerous than self-reactive substances type B;

-        the sample is packaged in accordance with packing method OP2 and the quantity per transport unit is limited to 10 kg;

-        the available data indicate that the control temperature, if any, is sufficiently low to prevent any dangerous decomposition and sufficiently high to prevent any dangerous phase separation.

Desensitization

2.2.41.1.16   In order to ensure safety during carriage, self-reactive substances are in many cases desensitized by use of a diluent. Where a percentage of a substance is stipulated, this refers to the percentage by mass, rounded to the nearest whole number. If a diluent is used, the self-reactive substance shall be tested with the diluent present in the concentration and form used in carriage. Diluents which may allow a self-reactive substance to concentrate to a dangerous extent in the event of leakage from a packaging shall not be used. Any diluent shall be compatible with the self-reactive substance. In this regard, compatible diluents are those solids or liquids which have no detrimental influence on the thermal stability and hazard type of the self-reactive substance. Liquid diluents in formulations requiring temperature control (see 2.2.41.1.14) shall have a boiling point of at least 60 °C and a flash-point not less than 5 °C. The boiling point of the liquid shall be at least 50 °C higher than the control temperature of the self-reactive substance.

Temperature control requirements

2.2.41.1.17   Certain self-reactive substances may only be carried under temperature controlled conditions. The control temperature is the maximum temperature at which the self-reactive substance can be safely carried. It is assumed that the temperature of the immediate surroundings of a package only exceeds 55 °C during carriage for a relatively short time in a 24 hour period. In the event of loss of temperature control, it may be necessary to implement emergency procedures. The emergency temperature is the temperature at which such procedures shall be implemented.

The control and emergency temperatures are derived from the SADT (see table 1). The SADT shall be determined in order to decide whether a substance shall be subjected to temperature control during carriage. Provisions for the determination of the SADT are given in the Manual of Tests and Criteria, Part II, Chapter 20 and Section 28.4.

Table 1: Derivation of control and emergency temperatures

Type of receptacle	SADT [a]	Control temperature	Emergency temperature
Single packagings and IBCs	20 °C or less	20 °C below SADT	10 °C below SADT
	over 20 °C to 35 °C	15 °C below SADT	10 °C below SADT
	over 35 °C	10 °C below SADT	5 °C below SADT
Tanks	not greater than 50 °C	10 °C below SADT	5 °C below SADT

[a]   *SADT of the substance as packaged for carriage.*

Self-reactive substances with an SADT not greater than 55 °C shall be subject to temperature control during carriage. Where applicable, control and emergency temperatures are listed in 2.2.41.4. The actual temperature during carriage may be lower than the control temperature but shall be selected so as to avoid dangerous separation of phases.

Solid desensitized explosives

2.2.41.1.18   Solid desensitized explosives are substances which are wetted with water or alcohols or are diluted with other substances to suppress their explosive properties. Such entries in Table A of Chapter 3.2 are: UN Nos. 1310, 1320, 1321, 1322, 1336, 1337, 1344, 1347, 1348, 1349, 1354, 1355, 1356, 1357, 1517, 1571, 2555, 2556, 2557, 2852, 2907, 3317, 3319, 3344, 3364, 3365, 3366, 3367, 3368, 3369, 3370, 3376, 3380 and 3474.

Substances related to self-reactive substances

2.2.41.1.19   Substances that:

(a)   have been provisionally accepted into Class 1 according to Test Series 1 and 2 but exempted from Class 1 by Test Series 6;

(b)   are not self-reactive substances of Class 4.1; and

(c)   are not substances of Classes 5.1 or 5.2;

are also assigned to Class 4.1. UN Nos. 2956, 3241, 3242 and 3251 are such entries.

Polymerizing substances

Definitions and properties

2.2.41.1.20   Polymerizing substances are substances which, without stabilization, are liable to undergo a strongly exothermic reaction resulting in the formation of larger molecules or resulting in the formation of polymers under conditions normally encountered in carriage. Such substances are considered to be polymerizing substances of Class 4.1 when:

(a)   their self-accelerating polymerization temperature (SAPT) is 75 °C or less under the conditions (with or without chemical stabilization as offered for carriage) and in the packaging, IBC or tank in which the substance or mixture is to be carried;

(b)   they exhibit a heat of reaction of more than 300 J/g; and

(c)   they do not meet any other criteria for inclusion in classes 1 to 8.

A mixture meeting the criteria of a polymerizing substance shall be classified as a polymerizing substance of Class 4.1.

Temperature control requirements

2.2.41.1.21   Polymerizing substances are subject to temperature control in carriage if their self-accelerating polymerization temperature (SAPT) is:

(a)   when offered for carriage in a packaging or IBC, 50 °C or less in the packaging or IBC in which the substance is to be carried; or

(b)   When offered for carriage in a tank, 45 °C or less in the tank in which the substance is to be carried.

2.2.41.2   *Substances not accepted for carriage*

2.2.41.2.1   The chemically unstable substances of Class 4.1 shall not be accepted for carriage unless the necessary steps have been taken to prevent their dangerous decomposition or polymerization during carriage. To this end, it shall in particular be ensured that receptacles and tanks do not contain any substance liable to promote these reactions.

2.2.41.2.2   Flammable solids, oxidizing, assigned to UN No. 3097 shall not be accepted for carriage unless they meet the requirements for Class 1 (see also 2.1.3.7).

2.2.41.2.3   The following substances shall not be accepted for carriage:

-    Self-reactive substances of type A (see Manual of Tests and Criteria, Part II, paragraph 20.4.2 (a));

-    Phosphorus sulphides which are not free from yellow and white phosphorus;

-    Solid densitized explosives other than those listed in Table A of Chapter 3.2;

-    Inorganic flammable substances in the molten form other than UN No. 2448 SULPHUR, MOLTEN.

2.2.41.3 List of collective entries

Flammable solids F	without subsidiary risk	organic	F1	3175 SOLIDS CONTAINING FLAMMABLE LIQUID, N.O.S. 1353 FIBRES IMPREGNATED WITH WEAKLY NITRATED NITROCELLULOSE, N.O.S. or 1353 FABRICS IMPREGNATED WITH WEAKLY NITRATED NITROCELLULOSE, N.O.S. 1325 FLAMMABLE SOLID, ORGANIC, N.O.S.
		organic molten	F2	3176 FLAMMABLE SOLID, ORGANIC, MOLTEN, N.O.S.
		inorganic	F3	3089 METAL POWDER, FLAMMABLE, N.O.S. [a] [b] 3181 METAL SALTS OF ORGANIC COMPOUNDS, FLAMMABLE, N.O.S. 3182 METAL HYDRIDES, FLAMMABLE,N.O.S. [c] 3178 FLAMMABLE SOLID, INORGANIC, N.O.S.
		articles	F4	3527 POLYESTER RESIN KIT, solid base material
	oxidizing		FO	3097 FLAMMABLE SOLID, OXIDIZING, N.O.S. (not allowed, see para. 2.2.41.2.2)
	toxic FT	organic	FT1	2926 FLAMMABLE SOLID, TOXIC, ORGANIC, N.O.S.
		inorganic	FT2	3179 FLAMMABLE SOLID, TOXIC, INORGANIC, N.O.S.
	corrosive FC	organic	FC1	2925 FLAMMABLE SOLID, TOXIC, INORGANIC, N.O.S.
		inorganic	FC2	3180 FLAMMABLE SOLID, CORROSIVE, INORGANIC, N.O.S.
Solid desensitized explosives	without subsidiary risk		D	3319 NITROGLYCERIN MIXTURE, DESENSITIZED, SOLID, N.O.S. with more than 2% but not more than 10% nitroglycerin by mass 3344 PENTAERYTHRITE TETRANITRATE (PENTAERYTHRITOL TETRANITRATE, PETN) MIXTURE, DESENSITIZED, SOLID, N.O.S. with more than 10% but not more than 20% PETN by mass 3380 DESENSITIZED EXPLOSIVE, SOLID, N.O.S.
	toxic		DT	Only substances listed in Table A of Chapter 3.2 are to be accepted for carriage as substances of Class 4.1
Self-reactive substances SR	Not requiring temperature control		SR1	SELF-REACTIVE LIQUID TYPE A — Not accepted for carriage, see 2.2.41.2.3 SELF-REACTIVE SOLID TYPE A 3221 SELF-REACTIVE LIQUID TYPE B 3222 SELF-REACTIVE SOLID TYPE B 3223 SELF-REACTIVE LIQUID TYPE C 3224 SELF-REACTIVE SOLID TYPE C 3225 SELF-REACTIVE LIQUID TYPE D 3226 SELF-REACTIVE SOLID TYPE D 3227 SELF-REACTIVE LIQUID TYPE E 3228 SELF-REACTIVE SOLID TYPE E 3229 SELF-REACTIVE LIQUID TYPE F 3230 SELF-REACTIVE SOLID TYPE F SELF-REACTIVE LIQUID TYPE G — Not subject to the provisions applicable to Class 4.1, see 2.2.41.1.11 SELF-REACTIVE SOLID TYPE G
	Requiring temperature control		SR2	3231 SELF-REACTIVE LIQUID TYPE B, TEMPERATURE CONTROLLED 3232 SELF-REACTIVE SOLID TYPE B, TEMPERATURE CONTROLLED 3233 SELF-REACTIVE LIQUID TYPE C, TEMPERATURE CONTROLLED 3234 SELF-REACTIVE SOLID TYPE C, TEMPERATURE CONTROLLED 3235 SELF-REACTIVE LIQUID TYPE D, TEMPERATURE CONTROLLED 3236 SELF-REACTIVE SOLID TYPE D, TEMPERATURE CONTROLLED 3237 SELF-REACTIVE LIQUID TYPE E, TEMPERATURE CONTROLLED 3238 SELF-REACTIVE SOLID TYPE E, TEMPERATURE CONTROLLED 3239 SELF-REACTIVE LIQUID TYPE F, TEMPERATURE CONTROLLED 3240 SELF-REACTIVE SOLID TYPE F, TEMPERATURE CONTROLLED

(cont'd on next page)

---

[a] *Metals and metal alloys in powdered or other flammable form, liable to spontaneous combustion, are substances of Class 4.2.*

[b] *Metals and metal alloys in powdered or other flammable form, which in contact with water, emit flammable gases, are substances of Class 4.3.*

[c] *Metals hydrides which, in contact with water, emit flammable gases, are substances of Class 4.3. Aluminium borohydride or aluminium borohydride in devices are substances of Class 4.2, UN No. 2870.*

2.2.41.3 *List of collective entries (cont'd)*

Polymerizing substances PM	not requiring temperature control	PM1	3531 3532	POLYMERIZING SUBSTANCE, SOLID, STABILIZED, N.O.S. POLYMERIZING SUBSTANCE, LIQUID, STABILIZED, N.O.S.
	requiring temperature control	PM2	3533 3534	POLYMERIZING SUBSTANCE, SOLID, TEMPERATURE CONTROLLED, N.O.S. POLYMERIZING SUBSTANCE, LIQUID, TEMPERATURE CONTROLLED, N.O.S.

2.2.41.4 *List of currently assigned self-reactive substances in packagings*

In the column "Packing Method" codes "OP1" to "OP8" refer to packing methods in 4.1.4.1, packing instruction P520 (see also 4.1.7.1). Self-reactive substances to be carried shall fulfil the classification and the control and emergency temperatures (derived from the SADT) as listed. For substances permitted in IBCs, see 4.1.4.2, packing instruction IBC520 and, for those permitted in tanks according to Chapter 4.2, see 4.2.5.2, portable tank instruction T23.

NOTE: The classification given in this table is based on the technically pure substance (except where a concentration of less than 100% is specified). For other concentrations, the substance may be classified differently following the procedures given in Part II of the Manual of Tests and Criteria and in 2.2.41.1.17.

SELF-REACTIVE SUBSTANCE	Concen-tration (%)	Packing method	Control tempera-ture (°C)	Emergency tempera-ture (°C)	UN generic entry	Remarks
ACETONE-PYROGALLOL COPOLYMER 2-DIAZO-1-NAPHTHOL-5-SULPHONATE	100	OP8			3228	
AZODICARBONAMIDE FORMULATION TYPE B, TEMPERATURE CONTROLLED	< 100	OP5			3232	(1) (2)
AZODICARBONAMIDE FORMULATION TYPE C	< 100	OP6			3224	(3)
AZODICARBONAMIDE FORMULATION TYPE C, TEMPERATURE CONTROLLED	< 100	OP6			3234	(4)
AZODICARBONAMIDE FORMULATION TYPE D	< 100	OP7			3226	(5)
AZODICARBONAMIDE FORMULATION TYPE D, TEMPERATURE CONTROLLED	< 100	OP7			3236	(6)
2,2' -AZODI(2,4-DIMETHYL- 4-METHOXY- VALERONITRILE)	100	OP7	-5	+5	3236	
2,2' -AZODI(2,4-DIMETHYL- VALERONITRILE)	100	OP7	+10	+15	3236	
2,2' -AZODI(ETHYL- 2-METHYL-PROPIONATE)	100	OP7	+20	+25	3235	
1,1-AZODI(HEXAHYDROBENZONITRILE)	100	OP7			3226	
2,2' -AZODI(ISOBUTYRONITRILE)	100	OP6	+40	+45	3234	
2,2' -AZODI(ISOBUTYRONITRILE) as a water based paste	≤ 50%	OP6			3224	
2,2' -AZODI(2-METHYLBUTYRONITRILE)	100	OP7	+35	+40	3236	
BENZENE-1,3-DISULPHONYL HYDRAZIDE, as a paste	52	OP7			3226	
BENZENE SULPHONYL HYDRAZIDE	100	OP7			3226	
4-(BENZYL(ETHYL)AMINO)-3-ETHOXY- BENZENEDIAZONIUM ZINC CHLORIDE	100	OP7			3226	
4-(BENZYL(METHYL)AMINO)-3-ETHOXY- BENZENEDIAZONIUM ZINC CHLORIDE	100	OP7	+40	+45	3236	
3-CHLORO-4-DIETHYLAMINOBENZENE- DIAZONIUM ZINC CHLORIDE	100	OP7			3226	
2-DIAZO-1-NAPHTHOL-4-SULPHONYL CHLORIDE	100	OP5			3222	(2)
2-DIAZO-1-NAPHTHOL-5-SULPHONYL CHLORIDE	100	OP5			3222	(2)

SELF-REACTIVE SUBSTANCE	Concen-tration (%)	Packing method	Control tempera-ture (°C)	Emergency tempera-ture (°C)	UN generic entry	Remarks
2-DIAZO-1-NAPHTHOL SULPHONIC ACID ESTER MIXTURE, TYPE D	< 100	OP7			3226	(9)
2,5-DIBUTOXY-4-(4-MORPHOLINYL)-BENZENEDIAZONIUM, TETRACHLOROZINCATE (2:1)	100	OP8			3228	
2,5-DIETHOXY-4-MORPHOLINO-BENZENEDIAZONIUM ZINC CHLORIDE	67-100	OP7	+35	+40	3236	
2,5-DIETHOXY-4-MORPHOLINO-BENZENEDIAZONIUM ZINC CHLORIDE	66	OP7	+40	+45	3236	
2,5-DIETHOXY-4-MORPHOLINO-BENZENEDIAZONIUM TETRAFLUOROBORATE	100	OP7	+30	+35	3236	
2,5-DIETHOXY-4-(4-MORPHOLINYL)-BENZENEDIAZONIUM SULPHATE	100	OP7			3226	
2,5-DIETHOXY-4-(PHENYLSULPHONYL)-BENZENEDIAZONIUM ZINC CHLORIDE	67	OP7	+40	+45	3236	
DIETHYLENEGLYCOL BIS (ALLYL CARBONATE) + DI- ISOPROPYL-PEROXYDICARBONATE	≥ 88+≤ 12	OP8	-10	0	3237	
2,5-DIMETHOXY-4-(4-METHYL-PHENYLSULPHONYL)BENZENE- DIAZONIUM ZINC CHLORIDE	79	OP7	+40	+45	3236	
4-(DIMETHYLAMINO)-BENZENE-DIAZONIUM TRICHLOROZINCATE (-1)	100	OP8			3228	
4-DIMETHYLAMINO-6-(2-DIMETHYL-AMINOETHOXY) TOLUENE- 2-DIAZONIUM ZINC CHLORIDE	100	OP7	+40	+45	3236	
N,N'-DINITROSO-N,N'- DIMETHYL TEREPHTHALAMIDE, as a paste	72	OP6			3224	
N,N'-DINITROSOPENTAMETHYLENE-TETRAMINE	82	OP6			3224	(7)
DIPHENYLOXIDE-4,4'-DISULPHONYL HYDRAZIDE	100	OP7			3226	
4-DIPROPYLAMINOBENZENE- DIAZONIUM ZINC CHLORIDE	100	OP7			3226	
2-(N,N-ETHOXYCARBONYL-PHENYLAMINO)-3-METHOXY-4-(N-METHYL-N-CYCLOHEXYLAMINO) BENZENEDIAZONIUM ZINC CHLORIDE	63-92	OP7	+ 40	+ 45	3236	
2-(N,N-ETHOXYCARBONYL-PHENYLAMINO)-3-METHOXY-4-(N-METHYL-N-CYCLOHEXYLAMINO) BENZENEDIAZONIUM ZINC CHLORIDE	62	OP7	+ 35	+ 40	3236	
N-FORMYL-2-(NITROMETHYLENE)-1,3-PERHYDROTHIAZINE	100	OP7	+45	+50	3236	
2-(2-HYDROXYETHOXY)-1-(PYRROLIDIN-1-YL)BENZENE-4- DIAZONIUM ZINC CHLORIDE	100	OP7	+ 45	+ 50	3236	
3-(2-HYDROXYETHOXY)-4- (PYRROLIDIN-1-YL) BENZENE DIAZONIUM ZINC CHLORIDE	100	OP7	+40	+45	3236	
2-(N,N-METHYLAMINOETHYLCARBONYL)-4-(3,4-DIMETHYLPHENYLSULPHONYL) BENZENEDIAZONIUM HYDROGEN SULPHATE	96	OP7	+45	+50	3236	
4-METHYLBENZENESULPHONYLHYDRAZIDE	100	OP7			3226	
3-METHYL-4-(PYRROLIDIN-1-YL) BENZENEDIAZONIUM TETRAFLUOROBORATE	95	OP6	+45	+50	3234	
4-NITROSOPHENOL	100	OP7	+35	+40	3236	

SELF-REACTIVE SUBSTANCE	Concen-tration (%)	Packing method	Control tempera-ture (°C)	Emergency tempera-ture (°C)	UN generic entry	Remarks
SELF-REACTIVE LIQUID, SAMPLE		OP2			3223	(8)
SELF-REACTIVE LIQUID, SAMPLE, TEMPERATURE CONTROLLED		OP2			3233	(8)
SELF-REACTIVE SOLID, SAMPLE		OP2			3224	(8)
SELF-REACTIVE SOLID, SAMPLE, TEMPERATURE CONTROLLED		OP2			3234	(8)
SODIUM 2-DIAZO-1-NAPHTHOL- 4-SULPHONATE	100	OP7			3226	
SODIUM 2-DIAZO-1-NAPHTHOL- 5-SULPHONATE	100	OP7			3226	
TETRAMINE PALLADIUM (II) NITRATE	100	OP6	+30	+35	3234	

Remarks

(1)   Azodicarbonamide formulations which fulfil the criteria of paragraph 20.4.2 (b) of the Manual of Tests and Criteria. The control and emergency temperatures shall be determined by the procedure given in 2.2.41.1.17.

(2)   "EXPLOSIVE" subsidiary risk label required (Model No. 1, see 5.2.2.2.2).

(3)   Azodicarbonamide formulations which fulfil the criteria of paragraph 20.4.2 (c) of the Manual of Tests and Criteria.

(4)   Azodicarbonamide formulations which fulfil the criteria of paragraph 20.4.2 (c) of the Manual of Tests and Criteria. The control and emergency temperatures shall be determined by the procedure given in 2.2.41.1.17.

(5)   Azodicarbonamide formulations which fulfil the criteria of paragraph 20.4.2 (d) of the Manual of Tests and Criteria.

(6)   Azodicarbonamide formulations which fulfil the criteria of paragraph 20.4.2 (d) of the Manual of Tests and Criteria. The control and emergency temperatures shall be determined by the procedure given in 2.2.41.1.17.

(7)   With a compatible diluent having a boiling point of not less than 150 °C.

(8)   See 2.2.41.1.15.

(9)   This entry applies to mixtures of esters of 2-diazo-1-naphthol-4-sulphonic acid and 2-diazo-1-naphthol-5-sulphonic acid which fulfil the criteria of paragraph 20.4.2 (d) of the *Manual of Test and Criteria*.

2.2.42   **Class 4.2 Substances liable to spontaneous combustion**

2.2.42.1   *Criteria*

2.2.42.1.1   The heading of Class 4.2 covers:

-      *Pyrophoric substances* which are substances, including mixtures and solutions (liquid or solid), which even in small quantities ignite on contact with air within five minutes. These are the Class 4.2 substances the most liable to spontaneous combustion; and

-      *Self-heating substances and articles* which are substances and articles, including mixtures and solutions, which, on contact with air, without energy supply, are liable to self-heating. These substances will ignite only in large amounts (kilograms) and after long periods of time (hours or days).

2.2.42.1.2   The substances and articles of Class 4.2 are subdivided as follows:

S   Substances liable to spontaneous combustion, without subsidiary risk:

S1   Organic, liquid;

S2   Organic, solid;

S3   Inorganic, liquid;

S4   Inorganic, solid;

S5   Organometallic;

SW   Substances liable to spontaneous combustion, which, in contact with water, emit flammable gases;

SO   Substances liable to spontaneous combustion, oxidizing;

ST   Substances liable to spontaneous combustion, toxic:

ST1   Organic, toxic, liquid;

ST2   Organic, toxic, solid;

ST3   Inorganic, toxic, liquid;

ST4   Inorganic, toxic, solid;

SC   Substances liable to spontaneous combustion, corrosive:

SC1   Organic, corrosive, liquid;

SC2   Organic, corrosive, solid;

SC3   Inorganic, corrosive, liquid;

SC4   Inorganic, corrosive, solid.

Properties

2.2.42.1.3   Self-heating of a substance is a process where the gradual reaction of that substance with oxygen (in air) generates heat. If the rate of heat production exceeds the rate of heat loss, then the temperature of the substance will rise which, after an induction time, may lead to self-ignition and combustion.

Classification

2.2.42.1.4   Substances and articles classified in Class 4.2 are listed in Table A of Chapter 3.2. The assignment of substances and articles not mentioned by name in Table A of Chapter 3.2 to the relevant specific N.O.S. entry of 2.2.42.3 in accordance with the provisions of Chapter 2.1 can be based on experience or the results of the test procedures in accordance with the Manual of Tests and Criteria, Part III, Section 33.3. Assignment to general N.O.S. entries of Class 4.2 shall be based on the results of the test procedures in accordance with the Manual of Tests and Criteria, Part III, section 33.3; experience shall also be taken into account when it leads to a more stringent assignment.

2.2.42.1.5   When substances or articles not mentioned by name are assigned to one of the entries listed in 2.2.42.3 on the basis of the test procedures in accordance with the Manual of Tests and Criteria, Part III, section 33.3, the following criteria shall apply:

(a)   Solids liable to spontaneous combustion (pyrophoric) shall be assigned to Class 4.2 when they ignite on falling from a height of 1 m or within five minutes;

(b)   Liquids liable to spontaneous combustion (pyrophoric) shall be assigned to Class 4.2 when:

(i)   on being poured on an inert carrier, they ignite within five minutes, or

(ii)   in the event of a negative result of the test according to (i), when poured on a dry, indented filter paper (Whatman No. 3 filter), they ignite or carbonize it within five minutes;

(c)   Substances in which, in a 10 cm sample cube, at 140 °C test temperature, spontaneous combustion or a rise in temperature to over 200 °C is observed within 24 hours shall be assigned to Class 4.2. This criterion is based on the temperature of the spontaneous combustion of charcoal, which is at 50 °C for a sample cube of 27 m^3. Substances with a temperature of spontaneous combustion higher than 50 °C for a volume of 27 m^3 are not to be assigned to Class 4.2.

NOTE 1: Substances carried in packages with a volume of not more than 3 m^3 are exempted from Class 4.2 if, tested with a 10 cm sample cube at 120 °C, no spontaneous combustion nor a rise in temperature to over 180 °C is observed within 24 hours.

NOTE 2: Substances carried in packages with a volume of not more than 450 litres are exempted from Class 4.2 if, tested with a 10 cm sample cube at 100 °C, no spontaneous combustion nor a rise in temperature to over 160 °C is observed within 24 hours.

NOTE 3: Since organometallic substances can be classified in Class 4.2 or 4.3 with additional subsidiary risks, depending on their properties, a specific classification flow chart for these substances is given in 2.3.5.

2.2.42.1.6   If substances of Class 4.2, as a result of admixtures, come into different categories of risk from those to which the substances mentioned by name in Table A of Chapter 3.2 belong, these mixtures shall be assigned to the entries to which they belong on the basis of their actual degree of danger.

NOTE: For the classification of solutions and mixtures (such as preparations and wastes), see also 2.1.3.

2.2.42.1.7   On the basis of the test procedure in the Manual of Tests and Criteria, Part III, section 33.3 and the criteria set out in 2.2.42.1.5, it may also be determined whether the nature of a substance mentioned by name is such that the substance is not subject to the provisions for this Class.

Assignment of packing groups

2.2.42.1.8   Substances and articles classified under the various entries in Table A of Chapter 3.2 shall be assigned to packing groups I, II or III on the basis of test procedures of the Manual of Tests and Criteria, Part III, section 33.3, in accordance with the following criteria:

(a)   Substances liable to spontaneous combustion (pyrophoric) shall be assigned to packing group I;

(b)   Self-heating substances and articles in which, in a 2.5 cm sample cube, at 140 °C test temperature, spontaneous combustion or a rise in temperature to over 200 °C is observed within 24 hours, shall be assigned to packing group II;

Substances with a temperature of spontaneous combustion higher than 50 °C for a volume of 450 litres are not to be assigned to packing group II;

(c)   Slightly self-heating substances in which, in a 2.5 cm sample cube, the phenomena referred to under (b) are not observed, in the given conditions, but in which in a 10 cm sample cube at 140 °C test temperature spontaneous combustion or a rise in temperature to over 200 °C is observed within 24 hours, shall be assigned to packing group III.

2.2.42.2   *Substances not accepted for carriage*

The following substances shall not be accepted for carriage:

-     UN No. 3255 tert-BUTYL HYPOCHLORITE; and

-     Self-heating solids, oxidizing, assigned to UN No. 3127 unless they meet the requirements for Class 1 (see 2.1.3.7).

2.2.42.3   *List of collective entries*

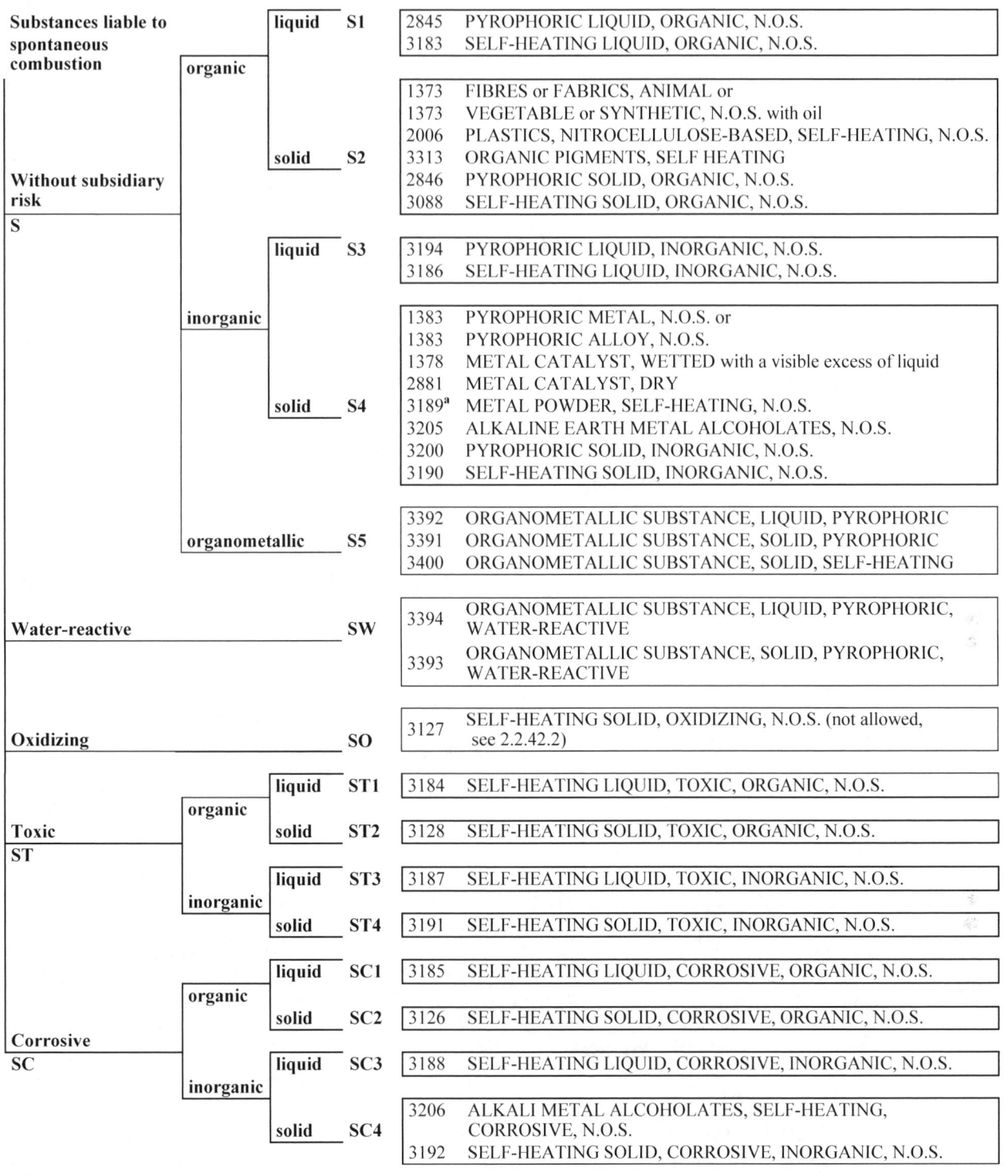

Substances liable to spontaneous combustion		liquid	S1	2845 3183	PYROPHORIC LIQUID, ORGANIC, N.O.S. SELF-HEATING LIQUID, ORGANIC, N.O.S.
	organic				
Without subsidiary risk **S**		solid	S2	1373 1373 2006 3313 2846 3088	FIBRES or FABRICS, ANIMAL or VEGETABLE or SYNTHETIC, N.O.S. with oil PLASTICS, NITROCELLULOSE-BASED, SELF-HEATING, N.O.S. ORGANIC PIGMENTS, SELF HEATING PYROPHORIC SOLID, ORGANIC, N.O.S. SELF-HEATING SOLID, ORGANIC, N.O.S.
		liquid	S3	3194 3186	PYROPHORIC LIQUID, INORGANIC, N.O.S. SELF-HEATING LIQUID, INORGANIC, N.O.S.
	inorganic	solid	S4	1383 1383 1378 2881 3189[a] 3205 3200 3190	PYROPHORIC METAL, N.O.S. or PYROPHORIC ALLOY, N.O.S. METAL CATALYST, WETTED with a visible excess of liquid METAL CATALYST, DRY METAL POWDER, SELF-HEATING, N.O.S. ALKALINE EARTH METAL ALCOHOLATES, N.O.S. PYROPHORIC SOLID, INORGANIC, N.O.S. SELF-HEATING SOLID, INORGANIC, N.O.S.
	organometallic		S5	3392 3391 3400	ORGANOMETALLIC SUBSTANCE, LIQUID, PYROPHORIC ORGANOMETALLIC SUBSTANCE, SOLID, PYROPHORIC ORGANOMETALLIC SUBSTANCE, SOLID, SELF-HEATING
Water-reactive			SW	3394 3393	ORGANOMETALLIC SUBSTANCE, LIQUID, PYROPHORIC, WATER-REACTIVE ORGANOMETALLIC SUBSTANCE, SOLID, PYROPHORIC, WATER-REACTIVE
Oxidizing			SO	3127	SELF-HEATING SOLID, OXIDIZING, N.O.S. (not allowed, see 2.2.42.2)
Toxic **ST**	organic	liquid	ST1	3184	SELF-HEATING LIQUID, TOXIC, ORGANIC, N.O.S.
		solid	ST2	3128	SELF-HEATING SOLID, TOXIC, ORGANIC, N.O.S.
	inorganic	liquid	ST3	3187	SELF-HEATING LIQUID, TOXIC, INORGANIC, N.O.S.
		solid	ST4	3191	SELF-HEATING SOLID, TOXIC, INORGANIC, N.O.S.
Corrosive **SC**	organic	liquid	SC1	3185	SELF-HEATING LIQUID, CORROSIVE, ORGANIC, N.O.S.
		solid	SC2	3126	SELF-HEATING SOLID, CORROSIVE, ORGANIC, N.O.S.
	inorganic	liquid	SC3	3188	SELF-HEATING LIQUID, CORROSIVE, INORGANIC, N.O.S.
		solid	SC4	3206 3192	ALKALI METAL ALCOHOLATES, SELF-HEATING, CORROSIVE, N.O.S. SELF-HEATING SOLID, CORROSIVE, INORGANIC, N.O.S.

---

[a]   *Dust and powder of metals, non toxic in a non-spontaneous combustible form which nevertheless, in contact with water, emit flammable gases, are substances of Class 4.3.*

2.2.43   **Class 4.3 Substances which, in contact with water, emit flammable gases**

2.2.43.1   *Criteria*

2.2.43.1.1   The heading of Class 4.3 covers substances which react with water to emit flammable gases liable to form explosive mixtures with air, and articles containing such substances.

2.2.43.1.2   Substances and articles of Class 4.3 are subdivided as follows:

W   Substances which, in contact with water, emit flammable gases, without subsidiary risk, and articles containing such substances:

        W1   Liquid;

        W2   Solid;

        W3   Articles;

WF1   Substances which, in contact with water, emit flammable gases, liquid, flammable;

WF2   Substances which, in contact with water, emit flammable gases, solid, flammable;

WS   Substances which, in contact with water, emit flammable gases, solid, self-heating;

WO   Substances which, in contact with water, emit flammable gases, oxidizing, solid;

WT   Substances which, in contact with water, emit flammable gases, toxic:

        WT1   Liquid;

        WT2   Solid;

WC   Substances which, in contact with water, emit flammable gases, corrosive:

        WC1   Liquid;

        WC2   Solid;

WFC   Substances which, in contact with water, emit flammable gases, flammable, corrosive.

Properties

2.2.43.1.3   Certain substances in contact with water may emit flammable gases that can form explosive mixtures with air. Such mixtures are easily ignited by all ordinary sources of ignition, for example naked lights, sparking handtools or unprotected lamps. The resulting blast wave and flames may endanger people and the environment. The test method referred to in 2.2.43.1.4 below is used to determine whether the reaction of a substance with water leads to the development of a dangerous amount of gases which may be flammable. This test method shall not be applied to pyrophoric substances.

Classification

2.2.43.1.4   Substances and articles classified in Class 4.3 are listed in Table A of Chapter 3.2. The assignment of substances and articles not mentioned by name in Table A of Chapter 3.2 to the relevant entry of 2.2.43.3 in accordance with the provisions of Chapter 2.1 shall be based on the results of the test procedure in accordance with the Manual of Tests and Criteria, Part III, Section 33.4; experience shall also be taken into account when it leads to a more stringent assignment.

2.2.43.1.5   When substances not mentioned by name are assigned to one of the entries listed in 2.2.43.3 on the basis of the test procedure in accordance with the Manual of Tests and Criteria, Part III, Section 33.4, the following criteria shall apply:

A substance shall be assigned to Class 4.3 if:

(a)   spontaneous ignition of the gas emitted takes place in any step of the test procedure; or

(b)   there is an evolution of flammable gas at a rate greater than 1 litre per kilogram of the substance to be tested per hour.

NOTE: Since organometallic substances can be classified in Class 4.2 or 4.3 with additional subsidiary risks, depending on their properties, a specific classification flow chart for these substances is given in 2.3.5.

2.2.43.1.6   If substances of Class 4.3, as a result of admixtures, come into different categories of risk from those to which the substances mentioned by name in Table A of Chapter 3.2 belong, these mixtures shall be assigned to the entries to which they belong on the basis of their actual degree of danger.

NOTE: For the classification of solutions and mixtures (such as preparations and wastes) see also 2.1.3.

2.2.43.1.7   On the basis of the test procedures in accordance with the Manual of Tests and Criteria, Part III, Section 33.4, and the criteria set out in paragraph 2.2.43.1.5, it may also be determined whether the nature of a substance mentioned by name is such that the substance is not subject to the provisions for this Class.

Assignment of packing groups

2.2.43.1.8   Substances and articles classified under the various entries in Table A of Chapter 3.2 shall be assigned to packing groups I, II or III on the basis of test procedures of the Manual of Tests and Criteria, Part III, section 33.4, in accordance with the following criteria:

(a)   Packing group I shall be assigned to any substance which reacts vigorously with water at ambient temperature and generally demonstrates a tendency for the gas produced to ignite spontaneously, or one which reacts readily with water at ambient temperatures such that the rate of evolution of flammable gas is equal to or greater than 10 litres per kilogram of substance over any one minute period;

(b)   Packing group II shall be assigned to any substance which reacts readily with water at ambient temperature such that the maximum rate of evolution of flammable gas is equal to or greater than 20 litres per kilogram of substance per hour, and which does not meet the criteria of packing group I;

(c)   Packing group III shall be assigned to any substance which reacts slowly with water at ambient temperature such that the maximum rate of evolution of flammable gas is greater than 1 litre per kilogram of substance per hour, and which does not meet the criteria of packing groups I or II.

2.2.43.2   ***Substances not accepted for carriage***

Water-reactive solids, oxidizing, assigned to UN No. 3133 shall not be accepted for carriage unless they meet the requirements for Class 1 (see also 2.1.3.7).

2.2.43.3 *List of collective entries*

Substances which, in contact with water, emit flammable gases			1389 ALKALI METAL AMALGAM, LIQUID 1391 ALKALI METAL DISPERSION or 1391 ALKALINE EARTH METAL DISPERSION 1392 ALKALINE EARTH METAL AMALGAM, LIQUID 1420 POTASSIUM METAL ALLOYS, LIQUID
	liquid	W1	1421 ALKALI METAL ALLOY, LIQUID, N.O.S. 1422 POTASSIUM SODIUM ALLOYS, LIQUID 3398 ORGANOMETALLIC SUBSTANCE, LIQUID, WATER-REACTIVE 3148 WATER-REACTIVE LIQUID, N.O.S.
Without subsidiary risk W	solid	W2 ª	1390 ALKALI METAL AMIDES 3401 ALKALI METAL AMALGAM, SOLID 3402 ALKALINE EARTH METAL AMALGAM, SOLID 3170 ALUMINIUM SMELTING BY-PRODUCTS or 3170 ALUMINIUM REMELTING BY-PRODUCTS 3403 POTASSIUM METAL ALLOYS, SOLID 3404 POTASSIUM SODIUM ALLOYS, SOLID 1393 ALKALINE EARTH METAL ALLOY, N.O.S. 1409 METAL HYDRIDES, WATER-REACTIVE, N.O.S. 3208 METALLIC SUBSTANCE, WATER-REACTIVE, N.O.S. 3395 ORGANOMETALLIC SUBSTANCE, SOLID, WATER-REACTIVE 2813 WATER-REACTIVE SOLID, N.O.S.
	articles	W3	3292 BATTERIES, CONTAINING SODIUM or 3292 CELLS, CONTAINING SODIUM
Liquid, flammable		WF1	3482 ALKALI METAL DISPERSION, FLAMMABLE or 3482 ALKALINE EARTH METAL DISPERSION, FLAMMABLE 3399 ORGANOMETALLIC SUBSTANCE, LIQUID, WATER-REACTIVE, FLAMMABLE
Solid, flammable		WF2	3396 ORGANOMETALLIC SUBSTANCE, SOLID, WATER-REACTIVE, FLAMMABLE 3132 WATER-REACTIVE SOLID, FLAMMABLE, N.O.S.
Solid, self-heating		WS ᵇ	3397 ORGANOMETALLIC SUBSTANCE, SOLID, WATER-REACTIVE, SELF-HEATING 3209 METALLIC SUBSTANCE, WATER-REACTIVE, SELF-HEATING, N.O.S. 3135 WATER-REACTIVE SOLID, SELF-HEATING, N.O.S.
Solid, oxidizing		WO	3133 WATER-REACTIVE SOLID, OXIDIZING, N.O.S. (not allowed, see 2.2.43.2)
Toxic WT	liquid	WT1	3130 WATER-REACTIVE LIQUID, TOXIC, N.O.S.
	solid	WT2	3134 WATER-REACTIVE SOLID, TOXIC, N.O.S.
Corrosive WC	liquid	WC1	3129 WATER-REACTIVE LIQUID, CORROSIVE, N.O.S.
	solid	WC2	3131 WATER-REACTIVE SOLID, CORROSIVE, N.O.S.
Flammable, corrosive		WFC ᶜ	2988 CHLOROSILANES, WATER-REACTIVE, FLAMMABLE, CORROSIVE, NO.S. (No other collective entry with this classification code available, if need be, classification under a collective entry with a classification code to be determined according to the table of precedence of hazard in 2.1.3.10.)

ª *Metals and metal alloys which, in contact with water, do not emit flammable gases and are not pyrophoric or self-heating, but which are readily flammable, are substances of Class 4.1. Alkaline-earth metals and alkaline-earth metal alloys in pyrophoric form are substances of Class 4.2 Dust and powders of metals in pyrophoric form are substances of Class 4.2. Metals and metal alloys in pyrophoric form are substances of Class 4.2. Compounds of phosphorus with heavy metals such as iron, copper, etc. are not subject to the provisions of ADR.*

ᵇ *Metals and metal alloys in pyrophoric form are substances of Class 4.2.*

ᶜ *Chlorosilanes, having a flash-point of less than 23 °C and which, in contact with water, do not emit flammable gases, are substances of Class 3. Chlorosilanes, having a flash-point equal to or greater than 23 °C and which, in contact with water, do not emit flammable gases, are substances of Class 8.*

2.2.51   **Class 5.1 Oxidizing substances**

2.2.51.1   *Criteria*

2.2.51.1.1   The heading of Class 5.1 covers substances which, while in themselves not necessarily combustible, may, generally by yielding oxygen, cause or contribute to the combustion of other material, and articles containing such substances.

2.2.51.1.2   The substances of Class 5.1 and articles containing such substances are subdivided as follows:

   O   Oxidizing substances without subsidiary risk or articles containing such substances:

      O1   Liquid;

      O2   Solid;

      O3   Articles;

   OF   Oxidizing substances, solid, flammable;

   OS   Oxidizing substances, solid, self-heating;

   OW   Oxidizing substances, solid which, in contact with water, emit flammable gases;

   OT   Oxidizing substances, toxic:

      OT1   Liquid;

      OT2   Solid;

   OC   Oxidizing substances, corrosive:

      OC1   Liquid;

      OC2   Solid;

   OTC   Oxidizing substances, toxic, corrosive.

2.2.51.1.3   Substances and articles classified in Class 5.1 are listed in Table A of Chapter 3.2. The assignment of substances and articles not mentioned by name in Table A of Chapter 3.2 to the relevant entry of 2.2.51.3 in accordance with the provisions of Chapter 2.1 can be based on the tests, methods and criteria in paragraphs 2.2.51.1.6-2.2.51.1.9 below and the Manual of Tests and Criteria, Part III, Section 34.4. In the event of divergence between test results and known experience, judgement based on known experience shall take precedence over test results.

2.2.51.1.4   If substances of Class 5.1, as a result of admixtures, come into different categories of risk from those to which the substances mentioned by name in Table A of Chapter 3.2 belong, these mixtures or solutions shall be assigned to the entries to which they belong on the basis of their actual degree of danger.

   NOTE: For the classification of solutions and mixtures (such as preparations and wastes), see also Section 2.1.3.

2.2.51.1.5   On the basis of the test procedures in the Manual of Tests and Criteria, Part III, Section 34.4 and the criteria set out in 2.2.51.1.6 to 2.2.51.1.9 it may also be determined whether the nature of a substance mentioned by name in Table A of Chapter 3.2 is such that the substance is not subject to the provisions for this class.

Oxidizing solids

Classification

2.2.51.1.6   When oxidizing solid substances not mentioned by name in Table A of Chapter 3.2 are assigned to one of the entries listed in 2.2.51.3 on the basis of the test procedure in accordance with the Manual of Tests and Criteria, Part III, sub-section 34.4.1 (test O.1) or alternatively, sub section 34.4.3 (test O.3), the following criteria shall apply:

(a)   In the test O.1, a solid substance shall be assigned to Class 5.1 if, in the 4:1 or the 1:1 sample-to-cellulose ratio (by mass) tested, it ignites or burns or exhibits mean burning times equal to or less than that of a 3:7 mixture (by mass) of potassium bromate and cellulose; or

(b)   In the test O.3, a solid substance shall be assigned to Class 5.1 if, in the 4:1 or the 1:1 sample-to-cellulose ratio (by mass) tested, it exhibits a mean burning rate equal to or greater than the mean burning rate of a 1:2 mixture (by mass) of calcium peroxide and cellulose.

Assignment of packing groups

2.2.51.1.7   Oxidizing solids classified under the various entries in Table A of Chapter 3.2 shall be assigned to packing groups I, II or III on the basis of test procedures of the Manual of Tests and Criteria, Part III, sub-section 34.4.1 (test O.1) or sub-section 34.4.3 (test O.3), in accordance with the following criteria:

(a)   Test O.1:

    (i)   Packing group I: any substance which, in the 4:1 or 1:1 sample-to-cellulose ratio (by mass) tested, exhibits a mean burning time less than the mean burning time of a 3:2 mixture, by mass, of potassium bromate and cellulose;

    (ii)   Packing group II: any substance which, in the 4:1 or 1:1 sample-to-cellulose ratio (by mass) tested, exhibits a mean burning time equal to or less than the mean burning time of a 2:3 mixture (by mass) of potassium bromate and cellulose and the criteria for packing group I are not met;

    (iii)   Packing group III: any substance which, in the 4:1 or 1:1 sample-to-cellulose ratio (by mass) tested, exhibits a mean burning time equal to or less than the mean burning time of a 3:7 mixture (by mass) of potassium bromate and cellulose and the criteria for packing groups I and II are not met;

(b)   Test O.3:

    (i)   Packing group I: any substance which, in the 4:1 or 1:1 sample-to-cellulose ratio (by mass) tested, exhibits a mean burning rate greater than the mean burning rate of a 3:1 mixture (by mass) of calcium peroxide and cellulose;

    (ii)   Packing group II: any substance which, in the 4:1 or 1:1 sample-to-cellulose ratio (by mass) tested, exhibits a mean burning rate equal to or greater than the mean burning rate of a 1:1 mixture (by mass) of calcium peroxide and cellulose, and the criteria for packing group I are not met;

    (iii)   Packing group III: any substance which, in the 4:1 or 1:1 sample-to-cellulose ratio (by mass) tested, exhibits a mean burning rate equal to or greater than the mean burning rate of a 1:2 mixture (by mass) of calcium peroxide and cellulose, and the criteria for packing groups I and II are not met.

Oxidizing liquids

Classification

2.2.51.1.8   When oxidizing liquid substances not mentioned by name in Table A of Chapter 3.2 are assigned to one of the entries listed in sub-section 2.2.51.3 on the basis of the test procedure in accordance with the Manual of Tests and Criteria, Part III, sub-section 34.4.2, the following criteria shall apply:

A liquid substance shall be assigned to Class 5.1 if, in the 1:1 mixture, by mass, of substance and cellulose tested, it exhibits a pressure rise of 2070 kPa gauge or more and a mean pressure rise time equal to or less than the mean pressure rise time of a 1:1 mixture, by mass, of 65% aqueous nitric acid and cellulose.

Assignment of packing groups

2.2.51.1.9   Oxidizing liquids classified under the various entries in Table A of Chapter 3.2 shall be assigned to packing groups I, II or III on the basis of test procedures of the Manual of Tests and Criteria, Part III, section 34.4.2, in accordance with the following criteria:

(a)   Packing group I: any substance which, in the 1:1 mixture, by mass, of substance and cellulose tested, spontaneously ignites; or the mean pressure rise time of a 1:1 mixture, by mass, of substance and cellulose is less than that of a 1:1 mixture, by mass, of 50% perchloric acid and cellulose;

(b)   Packing group II: any substance which, in the 1:1 mixture, by mass, of substance and cellulose tested, exhibits a mean pressure rise time less than or equal to the mean pressure rise time of a 1:1 mixture, by mass, of 40% aqueous sodium chlorate solution and cellulose; and the criteria for packing group I are not met;

(c)   Packing group III: any substance which, in the 1:1 mixture, by mass, of substance and cellulose tested, exhibits a mean pressure rise time less than or equal to the mean pressure rise time of a 1:1 mixture, by mass, of 65% aqueous nitric acid and cellulose; and the criteria for packing groups I and II are not met.

2.2.51.2   *Substances not accepted for carriage*

2.2.51.2.1   The chemically unstable substances of Class 5.1 shall not be accepted for carriage unless the necessary steps have been taken to prevent their dangerous decomposition or polymerization during carriage. To this end it shall in particular be ensured that receptacles and tanks do not contain any material liable to promote these reactions.

2.2.51.2.2   The following substances and mixtures shall not be accepted for carriage:

-    oxidizing solids, self-heating, assigned to UN No. 3100, oxidizing solids, water-reactive, assigned to UN No. 3121 and oxidizing solids, flammable, assigned to UN No. 3137, unless they meet the requirements for Class 1 (see also 2.1.3.7);

-    hydrogen peroxide, not stabilized or hydrogen peroxide, aqueous solutions, not stabilized containing more than 60% hydrogen peroxide;

-    tetranitromethane not free from combustible impurities;

-    perchloric acid solutions containing more than 72% (mass) acid, or mixtures of perchloric acid with any liquid other than water;

-    chloric acid solution containing more than 10% chloric acid or mixtures of chloric acid with any liquid other than water;

-    halogenated fluor compounds other than UN Nos. 1745 BROMINE PENTAFLUORIDE; 1746 BROMINE TRIFLUORIDE and 2495 IODINE PENTAFLUORIDE of Class 5.1 as well as UN Nos. 1749 CHLORINE TRIFLUORIDE and 2548 CHLORINE PENTAFLUORIDE of Class 2;

-    ammonium chlorate and its aqueous solutions and mixtures of a chlorate with an ammonium salt;

-    ammonium chlorite and its aqueous solutions and mixtures of a chlorite with an ammonium salt;

-    mixtures of a hypochlorite with an ammonium salt;

-    ammonium bromate and its aqueous solutions and mixtures of a bromate with an ammonium salt;

-    ammonium permanganate and its aqueous solutions and mixtures of a permanganate with an ammonium salt;

- ammonium nitrate containing more than 0.2% combustible substances (including any organic substance calculated as carbon) unless it is a constituent of a substance or article of Class 1;

- fertilizers having an ammonium nitrate content (in determining the ammonium nitrate content, all nitrate ions for which a molecular equivalent of ammonium ions is present in the mixture shall be calculated as ammonium nitrate) or a content in combustible substances exceeding the values specified in special provision 307 except under the conditions applicable to Class 1;

- ammonium nitrite and its aqueous solutions and mixtures of an inorganic nitrite with an ammonium salt;

- mixtures of potassium nitrate, sodium nitrite and an ammonium salt.

2.2.51.3   *List of collective entries*

Oxidizing substances and articles containing such substances	liquid	O1	3210 CHLORATES, INORGANIC, AQUEOUS SOLUTION, N.O.S. 3211 PERCHLORATES, INORGANIC, AQUEOUS SOLUTION, N.O.S. 3213 BROMATES, INORGANIC, AQUEOUS SOLUTION, N.O.S. 3214 PERMANGANATES, INORGANIC, AQUEOUS SOLUTION, N.O.S. 3216 PERSULPHATES, INORGANIC, AQUEOUS SOLUTION, N.O.S. 3218 NITRATES, INORGANIC, AQUEOUS SOLUTION, N.O.S. 3219 NITRITES, INORGANIC, AQUEOUS SOLUTION, N.O.S. 3139 OXIDIZING LIQUID, N.O.S.
Without subsidiary risk **O**	solid	O2	1450 BROMATES, INORGANIC, N.O.S 1461 CHLORATES, INORGANIC, N.O.S. 1462 CHLORITES, INORGANIC, N.O.S. 1477 NITRATES, INORGANIC, N.O.S 1481 PERCHLORATES, INORGANIC, N.O.S. 1482 PERMANGANATES, INORGANIC, N.O.S. 1483 PEROXIDES, INORGANIC, N.O.S 2627 NITRITES, INORGANIC, N.O.S. 3212 HYPOCHLORITES, INORGANIC, N.O.S. 3215 PERSULPHATES, INORGANIC, N.O.S. 1479 OXIDIZING SOLID, N.O.S.
	articles	O3	3356 OXYGEN GENERATOR, CHEMICAL
Solid, flammable		OF	3137 OXIDIZING SOLID, FLAMMABLE, N.O.S. (not allowed, see 2.2.51.2)
Solid, self-heating		OS	3100 OXIDIZING SOLID, SELF-HEATING, N.O.S. (not allowed, see 2.2.51.2)
Solid, water reactive		OW	3121 OXIDIZING SOLID, WATER REACTIVE, N.O.S. (not allowed, see 2.2.51.2)
Toxic **OT**	liquid	OT1	3099 OXIDIZING LIQUID, TOXIC, N.O.S.
	solid	OT2	3087 OXIDIZING SOLID, TOXIC, N.O.S.
Corrosive **OC**	liquid	OC1	3098 OXIDIZING LIQUID, CORROSIVE, N.O.S.
	solid	OC2	3085 OXIDIZING SOLID, CORROSIVE, N.O.S.
Toxic, corrosive		OTC	(No collective entry with this classification code available; if need be, classification under a collective entry with a classification code to be determined according to the table of precedence of hazard in 2.1.3.10.)

2.2.52   **Class 5.2 Organic peroxides**

2.2.52.1   *Criteria*

2.2.52.1.1   The heading of Class 5.2 covers organic peroxides and formulations of organic peroxides.

2.2.52.1.2   The substances of Class 5.2 are subdivided as follows:

P1   Organic peroxides, not requiring temperature control;

P2   Organic peroxides, requiring temperature control.

Definition

2.2.52.1.3   *Organic peroxides* are organic substances which contain the bivalent -O-O- structure and may be considered derivatives of hydrogen peroxide, where one or both of the hydrogen atoms have been replaced by organic radicals.

Properties

2.2.52.1.4   Organic peroxides are liable to exothermic decomposition at normal or elevated temperatures. The decomposition can be initiated by heat, contact with impurities (e.g. acids, heavy-metal compounds, amines), friction or impact. The rate of decomposition increases with temperature and varies with the organic peroxide formulation. Decomposition may result in the evolution of harmful, or flammable, gases or vapours. For certain organic peroxides the temperature shall be controlled during carriage. Some organic peroxides may decompose explosively, particularly if confined. This characteristic may be modified by the addition of diluents or by the use of appropriate packagings. Many organic peroxides burn vigorously. Contact of organic peroxides with the eyes is to be avoided. Some organic peroxides will cause serious injury to the cornea, even after brief contact, or will be corrosive to the skin.

NOTE: Test methods for determining the flammability of organic peroxides are set out in the Manual of Tests and Criteria, Part III, sub-section 32.4. Because organic peroxides may react vigorously when heated, it is recommended to determine their flash-point using small sample sizes such as described in ISO 3679:1983.

Classification

2.2.52.1.5   Any organic peroxide shall be considered for classification in Class 5.2 unless the organic peroxide formulation contains:

(a)   Not more than 1.0% available oxygen from the organic peroxides when containing not more than 1.0% hydrogen peroxide;

(b)   Not more than 0.5% available oxygen from the organic peroxides when containing more than 1.0% but not more than 7.0% hydrogen peroxide.

NOTE: The available oxygen content (%) of an organic peroxide formulation is given by the formula

$$16 \times \sum (n_i \times c_i/m_i)$$

where:
n_i =   number of peroxygen groups per molecule of organic peroxide i;
c_i =   concentration (mass %) of organic peroxide i; and
m_i =   molecular mass of organic peroxide i.

2.2.52.1.6   Organic peroxides are classified into seven types according to the degree of danger they present. The types of organic peroxide range from type A, which is not accepted for carriage in the packaging in which it is tested, to type G, which is not subject to the provisions of Class 5.2. The classification of types B to F is directly related to the maximum quantity allowed in one package. The principles to be applied to the classification of substances not listed in 2.2.52.4 are set out in the Manual of Tests and Criteria, Part II.

2.2.52.1.7 Organic peroxides which have already been classified and are already permitted for carriage in packagings are listed in 2.2.52.4, those already permitted for carriage in IBCs are listed in 4.1.4.2, packing instruction IBC520 and those already permitted for carriage in tanks in accordance with Chapters 4.2 and 4.3 are listed in 4.2.5.2, portable tank instruction T23. Each permitted substance listed is assigned to a generic entry of Table A of Chapter 3.2 (UN Nos. 3101 to 3120) and appropriate subsidiary risks and remarks providing relevant transport information are given.

These generic entries specify:

- the type (B to F) of organic peroxide (see 2.2.52.1.6 above);

- physical state (liquid/solid); and

- temperature control (when required), see 2.2.52.1.15 to 2.2.52.1.18.

Mixtures of these formulations may be classified as the same type of organic peroxide as that of the most dangerous component and be carried under the conditions of carriage given for this type. However, as two stable components can form a thermally less stable mixture, the self-accelerating decomposition temperature (SADT) of the mixture shall be determined and, if necessary, the control and emergency temperatures derived from the SADT in accordance with 2.2.52.1.16.

2.2.52.1.8 Classification of organic peroxides not listed in 2.2.52.4, 4.1.4.2 packing instruction IBC520 or 4.2.5.2, portable tank instruction T23, and assignment to a collective entry shall be made by the competent authority of the country of origin. The statement of approval shall contain the classification and the relevant conditions of carriage. If the country of origin is not a Contracting Party to ADR, the classification and conditions of carriage shall be recognized by the competent authority of the first country Contracting Party to ADR reached by the consignment.

2.2.52.1.9 Samples of organic peroxides or formulations of organic peroxides not listed in 2.2.52.4, for which a complete set of test results is not available and which are to be carried for further testing or evaluation, shall be assigned to one of the appropriate entries for organic peroxides type C provided the following conditions are met:

- the available data indicate that the sample would be no more dangerous than organic peroxides type B;

- the sample is packaged in accordance with packing method OP2 and the quantity per transport unit is limited to 10 kg;

- the available data indicate that the control temperature, if any, is sufficiently low to prevent any dangerous decomposition and sufficiently high to prevent any dangerous phase separation.

Desensitization of organic peroxides

2.2.52.1.10 In order to ensure safety during carriage, organic peroxides are in many cases desensitized by organic liquids or solids, inorganic solids or water. Where a percentage of a substance is stipulated, this refers to the percentage by mass, rounded to the nearest whole number. In general, desensitization shall be such that, in case of spillage, the organic peroxide will not concentrate to a dangerous extent.

2.2.52.1.11 Unless otherwise stated for the individual organic peroxide formulation, the following definition(s) shall apply to diluents used for desensitization:

- diluents type A are organic liquids which are compatible with the organic peroxide and which have a boiling point of not less than 150 °C. Type A diluents may be used for desensitizing all organic peroxides;

- diluents type B are organic liquids which are compatible with the organic peroxide and which have a boiling point of less than 150 °C but not less than 60 °C and a flash-point of not less than 5 °C.

Type B diluents may be used for desensitization of all organic peroxides provided that the boiling point of the liquid is at least 60 °C higher than the SADT in a 50 kg package.

2.2.52.1.12 Diluents, other than type A or type B, may be added to organic peroxide formulations as listed in 2.2.52.4 provided that they are compatible. However, replacement of all or part of a type A or type B diluent by another diluent with differing properties requires that the organic peroxide formulation be re-assessed in accordance with the normal acceptance procedure for Class 5.2.

2.2.52.1.13   Water may only be used for the desensitization of organic peroxides which are listed in 2.2.52.4 or in the competent authority decision according to 2.2.52.1.8 as being "with water" or "as a stable dispersion in water". Samples of organic peroxides or formulations of organic peroxides not listed in 2.2.52.4 may also be desensitized with water provided the requirements of 2.2.52.1.9 are met.

2.2.52.1.14   Organic and inorganic solids may be used for desensitization of organic peroxides provided that they are compatible. Compatible liquids and solids are those which have no detrimental influence on the thermal stability and hazard type of the organic peroxide formulation.

Temperature control requirements

2.2.52.1.15   Certain organic peroxides may only be carried under temperature-controlled conditions. The control temperature is the maximum temperature at which the organic peroxide can be safely carried. It is assumed that the temperature of the immediate surroundings of a package only exceeds 55 °C during carriage for a relatively short time in a 24 hour period. In the event of loss of temperature control, it may be necessary to implement emergency procedures. The emergency temperature is the temperature at which such procedures shall be implemented.

2.2.52.1.16   The control and emergency temperatures are derived from the SADT which is defined as the lowest temperature at which self-accelerating decomposition may occur with a substance in the packaging as used during carriage (see Table 1). The SADT shall be determined in order to decide whether a substance shall be subjected to temperature control during carriage. Provisions for the determination of the SADT are given in the Manual of Tests and Criteria, Part II, Sections 20 and 28.4.

Table 1: Derivation of control and emergency temperatures

Type of receptacle	SADT [a]	Control temperature	Emergency temperature
Single packagings and IBCs	20 °C or less	20 °C below SADT	10 °C below SADT
	over 20 °C to 35 °C	15 °C below SADT	10 °C below SADT
	over 35 °C	10 °C below SADT	5 °C below SADT
Tanks	not greater than 50 °C	10 °C below SADT	5 °C below SADT

[a]   *SADT of the substance as packaged for carriage*

2.2.52.1.17   The following organic peroxides shall be subject to temperature control during carriage:

-       organic peroxides types B and C with an SADT ≤ 50 °C;

-       organic peroxides type D showing a medium effect when heated under confinement with an SADT ≤ 50 °C or showing a low or no effect when heated under confinement with an SADT ≤ 45 °C; and

-       organic peroxides types E and F with an SADT ≤ 45 °C.

NOTE: *Provisions for the determination of the effects of heating under confinement are given in the Manual of Tests and Criteria, Part II, Section 20 and test series E in Section 25.*

2.2.52.1.18   Where applicable, control and emergency temperatures are listed in 2.2.52.4. The actual temperature during carriage may be lower than the control temperature but shall be selected so as to avoid dangerous separation of phases.

2.2.52.2   *Substances not accepted for carriage*

Organic peroxides, type A, shall not be accepted for carriage under the provisions of Class 5.2 (see Manual of Tests and Criteria, Part II, paragraph 20.4.3 (a)).

2.2.52.3   *List of collective entries*

Organic peroxides		ORGANIC PEROXIDE TYPE A, LIQUID ORGANIC PEROXIDE TYPE A, SOLID	} Not accepted for carriage, see 2.2.52.2
	3101	ORGANIC PEROXIDE TYPE B, LIQUID	
	3102	ORGANIC PEROXIDE TYPE B, SOLID	
	3103	ORGANIC PEROXIDE TYPE C, LIQUID	
	3104	ORGANIC PEROXIDE TYPE C, SOLID	
Not requiring temperature control **P1**	3105	ORGANIC PEROXIDE TYPE D, LIQUID	
	3106	ORGANIC PEROXIDE TYPE D, SOLID	
	3107	ORGANIC PEROXIDE TYPE E, LIQUID	
	3108	ORGANIC PEROXIDE TYPE E, SOLID	
	3109	ORGANIC PEROXIDE TYPE F, LIQUID	
	3110	ORGANIC PEROXIDE TYPE F, SOLID	
		ORGANIC PEROXIDE TYPE G, LIQUID ORGANIC PEROXIDE TYPE G, SOLID	} Not subject to the provisions applicable to Class 5.2, see 2.2.52.1.6

	3111	ORGANIC PEROXIDE TYPE B, LIQUID, TEMPERATURE CONTROLLED
	3112	ORGANIC PEROXIDE TYPE B, SOLID, TEMPERATURE CONTROLLED
	3113	ORGANIC PEROXIDE TYPE C, LIQUID, TEMPERATURE CONTROLLED
	3114	ORGANIC PEROXIDE TYPE C, SOLID, TEMPERATURE CONTROLLED
Requiring temperature control **P2**	3115	ORGANIC PEROXIDE TYPE D, LIQUID, TEMPERATURE CONTROLLED
	3116	ORGANIC PEROXIDE TYPE D, SOLID, TEMPERATURE CONTROLLED
	3117	ORGANIC PEROXIDE TYPE E, LIQUID, TEMPERATURE CONTROLLED
	3118	ORGANIC PEROXIDE TYPE E, SOLID, TEMPERATURE CONTROLLED
	3119	ORGANIC PEROXIDE TYPE F, LIQUID, TEMPERATURE CONTROLLED
	3120	ORGANIC PEROXIDE TYPE F, SOLID, TEMPERATURE CONTROLLED

2.2.52.4   *List of currently assigned organic peroxides in packagings*

In the column "Packing Method", codes "OP1" to "OP8" refer to packing methods in 4.1.4.1, packing instruction P520 (see also 4.1.7.1). Organic peroxides to be carried shall fulfil the classification and the control and emergency temperatures (derived from the SADT) as listed. For substances permitted in IBCs, see 4.1.4.2, packing instruction IBC520 and, for those permitted in tanks according to Chapters 4.2 and 4.3, see 4.2.5.2, portable tank instruction T23.

ORGANIC PEROXIDE	Concentration (%)	Diluent type A (%)	Diluent type B (%) 1)	Inert solid (%)	Water	Packing Method	Control temperature (°C)	Emergency temperature (°C)	Number (Generic entry)	Subsidiary risks and remarks
ACETYL ACETONE PEROXIDE	≤ 42	≥ 48			≥ 8	OP7			3105	2)
"	≤ 32 as a paste					OP7			3106	20)
ACETYL CYCLOHEXANESULPHONYL PEROXIDE	≤ 82				≥ 12	OP4	-10	0	3112	3)
"	≤ 32		≥ 68			OP7	-10	0	3115	
tert-AMYL HYDROPEROXIDE	≤ 88	≥ 6			≥ 6	OP8			3107	
tert-AMYL PEROXYACETATE	≤ 62	≥ 38				OP7			3105	
tert-AMYL PEROXYBENZOATE	≤ 100					OP5			3103	
tert-AMYL PEROXY-2-ETHYLHEXANOATE	≤ 100					OP7	+20	+25	3115	
tert-AMYL PEROXY-2-ETHYLHEXYL CARBONATE	≤ 100					OP7			3105	
tert-AMYL PEROXY ISOPROPYL CARBONATE	≤ 77	≥ 23				OP5			3103	
tert-AMYL PEROXYNEODECANOATE	≤ 77		≥ 23			OP7	0	+10	3115	
"	≤ 47	≥ 53				OP8	0	+10	3119	
tert-AMYL PEROXYPIVALATE	≤ 77		≥ 23			OP5	+10	+15	3113	
tert-AMYLPEROXY-3,5,5-TRIMETHYLHEXANOATE	≤ 100					OP7			3105	
tert-BUTYL CUMYL PEROXIDE	> 42 - 100					OP8			3109	
"	≤ 52			≥ 48		OP8			3108	
n-BUTYL-4,4-DI-(tert-BUTYLPEROXY)VALERATE	> 52 - 100					OP5			3103	
"	≤ 52			≥ 48		OP8			3108	
tert-BUTYL HYDROPEROXIDE	> 79 - 90				≥ 10	OP5			3103	13)
"	≤ 80	≥ 20				OP7			3105	4) 13)
"	≤ 79				> 14	OP8			3107	13) 23)
"	≤ 72				≥ 28	OP8			3109	13)
tert-BUTYL HYDROPEROXIDE + DI-tert-BUTYLPEROXIDE	< 82 + > 9				≥ 7	OP5			3103	13)

ORGANIC PEROXIDE	Concentration (%)	Diluent type A (%)	Diluent type B (%) 1)	Inert solid (%)	Water	Packing Method	Control temperature (°C)	Emergency temperature (°C)	Number (Generic entry)	Subsidiary risks and remarks
tert-BUTYL MONOPEROXYMALEATE	> 52 - 100					OP5			3102	3)
"	≤ 52	≥ 48				OP6			3103	
"	≤ 52			≥ 48		OP8			3108	
"	≤ 52 as a paste					OP8			3108	
tert-BUTYL PEROXYACETATE	> 52 - 77	≥ 23				OP5			3101	3)
"	> 32 - 52	≥ 48				OP6			3103	
"	≤ 32		≥ 68			OP8			3109	
tert-BUTYL PEROXYBENZOATE	> 77 - 100					OP5			3103	
"	> 52 - 77	≥ 23				OP7			3105	
"	≤ 52			≥ 48		OP7			3106	
tert-BUTYL PEROXYBUTYL FUMARATE	≤ 52	≥ 48				OP7			3105	
tert-BUTYL PEROXYCROTONATE	≤ 77	≥ 23				OP7			3105	
tert-BUTYL PEROXYDIETHYLACETATE	≤ 100					OP5	+20	+25	3113	
tert-BUTYL PEROXY-2-ETHYLHEXANOATE	> 52 - 100					OP6	+20	+25	3113	
"	> 32 - 52		≥ 48			OP8	+30	+35	3117	
"	≤ 52			≥ 48		OP8	+20	+25	3118	
"	≤ 32		≥ 68			OP8	+40	+45	3119	
tert-BUTYL PEROXY-2-ETHYLHEXANOATE + 2,2-DI-(tert-BUTYLPEROXY)BUTANE	≤ 12 + ≤ 14	≥ 14		≥ 60		OP7			3106	
"	≤ 31 + ≤ 36		≥ 33			OP7	+35	+40	3115	
tert-BUTYL PEROXY-2-ETHYLHEXYLCARBONATE	≤ 100					OP7			3105	
tert-BUTYL PEROXYISOBUTYRATE	> 52 - 77		≥ 23			OP5	+15	+20	3111	3)
"	≤ 52		≥ 48			OP7	+15	+20	3115	
tert-BUTYLPEROXY ISOPROPYLCARBONATE	≤ 77	≥ 23				OP5			3103	

ORGANIC PEROXIDE	Concentration (%)	Diluent type A (%)	Diluent type B (%) 1)	Inert solid (%)	Water	Packing Method	Control temperature (°C)	Emergency temperature (°C)	Number (Generic entry)	Subsidiary risks and remarks
1-(2-tert-BUTYLPEROXY ISOPROPYL)-3-ISOPROPENYLBENZENE	≤ 77	≥ 23				OP7			3105	
"	≤ 42			≥ 58		OP8			3108	
tert-BUTYL PEROXY-2-METHYLBENZOATE	≤ 100					OP5			3103	
tert-BUTYL PEROXYNEODECANOATE	> 77 - 100					OP7	-5	+5	3115	
"	≤ 77		≥ 23			OP7	0	+10	3115	
"	≤ 52 as a stable dispersion in water					OP8	0	+10	3119	
"	≤ 42 as a stable dispersion in water (frozen)					OP8	0	+10	3118	
"	≤ 32	≥ 68				OP8	0	+10	3119	
tert-BUTYL PEROXYNEOHEPTANOATE	≤ 77	≥ 23				OP7	0	+10	3115	
"	≤ 42 as a stable dispersion in water					OP8	0	+10	3117	
tert-BUTYL PEROXYPIVALATE	> 67 - 77	≥ 23				OP5	0	+10	3113	
"	> 27 - 67		≥ 33			OP7	0	+10	3115	
"	≤ 27		≥ 73			OP8	+30	+35	3119	
tert-BUTYLPEROXY STEARYLCARBONATE	≤ 100					OP7			3106	
tert-BUTYL PEROXY-3,5,5-TRIMETHYLHEXANOATE	> 37 - 100					OP7			3105	
"	≤ 42			≥ 58		OP7			3106	
"	≤ 37		≥ 63			OP8			3109	
3-CHLOROPEROXYBENZOIC ACID	> 57 - 86			≥ 14		OP1			3102	3)
"	≤ 57			≥ 3	≥ 40	OP7			3106	
"	≤ 77			≥ 6	≥ 17	OP7			3106	
CUMYL HYDROPEROXIDE	> 90 - 98	≤ 10				OP8			3107	13)
"	≤ 90	≥ 10				OP8			3109	13) 18)

ORGANIC PEROXIDE	Concentration (%)	Diluent type A (%)	Diluent type B (%) 1)	Inert solid (%)	Water	Packing Method	Control temperature (°C)	Emergency temperature (°C)	Number (Generic entry)	Subsidiary risks and remarks
CUMYL PEROXYNEODECANOATE	≤ 87	≥ 13				OP7	-10	0	3115	
"	≤ 77		≥ 23			OP7	-10	0	3115	
"	≤ 52 as a stable dispersion in water					OP8	-10	0	3119	
CUMYL PEROXYNEOHEPTANOATE	≤ 77	≥ 23				OP7	-10	0	3115	
CUMYL PEROXYPIVALATE	≤ 77		≥ 23			OP7	-5	+5	3115	
CYCLOHEXANONE PEROXIDE(S)	≤ 91				≥ 9	OP6			3104	13)
"	≤ 72	≥ 28				OP7			3105	5)
"	≤ 72 as a paste					OP7			3106	5) 20)
"	≤ 32			≥ 68					Exempt	29)
([3R-(3R,5aS,6S,8aS,9R,10R,12S,12aR**)]-DECAHYDRO-10-METHOXY-3,6,9-TRIMETHYL-3,12-EPOXY-12H-PYRANO[4,3-j]-1,2-BENZODIOXEPIN)	≤ 100					OP7			3106	
DIACETONE ALCOHOL PEROXIDES	≤ 57		≥ 26		≥ 8	OP7	+40	+45	3115	6)
DIACETYL PEROXIDE	≤ 27		≥ 73			OP7	+20	+25	3115	7) 13)
DI-tert-AMYL PEROXIDE	≤ 100					OP8			3107	
2,2-DI-(tert-AMYLPEROXY)BUTANE	≤ 57	≥ 43				OP7			3105	
1,1-DI-(tert-AMYLPEROXY)CYCLOHEXANE	≤ 82	≥ 18				OP6			3103	
DIBENZOYL PEROXIDE	> 52 - 100			≤ 48		OP2			3102	3)
"	> 77 - 94				≥ 6	OP4			3102	3)
"	≤ 77				≥ 23	OP6			3104	
"	≤ 62			≥ 28	≥ 10	OP7			3106	
"	> 52 – 62 as a paste					OP7			3106	20)
"	> 35 - 52			≥ 48		OP7			3106	
"	> 36 - 42	≥ 18			≤ 40	OP8			3107	
"	≤ 56.5 as a paste				≥ 15	OP8			3108	
"	≤ 52 as a paste					OP8			3108	20)
"	≤ 42 as a stable dispersion in water					OP8			3109	
"	≤ 35			≥ 65					Exempt	29)

ORGANIC PEROXIDE	Concentration (%)	Diluent type A (%)	Diluent type B (%) 1)	Inert solid (%)	Water	Packing Method	Control temperature (°C)	Emergency temperature (°C)	Number (Generic entry)	Subsidiary risks and remarks
DI-(4-tert-BUTYLCYCLOHEXYL) PEROXYDICARBONATE	≤ 100					OP6	+30	+35	3114	
"	≤ 42 as a stable dispersion in water					OP8	+30	+35	3119	
DI-tert-BUTYL PEROXIDE	> 52 - 100					OP8			3107	
"	≤ 52		≥ 48			OP8			3109	25)
DI-tert-BUTYL PEROXYAZELATE	≤ 52	≥ 48				OP7			3105	
2,2-DI-(tert-BUTYLPEROXY)BUTANE	≤ 52	≥ 48				OP6			3103	
1,6-Di-(tert-BUTYLPEROXYCARBONYLOXY) HEXANE	≤ 72	≥ 28				OP5			3103	
1,1-DI-(tert-BUTYLPEROXY) CYCLOHEXANE	> 80 - 100					OP5			3101	3)
"	≤ 72		≥ 28			OP5			3103	30)
"	> 52 - 80	≥ 20				OP5			3103	
"	> 42 - 52	≥ 48				OP7			3105	
"	≤ 42	≥ 13		≥ 45		OP7			3106	
"	≤ 42	≥ 58				OP8			3109	
"	≤ 27	≥ 25				OP8			3107	21)
"	≤ 13	≥ 13	≥ 74			OP8			3109	
1,1-DI-(tert-BUTYLPEROXY) CYCLOHEXANE + tert-BUTYL PEROXY-2-ETHYLHEXANOATE	≤ 43 + ≤ 16	≥ 41				OP 7			3105	
DI-n-BUTYL PEROXYDICARBONATE	> 27 - 52		≥ 48			OP7	-15	-5	3115	
"	≤ 27		≥ 73			OP8	-10	0	3117	
"	≤ 42 as a stable dispersion in water (frozen)					OP8	-15	-5	3118	
DI-sec-BUTYL PEROXYDICARBONATE	> 52 - 100					OP4	-20	-10	3113	
"	≤ 52		≥ 48			OP7	-15	-5	3115	

ORGANIC PEROXIDE	Concentration (%)	Diluent type A (%)	Diluent type B (%) 1)	Inert solid (%)	Water	Packing Method	Control temperature (°C)	Emergency temperature (°C)	Number (Generic entry)	Subsidiary risks and remarks
DI-(tert-BUTYLPEROXYISOPROPYL)BENZENE(S)	>42 - 100			≤57		OP7			3106	
"	≤42			≥58					Exempt	29)
DI-(tert-BUTYLPEROXY) PHTHALATE	>42 - 52	≥48				OP7			3105	
"	≤52 as a paste					OP7			3106	20)
"	≤42	≥58				OP8			3107	
2,2-DI-(tert-BUTYLPEROXY)PROPANE	≤52	≥48				OP7			3105	
"	≤42	≥13		≥45		OP7			3106	
1,1-DI-(tert-BUTYLPEROXY)-3,3,5-TRIMETHYLCYCLOHEXANE	>90 - 100					OP5			3101	3)
"	≤90		≥10			OP5			3103	30)
"	>57 - 90	≥10				OP5			3103	
"	≤77		≥23			OP5			3103	
"	≤57			≥43		OP8			3110	
"	≤57	≥43				OP8			3107	
"	≤32	≥26	≥42			OP8			3107	
DICETYL PEROXYDICARBONATE	≤100					OP8	+30	+35	3120	
"	≤42 as a stable dispersion in water					OP8	+30	+35	3119	
DI-4-CHLOROBENZOYL PEROXIDE	≤77				≥23	OP5			3102	3)
"	≤52 as a paste					OP7			3106	20)
"	≤32			≥68					Exempt	29)
DICUMYL PEROXIDE	>52 - 100					OP8			3110	12)
"	≤52			≥48					Exempt	29)
DICYCLOHEXYL PEROXYDICARBONATE	>91 - 100					OP3	+10	+15	3112	3)
"	≤91				≥9	OP5	+10	+15	3114	
"	≤42 as a stable dispersion in water					OP8	+15	+20	3119	

ORGANIC PEROXIDE	Concentration (%)	Diluent type A (%)	Diluent type B (%) 1)	Inert solid (%)	Water	Packing Method	Control temperature (°C)	Emergency temperature (°C)	Number (Generic entry)	Subsidiary risks and remarks
DIDECANOYL PEROXIDE	≤ 100					OP6	+30	+35	3114	
2,2-DI-(4,4-DI (tert-BUTYLPEROXY) CYCLOHEXYL) PROPANE	≤ 42			≥ 58		OP7			3106	
"	≤ 22		≥ 78			OP8			3107	
DI-2,4-DICHLOROBENZOYL PEROXIDE	≤ 77				≥ 23	OP5			3102	3)
"	≤ 52 as a paste					OP8	+ 20	+ 25	3118	
"	≤ 52 as a paste with silicon oil					OP7			3106	
DI-(2-ETHOXYETHYL) PEROXYDICARBONATE	≤ 52		≥ 48			OP7	-10	0	3115	
DI-(2-ETHYLHEXYL) PEROXYDICARBONATE	> 77 – 100					OP5	-20	-10	3113	
"	≤ 77		≥ 23			OP7	-15	-5	3115	
"	≤ 62 as a stable dispersion in water					OP8	-15	-5	3119	
"	≤ 52 as a stable dispersion in water (frozen)					OP8	-15	-5	3120	
2,2-DIHYDROPEROXYPROPANE	≤ 27			≥ 73		OP5			3102	3)
DI-(1-HYDROXYCYCLOHEXYL) PEROXIDE	≤ 100					OP7			3106	
DIISOBUTYRYL PEROXIDE	> 32 – 52		≥ 48			OP5	-20	-10	3111	3)
"	≤ 32		≥ 68			OP7	-20	-10	3115	
DIISOPROPYLBENZENE DIHYDROPEROXIDE	≤ 82	≥ 5			≥ 5	OP7			3106	24)
DIISOPROPYL PEROXYDICARBONATE	> 52-100					OP2	-15	-5	3112	3)
"	≤ 52		≥ 48			OP7	-20	-10	3115	
"	≤ 32	≥ 68				OP7	-15	-5	3115	
DILAUROYL PEROXIDE	≤ 100					OP7			3106	
"	≤ 42 as a stable dispersion in water					OP8			3109	
DI-(3-METHOXYBUTYL) PEROXYDICARBONATE	≤ 52		≥ 48			OP7	-5	+5	3115	
DI-(2-METHYLBENZOYL) PEROXIDE	≤ 87				≥ 13	OP5	+30	+35	3112	3)

ORGANIC PEROXIDE	Concentration (%)	Diluent type A (%)	Diluent type B (%) 1)	Inert solid (%)	Water	Packing Method	Control temperature (°C)	Emergency temperature (°C)	Number (Generic entry)	Subsidiary risks and remarks
DI-(3-METHYLBENZOYL) PEROXIDE + BENZOYL (3-METHYLBENZOYL) PEROXIDE + DIBENZOYL PEROXIDE	≤ 20 + ≤ 18 + ≤ 4		≥ 58			OP7	+35	+40	3115	
DI-(4-METHYLBENZOYL) PEROXIDE	≤ 52 as a paste with silicon oil					OP7			3106	
2,5-DIMETHYL-2,5-DI-(BENZOYLPEROXY)HEXANE	> 82-100					OP5			3102	3)
"	≤ 82			≥ 18		OP7			3106	
"	≤ 82				≥ 18	OP5			3104	
2,5-DIMETHYL-2,5-DI-(tert-BUTYLPEROXY)HEXANE	> 90 - 100					OP5			3103	
"	> 52 - 90	≥ 10				OP7			3105	
"	≤ 77			≥ 23		OP8			3108	
"	≤ 52	≥ 48				OP8			3109	
"	≤ 47 as a paste					OP8			3108	
2,5-DIMETHYL-2,5-DI-(tert-BUTYLPEROXY)HEXYNE-3	> 86-100					OP5			3101	3)
"	>52-86	≥ 14				OP5			3103	26)
"	≤ 52			≥ 48		OP7			3106	
2,5-DIMETHYL-2,5-DI-(2-ETHYLHEXANOYLPEROXY)HEXANE	≤ 100					OP5	+20	+25	3113	
2,5-DIMETHYL-2,5-DIHYDROPEROXYHEXANE	≤ 82				≥ 18	OP6			3104	
2,5-DIMETHYL-2,5-DI-(3,5,5-TRIMETHYLHEXANOYLPEROXY)HEXANE	≤ 77	≥ 23				OP7			3105	
1,1-DIMETHYL-3-HYDROXYBUTYL PEROXYNEOHEPTANOATE	≤ 52	≥ 48				OP8	0	+10	3117	
DIMYRISTYL PEROXYDICARBONATE	≤ 100					OP7	+20	+25	3116	
"	≤ 42 as a stable dispersion in water					OP8	+20	+25	3119	
DI-(2-NEODECANOYLPEROXYISOPROPYL) BENZENE	≤ 52					OP7	-10	0	3115	

ORGANIC PEROXIDE	Concentration (%)	Diluent type A (%)	Diluent type B (%) 1)	Inert solid (%)	Water	Packing Method	Control temperature (°C)	Emergency temperature (°C)	Number (Generic entry)	Subsidiary risks and remarks
DI-n-NONANOYL PEROXIDE	≤100					OP7	0	+10	3116	
DI-n-OCTANOYL PEROXIDE	≤100					OP5	+10	+15	3114	
DI-(2-PHENOXYETHYL) PEROXYDICARBONATE	>85-100					OP5			3102	3)
"	≤85				≥15	OP7			3106	
DIPROPIONYL PEROXIDE	≤27		≥73			OP8	+15	+20	3117	
DI-n-PROPYL PEROXYDICARBONATE	≤100					OP3	-25	-15	3113	
"	≤77		≥23			OP5	-20	-10	3113	
DISUCCINIC ACID PEROXIDE	>72-100					OP4			3102	3) 17)
"	≤72				≥28	OP7	+10	+15	3116	
DI-(3,5,5-TRIMETHYLHEXANOYL) PEROXIDE	>52-82	≥18				OP7	0	+10	3115	
"	≤52 as a stable dispersion in water					OP8	+10	+15	3119	
"	>38-52	≥48				OP8	+10	+15	3119	
"	≤38	≥62				OP8	+20	+25	3119	
ETHYL 3,3-DI-(tert-AMYLPEROXY)BUTYRATE	≤67	≥33				OP7			3105	
ETHYL 3,3-DI-(tert-BUTYLPEROXY)BUTYRATE	>77 - 100					OP5			3103	
"	≤77	≥23				OP7			3105	
"	≤52			≥48		OP7			3106	
1-(2-ETHYLHEXANOYLPEROXY)-1,3-DIMETHYLBUTYL PEROXYPIVALATE	≤52	≥45	≥10			OP7	-20	-10	3115	
tert-HEXYL PEROXYNEODECANOATE	≤71	≥29				OP7	0	+10	3115	
tert-HEXYL PEROXYPIVALATE	≤72		≥28			OP7	+10	+15	3115	
3-HYDROXY-1,1-DIMETHYLBUTYL PEROXYNEODECANOATE	≤77	≥23				OP 7	- 5	+ 5	3115	
"	≤52	≥48				OP 8	- 5	+ 5	3117	
"	≤52 as a stable dispersion in water					OP 8	- 5	+ 5	3119	

ORGANIC PEROXIDE	Concentration (%)	Diluent type A (%)	Diluent type B (%) 1)	Inert solid (%)	Water	Packing Method	Control temperature (°C)	Emergency temperature (°C)	Number (Generic entry)	Subsidiary risks and remarks
ISOPROPYL sec-BUTYL PEROXYDICARBONATE +DI-sec-BUTYL PEROXYDICARBONATE +DI-ISOPROPYL PEROXYDICARBONATE	≤32 + ≤15 – 18 ≤ 12 – 15	≥ 38				OP7	-20	-10	3115	
"	≤ 52 + ≤ 28 + ≤ 22					OP5	-20	-10	3111	3)
ISOPROPYLCUMYL HYDROPEROXIDE	≤ 72	≥ 28				OP8			3109	13)
p-MENTHYL HYDROPEROXIDE	> 72 - 100					OP7			3105	13)
"	≤ 72	≥ 28				OP8			3109	27)
METHYLCYCLOHEXANONE PEROXIDE(S)	≤ 67		≥ 33			OP7	+35	+40	3115	
METHYL ETHYL KETONE PEROXIDE(S)	see remark 8)	≥ 48				OP5			3101	3) 8) 13)
"	see remark 9)	≥ 55				OP7			3105	9)
"	see remark 10)	≥ 60				OP8			3107	10)
METHYL ISOBUTYL KETONE PEROXIDE(S)	≤ 62	≥ 19				OP7			3105	22)
METHYL ISOPROPYL KETONE PEROXIDE(S)	see remark 31)	≥ 70				OP8			3109	31)
ORGANIC PEROXIDE, LIQUID, SAMPLE						OP2			3103	11)
ORGANIC PEROXIDE, LIQUID, SAMPLE, TEMPERATURE CONTROLLED						OP2			3113	11)
ORGANIC PEROXIDE, SOLID, SAMPLE						OP2			3104	11)
ORGANIC PEROXIDE, SOLID, SAMPLE, TEMPERATURE CONTROLLED						OP2			3114	11)
3,3,5,7,7-PENTAMETHYL-1,2,4-TRIOXEPANE	≤ 100					OP8			3107	
PEROXYACETIC ACID, TYPE D, stabilized	≤ 43					OP7			3105	13) 14) 19)
PEROXYACETIC ACID, TYPE E, stabilized	≤ 43					OP8			3107	13) 15) 19)
PEROXYACETIC ACID, TYPE F, stabilized	≤ 43					OP8			3109	13) 16) 19)
PEROXYLAURIC ACID	≤ 100					OP8	+35	+40	3118	
PINANYL HYDROPEROXIDE	> 56 – 100					OP7			3105	13)
"	≤ 56	≥ 44				OP8			3109	
POLYETHER POLY-tert-BUTYLPEROXY-CARBONATE	≤ 52		≥ 48			OP8			3107	

ORGANIC PEROXIDE	Concentration (%)	Diluent type A (%)	Diluent type B (%) 1)	Inert solid (%)	Water	Packing Method	Control temperature (°C)	Emergency temperature (°C)	Number (Generic entry)	Subsidiary risks and remarks
1,1,3,3-TETRAMETHYLBUTYL HYDROPEROXIDE	≤100					OP7			3105	
1,1,3,3-TETRAMETHYLBUTYL PEROXY-2-ETHYLHEXANOATE	≤100					OP7	+15	+20	3115	
1,1,3,3- TETRAMETHYLBUTYL PEROXYNEODECANOATE	≤72		≥28			OP7	-5	+5	3115	
"	≤52 as a stable dispersion in water					OP8	-5	+5	3119	
1,1,3,3-TETRAMETHYLBUTYL PEROXYPIVALATE	≤77	≥23				OP7	0	+10	3115	
3,6,9-TRIETHYL-3,6,9-TRIMETHYL-1,4,7 TRIPEROXONANE	≤17	≥18		≥65		OP8			3110	
3,6,9-TRIETHYL-3,6,9-TRIMETHYL -1,4,7 TRIPEROXONANE	≤42	≥58				OP7			3105	28)

Remarks (refer to the last column of the Table in 2.2.52.4):

1)   *Diluent type B may always be replaced by diluent type A. The boiling point of diluent type B shall be at least 60°C higher than the SADT of the organic peroxide.*

2)   *Available oxygen ≤ 4.7%.*

3)   *"EXPLOSIVE" subsidiary risk label required (Model No.1, see 5.2.2.2.2).*

4)   *Diluent may be replaced by di-tert-butyl peroxide.*

5)   *Available oxygen ≤ 9%.*

6)   *With ≤ 9% hydrogen peroxide; available oxygen ≤ 10%.*

7)   *Only non-metallic packagings allowed.*

8)   *Available oxygen > 10% and ≤ 10.7%, with or without water.*

9)   *Available oxygen ≤ 10%, with or without water.*

10)   *Available oxygen ≤ 8.2%, with or without water.*

11)   *See 2.2.52.1.9.*

12)   *Up to 2000 kg per receptacle assigned to ORGANIC PEROXIDE TYPE F on the basis of large scale trials.*

13)   *"CORROSIVE" subsidiary risk label required (Model No.8, see 5.2.2.2.2).*

14)   *Peroxyacetic acid formulations which fulfil the criteria of the Manual of Tests and Criteria, paragraph 20.4.3 (d).*

15)   *Peroxyacetic acid formulations which fulfil the criteria of the Manual of Tests and Criteria, paragraph 20.4.3 (e).*

16)   *Peroxyacetic acid formulations which fulfil the criteria of the Manual of Tests and Criteria, paragraph 20.4.3 (f).*

17)   *Addition of water to this organic peroxide will decrease its thermal stability.*

18)   *No "CORROSIVE" subsidiary risk label (Model No.8, see 5.2.2.2.2) required for concentrations below 80%.*

19)   *Mixtures with hydrogen peroxide, water and acid(s).*

20)   *With diluent type A, with or without water.*

21)   *With ≥ 25% diluent type A by mass, and in addition ethylbenzene.*

22)   *With ≥ 19%, diluent type A by mass, and in addition methyl isobutyl ketone.*

23)   *With < 6% di-tert-butyl peroxide.*

24)   *With ≤ 8% 1-isopropylhydroperoxy-4-isopropylhydroxybenzene.*

25)   *Diluent type B with boiling point > 110 °C.*

26)   *With < 0.5% hydroperoxides content.*

27)   *For concentrations more than 56%, "CORROSIVE" subsidiary risk label required (Model No.8, see 5.2.2.2.2).*

28)   *Available active oxygen ≤ 7.6% in diluent type A having a 95% boil-off point in the range of 200 - 260 °C.*

29)   *Not subject to the requirements of ADR for Class 5.2.*

30)   *Diluent type B with boiling point > 130 °C.*

31)   *Active oxygen ≤ 6.7%.*

2.2.61   **Class 6.1 Toxic substances**

2.2.61.1   *Criteria*

2.2.61.1.1   The heading of Class 6.1 covers substances of which it is known by experience or regarding which it is presumed from experiments on animals that in relatively small quantities they are able by a single action or by action of short duration to cause damage to human health, or death, by inhalation, by cutaneous absorption or by ingestion.

NOTE: Genetically modified microorganisms and organisms shall be assigned to this Class if they meet the conditions for this Class.

2.2.61.1.2   Substances of Class 6.1 are subdivided as follows:

T   Toxic substances without subsidiary risk:

    T1   Organic, liquid;
    T2   Organic, solid;
    T3   Organometallic substances;
    T4   Inorganic, liquid;
    T5   Inorganic, solid;
    T6   Liquid, used as pesticides;
    T7   Solid, used as pesticides;
    T8   Samples;
    T9   Other toxic substances;

TF   Toxic substances, flammable:

    TF1   Liquid;
    TF2   Liquid, used as pesticides;
    TF3   Solid;

TS   Toxic substances, self-heating, solid;

TW   Toxic substances, which, in contact with water, emit flammable gases:

    TW1   Liquid;
    TW2   Solid;

TO   Toxic substances, oxidizing:

    TO1   Liquid;
    TO2   Solid;

TC   Toxic substances, corrosive:

    TC1   Organic, liquid;
    TC2   Organic, solid;
    TC3   Inorganic, liquid;
    TC4   Inorganic, solid;

TFC   Toxic substances, flammable, corrosive;

TFW   Toxic substances, flammable, which, in contact with water, emit flammable gases.

Definitions

2.2.61.1.3   For the purposes of ADR:

LD_{50} (median lethal dose) for acute oral toxicity is the statistically derived single dose of a substance that can be expected to cause death within 14 days in 50 per cent of young adult albino rats when administered by the oral route. The LD_{50} value is expressed in terms of mass of test substance per mass of test animal (mg/kg);

LD_{50} for acute dermal toxicity is that dose of the substance which, administered by continuous contact for 24 hours with the bare skin of albino rabbits, is most likely to cause death within 14 days in one half of the animals tested. The number of animals tested shall be sufficient to give a statistically significant result and be in conformity with good pharmacological practice. The result is expressed in milligrams per kg body mass;

LC_{50} for acute toxicity on inhalation is that concentration of vapour, mist or dust which, administered by continuous inhalation to both male and female young adult albino rats for one hour, is most likely to cause death within 14 days in one half of the animals tested. A solid substance shall be tested if at least 10% (by mass) of its total mass is likely to be dust in a respirable range, e.g. the aerodynamic diameter of that particle-fraction is 10 μm or less. A liquid substance shall be tested if a mist is likely to be generated in a leakage of the transport containment. Both for solid and liquid substances more than 90% (by mass) of a specimen prepared for inhalation toxicity shall be in the respirable range as defined above. The result is expressed in milligrams per litre of air for dusts and mists or in millilitres per cubic metre of air (parts per million) for vapours.

Classification and assignment of packing groups

2.2.61.1.4   Substances of Class 6.1 shall be classified in three packing groups according to the degree of danger they present for carriage, as follows:

Packing group I: highly toxic substances

Packing group II: toxic substances

Packing group III: slightly toxic substances.

2.2.61.1.5   Substances, mixtures, solutions and articles classified in Class 6.1 are listed in Table A of Chapter 3.2. The assignment of substances, mixtures and solutions not mentioned by name in Table A of Chapter 3.2 to the relevant entry of sub-section 2.2.61.3 and to the relevant packing group in accordance with the provisions of Chapter 2.1, shall be made according to the following criteria in 2.2.61.1.6 to 2.2.61.1.11.

2.2.61.1.6   To assess the degree of toxicity, account shall be taken of human experience of instances of accidental poisoning, as well as special properties possessed by any individual substances: liquid state, high volatility, any special likelihood of cutaneous absorption, and special biological effects.

2.2.61.1.7   In the absence of observations on humans, the degree of toxicity shall be assessed using the available data from animal experiments in accordance with the table below:

	Packing group	Oral toxicity LD_{50} (mg/kg)	Dermal toxicity LD_{50} (mg/kg)	Inhalation toxicity by dusts and mists LC_{50} (mg/l)
Highly toxic	I	≤ 5	≤ 50	≤ 0.2
Toxic	II	> 5 and ≤ 50	> 50 and ≤ 200	> 0.2 and ≤ 2
Slightly toxic	III [a]	> 50 and ≤ 300	> 200 and ≤ 1 000	> 2 and ≤ 4

[a]   *Tear gas substances shall be included in packing group II even if data concerning their toxicity correspond to packing group III criteria.*

2.2.61.1.7.1   Where a substance exhibits different degrees of toxicity for two or more kinds of exposure, it shall be classified under the highest such degree of toxicity.

2.2.61.1.7.2 Substances meeting the criteria of Class 8 and with an inhalation toxicity of dusts and mists (LC_{50}) leading to packing group I shall only be accepted for an allocation to Class 6.1 if the toxicity through oral ingestion or dermal contact is at least in the range of packing groups I or II. Otherwise an assignment to Class 8 shall be made if appropriate (see 2.2.8.1.5).

2.2.61.1.7.3 The criteria for inhalation toxicity of dusts and mists are based on LC_{50} data relating to 1-hour exposure, and where such information is available it shall be used. However, where only LC_{50} data relating to 4-hour exposure are available, such figures can be multiplied by four and the product substituted in the above criteria, i.e. LC_{50} value multiplied by four (4 hour) is considered the equivalent of LC_{50} (1 hour).

Inhalation toxicity of vapours

2.2.61.1.8 Liquids giving off toxic vapours shall be classified into the following groups where "V" is the saturated vapour concentration (in ml/m^3 of air) (volatility) at 20 °C and standard atmospheric pressure:

	Packing group	
Highly toxic	I	Where $V \geq 10\ LC_{50}$ and $LC_{50} \leq 1\ 000\ ml/m^3$
Toxic	II	Where $V \geq LC_{50}$ and $LC_{50} \leq 3\ 000\ ml/m^3$ and the criteria for packing group I are not met
Slightly toxic	III[a]	Where $V \geq 1/5\ LC_{50}$ and $LC_{50} \leq 5\ 000\ ml/m^3$ and the criteria for packing groups I and II are not met

[a] *Tear gas substances shall be included in packing group II even if data concerning their toxicity correspond to packing group III criteria.*

These criteria for inhalation toxicity of vapours are based on LC_{50} data relating to 1-hour exposure, and where such information is available, it shall be used.

However, where only LC_{50} data relating to 4-hour exposure to the vapours are available, such figures can be multiplied by two and the product substituted in the above criteria, i.e. LC_{50} (4 hour) × 2 is considered the equivalent of LC_{50} (1 hour).

In this figure, the criteria are expressed in graphical form, as an aid to easy classification. However, due to approximations inherent in the use of graphs, substances falling on or near group borderlines shall be checked using numerical criteria.

GROUP BORDERLINES INHALATION TOXICITY OF VAPOURS

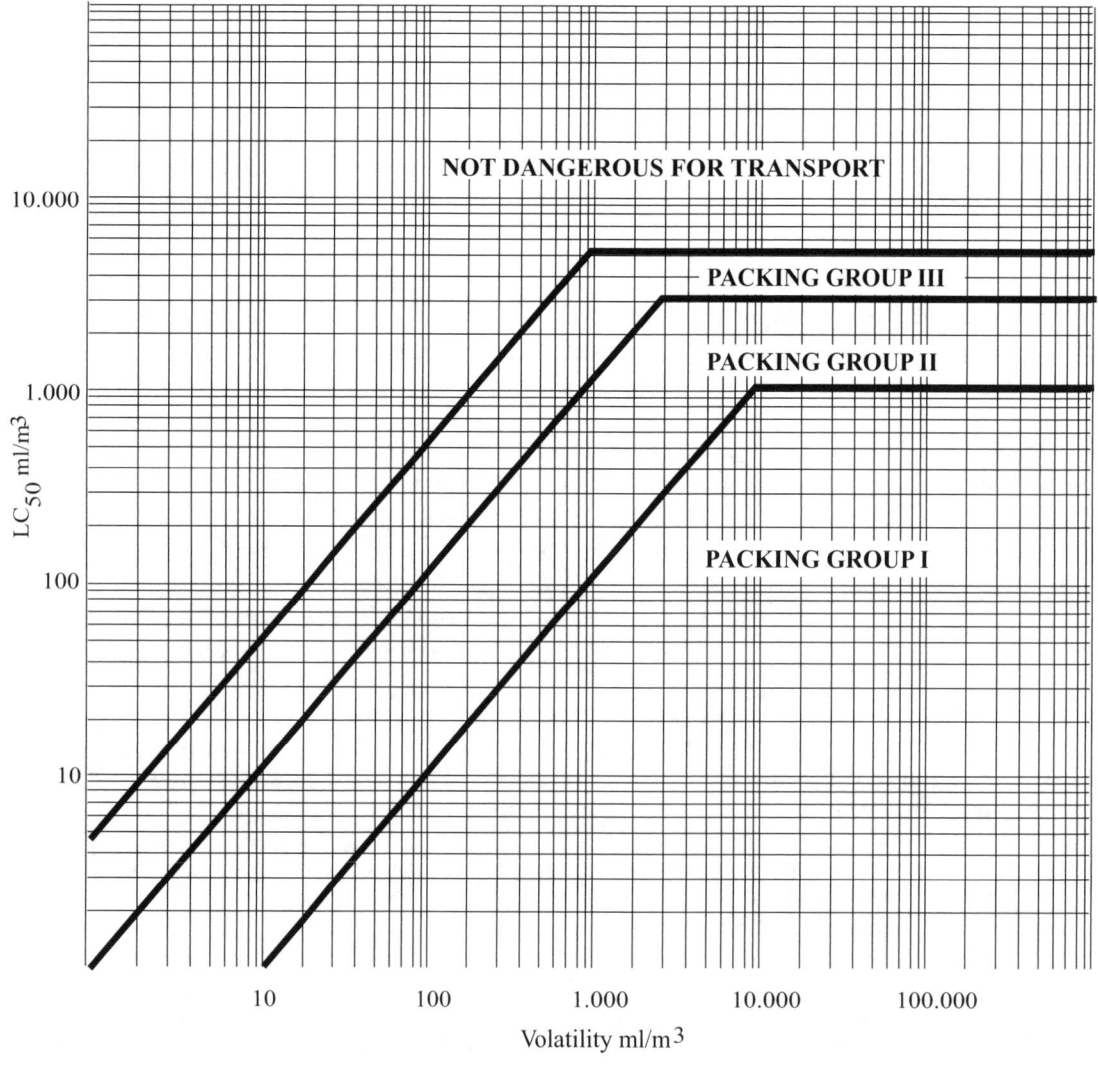

Mixtures of liquids

2.2.61.1.9   Mixtures of liquids which are toxic on inhalation shall be assigned to packing groups according to the following criteria:

2.2.61.1.9.1   If LC_{50} is known for each of the toxic substances constituting the mixture, the packing group may be determined as follows:

(a)   calculation of the LC_{50} of the mixture:

$$LC_{50} \text{ (mixture)} = \frac{1}{\sum_{i=1}^{1} \frac{f_i}{LC_{50i}}}$$

where f_i =   molar fraction of constituent i of the mixture;

LC_{50i} =   average lethal concentration of constituent i in ml/m^3.

(b)   calculation of volatility of each mixture constituent:

$$V_i = P_i \times \frac{10^6}{101.3} \, (ml/m^3)$$

where P_i =   partial pressure of constituent i in kPa at 20 °C and at standard atmospheric pressure.

- 184 -

(c)   calculation of the ratio of volatility to LC_{50}:

$$R = \sum_{i=1}^{n} \frac{V_i}{LC_{50i}}$$

(d)   the values calculated for LC_{50} (mixture) and R are then used to determine the packing group of the mixture:

Packing group I $R \geq 10$ and LC_{50} (mixture) $\leq 1\,000$ ml/m³;

Packing group II $R \geq 1$ and LC_{50} (mixture) $\leq 3\,000$ ml/m³, if the mixture does not meet the criteria for packing group I;

Packing group III $R \geq 1/5$ and LC_{50} (mixture) $\leq 5\,000$ ml/m³, if the mixture does not meet the criteria of packing groups I or II.

2.2.61.1.9.2   In the absence of LC_{50} data on the toxic constituent substances, the mixture may be assigned to a group based on the following simplified threshold toxicity tests. When these threshold tests are used, the most restrictive group shall be determined and used for carrying the mixture.

2.2.61.1.9.3   A mixture is assigned to packing group I only if it meets both of the following criteria:

(a)   A sample of the liquid mixture is vaporized and diluted with air to create a test atmosphere of 1 000 ml/m³ vaporized mixture in air. Ten albino rats (5 male and 5 female) are exposed to the test atmosphere for 1 hour and observed for 14 days. If five or more of the animals die within the 14-day observation period, the mixture is presumed to have an LC_{50} equal to or less than 1 000 ml/m³;

(b)   A sample of vapour in equilibrium with the liquid mixture is diluted with 9 equal volumes of air to form a test atmosphere. Ten albino rats (5 male and 5 female) are exposed to the test atmosphere for 1 hour and observed for 14 days. If five or more of the animals die within the 14-day observation period, the mixture is presumed to have a volatility equal to or greater than 10 times the mixture LC_{50}.

2.2.61.1.9.4   A mixture is assigned to packing group II only if it meets both of the following criteria, and does not meet the criteria for packing group I:

(a)   A sample of the liquid mixture is vaporized and diluted with air to create a test atmosphere of 3 000 ml/m³ vaporized mixture in air. Ten albino rats (5 male and 5 female) are exposed to the test atmosphere for 1 hour and observed for 14 days. If five or more of the animals die within the 14-day observation period, the mixture is presumed to have an LC_{50} equal to or less than 3 000 ml/m³;

(b)   A sample of the vapour in equilibrium with the liquid mixture is used to form a test atmosphere. Ten albino rats (5 male and 5 female) are exposed to the test atmosphere for 1 hour and observed for 14 days. If five or more of the animals die within the 14-day observation period, the mixture is presumed to have a volatility equal to or greater than the mixture LC_{50}.

2.2.61.1.9.5   A mixture is assigned to packing group III only if it meets both of the following criteria, and does not meet the criteria for packing groups I or II:

(a)   A sample of the liquid mixture is vaporized and diluted with air to create a test atmosphere of 5 000 ml/m³ vaporized mixture in air. Ten albino rats (5 male and 5 female) are exposed to the test atmosphere for 1 hour and observed for 14 days. If five or more of the animals die within the 14-day observation period, the mixture is presumed to have an LC_{50} equal to or less than 5 000 ml/m³;

(b)   The vapour concentration (volatility) of the liquid mixture is measured and if the vapour concentration is equal to or greater than 1 000 ml/m³, the mixture is presumed to have a volatility equal to or greater than 1/5 the mixture LC_{50}.

Methods for determining oral and dermal toxicity of mixtures

2.2.61.1.10   When classifying and assigning the appropriate packing group to mixtures in Class 6.1 in accordance with the oral and dermal toxicity criteria (see 2.2.61.1.3), it is necessary to determine the acute LD_{50} of the mixture.

2.2.61.1.10.1 If a mixture contains only one active substance, and the LD50 of that constituent is known, in the absence of reliable acute oral and dermal toxicity data on the actual mixture to be carried, the oral or dermal LD50 may be obtained by the following method:

$$LD_{50} \text{ value of preparation} = \frac{LD_{50} \text{ value of active substance} \times 100}{\text{percentage of active substance by mass}}$$

2.2.61.1.10.2 If a mixture contains more than one active constituent, there are three possible approaches that may be used to determine the oral or dermal LD_{50} of the mixture. The preferred method is to obtain reliable acute oral and dermal toxicity data on the actual mixture to be carried. If reliable, accurate data are not available, then either of the following methods may be performed:

(a)   Classify the formulation according to the most hazardous constituent of the mixture as if that constituent were present in the same concentration as the total concentration of all active constituents; or

(b)   Apply the formula:

$$\frac{C_A}{T_A} + \frac{C_B}{T_B} + ... + \frac{C_Z}{T_Z} = \frac{100}{T_M}$$

where:

C   =   the percentage concentration of constituent A, B, ..., Z in the mixture;

T   =   the oral LD_{50} values of constituent A, B, ... Z;

T_M =   the oral LD_{50} value of the mixture.

NOTE: This formula can also be used for dermal toxicities provided that this information is available on the same species for all constituents. The use of this formula does not take into account any potentiation or protective phenomena.

Classification of pesticides

2.2.61.1.11   All active pesticide substances and their preparations for which the LC_{50} and/or LD_{50} values are known and which are classified in Class 6.1 shall be classified under appropriate packing groups in accordance with the criteria given in 2.2.61.1.6 to 2.2.61.1.9. Substances and preparations which are characterized by subsidiary risks shall be classified according to the precedence of hazard Table in 2.1.3.10 with the assignment of appropriate packing groups.

2.2.61.1.11.1 If the oral or dermal LD_{50} value for a pesticide preparation is not known, but the LD_{50} value of its active substance(s) is known, the LD_{50} value for the preparation may be obtained by applying the procedures in 2.2.61.1.10.

NOTE: LD_{50} toxicity data for a number of common pesticides may be obtained from the most current edition of the document "The WHO Recommended Classification of Pesticides by Hazard and Guidelines to Classification" available from the International Programme on Chemical Safety, World Health Organisation (WHO), 1211 Geneva 27, Switzerland. While that document may be used as a source of LD_{50} data for pesticides, its classification system shall not be used for purposes of transport classification of, or assignment of packing groups to, pesticides, which shall be in accordance with the requirements of ADR.

2.2.61.1.11.2 The proper shipping name used in the carriage of the pesticide shall be selected on the basis of the active ingredient, of the physical state of the pesticide and any subsidiary risks it may exhibit (see 3.1.2).

2.2.61.1.12   If substances of Class 6.1, as a result of admixtures, come into categories of risk different from those to which the substances mentioned by name in Table A of Chapter 3.2 belong, these mixtures or solutions shall be assigned to the entries to which they belong on the basis of their actual degree of danger.

NOTE: For the classification of solutions and mixtures (such as preparations and wastes), see also 2.1.3.

2.2.61.1.13   On the basis of the criteria of 2.2.61.1.6 to 2.2.61.1.11, it may also be determined whether the nature of a solution or mixture mentioned by name or containing a substance mentioned by name is such that the solution or mixture is not subject to the requirements for this Class.

2.2.61.1.14   Substances, solutions and mixtures, with the exception of substances and preparations used as pesticides, which are not classified as acute toxic category 1, 2 or 3 according to Regulation (EC) No 1272/2008[3], may be considered as substances not belonging to class 6.1.

2.2.61.2   *Substances not accepted for carriage*

2.2.61.2.1   Chemically unstable substances of Class 6.1 shall not be accepted for carriage unless the necessary precautions have been taken to prevent the possibility of a dangerous decomposition or polymerization under normal conditions of carriage. For the precautions necessary to prevent polymerization, see special provision 386 of Chapter 3.3. To this end particular care shall be taken to ensure that receptacles and tanks do not contain any substances liable to promote these reactions.

2.2.61.2.2   The following substances and mixtures shall not be accepted for carriage:

-    Hydrogen cyanide, anhydrous or in solution, which do not meet the descriptions of UN Nos. 1051, 1613, 1614 and 3294;

-    metal carbonyls, having a flash-point below 23 °C, other than UN Nos. 1259 NICKEL CARBONYL and 1994 IRON PENTACARBONYL;

-    2,3,7,8-TETRACHLORODIBENZO-P-DIOXINE (TCDD) in concentrations considered highly toxic in accordance with the criteria in 2.2.61.1.7;

-    UN No. 2249 DICHLORODIMETHYL ETHER, SYMMETRICAL;

-    Preparations of phosphides without additives inhibiting the emission of toxic flammable gases.

---

[3]   *Regulation (EC) No 1272/2008 of the European Parliament and of the Council of 16 December 2008 on classification, labelling and packaging of substances and mixtures, amending and repealing Directive 67/548/EEC and 1999/45/EC; and amending Regulation (EC) No 1907/2006, published in the Official Journal of the European Union, L 353, 31 December 2008, p 1-1355.*

2.2.61.3 *List of collective entries*

Toxic substances without subsidiary risk(s)

Organic

liquid [a] **T1**

1583	CHLOROPICRIN MIXTURE, N.O.S.
1602	DYE, LIQUID, TOXIC, N.O.S., or
1602	DYE INTERMEDIATE, LIQUID, TOXIC, N.O.S.
1693	TEAR GAS SUBSTANCE, LIQUID, N.O.S.
1851	MEDICINE, LIQUID, TOXIC, N.O.S.
2206	ISOCYANATES, TOXIC, N.O.S. or
2206	ISOCYANATE SOLUTION, TOXIC, N.O.S.
3140	ALKALOIDS, LIQUID, N.O.S. or
3140	ALKALOID SALTS, LIQUID, N.O.S.
3142	DISINFECTANT, LIQUID, TOXIC, N.O.S.
3144	NICOTINE COMPOUND, LIQUID, N.O.S. or
3144	NICOTINE PREPARATION, LIQUID, N.O.S.
3172	TOXINS, EXTRACTED FROM LIVING SOURCES, LIQUID, N.O.S.
3276	NITRILES, LIQUID, TOXIC, N.O.S.
3278	ORGANOPHOSPHORUS COMPOUND, LIQUID, TOXIC, N.O.S.
3381	TOXIC BY INHALATION LIQUID, N.O.S. with an LC_{50} lower than or equal to 200 ml/m^3 and saturated vapour concentration greater than or equal to 500 LC_{50}
3382	TOXIC BY INHALATION LIQUID, N.O.S. with an LC_{50} lower than or equal to 1000 ml/m^3 and saturated vapour concentration greater than or equal to 10 LC_{50}
2810	TOXIC LIQUID, ORGANIC, N.O.S.

solid [a, b] **T2**

1544	ALKALOIDS, SOLID, N.O.S. or
1544	ALKALOID SALTS, SOLID, N.O.S.
1601	DISINFECTANT, SOLID, TOXIC, N.O.S.
1655	NICOTINE COMPOUND, SOLID, N.O.S., or
1655	NICOTINE PREPARATION, SOLID, N.O.S.
3448	TEAR GAS SUBSTANCE, SOLID, N.O.S.
3143	DYE, SOLID, TOXIC, N.O.S. or
3143	DYE INTERMEDIATE, SOLID, TOXIC, N.O.S.
3462	TOXINS, EXTRACTED FROM LIVING SOURCES, SOLID, N.O.S.
3249	MEDICINE, SOLID, TOXIC, N.O.S.
3464	ORGANOPHOSPHORUS COMPOUND, SOLID, TOXIC, N.O.S.
3439	NITRILES, SOLID, TOXIC, N.O.S.
2811	TOXIC SOLID, ORGANIC, N.O.S.

Organometallic [c, d] **T3**

2026	PHENYLMERCURIC COMPOUND, N.O.S.
2788	ORGANOTIN COMPOUND, LIQUID, N.O.S.
3146	ORGANOTIN COMPOUND, SOLID, N.O.S.
3280	ORGANOARSENIC COMPOUND, LIQUID, N.O.S.
3465	ORGANOARSENIC COMPOUND, SOLID, N.O.S.
3281	METAL CARBONYLS, LIQUID, N.O.S.
3466	METAL CARBONYLS, SOLID, N.O.S.
3282	ORGANOMETALLIC COMPOUND, LIQUID, TOXIC, N.O.S.
3467	ORGANOMETALLIC COMPOUND, SOLID, TOXIC, N.O.S.

(cont'd on next page)

[a] *Substances and preparations containing alkaloids or nicotine used as pesticides shall be classified under UN No. 2588 PESTICIDES, SOLID, TOXIC, N.O.S., UN No. 2902 PESTICIDES, LIQUID, TOXIC, N.O.S. or UN No. 2903 PESTICIDES, LIQUID, TOXIC, FLAMMABLE, N.O.S.*

[b] *Active substances and triturations or mixtures of substances intended for laboratories and experiments and for the manufacture of pharmaceutical products with other substances shall be classified according to their toxicity (see 2.2.61.1.7 to 2.2.61.1.11).*

[c] *Self-heating substances, slightly toxic and spontaneously combustible organometallic compounds, are substances of Class 4.2.*

[d] *Water-reactive substances, slightly toxic, and water-reactive organometallic compounds, are substances of Class 4.3.*

Toxic substances <u>without</u> subsidiary risk(s) *(cont'd)*

Inorganic	liquid ^{e}	T4	1556 ARSENIC COMPOUND, LIQUID, N.O.S., inorganic including: Arsenates, n.o.s., Arsenites, n.o.s.; and Arsenic sulphides, n.o.s.
			1935 CYANIDE SOLUTION, N.O.S.
			2024 MERCURY COMPOUND, LIQUID, N.O.S.
			3141 ANTIMONY COMPOUND, INORGANIC, LIQUID, N.O.S.
			3440 SELENIUM COMPOUND, LIQUID, N.O.S.
			3381 TOXIC BY INHALATION LIQUID, N.O.S. with an LC_{50} lower than or equal to 200 ml/m^3 and saturated vapour concentration greater than or equal to 500 LC_{50}
			3382 TOXIC BY INHALATION LIQUID, N.O.S. with an LC_{50} lower than or equal to 1000 ml/m^3 and saturated vapour concentration greater than or equal to 10 LC_{50}
			3287 TOXIC LIQUID, INORGANIC, N.O.S.
	solids ^{f,g}	T5	1549 ANTIMONY COMPOUND, INORGANIC, SOLID, N.O.S
			1557 ARSENIC COMPOUND, SOLID, N.O.S., including: Arsenates, n.o.s.; Arsenites, n.o.s.; and Arsenic sulphides, n.o.s.
			1564 BARIUM COMPOUND, N.O.S.
			1566 BERYLLIUM COMPOUND, N.O.S.
			1588 CYANIDES, INORGANIC, SOLID, N.O.S.
			1707 THALLIUM COMPOUND, N.O.S.
			2025 MERCURY COMPOUND, SOLID, N.O.S.
			2291 LEAD COMPOUND, SOLUBLE, N.O.S.
			2570 CADMIUM COMPOUND
			2630 SELENATES or
			2630 SELENITES
			2856 FLUOROSILICATES, N.O.S.
			3283 SELENIUM COMPOUND, SOLID, N.O.S.
			3284 TELLURIUM COMPOUND, N.O.S.
			3285 VANADIUM COMPOUND, N.O.S.
			3288 TOXIC SOLID, INORGANIC, N.O.S.
Pesticides	liquid ^{h}	T6	2992 CARBAMATE PESTICIDE, LIQUID, TOXIC
			2994 ARSENICAL PESTICIDE, LIQUID, TOXIC
			2996 ORGANOCHLORINE PESTICIDE, LIQUID, TOXIC
			2998 TRIAZINE PESTICIDE, LIQUID, TOXIC
			3006 THIOCARBAMATE PESTICIDE, LIQUID, TOXIC
			3010 COPPER BASED PESTICIDE, LIQUID, TOXIC
			3012 MERCURY BASED PESTICIDE, LIQUID, TOXIC
			3014 SUBSTITUTED NITROPHENOL PESTICIDE, LIQUID, TOXIC
			3016 BIPYRIDILIUM PESTICIDE, LIQUID, TOXIC
			3018 ORGANOPHOSPHORUS PESTICIDE, LIQUID, TOXIC
			3020 ORGANOTIN PESTICIDE, LIQUID, TOXIC
			3026 COUMARIN DERIVATIVE PESTICIDE, LIQUID, TOXIC
			3348 PHENOXYACETIC ACID DERIVATIVE PESTICIDE, LIQUID, TOXIC
			3352 PYRETHROID PESTICIDE, LIQUID, TOXIC
			2902 PESTICIDE, LIQUID, TOXIC, N.O.S.

(cont'd on next page)

---

^{e} *Mercury fulminate, wetted with not less than 20% water, or mixture of alcohol and water by mass is a substance of Class 1, UN No. 0135.*

^{f} *Ferricyanides, ferrocyanides, alkaline thiocyanates and ammonium thiocyanates are not subject to the provisions of ADR.*

^{g} *Lead salts and lead pigments which, when mixed in a ratio of 1:1,000 with 0.07M hydrochloric acid and stirred for one hour at a temperature of 23 °C ± 2 °C, exhibit a solubility of 5% or less, are not subject to the provisions of ADR.*

^{h} *Articles impregnated with this pesticide, such as fibreboard plates, paper strips, cotton-wool balls, sheets of plastics material, in hermetically closed wrappings, are not subject to the provisions of ADR.*

Toxic substances <u>without</u> subsidiary risk(s) *(cont'd)*

Pesticides *(cont'd)*				
	Solid[h]	T7	2757	CARBAMATE PESTICIDE, SOLID, TOXIC
			2759	ARSENICAL PESTICIDE, SOLID, TOXIC
			2761	ORGANOCHLORINE PESTICIDE, SOLID, TOXIC
			2763	TRIAZINE PESTICIDE, SOLID, TOXIC
			2771	THIOCARBAMATE PESTICIDE, SOLID, TOXIC
			2775	COPPER BASED PESTICIDE, SOLID, TOXIC
			2777	MERCURY BASED PESTICIDE, SOLID, TOXIC
			2779	SUBSTITUTED NITROPHENOL PESTICIDE, SOLID, TOXIC
			2781	BIPYRIDILIUM PESTICIDE, SOLID, TOXIC
			2783	ORGANOPHOSPHORUS PESTICIDE, SOLID, TOXIC
			2786	ORGANOTIN PESTICIDE, SOLID, TOXIC
			3027	COUMARIN DERIVATIVE PESTICIDE, SOLID, TOXIC
			3048	ALUMINIUM PHOSPHIDE PESTICIDE
			3345	PHENOXYACETIC ACID DERIVATIVE PESTICIDE, SOLID, TOXIC
			3349	PYRETHROID PESTICIDE, SOLID, TOXIC
			2588	PESTICIDE, SOLID, TOXIC, N.O.S.

Samples	T8	3315	CHEMICAL SAMPLE, TOXIC

Other toxic substances [i]	T9	3243	SOLIDS CONTAINING TOXIC LIQUID, N.O.S.

Toxic substances <u>with</u> subsidiary risk(s)

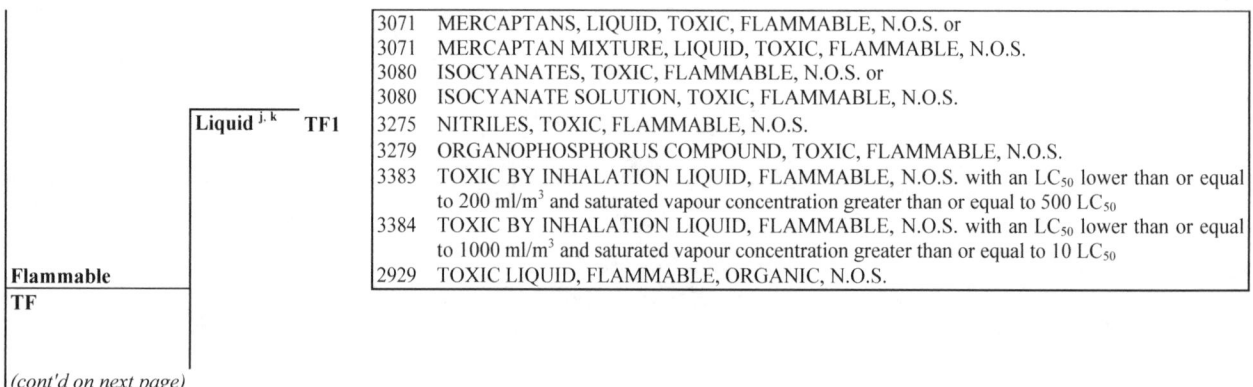

Flammable TF	Liquid [j, k]	TF1	3071	MERCAPTANS, LIQUID, TOXIC, FLAMMABLE, N.O.S. or
			3071	MERCAPTAN MIXTURE, LIQUID, TOXIC, FLAMMABLE, N.O.S.
			3080	ISOCYANATES, TOXIC, FLAMMABLE, N.O.S. or
			3080	ISOCYANATE SOLUTION, TOXIC, FLAMMABLE, N.O.S.
			3275	NITRILES, TOXIC, FLAMMABLE, N.O.S.
			3279	ORGANOPHOSPHORUS COMPOUND, TOXIC, FLAMMABLE, N.O.S.
			3383	TOXIC BY INHALATION LIQUID, FLAMMABLE, N.O.S. with an LC_{50} lower than or equal to 200 ml/m³ and saturated vapour concentration greater than or equal to 500 LC_{50}
			3384	TOXIC BY INHALATION LIQUID, FLAMMABLE, N.O.S. with an LC_{50} lower than or equal to 1000 ml/m³ and saturated vapour concentration greater than or equal to 10 LC_{50}
			2929	TOXIC LIQUID, FLAMMABLE, ORGANIC, N.O.S.

(cont'd on next page)

---

[h]   *Articles impregnated with this pesticide, such as fibreboard plates, paper strips, cotton-wool balls, sheets of plastics material, in hermetically closed wrappings, are not subject to the provisions of ADR.*

[i]   *Mixtures of solids which are not subject to the provisions of ADR and of toxic liquids may be carried under UN No. 3243 without first applying the classification criteria of Class 6.1, provided there is no free liquid visible at the time the substance is loaded or at the time the packaging, container or transport unit is closed. Each packaging shall correspond to a design type that has passed a leakproofness test at the packing group II level. This entry shall not be used for solids containing a packing group I liquid.*

[j]   *Highly toxic and toxic flammable liquids having a flash-point below 23 °C are substances of Class 3 except those which are highly toxic by inhalation, as defined in 2.2.61.1.4 to 2.2.61.1.9. Liquids which are highly toxic by inhalation are indicated as "toxic by inhalation" in their proper shipping name in Column (2) or by special provision 354 in Column (6) of Table A of Chapter 3.2.*

[k]   *Flammable liquids, slightly toxic, with the exception of substances and preparations used as pesticides, having a flash-point between 23 °C and 60 °C inclusive, are substances of Class 3.*

Toxic substances <u>with</u> subsidiary risk(s) (cont'd)

Flammable TF *(cont'd)*	pesticides, liquid (flash-point not less than 23 °C)	TF2	2991	CARBAMATE PESTICIDE, LIQUID, TOXIC, FLAMMABLE
			2993	ARSENICAL PESTICIDE, LIQUID, TOXIC, FLAMMABLE
			2995	ORGANOCHLORINE PESTICIDE, LIQUID, TOXIC, FLAMMABLE
			2997	TRIAZINE PESTICIDE, LIQUID, TOXIC, FLAMMABLE
			3005	THIOCARBAMATE PESTICIDE, LIQUID, TOXIC, FLAMMABLE
			3009	COPPER BASED PESTICIDE, LIQUID, TOXIC, FLAMMABLE
			3011	MERCURY BASED PESTICIDE, LIQUID, TOXIC, FLAMMABLE
			3013	SUBSTITUTED NITROPHENOL PESTICIDE, LIQUID, TOXIC, FLAMMABLE
			3015	BIPYRIDILIUM PESTICIDE, LIQUID, TOXIC, FLAMMABLE
			3017	ORGANOPHOSPHORUS PESTICIDE, LIQUID, TOXIC, FLAMMABLE
			3019	ORGANOTIN PESTICIDE, LIQUID, TOXIC, FLAMMABLE
			3025	COUMARIN DERIVATIVE PESTICIDE, LIQUID, TOXIC, FLAMMABLE
			3347	PHENOXYACETIC ACID DERIVATIVE PESTICIDE, LIQUID, TOXIC, FLAMMABLE
			3351	PYRETHROID PESTICIDE, LIQUID, TOXIC, FLAMMABLE
			2903	PESTICIDE, LIQUID, TOXIC, FLAMMABLE, N.O.S.
	solid	TF3	1700	TEAR GAS CANDLES
			2930	TOXIC SOLID, FLAMMABLE, ORGANIC, N.O.S.
Solid, self-heating [c] **TS**			3124	TOXIC SOLID, SELF-HEATING, N.O.S.
Water-reactive [d] **TW**	liquid	TW1	3385	TOXIC BY INHALATION LIQUID, WATER-REACTIVE, N.O.S. with an LC_{50} lower than or equal to 200 ml/m^3 and saturated vapour concentration greater than or equal to 500 LC_{50}
			3386	TOXIC BY INHALATION LIQUID, WATER-REACTIVE, N.O.S. with an LC_{50} lower than or equal to 1000 ml/m^3 and saturated vapour concentration greater than or equal to 10 LC_{50}
			3123	TOXIC LIQUID, WATER-REACTIVE, N.O.S.
	solid [n]	TW2	3125	TOXIC SOLID, WATER-REACTIVE, N.O.S.
Oxidizing [l] **TO**	liquid	TO1	3387	TOXIC BY INHALATION LIQUID, OXIDIZING, N.O.S. with an LC_{50} lower than or equal to 200 ml/m^3 and saturated vapour concentration greater than or equal to 500 LC_{50}
			3388	TOXIC BY INHALATION LIQUID, OXIDIZING, N.O.S. with an LC_{50} lower than or equal to 1000 ml/m^3 and saturated vapour concentration greater than or equal to 10 LC_{50}
			3122	TOXIC LIQUID, OXIDIZING, N.O.S.
	solid	TO2	3086	TOXIC SOLID, OXIDIZING, N.O.S.
Corro-sive [m] **TC**	organic — liquid	TC1	3277	CHLOROFORMATES, TOXIC, CORROSIVE, N.O.S.
			3361	CHLOROSILANES, TOXIC, CORROSIVE, N.O.S.
			3389	TOXIC BY INHALATION LIQUID, CORROSIVE, N.O.S. with an LC_{50} lower than or equal to 200 ml/m^3 and saturated vapour concentration greater than or equal to 500 LC_{50}
			3390	TOXIC BY INHALATION LIQUID, CORROSIVE, N.O.S. with an LC_{50} lower than or equal to 1000 ml/m^3 and saturated vapour concentration greater than or equal to 10 LC_{50}
			2927	TOXIC LIQUID, CORROSIVE, ORGANIC, N.O.S.
	organic — solid	TC2	2928	TOXIC SOLID, CORROSIVE, ORGANIC, N.O.S.

(cont'd on next page)

---

[c] *Self-heating substances, slightly toxic and spontaneously combustible organometallic compounds, are substances of Class 4.2.*

[d] *Water-reactive substances, slightly toxic, and water-reactive organometallic compounds, are substances of Class 4.3.*

[l] *Oxidizing substances, slightly toxic, are substances of Class 5.1.*

[m] *Substances slightly toxic and slightly corrosive, are substances of Class 8.*

[n] *Metal phosphides assigned to UN Nos. 1360, 1397, 1432, 1714, 2011 and 2013 are substances of Class 4.3.*

2.2.61.3 *List of collective entries (cont'd)*

Toxic substances <u>with</u> subsidiary risk(s) (cont'd)

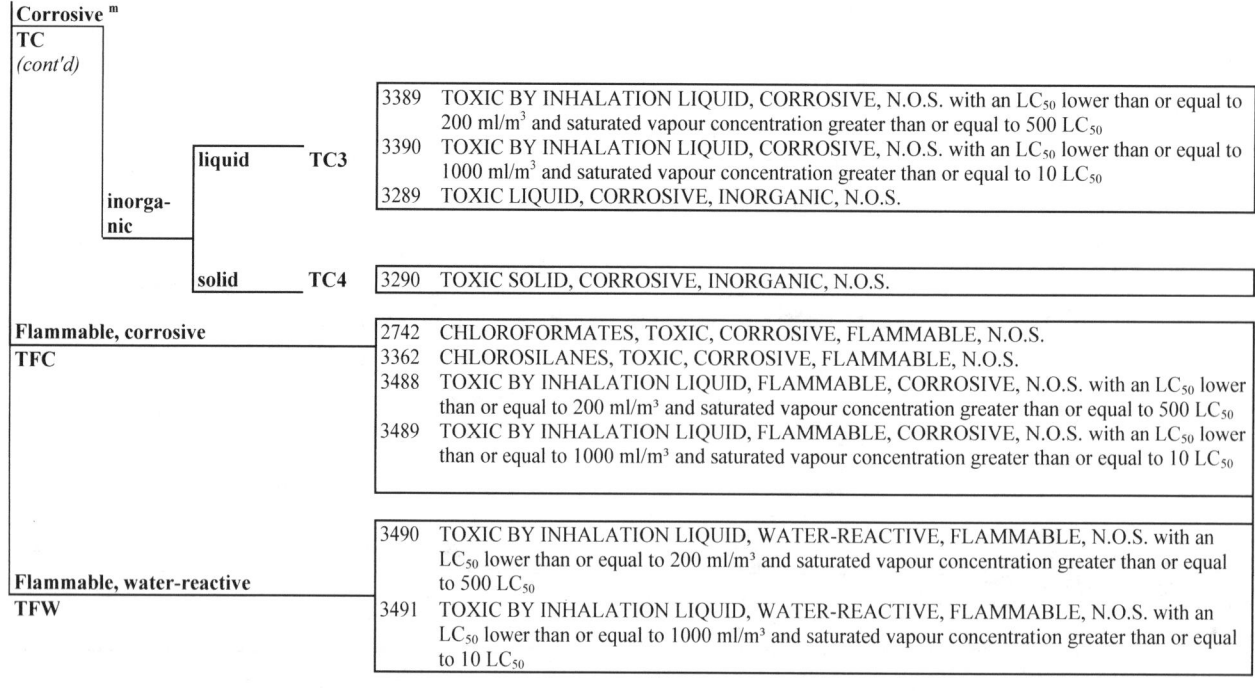

Corrosive [m]				
TC *(cont'd)*				
		liquid	TC3	3389 TOXIC BY INHALATION LIQUID, CORROSIVE, N.O.S. with an LC_{50} lower than or equal to 200 ml/m³ and saturated vapour concentration greater than or equal to 500 LC_{50}
	inorga-nic			3390 TOXIC BY INHALATION LIQUID, CORROSIVE, N.O.S. with an LC_{50} lower than or equal to 1000 ml/m³ and saturated vapour concentration greater than or equal to 10 LC_{50}
				3289 TOXIC LIQUID, CORROSIVE, INORGANIC, N.O.S.
		solid	TC4	3290 TOXIC SOLID, CORROSIVE, INORGANIC, N.O.S.

Flammable, corrosive	2742 CHLOROFORMATES, TOXIC, CORROSIVE, FLAMMABLE, N.O.S.
TFC	3362 CHLOROSILANES, TOXIC, CORROSIVE, FLAMMABLE, N.O.S.
	3488 TOXIC BY INHALATION LIQUID, FLAMMABLE, CORROSIVE, N.O.S. with an LC_{50} lower than or equal to 200 ml/m³ and saturated vapour concentration greater than or equal to 500 LC_{50}
	3489 TOXIC BY INHALATION LIQUID, FLAMMABLE, CORROSIVE, N.O.S. with an LC_{50} lower than or equal to 1000 ml/m³ and saturated vapour concentration greater than or equal to 10 LC_{50}

	3490 TOXIC BY INHALATION LIQUID, WATER-REACTIVE, FLAMMABLE, N.O.S. with an LC_{50} lower than or equal to 200 ml/m³ and saturated vapour concentration greater than or equal to 500 LC_{50}
Flammable, water-reactive	
TFW	3491 TOXIC BY INHALATION LIQUID, WATER-REACTIVE, FLAMMABLE, N.O.S. with an LC_{50} lower than or equal to 1000 ml/m³ and saturated vapour concentration greater than or equal to 10 LC_{50}

---

[m] *Substances slightly toxic and slightly corrosive, are substances of Class 8.*

2.2.62 **Class 6.2** **Infectious substances**

2.2.62.1 *Criteria*

2.2.62.1.1 The heading of Class 6.2 covers infectious substances. For the purposes of ADR, infectious substances are substances which are known or are reasonably expected to contain pathogens. Pathogens are defined as microorganisms (including bacteria, viruses, rickettsiae, parasites, fungi) and other agents such as prions, which can cause disease in humans or animals.

NOTE 1: Genetically modified microorganisms and organisms, biological products, diagnostic specimens and intentionally infected live animals shall be assigned to this Class if they meet the conditions for this Class.

The carriage of unintentionally or naturally infected live animals is subject only to the relevant rules and regulations of the respective countries of origin, transit and destination.

NOTE 2: Toxins from plant, animal or bacterial sources which do not contain any infectious substances or organisms or which are not contained in them are substances of Class 6.1, UN Nos. 3172 or 3462.

2.2.62.1.2 Substances of Class 6.2 are subdivided as follows:

I1   Infectious substances affecting humans;

I2   Infectious substances affecting animals only;

I3   Clinical waste;

I4   Biological substances.

Definitions

2.2.62.1.3 For the purposes of ADR,

"Biological products" are those products derived from living organisms which are manufactured and distributed in accordance with the requirements of appropriate national authorities, which may have special licensing requirements, and are used either for prevention, treatment, or diagnosis of disease in humans or animals, or for development, experimental or investigational purposes related thereto. They include, but are not limited to, finished or unfinished products such as vaccines;

"Cultures" are the result of a process by which pathogens are intentionally propagated. This definition does not include human or animal patient specimens as defined in this paragraph;

"Medical or clinical wastes" are wastes derived from the medical treatment of animals or humans or from bio-research;

"Patient specimens" are human or animal materials, collected directly from humans or animals, including, but not limited to, excreta, secreta, blood and its components, tissue and tissue fluid swabs, and body parts being carried for purposes such as research, diagnosis, investigational activities, disease treatment and prevention.

Classification

2.2.62.1.4 Infectious substances shall be classified in Class 6.2 and assigned to UN Nos. 2814, 2900, 3291 or 3373, as appropriate.

Infectious substances are divided into the following categories:

2.2.62.1.4.1   <u>Category A</u>: An infectious substance which is carried in a form that, when exposure to it occurs, is capable of causing permanent disability, life-threatening or fatal disease in otherwise healthy humans or animals. Indicative examples of substances that meet these criteria are given in the table in this paragraph.

NOTE: An exposure occurs when an infectious substance is released outside of the protective packaging, resulting in physical contact with humans or animals.

(a)   Infectious substances meeting these criteria which cause disease in humans or both in humans and animals shall be assigned to UN No. 2814. Infectious substances which cause disease only in animals shall be assigned to UN No. 2900;

(b)   Assignment to UN No. 2814 or UN No. 2900 shall be based on the known medical history and symptoms of the source human or animal, endemic local conditions, or professional judgement concerning individual circumstances of the source human or animal.

NOTE 1: The proper shipping name for UN No. 2814 is "INFECTIOUS SUBSTANCE, AFFECTING HUMANS". The proper shipping name for UN No. 2900 is "INFECTIOUS SUBSTANCE, AFFECTING ANIMALS only".

NOTE 2: The following table is not exhaustive. Infectious substances, including new or emerging pathogens, which do not appear in the table but which meet the same criteria shall be assigned to Category A. In addition, if there is doubt as to whether or not a substance meets the criteria it shall be included in Category A.

NOTE 3: In the following table, the microorganisms written in italics are bacteria, mycoplasmas, rickettsia or fungi.

INDICATIVE EXAMPLES OF INFECTIOUS SUBSTANCES INCLUDED IN CATEGORY A IN ANY FORM UNLESS OTHERWISE INDICATED (2.2.62.1.4.1)	
UN Number and name	**Microorganism**
UN No. 2814 Infectious substances affecting humans	*Bacillus anthracis (cultures only)* *Brucella abortus (cultures only)* *Brucella melitensis (cultures only)* *Brucella suis (cultures only)* *Burkholderia mallei - Pseudomonas mallei* – Glanders *(cultures only)* *Burkholderia pseudomallei – Pseudomonas pseudomallei (cultures only)* *Chlamydia psittaci* - avian strains (cultures only) *Clostridium botulinum (cultures only)* *Coccidioides immitis (cultures only)* *Coxiella burnetii (cultures only)* Crimean-Congo haemorrhagic fever virus Dengue virus (cultures only) Eastern equine encephalitis virus (cultures only) *Escherichia coli*, verotoxigenic (cultures only) [a] Ebola virus Flexal virus *Francisella tularensis (cultures only)* Guanarito virus Hantaan virus Hantavirus causing haemorrhagic fever with renal syndrome Hendra virus Hepatitis B virus (cultures only) Herpes B virus (cultures only) Human immunodeficiency virus (cultures only) Highly pathogenic avian influenza virus (cultures only) Japanese Encephalitis virus (cultures only) Junin virus Kyasanur Forest disease virus Lassa virus Machupo virus Marburg virus Monkeypox virus *Mycobacterium tuberculosis (cultures only)* [a] Nipah virus Omsk haemorrhagic fever virus Poliovirus (cultures only) Rabies virus (cultures only) *Rickettsia prowazekii (cultures only)* *Rickettsia rickettsii* (cultures only) Rift Valley fever virus (cultures only) Russian spring-summer encephalitis virus (cultures only) Sabia virus *Shigella dysenteriae type 1 (cultures only)* [a] Tick-borne encephalitis virus (cultures only) Variola virus Venezuelan equine encephalitis virus (cultures only) West Nile virus (cultures only) Yellow fever virus (cultures only) *Yersinia pestis* (cultures only)

---

[a]   *Nevertheless, when the cultures are intended for diagnostic or clinical purposes, they may be classified as infectious substances of Category B.*

INDICATIVE EXAMPLES OF INFECTIOUS SUBSTANCES INCLUDED IN CATEGORY A IN ANY FORM UNLESS OTHERWISE INDICATED (2.2.62.1.4.1)	
UN Number and name	**Microorganism**
UN No. 2900 Infectious substances affecting animals only	African swine fever virus (cultures only) Avian paramyxovirus Type 1 - Velogenic Newcastle disease virus (cultures only) Classical swine fever virus (cultures only) Foot and mouth disease virus (cultures only) Lumpy skin disease virus (cultures only) *Mycoplasma mycoides* - Contagious bovine pleuropneumonia (cultures only) Peste des petits ruminants virus (cultures only) Rinderpest virus (cultures only) Sheep-pox virus (cultures only) Goatpox virus (cultures only) Swine vesicular disease virus (cultures only) Vesicular stomatitis virus (cultures only)

2.2.62.1.4.2   <u>Category B</u>: An infectious substance which does not meet the criteria for inclusion in Category A. Infectious substances in Category B shall be assigned to UN No. 3373.

NOTE: The proper shipping name of UN No. 3373 is "BIOLOGICAL SUBSTANCE, CATEGORY B ".

2.2.62.1.5   *Exemptions*

2.2.62.1.5.1   Substances which do not contain infectious substances or substances which are unlikely to cause disease in humans or animals are not subject to the provisions of ADR unless they meet the criteria for inclusion in another class.

2.2.62.1.5.2   Substances containing microorganisms which are non-pathogenic to humans or animals are not subject to ADR unless they meet the criteria for inclusion in another class.

2.2.62.1.5.3   Substances in a form that any present pathogens have been neutralized or inactivated such that they no longer pose a health risk are not subject to ADR unless they meet the criteria for inclusion in another class.

NOTE: Medical equipment which has been drained of free liquid is deemed to meet the requirements of this paragraph and is not subject to the provisions of ADR.

2.2.62.1.5.4   Substances where the concentration of pathogens is at a level naturally encountered (including foodstuff and water samples) and which are not considered to pose a significant risk of infection are not subject to ADR unless they meet the criteria for inclusion in another class.

2.2.62.1.5.5   Dried blood spots, collected by applying a drop of blood onto absorbent material, are not subject to ADR.

2.2.62.1.5.6   Faecal occult blood screening samples are not subject to ADR.

2.2.62.1.5.7   Blood or blood components which have been collected for the purposes of transfusion or for the preparation of blood products to be used for transfusion or transplantation and any tissues or organs intended for use in transplantation as well as samples drawn in connection with such purposes are not subject to ADR.

2.2.62.1.5.8   Human or animal specimens for which there is minimal likelihood that pathogens are present are not subject to ADR if the specimen is carried in a packaging which will prevent any leakage and which is marked with the words "Exempt human specimen" or "Exempt animal specimen", as appropriate.

The packaging is deemed to comply with the above requirements if it meets the following conditions:

(a)   The packaging consists of three components:

(i)   a leak-proof primary receptacle(s);

(ii)   a leak-proof secondary packaging; and

(iii)   an outer packaging of adequate strength for its capacity, mass and intended use, and with at least one surface having minimum dimensions of 100 mm × 100 mm;

(b)   For liquids, absorbent material in sufficient quantity to absorb the entire contents is placed between the primary receptacle(s) and the secondary packaging so that, during carriage, any release or leak of a liquid substance will not reach the outer packaging and will not compromise the integrity of the cushioning material;

(c)   When multiple fragile primary receptacles are placed in a single secondary packaging, they are either individually wrapped or separated to prevent contact between them.

NOTE 1: An element of professional judgment is required to determine if a substance is exempt under this paragraph. That judgment should be based on the known medical history, symptoms and individual circumstances of the source, human or animal, and endemic local conditions. Examples of specimens which may be carried under this paragraph include the blood or urine tests to monitor cholesterol levels, blood glucose levels, hormone levels, or prostate specific antibodies (PSA); those required to monitor organ function such as heart, liver or kidney function for humans or animals with non-infectious diseases, or for therapeutic drug monitoring; those conducted for insurance or employment purposes and are intended to determine the presence of drugs or alcohol; pregnancy test; biopsies to detect cancer; and antibody detection in humans or animals in the absence of any concern for infection (e.g. evaluation of vaccine induced immunity, diagnosis of autoimmune disease, etc.).

NOTE 2: For air transport, packagings for specimens exempted under this paragraph shall meet the conditions in (a) to (c).

2.2.62.1.5.9   Except for:

(a)   Medical waste (UN No. 3291);

(b)   Medical devices or equipment contaminated with or containing infectious substances in Category A (UN No. 2814 or UN No. 2900); and

(c)   Medical devices or equipment contaminated with or containing other dangerous goods that meet the definition of another class,

medical devices or equipment potentially contaminated with or containing infectious substances which are being carried for disinfection, cleaning, sterilization, repair, or equipment evaluation are not subject to provisions of ADR other than those of this paragraph if packed in packagings designed and constructed in such a way that, under normal conditions of carriage, they cannot break, be punctured or leak their contents. Packagings shall be designed to meet the construction requirements listed in 6.1.4 or 6.6.4.

These packagings shall meet the general packing requirements of 4.1.1.1 and 4.1.1.2 and be capable of retaining the medical devices and equipment when dropped from a height of 1.2 m.

The packagings shall be marked "USED MEDICAL DEVICE" or "USED MEDICAL EQUIPMENT". When using overpacks, these shall be marked in the same way, except when the inscription remains visible.

2.2.62.1.6 to 2.2.62.1.8   *(Reserved)*

2.2.62.1.9   *Biological products*

For the purposes of ADR, biological products are divided into the following groups:

(a)   those which are manufactured and packaged in accordance with the requirements of appropriate national authorities and carried for the purposes of final packaging or distribution, and use for personal health care by medical professionals or individuals. Substances in this group are not subject to the provisions of ADR;

(b)   those which do not fall under paragraph (a) and are known or reasonably believed to contain infectious substances and which meet the criteria for inclusion in Category A or Category B. Substances in this group shall be assigned to UN Nos. 2814, 2900 or 3373, as appropriate.

NOTE: *Some licensed biological products may present a biohazard only in certain parts of the world. In that case, competent authorities may require these biological products to be in compliance with local requirements for infectious substances or may impose other restrictions.*

2.2.62.1.10   *Genetically modified microorganisms and organisms*

Genetically modified microorganisms not meeting the definition of infectious substance shall be classified according to section 2.2.9.

2.2.62.1.11   *Medical or clinical wastes*

2.2.62.1.11.1   Medical or clinical wastes containing Category A infectious substances shall be assigned to UN No. 2814 or UN No. 2900 as appropriate. Medical or clinical wastes containing infectious substances in Category B shall be assigned to UN No. 3291.

NOTE: *Medical or clinical wastes assigned to number 18 01 03 (Wastes from human or animal health care and/or related research – wastes from natal care, diagnosis, treatment or prevention of disease in humans – wastes whose collection and disposal is subject to special requirements in order to prevent infection) or 18 02 02 (Wastes from human or animal health care and/or related research – wastes from research, diagnosis, treatment or prevention of disease involving animals – wastes whose collection and disposal is subject to special requirements in order to prevent infection) according to the list of wastes annexed to the Commission Decision 2000/532/EC[4] as amended, shall be classified according to the provisions set out in this paragraph, based on the medical or veterinary diagnosis concerning the patient or the animal.*

2.2.62.1.11.2   Medical or clinical wastes which are reasonably believed to have a low probability of containing infectious substances shall be assigned to UN No. 3291. For the assignment, international, regional or national waste catalogues may be taken into account.

NOTE 1: *The proper shipping name for UN No. 3291 is "CLINICAL WASTE, UNSPECIFIED, N.O.S." or "(BIO) MEDICAL WASTE, N.O.S". or "REGULATED MEDICAL WASTE, N.O.S.".*

NOTE 2: *Notwithstanding the classification criteria set out above, medical or clinical wastes assigned to number 18 01 04 (Wastes from human or animal health care and/or related research – wastes from natal care, diagnosis, treatment or prevention of disease in humans – wastes whose collection and disposal is not subject to special requirements in order to prevent infection) or 18 02 03 (Wastes from human or animal health care and/or related research – wastes from research, diagnosis, treatment or prevention of disease involving animals – wastes whose collection and disposal is not subject to special requirements in order to prevent infection) according to the list of wastes annexed to the Commission Decision 2000/532/EC[5] as amended, are not subject to the provisions of ADR.*

2.2.62.1.11.3   Decontaminated medical or clinical wastes which previously contained infectious substances are not subject to the provisions of ADR unless they meet the criteria for inclusion in another class.

2.2.62.1.11.4   Medical or clinical wastes assigned to UN No. 3291 are assigned to packing group II.

2.2.62.1.12   *Infected animals*

2.2.62.1.12.1   Unless an infectious substance cannot be consigned by any other means, live animals shall not be used to consign such a substance. A live animal which has been intentionally infected and is known or suspected to contain an infectious substance shall only be carried under terms and conditions approved by the competent authority.

NOTE: *The approval of the competent authorities shall be issued on the basis of the relevant rules for the carriage of live animals, taking into consideration dangerous goods aspects. The authorities*

---

[4]   *Commission Decision 2000/532/EC of 3 May 2000 replacing Decision 94/3/EC establishing a list of wastes pursuant to Article 1(a) of Council Directive 75/442/EEC on waste (replaced by the Directive 2006/12/EC of the European Parliament and of the Council (Official Journal of the European Union No. L 114 of 27 April 2006, page 9)) and Council Decision 94/904/EC establishing a list of hazardous waste pursuant to Article 1(4) of Council Directive 91/689/EEC on hazardous waste (Official Journal of the European Communities No. L 226 of 6 September 2000, page 3).*

that are competent to lay down these conditions and rules for approval shall be regulated at national level.

If there is no approval by a competent authority of a Contracting Party to ADR, the competent authority of a Contracting Party to ADR may recognize an approval issued by the competent authority of a country that is not a Contracting Party to ADR.

Rules for the carriage of livestock are, for example, contained in Council Regulation (EC) No 1/2005 of 22 December 2004 on the protection of animals during transport (Official Journal of the European Union No L 3 of 5 January 2005) as amended.

2.2.62.1.12.2   Animal material affected by pathogens of Category A or by pathogens which would be assigned to Category A in cultures only, shall be assigned to UN 2814 or UN 2900 as appropriate. Animal material affected by pathogens of Category B, other than those which would be assigned to Category A if they were in cultures, shall be assigned to UN 3373.

2.2.62.2   *Substances not accepted for carriage*

Live vertebrate or invertebrate animals shall not be used to carry an infectious agent unless the agent cannot be carried by other means or unless this carriage has been approved by the competent authority (see 2.2.62.1.12.1).

2.2.62.3   *List of collective entries*

Effects on humans	I1	2814	INFECTIOUS SUBSTANCE, AFFECTING HUMANS
Effects on animals only	I2	2900	INFECTIOUS SUBSTANCE, AFFECTING ANIMALS only
Clinical waste	I3	3291 3291 3291	CLINICAL WASTE, UNSPECIFIED, N.O.S. or (BIO) MEDICAL WASTE, N.O.S. or REGULATED MEDICAL WASTE, N.O.S.
Biological substances	I4	3373	BIOLOGICAL SUBSTANCE, CATEGORY B

2.2.7 **Class 7 Radioactive material**

2.2.7.1 *Definitions*

2.2.7.1.1 *Radioactive material* means any material containing radionuclides where both the activity concentration and the total activity in the consignment exceed the values specified in 2.2.7.2.2.1 to 2.2.7.2.2.6.

2.2.7.1.2 *Contamination*

Contamination means the presence of a radioactive substance on a surface in quantities in excess of 0.4 Bq/cm^2 for beta and gamma emitters and low toxicity alpha emitters, or 0.04 Bq/cm^2 for all other alpha emitters.

Non-fixed contamination means contamination that can be removed from a surface during routine conditions of carriage.

Fixed contamination means contamination other than non-fixed contamination.

2.2.7.1.3 *Definitions of specific terms*

A_1 and A_2

A_1 means the activity value of special form radioactive material which is listed in the Table in 2.2.7.2.2.1 or derived in 2.2.7.2.2.2 and is used to determine the activity limits for the requirements of ADR.

A_2 means the activity value of radioactive material, other than special form radioactive material, which is listed in the Table in 2.2.7.2.2.1 or derived in 2.2.7.2.2.2 and is used to determine the activity limits for the requirements of ADR.

Fissile nuclides means uranium-233, uranium-235, plutonium-239 and plutonium-241. *Fissile material* means a material containing any of the fissile nuclides. Excluded from the definition of fissile material are the following:

(a) Natural uranium or depleted uranium which is unirradiated;

(b) Natural uranium or depleted uranium which has been irradiated in thermal reactors only;

(c) Material with fissile nuclides less than a total of 0.25 g;

(d) Any combination of (a), (b) and/or (c).

These exclusions are only valid if there is no other material with fissile nuclides in the package or in the consignment if shipped unpackaged.

Low dispersible radioactive material means either a solid radioactive material or a solid radioactive material in a sealed capsule, that has limited dispersibility and is not in powder form.

Low specific activity (LSA) material means radioactive material which by its nature has a limited specific activity, or radioactive material for which limits of estimated average specific activity apply. External shielding materials surrounding the LSA material shall not be considered in determining the estimated average specific activity.

Low toxicity alpha emitters are: natural uranium; depleted uranium; natural thorium; uranium-235 or uranium-238; thorium-232; thorium-228 and thorium-230 when contained in ores or physical and chemical concentrates; or alpha emitters with a half-life of less than 10 days.

Special form radioactive material means either:

(a) An indispersible solid radioactive material; or

(b) A sealed capsule containing radioactive material.

Specific activity of a radionuclide means the activity per unit mass of that nuclide. The specific activity of a material shall mean the activity per unit mass of the material in which the radionuclides are essentially uniformly distributed.

Surface contaminated object (SCO) means a solid object which is not itself radioactive but which has radioactive material distributed on its surface.

Unirradiated thorium means thorium containing not more than 10^{-7} g of uranium-233 per gram of thorium-232.

Unirradiated uranium means uranium containing not more than 2×10^3 Bq of plutonium per gram of uranium-235, not more than 9×10^6 Bq of fission products per gram of uranium-235 and not more than 5×10^{-3} g of uranium-236 per gram of uranium-235.

Uranium - natural, depleted, enriched means the following:

> *Natural uranium* means uranium (which may be chemically separated) containing the naturally occurring distribution of uranium isotopes (approximately 99.28% uranium-238, and 0.72% uranium-235 by mass).

> *Depleted uranium* means uranium containing a lesser mass percentage of uranium-235 than in natural uranium.

> *Enriched uranium* means uranium containing a greater mass percentage of uranium-235 than 0.72%.

> In all cases, a very small mass percentage of uranium-234 is present.

2.2.7.2   ***Classification***

2.2.7.2.1   *General provisions*

2.2.7.2.1.1   Radioactive material shall be assigned to one of the UN numbers specified in Table 2.2.7.2.1.1, in accordance with 2.2.7.2.4 and 2.2.7.2.5, taking into account the material characteristics determined in 2.2.7.2.3.

Table 2.2.7.2.1.1 Assignment of UN numbers

UN No.	Proper shipping name and description[a]
Excepted packages (1.7.1.5)	
UN 2908	RADIOACTIVE MATERIAL, EXCEPTED PACKAGE - EMPTY PACKAGING
UN 2909	RADIOACTIVE MATERIAL, EXCEPTED PACKAGE - ARTICLES MANUFACTURED FROM NATURAL URANIUM or DEPLETED URANIUM or NATURAL THORIUM
UN 2910	RADIOACTIVE MATERIAL, EXCEPTED PACKAGE - LIMITED QUANTITY OF MATERIAL
UN 2911	RADIOACTIVE MATERIAL, EXCEPTED PACKAGE - INSTRUMENTS or ARTICLES
UN 3507	URANIUM HEXAFLUORIDE, RADIOACTIVE MATERIAL, EXCEPTED PACKAGE less than 0.1 kg per package, non-fissile or fissile-excepted[b,c]
Low specific activity radioactive material (2.2.7.2.3.1)	
UN 2912	RADIOACTIVE MATERIAL, LOW SPECIFIC ACTIVITY (LSA-I), non fissile or fissile-excepted[b]
UN 3321	RADIOACTIVE MATERIAL, LOW SPECIFIC ACTIVITY (LSA-II), non fissile or fissile-excepted[b]
UN 3322	RADIOACTIVE MATERIAL, LOW SPECIFIC ACTIVITY (LSA-III), non fissile or fissile-excepted[b]
UN 3324	RADIOACTIVE MATERIAL, LOW SPECIFIC ACTIVITY (LSA-II), FISSILE
UN 3325	RADIOACTIVE MATERIAL, LOW SPECIFIC ACTIVITY, (LSA-III), FISSILE
Surface contaminated objects (2.2.7.2.3.2)	
UN 2913	RADIOACTIVE MATERIAL, SURFACE CONTAMINATED OBJECTS (SCO-I or SCO-II), non fissile or fissile-excepted[b]
UN 3326	RADIOACTIVE MATERIAL, SURFACE CONTAMINATED OBJECTS (SCO-I or SCO-II), FISSILE
Type A packages (2.2.7.2.4.4)	
UN 2915	RADIOACTIVE MATERIAL, TYPE A PACKAGE, non-special form, non fissile or fissile-excepted[b]
UN 3327	RADIOACTIVE MATERIAL, TYPE A PACKAGE, FISSILE, non-special form
UN 3332	RADIOACTIVE MATERIAL, TYPE A PACKAGE, SPECIAL FORM, non fissile or fissile-excepted[b]
UN 3333	RADIOACTIVE MATERIAL, TYPE A PACKAGE, SPECIAL FORM, FISSILE
Type B(U) packages (2.2.7.2.4.6)	
UN 2916	RADIOACTIVE MATERIAL, TYPE B(U) PACKAGE, non fissile or fissile-excepted[b]
UN 3328	RADIOACTIVE MATERIAL, TYPE B(U) PACKAGE, FISSILE
Type B(M) packages (2.2.7.2.4.6)	
UN 2917	RADIOACTIVE MATERIAL, TYPE B(M) PACKAGE, non fissile or fissile-excepted[b]
UN 3329	RADIOACTIVE MATERIAL, TYPE B(M) PACKAGE, FISSILE
Type C packages (2.2.7.2.4.6)	
UN 3323	RADIOACTIVE MATERIAL, TYPE C PACKAGE, non fissile or fissile-excepted[b]
UN 3330	RADIOACTIVE MATERIAL, TYPE C PACKAGE, FISSILE
Special arrangement (2.2.7.2.5)	
UN 2919	RADIOACTIVE MATERIAL, TRANSPORTED UNDER SPECIAL ARRANGEMENT, non fissile or fissile-excepted[b]
UN 3331	RADIOACTIVE MATERIAL, TRANSPORTED UNDER SPECIAL ARRANGEMENT, FISSILE
Uranium hexafluoride (2.2.7.2.4.5)	
UN 2977	RADIOACTIVE MATERIAL, URANIUM HEXAFLUORIDE, FISSILE
UN 2978	RADIOACTIVE MATERIAL, URANIUM HEXAFLUORIDE, non fissile or fissile-excepted[b]
UN 3507	URANIUM HEXAFLUORIDE, RADIOACTIVE MATERIAL, EXCEPTED PACKAGE less than 0.1 kg per package, non-fissile or fissile-excepted[b,c]

[a]   *The proper shipping name is found in the column "proper shipping name and description" and is restricted to that part shown in capital letters. In the cases of UN Nos. 2909, 2911, 2913 and 3326, where alternative proper shipping names are separated by the word "or" only the relevant proper shipping name shall be used.*

[b]   *The term "fissile-excepted" refers only to material excepted under 2.2.7.2.3.5.*

[c]   *For UN No. 3507, see also special provision 369 in Chapter 3.3.*

2.2.7.2.2   *Determination of basic radionuclide values*

2.2.7.2.2.1   The following basic values for individual radionuclides are given in Table 2.2.7.2.2.1:

(a) A_1 and A_2 in TBq;

(b)   Activity concentration limits for exempt material in Bq/g; and

(c)   Activity limits for exempt consignments in Bq.

Table 2.2.7.2.2.1: Basic radionuclides values for individual radionuclides

Radionuclide (atomic number)	A_1 (TBq)	A_2 (TBq)	Activity concentration limit for exempt material (Bq/g)	Activity limit for an exempt consignment (Bq)
Actinium (89)				
Ac-225 (a)	8×10^{-1}	6×10^{-3}	1×10^{1}	1×10^{4}
Ac-227 (a)	9×10^{-1}	9×10^{-5}	1×10^{-1}	1×10^{3}
Ac-228	6×10^{-1}	5×10^{-1}	1×10^{1}	1×10^{6}
Silver (47)				
Ag-105	2×10^{0}	2×10^{0}	1×10^{2}	1×10^{6}
Ag-108m (a)	7×10^{-1}	7×10^{-1}	1×10^{1} (b)	1×10^{6} (b)
Ag-110m (a)	4×10^{-1}	4×10^{-1}	1×10^{1}	1×10^{6}
Ag-111	2×10^{0}	6×10^{-1}	1×10^{3}	1×10^{6}
Aluminium (13)				
Al-26	1×10^{-1}	1×10^{-1}	1×10^{1}	1×10^{5}
Americium (95)				
Am-241	1×10^{1}	1×10^{-3}	1×10^{0}	1×10^{4}
Am-242m (a)	1×10^{1}	1×10^{-3}	1×10^{0} (b)	1×10^{4} (b)
Am-243 (a)	5×10^{0}	1×10^{-3}	1×10^{0} (b)	1×10^{3} (b)
Argon (18)				
Ar-37	4×10^{1}	4×10^{1}	1×10^{6}	1×10^{8}
Ar-39	4×10^{1}	2×10^{1}	1×10^{7}	1×10^{4}
Ar-41	3×10^{-1}	3×10^{-1}	1×10^{2}	1×10^{9}
Arsenic (33)				
As-72	3×10^{-1}	3×10^{-1}	1×10^{1}	1×10^{5}
As-73	4×10^{1}	4×10^{1}	1×10^{3}	1×10^{7}
As-74	1×10^{0}	9×10^{-1}	1×10^{1}	1×10^{6}
As-76	3×10^{-1}	3×10^{-1}	1×10^{2}	1×10^{5}
As-77	2×10^{1}	7×10^{-1}	1×10^{3}	1×10^{6}
Astatine (85)				
At-211 (a)	2×10^{1}	5×10^{-1}	1×10^{3}	1×10^{7}

Radionuclide (atomic number)	A_1 (TBq)	A_2 (TBq)	Activity concentration limit for exempt material (Bq/g)	Activity limit for an exempt consignment (Bq)
Gold (79)				
Au-193	7×10^0	2×10^0	1×10^2	1×10^7
Au-194	1×10^0	1×10^0	1×10^1	1×10^6
Au-195	1×10^1	6×10^0	1×10^2	1×10^7
Au-198	1×10^0	6×10^{-1}	1×10^2	1×10^6
Au-199	1×10^1	6×10^{-1}	1×10^2	1×10^6
Barium (56)				
Ba-131 (a)	2×10^0	2×10^0	1×10^2	1×10^6
Ba-133	3×10^0	3×10^0	1×10^2	1×10^6
Ba-133m	2×10^1	6×10^{-1}	1×10^2	1×10^6
Ba-140 (a)	5×10^{-1}	3×10^{-1}	1×10^1 (b)	1×10^5 (b)
Beryllium (4)				
Be-7	2×10^1	2×10^1	1×10^3	1×10^7
Be-10	4×10^1	6×10^{-1}	1×10^4	1×10^6
Bismuth (83)				
Bi-205	7×10^{-1}	7×10^{-1}	1×10^1	1×10^6
Bi-206	3×10^{-1}	3×10^{-1}	1×10^1	1×10^5
Bi-207	7×10^{-1}	7×10^{-1}	1×10^1	1×10^6
Bi-210	1×10^0	6×10^{-1}	1×10^3	1×10^6
Bi-210m (a)	6×10^{-1}	2×10^{-2}	1×10^1	1×10^5
Bi-212 (a)	7×10^{-1}	6×10^{-1}	1×10^1 (b)	1×10^5 (b)
Berkelium (97)				
Bk-247	8×10^0	8×10^{-4}	1×10^0	1×10^4
Bk-249 (a)	4×10^1	3×10^{-1}	1×10^3	1×10^6
Bromine (35)				
Br-76	4×10^{-1}	4×10^{-1}	1×10^1	1×10^5
Br-77	3×10^0	3×10^0	1×10^2	1×10^6
Br-82	4×10^{-1}	4×10^{-1}	1×10^1	1×10^6
Carbon (6)				
C-11	1×10^0	6×10^{-1}	1×10^1	1×10^6
C-14	4×10^1	3×10^0	1×10^4	1×10^7

Radionuclide (atomic number)	A_1 (TBq)	A_2 (TBq)	Activity concentration limit for exempt material (Bq/g)	Activity limit for an exempt consignment (Bq)
Calcium (20)				
Ca-41	Unlimited	Unlimited	1×10^5	1×10^7
Ca-45	4×10^1	1×10^0	1×10^4	1×10^7
Ca-47 (a)	3×10^0	3×10^{-1}	1×10^1	1×10^6
Cadmium (48)				
Cd-109	3×10^1	2×10^0	1×10^4	1×10^6
Cd-113m	4×10^1	5×10^{-1}	1×10^3	1×10^6
Cd-115 (a)	3×10^0	4×10^{-1}	1×10^2	1×10^6
Cd-115m	5×10^{-1}	5×10^{-1}	1×10^3	1×10^6
Cerium (58)				
Ce-139	7×10^0	2×10^0	1×10^2	1×10^6
Ce-141	2×10^1	6×10^{-1}	1×10^2	1×10^7
Ce-143	9×10^{-1}	6×10^{-1}	1×10^2	1×10^6
Ce-144 (a)	2×10^{-1}	2×10^{-1}	1×10^2 (b)	1×10^5 (b)
Californium (98)				
Cf-248	4×10^1	6×10^{-3}	1×10^1	1×10^4
Cf-249	3×10^0	8×10^{-4}	1×10^0	1×10^3
Cf-250	2×10^1	2×10^{-3}	1×10^1	1×10^4
Cf-251	7×10^0	7×10^{-4}	1×10^0	1×10^3
Cf-252	1×10^{-1}	3×10^{-3}	1×10^1	1×10^4
Cf-253 (a)	4×10^1	4×10^{-2}	1×10^2	1×10^5
Cf-254	1×10^{-3}	1×10^{-3}	1×10^0	1×10^3
Chlorine (17)				
Cl-36	1×10^1	6×10^{-1}	1×10^4	1×10^6
Cl-38	2×10^{-1}	2×10^{-1}	1×10^1	1×10^5
Curium (96)				
Cm-240	4×10^1	2×10^{-2}	1×10^2	1×10^5
Cm-241	2×10^0	1×10^0	1×10^2	1×10^6
Cm-242	4×10^1	1×10^{-2}	1×10^2	1×10^5
Cm-243	9×10^0	1×10^{-3}	1×10^0	1×10^4
Cm-244	2×10^1	2×10^{-3}	1×10^1	1×10^4
Cm-245	9×10^0	9×10^{-4}	1×10^0	1×10^3
Cm-246	9×10^0	9×10^{-4}	1×10^0	1×10^3

Radionuclide (atomic number)	A_1 (TBq)	A_2 (TBq)	Activity concentration limit for exempt material (Bq/g)	Activity limit for an exempt consignment (Bq)
Cm-247 (a)	3×10^0	1×10^{-3}	1×10^0	1×10^4
Cm-248	2×10^{-2}	3×10^{-4}	1×10^0	1×10^3
Cobalt (27)				
Co-55	5×10^{-1}	5×10^{-1}	1×10^1	1×10^6
Co-56	3×10^{-1}	3×10^{-1}	1×10^1	1×10^5
Co-57	1×10^1	1×10^1	1×10^2	1×10^6
Co-58	1×10^0	1×10^0	1×10^1	1×10^6
Co-58m	4×10^1	4×10^1	1×10^4	1×10^7
Co-60	4×10^{-1}	4×10^{-1}	1×10^1	1×10^5
Chromium (24)				
Cr-51	3×10^1	3×10^1	1×10^3	1×10^7
Caesium (55)				
Cs-129	4×10^0	4×10^0	1×10^2	1×10^5
Cs-131	3×10^1	3×10^1	1×10^3	1×10^6
Cs-132	1×10^0	1×10^0	1×10^1	1×10^5
Cs-134	7×10^{-1}	7×10^{-1}	1×10^1	1×10^4
Cs-134m	4×10^1	6×10^{-1}	1×10^3	1×10^5
Cs-135	4×10^1	1×10^0	1×10^4	1×10^7
Cs-136	5×10^{-1}	5×10^{-1}	1×10^1	1×10^5
Cs-137 (a)	2×10^0	6×10^{-1}	1×10^1 (b)	1×10^4 (b)
Copper (29)				
Cu-64	6×10^0	1×10^0	1×10^2	1×10^6
Cu-67	1×10^1	7×10^{-1}	1×10^2	1×10^6
Dysprosium (66)				
Dy-159	2×10^1	2×10^1	1×10^3	1×10^7
Dy-165	9×10^{-1}	6×10^{-1}	1×10^3	1×10^6
Dy-166 (a)	9×10^{-1}	3×10^{-1}	1×10^3	1×10^6
Erbium (68)				
Er-169	4×10^1	1×10^0	1×10^4	1×10^7
Er-171	8×10^{-1}	5×10^{-1}	1×10^2	1×10^6
Europium (63)				
Eu-147	2×10^0	2×10^0	1×10^2	1×10^6
Eu-148	5×10^{-1}	5×10^{-1}	1×10^1	1×10^6

Radionuclide (atomic number)	A_1 (TBq)	A_2 (TBq)	Activity concentration limit for exempt material (Bq/g)	Activity limit for an exempt consignment (Bq)
Eu-149	2×10^1	2×10^1	1×10^2	1×10^7
Eu-150(short lived)	2×10^0	7×10^{-1}	1×10^3	1×10^6
Eu-150(long lived)	7×10^{-1}	7×10^{-1}	1×10^1	1×10^6
Eu-152	1×10^0	1×10^0	1×10^1	1×10^6
Eu-152m	8×10^{-1}	8×10^{-1}	1×10^2	1×10^6
Eu-154	9×10^{-1}	6×10^{-1}	1×10^1	1×10^6
Eu-155	2×10^1	3×10^0	1×10^2	1×10^7
Eu-156	7×10^{-1}	7×10^{-1}	1×10^1	1×10^6
Fluorine (9)				
F-18	1×10^0	6×10^{-1}	1×10^1	1×10^6
Iron (26)				
Fe-52 (a)	3×10^{-1}	3×10^{-1}	1×10^1	1×10^6
Fe-55	4×10^1	4×10^1	1×10^4	1×10^6
Fe-59	9×10^{-1}	9×10^{-1}	1×10^1	1×10^6
Fe-60 (a)	4×10^1	2×10^{-1}	1×10^2	1×10^5
Gallium (31)				
Ga-67	7×10^0	3×10^0	1×10^2	1×10^6
Ga-68	5×10^{-1}	5×10^{-1}	1×10^1	1×10^5
Ga-72	4×10^{-1}	4×10^{-1}	1×10^1	1×10^5
Gadolinium (64)				
Gd-146 (a)	5×10^{-1}	5×10^{-1}	1×10^1	1×10^6
Gd-148	2×10^1	2×10^{-3}	1×10^1	1×10^4
Gd-153	1×10^1	9×10^0	1×10^2	1×10^7
Gd-159	3×10^0	6×10^{-1}	1×10^3	1×10^6
Germanium (32)				
Ge-68 (a)	5×10^{-1}	5×10^{-1}	1×10^1	1×10^5
Ge-71	4×10^1	4×10^1	1×10^4	1×10^8
Ge-77	3×10^{-1}	3×10^{-1}	1×10^1	1×10^5
Hafnium (72)				
Hf-172 (a)	6×10^{-1}	6×10^{-1}	1×10^1	1×10^6
Hf-175	3×10^0	3×10^0	1×10^2	1×10^6
Hf-181	2×10^0	5×10^{-1}	1×10^1	1×10^6
Hf-182	Unlimited	Unlimited	1×10^2	1×10^6

Radionuclide (atomic number)	A_1 (TBq)	A_2 (TBq)	Activity concentration limit for exempt material (Bq/g)	Activity limit for an exempt consignment (Bq)
Mercury (80)				
Hg-194 (a)	1×10^0	1×10^0	1×10^1	1×10^6
Hg-195m (a)	3×10^0	7×10^{-1}	1×10^2	1×10^6
Hg-197	2×10^1	1×10^1	1×10^2	1×10^7
Hg-197m	1×10^1	4×10^{-1}	1×10^2	1×10^6
Hg-203	5×10^0	1×10^0	1×10^2	1×10^5
Holmium (67)				
Ho-166	4×10^{-1}	4×10^{-1}	1×10^3	1×10^5
Ho-166m	6×10^{-1}	5×10^{-1}	1×10^1	1×10^6
Iodine (53)				
I-123	6×10^0	3×10^0	1×10^2	1×10^7
I-124	1×10^0	1×10^0	1×10^1	1×10^6
I-125	2×10^1	3×10^0	1×10^3	1×10^6
I-126	2×10^0	1×10^0	1×10^2	1×10^6
I-129	Unlimited	Unlimited	1×10^2	1×10^5
I-131	3×10^0	7×10^{-1}	1×10^2	1×10^6
I-132	4×10^{-1}	4×10^{-1}	1×10^1	1×10^5
I-133	7×10^{-1}	6×10^{-1}	1×10^1	1×10^6
I-134	3×10^{-1}	3×10^{-1}	1×10^1	1×10^5
I-135 (a)	6×10^{-1}	6×10^{-1}	1×10^1	1×10^6
Indium (49)				
In-111	3×10^0	3×10^0	1×10^2	1×10^6
In-113m	4×10^0	2×10^0	1×10^2	1×10^6
In-114m (a)	1×10^1	5×10^{-1}	1×10^2	1×10^6
In-115m	7×10^0	1×10^0	1×10^2	1×10^6
Iridium (77)				
Ir-189 (a)	1×10^1	1×10^1	1×10^2	1×10^7
Ir-190	7×10^{-1}	7×10^{-1}	1×10^1	1×10^6
Ir-192	1×10^0(c)	6×10^{-1}	1×10^1	1×10^4
Ir-194	3×10^{-1}	3×10^{-1}	1×10^2	1×10^5
Potassium (19)				
K-40	9×10^{-1}	9×10^{-1}	1×10^2	1×10^6
K-42	2×10^{-1}	2×10^{-1}	1×10^2	1×10^6

Radionuclide (atomic number)	A_1 (TBq)	A_2 (TBq)	Activity concentration limit for exempt material (Bq/g)	Activity limit for an exempt consignment (Bq)
K-43	7×10^{-1}	6×10^{-1}	1×10^1	1×10^6
Krypton (36)				
Kr-79	4×10^0	2×10^0	1×10^3	1×10^5
Kr-81	4×10^1	4×10^1	1×10^4	1×10^7
Kr-85	1×10^1	1×10^1	1×10^5	1×10^4
Kr-85m	8×10^0	3×10^0	1×10^3	1×10^{10}
Kr-87	2×10^{-1}	2×10^{-1}	1×10^2	1×10^9
Lanthanum (57)				
La-137	3×10^1	6×10^0	1×10^3	1×10^7
La-140	4×10^{-1}	4×10^{-1}	1×10^1	1×10^5
Lutetium (71)				
Lu-172	6×10^{-1}	6×10^{-1}	1×10^1	1×10^6
Lu-173	8×10^0	8×10^0	1×10^2	1×10^7
Lu-174	9×10^0	9×10^0	1×10^2	1×10^7
Lu-174m	2×10^1	1×10^1	1×10^2	1×10^7
Lu-177	3×10^1	7×10^{-1}	1×10^3	1×10^7
Magnesium (12)				
Mg-28 (a)	3×10^{-1}	3×10^{-1}	1×10^1	1×10^5
Manganese (25)				
Mn-52	3×10^{-1}	3×10^{-1}	1×10^1	1×10^5
Mn-53	Unlimited	Unlimited	1×10^4	1×10^9
Mn-54	1×10^0	1×10^0	1×10^1	1×10^6
Mn-56	3×10^{-1}	3×10^{-1}	1×10^1	1×10^5
Molybdenum (42)				
Mo-93	4×10^1	2×10^1	1×10^3	1×10^8
Mo-99 (a)	1×10^0	6×10^{-1}	1×10^2	1×10^6
Nitrogen (7)				
N-13	9×10^{-1}	6×10^{-1}	1×10^2	1×10^9
Sodium (11)				
Na-22	5×10^{-1}	5×10^{-1}	1×10^1	1×10^6
Na-24	2×10^{-1}	2×10^{-1}	1×10^1	1×10^5
Niobium (41)				
Nb-93m	4×10^1	3×10^1	1×10^4	1×10^7

Radionuclide (atomic number)	A_1 (TBq)	A_2 (TBq)	Activity concentration limit for exempt material (Bq/g)	Activity limit for an exempt consignment (Bq)
Nb-94	7×10^{-1}	7×10^{-1}	1×10^{1}	1×10^{6}
Nb-95	1×10^{0}	1×10^{0}	1×10^{1}	1×10^{6}
Nb-97	9×10^{-1}	6×10^{-1}	1×10^{1}	1×10^{6}
Neodymium (60)				
Nd-147	6×10^{0}	6×10^{-1}	1×10^{2}	1×10^{6}
Nd-149	6×10^{-1}	5×10^{-1}	1×10^{2}	1×10^{6}
Nickel (28)				
Ni-59	Unlimited	Unlimited	1×10^{4}	1×10^{8}
Ni-63	4×10^{1}	3×10^{1}	1×10^{5}	1×10^{8}
Ni-65	4×10^{-1}	4×10^{-1}	1×10^{1}	1×10^{6}
Neptunium (93)				
Np-235	4×10^{1}	4×10^{1}	1×10^{3}	1×10^{7}
Np-236(short-lived)	2×10^{1}	2×10^{0}	1×10^{3}	1×10^{7}
Np-236(long-lived)	9×10^{0}	2×10^{-2}	1×10^{2}	1×10^{5}
Np-237	2×10^{1}	2×10^{-3}	1×10^{0} (b)	1×10^{3} (b)
Np-239	7×10^{0}	4×10^{-1}	1×10^{2}	1×10^{7}
Osmium (76)				
Os-185	1×10^{0}	1×10^{0}	1×10^{1}	1×10^{6}
Os-191	1×10^{1}	2×10^{0}	1×10^{2}	1×10^{7}
Os-191m	4×10^{1}	3×10^{1}	1×10^{3}	1×10^{7}
Os-193	2×10^{0}	6×10^{-1}	1×10^{2}	1×10^{6}
Os-194 (a)	3×10^{-1}	3×10^{-1}	1×10^{2}	1×10^{5}
Phosphorus (15)				
P-32	5×10^{-1}	5×10^{-1}	1×10^{3}	1×10^{5}
P-33	4×10^{1}	1×10^{0}	1×10^{5}	1×10^{8}
Protactinium (91)				
Pa-230 (a)	2×10^{0}	7×10^{-2}	1×10^{1}	1×10^{6}
Pa-231	4×10^{0}	4×10^{-4}	1×10^{0}	1×10^{3}
Pa-233	5×10^{0}	7×10^{-1}	1×10^{2}	1×10^{7}
Lead (82)				
Pb-201	1×10^{0}	1×10^{0}	1×10^{1}	1×10^{6}
Pb-202	4×10^{1}	2×10^{1}	1×10^{3}	1×10^{6}
Pb-203	4×10^{0}	3×10^{0}	1×10^{2}	1×10^{6}

Radionuclide (atomic number)	A_1 (TBq)	A_2 (TBq)	Activity concentration limit for exempt material (Bq/g)	Activity limit for an exempt consignment (Bq)
Pb-205	Unlimited	Unlimited	1×10^4	1×10^7
Pb-210 (a)	1×10^0	5×10^{-2}	1×10^1 (b)	1×10^4 (b)
Pb-212 (a)	7×10^{-1}	2×10^{-1}	1×10^1 (b)	1×10^5 (b)
Palladium (46)				
Pd-103 (a)	4×10^1	4×10^1	1×10^3	1×10^8
Pd-107	Unlimited	Unlimited	1×10^5	1×10^8
Pd-109	2×10^0	5×10^{-1}	1×10^3	1×10^6
Promethium (61)				
Pm-143	3×10^0	3×10^0	1×10^2	1×10^6
Pm-144	7×10^{-1}	7×10^{-1}	1×10^1	1×10^6
Pm-145	3×10^1	1×10^1	1×10^3	1×10^7
Pm-147	4×10^1	2×10^0	1×10^4	1×10^7
Pm-148m (a)	8×10^{-1}	7×10^{-1}	1×10^1	1×10^6
Pm-149	2×10^0	6×10^{-1}	1×10^3	1×10^6
Pm-151	2×10^0	6×10^{-1}	1×10^2	1×10^6
Polonium (84)				
Po-210	4×10^1	2×10^{-2}	1×10^1	1×10^4
Praseodymium (59)				
Pr-142	4×10^{-1}	4×10^{-1}	1×10^2	1×10^5
Pr-143	3×10^0	6×10^{-1}	1×10^4	1×10^6
Platinum (78)				
Pt-188 (a)	1×10^0	8×10^{-1}	1×10^1	1×10^6
Pt-191	4×10^0	3×10^0	1×10^2	1×10^6
Pt-193	4×10^1	4×10^1	1×10^4	1×10^7
Pt-193m	4×10^1	5×10^{-1}	1×10^3	1×10^7
Pt-195m	1×10^1	5×10^{-1}	1×10^2	1×10^6
Pt-197	2×10^1	6×10^{-1}	1×10^3	1×10^6
Pt-197m	1×10^1	6×10^{-1}	1×10^2	1×10^6
Plutonium (94)				
Pu-236	3×10^1	3×10^{-3}	1×10^1	1×10^4
Pu-237	2×10^1	2×10^1	1×10^3	1×10^7
Pu-238	1×10^1	1×10^{-3}	1×10^0	1×10^4
Pu-239	1×10^1	1×10^{-3}	1×10^0	1×10^4

Radionuclide (atomic number)	A_1 (TBq)	A_2 (TBq)	Activity concentration limit for exempt material (Bq/g)	Activity limit for an exempt consignment (Bq)
Pu-240	1×10^1	1×10^{-3}	1×10^0	1×10^3
Pu-241 (a)	4×10^1	6×10^{-2}	1×10^2	1×10^5
Pu-242	1×10^1	1×10^{-3}	1×10^0	1×10^4
Pu-244 (a)	4×10^{-1}	1×10^{-3}	1×10^0	1×10^4
Radium (88)				
Ra-223 (a)	4×10^{-1}	7×10^{-3}	1×10^2 (b)	1×10^5 (b)
Ra-224 (a)	4×10^{-1}	2×10^{-2}	1×10^1 (b)	1×10^5 (b)
Ra-225 (a)	2×10^{-1}	4×10^{-3}	1×10^2	1×10^5
Ra-226 (a)	2×10^{-1}	3×10^{-3}	1×10^1 (b)	1×10^4 (b)
Ra-228 (a)	6×10^{-1}	2×10^{-2}	1×10^1 (b)	1×10^5 (b)
Rubidium (37)				
Rb-81	2×10^0	8×10^{-1}	1×10^1	1×10^6
Rb-83 (a)	2×10^0	2×10^0	1×10^2	1×10^6
Rb-84	1×10^0	1×10^0	1×10^1	1×10^6
Rb-86	5×10^{-1}	5×10^{-1}	1×10^2	1×10^5
Rb-87	Unlimited	Unlimited	1×10^4	1×10^7
Rb(nat)	Unlimited	Unlimited	1×10^4	1×10^7
Rhenium (75)				
Re-184	1×10^0	1×10^0	1×10^1	1×10^6
Re-184m	3×10^0	1×10^0	1×10^2	1×10^6
Re-186	2×10^0	6×10^{-1}	1×10^3	1×10^6
Re-187	Unlimited	Unlimited	1×10^6	1×10^9
Re-188	4×10^{-1}	4×10^{-1}	1×10^2	1×10^5
Re-189 (a)	3×10^0	6×10^{-1}	1×10^2	1×10^6
Re(nat)	Unlimited	Unlimited	1×10^6	1×10^9
Rhodium (45)				
Rh-99	2×10^0	2×10^0	1×10^1	1×10^6
Rh-101	4×10^0	3×10^0	1×10^2	1×10^7
Rh-102	5×10^{-1}	5×10^{-1}	1×10^1	1×10^6
Rh-102m	2×10^0	2×10^0	1×10^2	1×10^6
Rh-103m	4×10^1	4×10^1	1×10^4	1×10^8
Rh-105	1×10^1	8×10^{-1}	1×10^2	1×10^7
Radon (86)				

Radionuclide (atomic number)	A_1 (TBq)	A_2 (TBq)	Activity concentration limit for exempt material (Bq/g)	Activity limit for an exempt consignment (Bq)
Rn-222 (a)	3×10^{-1}	4×10^{-3}	1×10^1 (b)	1×10^8 (b)
Ruthenium (44)				
Ru-97	5×10^0	5×10^0	1×10^2	1×10^7
Ru-103 (a)	2×10^0	2×10^0	1×10^2	1×10^6
Ru-105	1×10^0	6×10^{-1}	1×10^1	1×10^6
Ru-106 (a)	2×10^{-1}	2×10^{-1}	1×10^2 (b)	1×10^5 (b)
Sulphur (16)				
S-35	4×10^1	3×10^0	1×10^5	1×10^8
Antimony (51)				
Sb-122	4×10^{-1}	4×10^{-1}	1×10^2	1×10^4
Sb-124	6×10^{-1}	6×10^{-1}	1×10^1	1×10^6
Sb-125	2×10^0	1×10^0	1×10^2	1×10^6
Sb-126	4×10^{-1}	4×10^{-1}	1×10^1	1×10^5
Scandium (21)				
Sc-44	5×10^{-1}	5×10^{-1}	1×10^1	1×10^5
Sc-46	5×10^{-1}	5×10^{-1}	1×10^1	1×10^6
Sc-47	1×10^1	7×10^{-1}	1×10^2	1×10^6
Sc-48	3×10^{-1}	3×10^{-1}	1×10^1	1×10^5
Selenium (34)				
Se-75	3×10^0	3×10^0	1×10^2	1×10^6
Se-79	4×10^1	2×10^0	1×10^4	1×10^7
Silicon (14)				
Si-31	6×10^{-1}	6×10^{-1}	1×10^3	1×10^6
Si-32	4×10^1	5×10^{-1}	1×10^3	1×10^6
Samarium (62)				
Sm-145	1×10^1	1×10^1	1×10^2	1×10^7
Sm-147	Unlimited	Unlimited	1×10^1	1×10^4
Sm-151	4×10^1	1×10^1	1×10^4	1×10^8
Sm-153	9×10^0	6×10^{-1}	1×10^2	1×10^6
Tin (50)				
Sn-113 (a)	4×10^0	2×10^0	1×10^3	1×10^7
Sn-117m	7×10^0	4×10^{-1}	1×10^2	1×10^6
Sn-119m	4×10^1	3×10^1	1×10^3	1×10^7

Radionuclide (atomic number)	A_1 (TBq)	A_2 (TBq)	Activity concentration limit for exempt material (Bq/g)	Activity limit for an exempt consignment (Bq)
Sn-121m (a)	4×10^1	9×10^{-1}	1×10^3	1×10^7
Sn-123	8×10^{-1}	6×10^{-1}	1×10^3	1×10^6
Sn-125	4×10^{-1}	4×10^{-1}	1×10^2	1×10^5
Sn-126 (a)	6×10^{-1}	4×10^{-1}	1×10^1	1×10^5
Strontium (38)				
Sr-82 (a)	2×10^{-1}	2×10^{-1}	1×10^1	1×10^5
Sr-85	2×10^0	2×10^0	1×10^2	1×10^6
Sr-85m	5×10^0	5×10^0	1×10^2	1×10^7
Sr-87m	3×10^0	3×10^0	1×10^2	1×10^6
Sr-89	6×10^{-1}	6×10^{-1}	1×10^3	1×10^6
Sr-90 (a)	3×10^{-1}	3×10^{-1}	1×10^2 (b)	1×10^4 (b)
Sr-91 (a)	3×10^{-1}	3×10^{-1}	1×10^1	1×10^5
Sr-92 (a)	1×10^0	3×10^{-1}	1×10^1	1×10^6
Tritium (1)				
T(H-3)	4×10^1	4×10^1	1×10^6	1×10^9
Tantalum (73)				
Ta-178(long-lived)	1×10^0	8×10^{-1}	1×10^1	1×10^6
Ta-179	3×10^1	3×10^1	1×10^3	1×10^7
Ta-182	9×10^{-1}	5×10^{-1}	1×10^1	1×10^4
Terbium (65)				
Tb-157	4×10^1	4×10^1	1×10^4	1×10^7
Tb-158	1×10^0	1×10^0	1×10^1	1×10^6
Tb-160	1×10^0	6×10^{-1}	1×10^1	1×10^6
Technetium (43)				
Tc-95m (a)	2×10^0	2×10^0	1×10^1	1×10^6
Tc-96	4×10^{-1}	4×10^{-1}	1×10^1	1×10^6
Tc-96m (a)	4×10^{-1}	4×10^{-1}	1×10^3	1×10^7
Tc-97	Unlimited	Unlimited	1×10^3	1×10^8
Tc-97m	4×10^1	1×10^0	1×10^3	1×10^7
Tc-98	8×10^{-1}	7×10^{-1}	1×10^1	1×10^6
Tc-99	4×10^1	9×10^{-1}	1×10^4	1×10^7
Tc-99m	1×10^1	4×10^0	1×10^2	1×10^7
Tellurium (52)				

Radionuclide (atomic number)	A_1 (TBq)	A_2 (TBq)	Activity concentration limit for exempt material (Bq/g)	Activity limit for an exempt consignment (Bq)
Te-121	2×10^0	2×10^0	1×10^1	1×10^6
Te-121m	5×10^0	3×10^0	1×10^2	1×10^6
Te-123m	8×10^0	1×10^0	1×10^2	1×10^7
Te-125m	2×10^1	9×10^{-1}	1×10^3	1×10^7
Te-127	2×10^1	7×10^{-1}	1×10^3	1×10^6
Te-127m (a)	2×10^1	5×10^{-1}	1×10^3	1×10^7
Te-129	7×10^{-1}	6×10^{-1}	1×10^2	1×10^6
Te-129m (a)	8×10^{-1}	4×10^{-1}	1×10^3	1×10^6
Te-131m (a)	7×10^{-1}	5×10^{-1}	1×10^1	1×10^6
Te-132 (a)	5×10^{-1}	4×10^{-1}	1×10^2	1×10^7
Thorium (90)				
Th-227	1×10^1	5×10^{-3}	1×10^1	1×10^4
Th-228 (a)	5×10^{-1}	1×10^{-3}	1×10^0 (b)	1×10^4 (b)
Th-229	5×10^0	5×10^{-4}	1×10^0 (b)	1×10^3 (b)
Th-230	1×10^1	1×10^{-3}	1×10^0	1×10^4
Th-231	4×10^1	2×10^{-2}	1×10^3	1×10^7
Th-232	Unlimited	Unlimited	1×10^1	1×10^4
Th-234 (a)	3×10^{-1}	3×10^{-1}	1×10^3 (b)	1×10^5 (b)
Th(nat)	Unlimited	Unlimited	1×10^0 (b)	1×10^3 (b)
Titanium (22)				
Ti-44 (a)	5×10^{-1}	4×10^{-1}	1×10^1	1×10^5
Thallium (81)				
Tl-200	9×10^{-1}	9×10^{-1}	1×10^1	1×10^6
Tl-201	1×10^1	4×10^0	1×10^2	1×10^6
Tl-202	2×10^0	2×10^0	1×10^2	1×10^6
Tl-204	1×10^1	7×10^{-1}	1×10^4	1×10^4
Thulium (69)				
Tm-167	7×10^0	8×10^{-1}	1×10^2	1×10^6
Tm-170	3×10^0	6×10^{-1}	1×10^3	1×10^6
Tm-171	4×10^1	4×10^1	1×10^4	1×10^8
Uranium (92)				
U-230 (fast lung absorption) (a)(d)	4×10^1	1×10^{-1}	1×10^1 (b)	1×10^5 (b)
U-230 (medium lung absorption) (a)(e)	4×10^1	4×10^{-3}	1×10^1	1×10^4

Radionuclide (atomic number)	A_1 (TBq)	A_2 (TBq)	Activity concentration limit for exempt material (Bq/g)	Activity limit for an exempt consignment (Bq)
U-230 (slow lung absorption) (a)(f)	3×10^1	3×10^{-3}	1×10^1	1×10^4
U-232 (fast lung absorption) (d)	4×10^1	1×10^{-2}	1×10^0 (b)	1×10^3 (b)
U-232 (medium lung absorption) (e)	4×10^1	7×10^{-3}	1×10^1	1×10^4
U-232 (slow lung absorption) (f)	1×10^1	1×10^{-3}	1×10^1	1×10^4
U-233 (fast lung absorption) (d)	4×10^1	9×10^{-2}	1×10^1	1×10^4
U-233 (medium lung absorption) (e)	4×10^1	2×10^{-2}	1×10^2	1×10^5
U-233 (slow lung absorption) (f)	4×10^1	6×10^{-3}	1×10^1	1×10^5
U-234 (fast lung absorption) (d)	4×10^1	9×10^{-2}	1×10^1	1×10^4
U-234 (medium lung absorption) (e)	4×10^1	2×10^{-2}	1×10^2	1×10^5
U-234 (slow lung absorption) (f)	4×10^1	6×10^{-3}	1×10^1	1×10^5
U-235 (all lung absorption types) (a)(d)(e)(f)	Unlimited	Unlimited	1×10^1 (b)	1×10^4 (b)
U-236 (fast lung absorption) (d)	Unlimited	Unlimited	1×10^1	1×10^4
U-236 (medium lung absorption) (e)	4×10^1	2×10^{-2}	1×10^2	1×10^5
U-236 (slow lung absorption) (f)	4×10^1	6×10^{-3}	1×10^1	1×10^4
U-238 (all lung absorption types) (d)(e)(f)	Unlimited	Unlimited	1×10^1 (b)	1×10^4 (b)
U (nat)	Unlimited	Unlimited	1×10^0 (b)	1×10^3 (b)
U (enriched to 20% or less) (g)	Unlimited	Unlimited	1×10^0	1×10^3
U (dep)	Unlimited	Unlimited	1×10^0	1×10^3
Vanadium (23)				
V-48	4×10^{-1}	4×10^{-1}	1×10^1	1×10^5
V-49	4×10^1	4×10^1	1×10^4	1×10^7
Tungsten (74)				
W-178 (a)	9×10^0	5×10^0	1×10^1	1×10^6
W-181	3×10^1	3×10^1	1×10^3	1×10^7
W-185	4×10^1	8×10^{-1}	1×10^4	1×10^7
W-187	2×10^0	6×10^{-1}	1×10^2	1×10^6
W-188 (a)	4×10^{-1}	3×10^{-1}	1×10^2	1×10^5
Xenon (54)				
Xe-122 (a)	4×10^{-1}	4×10^{-1}	1×10^2	1×10^9
Xe-123	2×10^0	7×10^{-1}	1×10^2	1×10^9
Xe-127	4×10^0	2×10^0	1×10^3	1×10^5
Xe-131m	4×10^1	4×10^1	1×10^4	1×10^4
Xe-133	2×10^1	1×10^1	1×10^3	1×10^4

Radionuclide (atomic number)	A_1 (TBq)	A_2 (TBq)	Activity concentration limit for exempt material (Bq/g)	Activity limit for an exempt consignment (Bq)
Xe-135	3×10^0	2×10^0	1×10^3	1×10^{10}
Yttrium (39)				
Y-87 (a)	1×10^0	1×10^0	1×10^1	1×10^6
Y-88	4×10^{-1}	4×10^{-1}	1×10^1	1×10^6
Y-90	3×10^{-1}	3×10^{-1}	1×10^3	1×10^5
Y-91	6×10^{-1}	6×10^{-1}	1×10^3	1×10^6
Y-91m	2×10^0	2×10^0	1×10^2	1×10^6
Y-92	2×10^{-1}	2×10^{-1}	1×10^2	1×10^5
Y-93	3×10^{-1}	3×10^{-1}	1×10^2	1×10^5
Ytterbium (70)				
Yb-169	4×10^0	1×10^0	1×10^2	1×10^7
Yb-175	3×10^1	9×10^{-1}	1×10^3	1×10^7
Zinc (30)				
Zn-65	2×10^0	2×10^0	1×10^1	1×10^6
Zn-69	3×10^0	6×10^{-1}	1×10^4	1×10^6
Zn-69m (a)	3×10^0	6×10^{-1}	1×10^2	1×10^6
Zirconium (40)				
Zr-88	3×10^0	3×10^0	1×10^2	1×10^6
Zr-93	Unlimited	Unlimited	1×10^3 (b)	1×10^7 (b)
Zr-95 (a)	2×10^0	8×10^{-1}	1×10^1	1×10^6
Zr-97 (a)	4×10^{-1}	4×10^{-1}	1×10^1 (b)	1×10^5 (b)

(a) A_1 and/or A_2 values for these parent radionuclides include contributions from their progeny with half-lives less than 10 days, as listed in the following:

Mg-28	Al-28
Ar-42	K-42
Ca-47	Sc-47
Ti-44	Sc-44
Fe-52	Mn-52m
Fe-60	Co-60m
Zn-69m	Zn-69
Ge-68	Ga-68
Rb-83	Kr-83m
Sr-82	Rb-82
Sr-90	Y-90
Sr-91	Y-91m
Sr-92	Y-92
Y-87	Sr-87m
Zr-95	Nb-95m
Zr-97	Nb-97m, Nb-97
Mo-99	Tc-99m
Tc-95m	Tc-95
Tc-96m	Tc-96

Ru-103	Rh-103m
Ru-106	Rh-106
Pd-103	Rh-103m
Ag-108m	Ag-108
Ag-110m	Ag-110
Cd-115	In-115m
In-114m	In-114
Sn-113	In-113m
Sn-121m	Sn-121
Sn-126	Sb-126m
Te-118	Sb-118
Te-127m	Te-127
Te-129m	Te-129
Te-131m	Te-131
Te-132	I-132
I-135	Xe-135m
Xe-122	I-122
Cs-137	Ba-137m
Ba-131	Cs-131
Ba-140	La-140
Ce-144	Pr-144m, Pr-144
Pm-148m	Pm-148
Gd-146	Eu-146
Dy-166	Ho-166
Hf-172	Lu-172
W-178	Ta-178
W-188	Re-188
Re-189	Os-189m
Os-194	Ir-194
Ir-189	Os-189m
Pt-188	Ir-188
Hg-194	Au-194
Hg-195m	Hg-195
Pb-210	Bi-210
Pb-212	Bi-212, Tl-208, Po-212
Bi-210m	Tl-206
Bi-212	Tl-208, Po-212
At-211	Po-211
Rn-222	Po-218, Pb-214, At-218, Bi-214, Po-214
Ra-223	Rn-219, Po-215, Pb-211, Bi-211, Po-211, Tl-207
Ra-224	Rn-220, Po-216, Pb-212, Bi-212, Tl-208, Po-212
Ra-225	Ac-225, Fr-221, At-217, Bi-213, Tl-209, Po-213, Pb-209
Ra-226	Rn-222, Po-218, Pb-214, At-218, Bi-214, Po-214
Ra-228	Ac-228
Ac-225	Fr-221, At-217, Bi-213, Tl-209, Po-213, Pb-209
Ac-227	Fr-223
Th-228	Ra-224, Rn-220, Po-216, Pb-212, Bi-212, Tl-208, Po-212
Th-234	Pa-234m, Pa-234
Pa-230	Ac-226, Th-226, Fr-222, Ra-222, Rn-218, Po-214
U-230	Th-226, Ra-222, Rn-218, Po-214
U-235	Th-231
Pu-241	U-237
Pu-244	U-240, Np-240m
Am-242m	Am-242, Np-238
Am-243	Np-239
Cm-247	Pu-243
Bk-249	Am-245
Cf-253	Cm-249

(b) Parent nuclides and their progeny included in secular equilibrium are listed in the following:

Sr-90	Y-90
Zr-93	Nb-93m
Zr-97	Nb-97
Ru-106	Rh-106
Ag-108m	Ag-108
Cs-137	Ba-137m
Ce-144	Pr-144
Ba-140	La-140
Bi-212	Tl-208 (0.36), Po-212 (0.64)
Pb-210	Bi-210, Po-210
Pb-212	Bi-212, Tl-208 (0.36), Po-212 (0.64)
Rn-222	Po-218, Pb-214, Bi-214, Po-214
Ra-223	Rn-219, Po-215, Pb-211, Bi-211, Tl-207
Ra-224	Rn-220, Po-216, Pb-212, Bi-212, Tl-208 (0.36), Po-212 (0.64)
Ra-226	Rn-222, Po-218, Pb-214, Bi-214, Po-214, Pb-210, Bi-210, Po-210
Ra-228	Ac-228
Th-228	Ra-224, Rn-220, Po-216, Pb212, Bi-212, Tl208 (0.36), Po-212 (0.64)
Th-229	Ra-225, Ac-225, Fr-221, At-217, Bi-213, Po-213, Pb-209
Th-nat	Ra-228, Ac-228, Th-228, Ra-224, Rn-220, Po-216, Pb-212, Bi-212, l208 (0.36), Po-212 (0.64)
Th-234	Pa-234m
U-230	Th-226, Ra-222, Rn-218, Po-214
U-232	Th-228, Ra-224, Rn-220, Po-216, Pb-212, Bi-212, Tl-208 (0.36), Po-212 (0.64)
U-235	Th-231
U-238	Th-234, Pa-234m
U-nat	Th-234, Pa-234m, U-234, Th-230, Ra-226, Rn-222, Po-218, Pb-214, Bi-214, Po-214, Pb-210, Bi-210, Po-210
Np-237	Pa-233
Am-242m	Am-242
Am-243	Np-239

(c) The quantity may be determined from a measurement of the rate of decay or a measurement of the radiation level at a prescribed distance from the source.

(d) These values apply only to compounds of uranium that take the chemical form of UF_6, UO_2F_2 and $UO_2(NO_3)_2$ in both normal and accident conditions of carriage.

(e) These values apply only to compounds of uranium that take the chemical form of UO_3, UF_4, UCl_4 and hexavalent compounds in both normal and accident conditions of carriage.

(f) These values apply to all compounds of uranium other than those specified in (d) and (e) above.

(g) These values apply to unirradiated uranium only.

2.2.7.2.2.2   For individual radionuclides:

(a) Which are not listed in Table 2.2.7.2.2.1 the determination of the basic radionuclide values referred to in 2.2.7.2.2.1 shall require multilateral approval. For these radionuclides, activity concentration limits for exempt material and activity limits for exempt consignments shall be calculated in accordance with the principles established in the International Basic Safety Standards for Protection against Ionizing Radiation and for the Safety of Radiation Sources, Safety Series No.115, IAEA, Vienna (1996). It is permissible to use an A_2 value calculated using a dose coefficient for the appropriate lung absorption type as recommended by the International Commission on Radiological Protection, if the chemical forms of each radionuclide under both normal and accident conditions of carriage are taken into consideration. Alternatively, the radionuclide values in Table 2.2.7.2.2.2 may be used without obtaining competent authority approval;

(b)	In instruments or articles in which the radioactive material is enclosed or is included as a component part of the instrument or other manufactured article and which meet 2.2.7.2.4.1.3 (c), alternative basic radionuclide values to those in Table 2.2.7.2.2.1 for the activity limit for an exempt consignment are permitted and shall require multilateral approval. Such alternative activity limits for an exempt consignment shall be calculated in accordance with the principles set out in the International Basic Safety Standards for Protection against Ionizing Radiation and for the Safety of Radiation Sources, Safety Series No.115, IAEA, Vienna (1996).

Table 2.2.7.2.2.2: Basic radionuclide values for unknown radionuclides or mixtures

Radioactive contents	A_1	A_2	Activity concentration limit for exempt material	Activity limit for exempt consignments
	(TBq)	(TBq)	(Bq/g)	(Bq)
Only beta or gamma emitting nuclides are known to be present	0.1	0.02	1×10^1	1×10^4
Alpha emitting nuclides but no neutron emitters are known to be present	0.2	9×10^{-5}	1×10^{-1}	1×10^3
Neutron emitting nuclides are known to be present or no relevant data are available	0.001	9×10^{-5}	1×10^{-1}	1×10^3

2.2.7.2.2.3	In the calculations of A_1 and A_2 for a radionuclide not in Table 2.2.7.2.2.1, a single radioactive decay chain in which the radionuclides are present in their naturally occurring proportions, and in which no daughter nuclide has a half-life either longer than 10 days or longer than that of the parent nuclide, shall be considered as a single radionuclide; and the activity to be taken into account and the A_1 or A_2 value to be applied shall be those corresponding to the parent nuclide of that chain. In the case of radioactive decay chains in which any daughter nuclide has a half-life either longer than 10 days or greater than that of the parent nuclide, the parent and such daughter nuclides shall be considered as mixtures of different nuclides.

2.2.7.2.2.4	For mixtures of radionuclides, the basic radionuclide values referred to in 2.2.7.2.2.1 may be determined as follows:

$$X_m = \frac{1}{\sum_i \frac{f(i)}{X(i)}}$$

where,

f(i)	is the fraction of activity or activity concentration of radionuclide i in the mixture;

X(i)	is the appropriate value of A_1 or A_2, or the activity concentration limit for exempt material or the activity limit for an exempt consignment as appropriate for the radionuclide i; and

X_m	is the derived value of A_1 or A_2, or the activity concentration limit for exempt material or the activity limit for an exempt consignment in the case of a mixture.

2.2.7.2.2.5	When the identity of each radionuclide is known but the individual activities of some of the radionuclides are not known, the radionuclides may be grouped and the lowest radionuclide value, as appropriate, for the radionuclides in each group may be used in applying the formulas in 2.2.7.2.2.4 and 2.2.7.2.4.4. Groups may be based on the total alpha activity and the total beta/gamma activity when these are known, using the lowest radionuclide values for the alpha emitters or beta/gamma emitters, respectively.

2.2.7.2.2.6	For individual radionuclides or for mixtures of radionuclides for which relevant data are not available, the values shown in Table 2.2.7.2.2.2 shall be used.

2.2.7.2.3	*Determination of other material characteristics*

2.2.7.2.3.1	Low specific activity (LSA) material

2.2.7.2.3.1.1   *(Reserved)*

2.2.7.2.3.1.2   LSA material shall be in one of three groups:

(a)   LSA-I

(i)   uranium and thorium ores and concentrates of such ores, and other ores containing naturally occurring radionuclides;

(ii)   natural uranium, depleted uranium, natural thorium or their compounds or mixtures, that are unirradiated and in solid or liquid form;

(iii)   radioactive material for which the A_2 value is unlimited. Fissile material may be included only if excepted under 2.2.7.2.3.5;

(iv)   other radioactive material in which the activity is distributed throughout and the estimated average specific activity does not exceed 30 times the values for activity concentration specified in 2.2.7.2.2.1 to 2.2.7.2.2.6. Fissile material may be included only if excepted under 2.2.7.2.3.5;

(b)   LSA-II

(i)   water with tritium concentration up to 0.8 TBq/l;

(ii)   other material in which the activity is distributed throughout and the estimated average specific activity does not exceed 10^{-4} A_2/g for solids and gases, and 10^{-5} A_2/g for liquids;

(c)   LSA-III - Solids (e.g. consolidated wastes, activated materials), excluding powders, that meet the requirements of 2.2.7.2.3.1.3, in which:

(i)   the radioactive material is distributed throughout a solid or a collection of solid objects, or is essentially uniformly distributed in a solid compact binding agent (such as concrete, bitumen and ceramic);

(ii)   the radioactive material is relatively insoluble, or it is intrinsically contained in a relatively insoluble matrix, so that, even under loss of packaging, the loss of radioactive material per package by leaching when placed in water for seven days would not exceed 0.1 A_2; and

(iii)   the estimated average specific activity of the solid, excluding any shielding material, does not exceed 2×10^{-3} A_2/g.

2.2.7.2.3.1.3   LSA-III material shall be a solid of such a nature that if the entire contents of a package were subjected to the test specified in 2.2.7.2.3.1.4 the activity in the water would not exceed 0.1 A_2.

2.2.7.2.3.1.4   LSA-III material shall be tested as follows:

A solid material sample representing the entire contents of the package shall be immersed for 7 days in water at ambient temperature. The volume of water to be used in the test shall be sufficient to ensure that at the end of the 7 day test period the free volume of the unabsorbed and unreacted water remaining shall be at least 10% of the volume of the solid test sample itself. The water shall have an initial pH of 6-8 and a maximum conductivity of 1 mS/m at 20 °C. The total activity of the free volume of water shall be measured following the 7 day immersion of the test sample.

2.2.7.2.3.1.5   Demonstration of compliance with the performance standards in 2.2.7.2.3.1.4 shall be in accordance with 6.4.12.1 and 6.4.12.2.

2.2.7.2.3.2   Surface contaminated object (SCO)

SCO is classified in one of two groups:

(a)   SCO-I: A solid object on which:

(i) the non-fixed contamination on the accessible surface averaged over 300 cm^2 (or the area of the surface if less than 300 cm^2) does not exceed 4 Bq/cm^2 for beta and gamma emitters and low toxicity alpha emitters, or 0.4 Bq/cm^2 for all other alpha emitters; and

(ii) the fixed contamination on the accessible surface averaged over 300 cm^2 (or the area of the surface if less than 300 cm^2) does not exceed 4×10^4 Bq/cm^2 for beta and gamma emitters and low toxicity alpha emitters, or 4×10^3 Bq/cm^2 for all other alpha emitters; and

(iii) the non-fixed contamination plus the fixed contamination on the inaccessible surface averaged over 300 cm^2 (or the area of the surface if less than 300 cm^2) does not exceed 4×10^4 Bq/cm^2 for beta and gamma emitters and low toxicity alpha emitters, or 4×10^3 Bq/cm^2 for all other alpha emitters;

(b) SCO-II: A solid object on which either the fixed or non-fixed contamination on the surface exceeds the applicable limits specified for SCO-I in (a) above and on which:

(i) the non-fixed contamination on the accessible surface averaged over 300 cm^2 (or the area of the surface if less than 300 cm^2) does not exceed 400 Bq/cm^2 for beta and gamma emitters and low toxicity alpha emitters, or 40 Bq/cm^2 for all other alpha emitters; and

(ii) the fixed contamination on the accessible surface, averaged over 300 cm^2 (or the area of the surface if less than 300 cm^2) does not exceed 8×10^5 Bq/cm^2 for beta and gamma emitters and low toxicity alpha emitters, or 8×10^4 Bq/cm^2 for all other alpha emitters; and

(iii) the non-fixed contamination plus the fixed contamination on the inaccessible surface averaged over 300 cm^2 (or the area of the surface if less than 300 cm^2) does not exceed 8×10^5 Bq/cm^2 for beta and gamma emitters and low toxicity alpha emitters, or 8×10^4 Bq/cm^2 for all other alpha emitters.

2.2.7.2.3.3   Special form radioactive material

2.2.7.2.3.3.1   Special form radioactive material shall have at least one dimension not less than 5 mm. When a sealed capsule constitutes part of the special form radioactive material, the capsule shall be so manufactured that it can be opened only by destroying it. The design for special form radioactive material requires unilateral approval.

2.2.7.2.3.3.2   Special form radioactive material shall be of such a nature or shall be so designed that if it is subjected to the tests specified in 2.2.7.2.3.3.4 to 2.2.7.2.3.3.8, it shall meet the following requirements:

(a) It would not break or shatter under the impact, percussion and bending tests 2.2.7.2.3.3.5 (a), (b), (c) and 2.2.7.2.3.3.6 (a) as applicable;

(b) It would not melt or disperse in the applicable heat test 2.2.7.2.3.3.5 (d) or 2.2.7.2.3.3.6 (b) as applicable; and

(c) The activity in the water from the leaching tests specified in 2.2.7.2.3.3.7 and 2.2.7.2.3.3.8 would not exceed 2 kBq; or alternatively for sealed sources, the leakage rate for the volumetric leakage assessment test specified in ISO 9978:1992 "Radiation Protection - Sealed Radioactive Sources - Leakage Test Methods", would not exceed the applicable acceptance threshold acceptable to the competent authority.

2.2.7.2.3.3.3   Demonstration of compliance with the performance standards in 2.2.7.2.3.3.2 shall be in accordance with 6.4.12.1 and 6.4.12.2.

2.2.7.2.3.3.4   Specimens that comprise or simulate special form radioactive material shall be subjected to the impact test, the percussion test, the bending test, and the heat test specified in 2.2.7.2.3.3.5 or alternative tests as authorized in 2.2.7.2.3.3.6. A different specimen may be used for each of the tests. Following each test, a leaching assessment or volumetric leakage test shall be performed on the specimen by a method no less sensitive than the methods given in 2.2.7.2.3.3.7 for indispersible solid material or 2.2.7.2.3.3.8 for encapsulated material.

2.2.7.2.3.3.5   The relevant test methods are:

(a)   Impact test: The specimen shall drop onto the target from a height of 9 m. The target shall be as defined in 6.4.14;

(b)   Percussion test: The specimen shall be placed on a sheet of lead which is supported by a smooth solid surface and struck by the flat face of a mild steel bar so as to cause an impact equivalent to that resulting from a free drop of 1.4 kg through 1 m. The lower part of the bar shall be 25 mm in diameter with the edges rounded off to a radius of (3.0 ± 0.3) mm. The lead, of hardness number 3.5 to 4.5 on the Vickers scale and not more than 25 mm thick, shall cover an area greater than that covered by the specimen. A fresh surface of lead shall be used for each impact. The bar shall strike the specimen so as to cause maximum damage;

(c)   Bending test: The test shall apply only to long, slender sources with both a minimum length of 10 cm and a length to minimum width ratio of not less than 10. The specimen shall be rigidly clamped in a horizontal position so that one half of its length protrudes from the face of the clamp. The orientation of the specimen shall be such that the specimen will suffer maximum damage when its free end is struck by the flat face of a steel bar. The bar shall strike the specimen so as to cause an impact equivalent to that resulting from a free vertical drop of 1.4 kg through 1 m. The lower part of the bar shall be 25 mm in diameter with the edges rounded off to a radius of (3.0 ± 0.3) mm;

(d)   Heat test: The specimen shall be heated in air to a temperature of 800 °C and held at that temperature for a period of 10 minutes and shall then be allowed to cool.

2.2.7.2.3.3.6   Specimens that comprise or simulate radioactive material enclosed in a sealed capsule may be excepted from:

(a)   The tests prescribed in 2.2.7.2.3.3.5 (a) and (b) provided that the specimens are alternatively subjected to the impact test prescribed in ISO 2919:2012: "Radiation Protection - Sealed Radioactive Sources - General requirements and classification":

(i)   The Class 4 impact test if the mass of the special form radioactive material is equal to or less than 200 g;

(ii)   The Class 5 impact test if the mass of the special form radioactive material is equal to or more than 200 g but is less than 500 g;

(b)   The test prescribed in 2.2.7.2.3.3.5 (d) provided they are alternatively subjected to the Class 6 temperature test specified in ISO 2919:2012 "Radiation protection - Sealed radioactive sources - General requirements and classification".

2.2.7.2.3.3.7   For specimens which comprise or simulate indispersible solid material, a leaching assessment shall be performed as follows:

(a)   The specimen shall be immersed for 7 days in water at ambient temperature. The volume of water to be used in the test shall be sufficient to ensure that at the end of the 7 day test period the free volume of the unabsorbed and unreacted water remaining shall be at least 10% of the volume of the solid test sample itself. The water shall have an initial pH of 6-8 and a maximum conductivity of 1 mS/m at 20 °C;

(b)   The water with specimen shall then be heated to a temperature of (50 ± 5) °C and maintained at this temperature for 4 hours;

(c)   The activity of the water shall then be determined;

(d)   The specimen shall then be kept for at least 7 days in still air at not less than 30 °C and relative humidity not less than 90%;

(e)   The specimen shall then be immersed in water of the same specification as in (a) above and the water with the specimen heated to (50 ± 5) °C and maintained at this temperature for 4 hours;

(f)   The activity of the water shall then be determined.

2.2.7.2.3.3.8 For specimens which comprise or simulate radioactive material enclosed in a sealed capsule, either a leaching assessment or a volumetric leakage assessment shall be performed as follows:

(a) The leaching assessment shall consist of the following steps:

(i) the specimen shall be immersed in water at ambient temperature. The water shall have an initial pH of 6-8 with a maximum conductivity of 1 mS/m at 20 °C;

(ii) the water and specimen shall be heated to a temperature of (50 ± 5) °C and maintained at this temperature for 4 hours;

(iii) the activity of the water shall then be determined;

(iv) the specimen shall then be kept for at least 7 days in still air at not less than 30 °C and relative humidity of not less than 90%;

(v) the process in (i), (ii) and (iii) shall be repeated;

(b) The alternative volumetric leakage assessment shall comprise any of the tests prescribed in ISO 9978:1992 "Radiation Protection - Sealed radioactive sources - Leakage test methods", provided that they are acceptable to the competent authority.

2.2.7.2.3.4 Low dispersible radioactive material

2.2.7.2.3.4.1 The design for low dispersible radioactive material shall require multilateral approval. Low dispersible radioactive material shall be such that the total amount of this radioactive material in a package, taking into account the provisions of 6.4.8.14, shall meet the following requirements:

(a) The radiation level at 3 m from the unshielded radioactive material does not exceed 10 mSv/h;

(b) If subjected to the tests specified in 6.4.20.3 and 6.4.20.4, the airborne release in gaseous and particulate forms of up to 100 μm aerodynamic equivalent diameter would not exceed 100 A_2. A separate specimen may be used for each test; and

(c) If subjected to the test specified in 2.2.7.2.3.1.4 the activity in the water would not exceed 100 A_2. In the application of this test, the damaging effects of the tests specified in (b) above shall be taken into account.

2.2.7.2.3.4.2 Low dispersible radioactive material shall be tested as follows:

A specimen that comprises or simulates low dispersible radioactive material shall be subjected to the enhanced thermal test specified in 6.4.20.3 and the impact test specified in 6.4.20.4. A different specimen may be used for each of the tests. Following each test, the specimen shall be subjected to the leach test specified in 2.2.7.2.3.1.4. After each test it shall be determined if the applicable requirements of 2.2.7.2.3.4.1 have been met.

2.2.7.2.3.4.3 Demonstration of compliance with the performance standards in 2.2.7.2.3.4.1 and 2.2.7.2.3.4.2 shall be in accordance with 6.4.12.1 and 6.4.12.2.

2.2.7.2.3.5 Fissile material

Fissile material and packages containing fissile material shall be classified under the relevant entry as "FISSILE" in accordance with Table 2.2.7.2.1.1 unless excepted by one of the provisions of sub-paragraphs (a) to (f) below and carried subject to the requirements of 7.5.11 CV33 (4.3). All provisions apply only to material in packages that meets the requirements of 6.4.7.2 unless unpackaged material is specifically allowed in the provision.

(a) Uranium enriched in uranium-235 to a maximum of 1% by mass, and with a total plutonium and uranium-233 content not exceeding 1% of the mass of uranium-235, provided that the fissile nuclides are distributed essentially homogeneously throughout the material. In addition, if uranium-235 is present in metallic, oxide or carbide forms, it shall not form a lattice arrangement;

(b)   Liquid solutions of uranyl nitrate enriched in uranium-235 to a maximum of 2% by mass, with a total plutonium and uranium-233 content not exceeding 0.002% of the mass of uranium, and with a minimum nitrogen to uranium atomic ratio (N/U) of 2;

(c)   Uranium with a maximum uranium enrichment of 5% by mass uranium-235 provided:

    (i)   There is no more than 3.5 g of uranium-235 per package;

    (ii)   The total plutonium and uranium-233 content does not exceed 1% of the mass of uranium-235 per package;

    (iii) Carriage of the package is subject to the consignment limit provided in 7.5.11 CV33 (4.3) (c);

(d)   Fissile nuclides with a total mass not greater than 2.0 g per package provided the package is carried subject to the consignment limit provided in 7.5.11 CV33 (4.3) (d);

(e)   Fissile nuclides with a total mass not greater than 45 g either packaged or unpackaged subject to limits provided in 7.5.11 CV33 (4.3) (e);

(f)   A fissile material that meets the requirements of 7.5.11 CV33 (4.3) (b), 2.2.7.2.3.6 and 5.1.5.2.1.

2.2.7.2.3.6   A fissile material excepted from classification as "FISSILE" under 2.2.7.2.3.5 (f) shall be subcritical without the need for accumulation control under the following conditions:

(a)   The conditions of 6.4.11.1 (a);

(b)   The conditions consistent with the assessment provisions stated in 6.4.11.12 (b) and 6.4.11.13 (b) for packages.

2.2.7.2.4   *Classification of packages or unpacked material*

The quantity of radioactive material in a package shall not exceed the relevant limits for the package type as specified below.

2.2.7.2.4.1   Classification as excepted package

2.2.7.2.4.1.1   A package may be classified as an excepted package if it meets one of the following   conditions:

(a)   It is an empty package having contained radioactive material;

(b)   It contains instruments or articles not exceeding the activity limits specified in columns (2) and (3) of Table 2.2.7.2.4.1.2;

(c)   It contains articles manufactured of natural uranium, depleted uranium or natural thorium;

(d)   It contains radioactive material not exceeding the activity limits specified in column (4) of Table 2.2.7.2.4.1.2; or

(e)   It contains less than 0.1 kg of uranium hexafluoride not exceeding the activity limits specified in column (4) of Table 2.2.7.2.4.1.2.

2.2.7.2.4.1.2   A package containing radioactive material may be classified as an excepted package provided that the radiation level at any point on its external surface does not exceed 5 µSv/h.

Table 2.2.7.2.4.1.2: Activity limits for excepted packages

| Physical state of contents | Instruments or articles | | Materials |
	Item limits [a]	Package limits [a]	Package limits [a]
(1)	(2)	(3)	(4)
Solids			
special form	$10^{-2}\,A_1$	A_1	$10^{-3}\,A_1$
other form	$10^{-2}\,A_2$	A_2	$10^{-3}\,A_2$
Liquids	$10^{-3}\,A_2$	$10^{-1}\,A_2$	$10^{-4}\,A_2$
Gases			
tritium	$2 \times 10^{-2}\,A_2$	$2 \times 10^{-1}\,A_2$	$2 \times 10^{-2}\,A_2$
special form	$10^{-3}\,A_1$	$10^{-2}\,A_1$	$10^{-3}\,A_1$
other forms	$10^{-3}\,A_2$	$10^{-2}\,A_2$	$10^{-3}\,A_2$

[a]   *For mixtures of radionuclides, see 2.2.7.2.2.4 to 2.2.7.2.2.6.*

2.2.7.2.4.1.3   Radioactive material which is enclosed in or is included as a component part of an instrument or other manufactured article may be classified under UN No. 2911 RADIOACTIVE MATERIAL, EXCEPTED PACKAGE - INSTRUMENTS or ARTICLES provided that:

(a)   The radiation level at 10 cm from any point on the external surface of any unpackaged instrument or article is not greater than 0.1 mSv/h;

(b)   Each instrument or manufactured article bears the mark "RADIOACTIVE" on its external surface except for the following:

(i)   radioluminescent time-pieces or devices;

(ii)   consumer products that have either received regulatory approval in accordance with 1.7.1.4 (e) or do not individually exceed the activity limit for an exempt consignment in Table 2.2.7.2.2.1 (column 5), provided such products are transported in a package that bears the mark "RADIOACTIVE" on its internal surface in such a manner that a warning of the presence of radioactive material is visible on opening the package; and

(iii)   other instruments or articles too small to bear the mark "RADIOACTIVE", provided that they are transported in a package that bears the mark "RADIOACTIVE" on its internal surface in such a manner that a warning of the presence of radioactive material is visible on opening the package;

(c)   The active material is completely enclosed by non-active components (a device performing the sole function of containing radioactive material shall not be considered to be an instrument or manufactured article); and

(d)   The limits specified in columns 2 and 3 of Table 2.2.7.2.4.1.2 are met for each individual item and each package, respectively.

2.2.7.2.4.1.4   Radioactive material in forms other than as specified in 2.2.7.2.4.1.3 and with an activity not exceeding the limits specified in column 4 of Table 2.2.7.2.4.1.2, may be classified under UN No. 2910 RADIOACTIVE MATERIAL, EXCEPTED PACKAGE - LIMITED QUANTITY OF MATERIAL provided that:

(a)   The package retains its radioactive contents under routine conditions of carriage; and

(b)   The package bears the mark "RADIOACTIVE" on either:

(i)   An internal surface in such a manner that a warning of the presence of radioactive material is visible on opening the package; or

(ii)   The outside of the package, where it is impractical to mark an internal surface.

2.2.7.2.4.1.5   Uranium hexafluoride not exceeding the limits specified in Column 4 of Table 2.2.7.2.4.1.2 may be classified under UN 3507 URANIUM HEXAFLUORIDE, RADIOACTIVE MATERIAL, EXCEPTED PACKAGE, less than 0.1 kg per package, non-fissile or fissile-excepted provided that:

(a)   The mass of uranium hexafluoride in the package is less than 0.1 kg;

(b)   The conditions of 2.2.7.2.4.5.2 and 2.2.7.2.4.1.4 (a) and (b) are met.

2.2.7.2.4.1.6   Articles manufactured of natural uranium, depleted uranium or natural thorium and articles in which the sole radioactive material is unirradiated natural uranium, unirradiated depleted uranium or unirradiated natural thorium may be classified under UN No. 2909 RADIOACTIVE MATERIAL, EXCEPTED PACKAGE - ARTICLES MANUFACTURED FROM NATURAL URANIUM or DEPLETED URANIUM or NATURAL THORIUM, provided that the outer surface of the uranium or thorium is enclosed in an inactive sheath made of metal or some other substantial material.

2.2.7.2.4.1.7   An empty packaging which had previously contained radioactive material may be classified under UN No. 2908 RADIOACTIVE MATERIAL, EXCEPTED PACKAGE - EMPTY PACKAGING, provided that:

(a)   It is in a well-maintained condition and securely closed;

(b)   The outer surface of any uranium or thorium in its structure is covered with an inactive sheath made of metal or some other substantial material;

(c)   The level of internal non-fixed contamination, when averaged over any 300 cm^2, does not exceed:

(i)   400 Bq/cm^2 for beta and gamma emitters and low toxicity alpha emitters; and

(ii)   40 Bq/cm^2 for all other alpha emitters; and

(d)   Any labels which may have been displayed on it in conformity with 5.2.2.1.11.1 are no longer visible.

2.2.7.2.4.2   Classification as Low specific activity (LSA) material

Radioactive material may only be classified as LSA material if the definition of LSA in 2.2.7.1.3 and the conditions of 2.2.7.2.3.1, 4.1.9.2 and 7.5.11 CV33 (2) are met.

2.2.7.2.4.3   Classification as Surface contaminated object (SCO)

Radioactive material may be classified as SCO if the definition of SCO in 2.2.7.1.3 and the conditions of 2.2.7.2.3.2, 4.1.9.2 and 7.5.11 CV33 (2) are met.

2.2.7.2.4.4   Classification as Type A package

Packages containing radioactive material may be classified as Type A packages provided that the following conditions are met:

Type A packages shall not contain activities greater than either of the following:

(a)   For special form radioactive material - A_1;

(b)   For all other radioactive material - A_2.

For mixtures of radionuclides whose identities and respective activities are known, the following condition shall apply to the radioactive contents of a Type A package:

$$\Sigma_i \frac{B(i)}{A_1(i)} + \Sigma_j \frac{C(j)}{A_2(j)} \leq 1$$

where $B(i)$ is the activity of radionuclide i as special form radioactive material;

         $A_1(i)$ is the A_1 value for radionuclide i;

         $C(j)$ is the activity of radionuclide j as other than special form radioactive material;

         $A_2(j)$ is the A_2 value for radionuclide j.

2.2.7.2.4.5   Classification of uranium hexafluoride

2.2.7.2.4.5.1   Uranium hexafluoride shall only be assigned to:

(a)   UN No. 2977, RADIOACTIVE MATERIAL, URANIUM HEXAFLUORIDE, FISSILE;

(b)   UN No. 2978, RADIOACTIVE MATERIAL, URANIUM HEXAFLUORIDE, non-fissile or fissile-excepted; or

(c)   UN No. 3507, URANIUM HEXAFLUORIDE, RADIOACTIVE MATERIAL, EXCEPTED PACKAGE less than 0.1 kg per package, non-fissile or fissile-excepted.

2.2.7.2.4.5.2   The contents of a package containing uranium hexafluoride shall comply with the following requirements:

(a)   For UN Nos. 2977 and 2978, the mass of uranium hexafluoride shall not be different from that allowed for the package design, and for UN No. 3507, the mass of uranium hexafluoride shall be less than 0.1 kg;

(b)   The mass of uranium hexafluoride shall not be greater than a value that would lead to an ullage smaller than 5% at the maximum temperature of the package as specified for the plant systems where the package shall be used; and

(c)   The uranium hexafluoride shall be in solid form and the internal pressure shall not be above atmospheric pressure when presented for carriage.

2.2.7.2.4.6   Classification as Type B(U), Type B(M) or Type C packages

2.2.7.2.4.6.1   Packages not otherwise classified in 2.2.7.2.4 (2.2.7.2.4.1 to 2.2.7.2.4.5) shall be classified in accordance with the competent authority certificate of approval for the package issued by the country of origin of design.

2.2.7.2.4.6.2   The contents of a Type B(U), Type B(M) or Type C package shall be as specified in the certificate of approval.

2.2.7.2.5   *Special arrangements*

Radioactive material shall be classified as transported under special arrangement when it is intended to be carried in accordance with 1.7.4.

2.2.8   **Class 8**   **Corrosive substances**

2.2.8.1   *Criteria*

2.2.8.1.1   The heading of Class 8 covers substances and articles containing substances of this class which by chemical action attack epithelial tissue - of skin or mucous membranes - with which they are in contact, or which in the event of leakage are capable of damaging or destroying other goods, or means of transport. The heading of this class also covers other substances which form a corrosive liquid only in the presence of water, or which produce corrosive vapour or mist in the presence of natural moisture of the air.

2.2.8.1.2   Substances and articles of Class 8 are subdivided as follows:

C1-C11 Corrosive substances without subsidiary risk and articles containing such substances:

C1-C4 Acid substances:

C1	Inorganic, liquid;
C2	Inorganic, solid;
C3	Organic, liquid;
C4	Organic, solid;

C5-C8 Basic substances:

C5	Inorganic, liquid;
C6	Inorganic, solid;
C7	Organic, liquid;
C8	Organic, solid;

C9-C10   Other corrosive substances:

C9	Liquid;
C10	Solid;

C11   Articles;

CF   Corrosive substances, flammable:

CF1   Liquid;
CF2   Solid;

CS   Corrosive substances, self-heating:

CS1   Liquid;
CS2   Solid;

CW   Corrosive substances which, in contact with water, emit flammable gases:

CW1   Liquid;
CW2   Solid;

CO   Corrosive substances, oxidizing:

CO1   Liquid;
CO2   Solid;

CT   Corrosive substances, toxic and articles containing such substances:

CT1   Liquid;
CT2   Solid;
CT3   Articles;

CFT   Corrosive substances, flammable, liquid, toxic;

COT   Corrosive substances, oxidizing, toxic.

Classification and assignment of packing groups

2.2.8.1.3   Substances of Class 8 shall be classified in three packing groups according to the degree of danger they present for carriage, as follows:

Packing group I: highly corrosive substances
Packing group II: corrosive substances
Packing group III: slightly corrosive substances.

2.2.8.1.4   Substances and articles classified in Class 8 are listed in Table A of Chapter 3.2. Allocation of substances to packing groups I, II and III has been made on the basis of experience taking into account such additional factors as inhalation risk (see 2.2.8.1.5) and reactivity with water (including the formation of dangerous decomposition products).

2.2.8.1.5   A substance or preparation meeting the criteria of Class 8 having an inhalation toxicity of dusts and mists (LC_{50}) in the range of packing group I, but toxicity through oral ingestion or dermal contact only in the range of packing group III or less, shall be allocated to Class 8.

2.2.8.1.6   Substances, including mixtures, not mentioned by name in Table A of Chapter 3.2 can be assigned to the relevant entry of sub-section 2.2.8.3, and to the relevant packing group on the basis of the length of time of contact necessary to produce full thickness destruction of human skin in accordance with the criteria of (a) to (c) below.

Liquids, and solids which may become liquid during carriage, which are judged not to cause full thickness destruction of human skin shall still be considered for their potential to cause corrosion to certain metal surfaces. In assigning the packing group, account shall be taken of human experience in instances of accidental exposure. In the absence of human experience, the grouping shall be based on data obtained from experiments in accordance with OECD Test Guideline 404[5] or 435[6]. A substance which is determined not to be corrosive in accordance with OECD Test Guideline 430[7] or 431[8] may be considered not to be corrosive to skin for the purposes of ADR without further testing.

(a)   Packing group I is assigned to substances that cause full thickness destruction of intact skin tissue within an observation period up to 60 minutes starting after the exposure time of 3 minutes or less;

(b)   Packing group II is assigned to substances that cause full thickness destruction of intact skin tissue within an observation period up to 14 days starting after the exposure time of more than 3 minutes but not more than 60 minutes;

(c)   Packing group III is assigned to substances that:

-    cause full thickness destruction of intact skin tissue within an observation period up to 14 days starting after the exposure time of more than 60 minutes but not more than 4 hours; or

-    are judged not to cause full thickness destruction of intact skin tissue, but which exhibit a corrosion rate on either steel or aluminium surfaces exceeding 6.25 mm a year at a test temperature of 55 °C when tested on both materials. For the purposes of testing steel, type S235JR+CR (1.0037 resp. St 37-2), S275J2G3+CR (1.0144 resp. St 44-3), ISO 3574, Unified Numbering System

(UNS) G10200 or SAE 1020, and for testing aluminium, non-clad, types 7075-T6 or AZ5GU-T6 shall be used. An acceptable test is prescribed in the Manual of Tests and Criteria, Part III, Section 37.

NOTE: Where an initial test on either steel or aluminium indicates the substance being tested is corrosive the follow up test on the other metal is not required.

---

[5]   *OECD Guideline for the testing of chemicals No. 404 "Acute Dermal Irritation/Corrosion" 2002.*

[6]   *OECD Guideline for the testing of chemicals No. 435 "In Vitro Membrane Barrier Test Method for Skin Corrosion" 2006.*

[7]   *OECD Guideline for the testing of chemicals No. 430 "In Vitro Skin Corrosion: Transcutaneous Electrical Resistance Test (TER)" 2004.*

[8]   *OECD Guideline for the testing of chemicals No. 431 "In Vitro Skin Corrosion: Human Skin Model Test" 2004.*

Table 2.2.8.1.6: Table summarizing the criteria in 2.2.8.1.6

Packing Group	Exposure Time	Observation Period	Effect
I	≤ 3 min	≤ 60 min	Full thickness destruction of intact skin
II	> 3 min ≤ 1 h	≤ 14 d	Full thickness destruction of intact skin
III	> 1 h ≤ 4 h	≤ 14 d	Full thickness destruction of intact skin
III	-	-	Corrosion rate on either steel or aluminium surfaces exceeding 6.25 mm a year at a test temperature of 55 °C when tested on both materials

2.2.8.1.7   If substances of Class 8, as a result of admixtures, come into categories of risk different from those to which the substances mentioned by name in Table A of Chapter 3.2 belong, these mixtures or solutions shall be assigned to the entries to which they belong, on the basis of their actual degree of danger.

NOTE: For the classification of solutions and mixtures (such as preparations and wastes), see also 2.1.3.

2.2.8.1.8   On the basis of the criteria set out in paragraph 2.2.8.1.6, it may also be determined whether the nature of a solution or mixture mentioned by name or containing a substance mentioned by name is such that the solution or mixture is not subject to the provisions for this class.

2.2.8.1.9   Substances, solutions and mixtures, which are not classified as corrosive to skin or metal of category 1 according to Regulation (EC) No 1272/2008[3] may be considered as substances not belonging to Class 8.

NOTE: UN No. 1910 calcium oxide and UN No. 2812 sodium aluminate, listed in the UN Model Regulations, are not subject to the provisions of ADR.

2.2.8.2   *Substances not accepted for carriage*

2.2.8.2.1   Chemically unstable substances of Class 8 shall not be accepted for carriage unless the necessary precautions have been taken to prevent the possibility of a dangerous decomposition or polymerization under normal conditions of carriage. For the precautions necessary to prevent polymerization, see special provision 386 of Chapter 3.3. To this end particular care shall be taken to ensure that receptacles and tanks do not contain any substances liable to promote these reactions.

2.2.8.2.2   The following substances shall not be accepted for carriage:

-    UN No. 1798 NITROHYDROCHLORIC ACID;

-    chemically unstable mixtures of spent sulphuric acid;

-    chemically unstable mixtures of nitrating acid or mixtures of residual sulphuric and nitric acids, not denitrated;

-    perchloric acid aqueous solution with more than 72% pure acid, by mass, or mixtures of perchloric acid with any liquid other than water.

---

[3]   *Regulation (EC) No 1272/2008 of the European Parliament and of the Council of 16 December 2008 on classification, labelling and packaging of substances and mixtures, amending and repealing Directive 67/548/EEC and 1999/45/EC; and amending Regulation (EC) No 1907/2006, published in the Official Journal of the European Union, L 353, 31 December 2008, p 1-1355.*

2.2.8.3   *List of collective entries*

Corrosive substances <u>without</u> subsidiary risk and articles containing such substances

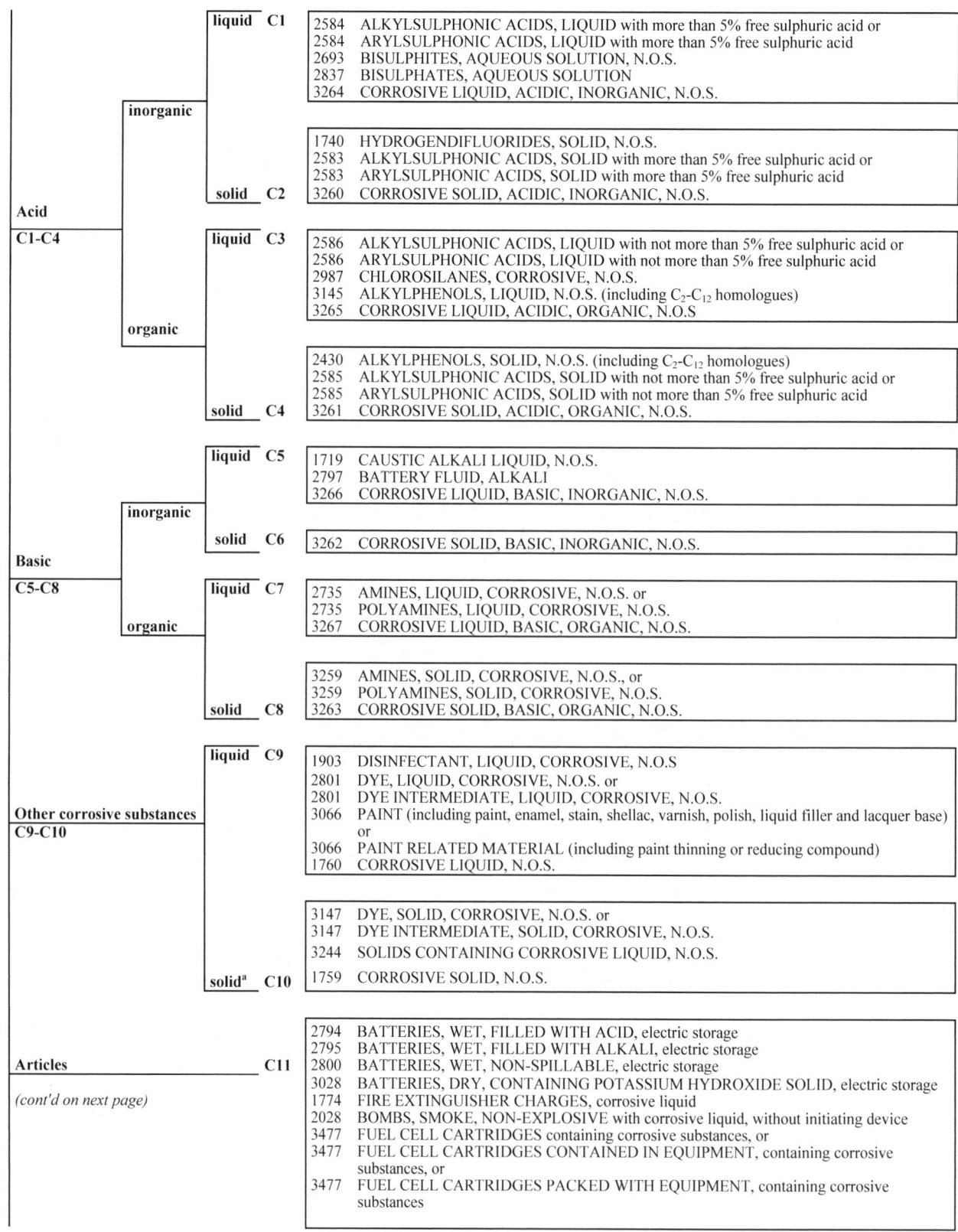

	liquid C1	2584 ALKYLSULPHONIC ACIDS, LIQUID with more than 5% free sulphuric acid or 2584 ARYLSULPHONIC ACIDS, LIQUID with more than 5% free sulphuric acid 2693 BISULPHITES, AQUEOUS SOLUTION, N.O.S. 2837 BISULPHATES, AQUEOUS SOLUTION 3264 CORROSIVE LIQUID, ACIDIC, INORGANIC, N.O.S.

inorganic

solid C2
1740 HYDROGENDIFLUORIDES, SOLID, N.O.S.
2583 ALKYLSULPHONIC ACIDS, SOLID with more than 5% free sulphuric acid or
2583 ARYLSULPHONIC ACIDS, SOLID with more than 5% free sulphuric acid
3260 CORROSIVE SOLID, ACIDIC, INORGANIC, N.O.S.

Acid
C1-C4

liquid C3
2586 ALKYLSULPHONIC ACIDS, LIQUID with not more than 5% free sulphuric acid or
2586 ARYLSULPHONIC ACIDS, LIQUID with not more than 5% free sulphuric acid
2987 CHLOROSILANES, CORROSIVE, N.O.S.
3145 ALKYLPHENOLS, LIQUID, N.O.S. (including C_2-C_{12} homologues)
3265 CORROSIVE LIQUID, ACIDIC, ORGANIC, N.O.S

organic

solid C4
2430 ALKYLPHENOLS, SOLID, N.O.S. (including C_2-C_{12} homologues)
2585 ALKYLSULPHONIC ACIDS, SOLID with not more than 5% free sulphuric acid or
2585 ARYLSULPHONIC ACIDS, SOLID with not more than 5% free sulphuric acid
3261 CORROSIVE SOLID, ACIDIC, ORGANIC, N.O.S.

liquid C5
1719 CAUSTIC ALKALI LIQUID, N.O.S.
2797 BATTERY FLUID, ALKALI
3266 CORROSIVE LIQUID, BASIC, INORGANIC, N.O.S.

inorganic

solid C6
3262 CORROSIVE SOLID, BASIC, INORGANIC, N.O.S.

Basic
C5-C8

liquid C7
2735 AMINES, LIQUID, CORROSIVE, N.O.S. or
2735 POLYAMINES, LIQUID, CORROSIVE, N.O.S.
3267 CORROSIVE LIQUID, BASIC, ORGANIC, N.O.S.

organic

solid C8
3259 AMINES, SOLID, CORROSIVE, N.O.S., or
3259 POLYAMINES, SOLID, CORROSIVE, N.O.S.
3263 CORROSIVE SOLID, BASIC, ORGANIC, N.O.S.

liquid C9
1903 DISINFECTANT, LIQUID, CORROSIVE, N.O.S
2801 DYE, LIQUID, CORROSIVE, N.O.S. or
2801 DYE INTERMEDIATE, LIQUID, CORROSIVE, N.O.S.
3066 PAINT (including paint, enamel, stain, shellac, varnish, polish, liquid filler and lacquer base) or
3066 PAINT RELATED MATERIAL (including paint thinning or reducing compound)
1760 CORROSIVE LIQUID, N.O.S.

Other corrosive substances
C9-C10

solid^a C10
3147 DYE, SOLID, CORROSIVE, N.O.S. or
3147 DYE INTERMEDIATE, SOLID, CORROSIVE, N.O.S.
3244 SOLIDS CONTAINING CORROSIVE LIQUID, N.O.S.
1759 CORROSIVE SOLID, N.O.S.

Articles   C11

(cont'd on next page)
2794 BATTERIES, WET, FILLED WITH ACID, electric storage
2795 BATTERIES, WET, FILLED WITH ALKALI, electric storage
2800 BATTERIES, WET, NON-SPILLABLE, electric storage
3028 BATTERIES, DRY, CONTAINING POTASSIUM HYDROXIDE SOLID, electric storage
1774 FIRE EXTINGUISHER CHARGES, corrosive liquid
2028 BOMBS, SMOKE, NON-EXPLOSIVE with corrosive liquid, without initiating device
3477 FUEL CELL CARTRIDGES containing corrosive substances, or
3477 FUEL CELL CARTRIDGES CONTAINED IN EQUIPMENT, containing corrosive substances, or
3477 FUEL CELL CARTRIDGES PACKED WITH EQUIPMENT, containing corrosive substances

---

^a   *Mixtures of solids which are not subject to the provisions of ADR and of corrosive liquids may be carried under UN No. 3244 without being subject to the classification criteria of Class 8, provided there is no free liquid visible at the time the substance is loaded or at the time the packaging, container or transport unit is closed. Each packaging shall correspond to a design type which has passed the leakproofness test for Packing group II level.*

Corrosive substances <u>with</u> subsidiary risk(s) and articles containing such substances

(cont'd)	liquid	CF1	3470	PAINT, CORROSIVE, FLAMMABLE (including paint, enamel, stain, shellac, varnish, polish, liquid filler and lacquer base) or
			3470	PAINT RELATED MATERIAL, CORROSIVE, FLAMMABLE (including paint thinning or reducing compound)
			2734	AMINES, LIQUID, CORROSIVE, FLAMMABLE, N.O.S. or
			2734	POLYAMINES, LIQUID, CORROSIVE, FLAMMABLE, N.O.S.
			2986	CHLOROSILANES, CORROSIVE, FLAMMABLE, N.O.S.
Flammable [b]			2920	CORROSIVE LIQUID, FLAMMABLE, N.O.S.
CF	solid	CF2	2921	CORROSIVE SOLID, FLAMMABLE, N.O.S.
	liquid	CS1	3301	CORROSIVE LIQUID, SELF-HEATING, N.O.S.
Self-heating **CS**	solid	CS2	3095	CORROSIVE SOLID, SELF-HEATING, N.O.S.
	liquid [b]	CW1	3094	CORROSIVE LIQUID, WATER-REACTIVE, N.O.S.
Water-reactive **CW**	solid	CW2	3096	CORROSIVE SOLID, WATER-REACTIVE, N.O.S.
	liquid	CO1	3093	CORROSIVE LIQUID, OXIDIZING, N.O.S.
Oxidizing **CO**	solid	CO2	3084	CORROSIVE SOLID, OXIDIZING, N.O.S.
	liquid [c]	CT1	3471	HYDROGENDIFLUORIDES SOLUTION, N.O.S.
			2922	CORROSIVE LIQUID, TOXIC, N.O.S.
Toxic [d] **CT**	solid [c]	CT2	2923	CORROSIVE SOLID, TOXIC, N.O.S.
	articles	CT3	3506	MERCURY CONTAINED IN MANUFACTURED ARTICLES
Flammable, liquid, toxic [d]		CFT	No collective entry with this classification code available; if need be, classification under a collective entry with a classification code to be determined according to table of precedence of hazard in 2.1.3.10.	
Oxidizing, toxic [d, e]		COT	No collective entry with this classification code available; if need be, classification under a collective entry with a classification code to be determined according to table of precedence of hazard in 2.1.3.10.	

---

[b]   *Chlorosilanes which, in contact with water or moist air, emit flammable gases, are substances of Class 4.3.*

[c]   *Chloroformates having predominantly toxic properties are substances of Class 6.1.*

[d]   *Corrosive substances which are highly toxic by inhalation, as defined in 2.2.61.1.4 to 2.2.61.1.9 are substances of Class 6.1.*

[e]   *UN No. 2505 AMMONIUM FLUORIDE, UN No. 1812 POTASSIUM FLUORIDE, SOLID, UN No. 1690 SODIUM FLUORIDE, SOLID, UN No. 2674 SODIUM FLUOROSILICATE, UN No. 2856 FLUOROSILICATES, N.O.S., UN No. 3415 SODIUM FLUORIDE SOLUTION and UN No. 3422 POTASSIUM FLUORIDE SOLUTION are substances of Class 6.1.*

2.2.9   **Class 9   Miscellaneous dangerous substances and articles**

2.2.9.1   *Criteria*

2.2.9.1.1   The heading of Class 9 covers substances and articles which, during carriage, present a danger not covered by the heading of other classes.

2.2.9.1.2   The substances and articles of Class 9 are subdivided as follows:

M1   Substances which, on inhalation as fine dust, may endanger health;

M2   Substances and articles which, in the event of fire, may form dioxins;

M3   Substances evolving flammable vapour;

M4   Lithium batteries;

M5   Life-saving appliances;

M6-M8   Environmentally hazardous substances:

M6   Pollutant to the aquatic environment, liquid;
M7   Pollutant to the aquatic environment, solid;
M8   Genetically modified microorganisms and organisms;

M9-M10   Elevated temperature substances:

M9   Liquid;
M10   Solid;

M11   Other substances and articles presenting a danger during carriage, but not meeting the definitions of another class.

Definitions and classification

2.2.9.1.3   Substances and articles classified in Class 9 are listed in Table A of Chapter 3.2. The assignment of substances and articles not mentioned by name in Table A of Chapter 3.2 to the relevant entry of that Table or of sub-section 2.2.9.3 shall be done in accordance with 2.2.9.1.4 to 2.2.9.1.14 below.

Substances which, on inhalation as fine dust, may endanger health

2.2.9.1.4   Substances which, on inhalation as fine dust, may endanger health include asbestos and mixtures containing asbestos.

Substances and articles which, in the event of fire, may form dioxins

2.2.9.1.5   Substances and articles which, in the event of fire, may form dioxins include polychlorinated biphenyls (PCBs) and terphenyls (PCTs) and polyhalogenated biphenyls and terphenyls and mixtures containing these substances, as well as articles such as transformers, condensers and articles containing those substances or mixtures.

NOTE: Mixtures with a PCB or PCT content of not more than 50 mg/kg are not subject to the provisions of ADR.

Substances evolving flammable vapour

2.2.9.1.6   Substances evolving flammable vapour include polymers containing flammable liquids with a flash-point not exceeding 55 °C.

Lithium batteries

2.2.9.1.7   Lithium batteries shall meet the following requirements, except when otherwise provided for in ADR (e.g. for prototype batteries and small production runs under special provision 310 or damaged batteries under special provision 376).

Cells and batteries, cells and batteries contained in equipment, or cells and batteries packed with equipment, containing lithium in any form shall be assigned to UN Nos. 3090, 3091, 3480 or 3481 as appropriate. They may be carried under these entries if they meet the following provisions:

(a) Each cell or battery is of the type proved to meet the requirements of each test of the Manual of Tests and Criteria, Part III, sub-section 38.3;

   NOTE: Batteries shall be of a type proved to meet the testing requirements of the Manual of Tests and Criteria, part III, sub-section 38.3, irrespective of whether the cells of which they are composed are of a tested type.

(b) Each cell and battery incorporates a safety venting device or is designed to preclude a violent rupture under normal conditions of carriage;

(c) Each cell and battery is equipped with an effective means of preventing external short circuits;

(d) Each battery containing cells or series of cells connected in parallel is equipped with effective means as necessary to prevent dangerous reverse current flow (e.g., diodes, fuses, etc.);

(e) Cells and batteries shall be manufactured under a quality management programme that includes:

   (i) A description of the organizational structure and responsibilities of personnel with regard to design and product quality;

   (ii) The relevant inspection and test, quality control, quality assurance, and process operation instructions that will be used;

   (iii) Process controls that should include relevant activities to prevent and detect internal short circuit failure during manufacture of cells;

   (iv) Quality records, such as inspection reports, test data, calibration data and certificates. Test data shall be kept and made available to the competent authority upon request;

   (v) Management reviews to ensure the effective operation of the quality management programme;

   (vi) A process for control of documents and their revision;

   (vii) A means for control of cells or batteries that are not conforming to the type tested as mentioned in (a) above;

   (viii) Training programmes and qualification procedures for relevant personnel; and

   (ix) Procedures to ensure that there is no damage to the final product.

   NOTE: In house quality management programmes may be accepted. Third party certification is not required, but the procedures listed in (i) to (ix) above shall be properly recorded and traceable. A copy of the quality management programme shall be made available to the competent authority upon request.

Lithium batteries are not subject to the provisions of ADR if they meet the requirements of special provision 188 of Chapter 3.3.

Life-saving appliances

2.2.9.1.8   Life-saving appliances include life-saving appliances and motor vehicle components which meet the descriptions of special provisions 235 or 296 of Chapter 3.3.

Environmentally hazardous substances

2.2.9.1.9   *(Deleted)*

Pollutants to the aquatic environment

2.2.9.1.10   *Environmentally hazardous substances (aquatic environment)*

2.2.9.1.10.1   General definitions

2.2.9.1.10.1.1 Environmentally hazardous substances include, inter alia, liquid or solid substances pollutant to the aquatic environment and solutions and mixtures of such substances (such as preparations and wastes).

For the purposes of 2.2.9.1.10, "*substance*" means chemical elements and their compounds in the natural state or obtained by any production process, including any additive necessary to preserve the stability of the product and any impurities deriving from the process used, but excluding any solvent which may be separated without affecting the stability of the substance or changing its composition.

2.2.9.1.10.1.2 The aquatic environment may be considered in terms of the aquatic organisms that live in the water, and the aquatic ecosystem of which they are part[9]. The basis, therefore, of the identification of hazard is the aquatic toxicity of the substance or mixture, although this may be modified by further information on the degradation and bioaccumulation behaviour.

2.2.9.1.10.1.3 While the following classification procedure is intended to apply to all substances and mixtures, it is recognised that in some cases, e.g. metals or poorly soluble inorganic compounds, special guidance will be necessary[10].

2.2.9.1.10.1.4 The following definitions apply for acronyms or terms used in this section:

-       BCF: Bioconcentration Factor;

-       BOD: Biochemical Oxygen Demand;

-       COD: Chemical Oxygen Demand;

-       GLP: Good Laboratory Practices;

-       EC_x: the concentration associated with x% response;

-       EC_{50}: the effective concentration of substance that causes 50% of the maximum response;

-       ErC_{50}: EC_{50} in terms of reduction of growth;

-       Kow: octanol/water partition coefficient;

-       LC_{50} (50% lethal concentration): the concentration of a substance in water which causes the death of 50% (one half) in a group of test animals;

-       $L(E)C_{50}$: LC_{50} or EC_{50};

-       NOEC (No Observed Effect Concentration): the test concentration immediately below the lowest tested concentration with statistically significant adverse effect. The NOEC has no statistically significant adverse effect compared to the control;

-       OECD Test Guidelines: Test guidelines published by the Organization for Economic Cooperation and Development (OECD).

---

[9]   *This does not address aquatic pollutants for which there may be a need to consider effects beyond the aquatic environment such as the impacts on human health etc.*

[10]   *This can be found in Annex 10 of the GHS.*

2.2.9.1.10.2   Definitions and data requirements

2.2.9.1.10.2.1   The basic elements for classification of environmentally hazardous substances (aquatic environment) are:

(a)   Acute aquatic toxicity;

(b)   Chronic aquatic toxicity;

(c)   Potential for or actual bioaccumulation; and

(d)   Degradation (biotic or abiotic) for organic chemicals.

2.2.9.1.10.2.2   While data from internationally harmonised test methods are preferred, in practice, data from national methods may also be used where they are considered as equivalent. In general, it has been agreed that freshwater and marine species toxicity data can be considered as equivalent data and are preferably to be derived using OECD Test Guidelines or equivalent according to the principles of Good Laboratory Practices (GLP). Where such data are not available, classification shall be based on the best available data.

2.2.9.1.10.2.3   *Acute aquatic toxicity* means the intrinsic property of a substance to be injurious to an organism in a short-term aquatic exposure to that substance.

Acute (short-term) hazard, for classification purposes, means the hazard of a chemical caused by its acute toxicity to an organism during short-term aquatic exposure to that chemical.

Acute aquatic toxicity shall normally be determined using a fish 96 hour LC_{50} (OECD Test Guideline 203 or equivalent), a crustacea species 48 hour EC_{50} (OECD Test Guideline 202 or equivalent) and/or an algal species 72 or 96 hour EC_{50} (OECD Test Guideline 201 or equivalent). These species are considered as surrogate for all aquatic organisms and data on other species such as Lemna may also be considered if the test methodology is suitable.

2.2.9.1.10.2.4   *Chronic aquatic toxicity* means the intrinsic property of a substance to cause adverse effects to aquatic organisms during aquatic exposures which are determined in relation to the life-cycle of the organism.

Long-term hazard, for classification purposes, means the hazard of a chemical caused by its chronic toxicity following long-term exposure in the aquatic environment.

Chronic toxicity data are less available than acute data and the range of testing procedures less standardised. Data generated according to the OECD Test Guidelines 210 (Fish Early Life Stage) or 211 (Daphnia Reproduction) and 201 (Algal Growth Inhibition) may be accepted. Other validated and internationally accepted tests may also be used. The NOECs or other equivalent EC_x shall be used.

2.2.9.1.10.2.5   *Bioaccumulation* means net result of uptake, transformation and elimination of a substance in an organism due to all routes of exposure (i.e. air, water, sediment/soil and food).

The potential for bioaccumulation shall normally be determined by using the octanol/water partition coefficient, usually reported as a log K_{ow} determined according to OECD Test Guideline 107 or 117. While this represents a potential to bioaccumulate, an experimentally determined Bioconcentration Factor (BCF) provides a better measure and shall be used in preference when available. A BCF shall be determined according to OECD Test Guideline 107, 117 or 123.

2.2.9.1.10.2.6   *Degradation* means the decomposition of organic molecules to smaller molecules and eventually to carbon dioxide, water and salts.

Environmental degradation may be biotic or abiotic (e.g. hydrolysis) and the criteria used reflect this fact. Ready biodegradation is most easily defined using the biodegradability tests (A-F) of OECD Test Guideline 301. A pass level in these tests may be considered as indicative of rapid degradation in most environments. These are freshwater tests and thus the use of the results from OECD Test Guideline 306, which is more suitable for marine environments, has also been included. Where such data are not available, a BOD(5 days)/COD ratio ≥ 0.5 is considered as indicative of rapid degradation.

Abiotic degradation such as hydrolysis, primary degradation, both abiotic and biotic, degradation in non-aquatic media and proven rapid degradation in the environment may all be considered in defining rapid degradability[11].

Substances are considered rapidly degradable in the environment if the following criteria are met:

(a)   In 28-day ready biodegradation studies, the following levels of degradation are achieved:

   (i)   Tests based on dissolved organic carbon: 70%;

   (ii)   Tests based on oxygen depletion or carbon dioxide generation: 60% of theoretical maxima;

   These levels of biodegradation shall be achieved within 10 days of the start of degradation which point is taken as the time when 10% of the substance has been degraded ", unless the substance is identified as a complex, multi-component substance with structurally similar constituents. In this case, and where there is sufficient justification, the 10-day window condition may be waived and the pass level applied at 28 days[12]; or

(b)   In those cases where only BOD and COD data are available, when the ratio of BOD5/COD is ≥ 0.5; or

(c)   If other convincing scientific evidence is available to demonstrate that the substance can be degraded (biotically and/or abiotically) in the aquatic environment to a level above 70% within a 28 day period.

2.2.9.1.10.3   Substance classification categories and criteria

2.2.9.1.10.3.1   Substances shall be classified as "environmentally hazardous substances (aquatic environment)", if they satisfy the criteria for Acute 1, Chronic 1 or Chronic 2, according to Table 2.2.9.1.10.3.1. These criteria describe in detail the classification categories. They are diagrammatically summarized in Table 2.2.9.1.10.3.2.

---

[11]   *Special guidance on data interpretation is provided in Chapter 4.1 and Annex 9 of the GHS.*
[12]   *See Chapter 4.1 and Annex 9, paragraph A9.4.2.2.3 of the GHS.*

Table 2.2.9.1.10.3.1: Categories for substances hazardous to the aquatic environment *(see Note 1)*

(a)   Acute (short-term) aquatic hazard

Category Acute 1: *(see Note 2)*	
96 hr LC_{50} (for fish)	≤ 1 mg/l and/or
48 hr EC_{50} (for crustacea)	≤ 1 mg/l and/or
72 or 96hr ErC_{50} (for algae or other aquatic plants)	≤ 1 mg/l *(see Note 3)*

(b)   Long-term aquatic hazard *(see also Figure 2.2.9.1.10.3.1)*

   (i)   Non-rapidly degradable substances (see Note 4) for which there are adequate chronic toxicity data available

Category Chronic 1: (see *Note 2*)	
Chronic NOEC or EC_x (for fish)	≤ 0.1 mg/l and/or
Chronic NOEC or EC_x (for crustacea)	≤ 0.1 mg/l and/or
Chronic NOEC or EC_x (for algae or other aquatic plants)	≤ 0.1 mg/l
Category Chronic 2:	
Chronic NOEC or EC_x (for fish)	≤ 1 mg/l and/or
Chronic NOEC or EC_x (for crustacea)	≤ 1 mg/l and/or
Chronic NOEC or EC_x (for algae or other aquatic plants)	≤ 1 mg/l

   (ii)   Rapidly degradable substances for which there are adequate chronic toxicity data available

Category Chronic 1: *(see Note 2)*	
Chronic NOEC or EC_x (for fish)	≤ 0.01 mg/l and/or
Chronic NOEC or EC_x (for crustacea)	≤ 0.01 mg/l and/or
Chronic NOEC or EC_x (for algae or other aquatic plants)	≤ 0.01 mg/l
Category Chronic 2:	
Chronic NOEC or EC_x (for fish)	≤ 0.1 mg/l and/or
Chronic NOEC or EC_x (for crustacea)	≤ 0.1 mg/l and/or
Chronic NOEC or EC_x (for algae or other aquatic plants)	≤ 0.1 mg/l

   (iii)   Substances for which adequate chronic toxicity data are not available

Category Chronic 1: *(see Note 2)*	
96 hr LC_{50} (for fish)	≤ 1 mg/l and/or
48 hr EC_{50} (for crustacea)	≤ 1 mg/l and/or
72 or 96hr ErC_{50} (for algae or other aquatic plants)	≤ 1 mg/l *(see Note 3)*
and the substance is not rapidly degradable and/or the experimentally determined BCF is ≥ 500 (or, if absent the log $K_{ow} \geq 4$) *(see Notes 4 and 5)*.	
Category Chronic 2:	
96 hr LC_{50} (for fish)	>1 but ≤ 10 mg/l and/or
48 hr EC_{50} (for crustacea)	>1 but ≤ 10 mg/l and/or
72 or 96hr ErC_{50} (for algae or other aquatic plants)	>1 but ≤ 10 mg/l *(see Note 3)*
and the substance is not rapidly degradable and/or the experimentally determined BCF is ≥ 500 (or, if absent the log $K_{ow} \geq 4$ *(see Notes 4 and 5)*.	

> *NOTE 1: The organisms fish, crustacea and algae are tested as surrogate species covering a range of trophic levels and taxa, and the test methods are highly standardized. Data on other organisms may also be considered, however, provided they represent equivalent species and test endpoints.*

> *NOTE 2: When classifying substances as Acute 1 and/or Chronic 1 it is necessary at the same time to indicate an appropriate M factor (see 2.9.1.10.4.6.4) to apply the summation method.*

> *NOTE 3: Where the algal toxicity ErC_{50} (= EC_{50} (growth rate)) falls more than 100 times below the next most sensitive species and results in a classification based solely on this effect, consideration shall be given to whether this toxicity is representative of the toxicity to aquatic plants. Where it can be shown that this is not the case, professional judgment shall be used in deciding if classification shall be applied. Classification shall be based on the ErC_{50}. In circumstances where the basis of the EC_{50} is not specified and no ErC_{50} is recorded, classification shall be based on the lowest EC_{50} available.*

NOTE 4: *Lack of rapid degradability is based on either a lack of ready biodegradability or other evidence of lack of rapid degradation. When no useful data on degradability are available, either experimentally determined or estimated data, the substance shall be regarded as not rapidly degradable.*

NOTE 5: *Potential to bioaccumulate, based on an experimentally derived BCF ≥ 500 or, if absent, a log $K_{ow} \geq 4$ provided log K_{ow} is an appropriate descriptor for the bioaccumulation potential of the substance. Measured log K_{ow} values take precedence over estimated values and measured BCF values take precedence over log K_{ow} values.*

Figure 2.2.9.1.10.3.1: Categories for substances long-term hazardous to the aquatic environment

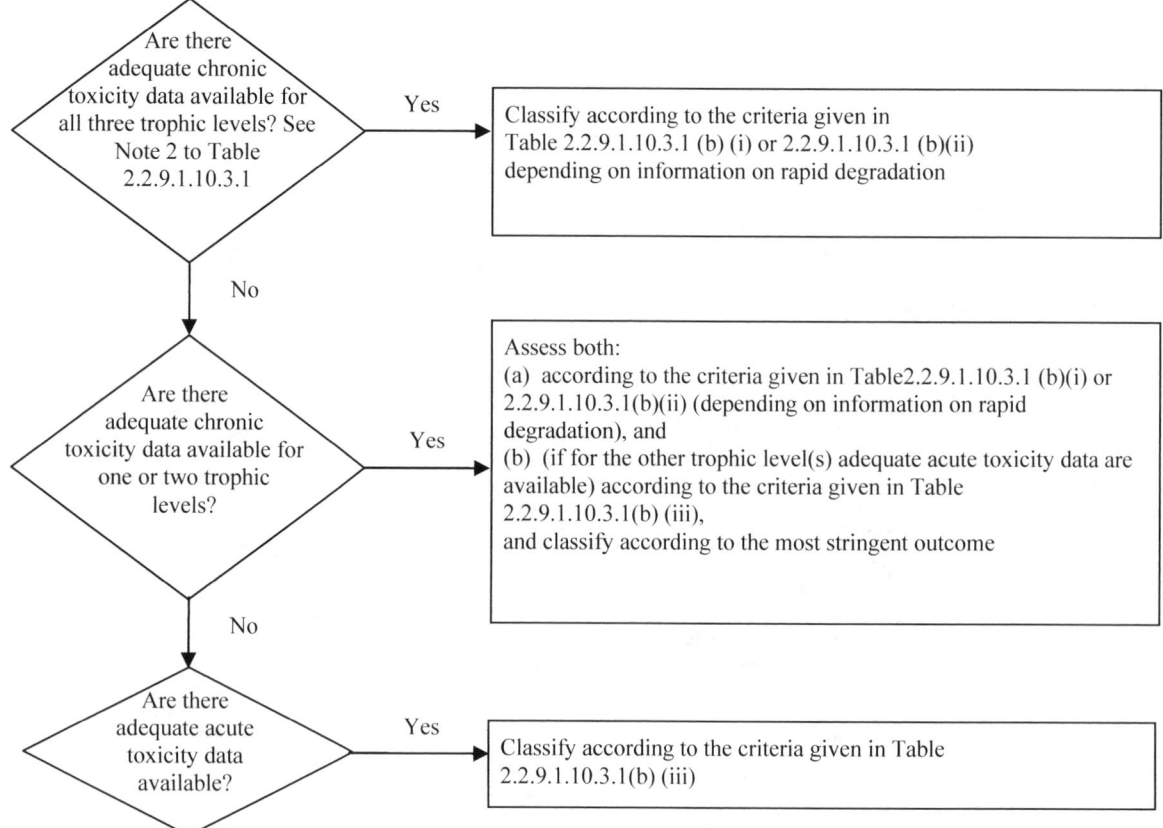

2.2.9.1.10.3.2 The classification scheme in Table 2.2.9.1.10.3.2 below summarizes the classification criteria for substances.

Table 2.2.9.1.10.3.2: Classification scheme for substances hazardous to the aquatic environment

Classification categories			
Acute hazard (*see Note 1*)	**Long-term hazard** (*see Note 2*)		
	Adequate chronic toxicity data available		**Adequate chronic toxicity data not available** (*see Note 1*)
	Non-rapidly degradable substances (*see Note 3*)	**Rapidly degradable substances** (*see Note 3*)	
Category: Acute 1	**Category: Chronic 1**	**Category: Chronic 1**	**Category: Chronic 1**
$L(E)C_{50} \leq 1.00$	NOEC or $EC_x \leq 0.1$	NOEC or $EC_x \leq 0.01$	$L(E)C_{50} \leq 1.00$ and lack of rapid degradability and/or BCF ≥ 500 or, if absent log $K_{ow} \geq 4$
	Category: Chronic 2	**Category: Chronic 2**	**Category: Chronic 2**
	$0.1 < $ NOEC or $EC_x \leq 1$	$0.01 < $ NOEC or $EC_x \leq 0.1$	$1.00 < L(E)C_{50} \leq 10.0$ and lack of rapid degradability and/or BCF ≥ 500 or, if absent log $K_{ow} \geq 4$

NOTE 1: *Acute toxicity band based on $L(E)C_{50}$ values in mg/l for fish, crustacea and/or algae or other aquatic plants (or Quantitative Structure Activity Relationships (QSAR) estimation if no experimental data[13]).*

NOTE 2: *Substances are classified in the various chronic categories unless there are adequate chronic toxicity data available for all three trophic levels above the water solubility or above 1 mg/l. ("Adequate" means that the data sufficiently cover the endpoint of concern. Generally this would mean measured test data, but in order to avoid unnecessary testing it can on a case by case basis also be estimated data, e.g. (Q)SAR, or for obvious cases expert judgment).*

NOTE 3: *Chronic toxicity band based on NOEC or equivalent EC_x values in mg/l for fish or crustacea or other recognized measures for chronic toxicity.*

2.2.9.1.10.4 Mixtures classification categories and criteria

2.2.9.1.10.4.1 The classification system for mixtures covers the classification categories which are used for substances, meaning categories Acute 1 and Chronic 1 and 2. In order to make use of all available data for purposes of classifying the aquatic environmental hazards of the mixture, the following assumption is made and is applied where appropriate:

The "relevant ingredients" of a mixture are those which are present in a concentration equal to or greater than 0.1% (by mass) for ingredients classified as Acute and/or Chronic 1 and equal to or greater than 1% for other ingredients, unless there is a presumption (e.g. in the case of highly toxic ingredients) that an ingredient present at less than 0.1% can still be relevant for classifying the mixture for aquatic environmental hazards.

2.2.9.1.10.4.2 The approach for classification of aquatic environmental hazards is tiered, and is dependent upon the type of information available for the mixture itself and for its ingredients. Elements of the tiered approach include:

(a) Classification based on tested mixtures;

(b) Classification based on bridging principles;

(c) The use of "summation of classified ingredients" and/or an "additivity formula".

Figure 2.2.9.1.10.4.2 below outlines the process to be followed.

---

[13] *Special guidance is provided in Chapter 4.1, paragraph 4.1.2.13 and Annex 9, Section A9.6 of the GHS.*

Figure 2.2.9.1.10.4.2: Tiered approach to classification of mixtures for acute and long-term aquatic environmental hazards

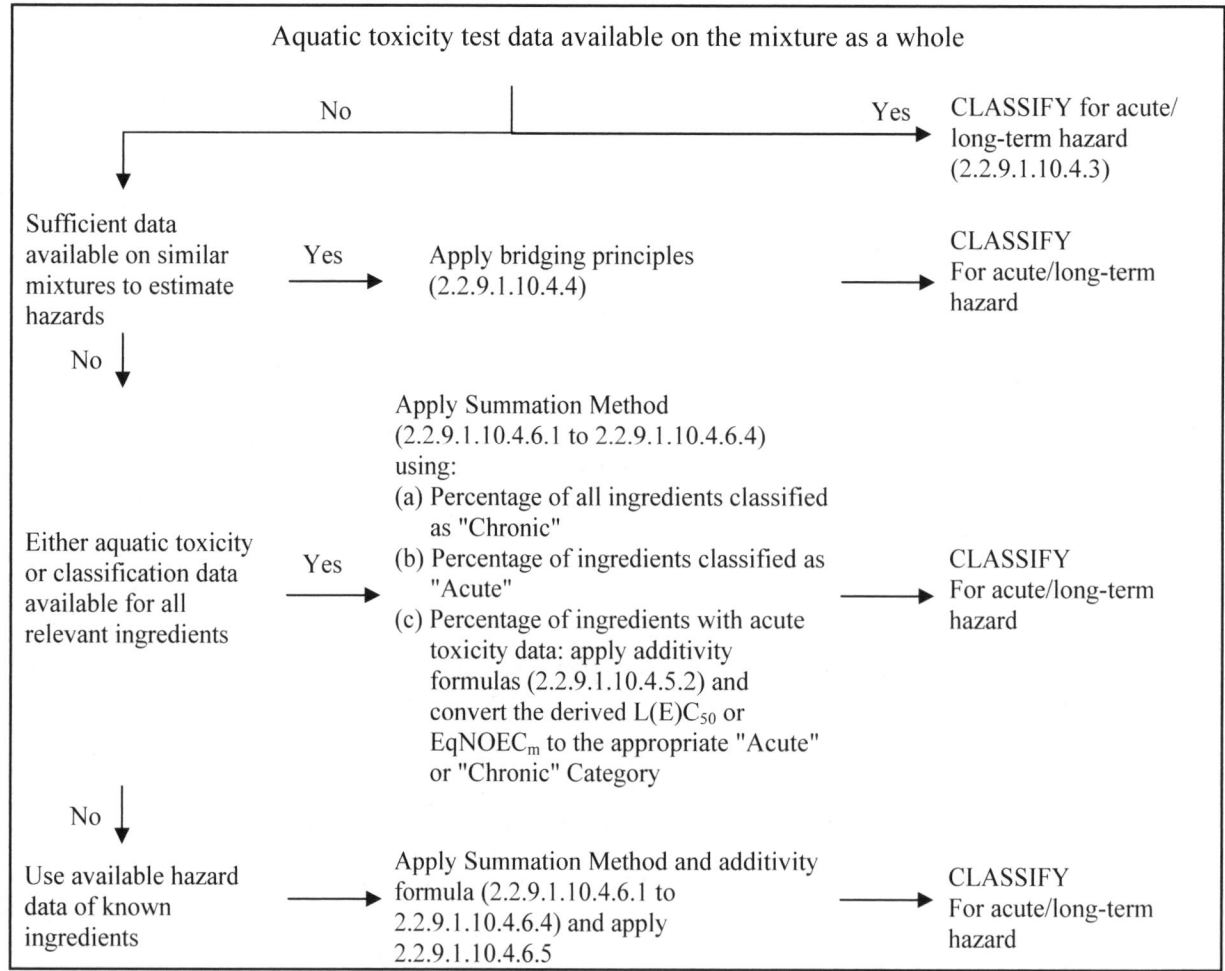

2.2.9.1.10.4.3   Classification of mixtures when toxicity data are available for the complete mixture

2.2.9.1.10.4.3.1   When the mixture as a whole has been tested to determine its aquatic toxicity, this information shall be used for classifying the mixture according to the criteria that have been agreed for substances. The classification is normally based on the data for fish, crustacea and algae/plants (see 2.2.9.1.10.2.3 and 2.2.9.1.10.2.4). When adequate acute or chronic data for the mixture as a whole are lacking, "bridging principles" or "summation method" shall be applied (see 2.2.9.1.10.4.4 to 2.2.9.1.10.4.6).

2.2.9.1.10.4.3.2   The long-term hazard classification of mixtures requires additional information on degradability and in certain cases bioaccumulation. There are no degradability and bioaccumulation data for mixtures as a whole. Degradability and bioaccumulation tests for mixtures are not used as they are usually difficult to interpret, and such tests may be meaningful only for single substances.

2.2.9.1.10.4.3.3   Classification for category Acute 1

(a)   When there are adequate acute toxicity test data (LC_{50} or EC_{50}) available for the mixture as a whole showing $L(E)C_{50} \leq 1$ mg/l:

Classify the mixture as Acute 1 in accordance with Table 2.2.9.1.10.3.1 (a);

(b)   When there are acute toxicity test data ($LC_{50}(s)$ or $EC_{50}(s)$) available for the mixture as a whole showing $L(E)C_{50}(s) > 1$ mg/l, or above the water solubility:

No need to classify for acute hazard under ADR.

2.2.9.1.10.4.3.4   Classification for categories Chronic 1 and 2

(a)   When there are adequate chronic toxicity data (EC_x or NOEC) available for the mixture as a whole showing EC_x or NOEC of the tested mixture $\leq$ 1mg/l:

(i)   classify the mixture as Chronic 1 or 2 in accordance with Table 2.2.9.1.10.3.1 (b) (ii) (rapidly degradable) if the available information allows the conclusion that all relevant ingredients of the mixture are rapidly degradable;

(ii)   classify the mixture as Chronic 1 or 2 in all other cases in accordance with Table 2.2.9.1.10.3.1 (b) (i) (non-rapidly degradable);

(b)   When there are adequate chronic toxicity data (EC_x or NOEC) available for the mixture as a whole showing EC_x(s) or NOEC(s) of the tested mixture > 1mg/l or above the water solubility:

No need to classify for long-term hazard under ADR.

2.2.9.1.10.4.4   Classification of mixtures when toxicity data are not available for the complete mixture: bridging principles

2.2.9.1.10.4.4.1   Where the mixture itself has not been tested to determine its aquatic environmental hazard, but there are sufficient data on the individual ingredients and similar tested mixtures to adequately characterise the hazards of the mixture, these data shall be used in accordance with the following agreed bridging rules. This ensures that the classification process uses the available data to the greatest extent possible in characterising the hazards of the mixture without the necessity for additional testing in animals.

2.2.9.1.10.4.4.2   Dilution

Where a new mixture is formed by diluting a tested mixture or a substance with a diluent which has an equivalent or lower aquatic hazard classification than the least toxic original ingredient and which is not expected to affect the aquatic hazards of other ingredients, then the resulting mixture shall be classified as equivalent to the original tested mixture or substance. Alternatively, the method explained in 2.2.9.1.10.4.5 may be applied.

2.2.9.1.10.4.4.3   Batching

The aquatic hazard classification of a tested production batch of a mixture shall be assumed to be substantially equivalent to that of another untested production batch of the same commercial product when produced by or under the control of the same manufacturer, unless there is reason to believe there is significant variation such that the aquatic hazard classification of the untested batch has changed. If the latter occurs, new classification is necessary.

2.2.9.1.10.4.4.4   Concentration of mixtures which are classified with the most severe classification categories (Chronic 1 and Acute 1)

If a tested mixture is classified as Chronic 1 and/or Acute 1, and the ingredients of the mixture which are classified as Chronic 1 and/or Acute 1 are further concentrated, the more concentrated untested mixture shall be classified with the same classification category as the original tested mixture without additional testing.

2.2.9.1.10.4.4.5   Interpolation within one toxicity category

For three mixtures (A, B and C) with identical ingredients, where mixtures A and B have been tested and are in the same toxicity category, and where untested mixture C has the same toxicologically active ingredients as mixtures A and B but has concentrations of toxicologically active ingredients intermediate to the concentrations in mixtures A and B, then mixture C is assumed to be in the same category as A and B.

2.2.9.1.10.4.4.6   Substantially similar mixtures

Given the following:

(a)   Two mixtures:

    (i)   A + B;

    (ii)   C + B;

(b)   The concentration of ingredient B is essentially the same in both mixtures;

(c)   The concentration of ingredient A in mixture (i) equals that of ingredient C in mixture (ii);

(d)   Data on aquatic hazards for A and C are available and are substantially equivalent, i.e. they are in the same hazard category and are not expected to affect the aquatic toxicity of B.

If mixture (i) or (ii) is already classified based on test data, then the other mixture can be assigned the same hazard category.

2.2.9.1.10.4.5   Classification of mixtures when toxicity data are available for all ingredients or only for some ingredients of the mixture

2.2.9.1.10.4.5.1   The classification of a mixture shall be based on summation of the concentrations of its classified ingredients. The percentage of ingredients classified as "Acute" or "Chronic" will feed straight into the summation method. Details of the summation method are described in 2.2.9.1.10.4.6.1 to 2.2.9.1.10.4.6.4.

2.2.9.1.10.4.5.2   Mixtures may be made of a combination of both ingredients that are classified (as Acute 1 and/or Chronic 1, 2) and those for which adequate toxicity test data are available. When adequate toxicity data are available for more than one ingredient in the mixture, the combined toxicity of those ingredients shall be calculated using the following additivity formulas (a) or (b), depending on the nature of the toxicity data:

(a)   Based on acute aquatic toxicity:

$$\frac{\sum C_i}{L(E)C_{50m}} = \sum_n \frac{C_i}{L(E)C_{50i}}$$

where:

C_i =   concentration of ingredient i (mass percentage);
$L(E)C_{50i}$ = LC_{50} or EC_{50} for ingredient i (mg/l);
n   =   number of ingredients, and i is running from 1 to n;
$L(E)C_{50m}$ = $L(E)C_{50}$ of the part of the mixture with test data;

The calculated toxicity shall be used to assign that portion of the mixture an acute hazard category which is then subsequently used in applying the summation method;

(b)   Based on chronic aquatic toxicity:

$$\frac{\sum C_i + \sum C_j}{EqNOEC_m} = \sum_n \frac{C_i}{NOEC_i} + \sum_n \frac{C_j}{0.1 \cdot NOEC_j}$$

where:

C_i =   concentration of ingredient i (mass percentage) covering the rapidly degradable ingredients;
C_j =   concentration of ingredient j (mass percentage) covering the non rapidly degradable ingredients;
$NOEC_i$ =   NOEC (or other recognized measures for chronic toxicity) for ingredient i covering the rapidly degradable ingredients, in mg/l;
$NOEC_j$ =   NOEC (or other recognized measures for chronic toxicity) for ingredient j covering the non-rapidly degradable ingredients, in mg/l;
n   =   number of ingredients, and i and j are running from 1 to n;

$EqNOEC_m$ = equivalent NOEC of the part of the mixture with test data;

The equivalent toxicity thus reflects the fact that non-rapidly degrading substances are classified one hazard category level more "severe" than rapidly degrading substances.

The calculated equivalent toxicity shall be used to assign that portion of the mixture a long-term hazard category, in accordance with the criteria for rapidly degradable substances (Table 2.2.9.1.10.3.1 (b) (ii)), which is then subsequently used in applying the summation method.

2.2.9.1.10.4.5.3 When applying the additivity formula for part of the mixture, it is preferable to calculate the toxicity of this part of the mixture using for each ingredient toxicity values that relate to the same taxonomic group (i.e. fish, crustacea or algae) and then to use the highest toxicity (lowest value) obtained (i.e. use the most sensitive of the three groups). However, when toxicity data for each ingredient are not available in the same taxonomic group, the toxicity value of each ingredient shall be selected in the same manner that toxicity values are selected for the classification of substances, i.e. the higher toxicity (from the most sensitive test organism) is used. The calculated acute and chronic toxicity shall then be used to classify this part of the mixture as Acute 1 and/or Chronic 1 or 2 using the same criteria described for substances.

2.2.9.1.10.4.5.4 If a mixture is classified in more than one way, the method yielding the more conservative result shall be used.

2.2.9.1.10.4.6 Summation method

2.2.9.1.10.4.6.1 Classification procedure

In general a more severe classification for mixtures overrides a less severe classification, e.g. a classification with Chronic 1 overrides a classification with Chronic 2. As a consequence the classification procedure is already completed if the results of the classification is Chronic 1. A more severe classification than Chronic 1 is not possible; therefore, it is not necessary to pursue the classification procedure further.

2.2.9.1.10.4.6.2 Classification for category Acute 1

2.2.9.1.10.4.6.2.1 First, all ingredients classified as Acute 1 are considered. If the sum of the concentrations (in %) of these ingredients is greater than or equal to 25% the whole mixture shall be classified as Acute 1. If the result of the calculation is a classification of the mixture as Acute 1, the classification process is completed.

2.2.9.1.10.4.6.2.2 The classification of mixtures for acute hazards based on this summation of the concentrations of classified ingredients is summarized in Table 2.2.9.1.10.4.6.2.2 below.

Table 2.2.9.1.10.4.6.2.2: Classification of a mixture for acute hazards based on summation of the concentrations of classified ingredients

Sum of the concentrations (in %) of ingredients classified as:	Mixture classified as:
Acute $1 \times M^a \geq 25\%$	Acute 1

[a] *For explanation of the M factor, see 2.2.9.1.10.4.6.4.*

2.2.9.1.10.4.6.3 Classification for categories Chronic 1 and 2

2.2.9.1.10.4.6.3.1 First, all ingredients classified as Chronic 1 are considered. If the sum of the concentrations (in %) of these ingredients is greater than or equal to 25% the mixture shall be classified as Chronic 1. If the result of the calculation is a classification of the mixture as Chronic 1 the classification procedure is completed.

2.2.9.1.10.4.6.3.2 In cases where the mixture is not classified as Chronic 1, classification of the mixture as Chronic 2 is considered. A mixture shall be classified as Chronic 2 if 10 times the sum of the concentrations (in %) of all ingredients classified as Chronic 1 plus the sum of the concentrations (in %) of all ingredients classified as Chronic 2 is greater than or equal to 25%. If the result of the calculation is classification of the mixture as Chronic 2, the classification process is completed.

2.2.9.1.10.4.6.3.3   The classification of mixtures for long-term hazards based on this summation of the concentrations of classified ingredients is summarized in Table 2.2.9.1.10.4.6.3.3 below.

Table 2.2.9.1.10.4.6.3.3: Classification of a mixture for long-term hazards based on summation of the concentrations of classified ingredients

Sum of the concentrations (in %) of ingredients classified as:		Mixture classified as:
Chronic 1 × M [a]	≥ 25%	Chronic 1
(M × 10 × Chronic 1) + Chronic 2	≥ 25%	Chronic 2

[a] *For explanation of the M factor, see 2.2.9.1.10.4.6.4.*

2.2.9.1.10.4.6.4   Mixtures with highly toxic ingredients

Acute 1 or Chronic 1 ingredients with acute toxicities well below 1 mg/l and/or chronic toxicities well below 0.1 mg/l (if non-rapidly degradable) and 0.01 mg/l (if rapidly degradable) may influence the toxicity of the mixture and are given increased weight in applying the summation method. When a mixture contains ingredients classified as acute or Chronic 1, the tiered approach described in 2.2.9.1.10.4.6.2 and 2.2.9.1.10.4.6.3 shall be applied using a weighted sum by multiplying the concentrations of Acute 1 and Chronic 1 ingredients by a factor, instead of merely adding up the percentages. This means that the concentration of "Acute 1" in the left column of Table 2.2.9.1.10.4.6.2.2 and the concentration of "Chronic 1" in the left column of Table 2.2.9.1.10.4.6.3.3 are multiplied by the appropriate multiplying factor. The multiplying factors to be applied to these ingredients are defined using the toxicity value, as summarised in Table 2.2.9.1.10.4.6.4 below. Therefore, in order to classify a mixture containing Acute 1 and/or Chronic 1 ingredients, the classifier needs to be informed of the value of the M factor in order to apply the summation method. Alternatively, the additivity formula (see 2.2.9.1.10.4.5.2) may be used when toxicity data are available for all highly toxic ingredients in the mixture and there is convincing evidence that all other ingredients, including those for which specific acute and/or chronic toxicity data are not available, are of low or no toxicity and do not significantly contribute to the environmental hazard of the mixture.

Table 2.2.9.1.10.4.6.4: Multiplying factors for highly toxic ingredients of mixtures

Acute toxicity	M factor	Chronic toxicity	M factor	
$L(E)C_{50}$ value		NOEC value	NRD [a] ingredients	RD [b] ingredients
$0.1 < L(E)C_{50} \le 1$	1	$0.01 < NOEC \le 0.1$	1	–
$0.01 < L(E)C_{50} \le 0.1$	10	$0.001 < NOEC \le 0.01$	10	1
$0.001 < L(E)C_{50} \le 0.01$	100	$0.0001 < NOEC \le 0.001$	100	10
$0.0001 < L(E)C_{50} \le 0.001$	1 000	$0.00001 < NOEC \le 0.0001$	1 000	100
$0.00001 < L(E)C_{50} \le 0.0001$	10 000	$0.000001 < NOEC \le 0.00001$	10 000	1 000
(continue in factor 10 intervals)		(continue in factor 10 intervals)		

[a] *Non-rapidly degradable.*
[b] *Rapidly degradable.*

2.2.9.1.10.4.6.5   Classification of mixtures with ingredients without any useable information

In the event that no useable information on acute and/or chronic aquatic toxicity is available for one or more relevant ingredients, it is concluded that the mixture cannot be attributed (a) definitive hazard category(ies). In this situation the mixture shall be classified based on the known ingredients only with the additional statement that: "x percent of the mixture consists of ingredient(s) of unknown hazard to the aquatic environment.

2.2.9.1.10.5 Substances or mixtures classified as environmentally hazardous substances (aquatic environment) on the basis of Regulation 1272/2008/EC[3]

If data for classification according to the criteria of 2.2.9.1.10.3 and 2.2.9.1.10.4 are not available, a substance or mixture:

(a)   Shall be classified as an environmentally hazardous substance (aquatic environment) if it has to be assigned category(ies) Aquatic Acute 1, Aquatic Chronic 1 or Aquatic Chronic 2 according to Regulation 1272/2008/EC[3];

(b)   May be regarded as not being an environmentally hazardous substance (aquatic environment) if it does not have to be assigned such a category according to the said Regulation.

2.2.9.1.10.6 Assignment of substances or mixtures classified as environmentally hazardous substances (aquatic environment) according to the provisions in 2.2.9.1.10.3, 2.2.9.1.10.4 or 2.2.9.1.10.5

Substances or mixtures classified as environmentally hazardous substances (aquatic environment), not otherwise classified under ADR shall be designated:

UN No. 3077 ENVIRONMENTALLY HAZARDOUS SUBSTANCE, SOLID, N.O.S.; or

UN No. 3082 ENVIRONMENTALLY HAZARDOUS SUBSTANCE, LIQUID, N.O.S.

They shall be assigned to packing group III.

Genetically modified microorganisms or organisms

2.2.9.1.11 Genetically modified microorganisms (GMMOs) and genetically modified organisms (GMOs) are microorganisms and organisms in which genetic material has been purposely altered through genetic engineering in a way that does not occur naturally. They are assigned to Class 9 (UN No. 3245) if they do not meet the definition of toxic substances or of infectious substances, but are capable of altering animals, plants or microbiological substances in a way not normally the result of natural reproduction.

NOTE 1:   GMMOs and GMOs which are infectious are substances of Class 6.2, UN Nos. 2814, 2900 or 3373.

NOTE 2:   GMMOs or GMOs are not subject to the provisions of ADR when authorized for use by the competent authorities of the countries of origin, transit and destination[14].

NOTE 3:   Genetically modified live animals which, in accordance with the current state of scientific knowledge, have no known pathogenic effect on humans, animals and plants and are carried in receptacles that are suitable for safely preventing both the escape of the animals and unauthorized access to them, are not subject to the provisions of ADR. The provisions specified by the International Air Transport Association (IATA) for air transport "Live Animals Regulations, LAR" can be drawn on as guidelines for suitable receptacles for the transport of live animals.

NOTE 4:   Live animals shall not be used to carry genetically modified microorganisms classified in Class 9 unless the substance can be carried no other way. Genetically modified live animals shall be carried under terms and conditions of the competent authorities of the countries of origin and destination.

2.2.9.1.12 *(Deleted)*

---

[3]   *Regulation (EC) No 1272/2008 of the European Parliament and of the Council of 16 December 2008 on classification, labelling and packaging of substances and mixtures, amending and repealing Directive 67/548/EEC and 1999/45/EC; and amending Regulation (EC) No 1907/2006, published in the Official Journal of the European Union, L 353, 31 December 2008, p 1-1355.*

[14]   *See Part C of Directive 2001/18/EC of the European Parliament and of the Council on the deliberate release into the environment of genetically modified organisms and repealing Council Directive 90/220/EEC (Official Journal of the European Communities, No. L 106, of 17 April 2001, pp 8-14) and Regulation (EC) No. 1829/2003 of the European Parliament and of the Council on genetically modified food and feed (Official Journal of the European Union, No. L 268, of 18 October 2003, pp 1-23), which set out the authorization procedures for the European Union.*

Elevated temperature substances

2.2.9.1.13 Elevated temperature substances include substances which are carried or handed over for carriage in the liquid state at or above 100 °C and, in the case of those with a flash-point, below their flash-point. They also include solids which are carried or handed over for carriage at or above 240 °C.

NOTE: *Elevated temperature substances may be assigned to Class 9 only if they do not meet the criteria of any other class.*

Other substances presenting a danger during carriage but not meeting the definitions of another class.

2.2.9.1.14 The following other miscellaneous substances not meeting the definitions of another class are assigned to Class 9:

Solid ammonia compounds having a flash-point below 60 °C;

Low hazard dithionites;

Highly volatile liquids;

Substances emitting noxious fumes;

Substances containing allergens;

Chemical kits and first aid kits;

Electric double layer capacitors (with an energy storage capacity greater than 0.3 Wh);

Vehicles, engines and machinery, internal combustion.

NOTE: *UN No. 1845 carbon dioxide, solid (dry ice)*[15] *UN No. 2071 ammonium nitrate fertilizers, UN No. 2216 fish meal (fish scrap), stabilized, UN No. 2807 magnetized material, UN No. 3334 aviation regulated liquid, n.o.s., UN No. 3335 aviation regulated solid, n.o.s. and UN No. 3363 dangerous goods in machinery or dangerous goods in apparatus listed in the UN Model Regulations, are not subject to the provisions of ADR.*

Assignment of the packing groups

2.2.9.1.15 When indicated in column (4) of Table A of Chapter 3.2, substances and articles of Class 9 are assigned to one of the following packing groups according to their degree of danger:

Packing group II: substances presenting medium danger;

Packing group III: substances presenting low danger.

2.2.9.2 Substances and articles not accepted for carriage

The following substances and articles shall not be accepted for carriage:

- Lithium batteries which do not meet the relevant conditions of special provisions 188, 230, 310 or 636 of Chapter 3.3;

- Uncleaned empty containment vessels for apparatus such as transformers, condensers and hydraulic apparatus containing substances assigned to UN Nos. 2315, 3151, 3152 or 3432.

---

[15] *For UN No. 1845 carbon dioxide, solid (dry ice) used as a coolant, see 5.5.3.*

2.2.9.3 *List of entries*

Substances which, on inhalation as fine dust, may endanger health		**M1**	2212 ASBESTOS, AMPHIBOLE (amosite, tremolite, actinolite, anthophyllite, crocidolite) 2590 ASBESTOS, CHRYSOTILE
Substances and articles which, in the event of fire, may form dioxins		**M2**	2315 POLYCHLORINATED BIPHENYLS, LIQUID 3432 POLYCHLORINATED BIPHENYLS, SOLID 3151 POLYHALOGENATED BIPHENYLS, LIQUID or 3151 HALOGENATED MONOMETHYLDIPHENYLMETHANES, LIQUID or 3151 POLYHALOGENATED TERPHENYLS, LIQUID 3152 POLYHALOGENATED BIPHENYLS, SOLID or 3152 HALOGENATED MONOMETHYLDIPHENYLMETHANES, SOLID or 3152 POLYHALOGENATED TERPHENYLS, SOLID
Substances evolving flammable vapour		**M3**	2211 POLYMERIC BEADS, EXPANDABLE, evolving flammable vapour 3314 PLASTICS MOULDING COMPOUND in dough, sheet or extruded rope form evolving flammable vapour
Lithium batteries		**M4**	3090 LITHIUM METAL BATTERIES (including lithium alloy batteries) 3091 LITHIUM METAL BATTERIES CONTAINED IN EQUIPMENT (including lithium alloy batteries) or 3091 LITHIUM METAL BATTERIES PACKED WITH EQUIPMENT (including lithium alloy batteries) 3480 LITHIUM ION BATTERIES (including lithium ion polymer batteries) 3481 LITHIUM ION BATTERIES CONTAINED IN EQUIPMENT (including lithium ion polymer batteries) or 3481 LITHIUM ION BATTERIES PACKED WITH EQUIPMENT (including lithium ion polymer batteries)
Live-saving appliances		**M5**	2990 LIFE-SAVING APPLIANCES, SELF-INFLATING 3072 LIFE-SAVING APPLIANCES NOT SELF-INFLATING containing dangerous goods as equipment 3268 SAFETY DEVICES, electrically initiated
Environmentally hazardous substances	**pollutant to the aquatic environment, liquid**	**M6**	3082 ENVIRONMENTALLY HAZARDOUS SUBSTANCE, LIQUID, N.O.S.
	pollutant to the aquatic environment, solid	**M7**	3077 ENVIRONMENTALLY HAZARDOUS SUBSTANCE, SOLID, N.O.S.
	genetically modified micro-organisms and organisms	**M8**	3245 GENETICALLY MODIFIED MICROORGANISMS or 3245 GENETICALLY MODIFIED ORGANISMS
Elevated temperature substances	**liquid**	**M9**	3257 ELEVATED TEMPERATURE LIQUID, N.O.S., at or above 100 °C and below its flash-point (including molten metal, molten salts, etc.)
	solid	**M10**	3258 ELEVATED TEMPERATURE SOLID, N.O.S., at or above 240 °C

(cont'd on next page)

Other substances and articles presenting a danger during carriage, but not meeting the definitions of another class **M11**	No collective entry available. Only substances and articles listed in Table A of Chapter 3.2 are subject to the provisions for Class 9 under this classification code, as follows: 1841 ACETALDEHYDE AMMONIA 1931 ZINC DITHIONITE (ZINC HYDROSULPHITE) 1941 DIBROMODIFLUOROMETHANE 1990 BENZALDEHYDE 2969 CASTOR BEANS, or 2969 CASTOR MEAL, or 2969 CASTOR POMACE, or 2969 CASTOR FLAKE 3166 VEHICLE, FLAMMABLE GAS POWERED or 3166 VEHICLE, FLAMMABLE LIQUID POWERED or 3166 VEHICLE, FUEL CELL, FLAMMABLE GAS POWERED or 3166 VEHICLE, FUEL CELL, FLAMMABLE LIQUID POWERED 3171 BATTERY POWERED VEHICLE or 3171 BATTERY POWERED EQUIPMENT 3316 CHEMICAL KIT, or 3316 FIRST AID KIT 3359 FUMIGATED UNIT 3499 CAPACITOR, ELECTRIC DOUBLE LAYER (with an energy storage capacity greater than 0.3Wh) 3508 CAPACITOR, ASYMMETRIC (with an energy storage capacity greater than 0.3Wh) 3509 PACKAGINGS, DISCARDED, EMPTY, UNCLEANED 3530 ENGINE, INTERNAL COMBUSTION or 3530 MACHINERY, INTERNAL COMBUSTION

CHAPTER 2.3

TEST METHODS

2.3.0   **General**

Unless otherwise provided for in Chapter 2.2 or in this Chapter, the test methods to be used for the classification of dangerous goods are those described in the Manual of Tests and Criteria.

2.3.1   **Exudation test for blasting explosives of Type A**

2.3.1.1   Blasting explosives of type A (UN No. 0081) shall, if they contain more than 40% liquid nitric ester, in addition to the testing specified in the Manual of Tests and Criteria, satisfy the following exudation test.

2.3.1.2   The apparatus for testing blasting explosive for exudation (figs. 1 to 3) consists of a hollow bronze cylinder. This cylinder, which is closed at one end by a plate of the same metal, has an internal diameter of 15.7 mm and a depth of 40 mm. It is pierced by 20 holes 0.5 mm in diameter (four sets of five holes) on the circumference. A bronze piston, cylindrically fashioned over a length of 48 mm and having a total length of 52 mm, slides into the vertically placed cylinder. The piston, whose diameter is 15.6 mm, is loaded with a mass of 2 220 g so that a pressure of 120 kPa (1.20 bar) is exerted on the base of the cylinder.

2.3.1.3   A small plug of blasting explosive weighing 5 to 8 g, 30 mm long and 15 mm in diameter, is wrapped in very fine gauze and placed in the cylinder; the piston and its loading mass are then placed on it so that the blasting explosive is subjected to a pressure of 120 kPa (1.20 bar). The time taken for the appearance of the first signs of oily droplets (nitroglycerine) at the outer orifices of the cylinder holes is noted.

2.3.1.4   The blasting explosive is considered satisfactory if the time elapsing before the appearance of the liquid exudations is more than five minutes, the test having been carried out at a temperature of 15 °C to 25 °C.

Test of blasting explosive for exudation

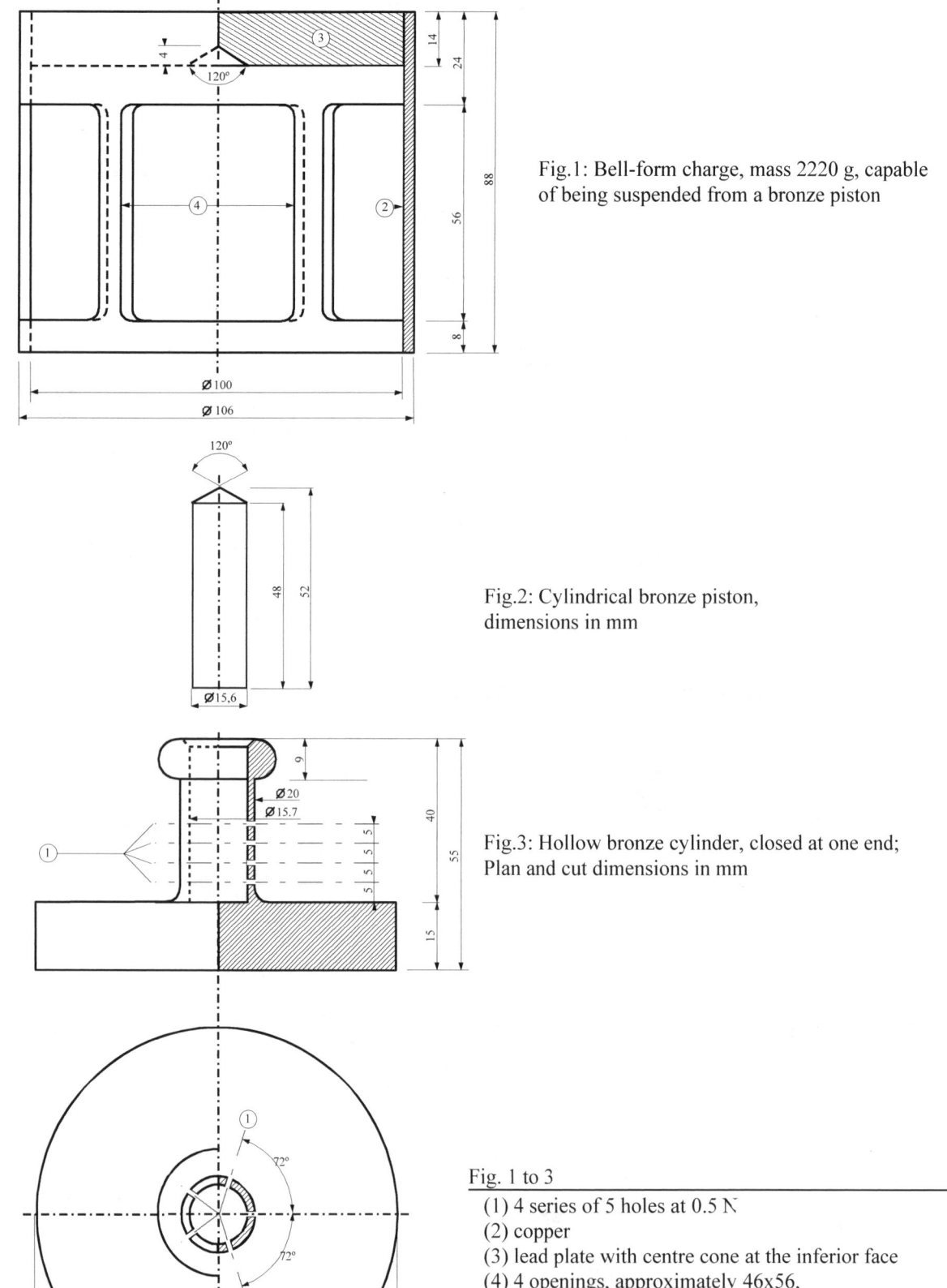

Fig.1: Bell-form charge, mass 2220 g, capable of being suspended from a bronze piston

Fig.2: Cylindrical bronze piston, dimensions in mm

Fig.3: Hollow bronze cylinder, closed at one end; Plan and cut dimensions in mm

Fig. 1 to 3

(1) 4 series of 5 holes at 0.5 N
(2) copper
(3) lead plate with centre cone at the inferior face
(4) 4 openings, approximately 46x56, set at even intervals on the periphery

2.3.2	**Tests relating to nitrated cellulose mixtures of Class 4.1**

2.3.2.1   Nitrocellulose heated for half an hour at 132 °C shall not give off visible yellowish-brown nitrous fumes (nitrous gases). The ignition temperature shall be above 180 °C. See 2.3.2.3 to 2.3.2.8, 2.3.2.9 (a) and 2.3.2.10 below.

2.3.2.2   3 g of plasticized nitrocellulose, heated for one hour at 132 °C, shall not give off visible yellowish-brown nitrous fumes (nitrous gases). The ignition temperature shall be above 170 °C. See 2.3.2.3 to 2.3.2.8, 2.3.2.9 (b) and 2.3.2.10 below.

2.3.2.3   The test procedures set out below are to be applied when differences of opinion arise as to the acceptability of substances for carriage by road.

2.3.2.4   If other methods or test procedures are used to verify the conditions of stability prescribed above in this section, those methods shall lead to the same findings as could be reached by the methods specified below.

2.3.2.5   In carrying out the stability tests by heating described below, the temperature of the oven containing the sample under test shall not deviate by more than 2 °C from the prescribed temperature; the prescribed duration of a 30-minute or 60-minute test shall be observed to within two minutes. The oven shall be such that the required temperature is restored not more than five minutes after insertion of the sample.

2.3.2.6   Before undergoing the tests in 2.3.2.9 and 2.3.2.10, the samples shall be dried for not less than 15 hours at the ambient temperature in a vacuum desiccator containing fused and granulated calcium chloride, the sample substance being spread in a thin layer; for this purpose, substances which are neither in powder form nor fibrous shall be ground, or grated, or cut into small pieces. The pressure in the desiccator shall be brought below 6.5 kPa (0.065 bar).

2.3.2.7   Before being dried as prescribed in 2.3.2.6 above, substances conforming to 2.3.2.2 shall undergo preliminary drying in a well-ventilated oven, with its temperature set at 70 °C, until the loss of mass per quarter-hour is less than 0.3% of the original mass.

2.3.2.8   Weakly nitrated nitrocellulose conforming to 2.3.2.1 shall first undergo preliminary drying as prescribed in 2.3.2.7 above; drying shall then be completed by keeping the nitrocellulose for at least 15 hours over concentrated sulphuric acid in a desiccator.

2.3.2.9   *Test of chemical stability under heat*

(a)   *Test of the substance listed in paragraph 2.3.2.1 above.*

(i)   In each of two glass test tubes having the following dimensions:

length	350	mm
internal diameter	16	mm
thickness of wall	1.5	mm

is placed 1 g of substance dried over calcium chloride (if necessary the drying shall be carried out after reducing the substance to pieces weighing not more than 0.05 g each).

Both test tubes, completely covered with loose-fitting closures, are then so placed in an oven that at least four-fifths of their length is visible, and are kept at a constant temperature of 132 °C for 30 minutes. It is observed whether nitrous gases in the form of yellowish-brown fumes clearly visible against a white background are given off during this time;

(ii)   In the absence of such fumes the substance is deemed to be stable;

(b)   *Test of plasticized nitrocellulose (see 2.3.2.2)*

(i)   3 g of plasticized nitrocellulose are placed in glass test tubes, similar to those referred to in (a), which are then placed in an oven kept at a constant temperature of 132 °C;

(ii)   The test tubes containing the plasticized nitrocellulose are kept in the oven for one hour. During this time no yellowish-brown nitrous fumes (nitrous gases) shall be visible. Observation and appraisal as in (a).

2.3.2.10   *Ignition temperature (see 2.3.2.1 and 2.3.2.2)*

(a)   The ignition temperature is determined by heating 0.2 g of substance enclosed in a glass test tube immersed in a Wood's alloy bath. The test tube is placed in the bath when the latter has reached 100 °C. The temperature of the bath is then progressively increased by 5 °C per minute;

(b)   The test tubes must have the following dimensions:

length	125 mm
internal diameter	15 mm
thickness of wall	0.5 mm

and shall be immersed to a depth of 20 mm;

(c)   The test shall be repeated three times, the temperature at which ignition of the substance occurs, i.e., slow or rapid combustion, deflagration or detonation, being noted each time;

(d)   The lowest temperature recorded in the three tests is the ignition temperature.

2.3.3   **Tests relating to flammable liquids of Classes 3, 6.1 and 8**

2.3.3.1   *Determination of flash-point*

2.3.3.1.1   The following methods for determining the flash-point of flammable liquids may be used:

International standards:

ISO 1516 (Determination of flash/no flash – Closed cup equilibrium method)

ISO 1523 (Determination of flash point – Closed cup equilibrium method)

ISO 2719 (Determination of flash point – Pensky-Martens closed cup method)

ISO 13736 (Determination of flash point – Abel closed-cup method)

ISO 3679 (Determination of flash point – Rapid equilibrium closed cup method)

ISO 3680 (Determination of flash/no flash – Rapid equilibrium closed cup method)

National standards:

American Society for Testing Materials International, 100 Barr Harbor Drive, PO Box C700, West Conshohocken, Pennsylvania, USA 19428-2959:

ASTM D3828-07a, Standard Test Methods for Flash Point by Small Scale Closed-Cup Tester

ASTM D56-05, Standard Test Method for Flash Point by Tag Closed-Cup Tester

ASTM D3278-96(2004)e1, Standard Test Methods for Flash Point of Liquids by Small Scale Closed-Cup Apparatus

ASTM D93-08, Standard Test Methods for Flash Point by Pensky-Martens Closed-Cup Tester

Association française de normalisation, AFNOR, 11, rue de Pressensé, F-93571 La Plaine Saint-Denis Cedex:

French standard NF M 07 - 019

French standards NF M 07 - 011 / NF T 30 - 050 / NF T 66 - 009

French standard NF M 07 - 036

Deutsches Institut für Normung, Burggrafenstr. 6, D-10787 Berlin:

Standard DIN 51755 (flash-points below 65 °C)

State Committee of the Council of Ministers for Standardization, RUS-113813, GSP, Moscow, M-49 Leninsky Prospect, 9:

GOST 12.1.044-84

2.3.3.1.2   To determine the flash-point of paints, gums and similar viscous products containing solvents, only apparatus and test methods suitable for determining the flash-point for viscous liquids shall be used, in accordance with the following standards:

(a)   International Standard ISO 3679: 1983;

(b)   International Standard ISO 3680: 1983;

(c)   International Standard ISO 1523: 1983;

(d)   International standards EN ISO 13736 and EN ISO 2719, Method B.

2.3.3.1.3   The standards listed in 2.3.3.1.1 shall only be used for flash-point ranges which are specified therein. The possibility of chemical reactions between the substance and the sample holder shall be considered when selecting the standard to be used. The apparatus shall, as far as is consistent with safety, be placed in a draught-free position. For safety, a method utilizing a small sample size, around 2 ml, shall be used for organic peroxides and self-reactive substances (also known as "energetic" substances), or for toxic substances.

2.3.3.1.4   When the flash-point, determined by a non-equilibrium method is found to be 23 ± 2 °C or 60 ± 2 °C, it shall be confirmed for each temperature range by an equilibrium method.

2.3.3.1.5   In the event of a dispute as to the classification of a flammable liquid, the classification proposed by the consignor shall be accepted if a check-test of the flash-point, yields a result not differing by more than 2 °C from the limits (23 °C and 60 °C respectively) stated in 2.2.3.1. If the difference is more than 2 °C, a second check-test shall be carried out, and the lowest figure of the flash-points obtained in either check-test shall be adopted.

2.3.3.2   *Determination of initial boiling point*

The following methods for determining the initial boiling point of flammable liquids may be used:

International standards:

ISO 3924 (Petroleum products – Determination of boiling range distribution – Gas chromatography method)

ISO 4626 (Volatile organic liquids – Determination of boiling range of organic solvents used as raw materials)

ISO 3405 (Petroleum products – Determination of distillation characteristics at atmospheric pressure)

National standards:

American Society for Testing Materials International, 100 Barr Harbor Drive, PO Box C700, West Conshohocken, Pennsylvania, USA 19428-2959:

ASTM D86-07a, Standard Test Method for Distillation of Petroleum Products at Atmospheric Pressure

ASTM D1078-05, Standard Test Method for Distillation Range of Volatile Organic Liquids

Further acceptable methods:

Method A.2 as described in Part A of the Annex to Commission Regulation (EC) No 440/2008[1].

---

[1]   *Commission Regulation (EC) No 440/2008 of 30 May 2008 laying down test methods pursuant to Regulation (EC) No 1907/2006 of the European Parliament and of the Council on the Registration, Evaluation, Authorisation and Restriction of Chemicals (REACH) (Official Journal of the European Union, No. L 142 of 31.05.2008, p.1-739 and No. L 143 of 03.06.2008, p.55).*

2.3.3.3   *Test for determining peroxide content*

To determine the peroxide content of a liquid, the procedure is as follows:

A quantity p (about 5 g, weighed to the nearest 0.01 g) of the liquid to be titrated is placed in an Erlenmeyer flask; 20 cm^3 of acetic anhydride and about 1 g of powdered solid potassium iodide are added; the flask is shaken and, after 10 minutes, heated for 3 minutes to about 60 °C. When it has been left to cool for 5 minutes, 25 cm^3 of water are added. After this, it is left standing for half an hour, then the liberated iodine is titrated with a decinormal solution of sodium thiosulphate, no indicator being added; complete discoloration indicates the end of the reaction. If n is the number of cm^3 of thiosulphate solution required, the percentage of peroxide (calculated as H_2O_2) present in the sample is obtained by the formula:

$$\frac{17n}{100p}$$

2.3.4   **Test for determining fluidity**

To determine the fluidity of liquid, viscous or pasty substances and mixtures, the following test method shall be used.

2.3.4.1   *Test apparatus*

Commercial penetrometer conforming to ISO 2137:1985, with a guide rod of 47.5 g ± 0.05 g; sieve disc of duralumin with conical bores and a mass of 102.5 g ± 0.05 g (see Figure 1); penetration vessel with an inside diameter of 72 mm to 80 mm for reception of the sample.

2.3.4.2   *Test procedure*

The sample is poured into the penetration vessel not less than half an hour before the measurement. The vessel is then hermetically closed and left standing until the measurement. The sample in the hermetically closed penetration vessel is heated to 35 °C ± 0.5 °C and is placed on the penetrometer table immediately prior to measurement (not more than two minutes). The point S of the sieve disc is then brought into contact with the surface of the liquid and the rate of penetration is measured.

2.3.4.3   *Evaluation of test results*

A substance is pasty if, after the centre S has been brought into contact with the surface of the sample, the penetration indicated by the dial gauge:

(a)   after a loading time of 5 s ± 0.1 s, is less than 15.0 mm ± 0.3 mm; or

(b)   after a loading time of 5 s ± 0.1 s, is greater than 15.0 mm ± 0.3 mm, but the additional penetration after another 55 s ± 0.5 s is less than 5.0 mm ± 0.5 mm.

NOTE: In the case of samples having a flow point, it is often impossible to produce a steady level surface in the penetration vessel and, hence, to establish satisfactory initial measuring conditions for the contact of the point S. Furthermore, with some samples, the impact of the sieve disc can cause an elastic deformation of the surface and, in the first few seconds, simulate a deeper penetration. In all these cases, it may be appropriate to make the evaluation in paragraph (b) above.

Figure 1 – Penetrometer

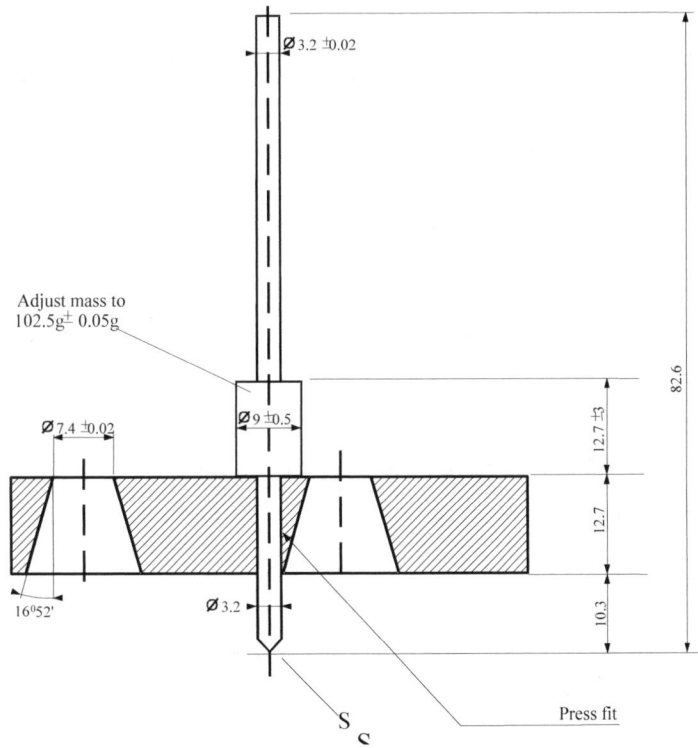

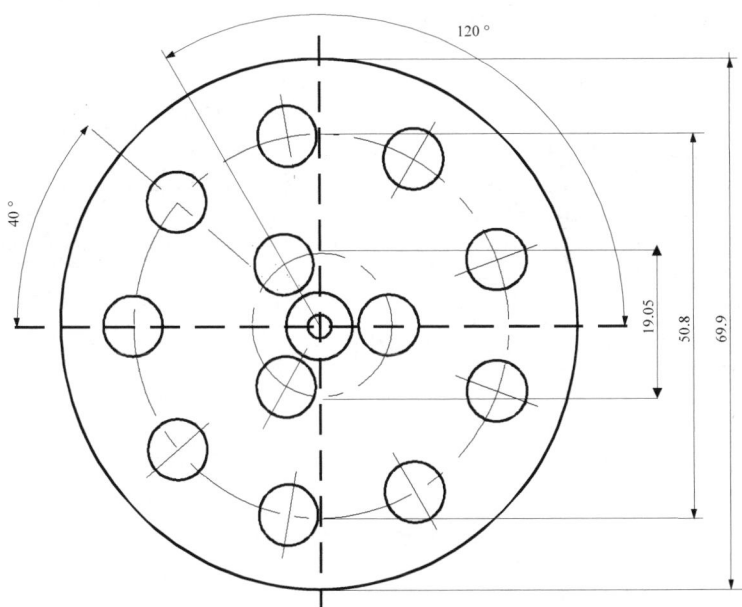

Tolerances not specified are ± 0.1 mm.

2.3.5   **Classification of organometallic substances in Classes 4.2 and 4.3**

Depending on their properties as determined in accordance with tests N.1 to N.5 of the Manual of Tests end Criteria, Part III, section 33, organometallic substances may be classified in Class 4.2 or 4.3, as appropriate, in accordance with the flowchart scheme given in Figure 2.3.5.

NOTE 1:   Depending on their other properties and on the precedence of hazard table (see 2.1.3.10), organometallic substances may have to be classified in other classes as appropriate.

NOTE 2:   Flammable solutions with organometallic compounds in concentrations which are not liable to spontaneous combustion or, in contact with water, do not emit flammable gases in dangerous quantities, are substances of Class 3.

**Figure 2.3.5: Flowchart scheme for the classification of organometallic substances
in Classes 4.2 and 4.3 [b]**

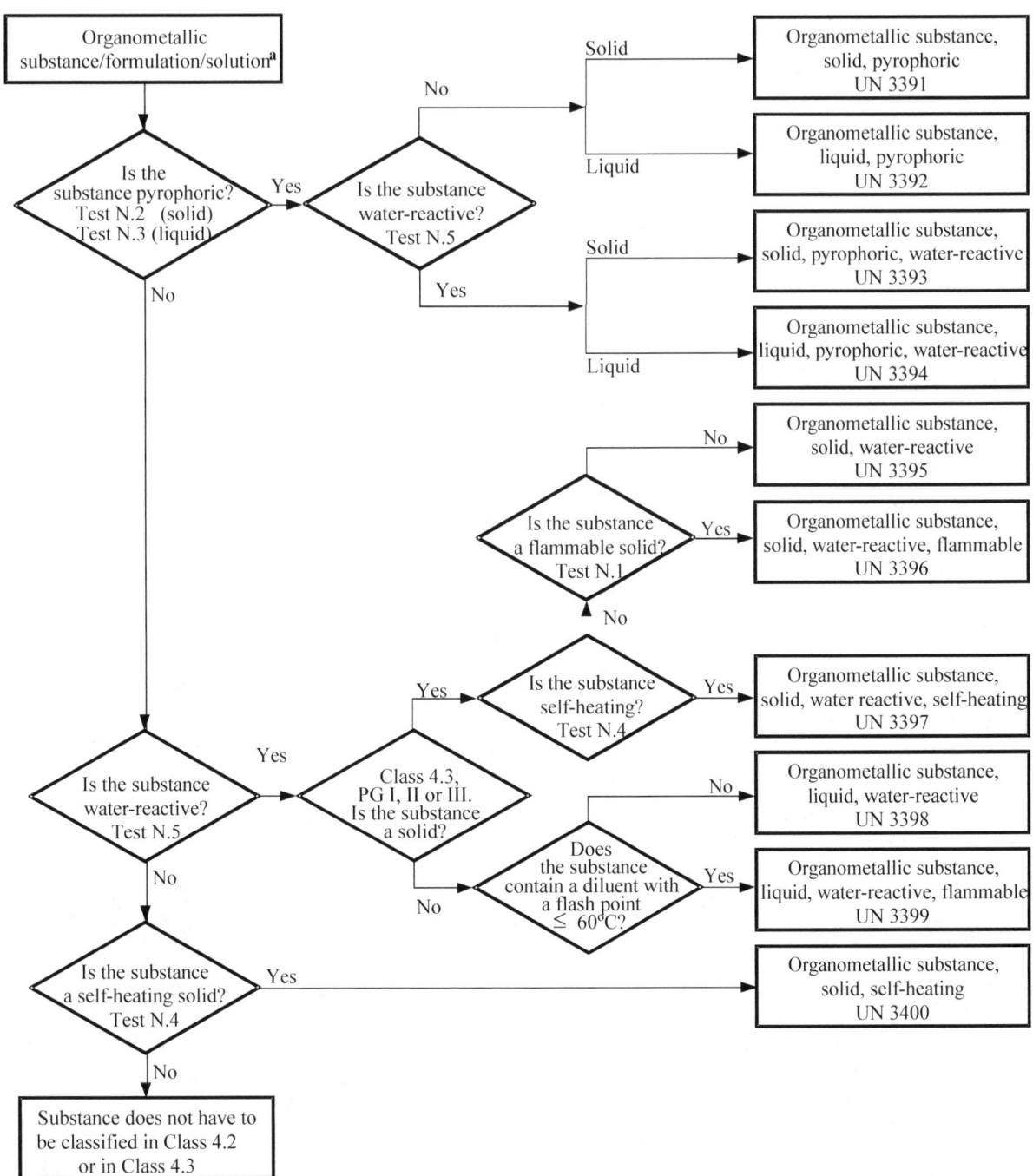

[a]   *If applicable and testing is relevant, taking into account reactivity properties, class 6.1 and 8 properties should be considered according to the precedence of hazard table of 2.1.3.10.*

[b]   *Test methods N.1 to N.5 can be found in the Manual of Tests and Criteria, Part III, Section 33.*

PART 3

Dangerous goods list, special provisions and exemptions related to limited and excepted quantities

CHAPTER 3.1

GENERAL

3.1.1   **Introduction**

In addition to the provisions referred to or given in the tables of this Part, the general requirements of each Part, Chapter and/or Section are to be observed. These general requirements are not given in the tables. When a general requirement is contradictory to a special provision, the special provision prevails.

3.1.2   **Proper shipping name**

NOTE: For proper shipping names used for the carriage of samples, see 2.1.4.1.

3.1.2.1   The proper shipping name is that portion of the entry most accurately describing the goods in Table A in Chapter 3.2, which is shown in upper case characters (plus any numbers, Greek letters, "sec", "tert", and the letters "m", "n", "o", "p", which form an integral part of the name). An alternative proper shipping name may be shown in brackets following the main proper shipping name [e.g., ETHANOL (ETHYL ALCOHOL)]. Portions of an entry appearing in lower case need not be considered as part of the proper shipping name.

3.1.2.2   When conjunctions such as "and" or "or" are in lower case or when segments of the name are punctuated by commas, the entire name of the entry need not necessarily be shown in the transport document or package marks. This is the case particularly when a combination of several distinct entries are listed under a single UN Number. Examples illustrating the selection of the proper shipping name for such entries are:

(a)   UN 1057 LIGHTERS or LIGHTER REFILLS - The proper shipping name is the most appropriate of the following combinations:

LIGHTERS
LIGHTER REFILLS;

(b)   UN 2793 FERROUS METAL BORINGS, SHAVINGS, TURNINGS or CUTTINGS in a form liable to self-heating. The proper shipping name is the most appropriate of the following combinations:

FERROUS METAL BORINGS
FERROUS METAL SHAVINGS
FERROUS METAL TURNINGS
FERROUS METAL CUTTINGS.

3.1.2.3   Proper shipping names may be used in the singular or plural as appropriate. In addition, when qualifying words are used as part of the proper shipping name, their sequence on documentation or package marks is optional. For instance, "DIMETHYLAMINE AQUEOUS SOLUTION" may alternatively be shown "AQUEOUS SOLUTION OF DIMETHYLAMINE". Commercial or military names for goods of Class 1 which contain the proper shipping name supplemented by additional descriptive text may be used.

3.1.2.4   Many substances have an entry for both the liquid and solid state (see definitions for liquid and solid in 1.2.1), or for the solid and solution. These are allocated separate UN numbers which are not necessarily adjacent to each other[1].

3.1.2.5   Unless it is already included in capital letters in the name indicated in Table A in Chapter 3.2, the qualifying word "MOLTEN" shall be added as part of the proper shipping name when a substance, which is a solid in accordance with the definition in 1.2.1, is offered for carriage in the molten state (e.g. ALKYLPHENOL, SOLID, N.O.S., MOLTEN).

---

[1]   *Details are provided in the alphabetical index (Table B of Chapter 3.2), e.g.:*

NITROXYLENES, LIQUID   6.1   1665;
NITROXYLENES, SOLID   6.1   3447.

3.1.2.6   Except for self-reactive substances and organic peroxides and unless it is already included in capital letters in the name indicated in Column (2) of Table A of Chapter 3.2, the word "STABILIZED" shall be added as part of the proper shipping name of a substance which without stabilization would be forbidden from carriage in accordance with paragraphs 2.2.X.2 due to it being liable to dangerously react under conditions normally encountered in carriage (e.g.: "TOXIC LIQUID, ORGANIC, N.O.S., STABILIZED").

When temperature control is used to stabilize such substances to prevent the development of any dangerous excess pressure, or the evolution of excessive heat, or when chemical stabilization is used in combination with temperature control, then:

(a)   For liquids and solids where the SAPT[2] (measured without or with inhibitor, when chemical stabilization is applied) is less than or equal to that prescribed in 2.2.41.1.21, the provisions of 2.2.41.1.17, special provision 386 of Chapter 3.3, special provision V8 of Chapter 7.2, special provision S4 of Chapter 8.5 and the requirements of Chapter 9.6 apply except that the term "SADT" as used in these paragraphs is understood to include also "SAPT" when the substance concerned reacts by polymerization;

(b)   For gases: the conditions of carriage shall be approved by the competent authority.

3.1.2.7   Hydrates may be carried under the proper shipping name for the anhydrous substance.

3.1.2.8   *Generic or "not otherwise specified" (N.O.S.) names*

3.1.2.8.1   Generic and "not otherwise specified" proper shipping names that are assigned to special provision 274 or 318 in Column (6) of Table A in Chapter 3.2 shall be supplemented with the technical name of the goods unless a national law or international convention prohibits its disclosure if it is a controlled substance. For explosives of Class 1, the dangerous goods description may be supplemented by additional descriptive text to indicate commercial or military names. Technical names shall be entered in brackets immediately following the proper shipping name. An appropriate modifier, such as "contains" or "containing" or other qualifying words such as "mixture", "solution", etc. and the percentage of the technical constituent may also be used. For example: "UN 1993 FLAMMABLE LIQUID, N.O.S. (CONTAINS XYLENE AND BENZENE), 3, II".

3.1.2.8.1.1   The technical name shall be a recognized chemical name or biological name, or other name currently used in scientific and technical handbooks, journals and texts. Trade names shall not be used for this purpose. In the case of pesticides, only ISO common name(s), other name(s) in the World Health Organization (WHO) Recommended Classification of Pesticides by Hazard and Guidelines to Classification, or the name(s) of the active substance(s) may be used.

3.1.2.8.1.2   When a mixture of dangerous goods is described by one of the "N.O.S." or "generic" entries to which special provision 274 has been allocated in Column (6) of Table A in Chapter 3.2, not more than the two constituents which most predominantly contribute to the hazard or hazards of a mixture need to be shown, excluding controlled substances when their disclosure is prohibited by national law or international convention. If a package containing a mixture is labelled with any subsidiary risk label, one of the two technical names shown in parentheses shall be the name of the constituent which compels the use of the subsidiary risk label.

NOTE: See 5.4.1.2.2.

3.1.2.8.1.3   Examples illustrating the selection of the proper shipping name supplemented with the technical name of goods for such N.O.S. entries are:

UN 2902   PESTICIDE, LIQUID, TOXIC, N.O.S. (drazoxolon);
UN 3394   ORGANOMETALLIC SUBSTANCE, LIQUID, PYROPHORIC, WATER-REACTIVE (trimethylgallium).

---

[2]   *For the definition of self-accelerating polymerization temperature (SAPT), see 1.2.1.*

3.1.3   **Solutions or mixtures**

NOTE: Where a substance is specifically mentioned by name in Table A of Chapter 3.2, it shall be identified in carriage by the proper shipping name in Column (2) of Table A of Chapter 3.2. Such substances may contain technical impurities (for example those deriving from the production process) or additives for stability or other purposes that do not affect its classification. However, a substance mentioned by name containing technical impurities or additives for stability or other purposes affecting its classification shall be considered a solution or mixture (see 2.1.3.3).

3.1.3.1   A solution or mixture is not subject to ADR if the characteristics, properties, form or physical state of the solution or mixture are such that it does not meet the criteria, including human experience criteria, for inclusion in any class.

3.1.3.2   A solution or mixture meeting the classification criteria of ADR composed of a single predominant substance mentioned by name in Table A of Chapter 3.2 and one or more substances not subject to ADR or traces of one or more substances mentioned by name in Table A of Chapter 3.2, shall be assigned the UN number and proper shipping name of the predominant substance mentioned by name in Table A of Chapter 3.2 unless:

(a)   The solution or mixture is mentioned by name in Table A of Chapter 3.2;

(b)   The name and description of the substance mentioned by name in Table A of Chapter 3.2 specifically indicate that they apply only to the pure substance;

(c)   The class, classification code, packing group, or physical state of the solution or mixture is different from that of the substance mentioned by name in Table A of Chapter 3.2; or

(d)   The hazard characteristics and properties of the solution or mixture necessitate emergency response measures that are different from those required for the substance mentioned by name in Table A of Chapter 3.2.

Qualifying words such as "SOLUTION" or "MIXTURE", as appropriate, shall be added as part of the proper shipping name, for example, "ACETONE SOLUTION". In addition, the concentration of the mixture or solution may also be indicated after the basic description of the mixture or solution, for example, "ACETONE 75% SOLUTION".

3.1.3.3   A solution or mixture meeting the classification criteria of ADR that is not mentioned by name in Table A of Chapter 3.2 and that is composed of two or more dangerous goods shall be assigned to an entry that has the proper shipping name, description, class, classification code and packing group that most precisely describe the solution or mixture.

CHAPTER 3.2

DANGEROUS GOODS LIST

3.2.1   **Table A: Dangerous Goods List**

Explanations

As a rule, each row of Table A of this Chapter deals with the substance(s) or article(s) covered by a specific UN number. However, when substances or articles belonging to the same UN number have different chemical properties, physical properties and/or carriage conditions, several consecutive rows may be used for that UN number.

Each column of Table A is dedicated to a specific subject as indicated in the explanatory notes below. The intersection of columns and rows (cell) contains information concerning the subject treated in that column, for the substance(s) or article(s) of that row:

-     The first four cells identify the substance(s) or article(s) belonging to that row (additional information in that respect may be given by the special provisions referred to in Column (6);

-     The following cells give the applicable special provisions, either in the form of complete information or in coded form. The codes cross-refer to detailed information that is to be found in the Part, Chapter, Section and/or Sub-section indicated in the explanatory notes below. An empty cell means either that there is no special provision and that only the general requirements apply, or that the carriage restriction indicated in the explanatory notes is in force. When used in this table, an alphanumeric code starting with the letters "SP" designates a special provision of Chapter 3.3.

The applicable general requirements are not referred to in the corresponding cells. The explanatory notes below indicate for every column the Part(s), Chapter(s), Section(s) and/or Sub-section(s) where these are to be found.

Explanatory notes for each column:

Column (1)   "UN No."

Contains the UN number:

-     of the dangerous substance or article if the substance or article has been assigned its own specific UN number; or

-     of the generic or n.o.s. entry to which the dangerous substances or articles not mentioned by name shall be assigned in accordance with the criteria ("decision trees") of Part 2.

Column (2)   "Name and description"

Contains, in upper case characters, the name of the substance or article, if the substance or article has been assigned its own specific UN number, or of the generic or n.o.s. entry to which it has been assigned in accordance with the criteria ("decision trees") of Part 2. This name shall be used as the proper shipping name or, when applicable, as part of the proper shipping name (see 3.1.2 for further details on the proper shipping name).

A descriptive text in lower case characters is added after the proper shipping name to clarify the scope of the entry if the classification and/or carriage conditions of the substance or article may be different under certain conditions.

Column (3a)   "Class"

Contains the number of the class, whose heading covers the dangerous substance or article. This class number is assigned in accordance with the procedures and criteria of Part 2.

Column (3b)	"Classification code"

Contains the classification code of the dangerous substance or article.

- For dangerous substances or articles of Class 1, the code consists of a division number and compatibility group letter, which are assigned in accordance with the procedures and criteria of 2.2.1.1.4;

- For dangerous substances or articles of Class 2, the code consists of a number and hazardous property group, which are explained in 2.2.2.1.2 and 2.2.2.1.3;

- For dangerous substances or articles of Classes 3, 4.1, 4.2, 4.3, 5.1, 5.2, 6.1, 6.2, 8 and 9, the codes are explained in 2.2.x.1.2 [1];

- Dangerous substances or articles of Class 7 do not have a classification code.

Column (4)	"Packing group"

Contains the packing group number(s) (I, II or III) assigned to the dangerous substance. These packing group numbers are assigned on the basis of the procedures and criteria of Part 2. Certain articles and substances are not assigned to packing groups.

Column (5)	"Labels"

Contains the model number of the labels/placards (see 5.2.2.2 and 5.3.1.7) that have to be affixed to packages, containers, tank-containers, portable tanks, MEGCs and vehicles. However, for substances or articles of Class 7, 7X means label model No.7A, 7B or 7C as appropriate according to the category (see 5.1.5.3.4 and 5.2.2.1.11.1) or placard No. 7D (see 5.3.1.1.3 and 5.3.1.7.2).

The general provisions on labelling/placarding (e.g. number of labels, their location) are to be found in 5.2.2.1 for packages, and in 5.3.1, for containers, tank-containers, MEGCs, portable tanks and vehicles.

NOTE: Special provisions, indicated in Column (6), may change the above labelling provisions.

Column (6)	"Special provisions"

Contains the numeric codes of special provisions that have to be met. These provisions concern a wide array of subjects, mainly connected with the contents of Columns (1) to (5) (e.g. carriage prohibitions, exemptions from requirements, explanations concerning the classification of certain forms of the dangerous goods concerned and additional labelling or marking provisions), and are listed in Chapter 3.3 in numerical order. If Column (6) is empty, no special provisions apply to the contents of Columns (1) to (5) for the dangerous goods concerned.

Column (7a)	"Limited Quantities"

Provides the maximum quantity per inner packaging or article for carrying dangerous goods as limited quantities in accordance with Chapter 3.4.

Column (7b)	"Excepted Quantities"

Contains an alphanumeric code with the following meaning:

- "E0" signifies that no exemption from the provisions of ADR exists for the dangerous goods packed in excepted quantities;

---

[1] *x = the class number of the dangerous substance or article, without dividing point if applicable.*

All the other alphanumerical codes starting with the letter "E" signify that the provisions of ADR are not applicable if the conditions indicated in Chapter 3.5 are fulfilled.

Column (8) "Packing instructions"

Contains the alphanumeric codes of the applicable packing instructions:

- Alphanumeric codes starting with the letter "P", which refers to packing instructions for packagings and receptacles (except IBCs and large packagings), or "R", which refers to packing instructions for light gauge metal packagings. These are listed in 4.1.4.1 in numerical order, and specify the packagings and receptacles that are authorized. They also indicate which of the general packing provisions of 4.1.1, 4.1.2 and 4.1.3, and which of the special packing provisions of 4.1.5, 4.1.6, 4.1.7, 4.1.8 and 4.1.9 have to be met. If Column (8) does not contain a code starting with the letters "P" or "R", the dangerous goods concerned may not be carried in packagings;

- Alphanumeric codes starting with the letters "IBC" refer to packing instructions for IBCs. These are listed in 4.1.4.2 in numerical order, and specify the IBCs that are authorized. They also indicate which of the general packing provisions of 4.1.1, 4.1.2 and 4.1.3, and which of the special packing provisions of 4.1.5, 4.1.6, 4.1.7, 4.1.8 and 4.1.9 have to be met. If Column (8) does not contain a code starting with the letters "IBC", the dangerous goods concerned may not be carried in IBCs;

- Alphanumeric codes starting with the letters "LP" refer to packing instructions for large packagings. These are listed in 4.1.4.3 in numerical order, and specify the large packagings that are authorized. They also indicate which of the general packing provisions of 4.1.1, 4.1.2 and 4.1.3, and which of the special packing provisions of 4.1.5, 4.1.6, 4.1.7, 4.1.8 and 4.1.9 have to be met. If Column (8) does not contain a code starting with the letters "LP", the dangerous goods concerned cannot be carried in large packagings;

NOTE: Special packing provisions, indicated in Column (9a), may change the above packing instructions.

Column (9a) "Special packing provisions"

Contains the alphanumeric codes of the applicable special packing provisions:

- Alphanumeric codes starting with the letters "PP" or "RR" refer to special packing provisions for packagings and receptacles (except IBCs and large packagings) that have additionally to be met. These are to be found in 4.1.4.1, at the end of the relevant packing instruction (with the letter "P" or "R") referred to in Column (8). If Column (9a) does not contain a code starting with the letters "PP" or "RR", none of the special packing provisions listed at the end of the relevant packing instruction apply;

- Alphanumeric codes starting with the letter "B" or the letters "BB" refer to special packing provisions for IBCs that have additionally to be met. These are to be found in 4.1.4.2, at the end of the relevant packing instruction (with the letters "IBC") referred to in Column (8). If Column (9a) does not contain a code starting with the letter "B" or the letters "BB", none of the special packing provisions listed at the end of the relevant packing instruction apply;

- Alphanumeric codes starting with the letter "L" refer to special packing provisions for large packagings that have additionally to be met. These are to be found in 4.1.4.3, at the end of the relevant packing instruction (with the letters "LP") referred to in Column (8). If Column (9a) does not contain a code starting with the letter "L", none of the special packing provisions listed at the end of the relevant packing instruction apply.

Column (9b)	"Mixed packing provisions"

Contains the alphanumeric codes starting with the letters "MP" of the applicable mixed packing provisions. These are listed in 4.1.10 in numerical order. If Column (9b) does not contain a code starting with the letters "MP", only the general requirements apply (see 4.1.1.5 and 4.1.1.6).

Column (10)	"Portable tank and bulk container instructions"

Contains an alphanumeric code assigned to a portable tank instruction, in accordance with 4.2.5.2.1 to 4.2.5.2.4 and 4.2.5.2.6. This portable tank instruction corresponds to the least stringent provisions that are acceptable for the carriage of the substance in portable tanks. The codes identifying the other portable tank instructions that are also permitted for the carriage of the substance are to be found in 4.2.5.2.5. If no code is given, carriage in portable tanks is not permitted unless a competent authority approval is granted as detailed in 6.7.1.3.

The general requirements for the design, construction, equipment, type approval, testing and marking of portable tanks are to be found in Chapter 6.7. The general requirements for the use (e.g. filling) are to be found in 4.2.1 to 4.2.4.

The indication of a "(M)" means that the substance may be carried in UN MEGCs.

NOTE: Special provisions, indicated in Column (11), may change the above requirements.

May also contain alphanumeric codes starting with the letters "BK" referring to types of bulk containers described in Chapter 6.11 which may be used for the carriage of bulk goods in accordance with 7.3.1.1 (a) and 7.3.2.

Column (11)	"Portable tank and bulk container special provisions"

Contains the alphanumeric codes of the portable tank special provisions that have additionally to be met. These codes, starting with the letters "TP" refer to special provisions for the construction or use of these portable tanks. They are to be found in 4.2.5.3.

NOTE: If technically relevant, these special provisions are not only applicable to the portable tanks specified in column (10), but also to the portable tanks that may be used according to the table in 4.2.5.2.5.

Column (12)	"Tank codes for ADR tanks"

Contains an alphanumeric code describing a tank type, in accordance with 4.3.3.1.1 (for gases of Class 2) or 4.3.4.1.1 (for substances of Classes 3 to 9). This tank type corresponds to the least stringent tank provisions that are acceptable for the carriage of the relevant substance in ADR tanks. The codes describing the other permitted tank types are to be found in 4.3.3.1.2 (for gases of Class 2) or 4.3.4.1.2 (for substances of Classes 3 to 9). If no code is given, carriage in ADR tanks is not permitted.

If in this column a tank code for solids (S) and for liquids (L) is indicated, this means that this substance may be offered for carriage in tanks in the solid or the liquid (molten) state. In general this provision is applicable to substances having melting points from 20 °C to 180 °C.

If for a solid, only a tank code for liquids (L) is indicated in this column, this means that this substance is only offered for carriage in tanks in the liquid (molten) state.

The general requirements for the construction, equipment, type approval, testing and marking that are not indicated in the tank code are to be found in 6.8.1, 6.8.2, 6.8.3 and 6.8.5. The general requirements for the use (e.g. maximum degree of filling, minimum test pressure) are to be found in 4.3.1 to 4.3.4.

The indication of a "(M)" after the tank code means that the substance can also be carried in battery-vehicles or MEGCs.

The indication of a (+) after the tank code means that the alternative use of the tanks is permitted only where this is specified in the certificate of type approval.

For fibre-reinforced plastic tanks, see 4.4.1 and Chapter 6.9; for vacuum operated waste tanks, see 4.5.1 and Chapter 6.10.

NOTE: Special provisions, indicated in Column (13), may change the above requirements.

Column (13)	"Special provisions for ADR tanks"

Contains the alphanumeric codes of the special provisions for ADR tanks that have additionally to be met:

- Alphanumeric codes starting with the letters "TU" refer to special provisions for the use of these tanks. These are to be found in 4.3.5;

- Alphanumeric codes starting with the letters "TC" refer to special provisions for the construction of these tanks. These are to be found in 6.8.4 (a);

- Alphanumeric codes starting with the letters "TE" refer to special provisions concerning the items of equipment of these tanks. These are to be found in 6.8.4 (b);

- Alphanumeric codes starting with the letters "TA" refer to special provisions for the type approval of these tanks. These are to be found in 6.8.4 (c);

- Alphanumeric codes starting with the letters "TT" refer to special provisions for the testing of these tanks. These are to be found in 6.8.4 (d);

- Alphanumeric codes starting with the letters "TM" refer to special provisions for the marking of these tanks. These are to be found in 6.8.4 (e).

NOTE: If technically relevant, these special provisions are not only applicable to the tanks specified in column (12), but also to the tanks that may be used according to the hierarchies in 4.3.3.1.2 and 4.3.4.1.2.

Column (14)	"Vehicle for tank carriage"

Contains a code designating the vehicle (including the drawing vehicle of trailers or semi-trailers) (see 9.1.1) to be used for the carriage of the substance in tank in accordance with 7.4.2. The requirements concerning the construction and approval of vehicles are to be found in Chapters 9.1, 9.2 and 9.7.

Column (15)	"Transport category / (Tunnel restriction code)"

Contains at the top of the cell a figure indicating the transport category to which the substance or article is assigned for the purposes of exemption related to quantities carried per transport unit (see 1.1.3.6).

Contains at the bottom of the cell, between brackets, the tunnel restriction code that refers to the applicable restriction for the passage of vehicles carrying the substance or article through road tunnels. These are to be found in Chapter 8.6. When no tunnel restriction code has been assigned, this is indicated by the mention '(—)'.

Column (16)	"Special provisions for carriage - Packages"

Contains the alphanumeric code(s), starting with letter "V", of the applicable special provisions (if any) for carriage in packages. These are listed in 7.2.4.

General provisions concerning the carriage in packages are to be found in Chapters 7.1 and 7.2.

NOTE: In addition, special provisions indicated in Column (18), concerning loading, unloading and handling, shall be observed.

Column (17) "Special provisions for carriage - Bulk"

Contains the alphanumeric code(s), starting with letters "VC", as well as the alphanumeric code(s) starting with letters "AP", of the applicable provisions for carriage in bulk. These are listed in 7.3.3. If no special provision, identified by the code "VC" or a reference to a specific paragraph, explicitly authorizing this mode of carriage is indicated in this column, and no special provision, identified by the code "BK" or a reference to a specific paragraph, explicitly authorizing this mode of carriage is indicated in column (10), carriage in bulk is not permitted. General and additional provisions concerning carriage in bulk are to be found in Chapters 7.1 and 7.3.

NOTE: In addition, special provisions indicated in Column (18), concerning loading, unloading and handling, shall be observed.

Column (18) "Special provisions for carriage – Loading, unloading and handling"

Contains the alphanumeric code(s), starting with letters "CV", of the applicable special provisions for loading, unloading and handling. These are listed in 7.5.11. If no code is given, only the general provisions apply (see 7.5.1 to 7.5.10).

Column (19) "Special provisions for carriage - Operation"

Contains the alphanumeric code(s), starting with letter "S", of the applicable special provisions for operation which are listed in Chapter 8.5. These provisions shall be applied in addition to the requirements of Chapters 8.1 to 8.4 but in the event of conflict with the requirements of Chapters 8.1 to 8.4, the special provisions shall take precedence.

Column (20) "Hazard identification number"

Contains a two or three figure number (preceded in certain cases by the letter "X") for substances and articles of classes 2 to 9, and for substances and articles of Class 1, the classification code (see column (3b)). In the cases described in 5.3.2.1, this number shall appear in the upper half of the orange-coloured plates. The meaning of the hazard identification numbers is explained in 5.3.2.3.

TABLE A

DANGEROUS GOODS LIST

UN No.	Name and description	Class	Classification code	Packing group	Labels	Special provisions	Limited and excepted quantities		Packaging			Portable tanks and bulk containers	
									Packing instructions	Special packing provisions	Mixed packing provisions	Instructions	Special provisions
3.1.2		2.2	2.2	2.1.1.3	5.2.2	3.3	3.4	3.5.1.2	4.1.4	4.1.4	4.1.10	4.2.5.2 7.3.2	4.2.5.3
(1)	(2)	(3a)	(3b)	(4)	(5)	(6)	(7a)	(7b)	(8)	(9a)	(9b)	(10)	(11)
0004	AMMONIUM PICRATE dry or wetted with less than 10% water, by mass	1	1.1D		1		0	E0	P112(a) P112(b) P112(c)	PP26	MP20		
0005	CARTRIDGES FOR WEAPONS with bursting charge	1	1.1F		1		0	E0	P130		MP23		
0006	CARTRIDGES FOR WEAPONS with bursting charge	1	1.1E		1		0	E0	P130 LP101	PP67 L1	MP21		
0007	CARTRIDGES FOR WEAPONS with bursting charge	1	1.2F		1		0	E0	P130		MP23		
0009	AMMUNITION, INCENDIARY with or without burster, expelling charge or propelling charge	1	1.2G		1		0	E0	P130 LP101	PP67 L1	MP23		
0010	AMMUNITION, INCENDIARY with or without burster, expelling charge or propelling charge	1	1.3G		1		0	E0	P130 LP101	PP67 L1	MP23		
0012	CARTRIDGES FOR WEAPONS, INERT PROJECTILE or CARTRIDGES, SMALL ARMS	1	1.4S		1.4	364	5 kg	E0	P130		MP23 MP24		
0014	CARTRIDGES FOR WEAPONS, BLANK or CARTRIDGES, SMALL ARMS, BLANK or CARTRIDGE FOR TOOLS, BLANK	1	1.4S		1.4	364	5 kg	E0	P130		MP23 MP24		
0015	AMMUNITION, SMOKE with or without burster, expelling charge or propelling charge	1	1.2G		1		0	E0	P130 LP101	PP67 L1	MP23		
0015	AMMUNITION, SMOKE with or without burster, expelling charge or propelling charge, containing corrosive substances	1	1.2G		1 +8		0	E0	P130 LP101	PP67 L1	MP23		
0015	AMMUNITION, SMOKE with or without burster, expelling charge or propelling charge, containing toxic by inhalation substances	1	1.2G		1 +6.1		0	E0	P130 LP101	PP67 L1	MP23		
0016	AMMUNITION, SMOKE with or without burster, expelling charge or propelling charge	1	1.3G		1		0	E0	P130 LP101	PP67 L1	MP23		
0016	AMMUNITION, SMOKE with or without burster, expelling charge or propelling charge, containing corrosive substances	1	1.3G		1 +8		0	E0	P130 LP101	PP67 L1	MP23		
0016	AMMUNITION, SMOKE with or without burster, expelling charge or propelling charge, containing toxic by inhalation substances	1	1.3G		1 +6.1		0	E0	P130 LP101	PP67 L1	MP23		
0018	AMMUNITION, TEAR-PRODUCING with burster, expelling charge or propelling charge	1	1.2G		1 +6.1 +8		0	E0	P130 LP101	PP67 L1	MP23		
0019	AMMUNITION, TEAR-PRODUCING with burster, expelling charge or propelling charge	1	1.3G		1 +6.1 +8		0	E0	P130 LP101	PP67 L1	MP23		
0020	AMMUNITION, TOXIC with burster, expelling charge or propelling charge	1	1.2K	CARRIAGE PROHIBITED										
0021	AMMUNITION, TOXIC with burster, expelling charge or propelling charge	1	1.3K	CARRIAGE PROHIBITED										
0027	BLACK POWDER (GUNPOWDER), granular or as a meal	1	1.1D		1		0	E0	P113	PP50	MP20 MP24		
0028	BLACK POWDER (GUNPOWDER), COMPRESSED or BLACK POWDER (GUNPOWDER), IN PELLETS	1	1.1D		1		0	E0	P113	PP51	MP20 MP24		
0029	DETONATORS, NON-ELECTRIC for blasting	1	1.1B		1		0	E0	P131	PP68	MP23		
0030	DETONATORS, ELECTRIC for blasting	1	1.1B		1		0	E0	P131		MP23		

ADR tank		Vehicle for tank carriage	Transport category (Tunnel restriction code)	Special provisions for carriage				Hazard identifi- cation No.	UN No.	Name and description
Tank code	Special provisions			Packages	Bulk	Loading, unloading and handling	Operation			
4.3	4.3.5, 6.8.4	9.1.1.2	1.1.3.6 (8.6)	7.2.4	7.3.3	7.5.11	8.5	5.3.2.3		3.1.2
(12)	(13)	(14)	(15)	(16)	(17)	(18)	(19)	(20)	(1)	(2)
			1 (B1000C)	V2 V3		CV1 CV2 CV3	S1		0004	AMMONIUM PICRATE dry or wetted with less than 10% water, by mass
			1 (B1000C)	V2		CV1 CV2 CV3	S1		0005	CARTRIDGES FOR WEAPONS with bursting charge
			1 (B1000C)	V2		CV1 CV2 CV3	S1		0006	CARTRIDGES FOR WEAPONS with bursting charge
			1 (B1000C)	V2		CV1 CV2 CV3	S1		0007	CARTRIDGES FOR WEAPONS with bursting charge
			1 (B1000C)	V2		CV1 CV2 CV3	S1		0009	AMMUNITION, INCENDIARY with or without burster, expelling charge or propelling charge
			1 (C5000D)	V2		CV1 CV2 CV3	S1		0010	AMMUNITION, INCENDIARY with or without burster, expelling charge or propelling charge
			4 (E)			CV1 CV2 CV3	S1		0012	CARTRIDGES FOR WEAPONS, INERT PROJECTILE or CARTRIDGES, SMALL ARMS
			4 (E)			CV1 CV2 CV3	S1		0014	CARTRIDGES FOR WEAPONS, BLANK or CARTRIDGES, SMALL ARMS, BLANK or CARTRIDGE FOR TOOLS, BLANK
			1 (B1000C)	V2		CV1 CV2 CV3	S1		0015	AMMUNITION, SMOKE with or without burster, expelling charge or propelling charge
			1 (B1000C)	V2		CV1 CV2 CV3	S1		0015	AMMUNITION, SMOKE with or without burster, expelling charge or propelling charge, containing corrosive substances
			1 (B1000C)	V2		CV1 CV2 CV3 CV28	S1		0015	AMMUNITION, SMOKE with or without burster, expelling charge or propelling charge, containing toxic by inhalation substances
			1 (C5000D)	V2		CV1 CV2 CV3	S1		0016	AMMUNITION, SMOKE with or without burster, expelling charge or propelling charge
			1 (C5000D)	V2		CV1 CV2 CV3	S1		0016	AMMUNITION, SMOKE with or without burster, expelling charge or propelling charge, containing corrosive substances
			1 (C5000D)	V2		CV1 CV2 CV3 CV28	S1		0016	AMMUNITION, SMOKE with or without burster, expelling charge or propelling charge, containing toxic by inhalation substances
			1 (B1000C)	V2		CV1 CV2 CV3 CV28	S1		0018	AMMUNITION, TEAR-PRODUCING with burster, expelling charge or propelling charge
			1 (C5000D)	V2		CV1 CV2 CV3 CV28	S1		0019	AMMUNITION, TEAR-PRODUCING with burster, expelling charge or propelling charge
CARRIAGE PROHIBITED									0020	AMMUNITION, TOXIC with burster, expelling charge or propelling charge
CARRIAGE PROHIBITED									0021	AMMUNITION, TOXIC with burster, expelling charge or propelling charge
			1 (B1000C)	V2 V3		CV1 CV2 CV3	S1		0027	BLACK POWDER (GUNPOWDER), granular or as a meal
			1 (B1000C)	V2		CV1 CV2 CV3	S1		0028	BLACK POWDER (GUNPOWDER), COMPRESSED or BLACK POWDER (GUNPOWDER), IN PELLETS
			1 (B1000C)	V2		CV1 CV2 CV3	S1		0029	DETONATORS, NON-ELECTRIC for blasting
			1 (B1000C)	V2		CV1 CV2 CV3	S1		0030	DETONATORS, ELECTRIC for blasting

UN No.	Name and description	Class	Classifi-cation code	Packing group	Labels	Special provi-sions	Limited and excepted quantities		Packaging			Portable tanks and bulk containers	
									Packing instruc-tions	Special packing provisions	Mixed packing provisions	Instruc-tions	Special provisions
	3.1.2	2.2	2.2	2.1.1.3	5.2.2	3.3	3.4	3.5.1.2	4.1.4	4.1.4	4.1.10	4.2.5.2 7.3.2	4.2.5.3
(1)	(2)	(3a)	(3b)	(4)	(5)	(6)	(7a)	(7b)	(8)	(9a)	(9b)	(10)	(11)
0033	BOMBS with bursting charge	1	1.1F		1		0	E0	P130		MP23		
0034	BOMBS with bursting charge	1	1.1D		1		0	E0	P130 LP101	PP67 L1	MP21		
0035	BOMBS with bursting charge	1	1.2D		1		0	E0	P130 LP101	PP67 L1	MP21		
0037	BOMBS, PHOTO-FLASH	1	1.1F		1		0	E0	P130		MP23		
0038	BOMBS, PHOTO-FLASH	1	1.1D		1		0	E0	P130 LP101	PP67 L1	MP21		
0039	BOMBS, PHOTO-FLASH	1	1.2G		1		0	E0	P130 LP101	PP67 L1	MP23		
0042	BOOSTERS without detonator	1	1.1D		1		0	E0	P132(a) P132(b)		MP21		
0043	BURSTERS, explosive	1	1.1D		1		0	E0	P133	PP69	MP21		
0044	PRIMERS, CAP TYPE	1	1.4S		1.4		0	E0	P133		MP23 MP24		
0048	CHARGES, DEMOLITION	1	1.1D		1		0	E0	P130 LP101	PP67 L1	MP21		
0049	CARTRIDGES, FLASH	1	1.1G		1		0	E0	P135		MP23		
0050	CARTRIDGES, FLASH	1	1.3G		1		0	E0	P135		MP23		
0054	CARTRIDGES, SIGNAL	1	1.3G		1		0	E0	P135		MP23 MP24		
0055	CASES, CARTRIDGE, EMPTY, WITH PRIMER	1	1.4S		1.4	364	5 kg	E0	P136		MP23		
0056	CHARGES, DEPTH	1	1.1D		1		0	E0	P130 LP101	PP67 L1	MP21		
0059	CHARGES, SHAPED without detonator	1	1.1D		1		0	E0	P137	PP70	MP21		
0060	CHARGES, SUPPLEMENTARY, EXPLOSIVE	1	1.1D		1		0	E0	P132(a) P132(b)		MP21		
0065	CORD, DETONATING, flexible	1	1.1D		1		0	E0	P139	PP71 PP72	MP21		
0066	CORD, IGNITER	1	1.4G		1.4		0	E0	P140		MP23		
0070	CUTTERS, CABLE, EXPLOSIVE	1	1.4S		1.4		0	E0	P134 LP102		MP23		
0072	CYCLOTRIMETHYLENE-TRINITRAMINE (CYCLONITE; HEXOGEN; RDX), WETTED with not less than 15% water, by mass	1	1.1D		1	266	0	E0	P112(a)	PP45	MP20		
0073	DETONATORS FOR AMMUNITION	1	1.1B		1		0	E0	P133		MP23		
0074	DIAZODINITROPHENOL, WETTED with not less than 40% water, or mixture of alcohol and water, by mass	1	1.1A		1	266	0	E0	P110(b)	PP42	MP20		

ADR tank		Vehicle for tank carriage	Transport category (Tunnel restriction code)	Special provisions for carriage				Hazard identifi-cation No.	UN No.	Name and description
Tank code	Special provisions			Packages	Bulk	Loading, unloading and handling	Operation			
4.3	4.3.5, 6.8.4	9.1.1.2	1.1.3.6 (8.6)	7.2.4	7.3.3	7.5.11	8.5	5.3.2.3		3.1.2
(12)	(13)	(14)	(15)	(16)	(17)	(18)	(19)	(20)	(1)	(2)
			1 (B1000C)	V2		CV1 CV2 CV3	S1		0033	BOMBS with bursting charge
			1 (B1000C)	V2		CV1 CV2 CV3	S1		0034	BOMBS with bursting charge
			1 (B1000C)	V2		CV1 CV2 CV3	S1		0035	BOMBS with bursting charge
			1 (B1000C)	V2		CV1 CV2 CV3	S1		0037	BOMBS, PHOTO-FLASH
			1 (B1000C)	V2		CV1 CV2 CV3	S1		0038	BOMBS, PHOTO-FLASH
			1 (B1000C)	V2		CV1 CV2 CV3	S1		0039	BOMBS, PHOTO-FLASH
			1 (B1000C)	V2		CV1 CV2 CV3	S1		0042	BOOSTERS without detonator
			1 (B1000C)	V2		CV1 CV2 CV3	S1		0043	BURSTERS, explosive
			4 (E)			CV1 CV2 CV3	S1		0044	PRIMERS, CAP TYPE
			1 (B1000C)	V2		CV1 CV2 CV3	S1		0048	CHARGES, DEMOLITION
			1 (B1000C)	V2		CV1 CV2 CV3	S1		0049	CARTRIDGES, FLASH
			1 (C5000D)	V2		CV1 CV2 CV3	S1		0050	CARTRIDGES, FLASH
			1 (C5000D)	V2		CV1 CV2 CV3	S1		0054	CARTRIDGES, SIGNAL
			4 (E)			CV1 CV2 CV3	S1		0055	CASES, CARTRIDGE, EMPTY, WITH PRIMER
			1 (B1000C)	V2		CV1 CV2 CV3	S1		0056	CHARGES, DEPTH
			1 (B1000C)	V2		CV1 CV2 CV3	S1		0059	CHARGES, SHAPED without detonator
			1 (B1000C)	V2		CV1 CV2 CV3	S1		0060	CHARGES, SUPPLEMENTARY, EXPLOSIVE
			1 (B1000C)	V2		CV1 CV2 CV3	S1		0065	CORD, DETONATING, flexible
			2 (E)	V2		CV1 CV2 CV3	S1		0066	CORD, IGNITER
			4 (E)			CV1 CV2 CV3	S1		0070	CUTTERS, CABLE, EXPLOSIVE
			1 (B1000C)	V2		CV1 CV2 CV3	S1		0072	CYCLOTRIMETHYLENE-TRINITRAMINE (CYCLONITE; HEXOGEN; RDX), WETTED with not less than 15% water, by mass
			1 (B1000C)	V2		CV1 CV2 CV3	S1		0073	DETONATORS FOR AMMUNITION
			0 (B)	V2		CV1 CV2 CV3	S1		0074	DIAZODINITROPHENOL, WETTED with not less than 40% water, or mixture of alcohol and water, by mass

UN No.	Name and description	Class	Classification code	Packing group	Labels	Special provisions	Limited and excepted quantities		Packaging			Portable tanks and bulk containers	
									Packing instructions	Special packing provisions	Mixed packing provisions	Instructions	Special provisions
	3.1.2	2.2	2.2	2.1.1.3	5.2.2	3.3	3.4	3.5.1.2	4.1.4	4.1.4	4.1.10	4.2.5.2 7.3.2	4.2.5.3
(1)	(2)	(3a)	(3b)	(4)	(5)	(6)	(7a)	(7b)	(8)	(9a)	(9b)	(10)	(11)
0075	DIETHYLENEGLYCOL DINITRATE, DESENSITIZED with not less than 25% non-volatile, water-insoluble phlegmatizer, by mass	1	1.1D		1	266	0	E0	P115	PP53 PP54 PP57 PP58	MP20		
0076	DINITROPHENOL, dry or wetted with less than 15% water, by mass	1	1.1D		1 +6.1		0	E0	P112(a) P112(b) P112(c)	PP26	MP20		
0077	DINITROPHENOLATES, alkali metals, dry or wetted with less than 15% water, by mass	1	1.3C		1 +6.1		0	E0	P114(a) P114(b)	PP26	MP20		
0078	DINITRORESORCINOL, dry or wetted with less than 15% water, by mass	1	1.1D		1		0	E0	P112(a) P112(b) P112(c)	PP26	MP20		
0079	HEXANITRODIPHENYL-AMINE (DIPICRYLAMINE; HEXYL)	1	1.1D		1		0	E0	P112(b) P112(c)		MP20		
0081	EXPLOSIVE, BLASTING, TYPE A	1	1.1D		1	616 617	0	E0	P116	PP63 PP66	MP20		
0082	EXPLOSIVE, BLASTING, TYPE B	1	1.1D		1	617	0	E0	P116 IBC100	PP61 PP62 B9	MP20		
0083	EXPLOSIVE, BLASTING, TYPE C	1	1.1D		1	267 617	0	E0	P116		MP20		
0084	EXPLOSIVE, BLASTING, TYPE D	1	1.1D		1	617	0	E0	P116		MP20		
0092	FLARES, SURFACE	1	1.3G		1		0	E0	P135		MP23		
0093	FLARES, AERIAL	1	1.3G		1		0	E0	P135		MP23		
0094	FLASH POWDER	1	1.1G		1		0	E0	P113	PP49	MP20		
0099	FRACTURING DEVICES, EXPLOSIVE without detonator, for oil wells	1	1.1D		1		0	E0	P134 LP102		MP21		
0101	FUSE, NON-DETONATING	1	1.3G		1		0	E0	P140	PP74 PP75	MP23		
0102	CORD (FUSE), DETONATING, metal clad	1	1.2D		1		0	E0	P139	PP71	MP21		
0103	FUSE, IGNITER, tubular, metal clad	1	1.4G		1.4		0	E0	P140		MP23		
0104	CORD (FUSE), DETONATING, MILD EFFECT, metal clad	1	1.4D		1.4		0	E0	P139	PP71	MP21		
0105	FUSE, SAFETY	1	1.4S		1.4		0	E0	P140	PP73	MP23		
0106	FUZES, DETONATING	1	1.1B		1		0	E0	P141		MP23		
0107	FUZES, DETONATING	1	1.2B		1		0	E0	P141		MP23		
0110	GRENADES, PRACTICE, hand or rifle	1	1.4S		1.4		0	E0	P141		MP23		
0113	GUANYLNITROSAMINO-GUANYLIDENE HYDRAZINE, WETTED with not less than 30% water, by mass	1	1.1A		1	266	0	E0	P110(b)	PP42	MP20		

ADR tank		Vehicle for tank carriage	Transport category (Tunnel restriction code)	Special provisions for carriage				Hazard identifi-cation No.	UN No.	Name and description
Tank code	Special provisions			Packages	Bulk	Loading, unloading and handling	Operation			
4.3	4.3.5, 6.8.4	9.1.1.2	1.1.3.6 (8.6)	7.2.4	7.3.3	7.5.11	8.5	5.3.2.3		3.1.2
(12)	(13)	(14)	(15)	(16)	(17)	(18)	(19)	(20)	(1)	(2)
			1 (B1000C)	V2		CV1 CV2 CV3	S1		0075	DIETHYLENEGLYCOL DINITRATE, DESENSITIZED with not less than 25% non-volatile, water-insoluble phlegmatizer, by mass
			1 (B1000C)	V2 V3		CV1 CV2 CV3 CV28	S1		0076	DINITROPHENOL, dry or wetted with less than 15% water, by mass
			1 (C5000D)	V2 V3		CV1 CV2 CV3 CV28	S1		0077	DINITROPHENOLATES, alkali metals, dry or wetted with less than 15% water, by mass
			1 (B1000C)	V2 V3		CV1 CV2 CV3	S1		0078	DINITRORESORCINOL, dry or wetted with less than 15% water, by mass
			1 (B1000C)	V2 V3		CV1 CV2 CV3	S1		0079	HEXANITRODIPHENYL-AMINE (DIPICRYLAMINE; HEXYL)
			1 (B1000C)	V2 V3		CV1 CV2 CV3	S1		0081	EXPLOSIVE, BLASTING, TYPE A
			1 (B1000C)	V2 V3 V12		CV1 CV2 CV3	S1		0082	EXPLOSIVE, BLASTING, TYPE B
			1 (B1000C)	V2 V3		CV1 CV2 CV3	S1		0083	EXPLOSIVE, BLASTING, TYPE C
			1 (B1000C)	V2		CV1 CV2 CV3	S1		0084	EXPLOSIVE, BLASTING, TYPE D
			1 (C5000D)	V2		CV1 CV2 CV3	S1		0092	FLARES, SURFACE
			1 (C5000D)	V2		CV1 CV2 CV3	S1		0093	FLARES, AERIAL
			1 (B1000C)	V2 V3		CV1 CV2 CV3	S1		0094	FLASH POWDER
			1 (B1000C)	V2		CV1 CV2 CV3	S1		0099	FRACTURING DEVICES, EXPLOSIVE without detonator, for oil wells
			1 (C5000D)	V2		CV1 CV2 CV3	S1		0101	FUSE, NON-DETONATING
			1 (B1000C)	V2		CV1 CV2 CV3	S1		0102	CORD (FUSE), DETONATING, metal clad
			2 (E)	V2		CV1 CV2 CV3	S1		0103	FUSE, IGNITER, tubular, metal clad
			2 (E)	V2		CV1 CV2 CV3	S1		0104	CORD (FUSE), DETONATING, MILD EFFECT, metal clad
			4 (E)			CV1 CV2 CV3	S1		0105	FUSE, SAFETY
			1 (B1000C)	V2		CV1 CV2 CV3	S1		0106	FUZES, DETONATING
			1 (B1000C)	V2		CV1 CV2 CV3	S1		0107	FUZES, DETONATING
			4 (E)			CV1 CV2 CV3	S1		0110	GRENADES, PRACTICE, hand or rifle
			0 (B)	V2		CV1 CV2 CV3	S1		0113	GUANYLNITROSAMINO-GUANYLIDENE HYDRAZINE, WETTED with not less than 30% water, by mass

UN No.	Name and description	Class	Classifi- cation code	Packing group	Labels	Special provi- sions	Limited and excepted quantities		Packaging			Portable tanks and bulk containers	
									Packing instruc- tions	Special packing provisions	Mixed packing provisions	Instruc- tions	Special provisions
	3.1.2	2.2	2.2	2.1.1.3	5.2.2	3.3	3.4	3.5.1.2	4.1.4	4.1.4	4.1.10	4.2.5.2 7.3.2	4.2.5.3
(1)	(2)	(3a)	(3b)	(4)	(5)	(6)	(7a)	(7b)	(8)	(9a)	(9b)	(10)	(11)
0114	GUANYLNITROSAMINO- GUANYLTETRAZENE (TETRAZENE), WETTED with not less than 30% water, or mixture of alcohol and water, by mass	1	1.1A		1	266	0	E0	P110(b)	PP42	MP20		
0118	HEXOLITE (HEXOTOL), dry or wetted with less than 15% water, by mass	1	1.1D		1		0	E0	P112(a) P112(b) P112(c)		MP20		
0121	IGNITERS	1	1.1G		1		0	E0	P142		MP23		
0124	JET PERFORATING GUNS, CHARGED, oil well, without detonator	1	1.1D		1		0	E0	P101		MP21		
0129	LEAD AZIDE, WETTED with not less than 20% water, or mixture of alcohol and water, by mass	1	1.1A		1	266	0	E0	P110(b)	PP42	MP20		
0130	LEAD STYPHNATE (LEAD TRINITRORESORCINATE), WETTED with not less than 20% water, or mixture of alcohol and water, by mass	1	1.1A		1	266	0	E0	P110(b)	PP42	MP20		
0131	LIGHTERS, FUSE	1	1.4S		1.4		0	E0	P142		MP23		
0132	DEFLAGRATING METAL SALTS OF AROMATIC NITRODERIVATIVES, N.O.S.	1	1.3C		1	274	0	E0	P114(a) P114(b)	PP26	MP2		
0133	MANNITOL HEXANITRATE (NITROMANNITE), WETTED with not less than 40% water, or mixture of alcohol and water, by mass	1	1.1D		1	266	0	E0	P112(a)		MP20		
0135	MERCURY FULMINATE, WETTED with not less than 20% water, or mixture of alcohol and water, by mass	1	1.1A		1	266	0	E0	P110(b)	PP42	MP20		
0136	MINES with bursting charge	1	1.1F		1		0	E0	P130		MP23		
0137	MINES with bursting charge	1	1.1D		1		0	E0	P130 LP101	PP67 L1	MP21		
0138	MINES with bursting charge	1	1.2D		1		0	E0	P130 LP101	PP67 L1	MP21		
0143	NITROGLYCERIN, DESENSITIZED with not less than 40% non-volatile water-insoluble phlegmatizer, by mass	1	1.1D		1 +6.1	266 271	0	E0	P115	PP53 PP54 PP57 PP58	MP20		
0144	NITROGLYCERIN SOLUTION IN ALCOHOL with more than 1% but not more than 10% nitroglycerin	1	1.1D		1	358	0	E0	P115	PP45 PP55 PP56 PP59 PP60	MP20		
0146	NITROSTARCH, dry or wetted with less than 20% water, by mass	1	1.1D		1		0	E0	P112(a) P112(b) P112(c)		MP20		
0147	NITRO UREA	1	1.1D		1		0	E0	P112(b)		MP20		
0150	PENTAERYTHRITE TETRANITRATE (PENTAERYTHRITOL TETRANITRATE; PETN), WETTED with not less than 25% water, by mass, or DESENSITIZED with not less than 15% phlegmatizer, by mass	1	1.1D		1	266	0	E0	P112(a) P112(b)		MP20		
0151	PENTOLITE, dry or wetted with less than 15% water, by mass	1	1.1D		1		0	E0	P112(a) P112(b) P112(c)		MP20		

ADR tank		Vehicle for tank carriage	Transport category (Tunnel restriction code)	Special provisions for carriage				Hazard identification No.	UN No.	Name and description
Tank code	Special provisions			Packages	Bulk	Loading, unloading and handling	Operation			
4.3	4.3.5, 6.8.4	9.1.1.2	1.1.3.6 (8.6)	7.2.4	7.3.3	7.5.11	8.5	5.3.2.3		3.1.2
(12)	(13)	(14)	(15)	(16)	(17)	(18)	(19)	(20)	(1)	(2)
			0 (B)	V2		CV1 CV2 CV3	S1		0114	GUANYLNITROSAMINO-GUANYLTETRAZENE (TETRAZENE), WETTED with not less than 30% water, or mixture of alcohol and water, by mass
			1 (B1000C)	V2 V3		CV1 CV2 CV3	S1		0118	HEXOLITE (HEXOTOL), dry or wetted with less than 15% water, by mass
			1 (B1000C)	V2		CV1 CV2 CV3	S1		0121	IGNITERS
			1 (B1000C)	V2		CV1 CV2 CV3	S1		0124	JET PERFORATING GUNS, CHARGED, oil well, without detonator
			0 (B)	V2		CV1 CV2 CV3	S1		0129	LEAD AZIDE, WETTED with not less than 20% water, or mixture of alcohol and water, by mass
			0 (B)	V2		CV1 CV2 CV3	S1		0130	LEAD STYPHNATE (LEAD TRINITRORESORCINATE), WETTED with not less than 20% water, or mixture of alcohol and water, by mass
			4 (E)			CV1 CV2 CV3	S1		0131	LIGHTERS, FUSE
			1 (C5000D)	V2 V3		CV1 CV2 CV3	S1		0132	DEFLAGRATING METAL SALTS OF AROMATIC NITRODERIVATIVES, N.O.S.
			1 (B1000C)	V2		CV1 CV2 CV3	S1		0133	MANNITOL HEXANITRATE (NITROMANNITE), WETTED with not less than 40% water, or mixture of alcohol and water, by mass
			0 (B)	V2		CV1 CV2 CV3	S1		0135	MERCURY FULMINATE, WETTED with not less than 20% water, or mixture of alcohol and water, by mass
			1 (B1000C)	V2		CV1 CV2 CV3	S1		0136	MINES with bursting charge
			1 (B1000C)	V2		CV1 CV2 CV3	S1		0137	MINES with bursting charge
			1 (B1000C)	V2		CV1 CV2 CV3	S1		0138	MINES with bursting charge
			1 (B1000C)	V2		CV1 CV2 CV3 CV28	S1		0143	NITROGLYCERIN, DESENSITIZED with not less than 40% non-volatile water-insoluble phlegmatizer, by mass
			1 (B1000C)	V2		CV1 CV2 CV3	S1		0144	NITROGLYCERIN SOLUTION IN ALCOHOL with more than 1% but not more than 10% nitroglycerin
			1 (B1000C)	V2 V3		CV1 CV2 CV3	S1		0146	NITROSTARCH, dry or wetted with less than 20% water, by mass
			1 (B1000C)	V2 V3		CV1 CV2 CV3	S1		0147	NITRO UREA
			1 (B1000C)	V2 V3		CV1 CV2 CV3	S1		0150	PENTAERYTHRITE TETRANITRATE (PENTAERYTHRITOL TETRANITRATE; PETN), WETTED with not less than 25% water, by mass, or DESENSITIZED with not less than 15% phlegmatizer, by mass
			1 (B1000C)	V2 V3		CV1 CV2 CV3	S1		0151	PENTOLITE, dry or wetted with less than 15% water, by mass

UN No.	Name and description	Class	Classifi-cation code	Packing group	Labels	Special provi-sions	Limited and excepted quantities		Packaging			Portable tanks and bulk containers	
									Packing instruc-tions	Special packing provisions	Mixed packing provisions	Instruc-tions	Special provisions
	3.1.2	2.2	2.2	2.1.1.3	5.2.2	3.3	3.4	3.5.1.2	4.1.4	4.1.4	4.1.10	4.2.5.2 7.3.2	4.2.5.3
(1)	(2)	(3a)	(3b)	(4)	(5)	(6)	(7a)	(7b)	(8)	(9a)	(9b)	(10)	(11)
0153	TRINITROANILINE (PICRAMIDE)	1	1.1D		1		0	E0	P112(b) P112(c)		MP20		
0154	TRINITROPHENOL (PICRIC ACID), dry or wetted with less than 30% water, by mass	1	1.1D		1		0	E0	P112(a) P112(b) P112(c)	PP26	MP20		
0155	TRINITROCHLORO-BENZENE (PICRYL CHLORIDE)	1	1.1D		1		0	E0	P112(b) P112(c)		MP20		
0159	POWDER CAKE (POWDER PASTE), WETTED with not less than 25% water, by mass	1	1.3C		1	266	0	E0	P111	PP43	MP20		
0160	POWDER, SMOKELESS	1	1.1C		1		0	E0	P114(b)	PP50 PP52	MP20 MP24		
0161	POWDER, SMOKELESS	1	1.3C		1		0	E0	P114(b)	PP50 PP52	MP20 MP24		
0167	PROJECTILES with bursting charge	1	1.1F		1		0	E0	P130		MP23		
0168	PROJECTILES with bursting charge	1	1.1D		1		0	E0	P130 LP101	PP67 L1	MP21		
0169	PROJECTILES with bursting charge	1	1.2D		1		0	E0	P130 LP101	PP67 L1	MP21		
0171	AMMUNITION, ILLUMINATING with or without burster, expelling charge or propelling charge	1	1.2G		1		0	E0	P130 LP101	PP67 L1	MP23		
0173	RELEASE DEVICES, EXPLOSIVE	1	1.4S		1.4		0	E0	P134 LP102		MP23		
0174	RIVETS, EXPLOSIVE	1	1.4S		1.4		0	E0	P134 LP102		MP23		
0180	ROCKETS with bursting charge	1	1.1F		1		0	E0	P130		MP23		
0181	ROCKETS with bursting charge	1	1.1E		1		0	E0	P130 LP101	PP67 L1	MP21		
0182	ROCKETS with bursting charge	1	1.2E		1		0	E0	P130 LP101	PP67 L1	MP21		
0183	ROCKETS with inert head	1	1.3C		1		0	E0	P130 LP101	PP67 L1	MP22		
0186	ROCKET MOTORS	1	1.3C		1		0	E0	P130 LP101	PP67 L1	MP22 MP24		
0190	SAMPLES, EXPLOSIVE, other than initiating explosive	1				16 274	0	E0	P101		MP2		
0191	SIGNAL DEVICES, HAND	1	1.4G		1.4		0	E0	P135		MP23 MP24		
0192	SIGNALS, RAILWAY TRACK, EXPLOSIVE	1	1.1G		1		0	E0	P135		MP23		
0193	SIGNALS, RAILWAY TRACK, EXPLOSIVE	1	1.4S		1.4		0	E0	P135		MP23		
0194	SIGNALS, DISTRESS, ship	1	1.1G		1		0	E0	P135		MP23 MP24		
0195	SIGNALS, DISTRESS, ship	1	1.3G		1		0	E0	P135		MP23 MP24		
0196	SIGNALS, SMOKE	1	1.1G		1		0	E0	P135		MP23		
0197	SIGNALS, SMOKE	1	1.4G		1.4		0	E0	P135		MP23 MP24		

ADR tank		Vehicle for tank carriage	Transport category (Tunnel restriction code)	Special provisions for carriage				Hazard identification No.	UN No.	Name and description
Tank code	Special provisions			Packages	Bulk	Loading, unloading and handling	Operation			
4.3	4.3.5, 6.8.4	9.1.1.2	1.1.3.6 (8.6)	7.2.4	7.3.3	7.5.11	8.5	5.3.2.3		3.1.2
(12)	(13)	(14)	(15)	(16)	(17)	(18)	(19)	(20)	(1)	(2)
			1 (B1000C)	V2 V3		CV1 CV2 CV3	S1		0153	TRINITROANILINE (PICRAMIDE)
			1 (B1000C)	V2 V3		CV1 CV2 CV3	S1		0154	TRINITROPHENOL (PICRIC ACID), dry or wetted with less than 30% water, by mass
			1 (B1000C)	V2 V3		CV1 CV2 CV3	S1		0155	TRINITROCHLORO-BENZENE (PICRYL CHLORIDE)
			1 (C5000D)	V2		CV1 CV2 CV3	S1		0159	POWDER CAKE (POWDER PASTE), WETTED with not less than 25% water, by mass
			1 (B1000C)	V2 V3		CV1 CV2 CV3	S1		0160	POWDER, SMOKELESS
			1 (C5000D)	V2 V3		CV1 CV2 CV3	S1		0161	POWDER, SMOKELESS
			1 (B1000C)	V2		CV1 CV2 CV3	S1		0167	PROJECTILES with bursting charge
			1 (B1000C)	V2		CV1 CV2 CV3	S1		0168	PROJECTILES with bursting charge
			1 (B1000C)	V2		CV1 CV2 CV3	S1		0169	PROJECTILES with bursting charge
			1 (B1000C)	V2		CV1 CV2 CV3	S1		0171	AMMUNITION, ILLUMINATING with or without burster, expelling charge or propelling charge
			4 (E)			CV1 CV2 CV3	S1		0173	RELEASE DEVICES, EXPLOSIVE
			4 (E)			CV1 CV2 CV3	S1		0174	RIVETS, EXPLOSIVE
			1 (B1000C)	V2		CV1 CV2 CV3	S1		0180	ROCKETS with bursting charge
			1 (B1000C)	V2		CV1 CV2 CV3	S1		0181	ROCKETS with bursting charge
			1 (B1000C)	V2		CV1 CV2 CV3	S1		0182	ROCKETS with bursting charge
			1 (C5000D)	V2		CV1 CV2 CV3	S1		0183	ROCKETS with inert head
			1 (C5000D)	V2		CV1 CV2 CV3	S1		0186	ROCKET MOTORS
			0 (E)	V2		CV1 CV2 CV3	S1		0190	SAMPLES, EXPLOSIVE, other than initiating explosive
			2 (E)	V2		CV1 CV2 CV3	S1		0191	SIGNAL DEVICES, HAND
			1 (B1000C)	V2		CV1 CV2 CV3	S1		0192	SIGNALS, RAILWAY TRACK, EXPLOSIVE
			4 (E)			CV1 CV2 CV3	S1		0193	SIGNALS, RAILWAY TRACK, EXPLOSIVE
			1 (B1000C)	V2		CV1 CV2 CV3	S1		0194	SIGNALS, DISTRESS, ship
			1 (C5000D)	V2		CV1 CV2 CV3	S1		0195	SIGNALS, DISTRESS, ship
			1 (B1000C)	V2		CV1 CV2 CV3	S1		0196	SIGNALS, SMOKE
			2 (E)	V2		CV1 CV2 CV3	S1		0197	SIGNALS, SMOKE

UN No.	Name and description	Class	Classification code	Packing group	Labels	Special provisions	Limited and excepted quantities		Packaging			Portable tanks and bulk containers	
									Packing instructions	Special packing provisions	Mixed packing provisions	Instructions	Special provisions
	3.1.2	2.2	2.2	2.1.1.3	5.2.2	3.3	3.4	3.5.1.2	4.1.4	4.1.4	4.1.10	4.2.5.2 7.3.2	4.2.5.3
(1)	(2)	(3a)	(3b)	(4)	(5)	(6)	(7a)	(7b)	(8)	(9a)	(9b)	(10)	(11)
0204	SOUNDING DEVICES, EXPLOSIVE	1	1.2F		1		0	E0	P134 LP102		MP23		
0207	TETRANITROANILINE	1	1.1D		1		0	E0	P112(b) P112(c)		MP20		
0208	TRINITROPHENYLMETHYL-NITRAMINE (TETRYL)	1	1.1D		1		0	E0	P112(b) P112(c)		MP20		
0209	TRINITROTOLUENE (TNT), dry or wetted with less than 30% water, by mass	1	1.1D		1		0	E0	P112(b) P112(c)	PP46	MP20		
0212	TRACERS FOR AMMUNITION	1	1.3G		1		0	E0	P133	PP69	MP23		
0213	TRINITROANISOLE	1	1.1D		1		0	E0	P112(b) P112(c)		MP20		
0214	TRINITROBENZENE, dry or wetted with less than 30% water, by mass	1	1.1D		1		0	E0	P112(a) P112(b) P112(c)		MP20		
0215	TRINITROBENZOIC ACID, dry or wetted with less than 30% water, by mass	1	1.1D		1		0	E0	P112(a) P112(b) P112(c)		MP20		
0216	TRINITRO-m-CRESOL	1	1.1D		1		0	E0	P112(b) P112(c)	PP26	MP20		
0217	TRINITRONAPHTHALENE	1	1.1D		1		0	E0	P112(b) P112(c)		MP20		
0218	TRINITROPHENETOLE	1	1.1D		1		0	E0	P112(b) P112(c)		MP20		
0219	TRINITRORESORCINOL (STYPHNIC ACID), dry or wetted with less than 20% water, or mixture of alcohol and water, by mass	1	1.1D		1		0	E0	P112(a) P112(b) P112(c)	PP26	MP20		
0220	UREA NITRATE, dry or wetted with less than 20% water, by mass	1	1.1D		1		0	E0	P112(a) P112(b) P112(c)		MP20		
0221	WARHEADS, TORPEDO with bursting charge	1	1.1D		1		0	E0	P130 LP101	PP67 L1	MP21		
0222	AMMONIUM NITRATE	1	1.1D		1	370	0	E0	P112(b) P112(c) IBC100	PP47 B3 B17	MP20		
0224	BARIUM AZIDE, dry or wetted with less than 50% water, by mass	1	1.1A		1 +6.1		0	E0	P110(b)	PP42	MP20		
0225	BOOSTERS WITH DETONATOR	1	1.1B		1		0	E0	P133	PP69	MP23		
0226	CYCLOTETRAMETHYLENE-TETRANITRAMINE (HMX; OCTOGEN), WETTED with not less than 15% water, by mass	1	1.1D		1	266	0	E0	P112(a)	PP45	MP20		
0234	SODIUM DINITRO-o-CRESOLATE, dry or wetted with less than 15% water, by mass	1	1.3C		1		0	E0	P114(a) P114(b)	PP26	MP20		
0235	SODIUM PICRAMATE, dry or wetted with less than 20% water, by mass	1	1.3C		1		0	E0	P114(a) P114(b)	PP26	MP20		
0236	ZIRCONIUM PICRAMATE, dry or wetted with less than 20% water, by mass	1	1.3C		1		0	E0	P114(a) P114(b)	PP26	MP20		
0237	CHARGES, SHAPED, FLEXIBLE, LINEAR	1	1.4D		1.4		0	E0	P138		MP21		
0238	ROCKETS, LINE-THROWING	1	1.2G		1		0	E0	P130		MP23 MP24		
0240	ROCKETS, LINE-THROWING	1	1.3G		1		0	E0	P130		MP23 MP24		

ADR tank		Vehicle for tank carriage	Transport category (Tunnel restriction code)	Special provisions for carriage				Hazard identifi-cation No.	UN No.	Name and description
Tank code	Special provisions			Packages	Bulk	Loading, unloading and handling	Operation			
4.3	4.3.5, 6.8.4	9.1.1.2	1.1.3.6 (8.6)	7.2.4	7.3.3	7.5.11	8.5	5.3.2.3		3.1.2
(12)	(13)	(14)	(15)	(16)	(17)	(18)	(19)	(20)	(1)	(2)
			1 (B1000C)	V2		CV1 CV2 CV3	S1		0204	SOUNDING DEVICES, EXPLOSIVE
			1 (B1000C)	V2 V3		CV1 CV2 CV3	S1		0207	TETRANITROANILINE
			1 (B1000C)	V2 V3		CV1 CV2 CV3	S1		0208	TRINITROPHENYLMETHYL-NITRAMINE (TETRYL)
			1 (B1000C)	V2 V3		CV1 CV2 CV3	S1		0209	TRINITROTOLUENE (TNT), dry or wetted with less than 30% water, by mass
			1 (C5000D)	V2		CV1 CV2 CV3	S1		0212	TRACERS FOR AMMUNITION
			1 (B1000C)	V2 V3		CV1 CV2 CV3	S1		0213	TRINITROANISOLE
			1 (B1000C)	V2 V3		CV1 CV2 CV3	S1		0214	TRINITROBENZENE, dry or wetted with less than 30% water, by mass
			1 (B1000C)	V2 V3		CV1 CV2 CV3	S1		0215	TRINITROBENZOIC ACID, dry or wetted with less than 30% water, by mass
			1 (B1000C)	V2 V3		CV1 CV2 CV3	S1		0216	TRINITRO-m-CRESOL
			1 (B1000C)	V2 V3		CV1 CV2 CV3	S1		0217	TRINITRONAPHTHALENE
			1 (B1000C)	V2 V3		CV1 CV2 CV3	S1		0218	TRINITROPHENETOLE
			1 (B1000C)	V2 V3		CV1 CV2 CV3	S1		0219	TRINITRORESORCINOL (STYPHNIC ACID), dry or wetted with less than 20% water, or mixture of alcohol and water, by mass
			1 (B1000C)	V2 V3		CV1 CV2 CV3	S1		0220	UREA NITRATE, dry or wetted with less than 20% water, by mass
			1 (B1000C)	V2		CV1 CV2 CV3	S1		0221	WARHEADS, TORPEDO with bursting charge
			1 (B1000C)	V2 V3		CV1 CV2 CV3	S1		0222	AMMONIUM NITRATE
			0 (B)	V2 V3		CV1 CV2 CV3 CV28	S1		0224	BARIUM AZIDE, dry or wetted with less than 50% water, by mass
			1 (B1000C)	V2		CV1 CV2 CV3	S1		0225	BOOSTERS WITH DETONATOR
			1 (B1000C)	V2		CV1 CV2 CV3	S1		0226	CYCLOTETRAMETHYLENE-TETRANITRAMINE (HMX; OCTOGEN), WETTED with not less than 15% water, by mass
			1 (C5000D)	V2 V3		CV1 CV2 CV3	S1		0234	SODIUM DINITRO-o-CRESOLATE, dry or wetted with less than 15% water, by mass
			1 (C5000D)	V2 V3		CV1 CV2 CV3	S1		0235	SODIUM PICRAMATE, dry or wetted with less than 20% water, by mass
			1 (C5000D)	V2 V3		CV1 CV2 CV3	S1		0236	ZIRCONIUM PICRAMATE, dry or wetted with less than 20% water, by mass
			2 (E)	V2		CV1 CV2 CV3	S1		0237	CHARGES, SHAPED, FLEXIBLE, LINEAR
			1 (B1000C)	V2		CV1 CV2 CV3	S1		0238	ROCKETS, LINE-THROWING
			1 (C5000D)	V2		CV1 CV2 CV3	S1		0240	ROCKETS, LINE-THROWING

UN No.	Name and description	Class	Classifi-cation code	Packing group	Labels	Special provi-sions	Limited and excepted quantities		Packaging			Portable tanks and bulk containers	
									Packing instruc-tions	Special packing provisions	Mixed packing provisions	Instruc-tions	Special provisions
	3.1.2	2.2	2.2	2.1.1.3	5.2.2	3.3	3.4	3.5.1.2	4.1.4	4.1.4	4.1.10	4.2.5.2 7.3.2	4.2.5.3
(1)	(2)	(3a)	(3b)	(4)	(5)	(6)	(7a)	(7b)	(8)	(9a)	(9b)	(10)	(11)
0241	EXPLOSIVE, BLASTING, TYPE E	1	1.1D		1	617	0	E0	P116 IBC100	PP61 PP62 B10	MP20		
0242	CHARGES, PROPELLING, FOR CANNON	1	1.3C		1		0	E0	P130		MP22		
0243	AMMUNITION, INCENDIARY, WHITE PHOSPHORUS with burster, expelling charge or propelling charge	1	1.2H		1		0	E0	P130 LP101	PP67 L1	MP23		
0244	AMMUNITION, INCENDIARY, WHITE PHOSPHORUS with burster, expelling charge or propelling charge	1	1.3H		1		0	E0	P130 LP101	PP67 L1	MP23		
0245	AMMUNITION, SMOKE, WHITE PHOSPHORUS with burster, expelling charge or propelling charge	1	1.2H		1		0	E0	P130 LP101	PP67 L1	MP23		
0246	AMMUNITION, SMOKE, WHITE PHOSPHORUS with burster, expelling charge or propelling charge	1	1.3H		1		0	E0	P130 LP101	PP67 L1	MP23		
0247	AMMUNITION, INCENDIARY, liquid or gel, with burster, expelling charge or propelling charge	1	1.3J		1		0	E0	P101		MP23		
0248	CONTRIVANCES, WATER-ACTIVATED with burster, expelling charge or propelling charge	1	1.2L		1	274	0	E0	P144	PP77	MP1		
0249	CONTRIVANCES, WATER-ACTIVATED with burster, expelling charge or propelling charge	1	1.3L		1	274	0	E0	P144	PP77	MP1		
0250	ROCKET MOTORS WITH HYPERGOLIC LIQUIDS with or without expelling charge	1	1.3L		1		0	E0	P101		MP1		
0254	AMMUNITION, ILLUMINATING with or without burster, expelling charge or propelling charge	1	1.3G		1		0	E0	P130 LP101	PP67 L1	MP23		
0255	DETONATORS, ELECTRIC for blasting	1	1.4B		1.4		0	E0	P131		MP23		
0257	FUZES, DETONATING	1	1.4B		1.4		0	E0	P141		MP23		
0266	OCTOLITE (OCTOL), dry or wetted with less than 15% water, by mass	1	1.1D		1		0	E0	P112(a) P112(b) P112(c)		MP20		
0267	DETONATORS, NON-ELECTRIC for blasting	1	1.4B		1.4		0	E0	P131	PP68	MP23		
0268	BOOSTERS WITH DETONATOR	1	1.2B		1		0	E0	P133	PP69	MP23		
0271	CHARGES, PROPELLING	1	1.1C		1		0	E0	P143	PP76	MP22		
0272	CHARGES, PROPELLING	1	1.3C		1		0	E0	P143	PP76	MP22		
0275	CARTRIDGES, POWER DEVICE	1	1.3C		1		0	E0	P134 LP102		MP22		
0276	CARTRIDGES, POWER DEVICE	1	1.4C		1.4		0	E0	P134 LP102		MP22		
0277	CARTRIDGES, OIL WELL	1	1.3C		1		0	E0	P134 LP102		MP22		
0278	CARTRIDGES, OIL WELL	1	1.4C		1.4		0	E0	P134 LP102		MP22		

ADR tank		Vehicle for tank carriage	Transport category (Tunnel restriction code)	Special provisions for carriage				Hazard identifi-cation No.	UN No.	Name and description
Tank code	Special provisions			Packages	Bulk	Loading, unloading and handling	Operation			
4.3	4.3.5, 6.8.4	9.1.1.2	1.1.3.6 (8.6)	7.2.4	7.3.3	7.5.11	8.5	5.3.2.3		3.1.2
(12)	(13)	(14)	(15)	(16)	(17)	(18)	(19)	(20)	(1)	(2)
			1 (B1000C)	V2 V12		CV1 CV2 CV3	S1		0241	EXPLOSIVE, BLASTING, TYPE E
			1 (C5000D)	V2		CV1 CV2 CV3	S1		0242	CHARGES, PROPELLING, FOR CANNON
			1 (B1000C)	V2		CV1 CV2 CV3	S1		0243	AMMUNITION, INCENDIARY, WHITE PHOSPHORUS with burster, expelling charge or propelling charge
			1 (C)	V2		CV1 CV2 CV3	S1		0244	AMMUNITION, INCENDIARY, WHITE PHOSPHORUS with burster, expelling charge or propelling charge
			1 (B1000C)	V2		CV1 CV2 CV3	S1		0245	AMMUNITION, SMOKE, WHITE PHOSPHORUS with burster, expelling charge or propelling charge
			1 (C)	V2		CV1 CV2 CV3	S1		0246	AMMUNITION, SMOKE, WHITE PHOSPHORUS with burster, expelling charge or propelling charge
			1 (C)	V2		CV1 CV2 CV3	S1		0247	AMMUNITION, INCENDIARY, liquid or gel, with burster, expelling charge or propelling charge
			0 (B)	V2		CV1 CV2 CV3 CV4	S1		0248	CONTRIVANCES, WATER-ACTIVATED with burster, expelling charge or propelling charge
			0 (B)	V2		CV1 CV2 CV3 CV4	S1		0249	CONTRIVANCES, WATER-ACTIVATED with burster, expelling charge or propelling charge
			0 (B)	V2		CV1 CV2 CV3 CV4	S1		0250	ROCKET MOTORS WITH HYPERGOLIC LIQUIDS with or without expelling charge
			1 (C5000D)	V2		CV1 CV2 CV3	S1		0254	AMMUNITION, ILLUMINATING with or without burster, expelling charge or propelling charge
			2 (E)	V2		CV1 CV2 CV3	S1		0255	DETONATORS, ELECTRIC for blasting
			2 (E)	V2		CV1 CV2 CV3	S1		0257	FUZES, DETONATING
			1 (B1000C)	V2 V3		CV1 CV2 CV3	S1		0266	OCTOLITE (OCTOL), dry or wetted with less than 15% water, by mass
			2 (E)	V2		CV1 CV2 CV3	S1		0267	DETONATORS, NON-ELECTRIC for blasting
			1 (B1000C)	V2		CV1 CV2 CV3	S1		0268	BOOSTERS WITH DETONATOR
			1 (B1000C)	V2		CV1 CV2 CV3	S1		0271	CHARGES, PROPELLING
			1 (C5000D)	V2		CV1 CV2 CV3	S1		0272	CHARGES, PROPELLING
			1 (C5000D)	V2		CV1 CV2 CV3	S1		0275	CARTRIDGES, POWER DEVICE
			2 (E)	V2		CV1 CV2 CV3	S1		0276	CARTRIDGES, POWER DEVICE
			1 (C5000D)	V2		CV1 CV2 CV3	S1		0277	CARTRIDGES, OIL WELL
			2 (E)	V2		CV1 CV2 CV3	S1		0278	CARTRIDGES, OIL WELL

UN No.	Name and description	Class	Classifi-cation code	Packing group	Labels	Special provi-sions	Limited and excepted quantities		Packaging			Portable tanks and bulk containers	
									Packing instruc-tions	Special packing provisions	Mixed packing provisions	Instruc-tions	Special provisions
	3.1.2	2.2	2.2	2.1.1.3	5.2.2	3.3	3.4	3.5.1.2	4.1.4	4.1.4	4.1.10	4.2.5.2 7.3.2	4.2.5.3
(1)	(2)	(3a)	(3b)	(4)	(5)	(6)	(7a)	(7b)	(8)	(9a)	(9b)	(10)	(11)
0279	CHARGES, PROPELLING, FOR CANNON	1	1.1C		1		0	E0	P130		MP22		
0280	ROCKET MOTORS	1	1.1C		1		0	E0	P130 LP101	PP67 L1	MP22		
0281	ROCKET MOTORS	1	1.2C		1		0	E0	P130 LP101	PP67 L1	MP22		
0282	NITROGUANIDINE (PICRITE), dry or wetted with less than 20% water, by mass	1	1.1D		1		0	E0	P112(a) P112(b) P112(c)		MP20		
0283	BOOSTERS without detonator	1	1.2D		1		0	E0	P132(a) P132(b)		MP21		
0284	GRENADES, hand or rifle, with bursting charge	1	1.1D		1		0	E0	P141		MP21		
0285	GRENADES, hand or rifle, with bursting charge	1	1.2D		1		0	E0	P141		MP21		
0286	WARHEADS, ROCKET with bursting charge	1	1.1D		1		0	E0	P130 LP101	PP67 L1	MP21		
0287	WARHEADS, ROCKET with bursting charge	1	1.2D		1		0	E0	P130 LP101	PP67 L1	MP21		
0288	CHARGES, SHAPED, FLEXIBLE, LINEAR	1	1.1D		1		0	E0	P138		MP21		
0289	CORD, DETONATING, flexible	1	1.4D		1.4		0	E0	P139	PP71 PP72	MP21		
0290	CORD (FUSE), DETONATING, metal clad	1	1.1D		1		0	E0	P139	PP71	MP21		
0291	BOMBS with bursting charge	1	1.2F		1		0	E0	P130		MP23		
0292	GRENADES, hand or rifle, with bursting charge	1	1.1F		1		0	E0	P141		MP23		
0293	GRENADES, hand or rifle, with bursting charge	1	1.2F		1		0	E0	P141		MP23		
0294	MINES with bursting charge	1	1.2F		1		0	E0	P130		MP23		
0295	ROCKETS with bursting charge	1	1.2F		1		0	E0	P130		MP23		
0296	SOUNDING DEVICES, EXPLOSIVE	1	1.1F		1		0	E0	P134 LP102		MP23		
0297	AMMUNITION, ILLUMINATING with or without burster, expelling charge or propelling charge	1	1.4G		1.4		0	E0	P130 LP101	PP67 L1	MP23		
0299	BOMBS, PHOTO-FLASH	1	1.3G		1		0	E0	P130 LP101	PP67 L1	MP23		
0300	AMMUNITION, INCENDIARY with or without burster, expelling charge or propelling charge	1	1.4G		1.4		0	E0	P130 LP101	PP67 L1	MP23		
0301	AMMUNITION, TEAR-PRODUCING with burster, expelling charge or propelling charge	1	1.4G		1.4 +6.1 +8		0	E0	P130 LP101	PP67 L1	MP23		
0303	AMMUNITION, SMOKE with or without burster, expelling charge or propelling charge	1	1.4G		1.4		0	E0	P130 LP101	PP67 L1	MP23		
0303	AMMUNITION, SMOKE with or without burster, expelling charge or propelling charge, containing corrosive substances	1	1.4G		1.4 +8		0	E0	P130 LP101	PP67 L1	MP23		

ADR tank		Vehicle for tank carriage	Transport category (Tunnel restriction code)	Special provisions for carriage				Hazard identifi-cation No.	UN No.	Name and description
Tank code	Special provisions			Packages	Bulk	Loading, unloading and handling	Operation			
4.3	4.3.5, 6.8.4	9.1.1.2	1.1.3.6 (8.6)	7.2.4	7.3.3	7.5.11	8.5	5.3.2.3		3.1.2
(12)	(13)	(14)	(15)	(16)	(17)	(18)	(19)	(20)	(1)	(2)
			1 (B1000C)	V2		CV1 CV2 CV3	S1		0279	CHARGES, PROPELLING, FOR CANNON
			1 (B1000C)	V2		CV1 CV2 CV3	S1		0280	ROCKET MOTORS
			1 (B1000C)	V2		CV1 CV2 CV3	S1		0281	ROCKET MOTORS
			1 (B1000C)	V2 V3		CV1 CV2 CV3	S1		0282	NITROGUANIDINE (PICRITE), dry or wetted with less than 20% water, by mass
			1 (B1000C)	V2		CV1 CV2 CV3	S1		0283	BOOSTERS without detonator
			1 (B1000C)	V2		CV1 CV2 CV3	S1		0284	GRENADES, hand or rifle, with bursting charge
			1 (B1000C)	V2		CV1 CV2 CV3	S1		0285	GRENADES, hand or rifle, with bursting charge
			1 (B1000C)	V2		CV1 CV2 CV3	S1		0286	WARHEADS, ROCKET with bursting charge
			1 (B1000C)	V2		CV1 CV2 CV3	S1		0287	WARHEADS, ROCKET with bursting charge
			1 (B1000C)	V2		CV1 CV2 CV3	S1		0288	CHARGES, SHAPED, FLEXIBLE, LINEAR
			2 (E)	V2		CV1 CV2 CV3	S1		0289	CORD, DETONATING, flexible
			1 (B1000C)	V2		CV1 CV2 CV3	S1		0290	CORD (FUSE), DETONATING, metal clad
			1 (B1000C)	V2		CV1 CV2 CV3	S1		0291	BOMBS with bursting charge
			1 (B1000C)	V2		CV1 CV2 CV3	S1		0292	GRENADES, hand or rifle, with bursting charge
			1 (B1000C)	V2		CV1 CV2 CV3	S1		0293	GRENADES, hand or rifle, with bursting charge
			1 (B1000C)	V2		CV1 CV2 CV3	S1		0294	MINES with bursting charge
			1 (B1000C)	V2		CV1 CV2 CV3	S1		0295	ROCKETS with bursting charge
			1 (B1000C)	V2		CV1 CV2 CV3	S1		0296	SOUNDING DEVICES, EXPLOSIVE
			2 (E)	V2		CV1 CV2 CV3	S1		0297	AMMUNITION, ILLUMINATING with or without burster, expelling charge or propelling charge
			1 (C5000D)	V2		CV1 CV2 CV3	S1		0299	BOMBS, PHOTO-FLASH
			2 (E)	V2		CV1 CV2 CV3	S1		0300	AMMUNITION, INCENDIARY with or without burster, expelling charge or propelling charge
			2 (E)	V2		CV1 CV2 CV3 CV28	S1		0301	AMMUNITION, TEAR-PRODUCING with burster, expelling charge or propelling charge
			2 (E)	V2		CV1 CV2 CV3	S1		0303	AMMUNITION, SMOKE with or without burster, expelling charge or propelling charge
			2 (E)	V2		CV1 CV2 CV3	S1		0303	AMMUNITION, SMOKE with or without burster, expelling charge or propelling charge, containing corrosive substances

UN No.	Name and description	Class	Classification code	Packing group	Labels	Special provisions	Limited and excepted quantities		Packaging			Portable tanks and bulk containers	
									Packing instructions	Special packing provisions	Mixed packing provisions	Instructions	Special provisions
	3.1.2	2.2	2.2	2.1.1.3	5.2.2	3.3	3.4	3.5.1.2	4.1.4	4.1.4	4.1.10	4.2.5.2 7.3.2	4.2.5.3
(1)	(2)	(3a)	(3b)	(4)	(5)	(6)	(7a)	(7b)	(8)	(9a)	(9b)	(10)	(11)
0303	AMMUNITION, SMOKE with or without burster, expelling charge or propelling charge, containing toxic by inhalation substances	1	1.4G		1.4 +6.1		0	E0	P130 LP101	PP67 L1	MP23		
0305	FLASH POWDER	1	1.3G		1		0	E0	P113	PP49	MP20		
0306	TRACERS FOR AMMUNITION	1	1.4G		1.4		0	E0	P133	PP69	MP23		
0312	CARTRIDGES, SIGNAL	1	1.4G		1.4		0	E0	P135		MP23 MP24		
0313	SIGNALS, SMOKE	1	1.2G		1		0	E0	P135		MP23		
0314	IGNITERS	1	1.2G		1		0	E0	P142		MP23		
0315	IGNITERS	1	1.3G		1		0	E0	P142		MP23		
0316	FUZES, IGNITING	1	1.3G		1		0	E0	P141		MP23		
0317	FUZES, IGNITING	1	1.4G		1.4		0	E0	P141		MP23		
0318	GRENADES, PRACTICE, hand or rifle	1	1.3G		1		0	E0	P141		MP23		
0319	PRIMERS, TUBULAR	1	1.3G		1		0	E0	P133		MP23		
0320	PRIMERS, TUBULAR	1	1.4G		1.4		0	E0	P133		MP23		
0321	CARTRIDGES FOR WEAPONS with bursting charge	1	1.2E		1		0	E0	P130 LP101	PP67 L1	MP21		
0322	ROCKET MOTORS WITH HYPERGOLIC LIQUIDS with or without expelling charge	1	1.2L		1		0	E0	P101		MP1		
0323	CARTRIDGES, POWER DEVICE	1	1.4S		1.4	347	0	E0	P134 LP102		MP23		
0324	PROJECTILES with bursting charge	1	1.2F		1		0	E0	P130		MP23		
0325	IGNITERS	1	1.4G		1.4		0	E0	P142		MP23		
0326	CARTRIDGES FOR WEAPONS, BLANK	1	1.1C		1		0	E0	P130		MP22		
0327	CARTRIDGES FOR WEAPONS, BLANK or CARTRIDGES, SMALL ARMS, BLANK	1	1.3C		1		0	E0	P130		MP22		
0328	CARTRIDGES FOR WEAPONS, INERT PROJECTILE	1	1.2C		1		0	E0	P130 LP101	PP67 L1	MP22		
0329	TORPEDOES with bursting charge	1	1.1E		1		0	E0	P130 LP101	PP67 L1	MP21		
0330	TORPEDOES with bursting charge	1	1.1F		1		0	E0	P130		MP23		
0331	EXPLOSIVE, BLASTING, TYPE B (AGENT, BLASTING, TYPE B)	1	1.5D		1.5	617	0	E0	P116 IBC100	PP61 PP62 PP64	MP20	T1	TP1 TP17 TP32

ADR tank		Vehicle for tank carriage	Transport category (Tunnel restriction code)	Special provisions for carriage				Hazard identifi-cation No.	UN No.	Name and description
Tank code	Special provisions			Packages	Bulk	Loading, unloading and handling	Operation			
4.3	4.3.5, 6.8.4	9.1.1.2	1.1.3.6 (8.6)	7.2.4	7.3.3	7.5.11	8.5	5.3.2.3		3.1.2
(12)	(13)	(14)	(15)	(16)	(17)	(18)	(19)	(20)	(1)	(2)
			2 (E)	V2		CV1 CV2 CV3 CV28	S1		0303	AMMUNITION, SMOKE with or without burster, expelling charge or propelling charge, containing toxic by inhalation substances
			1 (C5000D)	V2 V3		CV1 CV2 CV3	S1		0305	FLASH POWDER
			2 (E)	V2		CV1 CV2 CV3	S1		0306	TRACERS FOR AMMUNITION
			2 (E)	V2		CV1 CV2 CV3	S1		0312	CARTRIDGES, SIGNAL
			1 (B1000C)	V2		CV1 CV2 CV3	S1		0313	SIGNALS, SMOKE
			1 (B1000C)	V2		CV1 CV2 CV3	S1		0314	IGNITERS
			1 (C5000D)	V2		CV1 CV2 CV3	S1		0315	IGNITERS
			1 (C5000D)	V2		CV1 CV2 CV3	S1		0316	FUZES, IGNITING
			2 (E)	V2		CV1 CV2 CV3	S1		0317	FUZES, IGNITING
			1 (C5000D)	V2		CV1 CV2 CV3	S1		0318	GRENADES, PRACTICE, hand or rifle
			1 (C5000D)	V2		CV1 CV2 CV3	S1		0319	PRIMERS, TUBULAR
			2 (E)	V2		CV1 CV2 CV3	S1		0320	PRIMERS, TUBULAR
			1 (B1000C)	V2		CV1 CV2 CV3	S1		0321	CARTRIDGES FOR WEAPONS with bursting charge
			0 (B)	V2		CV1 CV2 CV3 CV4	S1		0322	ROCKET MOTORS WITH HYPERGOLIC LIQUIDS with or without expelling charge
			4 (E)			CV1 CV2 CV3	S1		0323	CARTRIDGES, POWER DEVICE
			1 (B1000C)	V2		CV1 CV2 CV3	S1		0324	PROJECTILES with bursting charge
			2 (E)	V2		CV1 CV2 CV3	S1		0325	IGNITERS
			1 (B1000C)	V2		CV1 CV2 CV3	S1		0326	CARTRIDGES FOR WEAPONS, BLANK
			1 (C5000D)	V2		CV1 CV2 CV3	S1		0327	CARTRIDGES FOR WEAPONS, BLANK or CARTRIDGES, SMALL ARMS, BLANK
			1 (B1000C)	V2		CV1 CV2 CV3	S1		0328	CARTRIDGES FOR WEAPONS, INERT PROJECTILE
			1 (B1000C)	V2		CV1 CV2 CV3	S1		0329	TORPEDOES with bursting charge
			1 (B1000C)	V2		CV1 CV2 CV3	S1		0330	TORPEDOES with bursting charge
S2.65AN(+)	TU3 TU12 TU41 TC8 TA1 TA5	EX/III	1 (B1000C)	V2 V12		CV1 CV2 CV3	S1	1.5D	0331	EXPLOSIVE, BLASTING, TYPE B (AGENT, BLASTING, TYPE B)

UN No.	Name and description	Class	Classifi-cation code	Packing group	Labels	Special provi-sions	Limited and excepted quantities		Packaging			Portable tanks and bulk containers	
									Packing instruc-tions	Special packing provisions	Mixed packing provisions	Instruc-tions	Special provisions
	3.1.2	2.2	2.2	2.1.1.3	5.2.2	3.3	3.4	3.5.1.2	4.1.4	4.1.4	4.1.10	4.2.5.2 7.3.2	4.2.5.3
(1)	(2)	(3a)	(3b)	(4)	(5)	(6)	(7a)	(7b)	(8)	(9a)	(9b)	(10)	(11)
0332	EXPLOSIVE, BLASTING, TYPE E (AGENT, BLASTING, TYPE E)	1	1.5D		1.5	617	0	E0	P116 IBC100	PP61 PP62	MP20	T1	TP1 TP17 TP32
0333	FIREWORKS	1	1.1G		1	645	0	E0	P135		MP23 MP24		
0334	FIREWORKS	1	1.2G		1	645	0	E0	P135		MP23 MP24		
0335	FIREWORKS	1	1.3G		1	645	0	E0	P135		MP23 MP24		
0336	FIREWORKS	1	1.4G		1.4	645 651	0	E0	P135		MP23 MP24		
0337	FIREWORKS	1	1.4S		1.4	645	0	E0	P135		MP23 MP24		
0338	CARTRIDGES FOR WEAPONS, BLANK or CARTRIDGES, SMALL ARMS, BLANK	1	1.4C		1.4		0	E0	P130		MP22		
0339	CARTRIDGES FOR WEAPONS, INERT PROJECTILE or CARTRIDGES, SMALL ARMS	1	1.4C		1.4		0	E0	P130		MP22		
0340	NITROCELLULOSE, dry or wetted with less than 25% water (or alcohol), by mass	1	1.1D		1		0	E0	P112(a) P112(b)		MP20		
0341	NITROCELLULOSE, unmodified or plasticized with less than 18% plasticizing substance, by mass	1	1.1D		1		0	E0	P112(b)		MP20		
0342	NITROCELLULOSE, WETTED with not less than 25% alcohol, by mass	1	1.3C		1	105	0	E0	P114(a)	PP43	MP20		
0343	NITROCELLULOSE, PLASTICIZED with not less than 18% plasticizing substance, by mass	1	1.3C		1	105	0	E0	P111		MP20		
0344	PROJECTILES with bursting charge	1	1.4D		1.4		0	E0	P130 LP101	PP67 L1	MP21		
0345	PROJECTILES, inert with tracer	1	1.4S		1.4		0	E0	P130 LP101	PP67 L1	MP23		
0346	PROJECTILES with burster or expelling charge	1	1.2D		1		0	E0	P130 LP101	PP67 L1	MP21		
0347	PROJECTILES with burster or expelling charge	1	1.4D		1.4		0	E0	P130 LP101	PP67 L1	MP21		
0348	CARTRIDGES FOR WEAPONS with bursting charge	1	1.4F		1.4		0	E0	P130		MP23		
0349	ARTICLES, EXPLOSIVE, N.O.S.	1	1.4S		1.4	178 274	0	E0	P101		MP2		
0350	ARTICLES, EXPLOSIVE, N.O.S.	1	1.4B		1.4	178 274	0	E0	P101		MP2		
0351	ARTICLES, EXPLOSIVE, N.O.S.	1	1.4C		1.4	178 274	0	E0	P101		MP2		
0352	ARTICLES, EXPLOSIVE, N.O.S.	1	1.4D		1.4	178 274	0	E0	P101		MP2		
0353	ARTICLES, EXPLOSIVE, N.O.S.	1	1.4G		1.4	178 274	0	E0	P101		MP2		
0354	ARTICLES, EXPLOSIVE, N.O.S.	1	1.1L		1	178 274	0	E0	P101		MP1		
0355	ARTICLES, EXPLOSIVE, N.O.S.	1	1.2L		1	178 274	0	E0	P101		MP1		

ADR tank		Vehicle for tank carriage	Transport category (Tunnel restriction code)	Special provisions for carriage				Hazard identifi- cation No.	UN No.	Name and description
Tank code	Special provisions			Packages	Bulk	Loading, unloading and handling	Operation			
4.3	4.3.5, 6.8.4	9.1.1.2	1.1.3.6 (8.6)	7.2.4	7.3.3	7.5.11	8.5	5.3.2.3		3.1.2
(12)	(13)	(14)	(15)	(16)	(17)	(18)	(19)	(20)	(1)	(2)
		EX/III	1 (B1000C)	V2 V12		CV1 CV2 CV3	S1	1.5D	0332	EXPLOSIVE, BLASTING, TYPE E (AGENT, BLASTING, TYPE E)
			1 (B1000C)	V2 V3		CV1 CV2 CV3	S1		0333	FIREWORKS
			1 (B1000C)	V2 V3		CV1 CV2 CV3	S1		0334	FIREWORKS
			1 (C5000D)	V2 V3		CV1 CV2 CV3	S1		0335	FIREWORKS
			2 (E)	V2		CV1 CV2 CV3	S1		0336	FIREWORKS
			4 (E)			CV1 CV2 CV3	S1		0337	FIREWORKS
			2 (E)	V2		CV1 CV2 CV3	S1		0338	CARTRIDGES FOR WEAPONS, BLANK or CARTRIDGES, SMALL ARMS, BLANK
			2 (E)	V2		CV1 CV2 CV3	S1		0339	CARTRIDGES FOR WEAPONS, INERT PROJECTILE or CARTRIDGES, SMALL ARMS
			1 (B1000C)	V2 V3		CV1 CV2 CV3	S1		0340	NITROCELLULOSE, dry or wetted with less than 25% water (or alcohol), by mass
			1 (B1000C)	V2 V3		CV1 CV2 CV3	S1		0341	NITROCELLULOSE, unmodified or plasticized with less than 18% plasticizing substance, by mass
			1 (C5000D)	V2		CV1 CV2 CV3	S1		0342	NITROCELLULOSE, WETTED with not less than 25% alcohol, by mass
			1 (C5000D)	V2		CV1 CV2 CV3	S1		0343	NITROCELLULOSE, PLASTICIZED with not less than 18% plasticizing substance, by mass
			2 (E)	V2		CV1 CV2 CV3	S1		0344	PROJECTILES with bursting charge
			4 (E)			CV1 CV2 CV3	S1		0345	PROJECTILES, inert with tracer
			1 (B1000C)	V2		CV1 CV2 CV3	S1		0346	PROJECTILES with burster or expelling charge
			2 (E)	V2		CV1 CV2 CV3	S1		0347	PROJECTILES with burster or expelling charge
			2 (E)	V2		CV1 CV2 CV3	S1		0348	CARTRIDGES FOR WEAPONS with bursting charge
			4 (E)			CV1 CV2 CV3	S1		0349	ARTICLES, EXPLOSIVE, N.O.S.
			2 (E)	V2		CV1 CV2 CV3	S1		0350	ARTICLES, EXPLOSIVE, N.O.S.
			2 (E)	V2		CV1 CV2 CV3	S1		0351	ARTICLES, EXPLOSIVE, N.O.S.
			2 (E)	V2		CV1 CV2 CV3	S1		0352	ARTICLES, EXPLOSIVE, N.O.S.
			2 (E)	V2		CV1 CV2 CV3	S1		0353	ARTICLES, EXPLOSIVE, N.O.S.
			0 (B)	V2		CV1 CV2 CV3 CV4	S1		0354	ARTICLES, EXPLOSIVE, N.O.S.
			0 (B)	V2		CV1 CV2 CV3 CV4	S1		0355	ARTICLES, EXPLOSIVE, N.O.S.

UN No.	Name and description	Class	Classification code	Packing group	Labels	Special provisions	Limited and excepted quantities		Packaging			Portable tanks and bulk containers	
									Packing instructions	Special packing provisions	Mixed packing provisions	Instructions	Special provisions
	3.1.2	2.2	2.2	2.1.1.3	5.2.2	3.3	3.4	3.5.1.2	4.1.4	4.1.4	4.1.10	4.2.5.2 7.3.2	4.2.5.3
(1)	(2)	(3a)	(3b)	(4)	(5)	(6)	(7a)	(7b)	(8)	(9a)	(9b)	(10)	(11)
0356	ARTICLES, EXPLOSIVE, N.O.S.	1	1.3L		1	178 274	0	E0	P101		MP1		
0357	SUBSTANCES, EXPLOSIVE, N.O.S.	1	1.1L		1	178 274	0	E0	P101		MP1		
0358	SUBSTANCES, EXPLOSIVE, N.O.S.	1	1.2L		1	178 274	0	E0	P101		MP1		
0359	SUBSTANCES, EXPLOSIVE, N.O.S.	1	1.3L		1	178 274	0	E0	P101		MP1		
0360	DETONATOR ASSEMBLIES, NON-ELECTRIC for blasting	1	1.1B		1		0	E0	P131		MP23		
0361	DETONATOR ASSEMBLIES, NON-ELECTRIC for blasting	1	1.4B		1.4		0	E0	P131		MP23		
0362	AMMUNITION, PRACTICE	1	1.4G		1.4		0	E0	P130 LP101	PP67 L1	MP23		
0363	AMMUNITION, PROOF	1	1.4G		1.4		0	E0	P130 LP101	PP67 L1	MP23		
0364	DETONATORS FOR AMMUNITION	1	1.2B		1		0	E0	P133		MP23		
0365	DETONATORS FOR AMMUNITION	1	1.4B		1.4		0	E0	P133		MP23		
0366	DETONATORS FOR AMMUNITION	1	1.4S		1.4	347	0	E0	P133		MP23		
0367	FUZES, DETONATING	1	1.4S		1.4		0	E0	P141		MP23		
0368	FUZES, IGNITING	1	1.4S		1.4		0	E0	P141		MP23		
0369	WARHEADS, ROCKET with bursting charge	1	1.1F		1		0	E0	P130		MP23		
0370	WARHEADS, ROCKET with burster or expelling charge	1	1.4D		1.4		0	E0	P130 LP101	PP67 L1	MP21		
0371	WARHEADS, ROCKET with burster or expelling charge	1	1.4F		1.4		0	E0	P130		MP23		
0372	GRENADES, PRACTICE, hand or rifle	1	1.2G		1		0	E0	P141		MP23		
0373	SIGNAL DEVICES, HAND	1	1.4S		1.4		0	E0	P135		MP23 MP24		
0374	SOUNDING DEVICES, EXPLOSIVE	1	1.1D		1		0	E0	P134 LP102		MP21		
0375	SOUNDING DEVICES, EXPLOSIVE	1	1.2D		1		0	E0	P134 LP102		MP21		
0376	PRIMERS, TUBULAR	1	1.4S		1.4		0	E0	P133		MP23		
0377	PRIMERS, CAP TYPE	1	1.1B		1		0	E0	P133		MP23		
0378	PRIMERS, CAP TYPE	1	1.4B		1.4		0	E0	P133		MP23		
0379	CASES, CARTRIDGE, EMPTY, WITH PRIMER	1	1.4C		1.4		0	E0	P136		MP22		

ADR tank		Vehicle for tank carriage	Transport category (Tunnel restriction code)	Special provisions for carriage				Hazard identification No.	UN No.	Name and description
Tank code	Special provisions			Packages	Bulk	Loading, unloading and handling	Operation			
4.3	4.3.5, 6.8.4	9.1.1.2	1.1.3.6 (8.6)	7.2.4	7.3.3	7.5.11	8.5	5.3.2.3		3.1.2
(12)	(13)	(14)	(15)	(16)	(17)	(18)	(19)	(20)	(1)	(2)
			0 (B)	V2		CV1 CV2 CV3 CV4	S1		0356	ARTICLES, EXPLOSIVE, N.O.S.
			0 (B)	V2		CV1 CV2 CV3 CV4	S1		0357	SUBSTANCES, EXPLOSIVE, N.O.S.
			0 (B)	V2		CV1 CV2 CV3 CV4	S1		0358	SUBSTANCES, EXPLOSIVE, N.O.S.
			0 (B)	V2		CV1 CV2 CV3 CV4	S1		0359	SUBSTANCES, EXPLOSIVE, N.O.S.
			1 (B1000C)	V2		CV1 CV2 CV3	S1		0360	DETONATOR ASSEMBLIES, NON-ELECTRIC for blasting
			2 (E)	V2		CV1 CV2 CV3	S1		0361	DETONATOR ASSEMBLIES, NON-ELECTRIC for blasting
			2 (E)	V2		CV1 CV2 CV3	S1		0362	AMMUNITION, PRACTICE
			2 (E)	V2		CV1 CV2 CV3	S1		0363	AMMUNITION, PROOF
			1 (B1000C)	V2		CV1 CV2 CV3	S1		0364	DETONATORS FOR AMMUNITION
			2 (E)	V2		CV1 CV2 CV3	S1		0365	DETONATORS FOR AMMUNITION
			4 (E)			CV1 CV2 CV3	S1		0366	DETONATORS FOR AMMUNITION
			4 (E)			CV1 CV2 CV3	S1		0367	FUZES, DETONATING
			4 (E)			CV1 CV2 CV3	S1		0368	FUZES, IGNITING
			1 (B1000C)	V2		CV1 CV2 CV3	S1		0369	WARHEADS, ROCKET with bursting charge
			2 (E)	V2		CV1 CV2 CV3	S1		0370	WARHEADS, ROCKET with burster or expelling charge
			2 (E)	V2		CV1 CV2 CV3	S1		0371	WARHEADS, ROCKET with burster or expelling charge
			1 (B1000C)	V2		CV1 CV2 CV3	S1		0372	GRENADES, PRACTICE, hand or rifle
			4 (E)			CV1 CV2 CV3	S1		0373	SIGNAL DEVICES, HAND
			1 (B1000C)	V2		CV1 CV2 CV3	S1		0374	SOUNDING DEVICES, EXPLOSIVE
			1 (B1000C)	V2		CV1 CV2 CV3	S1		0375	SOUNDING DEVICES, EXPLOSIVE
			4 (E)			CV1 CV2 CV3	S1		0376	PRIMERS, TUBULAR
			1 (B1000C)	V2		CV1 CV2 CV3	S1		0377	PRIMERS, CAP TYPE
			2 (E)	V2		CV1 CV2 CV3	S1		0378	PRIMERS, CAP TYPE
			2 (E)	V2		CV1 CV2 CV3	S1		0379	CASES, CARTRIDGE, EMPTY, WITH PRIMER

UN No.	Name and description	Class	Classifi-cation code	Packing group	Labels	Special provi-sions	Limited and excepted quantities		Packaging			Portable tanks and bulk containers	
									Packing instruc-tions	Special packing provisions	Mixed packing provisions	Instruc-tions	Special provisions
3.1.2	3.1.2	2.2	2.2	2.1.1.3	5.2.2	3.3	3.4	3.5.1.2	4.1.4	4.1.4	4.1.10	4.2.5.2 7.3.2	4.2.5.3
(1)	(2)	(3a)	(3b)	(4)	(5)	(6)	(7a)	(7b)	(8)	(9a)	(9b)	(10)	(11)
0380	ARTICLES, PYROPHORIC	1	1.2L		1		0	E0	P101		MP1		
0381	CARTRIDGES, POWER DEVICE	1	1.2C		1		0	E0	P134 LP102		MP22		
0382	COMPONENTS, EXPLOSIVE TRAIN, N.O.S.	1	1.2B		1	178 274	0	E0	P101		MP2		
0383	COMPONENTS, EXPLOSIVE TRAIN, N.O.S.	1	1.4B		1.4	178 274	0	E0	P101		MP2		
0384	COMPONENTS, EXPLOSIVE TRAIN, N.O.S.	1	1.4S		1.4	178 274	0	E0	P101		MP2		
0385	5-NITROBENZOTRIAZOL	1	1.1D		1		0	E0	P112(b) P112(c)		MP20		
0386	TRINITROBENZENE-SULPHONIC ACID	1	1.1D		1		0	E0	P112(b) P112(c)	PP26	MP20		
0387	TRINITROFLUORENONE	1	1.1D		1		0	E0	P112(b) P112(c)		MP20		
0388	TRINITROTOLUENE (TNT) AND TRINITROBENZENE MIXTURE or TRINITROTOLUENE (TNT) AND HEXANITROSTILBENE MIXTURE	1	1.1D		1		0	E0	P112(b) P112(c)		MP20		
0389	TRINITROTOLUENE (TNT) MIXTURE CONTAINING TRINITROBENZENE AND HEXANITROSTILBENE	1	1.1D		1		0	E0	P112(b) P112(c)		MP20		
0390	TRITONAL	1	1.1D		1		0	E0	P112(b) P112(c)		MP20		
0391	CYCLOTRIMETHYLENE-TRINITRAMINE (CYCLONITE; HEXOGEN; RDX) AND CYCLOTETRAMETHYLENE-TETRANITRAMINE (HMX; OCTOGEN) MIXTURE, WETTED with not less than 15% water, by mass or DESENSITIZED with not less than 10% phlegmatiser by mass	1	1.1D		1	266	0	E0	P112(a) P112(b)		MP20		
0392	HEXANITROSTILBENE	1	1.1D		1		0	E0	P112(b) P112(c)		MP20		
0393	HEXOTONAL	1	1.1D		1		0	E0	P112(b)		MP20		
0394	TRINITRORESORCINOL (STYPHNIC ACID), WETTED with not less than 20% water, or mixture of alcohol and water, by mass	1	1.1D		1		0	E0	P112(a)	PP26	MP20		
0395	ROCKET MOTORS, LIQUID FUELLED	1	1.2J		1		0	E0	P101		MP23		
0396	ROCKET MOTORS, LIQUID FUELLED	1	1.3J		1		0	E0	P101		MP23		
0397	ROCKETS, LIQUID FUELLED with bursting charge	1	1.1J		1		0	E0	P101		MP23		
0398	ROCKETS, LIQUID FUELLED with bursting charge	1	1.2J		1		0	E0	P101		MP23		
0399	BOMBS WITH FLAMMABLE LIQUID with bursting charge	1	1.1J		1		0	E0	P101		MP23		

ADR tank		Vehicle for tank carriage	Transport category (Tunnel restriction code)	Special provisions for carriage				Hazard identification No.	UN No.	Name and description
Tank code	Special provisions			Packages	Bulk	Loading, unloading and handling	Operation			
4.3	4.3.5, 6.8.4	9.1.1.2	1.1.3.6 (8.6)	7.2.4	7.3.3	7.5.11	8.5	5.3.2.3		3.1.2
(12)	(13)	(14)	(15)	(16)	(17)	(18)	(19)	(20)	(1)	(2)
			0 (B)	V2		CV1 CV2 CV3 CV4	S1		0380	ARTICLES, PYROPHORIC
			1 (B1000C)	V2		CV1 CV2 CV3	S1		0381	CARTRIDGES, POWER DEVICE
			1 (B1000C)	V2		CV1 CV2 CV3	S1		0382	COMPONENTS, EXPLOSIVE TRAIN, N.O.S.
			2 (E)	V2		CV1 CV2 CV3	S1		0383	COMPONENTS, EXPLOSIVE TRAIN, N.O.S.
			4 (E)			CV1 CV2 CV3	S1		0384	COMPONENTS, EXPLOSIVE TRAIN, N.O.S.
			1 (B1000C)	V2 V3		CV1 CV2 CV3	S1		0385	5-NITROBENZOTRIAZOL
			1 (B1000C)	V2 V3		CV1 CV2 CV3	S1		0386	TRINITROBENZENE-SULPHONIC ACID
			1 (B1000C)	V2 V3		CV1 CV2 CV3	S1		0387	TRINITROFLUORENONE
			1 (B1000C)	V2 V3		CV1 CV2 CV3	S1		0388	TRINITROTOLUENE (TNT) AND TRINITROBENZENE MIXTURE or TRINITROTOLUENE (TNT) AND HEXANITROSTILBENE MIXTURE
			1 (B1000C)	V2 V3		CV1 CV2 CV3	S1		0389	TRINITROTOLUENE (TNT) MIXTURE CONTAINING TRINITROBENZENE AND HEXANITROSTILBENE
			1 (B1000C)	V2 V3		CV1 CV2 CV3	S1		0390	TRITONAL
			1 (B1000C)	V2 V3		CV1 CV2 CV3	S1		0391	CYCLOTRIMETHYLENE-TRINITRAMINE (CYCLONITE; HEXOGEN; RDX) AND CYCLOTETRAMETHYLENE-TETRANITRAMINE (HMX; OCTOGEN) MIXTURE, WETTED with not less than 15% water, by mass or DESENSITIZED with not less than 10% phlegmatiser by mass
			1 (B1000C)	V2 V3		CV1 CV2 CV3	S1		0392	HEXANITROSTILBENE
			1 (B1000C)	V2 V3		CV1 CV2 CV3	S1		0393	HEXOTONAL
			1 (B1000C)	V2		CV1 CV2 CV3	S1		0394	TRINITRORESORCINOL (STYPHNIC ACID), WETTED with not less than 20% water, or mixture of alcohol and water, by mass
			1 (B1000C)	V2		CV1 CV2 CV3	S1		0395	ROCKET MOTORS, LIQUID FUELLED
			1 (C)	V2		CV1 CV2 CV3	S1		0396	ROCKET MOTORS, LIQUID FUELLED
			1 (B1000C)	V2		CV1 CV2 CV3	S1		0397	ROCKETS, LIQUID FUELLED with bursting charge
			1 (B1000C)	V2		CV1 CV2 CV3	S1		0398	ROCKETS, LIQUID FUELLED with bursting charge
			1 (B1000C)	V2		CV1 CV2 CV3	S1		0399	BOMBS WITH FLAMMABLE LIQUID with bursting charge

UN No.	Name and description	Class	Classification code	Packing group	Labels	Special provisions	Limited and excepted quantities		Packaging			Portable tanks and bulk containers	
									Packing instructions	Special packing provisions	Mixed packing provisions	Instructions	Special provisions
	3.1.2	2.2	2.2	2.1.1.3	5.2.2	3.3	3.4	3.5.1.2	4.1.4	4.1.4	4.1.10	4.2.5.2 7.3.2	4.2.5.3
(1)	(2)	(3a)	(3b)	(4)	(5)	(6)	(7a)	(7b)	(8)	(9a)	(9b)	(10)	(11)
0400	BOMBS WITH FLAMMABLE LIQUID with bursting charge	1	1.2J		1		0	E0	P101		MP23		
0401	DIPICRYL SULPHIDE, dry or wetted with less than 10% water, by mass	1	1.1D		1		0	E0	P112(a) P112(b) P112(c)		MP20		
0402	AMMONIUM PERCHLORATE	1	1.1D		1	152	0	E0	P112(b) P112(c)		MP20		
0403	FLARES, AERIAL	1	1.4G		1.4		0	E0	P135		MP23		
0404	FLARES, AERIAL	1	1.4S		1.4		0	E0	P135		MP23		
0405	CARTRIDGES, SIGNAL	1	1.4S		1.4		0	E0	P135		MP23 MP24		
0406	DINITROSOBENZENE	1	1.3C		1		0	E0	P114(b)		MP20		
0407	TETRAZOL-1-ACETIC ACID	1	1.4C		1.4		0	E0	P114(b)		MP20		
0408	FUZES, DETONATING with protective features	1	1.1D		1		0	E0	P141		MP21		
0409	FUZES, DETONATING with protective features	1	1.2D		1		0	E0	P141		MP21		
0410	FUZES, DETONATING with protective features	1	1.4D		1.4		0	E0	P141		MP21		
0411	PENTAERYTHRITE TETRANITRATE (PENTAERYTHRITOL TETRANITRATE; PETN) with not less than 7% wax, by mass	1	1.1D		1	131	0	E0	P112(b) P112(c)		MP20		
0412	CARTRIDGES FOR WEAPONS with bursting charge	1	1.4E		1.4		0	E0	P130 LP101	PP67 L1	MP21		
0413	CARTRIDGES FOR WEAPONS, BLANK	1	1.2C		1		0	E0	P130		MP22		
0414	CHARGES, PROPELLING, FOR CANNON	1	1.2C		1		0	E0	P130		MP22		
0415	CHARGES, PROPELLING	1	1.2C		1		0	E0	P143	PP76	MP22		
0417	CARTRIDGES FOR WEAPONS, INERT PROJECTILE or CARTRIDGES, SMALL ARMS	1	1.3C		1		0	E0	P130		MP22		
0418	FLARES, SURFACE	1	1.1G		1		0	E0	P135		MP23		
0419	FLARES, SURFACE	1	1.2G		1		0	E0	P135		MP23		
0420	FLARES, AERIAL	1	1.1G		1		0	E0	P135		MP23		
0421	FLARES, AERIAL	1	1.2G		1		0	E0	P135		MP23		
0424	PROJECTILES, inert with tracer	1	1.3G		1		0	E0	P130 LP101	PP67 L1	MP23		
0425	PROJECTILES, inert with tracer	1	1.4G		1.4		0	E0	P130 LP101	PP67 L1	MP23		
0426	PROJECTILES with burster or expelling charge	1	1.2F		1		0	E0	P130		MP23		

ADR tank		Vehicle for tank carriage	Transport category (Tunnel restriction code)	Special provisions for carriage				Hazard identification No.	UN No.	Name and description
Tank code	Special provisions			Packages	Bulk	Loading, unloading and handling	Operation			
4.3	4.3.5, 6.8.4	9.1.1.2	1.1.3.6 (8.6)	7.2.4	7.3.3	7.5.11	8.5	5.3.2.3		3.1.2
(12)	(13)	(14)	(15)	(16)	(17)	(18)	(19)	(20)	(1)	(2)
			1 (B1000C)	V2		CV1 CV2 CV3	S1		0400	BOMBS WITH FLAMMABLE LIQUID with bursting charge
			1 (B1000C)	V2 V3		CV1 CV2 CV3	S1		0401	DIPICRYL SULPHIDE, dry or wetted with less than 10% water, by mass
			1 (B1000C)	V2 V3		CV1 CV2 CV3	S1		0402	AMMONIUM PERCHLORATE
			2 (E)	V2		CV1 CV2 CV3	S1		0403	FLARES, AERIAL
			4 (E)			CV1 CV2 CV3	S1		0404	FLARES, AERIAL
			4 (E)			CV1 CV2 CV3	S1		0405	CARTRIDGES, SIGNAL
			1 (C5000D)	V2 V3		CV1 CV2 CV3	S1		0406	DINITROSOBENZENE
			2 (E)	V2		CV1 CV2 CV3	S1		0407	TETRAZOL-1-ACETIC ACID
			1 (B1000C)	V2		CV1 CV2 CV3	S1		0408	FUZES, DETONATING with protective features
			1 (B1000C)	V2		CV1 CV2 CV3	S1		0409	FUZES, DETONATING with protective features
			2 (E)	V2		CV1 CV2 CV3	S1		0410	FUZES, DETONATING with protective features
			1 (B1000C)	V2 V3		CV1 CV2 CV3	S1		0411	PENTAERYTHRITE TETRANITRATE (PENTAERYTHRITOL TETRANITRATE; PETN) with not less than 7% wax, by mass
			2 (E)	V2		CV1 CV2 CV3	S1		0412	CARTRIDGES FOR WEAPONS with bursting charge
			1 (B1000C)	V2		CV1 CV2 CV3	S1		0413	CARTRIDGES FOR WEAPONS, BLANK
			1 (B1000C)	V2		CV1 CV2 CV3	S1		0414	CHARGES, PROPELLING, FOR CANNON
			1 (B1000C)	V2		CV1 CV2 CV3	S1		0415	CHARGES, PROPELLING
			1 (C5000D)	V2		CV1 CV2 CV3	S1		0417	CARTRIDGES FOR WEAPONS, INERT PROJECTILE or CARTRIDGES, SMALL ARMS
			1 (B1000C)	V2		CV1 CV2 CV3	S1		0418	FLARES, SURFACE
			1 (B1000C)	V2		CV1 CV2 CV3	S1		0419	FLARES, SURFACE
			1 (B1000C)	V2		CV1 CV2 CV3	S1		0420	FLARES, AERIAL
			1 (B1000C)	V2		CV1 CV2 CV3	S1		0421	FLARES, AERIAL
			1 (C5000D)	V2		CV1 CV2 CV3	S1		0424	PROJECTILES, inert with tracer
			2 (E)	V2		CV1 CV2 CV3	S1		0425	PROJECTILES, inert with tracer
			1 (B1000C)	V2		CV1 CV2 CV3	S1		0426	PROJECTILES with burster or expelling charge

UN No.	Name and description	Class	Classification code	Packing group	Labels	Special provisions	Limited and excepted quantities		Packaging			Portable tanks and bulk containers	
									Packing instructions	Special packing provisions	Mixed packing provisions	Instructions	Special provisions
	3.1.2	2.2	2.2	2.1.1.3	5.2.2	3.3	3.4	3.5.1.2	4.1.4	4.1.4	4.1.10	4.2.5.2 7.3.2	4.2.5.3
(1)	(2)	(3a)	(3b)	(4)	(5)	(6)	(7a)	(7b)	(8)	(9a)	(9b)	(10)	(11)
0427	PROJECTILES with burster or expelling charge	1	1.4F		1.4		0	E0	P130		MP23		
0428	ARTICLES, PYROTECHNIC for technical purposes	1	1.1G		1		0	E0	P135		MP23 MP24		
0429	ARTICLES, PYROTECHNIC for technical purposes	1	1.2G		1		0	E0	P135		MP23 MP24		
0430	ARTICLES, PYROTECHNIC for technical purposes	1	1.3G		1		0	E0	P135		MP23 MP24		
0431	ARTICLES, PYROTECHNIC for technical purposes	1	1.4G		1.4		0	E0	P135		MP23 MP24		
0432	ARTICLES, PYROTECHNIC for technical purposes	1	1.4S		1.4		0	E0	P135		MP23 MP24		
0433	POWDER CAKE (POWDER PASTE), WETTED with not less than 17% alcohol, by mass	1	1.1C		1	266	0	E0	P111		MP20		
0434	PROJECTILES with burster or expelling charge	1	1.2G		1		0	E0	P130 LP101	PP67 L1	MP23		
0435	PROJECTILES with burster or expelling charge	1	1.4G		1.4		0	E0	P130 LP101	PP67 L1	MP23		
0436	ROCKETS with expelling charge	1	1.2C		1		0	E0	P130 LP101	PP67 L1	MP22		
0437	ROCKETS with expelling charge	1	1.3C		1		0	E0	P130 LP101	PP67 L1	MP22		
0438	ROCKETS with expelling charge	1	1.4C		1.4		0	E0	P130 LP101	PP67 L1	MP22		
0439	CHARGES, SHAPED, without detonator	1	1.2D		1		0	E0	P137	PP70	MP21		
0440	CHARGES, SHAPED, without detonator	1	1.4D		1.4		0	E0	P137	PP70	MP21		
0441	CHARGES, SHAPED, without detonator	1	1.4S		1.4	347	0	E0	P137	PP70	MP23		
0442	CHARGES, EXPLOSIVE, COMMERCIAL without detonator	1	1.1D		1		0	E0	P137		MP21		
0443	CHARGES, EXPLOSIVE, COMMERCIAL without detonator	1	1.2D		1		0	E0	P137		MP21		
0444	CHARGES, EXPLOSIVE, COMMERCIAL without detonator	1	1.4D		1.4		0	E0	P137		MP21		
0445	CHARGES, EXPLOSIVE, COMMERCIAL without detonator	1	1.4S		1.4	347	0	E0	P137		MP23		
0446	CASES, COMBUSTIBLE, EMPTY, WITHOUT PRIMER	1	1.4C		1.4		0	E0	P136		MP22		
0447	CASES, COMBUSTIBLE, EMPTY, WITHOUT PRIMER	1	1.3C		1		0	E0	P136		MP22		
0448	5-MERCAPTOTETRAZOL-1-ACETIC ACID	1	1.4C		1.4		0	E0	P114(b)		MP20		
0449	TORPEDOES, LIQUID FUELLED with or without bursting charge	1	1.1J		1		0	E0	P101		MP23		
0450	TORPEDOES, LIQUID FUELLED with inert head	1	1.3J		1		0	E0	P101		MP23		
0451	TORPEDOES with bursting charge	1	1.1D		1		0	E0	P130 LP101	PP67 L1	MP21		

ADR tank		Vehicle for tank carriage	Transport category (Tunnel restriction code)	Special provisions for carriage				Hazard identification No.	UN No.	Name and description
Tank code	Special provisions			Packages	Bulk	Loading, unloading and handling	Operation			
4.3	4.3.5, 6.8.4	9.1.1.2	1.1.3.6 (8.6)	7.2.4	7.3.3	7.5.11	8.5	5.3.2.3		3.1.2
(12)	(13)	(14)	(15)	(16)	(17)	(18)	(19)	(20)	(1)	(2)
			2 (E)	V2		CV1 CV2 CV3	S1		0427	PROJECTILES with burster or expelling charge
			1 (B1000C)	V2		CV1 CV2 CV3	S1		0428	ARTICLES, PYROTECHNIC for technical purposes
			1 (B1000C)	V2		CV1 CV2 CV3	S1		0429	ARTICLES, PYROTECHNIC for technical purposes
			1 (C5000D)	V2		CV1 CV2 CV3	S1		0430	ARTICLES, PYROTECHNIC for technical purposes
			2 (E)	V2		CV1 CV2 CV3	S1		0431	ARTICLES, PYROTECHNIC for technical purposes
			4 (E)			CV1 CV2 CV3	S1		0432	ARTICLES, PYROTECHNIC for technical purposes
			1 (B1000C)	V2		CV1 CV2 CV3	S1		0433	POWDER CAKE (POWDER PASTE), WETTED with not less than 17% alcohol, by mass
			1 (B1000C)	V2		CV1 CV2 CV3	S1		0434	PROJECTILES with burster or expelling charge
			2 (E)	V2		CV1 CV2 CV3	S1		0435	PROJECTILES with burster or expelling charge
			1 (B1000C)	V2		CV1 CV2 CV3	S1		0436	ROCKETS with expelling charge
			1 (C5000D)	V2		CV1 CV2 CV3	S1		0437	ROCKETS with expelling charge
			2 (E)	V2		CV1 CV2 CV3	S1		0438	ROCKETS with expelling charge
			1 (B1000C)	V2		CV1 CV2 CV3	S1		0439	CHARGES, SHAPED, without detonator
			2 (E)	V2		CV1 CV2 CV3	S1		0440	CHARGES, SHAPED, without detonator
			4 (E)			CV1 CV2 CV3	S1		0441	CHARGES, SHAPED, without detonator
			1 (B1000C)	V2		CV1 CV2 CV3	S1		0442	CHARGES, EXPLOSIVE, COMMERCIAL without detonator
			1 (B1000C)	V2		CV1 CV2 CV3	S1		0443	CHARGES, EXPLOSIVE, COMMERCIAL without detonator
			2 (E)	V2		CV1 CV2 CV3	S1		0444	CHARGES, EXPLOSIVE, COMMERCIAL without detonator
			4 (E)			CV1 CV2 CV3	S1		0445	CHARGES, EXPLOSIVE, COMMERCIAL without detonator
			2 (E)	V2		CV1 CV2 CV3	S1		0446	CASES, COMBUSTIBLE, EMPTY, WITHOUT PRIMER
			1 (C5000D)	V2		CV1 CV2 CV3	S1		0447	CASES, COMBUSTIBLE, EMPTY, WITHOUT PRIMER
			2 (E)	V2		CV1 CV2 CV3	S1		0448	5-MERCAPTOTETRAZOL-1-ACETIC ACID
			1 (B1000C)	V2		CV1 CV2 CV3	S1		0449	TORPEDOES, LIQUID FUELLED with or without bursting charge
			1 (C)	V2		CV1 CV2 CV3	S1		0450	TORPEDOES, LIQUID FUELLED with inert head
			1 (B1000C)	V2		CV1 CV2 CV3	S1		0451	TORPEDOES with bursting charge

UN No.	Name and description	Class	Classifi-cation code	Packing group	Labels	Special provi-sions	Limited and excepted quantities		Packaging			Portable tanks and bulk containers	
									Packing instruc-tions	Special packing provisions	Mixed packing provisions	Instruc-tions	Special provisions
	3.1.2	2.2	2.2	2.1.1.3	5.2.2	3.3	3.4	3.5.1.2	4.1.4	4.1.4	4.1.10	4.2.5.2 7.3.2	4.2.5.3
(1)	(2)	(3a)	(3b)	(4)	(5)	(6)	(7a)	(7b)	(8)	(9a)	(9b)	(10)	(11)
0452	GRENADES, PRACTICE, hand or rifle	1	1.4G		1.4		0	E0	P141		MP23		
0453	ROCKETS, LINE-THROWING	1	1.4G		1.4		0	E0	P130		MP23		
0454	IGNITERS	1	1.4S		1.4		0	E0	P142		MP23		
0455	DETONATORS, NON-ELECTRIC for blasting	1	1.4S		1.4	347	0	E0	P131	PP68	MP23		
0456	DETONATORS, ELECTRIC for blasting	1	1.4S		1.4	347	0	E0	P131		MP23		
0457	CHARGES, BURSTING, PLASTICS BONDED	1	1.1D		1		0	E0	P130		MP21		
0458	CHARGES, BURSTING, PLASTICS BONDED	1	1.2D		1		0	E0	P130		MP21		
0459	CHARGES, BURSTING, PLASTICS BONDED	1	1.4D		1.4		0	E0	P130		MP21		
0460	CHARGES, BURSTING, PLASTICS BONDED	1	1.4S		1.4	347	0	E0	P130		MP23		
0461	COMPONENTS, EXPLOSIVE TRAIN, N.O.S.	1	1.1B		1	178 274	0	E0	P101		MP2		
0462	ARTICLES, EXPLOSIVE, N.O.S.	1	1.1C		1	178 274	0	E0	P101		MP2		
0463	ARTICLES, EXPLOSIVE, N.O.S.	1	1.1D		1	178 274	0	E0	P101		MP2		
0464	ARTICLES, EXPLOSIVE, N.O.S.	1	1.1E		1	178 274	0	E0	P101		MP2		
0465	ARTICLES, EXPLOSIVE, N.O.S.	1	1.1F		1	178 274	0	E0	P101		MP2		
0466	ARTICLES, EXPLOSIVE, N.O.S.	1	1.2C		1	178 274	0	E0	P101		MP2		
0467	ARTICLES, EXPLOSIVE, N.O.S.	1	1.2D		1	178 274	0	E0	P101		MP2		
0468	ARTICLES, EXPLOSIVE, N.O.S.	1	1.2E		1	178 274	0	E0	P101		MP2		
0469	ARTICLES, EXPLOSIVE, N.O.S.	1	1.2F		1	178 274	0	E0	P101		MP2		
0470	ARTICLES, EXPLOSIVE, N.O.S.	1	1.3C		1	178 274	0	E0	P101		MP2		
0471	ARTICLES, EXPLOSIVE, N.O.S.	1	1.4E		1.4	178 274	0	E0	P101		MP2		
0472	ARTICLES, EXPLOSIVE, N.O.S.	1	1.4F		1.4	178 274	0	E0	P101		MP2		
0473	SUBSTANCES, EXPLOSIVE, N.O.S.	1	1.1A		1	178 274	0	E0	P101		MP2		
0474	SUBSTANCES, EXPLOSIVE, N.O.S.	1	1.1C		1	178 274	0	E0	P101		MP2		
0475	SUBSTANCES, EXPLOSIVE, N.O.S.	1	1.1D		1	178 274	0	E0	P101		MP2		
0476	SUBSTANCES, EXPLOSIVE, N.O.S.	1	1.1G		1	178 274	0	E0	P101		MP2		

ADR tank		Vehicle for tank carriage	Transport category (Tunnel restriction code)	Special provisions for carriage				Hazard identification No.	UN No.	Name and description
Tank code	Special provisions			Packages	Bulk	Loading, unloading and handling	Operation			
4.3	4.3.5, 6.8.4	9.1.1.2	1.1.3.6 (8.6)	7.2.4	7.3.3	7.5.11	8.5	5.3.2.3		3.1.2
(12)	(13)	(14)	(15)	(16)	(17)	(18)	(19)	(20)	(1)	(2)
			2 (E)	V2		CV1 CV2 CV3	S1		0452	GRENADES, PRACTICE, hand or rifle
			2 (E)	V2		CV1 CV2 CV3	S1		0453	ROCKETS, LINE-THROWING
			4 (E)			CV1 CV2 CV3	S1		0454	IGNITERS
			4 (E)			CV1 CV2 CV3	S1		0455	DETONATORS, NON-ELECTRIC for blasting
			4 (E)			CV1 CV2 CV3	S1		0456	DETONATORS, ELECTRIC for blasting
			1 (B1000C)	V2		CV1 CV2 CV3	S1		0457	CHARGES, BURSTING, PLASTICS BONDED
			1 (B1000C)	V2		CV1 CV2 CV3	S1		0458	CHARGES, BURSTING, PLASTICS BONDED
			2 (E)	V2		CV1 CV2 CV3	S1		0459	CHARGES, BURSTING, PLASTICS BONDED
			4 (E)			CV1 CV2 CV3	S1		0460	CHARGES, BURSTING, PLASTICS BONDED
			1 (B1000C)	V2		CV1 CV2 CV3	S1		0461	COMPONENTS, EXPLOSIVE TRAIN, N.O.S.
			1 (B1000C)	V2		CV1 CV2 CV3	S1		0462	ARTICLES, EXPLOSIVE, N.O.S.
			1 (B1000C)	V2		CV1 CV2 CV3	S1		0463	ARTICLES, EXPLOSIVE, N.O.S.
			1 (B1000C)	V2		CV1 CV2 CV3	S1		0464	ARTICLES, EXPLOSIVE, N.O.S.
			1 (B1000C)	V2		CV1 CV2 CV3	S1		0465	ARTICLES, EXPLOSIVE, N.O.S.
			1 (B1000C)	V2		CV1 CV2 CV3	S1		0466	ARTICLES, EXPLOSIVE, N.O.S.
			1 (B1000C)	V2		CV1 CV2 CV3	S1		0467	ARTICLES, EXPLOSIVE, N.O.S.
			1 (B1000C)	V2		CV1 CV2 CV3	S1		0468	ARTICLES, EXPLOSIVE, N.O.S.
			1 (B1000C)	V2		CV1 CV2 CV3	S1		0469	ARTICLES, EXPLOSIVE, N.O.S.
			1 (C5000D)	V2		CV1 CV2 CV3	S1		0470	ARTICLES, EXPLOSIVE, N.O.S.
			2 (E)	V2		CV1 CV2 CV3	S1		0471	ARTICLES, EXPLOSIVE, N.O.S.
			2 (E)	V2		CV1 CV2 CV3	S1		0472	ARTICLES, EXPLOSIVE, N.O.S.
			0 (B)	V2		CV1 CV2 CV3	S1		0473	SUBSTANCES, EXPLOSIVE, N.O.S.
			1 (B1000C)	V2 V3		CV1 CV2 CV3	S1		0474	SUBSTANCES, EXPLOSIVE, N.O.S.
			1 (B1000C)	V2 V3		CV1 CV2 CV3	S1		0475	SUBSTANCES, EXPLOSIVE, N.O.S.
			1 (B1000C)	V2 V3		CV1 CV2 CV3	S1		0476	SUBSTANCES, EXPLOSIVE, N.O.S.

UN No.	Name and description	Class	Classifi-cation code	Packing group	Labels	Special provi-sions	Limited and excepted quantities		Packaging			Portable tanks and bulk containers	
									Packing instruc-tions	Special packing provisions	Mixed packing provisions	Instruc-tions	Special provisions
	3.1.2	2.2	2.2	2.1.1.3	5.2.2	3.3	3.4	3.5.1.2	4.1.4	4.1.4	4.1.10	4.2.5.2 7.3.2	4.2.5.3
(1)	(2)	(3a)	(3b)	(4)	(5)	(6)	(7a)	(7b)	(8)	(9a)	(9b)	(10)	(11)
0477	SUBSTANCES, EXPLOSIVE, N.O.S.	1	1.3C		1	178 274	0	E0	P101		MP2		
0478	SUBSTANCES, EXPLOSIVE, N.O.S.	1	1.3G		1	178 274	0	E0	P101		MP2		
0479	SUBSTANCES, EXPLOSIVE, N.O.S.	1	1.4C		1.4	178 274	0	E0	P101		MP2		
0480	SUBSTANCES, EXPLOSIVE, N.O.S.	1	1.4D		1.4	178 274	0	E0	P101		MP2		
0481	SUBSTANCES, EXPLOSIVE, N.O.S.	1	1.4S		1.4	178 274	0	E0	P101		MP2		
0482	SUBSTANCES, EXPLOSIVE, VERY INSENSITIVE (SUBSTANCES, EVI), N.O.S.	1	1.5D		1.5	178 274	0	E0	P101		MP2		
0483	CYCLOTRIMETHYLENE-TRINITRAMINE (CYCLONITE; HEXOGEN; RDX), DESENSITIZED	1	1.1D		1		0	E0	P112(b) P112(c)		MP20		
0484	CYCLOTETRAMETHYLENE-TETRA-NITRAMINE (HMX; OCTOGEN), DESENSITIZED	1	1.1D		1		0	E0	P112(b) P112(c)		MP20		
0485	SUBSTANCES, EXPLOSIVE, N.O.S.	1	1.4G		1.4	178 274	0	E0	P101		MP2		
0486	ARTICLES, EXPLOSIVE, EXTREMELY INSENSITIVE (ARTICLES, EEI)	1	1.6N		1.6		0	E0	P101		MP23		
0487	SIGNALS, SMOKE	1	1.3G		1		0	E0	P135		MP23		
0488	AMMUNITION, PRACTICE	1	1.3G		1		0	E0	P130 LP101	PP67 L1	MP23		
0489	DINITROGLYCOLURIL (DINGU)	1	1.1D		1		0	E0	P112(b) P112(c)		MP20		
0490	NITROTRIAZOLONE (NTO)	1	1.1D		1		0	E0	P112(b) P112(c)		MP20		
0491	CHARGES, PROPELLING	1	1.4C		1.4		0	E0	P143	PP76	MP22		
0492	SIGNALS, RAILWAY TRACK, EXPLOSIVE	1	1.3G		1		0	E0	P135		MP23		
0493	SIGNALS, RAILWAY TRACK, EXPLOSIVE	1	1.4G		1.4		0	E0	P135		MP23		
0494	JET PERFORATING GUNS, CHARGED, oil well, without detonator	1	1.4D		1.4		0	E0	P101		MP21		
0495	PROPELLANT, LIQUID	1	1.3C		1	224	0	E0	P115	PP53 PP54 PP57 PP58	MP20		
0496	OCTONAL	1	1.1D		1		0	E0	P112(b) P112(c)		MP20		
0497	PROPELLANT, LIQUID	1	1.1C		1	224	0	E0	P115	PP53 PP54 PP57 PP58	MP20		
0498	PROPELLANT, SOLID	1	1.1C		1		0	E0	P114(b)		MP20		
0499	PROPELLANT, SOLID	1	1.3C		1		0	E0	P114(b)		MP20		
0500	DETONATOR ASSEMBLIES, NON-ELECTRIC for blasting	1	1.4S		1.4	347	0	E0	P131		MP23		

ADR tank		Vehicle for tank carriage	Transport category (Tunnel restriction code)	Special provisions for carriage				Hazard identification No.	UN No.	Name and description
Tank code	Special provisions			Packages	Bulk	Loading, unloading and handling	Operation			
4.3	4.3.5, 6.8.4	9.1.1.2	1.1.3.6 (8.6)	7.2.4	7.3.3	7.5.11	8.5	5.3.2.3		3.1.2
(12)	(13)	(14)	(15)	(16)	(17)	(18)	(19)	(20)	(1)	(2)
			1 (C5000D)	V2 V3		CV1 CV2 CV3	S1		0477	SUBSTANCES, EXPLOSIVE, N.O.S.
			1 (C5000D)	V2 V3		CV1 CV2 CV3	S1		0478	SUBSTANCES, EXPLOSIVE, N.O.S.
			2 (E)	V2		CV1 CV2 CV3	S1		0479	SUBSTANCES, EXPLOSIVE, N.O.S.
			2 (E)	V2		CV1 CV2 CV3	S1		0480	SUBSTANCES, EXPLOSIVE, N.O.S.
			4 (E)			CV1 CV2 CV3	S1		0481	SUBSTANCES, EXPLOSIVE, N.O.S.
			1 (B1000C)	V2		CV1 CV2 CV3	S1		0482	SUBSTANCES, EXPLOSIVE, VERY INSENSITIVE (SUBSTANCES, EVI), N.O.S.
			1 (B1000C)	V2 V3		CV1 CV2 CV3	S1		0483	CYCLOTRIMETHYLENE-TRINITRAMINE (CYCLONITE; HEXOGEN; RDX), DESENSITIZED
			1 (B1000C)	V2 V3		CV1 CV2 CV3	S1		0484	CYCLOTETRAMETHYLENE-TETRA-NITRAMINE (HMX; OCTOGEN), DESENSITIZED
			2 (E)	V2 V3		CV1 CV2 CV3	S1		0485	SUBSTANCES, EXPLOSIVE, N.O.S.
			2 (E)	V2		CV1 CV2 CV3	S1		0486	ARTICLES, EXPLOSIVE, EXTREMELY INSENSITIVE (ARTICLES, EEI)
			1 (C5000D)	V2		CV1 CV2 CV3	S1		0487	SIGNALS, SMOKE
			1 (C5000D)	V2		CV1 CV2 CV3	S1		0488	AMMUNITION, PRACTICE
			1 (B1000C)	V2 V3		CV1 CV2 CV3	S1		0489	DINITROGLYCOLURIL (DINGU)
			1 (B1000C)	V2 V3		CV1 CV2 CV3	S1		0490	NITROTRIAZOLONE (NTO)
			2 (E)	V2		CV1 CV2 CV3	S1		0491	CHARGES, PROPELLING
			1 (C5000D)	V2		CV1 CV2 CV3	S1		0492	SIGNALS, RAILWAY TRACK, EXPLOSIVE
			2 (E)	V2		CV1 CV2 CV3	S1		0493	SIGNALS, RAILWAY TRACK, EXPLOSIVE
			2 (E)	V2		CV1 CV2 CV3	S1		0494	JET PERFORATING GUNS, CHARGED, oil well, without detonator
			1 (C5000D)	V2		CV1 CV2 CV3	S1		0495	PROPELLANT, LIQUID
			1 (B1000C)	V2 V3		CV1 CV2 CV3	S1		0496	OCTONAL
			1 (B1000C)	V2		CV1 CV2 CV3	S1		0497	PROPELLANT, LIQUID
			1 (B1000C)	V2		CV1 CV2 CV3	S1		0498	PROPELLANT, SOLID
			1 (C5000D)	V2		CV1 CV2 CV3	S1		0499	PROPELLANT, SOLID
			4 (E)			CV1 CV2 CV3	S1		0500	DETONATOR ASSEMBLIES, NON-ELECTRIC for blasting

UN No.	Name and description	Class	Classifi-cation code	Packing group	Labels	Special provi-sions	Limited and excepted quantities		Packaging			Portable tanks and bulk containers	
									Packing instruc-tions	Special packing provisions	Mixed packing provisions	Instruc-tions	Special provisions
	3.1.2	2.2	2.2	2.1.1.3	5.2.2	3.3	3.4	3.5.1.2	4.1.4	4.1.4	4.1.10	4.2.5.2 7.3.2	4.2.5.3
(1)	(2)	(3a)	(3b)	(4)	(5)	(6)	(7a)	(7b)	(8)	(9a)	(9b)	(10)	(11)
0501	PROPELLANT, SOLID	1	1.4C		1.4		0	E0	P114(b)		MP20		
0502	ROCKETS with inert head	1	1.2C		1		0	E0	P130 LP101	PP67 L1	MP22		
0503	SAFETY DEVICES, PYROTECHNIC	1	1.4G		1.4	235 289	0	E0	P135		MP23		
0504	1H-TETRAZOLE	1	1.1D		1		0	E0	P112(c)	PP48	MP20		
0505	SIGNALS, DISTRESS, ship	1	1.4G		1.4		0	E0	P135		MP23 MP24		
0506	SIGNALS, DISTRESS, ship	1	1.4S		1.4		0	E0	P135		MP23 MP24		
0507	SIGNALS, SMOKE	1	1.4S		1.4		0	E0	P135		MP23 MP24		
0508	1-HYDROXY-BENZOTRIAZOLE, ANHYDROUS, dry or wetted with less than 20% water, by mass	1	1.3C		1		0	E0	P114(b)	PP48 PP50	MP20		
0509	POWDER, SMOKELESS	1	1.4C		1.4		0	E0	P114(b)	PP48	MP20		
0510	ROCKET MOTORS	1	1.4C		1.4		0	E0	P130 LP101	PP67 L1	MP22		
1001	ACETYLENE, DISSOLVED	2	4F		2.1	662	0	E0	P200		MP9		
1002	AIR, COMPRESSED	2	1A		2.2	655 662	120 ml	E1	P200		MP9	(M)	
1003	AIR, REFRIGERATED LIQUID	2	3O		2.2 +5.1		0	E0	P203		MP9	T75	TP5 TP22
1005	AMMONIA, ANHYDROUS	2	2TC		2.3 +8	23 379	0	E0	P200		MP9	(M) T50	
1006	ARGON, COMPRESSED	2	1A		2.2	378 653 662	120 ml	E1	P200		MP9	(M)	
1008	BORON TRIFLUORIDE	2	2TC		2.3 +8	373	0	E0	P200		MP9	(M)	
1009	BROMOTRIFLUORO-METHANE (REFRIGERANT GAS R 13B1)	2	2A		2.2	662	120 ml	E1	P200		MP9	(M) T50	
1010	BUTADIENES, STABILIZED or BUTADIENES AND HYDROCARBON MIXTURE, STABILIZED, having a vapour pressure at 70 °C not exceeding 1.1 Mpa (11 bar) and a density at 50 °C not lower than 0.525 kg/l	2	2F		2.1	386 618 662	0	E0	P200		MP9	(M) T50	
1011	BUTANE	2	2F		2.1	652 657 660 662	0	E0	P200		MP9	(M) T50	
1012	BUTYLENES MIXTURE or 1-BUTYLENE or cis-2-BUTYLENE or trans-2-BUTYLENE	2	2F		2.1	662	0	E0	P200		MP9	(M) T50	
1013	CARBON DIOXIDE	2	2A		2.2	378 584 653 662	120 ml	E1	P200		MP9	(M)	
1016	CARBON MONOXIDE, COMPRESSED	2	1TF		2.3 +2.1		0	E0	P200		MP9	(M)	

ADR tank		Vehicle for tank carriage	Transport category (Tunnel restriction code)	Special provisions for carriage				Hazard identifi-cation No.	UN No.	Name and description
Tank code	Special provisions			Packages	Bulk	Loading, unloading and handling	Operation			
4.3	4.3.5, 6.8.4	9.1.1.2	1.1.3.6 (8.6)	7.2.4	7.3.3	7.5.11	8.5	5.3.2.3		3.1.2
(12)	(13)	(14)	(15)	(16)	(17)	(18)	(19)	(20)	(1)	(2)
			2 (E)	V2		CV1 CV2 CV3	S1		0501	PROPELLANT, SOLID
			1 (B1000C)	V2		CV1 CV2 CV3	S1		0502	ROCKETS with inert head
			2 (E)	V2		CV1 CV2 CV3	S1		0503	SAFETY DEVICES, PYROTECHNIC
			1 (B1000C)	V2 V3		CV1 CV2 CV3	S1		0504	1H-TETRAZOLE
			2 (E)	V2		CV1 CV2 CV3	S1		0505	SIGNALS, DISTRESS, ship
			4 (E)			CV1 CV2 CV3	S1		0506	SIGNALS, DISTRESS, ship
			4 (E)			CV1 CV2 CV3	S1		0507	SIGNALS, SMOKE
			1 (C5000D)	V2 V3		CV1 CV2 CV3	S1		0508	1-HYDROXY-BENZOTRIAZOLE, ANHYDROUS, dry or wetted with less than 20% water, by mass
			2 (E)	V2		CV1 CV2 CV3	S1		0509	POWDER, SMOKELESS
			2 (E)	V2		CV1 CV2 CV3	S1		0510	ROCKET MOTORS
PxBN(M)	TU17 TA4 TT9	FL	2 (B/D)			CV9 CV10 CV36	S2	239	1001	ACETYLENE, DISSOLVED
CxBN(M)	TA4 TT9	AT	3 (E)			CV9 CV10		20	1002	AIR, COMPRESSED
RxBN	TU7 TU19 TA4 TT9	AT	3 (C/E)	V5		CV9 CV11 CV36	S20	225	1003	AIR, REFRIGERATED LIQUID
PxBH(M)	TA4 TT8 TT9	AT	1 (C/D)			CV9 CV10 CV36	S14	268	1005	AMMONIA, ANHYDROUS
CxBN(M)	TA4 TT9	AT	3 (E)			CV9 CV10 CV36		20	1006	ARGON, COMPRESSED
PxBH(M)	TA4 TT9 TT10	AT	1 (C/D)			CV9 CV10 CV36	S14	268	1008	BORON TRIFLUORIDE
PxBN(M)	TA4 TT9	AT	3 (C/E)			CV9 CV10 CV36		20	1009	BROMOTRIFLUORO-METHANE (REFRIGERANT GAS R 13B1)
PxBN(M)	TA4 TT9	FL	2 (B/D)	V8		CV9 CV10 CV36	S2 S4 S20	239	1010	BUTADIENES, STABILIZED or BUTADIENES AND HYDROCARBON MIXTURE, STABILIZED, having a vapour pressure at 70 °C not exceeding 1.1 Mpa (11 bar) and a density at 50 °C not lower than 0.525 kg/l
PxBN(M)	TA4 TT9 TT11	FL	2 (B/D)			CV9 CV10 CV36	S2 S20	23	1011	BUTANE
PxBN(M)	TA4 TT9	FL	2 (B/D)			CV9 CV10 CV36	S2 S20	23	1012	BUTYLENES MIXTURE or 1-BUTYLENE or cis-2-BUTYLENE or trans-2-BUTYLENE
PxBN(M)	TA4 TT9	AT	3 (C/E)			CV9 CV10 CV36		20	1013	CARBON DIOXIDE
CxBH(M)	TA4 TT9	FL	1 (B/D)			CV9 CV10 CV36	S2 S14	263	1016	CARBON MONOXIDE, COMPRESSED

UN No.	Name and description	Class	Classifi-cation code	Packing group	Labels	Special provi-sions	Limited and excepted quantities		Packaging			Portable tanks and bulk containers	
									Packing instruc-tions	Special packing provisions	Mixed packing provisions	Instruc-tions	Special provisions
3.1.2		2.2	2.2	2.1.1.3	5.2.2	3.3	3.4	3.5.1.2	4.1.4	4.1.4	4.1.10	4.2.5.2 7.3.2	4.2.5.3
(1)	(2)	(3a)	(3b)	(4)	(5)	(6)	(7a)	(7b)	(8)	(9a)	(9b)	(10)	(11)
1017	CHLORINE	2	2TOC		2.3 +5.1 +8		0	E0	P200		MP9	(M) T50	TP19
1018	CHLORODIFLUORO-METHANE (REFRIGERANT GAS R 22)	2	2A		2.2	662	120 ml	E1	P200		MP9	(M) T50	
1020	CHLOROPENTAFLUORO-ETHANE (REFRIGERANT GAS R 115)	2	2A		2.2	662	120 ml	E1	P200		MP9	(M) T50	
1021	1-CHLORO-1,2,2,2-TETRAFLUOROETHANE (REFRIGERANT GAS R 124)	2	2A		2.2	662	120 ml	E1	P200		MP9	(M) T50	
1022	CHLOROTRIFLUORO-METHANE (REFRIGERANT GAS R 13)	2	2A		2.2	662	120 ml	E1	P200		MP9	(M)	
1023	COAL GAS, COMPRESSED	2	1TF		2.3 +2.1		0	E0	P200		MP9	(M)	
1026	CYANOGEN	2	2TF		2.3 +2.1		0	E0	P200		MP9	(M)	
1027	CYCLOPROPANE	2	2F		2.1	662	0	E0	P200		MP9	(M) T50	
1028	DICHLORODIFLUORO-METHANE (REFRIGERANT GAS R 12)	2	2A		2.2	662	120 ml	E1	P200		MP9	(M) T50	
1029	DICHLOROFLUORO-METHANE (REFRIGERANT GAS R 21)	2	2A		2.2	662	120 ml	E1	P200		MP9	(M) T50	
1030	1,1-DIFLUOROETHANE (REFRIGERANT GAS R 152a)	2	2F		2.1	662	0	E0	P200		MP9	(M) T50	
1032	DIMETHYLAMINE, ANHYDROUS	2	2F		2.1	662	0	E0	P200		MP9	(M) T50	
1033	DIMETHYL ETHER	2	2F		2.1	662	0	E0	P200		MP9	(M) T50	
1035	ETHANE	2	2F		2.1	662	0	E0	P200		MP9	(M)	
1036	ETHYLAMINE	2	2F		2.1	662	0	E0	P200		MP9	(M) T50	
1037	ETHYL CHLORIDE	2	2F		2.1	662	0	E0	P200		MP9	(M) T50	
1038	ETHYLENE, REFRIGERATED LIQUID	2	3F		2.1		0	E0	P203		MP9	T75	TP5
1039	ETHYL METHYL ETHER	2	2F		2.1	662	0	E0	P200		MP9	(M)	
1040	ETHYLENE OXIDE	2	2TF		2.3 +2.1	342	0	E0	P200		MP9	(M)	
1040	ETHYLENE OXIDE WITH NITROGEN up to a total pressure of 1 MPa (10 bar) at 50 °C	2	2TF		2.3 +2.1	342	0	E0	P200		MP9	(M) T50	TP20
1041	ETHYLENE OXIDE AND CARBON DIOXIDE MIXTURE with more than 9% but not more than 87% ethylene oxide	2	2F		2.1	662	0	E0	P200		MP9	(M) T50	
1043	FERTILIZER AMMONIATING SOLUTION with free ammonia	2	4A		2.2	642							
1044	FIRE EXTINGUISHERS with compressed or liquefied gas	2	6A		2.2	225 594	120 ml	E0	P003	PP91	MP9		
1045	FLUORINE, COMPRESSED	2	1TOC		2.3 +5.1 +8		0	E0	P200		MP9		
1046	HELIUM, COMPRESSED	2	1A		2.2	378 653 662	120 ml	E1	P200		MP9	(M)	

ADR tank		Vehicle for tank carriage	Transport category (Tunnel restriction code)	Special provisions for carriage				Hazard identifi-cation No.	UN No.	Name and description
Tank code	Special provisions			Packages	Bulk	Loading, unloading and handling	Operation			
4.3	4.3.5, 6.8.4	9.1.1.2	1.1.3.6 (8.6)	7.2.4	7.3.3	7.5.11	8.5	5.3.2.3		3.1.2
(12)	(13)	(14)	(15)	(16)	(17)	(18)	(19)	(20)	(1)	(2)
P22DH(M)	TA4 TT9 TT10	AT	1 (C/D)			CV9 CV10 CV36	S14	265	1017	CHLORINE
PxBN(M)	TA4 TT9	AT	3 (C/E)			CV9 CV10 CV36		20	1018	CHLORODIFLUORO-METHANE (REFRIGERANT GAS R 22)
PxBN(M)	TA4 TT9	AT	3 (C/E)			CV9 CV10 CV36		20	1020	CHLOROPENTAFLUORO-ETHANE (REFRIGERANT GAS R 115)
PxBN(M)	TA4 TT9	AT	3 (C/E)			CV9 CV10 CV36		20	1021	1-CHLORO-1,2,2,2-TETRAFLUOROETHANE (REFRIGERANT GAS R 124)
PxBN(M)	TA4 TT9	AT	3 (C/E)			CV9 CV10 CV36		20	1022	CHLOROTRIFLUORO-METHANE (REFRIGERANT GAS R 13)
CxBH(M)	TA4 TT9	FL	1 (B/D)			CV9 CV10 CV36	S2 S14	263	1023	COAL GAS, COMPRESSED
PxBH(M)	TA4 TT9	FL	1 (B/D)			CV9 CV10 CV36	S2 S14	263	1026	CYANOGEN
PxBN(M)	TA4 TT9	FL	2 (B/D)			CV9 CV10 CV36	S2 S20	23	1027	CYCLOPROPANE
PxBN(M)	TA4 TT9	AT	3 (C/E)			CV9 CV10 CV36		20	1028	DICHLORODIFLUORO-METHANE (REFRIGERANT GAS R 12)
PxBN(M)	TA4 TT9	AT	3 (C/E)			CV9 CV10 CV36		20	1029	DICHLOROFLUORO-METHANE (REFRIGERANT GAS R 21)
PxBN(M)	TA4 TT9	FL	2 (B/D)			CV9 CV10 CV36	S2 S20	23	1030	1,1-DIFLUOROETHANE (REFRIGERANT GAS R 152a)
PxBN(M)	TA4 TT9	FL	2 (B/D)			CV9 CV10 CV36	S2 S20	23	1032	DIMETHYLAMINE, ANHYDROUS
PxBN(M)	TA4 TT9	FL	2 (B/D)			CV9 CV10 CV36	S2 S20	23	1033	DIMETHYL ETHER
PxBN(M)	TA4 TT9	FL	2 (B/D)			CV9 CV10 CV36	S2 S20	23	1035	ETHANE
PxBN(M)	TA4 TT9	FL	2 (B/D)			CV9 CV10 CV36	S2 S20	23	1036	ETHYLAMINE
PxBN(M)	TA4 TT9	FL	2 (B/D)			CV9 CV10 CV36	S2 S20	23	1037	ETHYL CHLORIDE
RxBN	TU18 TA4 TT9	FL	2 (B/D)	V5		CV9 CV11 CV36	S2 S17	223	1038	ETHYLENE, REFRIGERATED LIQUID
PxBN(M)	TA4 TT9	FL	2 (B/D)			CV9 CV10 CV36	S2 S20	23	1039	ETHYL METHYL ETHER
		FL	1 (B/D)			CV9 CV10 CV36	S2 S14	263	1040	ETHYLENE OXIDE
PxBH(M)	TA4 TT9	FL	1 (B/D)			CV9 CV10 CV36	S2 S14	263	1040	ETHYLENE OXIDE WITH NITROGEN up to a total pressure of 1 MPa (10 bar) at 50 °C
PxBN(M)	TA4 TT9	FL	2 (B/D)			CV9 CV10 CV36	S2 S20	239	1041	ETHYLENE OXIDE AND CARBON DIOXIDE MIXTURE with more than 9% but not more than 87% ethylene oxide
			(E)						1043	FERTILIZER AMMONIATING SOLUTION with free ammonia
			3 (E)			CV9			1044	FIRE EXTINGUISHERS with compressed or liquefied gas
			1 (D)			CV9 CV10 CV36	S14		1045	FLUORINE, COMPRESSED
CxBN(M)	TA4 TT9	AT	3 (E)			CV9 CV10 CV36		20	1046	HELIUM, COMPRESSED

UN No.	Name and description	Class	Classification code	Packing group	Labels	Special provisions	Limited and excepted quantities		Packaging			Portable tanks and bulk containers	
									Packing instructions	Special packing provisions	Mixed packing provisions	Instructions	Special provisions
	3.1.2	2.2	2.2	2.1.1.3	5.2.2	3.3	3.4	3.5.1.2	4.1.4	4.1.4	4.1.10	4.2.5.2 7.3.2	4.2.5.3
(1)	(2)	(3a)	(3b)	(4)	(5)	(6)	(7a)	(7b)	(8)	(9a)	(9b)	(10)	(11)
1048	HYDROGEN BROMIDE, ANHYDROUS	2	2TC		2.3 +8		0	E0	P200		MP9	(M)	
1049	HYDROGEN, COMPRESSED	2	1F		2.1	660 662	0	E0	P200		MP9	(M)	
1050	HYDROGEN CHLORIDE, ANHYDROUS	2	2TC		2.3 +8		0	E0	P200		MP9	(M)	
1051	HYDROGEN CYANIDE, STABILIZED containing less than 3% water	6.1	TF1	I	6.1 +3	386 603	0	E0	P200		MP2		
1052	HYDROGEN FLUORIDE, ANHYDROUS	8	CT1	I	8 +6.1		0	E0	P200		MP2	T10	TP2
1053	HYDROGEN SULPHIDE	2	2TF		2.3 +2.1		0	E0	P200		MP9	(M)	
1055	ISOBUTYLENE	2	2F		2.1	662	0	E0	P200		MP9	(M) T50	
1056	KRYPTON, COMPRESSED	2	1A		2.2	378 662	120 ml	E1	P200		MP9	(M)	
1057	LIGHTERS or LIGHTER REFILLS containing flammable gas	2	6F		2.1	201 654 658	0	E0	P002	PP84 RR5	MP9		
1058	LIQUEFIED GASES, non-flammable, charged with nitrogen, carbon dioxide or air	2	2A		2.2	662	120 ml	E1	P200		MP9	(M)	
1060	METHYLACETYLENE AND PROPADIENE MIXTURE, STABILIZED such as mixture P1 or mixture P2	2	2F		2.1	386 581 662	0	E0	P200		MP9	(M) T50	
1061	METHYLAMINE, ANHYDROUS	2	2F		2.1	662	0	E0	P200		MP9	(M) T50	
1062	METHYL BROMIDE with not more than 2% chloropicrin	2	2T		2.3	23	0	E0	P200		MP9	(M) T50	
1063	METHYL CHLORIDE (REFRIGERANT GAS R 40)	2	2F		2.1	662	0	E0	P200		MP9	(M) T50	
1064	METHYL MERCAPTAN	2	2TF		2.3 +2.1		0	E0	P200		MP9	(M) T50	
1065	NEON, COMPRESSED	2	1A		2.2	378 662	120 ml	E1	P200		MP9	(M)	
1066	NITROGEN, COMPRESSED	2	1A		2.2	378 653 662	120 ml	E1	P200		MP9	(M)	
1067	DINITROGEN TETROXIDE (NITROGEN DIOXIDE)	2	2TOC		2.3 +5.1 +8		0	E0	P200		MP9	T50	TP21
1069	NITROSYL CHLORIDE	2	2TC		2.3 +8		0	E0	P200		MP9		
1070	NITROUS OXIDE	2	2O		2.2 +5.1	584 662	0	E0	P200		MP9	(M)	
1071	OIL GAS, COMPRESSED	2	1TF		2.3 +2.1		0	E0	P200		MP9	(M)	
1072	OXYGEN, COMPRESSED	2	1O		2.2 +5.1	355 655 662	0	E0	P200		MP9	(M)	
1073	OXYGEN, REFRIGERATED LIQUID	2	3O		2.2 +5.1		0	E0	P203		MP9	T75	TP5 TP22
1075	PETROLEUM GASES, LIQUEFIED	2	2F		2.1	274 583 639 660 662	0	E0	P200		MP9	(M) T50	

ADR tank		Vehicle for tank carriage	Transport category (Tunnel restriction code)	Special provisions for carriage				Hazard identification No.	UN No.	Name and description
Tank code	Special provisions			Packages	Bulk	Loading, unloading and handling	Operation			
4.3	4.3.5, 6.8.4	9.1.1.2	1.1.3.6 (8.6)	7.2.4	7.3.3	7.5.11	8.5	5.3.2.3		3.1.2
(12)	(13)	(14)	(15)	(16)	(17)	(18)	(19)	(20)	(1)	(2)
PxBH(M)	TA4 TT9 TT10	AT	1 (C/D)			CV9 CV10 CV36	S14	268	1048	HYDROGEN BROMIDE, ANHYDROUS
CxBN(M)	TA4 TT9	FL	2 (B/D)			CV9 CV10 CV36	S2 S20	23	1049	HYDROGEN, COMPRESSED
PxBH(M)	TA4 TT9 TT10	AT	1 (C/D)			CV9 CV10 CV36	S14	268	1050	HYDROGEN CHLORIDE, ANHYDROUS
			0 (D)	V8		CV1 CV13 CV28	S2 S4 S9 S10 S14		1051	HYDROGEN CYANIDE, STABILIZED containing less than 3% water
L21DH(+)	TU14 TU34 TC1 TE21 TA4 TT9 TM3	AT	1 (C/D)			CV13 CV28 CV34	S14	886	1052	HYDROGEN FLUORIDE, ANHYDROUS
PxDH(M)	TA4 TT9 TT10	FL	1 (B/D)			CV9 CV10 CV36	S2 S14	263	1053	HYDROGEN SULPHIDE
PxBN(M)	TA4 TT9	FL	2 (B/D)			CV9 CV10 CV36	S2 S20	23	1055	ISOBUTYLENE
CxBN(M)	TA4 TT9	AT	3 (E)			CV9 CV10 CV36		20	1056	KRYPTON, COMPRESSED
			2 (D)			CV9	S2		1057	LIGHTERS or LIGHTER REFILLS containing flammable gas
PxBN(M)	TA4 TT9	AT	3 (C/E)			CV9 CV10 CV36		20	1058	LIQUEFIED GASES, non-flammable, charged with nitrogen, carbon dioxide or air
PxBN(M)	TA4 TT9	FL	2 (B/D)	V8		CV9 CV10 CV36	S2 S4 S20	239	1060	METHYLACETYLENE AND PROPADIENE MIXTURE, STABILIZED such as mixture P1 or mixture P2
PxBN(M)	TA4 TT9	FL	2 (B/D)			CV9 CV10 CV36	S2 S20	23	1061	METHYLAMINE, ANHYDROUS
PxBH(M)	TA4 TT9	AT	1 (C/D)			CV9 CV10 CV36	S14	26	1062	METHYL BROMIDE with not more than 2% chloropicrin
PxBN(M)	TA4 TT9	FL	2 (B/D)			CV9 CV10 CV36	S2 S20	23	1063	METHYL CHLORIDE (REFRIGERANT GAS R 40)
PxDH(M)	TA4 TT9	FL	1 (B/D)			CV9 CV10 CV36	S2 S14	263	1064	METHYL MERCAPTAN
CxBN(M)	TA4 TT9	AT	3 (E)			CV9 CV10 CV36		20	1065	NEON, COMPRESSED
CxBN(M)	TA4 TT9	AT	3 (E)			CV9 CV10 CV36		20	1066	NITROGEN, COMPRESSED
PxBH(M)	TU17 TA4 TT9	AT	1 (C/D)			CV9 CV10 CV36	S14	265	1067	DINITROGEN TETROXIDE (NITROGEN DIOXIDE)
			1 (D)			CV9 CV10 CV36	S14		1069	NITROSYL CHLORIDE
PxBN(M)	TA4 TT9	AT	3 (C/E)			CV9 CV10 CV36		25	1070	NITROUS OXIDE
CxBH(M)	TA4 TT9	FL	1 (B/D)			CV9 CV10 CV36	S2 S14	263	1071	OIL GAS, COMPRESSED
CxBN(M)	TA4 TT9	AT	3 (E)			CV9 CV10 CV36		25	1072	OXYGEN, COMPRESSED
RxBN	TU7 TU19 TA4 TT9	AT	3 (C/E)	V5		CV9 CV11 CV36	S20	225	1073	OXYGEN, REFRIGERATED LIQUID
PxBN(M)	TA4 TT9 TT11	FL	2 (B/D)			CV9 CV10 CV36	S2 S20	23	1075	PETROLEUM GASES, LIQUEFIED

UN No.	Name and description	Class	Classification code	Packing group	Labels	Special provisions	Limited and excepted quantities		Packaging			Portable tanks and bulk containers	
									Packing instructions	Special packing provisions	Mixed packing provisions	Instructions	Special provisions
	3.1.2	2.2	2.2	2.1.1.3	5.2.2	3.3	3.4	3.5.1.2	4.1.4	4.1.4	4.1.10	4.2.5.2 7.3.2	4.2.5.3
(1)	(2)	(3a)	(3b)	(4)	(5)	(6)	(7a)	(7b)	(8)	(9a)	(9b)	(10)	(11)
1076	PHOSGENE	2	2TC		2.3 +8		0	E0	P200		MP9		
1077	PROPYLENE	2	2F		2.1	662	0	E0	P200		MP9	(M) T50	
1078	REFRIGERANT GAS, N.O.S., such as mixture F1, mixture F2 or mixture F3	2	2A		2.2	274 582 662	120 ml	E1	P200		MP9	(M) T50	
1079	SULPHUR DIOXIDE	2	2TC		2.3 +8		0	E0	P200		MP9	(M) T50	TP19
1080	SULPHUR HEXAFLUORIDE	2	2A		2.2	662	120 ml	E1	P200		MP9	(M)	
1081	TETRAFLUOROETHYLENE, STABILIZED	2	2F		2.1	386 662	0	E0	P200		MP9	(M)	
1082	TRIFLUOROCHLORO-ETHYLENE, STABILIZED (REFRIGERANT GAS R 1113)	2	2TF		2.3 +2.1	386	0	E0	P200		MP9	(M) T50	
1083	TRIMETHYLAMINE, ANHYDROUS	2	2F		2.1	662	0	E0	P200		MP9	(M) T50	
1085	VINYL BROMIDE, STABILIZED	2	2F		2.1	386 662	0	E0	P200		MP9	(M) T50	
1086	VINYL CHLORIDE, STABILIZED	2	2F		2.1	386 662	0	E0	P200		MP9	(M) T50	
1087	VINYL METHYL ETHER, STABILIZED	2	2F		2.1	386 662	0	E0	P200		MP9	(M) T50	
1088	ACETAL	3	F1	II	3		1 L	E2	P001 IBC02 R001		MP19	T4	TP1
1089	ACETALDEHYDE	3	F1	I	3		0	E0	P001		MP7 MP17	T11	TP2 TP7
1090	ACETONE	3	F1	II	3		1 L	E2	P001 IBC02 R001		MP19	T4	TP1
1091	ACETONE OILS	3	F1	II	3		1 L	E2	P001 IBC02 R001		MP19	T4	TP1 TP8
1092	ACROLEIN, STABILIZED	6.1	TF1	I	6.1 +3	354 386	0	E0	P601		MP8 MP17	T22	TP2 TP7
1093	ACRYLONITRILE, STABILIZED	3	FT1	I	3 +6.1	386	0	E0	P001		MP7 MP17	T14	TP2
1098	ALLYL ALCOHOL	6.1	TF1	I	6.1 +3	354	0	E0	P602		MP8 MP17	T20	TP2
1099	ALLYL BROMIDE	3	FT1	I	3 +6.1		0	E0	P001		MP7 MP17	T14	TP2
1100	ALLYL CHLORIDE	3	FT1	I	3 +6.1		0	E0	P001		MP7 MP17	T14	TP2
1104	AMYL ACETATES	3	F1	III	3		5 L	E1	P001 IBC03 LP01 R001		MP19	T2	TP1
1105	PENTANOLS	3	F1	II	3		1 L	E2	P001 IBC02 R001		MP19	T4	TP1 TP29
1105	PENTANOLS	3	F1	III	3		5 L	E1	P001 IBC03 LP01 R001		MP19	T2	TP1
1106	AMYLAMINE	3	FC	II	3 +8		1 L	E2	P001 IBC02		MP19	T7	TP1
1106	AMYLAMINE	3	FC	III	3 +8		5 L	E1	P001 IBC03 R001		MP19	T4	TP1
1107	AMYL CHLORIDE	3	F1	II	3		1 L	E2	P001 IBC02 R001		MP19	T4	TP1

ADR tank		Vehicle for tank carriage	Transport category (Tunnel restriction code)	Special provisions for carriage				Hazard identifi-cation No.	UN No.	Name and description
Tank code	Special provisions			Packages	Bulk	Loading, unloading and handling	Operation			
4.3	4.3.5, 6.8.4	9.1.1.2	1.1.3.6 (8.6)	7.2.4	7.3.3	7.5.11	8.5	5.3.2.3		3.1.2
(12)	(13)	(14)	(15)	(16)	(17)	(18)	(19)	(20)	(1)	(2)
P22DH(M)	TU17 TA4 TT9	AT	1 (C/D)			CV9 CV10 CV36	S14	268	1076	PHOSGENE
PxBN(M)	TA4 TT9	FL	2 (B/D)			CV9 CV10 CV36	S2 S20	23	1077	PROPYLENE
PxBN(M)	TA4 TT9	AT	3 (C/E)			CV9 CV10 CV36		20	1078	REFRIGERANT GAS, N.O.S., such as mixture F1, mixture F2 or mixture F3
PxDH(M)	TA4 TT9 TT10	AT	1 (C/D)			CV9 CV10 CV36	S14	268	1079	SULPHUR DIOXIDE
PxBN(M)	TA4 TT9	AT	3 (C/E)			CV9 CV10 CV36		20	1080	SULPHUR HEXAFLUORIDE
PxBN(M)	TU40 TA4 TT9	FL	2 (B/D)	V8		CV9 CV10 CV36	S2 S4 S20	239	1081	TETRAFLUOROETHYLENE, STABILIZED
PxBH(M)	TA4 TT9	FL	1 (B/D)	V8		CV9 CV10 CV36	S2 S4 S14	263	1082	TRIFLUOROCHLORO-ETHYLENE, STABILIZED (REFRIGERANT GAS R 1113)
PxBN(M)	TA4 TT9	FL	2 (B/D)			CV9 CV10 CV36	S2 S20	23	1083	TRIMETHYLAMINE, ANHYDROUS
PxBN(M)	TA4 TT9	FL	2 (B/D)	V8		CV9 CV10 CV36	S2 S4 S20	239	1085	VINYL BROMIDE, STABILIZED
PxBN(M)	TA4 TT9	FL	2 (B/D)	V8		CV9 CV10 CV36	S2 S4 S20	239	1086	VINYL CHLORIDE, STABILIZED
PxBN(M)	TA4 TT9	FL	2 (B/D)	V8		CV9 CV10 CV36	S2 S4 S20	239	1087	VINYL METHYL ETHER, STABILIZED
LGBF		FL	2 (D/E)				S2 S20	33	1088	ACETAL
L4BN	TU8	FL	1 (D/E)				S2 S20	33	1089	ACETALDEHYDE
LGBF		FL	2 (D/E)				S2 S20	33	1090	ACETONE
LGBF		FL	2 (D/E)				S2 S20	33	1091	ACETONE OILS
L15CH	TU14 TU15 TE19 TE21	FL	1 (C/D)	V8		CV1 CV13 CV28	S2 S4 S9 S14	663	1092	ACROLEIN, STABILIZED
L10CH	TU14 TU15 TE21	FL	1 (C/E)	V8		CV13 CV28	S2 S4 S22	336	1093	ACRYLONITRILE, STABILIZED
L10CH	TU14 TU15 TE19 TE21	FL	1 (C/D)			CV1 CV13 CV28	S2 S9 S14	663	1098	ALLYL ALCOHOL
L10CH	TU14 TU15 TE21	FL	1 (C/E)			CV13 CV28	S2 S22	336	1099	ALLYL BROMIDE
L10CH	TU14 TU15 TE21	FL	1 (C/E)			CV13 CV28	S2 S22	336	1100	ALLYL CHLORIDE
LGBF		FL	3 (D/E)	V12			S2	30	1104	AMYL ACETATES
LGBF		FL	2 (D/E)				S2 S20	33	1105	PENTANOLS
LGBF		FL	3 (D/E)	V12			S2	30	1105	PENTANOLS
L4BH		FL	2 (D/E)				S2 S20	338	1106	AMYLAMINE
L4BN		FL	3 (D/E)	V12			S2	38	1106	AMYLAMINE
LGBF		FL	2 (D/E)				S2 S20	33	1107	AMYL CHLORIDE

UN No.	Name and description	Class	Classification code	Packing group	Labels	Special provisions	Limited and excepted quantities		Packaging			Portable tanks and bulk containers	
									Packing instructions	Special packing provisions	Mixed packing provisions	Instructions	Special provisions
	3.1.2	2.2	2.2	2.1.1.3	5.2.2	3.3	3.4	3.5.1.2	4.1.4	4.1.4	4.1.10	4.2.5.2 7.3.2	4.2.5.3
(1)	(2)	(3a)	(3b)	(4)	(5)	(6)	(7a)	(7b)	(8)	(9a)	(9b)	(10)	(11)
1108	1-PENTENE (n-AMYLENE)	3	F1	I	3		0	E3	P001		MP7 MP17	T11	TP2
1109	AMYL FORMATES	3	F1	III	3		5 L	E1	P001 IBC03 LP01 R001		MP19	T2	TP1
1110	n-AMYL METHYL KETONE	3	F1	III	3		5 L	E1	P001 IBC03 LP01 R001		MP19	T2	TP1
1111	AMYL MERCAPTAN	3	F1	II	3		1 L	E2	P001 IBC02 R001		MP19	T4	TP1
1112	AMYL NITRATE	3	F1	III	3		5 L	E1	P001 IBC03 LP01 R001		MP19	T2	TP1
1113	AMYL NITRITE	3	F1	II	3		1 L	E2	P001 IBC02 R001		MP19	T4	TP1
1114	BENZENE	3	F1	II	3		1 L	E2	P001 IBC02 R001		MP19	T4	TP1
1120	BUTANOLS	3	F1	II	3		1 L	E2	P001 IBC02 R001		MP19	T4	TP1 TP29
1120	BUTANOLS	3	F1	III	3		5 L	E1	P001 IBC03 LP01 R001		MP19	T2	TP1
1123	BUTYL ACETATES	3	F1	II	3		1 L	E2	P001 IBC02 R001		MP19	T4	TP1
1123	BUTYL ACETATES	3	F1	III	3		5 L	E1	P001 IBC03 LP01 R001		MP19	T2	TP1
1125	n-BUTYLAMINE	3	FC	II	3 +8		1 L	E2	P001 IBC02		MP19	T7	TP1
1126	1-BROMOBUTANE	3	F1	II	3		1 L	E2	P001 IBC02 R001		MP19	T4	TP1
1127	CHLOROBUTANES	3	F1	II	3		1 L	E2	P001 IBC02 R001		MP19	T4	TP1
1128	n-BUTYL FORMATE	3	F1	II	3		1 L	E2	P001 IBC02 R001		MP19	T4	TP1
1129	BUTYRALDEHYDE	3	F1	II	3		1 L	E2	P001 IBC02 R001		MP19	T4	TP1
1130	CAMPHOR OIL	3	F1	III	3		5 L	E1	P001 IBC03 LP01 R001		MP19	T2	TP1
1131	CARBON DISULPHIDE	3	FT1	I	3 +6.1		0	E0	P001	PP31	MP7 MP17	T14	TP2 TP7
1133	ADHESIVES containing flammable liquid	3	F1	I	3		500 ml	E3	P001		MP7 MP17	T11	TP1 TP8 TP27
1133	ADHESIVES containing flammable liquid (vapour pressure at 50 °C more than 110 kPa)	3	F1	II	3	640C	5 L	E2	P001	PP1	MP19	T4	TP1 TP8
1133	ADHESIVES containing flammable liquid (vapour pressure at 50 °C not more than 110 kPa)	3	F1	II	3	640D	5 L	E2	P001 IBC02 R001	PP1	MP19	T4	TP1 TP8
1133	ADHESIVES containing flammable liquid	3	F1	III	3		5 L	E1	P001 IBC03 LP01 R001	PP1	MP19	T2	TP1

ADR tank		Vehicle for tank carriage	Transport category (Tunnel restriction code)	Special provisions for carriage				Hazard identifi-cation No.	UN No.	Name and description
Tank code	Special provisions			Packages	Bulk	Loading, unloading and handling	Operation			
4.3	4.3.5, 6.8.4	9.1.1.2	1.1.3.6 (8.6)	7.2.4	7.3.3	7.5.11	8.5	5.3.2.3		3.1.2
(12)	(13)	(14)	(15)	(16)	(17)	(18)	(19)	(20)	(1)	(2)
L4BN		FL	1 (D/E)				S2 S20	33	1108	1-PENTENE (n-AMYLENE)
LGBF		FL	3 (D/E)	V12			S2	30	1109	AMYL FORMATES
LGBF		FL	3 (D/E)	V12			S2	30	1110	n-AMYL METHYL KETONE
LGBF		FL	2 (D/E)				S2 S20	33	1111	AMYL MERCAPTAN
LGBF		FL	3 (D/E)	V12			S2	30	1112	AMYL NITRATE
LGBF		FL	2 (D/E)				S2 S20	33	1113	AMYL NITRITE
LGBF		FL	2 (D/E)				S2 S20	33	1114	BENZENE
LGBF		FL	2 (D/E)				S2 S20	33	1120	BUTANOLS
LGBF		FL	3 (D/E)	V12			S2	30	1120	BUTANOLS
LGBF		FL	2 (D/E)				S2 S20	33	1123	BUTYL ACETATES
LGBF		FL	3 (D/E)	V12			S2	30	1123	BUTYL ACETATES
L4BH		FL	2 (D/E)				S2 S20	338	1125	n-BUTYLAMINE
LGBF		FL	2 (D/E)				S2 S20	33	1126	1-BROMOBUTANE
LGBF		FL	2 (D/E)				S2 S20	33	1127	CHLOROBUTANES
LGBF		FL	2 (D/E)				S2 S20	33	1128	n-BUTYL FORMATE
LGBF		FL	2 (D/E)				S2 S20	33	1129	BUTYRALDEHYDE
LGBF		FL	3 (D/E)	V12			S2	30	1130	CAMPHOR OIL
L10CH	TU2 TU14 TU15 TE21	FL	1 (C/E)			CV13 CV28	S2 S22	336	1131	CARBON DISULPHIDE
L4BN		FL	1 (D/E)				S2 S20	33	1133	ADHESIVES containing flammable liquid
L1.5BN		FL	2 (D/E)				S2 S20	33	1133	ADHESIVES containing flammable liquid (vapour pressure at 50 °C more than 110 kPa)
LGBF		FL	2 (D/E)				S2 S20	33	1133	ADHESIVES containing flammable liquid (vapour pressure at 50 °C not more than 110 kPa)
LGBF		FL	3 (D/E)	V12			S2	30	1133	ADHESIVES containing flammable liquid

UN No.	Name and description	Class	Classification code	Packing group	Labels	Special provisions	Limited and excepted quantities		Packaging			Portable tanks and bulk containers	
									Packing instructions	Special packing provisions	Mixed packing provisions	Instructions	Special provisions
	3.1.2	2.2	2.2	2.1.1.3	5.2.2	3.3	3.4	3.5.1.2	4.1.4	4.1.4	4.1.10	4.2.5.2 7.3.2	4.2.5.3
(1)	(2)	(3a)	(3b)	(4)	(5)	(6)	(7a)	(7b)	(8)	(9a)	(9b)	(10)	(11)
1133	ADHESIVES containing flammable liquid (having a flash-point below 23 °C and viscous according to 2.2.3.1.4) (vapour pressure at 50 °C more than 110 kPa)	3	F1	III	3		5 L	E1	P001 R001	PP1	MP19		
1133	ADHESIVES containing flammable liquid (having a flash-point below 23 °C and viscous according to 2.2.3.1.4) (vapour pressure at 50 °C not more than 110 kPa)	3	F1	III	3		5 L	E1	P001 IBC02 R001	PP1 BB4	MP19		
1134	CHLOROBENZENE	3	F1	III	3		5 L	E1	P001 IBC03 LP01 R001		MP19	T2	TP1
1135	ETHYLENE CHLOROHYDRIN	6.1	TF1	I	6.1 +3	354	0	E0	P602		MP8 MP17	T20	TP2
1136	COAL TAR DISTILLATES, FLAMMABLE	3	F1	II	3		1 L	E2	P001 IBC02 R001		MP19	T4	TP1
1136	COAL TAR DISTILLATES, FLAMMABLE	3	F1	III	3		5 L	E1	P001 IBC03 LP01 R001		MP19	T4	TP1 TP29
1139	COATING SOLUTION (includes surface treatments or coatings used for industrial or other purposes such as vehicle under coating, drum or barrel lining)	3	F1	I	3		500 ml	E3	P001		MP7 MP17	T11	TP1 TP8 TP27
1139	COATING SOLUTION (includes surface treatments or coatings used for industrial or other purposes such as vehicle under coating, drum or barrel lining) (vapour pressure at 50 °C more than 110 kPa)	3	F1	II	3	640C	5 L	E2	P001		MP19	T4	TP1 TP8
1139	COATING SOLUTION (includes surface treatments or coatings used for industrial or other purposes such as vehicle under coating, drum or barrel lining) (vapour pressure at 50 °C not more than 110 kPa)	3	F1	II	3	640D	5 L	E2	P001 IBC02 R001		MP19	T4	TP1 TP8
1139	COATING SOLUTION (includes surface treatments or coatings used for industrial or other purposes such as vehicle under coating, drum or barrel lining)	3	F1	III	3		5 L	E1	P001 IBC03 LP01 R001		MP19	T2	TP1
1139	COATING SOLUTION (includes surface treatments or coatings used for industrial or other purposes such as vehicle under coating, drum or barrel lining) (having a flash-point below 23 °C and viscous according to 2.2.3.1.4) (vapour pressure at 50 °C more than 110 kPa)	3	F1	III	3		5 L	E1	P001 R001		MP19		
1139	COATING SOLUTION (includes surface treatments or coatings used for industrial or other purposes such as vehicle under coating, drum or barrel lining) (having a flash-point below 23 °C and viscous according to 2.2.3.1.4) (vapour pressure at 50 °C not more than 110 kPa)	3	F1	III	3		5 L	E1	P001 IBC02 R001	BB4	MP19		
1143	CROTONALDEHYDE or CROTONALDEHYDE, STABILIZED	6.1	TF1	I	6.1 +3	324 354 386	0	E0	P602		MP8 MP17	T20	TP2
1144	CROTONYLENE	3	F1	I	3		0	E3	P001		MP7 MP17	T11	TP2
1145	CYCLOHEXANE	3	F1	II	3		1 L	E2	P001 IBC02 R001		MP19	T4	TP1

ADR tank		Vehicle for tank carriage	Transport category (Tunnel restriction code)	Special provisions for carriage				Hazard identifi-cation No.	UN No.	Name and description
Tank code	Special provisions			Packages	Bulk	Loading, unloading and handling	Operation			
4.3	4.3.5, 6.8.4	9.1.1.2	1.1.3.6 (8.6)	7.2.4	7.3.3	7.5.11	8.5	5.3.2.3		3.1.2
(12)	(13)	(14)	(15)	(16)	(17)	(18)	(19)	(20)	(1)	(2)
			3 (E)				S2		1133	ADHESIVES containing flammable liquid (having a flash-point below 23 °C and viscous according to 2.2.3.1.4) (vapour pressure at 50 °C more than 110 kPa)
			3 (E)				S2		1133	ADHESIVES containing flammable liquid (having a flash-point below 23 °C and viscous according to 2.2.3.1.4) (vapour pressure at 50 °C not more than 110 kPa)
LGBF		FL	3 (D/E)	V12			S2	30	1134	CHLOROBENZENE
L10CH	TU14 TU15 TE19 TE21	FL	1 (C/D)			CV1 CV13 CV28	S2 S9 S14	663	1135	ETHYLENE CHLOROHYDRIN
LGBF		FL	2 (D/E)				S2 S20	33	1136	COAL TAR DISTILLATES, FLAMMABLE
LGBF		FL	3 (D/E)	V12			S2	30	1136	COAL TAR DISTILLATES, FLAMMABLE
L4BN		FL	1 (D/E)				S2 S20	33	1139	COATING SOLUTION (includes surface treatments or coatings used for industrial or other purposes such as vehicle under coating, drum or barrel lining)
L1.5BN		FL	2 (D/E)				S2 S20	33	1139	COATING SOLUTION (includes surface treatments or coatings used for industrial or other purposes such as vehicle under coating, drum or barrel lining) (vapour pressure at 50 °C more than 110 kPa)
LGBF		FL	2 (D/E)				S2 S20	33	1139	COATING SOLUTION (includes surface treatments or coatings used for industrial or other purposes such as vehicle under coating, drum or barrel lining) (vapour pressure at 50 °C not more than 110 kPa)
LGBF		FL	3 (D/E)	V12			S2	30	1139	COATING SOLUTION (includes surface treatments or coatings used for industrial or other purposes such as vehicle under coating, drum or barrel lining)
			3 (E)				S2		1139	COATING SOLUTION (includes surface treatments or coatings used for industrial or other purposes such as vehicle under coating, drum or barrel lining) (having a flash-point below 23 °C and viscous according to 2.2.3.1.4) (vapour pressure at 50 °C more than 110 kPa)
			3 (E)				S2		1139	COATING SOLUTION (includes surface treatments or coatings used for industrial or other purposes such as vehicle under coating, drum or barrel lining) (having a flash-point below 23 °C and viscous according to 2.2.3.1.4) (vapour pressure at 50 °C not more than 110 kPa)
L10CH	TU14 TU15 TE19 TE21	FL	1 (C/D)	V8		CV1 CV13 CV28	S2 S4 S9 S14	663	1143	CROTONALDEHYDE or CROTONALDEHYDE, STABILIZED
L4BN		FL	1 (D/E)				S2 S20	339	1144	CROTONYLENE
LGBF		FL	2 (D/E)				S2 S20	33	1145	CYCLOHEXANE

UN No.	Name and description	Class	Classifi-cation code	Packing group	Labels	Special provisions	Limited and excepted quantities		Packaging			Portable tanks and bulk containers	
									Packing instruc-tions	Special packing provisions	Mixed packing provisions	Instruc-tions	Special provisions
	3.1.2	2.2	2.2	2.1.1.3	5.2.2	3.3	3.4	3.5.1.2	4.1.4	4.1.4	4.1.10	4.2.5.2 7.3.2	4.2.5.3
(1)	(2)	(3a)	(3b)	(4)	(5)	(6)	(7a)	(7b)	(8)	(9a)	(9b)	(10)	(11)
1146	CYCLOPENTANE	3	F1	II	3		1 L	E2	P001 IBC02 R001		MP19	T7	TP1
1147	DECAHYDRO-NAPHTHALENE	3	F1	III	3		5 L	E1	P001 IBC03 LP01 R001		MP19	T2	TP1
1148	DIACETONE ALCOHOL	3	F1	II	3		1 L	E2	P001 IBC02 R001		MP19	T4	TP1
1148	DIACETONE ALCOHOL	3	F1	III	3		5 L	E1	P001 IBC03 LP01 R001		MP19	T2	TP1
1149	DIBUTYL ETHERS	3	F1	III	3		5 L	E1	P001 IBC03 LP01 R001		MP19	T2	TP1
1150	1,2-DICHLOROETHYLENE	3	F1	II	3		1 L	E2	P001 IBC02 R001		MP19	T7	TP2
1152	DICHLOROPENTANES	3	F1	III	3		5 L	E1	P001 IBC03 LP01 R001		MP19	T2	TP1
1153	ETHYLENE GLYCOL DIETHYL ETHER	3	F1	II	3		1 L	E2	P001 IBC02 R001		MP19	T4	TP1
1153	ETHYLENE GLYCOL DIETHYL ETHER	3	F1	III	3		5 L	E1	P001 IBC03 LP01 R001		MP19	T2	TP1
1154	DIETHYLAMINE	3	FC	II	3 +8		1 L	E2	P001 IBC02		MP19	T7	TP1
1155	DIETHYL ETHER (ETHYL ETHER)	3	F1	I	3		0	E3	P001		MP7 MP17	T11	TP2
1156	DIETHYL KETONE	3	F1	II	3		1 L	E2	P001 IBC02 R001		MP19	T4	TP1
1157	DIISOBUTYL KETONE	3	F1	III	3		5 L	E1	P001 IBC03 LP01 R001		MP19	T2	TP1
1158	DIISOPROPYLAMINE	3	FC	II	3 +8		1 L	E2	P001 IBC02		MP19	T7	TP1
1159	DIISOPROPYL ETHER	3	F1	II	3		1 L	E2	P001 IBC02 R001		MP19	T4	TP1
1160	DIMETHYLAMINE AQUEOUS SOLUTION	3	FC	II	3 +8		1 L	E2	P001 IBC02		MP19	T7	TP1
1161	DIMETHYL CARBONATE	3	F1	II	3		1 L	E2	P001 IBC02 R001		MP19	T4	TP1
1162	DIMETHYLDICHLORO-SILANE	3	FC	II	3 +8		0	E0	P010		MP19	T10	TP2 TP7
1163	DIMETHYLHYDRAZINE, UNSYMMETRICAL	6.1	TFC	I	6.1 +3 +8	354	0	E0	P602		MP8 MP17	T20	TP2
1164	DIMETHYL SULPHIDE	3	F1	II	3		1 L	E2	P001 IBC02	B8	MP19	T7	TP2
1165	DIOXANE	3	F1	II	3		1 L	E2	P001 IBC02 R001		MP19	T4	TP1
1166	DIOXOLANE	3	F1	II	3		1 L	E2	P001 IBC02 R001		MP19	T4	TP1
1167	DIVINYL ETHER, STABILIZED	3	F1	I	3	386	0	E3	P001		MP7 MP17	T11	TP2
1169	EXTRACTS, AROMATIC, LIQUID (vapour pressure at 50 °C more than 110 kPa)	3	F1	II	3	601 640C	5 L	E2	P001		MP19	T4	TP1 TP8
1169	EXTRACTS, AROMATIC, LIQUID (vapour pressure at 50 °C not more than 110 kPa)	3	F1	II	3	601 640D	5 L	E2	P001 IBC02 R001		MP19	T4	TP1 TP8

ADR tank		Vehicle for tank carriage	Transport category (Tunnel restriction code)	Special provisions for carriage				Hazard identification No.	UN No.	Name and description
Tank code	Special provisions			Packages	Bulk	Loading, unloading and handling	Operation			
4.3	4.3.5, 6.8.4	9.1.1.2	1.1.3.6 (8.6)	7.2.4	7.3.3	7.5.11	8.5	5.3.2.3		3.1.2
(12)	(13)	(14)	(15)	(16)	(17)	(18)	(19)	(20)	(1)	(2)
LGBF		FL	2 (D/E)				S2 S20	33	1146	CYCLOPENTANE
LGBF		FL	3 (D/E)	V12			S2	30	1147	DECAHYDRO-NAPHTHALENE
LGBF		FL	2 (D/E)				S2 S20	33	1148	DIACETONE ALCOHOL
LGBF		FL	3 (D/E)	V12			S2	30	1148	DIACETONE ALCOHOL
LGBF		FL	3 (D/E)	V12			S2	30	1149	DIBUTYL ETHERS
LGBF		FL	2 (D/E)				S2 S20	33	1150	1,2-DICHLOROETHYLENE
LGBF		FL	3 (D/E)	V12			S2	30	1152	DICHLOROPENTANES
LGBF		FL	2 (D/E)				S2 S20	33	1153	ETHYLENE GLYCOL DIETHYL ETHER
LGBF		FL	3 (D/E)	V12			S2	30	1153	ETHYLENE GLYCOL DIETHYL ETHER
L4BH		FL	2 (D/E)				S2 S20	338	1154	DIETHYLAMINE
L4BN		FL	1 (D/E)				S2 S20	33	1155	DIETHYL ETHER (ETHYL ETHER)
LGBF		FL	2 (D/E)				S2 S20	33	1156	DIETHYL KETONE
LGBF		FL	3 (D/E)	V12			S2	30	1157	DIISOBUTYL KETONE
L4BH		FL	2 (D/E)				S2 S20	338	1158	DIISOPROPYLAMINE
LGBF		FL	2 (D/E)				S2 S20	33	1159	DIISOPROPYL ETHER
L4BH		FL	2 (D/E)				S2 S20	338	1160	DIMETHYLAMINE AQUEOUS SOLUTION
LGBF		FL	2 (D/E)				S2 S20	33	1161	DIMETHYL CARBONATE
L4BH		FL	2 (D/E)				S2 S20	X338	1162	DIMETHYLDICHLORO-SILANE
L10CH	TU14 TU15 TE19 TE21	FL	1 (C/D)			CV1 CV13 CV28	S2 S9 S14	663	1163	DIMETHYLHYDRAZINE, UNSYMMETRICAL
L1.5BN		FL	2 (D/E)				S2 S20	33	1164	DIMETHYL SULPHIDE
LGBF		FL	2 (D/E)				S2 S20	33	1165	DIOXANE
LGBF		FL	2 (D/E)				S2 S20	33	1166	DIOXOLANE
L4BN		FL	1 (D/E)	V8			S2 S4 S20	339	1167	DIVINYL ETHER, STABILIZED
L1.5BN		FL	2 (D/E)				S2 S20	33	1169	EXTRACTS, AROMATIC, LIQUID (vapour pressure at 50 °C more than 110 kPa)
LGBF		FL	2 (D/E)				S2 S20	33	1169	EXTRACTS, AROMATIC, LIQUID (vapour pressure at 50 °C not more than 110 kPa)

UN No.	Name and description	Class	Classifi-cation code	Packing group	Labels	Special provi-sions	Limited and excepted quantities		Packaging			Portable tanks and bulk containers	
									Packing instruc-tions	Special packing provisions	Mixed packing provisions	Instruc-tions	Special provisions
3.1.2		2.2	2.2	2.1.1.3	5.2.2	3.3	3.4	3.5.1.2	4.1.4	4.1.4	4.1.10	4.2.5.2 7.3.2	4.2.5.3
(1)	(2)	(3a)	(3b)	(4)	(5)	(6)	(7a)	(7b)	(8)	(9a)	(9b)	(10)	(11)
1169	EXTRACTS, AROMATIC, LIQUID	3	F1	III	3	601	5 L	E1	P001 IBC03 LP01 R001		MP19	T2	TP1
1169	EXTRACTS, AROMATIC, LIQUID (having a flash-point below 23 °C and viscous according to 2.2.3.1.4) (vapour pressure at 50 °C more than 110 kPa)	3	F1	III	3	601	5 L	E1	P001 R001		MP19		
1169	EXTRACTS, AROMATIC, LIQUID (having a flash-point below 23 °C and viscous according to 2.2.3.1.4) (vapour pressure at 50 °C not more than 110 kPa)	3	F1	III	3	601	5 L	E1	P001 IBC02 R001	BB4	MP19		
1170	ETHANOL (ETHYL ALCOHOL) or ETHANOL SOLUTION (ETHYL ALCOHOL SOLUTION)	3	F1	II	3	144 601	1 L	E2	P001 IBC02 R001		MP19	T4	TP1
1170	ETHANOL SOLUTION (ETHYL ALCOHOL SOLUTION)	3	F1	III	3	144 601	5 L	E1	P001 IBC03 LP01 R001		MP19	T2	TP1
1171	ETHYLENE GLYCOL MONOETHYL ETHER	3	F1	III	3		5 L	E1	P001 IBC03 LP01 R001		MP19	T2	TP1
1172	ETHYLENE GLYCOL MONOETHYL ETHER ACETATE	3	F1	III	3		5 L	E1	P001 IBC03 LP01 R001		MP19	T2	TP1
1173	ETHYL ACETATE	3	F1	II	3		1 L	E2	P001 IBC02 R001		MP19	T4	TP1
1175	ETHYLBENZENE	3	F1	II	3		1 L	E2	P001 IBC02 R001		MP19	T4	TP1
1176	ETHYL BORATE	3	F1	II	3		1 L	E2	P001 IBC02 R001		MP19	T4	TP1
1177	2-ETHYLBUTYL ACETATE	3	F1	III	3		5 L	E1	P001 IBC03 LP01 R001		MP19	T2	TP1
1178	2-ETHYLBUTYRALDEHYDE	3	F1	II	3		1 L	E2	P001 IBC02 R001		MP19	T4	TP1
1179	ETHYL BUTYL ETHER	3	F1	II	3		1 L	E2	P001 IBC02 R001		MP19	T4	TP1
1180	ETHYL BUTYRATE	3	F1	III	3		5 L	E1	P001 IBC03 LP01 R001		MP19	T2	TP1
1181	ETHYL CHLOROACETATE	6.1	TF1	II	6.1 +3		100 ml	E4	P001 IBC02		MP15	T7	TP2
1182	ETHYL CHLOROFORMATE	6.1	TFC	I	6.1 +3 +8	354	0	E0	P602		MP8 MP17	T20	TP2
1183	ETHYLDICHLOROSILANE	4.3	WFC	I	4.3 +3 +8		0	E0	P401	RR7	MP2	T14	TP2 TP7
1184	ETHYLENE DICHLORIDE	3	FT1	II	3 +6.1		1 L	E2	P001 IBC02		MP19	T7	TP1
1185	ETHYLENEIMINE, STABILIZED	6.1	TF1	I	6.1 +3	354 386	0	E0	P601		MP2	T22	TP2
1188	ETHYLENE GLYCOL MONOMETHYL ETHER	3	F1	III	3		5 L	E1	P001 IBC03 LP01 R001		MP19	T2	TP1
1189	ETHYLENE GLYCOL MONOMETHYL ETHER ACETATE	3	F1	III	3		5 L	E1	P001 IBC03 LP01 R001		MP19	T2	TP1

ADR tank		Vehicle for tank carriage	Transport category (Tunnel restriction code)	Special provisions for carriage				Hazard identifi- cation No.	UN No.	Name and description
Tank code	Special provisions			Packages	Bulk	Loading, unloading and handling	Operation			
4.3	4.3.5, 6.8.4	9.1.1.2	1.1.3.6 (8.6)	7.2.4	7.3.3	7.5.11	8.5	5.3.2.3		3.1.2
(12)	(13)	(14)	(15)	(16)	(17)	(18)	(19)	(20)	(1)	(2)
LGBF		FL	3 (D/E)	V12			S2	30	1169	EXTRACTS, AROMATIC, LIQUID
			3 (E)				S2		1169	EXTRACTS, AROMATIC, LIQUID (having a flash-point below 23 °C and viscous according to 2.2.3.1.4) (vapour pressure at 50 °C more than 110 kPa)
			3 (E)				S2		1169	EXTRACTS, AROMATIC, LIQUID (having a flash-point below 23 °C and viscous according to 2.2.3.1.4) (vapour pressure at 50 °C not more than 110 kPa)
LGBF		FL	2 (D/E)				S2 S20	33	1170	ETHANOL (ETHYL ALCOHOL) or ETHANOL SOLUTION (ETHYL ALCOHOL SOLUTION)
LGBF		FL	3 (D/E)	V12			S2	30	1170	ETHANOL SOLUTION (ETHYL ALCOHOL SOLUTION)
LGBF		FL	3 (D/E)	V12			S2	30	1171	ETHYLENE GLYCOL MONOETHYL ETHER
LGBF		FL	3 (D/E)	V12			S2	30	1172	ETHYLENE GLYCOL MONOETHYL ETHER ACETATE
LGBF		FL	2 (D/E)				S2 S20	33	1173	ETHYL ACETATE
LGBF		FL	2 (D/E)				S2 S20	33	1175	ETHYLBENZENE
LGBF		FL	2 (D/E)				S2 S20	33	1176	ETHYL BORATE
LGBF		FL	3 (D/E)	V12			S2	30	1177	2-ETHYLBUTYL ACETATE
LGBF		FL	2 (D/E)				S2 S20	33	1178	2-ETHYLBUTYRALDEHYDE
LGBF		FL	2 (D/E)				S2 S20	33	1179	ETHYL BUTYL ETHER
LGBF		FL	3 (D/E)	V12			S2	30	1180	ETHYL BUTYRATE
L4BH	TU15 TE19	FL	2 (D/E)			CV13 CV28	S2 S9 S19	63	1181	ETHYL CHLOROACETATE
L10CH	TU14 TU15 TE19 TE21	FL	1 (C/D)			CV1 CV13 CV28	S2 S9 S14	663	1182	ETHYL CHLOROFORMATE
L10DH	TU14 TU23 TE21 TM2 TM3	FL	0 (B/E)	V1		CV23	S2 S20	X338	1183	ETHYLDICHLOROSILANE
L4BH	TU15	FL	2 (D/E)			CV13 CV28	S2 S19	336	1184	ETHYLENE DICHLORIDE
L15CH	TU14 TU15 TE19 TE21	FL	1 (C/D)	V8		CV1 CV13 CV28	S2 S4 S9 S14	663	1185	ETHYLENEIMINE, STABILIZED
LGBF		FL	3 (D/E)	V12			S2	30	1188	ETHYLENE GLYCOL MONOMETHYL ETHER
LGBF		FL	3 (D/E)	V12			S2	30	1189	ETHYLENE GLYCOL MONOMETHYL ETHER ACETATE

UN No.	Name and description	Class	Classifi-cation code	Packing group	Labels	Special provi-sions	Limited and excepted quantities		Packaging			Portable tanks and bulk containers	
									Packing instruc-tions	Special packing provisions	Mixed packing provisions	Instruc-tions	Special provisions
	3.1.2	2.2	2.2	2.1.1.3	5.2.2	3.3	3.4	3.5.1.2	4.1.4	4.1.4	4.1.10	4.2.5.2 7.3.2	4.2.5.3
(1)	(2)	(3a)	(3b)	(4)	(5)	(6)	(7a)	(7b)	(8)	(9a)	(9b)	(10)	(11)
1190	ETHYL FORMATE	3	F1	II	3		1 L	E2	P001 IBC02 R001		MP19	T4	TP1
1191	OCTYL ALDEHYDES	3	F1	III	3		5 L	E1	P001 IBC03 LP01 R001		MP19	T2	TP1
1192	ETHYL LACTATE	3	F1	III	3		5 L	E1	P001 IBC03 LP01 R001		MP19	T2	TP1
1193	ETHYL METHYL KETONE (METHYL ETHYL KETONE)	3	F1	II	3		1 L	E2	P001 IBC02 R001		MP19	T4	TP1
1194	ETHYL NITRITE SOLUTION	3	FT1	I	3 +6.1		0	E0	P001		MP7 MP17		
1195	ETHYL PROPIONATE	3	F1	II	3		1 L	E2	P001 IBC02 R001		MP19	T4	TP1
1196	ETHYLTRICHLOROSILANE	3	FC	II	3 +8		0	E0	P010		MP19	T10	TP2 TP7
1197	EXTRACTS, FLAVOURING, LIQUID (vapour pressure at 50 °C more than 110 kPa)	3	F1	II	3	601 640C	5 L	E2	P001		MP19	T4	TP1 TP8
1197	EXTRACTS, FLAVOURING, LIQUID (vapour pressure at 50 °C not more than 110 kPa)	3	F1	II	3	601 640D	5 L	E2	P001 IBC02 R001		MP19	T4	TP1 TP8
1197	EXTRACTS, FLAVOURING, LIQUID	3	F1	III	3	601	5 L	E1	P001 IBC03 LP01 R001		MP19	T2	TP1
1197	EXTRACTS, FLAVOURING, LIQUID (having a flash-point below 23 °C and viscous according to 2.2.3.1.4) (vapour pressure at 50 °C more than 110 kPa)	3	F1	III	3	601	5 L	E1	P001 R001		MP19		
1197	EXTRACTS, FLAVOURING, LIQUID (having a flash-point below 23 °C and viscous according to 2.2.3.1.4) (vapour pressure at 50 °C not more than 110 kPa)	3	F1	III	3	601	5 L	E1	P001 IBC02 R001	BB4	MP19		
1198	FORMALDEHYDE SOLUTION, FLAMMABLE	3	FC	III	3 +8		5 L	E1	P001 IBC03 R001		MP19	T4	TP1
1199	FURALDEHYDES	6.1	TF1	II	6.1 +3		100 ml	E4	P001 IBC02		MP15	T7	TP2
1201	FUSEL OIL	3	F1	II	3		1 L	E2	P001 IBC02 R001		MP19	T4	TP1
1201	FUSEL OIL	3	F1	III	3		5 L	E1	P001 IBC03 LP01 R001		MP19	T2	TP1
1202	GAS OIL or DIESEL FUEL or HEATING OIL, LIGHT (flash-point not more than 60 °C)	3	F1	III	3	640K 664	5 L	E1	P001 IBC03 LP01 R001		MP19	T2	TP1
1202	DIESEL FUEL complying with standard EN 590:2013 + AC:2014 or GAS OIL or HEATING OIL, LIGHT with a flash-point as specified in EN 590:2013 + AC:2014	3	F1	III	3	640L 664	5 L	E1	P001 IBC03 LP01 R001		MP19	T2	TP1
1202	GAS OIL or DIESEL FUEL or HEATING OIL, LIGHT (flash-point more than 60 °C and not more than 100 °C)	3	F1	III	3	640M 664	5 L	E1	P001 IBC03 LP01 R001		MP19	T2	TP1
1203	MOTOR SPIRIT or GASOLINE or PETROL	3	F1	II	3	243 534 664	1 L	E2	P001 IBC02 R001	BB2	MP19	T4	TP1
1204	NITROGLYCERIN SOLUTION IN ALCOHOL with not more than 1% nitroglycerin	3	D	II	3	601	1 L	E0	P001 IBC02	PP5	MP2		

ADR tank		Vehicle for tank carriage	Transport category (Tunnel restriction code)	Special provisions for carriage				Hazard identifi-cation No.	UN No.	Name and description
Tank code	Special provisions			Packages	Bulk	Loading, unloading and handling	Operation			
4.3	4.3.5, 6.8.4	9.1.1.2	1.1.3.6 (8.6)	7.2.4	7.3.3	7.5.11	8.5	5.3.2.3		3.1.2
(12)	(13)	(14)	(15)	(16)	(17)	(18)	(19)	(20)	(1)	(2)
LGBF		FL	2 (D/E)				S2 S20	33	1190	ETHYL FORMATE
LGBF		FL	3 (D/E)	V12			S2	30	1191	OCTYL ALDEHYDES
LGBF		FL	3 (D/E)	V12			S2	30	1192	ETHYL LACTATE
LGBF		FL	2 (D/E)				S2 S20	33	1193	ETHYL METHYL KETONE (METHYL ETHYL KETONE)
L10CH	TU14 TU15 TE21	FL	1 (C/E)			CV13 CV28	S2 S22	336	1194	ETHYL NITRITE SOLUTION
LGBF		FL	2 (D/E)				S2 S20	33	1195	ETHYL PROPIONATE
L4BH		FL	2 (D/E)				S2 S20	X338	1196	ETHYLTRICHLOROSILANE
L1.5BN		FL	2 (D/E)				S2 S20	33	1197	EXTRACTS, FLAVOURING, LIQUID (vapour pressure at 50 °C more than 110 kPa)
LGBF		FL	2 (D/E)				S2 S20	33	1197	EXTRACTS, FLAVOURING, LIQUID (vapour pressure at 50 °C not more than 110 kPa)
LGBF		FL	3 (D/E)	V12			S2	30	1197	EXTRACTS, FLAVOURING, LIQUID
			3 (E)				S2		1197	EXTRACTS, FLAVOURING, LIQUID (having a flash-point below 23 °C and viscous according to 2.2.3.1.4) (vapour pressure at 50 °C more than 110 kPa)
			3 (E)				S2		1197	EXTRACTS, FLAVOURING, LIQUID (having a flash-point below 23 °C and viscous according to 2.2.3.1.4) (vapour pressure at 50 °C not more than 110 kPa)
L4BN		FL	3 (D/E)	V12			S2	38	1198	FORMALDEHYDE SOLUTION, FLAMMABLE
L4BH	TU15 TE19	FL	2 (D/E)			CV13 CV28	S2 S9 S19	63	1199	FURALDEHYDES
LGBF		FL	2 (D/E)				S2 S20	33	1201	FUSEL OIL
LGBF		FL	3 (D/E)	V12			S2	30	1201	FUSEL OIL
LGBF		FL	3 (D/E)	V12			S2	30	1202	GAS OIL or DIESEL FUEL or HEATING OIL, LIGHT (flash-point not more than 60 °C)
LGBF		AT	3 (D/E)	V12			S2	30	1202	DIESEL FUEL complying with standard EN 590:2013 + AC:2014 or GAS OIL or HEATING OIL, LIGHT with a flash-point as specified in EN 590:2013 + AC:2014
LGBV		AT	3 (D/E)	V12				30	1202	GAS OIL or DIESEL FUEL or HEATING OIL, LIGHT (flash-point more than 60 °C and not more than 100 °C)
LGBF	TU9	FL	2 (D/E)				S2 S20	33	1203	MOTOR SPIRIT or GASOLINE or PETROL
			2 (B)				S2 S14		1204	NITROGLYCERIN SOLUTION IN ALCOHOL with not more than 1% nitroglycerin

UN No.	Name and description	Class	Classifi- cation code	Packing group	Labels	Special provi- sions	Limited and excepted quantities		Packaging			Portable tanks and bulk containers	
									Packing instruc- tions	Special packing provisions	Mixed packing provisions	Instruc- tions	Special provisions
	3.1.2	2.2	2.2	2.1.1.3	5.2.2	3.3	3.4	3.5.1.2	4.1.4	4.1.4	4.1.10	4.2.5.2 7.3.2	4.2.5.3
(1)	(2)	(3a)	(3b)	(4)	(5)	(6)	(7a)	(7b)	(8)	(9a)	(9b)	(10)	(11)
1206	HEPTANES	3	F1	II	3		1 L	E2	P001 IBC02 R001		MP19	T4	TP1
1207	HEXALDEHYDE	3	F1	III	3		5 L	E1	P001 IBC03 LP01 R001		MP19	T2	TP1
1208	HEXANES	3	F1	II	3		1 L	E2	P001 IBC02 R001		MP19	T4	TP1
1210	PRINTING INK, flammable or PRINTING INK RELATED MATERIAL (including printing ink thinning or reducing compound), flammable	3	F1	I	3	163 367	500 ml	E3	P001		MP7 MP17	T11	TP1 TP8
1210	PRINTING INK, flammable or PRINTING INK RELATED MATERIAL (including printing ink thinning or reducing compound), flammable (vapour pressure at 50 °C more than 110 kPa)	3	F1	II	3	163 367 640C	5 L	E2	P001	PP1	MP19	T4	TP1 TP8
1210	PRINTING INK, flammable or PRINTING INK RELATED MATERIAL (including printing ink thinning or reducing compound), flammable (vapour pressure at 50 °C not more than 110 kPa)	3	F1	II	3	163 367 640D	5 L	E2	P001 IBC02 R001	PP1	MP19	T4	TP1 TP8
1210	PRINTING INK, flammable or PRINTING INK RELATED MATERIAL (including printing ink thinning or reducing compound), flammable	3	F1	III	3	163 367	5 L	E1	P001 IBC03 LP01 R001	PP1	MP19	T2	TP1
1210	PRINTING INK, flammable or PRINTING INK RELATED MATERIAL (including printing ink thinning or reducing compound), flammable (having a flash-point below 23 °C and viscous according to 2.2.3.1.4) (vapour pressure at 50 °C more than 110 kPa)	3	F1	III	3	163 367	5 L	E1	P001 R001	PP1	MP19		
1210	PRINTING INK, flammable or PRINTING INK RELATED MATERIAL (including printing ink thinning or reducing compound), flammable (having a flash-point below 23 °C and viscous according to 2.2.3.1.4) (vapour pressure at 50 °C not more than 110 kPa)	3	F1	III	3	163 367	5 L	E1	P001 IBC02 R001	PP1 BB4	MP19		
1212	ISOBUTANOL (ISOBUTYL ALCOHOL)	3	F1	III	3		5 L	E1	P001 IBC03 LP01 R001		MP19	T2	TP1
1213	ISOBUTYL ACETATE	3	F1	II	3		1 L	E2	P001 IBC02 R001		MP19	T4	TP1
1214	ISOBUTYLAMINE	3	FC	II	3 +8		1 L	E2	P001 IBC02		MP19	T7	TP1
1216	ISOOCTENES	3	F1	II	3		1 L	E2	P001 IBC02 R001		MP19	T4	TP1
1218	ISOPRENE, STABILIZED	3	F1	I	3	386	0	E3	P001		MP7 MP17	T11	TP2
1219	ISOPROPANOL (ISOPROPYL ALCOHOL)	3	F1	II	3	601	1 L	E2	P001 IBC02 R001		MP19	T4	TP1
1220	ISOPROPYL ACETATE	3	F1	II	3		1 L	E2	P001 IBC02 R001		MP19	T4	TP1
1221	ISOPROPYLAMINE	3	FC	I	3 +8		0	E0	P001		MP7 MP17	T11	TP2

ADR tank		Vehicle for tank carriage	Transport category (Tunnel restriction code)	Special provisions for carriage				Hazard identifi-cation No.	UN No.	Name and description
Tank code	Special provisions			Packages	Bulk	Loading, unloading and handling	Operation			
4.3	4.3.5, 6.8.4	9.1.1.2	1.1.3.6 (8.6)	7.2.4	7.3.3	7.5.11	8.5	5.3.2.3		3.1.2
(12)	(13)	(14)	(15)	(16)	(17)	(18)	(19)	(20)	(1)	(2)
LGBF		FL	2 (D/E)				S2 S20	33	1206	HEPTANES
LGBF		FL	3 (D/E)	V12			S2	30	1207	HEXALDEHYDE
LGBF		FL	2 (D/E)				S2 S20	33	1208	HEXANES
L4BN		FL	1 (D/E)				S2 S20	33	1210	PRINTING INK, flammable or PRINTING INK RELATED MATERIAL (including printing ink thinning or reducing compound), flammable
L1.5BN		FL	2 (D/E)				S2 S20	33	1210	PRINTING INK, flammable or PRINTING INK RELATED MATERIAL (including printing ink thinning or reducing compound), flammable (vapour pressure at 50 °C more than 110 kPa)
LGBF		FL	2 (D/E)				S2 S20	33	1210	PRINTING INK, flammable or PRINTING INK RELATED MATERIAL (including printing ink thinning or reducing compound), flammable (vapour pressure at 50 °C not more than 110 kPa)
LGBF		FL	3 (D/E)	V12			S2	30	1210	PRINTING INK, flammable or PRINTING INK RELATED MATERIAL (including printing ink thinning or reducing compound), flammable
			3 (E)				S2		1210	PRINTING INK, flammable or PRINTING INK RELATED MATERIAL (including printing ink thinning or reducing compound), flammable (having a flash-point below 23 °C and viscous according to 2.2.3.1.4) (vapour pressure at 50 °C more than 110 kPa)
			3 (E)				S2		1210	PRINTING INK, flammable or PRINTING INK RELATED MATERIAL (including printing ink thinning or reducing compound), flammable (having a flash-point below 23 °C and viscous according to 2.2.3.1.4) (vapour pressure at 50 °C not more than 110 kPa)
LGBF		FL	3 (D/E)	V12			S2	30	1212	ISOBUTANOL (ISOBUTYL ALCOHOL)
LGBF		FL	2 (D/E)				S2 S20	33	1213	ISOBUTYL ACETATE
L4BH		FL	2 (D/E)				S2 S20	338	1214	ISOBUTYLAMINE
LGBF		FL	2 (D/E)				S2 S20	33	1216	ISOOCTENES
L4BN		FL	1 (D/E)	V8			S2 S4 S20	339	1218	ISOPRENE, STABILIZED
LGBF		FL	2 (D/E)				S2 S20	33	1219	ISOPROPANOL (ISOPROPYL ALCOHOL)
LGBF		FL	2 (D/E)				S2 S20	33	1220	ISOPROPYL ACETATE
L10CH	TU14 TE21	FL	1 (C/E)				S2 S20	338	1221	ISOPROPYLAMINE

UN No.	Name and description	Class	Classification code	Packing group	Labels	Special provisions	Limited and excepted quantities		Packaging			Portable tanks and bulk containers	
									Packing instructions	Special packing provisions	Mixed packing provisions	Instructions	Special provisions
	3.1.2	2.2	2.2	2.1.1.3	5.2.2	3.3	3.4	3.5.1.2	4.1.4	4.1.4	4.1.10	4.2.5.2 7.3.2	4.2.5.3
(1)	(2)	(3a)	(3b)	(4)	(5)	(6)	(7a)	(7b)	(8)	(9a)	(9b)	(10)	(11)
1222	ISOPROPYL NITRATE	3	F1	II	3		1 L	E2	P001 IBC02 R001	B7	MP19		
1223	KEROSENE	3	F1	III	3	664	5 L	E1	P001 IBC03 LP01 R001		MP19	T2	TP2
1224	KETONES, LIQUID, N.O.S. (vapour pressure at 50 °C more than 110 kPa)	3	F1	II	3	274 640C	1 L	E2	P001		MP19	T7	TP1 TP8 TP28
1224	KETONES, LIQUID, N.O.S. (vapour pressure at 50 °C not more than 110 kPa)	3	F1	II	3	274 640D	1 L	E2	P001 IBC02 R001		MP19	T7	TP1 TP8 TP28
1224	KETONES, LIQUID, N.O.S.	3	F1	III	3	274	5 L	E1	P001 IBC03 LP01 R001		MP19	T4	TP1 TP29
1228	MERCAPTANS, LIQUID, FLAMMABLE, TOXIC, N.O.S. or MERCAPTAN MIXTURE, LIQUID, FLAMMABLE, TOXIC, N.O.S.	3	FT1	II	3 +6.1	274	1 L	E0	P001 IBC02		MP19	T11	TP2 TP27
1228	MERCAPTANS, LIQUID, FLAMMABLE, TOXIC, N.O.S. or MERCAPTAN MIXTURE, LIQUID, FLAMMABLE, TOXIC, N.O.S.	3	FT1	III	3 +6.1	274	5 L	E1	P001 IBC03 R001		MP19	T7	TP1 TP28
1229	MESITYL OXIDE	3	F1	III	3		5 L	E1	P001 IBC03 LP01 R001		MP19	T2	TP1
1230	METHANOL	3	FT1	II	3 +6.1	279	1 L	E2	P001 IBC02		MP19	T7	TP2
1231	METHYL ACETATE	3	F1	II	3		1 L	E2	P001 IBC02 R001		MP19	T4	TP1
1233	METHYLAMYL ACETATE	3	F1	III	3		5 L	E1	P001 IBC03 LP01 R001		MP19	T2	TP1
1234	METHYLAL	3	F1	II	3		1 L	E2	P001 IBC02	B8	MP19	T7	TP2
1235	METHYLAMINE, AQUEOUS SOLUTION	3	FC	II	3 +8		1 L	E2	P001 IBC02		MP19	T7	TP1
1237	METHYL BUTYRATE	3	F1	II	3		1 L	E2	P001 IBC02 R001		MP19	T4	TP1
1238	METHYL CHLOROFORMATE	6.1	TFC	I	6.1 +3 +8	354	0	E0	P602		MP8 MP17	T22	TP2
1239	METHYL CHLORO-METHYL ETHER	6.1	TF1	I	6.1 +3	354	0	E0	P602		MP8 MP17	T22	TP2
1242	METHYLDICHLOROSILANE	4.3	WFC	I	4.3 +3 +8		0	E0	P401	RR7	MP2	T14	TP2 TP7
1243	METHYL FORMATE	3	F1	I	3		0	E3	P001		MP7 MP17	T11	TP2
1244	METHYLHYDRAZINE	6.1	TFC	I	6.1 +3 +8	354	0	E0	P602		MP8 MP17	T22	TP2
1245	METHYL ISOBUTYL KETONE	3	F1	II	3		1 L	E2	P001 IBC02 R001		MP19	T4	TP1
1246	METHYL ISOPROPENYL KETONE, STABILIZED	3	F1	II	3	386	1 L	E2	P001 IBC02 R001		MP19	T4	TP1
1247	METHYL METHACRYLATE MONOMER, STABILIZED	3	F1	II	3	386	1 L	E2	P001 IBC02 R001		MP19	T4	TP1
1248	METHYL PROPIONATE	3	F1	II	3		1 L	E2	P001 IBC02 R001		MP19	T4	TP1
1249	METHYL PROPYL KETONE	3	F1	II	3		1 L	E2	P001 IBC02 R001		MP19	T4	TP1

ADR tank		Vehicle for tank carriage	Transport category (Tunnel restriction code)	Special provisions for carriage				Hazard identification No.	UN No.	Name and description
Tank code	Special provisions			Packages	Bulk	Loading, unloading and handling	Operation			
4.3	4.3.5, 6.8.4	9.1.1.2	1.1.3.6 (8.6)	7.2.4	7.3.3	7.5.11	8.5	5.3.2.3		3.1.2
(12)	(13)	(14)	(15)	(16)	(17)	(18)	(19)	(20)	(1)	(2)
			2 (E)				S2 S20		1222	ISOPROPYL NITRATE
LGBF		FL	3 (D/E)	V12			S2	30	1223	KEROSENE
L1.5BN		FL	2 (D/E)				S2 S20	33	1224	KETONES, LIQUID, N.O.S. (vapour pressure at 50 °C more than 110 kPa)
LGBF		FL	2 (D/E)				S2 S20	33	1224	KETONES, LIQUID, N.O.S. (vapour pressure at 50 °C not more than 110 kPa)
LGBF		FL	3 (D/E)	V12			S2	30	1224	KETONES, LIQUID, N.O.S.
L4BH	TU15	FL	2 (D/E)			CV13 CV28	S2 S19	336	1228	MERCAPTANS, LIQUID, FLAMMABLE, TOXIC, N.O.S. or MERCAPTAN MIXTURE, LIQUID, FLAMMABLE, TOXIC, N.O.S.
L4BH	TU15	FL	3 (D/E)	V12		CV13 CV28	S2	36	1228	MERCAPTANS, LIQUID, FLAMMABLE, TOXIC, N.O.S. or MERCAPTAN MIXTURE, LIQUID, FLAMMABLE, TOXIC, N.O.S.
LGBF		FL	3 (D/E)	V12			S2	30	1229	MESITYL OXIDE
L4BH	TU15	FL	2 (D/E)			CV13 CV28	S2 S19	336	1230	METHANOL
LGBF		FL	2 (D/E)				S2 S20	33	1231	METHYL ACETATE
LGBF		FL	3 (D/E)	V12			S2	30	1233	METHYLAMYL ACETATE
L1.5BN		FL	2 (D/E)				S2 S20	33	1234	METHYLAL
L4BH		FL	2 (D/E)				S2 S20	338	1235	METHYLAMINE, AQUEOUS SOLUTION
LGBF		FL	2 (D/E)				S2 S20	33	1237	METHYL BUTYRATE
L15CH	TU14 TU15 TE19 TE21	FL	1 (C/D)			CV1 CV13 CV28	S2 S9 S14	663	1238	METHYL CHLOROFORMATE
L15CH	TU14 TU15 TE19 TE21	FL	1 (C/D)			CV1 CV13 CV28	S2 S9 S14	663	1239	METHYL CHLORO-METHYL ETHER
L10DH	TU14 TU24 TE21 TM2 TM3	FL	0 (B/E)	V1		CV23	S2 S20	X338	1242	METHYLDICHLOROSILANE
L4BN		FL	1 (D/E)				S2 S20	33	1243	METHYL FORMATE
L15CH	TU14 TU15 TE19 TE21	FL	1 (C/D)			CV1 CV13 CV28	S2 S9 S14	663	1244	METHYLHYDRAZINE
LGBF		FL	2 (D/E)				S2 S20	33	1245	METHYL ISOBUTYL KETONE
LGBF		FL	2 (D/E)	V8			S2 S4 S20	339	1246	METHYL ISOPROPENYL KETONE, STABILIZED
LGBF		FL	2 (D/E)	V8			S2 S4 S20	339	1247	METHYL METHACRYLATE MONOMER, STABILIZED
LGBF		FL	2 (D/E)				S2 S20	33	1248	METHYL PROPIONATE
LGBF		FL	2 (D/E)				S2 S20	33	1249	METHYL PROPYL KETONE

UN No.	Name and description	Class	Classifi-cation code	Packing group	Labels	Special provi-sions	Limited and excepted quantities		Packaging			Portable tanks and bulk containers	
									Packing instruc-tions	Special packing provisions	Mixed packing provisions	Instruc-tions	Special provisions
	3.1.2	2.2	2.2	2.1.1.3	5.2.2	3.3	3.4	3.5.1.2	4.1.4	4.1.4	4.1.10	4.2.5.2 7.3.2	4.2.5.3
(1)	(2)	(3a)	(3b)	(4)	(5)	(6)	(7a)	(7b)	(8)	(9a)	(9b)	(10)	(11)
1250	METHYLTRICHLORO-SILANE	3	FC	II	3 +8		0	E0	P010		MP19	T10	TP2 TP7
1251	METHYL VINYL KETONE, STABILIZED	6.1	TFC	I	6.1 +3 +8	354 386	0	E0	P601	RR7	MP8 MP17	T22	TP2
1259	NICKEL CARBONYL	6.1	TF1	I	6.1 +3		0	E0	P601		MP2		
1261	NITROMETHANE	3	F1	II	3		1 L	E0	P001 R001	RR2	MP19		
1262	OCTANES	3	F1	II	3		1 L	E2	P001 IBC02 R001		MP19	T4	TP1
1263	PAINT (including paint, lacquer, enamel, stain, shellac, varnish, polish, liquid filler and liquid lacquer base) or PAINT RELATED MATERIAL (including paint thinning and reducing compound)	3	F1	I	3	163 367 650	500 ml	E3	P001		MP7 MP17	T11	TP1 TP8 TP27
1263	PAINT (including paint, lacquer, enamel, stain, shellac, varnish, polish, liquid filler and liquid lacquer base) or PAINT RELATED MATERIAL (including paint thinning and reducing compound) (vapour pressure at 50 °C more than 110 kPa)	3	F1	II	3	163 367 640C 650	5 L	E2	P001	PP1	MP19	T4	TP1 TP8 TP28
1263	PAINT (including paint, lacquer, enamel, stain, shellac, varnish, polish, liquid filler and liquid lacquer base) or PAINT RELATED MATERIAL (including paint thinning and reducing compound) (vapour pressure at 50 °C not more than 110 kPa)	3	F1	II	3	163 367 640D 650	5 L	E2	P001 IBC02 R001	PP1	MP19	T4	TP1 TP8 TP28
1263	PAINT (including paint, lacquer, enamel, stain, shellac, varnish, polish, liquid filler and liquid lacquer base) or PAINT RELATED MATERIAL (including paint thinning and reducing compound)	3	F1	III	3	163 367 650	5 L	E1	P001 IBC03 LP01 R001	PP1	MP19	T2	TP1 TP29
1263	PAINT (including paint, lacquer, enamel, stain, shellac, varnish, polish, liquid filler and liquid lacquer base) or PAINT RELATED MATERIAL (including paint thinning and reducing compound) (having a flash-point below 23 °C and viscous according to 2.2.3.1.4) (vapour pressure at 50 °C more than 110 kPa)	3	F1	III	3	163 367 650	5 L	E1	P001 R001	PP1	MP19		
1263	PAINT (including paint, lacquer, enamel, stain, shellac, varnish, polish, liquid filler and liquid lacquer base) or PAINT RELATED MATERIAL (including paint thinning and reducing compound) (having a flash-point below 23 °C and viscous according to 2.2.3.1.4) (vapour pressure at 50 °C not more than 110 kPa)	3	F1	III	3	163 367 650	5 L	E1	P001 IBC02 R001	PP1 BB4	MP19		
1264	PARALDEHYDE	3	F1	III	3		5 L	E1	P001 IBC03 LP01 R001		MP19	T2	TP1
1265	PENTANES, liquid	3	F1	I	3		0	E3	P001		MP7 MP17	T11	TP2
1265	PENTANES, liquid	3	F1	II	3		1 L	E2	P001 IBC02	B8	MP19	T4	TP1

ADR tank		Vehicle for tank carriage	Transport category (Tunnel restriction code)	Special provisions for carriage				Hazard identifi- cation No.	UN No.	Name and description
Tank code	Special provisions			Packages	Bulk	Loading, unloading and handling	Operation			
4.3	4.3.5, 6.8.4	9.1.1.2	1.1.3.6 (8.6)	7.2.4	7.3.3	7.5.11	8.5	5.3.2.3		3.1.2
(12)	(13)	(14)	(15)	(16)	(17)	(18)	(19)	(20)	(1)	(2)
L4BH		FL	2 (D/E)				S2 S20	X338	1250	METHYLTRICHLORO-SILANE
L15CH	TU14 TU15 TE19 TE21	FL	1 (C/D)	V8		CV1 CV13 CV28	S2 S4 S9 S14	639	1251	METHYL VINYL KETONE, STABILIZED
L15CH	TU14 TU15 TU31 TE19 TE21 TM3	FL	1 (C/D)			CV1 CV13 CV28	S2 S9 S14	663	1259	NICKEL CARBONYL
			2 (E)				S2 S20		1261	NITROMETHANE
LGBF		FL	2 (D/E)				S2 S20	33	1262	OCTANES
L4BN		FL	1 (D/E)				S2 S20	33	1263	PAINT (including paint, lacquer, enamel, stain, shellac, varnish, polish, liquid filler and liquid lacquer base) or PAINT RELATED MATERIAL (including paint thinning and reducing compound)
L1.5BN		FL	2 (D/E)				S2 S20	33	1263	PAINT (including paint, lacquer, enamel, stain, shellac, varnish, polish, liquid filler and liquid lacquer base) or PAINT RELATED MATERIAL (including paint thinning and reducing compound) (vapour pressure at 50 °C more than 110 kPa)
LGBF		FL	2 (D/E)				S2 S20	33	1263	PAINT (including paint, lacquer, enamel, stain, shellac, varnish, polish, liquid filler and liquid lacquer base) or PAINT RELATED MATERIAL (including paint thinning and reducing compound) (vapour pressure at 50 °C not more than 110 kPa)
LGBF		FL	3 (D/E)	V12			S2	30	1263	PAINT (including paint, lacquer, enamel, stain, shellac, varnish, polish, liquid filler and liquid lacquer base) or PAINT RELATED MATERIAL (including paint thinning and reducing compound)
			3 (E)				S2		1263	PAINT (including paint, lacquer, enamel, stain, shellac, varnish, polish, liquid filler and liquid lacquer base) or PAINT RELATED MATERIAL (including paint thinning and reducing compound) (having a flash-point below 23 °C and viscous according to 2.2.3.1.4) (vapour pressure at 50 °C more than 110 kPa)
			3 (E)				S2		1263	PAINT (including paint, lacquer, enamel, stain, shellac, varnish, polish, liquid filler and liquid lacquer base) or PAINT RELATED MATERIAL (including paint thinning and reducing compound) (having a flash-point below 23 °C and viscous according to 2.2.3.1.4) (vapour pressure at 50 °C not more than 110 kPa)
LGBF		FL	3 (D/E)	V12			S2	30	1264	PARALDEHYDE
L4BN		FL	1 (D/E)				S2 S20	33	1265	PENTANES, liquid
L1.5BN		FL	2 (D/E)				S2 S20	33	1265	PENTANES, liquid

UN No.	Name and description	Class	Classification code	Packing group	Labels	Special provisions	Limited and excepted quantities		Packaging			Portable tanks and bulk containers	
									Packing instructions	Special packing provisions	Mixed packing provisions	Instructions	Special provisions
	3.1.2	2.2	2.2	2.1.1.3	5.2.2	3.3	3.4	3.5.1.2	4.1.4	4.1.4	4.1.10	4.2.5.2 7.3.2	4.2.5.3
(1)	(2)	(3a)	(3b)	(4)	(5)	(6)	(7a)	(7b)	(8)	(9a)	(9b)	(10)	(11)
1266	PERFUMERY PRODUCTS with flammable solvents (vapour pressure at 50 °C more than 110 kPa)	3	F1	II	3	163 640C	5 L	E2	P001		MP19	T4	TP1 TP8
1266	PERFUMERY PRODUCTS with flammable solvents (vapour pressure at 50 °C not more than 110 kPa)	3	F1	II	3	163 640D	5 L	E2	P001 IBC02 R001		MP19	T4	TP1 TP8
1266	PERFUMERY PRODUCTS with flammable solvents	3	F1	III	3	163	5 L	E1	P001 IBC03 LP01 R001		MP19	T2	TP1
1266	PERFUMERY PRODUCTS with flammable solvents (having a flash-point below 23 °C and viscous according to 2.2.3.1.4) (vapour pressure at 50 °C more than 110 kPa)	3	F1	III	3	163	5 L	E1	P001 R001		MP19		
1266	PERFUMERY PRODUCTS with flammable solvents (having a flash-point below 23 °C and viscous according to 2.2.3.1.4) (vapour pressure at 50 °C not more than 110 kPa)	3	F1	III	3	163	5 L	E1	P001 IBC02 R001	BB4	MP19		
1267	PETROLEUM CRUDE OIL	3	F1	I	3	357	500 ml	E3	P001		MP7 MP17	T11	TP1 TP8
1267	PETROLEUM CRUDE OIL (vapour pressure at 50 °C more than 110 kPa)	3	F1	II	3	357 640C	1 L	E2	P001		MP19	T4	TP1 TP8
1267	PETROLEUM CRUDE OIL (vapour pressure at 50 °C not more than 110 kPa)	3	F1	II	3	357 640D	1 L	E2	P001 IBC02 R001		MP19	T4	TP1 TP8
1267	PETROLEUM CRUDE OIL	3	F1	III	3	357	5 L	E1	P001 IBC03 LP01 R001		MP19	T2	TP1
1268	PETROLEUM DISTILLATES, N.O.S. or PETROLEUM PRODUCTS, N.O.S.	3	F1	I	3	664	500 ml	E3	P001		MP7 MP17	T11	TP1 TP8
1268	PETROLEUM DISTILLATES, N.O.S. or PETROLEUM PRODUCTS, N.O.S. (vapour pressure at 50 °C more than 110 kPa)	3	F1	II	3	640C 664	1 L	E2	P001		MP19	T7	TP1 TP8 TP28
1268	PETROLEUM DISTILLATES, N.O.S. or PETROLEUM PRODUCTS, N.O.S. (vapour pressure at 50 °C not more than 110 kPa)	3	F1	II	3	640D 664	1 L	E2	P001 IBC02 R001		MP19	T7	TP1 TP8 TP28
1268	PETROLEUM DISTILLATES, N.O.S. or PETROLEUM PRODUCTS, N.O.S.	3	F1	III	3	664	5 L	E1	P001 IBC03 LP01 R001		MP19	T4	TP1 TP29
1272	PINE OIL	3	F1	III	3		5 L	E1	P001 IBC03 LP01 R001		MP19	T2	TP1
1274	n-PROPANOL (PROPYL ALCOHOL, NORMAL)	3	F1	II	3		1 L	E2	P001 IBC02 R001		MP19	T4	TP1
1274	n-PROPANOL (PROPYL ALCOHOL, NORMAL)	3	F1	III	3		5 L	E1	P001 IBC03 LP01 R001		MP19	T2	TP1
1275	PROPIONALDEHYDE	3	F1	II	3		1 L	E2	P001 IBC02 R001		MP19	T7	TP1
1276	n-PROPYL ACETATE	3	F1	II	3		1 L	E2	P001 IBC02 R001		MP19	T4	TP1
1277	PROPYLAMINE	3	FC	II	3 +8		1 L	E2	P001 IBC02		MP19	T7	TP1
1278	1-CHLOROPROPANE	3	F1	II	3		1 L	E0	P001 IBC02	B8	MP19	T7	TP2

ADR tank		Vehicle for tank carriage	Transport category (Tunnel restriction code)	Special provisions for carriage				Hazard identifi-cation No.	UN No.	Name and description
Tank code	Special provisions			Packages	Bulk	Loading, unloading and handling	Operation			
4.3	4.3.5, 6.8.4	9.1.1.2	1.1.3.6 (8.6)	7.2.4	7.3.3	7.5.11	8.5	5.3.2.3		3.1.2
(12)	(13)	(14)	(15)	(16)	(17)	(18)	(19)	(20)	(1)	(2)
L1.5BN		FL	2 (D/E)				S2 S20	33	1266	PERFUMERY PRODUCTS with flammable solvents (vapour pressure at 50 °C more than 110 kPa)
LGBF		FL	2 (D/E)				S2 S20	33	1266	PERFUMERY PRODUCTS with flammable solvents (vapour pressure at 50 °C not more than 110 kPa)
LGBF		FL	3 (D/E)	V12			S2	30	1266	PERFUMERY PRODUCTS with flammable solvents
			3 (E)				S2		1266	PERFUMERY PRODUCTS with flammable solvents (having a flash-point below 23 °C and viscous according to 2.2.3.1.4) (vapour pressure at 50 °C more than 110 kPa)
			3 (E)				S2		1266	PERFUMERY PRODUCTS with flammable solvents (having a flash-point below 23 °C and viscous according to 2.2.3.1.4) (vapour pressure at 50 °C not more than 110 kPa)
L4BN		FL	1 (D/E)				S2 S20	33	1267	PETROLEUM CRUDE OIL
L1.5BN		FL	2 (D/E)				S2 S20	33	1267	PETROLEUM CRUDE OIL (vapour pressure at 50 °C more than 110 kPa)
LGBF		FL	2 (D/E)				S2 S20	33	1267	PETROLEUM CRUDE OIL (vapour pressure at 50 °C not more than 110 kPa)
LGBF		FL	3 (D/E)	V12			S2	30	1267	PETROLEUM CRUDE OIL
L4BN		FL	1 (D/E)				S2 S20	33	1268	PETROLEUM DISTILLATES, N.O.S. or PETROLEUM PRODUCTS, N.O.S.
L1.5BN		FL	2 (D/E)				S2 S20	33	1268	PETROLEUM DISTILLATES, N.O.S. or PETROLEUM PRODUCTS, N.O.S. (vapour pressure at 50 °C more than 110 kPa)
LGBF		FL	2 (D/E)				S2 S20	33	1268	PETROLEUM DISTILLATES, N.O.S. or PETROLEUM PRODUCTS, N.O.S. (vapour pressure at 50 °C not more than 110 kPa)
LGBF		FL	3 (D/E)	V12			S2	30	1268	PETROLEUM DISTILLATES, N.O.S. or PETROLEUM PRODUCTS, N.O.S.
LGBF		FL	3 (D/E)	V12			S2	30	1272	PINE OIL
LGBF		FL	2 (D/E)				S2 S20	33	1274	n-PROPANOL (PROPYL ALCOHOL, NORMAL)
LGBF		FL	3 (D/E)	V12			S2	30	1274	n-PROPANOL (PROPYL ALCOHOL, NORMAL)
LGBF		FL	2 (D/E)				S2 S20	33	1275	PROPIONALDEHYDE
LGBF		FL	2 (D/E)				S2 S20	33	1276	n-PROPYL ACETATE
L4BH		FL	2 (D/E)				S2 S20	338	1277	PROPYLAMINE
L1.5BN		FL	2 (D/E)				S2 S20	33	1278	1-CHLOROPROPANE

UN No.	Name and description	Class	Classification code	Packing group	Labels	Special provisions	Limited and excepted quantities		Packaging			Portable tanks and bulk containers	
									Packing instructions	Special packing provisions	Mixed packing provisions	Instructions	Special provisions
	3.1.2	2.2	2.2	2.1.1.3	5.2.2	3.3	3.4	3.5.1.2	4.1.4	4.1.4	4.1.10	4.2.5.2 7.3.2	4.2.5.3
(1)	(2)	(3a)	(3b)	(4)	(5)	(6)	(7a)	(7b)	(8)	(9a)	(9b)	(10)	(11)
1279	1,2-DICHLOROPROPANE	3	F1	II	3		1 L	E2	P001 IBC02 R001		MP19	T4	TP1
1280	PROPYLENE OXIDE	3	F1	I	3		0	E3	P001		MP7 MP17	T11	TP2 TP7
1281	PROPYL FORMATES	3	F1	II	3		1 L	E2	P001 IBC02 R001		MP19	T4	TP1
1282	PYRIDINE	3	F1	II	3		1 L	E2	P001 IBC02 R001		MP19	T4	TP2
1286	ROSIN OIL (vapour pressure at 50 °C more than 110 kPa)	3	F1	II	3	640C	5 L	E2	P001		MP19	T4	TP1
1286	ROSIN OIL (vapour pressure at 50 °C not more than 110 kPa)	3	F1	II	3	640D	5 L	E2	P001 IBC02 R001		MP19	T4	TP1
1286	ROSIN OIL	3	F1	III	3		5 L	E1	P001 IBC03 LP01 R001		MP19	T2	TP1
1286	ROSIN OIL (having a flash-point below 23 °C and viscous according to 2.2.3.1.4) (vapour pressure at 50 °C more than 110 kPa)	3	F1	III	3		5 L	E1	P001 R001		MP19		
1286	ROSIN OIL (having a flash-point below 23 °C and viscous according to 2.2.3.1.4) (vapour pressure at 50 °C not more than 110 kPa)	3	F1	III	3		5 L	E1	P001 IBC02 R001	BB4	MP19		
1287	RUBBER SOLUTION (vapour pressure at 50 °C more than 110 kPa)	3	F1	II	3	640C	5 L	E2	P001		MP19	T4	TP1 TP8
1287	RUBBER SOLUTION (vapour pressure at 50 °C not more than 110 kPa)	3	F1	II	3	640D	5 L	E2	P001 IBC02 R001		MP19	T4	TP1 TP8
1287	RUBBER SOLUTION	3	F1	III	3		5 L	E1	P001 IBC03 LP01 R001		MP19	T2	TP1
1287	RUBBER SOLUTION (having a flash point below 23 °C and viscous according to 2.2.3.1.4) (vapour pressure at 50 °C more than 110 kPa)	3	F1	III	3		5 L	E1	P001 R001		MP19		
1287	RUBBER SOLUTION (having a flash point below 23 °C and viscous according to 2.2.3.1.4) (vapour pressure at 50 °C not more than 110 kPa)	3	F1	III	3		5 L	E1	P001 IBC02 R001	BB4	MP19		
1288	SHALE OIL	3	F1	II	3		1 L	E2	P001 IBC02 R001		MP19	T4	TP1 TP8
1288	SHALE OIL	3	F1	III	3		5 L	E1	P001 IBC03 LP01 R001		MP19	T2	TP1
1289	SODIUM METHYLATE SOLUTION in alcohol	3	FC	II	3 +8		1 L	E2	P001 IBC02		MP19	T7	TP1 TP8
1289	SODIUM METHYLATE SOLUTION in alcohol	3	FC	III	3 +8		5 L	E1	P001 IBC02 R001		MP19	T4	TP1
1292	TETRAETHYL SILICATE	3	F1	III	3		5 L	E1	P001 IBC03 LP01 R001		MP19	T2	TP1
1293	TINCTURES, MEDICINAL	3	F1	II	3	601	1 L	E2	P001 IBC02 R001		MP19	T4	TP1 TP8
1293	TINCTURES, MEDICINAL	3	F1	III	3	601	5 L	E1	P001 IBC03 LP01 R001		MP19	T2	TP1
1294	TOLUENE	3	F1	II	3		1 L	E2	P001 IBC02 R001		MP19	T4	TP1

ADR tank		Vehicle for tank carriage	Transport category (Tunnel restriction code)	Special provisions for carriage				Hazard identification No.	UN No.	Name and description
Tank code	Special provisions			Packages	Bulk	Loading, unloading and handling	Operation			
4.3	4.3.5, 6.8.4	9.1.1.2	1.1.3.6 (8.6)	7.2.4	7.3.3	7.5.11	8.5	5.3.2.3		3.1.2
(12)	(13)	(14)	(15)	(16)	(17)	(18)	(19)	(20)	(1)	(2)
LGBF		FL	2 (D/E)				S2 S20	33	1279	1,2-DICHLOROPROPANE
L4BN		FL	1 (D/E)				S2 S20	33	1280	PROPYLENE OXIDE
LGBF		FL	2 (D/E)				S2 S20	33	1281	PROPYL FORMATES
LGBF		FL	2 (D/E)				S2 S20	33	1282	PYRIDINE
L1.5BN		FL	2 (D/E)				S2 S20	33	1286	ROSIN OIL (vapour pressure at 50 °C more than 110 kPa)
LGBF		FL	2 (D/E)				S2 S20	33	1286	ROSIN OIL (vapour pressure at 50 °C not more than 110 kPa)
LGBF		FL	3 (D/E)	V12			S2	30	1286	ROSIN OIL
			3 (E)				S2		1286	ROSIN OIL (having a flash-point below 23 °C and viscous according to 2.2.3.1.4) (vapour pressure at 50 °C more than 110 kPa)
			3 (E)				S2		1286	ROSIN OIL (having a flash-point below 23 °C and viscous according to 2.2.3.1.4) (vapour pressure at 50 °C not more than 110 kPa)
L1.5BN		FL	2 (D/E)				S2 S20	33	1287	RUBBER SOLUTION (vapour pressure at 50 °C more than 110 kPa)
LGBF		FL	2 (D/E)				S2 S20	33	1287	RUBBER SOLUTION (vapour pressure at 50 °C not more than 110 kPa)
LGBF		FL	3 (D/E)	V12			S2	30	1287	RUBBER SOLUTION
			3 (E)				S2		1287	RUBBER SOLUTION (having a flash point below 23 °C and viscous according to 2.2.3.1.4) (vapour pressure at 50 °C more than 110 kPa)
			3 (E)				S2		1287	RUBBER SOLUTION (having a flash point below 23 °C and viscous according to 2.2.3.1.4) (vapour pressure at 50 °C not more than 110 kPa)
LGBF		FL	2 (D/E)				S2 S20	33	1288	SHALE OIL
LGBF		FL	3 (D/E)	V12			S2	30	1288	SHALE OIL
L4BH		FL	2 (D/E)				S2 S20	338	1289	SODIUM METHYLATE SOLUTION in alcohol
L4BN		FL	3 (D/E)				S2	38	1289	SODIUM METHYLATE SOLUTION in alcohol
LGBF		FL	3 (D/E)	V12			S2	30	1292	TETRAETHYL SILICATE
LGBF		FL	2 (D/E)				S2 S20	33	1293	TINCTURES, MEDICINAL
LGBF		FL	3 (D/E)	V12			S2	30	1293	TINCTURES, MEDICINAL
LGBF		FL	2 (D/E)				S2 S20	33	1294	TOLUENE

UN No.	Name and description	Class	Classifi-cation code	Packing group	Labels	Special provi-sions	Limited and excepted quantities		Packaging			Portable tanks and bulk containers	
									Packing instruc-tions	Special packing provisions	Mixed packing provisions	Instruc-tions	Special provisions
3.1.2	3.1.2	2.2	2.2	2.1.1.3	5.2.2	3.3	3.4	3.5.1.2	4.1.4	4.1.4	4.1.10	4.2.5.2 7.3.2	4.2.5.3
(1)	(2)	(3a)	(3b)	(4)	(5)	(6)	(7a)	(7b)	(8)	(9a)	(9b)	(10)	(11)
1295	TRICHLOROSILANE	4.3	WFC	I	4.3 +3 +8		0	E0	P401	RR7	MP2	T14	TP2 TP7
1296	TRIETHYLAMINE	3	FC	II	3 +8		1 L	E2	P001 IBC02		MP19	T7	TP1
1297	TRIMETHYLAMINE, AQUEOUS SOLUTION, not more than 50% trimethylamine, by mass	3	FC	I	3 +8		0	E0	P001		MP7 MP17	T11	TP1
1297	TRIMETHYLAMINE, AQUEOUS SOLUTION, not more than 50% trimethylamine, by mass	3	FC	II	3 +8		1 L	E2	P001 IBC02		MP19	T7	TP1
1297	TRIMETHYLAMINE, AQUEOUS SOLUTION, not more than 50% trimethylamine, by mass	3	FC	III	3 +8		5 L	E1	P001 IBC03 R001		MP19	T7	TP1
1298	TRIMETHYLCHLORO-SILANE	3	FC	II	3 +8		0	E0	P010		MP19	T10	TP2 TP7
1299	TURPENTINE	3	F1	III	3		5 L	E1	P001 IBC03 LP01 R001		MP19	T2	TP1
1300	TURPENTINE SUBSTITUTE	3	F1	II	3		1 L	E2	P001 IBC02 R001		MP19	T4	TP1
1300	TURPENTINE SUBSTITUTE	3	F1	III	3		5 L	E1	P001 IBC03 LP01 R001		MP19	T2	TP1
1301	VINYL ACETATE, STABILIZED	3	F1	II	3	386	1 L	E2	P001 IBC02 R001		MP19	T4	TP1
1302	VINYL ETHYL ETHER, STABILIZED	3	F1	I	3	386	0	E3	P001		MP7 MP17	T11	TP2
1303	VINYLIDENE CHLORIDE, STABILIZED	3	F1	I	3	386	0	E3	P001		MP7 MP17	T12	TP2 TP7
1304	VINYL ISOBUTYL ETHER, STABILIZED	3	F1	II	3	386	1 L	E2	P001 IBC02 R001		MP19	T4	TP1
1305	VINYLTRICHLOROSILANE	3	FC	II	3 +8		0	E0	P010		MP19	T10	TP2 TP7
1306	WOOD PRESERVATIVES, LIQUID (vapour pressure at 50 °C more than 110 kPa)	3	F1	II	3	640C	5 L	E2	P001		MP19	T4	TP1 TP8
1306	WOOD PRESERVATIVES, LIQUID (vapour pressure at 50 °C not more than 110 kPa)	3	F1	II	3	640D	5 L	E2	P001 IBC02 R001		MP19	T4	TP1 TP8
1306	WOOD PRESERVATIVES, LIQUID	3	F1	III	3		5 L	E1	P001 IBC03 LP01 R001		MP19	T2	TP1
1306	WOOD PRESERVATIVES, LIQUID (having a flash-point below 23 °C and viscous according to 2.2.3.1.4) (vapour pressure at 50 °C more than 110 kPa)	3	F1	III	3		5 L	E1	P001 R001		MP19		
1306	WOOD PRESERVATIVES, LIQUID (having a flash-point below 23 °C and viscous according to 2.2.3.1.4) (vapour pressure at 50 °C not more than 110 kPa)	3	F1	III	3		5 L	E1	P001 IBC02 R001	BB4	MP19		
1307	XYLENES	3	F1	II	3		1 L	E2	P001 IBC02 R001		MP19	T4	TP1
1307	XYLENES	3	F1	III	3		5 L	E1	P001 IBC03 LP01 R001		MP19	T2	TP1
1308	ZIRCONIUM SUSPENDED IN A FLAMMABLE LIQUID	3	F1	I	3		0	E0	P001	PP33	MP7 MP17		

ADR tank Tank code (4.3) (12)	ADR tank Special provisions (4.3.5, 6.8.4) (13)	Vehicle for tank carriage (9.1.1.2) (14)	Transport category (Tunnel restriction code) (1.1.3.6) (8.6) (15)	Special provisions for carriage Packages (7.2.4) (16)	Bulk (7.3.3) (17)	Loading, unloading and handling (7.5.11) (18)	Operation (8.5) (19)	Hazard identification No. (5.3.2.3) (20)	UN No. (1)	Name and description (3.1.2) (2)
L10DH	TU14 TU25 TE21 TM2 TM3	FL	0 (B/E)	V1		CV23	S2 S20	X338	1295	TRICHLOROSILANE
L4BH		FL	2 (D/E)				S2 S20	338	1296	TRIETHYLAMINE
L10CH	TU14 TE21	FL	1 (C/E)				S2 S20	338	1297	TRIMETHYLAMINE, AQUEOUS SOLUTION, not more than 50% trimethylamine, by mass
L4BH		FL	2 (D/E)				S2 S20	338	1297	TRIMETHYLAMINE, AQUEOUS SOLUTION, not more than 50% trimethylamine, by mass
L4BN		FL	3 (D/E)	V12			S2	38	1297	TRIMETHYLAMINE, AQUEOUS SOLUTION, not more than 50% trimethylamine, by mass
L4BH		FL	2 (D/E)				S2 S20	X338	1298	TRIMETHYLCHLORO-SILANE
LGBF		FL	3 (D/E)	V12			S2	30	1299	TURPENTINE
LGBF		FL	2 (D/E)				S2 S20	33	1300	TURPENTINE SUBSTITUTE
LGBF		FL	3 (D/E)	V12			S2	30	1300	TURPENTINE SUBSTITUTE
LGBF		FL	2 (D/E)	V8			S2 S4 S20	339	1301	VINYL ACETATE, STABILIZED
L4BN		FL	1 (D/E)	V8			S2 S4 S20	339	1302	VINYL ETHYL ETHER, STABILIZED
L4BN		FL	1 (D/E)	V8			S2 S4 S20	339	1303	VINYLIDENE CHLORIDE, STABILIZED
LGBF		FL	2 (D/E)	V8			S2 S4 S20	339	1304	VINYL ISOBUTYL ETHER, STABILIZED
L4BH		FL	2 (D/E)				S2 S20	X338	1305	VINYLTRICHLOROSILANE
L1.5BN		FL	2 (D/E)				S2 S20	33	1306	WOOD PRESERVATIVES, LIQUID (vapour pressure at 50 °C more than 110 kPa)
LGBF		FL	2 (D/E)				S2 S20	33	1306	WOOD PRESERVATIVES, LIQUID (vapour pressure at 50 °C not more than 110 kPa)
LGBF		FL	3 (D/E)	V12			S2	30	1306	WOOD PRESERVATIVES, LIQUID
			3 (E)				S2		1306	WOOD PRESERVATIVES, LIQUID (having a flash-point below 23 °C and viscous according to 2.2.3.1.4) (vapour pressure at 50 °C more than 110 kPa)
			3 (E)				S2		1306	WOOD PRESERVATIVES, LIQUID (having a flash-point below 23 °C and viscous according to 2.2.3.1.4) (vapour pressure at 50 °C not more than 110 kPa)
LGBF		FL	2 (D/E)				S2 S20	33	1307	XYLENES
LGBF		FL	3 (D/E)	V12			S2	30	1307	XYLENES
L4BN		FL	1 (D/E)				S2 S20	33	1308	ZIRCONIUM SUSPENDED IN A FLAMMABLE LIQUID

UN No.	Name and description	Class	Classifi-cation code	Packing group	Labels	Special provi-sions	Limited and excepted quantities		Packaging			Portable tanks and bulk containers	
									Packing instruc-tions	Special packing provisions	Mixed packing provisions	Instruc-tions	Special provisions
	3.1.2	2.2	2.2	2.1.1.3	5.2.2	3.3	3.4	3.5.1.2	4.1.4	4.1.4	4.1.10	4.2.5.2 7.3.2	4.2.5.3
(1)	(2)	(3a)	(3b)	(4)	(5)	(6)	(7a)	(7b)	(8)	(9a)	(9b)	(10)	(11)
1308	ZIRCONIUM SUSPENDED IN A FLAMMABLE LIQUID (vapour pressure at 50 °C more than 110 kPa)	3	F1	II	3	640C	1 L	E2	P001 R001	PP33	MP19		
1308	ZIRCONIUM SUSPENDED IN A FLAMMABLE LIQUID (vapour pressure at 50 °C not more than 110 kPa)	3	F1	II	3	640D	1 L	E2	P001 R001	PP33	MP19		
1308	ZIRCONIUM SUSPENDED IN A FLAMMABLE LIQUID	3	F1	III	3		5 L	E1	P001 R001		MP19		
1309	ALUMINIUM POWDER, COATED	4.1	F3	II	4.1		1 kg	E2	P002 IBC08	PP38 B4	MP11	T3	TP33
1309	ALUMINIUM POWDER, COATED	4.1	F3	III	4.1		5 kg	E1	P002 IBC08 LP02 R001	PP11 B3	MP11	T1	TP33
1310	AMMONIUM PICRATE, WETTED with not less than 10% water, by mass	4.1	D	I	4.1		0	E0	P406	PP26	MP2		
1312	BORNEOL	4.1	F1	III	4.1		5 kg	E1	P002 IBC08 LP02 R001	B3	MP10	T1	TP33
1313	CALCIUM RESINATE	4.1	F3	III	4.1		5 kg	E1	P002 IBC06 R001		MP11	T1	TP33
1314	CALCIUM RESINATE, FUSED	4.1	F3	III	4.1		5 kg	E1	P002 IBC04 R001		MP11	T1	TP33
1318	COBALT RESINATE, PRECIPITATED	4.1	F3	III	4.1		5 kg	E1	P002 IBC06 R001		MP11	T1	TP33
1320	DINITROPHENOL, WETTED with not less than 15% water, by mass	4.1	DT	I	4.1 +6.1		0	E0	P406	PP26	MP2		
1321	DINITROPHENOLATES, WETTED with not less than 15% water, by mass	4.1	DT	I	4.1 +6.1		0	E0	P406	PP26	MP2		
1322	DINITRORESORCINOL, WETTED with not less than 15% water, by mass	4.1	D	I	4.1		0	E0	P406	PP26	MP2		
1323	FERROCERIUM	4.1	F3	II	4.1	249	1 kg	E2	P002 IBC08	B4	MP11	T3	TP33
1324	FILMS, NITROCELLULOSE BASE, gelatin coated, except scrap	4.1	F1	III	4.1		5 kg	E1	P002 R001	PP15	MP11		
1325	FLAMMABLE SOLID, ORGANIC, N.O.S.	4.1	F1	II	4.1	274	1 kg	E2	P002 IBC08	B4	MP10	T3	TP33
1325	FLAMMABLE SOLID, ORGANIC, N.O.S.	4.1	F1	III	4.1	274	5 kg	E1	P002 IBC08 LP02 R001	B3	MP10	T1	TP33
1326	HAFNIUM POWDER, WETTED with not less than 25% water	4.1	F3	II	4.1	586	1 kg	E2	P410 IBC06	PP40	MP11	T3	TP33
1327	Hay, Straw or Bhusa	4.1	F1				NOT SUBJECT TO ADR						
1328	HEXAMETHYLENETE-TRAMINE	4.1	F1	III	4.1		5 kg	E1	P002 IBC08 R001	B3	MP10	T1	TP33
1330	MANGANESE RESINATE	4.1	F3	III	4.1		5 kg	E1	P002 IBC06 R001		MP11	T1	TP33
1331	MATCHES, 'STRIKE ANYWHERE'	4.1	F1	III	4.1	293	5 kg	E0	P407	PP27	MP12		
1332	METALDEHYDE	4.1	F1	III	4.1		5 kg	E1	P002 IBC08 LP02 R001	B3	MP10	T1	TP33
1333	CERIUM, slabs, ingots or rods	4.1	F3	II	4.1		1 kg	E2	P002 IBC08	B4	MP11		
1334	NAPHTHALENE, CRUDE or NAPHTHALENE, REFINED	4.1	F1	III	4.1	501	5 kg	E1	P002 IBC08 LP02 R001	B3	MP10	T1 BK1 BK2 BK3	TP33
1336	NITROGUANIDINE (PICRITE), WETTED with not less than 20% water, by mass	4.1	D	I	4.1		0	E0	P406		MP2		

ADR tank		Vehicle for tank carriage	Transport category (Tunnel restriction code)	Special provisions for carriage				Hazard identification No.	UN No.	Name and description
Tank code	Special provisions			Packages	Bulk	Loading, unloading and handling	Operation			
4.3	4.3.5, 6.8.4	9.1.1.2	1.1.3.6 (8.6)	7.2.4	7.3.3	7.5.11	8.5	5.3.2.3		3.1.2
(12)	(13)	(14)	(15)	(16)	(17)	(18)	(19)	(20)	(1)	(2)
L1.5BN		FL	2 (D/E)				S2 S20	33	1308	ZIRCONIUM SUSPENDED IN A FLAMMABLE LIQUID (vapour pressure at 50 °C more than 110 kPa)
LGBF		FL	2 (D/E)				S2 S20	33	1308	ZIRCONIUM SUSPENDED IN A FLAMMABLE LIQUID (vapour pressure at 50 °C not more than 110 kPa)
LGBF		FL	3 (D/E)				S2	30	1308	ZIRCONIUM SUSPENDED IN A FLAMMABLE LIQUID
SGAN		AT	2 (E)	V11				40	1309	ALUMINIUM POWDER, COATED
SGAV		AT	3 (E)		VC1 VC2			40	1309	ALUMINIUM POWDER, COATED
			1 (B)				S14		1310	AMMONIUM PICRATE, WETTED with not less than 10% water, by mass
SGAV		AT	3 (E)		VC1 VC2			40	1312	BORNEOL
SGAV		AT	3 (E)		VC1 VC2			40	1313	CALCIUM RESINATE
SGAV		AT	3 (E)		VC1 VC2			40	1314	CALCIUM RESINATE, FUSED
SGAV		AT	3 (E)		VC1 VC2			40	1318	COBALT RESINATE, PRECIPITATED
			1 (B)			CV28	S14		1320	DINITROPHENOL, WETTED with not less than 15% water, by mass
			1 (B)			CV28	S14		1321	DINITROPHENOLATES, WETTED with not less than 15% water, by mass
			1 (B)				S14		1322	DINITRORESORCINOL, WETTED with not less than 15% water, by mass
SGAN		AT	2 (E)	V11				40	1323	FERROCERIUM
			3 (E)						1324	FILMS, NITROCELLULOSE BASE, gelatin coated, except scrap
SGAN		AT	2 (E)	V11				40	1325	FLAMMABLE SOLID, ORGANIC, N.O.S.
SGAV		AT	3 (E)		VC1 VC2			40	1325	FLAMMABLE SOLID, ORGANIC, N.O.S.
SGAN		AT	2 (E)	V11				40	1326	HAFNIUM POWDER, WETTED with not less than 25% water
NOT SUBJECT TO ADR									1327	Hay, Straw or Bhusa
SGAV		AT	3 (E)		VC1 VC2			40	1328	HEXAMETHYLENETE-TRAMINE
SGAV		AT	3 (E)		VC1 VC2			40	1330	MANGANESE RESINATE
			4 (E)						1331	MATCHES, 'STRIKE ANYWHERE'
SGAV		AT	3 (E)		VC1 VC2			40	1332	METALDEHYDE
			2 (E)	V11					1333	CERIUM, slabs, ingots or rods
SGAV		AT	3 (E)		VC1 VC2 AP1			40	1334	NAPHTHALENE, CRUDE or NAPHTHALENE, REFINED
			1 (B)				S14		1336	NITROGUANIDINE (PICRITE), WETTED with not less than 20% water, by mass

UN No.	Name and description	Class	Classifi-cation code	Packing group	Labels	Special provi-sions	Limited and excepted quantities		Packaging			Portable tanks and bulk containers	
									Packing instruc-tions	Special packing provisions	Mixed packing provisions	Instruc-tions	Special provisions
	3.1.2	2.2	2.2	2.1.1.3	5.2.2	3.3	3.4	3.5.1.2	4.1.4	4.1.4	4.1.10	4.2.5.2 7.3.2	4.2.5.3
(1)	(2)	(3a)	(3b)	(4)	(5)	(6)	(7a)	(7b)	(8)	(9a)	(9b)	(10)	(11)
1337	NITROSTARCH, WETTED with not less than 20% water, by mass	4.1	D	I	4.1		0	E0	P406		MP2		
1338	PHOSPHORUS, AMORPHOUS	4.1	F3	III	4.1		5 kg	E1	P410 IBC08 R001	B3	MP11	T1	TP33
1339	PHOSPHORUS HEPTASULPHIDE, free from yellow and white phosphorus	4.1	F3	II	4.1	602	1 kg	E2	P410 IBC04		MP11	T3	TP33
1340	PHOSPHORUS PENTASULPHIDE, free from yellow and white phosphorus	4.3	WF2	II	4.3 +4.1	602	500 g	E2	P410 IBC04		MP14	T3	TP33
1341	PHOSPHORUS SESQUISULPHIDE, free from yellow and white phosphorus	4.1	F3	II	4.1	602	1 kg	E2	P410 IBC04		MP11	T3	TP33
1343	PHOSPHORUS TRISULPHIDE, free from yellow and white phosphorus	4.1	F3	II	4.1	602	1 kg	E2	P410 IBC04		MP11	T3	TP33
1344	TRINITROPHENOL (PICRIC ACID), WETTED with not less than 30% water, by mass	4.1	D	I	4.1		0	E0	P406	PP26	MP2		
1345	RUBBER SCRAP or RUBBER SHODDY, powdered or granulated	4.1	F1	II	4.1		1 kg	E2	P002 IBC08	B4	MP11	T3	TP33
1346	SILICON POWDER, AMORPHOUS	4.1	F3	III	4.1	32	5 kg	E1	P002 IBC08 LP02 R001	B3	MP11	T1	TP33
1347	SILVER PICRATE, WETTED with not less than 30% water, by mass	4.1	D	I	4.1		0	E0	P406	PP25 PP26	MP2		
1348	SODIUM DINITRO-o-CRESOLATE, WETTED with not less than 15% water, by mass	4.1	DT	I	4.1 +6.1		0	E0	P406	PP26	MP2		
1349	SODIUM PICRAMATE, WETTED with not less than 20% water, by mass	4.1	D	I	4.1		0	E0	P406	PP26	MP2		
1350	SULPHUR	4.1	F3	III	4.1	242	5 kg	E1	P002 IBC08 LP02 R001	B3	MP11	T1 BK1 BK2 BK3	TP33
1352	TITANIUM POWDER, WETTED with not less than 25% water	4.1	F3	II	4.1	586	1 kg	E2	P410 IBC06	PP40	MP11	T3	TP33
1353	FIBRES or FABRICS IMPREGNATED WITH WEAKLY NITRATED NITROCELLULOSE, N.O.S.	4.1	F1	III	4.1	502	5 kg	E1	P410 IBC08 R001	B3	MP11		
1354	TRINITROBENZENE, WETTED with not less than 30% water, by mass	4.1	D	I	4.1		0	E0	P406		MP2		
1355	TRINITROBENZOIC ACID, WETTED with not less than 30% water, by mass	4.1	D	I	4.1		0	E0	P406		MP2		
1356	TRINITROTOLUENE (TNT), WETTED with not less than 30% water, by mass	4.1	D	I	4.1		0	E0	P406		MP2		
1357	UREA NITRATE, WETTED with not less than 20% water, by mass	4.1	D	I	4.1	227	0	E0	P406		MP2		
1358	ZIRCONIUM POWDER, WETTED with not less than 25% water	4.1	F3	II	4.1	586	1 kg	E2	P410 IBC06	PP40	MP11	T3	TP33
1360	CALCIUM PHOSPHIDE	4.3	WT2	I	4.3 +6.1		0	E0	P403		MP2		
1361	CARBON, animal or vegetable origin	4.2	S2	II	4.2		0	E0	P002 IBC06	PP12	MP14	T3	TP33
1361	CARBON, animal or vegetable origin	4.2	S2	III	4.2	665	0	E0	P002 IBC08 LP02 R001	PP12 B3	MP14	T1	TP33
1362	CARBON, ACTIVATED	4.2	S2	III	4.2	646	0	E1	P002 IBC08 LP02 R001	PP11 B3	MP14	T1	TP33
1363	COPRA	4.2	S2	III	4.2		0	E0	P003 IBC08 LP02 R001	PP20 B3 B6	MP14		

ADR tank		Vehicle for tank carriage	Transport category (Tunnel restriction code)	Special provisions for carriage				Hazard identification No.	UN No.	Name and description
Tank code	Special provisions			Packages	Bulk	Loading, unloading and handling	Operation			
4.3	4.3.5, 6.8.4	9.1.1.2	1.1.3.6 (8.6)	7.2.4	7.3.3	7.5.11	8.5	5.3.2.3		3.1.2
(12)	(13)	(14)	(15)	(16)	(17)	(18)	(19)	(20)	(1)	(2)
			1 (B)				S14		1337	NITROSTARCH, WETTED with not less than 20% water, by mass
SGAV		AT	3 (E)		VC1 VC2			40	1338	PHOSPHORUS, AMORPHOUS
SGAN		AT	2 (E)					40	1339	PHOSPHORUS HEPTASULPHIDE, free from yellow and white phosphorus
SGAN		AT	0 (D/E)	V1		CV23		423	1340	PHOSPHORUS PENTASULPHIDE, free from yellow and white phosphorus
SGAN		AT	2 (E)					40	1341	PHOSPHORUS SESQUISULPHIDE, free from yellow and white phosphorus
SGAN		AT	2 (E)					40	1343	PHOSPHORUS TRISULPHIDE, free from yellow and white phosphorus
			1 (B)				S14		1344	TRINITROPHENOL (PICRIC ACID), WETTED with not less than 30% water, by mass
SGAN		AT	4 (E)	V11				40	1345	RUBBER SCRAP or RUBBER SHODDY, powdered or granulated
SGAV		AT	3 (E)		VC1 VC2			40	1346	SILICON POWDER, AMORPHOUS
			1 (B)				S14		1347	SILVER PICRATE, WETTED with not less than 30% water, by mass
			1 (B)			CV28	S14		1348	SODIUM DINITRO-o-CRESOLATE, WETTED with not less than 15% water, by mass
			1 (B)				S14		1349	SODIUM PICRAMATE, WETTED with not less than 20% water, by mass
SGAV		AT	3 (E)		VC1 VC2			40	1350	SULPHUR
SGAN		AT	2 (E)	V11				40	1352	TITANIUM POWDER, WETTED with not less than 25% water
			3 (E)						1353	FIBRES or FABRICS IMPREGNATED WITH WEAKLY NITRATED NITROCELLULOSE, N.O.S.
			1 (B)				S14		1354	TRINITROBENZENE, WETTED with not less than 30% water, by mass
			1 (B)				S14		1355	TRINITROBENZOIC ACID, WETTED with not less than 30% water, by mass
			1 (B)				S14		1356	TRINITROTOLUENE (TNT), WETTED with not less than 30% water, by mass
			1 (B)				S14		1357	UREA NITRATE, WETTED with not less than 20% water, by mass
SGAN		AT	2 (E)	V11				40	1358	ZIRCONIUM POWDER, WETTED with not less than 25% water
			1 (E)	V1		CV23 CV28	S20		1360	CALCIUM PHOSPHIDE
SGAN	TU11	AT	2 (D/E)	V1 V13				40	1361	CARBON, animal or vegetable origin
SGAV		AT	4 (E)	V1 V13	VC1 VC2 AP1			40	1361	CARBON, animal or vegetable origin
SGAV		AT	4 (E)	V1	VC1 VC2 AP1			40	1362	CARBON, ACTIVATED
			3 (E)	V1	VC1 VC2 AP1			40	1363	COPRA

UN No.	Name and description	Class	Classification code	Packing group	Labels	Special provisions	Limited and excepted quantities		Packaging			Portable tanks and bulk containers	
									Packing instructions	Special packing provisions	Mixed packing provisions	Instructions	Special provisions
3.1.2		2.2	2.2	2.1.1.3	5.2.2	3.3	3.4	3.5.1.2	4.1.4	4.1.4	4.1.10	4.2.5.2 7.3.2	4.2.5.3
(1)	(2)	(3a)	(3b)	(4)	(5)	(6)	(7a)	(7b)	(8)	(9a)	(9b)	(10)	(11)
1364	COTTON WASTE, OILY	4.2	S2	III	4.2		0	E0	P003 IBC08 LP02 R001	PP19 B3 B6	MP14		
1365	COTTON, WET	4.2	S2	III	4.2		0	E0	P003 IBC08 LP02 R001	PP19 B3 B6	MP14		
1369	p-NITROSODIMETHYL-ANILINE	4.2	S2	II	4.2		0	E2	P410 IBC06		MP14	T3	TP33
1372	Fibres, animal or fibres, vegetable burnt, wet or damp	4.2	S2				NOT SUBJECT TO ADR						
1373	FIBRES or FABRICS, ANIMAL or VEGETABLE or SYNTHETIC, N.O.S. with oil	4.2	S2	III	4.2		0	E0	P410 IBC08 R001	B3	MP14	T1	TP33
1374	FISH MEAL (FISH SCRAP), UNSTABILIZED	4.2	S2	II	4.2	300	0	E2	P410 IBC08	B4	MP14	T3	TP33
1376	IRON OXIDE, SPENT or IRON SPONGE, SPENT obtained from coal gas purification	4.2	S4	III	4.2	592	0	E0	P002 IBC08 LP02 R001	B3	MP14	T1 BK2	TP33
1378	METAL CATALYST, WETTED with a visible excess of liquid	4.2	S4	II	4.2	274	0	E0	P410 IBC01	PP39	MP14	T3	TP33
1379	PAPER, UNSATURATED OIL TREATED, incompletely dried (including carbon paper)	4.2	S2	III	4.2		0	E0	P410 IBC08 R001	B3	MP14		
1380	PENTABORANE	4.2	ST3	I	4.2 +6.1		0	E0	P601		MP2		
1381	PHOSPHORUS, WHITE or YELLOW, UNDER WATER or IN SOLUTION	4.2	ST3	I	4.2 +6.1	503	0	E0	P405		MP2	T9	TP3 TP31
1381	PHOSPHORUS, WHITE or YELLOW, DRY	4.2	ST4	I	4.2 +6.1	503	0	E0	P405		MP2	T9	TP3 TP31
1382	POTASSIUM SULPHIDE, ANHYDROUS or POTASSIUM SULPHIDE with less than 30% water of crystallization	4.2	S4	II	4.2	504	0	E2	P410 IBC06		MP14	T3	TP33
1383	PYROPHORIC METAL, N.O.S. or PYROPHORIC ALLOY, N.O.S.	4.2	S4	I	4.2	274	0	E0	P404		MP13	T21	TP7 TP33
1384	SODIUM DITHIONITE (SODIUM HYDROSULPHITE)	4.2	S4	II	4.2		0	E2	P410 IBC06		MP14	T3	TP33
1385	SODIUM SULPHIDE, ANHYDROUS or SODIUM SULPHIDE with less than 30% water of crystallization	4.2	S4	II	4.2	504	0	E2	P410 IBC06		MP14	T3	TP33
1386	SEED CAKE with more than 1.5% oil and not more than 11% moisture	4.2	S2	III	4.2		0	E0	P003 IBC08 LP02 R001	PP20 B3 B6	MP14		
1387	Wool waste, wet	4.2	S2				NOT SUBJECT TO ADR						
1389	ALKALI METAL AMALGAM, LIQUID	4.3	W1	I	4.3	182	0	E0	P402	RR8	MP2		
1390	ALKALI METAL AMIDES	4.3	W2	II	4.3	182 505	500 g	E2	P410 IBC07		MP14	T3	TP33
1391	ALKALI METAL DISPERSION or ALKALINE EARTH METAL DISPERSION	4.3	W1	I	4.3	182 183 506	0	E0	P402	RR8	MP2		
1392	ALKALINE EARTH METAL AMALGAM, LIQUID	4.3	W1	I	4.3	183 506	0	E0	P402		MP2		
1393	ALKALINE EARTH METAL ALLOY, N.O.S.	4.3	W2	II	4.3	183 506	500 g	E2	P410 IBC07		MP14	T3	TP33
1394	ALUMINIUM CARBIDE	4.3	W2	II	4.3		500 g	E2	P410 IBC07		MP14	T3	TP33
1395	ALUMINIUM FERROSILICON POWDER	4.3	WT2	II	4.3 +6.1		500 g	E2	P410 IBC05	PP40	MP14	T3	TP33
1396	ALUMINIUM POWDER, UNCOATED	4.3	W2	II	4.3		500 g	E2	P410 IBC07	PP40	MP14	T3	TP33
1396	ALUMINIUM POWDER, UNCOATED	4.3	W2	III	4.3		1 kg	E1	P410 IBC08 R001	B4	MP14	T1	TP33
1397	ALUMINIUM PHOSPHIDE	4.3	WT2	I	4.3 +6.1	507	0	E0	P403		MP2		

ADR tank		Vehicle for tank carriage	Transport category (Tunnel restriction code)	Special provisions for carriage				Hazard identifi-cation No.	UN No.	Name and description
Tank code	Special provisions			Packages	Bulk	Loading, unloading and handling	Operation			
4.3	4.3.5, 6.8.4	9.1.1.2	1.1.3.6 (8.6)	7.2.4	7.3.3	7.5.11	8.5	5.3.2.3		3.1.2
(12)	(13)	(14)	(15)	(16)	(17)	(18)	(19)	(20)	(1)	(2)
			3 (E)	V1	VC1 VC2 AP1			40	1364	COTTON WASTE, OILY
			3 (E)	V1	VC1 VC2 AP1			40	1365	COTTON, WET
SGAN		AT	2 (D/E)	V1				40	1369	p-NITROSODIMETHYL-ANILINE
NOT SUBJECT TO ADR									1372	Fibres, animal or fibres, vegetable burnt, wet or damp
		AT	3 (E)	V1	VC1 VC2 AP1			40	1373	FIBRES or FABRICS, ANIMAL or VEGETABLE or SYNTHETIC, N.O.S. with oil
		AT	2 (D/E)	V1				40	1374	FISH MEAL (FISH SCRAP), UNSTABILIZED
SGAV		AT	3 (E)	V1	VC1 VC2 AP1			40	1376	IRON OXIDE, SPENT or IRON SPONGE, SPENT obtained from coal gas purification
SGAN		AT	2 (D/E)	V1				40	1378	METAL CATALYST, WETTED with a visible excess of liquid
			3 (E)	V1	VC1 VC2 AP1			40	1379	PAPER, UNSATURATED OIL TREATED, incompletely dried (including carbon paper)
L21DH	TU14 TC1 TE21 TM1	AT	0 (B/E)	V1		CV28	S20	333	1380	PENTABORANE
L10DH(+)	TU14 TU16 TU21 TE3 TE21	AT	0 (B/E)	V1		CV28	S20	46	1381	PHOSPHORUS, WHITE or YELLOW, UNDER WATER or IN SOLUTION
L10DH(+)	TU14 TU16 TU21 TE3 TE21	AT	0 (B/E)	V1		CV28	S20	46	1381	PHOSPHORUS, WHITE or YELLOW, DRY
SGAN		AT	2 (D/E)	V1				40	1382	POTASSIUM SULPHIDE, ANHYDROUS or POTASSIUM SULPHIDE with less than 30% water of crystallization
		AT	0 (B/E)	V1			S20	43	1383	PYROPHORIC METAL, N.O.S. or PYROPHORIC ALLOY, N.O.S.
SGAN		AT	2 (D/E)	V1				40	1384	SODIUM DITHIONITE (SODIUM HYDROSULPHITE)
SGAN		AT	2 (D/E)	V1				40	1385	SODIUM SULPHIDE, ANHYDROUS or SODIUM SULPHIDE with less than 30% water of crystallization
			3 (E)	V1	VC1 VC2 AP1			40	1386	SEED CAKE with more than 1.5% oil and not more than 11% moisture
NOT SUBJECT TO ADR									1387	Wool waste, wet
L10BN(+)	TU1 TE5 TT3 TM2	AT	1 (B/E)	V1		CV23	S20	X323	1389	ALKALI METAL AMALGAM, LIQUID
SGAN		AT	0 (D/E)	V1		CV23		423	1390	ALKALI METAL AMIDES
L10BN(+)	TU1 TE5 TT3 TM2	AT	1 (B/E)	V1		CV23	S20	X323	1391	ALKALI METAL DISPERSION or ALKALINE EARTH METAL DISPERSION
L10BN(+)	TU1 TE5 TT3 TM2	AT	1 (B/E)	V1		CV23	S20	X323	1392	ALKALINE EARTH METAL AMALGAM, LIQUID
SGAN		AT	2 (D/E)	V1		CV23		423	1393	ALKALINE EARTH METAL ALLOY, N.O.S.
SGAN		AT	2 (D/E)	V1	VC1 VC2 AP3 AP4 AP5	CV23		423	1394	ALUMINIUM CARBIDE
SGAN		AT	2 (D/E)	V1		CV23 CV28		462	1395	ALUMINIUM FERROSILICON POWDER
SGAN		AT	2 (D/E)	V1		CV23		423	1396	ALUMINIUM POWDER, UNCOATED
SGAN		AT	3 (E)	V1	VC2 AP4 AP5	CV23		423	1396	ALUMINIUM POWDER, UNCOATED
			1 (E)	V1		CV23 CV28	S20		1397	ALUMINIUM PHOSPHIDE

UN No.	Name and description	Class	Classification code	Packing group	Labels	Special provisions	Limited and excepted quantities		Packaging			Portable tanks and bulk containers	
									Packing instructions	Special packing provisions	Mixed packing provisions	Instructions	Special provisions
3.1.2	3.1.2	2.2	2.2	2.1.1.3	5.2.2	3.3	3.4	3.5.1.2	4.1.4	4.1.4	4.1.10	4.2.5.2 7.3.2	4.2.5.3
(1)	(2)	(3a)	(3b)	(4)	(5)	(6)	(7a)	(7b)	(8)	(9a)	(9b)	(10)	(11)
1398	ALUMINIUM SILICON POWDER, UNCOATED	4.3	W2	III	4.3	37	1 kg	E1	P410 IBC08 R001	B4	MP14	T1	TP33
1400	BARIUM	4.3	W2	II	4.3		500 g	E2	P410 IBC07		MP14	T3	TP33
1401	CALCIUM	4.3	W2	II	4.3		500 g	E2	P410 IBC07		MP14	T3	TP33
1402	CALCIUM CARBIDE	4.3	W2	I	4.3		0	E0	P403 IBC04		MP2	T9	TP7 TP33
1402	CALCIUM CARBIDE	4.3	W2	II	4.3		500 g	E2	P410 IBC07		MP14	T3	TP33
1403	CALCIUM CYANAMIDE with more than 0.1% calcium carbide	4.3	W2	III	4.3	38	1 kg	E1	P410 IBC08 R001	B4	MP14	T1	TP33
1404	CALCIUM HYDRIDE	4.3	W2	I	4.3		0	E0	P403		MP2		
1405	CALCIUM SILICIDE	4.3	W2	II	4.3		500 g	E2	P410 IBC07		MP14	T3	TP33
1405	CALCIUM SILICIDE	4.3	W2	III	4.3		1 kg	E1	P410 IBC08 R001	B4	MP14	T1	TP33
1407	CAESIUM	4.3	W2	I	4.3		0	E0	P403 IBC04		MP2		
1408	FERROSILICON with 30% or more but less than 90% silicon	4.3	WT2	III	4.3 +6.1	39	1 kg	E1	P003 IBC08 R001	PP20 B4 B6	MP14	T1 BK2	TP33
1409	METAL HYDRIDES, WATER-REACTIVE, N.O.S.	4.3	W2	I	4.3	274 508	0	E0	P403		MP2		
1409	METAL HYDRIDES, WATER-REACTIVE, N.O.S.	4.3	W2	II	4.3	274 508	500 g	E2	P410 IBC04		MP14	T3	TP33
1410	LITHIUM ALUMINIUM HYDRIDE	4.3	W2	I	4.3		0	E0	P403		MP2		
1411	LITHIUM ALUMINIUM HYDRIDE, ETHEREAL	4.3	WF1	I	4.3 +3		0	E0	P402	RR8	MP2		
1413	LITHIUM BOROHYDRIDE	4.3	W2	I	4.3		0	E0	P403		MP2		
1414	LITHIUM HYDRIDE	4.3	W2	I	4.3		0	E0	P403		MP2		
1415	LITHIUM	4.3	W2	I	4.3		0	E0	P403 IBC04		MP2	T9	TP7 TP33
1417	LITHIUM SILICON	4.3	W2	II	4.3		500 g	E2	P410 IBC07		MP14	T3	TP33
1418	MAGNESIUM POWDER or MAGNESIUM ALLOYS POWDER	4.3	WS	I	4.3 +4.2		0	E0	P403		MP2		
1418	MAGNESIUM POWDER or MAGNESIUM ALLOYS POWDER	4.3	WS	II	4.3 +4.2		0	E2	P410 IBC05		MP14	T3	TP33
1418	MAGNESIUM POWDER or MAGNESIUM ALLOYS POWDER	4.3	WS	III	4.3 +4.2		0	E1	P410 IBC08 R001	B4	MP14	T1	TP33
1419	MAGNESIUM ALUMINIUM PHOSPHIDE	4.3	WT2	I	4.3 +6.1		0	E0	P403		MP2		
1420	POTASSIUM METAL ALLOYS, LIQUID	4.3	W1	I	4.3		0	E0	P402		MP2		
1421	ALKALI METAL ALLOY, LIQUID, N.O.S.	4.3	W1	I	4.3	182	0	E0	P402	RR8	MP2		
1422	POTASSIUM SODIUM ALLOYS, LIQUID	4.3	W1	I	4.3		0	E0	P402		MP2	T9	TP3 TP7 TP31
1423	RUBIDIUM	4.3	W2	I	4.3		0	E0	P403 IBC04		MP2		
1426	SODIUM BOROHYDRIDE	4.3	W2	I	4.3		0	E0	P403		MP2		
1427	SODIUM HYDRIDE	4.3	W2	I	4.3		0	E0	P403		MP2		
1428	SODIUM	4.3	W2	I	4.3		0	E0	P403 IBC04		MP2	T9	TP7 TP33
1431	SODIUM METHYLATE	4.2	SC4	II	4.2 +8		0	E2	P410 IBC05		MP14	T3	TP33

ADR tank		Vehicle for tank carriage	Transport category (Tunnel restriction code)	Special provisions for carriage				Hazard identifi- cation No.	UN No.	Name and description
Tank code	Special provisions			Packages	Bulk	Loading, unloading and handling	Operation			
4.3	4.3.5, 6.8.4	9.1.1.2	1.1.3.6 (8.6)	7.2.4	7.3.3	7.5.11	8.5	5.3.2.3		3.1.2
(12)	(13)	(14)	(15)	(16)	(17)	(18)	(19)	(20)	(1)	(2)
SGAN		AT	3 (E)	V1	VC2 AP4 AP5	CV23		423	1398	ALUMINIUM SILICON POWDER, UNCOATED
SGAN		AT	2 (D/E)	V1		CV23		423	1400	BARIUM
SGAN		AT	2 (D/E)	V1		CV23		423	1401	CALCIUM
S2.65AN(+)	TU4 TU22 TM2 TA5	AT	1 (B/E)	V1		CV23	S20	X423	1402	CALCIUM CARBIDE
SGAN		AT	2 (D/E)	V1	VC1 VC2 AP3 AP4 AP5	CV23		423	1402	CALCIUM CARBIDE
SGAN		AT	0 (E)	V1		CV23		423	1403	CALCIUM CYANAMIDE with more than 0.1% calcium carbide
			1 (E)	V1		CV23	S20		1404	CALCIUM HYDRIDE
SGAN		AT	2 (D/E)	V1	VC1 VC2 AP3 AP4 AP5	CV23		423	1405	CALCIUM SILICIDE
SGAN		AT	3 (E)	V1	VC1 VC2 AP3 AP4 AP5	CV23		423	1405	CALCIUM SILICIDE
L10CH(+)	TU2 TU14 TE5 TE21 TT3 TM2	AT	1 (B/E)	V1		CV23	S20	X423	1407	CAESIUM
SGAN		AT	3 (E)	V1	VC1 VC2 AP3 AP4 AP5	CV23 CV28		462	1408	FERROSILICON with 30% or more but less than 90% silicon
			1 (E)	V1		CV23	S20		1409	METAL HYDRIDES, WATER-REACTIVE, N.O.S.
SGAN		AT	2 (D/E)	V1		CV23		423	1409	METAL HYDRIDES, WATER-REACTIVE, N.O.S.
			1 (E)	V1		CV23	S20		1410	LITHIUM ALUMINIUM HYDRIDE
			1 (E)	V1		CV23	S2 S20		1411	LITHIUM ALUMINIUM HYDRIDE, ETHEREAL
			1 (E)	V1		CV23	S20		1413	LITHIUM BOROHYDRIDE
			1 (E)	V1		CV23	S20		1414	LITHIUM HYDRIDE
L10BN(+)	TU1 TE5 TT3 TM2	AT	1 (B/E)	V1		CV23	S20	X423	1415	LITHIUM
SGAN		AT	2 (D/E)	V1		CV23		423	1417	LITHIUM SILICON
			1 (E)	V1		CV23	S20		1418	MAGNESIUM POWDER or MAGNESIUM ALLOYS POWDER
SGAN		AT	2 (D/E)	V1		CV23		423	1418	MAGNESIUM POWDER or MAGNESIUM ALLOYS POWDER
SGAN		AT	3 (E)	V1	VC2 AP4 AP5	CV23		423	1418	MAGNESIUM POWDER or MAGNESIUM ALLOYS POWDER
			1 (E)	V1		CV23 CV28	S20		1419	MAGNESIUM ALUMINIUM PHOSPHIDE
L10BN(+)	TU1 TE5 TT3 TM2	AT	1 (B/E)	V1		CV23	S20	X323	1420	POTASSIUM METAL ALLOYS, LIQUID
L10BN(+)	TU1 TE5 TT3 TM2	AT	1 (B/E)	V1		CV23	S20	X323	1421	ALKALI METAL ALLOY, LIQUID, N.O.S.
L10BN(+)	TU1 TE5 TT3 TM2	AT	1 (B/E)	V1		CV23	S20	X323	1422	POTASSIUM SODIUM ALLOYS, LIQUID
L10CH(+)	TU2 TU14 TE5 TE21 TT3 TM2	AT	1 (B/E)	V1		CV23	S20	X423	1423	RUBIDIUM
			1 (E)	V1		CV23	S20		1426	SODIUM BOROHYDRIDE
			1 (E)	V1		CV23	S20		1427	SODIUM HYDRIDE
L10BN(+)	TU1 TE5 TT3 TM2	AT	1 (B/E)	V1		CV23	S20	X423	1428	SODIUM
SGAN		AT	2 (D/E)	V1				48	1431	SODIUM METHYLATE

UN No.	Name and description	Class	Classifi-cation code	Packing group	Labels	Special provi-sions	Limited and excepted quantities		Packaging			Portable tanks and bulk containers	
									Packing instruc-tions	Special packing provisions	Mixed packing provisions	Instruc-tions	Special provisions
	3.1.2	2.2	2.2	2.1.1.3	5.2.2	3.3	3.4	3.5.1.2	4.1.4	4.1.4	4.1.10	4.2.5.2 7.3.2	4.2.5.3
(1)	(2)	(3a)	(3b)	(4)	(5)	(6)	(7a)	(7b)	(8)	(9a)	(9b)	(10)	(11)
1432	SODIUM PHOSPHIDE	4.3	WT2	I	4.3 +6.1		0	E0	P403		MP2		
1433	STANNIC PHOSPHIDES	4.3	WT2	I	4.3 +6.1		0	E0	P403		MP2		
1435	ZINC ASHES	4.3	W2	III	4.3		1 kg	E1	P002 IBC08 R001	B4	MP14	T1	TP33
1436	ZINC POWDER or ZINC DUST	4.3	WS	I	4.3 +4.2		0	E0	P403		MP2		
1436	ZINC POWDER or ZINC DUST	4.3	WS	II	4.3 +4.2		0	E2	P410 IBC07	PP40	MP14	T3	TP33
1436	ZINC POWDER or ZINC DUST	4.3	WS	III	4.3 +4.2		0	E1	P410 IBC08 R001	B4	MP14	T1	TP33
1437	ZIRCONIUM HYDRIDE	4.1	F3	II	4.1		1 kg	E2	P410 IBC04	PP40	MP11	T3	TP33
1438	ALUMINIUM NITRATE	5.1	O2	III	5.1		5 kg	E1	P002 IBC08 LP02 R001	B3	MP10	T1 BK1 BK2	TP33
1439	AMMONIUM DICHROMATE	5.1	O2	II	5.1		1 kg	E2	P002 IBC08	B4	MP2	T3	TP33
1442	AMMONIUM PERCHLORATE	5.1	O2	II	5.1	152	1 kg	E2	P002 IBC06		MP2	T3	TP33
1444	AMMONIUM PERSULPHATE	5.1	O2	III	5.1		5 kg	E1	P002 IBC08 LP02 R001	B3	MP10	T1	TP33
1445	BARIUM CHLORATE, SOLID	5.1	OT2	II	5.1 +6.1		1 kg	E2	P002 IBC06		MP2	T3	TP33
1446	BARIUM NITRATE	5.1	OT2	II	5.1 +6.1		1 kg	E2	P002 IBC08	B4	MP2	T3	TP33
1447	BARIUM PERCHLORATE, SOLID	5.1	OT2	II	5.1 +6.1		1 kg	E2	P002 IBC06		MP2	T3	TP33
1448	BARIUM PERMANGANATE	5.1	OT2	II	5.1 +6.1		1 kg	E2	P002 IBC06		MP2	T3	TP33
1449	BARIUM PEROXIDE	5.1	OT2	II	5.1 +6.1		1 kg	E2	P002 IBC06		MP2	T3	TP33
1450	BROMATES, INORGANIC, N.O.S.	5.1	O2	II	5.1	274 350	1 kg	E2	P002 IBC08	B4	MP2	T3	TP33
1451	CAESIUM NITRATE	5.1	O2	III	5.1		5 kg	E1	P002 IBC08 LP02 R001	B3	MP10	T1	TP33
1452	CALCIUM CHLORATE	5.1	O2	II	5.1		1 kg	E2	P002 IBC08	B4	MP2	T3	TP33
1453	CALCIUM CHLORITE	5.1	O2	II	5.1		1 kg	E2	P002 IBC08	B4	MP2	T3	TP33
1454	CALCIUM NITRATE	5.1	O2	III	5.1	208	5 kg	E1	P002 IBC08 LP02 R001	B3	MP10	T1 BK1 BK2 BK3	TP33
1455	CALCIUM PERCHLORATE	5.1	O2	II	5.1		1 kg	E2	P002 IBC06		MP2	T3	TP33
1456	CALCIUM PERMANGANATE	5.1	O2	II	5.1		1 kg	E2	P002 IBC06		MP2	T3	TP33
1457	CALCIUM PEROXIDE	5.1	O2	II	5.1		1 kg	E2	P002 IBC06		MP2	T3	TP33
1458	CHLORATE AND BORATE MIXTURE	5.1	O2	II	5.1		1 kg	E2	P002 IBC08	B4	MP2	T3	TP33
1458	CHLORATE AND BORATE MIXTURE	5.1	O2	III	5.1		5 kg	E1	P002 IBC08 LP02 R001	B3	MP2	T1	TP33
1459	CHLORATE AND MAGNESIUM CHLORIDE MIXTURE, SOLID	5.1	O2	II	5.1		1 kg	E2	P002 IBC08	B4	MP2	T3	TP33
1459	CHLORATE AND MAGNESIUM CHLORIDE MIXTURE, SOLID	5.1	O2	III	5.1		5 kg	E1	P002 IBC08 LP02 R001	B3	MP2	T1	TP33
1461	CHLORATES, INORGANIC, N.O.S.	5.1	O2	II	5.1	274 351	1 kg	E2	P002 IBC06		MP2	T3	TP33
1462	CHLORITES, INORGANIC, N.O.S.	5.1	O2	II	5.1	274 352 509	1 kg	E2	P002 IBC06		MP2	T3	TP33

ADR tank		Vehicle for tank carriage	Transport category (Tunnel restriction code)	Special provisions for carriage				Hazard identification No.	UN No.	Name and description
Tank code	Special provisions			Packages	Bulk	Loading, unloading and handling	Operation			
4.3	4.3.5, 6.8.4	9.1.1.2	1.1.3.6 (8.6)	7.2.4	7.3.3	7.5.11	8.5	5.3.2.3		3.1.2
(12)	(13)	(14)	(15)	(16)	(17)	(18)	(19)	(20)	(1)	(2)
			1 (E)	V1		CV23 CV28	S20		1432	SODIUM PHOSPHIDE
			1 (E)	V1		CV23 CV28	S20		1433	STANNIC PHOSPHIDES
SGAN		AT	3 (E)	V1	VC1 VC2 AP3 AP4 AP5	CV23		423	1435	ZINC ASHES
			1 (E)	V1		CV23	S20		1436	ZINC POWDER or ZINC DUST
SGAN		AT	2 (D/E)	V1		CV23		423	1436	ZINC POWDER or ZINC DUST
SGAN		AT	3 (E)	V1	VC2 AP4 AP5	CV23		423	1436	ZINC POWDER or ZINC DUST
SGAN		AT	2 (E)					40	1437	ZIRCONIUM HYDRIDE
SGAV	TU3	AT	3 (E)		VC1 VC2 AP6 AP7	CV24		50	1438	ALUMINIUM NITRATE
SGAN	TU3	AT	2 (E)	V11		CV24		50	1439	AMMONIUM DICHROMATE
		AT	2 (E)	V11	VC1 VC2 AP6 AP7	CV24	S23	50	1442	AMMONIUM PERCHLORATE
SGAV	TU3	AT	3 (E)		VC1 VC2 AP6 AP7	CV24		50	1444	AMMONIUM PERSULPHATE
SGAN	TU3	AT	2 (E)	V11		CV24 CV28		56	1445	BARIUM CHLORATE, SOLID
SGAN	TU3	AT	2 (E)	V11		CV24 CV28		56	1446	BARIUM NITRATE
SGAN	TU3	AT	2 (E)	V11		CV24 CV28	S23	56	1447	BARIUM PERCHLORATE, SOLID
SGAN	TU3	AT	2 (E)	V11		CV24 CV28		56	1448	BARIUM PERMANGANATE
SGAN	TU3	AT	2 (E)	V11		CV24 CV28		56	1449	BARIUM PEROXIDE
SGAV	TU3	AT	2 (E)	V11	VC1 VC2 AP6 AP7	CV24		50	1450	BROMATES, INORGANIC, N.O.S.
SGAV	TU3	AT	3 (E)		VC1 VC2 AP6 AP7	CV24		50	1451	CAESIUM NITRATE
SGAV	TU3	AT	2 (E)	V11	VC1 VC2 AP6 AP7	CV24		50	1452	CALCIUM CHLORATE
SGAN	TU3	AT	2 (E)	V11		CV24		50	1453	CALCIUM CHLORITE
SGAV	TU3	AT	3 (E)		VC1 VC2 AP6 AP7	CV24		50	1454	CALCIUM NITRATE
SGAV	TU3	AT	2 (E)	V11	VC1 VC2 AP6 AP7	CV24	S23	50	1455	CALCIUM PERCHLORATE
SGAN	TU3	AT	2 (E)	V11		CV24		50	1456	CALCIUM PERMANGANATE
SGAN	TU3	AT	2 (E)	V11		CV24		50	1457	CALCIUM PEROXIDE
SGAV	TU3	AT	2 (E)	V11	VC1 VC2 AP6 AP7	CV24		50	1458	CHLORATE AND BORATE MIXTURE
SGAV	TU3	AT	3 (E)		VC1 VC2 AP6 AP7	CV24		50	1458	CHLORATE AND BORATE MIXTURE
SGAV	TU3	AT	2 (E)	V11	VC1 VC2 AP6 AP7	CV24		50	1459	CHLORATE AND MAGNESIUM CHLORIDE MIXTURE, SOLID
SGAV	TU3	AT	3 (E)		VC1 VC2 AP6 AP7	CV24		50	1459	CHLORATE AND MAGNESIUM CHLORIDE MIXTURE, SOLID
SGAV	TU3	AT	2 (E)	V11	VC1 VC2 AP6 AP7	CV24		50	1461	CHLORATES, INORGANIC, N.O.S.
SGAN	TU3	AT	2 (E)	V11		CV24		50	1462	CHLORITES, INORGANIC, N.O.S.

UN No.	Name and description	Class	Classifi-cation code	Packing group	Labels	Special provi-sions	Limited and excepted quantities		Packaging			Portable tanks and bulk containers	
									Packing instruc-tions	Special packing provisions	Mixed packing provisions	Instruc-tions	Special provisions
	3.1.2	2.2	2.2	2.1.1.3	5.2.2	3.3	3.4	3.5.1.2	4.1.4	4.1.4	4.1.10	4.2.5.2 7.3.2	4.2.5.3
(1)	(2)	(3a)	(3b)	(4)	(5)	(6)	(7a)	(7b)	(8)	(9a)	(9b)	(10)	(11)
1463	CHROMIUM TRIOXIDE, ANHYDROUS	5.1	OTC	II	5.1 +6.1 +8	510	1 kg	E2	P002 IBC08	B4	MP2	T3	TP33
1465	DIDYMIUM NITRATE	5.1	O2	III	5.1		5 kg	E1	P002 IBC08 LP02 R001	B3	MP10	T1	TP33
1466	FERRIC NITRATE	5.1	O2	III	5.1		5 kg	E1	P002 IBC08 LP02 R001	B3	MP10	T1	TP33
1467	GUANIDINE NITRATE	5.1	O2	III	5.1		5 kg	E1	P002 IBC08 LP02 R001	B3	MP10	T1	TP33
1469	LEAD NITRATE	5.1	OT2	II	5.1 +6.1		1 kg	E2	P002 IBC08	B4	MP2	T3	TP33
1470	LEAD PERCHLORATE, SOLID	5.1	OT2	II	5.1 +6.1		1 kg	E2	P002 IBC06		MP2	T3	TP33
1471	LITHIUM HYPOCHLORITE, DRY or LITHIUM HYPOCHLORITE MIXTURE	5.1	O2	II	5.1		1 kg	E2	P002 IBC08	B4	MP10		
1471	LITHIUM HYPOCHLORITE, DRY or LITHIUM HYPOCHLORITE MIXTURE	5.1	O2	III	5.1		5 kg	E1	P002 IBC08 LP02 R001	B3	MP10	T1	TP33
1472	LITHIUM PEROXIDE	5.1	O2	II	5.1		1 kg	E2	P002 IBC06		MP2	T3	TP33
1473	MAGNESIUM BROMATE	5.1	O2	II	5.1		1 kg	E2	P002 IBC08	B4	MP2	T3	TP33
1474	MAGNESIUM NITRATE	5.1	O2	III	5.1	332	5 kg	E1	P002 IBC08 LP02 R001	B3	MP10	T1 BK1 BK2 BK3	TP33
1475	MAGNESIUM PERCHLORATE	5.1	O2	II	5.1		1 kg	E2	P002 IBC06		MP2	T3	TP33
1476	MAGNESIUM PEROXIDE	5.1	O2	II	5.1		1 kg	E2	P002 IBC06		MP2	T3	TP33
1477	NITRATES, INORGANIC, N.O.S.	5.1	O2	II	5.1	511	1 kg	E2	P002 IBC08	B4	MP10	T3	TP33
1477	NITRATES, INORGANIC, N.O.S.	5.1	O2	III	5.1	511	5 kg	E1	P002 IBC08 LP02 R001	B3	MP10	T1	TP33
1479	OXIDIZING SOLID, N.O.S.	5.1	O2	I	5.1	274	0	E0	P503 IBC05		MP2		
1479	OXIDIZING SOLID, N.O.S.	5.1	O2	II	5.1	274	1 kg	E2	P002 IBC08	B4	MP2	T3	TP33
1479	OXIDIZING SOLID, N.O.S.	5.1	O2	III	5.1	274	5 kg	E1	P002 IBC08 LP02 R001	B3	MP2	T1	TP33
1481	PERCHLORATES, INORGANIC, N.O.S.	5.1	O2	II	5.1		1 kg	E2	P002 IBC06		MP2	T3	TP33
1481	PERCHLORATES, INORGANIC, N.O.S.	5.1	O2	III	5.1		5 kg	E1	P002 IBC08 LP02 R001	B3	MP2	T1	TP33
1482	PERMANGANATES, INORGANIC, N.O.S.	5.1	O2	II	5.1	274 353	1 kg	E2	P002 IBC06		MP2	T3	TP33
1482	PERMANGANATES, INORGANIC, N.O.S.	5.1	O2	III	5.1	274 353	5 kg	E1	P002 IBC08 LP02 R001	B3	MP2	T1	TP33
1483	PEROXIDES, INORGANIC, N.O.S.	5.1	O2	II	5.1		1 kg	E2	P002 IBC06		MP2	T3	TP33
1483	PEROXIDES, INORGANIC, N.O.S.	5.1	O2	III	5.1		5 kg	E1	P002 IBC08 LP02 R001	B3	MP2	T1	TP33
1484	POTASSIUM BROMATE	5.1	O2	II	5.1		1 kg	E2	P002 IBC08	B4	MP2	T3	TP33
1485	POTASSIUM CHLORATE	5.1	O2	II	5.1		1 kg	E2	P002 IBC08	B4	MP2	T3	TP33

ADR tank		Vehicle for tank carriage	Transport category (Tunnel restriction code)	Special provisions for carriage				Hazard identification No.	UN No.	Name and description
Tank code	Special provisions			Packages	Bulk	Loading, unloading and handling	Operation			
4.3	4.3.5, 6.8.4	9.1.1.2	1.1.3.6 (8.6)	7.2.4	7.3.3	7.5.11	8.5	5.3.2.3		3.1.2
(12)	(13)	(14)	(15)	(16)	(17)	(18)	(19)	(20)	(1)	(2)
SGAN	TU3	AT	2 (E)	V11		CV24 CV28		568	1463	CHROMIUM TRIOXIDE, ANHYDROUS
SGAV	TU3	AT	3 (E)		VC1 VC2 AP6 AP7	CV24		50	1465	DIDYMIUM NITRATE
SGAV	TU3	AT	3 (E)		VC1 VC2 AP6 AP7	CV24		50	1466	FERRIC NITRATE
SGAV	TU3	AT	3 (E)		VC1 VC2 AP6 AP7	CV24		50	1467	GUANIDINE NITRATE
SGAN	TU3	AT	2 (E)	V11		CV24 CV28		56	1469	LEAD NITRATE
SGAN	TU3	AT	2 (E)	V11		CV24 CV28	S23	56	1470	LEAD PERCHLORATE, SOLID
SGAN	TU3	AT	2 (E)	V11		CV24		50	1471	LITHIUM HYPOCHLORITE, DRY or LITHIUM HYPOCHLORITE MIXTURE
SGAV	TU3	AT	3 (E)			CV24		50	1471	LITHIUM HYPOCHLORITE, DRY or LITHIUM HYPOCHLORITE MIXTURE
SGAN	TU3	AT	2 (E)	V11		CV24		50	1472	LITHIUM PEROXIDE
SGAV	TU3	AT	2 (E)	V11	VC1 VC2 AP6 AP7	CV24		50	1473	MAGNESIUM BROMATE
SGAV	TU3	AT	3 (E)		VC1 VC2 AP6 AP7	CV24		50	1474	MAGNESIUM NITRATE
SGAV	TU3	AT	2 (E)	V11	VC1 VC2 AP6 AP7	CV24	S23	50	1475	MAGNESIUM PERCHLORATE
SGAN	TU3	AT	2 (E)	V11		CV24		50	1476	MAGNESIUM PEROXIDE
SGAN	TU3	AT	2 (E)	V11		CV24		50	1477	NITRATES, INORGANIC, N.O.S.
SGAV	TU3	AT	3 (E)		VC1 VC2 AP6 AP7	CV24		50	1477	NITRATES, INORGANIC, N.O.S.
			1 (E)	V10		CV24	S20		1479	OXIDIZING SOLID, N.O.S.
SGAN	TU3	AT	2 (E)	V11		CV24		50	1479	OXIDIZING SOLID, N.O.S.
SGAN	TU3	AT	3 (E)			CV24		50	1479	OXIDIZING SOLID, N.O.S.
SGAV	TU3	AT	2 (E)	V11	VC1 VC2 AP6 AP7	CV24	S23	50	1481	PERCHLORATES, INORGANIC, N.O.S.
SGAV	TU3	AT	3 (E)		VC1 VC2 AP6 AP7	CV24	S23	50	1481	PERCHLORATES, INORGANIC, N.O.S.
SGAN	TU3	AT	2 (E)	V11		CV24		50	1482	PERMANGANATES, INORGANIC, N.O.S.
SGAN	TU3	AT	3 (E)			CV24		50	1482	PERMANGANATES, INORGANIC, N.O.S.
SGAN	TU3	AT	2 (E)	V11		CV24		50	1483	PEROXIDES, INORGANIC, N.O.S.
SGAN	TU3	AT	3 (E)			CV24		50	1483	PEROXIDES, INORGANIC, N.O.S.
SGAV	TU3	AT	2 (E)	V11	VC1 VC2 AP6 AP7	CV24		50	1484	POTASSIUM BROMATE
SGAV	TU3	AT	2 (E)	V11	VC1 VC2 AP6 AP7	CV24		50	1485	POTASSIUM CHLORATE

UN No.	Name and description	Class	Classifi-cation code	Packing group	Labels	Special provi-sions	Limited and excepted quantities		Packaging			Portable tanks and bulk containers	
									Packing instruc-tions	Special packing provisions	Mixed packing provisions	Instruc-tions	Special provisions
	3.1.2	2.2	2.2	2.1.1.3	5.2.2	3.3	3.4	3.5.1.2	4.1.4	4.1.4	4.1.10	4.2.5.2 7.3.2	4.2.5.3
(1)	(2)	(3a)	(3b)	(4)	(5)	(6)	(7a)	(7b)	(8)	(9a)	(9b)	(10)	(11)
1486	POTASSIUM NITRATE	5.1	O2	III	5.1		5 kg	E1	P002 IBC08 LP02 R001	B3	MP10	T1 BK1 BK2 BK3	TP33
1487	POTASSIUM NITRATE AND SODIUM NITRITE MIXTURE	5.1	O2	II	5.1	607	1 kg	E2	P002 IBC08	B4	MP10	T3	TP33
1488	POTASSIUM NITRITE	5.1	O2	II	5.1		1 kg	E2	P002 IBC08	B4	MP10	T3	TP33
1489	POTASSIUM PERCHLORATE	5.1	O2	II	5.1		1 kg	E2	P002 IBC06		MP2	T3	TP33
1490	POTASSIUM PERMANGANATE	5.1	O2	II	5.1		1 kg	E2	P002 IBC08	B4	MP2	T3	TP33
1491	POTASSIUM PEROXIDE	5.1	O2	I	5.1		0	E0	P503 IBC06		MP2		
1492	POTASSIUM PERSULPHATE	5.1	O2	III	5.1		5 kg	E1	P002 IBC08 LP02 R001	B3	MP10	T1	TP33
1493	SILVER NITRATE	5.1	O2	II	5.1		1 kg	E2	P002 IBC08	B4	MP10	T3	TP33
1494	SODIUM BROMATE	5.1	O2	II	5.1		1 kg	E2	P002 IBC08	B4	MP2	T3	TP33
1495	SODIUM CHLORATE	5.1	O2	II	5.1		1 kg	E2	P002 IBC08	B4	MP2	T3 BK1 BK2	TP33
1496	SODIUM CHLORITE	5.1	O2	II	5.1		1 kg	E2	P002 IBC08	B4	MP2	T3	TP33
1498	SODIUM NITRATE	5.1	O2	III	5.1		5 kg	E1	P002 IBC08 LP02 R001	B3	MP10	T1 BK1 BK2 BK3	TP33
1499	SODIUM NITRATE AND POTASSIUM NITRATE MIXTURE	5.1	O2	III	5.1		5 kg	E1	P002 IBC08 LP02 R001	B3	MP10	T1 BK1 BK2 BK3	TP33
1500	SODIUM NITRITE	5.1	OT2	III	5.1 +6.1		5 kg	E1	P002 IBC08 R001	B3	MP10	T1	TP33
1502	SODIUM PERCHLORATE	5.1	O2	II	5.1		1 kg	E2	P002 IBC06		MP2	T3	TP33
1503	SODIUM PERMANGANATE	5.1	O2	II	5.1		1 kg	E2	P002 IBC06		MP2	T3	TP33
1504	SODIUM PEROXIDE	5.1	O2	I	5.1		0	E0	P503 IBC05		MP2		
1505	SODIUM PERSULPHATE	5.1	O2	III	5.1		5 kg	E1	P002 IBC08 LP02 R001	B3	MP10	T1	TP33
1506	STRONTIUM CHLORATE	5.1	O2	II	5.1		1 kg	E2	P002 IBC08	B4	MP2	T3	TP33
1507	STRONTIUM NITRATE	5.1	O2	III	5.1		5 kg	E1	P002 IBC08 LP02 R001	B3	MP10	T1	TP33
1508	STRONTIUM PERCHLORATE	5.1	O2	II	5.1		1 kg	E2	P002 IBC06		MP2	T3	TP33
1509	STRONTIUM PEROXIDE	5.1	O2	II	5.1		1 kg	E2	P002 IBC06		MP2	T3	TP33
1510	TETRANITROMETHANE	6.1	TO1	I	6.1 +5.1	354 609	0	E0	P602		MP8 MP17		
1511	UREA HYDROGEN PEROXIDE	5.1	OC2	III	5.1 +8		5 kg	E1	P002 IBC08 R001	B3	MP2	T1	TP33
1512	ZINC AMMONIUM NITRITE	5.1	O2	II	5.1		1 kg	E2	P002 IBC08	B4	MP10	T3	TP33
1513	ZINC CHLORATE	5.1	O2	II	5.1		1 kg	E2	P002 IBC08	B4	MP2	T3	TP33
1514	ZINC NITRATE	5.1	O2	II	5.1		1 kg	E2	P002 IBC08	B4	MP10	T3	TP33
1515	ZINC PERMANGANATE	5.1	O2	II	5.1		1 kg	E2	P002 IBC06		MP2	T3	TP33
1516	ZINC PEROXIDE	5.1	O2	II	5.1		1 kg	E2	P002 IBC06		MP2	T3	TP33

ADR tank		Vehicle for tank carriage	Transport category (Tunnel restriction code)	Special provisions for carriage				Hazard identifi-cation No.	UN No.	Name and description
Tank code	Special provisions			Packages	Bulk	Loading, unloading and handling	Operation			
4.3	4.3.5, 6.8.4	9.1.1.2	1.1.3.6 (8.6)	7.2.4	7.3.3	7.5.11	8.5	5.3.2.3		3.1.2
(12)	(13)	(14)	(15)	(16)	(17)	(18)	(19)	(20)	(1)	(2)
SGAV	TU3	AT	3 (E)		VC1 VC2 AP6 AP7	CV24		50	1486	POTASSIUM NITRATE
SGAV	TU3	AT	2 (E)	V11	VC1 VC2 AP6 AP7	CV24		50	1487	POTASSIUM NITRATE AND SODIUM NITRITE MIXTURE
SGAV	TU3	AT	2 (E)	V11	VC1 VC2 AP6 AP7	CV24		50	1488	POTASSIUM NITRITE
SGAV	TU3	AT	2 (E)	V11	VC1 VC2 AP6 AP7	CV24	S23	50	1489	POTASSIUM PERCHLORATE
SGAN	TU3	AT	2 (E)	V11		CV24		50	1490	POTASSIUM PERMANGANATE
			1 (E)	V10		CV24	S20		1491	POTASSIUM PEROXIDE
SGAV	TU3	AT	3 (E)		VC1 VC2 AP6 AP7	CV24		50	1492	POTASSIUM PERSULPHATE
SGAV	TU3	AT	2 (E)	V11	VC1 VC2 AP6 AP7	CV24		50	1493	SILVER NITRATE
SGAV	TU3	AT	2 (E)	V11	VC1 VC2 AP6 AP7	CV24		50	1494	SODIUM BROMATE
SGAV	TU3	AT	2 (E)	V11	VC1 VC2 AP6 AP7	CV24		50	1495	SODIUM CHLORATE
SGAN	TU3	AT	2 (E)	V11		CV24		50	1496	SODIUM CHLORITE
SGAV	TU3	AT	3 (E)		VC1 VC2 AP6 AP7	CV24		50	1498	SODIUM NITRATE
SGAV	TU3	AT	3 (E)		VC1 VC2 AP6 AP7	CV24		50	1499	SODIUM NITRATE AND POTASSIUM NITRATE MIXTURE
SGAN	TU3	AT	3 (E)			CV24 CV28		56	1500	SODIUM NITRITE
SGAV	TU3	AT	2 (E)	V11	VC1 VC2 AP6 AP7	CV24	S23	50	1502	SODIUM PERCHLORATE
SGAN	TU3	AT	2 (E)	V11		CV24		50	1503	SODIUM PERMANGANATE
			1 (E)	V10		CV24	S20		1504	SODIUM PEROXIDE
SGAV	TU3	AT	3 (E)		VC1 VC2 AP6 AP7	CV24		50	1505	SODIUM PERSULPHATE
SGAV	TU3	AT	2 (E)	V11	VC1 VC2 AP6 AP7	CV24		50	1506	STRONTIUM CHLORATE
SGAV	TU3	AT	3 (E)		VC1 VC2 AP6 AP7	CV24		50	1507	STRONTIUM NITRATE
SGAV	TU3	AT	2 (E)	V11	VC1 VC2 AP6 AP7	CV24	S23	50	1508	STRONTIUM PERCHLORATE
SGAN	TU3	AT	2 (E)	V11		CV24		50	1509	STRONTIUM PEROXIDE
L10CH	TU14 TU15 TE19 TE21	AT	1 (B/D)			CV1 CV13 CV28	S9 S14	665	1510	TETRANITROMETHANE
SGAN	TU3	AT	3 (E)			CV24		58	1511	UREA HYDROGEN PEROXIDE
SGAN	TU3	AT	2 (E)	V11		CV24		50	1512	ZINC AMMONIUM NITRITE
SGAV	TU3	AT	2 (E)	V11	VC1 VC2 AP6 AP7	CV24		50	1513	ZINC CHLORATE
SGAN	TU3	AT	2 (E)	V11		CV24		50	1514	ZINC NITRATE
SGAN	TU3	AT	2 (E)	V11		CV24		50	1515	ZINC PERMANGANATE
SGAN	TU3	AT	2 (E)	V11		CV24		50	1516	ZINC PEROXIDE

UN No.	Name and description	Class	Classifi-cation code	Packing group	Labels	Special provi-sions	Limited and excepted quantities		Packaging			Portable tanks and bulk containers	
									Packing instruc-tions	Special packing provisions	Mixed packing provisions	Instruc-tions	Special provisions
	3.1.2	2.2	2.2	2.1.1.3	5.2.2	3.3	3.4	3.5.1.2	4.1.4	4.1.4	4.1.10	4.2.5.2 7.3.2	4.2.5.3
(1)	(2)	(3a)	(3b)	(4)	(5)	(6)	(7a)	(7b)	(8)	(9a)	(9b)	(10)	(11)
1517	ZIRCONIUM PICRAMATE, WETTED with not less than 20% water, by mass	4.1	D	I	4.1		0	E0	P406	PP26	MP2		
1541	ACETONE CYANOHYDRIN, STABILIZED	6.1	T1	I	6.1	354	0	E0	P602		MP8 MP17	T20	TP2
1544	ALKALOIDS, SOLID, N.O.S. or ALKALOID SALTS, SOLID, N.O.S.	6.1	T2	I	6.1	43 274	0	E5	P002 IBC07		MP18	T6	TP33
1544	ALKALOIDS, SOLID, N.O.S. or ALKALOID SALTS, SOLID, N.O.S.	6.1	T2	II	6.1	43 274	500 g	E4	P002 IBC08	B4	MP10	T3	TP33
1544	ALKALOIDS, SOLID, N.O.S. or ALKALOID SALTS, SOLID, N.O.S.	6.1	T2	III	6.1	43 274	5 kg	E1	P002 IBC08 LP02 R001	B3	MP10	T1	TP33
1545	ALLYL ISOTHIOCYANATE, STABILIZED	6.1	TF1	II	6.1 +3	386	100 ml	E0	P001 IBC02		MP15	T7	TP2
1546	AMMONIUM ARSENATE	6.1	T5	II	6.1		500 g	E4	P002 IBC08	B4	MP10	T3	TP33
1547	ANILINE	6.1	T1	II	6.1	279	100 ml	E4	P001 IBC02		MP15	T7	TP2
1548	ANILINE HYDROCHLORIDE	6.1	T2	III	6.1		5 kg	E1	P002 IBC08 LP02 R001	B3	MP10	T1	TP33
1549	ANTIMONY COMPOUND, INORGANIC, SOLID, N.O.S.	6.1	T5	III	6.1	45 274 512	5 kg	E1	P002 IBC08 LP02 R001	B3	MP10	T1	TP33
1550	ANTIMONY LACTATE	6.1	T5	III	6.1		5 kg	E1	P002 IBC08 LP02 R001	B3	MP10	T1	TP33
1551	ANTIMONY POTASSIUM TARTRATE	6.1	T5	III	6.1		5 kg	E1	P002 IBC08 LP02 R001	B3	MP10	T1	TP33
1553	ARSENIC ACID, LIQUID	6.1	T4	I	6.1		0	E5	P001		MP8 MP17	T20	TP2 TP7
1554	ARSENIC ACID, SOLID	6.1	T5	II	6.1		500 g	E4	P002 IBC08	B4	MP10	T3	TP33
1555	ARSENIC BROMIDE	6.1	T5	II	6.1		500 g	E4	P002 IBC08	B4	MP10	T3	TP33
1556	ARSENIC COMPOUND, LIQUID, N.O.S., inorganic, including: Arsenates, n.o.s., Arsenites, n.o.s.; and Arsenic sulphides, n.o.s.	6.1	T4	I	6.1	43 274	0	E5	P001		MP8 MP17	T14	TP2 TP27
1556	ARSENIC COMPOUND, LIQUID, N.O.S., inorganic, including: Arsenates, n.o.s., Arsenites, n.o.s.; and Arsenic sulphides, n.o.s.	6.1	T4	II	6.1	43 274	100 ml	E4	P001 IBC02		MP15	T11	TP2 TP27
1556	ARSENIC COMPOUND, LIQUID, N.O.S., inorganic, including: Arsenates, n.o.s., Arsenites, n.o.s.; and Arsenic sulphides, n.o.s.	6.1	T4	III	6.1	43 274	5 L	E1	P001 IBC03 LP01 R001		MP19	T7	TP2 TP28
1557	ARSENIC COMPOUND, SOLID, N.O.S., inorganic, including: Arsenates, n.o.s.; Arsenites, n.o.s.; and Arsenic sulphides, n.o.s.	6.1	T5	I	6.1	43 274	0	E5	P002 IBC07		MP18	T6	TP33
1557	ARSENIC COMPOUND, SOLID, N.O.S., inorganic, including: Arsenates, n.o.s.; Arsenites, n.o.s.; and Arsenic sulphides, n.o.s.	6.1	T5	II	6.1	43 274	500 g	E4	P002 IBC08	B4	MP10	T3	TP33
1557	ARSENIC COMPOUND, SOLID, N.O.S., inorganic, including: Arsenates, n.o.s.; Arsenites, n.o.s.; and Arsenic sulphides, n.o.s.	6.1	T5	III	6.1	43 274	5 kg	E1	P002 IBC08 LP02 R001	B3	MP10	T1	TP33

ADR tank		Vehicle for tank carriage	Transport category (Tunnel restriction code)	Special provisions for carriage				Hazard identification No.	UN No.	Name and description
Tank code	Special provisions			Packages	Bulk	Loading, unloading and handling	Operation			
4.3	4.3.5, 6.8.4	9.1.1.2	1.1.3.6 (8.6)	7.2.4	7.3.3	7.5.11	8.5	5.3.2.3		3.1.2
(12)	(13)	(14)	(15)	(16)	(17)	(18)	(19)	(20)	(1)	(2)
			1 (B)				S14		1517	ZIRCONIUM PICRAMATE, WETTED with not less than 20% water, by mass
L10CH	TU14 TU15 TE19 TE21	AT	1 (C/D)			CV1 CV13 CV28	S9 S14	669	1541	ACETONE CYANOHYDRIN, STABILIZED
S10AH	TU15 TE19	AT	1 (C/E)	V10		CV1 CV13 CV28	S9 S14	66	1544	ALKALOIDS, SOLID, N.O.S. or ALKALOID SALTS, SOLID, N.O.S.
SGAH L4BH	TU15 TE19	AT	2 (D/E)	V11		CV13 CV28	S9 S19	60	1544	ALKALOIDS, SOLID, N.O.S. or ALKALOID SALTS, SOLID, N.O.S.
SGAH L4BH	TU15 TE19	AT	2 (E)		VC1 VC2 AP7	CV13 CV28	S9	60	1544	ALKALOIDS, SOLID, N.O.S. or ALKALOID SALTS, SOLID, N.O.S.
L4BH	TU15 TE19	FL	2 (D/E)	V8		CV13 CV28	S2 S4 S9 S19	639	1545	ALLYL ISOTHIOCYANATE, STABILIZED
SGAH	TU15 TE19	AT	2 (D/E)	V11		CV13 CV28	S9 S19	60	1546	AMMONIUM ARSENATE
L4BH	TU15 TE19	AT	2 (D/E)			CV13 CV28	S9 S19	60	1547	ANILINE
SGAH	TU15 TE19	AT	2 (E)		VC1 VC2 AP7	CV13 CV28	S9	60	1548	ANILINE HYDROCHLORIDE
SGAH L4BH	TU15 TE19	AT	2 (E)		VC1 VC2 AP7	CV13 CV28	S9	60	1549	ANTIMONY COMPOUND, INORGANIC, SOLID, N.O.S.
SGAH L4BH	TU15 TE19	AT	2 (E)		VC1 VC2 AP7	CV13 CV28	S9	60	1550	ANTIMONY LACTATE
SGAH L4BH	TU15 TE19	AT	2 (E)		VC1 VC2 AP7	CV13 CV28	S9	60	1551	ANTIMONY POTASSIUM TARTRATE
L10CH	TU14 TU15 TE19 TE21	AT	1 (C/E)			CV1 CV13 CV28	S9 S14	66	1553	ARSENIC ACID, LIQUID
SGAH L4BH	TU15 TE19	AT	2 (D/E)	V11		CV13 CV28	S9 S19	60	1554	ARSENIC ACID, SOLID
SGAH L4BH	TU15 TE19	AT	2 (D/E)	V11		CV13 CV28	S9 S19	60	1555	ARSENIC BROMIDE
L10CH	TU14 TU15 TE19 TE21	AT	1 (C/E)			CV1 CV13 CV28	S9 S14	66	1556	ARSENIC COMPOUND, LIQUID, N.O.S., inorganic, including: Arsenates, n.o.s., Arsenites, n.o.s.; and Arsenic sulphides, n.o.s.
L4BH	TU15 TE19	AT	2 (D/E)			CV13 CV28	S9 S19	60	1556	ARSENIC COMPOUND, LIQUID, N.O.S., inorganic, including: Arsenates, n.o.s., Arsenites, n.o.s.; and Arsenic sulphides, n.o.s.
L4BH	TU15 TE19	AT	2 (E)	V12		CV13 CV28	S9	60	1556	ARSENIC COMPOUND, LIQUID, N.O.S., inorganic, including: Arsenates, n.o.s., Arsenites, n.o.s.; and Arsenic sulphides, n.o.s.
S10AH L10CH	TU15 TE19	AT	1 (C/E)	V10		CV1 CV13 CV28	S9 S14	66	1557	ARSENIC COMPOUND, SOLID, N.O.S., inorganic, including: Arsenates, n.o.s.; Arsenites, n.o.s.; and Arsenic sulphides, n.o.s.
SGAH L4BH	TU15 TE19	AT	2 (D/E)	V11		CV13 CV28	S9 S19	60	1557	ARSENIC COMPOUND, SOLID, N.O.S., inorganic, including: Arsenates, n.o.s.; Arsenites, n.o.s.; and Arsenic sulphides, n.o.s.
SGAH L4BH	TU15 TE19	AT	2 (E)		VC1 VC2 AP7	CV13 CV28	S9	60	1557	ARSENIC COMPOUND, SOLID, N.O.S., inorganic, including: Arsenates, n.o.s.; Arsenites, n.o.s.; and Arsenic sulphides, n.o.s.

UN No.	Name and description	Class	Classification code	Packing group	Labels	Special provisions	Limited and excepted quantities		Packaging			Portable tanks and bulk containers	
									Packing instructions	Special packing provisions	Mixed packing provisions	Instructions	Special provisions
3.1.2	3.1.2	2.2	2.2	2.1.1.3	5.2.2	3.3	3.4	3.5.1.2	4.1.4	4.1.4	4.1.10	4.2.5.2 7.3.2	4.2.5.3
(1)	(2)	(3a)	(3b)	(4)	(5)	(6)	(7a)	(7b)	(8)	(9a)	(9b)	(10)	(11)
1558	ARSENIC	6.1	T5	II	6.1		500 g	E4	P002 IBC08	B4	MP10	T3	TP33
1559	ARSENIC PENTOXIDE	6.1	T5	II	6.1		500 g	E4	P002 IBC08	B4	MP10	T3	TP33
1560	ARSENIC TRICHLORIDE	6.1	T4	I	6.1		0	E0	P602		MP8 MP17	T14	TP2
1561	ARSENIC TRIOXIDE	6.1	T5	II	6.1		500 g	E4	P002 IBC08	B4	MP10	T3	TP33
1562	ARSENICAL DUST	6.1	T5	II	6.1		500 g	E4	P002 IBC08	B4	MP10	T3	TP33
1564	BARIUM COMPOUND, N.O.S.	6.1	T5	II	6.1	177 274 513 587	500 g	E4	P002 IBC08	B4	MP10	T3	TP33
1564	BARIUM COMPOUND, N.O.S.	6.1	T5	III	6.1	177 274 513 587	5 kg	E1	P002 IBC08 LP02 R001	B3	MP10	T1	TP33
1565	BARIUM CYANIDE	6.1	T5	I	6.1		0	E5	P002 IBC07		MP18	T6	TP33
1566	BERYLLIUM COMPOUND, N.O.S.	6.1	T5	II	6.1	274 514	500 g	E4	P002 IBC08	B4	MP10	T3	TP33
1566	BERYLLIUM COMPOUND, N.O.S.	6.1	T5	III	6.1	274 514	5 kg	E1	P002 IBC08 LP02 R001	B3	MP10	T1	TP33
1567	BERYLLIUM POWDER	6.1	TF3	II	6.1 +4.1		500 g	E4	P002 IBC08	B4	MP10	T3	TP33
1569	BROMOACETONE	6.1	TF1	II	6.1 +3		0	E0	P602		MP15	T20	TP2
1570	BRUCINE	6.1	T2	I	6.1	43	0	E5	P002 IBC07		MP18	T6	TP33
1571	BARIUM AZIDE, WETTED with not less than 50% water, by mass	4.1	DT	I	4.1 +6.1	568	0	E0	P406		MP2		
1572	CACODYLIC ACID	6.1	T5	II	6.1		500 g	E4	P002 IBC08	B4	MP10	T3	TP33
1573	CALCIUM ARSENATE	6.1	T5	II	6.1		500 g	E4	P002 IBC08	B4	MP10	T3	TP33
1574	CALCIUM ARSENATE AND CALCIUM ARSENITE MIXTURE, SOLID	6.1	T5	II	6.1		500 g	E4	P002 IBC08	B4	MP10	T3	TP33
1575	CALCIUM CYANIDE	6.1	T5	I	6.1		0	E5	P002 IBC07		MP18	T6	TP33
1577	CHLORODINITRO-BENZENES, LIQUID	6.1	T1	II	6.1	279	100 ml	E4	P001 IBC02		MP15	T7	TP2
1578	CHLORONITROBENZENES, SOLID	6.1	T2	II	6.1	279	500 g	E4	P002 IBC08	B4	MP10	T3	TP33
1579	4-CHLORO-o-TOLUIDINE HYDROCHLORIDE, SOLID	6.1	T2	III	6.1		5 kg	E1	P002 IBC08 LP02 R001	B3	MP10	T1	TP33
1580	CHLOROPICRIN	6.1	T1	I	6.1	354	0	E0	P601		MP8 MP17	T22	TP2
1581	CHLOROPICRIN AND METHYL BROMIDE MIXTURE with more than 2% chloropicrin	2	2T		2.3		0	E0	P200		MP9	(M) T50	
1582	CHLOROPICRIN AND METHYL CHLORIDE MIXTURE	2	2T		2.3		0	E0	P200		MP9	(M) T50	
1583	CHLOROPICRIN MIXTURE, N.O.S.	6.1	T1	I	6.1	274 315 515	0	E0	P602		MP8 MP17		
1583	CHLOROPICRIN MIXTURE, N.O.S.	6.1	T1	II	6.1	274 515	100 ml	E0	P001 IBC02		MP15		
1583	CHLOROPICRIN MIXTURE, N.O.S.	6.1	T1	III	6.1	274 515	5 L	E0	P001 IBC03 LP01 R001		MP19		

ADR tank		Vehicle for tank carriage	Transport category (Tunnel restriction code)	Special provisions for carriage				Hazard identifi-cation No.	UN No.	Name and description
Tank code	Special provisions			Packages	Bulk	Loading, unloading and handling	Operation			
4.3	4.3.5, 6.8.4	9.1.1.2	1.1.3.6 (8.6)	7.2.4	7.3.3	7.5.11	8.5	5.3.2.3		3.1.2
(12)	(13)	(14)	(15)	(16)	(17)	(18)	(19)	(20)	(1)	(2)
SGAH	TU15 TE19	AT	2 (D/E)	V11		CV13 CV28	S9 S19	60	1558	ARSENIC
SGAH	TU15 TE19	AT	2 (D/E)	V11		CV13 CV28	S9 S19	60	1559	ARSENIC PENTOXIDE
L10CH	TU14 TU15 TE19 TE21	AT	1 (C/E)			CV1 CV13 CV28	S9 S14	66	1560	ARSENIC TRICHLORIDE
SGAH	TU15 TE19	AT	2 (D/E)	V11		CV13 CV28	S9 S19	60	1561	ARSENIC TRIOXIDE
SGAH	TU15 TE19	AT	2 (D/E)	V11		CV13 CV28	S9 S19	60	1562	ARSENICAL DUST
SGAH L4BH	TU15 TE19	AT	2 (D/E)	V11		CV13 CV28	S9 S19	60	1564	BARIUM COMPOUND, N.O.S.
SGAH L4BH	TU15 TE19	AT	2 (E)		VC1 VC2 AP7	CV13 CV28	S9	60	1564	BARIUM COMPOUND, N.O.S.
S10AH	TU15 TE19	AT	1 (C/E)	V10		CV1 CV13 CV28	S9 S14	66	1565	BARIUM CYANIDE
SGAH L4BH	TU15 TE19	AT	2 (D/E)	V11		CV13 CV28	S9 S19	60	1566	BERYLLIUM COMPOUND, N.O.S.
SGAH L4BH	TU15 TE19	AT	2 (E)		VC1 VC2 AP7	CV13 CV28	S9	60	1566	BERYLLIUM COMPOUND, N.O.S.
SGAH	TU15 TE19	AT	2 (D/E)	V11		CV13 CV28	S9 S19	64	1567	BERYLLIUM POWDER
L4BH	TU15 TE19	FL	2 (D/E)			CV13 CV28	S2 S9 S19	63	1569	BROMOACETONE
S10AH L10CH	TU14 TU15 TE19 TE21	AT	1 (C/E)	V10		CV1 CV13 CV28	S9 S14	66	1570	BRUCINE
			1 (B)			CV28	S14		1571	BARIUM AZIDE, WETTED with not less than 50% water, by mass
SGAH	TU15 TE19	AT	2 (D/E)	V11		CV13 CV28	S9 S19	60	1572	CACODYLIC ACID
SGAH	TU15 TE19	AT	2 (D/E)	V11		CV13 CV28	S9 S19	60	1573	CALCIUM ARSENATE
SGAH	TU15 TE19	AT	2 (D/E)	V11		CV13 CV28	S9 S19	60	1574	CALCIUM ARSENATE AND CALCIUM ARSENITE MIXTURE, SOLID
S10AH	TU15 TE19	AT	1 (C/E)	V10		CV1 CV13 CV28	S9 S14	66	1575	CALCIUM CYANIDE
L4BH	TU15 TE19	AT	2 (D/E)			CV13 CV28	S9 S19	60	1577	CHLORODINITRO-BENZENES, LIQUID
SGAH	TU15 TE19	AT	2 (D/E)	V11		CV13 CV28	S9 S19	60	1578	CHLORONITROBENZENES, SOLID
SGAH L4BH	TU15 TE19	AT	2 (E)		VC1 VC2 AP7	CV13 CV28	S9	60	1579	4-CHLORO-o-TOLUIDINE HYDROCHLORIDE, SOLID
L15CH	TU14 TU15 TE19 TE21	AT	1 (C/D)			CV1 CV13 CV28	S9 S14	66	1580	CHLOROPICRIN
PxBH(M)	TA4 TT9	AT	1 (C/D)			CV9 CV10 CV36	S14	26	1581	CHLOROPICRIN AND METHYL BROMIDE MIXTURE with more than 2% chloropicrin
PxBH(M)	TA4 TT9	AT	1 (C/D)			CV9 CV10 CV36	S14	26	1582	CHLOROPICRIN AND METHYL CHLORIDE MIXTURE
L10CH	TU14 TU15 TE19 TE21	AT	1 (C/E)			CV1 CV13 CV28	S9 S14	66	1583	CHLOROPICRIN MIXTURE, N.O.S.
L4BH	TU15 TE19	AT	2 (D/E)			CV13 CV28	S9 S19	60	1583	CHLOROPICRIN MIXTURE, N.O.S.
L4BH	TU15 TE19	AT	2 (E)	V12		CV13 CV28	S9	60	1583	CHLOROPICRIN MIXTURE, N.O.S.

UN No.	Name and description	Class	Classifi-cation code	Packing group	Labels	Special provi-sions	Limited and excepted quantities		Packaging			Portable tanks and bulk containers	
									Packing instruc-tions	Special packing provisions	Mixed packing provisions	Instruc-tions	Special provisions
	3.1.2	2.2	2.2	2.1.1.3	5.2.2	3.3	3.4	3.5.1.2	4.1.4	4.1.4	4.1.10	4.2.5.2 7.3.2	4.2.5.3
(1)	(2)	(3a)	(3b)	(4)	(5)	(6)	(7a)	(7b)	(8)	(9a)	(9b)	(10)	(11)
1585	COPPER ACETOARSENITE	6.1	T5	II	6.1		500 g	E4	P002 IBC08	B4	MP10	T3	TP33
1586	COPPER ARSENITE	6.1	T5	II	6.1		500 g	E4	P002 IBC08	B4	MP10	T3	TP33
1587	COPPER CYANIDE	6.1	T5	II	6.1		500 g	E4	P002 IBC08	B4	MP10	T3	TP33
1588	CYANIDES, INORGANIC, SOLID, N.O.S.	6.1	T5	I	6.1	47 274	0	E5	P002 IBC07		MP18	T6	TP33
1588	CYANIDES, INORGANIC, SOLID, N.O.S.	6.1	T5	II	6.1	47 274	500 g	E4	P002 IBC08	B4	MP10	T3	TP33
1588	CYANIDES, INORGANIC, SOLID, N.O.S.	6.1	T5	III	6.1	47 274	5 kg	E1	P002 IBC08 LP02 R001	B3	MP10	T1	TP33
1589	CYANOGEN CHLORIDE, STABILIZED	2	2TC		2.3 +8	386	0	E0	P200		MP9		
1590	DICHLOROANILINES, LIQUID	6.1	T1	II	6.1	279	100 ml	E4	P001 IBC02		MP15	T7	TP2
1591	o-DICHLOROBENZENE	6.1	T1	III	6.1	279	5 L	E1	P001 IBC03 LP01 R001		MP19	T4	TP1
1593	DICHLOROMETHANE	6.1	T1	III	6.1	516	5 L	E1	P001 IBC03 LP01 R001	B8	MP19	T7	TP2
1594	DIETHYL SULPHATE	6.1	T1	II	6.1		100 ml	E4	P001 IBC02		MP15	T7	TP2
1595	DIMETHYL SULPHATE	6.1	TC1	I	6.1 +8	354	0	E0	P602		MP8 MP17	T20	TP2
1596	DINITROANILINES	6.1	T2	II	6.1		500 g	E4	P002 IBC08	B4	MP10	T3	TP33
1597	DINITROBENZENES, LIQUID	6.1	T1	II	6.1		100 ml	E4	P001 IBC02		MP15	T7	TP2
1597	DINITROBENZENES, LIQUID	6.1	T1	III	6.1		5 L	E1	P001 IBC03 LP01 R001		MP19	T7	TP2
1598	DINITRO-o-CRESOL	6.1	T2	II	6.1	43	500 g	E4	P002 IBC08	B4	MP10	T3	TP33
1599	DINITROPHENOL SOLUTION	6.1	T1	II	6.1		100 ml	E4	P001 IBC02		MP15	T7	TP2
1599	DINITROPHENOL SOLUTION	6.1	T1	III	6.1		5 L	E1	P001 IBC03 LP01 R001		MP19	T4	TP1
1600	DINITROTOLUENES, MOLTEN	6.1	T1	II	6.1		0	E0				T7	TP3
1601	DISINFECTANT, SOLID, TOXIC, N.O.S.	6.1	T2	I	6.1	274	0	E5	P002 IBC07		MP18	T6	TP33
1601	DISINFECTANT, SOLID, TOXIC, N.O.S.	6.1	T2	II	6.1	274	500 g	E4	P002 IBC08	B4	MP10	T3	TP33
1601	DISINFECTANT, SOLID, TOXIC, N.O.S.	6.1	T2	III	6.1	274	5 kg	E1	P002 IBC08 LP02 R001	B3	MP10	T1	TP33
1602	DYE, LIQUID, TOXIC, N.O.S. or DYE INTERMEDIATE, LIQUID, TOXIC, N.O.S.	6.1	T1	I	6.1	274	0	E5	P001		MP8 MP17		
1602	DYE, LIQUID, TOXIC, N.O.S. or DYE INTERMEDIATE, LIQUID, TOXIC, N.O.S.	6.1	T1	II	6.1	274	100 ml	E4	P001 IBC02		MP15		
1602	DYE, LIQUID, TOXIC, N.O.S. or DYE INTERMEDIATE, LIQUID, TOXIC, N.O.S.	6.1	T1	III	6.1	274	5 L	E1	P001 IBC03 LP01 R001		MP19		
1603	ETHYL BROMOACETATE	6.1	TF1	II	6.1 +3		100 ml	E0	P001 IBC02		MP15	T7	TP2
1604	ETHYLENEDIAMINE	8	CF1	II	8 +3		1 L	E2	P001 IBC02		MP15	T7	TP2

ADR tank		Vehicle for tank carriage	Transport category (Tunnel restriction code)	Special provisions for carriage				Hazard identifi- cation No.	UN No.	Name and description
Tank code	Special provisions			Packages	Bulk	Loading, unloading and handling	Operation			
4.3	4.3.5, 6.8.4	9.1.1.2	1.1.3.6 (8.6)	7.2.4	7.3.3	7.5.11	8.5	5.3.2.3		3.1.2
(12)	(13)	(14)	(15)	(16)	(17)	(18)	(19)	(20)	(1)	(2)
SGAH	TU15 TE19	AT	2 (D/E)	V11		CV13 CV28	S9 S19	60	1585	COPPER ACETOARSENITE
SGAH	TU15 TE19	AT	2 (D/E)	V11		CV13 CV28	S9 S19	60	1586	COPPER ARSENITE
SGAH	TU15 TE19	AT	2 (D/E)	V11		CV13 CV28	S9 S19	60	1587	COPPER CYANIDE
S10AH	TU15 TE19	AT	1 (C/E)	V10		CV1 CV13 CV28	S9 S14	66	1588	CYANIDES, INORGANIC, SOLID, N.O.S.
SGAH	TU15 TE19	AT	2 (D/E)	V11		CV13 CV28	S9 S19	60	1588	CYANIDES, INORGANIC, SOLID, N.O.S.
SGAH	TU15 TE19	AT	2 (E)		VC1 VC2 AP7	CV13 CV28	S9	60	1588	CYANIDES, INORGANIC, SOLID, N.O.S.
			1 (D)	V8		CV9 CV10 CV36	S4 S14		1589	CYANOGEN CHLORIDE, STABILIZED
L4BH	TU15 TE19	AT	2 (D/E)			CV13 CV28	S9 S19	60	1590	DICHLOROANILINES, LIQUID
L4BH	TU15 TE19	AT	2 (E)	V12		CV13 CV28	S9	60	1591	o-DICHLOROBENZENE
L4BH	TU15 TE19	AT	2 (E)	V12		CV13 CV28	S9	60	1593	DICHLOROMETHANE
L4BH	TU15 TE19	AT	2 (D/E)			CV13 CV28	S9 S19	60	1594	DIETHYL SULPHATE
L10CH	TU14 TU15 TE19 TE21	AT	1 (C/D)			CV1 CV13 CV28	S9 S14	668	1595	DIMETHYL SULPHATE
SGAH L4BH	TU15 TE19	AT	2 (D/E)	V11		CV13 CV28	S9 S19	60	1596	DINITROANILINES
L4BH	TU15 TE19	AT	2 (D/E)			CV13 CV28	S9 S19	60	1597	DINITROBENZENES, LIQUID
L4BH	TU15 TE19	AT	2 (E)	V12		CV13 CV28	S9	60	1597	DINITROBENZENES, LIQUID
SGAH L4BH	TU15 TE19	AT	2 (D/E)	V11		CV13 CV28	S9 S19	60	1598	DINITRO-o-CRESOL
L4BH	TU15 TE19	AT	2 (D/E)			CV13 CV28	S9 S19	60	1599	DINITROPHENOL SOLUTION
L4BH	TU15 TE19	AT	2 (E)	V12		CV13 CV28	S9	60	1599	DINITROPHENOL SOLUTION
L4BH	TU15 TE19	AT	0 (D/E)			CV13	S9 S19	60	1600	DINITROTOLUENES, MOLTEN
S10AH L10CH	TU15 TE19	AT	1 (C/E)	V10		CV1 CV13 CV28	S9 S14	66	1601	DISINFECTANT, SOLID, TOXIC, N.O.S.
SGAH L4BH	TU15 TE19	AT	2 (D/E)	V11		CV13 CV28	S9 S19	60	1601	DISINFECTANT, SOLID, TOXIC, N.O.S.
SGAH L4BH	TU15 TE19	AT	2 (E)		VC1 VC2 AP7	CV13 CV28	S9	60	1601	DISINFECTANT, SOLID, TOXIC, N.O.S.
L10CH	TU14 TU15 TE19 TE21	AT	1 (C/E)			CV1 CV13 CV28	S9 S14	66	1602	DYE, LIQUID, TOXIC, N.O.S. or DYE INTERMEDIATE, LIQUID, TOXIC, N.O.S.
L4BH	TU15 TE19	AT	2 (D/E)			CV13 CV28	S9 S19	60	1602	DYE, LIQUID, TOXIC, N.O.S. or DYE INTERMEDIATE, LIQUID, TOXIC, N.O.S.
L4BH	TU15 TE19	AT	2 (E)	V12		CV13 CV28	S9	60	1602	DYE, LIQUID. TOXIC, N.O.S. or DYE INTERMEDIATE, LIQUID, TOXIC, N.O.S.
L4BH	TU15 TE19	FL	2 (D/E)			CV13 CV28	S2 S9 S19	63	1603	ETHYL BROMOACETATE
L4BN		FL	2 (D/E)				S2	83	1604	ETHYLENEDIAMINE

UN No.	Name and description	Class	Classifi-cation code	Packing group	Labels	Special provi-sions	Limited and excepted quantities		Packaging			Portable tanks and bulk containers	
									Packing instruc-tions	Special packing provisions	Mixed packing provisions	Instruc-tions	Special provisions
	3.1.2	2.2	2.2	2.1.1.3	5.2.2	3.3	3.4	3.5.1.2	4.1.4	4.1.4	4.1.10	4.2.5.2 7.3.2	4.2.5.3
(1)	(2)	(3a)	(3b)	(4)	(5)	(6)	(7a)	(7b)	(8)	(9a)	(9b)	(10)	(11)
1605	ETHYLENE DIBROMIDE	6.1	T1	I	6.1	354	0	E0	P602		MP8 MP17	T20	TP2
1606	FERRIC ARSENATE	6.1	T5	II	6.1		500 g	E4	P002 IBC08	B4	MP10	T3	TP33
1607	FERRIC ARSENITE	6.1	T5	II	6.1		500 g	E4	P002 IBC08	B4	MP10	T3	TP33
1608	FERROUS ARSENATE	6.1	T5	II	6.1		500 g	E4	P002 IBC08	B4	MP10	T3	TP33
1611	HEXAETHYL TETRAPHOSPHATE	6.1	T1	II	6.1		100 ml	E4	P001 IBC02		MP15	T7	TP2
1612	HEXAETHYL TETRAPHOSPHATE AND COMPRESSED GAS MIXTURE	2	1T		2.3		0	E0	P200		MP9	(M)	
1613	HYDROCYANIC ACID, AQUEOUS SOLUTION (HYDROGEN CYANIDE, AQUEOUS SOLUTION) with not more than 20% hydrogen cyanide	6.1	TF1	I	6.1 +3	48	0	E0	P601		MP8 MP17	T14	TP2
1614	HYDROGEN CYANIDE, STABILIZED, containing less than 3% water and absorbed in a porous inert material	6.1	TF1	I	6.1 +3	386 603	0	E0	P099 P601	RR10	MP2		
1616	LEAD ACETATE	6.1	T5	III	6.1		5 kg	E1	P002 IBC08 LP02 R001	B3	MP10	T1	TP33
1617	LEAD ARSENATES	6.1	T5	II	6.1		500 g	E4	P002 IBC08	B4	MP10	T3	TP33
1618	LEAD ARSENITES	6.1	T5	II	6.1		500 g	E4	P002 IBC08	B4	MP10	T3	TP33
1620	LEAD CYANIDE	6.1	T5	II	6.1		500 g	E4	P002 IBC08	B4	MP10	T3	TP33
1621	LONDON PURPLE	6.1	T5	II	6.1	43	500 g	E4	P002 IBC08	B4	MP10	T3	TP33
1622	MAGNESIUM ARSENATE	6.1	T5	II	6.1		500 g	E4	P002 IBC08	B4	MP10	T3	TP33
1623	MERCURIC ARSENATE	6.1	T5	II	6.1		500 g	E4	P002 IBC08	B4	MP10	T3	TP33
1624	MERCURIC CHLORIDE	6.1	T5	II	6.1		500 g	E4	P002 IBC08	B4	MP10	T3	TP33
1625	MERCURIC NITRATE	6.1	T5	II	6.1		500 g	E4	P002 IBC08	B4	MP10	T3	TP33
1626	MERCURIC POTASSIUM CYANIDE	6.1	T5	I	6.1		0	E5	P002 IBC07		MP18	T6	TP33
1627	MERCUROUS NITRATE	6.1	T5	II	6.1		500 g	E4	P002 IBC08	B4	MP10	T3	TP33
1629	MERCURY ACETATE	6.1	T5	II	6.1		500 g	E4	P002 IBC08	B4	MP10	T3	TP33
1630	MERCURY AMMONIUM CHLORIDE	6.1	T5	II	6.1		500 g	E4	P002 IBC08	B4	MP10	T3	TP33
1631	MERCURY BENZOATE	6.1	T5	II	6.1		500 g	E4	P002 IBC08	B4	MP10	T3	TP33
1634	MERCURY BROMIDES	6.1	T5	II	6.1		500 g	E4	P002 IBC08	B4	MP10	T3	TP33
1636	MERCURY CYANIDE	6.1	T5	II	6.1		500 g	E4	P002 IBC08	B4	MP10	T3	TP33
1637	MERCURY GLUCONATE	6.1	T5	II	6.1		500 g	E4	P002 IBC08	B4	MP10	T3	TP33
1638	MERCURY IODIDE	6.1	T5	II	6.1		500 g	E4	P002 IBC08	B4	MP10	T3	TP33
1639	MERCURY NUCLEATE	6.1	T5	II	6.1		500 g	E4	P002 IBC08	B4	MP10	T3	TP33
1640	MERCURY OLEATE	6.1	T5	II	6.1		500 g	E4	P002 IBC08	B4	MP10	T3	TP33
1641	MERCURY OXIDE	6.1	T5	II	6.1		500 g	E4	P002 IBC08	B4	MP10	T3	TP33
1642	MERCURY OXYCYANIDE, DESENSITIZED	6.1	T5	II	6.1		500 g	E4	P002 IBC08	B4	MP10	T3	TP33
1643	MERCURY POTASSIUM IODIDE	6.1	T5	II	6.1		500 g	E4	P002 IBC08	B4	MP10	T3	TP33
1644	MERCURY SALICYLATE	6.1	T5	II	6.1		500 g	E4	P002 IBC08	B4	MP10	T3	TP33

ADR tank		Vehicle for tank carriage	Transport category (Tunnel restriction code)	Special provisions for carriage				Hazard identification No.	UN No.	Name and description
Tank code	Special provisions			Packages	Bulk	Loading, unloading and handling	Operation			
4.3	4.3.5, 6.8.4	9.1.1.2	1.1.3.6 (8.6)	7.2.4	7.3.3	7.5.11	8.5	5.3.2.3		3.1.2
(12)	(13)	(14)	(15)	(16)	(17)	(18)	(19)	(20)	(1)	(2)
L10CH	TU14 TU15 TE19 TE21	AT	1 (C/D)			CV1 CV13 CV28	S9 S14	66	1605	ETHYLENE DIBROMIDE
SGAH	TU15 TE19	AT	2 (D/E)	V11		CV13 CV28	S9 S19	60	1606	FERRIC ARSENATE
SGAH	TU15 TE19	AT	2 (D/E)	V11		CV13 CV28	S9 S19	60	1607	FERRIC ARSENITE
SGAH	TU15 TE19	AT	2 (D/E)	V11		CV13 CV28	S9 S19	60	1608	FERROUS ARSENATE
L4BH	TU15 TE19	AT	2 (D/E)			CV13 CV28	S9 S19	60	1611	HEXAETHYL TETRAPHOSPHATE
CxBH(M)	TA4 TT9	AT	1 (C/D)			CV9 CV10 CV36	S14	26	1612	HEXAETHYL TETRAPHOSPHATE AND COMPRESSED GAS MIXTURE
L15DH(+)	TU14 TU15 TE19 TE21	FL	0 (C/D)			CV1 CV13 CV28	S2 S9 S14	663	1613	HYDROCYANIC ACID, AQUEOUS SOLUTION (HYDROGEN CYANIDE, AQUEOUS SOLUTION) with not more than 20% hydrogen cyanide
			0 (D)	V8		CV1 CV13 CV28	S2 S4 S9 S10 S14		1614	HYDROGEN CYANIDE, STABILIZED, containing less than 3% water and absorbed in a porous inert material
SGAH L4BH	TU15 TE19	AT	2 (E)		VC1 VC2 AP7	CV13 CV28	S9	60	1616	LEAD ACETATE
SGAH	TU15 TE19	AT	2 (D/E)	V11		CV13 CV28	S9 S19	60	1617	LEAD ARSENATES
SGAH	TU15 TE19	AT	2 (D/E)	V11		CV13 CV28	S9 S19	60	1618	LEAD ARSENITES
SGAH	TU15 TE19	AT	2 (D/E)	V11		CV13 CV28	S9 S19	60	1620	LEAD CYANIDE
SGAH	TU15 TE19	AT	2 (D/E)	V11		CV13 CV28	S9 S19	60	1621	LONDON PURPLE
SGAH	TU15 TE19	AT	2 (D/E)	V11		CV13 CV28	S9 S19	60	1622	MAGNESIUM ARSENATE
SGAH	TU15 TE19	AT	2 (D/E)	V11		CV13 CV28	S9 S19	60	1623	MERCURIC ARSENATE
SGAH	TU15 TE19	AT	2 (D/E)	V11		CV13 CV28	S9 S19	60	1624	MERCURIC CHLORIDE
SGAH	TU15 TE19	AT	2 (D/E)	V11		CV13 CV28	S9 S19	60	1625	MERCURIC NITRATE
S10AH	TU15 TE19	AT	1 (C/E)	V10		CV1 CV13 CV28	S9 S14	66	1626	MERCURIC POTASSIUM CYANIDE
SGAH	TU15 TE19	AT	2 (D/E)	V11		CV13 CV28	S9 S19	60	1627	MERCUROUS NITRATE
SGAH	TU15 TE19	AT	2 (D/E)	V11		CV13 CV28	S9 S19	60	1629	MERCURY ACETATE
SGAH	TU15 TE19	AT	2 (D/E)	V11		CV13 CV28	S9 S19	60	1630	MERCURY AMMONIUM CHLORIDE
SGAH	TU15 TE19	AT	2 (D/E)	V11		CV13 CV28	S9 S19	60	1631	MERCURY BENZOATE
SGAH	TU15 TE19	AT	2 (D/E)	V11		CV13 CV28	S9 S19	60	1634	MERCURY BROMIDES
SGAH	TU15 TE19	AT	2 (D/E)	V11		CV13 CV28	S9 S19	60	1636	MERCURY CYANIDE
SGAH	TU15 TE19	AT	2 (D/E)	V11		CV13 CV28	S9 S19	60	1637	MERCURY GLUCONATE
SGAH	TU15 TE19	AT	2 (D/E)	V11		CV13 CV28	S9 S19	60	1638	MERCURY IODIDE
SGAH	TU15 TE19	AT	2 (D/E)	V11		CV13 CV28	S9 S19	60	1639	MERCURY NUCLEATE
SGAH	TU15 TE19	AT	2 (D/E)	V11		CV13 CV28	S9 S19	60	1640	MERCURY OLEATE
SGAH	TU15 TE19	AT	2 (D/E)	V11		CV13 CV28	S9 S19	60	1641	MERCURY OXIDE
SGAH	TU15 TE19	AT	2 (D/E)	V11		CV13 CV28	S9 S19	60	1642	MERCURY OXYCYANIDE, DESENSITIZED
SGAH	TU15 TE19	AT	2 (D/E)	V11		CV13 CV28	S9 S19	60	1643	MERCURY POTASSIUM IODIDE
SGAH	TU15 TE19	AT	2 (D/E)	V11		CV13 CV28	S9 S19	60	1644	MERCURY SALICYLATE

UN No.	Name and description	Class	Classifi-cation code	Packing group	Labels	Special provi-sions	Limited and excepted quantities		Packaging			Portable tanks and bulk containers	
									Packing instruc-tions	Special packing provisions	Mixed packing provisions	Instruc-tions	Special provisions
3.1.2		2.2	2.2	2.1.1.3	5.2.2	3.3	3.4	3.5.1.2	4.1.4	4.1.4	4.1.10	4.2.5.2 7.3.2	4.2.5.3
(1)	(2)	(3a)	(3b)	(4)	(5)	(6)	(7a)	(7b)	(8)	(9a)	(9b)	(10)	(11)
1645	MERCURY SULPHATE	6.1	T5	II	6.1		500 g	E4	P002 IBC08	B4	MP10	T3	TP33
1646	MERCURY THIOCYANATE	6.1	T5	II	6.1		500 g	E4	P002 IBC08	B4	MP10	T3	TP33
1647	METHYL BROMIDE AND ETHYLENE DIBROMIDE MIXTURE, LIQUID	6.1	T1	I	6.1	354	0	E0	P602		MP8 MP17	T20	TP2
1648	ACETONITRILE	3	F1	II	3		1 L	E2	P001 IBC02 R001		MP19	T7	TP2
1649	MOTOR FUEL ANTI-KNOCK MIXTURE	6.1	T3	I	6.1		0	E0	P602		MP8 MP17	T14	TP2
1650	beta-NAPHTHYLAMINE, SOLID	6.1	T2	II	6.1		500 g	E4	P002 IBC08	B4	MP10	T3	TP33
1651	NAPHTHYLTHIOUREA	6.1	T2	II	6.1	43	500 g	E4	P002 IBC08	B4	MP10	T3	TP33
1652	NAPHTHYLUREA	6.1	T2	II	6.1		500 g	E4	P002 IBC08	B4	MP10	T3	TP33
1653	NICKEL CYANIDE	6.1	T5	II	6.1		500 g	E4	P002 IBC08	B4	MP10	T3	TP33
1654	NICOTINE	6.1	T1	II	6.1		100 ml	E4	P001 IBC02		MP15		
1655	NICOTINE COMPOUND, SOLID, N.O.S. or NICOTINE PREPARATION, SOLID, N.O.S.	6.1	T2	I	6.1	43 274	0	E5	P002 IBC07		MP18	T6	TP33
1655	NICOTINE COMPOUND, SOLID, N.O.S. or NICOTINE PREPARATION, SOLID, N.O.S.	6.1	T2	II	6.1	43 274	500 g	E4	P002 IBC08	B4	MP10	T3	TP33
1655	NICOTINE COMPOUND, SOLID, N.O.S. or NICOTINE PREPARATION, SOLID, N.O.S.	6.1	T2	III	6.1	43 274	5 kg	E1	P002 IBC08 LP02 R001	B3	MP10	T1	TP33
1656	NICOTINE HYDROCHLORIDE, LIQUID or SOLUTION	6.1	T1	II	6.1	43	100 ml	E4	P001 IBC02		MP15		
1656	NICOTINE HYDROCHLORIDE, LIQUID or SOLUTION	6.1	T1	III	6.1	43	5 L	E1	P001 IBC03 LP01 R001		MP19		
1657	NICOTINE SALICYLATE	6.1	T2	II	6.1		500 g	E4	P002 IBC08	B4	MP10	T3	TP33
1658	NICOTINE SULPHATE, SOLUTION	6.1	T1	II	6.1		100 ml	E4	P001 IBC02		MP15	T7	TP2
1658	NICOTINE SULPHATE, SOLUTION	6.1	T1	III	6.1		5 L	E1	P001 IBC03 LP01 R001		MP19	T7	TP2
1659	NICOTINE TARTRATE	6.1	T2	II	6.1		500 g	E4	P002 IBC08	B4	MP10	T3	TP33
1660	NITRIC OXIDE, COMPRESSED	2	1TOC		2.3 +5.1 +8		0	E0	P200		MP9		
1661	NITROANILINES (o-, m-, p-)	6.1	T2	II	6.1	279	500 g	E4	P002 IBC08	B4	MP10	T3	TP33
1662	NITROBENZENE	6.1	T1	II	6.1	279	100 ml	E4	P001 IBC02		MP15	T7	TP2
1663	NITROPHENOLS (o-, m-, p-)	6.1	T2	III	6.1	279	5 kg	E1	P002 IBC08 LP02 R001	B3	MP10	T1	TP33
1664	NITROTOLUENES, LIQUID	6.1	T1	II	6.1		100 ml	E4	P001 IBC02		MP15	T7	TP2
1665	NITROXYLENES, LIQUID	6.1	T1	II	6.1		100 ml	E4	P001 IBC02		MP15	T7	TP2
1669	PENTACHLOROETHANE	6.1	T1	II	6.1		100 ml	E4	P001 IBC02		MP15	T7	TP2
1670	PERCHLOROMETHYL MERCAPTAN	6.1	T1	I	6.1	354	0	E0	P602		MP8 MP17	T20	TP2
1671	PHENOL, SOLID	6.1	T2	II	6.1	279	500 g	E4	P002 IBC08	B4	MP10	T3	TP33
1672	PHENYLCARBYLAMINE CHLORIDE	6.1	T1	I	6.1		0	E0	P602		MP8 MP17	T14	TP2

ADR tank		Vehicle for tank carriage	Transport category (Tunnel restriction code)	Special provisions for carriage				Hazard identification No.	UN No.	Name and description
Tank code	Special provisions			Packages	Bulk	Loading, unloading and handling	Operation			
4.3	4.3.5, 6.8.4	9.1.1.2	1.1.3.6 (8.6)	7.2.4	7.3.3	7.5.11	8.5	5.3.2.3		3.1.2
(12)	(13)	(14)	(15)	(16)	(17)	(18)	(19)	(20)	(1)	(2)
SGAH	TU15 TE19	AT	2 (D/E)	V11		CV13 CV28	S9 S19	60	1645	MERCURY SULPHATE
SGAH	TU15 TE19	AT	2 (D/E)	V11		CV13 CV28	S9 S19	60	1646	MERCURY THIOCYANATE
L10CH	TU14 TU15 TE19 TE21	AT	1 (C/D)			CV1 CV13 CV28	S9 S14	66	1647	METHYL BROMIDE AND ETHYLENE DIBROMIDE MIXTURE, LIQUID
LGBF		FL	2 (D/E)				S2 S20	33	1648	ACETONITRILE
L10CH	TU14 TU15 TE19 TE21 TT6	AT	1 (C/E)			CV1 CV13 CV28	S9 S14	66	1649	MOTOR FUEL ANTI-KNOCK MIXTURE
SGAH L4BH	TU15 TE19	AT	2 (D/E)	V11		CV13 CV28	S9 S19	60	1650	beta-NAPHTHYLAMINE, SOLID
SGAH	TU15 TE19	AT	2 (D/E)	V11		CV13 CV28	S9 S19	60	1651	NAPHTHYLTHIOUREA
SGAH	TU15 TE19	AT	2 (D/E)	V11		CV13 CV28	S9 S19	60	1652	NAPHTHYLUREA
SGAH L4BH	TU15 TE19	AT	2 (D/E)	V11		CV13 CV28	S9 S19	60	1653	NICKEL CYANIDE
L4BH	TU15 TE19	AT	2 (D/E)			CV13 CV28	S9 S19	60	1654	NICOTINE
S10AH L10CH	TU15 TE19	AT	1 (C/E)	V10		CV1 CV13 CV28	S9 S14	66	1655	NICOTINE COMPOUND, SOLID, N.O.S. or NICOTINE PREPARATION, SOLID, N.O.S.
SGAH L4BH	TU15 TE19	AT	2 (D/E)	V11		CV13 CV28	S9 S19	60	1655	NICOTINE COMPOUND, SOLID, N.O.S. or NICOTINE PREPARATION, SOLID, N.O.S.
SGAH L4BH	TU15 TE19	AT	2 (E)		VC1 VC2 AP7	CV13 CV28	S9	60	1655	NICOTINE COMPOUND, SOLID, N.O.S. or NICOTINE PREPARATION, SOLID, N.O.S.
L4BH	TU15 TE19	AT	2 (D/E)			CV13 CV28	S9 S19	60	1656	NICOTINE HYDROCHLORIDE, LIQUID or SOLUTION
L4BH	TU15 TE19	AT	2 (E)	V12		CV13 CV28	S9	60	1656	NICOTINE HYDROCHLORIDE, LIQUID or SOLUTION
SGAH L4BH	TU15 TE19	AT	2 (D/E)	V11		CV13 CV28	S9 S19	60	1657	NICOTINE SALICYLATE
L4BH	TU15 TE19	AT	2 (D/E)			CV13 CV28	S9 S19	60	1658	NICOTINE SULPHATE, SOLUTION
L4BH	TU15 TE19	AT	2 (E)	V12		CV13 CV28	S9	60	1658	NICOTINE SULPHATE, SOLUTION
SGAH L4BH	TU15 TE19	AT	2 (D/E)	V11		CV13 CV28	S9 S19	60	1659	NICOTINE TARTRATE
			1 (D)			CV9 CV10 CV36	S14		1660	NITRIC OXIDE, COMPRESSED
SGAH L4BH	TU15 TE19	AT	2 (D/E)	V11		CV13 CV28	S9 S19	60	1661	NITROANILINES (o-, m-, p-)
L4BH	TU15 TE19	AT	2 (D/E)			CV13 CV28	S9 S19	60	1662	NITROBENZENE
SGAH L4BH	TU15 TE19	AT	2 (E)		VC1 VC2 AP7	CV13 CV28	S9	60	1663	NITROPHENOLS (o-, m-, p-)
L4BH	TU15 TE19	AT	2 (D/E)			CV13 CV28	S9 S19	60	1664	NITROTOLUENES, LIQUID
L4BH	TU15 TE19	AT	2 (D/E)			CV13 CV28	S9 S19	60	1665	NITROXYLENES, LIQUID
L4BH	TU15 TE19	AT	2 (D/E)			CV13 CV28	S9 S19	60	1669	PENTACHLOROETHANE
L10CH	TU14 TU15 TE19 TE21	AT	1 (C/D)			CV1 CV13 CV28	S9 S14	66	1670	PERCHLOROMETHYL MERCAPTAN
SGAH	TU15 TE19	AT	2 (D/E)	V11		CV13 CV28	S9 S19	60	1671	PHENOL, SOLID
L10CH	TU14 TU15 TE19 TE21	AT	1 (C/E)			CV1 CV13 CV28	S9 S14	66	1672	PHENYLCARBYLAMINE CHLORIDE

UN No.	Name and description	Class	Classifi-cation code	Packing group	Labels	Special provi-sions	Limited and excepted quantities		Packaging			Portable tanks and bulk containers	
									Packing instruc-tions	Special packing provisions	Mixed packing provisions	Instruc-tions	Special provisions
	3.1.2	2.2	2.2	2.1.1.3	5.2.2	3.3	3.4	3.5.1.2	4.1.4	4.1.4	4.1.10	4.2.5.2 7.3.2	4.2.5.3
(1)	(2)	(3a)	(3b)	(4)	(5)	(6)	(7a)	(7b)	(8)	(9a)	(9b)	(10)	(11)
1673	PHENYLENEDIAMINES (o-, m-, p-)	6.1	T2	III	6.1	279	5 kg	E1	P002 IBC08 LP02 R001	B3	MP10	T1	TP33
1674	PHENYLMERCURIC ACETATE	6.1	T3	II	6.1	43	500 g	E4	P002 IBC08	B4	MP10	T3	TP33
1677	POTASSIUM ARSENATE	6.1	T5	II	6.1		500 g	E4	P002 IBC08	B4	MP10	T3	TP33
1678	POTASSIUM ARSENITE	6.1	T5	II	6.1		500 g	E4	P002 IBC08	B4	MP10	T3	TP33
1679	POTASSIUM CUPROCYANIDE	6.1	T5	II	6.1		500 g	E4	P002 IBC08	B4	MP10	T3	TP33
1680	POTASSIUM CYANIDE, SOLID	6.1	T5	I	6.1		0	E5	P002 IBC07		MP18	T6	TP33
1683	SILVER ARSENITE	6.1	T5	II	6.1		500 g	E4	P002 IBC08	B4	MP10	T3	TP33
1684	SILVER CYANIDE	6.1	T5	II	6.1		500 g	E4	P002 IBC08	B4	MP10	T3	TP33
1685	SODIUM ARSENATE	6.1	T5	II	6.1		500 g	E4	P002 IBC08	B4	MP10	T3	TP33
1686	SODIUM ARSENITE, AQUEOUS SOLUTION	6.1	T4	II	6.1	43	100 ml	E4	P001 IBC02		MP15	T7	TP2
1686	SODIUM ARSENITE, AQUEOUS SOLUTION	6.1	T4	III	6.1	43	5 L	E1	P001 IBC03 LP01 R001		MP19	T4	TP2
1687	SODIUM AZIDE	6.1	T5	II	6.1		500 g	E4	P002 IBC08	B4	MP10		
1688	SODIUM CACODYLATE	6.1	T5	II	6.1		500 g	E4	P002 IBC08	B4	MP10	T3	TP33
1689	SODIUM CYANIDE, SOLID	6.1	T5	I	6.1		0	E5	P002 IBC07		MP18	T6	TP33
1690	SODIUM FLUORIDE, SOLID	6.1	T5	III	6.1		5 kg	E1	P002 IBC08 LP02 R001	B3	MP10	T1	TP33
1691	STRONTIUM ARSENITE	6.1	T5	II	6.1		500 g	E4	P002 IBC08	B4	MP10	T3	TP33
1692	STRYCHNINE or STRYCHNINE SALTS	6.1	T2	I	6.1		0	E5	P002 IBC07		MP18	T6	TP33
1693	TEAR GAS SUBSTANCE, LIQUID, N.O.S.	6.1	T1	I	6.1	274	0	E0	P001		MP8 MP17		
1693	TEAR GAS SUBSTANCE, LIQUID, N.O.S.	6.1	T1	II	6.1	274	0	E0	P001 IBC02		MP15		
1694	BROMOBENZYL CYANIDES, LIQUID	6.1	T1	I	6.1	138	0	E0	P001		MP8 MP17	T14	TP2
1695	CHLOROACETONE, STABILIZED	6.1	TFC	I	6.1 +3 +8	354	0	E0	P602		MP8 MP17	T20	TP2
1697	CHLOROACETOPHENONE, SOLID	6.1	T2	II	6.1		0	E0	P002 IBC08	B4	MP10	T3	TP33
1698	DIPHENYLAMINE CHLOROARSINE	6.1	T3	I	6.1		0	E0	P002		MP18	T6	TP33
1699	DIPHENYLCHLORO-ARSINE, LIQUID	6.1	T3	I	6.1		0	E0	P001		MP8 MP17		
1700	TEAR GAS CANDLES	6.1	TF3		6.1 +4.1		0	E0	P600				
1701	XYLYL BROMIDE, LIQUID	6.1	T1	II	6.1		0	E0	P001 IBC02		MP15	T7	TP2
1702	1,1,2,2-TETRACHLOROETHANE	6.1	T1	II	6.1		100 ml	E4	P001 IBC02		MP15	T7	TP2
1704	TETRAETHYL DITHIOPYROPHOSPHATE	6.1	T1	II	6.1	43	100 ml	E4	P001 IBC02		MP15	T7	TP2
1707	THALLIUM COMPOUND, N.O.S.	6.1	T5	II	6.1	43 274	500 g	E4	P002 IBC08	B4	MP10	T3	TP33
1708	TOLUIDINES, LIQUID	6.1	T1	II	6.1	279	100 ml	E4	P001 IBC02		MP15	T7	TP2

ADR tank		Vehicle for tank carriage	Transport category (Tunnel restriction code)	Special provisions for carriage				Hazard identification No.	UN No.	Name and description
Tank code	Special provisions			Packages	Bulk	Loading, unloading and handling	Operation			
4.3	4.3.5, 6.8.4	9.1.1.2	1.1.3.6 (8.6)	7.2.4	7.3.3	7.5.11	8.5	5.3.2.3		3.1.2
(12)	(13)	(14)	(15)	(16)	(17)	(18)	(19)	(20)	(1)	(2)
SGAH L4BH	TU15 TE19	AT	2 (E)		VC1 VC2 AP7	CV13 CV28	S9	60	1673	PHENYLENEDIAMINES (o-, m-, p-)
SGAH L4BH	TU15 TE19	AT	2 (D/E)	V11		CV13 CV28	S9 S19	60	1674	PHENYLMERCURIC ACETATE
SGAH	TU15 TE19	AT	2 (D/E)	V11		CV13 CV28	S9 S19	60	1677	POTASSIUM ARSENATE
SGAH	TU15 TE19	AT	2 (D/E)	V11		CV13 CV28	S9 S19	60	1678	POTASSIUM ARSENITE
SGAH	TU15 TE19	AT	2 (D/E)	V11		CV13 CV28	S9 S19	60	1679	POTASSIUM CUPROCYANIDE
S10AH	TU15 TE19	AT	1 (C/E)	V10		CV1 CV13 CV28	S9 S14	66	1680	POTASSIUM CYANIDE, SOLID
SGAH	TU15 TE19	AT	2 (D/E)	V11		CV13 CV28	S9 S19	60	1683	SILVER ARSENITE
SGAH	TU15 TE19	AT	2 (D/E)	V11		CV13 CV28	S9 S19	60	1684	SILVER CYANIDE
SGAH	TU15 TE19	AT	2 (D/E)	V11		CV13 CV28	S9 S19	60	1685	SODIUM ARSENATE
L4BH	TU15 TE19	AT	2 (D/E)			CV13 CV28	S9 S19	60	1686	SODIUM ARSENITE, AQUEOUS SOLUTION
L4BH	TU15 TE19	AT	2 (E)	V12		CV13 CV28	S9	60	1686	SODIUM ARSENITE, AQUEOUS SOLUTION
			2 (E)	V11		CV13 CV28	S9 S19		1687	SODIUM AZIDE
SGAH	TU15 TE19	AT	2 (D/E)	V11		CV13 CV28	S9 S19	60	1688	SODIUM CACODYLATE
S10AH	TU15 TE19	AT	1 (C/E)	V10		CV1 CV13 CV28	S9 S14	66	1689	SODIUM CYANIDE, SOLID
SGAH	TU15 TE19	AT	2 (E)		VC1 VC2 AP7	CV13 CV28	S9	60	1690	SODIUM FLUORIDE, SOLID
SGAH	TU15 TE19	AT	2 (D/E)	V11		CV13 CV28	S9 S19	60	1691	STRONTIUM ARSENITE
S10AH	TU15 TE19	AT	1 (C/E)	V10		CV1 CV13 CV28	S9 S14	66	1692	STRYCHNINE or STRYCHNINE SALTS
L10CH	TU14 TU15 TE19 TE21	AT	1 (C/E)			CV1 CV13 CV28	S9 S14	66	1693	TEAR GAS SUBSTANCE, LIQUID, N.O.S.
L4BH	TU15 TE19	AT	2 (D/E)			CV13 CV28	S9 S19	60	1693	TEAR GAS SUBSTANCE, LIQUID, N.O.S.
L10CH	TU14 TU15 TE19 TE21	AT	1 (C/E)			CV1 CV13 CV28	S9 S14	66	1694	BROMOBENZYL CYANIDES, LIQUID
L10CH	TU14 TU15 TE19 TE21	FL	1 (C/D)			CV1 CV13 CV28	S2 S9 S14	663	1695	CHLOROACETONE, STABILIZED
SGAH L4BH	TU15 TE19	AT	2 (D/E)	V11		CV13 CV28	S9 S19	60	1697	CHLOROACETOPHENONE, SOLID
S10AH	TU15 TE19	AT	1 (C/E)			CV1 CV13 CV28	S9 S14	66	1698	DIPHENYLAMINE CHLOROARSINE
L10CH	TU14 TU15 TE19 TE21	AT	1 (C/E)			CV1 CV13 CV28	S9 S14	66	1699	DIPHENYLCHLORO-ARSINE, LIQUID
			2 (E)			CV13 CV28	S9 S19		1700	TEAR GAS CANDLES
L4BH	TU15 TE19	AT	2 (D/E)			CV13 CV28	S9 S19	60	1701	XYLYL BROMIDE, LIQUID
L4BH	TU15 TE19	AT	2 (D/E)			CV13 CV28	S9 S19	60	1702	1,1,2,2-TETRACHLOROETHANE
L4BH	TU15 TE19	AT	2 (D/E)			CV13 CV28	S9 S19	60	1704	TETRAETHYL DITHIOPYROPHOSPHATE
SGAH L4BH	TU15 TE19	AT	2 (D/E)	V11		CV13 CV28	S9 S19	60	1707	THALLIUM COMPOUND, N.O.S.
L4BH	TU15 TE19	AT	2 (D/E)			CV13 CV28	S9 S19	60	1708	TOLUIDINES, LIQUID

UN No.	Name and description	Class	Classification code	Packing group	Labels	Special provisions	Limited and excepted quantities		Packaging			Portable tanks and bulk containers	
									Packing instructions	Special packing provisions	Mixed packing provisions	Instructions	Special provisions
	3.1.2	2.2	2.2	2.1.1.3	5.2.2	3.3	3.4	3.5.1.2	4.1.4	4.1.4	4.1.10	4.2.5.2 7.3.2	4.2.5.3
(1)	(2)	(3a)	(3b)	(4)	(5)	(6)	(7a)	(7b)	(8)	(9a)	(9b)	(10)	(11)
1709	2,4-TOLUYLENEDIAMINE, SOLID	6.1	T2	III	6.1		5 kg	E1	P002 IBC08 LP02 R001	B3	MP10	T1	TP33
1710	TRICHLOROETHYLENE	6.1	T1	III	6.1		5 L	E1	P001 IBC03 LP01 R001		MP19	T4	TP1
1711	XYLIDINES, LIQUID	6.1	T1	II	6.1		100 ml	E4	P001 IBC02		MP15	T7	TP2
1712	ZINC ARSENATE, ZINC ARSENITE or ZINC ARSENATE AND ZINC ARSENITE MIXTURE	6.1	T5	II	6.1		500 g	E4	P002 IBC08	B4	MP10	T3	TP33
1713	ZINC CYANIDE	6.1	T5	I	6.1		0	E5	P002 IBC07		MP18	T6	TP33
1714	ZINC PHOSPHIDE	4.3	WT2	I	4.3 +6.1		0	E0	P403		MP2		
1715	ACETIC ANHYDRIDE	8	CF1	II	8 +3		1 L	E2	P001 IBC02		MP15	T7	TP2
1716	ACETYL BROMIDE	8	C3	II	8		1 L	E2	P001 IBC02		MP15	T8	TP2
1717	ACETYL CHLORIDE	3	FC	II	3 +8		1 L	E2	P001 IBC02		MP19	T8	TP2
1718	BUTYL ACID PHOSPHATE	8	C3	III	8		5 L	E1	P001 IBC03 LP01 R001		MP19	T4	TP1
1719	CAUSTIC ALKALI LIQUID, N.O.S.	8	C5	II	8	274	1 L	E2	P001 IBC02		MP15	T11	TP2 TP27
1719	CAUSTIC ALKALI LIQUID, N.O.S.	8	C5	III	8	274	5 L	E1	P001 IBC03 R001		MP19	T7	TP1 TP28
1722	ALLYL CHLOROFORMATE	6.1	TFC	I	6.1 +3 +8		0	E0	P001		MP8 MP17	T14	TP2
1723	ALLYL IODIDE	3	FC	II	3 +8		1 L	E2	P001 IBC02		MP19	T7	TP2
1724	ALLYLTRICHLOROSILANE, STABILIZED	8	CF1	II	8 +3	386	0	E0	P010		MP15	T10	TP2 TP7
1725	ALUMINIUM BROMIDE, ANHYDROUS	8	C2	II	8	588	1 kg	E2	P002 IBC08	B4	MP10	T3	TP33
1726	ALUMINIUM CHLORIDE, ANHYDROUS	8	C2	II	8	588	1 kg	E2	P002 IBC08	B4	MP10	T3	TP33
1727	AMMONIUM HYDROGENDIFLUORIDE, SOLID	8	C2	II	8		1 kg	E2	P002 IBC08	B4	MP10	T3	TP33
1728	AMYLTRICHLOROSILANE	8	C3	II	8		0	E0	P010		MP15	T10	TP2 TP7
1729	ANISOYL CHLORIDE	8	C4	II	8		1 kg	E2	P002 IBC08	B4	MP10	T3	TP33
1730	ANTIMONY PENTACHLORIDE, LIQUID	8	C1	II	8		1 L	E2	P001 IBC02		MP15	T7	TP2
1731	ANTIMONY PENTACHLORIDE SOLUTION	8	C1	II	8		1 L	E2	P001 IBC02		MP15	T7	TP2
1731	ANTIMONY PENTACHLORIDE SOLUTION	8	C1	III	8		5 L	E1	P001 IBC03 LP01 R001		MP19	T4	TP1
1732	ANTIMONY PENTAFLUORIDE	8	CT1	II	8 +6.1		1 L	E0	P001 IBC02		MP15	T7	TP2
1733	ANTIMONY TRICHLORIDE	8	C2	II	8		1 kg	E2	P002 IBC08	B4	MP10	T3	TP33
1736	BENZOYL CHLORIDE	8	C3	II	8		1 L	E2	P001 IBC02		MP15	T8	TP2
1737	BENZYL BROMIDE	6.1	TC1	II	6.1 +8		0	E4	P001 IBC02		MP15	T8	TP2
1738	BENZYL CHLORIDE	6.1	TC1	II	6.1 +8		0	E4	P001 IBC02		MP15	T8	TP2
1739	BENZYL CHLOROFORMATE	8	C9	I	8		0	E0	P001		MP8 MP17	T10	TP2
1740	HYDROGENDIFLUORIDES, SOLID, N.O.S.	8	C2	II	8	517	1 kg	E2	P002 IBC08	B4	MP10	T3	TP33

ADR tank		Vehicle for tank carriage	Transport category (Tunnel restriction code)	Special provisions for carriage				Hazard identifi-cation No.	UN No.	Name and description
Tank code	Special provisions			Packages	Bulk	Loading, unloading and handling	Operation			
4.3	4.3.5, 6.8.4	9.1.1.2	1.1.3.6 (8.6)	7.2.4	7.3.3	7.5.11	8.5	5.3.2.3		3.1.2
(12)	(13)	(14)	(15)	(16)	(17)	(18)	(19)	(20)	(1)	(2)
SGAH L4BH	TU15 TE19	AT	2 (E)		VC1 VC2 AP7	CV13 CV28	S9	60	1709	2,4-TOLUYLENEDIAMINE, SOLID
L4BH	TU15 TE19	AT	2 (E)	V12		CV13 CV28	S9	60	1710	TRICHLOROETHYLENE
L4BH	TU15 TE19	AT	2 (D/E)			CV13 CV28	S9 S19	60	1711	XYLIDINES, LIQUID
SGAH	TU15 TE19	AT	2 (D/E)	V11		CV13 CV28	S9 S19	60	1712	ZINC ARSENATE, ZINC ARSENITE or ZINC ARSENATE AND ZINC ARSENITE MIXTURE
S10AH	TU15 TE19	AT	1 (C/E)	V10		CV1 CV13 CV28	S9 S14	66	1713	ZINC CYANIDE
			1 (E)	V1		CV23 CV28	S14		1714	ZINC PHOSPHIDE
L4BN		FL	2 (D/E)				S2	83	1715	ACETIC ANHYDRIDE
L4BN		AT	2 (E)					80	1716	ACETYL BROMIDE
L4BH		FL	2 (D/E)				S2 S20	X338	1717	ACETYL CHLORIDE
L4BN		AT	3 (E)	V12				80	1718	BUTYL ACID PHOSPHATE
L4BN		AT	2 (E)					80	1719	CAUSTIC ALKALI LIQUID, N.O.S.
L4BN		AT	3 (E)	V12				80	1719	CAUSTIC ALKALI LIQUID, N.O.S.
L10CH	TU14 TU15 TE19 TE21	FL	1 (C/D)			CV1 CV13 CV28	S2 S9 S14	668	1722	ALLYL CHLOROFORMATE
L4BH		FL	2 (D/E)				S2 S20	338	1723	ALLYL IODIDE
L4BN		FL	2 (D/E)	V8			S2 S4	X839	1724	ALLYLTRICHLOROSILANE, STABILIZED
SGAN		AT	2 (E)	V11				80	1725	ALUMINIUM BROMIDE, ANHYDROUS
SGAN		AT	2 (E)	V11				80	1726	ALUMINIUM CHLORIDE, ANHYDROUS
SGAN		AT	2 (E)	V11				80	1727	AMMONIUM HYDROGENDIFLUORIDE, SOLID
L4BN		AT	2 (E)					X80	1728	AMYLTRICHLOROSILANE
SGAN L4BN		AT	2 (E)	V11				80	1729	ANISOYL CHLORIDE
L4BN		AT	2 (E)					X80	1730	ANTIMONY PENTACHLORIDE, LIQUID
L4BN		AT	2 (E)					80	1731	ANTIMONY PENTACHLORIDE SOLUTION
L4BN		AT	3 (E)	V12				80	1731	ANTIMONY PENTACHLORIDE SOLUTION
L4BN		AT	2 (E)			CV13 CV28		86	1732	ANTIMONY PENTAFLUORIDE
SGAN L4BN		AT	2 (E)	V11				80	1733	ANTIMONY TRICHLORIDE
L4BN		AT	2 (E)					80	1736	BENZOYL CHLORIDE
L4BH	TU15 TE19	AT	2 (D/E)			CV13 CV28	S9 S19	68	1737	BENZYL BROMIDE
L4BH	TU15 TE19	AT	2 (D/E)			CV13 CV28	S9 S19	68	1738	BENZYL CHLORIDE
L10BH		AT	1 (E)				S20	88	1739	BENZYL CHLOROFORMATE
SGAN		AT	2 (E)	V11				80	1740	HYDROGENDIFLUORIDES, SOLID, N.O.S.

UN No.	Name and description	Class	Classi-fication code	Packing group	Labels	Special provi-sions	Limited and excepted quantities		Packaging			Portable tanks and bulk containers	
									Packing instruc-tions	Special packing provisions	Mixed packing provisions	Instruc-tions	Special provisions
	3.1.2	2.2	2.2	2.1.1.3	5.2.2	3.3	3.4	3.5.1.2	4.1.4	4.1.4	4.1.10	4.2.5.2 7.3.2	4.2.5.3
(1)	(2)	(3a)	(3b)	(4)	(5)	(6)	(7a)	(7b)	(8)	(9a)	(9b)	(10)	(11)
1740	HYDROGENDIFLUORIDES, SOLID, N.O.S.	8	C2	III	8	517	5 kg	E1	P002 IBC08 LP02 R001	B3	MP10	T1	TP33
1741	BORON TRICHLORIDE	2	2TC		2.3 +8		0	E0	P200		MP9	(M)	
1742	BORON TRIFLUORIDE ACETIC ACID COMPLEX, LIQUID	8	C3	II	8		1 L	E2	P001 IBC02		MP15	T8	TP2
1743	BORON TRIFLUORIDE PROPIONIC ACID COMPLEX, LIQUID	8	C3	II	8		1 L	E2	P001 IBC02		MP15	T8	TP2
1744	BROMINE or BROMINE SOLUTION	8	CT1	I	8 +6.1		0	E0	P804		MP2	T22	TP2 TP10
1745	BROMINE PENTAFLUORIDE	5.1	OTC	I	5.1 +6.1 +8		0	E0	P200		MP2	T22	TP2
1746	BROMINE TRIFLUORIDE	5.1	OTC	I	5.1 +6.1 +8		0	E0	P200		MP2	T22	TP2
1747	BUTYLTRICHLOROSILANE	8	CF1	II	8 +3		0	E0	P010		MP15	T10	TP2 TP7
1748	CALCIUM HYPOCHLORITE, DRY or CALCIUM HYPOCHLORITE MIXTURE, DRY with more than 39% available chlorine (8.8% available oxygen)	5.1	O2	II	5.1	314	1 kg	E2	P002 IBC08	B4 B13	MP10		
1748	CALCIUM HYPOCHLORITE, DRY or CALCIUM HYPOCHLORITE MIXTURE, DRY with more than 39% available chlorine (8.8% available oxygen)	5.1	O2	III	5.1	316	5 kg	E1	P002 IBC08 R001	B4 B13	MP10		
1749	CHLORINE TRIFLUORIDE	2	2TOC		2.3 +5.1 +8		0	E0	P200		MP9	(M)	
1750	CHLOROACETIC ACID SOLUTION	6.1	TC1	II	6.1 +8		100 ml	E4	P001 IBC02		MP15	T7	TP2
1751	CHLOROACETIC ACID, SOLID	6.1	TC2	II	6.1 +8		500 g	E4	P002 IBC08	B4	MP10	T3	TP33
1752	CHLOROACETYL CHLORIDE	6.1	TC1	I	6.1 +8	354	0	E0	P602		MP8 MP17	T20	TP2
1753	CHLOROPHENYL-TRICHLOROSILANE	8	C3	II	8		0	E0	P010		MP15	T10	TP2 TP7
1754	CHLOROSULPHONIC ACID (with or without sulphur trioxide)	8	C1	I	8		0	E0	P001		MP8 MP17	T20	TP2
1755	CHROMIC ACID SOLUTION	8	C1	II	8	518	1 L	E2	P001 IBC02		MP15	T8	TP2
1755	CHROMIC ACID SOLUTION	8	C1	III	8	518	5 L	E1	P001 IBC02 LP01 R001		MP19	T4	TP1
1756	CHROMIC FLUORIDE, SOLID	8	C2	II	8		1 kg	E2	P002 IBC08	B4	MP10	T3	TP33
1757	CHROMIC FLUORIDE SOLUTION	8	C1	II	8		1 L	E2	P001 IBC02		MP15	T7	TP2
1757	CHROMIC FLUORIDE SOLUTION	8	C1	III	8		5 L	E1	P001 IBC03 LP01 R001		MP19	T4	TP1
1758	CHROMIUM OXYCHLORIDE	8	C1	I	8		0	E0	P001		MP8 MP17	T10	TP2
1759	CORROSIVE SOLID, N.O.S.	8	C10	I	8	274	0	E0	P002 IBC07		MP18	T6	TP33
1759	CORROSIVE SOLID, N.O.S.	8	C10	II	8	274	1 kg	E2	P002 IBC08	B4	MP10	T3	TP33
1759	CORROSIVE SOLID, N.O.S.	8	C10	III	8	274	5 kg	E1	P002 IBC08 LP02 R001	B3	MP10	T1	TP33
1760	CORROSIVE LIQUID, N.O.S.	8	C9	I	8	274	0	E0	P001		MP8 MP17	T14	TP2 TP27

ADR tank		Vehicle for tank carriage	Transport category (Tunnel restriction code)	Special provisions for carriage				Hazard identification No.	UN No.	Name and description
Tank code	Special provisions			Packages	Bulk	Loading, unloading and handling	Operation			
4.3	4.3.5, 6.8.4	9.1.1.2	1.1.3.6 (8.6)	7.2.4	7.3.3	7.5.11	8.5	5.3.2.3		3.1.2
(12)	(13)	(14)	(15)	(16)	(17)	(18)	(19)	(20)	(1)	(2)
SGAV		AT	3 (E)		VC1 VC2 AP7			80	1740	HYDROGENDIFLUORIDES, SOLID, N.O.S.
		AT	1 (C/D)			CV9 CV10 CV36	S14	268	1741	BORON TRICHLORIDE
L4BN		AT	2 (E)					80	1742	BORON TRIFLUORIDE ACETIC ACID COMPLEX, LIQUID
L4BN		AT	2 (E)					80	1743	BORON TRIFLUORIDE PROPIONIC ACID COMPLEX, LIQUID
L21DH(+)	TU14 TU33 TC5 TE21 TT2 TM3 TM5	AT	1 (C/D)			CV13 CV28	S14	886	1744	BROMINE or BROMINE SOLUTION
L10DH	TU3	AT	1 (B/E)			CV24 CV28	S14	568	1745	BROMINE PENTAFLUORIDE
L10DH	TU3	AT	1 (B/E)			CV24 CV28	S14	568	1746	BROMINE TRIFLUORIDE
L4BN		FL	2 (D/E)				S2	X83	1747	BUTYLTRICHLOROSILANE
SGAN	TU3	AT	2 (E)	V11		CV24 CV35		50	1748	CALCIUM HYPOCHLORITE, DRY or CALCIUM HYPOCHLORITE MIXTURE, DRY with more than 39% available chlorine (8.8% available oxygen)
SGAV	TU3	AT	3 (E)			CV24 CV35		50	1748	CALCIUM HYPOCHLORITE, DRY or CALCIUM HYPOCHLORITE MIXTURE, DRY with more than 39% available chlorine (8.8% available oxygen)
PxBH(M)	TA4 TT9	AT	1 (C/D)			CV9 CV10 CV36	S14	265	1749	CHLORINE TRIFLUORIDE
L4BH	TU15 TE19	AT	2 (D/E)			CV13 CV28	S9 S19	68	1750	CHLOROACETIC ACID SOLUTION
SGAH	TU15 TE19	AT	2 (D/E)	V11		CV13 CV28	S9 S19	68	1751	CHLOROACETIC ACID, SOLID
L10CH	TU14 TU15 TE19 TE21	AT	1 (C/D)			CV1 CV13 CV28	S9 S14	668	1752	CHLOROACETYL CHLORIDE
L4BN		AT	2 (E)					X80	1753	CHLOROPHENYL-TRICHLOROSILANE
L10BH		AT	1 (E)				S20	X88	1754	CHLOROSULPHONIC ACID (with or without sulphur trioxide)
L4BN		AT	2 (E)					80	1755	CHROMIC ACID SOLUTION
L4BN		AT	3 (E)					80	1755	CHROMIC ACID SOLUTION
SGAN		AT	2 (E)	V11				80	1756	CHROMIC FLUORIDE, SOLID
L4BN		AT	2 (E)					80	1757	CHROMIC FLUORIDE SOLUTION
L4BN		AT	3 (E)	V12				80	1757	CHROMIC FLUORIDE SOLUTION
L10BH		AT	1 (E)				S20	X88	1758	CHROMIUM OXYCHLORIDE
S10AN L10BH		AT	1 (E)	V10			S20	88	1759	CORROSIVE SOLID, N.O.S.
SGAN L4BN		AT	2 (E)	V11				80	1759	CORROSIVE SOLID, N.O.S.
SGAV L4BN		AT	3 (E)		VC1 VC2 AP7			80	1759	CORROSIVE SOLID, N.O.S.
L10BH		AT	1 (E)				S20	88	1760	CORROSIVE LIQUID, N.O.S.

UN No.	Name and description	Class	Classifi-cation code	Packing group	Labels	Special provi-sions	Limited and excepted quantities		Packaging			Portable tanks and bulk containers	
									Packing instruc-tions	Special packing provisions	Mixed packing provisions	Instruc-tions	Special provisions
	3.1.2	2.2	2.2	2.1.1.3	5.2.2	3.3	3.4	3.5.1.2	4.1.4	4.1.4	4.1.10	4.2.5.2 7.3.2	4.2.5.3
(1)	(2)	(3a)	(3b)	(4)	(5)	(6)	(7a)	(7b)	(8)	(9a)	(9b)	(10)	(11)
1760	CORROSIVE LIQUID, N.O.S.	8	C9	II	8	274	1 L	E2	P001 IBC02		MP15	T11	TP2 TP27
1760	CORROSIVE LIQUID, N.O.S.	8	C9	III	8	274	5 L	E1	P001 IBC03 LP01 R001		MP19	T7	TP1 TP28
1761	CUPRIETHYLENEDIAMINE SOLUTION	8	CT1	II	8 +6.1		1 L	E2	P001 IBC02		MP15	T7	TP2
1761	CUPRIETHYLENEDIAMINE SOLUTION	8	CT1	III	8 +6.1		5 L	E1	P001 IBC03 R001		MP19	T7	TP1 TP28
1762	CYCLOHEXENYLTRICHLORO-SILANE	8	C3	II	8		0	E0	P010		MP15	T10	TP2 TP7
1763	CYCLOHEXYLTRICHLORO-SILANE	8	C3	II	8		0	E0	P010		MP15	T10	TP2 TP7
1764	DICHLOROACETIC ACID	8	C3	II	8		1 L	E2	P001 IBC02		MP15	T8	TP2
1765	DICHLOROACETYL CHLORIDE	8	C3	II	8		1 L	E2	P001 IBC02		MP15	T7	TP2
1766	DICHLOROPHENYL-TRICHLOROSILANE	8	C3	II	8		0	E0	P010		MP15	T10	TP2 TP7
1767	DIETHYLDICHLORO-SILANE	8	CF1	II	8 +3		0	E0	P010		MP15	T10	TP2 TP7
1768	DIFLUOROPHOSPHORIC ACID, ANHYDROUS	8	C1	II	8		1 L	E2	P001 IBC02		MP15	T8	TP2
1769	DIPHENYLDICHLORO-SILANE	8	C3	II	8		0	E0	P010		MP15	T10	TP2 TP7
1770	DIPHENYLMETHYL BROMIDE	8	C10	II	8		1 kg	E2	P002 IBC08	B4	MP10	T3	TP33
1771	DODECYLTRICHLORO-SILANE	8	C3	II	8		0	E0	P010		MP15	T10	TP2 TP7
1773	FERRIC CHLORIDE, ANHYDROUS	8	C2	III	8	590	5 kg	E1	P002 IBC08 LP02 R001	B3	MP10	T1	TP33
1774	FIRE EXTINGUISHER CHARGES, corrosive liquid	8	C11	II	8		1 L	E0	P001	PP4			
1775	FLUOROBORIC ACID	8	C1	II	8		1 L	E2	P001 IBC02		MP15	T7	TP2
1776	FLUOROPHOSPHORIC ACID, ANHYDROUS	8	C1	II	8		1 L	E2	P001 IBC02		MP15	T8	TP2
1777	FLUOROSULPHONIC ACID	8	C1	I	8		0	E0	P001		MP8 MP17	T10	TP2
1778	FLUOROSILICIC ACID	8	C1	II	8		1 L	E2	P001 IBC02		MP15	T8	TP2
1779	FORMIC ACID with more than 85% acid by mass	8	CF1	II	8 +3		1 L	E2	P001 IBC02		MP15	T7	TP2
1780	FUMARYL CHLORIDE	8	C3	II	8		1 L	E2	P001 IBC02		MP15	T7	TP2
1781	HEXADECYLTRICHLORO-SILANE	8	C3	II	8		0	E0	P010		MP15	T10	TP2 TP7
1782	HEXAFLUORO-PHOSPHORIC ACID	8	C1	II	8		1 L	E2	P001 IBC02		MP15	T8	TP2
1783	HEXAMETHYLENE-DIAMINE SOLUTION	8	C7	II	8		1 L	E2	P001 IBC02		MP15	T7	TP2
1783	HEXAMETHYLENE-DIAMINE SOLUTION	8	C7	III	8		5 L	E1	P001 IBC03 LP01 R001		MP19	T4	TP1
1784	HEXYLTRICHLOROSILANE	8	C3	II	8		0	E0	P010		MP15	T10	TP2 TP7
1786	HYDROFLUORIC ACID AND SULPHURIC ACID MIXTURE	8	CT1	I	8 +6.1		0	E0	P001		MP8 MP17	T10	TP2
1787	HYDRIODIC ACID	8	C1	II	8		1 L	E2	P001 IBC02		MP15	T7	TP2
1787	HYDRIODIC ACID	8	C1	III	8		5 L	E1	P001 IBC03 LP01 R001		MP19	T4	TP1
1788	HYDROBROMIC ACID	8	C1	II	8	519	1 L	E2	P001 IBC02		MP15	T7	TP2
1788	HYDROBROMIC ACID	8	C1	III	8	519	5 L	E1	P001 IBC03 LP01 R001		MP19	T4	TP1

ADR tank		Vehicle for tank carriage	Transport category (Tunnel restriction code)	Special provisions for carriage				Hazard identification No.	UN No.	Name and description
Tank code	Special provisions			Packages	Bulk	Loading, unloading and handling	Operation			
4.3	4.3.5, 6.8.4	9.1.1.2	1.1.3.6 (8.6)	7.2.4	7.3.3	7.5.11	8.5	5.3.2.3		3.1.2
(12)	(13)	(14)	(15)	(16)	(17)	(18)	(19)	(20)	(1)	(2)
L4BN		AT	2 (E)					80	1760	CORROSIVE LIQUID, N.O.S.
L4BN		AT	3 (E)	V12				80	1760	CORROSIVE LIQUID, N.O.S.
L4BN		AT	2 (E)			CV13 CV28		86	1761	CUPRIETHYLENEDIAMINE SOLUTION
L4BN		AT	3 (E)	V12		CV13 CV28		86	1761	CUPRIETHYLENEDIAMINE SOLUTION
L4BN		AT	2 (E)					X80	1762	CYCLOHEXENYLTRICHLORO-SILANE
L4BN		AT	2 (E)					X80	1763	CYCLOHEXYLTRICHLORO-SILANE
L4BN		AT	2 (E)					80	1764	DICHLOROACETIC ACID
L4BN		AT	2 (E)					X80	1765	DICHLOROACETYL CHLORIDE
L4BN		AT	2 (E)					X80	1766	DICHLOROPHENYL-TRICHLOROSILANE
L4BN		FL	2 (D/E)				S2	X83	1767	DIETHYLDICHLORO-SILANE
L4BN		AT	2 (E)					80	1768	DIFLUOROPHOSPHORIC ACID, ANHYDROUS
L4BN		AT	2 (E)					X80	1769	DIPHENYLDICHLORO-SILANE
SGAN L4BN		AT	2 (E)	V11				80	1770	DIPHENYLMETHYL BROMIDE
L4BN		AT	2 (E)					X80	1771	DODECYLTRICHLORO-SILANE
SGAV		AT	3 (E)		VC1 VC2 AP7			80	1773	FERRIC CHLORIDE, ANHYDROUS
			2 (E)						1774	FIRE EXTINGUISHER CHARGES, corrosive liquid
L4BN		AT	2 (E)					80	1775	FLUOROBORIC ACID
L4BN		AT	2 (E)					80	1776	FLUOROPHOSPHORIC ACID, ANHYDROUS
L10BH		AT	1 (E)				S20	88	1777	FLUOROSULPHONIC ACID
L4BN		AT	2 (E)					80	1778	FLUOROSILICIC ACID
L4BN		FL	2 (D/E)				S2	83	1779	FORMIC ACID with more than 85% acid by mass
L4BN		AT	2 (E)					80	1780	FUMARYL CHLORIDE
L4BN		AT	2 (E)					X80	1781	HEXADECYLTRICHLORO-SILANE
L4BN		AT	2 (E)					80	1782	HEXAFLUORO-PHOSPHORIC ACID
L4BN		AT	2 (E)					80	1783	HEXAMETHYLENE-DIAMINE SOLUTION
L4BN		AT	3 (E)	V12				80	1783	HEXAMETHYLENE-DIAMINE SOLUTION
L4BN		AT	2 (E)					X80	1784	HEXYLTRICHLOROSILANE
L10DH	TU14 TE21	AT	1 (C/D)			CV13 CV28	S14	886	1786	HYDROFLUORIC ACID AND SULPHURIC ACID MIXTURE
L4BN		AT	2 (E)					80	1787	HYDRIODIC ACID
L4BN		AT	3 (E)	V12				80	1787	HYDRIODIC ACID
L4BN		AT	2 (E)					80	1788	HYDROBROMIC ACID
L4BN		AT	3 (E)	V12				80	1788	HYDROBROMIC ACID

UN No.	Name and description	Class	Classifi-cation code	Packing group	Labels	Special provi-sions	Limited and excepted quantities		Packaging			Portable tanks and bulk containers	
									Packing instruc-tions	Special packing provisions	Mixed packing provisions	Instruc-tions	Special provisions
3.1.2	3.1.2	2.2	2.2	2.1.1.3	5.2.2	3.3	3.4	3.5.1.2	4.1.4	4.1.4	4.1.10	4.2.5.2 7.3.2	4.2.5.3
(1)	(2)	(3a)	(3b)	(4)	(5)	(6)	(7a)	(7b)	(8)	(9a)	(9b)	(10)	(11)
1789	HYDROCHLORIC ACID	8	C1	II	8	520	1 L	E2	P001 IBC02		MP15	T8	TP2
1789	HYDROCHLORIC ACID	8	C1	III	8	520	5 L	E1	P001 IBC03 LP01 R001		MP19	T4	TP1
1790	HYDROFLUORIC ACID with more than 85% hydrogen fluoride	8	CT1	I	8 +6.1	640I	0	E0	P802		MP2	T10	TP2
1790	HYDROFLUORIC ACID with more than 60% but not more than 85% hydrogen fluoride	8	CT1	I	8 +6.1	640J	0	E0	P001	PP81	MP8 MP17	T10	TP2
1790	HYDROFLUORIC ACID with not more than 60% hydrogen fluoride	8	CT1	II	8 +6.1		1 L	E2	P001 IBC02		MP15	T8	TP2
1791	HYPOCHLORITE SOLUTION	8	C9	II	8	521	1 L	E2	P001 IBC02	PP10 B5	MP15	T7	TP2 TP24
1791	HYPOCHLORITE SOLUTION	8	C9	III	8	521	5 L	E1	P001 IBC02 LP01 R001	B5	MP19	T4	TP2 TP24
1792	IODINE MONOCHLORIDE, SOLID	8	C2	II	8		1 kg	E0	P002 IBC08	B4	MP10	T7	TP2
1793	ISOPROPYL ACID PHOSPHATE	8	C3	III	8		5 L	E1	P001 IBC02 LP01 R001		MP19	T4	TP1
1794	LEAD SULPHATE with more than 3% free acid	8	C2	II	8	591	1 kg	E2	P002 IBC08	B4	MP10	T3	TP33
1796	NITRATING ACID MIXTURE with more than 50% nitric acid	8	CO1	I	8 +5.1		0	E0	P001		MP8 MP17	T10	TP2
1796	NITRATING ACID MIXTURE with not more than 50% nitric acid	8	C1	II	8 +5.1		1 L	E0	P001 IBC02		MP15	T8	TP2
1798	NITROHYDROCHLORIC ACID	8	COT				CARRIAGE PROHIBITED							
1799	NONYLTRICHLOROSILANE	8	C3	II	8		0	E0	P010		MP15	T10	TP2 TP7
1800	OCTADECYLTRICHLORO-SILANE	8	C3	II	8		0	E0	P010		MP15	T10	TP2 TP7
1801	OCTYLTRICHLOROSILANE	8	C3	II	8		0	E0	P010		MP15	T10	TP2 TP7
1802	PERCHLORIC ACID with not more than 50% acid, by mass	8	CO1	II	8 +5.1	522	1 L	E0	P001 IBC02		MP3	T7	TP2
1803	PHENOLSULPHONIC ACID, LIQUID	8	C3	II	8		1 L	E2	P001 IBC02		MP15	T7	TP2
1804	PHENYLTRICHLORO-SILANE	8	C3	II	8		0	E0	P010		MP15	T10	TP2 TP7
1805	PHOSPHORIC ACID, SOLUTION	8	C1	III	8		5 L	E1	P001 IBC03 LP01 R001		MP19	T4	TP1
1806	PHOSPHORUS PENTACHLORIDE	8	C2	II	8		1 kg	E0	P002 IBC08	B4	MP10	T3	TP33
1807	PHOSPHORUS PENTOXIDE	8	C2	II	8		1 kg	E2	P002 IBC08	B4	MP10	T3	TP33
1808	PHOSPHORUS TRIBROMIDE	8	C1	II	8		1 L	E0	P001 IBC02		MP15	T7	TP2
1809	PHOSPHORUS TRICHLORIDE	6.1	TC3	I	6.1 +8	354	0	E0	P602		MP8 MP17	T20	TP2
1810	PHOSPHORUS OXYCHLORIDE	6.1	TC3	I	6.1 +8	354	0	E0	P602		MP8 MP17	T20	TP2
1811	POTASSIUM HYDROGENDIFLUORIDE, SOLID	8	CT2	II	8 +6.1		1 kg	E2	P002 IBC08	B4	MP10	T3	TP33
1812	POTASSIUM FLUORIDE, SOLID	6.1	T5	III	6.1		5 kg	E1	P002 IBC08 LP02 R001	B3	MP10	T1	TP33
1813	POTASSIUM HYDROXIDE, SOLID	8	C6	II	8		1 kg	E2	P002 IBC08	B4	MP10	T3	TP33
1814	POTASSIUM HYDROXIDE SOLUTION	8	C5	II	8		1 L	E2	P001 IBC02		MP15	T7	TP2

ADR tank		Vehicle for tank carriage	Transport category (Tunnel restriction code)	Special provisions for carriage				Hazard identifi-cation No.	UN No.	Name and description
Tank code	Special provisions			Packages	Bulk	Loading, unloading and handling	Operation			
4.3	4.3.5, 6.8.4	9.1.1.2	1.1.3.6 (8.6)	7.2.4	7.3.3	7.5.11	8.5	5.3.2.3		3.1.2
(12)	(13)	(14)	(15)	(16)	(17)	(18)	(19)	(20)	(1)	(2)
L4BN		AT	2 (E)					80	1789	HYDROCHLORIC ACID
L4BN		AT	3 (E)	V12				80	1789	HYDROCHLORIC ACID
L21DH(+)	TU14 TU34 TC1 TE21 TA4 TT9 TM3	AT	1 (C/D)			CV13 CV28	S14	886	1790	HYDROFLUORIC ACID with more than 85% hydrogen fluoride
L10DH	TU14 TE21	AT	1 (C/D)			CV13 CV28	S14	886	1790	HYDROFLUORIC ACID with more than 60% but not more than 85% hydrogen fluoride
L4DH	TU14 TE21	AT	2 (E)			CV13 CV28		86	1790	HYDROFLUORIC ACID with not more than 60% hydrogen fluoride
L4BV(+)	TE11	AT	2 (E)					80	1791	HYPOCHLORITE SOLUTION
L4BV(+)	TE11	AT	3 (E)					80	1791	HYPOCHLORITE SOLUTION
SGAN L4BN		AT	2 (E)	V11				80	1792	IODINE MONOCHLORIDE, SOLID
L4BN		AT	3 (E)					80	1793	ISOPROPYL ACID PHOSPHATE
SGAN		AT	2 (E)	V11	VC1 VC2 AP7			80	1794	LEAD SULPHATE with more than 3% free acid
L10BH	TC6 TT1	AT	1 (E)			CV24	S14	885	1796	NITRATING ACID MIXTURE with more than 50% nitric acid
L4BN		AT	2 (E)					80	1796	NITRATING ACID MIXTURE with not more than 50% nitric acid
CARRIAGE PROHIBITED									1798	NITROHYDROCHLORIC ACID
L4BN		AT	2 (E)					X80	1799	NONYLTRICHLOROSILANE
L4BN		AT	2 (E)					X80	1800	OCTADECYLTRICHLORO-SILANE
L4BN		AT	2 (E)					X80	1801	OCTYLTRICHLOROSILANE
L4BN		AT	2 (E)			CV24		85	1802	PERCHLORIC ACID with not more than 50% acid, by mass
L4BN		AT	2 (E)					80	1803	PHENOLSULPHONIC ACID, LIQUID
L4BN		AT	2 (E)					X80	1804	PHENYLTRICHLORO-SILANE
L4BN		AT	3 (E)	V12				80	1805	PHOSPHORIC ACID, SOLUTION
SGAN		AT	2 (E)	V11				80	1806	PHOSPHORUS PENTACHLORIDE
SGAN		AT	2 (E)	V11				80	1807	PHOSPHORUS PENTOXIDE
L4BN		AT	2 (E)					X80	1808	PHOSPHORUS TRIBROMIDE
L10CH	TU14 TU15 TE19 TE21	AT	1 (C/D)			CV1 CV13 CV28	S9 S14	668	1809	PHOSPHORUS TRICHLORIDE
L10CH	TU14 TU15 TE19 TE21	AT	1 (C/D)			CV1 CV13 CV28	S9 S14	X668	1810	PHOSPHORUS OXYCHLORIDE
SGAN		AT	2 (E)	V11		CV13 CV28		86	1811	POTASSIUM HYDROGENDIFLUORIDE, SOLID
SGAH	TU15 TE19	AT	2 (E)		VC1 VC2 AP7	CV13 CV28	S9	60	1812	POTASSIUM FLUORIDE, SOLID
SGAN		AT	2 (E)	V11				80	1813	POTASSIUM HYDROXIDE, SOLID
L4BN		AT	2 (E)					80	1814	POTASSIUM HYDROXIDE SOLUTION

UN No.	Name and description	Class	Classifi-cation code	Packing group	Labels	Special provi-sions	Limited and excepted quantities		Packaging			Portable tanks and bulk containers	
									Packing instruc-tions	Special packing provisions	Mixed packing provisions	Instruc-tions	Special provisions
	3.1.2	2.2	2.2	2.1.1.3	5.2.2	3.3	3.4	3.5.1.2	4.1.4	4.1.4	4.1.10	4.2.5.2 7.3.2	4.2.5.3
(1)	(2)	(3a)	(3b)	(4)	(5)	(6)	(7a)	(7b)	(8)	(9a)	(9b)	(10)	(11)
1814	POTASSIUM HYDROXIDE SOLUTION	8	C5	III	8		5 L	E1	P001 IBC03 LP01 R001		MP19	T4	TP1
1815	PROPIONYL CHLORIDE	3	FC	II	3 +8		1 L	E2	P001 IBC02		MP19	T7	TP1
1816	PROPYLTRICHLORO-SILANE	8	CF1	II	8 +3		0	E0	P010		MP15	T10	TP2 TP7
1817	PYROSULPHURYL CHLORIDE	8	C1	II	8		1 L	E2	P001 IBC02		MP15	T8	TP2
1818	SILICON TETRACHLORIDE	8	C1	II	8		0	E0	P010		MP15	T10	TP2 TP7
1819	SODIUM ALUMINATE SOLUTION	8	C5	II	8		1 L	E2	P001 IBC02		MP15	T7	TP2
1819	SODIUM ALUMINATE SOLUTION	8	C5	III	8		5 L	E1	P001 IBC03 LP01 R001		MP19	T4	TP1
1823	SODIUM HYDROXIDE, SOLID	8	C6	II	8		1 kg	E2	P002 IBC08	B4	MP10	T3	TP33
1824	SODIUM HYDROXIDE SOLUTION	8	C5	II	8		1 L	E2	P001 IBC02		MP15	T7	TP2
1824	SODIUM HYDROXIDE SOLUTION	8	C5	III	8		5 L	E1	P001 IBC03 LP01 R001		MP19	T4	TP1
1825	SODIUM MONOXIDE	8	C6	II	8		1 kg	E2	P002 IBC08	B4	MP10	T3	TP33
1826	NITRATING ACID MIXTURE, SPENT, with more than 50% nitric acid	8	CO1	I	8 +5.1	113	0	E0	P001		MP8 MP17	T10	TP2
1826	NITRATING ACID MIXTURE, SPENT, with not more than 50% nitric acid	8	C1	II	8	113	1 L	E0	P001 IBC02		MP15	T8	TP2
1827	STANNIC CHLORIDE, ANHYDROUS	8	C1	II	8		1 L	E2	P001 IBC02		MP15	T7	TP2
1828	SULPHUR CHLORIDES	8	C1	I	8		0	E0	P602		MP8 MP17	T20	TP2
1829	SULPHUR TRIOXIDE, STABILIZED	8	C1	I	8	386 623	0	E0	P001		MP8 MP17	T20	TP4 TP25 TP26
1830	SULPHURIC ACID with more than 51% acid	8	C1	II	8		1 L	E2	P001 IBC02		MP15	T8	TP2
1831	SULPHURIC ACID, FUMING	8	CT1	I	8 +6.1		0	E0	P602		MP8 MP17	T20	TP2
1832	SULPHURIC ACID, SPENT	8	C1	II	8	113	1 L	E0	P001 IBC02		MP15	T8	TP2
1833	SULPHUROUS ACID	8	C1	II	8		1 L	E2	P001 IBC02		MP15	T7	TP2
1834	SULPHURYL CHLORIDE	6.1	TC3	I	6.1 +8	354	0	E0	P602		MP8 MP17	T20	TP2
1835	TETRAMETHYL-AMMONIUM HYDROXIDE SOLUTION	8	C7	II	8		1 L	E2	P001 IBC02		MP15	T7	TP2
1835	TETRAMETHYL-AMMONIUM HYDROXIDE SOLUTION	8	C7	III	8		5 L	E1	P001 IBC03 LP01 R001		MP19	T7	TP2
1836	THIONYL CHLORIDE	8	C1	I	8		0	E0	P802		MP8 MP17	T10	TP2
1837	THIOPHOSPHORYL CHLORIDE	8	C1	II	8		1 L	E0	P001 IBC02		MP15	T7	TP2
1838	TITANIUM TETRACHLORIDE	6.1	TC3	I	6.1 +8	354	0	E0	P602		MP8 MP17	T20	TP2
1839	TRICHLOROACETIC ACID	8	C4	II	8		1 kg	E2	P002 IBC08	B4	MP10	T3	TP33
1840	ZINC CHLORIDE SOLUTION	8	C1	III	8		5 L	E1	P001 IBC03 LP01 R001		MP19	T4	TP1
1841	ACETALDEHYDE AMMONIA	9	M11	III	9		5 kg	E1	P002 IBC08 LP02 R001	B3 B6	MP10	T1	TP33

ADR tank		Vehicle for tank carriage	Transport category (Tunnel restriction code)	Special provisions for carriage				Hazard identifi-cation No.	UN No.	Name and description
Tank code	Special provisions			Packages	Bulk	Loading, unloading and handling	Operation			
4.3	4.3.5, 6.8.4	9.1.1.2	1.1.3.6 (8.6)	7.2.4	7.3.3	7.5.11	8.5	5.3.2.3		3.1.2
(12)	(13)	(14)	(15)	(16)	(17)	(18)	(19)	(20)	(1)	(2)
L4BN		AT	3 (E)	V12				80	1814	POTASSIUM HYDROXIDE SOLUTION
L4BH		FL	2 (D/E)				S2 S20	338	1815	PROPIONYL CHLORIDE
L4BN		FL	2 (D/E)				S2	X83	1816	PROPYLTRICHLORO-SILANE
L4BN		AT	2 (E)					X80	1817	PYROSULPHURYL CHLORIDE
L4BN		AT	2 (E)					X80	1818	SILICON TETRACHLORIDE
L4BN		AT	2 (E)					80	1819	SODIUM ALUMINATE SOLUTION
L4BN		AT	3 (E)	V12				80	1819	SODIUM ALUMINATE SOLUTION
SGAN		AT	2 (E)	V11				80	1823	SODIUM HYDROXIDE, SOLID
L4BN		AT	2 (E)					80	1824	SODIUM HYDROXIDE SOLUTION
L4BN		AT	3 (E)	V12				80	1824	SODIUM HYDROXIDE SOLUTION
SGAN		AT	2 (E)	V11				80	1825	SODIUM MONOXIDE
L10BH		AT	1 (E)			CV24	S14	885	1826	NITRATING ACID MIXTURE, SPENT, with more than 50% nitric acid
L4BN		AT	2 (E)					80	1826	NITRATING ACID MIXTURE, SPENT, with not more than 50% nitric acid
L4BN		AT	2 (E)					X80	1827	STANNIC CHLORIDE, ANHYDROUS
L10BH		AT	1 (E)				S20	X88	1828	SULPHUR CHLORIDES
L10BH	TU32 TE13 TT5 TM3	AT	1 (E)	V8			S4 S20	X88	1829	SULPHUR TRIOXIDE, STABILIZED
L4BN		AT	2 (E)					80	1830	SULPHURIC ACID with more than 51% acid
L10BH		AT	1 (C/D)			CV13 CV28	S14	X886	1831	SULPHURIC ACID, FUMING
L4BN		AT	2 (E)					80	1832	SULPHURIC ACID, SPENT
L4BN		AT	2 (E)					80	1833	SULPHUROUS ACID
L10CH	TU14 TU15 TE19 TE21	AT	1 (C/D)			CV1 CV13 CV28	S9 S14	X668	1834	SULPHURYL CHLORIDE
L4BN		AT	2 (E)					80	1835	TETRAMETHYL-AMMONIUM HYDROXIDE SOLUTION
L4BN		AT	3 (E)	V12				80	1835	TETRAMETHYL-AMMONIUM HYDROXIDE SOLUTION
L10BH		AT	1 (E)				S20	X88	1836	THIONYL CHLORIDE
L4BN		AT	2 (E)					X80	1837	THIOPHOSPHORYL CHLORIDE
L10CH	TU14 TU15 TE19 TE21	AT	1 (C/D)			CV1 CV13 CV28	S9 S14	X668	1838	TITANIUM TETRACHLORIDE
SGAN L4BN		AT	2 (E)	V11				80	1839	TRICHLOROACETIC ACID
L4BN		AT	3 (E)	V12				80	1840	ZINC CHLORIDE SOLUTION
SGAV		AT	3 (E)		VC1 VC2			90	1841	ACETALDEHYDE AMMONIA

UN No.	Name and description	Class	Classification code	Packing group	Labels	Special provisions	Limited and excepted quantities		Packaging			Portable tanks and bulk containers	
									Packing instructions	Special packing provisions	Mixed packing provisions	Instructions	Special provisions
3.1.2	3.1.2	2.2	2.2	2.1.1.3	5.2.2	3.3	3.4	3.5.1.2	4.1.4	4.1.4	4.1.10	4.2.5.2 7.3.2	4.2.5.3
(1)	(2)	(3a)	(3b)	(4)	(5)	(6)	(7a)	(7b)	(8)	(9a)	(9b)	(10)	(11)
1843	AMMONIUM DINITRO-o-CRESOLATE, SOLID	6.1	T2	II	6.1		500 g	E4	P002 IBC08	B4	MP10	T3	TP33
1845	Carbon dioxide, solid (Dry ice)	9	M11				NOT SUBJECT TO ADR except for 5.5.3						
1846	CARBON TETRACHLORIDE	6.1	T1	II	6.1		100 ml	E4	P001 IBC02		MP15	T7	TP2
1847	POTASSIUM SULPHIDE, HYDRATED with not less than 30% water of crystallization	8	C6	II	8	523	1 kg	E2	P002 IBC08	B4	MP10	T3	TP33
1848	PROPIONIC ACID with not less than 10% and less than 90% acid by mass	8	C3	III	8		5 L	E1	P001 IBC03 LP01 R001		MP19	T4	TP1
1849	SODIUM SULPHIDE, HYDRATED with not less than 30% water	8	C6	II	8	523	1 kg	E2	P002 IBC08	B4	MP10	T3	TP33
1851	MEDICINE, LIQUID, TOXIC, N.O.S.	6.1	T1	II	6.1	221 601	100 ml	E4	P001		MP15		
1851	MEDICINE, LIQUID, TOXIC, N.O.S.	6.1	T1	III	6.1	221 601	5 L	E1	P001 LP01 R001		MP19		
1854	BARIUM ALLOYS, PYROPHORIC	4.2	S4	I	4.2		0	E0	P404		MP13	T21	TP7 TP33
1855	CALCIUM, PYROPHORIC or CALCIUM ALLOYS, PYROPHORIC	4.2	S4	I	4.2		0	E0	P404		MP13		
1856	Rags, oily	4.2	S2				NOT SUBJECT TO ADR						
1857	Textile waste, wet	4.2	S2				NOT SUBJECT TO ADR						
1858	HEXAFLUOROPROPYLENE (REFRIGERANT GAS R 1216)	2	2A		2.2	662	120 ml	E1	P200		MP9	(M) T50	
1859	SILICON TETRAFLUORIDE	2	2TC		2.3 +8		0	E0	P200		MP9	(M)	
1860	VINYL FLUORIDE, STABILIZED	2	2F		2.1	386 662	0	E0	P200		MP9	(M)	
1862	ETHYL CROTONATE	3	F1	II	3		1 L	E2	P001 IBC02 R001		MP19	T4	TP2
1863	FUEL, AVIATION, TURBINE ENGINE	3	F1	I	3	664	500 ml	E3	P001		MP7 MP17	T11	TP1 TP8 TP28
1863	FUEL, AVIATION, TURBINE ENGINE (vapour pressure at 50 °C more than 110 kPa)	3	F1	II	3	640C 664	1 L	E2	P001		MP19	T4	TP1 TP8
1863	FUEL, AVIATION, TURBINE ENGINE (vapour pressure at 50 °C not more than 110 kPa)	3	F1	II	3	640D 664	1 L	E2	P001 IBC02 R001		MP19	T4	TP1 TP8
1863	FUEL, AVIATION, TURBINE ENGINE	3	F1	III	3	664	5 L	E1	P001 IBC03 LP01 R001		MP19	T2	TP1
1865	n-PROPYL NITRATE	3	F1	II	3		1 L	E2	P001 IBC02 R001	B7	MP19		
1866	RESIN SOLUTION, flammable	3	F1	I	3		500 ml	E3	P001		MP7 MP17	T11	TP1 TP8 TP28
1866	RESIN SOLUTION, flammable (vapour pressure at 50 °C more than 110 kPa)	3	F1	II	3	640C	5 L	E2	P001	PP1	MP19	T4	TP1 TP8
1866	RESIN SOLUTION, flammable (vapour pressure at 50 °C not more than 110 kPa)	3	F1	II	3	640D	5 L	E2	P001 IBC02 R001	PP1	MP19	T4	TP1 TP8
1866	RESIN SOLUTION, flammable	3	F1	III	3		5 L	E1	P001 IBC03 LP01 R001	PP1	MP19	T2	TP1
1866	RESIN SOLUTION, flammable (having a flash-point below 23 °C and viscous according to 2.2.3.1.4) (vapour pressure at 50 °C more than 110 kPa)	3	F1	III	3		5 L	E1	P001 R001	PP1	MP19		

ADR tank		Vehicle for tank carriage	Transport category (Tunnel restriction code)	Special provisions for carriage				Hazard identifi-cation No.	UN No.	Name and description
Tank code	Special provisions			Packages	Bulk	Loading, unloading and handling	Operation			
4.3	4.3.5, 6.8.4	9.1.1.2	1.1.3.6 (8.6)	7.2.4	7.3.3	7.5.11	8.5	5.3.2.3		3.1.2
(12)	(13)	(14)	(15)	(16)	(17)	(18)	(19)	(20)	(1)	(2)
SGAH	TU15 TE19	AT	2 (D/E)	V11		CV13 CV28	S9 S19	60	1843	AMMONIUM DINITRO-o-CRESOLATE, SOLID
NOT SUBJECT TO ADR except for 5.5.3									1845	Carbon dioxide, solid (Dry ice)
L4BH	TU15 TE19	AT	2 (D/E)			CV13 CV28	S9 S19	60	1846	CARBON TETRACHLORIDE
SGAN L4BN		AT	2 (E)	V11				80	1847	POTASSIUM SULPHIDE, HYDRATED with not less than 30% water of crystallization
L4BN		AT	3 (E)	V12				80	1848	PROPIONIC ACID with not less than 10% and less than 90% acid by mass
SGAN L4BN		AT	2 (E)	V11				80	1849	SODIUM SULPHIDE, HYDRATED with not less than 30% water
L4BH	TU15 TE19	AT	2 (D/E)			CV13 CV28	S9 S19	60	1851	MEDICINE, LIQUID, TOXIC, N.O.S.
L4BH	TU15 TE19	AT	2 (E)			CV13 CV28	S9	60	1851	MEDICINE, LIQUID, TOXIC, N.O.S.
		AT	0 (B/E)	V1			S20	43	1854	BARIUM ALLOYS, PYROPHORIC
			0 (E)	V1			S20		1855	CALCIUM, PYROPHORIC or CALCIUM ALLOYS, PYROPHORIC
NOT SUBJECT TO ADR									1856	Rags, oily
NOT SUBJECT TO ADR									1857	Textile waste, wet
PxBN(M)	TA4 TT9	AT	3 (C/E)			CV9 CV10 CV36		20	1858	HEXAFLUOROPROPYLENE (REFRIGERANT GAS R 1216)
PxBH(M)	TA4 TT9	AT	1 (C/D)			CV9 CV10 CV36	S14	268	1859	SILICON TETRAFLUORIDE
PxBN(M)	TA4 TT9	FL	2 (B/D)	V8		CV9 CV10 CV36	S2 S4 S20	239	1860	VINYL FLUORIDE, STABILIZED
LGBF		FL	2 (D/E)				S2 S20	33	1862	ETHYL CROTONATE
L4BN		FL	1 (D/E)				S2 S20	33	1863	FUEL, AVIATION, TURBINE ENGINE
L1.5BN		FL	2 (D/E)				S2 S20	33	1863	FUEL, AVIATION, TURBINE ENGINE (vapour pressure at 50 °C more than 110 kPa)
LGBF		FL	2 (D/E)				S2 S20	33	1863	FUEL, AVIATION, TURBINE ENGINE (vapour pressure at 50 °C not more than 110 kPa)
LGBF		FL	3 (D/E)	V12			S2	30	1863	FUEL, AVIATION, TURBINE ENGINE
			2 (E)				S2 S20		1865	n-PROPYL NITRATE
L4BN		FL	1 (D/E)				S2 S20	33	1866	RESIN SOLUTION, flammable
L1.5BN		FL	2 (D/E)				S2 S20	33	1866	RESIN SOLUTION, flammable (vapour pressure at 50 °C more than 110 kPa)
LGBF		FL	2 (D/E)				S2 S20	33	1866	RESIN SOLUTION, flammable (vapour pressure at 50 °C not more than 110 kPa)
LGBF		FL	3 (D/E)	V12			S2	30	1866	RESIN SOLUTION, flammable
			3 (E)				S2		1866	RESIN SOLUTION, flammable (having a flash-point below 23 °C and viscous according to 2.2.3.1.4) (vapour pressure at 50 °C more than 110 kPa)

UN No.	Name and description	Class	Classification code	Packing group	Labels	Special provisions	Limited and excepted quantities		Packaging			Portable tanks and bulk containers	
									Packing instructions	Special packing provisions	Mixed packing provisions	Instructions	Special provisions
	3.1.2	2.2	2.2	2.1.1.3	5.2.2	3.3	3.4	3.5.1.2	4.1.4	4.1.4	4.1.10	4.2.5.2 7.3.2	4.2.5.3
(1)	(2)	(3a)	(3b)	(4)	(5)	(6)	(7a)	(7b)	(8)	(9a)	(9b)	(10)	(11)
1866	RESIN SOLUTION, flammable (having a flash-point below 23 °C and viscous according to 2.2.3.1.4) (vapour pressure at 50 °C not more than 110 kPa)	3	F1	III	3		5 L	E1	P001 IBC02 R001	PP1 BB4	MP19		
1868	DECABORANE	4.1	FT2	II	4.1 +6.1		1 kg	E0	P002 IBC06		MP10	T3	TP33
1869	MAGNESIUM or MAGNESIUM ALLOYS with more than 50% magnesium in pellets, turnings or ribbons	4.1	F3	III	4.1	59	5 kg	E1	P002 IBC08 LP02 R001	B3	MP11	T1	TP33
1870	POTASSIUM BOROHYDRIDE	4.3	W2	I	4.3		0	E0	P403		MP2		
1871	TITANIUM HYDRIDE	4.1	F3	II	4.1		1 kg	E2	P410 IBC04	PP40	MP11	T3	TP33
1872	LEAD DIOXIDE	5.1	OT2	III	5.1 +6.1		5 kg	E1	P002 IBC08 LP02 R001	B3	MP2	T1	TP33
1873	PERCHLORIC ACID with more than 50% but not more than 72% acid, by mass	5.1	OC1	I	5.1 +8	60	0	E0	P502	PP28	MP3	T10	TP1
1884	BARIUM OXIDE	6.1	T5	III	6.1		5 kg	E1	P002 IBC08 LP02 R001	B3	MP10	T1	TP33
1885	BENZIDINE	6.1	T2	II	6.1		500 g	E4	P002 IBC08	B4	MP10	T3	TP33
1886	BENZYLIDENE CHLORIDE	6.1	T1	II	6.1		100 ml	E4	P001 IBC02		MP15	T7	TP2
1887	BROMOCHLOROMETHANE	6.1	T1	III	6.1		5 L	E1	P001 IBC03 LP01 R001		MP19	T4	TP1
1888	CHLOROFORM	6.1	T1	III	6.1		5 L	E1	P001 IBC03 LP01 R001		MP19	T7	TP2
1889	CYANOGEN BROMIDE	6.1	TC2	I	6.1 +8		0	E0	P002		MP18	T6	TP33
1891	ETHYL BROMIDE	6.1	T1	II	6.1		100 ml	E4	P001 IBC02	B8	MP15	T7	TP2
1892	ETHYLDICHLOROARSINE	6.1	T3	I	6.1	354	0	E0	P602		MP8 MP17	T20	TP2
1894	PHENYLMERCURIC HYDROXIDE	6.1	T3	II	6.1		500 g	E4	P002 IBC08	B4	MP10	T3	TP33
1895	PHENYLMERCURIC NITRATE	6.1	T3	II	6.1		500 g	E4	P002 IBC08	B4	MP10	T3	TP33
1897	TETRACHLOROETHYLENE	6.1	T1	III	6.1		5 L	E1	P001 IBC03 LP01 R001		MP19	T4	TP1
1898	ACETYL IODIDE	8	C3	II	8		1 L	E2	P001 IBC02		MP15	T7	TP2
1902	DIISOOCTYL ACID PHOSPHATE	8	C3	III	8		5 L	E1	P001 IBC03 LP01 R001		MP19	T4	TP1
1903	DISINFECTANT, LIQUID, CORROSIVE, N.O.S.	8	C9	I	8	274	0	E0	P001		MP8 MP17		
1903	DISINFECTANT, LIQUID, CORROSIVE, N.O.S.	8	C9	II	8	274	1 L	E2	P001 IBC02		MP15		
1903	DISINFECTANT, LIQUID, CORROSIVE, N.O.S.	8	C9	III	8	274	5 L	E1	P001 IBC03 LP01 R001		MP19		
1905	SELENIC ACID	8	C2	I	8		0	E0	P002 IBC07		MP18	T6	TP33
1906	SLUDGE ACID	8	C1	II	8		1 L	E0	P001 IBC02		MP15	T8	TP2 TP28
1907	SODA LIME with more than 4% sodium hydroxide	8	C6	III	8	62	5 kg	E1	P002 IBC08 LP02 R001	B3	MP10	T1	TP33

Tank code	Special provisions	Vehicle for tank carriage	Transport category (Tunnel restriction code)	Packages	Bulk	Loading, unloading and handling	Operation	Hazard identification No.	UN No.	Name and description
4.3	4.3.5, 6.8.4	9.1.1.2	1.1.3.6 (8.6)	7.2.4	7.3.3	7.5.11	8.5	5.3.2.3		3.1.2
(12)	(13)	(14)	(15)	(16)	(17)	(18)	(19)	(20)	(1)	(2)
			3 (E)				S2		1866	RESIN SOLUTION, flammable (having a flash-point below 23 °C and viscous according to 2.2.3.1.4) (vapour pressure at 50 °C not more than 110 kPa)
SGAN		AT	2 (E)	V11		CV28		46	1868	DECABORANE
SGAV		AT	3 (E)		VC1 VC2			40	1869	MAGNESIUM or MAGNESIUM ALLOYS with more than 50% magnesium in pellets, turnings or ribbons
			1 (E)	V1		CV23	S20		1870	POTASSIUM BOROHYDRIDE
SGAN		AT	2 (E)					40	1871	TITANIUM HYDRIDE
SGAN	TU3	AT	3 (E)			CV24 CV28		56	1872	LEAD DIOXIDE
L4DN(+)	TU3 TU28	AT	1 (B/E)			CV24	S20	558	1873	PERCHLORIC ACID with more than 50% but not more than 72% acid, by mass
SGAH L4BH	TU15 TE19	AT	2 (E)		VC1 VC2 AP7	CV13 CV28	S9	60	1884	BARIUM OXIDE
SGAH L4BH	TU15 TE19	AT	2 (D/E)	V11		CV13 CV28	S9 S19	60	1885	BENZIDINE
L4BH	TU15 TE19	AT	2 (D/E)			CV13 CV28	S9 S19	60	1886	BENZYLIDENE CHLORIDE
L4BH	TU15 TE19	AT	2 (E)	V12		CV13 CV28	S9	60	1887	BROMOCHLOROMETHANE
L4BH	TU15 TE19	AT	2 (E)	V12		CV13 CV28	S9	60	1888	CHLOROFORM
S10AH L10CH	TU14 TU15 TE19 TE21	AT	1 (C/E)			CV1 CV13 CV28	S9 S14	668	1889	CYANOGEN BROMIDE
L4BH	TU15 TE19	AT	2 (D/E)			CV13 CV28	S9 S19	60	1891	ETHYL BROMIDE
L10CH	TU14 TU15 TE19 TE21	AT	1 (C/D)			CV1 CV13 CV28	S9 S14	66	1892	ETHYLDICHLOROARSINE
SGAH	TU15 TE19	AT	2 (D/E)	V11		CV13 CV28	S9 S19	60	1894	PHENYLMERCURIC HYDROXIDE
SGAH	TU15 TE19	AT	2 (D/E)	V11		CV13 CV28	S9 S19	60	1895	PHENYLMERCURIC NITRATE
L4BH	TU15 TE19	AT	2 (E)	V12		CV13 CV28	S9	60	1897	TETRACHLOROETHYLENE
L4BN		AT	2 (E)					80	1898	ACETYL IODIDE
L4BN		AT	3 (E)	V12				80	1902	DIISOOCTYL ACID PHOSPHATE
L10BH		AT	1 (E)				S20	88	1903	DISINFECTANT, LIQUID, CORROSIVE, N.O.S.
L4BN		AT	2 (E)					80	1903	DISINFECTANT, LIQUID, CORROSIVE, N.O.S.
L4BN		AT	3 (E)	V12				80	1903	DISINFECTANT, LIQUID, CORROSIVE, N.O.S.
S10AN		AT	1 (E)	V10			S20	88	1905	SELENIC ACID
L4BN		AT	2 (E)					80	1906	SLUDGE ACID
SGAV		AT	3 (E)		VC1 VC2 AP7			80	1907	SODA LIME with more than 4% sodium hydroxide

UN No.	Name and description	Class	Classification code	Packing group	Labels	Special provisions	Limited quantities	Excepted quantities	Packing instructions	Special packing provisions	Mixed packing provisions	Instructions	Special provisions
													Portable tanks and bulk containers
3.1.2		2.2	2.2	2.1.1.3	5.2.2	3.3	3.4	3.5.1.2	4.1.4	4.1.4	4.1.10	4.2.5.2 7.3.2	4.2.5.3
(1)	(2)	(3a)	(3b)	(4)	(5)	(6)	(7a)	(7b)	(8)	(9a)	(9b)	(10)	(11)
1908	CHLORITE SOLUTION	8	C9	II	8	521	1 L	E2	P001 IBC02		MP15	T7	TP2 TP24
1908	CHLORITE SOLUTION	8	C9	III	8	521	5 L	E1	P001 IBC03 LP01 R001		MP19	T4	TP2 TP24
1910	Calcium oxide	8	C6					NOT SUBJECT TO ADR					
1911	DIBORANE	2	2TF		2.3 +2.1		0	E0	P200		MP9		
1912	METHYL CHLORIDE AND METHYLENE CHLORIDE MIXTURE	2	2F		2.1	228 662	0	E0	P200		MP9	(M) T50	
1913	NEON, REFRIGERATED LIQUID	2	3A		2.2	593	120 ml	E1	P203		MP9	T75	TP5
1914	BUTYL PROPIONATES	3	F1	III	3		5 L	E1	P001 IBC03 LP01 R001		MP19	T2	TP1
1915	CYCLOHEXANONE	3	F1	III	3		5 L	E1	P001 IBC03 LP01 R001		MP19	T2	TP1
1916	2,2'-DICHLORODIETHYL ETHER	6.1	TF1	II	6.1 +3		100 ml	E4	P001 IBC02		MP15	T7	TP2
1917	ETHYL ACRYLATE, STABILIZED	3	F1	II	3	386	1 L	E2	P001 IBC02 R001		MP19	T4	TP1
1918	ISOPROPYLBENZENE	3	F1	III	3		5 L	E1	P001 IBC03 LP01 R001		MP19	T2	TP1
1919	METHYL ACRYLATE, STABILIZED	3	F1	II	3	386	1 L	E2	P001 IBC02 R001		MP19	T4	TP1
1920	NONANES	3	F1	III	3		5 L	E1	P001 IBC03 LP01 R001		MP19	T2	TP1
1921	PROPYLENEIMINE, STABILIZED	3	FT1	I	3 +6.1	386	0	E0	P001		MP2	T14	TP2
1922	PYRROLIDINE	3	FC	II	3 +8		1 L	E2	P001 IBC02		MP19	T7	TP1
1923	CALCIUM DITHIONITE (CALCIUM HYDROSULPHITE)	4.2	S4	II	4.2		0	E2	P410 IBC06		MP14	T3	TP33
1928	METHYL MAGNESIUM BROMIDE IN ETHYL ETHER	4.3	WF1	I	4.3 +3		0	E0	P402	RR8	MP2		
1929	POTASSIUM DITHIONITE (POTASSIUM HYDROSULPHITE)	4.2	S4	II	4.2		0	E2	P410 IBC06		MP14	T3	TP33
1931	ZINC DITHIONITE (ZINC HYDROSULPHITE)	9	M11	III	9		5 kg	E1	P002 IBC08 LP02 R001	B3	MP10	T1	TP33
1932	ZIRCONIUM SCRAP	4.2	S4	III	4.2	524 592	0	E0	P002 IBC08 LP02 R001	B3	MP14	T1	TP33
1935	CYANIDE SOLUTION, N.O.S.	6.1	T4	I	6.1	274 525	0	E5	P001		MP8 MP17	T14	TP2 TP27
1935	CYANIDE SOLUTION, N.O.S.	6.1	T4	II	6.1	274 525	100 ml	E4	P001 IBC02		MP15	T11	TP2 TP27
1935	CYANIDE SOLUTION, N.O.S.	6.1	T4	III	6.1	274 525	5 L	E1	P001 IBC03 LP01 R001		MP19	T7	TP2 TP28
1938	BROMOACETIC ACID SOLUTION	8	C3	II	8		1 L	E2	P001 IBC02		MP15	T7	TP2
1938	BROMOACETIC ACID SOLUTION	8	C3	III	8		5 L	E1	P001 IBC02 LP01 R001		MP19	T7	TP2

ADR tank		Vehicle for tank carriage	Transport category (Tunnel restriction code)	Special provisions for carriage				Hazard identification No.	UN No.	Name and description
Tank code	Special provisions			Packages	Bulk	Loading, unloading and handling	Operation			
4.3	4.3.5, 6.8.4	9.1.1.2	1.1.3.6 (8.6)	7.2.4	7.3.3	7.5.11	8.5	5.3.2.3		3.1.2
(12)	(13)	(14)	(15)	(16)	(17)	(18)	(19)	(20)	(1)	(2)
L4BV(+)	TE11	AT	2 (E)					80	1908	CHLORITE SOLUTION
L4BV(+)	TE11	AT	3 (E)	V12				80	1908	CHLORITE SOLUTION
		NOT SUBJECT TO ADR							1910	Calcium oxide
			1 (D)			CV9 CV10 CV36	S2 S14		1911	DIBORANE
PxBN(M)	TA4 TT9	FL	2 (B/D)			CV9 CV10 CV36	S2 S20	23	1912	METHYL CHLORIDE AND METHYLENE CHLORIDE MIXTURE
RxBN	TU19 TA4 TT9	AT	3 (C/E)	V5		CV9 CV11 CV36	S20	22	1913	NEON, REFRIGERATED LIQUID
LGBF		FL	3 (D/E)	V12			S2	30	1914	BUTYL PROPIONATES
LGBF		FL	3 (D/E)	V12			S2	30	1915	CYCLOHEXANONE
L4BH	TU15 TE19	FL	2 (D/E)			CV13 CV28	S2 S9 S19	63	1916	2,2'-DICHLORODIETHYL ETHER
LGBF		FL	2 (D/E)	V8			S2 S4 S20	339	1917	ETHYL ACRYLATE, STABILIZED
LGBF		FL	3 (D/E)	V12			S2	30	1918	ISOPROPYLBENZENE
LGBF		FL	2 (D/E)	V8			S2 S4 S20	339	1919	METHYL ACRYLATE, STABILIZED
LGBF		FL	3 (D/E)	V12			S2	30	1920	NONANES
L15CH	TU14 TU15 TE21	FL	1 (C/E)	V8		CV13 CV28	S2 S4 S22	336	1921	PROPYLENEIMINE, STABILIZED
L4BH		FL	2 (D/E)				S2 S20	338	1922	PYRROLIDINE
SGAN		AT	2 (D/E)	V1				40	1923	CALCIUM DITHIONITE (CALCIUM HYDROSULPHITE)
L10DH	TU4 TU14 TU22 TE21 TM2	FL	0 (B/E)	V1		CV23	S2 S20	X323	1928	METHYL MAGNESIUM BROMIDE IN ETHYL ETHER
SGAN		AT	2 (D/E)	V1				40	1929	POTASSIUM DITHIONITE (POTASSIUM HYDROSULPHITE)
SGAV		AT	3 (E)		VC1 VC2			90	1931	ZINC DITHIONITE (ZINC HYDROSULPHITE)
SGAN		AT	3 (E)	V1	VC1 VC2 AP1			40	1932	ZIRCONIUM SCRAP
L10CH	TU14 TU15 TE19 TE21	AT	1 (C/E)			CV1 CV13 CV28	S9 S14	66	1935	CYANIDE SOLUTION, N.O.S.
L4BH	TU15 TE19	AT	2 (D/E)			CV13 CV28	S9 S19	60	1935	CYANIDE SOLUTION, N.O.S.
L4BH	TU15 TE19	AT	2 (E)	V12		CV13 CV28	S9	60	1935	CYANIDE SOLUTION, N.O.S.
L4BN		AT	2 (E)					80	1938	BROMOACETIC ACID SOLUTION
L4BN		AT	3 (E)					80	1938	BROMOACETIC ACID SOLUTION

UN No.	Name and description	Class	Classification code	Packing group	Labels	Special provisions	Limited and excepted quantities		Packaging			Portable tanks and bulk containers	
							Limited	Excepted	Packing instructions	Special packing provisions	Mixed packing provisions	Instructions	Special provisions
	3.1.2	2.2	2.2	2.1.1.3	5.2.2	3.3	3.4	3.5.1.2	4.1.4	4.1.4	4.1.10	4.2.5.2 7.3.2	4.2.5.3
(1)	(2)	(3a)	(3b)	(4)	(5)	(6)	(7a)	(7b)	(8)	(9a)	(9b)	(10)	(11)
1939	PHOSPHORUS OXYBROMIDE	8	C2	II	8		1 kg	E0	P002 IBC08	B4	MP10	T3	TP33
1940	THIOGLYCOLIC ACID	8	C3	II	8		1 L	E2	P001 IBC02		MP15	T7	TP2
1941	DIBROMODIFLUORO-METHANE	9	M11	III	9		5 L	E1	P001 LP01 R001		MP15	T11	TP2
1942	AMMONIUM NITRATE with not more than 0.2% combustible substances, including any organic substance calculated as carbon, to the exclusion of any other added substance	5.1	O2	III	5.1	306 611	5 kg	E1	P002 IBC08 LP02 R001	B3	MP10	T1 BK1 BK2 BK3	TP33
1944	MATCHES, SAFETY (book, card or strike on box)	4.1	F1	III	4.1	293	5 kg	E1	P407 R001		MP11		
1945	MATCHES, WAX 'VESTA'	4.1	F1	III	4.1	293	5 kg	E1	P407 R001		MP11		
1950	AEROSOLS, asphyxiant	2	5A		2.2	190 327 344 625	1 L	E0	P207 LP200	PP87 RR6 L2	MP9		
1950	AEROSOLS, corrosive	2	5C		2.2 +8	190 327 344 625	1 L	E0	P207 LP200	PP87 RR6 L2	MP9		
1950	AEROSOLS, corrosive, oxidizing	2	5CO		2.2 +5.1 +8	190 327 344 625	1 L	E0	P207 LP200	PP87 RR6 L2	MP9		
1950	AEROSOLS, flammable	2	5F		2.1	190 327 344 625	1 L	E0	P207 LP200	PP87 RR6 L2	MP9		
1950	AEROSOLS, flammable, corrosive	2	5FC		2.1 +8	190 327 344 625	1 L	E0	P207 LP200	PP87 RR6 L2	MP9		
1950	AEROSOLS, oxidizing	2	5O		2.2 +5.1	190 327 344 625	1 L	E0	P207 LP200	PP87 RR6 L2	MP9		
1950	AEROSOLS, toxic	2	5T		2.2 +6.1	190 327 344 625	120 ml	E0	P207 LP200	PP87 RR6 L2	MP9		
1950	AEROSOLS, toxic, corrosive	2	5TC		2.2 +6.1 +8	190 327 344 625	120 ml	E0	P207 LP200	PP87 RR6 L2	MP9		
1950	AEROSOLS, toxic, flammable	2	5TF		2.1 +6.1	190 327 344 625	120 ml	E0	P207 LP200	PP87 RR6 L2	MP9		
1950	AEROSOLS, toxic, flammable, corrosive	2	5TFC		2.1 +6.1 +8	190 327 344 625	120 ml	E0	P207 LP200	PP87 RR6 L2	MP9		
1950	AEROSOLS, toxic, oxidizing	2	5TO		2.2 +5.1 +6.1	190 327 344 625	120 ml	E0	P207 LP200	PP87 RR6 L2	MP9		
1950	AEROSOLS, toxic, oxidizing, corrosive	2	5TOC		2.2 +5.1 +6.1 +8	190 327 344 625	120 ml	E0	P207 LP200	PP87 RR6 L2	MP9		
1951	ARGON, REFRIGERATED LIQUID	2	3A		2.2	593	120 ml	E1	P203		MP9	T75	TP5
1952	ETHYLENE OXIDE AND CARBON DIOXIDE MIXTURE with not more than 9% ethylene oxide	2	2A		2.2	662	120 ml	E1	P200		MP9	(M)	
1953	COMPRESSED GAS, TOXIC, FLAMMABLE, N.O.S.	2	1TF		2.3 +2.1	274	0	E0	P200		MP9	(M)	

ADR tank		Vehicle for tank carriage	Transport category (Tunnel restriction code)	Special provisions for carriage				Hazard identification No.	UN No.	Name and description
Tank code	Special provisions			Packages	Bulk	Loading, unloading and handling	Operation			
4.3	4.3.5, 6.8.4	9.1.1.2	1.1.3.6 (8.6)	7.2.4	7.3.3	7.5.11	8.5	5.3.2.3		3.1.2
(12)	(13)	(14)	(15)	(16)	(17)	(18)	(19)	(20)	(1)	(2)
SGAN		AT	2 (E)	V11				80	1939	PHOSPHORUS OXYBROMIDE
L4BN		AT	2 (E)					80	1940	THIOGLYCOLIC ACID
L4BN		AT	3 (E)					90	1941	DIBROMODIFLUORO-METHANE
SGAV	TU3	AT	3 (E)		VC1 VC2 AP6 AP7	CV24	S23	50	1942	AMMONIUM NITRATE with not more than 0.2% combustible substances, including any organic substance calculated as carbon, to the exclusion of any other added substance
			4 (E)						1944	MATCHES, SAFETY (book, card or strike on box)
			4 (E)						1945	MATCHES, WAX 'VESTA'
			3 (E)	V14		CV9 CV12			1950	AEROSOLS, asphyxiant
			1 (E)	V14		CV9 CV12			1950	AEROSOLS, corrosive
			1 (E)	V14		CV9 CV12			1950	AEROSOLS, corrosive, oxidizing
			2 (D)	V14		CV9 CV12	S2		1950	AEROSOLS, flammable
			1 (D)	V14		CV9 CV12	S2		1950	AEROSOLS, flammable, corrosive
			3 (E)	V14		CV9 CV12			1950	AEROSOLS, oxidizing
			1 (D)	V14		CV9 CV12 CV28			1950	AEROSOLS, toxic
			1 (D)	V14		CV9 CV12 CV28			1950	AEROSOLS, toxic, corrosive
			1 (D)	V14		CV9 CV12 CV28	S2		1950	AEROSOLS, toxic, flammable
			1 (D)	V14		CV9 CV12 CV28	S2		1950	AEROSOLS, toxic, flammable, corrosive
			1 (D)	V14		CV9 CV12 CV28			1950	AEROSOLS, toxic, oxidizing
			1 (D)	V14		CV9 CV12 CV28			1950	AEROSOLS, toxic, oxidizing, corrosive
RxBN	TU19 TA4 TT9	AT	3 (C/E)	V5		CV9 CV11 CV36	S20	22	1951	ARGON, REFRIGERATED LIQUID
PxBN(M)	TA4 TT9	AT	3 (C/E)			CV9 CV10 CV36		20	1952	ETHYLENE OXIDE AND CARBON DIOXIDE MIXTURE with not more than 9% ethylene oxide
CxBH(M)	TU6 TA4 TT9	FL	1 (B/D)			CV9 CV10 CV36	S2 S14	263	1953	COMPRESSED GAS, TOXIC, FLAMMABLE, N.O.S.

UN No.	Name and description	Class	Classification code	Packing group	Labels	Special provisions	Limited and excepted quantities		Packaging			Portable tanks and bulk containers	
									Packing instructions	Special packing provisions	Mixed packing provisions	Instructions	Special provisions
	3.1.2	2.2	2.2	2.1.1.3	5.2.2	3.3	3.4	3.5.1.2	4.1.4	4.1.4	4.1.10	4.2.5.2 7.3.2	4.2.5.3
(1)	(2)	(3a)	(3b)	(4)	(5)	(6)	(7a)	(7b)	(8)	(9a)	(9b)	(10)	(11)
1954	COMPRESSED GAS, FLAMMABLE, N.O.S.	2	1F		2.1	274 660 662	0	E0	P200		MP9	(M)	
1955	COMPRESSED GAS, TOXIC, N.O.S.	2	1T		2.3	274	0	E0	P200		MP9	(M)	
1956	COMPRESSED GAS, N.O.S.	2	1A		2.2	274 378 655 662	120 ml	E1	P200		MP9	(M)	
1957	DEUTERIUM, COMPRESSED	2	1F		2.1	662	0	E0	P200		MP9	(M)	
1958	1,2-DICHLORO-1,1,2,2-TETRAFLUOROETHANE (REFRIGERANT GAS R 114)	2	2A		2.2	662	120 ml	E1	P200		MP9	(M) T50	
1959	1,1-DIFLUOROETHYLENE (REFRIGERANT GAS R 1132a)	2	2F		2.1	662	0	E0	P200		MP9	(M)	
1961	ETHANE, REFRIGERATED LIQUID	2	3F		2.1		0	E0	P203		MP9	T75	TP5
1962	ETHYLENE	2	2F		2.1	662	0	E0	P200		MP9	(M)	
1963	HELIUM, REFRIGERATED LIQUID	2	3A		2.2	593	120 ml	E1	P203		MP9	T75	TP5 TP34
1964	HYDROCARBON GAS MIXTURE, COMPRESSED, N.O.S.	2	1F		2.1	274 662	0	E0	P200		MP9	(M)	
1965	HYDROCARBON GAS MIXTURE, LIQUEFIED, N.O.S. such as mixtures A, A01, A02, A0, A1, B1, B2, B or C	2	2F		2.1	274 583 652 660 662	0	E0	P200		MP9	(M) T50	
1966	HYDROGEN, REFRIGERATED LIQUID	2	3F		2.1		0	E0	P203		MP9	T75	TP5 TP34
1967	INSECTICIDE GAS, TOXIC, N.O.S.	2	2T		2.3	274	0	E0	P200		MP9	(M)	
1968	INSECTICIDE GAS, N.O.S.	2	2A		2.2	274 662	120 ml	E1	P200		MP9	(M)	
1969	ISOBUTANE	2	2F		2.1	657 660 662	0	E0	P200		MP9	(M) T50	
1970	KRYPTON, REFRIGERATED LIQUID	2	3A		2.2	593	120 ml	E1	P203		MP9	T75	TP5
1971	METHANE, COMPRESSED or NATURAL GAS, COMPRESSED with high methane content	2	1F		2.1	660 662	0	E0	P200		MP9	(M)	
1972	METHANE, REFRIGERATED LIQUID or NATURAL GAS, REFRIGERATED LIQUID with high methane content	2	3F		2.1	660	0	E0	P203		MP9	T75	TP5
1973	CHLORODIFLUORO-METHANE AND CHLOROPENTAFLUORO-ETHANE MIXTURE with fixed boiling point, with approximately 49% chlorodifluoromethane (REFRIGERANT GAS R 502)	2	2A		2.2	662	120 ml	E1	P200		MP9	(M) T50	
1974	CHLORODIFLUOROBROMO-METHANE (REFRIGERANT GAS R 12B1)	2	2A		2.2	662	120 ml	E1	P200		MP9	(M) T50	
1975	NITRIC OXIDE AND DINITROGEN TETROXIDE MIXTURE (NITRIC OXIDE AND NITROGEN DIOXIDE MIXTURE)	2	2TOC		2.3 +5.1 +8		0	E0	P200		MP9		
1976	OCTAFLUOROCYCLO-BUTANE (REFRIGERANT GAS RC 318)	2	2A		2.2	662	120 ml	E1	P200		MP9	(M) T50	

Tank code	Special provisions	Vehicle for tank carriage	Transport category (Tunnel restriction code)	Packages	Bulk	Loading, unloading and handling	Operation	Hazard identification No.	UN No.	Name and description
4.3	4.3.5, 6.8.4	9.1.1.2	1.1.3.6 (8.6)	7.2.4	7.3.3	7.5.11	8.5	5.3.2.3		3.1.2
(12)	(13)	(14)	(15)	(16)	(17)	(18)	(19)	(20)	(1)	(2)
CxBN(M)	TA4 TT9	FL	2 (B/D)			CV9 CV10 CV36	S2 S20	23	1954	COMPRESSED GAS, FLAMMABLE, N.O.S.
CxBH(M)	TU6 TA4 TT9	AT	1 (C/D)			CV9 CV10 CV36	S14	26	1955	COMPRESSED GAS, TOXIC, N.O.S.
CxBN(M)	TA4 TT9	AT	3 (E)			CV9 CV10 CV36		20	1956	COMPRESSED GAS, N.O.S.
CxBN(M)	TA4 TT9	FL	2 (B/D)			CV9 CV10 CV36	S2 S20	23	1957	DEUTERIUM, COMPRESSED
PxBN(M)	TA4 TT9	AT	3 (C/E)			CV9 CV10 CV36		20	1958	1,2-DICHLORO-1,1,2,2-TETRAFLUOROETHANE (REFRIGERANT GAS R 114)
PxBN(M)	TA4 TT9	FL	2 (B/D)			CV9 CV10 CV36	S2 S20	239	1959	1,1-DIFLUOROETHYLENE (REFRIGERANT GAS R 1132a)
RxBN	TU18 TA4 TT9	FL	2 (B/D)	V5		CV9 CV11 CV36	S2 S17	223	1961	ETHANE, REFRIGERATED LIQUID
PxBN(M)	TA4 TT9	FL	2 (B/D)			CV9 CV10 CV36	S2 S20	23	1962	ETHYLENE
RxBN	TU19 TA4 TT9	AT	3 (C/E)	V5		CV9 CV11 CV36	S20	22	1963	HELIUM, REFRIGERATED LIQUID
CxBN(M)	TA4 TT9	FL	2 (B/D)			CV9 CV10 CV36	S2 S20	23	1964	HYDROCARBON GAS MIXTURE, COMPRESSED, N.O.S.
PxBN(M)	TA4 TT9 TT11	FL	2 (B/D)			CV9 CV10 CV36	S2 S20	23	1965	HYDROCARBON GAS MIXTURE, LIQUEFIED, N.O.S. such as mixtures A, A01, A02, A0, A1, B1, B2, B or C
RxBN	TU18 TA4 TT9	FL	2 (B/D)	V5		CV9 CV11 CV36	S2 S17	223	1966	HYDROGEN, REFRIGERATED LIQUID
PxBH(M)	TU6 TA4 TT9	AT	1 (C/D)			CV9 CV10 CV36	S14	26	1967	INSECTICIDE GAS, TOXIC, N.O.S.
PxBN(M)	TA4 TT9	AT	3 (C/E)			CV9 CV10 CV36		20	1968	INSECTICIDE GAS, N.O.S.
PxBN(M)	TA4 TT9 TT11	FL	2 (B/D)			CV9 CV10 CV36	S2 S20	23	1969	ISOBUTANE
RxBN	TU19 TA4 TT9	AT	3 (C/E)	V5		CV9 CV11 CV36	S20	22	1970	KRYPTON, REFRIGERATED LIQUID
CxBN(M)	TA4 TT9	FL	2 (B/D)			CV9 CV10 CV36	S2 S20	23	1971	METHANE, COMPRESSED or NATURAL GAS, COMPRESSED with high methane content
RxBN	TU18 TA4 TT9	FL	2 (B/D)	V5		CV9 CV11 CV36	S2 S17	223	1972	METHANE, REFRIGERATED LIQUID or NATURAL GAS, REFRIGERATED LIQUID with high methane content
PxBN(M)	TA4 TT9	AT	3 (C/E)			CV9 CV10 CV36		20	1973	CHLORODIFLUORO-METHANE AND CHLOROPENTAFLUORO-ETHANE MIXTURE with fixed boiling point, with approximately 49% chlorodifluoromethane (REFRIGERANT GAS R 502)
PxBN(M)	TA4 TT9	AT	3 (C/E)			CV9 CV10 CV36		20	1974	CHLORODIFLUOROBROMO-METHANE (REFRIGERANT GAS R 12B1)
			1 (D)			CV9 CV10 CV36	S14		1975	NITRIC OXIDE AND DINITROGEN TETROXIDE MIXTURE (NITRIC OXIDE AND NITROGEN DIOXIDE MIXTURE)
PxBN(M)	TA4 TT9	AT	3 (C/E)			CV9 CV10 CV36		20	1976	OCTAFLUOROCYCLO-BUTANE (REFRIGERANT GAS RC 318)

UN No.	Name and description	Class	Classification code	Packing group	Labels	Special provisions	Limited and excepted quantities		Packaging			Portable tanks and bulk containers	
									Packing instructions	Special packing provisions	Mixed packing provisions	Instructions	Special provisions
	3.1.2	2.2	2.2	2.1.1.3	5.2.2	3.3	3.4	3.5.1.2	4.1.4	4.1.4	4.1.10	4.2.5.2 7.3.2	4.2.5.3
(1)	(2)	(3a)	(3b)	(4)	(5)	(6)	(7a)	(7b)	(8)	(9a)	(9b)	(10)	(11)
1977	NITROGEN, REFRIGERATED LIQUID	2	3A		2.2	345 346 593	120 ml	E1	P203		MP9	T75	TP5
1978	PROPANE	2	2F		2.1	652 657 660 662	0	E0	P200		MP9	(M) T50	
1982	TETRAFLUOROMETHANE (REFRIGERANT GAS R 14)	2	2A		2.2	662	120 ml	E1	P200		MP9	(M)	
1983	1-CHLORO-2,2,2-TRIFLUOROETHANE (REFRIGERANT GAS R 133a)	2	2A		2.2	662	120 ml	E1	P200		MP9	(M) T50	
1984	TRIFLUOROMETHANE (REFRIGERANT GAS R 23)	2	2A		2.2	662	120 ml	E1	P200		MP9	(M)	
1986	ALCOHOLS, FLAMMABLE, TOXIC, N.O.S.	3	FT1	I	3 +6.1	274	0	E0	P001		MP7 MP17	T14	TP2 TP27
1986	ALCOHOLS, FLAMMABLE, TOXIC, N.O.S.	3	FT1	II	3 +6.1	274	1 L	E2	P001 IBC02		MP19	T11	TP2 TP27
1986	ALCOHOLS, FLAMMABLE, TOXIC, N.O.S.	3	FT1	III	3 +6.1	274	5 L	E1	P001 IBC03 R001		MP19	T7	TP1 TP28
1987	ALCOHOLS, N.O.S. (vapour pressure at 50 °C more than 110 kPa)	3	F1	II	3	274 601 640C	1 L	E2	P001		MP19	T7	TP1 TP8 TP28
1987	ALCOHOLS, N.O.S. (vapour pressure at 50 °C not more than 110 kPa)	3	F1	II	3	274 601 640D	1 L	E2	P001 IBC02 R001		MP19	T7	TP1 TP8 TP28
1987	ALCOHOLS, N.O.S.	3	F1	III	3	274 601	5 L	E1	P001 IBC03 LP01 R001		MP19	T4	TP1 TP29
1988	ALDEHYDES, FLAMMABLE, TOXIC, N.O.S.	3	FT1	I	3 +6.1	274	0	E0	P001		MP7 MP17	T14	TP2 TP27
1988	ALDEHYDES, FLAMMABLE, TOXIC, N.O.S.	3	FT1	II	3 +6.1	274	1 L	E2	P001 IBC02		MP19	T11	TP2 TP27
1988	ALDEHYDES, FLAMMABLE, TOXIC, N.O.S.	3	FT1	III	3 +6.1	274	5 L	E1	P001 IBC03 R001		MP19	T7	TP1 TP28
1989	ALDEHYDES, N.O.S.	3	F1	I	3	274	0	E3	P001		MP7 MP17	T11	TP1 TP27
1989	ALDEHYDES, N.O.S. (vapour pressure at 50 °C more than 110 kPa)	3	F1	II	3	274 640C	1 L	E2	P001		MP19	T7	TP1 TP8 TP28
1989	ALDEHYDES, N.O.S. (vapour pressure at 50 °C not more than 110 kPa)	3	F1	II	3	274 640D	1 L	E2	P001 IBC02 R001		MP19	T7	TP1 TP8 TP28
1989	ALDEHYDES, N.O.S.	3	F1	III	3	274	5 L	E1	P001 IBC03 LP01 R001		MP19	T4	TP1 TP29
1990	BENZALDEHYDE	9	M11	III	9		5 L	E1	P001 IBC03 LP01 R001		MP15	T2	TP1
1991	CHLOROPRENE, STABILIZED	3	FT1	I	3 +6.1	386	0	E0	P001		MP7 MP17	T14	TP2 TP6
1992	FLAMMABLE LIQUID, TOXIC, N.O.S.	3	FT1	I	3 +6.1	274	0	E0	P001		MP7 MP17	T14	TP2 TP27
1992	FLAMMABLE LIQUID, TOXIC, N.O.S.	3	FT1	II	3 +6.1	274	1 L	E2	P001 IBC02		MP19	T7	TP2
1992	FLAMMABLE LIQUID, TOXIC, N.O.S.	3	FT1	III	3 +6.1	274	5 L	E1	P001 IBC03 R001		MP19	T7	TP1 TP28
1993	FLAMMABLE LIQUID, N.O.S.	3	F1	I	3	274	0	E3	P001		MP7 MP17	T11	TP1 TP27
1993	FLAMMABLE LIQUID, N.O.S. (vapour pressure at 50 °C more than 110 kPa)	3	F1	II	3	274 601 640C	1 L	E2	P001		MP19	T7	TP1 TP8 TP28
1993	FLAMMABLE LIQUID, N.O.S. (vapour pressure at 50 °C not more than 110 kPa)	3	F1	II	3	274 601 640D	1 L	E2	P001 IBC02 R001		MP19	T7	TP1 TP8 TP28

ADR tank		Vehicle for tank carriage	Transport category (Tunnel restriction code)	Special provisions for carriage				Hazard identifi-cation No.	UN No.	Name and description
Tank code	Special provisions			Packages	Bulk	Loading, unloading and handling	Operation			
4.3	4.3.5, 6.8.4	9.1.1.2	1.1.3.6 (8.6)	7.2.4	7.3.3	7.5.11	8.5	5.3.2.3		3.1.2
(12)	(13)	(14)	(15)	(16)	(17)	(18)	(19)	(20)	(1)	(2)
RxBN	TU19 TA4 TT9	AT	3 (C/E)	V5		CV9 CV11 CV36	S20	22	1977	NITROGEN, REFRIGERATED LIQUID
PxBN(M)	TA4 TT9 TT11	FL	2 (B/D)			CV9 CV10 CV36	S2 S20	23	1978	PROPANE
PxBN(M)	TA4 TT9	AT	3 (C/E)			CV9 CV10 CV36		20	1982	TETRAFLUOROMETHANE (REFRIGERANT GAS R 14)
PxBN(M)	TA4 TT9	AT	3 (C/E)			CV9 CV10 CV36		20	1983	1-CHLORO-2,2,2-TRIFLUOROETHANE (REFRIGERANT GAS R 133a)
PxBN(M)	TA4 TT9	AT	3 (C/E)			CV9 CV10 CV36		20	1984	TRIFLUOROMETHANE (REFRIGERANT GAS R 23)
L10CH	TU14 TU15 TE21	FL	1 (C/E)			CV13 CV28	S2 S22	336	1986	ALCOHOLS, FLAMMABLE, TOXIC, N.O.S.
L4BH	TU15	FL	2 (D/E)			CV13 CV28	S2 S22	336	1986	ALCOHOLS, FLAMMABLE, TOXIC, N.O.S.
L4BH	TU15	FL	3 (D/E)	V12		CV13 CV28	S2	36	1986	ALCOHOLS, FLAMMABLE, TOXIC, N.O.S.
L1.5BN		FL	2 (D/E)				S2 S20	33	1987	ALCOHOLS, N.O.S. (vapour pressure at 50 °C more than 110 kPa)
LGBF		FL	2 (D/E)				S2 S20	33	1987	ALCOHOLS, N.O.S. (vapour pressure at 50 °C not more than 110 kPa)
LGBF		FL	3 (D/E)	V12			S2	30	1987	ALCOHOLS, N.O.S.
L10CH	TU14 TU15 TE21	FL	1 (C/E)			CV13 CV28	S2 S22	336	1988	ALDEHYDES, FLAMMABLE, TOXIC, N.O.S.
L4BH	TU15	FL	2 (D/E)			CV13 CV28	S2 S22	336	1988	ALDEHYDES, FLAMMABLE, TOXIC, N.O.S.
L4BH	TU15	FL	3 (D/E)	V12		CV13 CV28	S2	36	1988	ALDEHYDES, FLAMMABLE, TOXIC, N.O.S.
L4BN		FL	1 (D/E)				S2 S20	33	1989	ALDEHYDES, N.O.S.
L1.5BN		FL	2 (D/E)				S2 S20	33	1989	ALDEHYDES, N.O.S. (vapour pressure at 50 °C more than 110 kPa)
LGBF		FL	2 (D/E)				S2 S20	33	1989	ALDEHYDES, N.O.S. (vapour pressure at 50 °C not more than 110 kPa)
LGBF		FL	3 (D/E)	V12			S2	30	1989	ALDEHYDES, N.O.S.
LGBV		AT	3 (E)	V12				90	1990	BENZALDEHYDE
L10CH	TU14 TU15 TE21	FL	1 (C/E)	V8		CV13 CV28	S2 S4 S22	336	1991	CHLOROPRENE, STABILIZED
L10CH	TU14 TU15 TE21	FL	1 (C/E)			CV13 CV28	S2 S22	336	1992	FLAMMABLE LIQUID, TOXIC, N.O.S.
L4BH	TU15	FL	2 (D/E)			CV13 CV28	S2 S22	336	1992	FLAMMABLE LIQUID, TOXIC, N.O.S.
L4BH	TU15	FL	3 (D/E)	V12		CV13 CV28	S2	36	1992	FLAMMABLE LIQUID, TOXIC, N.O.S.
L4BN		FL	1 (D/E)				S2 S20	33	1993	FLAMMABLE LIQUID, N.O.S.
L1.5BN		FL	2 (D/E)				S2 S20	33	1993	FLAMMABLE LIQUID, N.O.S. (vapour pressure at 50 °C more than 110 kPa)
LGBF		FL	2 (D/E)				S2 S20	33	1993	FLAMMABLE LIQUID, N.O.S. (vapour pressure at 50 °C not more than 110 kPa)

UN No.	Name and description	Class	Classification code	Packing group	Labels	Special provisions	Limited and excepted quantities		Packaging			Portable tanks and bulk containers	
									Packing instructions	Special packing provisions	Mixed packing provisions	Instructions	Special provisions
	3.1.2	2.2	2.2	2.1.1.3	5.2.2	3.3	3.4	3.5.1.2	4.1.4	4.1.4	4.1.10	4.2.5.2 7.3.2	4.2.5.3
(1)	(2)	(3a)	(3b)	(4)	(5)	(6)	(7a)	(7b)	(8)	(9a)	(9b)	(10)	(11)
1993	FLAMMABLE LIQUID, N.O.S.	3	F1	III	3	274 601	5 L	E1	P001 IBC03 LP01 R001		MP19	T4	TP1 TP29
1993	FLAMMABLE LIQUID, N.O.S. (having a flash-point below 23 °C and viscous according to 2.2.3.1.4) (vapour pressure at 50 °C more than 110 kPa)	3	F1	III	3	274 601	5 L	E1	P001 R001		MP19		
1993	FLAMMABLE LIQUID, N.O.S. (having a flash-point below 23 °C and viscous according to 2.2.3.1.4) (vapour pressure at 50 °C not more than 110 kPa)	3	F1	III	3	274 601	5 L	E1	P001 IBC02 R001	BB4	MP19		
1994	IRON PENTACARBONYL	6.1	TF1	I	6.1 +3	354	0	E0	P601		MP2	T22	TP2
1999	TARS, LIQUID, including road oils, and cutback bitumens (vapour pressure at 50 °C more than 110 kPa)	3	F1	II	3	640C	5 L	E2	P001		MP19	T3	TP3 TP29
1999	TARS, LIQUID, including road oils, and cutback bitumens (vapour pressure at 50 °C not more than 110 kPa)	3	F1	II	3	640D	5 L	E2	P001 IBC02 R001		MP19	T3	TP3 TP29
1999	TARS, LIQUID, including road oils, and cutback bitumens	3	F1	III	3		5 L	E1	P001 IBC03 LP01 R001		MP19	T1	TP3
1999	TARS, LIQUID, including road oils, and cutback bitumens (having a flash-point below 23 °C and viscous according to 2.2.3.1.4) (vapour pressure at 50 °C more than 110 kPa)	3	F1	III	3		5 L	E1	P001 R001		MP19		
1999	TARS, LIQUID, including road oils, and cutback bitumens (having a flash-point below 23 °C and viscous according to 2.2.3.1.4) (vapour pressure at 50 °C not more than 110 kPa)	3	F1	III	3		5 L	E1	P001 IBC02 R001	BB4	MP19		
2000	CELLULOID in block, rods, rolls, sheets, tubes, etc., except scrap	4.1	F1	III	4.1	383 502	5 kg	E1	P002 LP02 R001	PP7	MP11		
2001	COBALT NAPHTHENATES, POWDER	4.1	F3	III	4.1		5 kg	E1	P002 IBC08 LP02 R001	B3	MP11	T1	TP33
2002	CELLULOID, SCRAP	4.2	S2	III	4.2	526 592	0	E0	P002 IBC08 LP02 R001	PP8 B3	MP14		
2004	MAGNESIUM DIAMIDE	4.2	S4	II	4.2		0	E2	P410 IBC06		MP14	T3	TP33
2006	PLASTICS, NITROCELLULOSE-BASED, SELF-HEATING, N.O.S.	4.2	S2	III	4.2	274 528	0	E0	P002 R001		MP14		
2008	ZIRCONIUM POWDER, DRY	4.2	S4	I	4.2	524 540	0	E0	P404		MP13	T21	TP7 TP33
2008	ZIRCONIUM POWDER, DRY	4.2	S4	II	4.2	524 540	0	E2	P410 IBC06		MP14	T3	TP33
2008	ZIRCONIUM POWDER, DRY	4.2	S4	III	4.2	524 540	0	E1	P002 IBC08 LP02 R001	B3	MP14	T1	TP33
2009	ZIRCONIUM, DRY, finished sheets, strip or coiled wire	4.2	S4	III	4.2	524 592	0	E1	P002 LP02 R001		MP14		
2010	MAGNESIUM HYDRIDE	4.3	W2	I	4.3		0	E0	P403		MP2		
2011	MAGNESIUM PHOSPHIDE	4.3	WT2	I	4.3 +6.1		0	E0	P403		MP2		
2012	POTASSIUM PHOSPHIDE	4.3	WT2	I	4.3 +6.1		0	E0	P403		MP2		
2013	STRONTIUM PHOSPHIDE	4.3	WT2	I	4.3 +6.1		0	E0	P403		MP2		

ADR tank		Vehicle for tank carriage	Transport category (Tunnel restriction code)	Special provisions for carriage				Hazard identification No.	UN No.	Name and description
Tank code	Special provisions			Packages	Bulk	Loading, unloading and handling	Operation			
4.3	4.3.5, 6.8.4	9.1.1.2	1.1.3.6 (8.6)	7.2.4	7.3.3	7.5.11	8.5	5.3.2.3		3.1.2
(12)	(13)	(14)	(15)	(16)	(17)	(18)	(19)	(20)	(1)	(2)
LGBF		FL	3 (D/E)	V12			S2	30	1993	FLAMMABLE LIQUID, N.O.S.
			3 (E)				S2		1993	FLAMMABLE LIQUID, N.O.S. (having a flash-point below 23 °C and viscous according to 2.2.3.1.4) (vapour pressure at 50 °C more than 110 kPa)
			3 (E)				S2		1993	FLAMMABLE LIQUID, N.O.S. (having a flash-point below 23 °C and viscous according to 2.2.3.1.4) (vapour pressure at 50 °C not more than 110 kPa)
L15CH	TU14 TU15 TU31 TE19 TE21 TM3	FL	1 (C/D)			CV1 CV13 CV28	S2 S9 S14	663	1994	IRON PENTACARBONYL
L1.5BN		FL	2 (D/E)				S2 S20	33	1999	TARS, LIQUID, including road oils, and cutback bitumens (vapour pressure at 50 °C more than 110 kPa)
LGBF		FL	2 (D/E)				S2 S20	33	1999	TARS, LIQUID, including road oils, and cutback bitumens (vapour pressure at 50 °C not more than 110 kPa)
LGBF		FL	3 (D/E)	V12			S2	30	1999	TARS, LIQUID, including road oils, and cutback bitumens
			3 (E)				S2		1999	TARS, LIQUID, including road oils, and cutback bitumens (having a flash-point below 23 °C and viscous according to 2.2.3.1.4) (vapour pressure at 50 °C more than 110 kPa)
			3 (E)				S2		1999	TARS, LIQUID, including road oils, and cutback bitumens (having a flash-point below 23 °C and viscous according to 2.2.3.1.4) (vapour pressure at 50 °C not more than 110 kPa)
			3 (E)						2000	CELLULOID in block, rods, rolls, sheets, tubes, etc., except scrap
SGAV		AT	3 (E)		VC1 VC2			40	2001	COBALT NAPHTHENATES, POWDER
			3 (E)	V1					2002	CELLULOID, SCRAP
SGAN		AT	2 (D/E)	V1				40	2004	MAGNESIUM DIAMIDE
			3 (E)	V1					2006	PLASTICS, NITROCELLULOSE-BASED, SELF-HEATING, N.O.S.
		AT	0 (B/E)	V1			S20	43	2008	ZIRCONIUM POWDER, DRY
SGAN		AT	2 (D/E)	V1				40	2008	ZIRCONIUM POWDER, DRY
SGAN		AT	3 (E)	V1	VC1 VC2 AP1			40	2008	ZIRCONIUM POWDER, DRY
			3 (E)	V1	VC1 VC2 AP1			40	2009	ZIRCONIUM, DRY, finished sheets, strip or coiled wire
			1 (E)	V1		CV23	S20		2010	MAGNESIUM HYDRIDE
			1 (E)	V1		CV23 CV28	S20		2011	MAGNESIUM PHOSPHIDE
			1 (E)	V1		CV23 CV28	S20		2012	POTASSIUM PHOSPHIDE
			1 (E)	V1		CV23 CV28	S20		2013	STRONTIUM PHOSPHIDE

UN No.	Name and description	Class	Classifi- cation code	Packing group	Labels	Special provi- sions	Limited and excepted quantities		Packaging			Portable tanks and bulk containers	
									Packing instruc- tions	Special packing provisions	Mixed packing provisions	Instruc- tions	Special provisions
	3.1.2	2.2	2.2	2.1.1.3	5.2.2	3.3	3.4	3.5.1.2	4.1.4	4.1.4	4.1.10	4.2.5.2 7.3.2	4.2.5.3
(1)	(2)	(3a)	(3b)	(4)	(5)	(6)	(7a)	(7b)	(8)	(9a)	(9b)	(10)	(11)
2014	HYDROGEN PEROXIDE, AQUEOUS SOLUTION with not less than 20% but not more than 60% hydrogen peroxide (stabilized as necessary)	5.1	OC1	II	5.1 +8		1 L	E2	P504 IBC02	PP10 B5	MP15	T7	TP2 TP6 TP24
2015	HYDROGEN PEROXIDE, AQUEOUS SOLUTION, STABILIZED with more than 70% hydrogen peroxide	5.1	OC1	I	5.1 +8	640N	0	E0	P501		MP2	T9	TP2 TP6 TP24
2015	HYDROGEN PEROXIDE, AQUEOUS SOLUTION, STABILIZED with more than 60% hydrogen peroxide and not more than 70% hydrogen peroxide	5.1	OC1	I	5.1 +8	640O	0	E0	P501		MP2	T9	TP2 TP6 TP24
2016	AMMUNITION, TOXIC, NON-EXPLOSIVE without burster or expelling charge, non-fuzed	6.1	T2		6.1		0	E0	P600		MP10		
2017	AMMUNITION, TEAR-PRODUCING, NON-EXPLOSIVE without burster or expelling charge, non-fuzed	6.1	TC2		6.1 +8		0	E0	P600				
2018	CHLOROANILINES, SOLID	6.1	T2	II	6.1		500 g	E4	P002 IBC08	B4	MP10	T3	TP33
2019	CHLOROANILINES, LIQUID	6.1	T1	II	6.1		100 ml	E4	P001 IBC02		MP15	T7	TP2
2020	CHLOROPHENOLS, SOLID	6.1	T2	III	6.1	205	5 kg	E1	P002 IBC08 LP02 R001	B3	MP10	T1	TP33
2021	CHLOROPHENOLS, LIQUID	6.1	T1	III	6.1		5 L	E1	P001 IBC03 LP01 R001		MP19	T4	TP1
2022	CRESYLIC ACID	6.1	TC1	II	6.1 +8		100 ml	E4	P001 IBC02		MP15	T7	TP2
2023	EPICHLOROHYDRIN	6.1	TF1	II	6.1 +3	279	100 ml	E4	P001 IBC02		MP15	T7	TP2
2024	MERCURY COMPOUND, LIQUID, N.O.S.	6.1	T4	I	6.1	43 274	0	E5	P001		MP8 MP17		
2024	MERCURY COMPOUND, LIQUID, N.O.S.	6.1	T4	II	6.1	43 274	100 ml	E4	P001 IBC02		MP15		
2024	MERCURY COMPOUND, LIQUID, N.O.S.	6.1	T4	III	6.1	43 274	5 L	E1	P001 IBC03 LP01 R001		MP19		
2025	MERCURY COMPOUND, SOLID, N.O.S.	6.1	T5	I	6.1	43 66 274 529	0	E5	P002 IBC07		MP18	T6	TP33
2025	MERCURY COMPOUND, SOLID, N.O.S.	6.1	T5	II	6.1	43 66 274 529	500 g	E4	P002 IBC08	B4	MP10	T3	TP33
2025	MERCURY COMPOUND, SOLID, N.O.S.	6.1	T5	III	6.1	43 66 274 529	5 kg	E1	P002 IBC08 LP02 R001	B3	MP10	T1	TP33
2026	PHENYLMERCURIC COMPOUND, N.O.S.	6.1	T3	I	6.1	43 274	0	E5	P002 IBC07		MP18	T6	TP33
2026	PHENYLMERCURIC COMPOUND, N.O.S.	6.1	T3	II	6.1	43 274	500 g	E4	P002 IBC08	B4	MP10	T3	TP33
2026	PHENYLMERCURIC COMPOUND, N.O.S.	6.1	T3	III	6.1	43 274	5 kg	E1	P002 IBC08 LP02 R001	B3	MP10	T1	TP33
2027	SODIUM ARSENITE, SOLID	6.1	T5	II	6.1	43	500 g	E4	P002 IBC08	B4	MP10	T3	TP33
2028	BOMBS, SMOKE, NON-EXPLOSIVE with corrosive liquid, without initiating device	8	C11	II	8		0	E0	P803				
2029	HYDRAZINE, ANHYDROUS	8	CFT	I	8 +3 +6.1		0	E0	P001		MP8 MP17		

ADR tank		Vehicle for tank carriage	Transport category (Tunnel restriction code)	Special provisions for carriage				Hazard identifi-cation No.	UN No.	Name and description
Tank code	Special provisions			Packages	Bulk	Loading, unloading and handling	Operation			
4.3	4.3.5, 6.8.4	9.1.1.2	1.1.3.6 (8.6)	7.2.4	7.3.3	7.5.11	8.5	5.3.2.3		3.1.2
(12)	(13)	(14)	(15)	(16)	(17)	(18)	(19)	(20)	(1)	(2)
L4BV(+)	TU3 TC2 TE8 TE11 TT1	AT	2 (E)			CV24		58	2014	HYDROGEN PEROXIDE, AQUEOUS SOLUTION with not less than 20% but not more than 60% hydrogen peroxide (stabilized as necessary)
L4DV(+)	TU3 TU28 TC2 TE8 TE9 TT1	FL	1 (B/E)	V5		CV24	S20	559	2015	HYDROGEN PEROXIDE, AQUEOUS SOLUTION, STABILIZED with more than 70% hydrogen peroxide
L4BV(+)	TU3 TU28 TC2 TE7 TE8 TE9 TT1	FL	1 (B/E)	V5		CV24	S20	559	2015	HYDROGEN PEROXIDE, AQUEOUS SOLUTION, STABILIZED with more than 60% hydrogen peroxide and not more than 70% hydrogen peroxide
			2 (E)			CV13 CV28	S9 S19		2016	AMMUNITION, TOXIC, NON-EXPLOSIVE without burster or expelling charge, non-fuzed
			2 (E)			CV13 CV28	S9 S19		2017	AMMUNITION, TEAR-PRODUCING, NON-EXPLOSIVE without burster or expelling charge, non-fuzed
SGAH L4BH	TU15 TE19	AT	2 (D/E)	V11		CV13 CV28	S9 S19	60	2018	CHLOROANILINES, SOLID
L4BH	TU15 TE19	AT	2 (D/E)			CV13 CV28	S9 S19	60	2019	CHLOROANILINES, LIQUID
SGAH	TU15 TE19	AT	2 (E)		VC1 VC2 AP7	CV13 CV28	S9	60	2020	CHLOROPHENOLS, SOLID
L4BH	TU15 TE19	AT	2 (E)	V12		CV13 CV28	S9	60	2021	CHLOROPHENOLS, LIQUID
L4BH	TU15 TE19	AT	2 (D/E)			CV13 CV28	S9 S19	68	2022	CRESYLIC ACID
L4BH	TU15 TE19	FL	2 (D/E)			CV13 CV28	S2 S9 S19	63	2023	EPICHLOROHYDRIN
L10CH	TU14 TU15 TE19 TE21	AT	1 (C/E)			CV1 CV13 CV28	S9 S14	66	2024	MERCURY COMPOUND, LIQUID, N.O.S.
L4BH	TU15 TE19	AT	2 (D/E)			CV13 CV28	S9 S19	60	2024	MERCURY COMPOUND, LIQUID, N.O.S.
L4BH	TU15 TE19	AT	2 (E)	V12		CV13 CV28	S9	60	2024	MERCURY COMPOUND, LIQUID, N.O.S.
S10AH	TU15 TE19	AT	1 (C/E)	V10		CV1 CV13 CV28	S9 S14	66	2025	MERCURY COMPOUND, SOLID, N.O.S.
SGAH	TU15 TE19	AT	2 (D/E)	V11		CV13 CV28	S9 S19	60	2025	MERCURY COMPOUND, SOLID, N.O.S.
SGAH	TU15 TE19	AT	2 (E)		VC1 VC2 AP7	CV13 CV28	S9	60	2025	MERCURY COMPOUND, SOLID, N.O.S.
S10AH L10CH	TU14 TU15 TE19 TE21	AT	1 (C/E)	V10		CV1 CV13 CV28	S9 S14	66	2026	PHENYLMERCURIC COMPOUND, N.O.S.
SGAH L4BH	TU15 TE19	AT	2 (D/E)	V11		CV13 CV28	S9 S19	60	2026	PHENYLMERCURIC COMPOUND, N.O.S.
SGAH L4BH	TU15 TE19	AT	2 (E)		VC1 VC2 AP7	CV13 CV28	S9	60	2026	PHENYLMERCURIC COMPOUND, N.O.S.
SGAH	TU15 TE19	AT	2 (D/E)	V11		CV13 CV28	S9 S19	60	2027	SODIUM ARSENITE, SOLID
			2 (E)						2028	BOMBS, SMOKE, NON-EXPLOSIVE with corrosive liquid, without initiating device
			1 (E)			CV13 CV28	S2 S14		2029	HYDRAZINE, ANHYDROUS

UN No.	Name and description	Class	Classification code	Packing group	Labels	Special provisions	Limited quantities	Excepted quantities	Packing instructions	Special packing provisions	Mixed packing provisions	Portable tanks Instructions	Special provisions
	3.1.2	2.2	2.2	2.1.1.3	5.2.2	3.3	3.4	3.5.1.2	4.1.4	4.1.4	4.1.10	4.2.5.2 7.3.2	4.2.5.3
(1)	(2)	(3a)	(3b)	(4)	(5)	(6)	(7a)	(7b)	(8)	(9a)	(9b)	(10)	(11)
2030	HYDRAZINE AQUEOUS SOLUTION, with more than 37% hydrazine by mass	8	CT1	I	8 +6.1	530	0	E0	P001		MP8 MP17	T10	TP2
2030	HYDRAZINE AQUEOUS SOLUTION, with more than 37% hydrazine by mass	8	CT1	II	8 +6.1	530	1 L	E0	P001 IBC02		MP15	T7	TP2
2030	HYDRAZINE AQUEOUS SOLUTION, with more than 37% hydrazine by mass	8	CT1	III	8 +6.1	530	5 L	E1	P001 IBC03 LP01 R001		MP19	T4	TP1
2031	NITRIC ACID, other than red fuming, with more than 70% nitric acid	8	CO1	I	8 +5.1		0	E0	P001	PP81	MP8 MP17	T10	TP2
2031	NITRIC ACID, other than red fuming, with at least 65%, but not more than 70% nitric acid	8	CO1	II	8 +5.1		1 L	E2	P001 IBC02	PP81 B15	MP15	T8	TP2
2031	NITRIC ACID, other than red fuming, with less than 65% nitric acid	8	C1	II	8		1 L	E2	P001 IBC02	PP81 B15	MP15	T8	TP2
2032	NITRIC ACID, RED FUMING	8	COT	I	8 +5.1 +6.1		0	E0	P602		MP8 MP17	T20	TP2
2033	POTASSIUM MONOXIDE	8	C6	II	8		1 kg	E2	P002 IBC08	B4	MP10	T3	TP33
2034	HYDROGEN AND METHANE MIXTURE, COMPRESSED	2	1F		2.1	662	0	E0	P200		MP9	(M)	
2035	1,1,1-TRIFLUOROETHANE (REFRIGERANT GAS R 143a)	2	2F		2.1	662	0	E0	P200		MP9	(M) T50	
2036	XENON	2	2A		2.2	378 662	120 ml	E1	P200		MP9	(M)	
2037	RECEPTACLES, SMALL, CONTAINING GAS (GAS CARTRIDGES) without a release device, non-refillable	2	5A		2.2	191 303 344	1 L	E0	P003	PP17 RR6	MP9		
2037	RECEPTACLES, SMALL, CONTAINING GAS (GAS CARTRIDGES) without a release device, non-refillable	2	5F		2.1	191 303 344	1 L	E0	P003	PP17 RR6	MP9		
2037	RECEPTACLES, SMALL, CONTAINING GAS (GAS CARTRIDGES) without a release device, non-refillable	2	5O		2.2 +5.1	191 303 344	1 L	E0	P003	PP17 RR6	MP9		
2037	RECEPTACLES, SMALL, CONTAINING GAS (GAS CARTRIDGES) without a release device, non-refillable	2	5T		2.3	303 344	120 ml	E0	P003	PP17 RR6	MP9		
2037	RECEPTACLES, SMALL, CONTAINING GAS (GAS CARTRIDGES) without a release device, non-refillable	2	5TC		2.3 +8	303 344	120 ml	E0	P003	PP17 RR6	MP9		
2037	RECEPTACLES, SMALL, CONTAINING GAS (GAS CARTRIDGES) without a release device, non-refillable	2	5TF		2.3 +2.1	303 344	120 ml	E0	P003	PP17 RR6	MP9		
2037	RECEPTACLES, SMALL, CONTAINING GAS (GAS CARTRIDGES) without a release device, non-refillable	2	5TFC		2.3 +2.1 +8	303 344	120 ml	E0	P003	PP17 RR6	MP9		
2037	RECEPTACLES, SMALL, CONTAINING GAS (GAS CARTRIDGES) without a release device, non-refillable	2	5TO		2.3 +5.1	303 344	120 ml	E0	P003	PP17 RR6	MP9		
2037	RECEPTACLES, SMALL, CONTAINING GAS (GAS CARTRIDGES) without a release device, non-refillable	2	5TOC		2.3 +5.1 +8	303 344	120 ml	E0	P003	PP17 RR6	MP9		
2038	DINITROTOLUENES, LIQUID	6.1	T1	II	6.1		100 ml	E4	P001 IBC02		MP15	T7	TP2
2044	2,2-DIMETHYLPROPANE	2	2F		2.1	662	0	E0	P200		MP9	(M)	

ADR tank		Vehicle for tank carriage	Transport category (Tunnel restriction code)	Special provisions for carriage				Hazard identification No.	UN No.	Name and description
Tank code	Special provisions			Packages	Bulk	Loading, unloading and handling	Operation			
4.3	4.3.5, 6.8.4	9.1.1.2	1.1.3.6 (8.6)	7.2.4	7.3.3	7.5.11	8.5	5.3.2.3		3.1.2
(12)	(13)	(14)	(15)	(16)	(17)	(18)	(19)	(20)	(1)	(2)
L10BH		AT	1 (C/D)			CV13 CV28	S14	886	2030	HYDRAZINE AQUEOUS SOLUTION, with more than 37% hydrazine by mass
L4BN		AT	2 (E)			CV13 CV28		86	2030	HYDRAZINE AQUEOUS SOLUTION, with more than 37% hydrazine by mass
L4BN		AT	3 (E)	V12		CV13 CV28		86	2030	HYDRAZINE AQUEOUS SOLUTION, with more than 37% hydrazine by mass
L10BH	TC6 TT1	AT	1 (E)			CV24	S14	885	2031	NITRIC ACID, other than red fuming, with more than 70% nitric acid
L4BN		AT	2 (E)					85	2031	NITRIC ACID, other than red fuming, with at least 65%, but not more than 70% nitric acid
L4BN		AT	2 (E)					80	2031	NITRIC ACID, other than red fuming, with less than 65% nitric acid
L10BH	TC6 TT1	AT	1 (C/D)			CV13 CV24 CV28	S14	856	2032	NITRIC ACID, RED FUMING
SGAN		AT	2 (E)	V11				80	2033	POTASSIUM MONOXIDE
CxBN(M)	TA4 TT9	FL	2 (B/D)			CV9 CV10 CV36	S2 S20	23	2034	HYDROGEN AND METHANE MIXTURE, COMPRESSED
PxBN(M)	TA4 TT9	FL	2 (B/D)			CV9 CV10 CV36	S2 S20	23	2035	1,1,1-TRIFLUOROETHANE (REFRIGERANT GAS R 143a)
PxBN(M)	TA4 TT9	AT	3 (C/E)			CV9 CV10 CV36		20	2036	XENON
			3 (E)			CV9 CV12			2037	RECEPTACLES, SMALL, CONTAINING GAS (GAS CARTRIDGES) without a release device, non-refillable
			2 (D)			CV9 CV12	S2		2037	RECEPTACLES, SMALL, CONTAINING GAS (GAS CARTRIDGES) without a release device, non-refillable
			3 (E)			CV9 CV12			2037	RECEPTACLES, SMALL, CONTAINING GAS (GAS CARTRIDGES) without a release device, non-refillable
			1 (D)			CV9 CV12			2037	RECEPTACLES, SMALL, CONTAINING GAS (GAS CARTRIDGES) without a release device, non-refillable
			1 (D)			CV9 CV12			2037	RECEPTACLES, SMALL, CONTAINING GAS (GAS CARTRIDGES) without a release device, non-refillable
			1 (D)			CV9 CV12	S2		2037	RECEPTACLES, SMALL, CONTAINING GAS (GAS CARTRIDGES) without a release device, non-refillable
			1 (D)			CV9 CV12	S2		2037	RECEPTACLES, SMALL, CONTAINING GAS (GAS CARTRIDGES) without a release device, non-refillable
			1 (D)			CV9 CV12			2037	RECEPTACLES, SMALL, CONTAINING GAS (GAS CARTRIDGES) without a release device, non-refillable
			1 (D)			CV9 CV12			2037	RECEPTACLES, SMALL, CONTAINING GAS (GAS CARTRIDGES) without a release device, non-refillable
L4BH	TU15 TE19	AT	2 (D/E)			CV13 CV28	S9 S19	60	2038	DINITROTOLUENES, LIQUID
PxBN(M)	TA4 TT9	FL	2 (B/D)			CV9 CV10 CV36	S2 S20	23	2044	2,2-DIMETHYLPROPANE

UN No.	Name and description	Class	Classifi-cation code	Packing group	Labels	Special provi-sions	Limited and excepted quantities		Packaging			Portable tanks and bulk containers	
									Packing instruc-tions	Special packing provisions	Mixed packing provisions	Instruc-tions	Special provisions
	3.1.2	2.2	2.2	2.1.1.3	5.2.2	3.3	3.4	3.5.1.2	4.1.4	4.1.4	4.1.10	4.2.5.2 7.3.2	4.2.5.3
(1)	(2)	(3a)	(3b)	(4)	(5)	(6)	(7a)	(7b)	(8)	(9a)	(9b)	(10)	(11)
2045	ISOBUTYRALDEHYDE (ISOBUTYL ALDEHYDE)	3	F1	II	3		1 L	E2	P001 IBC02 R001		MP19	T4	TP1
2046	CYMENES	3	F1	III	3		5 L	E1	P001 IBC03 LP01 R001		MP19	T2	TP1
2047	DICHLOROPROPENES	3	F1	II	3		1 L	E2	P001 IBC02 R001		MP19	T4	TP1
2047	DICHLOROPROPENES	3	F1	III	3		5 L	E1	P001 IBC03 LP01 R001		MP19	T2	TP1
2048	DICYCLOPENTADIENE	3	F1	III	3		5 L	E1	P001 IBC03 LP01 R001		MP19	T2	TP1
2049	DIETHYLBENZENE	3	F1	III	3		5 L	E1	P001 IBC03 LP01 R001		MP19	T2	TP1
2050	DIISOBUTYLENE, ISOMERIC COMPOUNDS	3	F1	II	3		1 L	E2	P001 IBC02 R001		MP19	T4	TP1
2051	2-DIMETHYLAMINO-ETHANOL	8	CF1	II	8 +3		1 L	E2	P001 IBC02		MP15	T7	TP2
2052	DIPENTENE	3	F1	III	3		5 L	E1	P001 IBC03 LP01 R001		MP19	T2	TP1
2053	METHYL ISOBUTYL CARBINOL	3	F1	III	3		5 L	E1	P001 IBC03 LP01 R001		MP19	T2	TP1
2054	MORPHOLINE	8	CF1	I	8 +3		0	E0	P001		MP8 MP17	T10	TP2
2055	STYRENE MONOMER, STABILIZED	3	F1	III	3	386	5 L	E1	P001 IBC03 LP01 R001		MP19	T2	TP1
2056	TETRAHYDROFURAN	3	F1	II	3		1 L	E2	P001 IBC02 R001		MP19	T4	TP1
2057	TRIPROPYLENE	3	F1	II	3		1 L	E2	P001 IBC02 R001		MP19	T4	TP1
2057	TRIPROPYLENE	3	F1	III	3		5 L	E1	P001 IBC03 LP01 R001		MP19	T2	TP1
2058	VALERALDEHYDE	3	F1	II	3		1 L	E2	P001 IBC02 R001		MP19	T4	TP1
2059	NITROCELLULOSE SOLUTION, FLAMMABLE with not more than 12.6% nitrogen, by dry mass, and not more than 55% nitrocellulose	3	D	I	3	198 531	0	E0	P001		MP7 MP17	T11	TP1 TP8 TP27
2059	NITROCELLULOSE SOLUTION, FLAMMABLE with not more than 12.6% nitrogen, by dry mass, and not more than 55% nitrocellulose (vapour pressure at 50 °C more than 110 kPa)	3	D	II	3	198 531 640C	1 L	E0	P001 IBC02		MP19	T4	TP1 TP8
2059	NITROCELLULOSE SOLUTION, FLAMMABLE with not more than 12.6% nitrogen, by dry mass, and not more than 55% nitrocellulose (vapour pressure at 50 °C not more than 110 kPa)	3	D	II	3	198 531 640D	1 L	E0	P001 IBC02 R001		MP19	T4	TP1 TP8
2059	NITROCELLULOSE SOLUTION, FLAMMABLE with not more than 12.6% nitrogen, by dry mass, and not more than 55% nitrocellulose	3	D	III	3	198 531	5 L	E0	P001 IBC03 LP01 R001		MP19	T2	TP1

ADR tank		Vehicle for tank carriage	Transport category (Tunnel restriction code)	Special provisions for carriage				Hazard identifi-cation No.	UN No.	Name and description
Tank code	Special provisions			Packages	Bulk	Loading, unloading and handling	Operation			
4.3	4.3.5, 6.8.4	9.1.1.2	1.1.3.6 (8.6)	7.2.4	7.3.3	7.5.11	8.5	5.3.2.3		3.1.2
(12)	(13)	(14)	(15)	(16)	(17)	(18)	(19)	(20)	(1)	(2)
LGBF		FL	2 (D/E)				S2 S20	33	2045	ISOBUTYRALDEHYDE (ISOBUTYL ALDEHYDE)
LGBF		FL	3 (D/E)	V12			S2	30	2046	CYMENES
LGBF		FL	2 (D/E)				S2 S20	33	2047	DICHLOROPROPENES
LGBF		FL	3 (D/E)	V12			S2	30	2047	DICHLOROPROPENES
LGBF		FL	3 (D/E)	V12			S2	30	2048	DICYCLOPENTADIENE
LGBF		FL	3 (D/E)	V12			S2	30	2049	DIETHYLBENZENE
LGBF		FL	2 (D/E)				S2 S20	33	2050	DIISOBUTYLFNE, ISOMERIC COMPOUNDS
L4BN		FL	2 (D/E)				S2	83	2051	2-DIMETHYLAMINO-ETHANOL
LGBF		FL	3 (D/E)	V12			S2	30	2052	DIPENTENE
LGBF		FL	3 (D/E)	V12			S2	30	2053	METHYL ISOBUTYL CARBINOL
L10BH		FL	1 (D/E)				S2 S14	883	2054	MORPHOLINE
LGBF		FL	3 (D/E)	V8 V12			S2 S4	39	2055	STYRENE MONOMER, STABILIZED
LGBF		FL	2 (D/E)				S2 S20	33	2056	TETRAHYDROFURAN
LGBF		FL	2 (D/E)				S2 S20	33	2057	TRIPROPYLENE
LGBF		FL	3 (D/E)	V12			S2	30	2057	TRIPROPYLENE
LGBF		FL	2 (D/E)				S2 S20	33	2058	VALERALDEHYDE
L4BN		FL	1 (B)				S2 S14	33	2059	NITROCELLULOSE SOLUTION, FLAMMABLE with not more than 12.6% nitrogen, by dry mass, and not more than 55% nitrocellulose
L1.5BN		FL	2 (B)				S2 S14	33	2059	NITROCELLULOSE SOLUTION, FLAMMABLE with not more than 12.6% nitrogen, by dry mass, and not more than 55% nitrocellulose (vapour pressure at 50 °C more than 110 kPa)
LGBF		FL	2 (B)				S2 S14	33	2059	NITROCELLULOSE SOLUTION, FLAMMABLE with not more than 12.6% nitrogen, by dry mass, and not more than 55% nitrocellulose (vapour pressure at 50 °C not more than 110 kPa)
LGBF		FL	3 (B)	V12			S2 S14	30	2059	NITROCELLULOSE SOLUTION, FLAMMABLE with not more than 12.6% nitrogen, by dry mass, and not more than 55% nitrocellulose

UN No.	Name and description	Class	Classifi-cation code	Packing group	Labels	Special provi-sions	Limited and excepted quantities		Packaging			Portable tanks and bulk containers	
									Packing instruc-tions	Special packing provisions	Mixed packing provisions	Instruc-tions	Special provisions
3.1.2	3.1.2	2.2	2.2	2.1.1.3	5.2.2	3.3	3.4	3.5.1.2	4.1.4	4.1.4	4.1.10	4.2.5.2 7.3.2	4.2.5.3
(1)	(2)	(3a)	(3b)	(4)	(5)	(6)	(7a)	(7b)	(8)	(9a)	(9b)	(10)	(11)
2067	AMMONIUM NITRATE BASED FERTILIZER	5.1	O2	III	5.1	186 306 307	5 kg	E1	P002 IBC08 LP02 R001	B3	MP10	T1 BK1 BK2 BK3	TP33
2071	Ammonium nitrate based fertilizer, uniform mixtures of the nitrogen/phosphate, nitrogen/potash or nitrogen/phosphate/potash type, containing not more than 70% ammonium nitrate and not more than 0.4% total combustible/organic material calculated as carbon or with not more than 45% ammonium nitrate and unrestricted combustible material	9	M11	NOT SUBJECT TO ADR									
2073	AMMONIA SOLUTION, relative density less than 0.880 at 15 °C in water, with more than 35% but not more than 50% ammonia	2	4A		2.2	532	120 ml	E0	P200		MP9	(M)	
2074	ACRYLAMIDE, SOLID	6.1	T2	III	6.1		5 kg	E1	P002 IBC08 LP02 R001	B3	MP10	T1	TP33
2075	CHLORAL, ANHYDROUS, STABILIZED	6.1	T1	II	6.1		100 ml	E4	P001 IBC02		MP15	T7	TP2
2076	CRESOLS, LIQUID	6.1	TC1	II	6.1 +8		100 ml	E4	P001 IBC02		MP15	T7	TP2
2077	alpha-NAPHTHYLAMINE	6.1	T2	III	6.1		5 kg	E1	P002 IBC08 LP02 R001	B3	MP10	T1	TP33
2078	TOLUENE DIISOCYANATE	6.1	T1	II	6.1	279	100 ml	E4	P001 IBC02		MP15	T7	TP2
2079	DIETHYLENETRIAMINE	8	C7	II	8		1 L	E2	P001 IBC02		MP15	T7	TP2
2186	HYDROGEN CHLORIDE, REFRIGERATED LIQUID	2	3TC	CARRIAGE PROHIBITED									
2187	CARBON DIOXIDE, REFRIGERATED LIQUID	2	3A		2.2		120 ml	E1	P203		MP9	T75	TP5
2188	ARSINE	2	2TF		2.3 +2.1		0	E0	P200		MP9		
2189	DICHLOROSILANE	2	2TFC		2.3 +2.1 +8		0	E0	P200		MP9	(M)	
2190	OXYGEN DIFLUORIDE, COMPRESSED	2	1TOC		2.3 +5.1 +8		0	E0	P200		MP9		
2191	SULPHURYL FLUORIDE	2	2T		2.3		0	E0	P200		MP9	(M)	
2192	GERMANE	2	2TF		2.3 +2.1	632	0	E0	P200		MP9	(M)	
2193	HEXAFLUOROETHANE (REFRIGERANT GAS R 116)	2	2A		2.2	662	120 ml	E1	P200		MP9	(M)	
2194	SELENIUM HEXAFLUORIDE	2	2TC		2.3 +8		0	E0	P200		MP9		
2195	TELLURIUM HEXAFLUORIDE	2	2TC		2.3 +8		0	E0	P200		MP9		
2196	TUNGSTEN HEXAFLUORIDE	2	2TC		2.3 +8		0	E0	P200		MP9		
2197	HYDROGEN IODIDE, ANHYDROUS	2	2TC		2.3 +8		0	E0	P200		MP9	(M)	
2198	PHOSPHORUS PENTAFLUORIDE	2	2TC		2.3 +8		0	E0	P200		MP9		

ADR tank		Vehicle for tank carriage	Transport category (Tunnel restriction code)	Special provisions for carriage				Hazard identification No.	UN No.	Name and description
Tank code	Special provisions			Packages	Bulk	Loading, unloading and handling	Operation			
4.3	4.3.5, 6.8.4	9.1.1.2	1.1.3.6 (8.6)	7.2.4	7.3.3	7.5.11	8.5	5.3.2.3		3.1.2
(12)	(13)	(14)	(15)	(16)	(17)	(18)	(19)	(20)	(1)	(2)
SGAV	TU3	AT	3 (E)		VC1 VC2 AP6 AP7	CV24	S23	50	2067	AMMONIUM NITRATE BASED FERTILIZER
NOT SUBJECT TO ADR									2071	Ammonium nitrate based fertilizer, uniform mixtures of the nitrogen/phosphate, nitrogen/potash or nitrogen/phosphate/potash type, containing not more than 70% ammonium nitrate and not more than 0.4% total combustible/organic material calculated as carbon or with not more than 45% ammonium nitrate and unrestricted combustible material
PxBN(M)	TA4 TT9	AT	3 (E)			CV9 CV10		20	2073	AMMONIA SOLUTION, relative density less than 0.880 at 15 °C in water, with more than 35% but not more than 50% ammonia
SGAH L4BH	TU15 TE19	AT	2 (E)		VC1 VC2 AP7	CV13 CV28	S9	60	2074	ACRYLAMIDE, SOLID
L4BH	TU15 TE19	AT	2 (D/E)			CV13 CV28	S9 S19	69	2075	CHLORAL, ANHYDROUS, STABILIZED
L4BH	TU15 TE19	AT	2 (D/E)			CV13 CV28	S9 S19	68	2076	CRESOLS, LIQUID
SGAH L4BH	TU15 TE19	AT	2 (E)		VC1 VC2 AP7	CV13 CV28	S9	60	2077	alpha-NAPHTHYLAMINE
L4BH	TU15 TE19	AT	2 (D/E)			CV13 CV28	S9 S19	60	2078	TOLUENE DIISOCYANATE
L4BN		AT	2 (E)					80	2079	DIETHYLENETRIAMINE
CARRIAGE PROHIBITED									2186	HYDROGEN CHLORIDE, REFRIGERATED LIQUID
RxBN	TU19 TA4 TT9	AT	3 (C/E)	V5		CV9 CV11 CV36	S20	22	2187	CARBON DIOXIDE, REFRIGERATED LIQUID
			1 (D)			CV9 CV10 CV36	S2 S14		2188	ARSINE
PxBH(M)	TA4 TT9	FL	1 (B/D)			CV9 CV10 CV36	S2 S14	263	2189	DICHLOROSILANE
			1 (D)			CV9 CV10 CV36	S14		2190	OXYGEN DIFLUORIDE, COMPRESSED
PxBH(M)	TA4 TT9	AT	1 (C/D)			CV9 CV10 CV36	S14	26	2191	SULPHURYL FLUORIDE
		FL	1 (B/D)			CV9 CV10 CV36	S2 S14	263	2192	GERMANE
PxBN(M)	TA4 TT9	AT	3 (C/E)			CV9 CV10 CV36		20	2193	HEXAFLUOROETHANE (REFRIGERANT GAS R 116)
			1 (D)			CV9 CV10 CV36	S14		2194	SELENIUM HEXAFLUORIDE
			1 (D)			CV9 CV10 CV36	S14		2195	TELLURIUM HEXAFLUORIDE
			1 (D)			CV9 CV10 CV36	S14		2196	TUNGSTEN HEXAFLUORIDE
PxBH(M)	TA4 TT9	AT	1 (C/D)			CV9 CV10 CV36	S14	268	2197	HYDROGEN IODIDE, ANHYDROUS
			1 (D)			CV9 CV10 CV36	S14		2198	PHOSPHORUS PENTAFLUORIDE

UN No.	Name and description	Class	Classifi-cation code	Packing group	Labels	Special provi-sions	Limited quantities	Excepted quantities	Packing instruc-tions	Special packing provisions	Mixed packing provisions	Instruc-tions	Special provisions
	3.1.2	2.2	2.2	2.1.1.3	5.2.2	3.3	3.4	3.5.1.2	4.1.4	4.1.4	4.1.10	4.2.5.2 7.3.2	4.2.5.3
(1)	(2)	(3a)	(3b)	(4)	(5)	(6)	(7a)	(7b)	(8)	(9a)	(9b)	(10)	(11)
2199	PHOSPHINE	2	2TF		2.3 +2.1	632	0	E0	P200		MP9		
2200	PROPADIENE, STABILIZED	2	2F		2.1	386 662	0	E0	P200		MP9	(M)	
2201	NITROUS OXIDE, REFRIGERATED LIQUID	2	3O		2.2 +5.1		0	E0	P203		MP9	T75	TP5 TP22
2202	HYDROGEN SELENIDE, ANHYDROUS	2	2TF		2.3 +2.1		0	E0	P200		MP9		
2203	SILANE	2	2F		2.1	632 662	0	E0	P200		MP9	(M)	
2204	CARBONYL SULPHIDE	2	2TF		2.3 +2.1		0	E0	P200		MP9	(M)	
2205	ADIPONITRILE	6.1	T1	III	6.1		5 L	E1	P001 IBC03 LP01 R001		MP19	T3	TP1
2206	ISOCYANATES, TOXIC, N.O.S. or ISOCYANATE SOLUTION, TOXIC, N.O.S.	6.1	T1	II	6.1	274 551	100 ml	E4	P001 IBC02		MP15	T11	TP2 TP27
2206	ISOCYANATES, TOXIC, N.O.S. or ISOCYANATE SOLUTION, TOXIC, N.O.S.	6.1	T1	III	6.1	274 551	5 L	E1	P001 IBC03 LP01 R001		MP19	T7	TP1 TP28
2208	CALCIUM HYPOCHLORITE MIXTURE, DRY with more than 10% but not more than 39% available chlorine	5.1	O2	III	5.1	314	5 kg	E1	P002 IBC08 LP02 R001	B3 B13 L3	MP10		
2209	FORMALDEHYDE SOLUTION with not less than 25% formaldehyde	8	C9	III	8	533	5 L	E1	P001 IBC03 LP01 R001		MP19	T4	TP1
2210	MANEB or MANEB PREPARATION with not less than 60% maneb	4.2	SW	III	4.2 +4.3	273	0	E1	P002 IBC06 R001		MP14	T1	TP33
2211	POLYMERIC BEADS, EXPANDABLE, evolving flammable vapour	9	M3	III	None	382 633	5 kg	E1	P002 IBC08 R001	PP14 B3 B6	MP10	T1	TP33
2212	ASBESTOS, AMPHIBOLE (amosite, tremolite, actinolite, anthophyllite, crocidolite)	9	M1	II	9	168 274 542	1 kg	E0	P002 IBC08	PP37 B4	MP10	T3	TP33
2213	PARAFORMALDEHYDE	4.1	F1	III	4.1		5 kg	E1	P002 IBC08 LP02 R001	PP12 B3	MP10	T1 BK1 BK2 BK3	TP33
2214	PHTHALIC ANHYDRIDE with more than 0.05% of maleic anhydride	8	C4	III	8	169	5 kg	E1	P002 IBC08 LP02 R001	B3	MP10	T1	TP33
2215	MALEIC ANHYDRIDE, MOLTEN	8	C3	III	8		0	E0				T4	TP3
2215	MALEIC ANHYDRIDE	8	C4	III	8		5 kg	E1	P002 IBC08 R001	B3	MP10	T1	TP33
2216	Fish meal (Fish scrap), stabilized	9	M11				NOT SUBJECT TO ADR							
2217	SEED CAKE with not more than 1.5% oil and not more than 11% moisture	4.2	S2	III	4.2	142	0	E0	P002 IBC08 LP02 R001	PP20 B3 B6	MP14		
2218	ACRYLIC ACID, STABILIZED	8	CF1	II	8 +3	386	1 L	E2	P001 IBC02		MP15	T7	TP2
2219	ALLYL GLYCIDYL ETHER	3	F1	III	3		5 L	E1	P001 IBC03 LP01 R001		MP19	T2	TP1
2222	ANISOLE	3	F1	III	3		5 L	E1	P001 IBC03 LP01 R001		MP19	T2	TP1
2224	BENZONITRILE	6.1	T1	II	6.1		100 ml	E4	P001 IBC02		MP15	T7	TP2

ADR tank		Vehicle for tank carriage	Transport category (Tunnel restriction code)	Special provisions for carriage				Hazard identification No.	UN No.	Name and description
Tank code	Special provisions			Packages	Bulk	Loading, unloading and handling	Operation			
4.3	4.3.5, 6.8.4	9.1.1.2	1.1.3.6 (8.6)	7.2.4	7.3.3	7.5.11	8.5	5.3.2.3		3.1.2
(12)	(13)	(14)	(15)	(16)	(17)	(18)	(19)	(20)	(1)	(2)
			1 (D)			CV9 CV10 CV36	S2 S14		2199	PHOSPHINE
PxBN(M)	TA4 TT9	FL	2 (B/D)	V8		CV9 CV10 CV36	S2 S4 S20	239	2200	PROPADIENE, STABILIZED
RxBN	TU7 TU19 TA4 TT9	AT	3 (C/E)	V5		CV9 CV11 CV36	S20	225	2201	NITROUS OXIDE, REFRIGERATED LIQUID
			1 (D)			CV9 CV10 CV36	S2 S14		2202	HYDROGEN SELENIDE, ANHYDROUS
PxBN(M)	TA4 TT9	FL	2 (B/D)			CV9 CV10 CV36	S2 S20	23	2203	SILANE
PxBH(M)	TA4 TT9	FL	1 (B/D)			CV9 CV10 CV36	S2 S14	263	2204	CARBONYL SULPHIDE
L4BH	TU15 TE19	AT	2 (E)	V12		CV13 CV28	S9	60	2205	ADIPONITRILE
L4BH	TU15 TE19	AT	2 (D/E)			CV13 CV28	S9 S19	60	2206	ISOCYANATES, TOXIC, N.O.S. or ISOCYANATE SOLUTION, TOXIC, N.O.S.
L4BH	TU15 TE19	AT	2 (E)	V12		CV13 CV28	S9	60	2206	ISOCYANATES, TOXIC, N.O.S. or ISOCYANATE SOLUTION, TOXIC, N.O.S.
SGAN	TU3	AT	3 (E)			CV24 CV35		50	2208	CALCIUM HYPOCHLORITE MIXTURE, DRY with more than 10% but not more than 39% available chlorine
L4BN		AT	3 (E)	V12				80	2209	FORMALDEHYDE SOLUTION with not less than 25% formaldehyde
SGAN		AT	3 (E)	V1	VC1 VC2 AP1			40	2210	MANEB or MANEB PREPARATION with not less than 60% maneb
SGAN	TE20	AT	3 (D/E)		VC1 VC2 AP2	CV36		90	2211	POLYMERIC BEADS, EXPANDABLE, evolving flammable vapour
SGAH	TU15	AT	2 (E)	V11		CV1 CV13 CV28	S19	90	2212	ASBESTOS, AMPHIBOLE (amosite, tremolite, actinolite, anthophyllite, crocidolite)
SGAV		AT	3 (E)	V13	VC1 VC2			40	2213	PARAFORMALDEHYDE
SGAV L4BN		AT	3 (E)		VC1 VC2 AP7			80	2214	PHTHALIC ANHYDRIDE with more than 0.05% of maleic anhydride
L4BN		AT	0 (E)					80	2215	MALEIC ANHYDRIDE, MOLTEN
SGAV		AT	3 (E)		VC1 VC2 AP7			80	2215	MALEIC ANHYDRIDE
NOT SUBJECT TO ADR									2216	Fish meal (Fish scrap), stabilized
			3 (E)	V1	VC1 VC2 AP1			40	2217	SEED CAKE with not more than 1.5% oil and not more than 11% moisture
L4BN		FL	2 (D/E)	V8			S2 S4	839	2218	ACRYLIC ACID, STABILIZED
LGBF		FL	3 (D/E)	V12			S2	30	2219	ALLYL GLYCIDYL ETHER
LGBF		FL	3 (D/E)	V12			S2	30	2222	ANISOLE
L4BH	TU15 TE19	AT	2 (D/E)			CV13 CV28	S9 S19	60	2224	BENZONITRILE

UN No.	Name and description	Class	Classification code	Packing group	Labels	Special provisions	Limited and excepted quantities		Packaging			Portable tanks and bulk containers	
									Packing instructions	Special packing provisions	Mixed packing provisions	Instructions	Special provisions
	3.1.2	2.2	2.2	2.1.1.3	5.2.2	3.3	3.4	3.5.1.2	4.1.4	4.1.4	4.1.10	4.2.5.2 7.3.2	4.2.5.3
(1)	(2)	(3a)	(3b)	(4)	(5)	(6)	(7a)	(7b)	(8)	(9a)	(9b)	(10)	(11)
2225	BENZENESULPHONYL CHLORIDE	8	C3	III	8		5 L	E1	P001 IBC03 LP01 R001		MP19	T4	TP1
2226	BENZOTRICHLORIDE	8	C9	II	8		1 L	E2	P001 IBC02		MP15	T7	TP2
2227	n-BUTYL METHACRYLATE, STABILIZED	3	F1	III	3	386	5 L	E1	P001 IBC03 LP01 R001		MP19	T2	TP1
2232	2-CHLOROETHANAL	6.1	T1	I	6.1	354	0	E0	P602		MP8 MP17	T20	TP2
2233	CHLOROANISIDINES	6.1	T2	III	6.1		5 kg	E1	P002 IBC08 LP02 R001	B3	MP10	T1	TP33
2234	CHLOROBENZOTRI-FLUORIDES	3	F1	III	3		5 L	E1	P001 IBC03 LP01 R001		MP19	T2	TP1
2235	CHLOROBENZYL CHLORIDES, LIQUID	6.1	T1	III	6.1		5 L	E1	P001 IBC03 LP01 R001		MP19	T4	TP1
2236	3-CHLORO-4-METHYLPHENYL ISOCYANATE, LIQUID	6.1	T1	II	6.1		100 ml	E4	P001 IBC02		MP15		
2237	CHLORONITROANILINES	6.1	T2	III	6.1		5 kg	E1	P002 IBC08 LP02 R001	B3	MP10	T1	TP33
2238	CHLOROTOLUENES	3	F1	III	3		5 L	E1	P001 IBC03 LP01 R001		MP19	T2	TP1
2239	CHLOROTOLUIDINES, SOLID	6.1	T2	III	6.1		5 kg	E1	P002 IBC08 LP02 R001	B3	MP10	T1	TP33
2240	CHROMOSULPHURIC ACID	8	C1	I	8		0	E0	P001		MP8 MP17	T10	TP2
2241	CYCLOHEPTANE	3	F1	II	3		1 L	E2	P001 IBC02 R001		MP19	T4	TP1
2242	CYCLOHEPTENE	3	F1	II	3		1 L	E2	P001 IBC02 R001		MP19	T4	TP1
2243	CYCLOHEXYL ACETATE	3	F1	III	3		5 L	E1	P001 IBC03 LP01 R001		MP19	T2	TP1
2244	CYCLOPENTANOL	3	F1	III	3		5 L	E1	P001 IBC03 LP01 R001		MP19	T2	TP1
2245	CYCLOPENTANONE	3	F1	III	3		5 L	E1	P001 IBC03 LP01 R001		MP19	T2	TP1
2246	CYCLOPENTENE	3	F1	II	3		1 L	E2	P001 IBC02	B8	MP19	T7	TP2
2247	n-DECANE	3	F1	III	3		5 L	E1	P001 IBC03 LP01 R001		MP19	T2	TP1
2248	DI-n-BUTYLAMINE	8	CF1	II	8 +3		1 L	E2	P001 IBC02		MP15	T7	TP2
2249	DICHLORODIMETHYL ETHER, SYMMETRICAL	6.1	TF1				CARRIAGE PROHIBITED						
2250	DICHLOROPHENYL ISOCYANATES	6.1	T2	II	6.1		500 g	E4	P002 IBC08	B4	MP10	T3	TP33
2251	BICYCLO[2.2.1]HEPTA-2,5-DIENE, STABILIZED (2,5-NORBORNADIENE, STABILIZED)	3	F1	II	3	386	1 L	E2	P001 IBC02 R001		MP19	T7	TP2

ADR tank		Vehicle for tank carriage	Transport category (Tunnel restriction code)	Special provisions for carriage				Hazard identification No.	UN No.	Name and description
Tank code	Special provisions			Packages	Bulk	Loading, unloading and handling	Operation			
4.3	4.3.5, 6.8.4	9.1.1.2	1.1.3.6 (8.6)	7.2.4	7.3.3	7.5.11	8.5	5.3.2.3		3.1.2
(12)	(13)	(14)	(15)	(16)	(17)	(18)	(19)	(20)	(1)	(2)
L4BN		AT	3 (E)	V12				80	2225	BENZENESULPHONYL CHLORIDE
L4BN		AT	2 (E)					80	2226	BENZOTRICHLORIDE
LGBF		FL	3 (D/E)	V8 V12			S2 S4	39	2227	n-BUTYL METHACRYLATE, STABILIZED
L10CH	TU14 TU15 TE19 TE21	AT	1 (C/D)			CV1 CV13 CV28	S9 S14	66	2232	2-CHLOROETHANAL
SGAH L4BH	TU15 TE19	AT	2 (E)		VC1 VC2 AP7	CV13 CV28	S9	60	2233	CHLOROANISIDINES
LGBF		FL	3 (D/E)	V12			S2	30	2234	CHLOROBENZOTRI-FLUORIDES
L4BH	TU15 TE19	AT	2 (E)	V12		CV13 CV28	S9	60	2235	CHLOROBENZYL CHLORIDES, LIQUID
L4BH	TU15 TE19	AT	2 (D/E)			CV13 CV28	S9 S19	60	2236	3-CHLORO-4-METHYLPHENYL ISOCYANATE, LIQUID
SGAH L4BH	TU15 TE19	AT	2 (E)		VC1 VC2 AP7	CV13 CV28	S9	60	2237	CHLORONITROANILINES
LGBF		FL	3 (D/E)	V12			S2	30	2238	CHLOROTOLUENES
SGAH L4BH	TU15 TE19	AT	2 (E)		VC1 VC2 AP7	CV13 CV28	S9	60	2239	CHLOROTOLUIDINES, SOLID
L10BH		AT	1 (E)				S20	88	2240	CHROMOSULPHURIC ACID
LGBF		FL	2 (D/E)				S2 S20	33	2241	CYCLOHEPTANE
LGBF		FL	2 (D/E)				S2 S20	33	2242	CYCLOHEPTENE
LGBF		FL	3 (D/E)	V12			S2	30	2243	CYCLOHEXYL ACETATE
LGBF		FL	3 (D/E)	V12			S2	30	2244	CYCLOPENTANOL
LGBF		FL	3 (D/E)	V12			S2	30	2245	CYCLOPENTANONE
L1.5BN		FL	2 (D/E)				S2 S20	33	2246	CYCLOPENTENE
LGBF		FL	3 (D/E)	V12			S2	30	2247	n-DECANE
L4BN		FL	2 (D/E)				S2	83	2248	DI-n-BUTYLAMINE
CARRIAGE PROHIBITED									2249	DICHLORODIMETHYL ETHER, SYMMETRICAL
SGAH L4BH	TU15 TE19	AT	2 (D/E)	V11		CV13 CV28	S9 S19	60	2250	DICHLOROPHENYL ISOCYANATES
LGBF		FL	2 (D/E)	V8			S2 S4 S20	339	2251	BICYCLO[2.2.1]HEPTA-2,5-DIENE, STABILIZED (2,5-NORBORNADIENE, STABILIZED)

UN No.	Name and description	Class	Classification code	Packing group	Labels	Special provisions	Limited and excepted quantities		Packaging			Portable tanks and bulk containers	
									Packing instructions	Special packing provisions	Mixed packing provisions	Instructions	Special provisions
	3.1.2	2.2	2.2	2.1.1.3	5.2.2	3.3	3.4	3.5.1.2	4.1.4	4.1.4	4.1.10	4.2.5.2 7.3.2	4.2.5.3
(1)	(2)	(3a)	(3b)	(4)	(5)	(6)	(7a)	(7b)	(8)	(9a)	(9b)	(10)	(11)
2252	1,2-DIMETHOXYETHANE	3	F1	II	3		1 L	E2	P001 IBC02 R001		MP19	T4	TP1
2253	N,N-DIMETHYLANILINE	6.1	T1	II	6.1		100 ml	E4	P001 IBC02		MP15	T7	TP2
2254	MATCHES, FUSEE	4.1	F1	III	4.1	293	5 kg	E0	P407 R001		MP11		
2256	CYCLOHEXENE	3	F1	II	3		1 L	E2	P001 IBC02 R001		MP19	T4	TP1
2257	POTASSIUM	4.3	W2	I	4.3		0	E0	P403 IBC04		MP2	T9	TP7 TP33
2258	1,2-PROPYLENEDIAMINE	8	CF1	II	8 +3		1 L	E2	P001 IBC02		MP15	T7	TP2
2259	TRIETHYLENETETRAMINE	8	C7	II	8		1 L	E2	P001 IBC02		MP15	T7	TP2
2260	TRIPROPYLAMINE	3	FC	III	3 +8		5 L	E1	P001 IBC03 R001		MP19	T4	TP1
2261	XYLENOLS, SOLID	6.1	T2	II	6.1		500 g	E4	P002 IBC08	B4	MP10	T3	TP33
2262	DIMETHYLCARBAMOYL CHLORIDE	8	C3	II	8		1 L	E2	P001 IBC02		MP15	T7	TP2
2263	DIMETHYL-CYCLOHEXANES	3	F1	II	3		1 L	E2	P001 IBC02 R001		MP19	T4	TP1
2264	N,N-DIMETHYL-CYCLOHEXYLAMINE	8	CF1	II	8 +3		1 L	E2	P001 IBC02		MP15	T7	TP2
2265	N,N-DIMETHYL-FORMAMIDE	3	F1	III	3		5 L	E1	P001 IBC03 LP01 R001		MP19	T2	TP2
2266	DIMETHYL-N-PROPYLAMINE	3	FC	II	3 +8		1 L	E2	P001 IBC02		MP19	T7	TP2
2267	DIMETHYL THIOPHOSPHORYL CHLORIDE	6.1	TC1	II	6.1 +8		100 ml	E4	P001 IBC02		MP15	T7	TP2
2269	3,3'-IMINODIPROPYLAMINE	8	C7	III	8		5 L	E1	P001 IBC03 LP01 R001		MP19	T4	TP2
2270	ETHYLAMINE, AQUEOUS SOLUTION with not less than 50% but not more than 70% ethylamine	3	FC	II	3 +8		1 L	E2	P001 IBC02		MP19	T7	TP1
2271	ETHYL AMYL KETONE	3	F1	III	3		5 L	E1	P001 IBC03 LP01 R001		MP19	T2	TP1
2272	N-ETHYLANILINE	6.1	T1	III	6.1		5 L	E1	P001 IBC03 LP01 R001		MP19	T4	TP1
2273	2-ETHYLANILINE	6.1	T1	III	6.1		5 L	E1	P001 IBC03 LP01 R001		MP19	T4	TP1
2274	N-ETHYL-N-BENZYLANILINE	6.1	T1	III	6.1		5 L	E1	P001 IBC03 LP01 R001		MP19	T4	TP1
2275	2-ETHYLBUTANOL	3	F1	III	3		5 L	E1	P001 IBC03 LP01 R001		MP19	T2	TP1
2276	2-ETHYLHEXYLAMINE	3	FC	III	3 +8		5 L	E1	P001 IBC03 R001		MP19	T4	TP1
2277	ETHYL METHACRYLATE, STABILIZED	3	F1	II	3	386	1 L	E2	P001 IBC02 R001		MP19	T4	TP1
2278	n-HEPTENE	3	F1	II	3		1 L	E2	P001 IBC02 R001		MP19	T4	TP1

ADR tank		Vehicle for tank carriage	Transport category (Tunnel restriction code)	Special provisions for carriage				Hazard identification No.	UN No.	Name and description
Tank code	Special provisions			Packages	Bulk	Loading, unloading and handling	Operation			
4.3	4.3.5, 6.8.4	9.1.1.2	1.1.3.6 (8.6)	7.2.4	7.3.3	7.5.11	8.5	5.3.2.3		3.1.2
(12)	(13)	(14)	(15)	(16)	(17)	(18)	(19)	(20)	(1)	(2)
LGBF		FL	2 (D/E)				S2 S20	33	2252	1,2-DIMETHOXYETHANE
L4BH	TU15 TE19	AT	2 (D/E)			CV13 CV28	S9 S19	60	2253	N,N-DIMETHYLANILINE
			4 (E)						2254	MATCHES, FUSEE
LGBF		FL	2 (D/E)				S2 S20	33	2256	CYCLOHEXENE
L10BN(+)	TU1 TE5 TT3 TM2	AT	1 (B/E)	V1		CV23	S20	X423	2257	POTASSIUM
L4BN		FL	2 (D/E)				S2	83	2258	1,2-PROPYLENEDIAMINE
L4BN		AT	2 (E)					80	2259	TRIETHYLENETETRAMINE
L4BN		FL	3 (D/E)	V12			S2	38	2260	TRIPROPYLAMINE
SGAH L4BH	TU15 TE19	AT	2 (D/E)	V11		CV13 CV28	S9 S19	60	2261	XYLENOLS, SOLID
L4BN		AT	2 (E)					80	2262	DIMETHYLCARBAMOYL CHLORIDE
LGBF		FL	2 (D/E)				S2 S20	33	2263	DIMETHYL-CYCLOHEXANES
L4BN		FL	2 (D/E)				S2	83	2264	N,N-DIMETHYL-CYCLOHEXYLAMINE
LGBF		FL	3 (D/E)	V12			S2	30	2265	N,N-DIMETHYL-FORMAMIDE
L4BH		FL	2 (D/E)				S2 S20	338	2266	DIMETHYL-N-PROPYLAMINE
L4BH	TU15 TE19	AT	2 (D/E)			CV13 CV28	S9 S19	68	2267	DIMETHYL THIOPHOSPHORYL CHLORIDE
L4BN		AT	3 (E)	V12				80	2269	3,3'-IMINODIPROPYLAMINE
L4BH		FL	2 (D/E)				S2 S20	338	2270	ETHYLAMINE, AQUEOUS SOLUTION with not less than 50% but not more than 70% ethylamine
LGBF		FL	3 (D/E)	V12			S2	30	2271	ETHYL AMYL KETONE
L4BH	TU15 TE19	AT	2 (E)	V12		CV13 CV28	S9	60	2272	N-ETHYLANILINE
L4BH	TU15 TE19	AT	2 (E)	V12		CV13 CV28	S9	60	2273	2-ETHYLANILINE
L4BH	TU15 TE19	AT	2 (E)	V12		CV13 CV28	S9	60	2274	N-ETHYL-N-BENZYLANILINE
LGBF		FL	3 (D/E)	V12			S2	30	2275	2-ETHYLBUTANOL
L4BN		FL	3 (D/E)	V12			S2	38	2276	2-ETHYLHEXYLAMINE
LGBF		FL	2 (D/E)	V8			S2 S4 S20	339	2277	ETHYL METHACRYLATE, STABILIZED
LGBF		FL	2 (D/E)				S2 S20	33	2278	n-HEPTENE

UN No.	Name and description	Class	Classification code	Packing group	Labels	Special provisions	Limited and excepted quantities		Packaging			Portable tanks and bulk containers	
									Packing instructions	Special packing provisions	Mixed packing provisions	Instructions	Special provisions
	3.1.2	2.2	2.2	2.1.1.3	5.2.2	3.3	3.4	3.5.1.2	4.1.4	4.1.4	4.1.10	4.2.5.2 7.3.2	4.2.5.3
(1)	(2)	(3a)	(3b)	(4)	(5)	(6)	(7a)	(7b)	(8)	(9a)	(9b)	(10)	(11)
2279	HEXACHLOROBUTADIENE	6.1	T1	III	6.1		5 L	E1	P001 IBC03 LP01 R001		MP19	T4	TP1
2280	HEXAMETHYLENE-DIAMINE, SOLID	8	C8	III	8		5 kg	E1	P002 IBC08 LP02 R001	B3	MP10	T1	TP33
2281	HEXAMETHYLENE DIISOCYANATE	6.1	T1	II	6.1		100 ml	E4	P001 IBC02		MP15	T7	TP2
2282	HEXANOLS	3	F1	III	3		5 L	E1	P001 IBC03 LP01 R001		MP19	T2	TP1
2283	ISOBUTYL METHACRYLATE, STABILIZED	3	F1	III	3	386	5 L	E1	P001 IBC03 LP01 R001		MP19	T2	TP1
2284	ISOBUTYRONITRILE	3	FT1	II	3 +6.1		1 L	E2	P001 IBC02		MP19	T7	TP2
2285	ISOCYANATOBENZO-TRIFLUORIDES	6.1	TF1	II	6.1 +3		100 ml	E4	P001 IBC02		MP15	T7	TP2
2286	PENTAMETHYLHEPTANE	3	F1	III	3		5 L	E1	P001 IBC03 LP01 R001		MP19	T2	TP1
2287	ISOHEPTENE	3	F1	II	3		1 L	E2	P001 IBC02 R001		MP19	T4	TP1
2288	ISOHEXENE	3	F1	II	3		1 L	E2	P001 IBC02 R001	B8	MP19	T11	TP1
2289	ISOPHORONEDIAMINE	8	C7	III	8		5 L	E1	P001 IBC03 LP01 R001		MP19	T4	TP1
2290	ISOPHORONE DIISOCYANATE	6.1	T1	III	6.1		5 L	E1	P001 IBC03 LP01 R001		MP19	T4	TP2
2291	LEAD COMPOUND, SOLUBLE, N.O.S.	6.1	T5	III	6.1	199 274 535	5 kg	E1	P002 IBC08 LP02 R001	B3	MP10	T1	TP33
2293	4-METHOXY-4-METHYLPENTAN-2-ONE	3	F1	III	3		5 L	E1	P001 IBC03 LP01 R001		MP19	T2	TP1
2294	N-METHYLANILINE	6.1	T1	III	6.1		5 L	E1	P001 IBC03 LP01 R001		MP19	T4	TP1
2295	METHYL CHLOROACETATE	6.1	TF1	I	6.1 +3		0	E0	P001		MP8 MP17	T14	TP2
2296	METHYLCYCLOHEXANE	3	F1	II	3		1 L	E2	P001 IBC02 R001		MP19	T4	TP1
2297	METHYLCYCLO-HEXANONE	3	F1	III	3		5 L	E1	P001 IBC03 LP01 R001		MP19	T2	TP1
2298	METHYLCYCLOPENTANE	3	F1	II	3		1 L	E2	P001 IBC02 R001		MP19	T4	TP1
2299	METHYL DICHLOROACETATE	6.1	T1	III	6.1		5 L	E1	P001 IBC03 LP01 R001		MP19	T4	TP1
2300	2-METHYL-5-ETHYLPYRIDINE	6.1	T1	III	6.1		5 L	E1	P001 IBC03 LP01 R001		MP19	T4	TP1
2301	2-METHYLFURAN	3	F1	II	3		1 L	E2	P001 IBC02 R001		MP19	T4	TP1

ADR tank		Vehicle for tank carriage	Transport category (Tunnel restriction code)	Special provisions for carriage				Hazard identifi-cation No.	UN No.	Name and description
Tank code	Special provisions			Packages	Bulk	Loading, unloading and handling	Operation			
4.3	4.3.5, 6.8.4	9.1.1.2	1.1.3.6 (8.6)	7.2.4	7.3.3	7.5.11	8.5	5.3.2.3		3.1.2
(12)	(13)	(14)	(15)	(16)	(17)	(18)	(19)	(20)	(1)	(2)
L4BH	TU15 TE19	AT	2 (E)	V12		CV13 CV28	S9	60	2279	HEXACHLOROBUTADIENE
SGAV L4BN		AT	3 (E)		VC1 VC2 AP7			80	2280	HEXAMETHYLENE-DIAMINE, SOLID
L4BH	TU15 TE19	AT	2 (D/E)			CV13 CV28	S9 S19	60	2281	HEXAMETHYLENE DIISOCYANATE
LGBF		FL	3 (D/E)	V12			S2	30	2282	HEXANOLS
LGBF		FL	3 (D/E)	V8 V12			S2 S4	39	2283	ISOBUTYL METHACRYLATE, STABILIZED
L4BH	TU15	FL	2 (D/E)			CV13 CV28	S2 S19	336	2284	ISOBUTYRONITRILE
L4BH	TU15 TE19	FL	2 (D/E)			CV13 CV28	S2 S9 S19	63	2285	ISOCYANATOBENZO-TRIFLUORIDES
LGBF		FL	3 (D/E)	V12			S2	30	2286	PENTAMETHYLHEPTANE
LGBF		FL	2 (D/E)				S2 S20	33	2287	ISOHEPTENE
LGBF		FL	2 (D/E)				S2 S20	33	2288	ISOHEXENE
L4BN		AT	3 (E)	V12				80	2289	ISOPHORONEDIAMINE
L4BH	TU15 TE19	AT	2 (E)	V12		CV13 CV28	S9	60	2290	ISOPHORONE DIISOCYANATE
SGAH L4BH	TU15 TE19	AT	2 (E)		VC1 VC2 AP7	CV13 CV28	S9	60	2291	LEAD COMPOUND, SOLUBLE, N.O.S.
LGBF		FL	3 (D/E)	V12			S2	30	2293	4-METHOXY-4-METHYLPENTAN-2-ONE
L4BH	TU15 TE19	AT	2 (E)	V12		CV13 CV28	S9	60	2294	N-METHYLANILINE
L10CH	TU14 TU15 TE19 TE21	FL	1 (C/D)			CV1 CV13 CV28	S2 S9 S14	663	2295	METHYL CHLOROACETATE
LGBF		FL	2 (D/E)				S2 S20	33	2296	METHYLCYCLOHEXANE
LGBF		FL	3 (D/E)	V12			S2	30	2297	METHYLCYCLO-HEXANONE
LGBF		FL	2 (D/E)				S2 S20	33	2298	METHYLCYCLOPENTANE
L4BH	TU15 TE19	AT	2 (E)	V12		CV13 CV28	S9	60	2299	METHYL DICHLOROACETATE
L4BH	TU15 TE19	AT	2 (E)	V12		CV13 CV28	S9	60	2300	2-METHYL-5-ETHYLPYRIDINE
LGBF		FL	2 (D/E)				S2 S20	33	2301	2-METHYLFURAN

UN No.	Name and description	Class	Classification code	Packing group	Labels	Special provisions	Limited and excepted quantities		Packaging			Portable tanks and bulk containers	
									Packing instructions	Special packing provisions	Mixed packing provisions	Instructions	Special provisions
	3.1.2	2.2	2.2	2.1.1.3	5.2.2	3.3	3.4	3.5.1.2	4.1.4	4.1.4	4.1.10	4.2.5.2 7.3.2	4.2.5.3
(1)	(2)	(3a)	(3b)	(4)	(5)	(6)	(7a)	(7b)	(8)	(9a)	(9b)	(10)	(11)
2302	5-METHYLHEXAN-2-ONE	3	F1	III	3		5 L	E1	P001 IBC03 LP01 R001		MP19	T2	TP1
2303	ISOPROPENYLBENZENE	3	F1	III	3		5 L	E1	P001 IBC03 LP01 R001		MP19	T2	TP1
2304	NAPHTHALENE, MOLTEN	4.1	F2	III	4.1	536	0	E0				T1	TP3
2305	NITROBENZENE-SULPHONIC ACID	8	C4	II	8		1 kg	E2	P002 IBC08	B4	MP10	T3	TP33
2306	NITROBENZOTRI-FLUORIDES, LIQUID	6.1	T1	II	6.1		100 ml	E4	P001 IBC02		MP15	T7	TP2
2307	3-NITRO-4-CHLORO-BENZOTRIFLUORIDE	6.1	T1	II	6.1		100 ml	E4	P001 IBC02		MP10	T7	TP2
2308	NITROSYLSULPHURIC ACID, LIQUID	8	C1	II	8		1 L	E2	P001 IBC02		MP15	T8	TP2
2309	OCTADIENES	3	F1	II	3		1 L	E2	P001 IBC02 R001		MP19	T4	TP1
2310	PENTANE-2,4-DIONE	3	FT1	III	3 +6.1		5 L	E1	P001 IBC03 R001		MP19	T4	TP1
2311	PHENETIDINES	6.1	T1	III	6.1	279	5 L	E1	P001 IBC03 LP01 R001		MP19	T4	TP1
2312	PHENOL, MOLTEN	6.1	T1	II	6.1		0	E0				T7	TP3
2313	PICOLINES	3	F1	III	3		5 L	E1	P001 IBC03 LP01 R001		MP19	T4	TP1
2315	POLYCHLORINATED BIPHENYLS, LIQUID	9	M2	II	9	305	1 L	E2	P906 IBC02		MP15	T4	TP1
2316	SODIUM CUPROCYANIDE, SOLID	6.1	T5	I	6.1		0	E5	P002 IBC07		MP18	T6	TP33
2317	SODIUM CUPROCYANIDE SOLUTION	6.1	T4	I	6.1		0	E5	P001		MP8 MP17	T14	TP2
2318	SODIUM HYDROSULPHIDE with less than 25% water of crystallization	4.2	S4	II	4.2	504	0	E2	P410 IBC06		MP14	T3	TP33
2319	TERPENE HYDROCARBONS, N.O.S.	3	F1	III	3		5 L	E1	P001 IBC03 LP01 R001		MP19	T4	TP1 TP29
2320	TETRAETHYLENE-PENTAMINE	8	C7	III	8		5 L	E1	P001 IBC03 LP01 R001		MP19	T4	TP1
2321	TRICHLOROBENZENES, LIQUID	6.1	T1	III	6.1		5 L	E1	P001 IBC03 LP01 R001		MP19	T4	TP1
2322	TRICHLOROBUTENE	6.1	T1	II	6.1		100 ml	E4	P001 IBC02		MP15	T7	TP2
2323	TRIETHYL PHOSPHITE	3	F1	III	3		5 L	E1	P001 IBC03 LP01 R001		MP19	T2	TP1
2324	TRIISOBUTYLENE	3	F1	III	3		5 L	E1	P001 IBC03 LP01 R001		MP19	T4	TP1
2325	1,3,5-TRIMETHYLBENZENE	3	F1	III	3		5 L	E1	P001 IBC03 LP01 R001		MP19	T2	TP1
2326	TRIMETHYLCYCLO-HEXYLAMINE	8	C7	III	8		5 L	E1	P001 IBC03 LP01 R001		MP19	T4	TP1

(12)	(13)	(14)	(15)	(16)	(17)	(18)	(19)	(20)	(1)	(2)
Tank code	Special provisions	Vehicle for tank carriage	Transport category (Tunnel restriction code)	Packages	Bulk	Loading, unloading and handling	Operation	Hazard identification No.	UN No.	Name and description
4.3	4.3.5, 6.8.4	9.1.1.2	1.1.3.6 (8.6)	7.2.4	7.3.3	7.5.11	8.5	5.3.2.3		3.1.2
LGBF		FL	3 (D/E)	V12			S2	30	2302	5-METHYLHEXAN-2-ONE
LGBF		FL	3 (D/E)	V12			S2	30	2303	ISOPROPENYLBENZENE
LGBV	TU27 TE4 TE6	AT	3 (E)					44	2304	NAPHTHALENE, MOLTEN
SGAN L4BN		AT	2 (E)	V11				80	2305	NITROBENZENE-SULPHONIC ACID
L4BH	TU15 TE19	AT	2 (D/E)			CV13 CV28	S9 S19	60	2306	NITROBENZOTRI-FLUORIDES, LIQUID
L4BH	TU15 TE19	AT	2 (D/E)			CV13 CV28	S9 S19	60	2307	3-NITRO-4-CHLORO-BENZOTRIFLUORIDE
L4BN		AT	2 (E)					X80	2308	NITROSYLSULPHURIC ACID, LIQUID
LGBF		FL	2 (D/E)				S2 S20	33	2309	OCTADIENES
L4BH	TU15	FL	3 (D/E)	V12		CV13 CV28	S2	36	2310	PENTANE-2,4-DIONE
L4BH	TU15 TE19	AT	2 (E)	V12		CV13 CV28	S9	60	2311	PHENETIDINES
L4BH	TU15 TE19	AT	0 (D/E)			CV13	S9 S19	60	2312	PHENOL, MOLTEN
LGBF		FL	3 (D/E)	V12			S2	30	2313	PICOLINES
L4BH	TU15	AT	0 (D/E)		VC1 VC2 AP9	CV1 CV13 CV28	S19	90	2315	POLYCHLORINATED BIPHENYLS, LIQUID
S10AH	TU15 TE19	AT	1 (C/E)	V10		CV1 CV13 CV28	S9 S14	66	2316	SODIUM CUPROCYANIDE, SOLID
L10CH	TU14 TU15 TE19 TE21	AT	1 (C/E)			CV1 CV13 CV28	S9 S14	66	2317	SODIUM CUPROCYANIDE SOLUTION
SGAN		AT	2 (D/E)	V1				40	2318	SODIUM HYDROSULPHIDE with less than 25% water of crystallization
LGBF		FL	3 (D/E)	V12			S2	30	2319	TERPENE HYDROCARBONS, N.O.S.
L4BN		AT	3 (E)	V12				80	2320	TETRAETHYLENE-PENTAMINE
L4BH	TU15 TE19	AT	2 (E)	V12		CV13 CV28	S9	60	2321	TRICHLOROBENZENES, LIQUID
L4BH	TU15 TE19	AT	2 (D/E)			CV13 CV28	S9 S19	60	2322	TRICHLOROBUTENE
LGBF		FL	3 (D/E)	V12			S2	30	2323	TRIETHYL PHOSPHITE
LGBF		FL	3 (D/E)	V12			S2	30	2324	TRIISOBUTYLENE
LGBF		FL	3 (D/E)	V12			S2	30	2325	1,3,5-TRIMETHYLBENZENE
L4BN		AT	3 (E)	V12				80	2326	TRIMETHYLCYCLO-HEXYLAMINE

UN No.	Name and description	Class	Classifi-cation code	Packing group	Labels	Special provi-sions	Limited and excepted quantities		Packaging			Portable tanks and bulk containers	
									Packing instruc-tions	Special packing provisions	Mixed packing provisions	Instruc-tions	Special provisions
3.1.2		2.2	2.2	2.1.1.3	5.2.2	3.3	3.4	3.5.1.2	4.1.4	4.1.4	4.1.10	4.2.5.2 7.3.2	4.2.5.3
(1)	(2)	(3a)	(3b)	(4)	(5)	(6)	(7a)	(7b)	(8)	(9a)	(9b)	(10)	(11)
2327	TRIMETHYLHEXA-METHYLENEDIAMINES	8	C7	III	8		5 L	E1	P001 IBC03 LP01 R001		MP19	T4	TP1
2328	TRIMETHYLHEXA-METHYLENE DIISOCYANATE	6.1	T1	III	6.1		5 L	E1	P001 IBC03 LP01 R001		MP19	T4	TP2
2329	TRIMETHYL PHOSPHITE	3	F1	III	3		5 L	E1	P001 IBC03 LP01 R001		MP19	T2	TP1
2330	UNDECANE	3	F1	III	3		5 L	E1	P001 IBC03 LP01 R001		MP19	T2	TP1
2331	ZINC CHLORIDE, ANHYDROUS	8	C2	III	8		5 kg	E1	P002 IBC08 LP02 R001	B3	MP10	T1	TP33
2332	ACETALDEHYDE OXIME	3	F1	III	3		5 L	E1	P001 IBC03 LP01 R001		MP19	T4	TP1
2333	ALLYL ACETATE	3	FT1	II	3 +6.1		1 L	E2	P001 IBC02		MP19	T7	TP1
2334	ALLYLAMINE	6.1	TF1	I	6.1 +3	354	0	E0	P602		MP8 MP17	T20	TP2
2335	ALLYL ETHYL ETHER	3	FT1	II	3 +6.1		1 L	E2	P001 IBC02		MP19	T7	TP1
2336	ALLYL FORMATE	3	FT1	I	3 +6.1		0	E0	P001		MP7 MP17	T14	TP2
2337	PHENYL MERCAPTAN	6.1	TF1	I	6.1 +3	354	0	E0	P602		MP8 MP17	T20	TP2
2338	BENZOTRIFLUORIDE	3	F1	II	3		1 L	E2	P001 IBC02 R001		MP19	T4	TP1
2339	2-BROMOBUTANE	3	F1	II	3		1 L	E2	P001 IBC02 R001		MP19	T4	TP1
2340	2-BROMOETHYL ETHYL ETHER	3	F1	II	3		1 L	E2	P001 IBC02 R001		MP19	T4	TP1
2341	1-BROMO-3-METHYLBUTANE	3	F1	III	3		5 L	E1	P001 IBC03 LP01 R001		MP19	T2	TP1
2342	BROMOMETHYL-PROPANES	3	F1	II	3		1 L	E2	P001 IBC02 R001		MP19	T4	TP1
2343	2-BROMOPENTANE	3	F1	II	3		1 L	E2	P001 IBC02 R001		MP19	T4	TP1
2344	BROMOPROPANES	3	F1	II	3		1 L	E2	P001 IBC02 R001		MP19	T4	TP1
2344	BROMOPROPANES	3	F1	III	3		5 L	E1	P001 IBC03 LP01 R001		MP19	T2	TP1
2345	3-BROMOPROPYNE	3	F1	II	3		1 L	E2	P001 IBC02 R001		MP19	T4	TP1
2346	BUTANEDIONE	3	F1	II	3		1 L	E2	P001 IBC02 R001		MP19	T4	TP1
2347	BUTYL MERCAPTAN	3	F1	II	3		1 L	E2	P001 IBC02 R001		MP19	T4	TP1
2348	BUTYL ACRYLATES, STABILIZED	3	F1	III	3	386	5 L	E1	P001 IBC03 LP01 R001		MP19	T2	TP1

ADR tank		Vehicle for tank carriage	Transport category (Tunnel restriction code)	Special provisions for carriage				Hazard identifi-cation No.	UN No.	Name and description
Tank code	Special provisions			Packages	Bulk	Loading, unloading and handling	Operation			
4.3	4.3.5, 6.8.4	9.1.1.2	1.1.3.6 (8.6)	7.2.4	7.3.3	7.5.11	8.5	5.3.2.3		3.1.2
(12)	(13)	(14)	(15)	(16)	(17)	(18)	(19)	(20)	(1)	(2)
L4BN		AT	3 (E)	V12				80	2327	TRIMETHYLHEXA-METHYLENEDIAMINES
L4BH	TU15 TE19	AT	2 (E)	V12		CV13 CV28	S9	60	2328	TRIMETHYLHEXA-METHYLENE DIISOCYANATE
LGBF		FL	3 (D/E)	V12			S2	30	2329	TRIMETHYL PHOSPHITE
LGBF		FL	3 (D/E)	V12			S2	30	2330	UNDECANE
SGAV		AT	3 (E)		VC1 VC2 AP7			80	2331	ZINC CHLORIDE, ANHYDROUS
LGBF		FL	3 (D/E)	V12			S2	30	2332	ACETALDEHYDE OXIME
L4BH	TU15	FL	2 (D/E)			CV13 CV28	S2 S19	336	2333	ALLYL ACETATE
L10CH	TU14 TU15 TE19 TE21	FL	1 (C/D)			CV1 CV13 CV28	S2 S9 S14	663	2334	ALLYLAMINE
L4BH	TU15	FL	2 (D/E)			CV13 CV28	S2 S19	336	2335	ALLYL ETHYL ETHER
L10CH	TU14 TU15 TE21	FL	1 (C/E)			CV13 CV28	S2 S22	336	2336	ALLYL FORMATE
L10CH	TU14 TU15 TE19 TE21	FL	1 (C/D)			CV1 CV13 CV28	S2 S9 S14	663	2337	PHENYL MERCAPTAN
LGBF		FL	2 (D/E)				S2 S20	33	2338	BENZOTRIFLUORIDE
LGBF		FL	2 (D/E)				S2 S20	33	2339	2-BROMOBUTANE
LGBF		FL	2 (D/E)				S2 S20	33	2340	2-BROMOETHYL ETHYL ETHER
LGBF		FL	3 (D/E)	V12			S2	30	2341	1-BROMO-3-METHYLBUTANE
LGBF		FL	2 (D/E)				S2 S20	33	2342	BROMOMETHYL-PROPANES
LGBF		FL	2 (D/E)				S2 S20	33	2343	2-BROMOPENTANE
LGBF		FL	2 (D/E)				S2 S20	33	2344	BROMOPROPANES
LGBF		FL	3 (D/E)	V12			S2	30	2344	BROMOPROPANES
LGBF		FL	2 (D/E)				S2 S20	33	2345	3-BROMOPROPYNE
LGBF		FL	2 (D/E)				S2 S20	33	2346	BUTANEDIONE
LGBF		FL	2 (D/E)				S2 S20	33	2347	BUTYL MERCAPTAN
LGBF		FL	3 (D/E)	V8 V12			S2 S4	39	2348	BUTYL ACRYLATES, STABILIZED

UN No.	Name and description	Class	Classification code	Packing group	Labels	Special provisions	Limited and excepted quantities		Packaging			Portable tanks and bulk containers	
									Packing instructions	Special packing provisions	Mixed packing provisions	Instructions	Special provisions
	3.1.2	2.2	2.2	2.1.1.3	5.2.2	3.3	3.4	3.5.1.2	4.1.4	4.1.4	4.1.10	4.2.5.2 7.3.2	4.2.5.3
(1)	(2)	(3a)	(3b)	(4)	(5)	(6)	(7a)	(7b)	(8)	(9a)	(9b)	(10)	(11)
2350	BUTYL METHYL ETHER	3	F1	II	3		1 L	E2	P001 IBC02 R001		MP19	T4	TP1
2351	BUTYL NITRITES	3	F1	II	3		1 L	E2	P001 IBC02 R001		MP19	T4	TP1
2351	BUTYL NITRITES	3	F1	III	3		5 L	E1	P001 IBC03 LP01 R001		MP19	T2	TP1
2352	BUTYL VINYL ETHER, STABILIZED	3	F1	II	3	386	1 L	E2	P001 IBC02 R001		MP19	T4	TP1
2353	BUTYRYL CHLORIDE	3	FC	II	3 +8		1 L	E2	P001 IBC02		MP19	T8	TP2
2354	CHLOROMETHYL ETHYL ETHER	3	FT1	II	3 +6.1		1 L	E2	P001 IBC02		MP19	T7	TP1
2356	2-CHLOROPROPANE	3	F1	I	3		0	E3	P001		MP7 MP17	T11	TP2
2357	CYCLOHEXYLAMINE	8	CF1	II	8 +3		1 L	E2	P001 IBC02		MP15	T7	TP2
2358	CYCLOOCTATETRAENE	3	F1	II	3		1 L	E2	P001 IBC02 R001		MP19	T4	TP1
2359	DIALLYLAMINE	3	FTC	II	3 +6.1 +8		1 L	E2	P001 IBC02		MP19	T7	TP1
2360	DIALLYL ETHER	3	FT1	II	3 +6.1		1 L	E2	P001 IBC02		MP19	T7	TP1
2361	DIISOBUTYLAMINE	3	FC	III	3 +8		5 L	E1	P001 IBC03 R001		MP19	T4	TP1
2362	1,1-DICHLOROETHANE	3	F1	II	3		1 L	E2	P001 IBC02 R001		MP19	T4	TP1
2363	ETHYL MERCAPTAN	3	F1	I	3		0	E0	P001		MP7 MP17	T11	TP2
2364	n-PROPYLBENZENE	3	F1	III	3		5 L	E1	P001 IBC03 LP01 R001		MP19	T2	TP1
2366	DIETHYL CARBONATE	3	F1	III	3		5 L	E1	P001 IBC03 LP01 R001		MP19	T2	TP1
2367	alpha-METHYL-VALERALDEHYDE	3	F1	II	3		1 L	E2	P001 IBC02 R001		MP19	T4	TP1
2368	alpha-PINENE	3	F1	III	3		5 L	E1	P001 IBC03 LP01 R001		MP19	T2	TP1
2370	1-HEXENE	3	F1	II	3		1 L	E2	P001 IBC02 R001		MP19	T4	TP1
2371	ISOPENTENES	3	F1	I	3		0	E3	P001		MP7 MP17	T11	TP2
2372	1,2-DI-(DIMETHYLAMINO) ETHANE	3	F1	II	3		1 L	E2	P001 IBC02 R001		MP19	T4	TP1
2373	DIETHOXYMETHANE	3	F1	II	3		1 L	E2	P001 IBC02 R001		MP19	T4	TP1
2374	3,3-DIETHOXYPROPENE	3	F1	II	3		1 L	E2	P001 IBC02 R001		MP19	T4	TP1
2375	DIETHYL SULPHIDE	3	F1	II	3		1 L	E2	P001 IBC02 R001		MP19	T7	TP1
2376	2,3-DIHYDROPYRAN	3	F1	II	3		1 L	E2	P001 IBC02 R001		MP19	T4	TP1
2377	1,1-DIMETHOXYETHANE	3	F1	II	3		1 L	E2	P001 IBC02 R001		MP19	T7	TP1

ADR tank		Vehicle for tank carriage	Transport category (Tunnel restriction code)	Special provisions for carriage				Hazard identifi-cation No.	UN No.	Name and description
Tank code	Special provisions			Packages	Bulk	Loading, unloading and handling	Operation			
4.3	4.3.5, 6.8.4	9.1.1.2	1.1.3.6 (8.6)	7.2.4	7.3.3	7.5.11	8.5	5.3.2.3		3.1.2
(12)	(13)	(14)	(15)	(16)	(17)	(18)	(19)	(20)	(1)	(2)
LGBF		FL	2 (D/E)				S2 S20	33	2350	BUTYL METHYL ETHER
LGBF		FL	2 (D/E)				S2 S20	33	2351	BUTYL NITRITES
LGBF		FL	3 (D/E)	V12			S2	30	2351	BUTYL NITRITES
LGBF		FL	2 (D/E)	V8			S2 S4 S20	339	2352	BUTYL VINYL ETHER, STABILIZED
L4BH		FL	2 (D/E)				S2 S20	338	2353	BUTYRYL CHLORIDE
L4BH	TU15	FL	2 (D/E)			CV13 CV28	S2 S19	336	2354	CHLOROMETHYL ETHYL ETHER
L4BN		FL	1 (D/E)				S2 S20	33	2356	2-CHLOROPROPANE
L4BN		FL	2 (D/E)				S2	83	2357	CYCLOHEXYLAMINE
LGBF		FL	2 (D/E)				S2 S20	33	2358	CYCLOOCTATETRAENE
L4BH	TU15	FL	2 (D/E)			CV13 CV28	S2 S19	338	2359	DIALLYLAMINE
L4BH	TU15	FL	2 (D/E)			CV13 CV28	S2 S19	336	2360	DIALLYL ETHER
L4BN		FL	3 (D/E)	V12			S2	38	2361	DIISOBUTYLAMINE
LGBF		FL	2 (D/E)				S2 S20	33	2362	1,1-DICHLOROETHANE
L4BN		FL	1 (D/E)				S2 S20	33	2363	ETHYL MERCAPTAN
LGBF		FL	3 (D/E)	V12			S2	30	2364	n-PROPYLBENZENE
LGBF		FL	3 (D/E)	V12			S2	30	2366	DIETHYL CARBONATE
LGBF		FL	2 (D/E)				S2 S20	33	2367	alpha-METHYL-VALERALDEHYDE
LGBF		FL	3 (D/E)	V12			S2	30	2368	alpha-PINENE
LGBF		FL	2 (D/E)				S2 S20	33	2370	1-HEXENE
L4BN		FL	1 (D/E)				S2 S20	33	2371	ISOPENTENES
LGBF		FL	2 (D/E)				S2 S20	33	2372	1,2-DI-(DIMETHYLAMINO) ETHANE
LGBF		FL	2 (D/E)				S2 S20	33	2373	DIETHOXYMETHANE
LGBF		FL	2 (D/E)				S2 S20	33	2374	3,3-DIETHOXYPROPENE
LGBF		FL	2 (D/E)				S2 S20	33	2375	DIETHYL SULPHIDE
LGBF		FL	2 (D/E)				S2 S20	33	2376	2,3-DIHYDROPYRAN
LGBF		FL	2 (D/E)				S2 S20	33	2377	1,1-DIMETHOXYETHANE

UN No.	Name and description	Class	Classification code	Packing group	Labels	Special provisions	Limited and excepted quantities		Packaging			Portable tanks and bulk containers	
									Packing instructions	Special packing provisions	Mixed packing provisions	Instructions	Special provisions
	3.1.2	2.2	2.2	2.1.1.3	5.2.2	3.3	3.4	3.5.1.2	4.1.4	4.1.4	4.1.10	4.2.5.2 7.3.2	4.2.5.3
(1)	(2)	(3a)	(3b)	(4)	(5)	(6)	(7a)	(7b)	(8)	(9a)	(9b)	(10)	(11)
2378	2-DIMETHYLAMINO-ACETONITRILE	3	FT1	II	3 +6.1		1 L	E2	P001 IBC02		MP19	T7	TP1
2379	1,3-DIMETHYLBUTYLAMINE	3	FC	II	3 +8		1 L	E2	P001 IBC02		MP19	T7	TP1
2380	DIMETHYLDIETHOXY-SILANE	3	F1	II	3		1 L	E2	P001 IBC02 R001		MP19	T4	TP1
2381	DIMETHYL DISULPHIDE	3	FT1	II	3 +6.1		1 L	E0	P001 IBC02		MP19	T7	TP2 TP39
2382	DIMETHYLHYDRAZINE, SYMMETRICAL	6.1	TF1	I	6.1 +3	354	0	E0	P602		MP8 MP17	T20	TP2
2383	DIPROPYLAMINE	3	FC	II	3 +8	386	1 L	E2	P001 IBC02		MP19	T7	TP1
2384	DI-n-PROPYL ETHER	3	F1	II	3		1 L	E2	P001 IBC02 R001		MP19	T4	TP1
2385	ETHYL ISOBUTYRATE	3	F1	II	3		1 L	E2	P001 IBC02 R001		MP19	T4	TP1
2386	1-ETHYLPIPERIDINE	3	FC	II	3 +8		1 L	E2	P001 IBC02		MP19	T7	TP1
2387	FLUOROBENZENE	3	F1	II	3		1 L	E2	P001 IBC02 R001		MP19	T4	TP1
2388	FLUOROTOLUENES	3	F1	II	3		1 L	E2	P001 IBC02 R001		MP19	T4	TP1
2389	FURAN	3	F1	I	3		0	E3	P001		MP7 MP17	T12	TP2
2390	2-IODOBUTANE	3	F1	II	3		1 L	E2	P001 IBC02 R001		MP19	T4	TP1
2391	IODOMETHYLPROPANES	3	F1	II	3		1 L	E2	P001 IBC02 R001		MP19	T4	TP1
2392	IODOPROPANES	3	F1	III	3		5 L	E1	P001 IBC03 LP01 R001		MP19	T2	TP1
2393	ISOBUTYL FORMATE	3	F1	II	3		1 L	E2	P001 IBC02 R001		MP19	T4	TP1
2394	ISOBUTYL PROPIONATE	3	F1	III	3		5 L	E1	P001 IBC03 LP01 R001		MP19	T2	TP1
2395	ISOBUTYRYL CHLORIDE	3	FC	II	3 +8		1 L	E2	P001 IBC02		MP19	T7	TP2
2396	METHACRYLALDEHYDE, STABILIZED	3	FT1	II	3 +6.1	386	1 L	E2	P001 IBC02		MP19	T7	TP1
2397	3-METHYLBUTAN-2-ONE	3	F1	II	3		1 L	E2	P001 IBC02 R001		MP19	T4	TP1
2398	METHYL tert-BUTYL ETHER	3	F1	II	3		1 L	E2	P001 IBC02 R001		MP19	T7	TP1
2399	1-METHYLPIPERIDINE	3	FC	II	3 +8		1 L	E2	P001 IBC02		MP19	T7	TP1
2400	METHYL ISOVALERATE	3	F1	II	3		1 L	E2	P001 IBC02 R001		MP19	T4	TP1
2401	PIPERIDINE	8	CF1	I	8 +3		0	E0	P001		MP8 MP17	T10	TP2
2402	PROPANETHIOLS	3	F1	II	3		1 L	E2	P001 IBC02 R001		MP19	T4	TP1
2403	ISOPROPENYL ACETATE	3	F1	II	3		1 L	E2	P001 IBC02 R001		MP19	T4	TP1
2404	PROPIONITRILE	3	FT1	II	3 +6.1		1 L	E0	P001 IBC02		MP19	T7	TP1

ADR tank		Vehicle for tank carriage	Transport category (Tunnel restriction code)	Special provisions for carriage				Hazard identifi-cation No.	UN No.	Name and description
Tank code	Special provisions			Packages	Bulk	Loading, unloading and handling	Operation			
4.3	4.3.5, 6.8.4	9.1.1.2	1.1.3.6 (8.6)	7.2.4	7.3.3	7.5.11	8.5	5.3.2.3		3.1.2
(12)	(13)	(14)	(15)	(16)	(17)	(18)	(19)	(20)	(1)	(2)
L4BH	TU15	FL	2 (D/E)			CV13 CV28	S2 S19	336	2378	2-DIMETHYLAMINO-ACETONITRILE
L4BH		FL	2 (D/E)				S2 S20	338	2379	1,3-DIMETHYLBUTYLAMINE
LGBF		FL	2 (D/E)				S2 S20	33	2380	DIMETHYLDIETHOXY-SILANE
L4BH	TU15	FL	2 (D/E)			CV13 CV28	S2 S22	336	2381	DIMETHYL DISULPHIDE
L10CH	TU14 TU15 TE19 TE21	FL	1 (C/D)			CV1 CV13 CV28	S2 S9 S14	663	2382	DIMETHYLHYDRAZINE, SYMMETRICAL
L4BH		FL	2 (D/E)	V8			S2 S4 S20	338	2383	DIPROPYLAMINE
LGBF		FL	2 (D/E)				S2 S20	33	2384	DI-n-PROPYL ETHER
LGBF		FL	2 (D/E)				S2 S20	33	2385	ETHYL ISOBUTYRATE
L4BH		FL	2 (D/E)				S2 S20	338	2386	1-ETHYLPIPERIDINE
LGBF		FL	2 (D/E)				S2 S20	33	2387	FLUOROBENZENE
LGBF		FL	2 (D/E)				S2 S20	33	2388	FLUOROTOLUENES
L4BN		FL	1 (D/E)				S2 S20	33	2389	FURAN
LGBF		FL	2 (D/E)				S2 S20	33	2390	2-IODOBUTANE
LGBF		FL	2 (D/E)				S2 S20	33	2391	IODOMETHYLPROPANES
LGBF		FL	3 (D/E)	V12			S2	30	2392	IODOPROPANES
LGBF		FL	2 (D/E)				S2 S20	33	2393	ISOBUTYL FORMATE
LGBF		FL	3 (D/E)	V12			S2	30	2394	ISOBUTYL PROPIONATE
L4BH		FL	2 (D/E)				S2 S20	338	2395	ISOBUTYRYL CHLORIDE
L4BH	TU15	FL	2 (D/E)	V8		CV13 CV28	S2 S4 S19	336	2396	METHACRYLALDEHYDE, STABILIZED
LGBF		FL	2 (D/E)				S2 S20	33	2397	3-METHYLBUTAN-2-ONE
LGBF		FL	2 (D/E)				S2 S20	33	2398	METHYL tert-BUTYL ETHER
L4BH		FL	2 (D/E)				S2 S20	338	2399	1-METHYLPIPERIDINE
LGBF		FL	2 (D/E)				S2 S20	33	2400	METHYL ISOVALERATE
L10BH		FL	1 (D/E)				S2 S14	883	2401	PIPERIDINE
LGBF		FL	2 (D/E)				S2 S20	33	2402	PROPANETHIOLS
LGBF		FL	2 (D/E)				S2 S20	33	2403	ISOPROPENYL ACETATE
L4BH	TU15	FL	2 (D/E)			CV13 CV28	S2 S19	336	2404	PROPIONITRILE

UN No.	Name and description	Class	Classifi-cation code	Packing group	Labels	Special provi-sions	Limited and excepted quantities		Packaging			Portable tanks and bulk containers	
									Packing instruc-tions	Special packing provisions	Mixed packing provisions	Instruc-tions	Special provisions
3.1.2		2.2	2.2	2.1.1.3	5.2.2	3.3	3.4	3.5.1.2	4.1.4	4.1.4	4.1.10	4.2.5.2 7.3.2	4.2.5.3
(1)	(2)	(3a)	(3b)	(4)	(5)	(6)	(7a)	(7b)	(8)	(9a)	(9b)	(10)	(11)
2405	ISOPROPYL BUTYRATE	3	F1	III	3		5 L	E1	P001 IBC03 LP01 R001		MP19	T2	TP1
2406	ISOPROPYL ISOBUTYRATE	3	F1	II	3		1 L	E2	P001 IBC02 R001		MP19	T4	TP1
2407	ISOPROPYL CHLOROFORMATE	6.1	TFC	I	6.1 +3 +8	354	0	E0	P602		MP8 MP17		
2409	ISOPROPYL PROPIONATE	3	F1	II	3		1 L	E2	P001 IBC02 R001		MP19	T4	TP1
2410	1,2,3,6-TETRAHYDROPYRIDINE	3	F1	II	3		1 L	E2	P001 IBC02 R001		MP19	T4	TP1
2411	BUTYRONITRILE	3	FT1	II	3 +6.1		1 L	E2	P001 IBC02		MP19	T7	TP1
2412	TETRAHYDROTHIOPHENE	3	F1	II	3		1 L	E2	P001 IBC02 R001		MP19	T4	TP1
2413	TETRAPROPYL ORTHOTITANATE	3	F1	III	3		5 L	E1	P001 IBC03 LP01 R001		MP19	T4	TP1
2414	THIOPHENE	3	F1	II	3		1 L	E2	P001 IBC02 R001		MP19	T4	TP1
2416	TRIMETHYL BORATE	3	F1	II	3		1 L	E2	P001 IBC02 R001		MP19	T7	TP1
2417	CARBONYL FLUORIDE	2	2TC		2.3 +8		0	E0	P200		MP9	(M)	
2418	SULPHUR TETRAFLUORIDE	2	2TC		2.3 +8		0	E0	P200		MP9		
2419	BROMOTRIFLUORO-ETHYLENE	2	2F		2.1	662	0	E0	P200		MP9	(M)	
2420	HEXAFLUOROACETONE	2	2TC		2.3 +8		0	E0	P200		MP9	(M)	
2421	NITROGEN TRIOXIDE	2	2TOC				CARRIAGE PROHIBITED						
2422	OCTAFLUOROBUT-2-ENE (REFRIGERANT GAS R 1318)	2	2A		2.2	662	120 ml	E1	P200		MP9	(M)	
2424	OCTAFLUOROPROPANE (REFRIGERANT GAS R 218)	2	2A		2.2	662	120 ml	E1	P200		MP9	(M) T50	
2426	AMMONIUM NITRATE, LIQUID, hot concentrated solution, in a concentration of more than 80% but not more than 93%	5.1	O1		5.1	252 644	0	E0				T7	TP1 TP16 TP17
2427	POTASSIUM CHLORATE, AQUEOUS SOLUTION	5.1	O1	II	5.1		1 L	E2	P504 IBC02		MP2	T4	TP1
2427	POTASSIUM CHLORATE, AQUEOUS SOLUTION	5.1	O1	III	5.1		5 L	E1	P504 IBC02 R001		MP2	T4	TP1
2428	SODIUM CHLORATE, AQUEOUS SOLUTION	5.1	O1	II	5.1		1 L	E2	P504 IBC02		MP2	T4	TP1
2428	SODIUM CHLORATE, AQUEOUS SOLUTION	5.1	O1	III	5.1		5 L	E1	P504 IBC02 R001		MP2	T4	TP1
2429	CALCIUM CHLORATE, AQUEOUS SOLUTION	5.1	O1	II	5.1		1 L	E2	P504 IBC02		MP2	T4	TP1
2429	CALCIUM CHLORATE, AQUEOUS SOLUTION	5.1	O1	III	5.1		5 L	E1	P504 IBC02 R001		MP2	T4	TP1
2430	ALKYLPHENOLS, SOLID, N.O.S. (including C2-C12 homologues)	8	C4	I	8		0	E0	P002 IBC07		MP18	T6	TP33
2430	ALKYLPHENOLS, SOLID, N.O.S. (including C2-C12 homologues)	8	C4	II	8		1 kg	E2	P002 IBC08	B4	MP10	T3	TP33

ADR tank		Vehicle for tank carriage	Transport category (Tunnel restriction code)	Special provisions for carriage				Hazard identification No.	UN No.	Name and description
Tank code	Special provisions			Packages	Bulk	Loading, unloading and handling	Operation			
4.3	4.3.5, 6.8.4	9.1.1.2	1.1.3.6 (8.6)	7.2.4	7.3.3	7.5.11	8.5	5.3.2.3		3.1.2
(12)	(13)	(14)	(15)	(16)	(17)	(18)	(19)	(20)	(1)	(2)
LGBF		FL	3 (D/E)	V12			S2	30	2405	ISOPROPYL BUTYRATE
LGBF		FL	2 (D/E)				S2 S20	33	2406	ISOPROPYL ISOBUTYRATE
			1 (D)			CV1 CV13 CV28	S2 S9 S14		2407	ISOPROPYL CHLOROFORMATE
LGBF		FL	2 (D/E)				S2 S20	33	2409	ISOPROPYL PROPIONATE
LGBF		FL	2 (D/E)				S2 S20	33	2410	1,2,3,6-TETRAHYDROPYRIDINE
L4BH	TU15	FL	2 (D/E)			CV13 CV28	S2 S19	336	2411	BUTYRONITRILE
LGBF		FL	2 (D/E)				S2 S20	33	2412	TETRAHYDROTHIOPHENE
LGBF		FL	3 (D/E)	V12			S2	30	2413	TETRAPROPYL ORTHOTITANATE
LGBF		FL	2 (D/E)				S2 S20	33	2414	THIOPHENE
LGBF		FL	2 (D/E)				S2 S20	33	2416	TRIMETHYL BORATE
PxBH(M)	TA4 TT9	AT	1 (C/D)			CV9 CV10 CV36	S14	268	2417	CARBONYL FLUORIDE
			1 (D)			CV9 CV10 CV36	S14		2418	SULPHUR TETRAFLUORIDE
PxBN(M)	TA4 TT9	FL	2 (B/D)			CV9 CV10 CV36	S2 S20	23	2419	BROMOTRIFLUORO-ETHYLENE
PxBH(M)	TA4 TT9	AT	1 (C/D)			CV9 CV10 CV36	S14	268	2420	HEXAFLUOROACETONE
CARRIAGE PROHIBITED									2421	NITROGEN TRIOXIDE
PxBN(M)	TA4 TT9	AT	3 (C/E)			CV9 CV10 CV36		20	2422	OCTAFLUOROBUT-2-ENE (REFRIGERANT GAS R 1318)
PxBN(M)	TA4 TT9	AT	3 (C/E)			CV9 CV10 CV36		20	2424	OCTAFLUOROPROPANE (REFRIGERANT GAS R 218)
L4BV(+)	TU3 TU12 TU29 TC3 TE9 TE10 TA1	AT	0 (E)				S23	59	2426	AMMONIUM NITRATE, LIQUID, hot concentrated solution, in a concentration of more than 80% but not more than 93%
L4BN	TU3	AT	2 (E)			CV24		50	2427	POTASSIUM CHLORATE, AQUEOUS SOLUTION
LGBV	TU3	AT	3 (E)			CV24		50	2427	POTASSIUM CHLORATE, AQUEOUS SOLUTION
L4BN	TU3	AT	2 (E)			CV24		50	2428	SODIUM CHLORATE, AQUEOUS SOLUTION
LGBV	TU3	AT	3 (E)			CV24		50	2428	SODIUM CHLORATE, AQUEOUS SOLUTION
L4BN	TU3	AT	2 (E)			CV24		50	2429	CALCIUM CHLORATE, AQUEOUS SOLUTION
LGBV	TU3	AT	3 (E)			CV24		50	2429	CALCIUM CHLORATE, AQUEOUS SOLUTION
S10AN L10BH		AT	1 (E)	V10			S20	88	2430	ALKYLPHENOLS, SOLID, N.O.S. (including C2-C12 homologues)
SGAN L4BN		AT	2 (E)	V11				80	2430	ALKYLPHENOLS, SOLID, N.O.S. (including C2-C12 homologues)

UN No.	Name and description	Class	Classification code	Packing group	Labels	Special provisions	Limited quantities	Excepted quantities	Packing instructions	Special packing provisions	Mixed packing provisions	Instructions	Special provisions
3.1.2	3.1.2	2.2	2.2	2.1.1.3	5.2.2	3.3	3.4	3.5.1.2	4.1.4	4.1.4	4.1.10	4.2.5.2 7.3.2	4.2.5.3
(1)	(2)	(3a)	(3b)	(4)	(5)	(6)	(7a)	(7b)	(8)	(9a)	(9b)	(10)	(11)
2430	ALKYLPHENOLS, SOLID, N.O.S. (including C2-C12 homologues)	8	C4	III	8		5 kg	E1	P002 IBC08 LP02 R001	B3	MP10	T1	TP33
2431	ANISIDINES	6.1	T1	III	6.1		5 L	E1	P001 IBC03 LP01 R001		MP19	T4	TP1
2432	N,N-DIETHYLANILINE	6.1	T1	III	6.1	279	5 L	E1	P001 IBC03 LP01 R001		MP19	T4	TP1
2433	CHLORONITROTOLUENES, LIQUID	6.1	T1	III	6.1		5 L	E1	P001 IBC03 LP01 R001		MP19	T4	TP1
2434	DIBENZYL-DICHLOROSILANE	8	C3	II	8		0	E0	P010		MP15	T10	TP2 TP7
2435	ETHYLPHENYL-DICHLOROSILANE	8	C3	II	8		0	E0	P010		MP15	T10	TP2 TP7
2436	THIOACETIC ACID	3	F1	II	3		1 L	E2	P001 IBC02 R001		MP19	T4	TP1
2437	METHYLPHENYL-DICHLOROSILANE	8	C3	II	8		0	E0	P010		MP15	T10	TP2 TP7
2438	TRIMETHYLACETYL CHLORIDE	6.1	TFC	I	6.1 +3 +8		0	E0	P001		MP8 MP17	T14	TP2
2439	SODIUM HYDROGENDIFLUORIDE	8	C2	II	8		1 kg	E2	P002 IBC08	B4	MP10	T3	TP33
2440	STANNIC CHLORIDE PENTAHYDRATE	8	C2	III	8		5 kg	E1	P002 IBC08 LP02 R001	B3	MP10	T1	TP33
2441	TITANIUM TRICHLORIDE, PYROPHORIC or TITANIUM TRICHLORIDE MIXTURE, PYROPHORIC	4.2	SC4	I	4.2 +8	537	0	E0	P404		MP13		
2442	TRICHLOROACETYL CHLORIDE	8	C3	II	8		0	E0	P001		MP15	T7	TP2
2443	VANADIUM OXYTRICHLORIDE	8	C1	II	8		1 L	E0	P001 IBC02		MP15	T7	TP2
2444	VANADIUM TETRACHLORIDE	8	C1	I	8		0	E0	P802		MP8 MP17	T10	TP2
2446	NITROCRESOLS, SOLID	6.1	T2	III	6.1		5 kg	E1	P002 IBC08 LP02 R001	B3	MP10	T1	TP33
2447	PHOSPHORUS, WHITE, MOLTEN	4.2	ST3	I	4.2 +6.1		0	E0				T21	TP3 TP7 TP26
2448	SULPHUR, MOLTEN	4.1	F3	III	4.1	538	0	E0				T1	TP3
2451	NITROGEN TRIFLUORIDE	2	2O		2.2 +5.1	662	0	E0	P200		MP9	(M)	
2452	ETHYLACETYLENE, STABILIZED	2	2F		2.1	386 662	0	E0	P200		MP9	(M)	
2453	ETHYL FLUORIDE (REFRIGERANT GAS R 161)	2	2F		2.1	662	0	E0	P200		MP9	(M)	
2454	METHYL FLUORIDE (REFRIGERANT GAS R 41)	2	2F		2.1	662	0	E0	P200		MP9	(M)	
2455	METHYL NITRITE	2	2A	CARRIAGE PROHIBITED									
2456	2-CHLOROPROPENE	3	F1	I	3		0	E3	P001		MP7 MP17	T11	TP2
2457	2,3-DIMETHYLBUTANE	3	F1	II	3		1 L	E2	P001 IBC02 R001		MP19	T7	TP1
2458	HEXADIENES	3	F1	II	3		1 L	E2	P001 IBC02 R001		MP19	T4	TP1
2459	2-METHYL-1-BUTENE	3	F1	I	3		0	E3	P001		MP7 MP17	T11	TP2

Tank code (4.3)	Special provisions (4.3.5, 6.8.4)	Vehicle for tank carriage (9.1.1.2)	Transport category (Tunnel restriction code) (1.1.3.6) (8.6)	Packages (7.2.4)	Bulk (7.3.3)	Loading, unloading and handling (7.5.11)	Operation (8.5)	Hazard identification No. (5.3.2.3)	UN No. (1)	Name and description (3.1.2)
(12)	(13)	(14)	(15)	(16)	(17)	(18)	(19)	(20)	(1)	(2)
SGAV L4BN		AT	3 (E)		VC1 VC2 AP7			80	2430	ALKYLPHENOLS, SOLID, N.O.S. (including C2-C12 homologues)
L4BH	TU15 TE19	AT	2 (E)	V12		CV13 CV28	S9	60	2431	ANISIDINES
L4BH	TU15 TE19	AT	2 (E)	V12		CV13 CV28	S9	60	2432	N,N-DIETHYLANILINE
L4BH	TU15 TE19	AT	2 (E)	V12		CV13 CV28	S9	60	2433	CHLORONITROTOLUENES, LIQUID
L4BN		AT	2 (E)					X80	2434	DIBENZYL-DICHLOROSILANE
L4BN		AT	2 (E)					X80	2435	ETHYLPHENYL-DICHLOROSILANE
LGBF		FL	2 (D/E)				S2 S20	33	2436	THIOACETIC ACID
L4BN		AT	2 (E)					X80	2437	METHYLPHENYL-DICHLOROSILANE
L10CH	TU14 TU15 TE19 TE21	FL	1 (C/D)			CV1 CV13 CV28	S2 S9 S14	663	2438	TRIMETHYLACETYL CHLORIDE
SGAN		AT	2 (E)	V11				80	2439	SODIUM HYDROGENDIFLUORIDE
SGAV		AT	3 (E)		VC1 VC2 AP7			80	2440	STANNIC CHLORIDE PENTAHYDRATE
			0 (E)	V1			S20		2441	TITANIUM TRICHLORIDE, PYROPHORIC or TITANIUM TRICHLORIDE MIXTURE, PYROPHORIC
L4BN		AT	2 (E)					X80	2442	TRICHLOROACETYL CHLORIDE
L4BN		AT	2 (E)					80	2443	VANADIUM OXYTRICHLORIDE
L10BH		AT	1 (E)				S20	X88	2444	VANADIUM TETRACHLORIDE
SGAH L4BH	TU15 TE19	AT	2 (E)		VC1 VC2 AP7	CV13 CV28	S9	60	2446	NITROCRESOLS, SOLID
L10DH(+)	TU14 TU16 TU21 TE3 TE21	AT	0 (B/E)				S20	446	2447	PHOSPHORUS, WHITE, MOLTEN
LGBV(+)	TU27 TE4 TE6	AT	3 (E)					44	2448	SULPHUR, MOLTEN
PxBN(M)	TA4 TT9	AT	3 (C/E)			CV9 CV10 CV36		25	2451	NITROGEN TRIFLUORIDE
PxBN(M)	TA4 TT9	FL	2 (B/D)	V8		CV9 CV10 CV36	S2 S4 S20	239	2452	ETHYLACETYLENE, STABILIZED
PxBN(M)	TA4 TT9	FL	2 (B/D)			CV9 CV10 CV36	S2 S20	23	2453	ETHYL FLUORIDE (REFRIGERANT GAS R 161)
PxBN(M)	TA4 TT9	FL	2 (B/D)			CV9 CV10 CV36	S2 S20	23	2454	METHYL FLUORIDE (REFRIGERANT GAS R 41)
CARRIAGE PROHIBITED									2455	METHYL NITRITE
L4BN		FL	1 (D/E)				S2 S20	33	2456	2-CHLOROPROPENE
LGBF		FL	2 (D/E)				S2 S20	33	2457	2,3-DIMETHYLBUTANE
LGBF		FL	2 (D/E)				S2 S20	33	2458	HEXADIENES
L4BN		FL	1 (D/E)				S2 S20	33	2459	2-METHYL-1-BUTENE

UN No.	Name and description	Class	Classification code	Packing group	Labels	Special provisions	Limited and excepted quantities		Packaging			Portable tanks and bulk containers	
									Packing instructions	Special packing provisions	Mixed packing provisions	Instructions	Special provisions
	3.1.2	2.2	2.2	2.1.1.3	5.2.2	3.3	3.4	3.5.1.2	4.1.4	4.1.4	4.1.10	4.2.5.2 7.3.2	4.2.5.3
(1)	(2)	(3a)	(3b)	(4)	(5)	(6)	(7a)	(7b)	(8)	(9a)	(9b)	(10)	(11)
2460	2-METHYL-2-BUTENE	3	F1	II	3		1 L	E2	P001 IBC02	B8	MP19	T7	TP1
2461	METHYLPENTADIENE	3	F1	II	3		1 L	E2	P001 IBC02 R001		MP19	T4	TP1
2463	ALUMINIUM HYDRIDE	4.3	W2	I	4.3		0	E0	P403		MP2		
2464	BERYLLIUM NITRATE	5.1	OT2	II	5.1 +6.1		1 kg	E2	P002 IBC08	B4	MP2	T3	TP33
2465	DICHLOROISOCYANURIC ACID, DRY or DICHLOROISOCYANURIC ACID SALTS	5.1	O2	II	5.1	135	1 kg	E2	P002 IBC08	B4	MP10	T3	TP33
2466	POTASSIUM SUPEROXIDE	5.1	O2	I	5.1		0	E0	P503 IBC06		MP2		
2468	TRICHLOROISOCYANURIC ACID, DRY	5.1	O2	II	5.1		1 kg	E2	P002 IBC08	B4	MP10	T3	TP33
2469	ZINC BROMATE	5.1	O2	III	5.1		5 kg	E1	P002 IBC08 LP02 R001	B3	MP10	T1	TP33
2470	PHENYLACETONITRILE, LIQUID	6.1	T1	III	6.1		5 L	E1	P001 IBC03 LP01 R001		MP19	T4	TP1
2471	OSMIUM TETROXIDE	6.1	T5	I	6.1		0	E5	P002 IBC07	PP30	MP18	T6	TP33
2473	SODIUM ARSANILATE	6.1	T3	III	6.1		5 kg	E1	P002 IBC08 LP02 R001	B3	MP10	T1	TP33
2474	THIOPHOSGENE	6.1	T1	I	6.1	279 354	0	E0	P602		MP8 MP17	T20	TP2
2475	VANADIUM TRICHLORIDE	8	C2	III	8		5 kg	E1	P002 IBC08 LP02 R001	B3	MP10	T1	TP33
2477	METHYL ISOTHIOCYANATE	6.1	TF1	I	6.1 +3	354	0	E0	P602		MP8 MP17	T20	TP2
2478	ISOCYANATES, FLAMMABLE, TOXIC, N.O.S. or ISOCYANATE SOLUTION, FLAMMABLE, TOXIC, N.O.S.	3	FT1	II	3 +6.1	274 539	1 L	E2	P001 IBC02		MP19	T11	TP2 TP27
2478	ISOCYANATES, FLAMMABLE, TOXIC, N.O.S. or ISOCYANATE SOLUTION, FLAMMABLE, TOXIC, N.O.S.	3	FT1	III	3 +6.1	274	5 L	E1	P001 IBC03 R001		MP19	T7	TP1 TP28
2480	METHYL ISOCYANATE	6.1	TF1	I	6.1 +3	354	0	E0	P601		MP2	T22	TP2
2481	ETHYL ISOCYANATE	6.1	TF1	I	6.1 +3	354	0	E0	P602		MP8 MP17	T20	TP2
2482	n-PROPYL ISOCYANATE	6.1	TF1	I	6.1 +3	354	0	E0	P602		MP8 MP17	T20	TP2
2483	ISOPROPYL ISOCYANATE	6.1	TF1	I	6.1 +3	354	0	E0	P602		MP8 MP17	T20	TP2
2484	tert-BUTYL ISOCYANATE	6.1	TF1	I	6.1 +3	354	0	E0	P602		MP8 MP17	T20	TP2
2485	n-BUTYL ISOCYANATE	6.1	TF1	I	6.1 +3	354	0	E0	P602		MP8 MP17	T20	TP2
2486	ISOBUTYL ISOCYANATE	6.1	TF1	I	6.1 +3	354	0	E0	P602		MP8 MP17	T20	TP2
2487	PHENYL ISOCYANATE	6.1	TF1	I	6.1 +3	354	0	E0	P602		MP8 MP17	T20	TP2

ADR tank		Vehicle for tank carriage	Transport category (Tunnel restriction code)	Special provisions for carriage				Hazard identification No.	UN No.	Name and description
Tank code	Special provisions			Packages	Bulk	Loading, unloading and handling	Operation			
4.3	4.3.5, 6.8.4	9.1.1.2	1.1.3.6 (8.6)	7.2.4	7.3.3	7.5.11	8.5	5.3.2.3		3.1.2
(12)	(13)	(14)	(15)	(16)	(17)	(18)	(19)	(20)	(1)	(2)
L1.5BN		FL	2 (D/E)				S2 S20	33	2460	2-METHYL-2-BUTENE
LGBF		FL	2 (D/E)				S2 S20	33	2461	METHYLPENTADIENE
			1 (E)	V1		CV23	S20		2463	ALUMINIUM HYDRIDE
SGAN	TU3	AT	2 (E)	V11		CV24 CV28		56	2464	BERYLLIUM NITRATE
SGAN	TU3	AT	2 (E)	V11		CV24		50	2465	DICHLOROISOCYANURIC ACID, DRY or DICHLOROISOCYANURIC ACID SALTS
			1 (E)	V10		CV24	S20		2466	POTASSIUM SUPEROXIDE
SGAN	TU3	AT	2 (E)	V11		CV24		50	2468	TRICHLOROISOCYANURIC ACID, DRY
SGAV	TU3	AT	3 (E)		VC1 VC2 AP6 AP7	CV24		50	2469	ZINC BROMATE
L4BH	TU15 TE19	AT	2 (E)	V12		CV13 CV28	S9	60	2470	PHENYLACETONITRILE, LIQUID
S10AH	TU15 TE19	AT	1 (C/E)	V10		CV1 CV13 CV28	S9 S14	66	2471	OSMIUM TETROXIDE
SGAH L4BH	TU15 TE19	AT	2 (E)		VC1 VC2 AP7	CV13 CV28	S9	60	2473	SODIUM ARSANILATE
L10CH	TU14 TU15 TE19 TE21	AT	1 (C/D)			CV1 CV13 CV28	S9 S14	66	2474	THIOPHOSGENE
SGAV		AT	3 (E)		VC1 VC2 AP7			80	2475	VANADIUM TRICHLORIDE
L10CH	TU14 TU15 TE19 TE21	FL	1 (C/D)			CV1 CV13 CV28	S2 S9 S14	663	2477	METHYL ISOTHIOCYANATE
L4BH	TU15	FL	2 (D/E)			CV13 CV28	S2 S19	336	2478	ISOCYANATES, FLAMMABLE, TOXIC, N.O.S. or ISOCYANATE SOLUTION, FLAMMABLE, TOXIC, N.O.S.
L4BH	TU15	FL	3 (D/E)	V12		CV13 CV28	S2	36	2478	ISOCYANATES, FLAMMABLE, TOXIC, N.O.S. or ISOCYANATE SOLUTION, FLAMMABLE, TOXIC, N.O.S.
L15CH	TU14 TU15 TE19 TE21	FL	1 (C/D)			CV1 CV13 CV28	S2 S9 S14	663	2480	METHYL ISOCYANATE
L15CH	TU14 TU15 TE19 TE21	FL	1 (C/D)			CV1 CV13 CV28	S2 S9 S14	663	2481	ETHYL ISOCYANATE
L10CH	TU14 TU15 TE19 TE21	FL	1 (C/D)			CV1 CV13 CV28	S2 S9 S14	663	2482	n-PROPYL ISOCYANATE
L10CH	TU14 TU15 TE19 TE21	FL	1 (C/D)			CV1 CV13 CV28	S2 S9 S14	663	2483	ISOPROPYL ISOCYANATE
L10CH	TU14 TU15 TE19 TE21	FL	1 (C/D)			CV1 CV13 CV28	S2 S9 S14	663	2484	tert-BUTYL ISOCYANATE
L10CH	TU14 TU15 TE19 TE21	FL	1 (C/D)			CV1 CV13 CV28	S2 S9 S14	663	2485	n-BUTYL ISOCYANATE
L10CH	TU14 TU15 TE19 TE21	FL	1 (C/D)			CV1 CV13 CV28	S2 S9 S14	663	2486	ISOBUTYL ISOCYANATE
L10CH	TU14 TU15 TE19 TE21	FL	1 (C/D)			CV1 CV13 CV28	S2 S9 S14	663	2487	PHENYL ISOCYANATE

UN No.	Name and description	Class	Classifi-cation code	Packing group	Labels	Special provi-sions	Limited and excepted quantities		Packaging			Portable tanks and bulk containers	
									Packing instruc-tions	Special packing provisions	Mixed packing provisions	Instruc-tions	Special provisions
	3.1.2	2.2	2.2	2.1.1.3	5.2.2	3.3	3.4	3.5.1.2	4.1.4	4.1.4	4.1.10	4.2.5.2 7.3.2	4.2.5.3
(1)	(2)	(3a)	(3b)	(4)	(5)	(6)	(7a)	(7b)	(8)	(9a)	(9b)	(10)	(11)
2488	CYCLOHEXYL ISOCYANATE	6.1	TF1	I	6.1 +3	354	0	E0	P602		MP8 MP17	T20	TP2
2490	DICHLOROISOPROPYL ETHER	6.1	T1	II	6.1		100 ml	E4	P001 IBC02		MP15	T7	TP2
2491	ETHANOLAMINE or ETHANOLAMINE SOLUTION	8	C7	III	8		5 L	E1	P001 IBC03 LP01 R001		MP19	T4	TP1
2493	HEXAMETHYLENEIMINE	3	FC	II	3 +8		1 L	E2	P001 IBC02		MP19	T7	TP1
2495	IODINE PENTAFLUORIDE	5.1	OTC	I	5.1 +6.1 +8		0	E0	P200		MP2		
2496	PROPIONIC ANHYDRIDE	8	C3	III	8		5 L	E1	P001 IBC03 LP01 R001		MP19	T4	TP1
2498	1,2,3,6-TETRAHYDROBENZAL-DEHYDE	3	F1	III	3		5 L	E1	P001 IBC03 LP01 R001		MP19	T2	TP1
2501	TRIS-(1-AZIRIDINYL) PHOSPHINE OXIDE SOLUTION	6.1	T1	II	6.1		100 ml	E4	P001 IBC02		MP15	T7	TP2
2501	TRIS-(1-AZIRIDINYL) PHOSPHINE OXIDE SOLUTION	6.1	T1	III	6.1		5 L	E1	P001 IBC03 LP01 R001		MP19	T4	TP1
2502	VALERYL CHLORIDE	8	CF1	II	8 +3		1 L	E2	P001 IBC02		MP15	T7	TP2
2503	ZIRCONIUM TETRACHLORIDE	8	C2	III	8		5 kg	E1	P002 IBC08 LP02 R001	B3	MP10	T1	TP33
2504	TETRABROMOETHANE	6.1	T1	III	6.1		5 L	E1	P001 IBC03 LP01 R001		MP19	T4	TP1
2505	AMMONIUM FLUORIDE	6.1	T5	III	6.1		5 kg	E1	P002 IBC08 LP02 R001	B3	MP10	T1	TP33
2506	AMMONIUM HYDROGEN SULPHATE	8	C2	II	8		1 kg	E2	P002 IBC08	B4	MP10	T3	TP33
2507	CHLOROPLATINIC ACID, SOLID	8	C2	III	8		5 kg	E1	P002 IBC08 LP02 R001	B3	MP10	T1	TP33
2508	MOLYBDENUM PENTACHLORIDE	8	C2	III	8		5 kg	E1	P002 IBC08 LP02 R001	B3	MP10	T1	TP33
2509	POTASSIUM HYDROGEN SULPHATE	8	C2	II	8		1 kg	E2	P002 IBC08	B4	MP10	T3	TP33
2511	2-CHLOROPROPIONIC ACID	8	C3	III	8		5 L	E1	P001 IBC03 LP01 R001		MP19	T4	TP2
2512	AMINOPHENOLS (o-, m-, p-)	6.1	T2	III	6.1	279	5 kg	E1	P002 IBC08 LP02 R001	B3	MP10	T1	TP33
2513	BROMOACETYL BROMIDE	8	C3	II	8		1 L	E2	P001 IBC02		MP15	T8	TP2
2514	BROMOBENZENE	3	F1	III	3		5 L	E1	P001 IBC03 LP01 R001		MP19	T2	TP1
2515	BROMOFORM	6.1	T1	III	6.1		5 L	E1	P001 IBC03 LP01 R001		MP19	T4	TP1
2516	CARBON TETRABROMIDE	6.1	T2	III	6.1		5 kg	E1	P002 IBC08 LP02 R001	B3	MP10	T1	TP33

ADR tank		Vehicle for tank carriage	Transport category (Tunnel restriction code)	Special provisions for carriage				Hazard identification No.	UN No.	Name and description
Tank code	Special provisions			Packages	Bulk	Loading, unloading and handling	Operation			
4.3	4.3.5, 6.8.4	9.1.1.2	1.1.3.6 (8.6)	7.2.4	7.3.3	7.5.11	8.5	5.3.2.3		3.1.2
(12)	(13)	(14)	(15)	(16)	(17)	(18)	(19)	(20)	(1)	(2)
L10CH	TU14 TU15 TE19 TE21	FL	1 (C/D)			CV1 CV13 CV28	S2 S9 S14	663	2488	CYCLOHEXYL ISOCYANATE
L4BH	TU15 TE19	AT	2 (D/E)			CV13 CV28	S9 S19	60	2490	DICHLOROISOPROPYL ETHER
L4BN		AT	3 (E)	V12				80	2491	ETHANOLAMINE or ETHANOLAMINE SOLUTION
L4BH		FL	2 (D/E)				S2 S20	338	2493	HEXAMETHYLENEIMINE
L10DH	TU3	AT	1 (B/E)			CV24 CV28	S20	568	2495	IODINE PENTAFLUORIDE
L4BN		AT	3 (E)	V12				80	2496	PROPIONIC ANHYDRIDE
LGBF		FL	3 (D/E)	V12			S2	30	2498	1,2,3,6-TETRAHYDROBENZAL-DEHYDE
L4BH	TU15 TE19	AT	2 (D/E)			CV13 CV28	S9 S19	60	2501	TRIS-(1-AZIRIDINYL) PHOSPHINE OXIDE SOLUTION
L4BH	TU15 TE19	AT	2 (E)	V12		CV13 CV28	S9	60	2501	TRIS-(1-AZIRIDINYL) PHOSPHINE OXIDE SOLUTION
L4BN		FL	2 (D/E)				S2	83	2502	VALERYL CHLORIDE
SGAV		AT	3 (E)		VC1 VC2 AP7			80	2503	ZIRCONIUM TETRACHLORIDE
L4BH	TU15 TE19	AT	2 (E)	V12		CV13 CV28	S9	60	2504	TETRABROMOETHANE
SGAH	TU15 TE19	AT	2 (E)		VC1 VC2 AP7	CV13 CV28	S9	60	2505	AMMONIUM FLUORIDE
SGAV		AT	2 (E)	V11	VC1 VC2 AP7			80	2506	AMMONIUM HYDROGEN SULPHATE
SGAV		AT	3 (E)		VC1 VC2 AP7			80	2507	CHLOROPLATINIC ACID, SOLID
SGAV		AT	3 (E)		VC1 VC2 AP7			80	2508	MOLYBDENUM PENTACHLORIDE
SGAV		AT	2 (E)	V11	VC1 VC2 AP7			80	2509	POTASSIUM HYDROGEN SULPHATE
L4BN		AT	3 (E)	V12				80	2511	2-CHLOROPROPIONIC ACID
SGAH L4BH	TU15 TE19	AT	2 (E)		VC1 VC2 AP7	CV13 CV28	S9	60	2512	AMINOPHENOLS (o-, m-, p-)
L4BN		AT	2 (E)					X80	2513	BROMOACETYL BROMIDE
LGBF		FL	3 (D/E)	V12			S2	30	2514	BROMOBENZENE
L4BH	TU15 TE19	AT	2 (E)	V12		CV13 CV28	S9	60	2515	BROMOFORM
SGAH L4BH	TU15 TE19	AT	2 (E)		VC1 VC2 AP7	CV13 CV28	S9	60	2516	CARBON TETRABROMIDE

UN No.	Name and description	Class	Classifi- cation code	Packing group	Labels	Special provi- sions	Limited and excepted quantities		Packaging			Portable tanks and bulk containers	
									Packing instruc- tions	Special packing provisions	Mixed packing provisions	Instruc- tions	Special provisions
	3.1.2	2.2	2.2	2.1.1.3	5.2.2	3.3	3.4	3.5.1.2	4.1.4	4.1.4	4.1.10	4.2.5.2 7.3.2	4.2.5.3
(1)	(2)	(3a)	(3b)	(4)	(5)	(6)	(7a)	(7b)	(8)	(9a)	(9b)	(10)	(11)
2517	1-CHLORO-1,1- DIFLUOROETHANE (REFRIGERANT GAS R 142b)	2	2F		2.1	662	0	E0	P200		MP9	(M) T50	
2518	1,5,9-CYCLODODECATRIENE	6.1	T1	III	6.1		5 L	E1	P001 IBC03 LP01 R001		MP19	T4	TP1
2520	CYCLOOCTADIENES	3	F1	III	3		5 L	E1	P001 IBC03 LP01 R001		MP19	T2	TP1
2521	DIKETENE, STABILIZED	6.1	TF1	I	6.1 +3	354 386	0	E0	P602		MP8 MP17	T20	TP2
2522	2-DIMETHYLAMINOETHYL METHACRYLATE	6.1	T1	II	6.1		100 ml	E4	P001 IBC02		MP15	T7	TP2
2524	ETHYL ORTHOFORMATE	3	F1	III	3		5 L	E1	P001 IBC03 LP01 R001		MP19	T2	TP1
2525	ETHYL OXALATE	6.1	T1	III	6.1		5 L	E1	P001 IBC03 LP01 R001		MP19	T4	TP1
2526	FURFURYLAMINE	3	FC	III	3 +8		5 L	E1	P001 IBC03 R001		MP19	T4	TP1
2527	ISOBUTYL ACRYLATE, STABILIZED	3	F1	III	3	386	5 L	E1	P001 IBC03 LP01 R001		MP19	T2	TP1
2528	ISOBUTYL ISOBUTYRATE	3	F1	III	3		5 L	E1	P001 IBC03 LP01 R001		MP19	T2	TP1
2529	ISOBUTYRIC ACID	3	FC	III	3 +8		5 L	E1	P001 IBC03 R001		MP19	T4	TP1
2531	METHACRYLIC ACID, STABILIZED	8	C3	II	8	386	1 L	E2	P001 IBC02 LP01		MP15	T7	TP2 TP18 TP30
2533	METHYL TRICHLOROACETATE	6.1	T1	III	6.1		5 L	E1	P001 IBC03 LP01 R001		MP19	T4	TP1
2534	METHYLCHLOROSILANE	2	2TFC		2.3 +2.1 +8		0	E0	P200		MP9	(M)	
2535	4-METHYLMORPHOLINE (N-METHYLMORPHOLINE)	3	FC	II	3 +8		1 L	E2	P001 IBC02		MP19	T7	TP1
2536	METHYLTETRAHYDRO-FURAN	3	F1	II	3		1 L	E2	P001 IBC02 R001		MP19	T4	TP1
2538	NITRONAPHTHALENE	4.1	F1	III	4.1		5 kg	E1	P002 IBC08 LP02 R001	B3	MP10	T1	TP33
2541	TERPINOLENE	3	F1	III	3		5 L	E1	P001 IBC03 LP01 R001		MP19	T2	TP1
2542	TRIBUTYLAMINE	6.1	T1	II	6.1		100 ml	E4	P001 IBC02		MP15	T7	TP2
2545	HAFNIUM POWDER, DRY	4.2	S4	I	4.2	540	0	E0	P404		MP13		
2545	HAFNIUM POWDER, DRY	4.2	S4	II	4.2	540	0	E2	P410 IBC06		MP14	T3	TP33
2545	HAFNIUM POWDER, DRY	4.2	S4	III	4.2	540	0	E1	P002 IBC08 LP02 R001	B3	MP14	T1	TP33
2546	TITANIUM POWDER, DRY	4.2	S4	I	4.2	540	0	E0	P404		MP13		
2546	TITANIUM POWDER, DRY	4.2	S4	II	4.2	540	0	E2	P410 IBC06		MP14	T3	TP33

ADR tank		Vehicle for tank carriage	Transport category (Tunnel restriction code)	Special provisions for carriage				Hazard identifi-cation No.	UN No.	Name and description
Tank code	Special provisions			Packages	Bulk	Loading, unloading and handling	Operation			
4.3	4.3.5, 6.8.4	9.1.1.2	1.1.3.6 (8.6)	7.2.4	7.3.3	7.5.11	8.5	5.3.2.3		3.1.2
(12)	(13)	(14)	(15)	(16)	(17)	(18)	(19)	(20)	(1)	(2)
PxBN(M)	TA4 TT9	FL	2 (B/D)			CV9 CV10 CV36	S2 S20	23	2517	1-CHLORO-1,1-DIFLUOROETHANE (REFRIGERANT GAS R 142b)
L4BH	TU15 TE19	AT	2 (E)	V12		CV13 CV28	S9	60	2518	1,5,9-CYCLODODECATRIENE
LGBF		FL	3 (D/E)	V12			S2	30	2520	CYCLOOCTADIENES
L10CH	TU14 TU15 TE19 TE21	FL	1 (C/D)	V8		CV1 CV13 CV28	S2 S4 S9 S14	663	2521	DIKETENE, STABILIZED
L4BH	TU15 TE19	AT	2 (D/E)			CV13 CV28	S9 S19	69	2522	2-DIMETHYLAMINOETHYL METHACRYLATE
LGBF		FL	3 (D/E)	V12			S2	30	2524	ETHYL ORTHOFORMATE
L4BH	TU15 TE19	AT	2 (E)	V12		CV13 CV28	S9	60	2525	ETHYL OXALATE
L4BN		FL	3 (D/E)	V12			S2	38	2526	FURFURYLAMINE
LGBF		FL	3 (D/E)	V8 V12			S2 S4	39	2527	ISOBUTYL ACRYLATE, STABILIZED
LGBF		FL	3 (D/E)	V12			S2	30	2528	ISOBUTYL ISOBUTYRATE
L4BN		FL	3 (D/E)	V12			S2	38	2529	ISOBUTYRIC ACID
L4BN		AT	2 (E)	V8			S4	89	2531	METHACRYLIC ACID, STABILIZED
L4BH	TU15 TE19	AT	2 (E)	V12		CV13 CV28	S9	60	2533	METHYL TRICHLOROACETATE
		FL	1 (B/D)			CV9 CV10 CV36	S2 S14	263	2534	METHYLCHLOROSILANE
L4BH		FL	2 (D/E)				S2 S20	338	2535	4-METHYLMORPHOLINE (N-METHYLMORPHOLINE)
LGBF		FL	2 (D/E)				S2 S20	33	2536	METHYLTETRAHYDRO-FURAN
SGAV		AT	3 (E)		VC1 VC2			40	2538	NITRONAPHTHALENE
LGBF		FL	3 (D/E)	V12			S2	30	2541	TERPINOLENE
L4BH	TU15 TE19	AT	2 (D/E)			CV13 CV28	S9 S19	60	2542	TRIBUTYLAMINE
			0 (E)	V1			S20		2545	HAFNIUM POWDER, DRY
SGAN		AT	2 (D/E)	V1				40	2545	HAFNIUM POWDER, DRY
SGAN		AT	3 (E)	V1	VC1 VC2 AP1			40	2545	HAFNIUM POWDER, DRY
			0 (E)	V1			S20		2546	TITANIUM POWDER, DRY
SGAN		AT	2 (D/E)	V1				40	2546	TITANIUM POWDER, DRY

UN No.	Name and description	Class	Classifi-cation code	Packing group	Labels	Special provi-sions	Limited and excepted quantities		Packaging			Portable tanks and bulk containers	
									Packing instruc-tions	Special packing provisions	Mixed packing provisions	Instruc-tions	Special provisions
	3.1.2	2.2	2.2	2.1.1.3	5.2.2	3.3	3.4	3.5.1.2	4.1.4	4.1.4	4.1.10	4.2.5.2 7.3.2	4.2.5.3
(1)	(2)	(3a)	(3b)	(4)	(5)	(6)	(7a)	(7b)	(8)	(9a)	(9b)	(10)	(11)
2546	TITANIUM POWDER, DRY	4.2	S4	III	4.2	540	0	E1	P002 IBC08 LP02 R001	B3	MP14	T1	TP33
2547	SODIUM SUPEROXIDE	5.1	O2	I	5.1		0	E0	P503 IBC06		MP2		
2548	CHLORINE PENTAFLUORIDE	2	2TOC		2.3 +5.1 +8		0	E0	P200		MP9		
2552	HEXAFLUOROACETONE HYDRATE, LIQUID	6.1	T1	II	6.1		100 ml	E4	P001 IBC02		MP15	T7	TP2
2554	METHYLALLYL CHLORIDE	3	F1	II	3		1 L	E2	P001 IBC02 R001		MP19	T4	TP1
2555	NITROCELLULOSE WITH WATER (not less than 25% water, by mass)	4.1	D	II	4.1	541	0	E0	P406		MP2		
2556	NITROCELLULOSE WITH ALCOHOL (not less than 25% alcohol, by mass, and not more than 12.6% nitrogen, by dry mass)	4.1	D	II	4.1	541	0	E0	P406		MP2		
2557	NITROCELLULOSE, with not more than 12.6% nitrogen, by dry mass, MIXTURE WITH or WITHOUT PLASTICIZER, WITH or WITHOUT PIGMENT	4.1	D	II	4.1	241 541	0	E0	P406		MP2		
2558	EPIBROMOHYDRIN	6.1	TF1	I	6.1 +3		0	E0	P001		MP8 MP17	T14	TP2
2560	2-METHYLPENTAN-2-OL	3	F1	III	3		5 L	E1	P001 IBC03 LP01 R001		MP19	T2	TP1
2561	3-METHYL-1-BUTENE	3	F1	I	3		0	E3	P001		MP7 MP17	T11	TP2
2564	TRICHLOROACETIC ACID SOLUTION	8	C3	II	8		1 L	E2	P001 IBC02		MP15	T7	TP2
2564	TRICHLOROACETIC ACID SOLUTION	8	C3	III	8		5 L	E1	P001 IBC03 LP01 R001		MP19	T4	TP1
2565	DICYCLOHEXYLAMINE	8	C7	III	8		5 L	E1	P001 IBC03 LP01 R001		MP19	T4	TP1
2567	SODIUM PENTACHLOROPHENATE	6.1	T2	II	6.1		500 g	E4	P002 IBC08	B4	MP10	T3	TP33
2570	CADMIUM COMPOUND	6.1	T5	I	6.1	274 596	0	E5	P002 IBC07		MP18	T6	TP33
2570	CADMIUM COMPOUND	6.1	T5	II	6.1	274 596	500 g	E4	P002 IBC08	B4	MP10	T3	TP33
2570	CADMIUM COMPOUND	6.1	T5	III	6.1	274 596	5 kg	E1	P002 IBC08 LP02 R001	B3	MP10	T1	TP33
2571	ALKYLSULPHURIC ACIDS	8	C3	II	8		1 L	E2	P001 IBC02		MP15	T8	TP2 TP28
2572	PHENYLHYDRAZINE	6.1	T1	II	6.1		100 ml	E4	P001 IBC02		MP15	T7	TP2
2573	THALLIUM CHLORATE	5.1	OT2	II	5.1 +6.1		1 kg	E2	P002 IBC06		MP2	T3	TP33
2574	TRICRESYL PHOSPHATE with more than 3% ortho isomer	6.1	T1	II	6.1		100 ml	E4	P001 IBC02		MP15	T7	TP2
2576	PHOSPHORUS OXYBROMIDE, MOLTEN	8	C1	II	8		0	E0				T7	TP3
2577	PHENYLACETYL CHLORIDE	8	C3	II	8		1 L	E2	P001 IBC02		MP15	T7	TP2
2578	PHOSPHORUS TRIOXIDE	8	C2	III	8		5 kg	E1	P002 IBC08 LP02 R001	B3	MP10	T1	TP33

ADR tank		Vehicle for tank carriage	Transport category (Tunnel restriction code)	Special provisions for carriage				Hazard identifi-cation No.	UN No.	Name and description
Tank code	Special provisions			Packages	Bulk	Loading, unloading and handling	Operation			
4.3	4.3.5, 6.8.4	9.1.1.2	1.1.3.6 (8.6)	7.2.4	7.3.3	7.5.11	8.5	5.3.2.3		3.1.2
(12)	(13)	(14)	(15)	(16)	(17)	(18)	(19)	(20)	(1)	(2)
SGAN		AT	3 (E)	V1	VC1 VC2 AP1			40	2546	TITANIUM POWDER, DRY
			1 (E)	V10		CV24	S20		2547	SODIUM SUPEROXIDE
			1 (D)			CV9 CV10 CV36	S14		2548	CHLORINE PENTAFLUORIDE
L4BH	TU15 TE19	AT	2 (D/E)			CV13 CV28	S9 S19	60	2552	HEXAFLUOROACETONE HYDRATE, LIQUID
LGBF		FL	2 (D/E)				S2 S20	33	2554	METHYLALLYL CHLORIDE
			2 (B)				S14		2555	NITROCELLULOSE WITH WATER (not less than 25% water, by mass)
			2 (B)				S14		2556	NITROCELLULOSE WITH ALCOHOL (not less than 25% alcohol, by mass, and not more than 12.6% nitrogen, by dry mass)
			2 (B)				S14		2557	NITROCELLULOSE, with not more than 12.6% nitrogen, by dry mass, MIXTURE WITH or WITHOUT PLASTICIZER, WITH or WITHOUT PIGMENT
L10CH	TU14 TU15 TE19 TE21	FL	1 (C/D)			CV1 CV13 CV28	S2 S9 S14	663	2558	EPIBROMOHYDRIN
LGBF		FL	3 (D/E)	V12			S2	30	2560	2-METHYLPENTAN-2-OL
L4BN		FL	1 (D/E)				S2 S20	33	2561	3-METHYL-1-BUTENE
L4BN		AT	2 (E)					80	2564	TRICHLOROACETIC ACID SOLUTION
L4BN		AT	3 (E)	V12				80	2564	TRICHLOROACETIC ACID SOLUTION
L4BN		AT	3 (E)	V12				80	2565	DICYCLOHEXYLAMINE
SGAH	TU15 TE19	AT	2 (D/E)	V11		CV13 CV28	S9 S19	60	2567	SODIUM PENTACHLOROPHENATE
S10AH L10CH	TU14 TU15 TE19 TE21	AT	1 (C/E)	V10		CV1 CV13 CV28	S9 S14	66	2570	CADMIUM COMPOUND
SGAH L4BH	TU15 TE19	AT	2 (D/E)	V11		CV13 CV28	S9 S19	60	2570	CADMIUM COMPOUND
SGAH L4BH	TU15 TE19	AT	2 (E)		VC1 VC2 AP7	CV13 CV28	S9	60	2570	CADMIUM COMPOUND
L4BN		AT	2 (E)					80	2571	ALKYLSULPHURIC ACIDS
L4BH	TU15 TE19	AT	2 (D/E)			CV13 CV28	S9 S19	60	2572	PHENYLHYDRAZINE
SGAN	TU3	AT	2 (E)	V11		CV24 CV28		56	2573	THALLIUM CHLORATE
L4BH	TU15 TE19	AT	2 (D/E)			CV13 CV28	S9 S19	60	2574	TRICRESYL PHOSPHATE with more than 3% ortho isomer
L4BN		AT	2 (E)					80	2576	PHOSPHORUS OXYBROMIDE, MOLTEN
L4BN		AT	2 (E)					80	2577	PHENYLACETYL CHLORIDE
SGAV		AT	3 (E)		VC1 VC2 AP7			80	2578	PHOSPHORUS TRIOXIDE

UN No.	Name and description	Class	Classifi-cation code	Packing group	Labels	Special provi-sions	Limited and excepted quantities		Packaging			Portable tanks and bulk containers	
									Packing instruc-tions	Special packing provisions	Mixed packing provisions	Instruc-tions	Special provisions
	3.1.2	2.2	2.2	2.1.1.3	5.2.2	3.3	3.4	3.5.1.2	4.1.4	4.1.4	4.1.10	4.2.5.2 7.3.2	4.2.5.3
(1)	(2)	(3a)	(3b)	(4)	(5)	(6)	(7a)	(7b)	(8)	(9a)	(9b)	(10)	(11)
2579	PIPERAZINE	8	C8	III	8		5 kg	E1	P002 IBC08 LP02 R001	B3	MP10	T1	TP33
2580	ALUMINIUM BROMIDE SOLUTION	8	C1	III	8		5 L	E1	P001 IBC03 LP01 R001		MP19	T4	TP1
2581	ALUMINIUM CHLORIDE SOLUTION	8	C1	III	8		5 L	E1	P001 IBC03 LP01 R001		MP19	T4	TP1
2582	FERRIC CHLORIDE SOLUTION	8	C1	III	8		5 L	E1	P001 IBC03 LP01 R001		MP19	T4	TP1
2583	ALKYLSULPHONIC ACIDS, SOLID or ARYLSULPHONIC ACIDS, SOLID with more than 5% free sulphuric acid	8	C2	II	8		1 kg	E2	P002 IBC08	B4	MP10	T3	TP33
2584	ALKYLSULPHONIC ACIDS, LIQUID or ARYLSULPHONIC ACIDS, LIQUID with more than 5% free sulphuric acid	8	C1	II	8		1 L	E2	P001 IBC02		MP15	T8	TP2
2585	ALKYLSULPHONIC ACIDS, SOLID or ARYLSULPHONIC ACIDS, SOLID with not more than 5% free sulphuric acid	8	C4	III	8		5 kg	E1	P002 IBC08 LP02 R001	B3	MP10	T1	TP33
2586	ALKYLSULPHONIC ACIDS, LIQUID or ARYLSULPHONIC ACIDS, LIQUID with not more than 5% free sulphuric acid	8	C3	III	8		5 L	E1	P001 IBC03 LP01 R001		MP19	T4	TP1
2587	BENZOQUINONE	6.1	T2	II	6.1		500 g	E4	P002 IBC08	B4	MP10	T3	TP33
2588	PESTICIDE, SOLID, TOXIC, N.O.S.	6.1	T7	I	6.1	61 274 648	0	E5	P002 IBC02		MP18	T6	TP33
2588	PESTICIDE, SOLID, TOXIC, N.O.S.	6.1	T7	II	6.1	61 274 648	500 g	E4	P002 IBC08	B4	MP10	T3	TP33
2588	PESTICIDE, SOLID, TOXIC, N.O.S.	6.1	T7	III	6.1	61 274 648	5 kg	E1	P002 IBC08 LP02 R001	B3	MP10	T1	TP33
2589	VINYL CHLOROACETATE	6.1	TF1	II	6.1 +3		100 ml	E4	P001 IBC02		MP15	T7	TP2
2590	ASBESTOS, CHRYSOTILE	9	M1	III	9	168 542	5 kg	E1	P002 IBC08 R001	PP37 B4	MP10	T1	TP33
2591	XENON, REFRIGERATED LIQUID	2	3A		2.2	593	120 ml	E1	P203		MP9	T75	TP5
2599	CHLOROTRIFLUORO-METHANE AND TRIFLUOROMETHANE AZEOTROPIC MIXTURE with approximately 60% chlorotrifluoromethane (REFRIGERANT GAS R 503)	2	2A		2.2	662	120 ml	E1	P200		MP9	(M)	
2601	CYCLOBUTANE	2	2F		2.1	662	0	E0	P200		MP9	(M)	
2602	DICHLORODIFLUORO-METHANE AND 1,1-DIFLUOROETHANE AZEOTROPIC MIXTURE with approximately 74% dichlorodifluoromethane (REFRIGERANT GAS R 500)	2	2A		2.2	662	120 ml	E1	P200		MP9	(M) T50	
2603	CYCLOHEPTATRIENE	3	FT1	II	3 +6.1		1 L	E2	P001 IBC02		MP19	T7	TP1
2604	BORON TRIFLUORIDE DIETHYL ETHERATE	8	CF1	I	8 +3		0	E0	P001		MP8 MP17	T10	TP2
2605	METHOXYMETHYL ISOCYANATE	6.1	TF1	I	6.1 +3	354	0	E0	P602		MP8 MP17	T20	TP2

ADR tank		Vehicle for tank carriage	Transport category (Tunnel restriction code)	Special provisions for carriage				Hazard identifi-cation No.	UN No.	Name and description
Tank code	Special provisions			Packages	Bulk	Loading, unloading and handling	Operation			
4.3	4.3.5, 6.8.4	9.1.1.2	1.1.3.6 (8.6)	7.2.4	7.3.3	7.5.11	8.5	5.3.2.3		3.1.2
(12)	(13)	(14)	(15)	(16)	(17)	(18)	(19)	(20)	(1)	(2)
SGAV L4BN		AT	3 (E)		VC1 VC2 AP7			80	2579	PIPERAZINE
L4BN		AT	3 (E)	V12				80	2580	ALUMINIUM BROMIDE SOLUTION
L4BN		AT	3 (E)	V12				80	2581	ALUMINIUM CHLORIDE SOLUTION
L4BN		AT	3 (E)	V12				80	2582	FERRIC CHLORIDE SOLUTION
SGAN L4BN		AT	2 (E)	V11				80	2583	ALKYLSULPHONIC ACIDS, SOLID or ARYLSULPHONIC ACIDS, SOLID with more than 5% free sulphuric acid
L4BN		AT	2 (E)					80	2584	ALKYLSULPHONIC ACIDS, LIQUID or ARYLSULPHONIC ACIDS, LIQUID with more than 5% free sulphuric acid
SGAV		AT	3 (E)		VC1 VC2 AP7			80	2585	ALKYLSULPHONIC ACIDS, SOLID or ARYLSULPHONIC ACIDS, SOLID with not more than 5% free sulphuric acid
L4BN		AT	3 (E)	V12				80	2586	ALKYLSULPHONIC ACIDS, LIQUID or ARYLSULPHONIC ACIDS, LIQUID with not more than 5% free sulphuric acid
SGAH L4BH	TU15 TE19	AT	2 (D/E)	V11		CV13 CV28	S9 S19	60	2587	BENZOQUINONE
S10AH L10CH	TU14 TU15 TE19 TE21	AT	1 (C/E)			CV1 CV13 CV28	S9 S14	66	2588	PESTICIDE, SOLID, TOXIC, N.O.S.
SGAH L4BH	TU15 TE19	AT	2 (D/E)	V11		.CV13 CV28	S9 S19	60	2588	PESTICIDE, SOLID, TOXIC, N.O.S.
SGAH L4BH	TU15 TE19	AT	2 (E)		VC1 VC2 AP7	CV13 CV28	S9	60	2588	PESTICIDE, SOLID, TOXIC, N.O.S.
L4BH	TU15 TE19	FL	2 (D/E)			CV13 CV28	S2 S9 S19	63	2589	VINYL CHLOROACETATE
SGAH	TU15	AT	3 (E)	V11		CV13 CV28		90	2590	ASBESTOS, CHRYSOTILE
RxBN	TU19 TA4 TT9	AT	3 (C/E)	V5		CV9 CV11 CV36	S20	22	2591	XENON, REFRIGERATED LIQUID
PxBN(M)	TA4 TT9	AT	3 (C/E)			CV9 CV10 CV36		20	2599	CHLOROTRIFLUORO-METHANE AND TRIFLUOROMETHANE AZEOTROPIC MIXTURE with approximately 60% chlorotrifluoromethane (REFRIGERANT GAS R 503)
PxBN(M)	TA4 TT9	FL	2 (B/D)			CV9 CV10 CV36	S2 S20	23	2601	CYCLOBUTANE
PxBN(M)	TA4 TT9	AT	3 (C/E)			CV9 CV10 CV36		20	2602	DICHLORODIFLUORO-METHANE AND 1,1-DIFLUOROETHANE AZEOTROPIC MIXTURE with approximately 74% dichlorodifluoromethane (REFRIGERANT GAS R 500)
L4BH	TU15	FL	2 (D/E)			CV13 CV28	S2 S19	336	2603	CYCLOHEPTATRIENE
L10BH		FL	1 (D/E)				S2 S14	883	2604	BORON TRIFLUORIDE DIETHYL ETHERATE
L10CH	TU14 TU15 TE19 TE21	FL	1 (C/D)			CV1 CV13 CV28	S2 S9 S14	663	2605	METHOXYMETHYL ISOCYANATE

UN No.	Name and description	Class	Classifi- cation code	Packing group	Labels	Special provi- sions	Limited and excepted quantities		Packaging			Portable tanks and bulk containers	
									Packing instruc- tions	Special packing provisions	Mixed packing provisions	Instruc- tions	Special provisions
	3.1.2	2.2	2.2	2.1.1.3	5.2.2	3.3	3.4	3.5.1.2	4.1.4	4.1.4	4.1.10	4.2.5.2 7.3.2	4.2.5.3
(1)	(2)	(3a)	(3b)	(4)	(5)	(6)	(7a)	(7b)	(8)	(9a)	(9b)	(10)	(11)
2606	METHYL ORTHOSILICATE	6.1	TF1	I	6.1 +3	354	0	E0	P602		MP8 MP17	T20	TP2
2607	ACROLEIN DIMER, STABILIZED	3	F1	III	3	386	5 L	E1	P001 IBC03 LP01 R001		MP19	T2	TP1
2608	NITROPROPANES	3	F1	III	3		5 L	E1	P001 IBC03 LP01 R001		MP19	T2	TP1
2609	TRIALLYL BORATE	6.1	T1	III	6.1		5 L	E1	P001 IBC03 LP01 R001		MP19		
2610	TRIALLYLAMINE	3	FC	III	3 +8		5 L	E1	P001 IBC03 R001		MP19	T4	TP1
2611	PROPYLENE CHLOROHYDRIN	6.1	TF1	II	6.1 +3		100 ml	E4	P001 IBC02		MP15	T7	TP2
2612	METHYL PROPYL ETHER	3	F1	II	3		1 L	E2	P001 IBC02	B8	MP19	T7	TP2
2614	METHALLYL ALCOHOL	3	F1	III	3		5 L	E1	P001 IBC03 LP01 R001		MP19	T2	TP1
2615	ETHYL PROPYL ETHER	3	F1	II	3		1 L	E2	P001 IBC02 R001		MP19	T4	TP1
2616	TRIISOPROPYL BORATE	3	F1	II	3		1 L	E2	P001 IBC02 R001		MP19	T4	TP1
2616	TRIISOPROPYL BORATE	3	F1	III	3		5 L	E1	P001 IBC03 LP01 R001		MP19	T2	TP1
2617	METHYLCYCLO-HEXANOLS, flammable	3	F1	III	3		5 L	E1	P001 IBC03 LP01 R001		MP19	T2	TP1
2618	VINYLTOLUENES, STABILIZED	3	F1	III	3	386	5 L	E1	P001 IBC03 LP01 R001		MP19	T2	TP1
2619	BENZYLDIMETHYLAMINE	8	CF1	II	8 +3		1 L	E2	P001 IBC02		MP15	T7	TP2
2620	AMYL BUTYRATES	3	F1	III	3		5 L	E1	P001 IBC03 LP01 R001		MP19	T2	TP1
2621	ACETYL METHYL CARBINOL	3	F1	III	3		5 L	E1	P001 IBC03 LP01 R001		MP19	T2	TP1
2622	GLYCIDALDEHYDE	3	FT1	II	3 +6.1		1 L	E2	P001 IBC02	B8	MP19	T7	TP1
2623	FIRELIGHTERS, SOLID with flammable liquid	4.1	F1	III	4.1		5 kg	E1	P002 LP02 R001	PP15	MP11		
2624	MAGNESIUM SILICIDE	4.3	W2	II	4.3		500 g	E2	P410 IBC07		MP14	T3	TP33
2626	CHLORIC ACID, AQUEOUS SOLUTION with not more than 10% chloric acid	5.1	O1	II	5.1	613	1 L	E0	P504 IBC02		MP2	T4	TP1
2627	NITRITES, INORGANIC, N.O.S.	5.1	O2	II	5.1	103 274	1 kg	E2	P002 IBC08	B4	MP10	T3	TP33
2628	POTASSIUM FLUOROACETATE	6.1	T2	I	6.1		0	E5	P002 IBC07		MP18	T6	TP33
2629	SODIUM FLUOROACETATE	6.1	T2	I	6.1		0	E5	P002 IBC07		MP18	T6	TP33
2630	SELENATES or SELENITES	6.1	T5	I	6.1	274	0	E5	P002 IBC07		MP18	T6	TP33

ADR tank		Vehicle for tank carriage	Transport category (Tunnel restriction code)	Special provisions for carriage				Hazard identifi-cation No.	UN No.	Name and description
Tank code	Special provisions			Packages	Bulk	Loading, unloading and handling	Operation			
4.3	4.3.5, 6.8.4	9.1.1.2	1.1.3.6 (8.6)	7.2.4	7.3.3	7.5.11	8.5	5.3.2.3		3.1.2
(12)	(13)	(14)	(15)	(16)	(17)	(18)	(19)	(20)	(1)	(2)
L10CH	TU14 TU15 TE19 TE21	FL	1 (C/D)			CV1 CV13 CV28	S2 S9 S14	663	2606	METHYL ORTHOSILICATE
LGBF		FL	3 (D/E)	V8			S2 S4	39	2607	ACROLEIN DIMER, STABILIZED
LGBF		FL	3 (D/E)	V12			S2	30	2608	NITROPROPANES
L4BH	TU15 TE19	AT	2 (E)	V12		CV13 CV28	S9	60	2609	TRIALLYL BORATE
L4BN		FL	3 (D/E)	V12			S2	38	2610	TRIALLYLAMINE
L4BH	TU15 TE19	FL	2 (D/E)			CV13 CV28	S2 S9 S19	63	2611	PROPYLENE CHLOROHYDRIN
L1.5BN		FL	2 (D/E)				S2 S20	33	2612	METHYL PROPYL ETHER
LGBF		FL	3 (D/E)	V12			S2	30	2614	METHALLYL ALCOHOL
LGBF		FL	2 (D/E)				S2 S20	33	2615	ETHYL PROPYL ETHER
LGBF		FL	2 (D/E)				S2 S20	33	2616	TRIISOPROPYL BORATE
LGBF		FL	3 (D/E)	V12			S2	30	2616	TRIISOPROPYL BORATE
LGBF		FL	3 (D/E)	V12			S2	30	2617	METHYLCYCLO-HEXANOLS, flammable
LGBF		FL	3 (D/E)	V8 V12			S2 S4	39	2618	VINYLTOLUENES, STABILIZED
L4BN		FL	2 (D/E)				S2	83	2619	BENZYLDIMETHYLAMINE
LGBF		FL	3 (D/E)	V12			S2	30	2620	AMYL BUTYRATES
LGBF		FL	3 (D/E)	V12			S2	30	2621	ACETYL METHYL CARBINOL
L4BH	TU15	FL	2 (D/E)			CV13 CV28	S2 S19	336	2622	GLYCIDALDEHYDE
			4 (E)						2623	FIRELIGHTERS, SOLID with flammable liquid
SGAN		AT	2 (D/E)	V1		CV23		423	2624	MAGNESIUM SILICIDE
L4BN	TU3	AT	2 (E)			CV24		50	2626	CHLORIC ACID, AQUEOUS SOLUTION with not more than 10% chloric acid
SGAN	TU3	AT	2 (E)	V11		CV24		50	2627	NITRITES, INORGANIC, N.O.S.
S10AH	TU15 TE19	AT	1 (C/E)	V10		CV1 CV13 CV28	S9 S14	66	2628	POTASSIUM FLUOROACETATE
S10AH	TU15 TE19	AT	1 (C/E)	V10		CV1 CV13 CV28	S9 S14	66	2629	SODIUM FLUOROACETATE
S10AH L10CH	TU14 TU15 TE19 TE21	AT	1 (C/E)	V10		CV1 CV13 CV28	S9 S14	66	2630	SELENATES or SELENITES

UN No.	Name and description	Class	Classifi-cation code	Packing group	Labels	Special provi-sions	Limited and excepted quantities		Packaging			Portable tanks and bulk containers	
									Packing instruc-tions	Special packing provisions	Mixed packing provisions	Instruc-tions	Special provisions
	3.1.2	2.2	2.2	2.1.1.3	5.2.2	3.3	3.4	3.5.1.2	4.1.4	4.1.4	4.1.10	4.2.5.2 7.3.2	4.2.5.3
(1)	(2)	(3a)	(3b)	(4)	(5)	(6)	(7a)	(7b)	(8)	(9a)	(9b)	(10)	(11)
2642	FLUOROACETIC ACID	6.1	T2	I	6.1		0	E5	P002 IBC07		MP18	T6	TP33
2643	METHYL BROMOACETATE	6.1	T1	II	6.1		100 ml	E4	P001 IBC02		MP15	T7	TP2
2644	METHYL IODIDE	6.1	T1	I	6.1	354	0	E0	P602		MP8 MP17	T20	TP2
2645	PHENACYL BROMIDE	6.1	T2	II	6.1		500 g	E4	P002 IBC08	B4	MP10	T3	TP33
2646	HEXACHLOROCYCLO-PENTADIENE	6.1	T1	I	6.1	354	0	E0	P602		MP8 MP17	T20	TP2
2647	MALONONITRILE	6.1	T2	II	6.1		500 g	E4	P002 IBC08	B4	MP10	T3	TP33
2648	1,2-DIBROMOBUTAN-3-ONE	6.1	T1	II	6.1		100 ml	E4	P001 IBC02		MP15		
2649	1,3-DICHLOROACETONE	6.1	T2	II	6.1		500 g	E4	P002 IBC08	B4	MP10	T3	TP33
2650	1,1-DICHLORO-1-NITROETHANE	6.1	T1	II	6.1		100 ml	E4	P001 IBC02		MP15	T7	TP2
2651	4,4'-DIAMINODIPHENYL-METHANE	6.1	T2	III	6.1		5 kg	E1	P002 IBC08 LP02 R001	B3	MP10	T1	TP33
2653	BENZYL IODIDE	6.1	T1	II	6.1		100 ml	E4	P001 IBC02		MP15	T7	TP2
2655	POTASSIUM FLUOROSILICATE	6.1	T5	III	6.1		5 kg	E1	P002 IBC08 LP02 R001	B3	MP10	T1	TP33
2656	QUINOLINE	6.1	T1	III	6.1		5 L	E1	P001 IBC03 LP01 R001		MP19	T4	TP1
2657	SELENIUM DISULPHIDE	6.1	T5	II	6.1		500 g	E4	P002 IBC08	B4	MP10	T3	TP33
2659	SODIUM CHLOROACETATE	6.1	T2	III	6.1		5 kg	E1	P002 IBC08 LP02 R001	B3	MP10	T1	TP33
2660	NITROTOLUIDINES (MONO)	6.1	T2	III	6.1		5 kg	E1	P002 IBC08 LP02 R001	B3	MP10	T1	TP33
2661	HEXACHLOROACETONE	6.1	T1	III	6.1		5 L	E1	P001 IBC03 LP01 R001		MP19	T4	TP1
2664	DIBROMOMETHANE	6.1	T1	III	6.1		5 L	E1	P001 IBC03 LP01 R001		MP19	T4	TP1
2667	BUTYLTOLUENES	6.1	T1	III	6.1		5 L	E1	P001 IBC03 LP01 R001		MP19	T4	TP1
2668	CHLOROACETONITRILE	6.1	TF1	I	6.1 +3	354	0	E0	P602		MP8 MP17	T20	TP2
2669	CHLOROCRESOLS SOLUTION	6.1	T1	II	6.1		100 ml	E4	P001 IBC02		MP15	T7	TP2
2669	CHLOROCRESOLS SOLUTION	6.1	T1	III	6.1		5 L	E1	P001 IBC03 LP01 R001		MP19	T7	TP2
2670	CYANURIC CHLORIDE	8	C4	II	8		1 kg	E2	P002 IBC08	B4	MP10	T3	TP33
2671	AMINOPYRIDINES (o-, m-, p-)	6.1	T2	II	6.1		500 g	E4	P002 IBC08	B4	MP10	T3	TP33
2672	AMMONIA SOLUTION, relative density between 0.880 and 0.957 at 15 °C in water, with more than 10% but not more than 35% ammonia	8	C5	III	8	543	5 L	E1	P001 IBC03 LP01 R001		MP19	T7	TP1

ADR tank		Vehicle for tank carriage	Transport category (Tunnel restriction code)	Special provisions for carriage				Hazard identification No.	UN No.	Name and description
Tank code	Special provisions			Packages	Bulk	Loading, unloading and handling	Operation			
4.3	4.3.5, 6.8.4	9.1.1.2	1.1.3.6 (8.6)	7.2.4	7.3.3	7.5.11	8.5	5.3.2.3		3.1.2
(12)	(13)	(14)	(15)	(16)	(17)	(18)	(19)	(20)	(1)	(2)
S10AH L10CH	TU14 TU15 TE19 TE21	AT	1 (C/E)	V10		CV1 CV13 CV28	S9 S14	66	2642	FLUOROACETIC ACID
L4BH	TU15 TE19	AT	2 (D/E)			CV13 CV28	S9 S19	60	2643	METHYL BROMOACETATE
L10CH	TU14 TU15 TE19 TE21	AT	1 (C/D)			CV1 CV13 CV28	S9 S14	66	2644	METHYL IODIDE
SGAH L4BH	TU15 TE19	AT	2 (D/E)	V11		CV13 CV28	S9 S19	60	2645	PHENACYL BROMIDE
L10CH	TU14 TU15 TE19 TE21	AT	1 (C/D)			CV1 CV13 CV28	S9 S14	66	2646	HEXACHLOROCYCLO-PENTADIENE
SGAH L4BH	TU15 TE19	AT	2 (D/E)	V11		CV13 CV28	S9 S19	60	2647	MALONONITRILE
L4BH	TU15 TE19	AT	2 (D/E)			CV13 CV28	S9 S19	60	2648	1,2-DIBROMOBUTAN-3-ONE
SGAH L4BH	TU15 TE19	AT	2 (D/E)	V11		CV13 CV28	S9 S19	60	2649	1,3-DICHLOROACETONE
L4BH	TU15 TE19	AT	2 (D/E)			CV13 CV28	S9 S19	60	2650	1,1-DICHLORO-1-NITROETHANE
SGAH L4BH	TU15 TE19	AT	2 (E)		VC1 VC2 AP7	CV13 CV28	S9	60	2651	4,4'-DIAMINODIPHENYL-METHANE
L4BH	TU15 TE19	AT	2 (D/E)			CV13 CV28	S9 S19	60	2653	BENZYL IODIDE
SGAH L4BH	TU15 TE19	AT	2 (E)		VC1 VC2 AP7	CV13 CV28	S9	60	2655	POTASSIUM FLUOROSILICATE
L4BH	TU15 TE19	AT	2 (E)	V12		CV13 CV28	S9	60	2656	QUINOLINE
SGAH L4BH	TU15 TE19	AT	2 (D/E)	V11		CV13 CV28	S9 S19	60	2657	SELENIUM DISULPHIDE
SGAH	TU15 TE19	AT	2 (E)		VC1 VC2 AP7	CV13 CV28	S9	60	2659	SODIUM CHLOROACETATE
SGAH L4BH	TU15 TE19	AT	2 (E)		VC1 VC2 AP7	CV13 CV28	S9	60	2660	NITROTOLUIDINES (MONO)
L4BH	TU15 TE19	AT	2 (E)	V12		CV13 CV28	S9	60	2661	HEXACHLOROACETONE
L4BH	TU15 TE19	AT	2 (E)	V12		CV13 CV28	S9	60	2664	DIBROMOMETHANE
L4BH	TU15 TE19	AT	2 (E)	V12		CV13 CV28	S9	60	2667	BUTYLTOLUENES
L10CH	TU14 TU15 TE19 TE21	FL	1 (C/D)			CV1 CV13 CV28	S2 S9 S14	663	2668	CHLOROACETONITRILE
L4BH	TU15 TE19	AT	2 (D/E)			CV13 CV28	S9 S19	60	2669	CHLOROCRESOLS SOLUTION
L4BH	TU15 TE19	AT	2 (E)	V12		CV13 CV28	S9	60	2669	CHLOROCRESOLS SOLUTION
SGAN L4BN		AT	2 (E)	V11				80	2670	CYANURIC CHLORIDE
SGAH L4BH	TU15 TE19	AT	2 (D/E)	V11		CV13 CV28	S9 S19	60	2671	AMINOPYRIDINES (o-, m-, p-)
L4BN		AT	3 (E)	V12				80	2672	AMMONIA SOLUTION, relative density between 0.880 and 0.957 at 15 °C in water, with more than 10% but not more than 35% ammonia

UN No.	Name and description	Class	Classification code	Packing group	Labels	Special provisions	Limited and excepted quantities		Packaging			Portable tanks and bulk containers	
									Packing instructions	Special packing provisions	Mixed packing provisions	Instructions	Special provisions
3.1.2	3.1.2	2.2	2.2	2.1.1.3	5.2.2	3.3	3.4	3.5.1.2	4.1.4	4.1.4	4.1.10	4.2.5.2 7.3.2	4.2.5.3
(1)	(2)	(3a)	(3b)	(4)	(5)	(6)	(7a)	(7b)	(8)	(9a)	(9b)	(10)	(11)
2673	2-AMINO-4-CHLOROPHENOL	6.1	T2	II	6.1		500 g	E4	P002 IBC08	B4	MP10	T3	TP33
2674	SODIUM FLUOROSILICATE	6.1	T5	III	6.1		5 kg	E1	P002 IBC08 LP02 R001	B3	MP10	T1	TP33
2676	STIBINE	2	2TF		2.3 +2.1		0	E0	P200		MP9		
2677	RUBIDIUM HYDROXIDE SOLUTION	8	C5	II	8		1 L	E2	P001 IBC02		MP15	T7	TP2
2677	RUBIDIUM HYDROXIDE SOLUTION	8	C5	III	8		5 L	E1	P001 IBC03 LP01 R001		MP19	T4	TP1
2678	RUBIDIUM HYDROXIDE	8	C6	II	8		1 kg	E2	P002 IBC08	B4	MP10	T3	TP33
2679	LITHIUM HYDROXIDE SOLUTION	8	C5	II	8		1 L	E2	P001 IBC02		MP15	T7	TP2
2679	LITHIUM HYDROXIDE SOLUTION	8	C5	III	8		5 L	E1	P001 IBC03 LP01 R001		MP19	T4	TP2
2680	LITHIUM HYDROXIDE	8	C6	II	8		1 kg	E2	P002 IBC08	B4	MP10	T3	TP33
2681	CAESIUM HYDROXIDE SOLUTION	8	C5	II	8		1 L	E2	P001 IBC02		MP15	T7	TP2
2681	CAESIUM HYDROXIDE SOLUTION	8	C5	III	8		5 L	E1	P001 IBC03 LP01 R001		MP19	T4	TP1
2682	CAESIUM HYDROXIDE	8	C6	II	8		1 kg	E2	P002 IBC08	B4	MP10	T3	TP33
2683	AMMONIUM SULPHIDE SOLUTION	8	CFT	II	8 +3 +6.1		1 L	E2	P001 IBC01		MP15	T7	TP2
2684	3-DIETHYLAMINOPROPYL-AMINE	3	FC	III	3 +8		5 L	E1	P001 IBC03 R001		MP19	T4	TP1
2685	N,N-DIETHYLETHYLENE-DIAMINE	8	CF1	II	8 +3		1 L	E2	P001 IBC02		MP15	T7	TP2
2686	2-DIETHYLAMINO-ETHANOL	8	CF1	II	8 +3		1 L	E2	P001 IBC02		MP15	T7	TP2
2687	DICYCLOHEXYL-AMMONIUM NITRITE	4.1	F3	III	4.1		5 kg	E1	P002 IBC08 LP02 R001	B3	MP11	T1	TP33
2688	1-BROMO-3-CHLOROPROPANE	6.1	T1	III	6.1		5 L	E1	P001 IBC03 LP01 R001		MP19	T4	TP1
2689	GLYCEROL alpha-MONOCHLOROHYDRIN	6.1	T1	III	6.1		5 L	E1	P001 IBC03 LP01 R001		MP19	T4	TP1
2690	N,n-BUTYLIMIDAZOLE	6.1	T1	II	6.1		100 ml	E4	P001 IBC02		MP15	T7	TP2
2691	PHOSPHORUS PENTABROMIDE	8	C2	II	8		1 kg	E0	P002 IBC08	B4	MP10	T3	TP33
2692	BORON TRIBROMIDE	8	C1	I	8		0	E0	P602		MP8 MP17	T20	TP2
2693	BISULPHITES, AQUEOUS SOLUTION, N.O.S.	8	C1	III	8	274	5 L	E1	P001 IBC03 LP01 R001		MP19	T7	TP1 TP28
2698	TETRAHYDROPHTHALIC ANHYDRIDES with more than 0.05% of maleic anhydride	8	C4	III	8	169	5 kg	E1	P002 IBC08 LP02 R001	PP14 B3	MP10	T1	TP33
2699	TRIFLUOROACETIC ACID	8	C3	I	8		0	E0	P001		MP8 MP17	T10	TP2
2705	1-PENTOL	8	C9	II	8		1 L	E2	P001 IBC02		MP15	T7	TP2
2707	DIMETHYLDIOXANES	3	F1	II	3		1 L	E2	P001 IBC02 R001		MP19	T4	TP1

ADR tank		Vehicle for tank carriage	Transport category (Tunnel restriction code)	Special provisions for carriage				Hazard identification No.	UN No.	Name and description
Tank code	Special provisions			Packages	Bulk	Loading, unloading and handling	Operation			
4.3	4.3.5, 6.8.4	9.1.1.2	1.1.3.6 (8.6)	7.2.4	7.3.3	7.5.11	8.5	5.3.2.3		3.1.2
(12)	(13)	(14)	(15)	(16)	(17)	(18)	(19)	(20)	(1)	(2)
SGAH L4BH	TU15 TE19	AT	2 (D/E)	V11		CV13 CV28	S9 S19	60	2673	2-AMINO-4-CHLOROPHENOL
SGAH L4BH	TU15 TE19	AT	2 (E)		VC1 VC2 AP7	CV13 CV28	S9	60	2674	SODIUM FLUOROSILICATE
			1 (D)			CV9 CV10 CV36	S2 S14		2676	STIBINE
L4BN		AT	2 (E)					80	2677	RUBIDIUM HYDROXIDE SOLUTION
L4BN		AT	3 (E)	V12				80	2677	RUBIDIUM HYDROXIDE SOLUTION
SGAN		AT	2 (E)	V11				80	2678	RUBIDIUM HYDROXIDE
L4BN		AT	2 (E)					80	2679	LITHIUM HYDROXIDE SOLUTION
L4BN		AT	3 (E)	V12				80	2679	LITHIUM HYDROXIDE SOLUTION
SGAN		AT	2 (E)	V11				80	2680	LITHIUM HYDROXIDE
L4BN		AT	2 (E)					80	2681	CAESIUM HYDROXIDE SOLUTION
L4BN		AT	3 (E)	V12				80	2681	CAESIUM HYDROXIDE SOLUTION
SGAN		AT	2 (E)	V11				80	2682	CAESIUM HYDROXIDE
L4BN		FL	2 (D/E)			CV13 CV28	S2	86	2683	AMMONIUM SULPHIDE SOLUTION
L4BN		FL	3 (D/E)	V12			S2	38	2684	3-DIETHYLAMINOPROPYL-AMINE
L4BN		FL	2 (D/E)				S2	83	2685	N,N-DIETHYLETHYLENE-DIAMINE
L4BN		FL	2 (D/E)				S2	83	2686	2-DIETHYLAMINO-ETHANOL
SGAV		AT	3 (E)		VC1 VC2			40	2687	DICYCLOHEXYL-AMMONIUM NITRITE
L4BH	TU15 TE19	AT	2 (E)	V12		CV13 CV28	S9	60	2688	1-BROMO-3-CHLOROPROPANE
L4BH	TU15 TE19	AT	2 (E)	V12		CV13 CV28	S9	60	2689	GLYCEROL alpha-MONOCHLOROHYDRIN
L4BH	TU15 TE19	AT	2 (D/E)			CV13 CV28	S9 S19	60	2690	N,n-BUTYLIMIDAZOLE
SGAN		AT	2 (E)	V11				80	2691	PHOSPHORUS PENTABROMIDE
L10BH		AT	1 (E)				S20	X88	2692	BORON TRIBROMIDE
L4BN		AT	3 (E)	V12				80	2693	BISULPHITES, AQUEOUS SOLUTION, N.O.S.
SGAV L4BN		AT	3 (E)		VC1 VC2 AP7			80	2698	TETRAHYDROPHTHALIC ANHYDRIDES with more than 0.05% of maleic anhydride
L10BH		AT	1 (E)				S20	88	2699	TRIFLUOROACETIC ACID
L4BN		AT	2 (E)					80	2705	1-PENTOL
LGBF		FL	2 (D/E)				S2 S20	33	2707	DIMETHYLDIOXANES

UN No.	Name and description	Class	Classifi-cation code	Packing group	Labels	Special provi-sions	Limited and excepted quantities		Packaging			Portable tanks and bulk containers	
									Packing instruc-tions	Special packing provisions	Mixed packing provisions	Instruc-tions	Special provisions
	3.1.2	2.2	2.2	2.1.1.3	5.2.2	3.3	3.4	3.5.1.2	4.1.4	4.1.4	4.1.10	4.2.5.2 7.3.2	4.2.5.3
(1)	(2)	(3a)	(3b)	(4)	(5)	(6)	(7a)	(7b)	(8)	(9a)	(9b)	(10)	(11)
2707	DIMETHYLDIOXANES	3	F1	III	3		5 L	E1	P001 IBC03 LP01 R001		MP19	T2	TP1
2709	BUTYLBENZENES	3	F1	III	3		5 L	E1	P001 IBC03 LP01 R001		MP19	T2	TP1
2710	DIPROPYL KETONE	3	F1	III	3		5 L	E1	P001 IBC03 LP01 R001		MP19	T2	TP1
2713	ACRIDINE	6.1	T2	III	6.1		5 kg	E1	P002 IBC08 LP02 R001	B3	MP10	T1	TP33
2714	ZINC RESINATE	4.1	F3	III	4.1		5 kg	E1	P002 IBC06 R001		MP11	T1	TP33
2715	ALUMINIUM RESINATE	4.1	F3	III	4.1		5 kg	E1	P002 IBC06 R001		MP11	T1	TP33
2716	1,4-BUTYNEDIOL	6.1	T2	III	6.1		5 kg	E1	P002 IBC08 LP02 R001	B3	MP10	T1	TP33
2717	CAMPHOR, synthetic	4.1	F1	III	4.1		5 kg	E1	P002 IBC08 LP02 R001	B3	MP10	T1	TP33
2719	BARIUM BROMATE	5.1	OT2	II	5.1 +6.1		1 kg	E2	P002 IBC08	B4	MP2	T3	TP33
2720	CHROMIUM NITRATE	5.1	O2	III	5.1		5 kg	E1	P002 IBC08 LP02 R001	B3	MP10	T1	TP33
2721	COPPER CHLORATE	5.1	O2	II	5.1		1 kg	E2	P002 IBC08	B4	MP2	T3	TP33
2722	LITHIUM NITRATE	5.1	O2	III	5.1		5 kg	E1	P002 IBC08 LP02 R001	B3	MP10	T1	TP33
2723	MAGNESIUM CHLORATE	5.1	O2	II	5.1		1 kg	E2	P002 IBC08	B4	MP2	T3	TP33
2724	MANGANESE NITRATE	5.1	O2	III	5.1		5 kg	E1	P002 IBC08 LP02 R001	B3	MP10	T1	TP33
2725	NICKEL NITRATE	5.1	O2	III	5.1		5 kg	E1	P002 IBC08 LP02 R001	B3	MP10	T1	TP33
2726	NICKEL NITRITE	5.1	O2	III	5.1		5 kg	E1	P002 IBC08 LP02 R001	B3	MP10	T1	TP33
2727	THALLIUM NITRATE	6.1	TO2	II	6.1 +5.1		500 g	E4	P002 IBC06		MP10	T3	TP33
2728	ZIRCONIUM NITRATE	5.1	O2	III	5.1		5 kg	E1	P002 IBC08 LP02 R001	B3	MP10	T1	TP33
2729	HEXACHLOROBENZENE	6.1	T2	III	6.1		5 kg	E1	P002 IBC08 LP02 R001	B3	MP10	T1	TP33
2730	NITROANISOLES, LIQUID	6.1	T1	III	6.1	279	5 L	E1	P001 IBC03 LP01 R001		MP19	T4	TP1
2732	NITROBROMOBENZENES, LIQUID	6.1	T1	III	6.1		5 L	E1	P001 IBC03 LP01 R001		MP19	T4	TP1

ADR tank		Vehicle for tank carriage	Transport category (Tunnel restriction code)	Special provisions for carriage				Hazard identification No.	UN No.	Name and description
Tank code	Special provisions			Packages	Bulk	Loading, unloading and handling	Operation			
4.3	4.3.5, 6.8.4	9.1.1.2	1.1.3.6 (8.6)	7.2.4	7.3.3	7.5.11	8.5	5.3.2.3		3.1.2
(12)	(13)	(14)	(15)	(16)	(17)	(18)	(19)	(20)	(1)	(2)
LGBF		FL	3 (D/E)	V12			S2	30	2707	DIMETHYLDIOXANES
LGBF		FL	3 (D/E)	V12			S2	30	2709	BUTYLBENZENES
LGBF		FL	3 (D/E)	V12			S2	30	2710	DIPROPYL KETONE
SGAH L4BH	TU15 TE19	AT	2 (E)		VC1 VC2 AP7	CV13 CV28	S9	60	2713	ACRIDINE
SGAV		AT	3 (E)		VC1 VC2			40	2714	ZINC RESINATE
SGAV		AT	3 (E)		VC1 VC2			40	2715	ALUMINIUM RESINATE
SGAH L4BH	TU15 TE19	AT	2 (E)		VC1 VC2 AP7	CV13 CV28	S9	60	2716	1,4-BUTYNEDIOL
SGAV		AT	3 (E)		VC1 VC2			40	2717	CAMPHOR, synthetic
SGAN	TU3	AT	2 (E)	V11		CV24 CV28		56	2719	BARIUM BROMATE
SGAV	TU3	AT	3 (E)		VC1 VC2 AP6 AP7	CV24		50	2720	CHROMIUM NITRATE
SGAV	TU3	AT	2 (E)	V11	VC1 VC2 AP6 AP7	CV24		50	2721	COPPER CHLORATE
SGAV	TU3	AT	3 (E)		VC1 VC2 AP6 AP7	CV24		50	2722	LITHIUM NITRATE
SGAV	TU3	AT	2 (E)	V11	VC1 VC2 AP6 AP7	CV24		50	2723	MAGNESIUM CHLORATE
SGAV	TU3	AT	3 (E)		VC1 VC2 AP6 AP7	CV24		50	2724	MANGANESE NITRATE
SGAV	TU3	AT	3 (E)		VC1 VC2 AP6 AP7	CV24		50	2725	NICKEL NITRATE
SGAV	TU3	AT	3 (E)		VC1 VC2 AP6 AP7	CV24		50	2726	NICKEL NITRITE
SGAH	TU15 TE19	AT	2 (D/E)	V11		CV13 CV28	S9 S19	65	2727	THALLIUM NITRATE
SGAV	TU3	AT	3 (E)		VC1 VC2 AP6 AP7	CV24		50	2728	ZIRCONIUM NITRATE
SGAH	TU15 TE19	AT	2 (E)		VC1 VC2 AP7	CV13 CV28	S9	60	2729	HEXACHLOROBENZENE
L4BH	TU15 TE19	AT	2 (E)	V12		CV13 CV28	S9	60	2730	NITROANISOLES, LIQUID
L4BH	TU15 TE19	AT	2 (E)	V12		CV13 CV28	S9	60	2732	NITROBROMOBENZENES, LIQUID

UN No.	Name and description	Class	Classifi-cation code	Packing group	Labels	Special provi-sions	Limited and excepted quantities		Packaging			Portable tanks and bulk containers	
									Packing instruc-tions	Special packing provisions	Mixed packing provisions	Instruc-tions	Special provisions
3.1.2		2.2	2.2	2.1.1.3	5.2.2	3.3	3.4	3.5.1.2	4.1.4	4.1.4	4.1.10	4.2.5.2 7.3.2	4.2.5.3
(1)	(2)	(3a)	(3b)	(4)	(5)	(6)	(7a)	(7b)	(8)	(9a)	(9b)	(10)	(11)
2733	AMINES, FLAMMABLE, CORROSIVE, N.O.S. or POLYAMINES, FLAMMABLE, CORROSIVE, N.O.S.	3	FC	I	3 +8	274 544	0	E0	P001		MP7 MP17	T14	TP1 TP27
2733	AMINES, FLAMMABLE, CORROSIVE, N.O.S. or POLYAMINES, FLAMMABLE, CORROSIVE, N.O.S.	3	FC	II	3 +8	274 544	1 L	E2	P001 IBC02		MP19	T11	TP1 TP27
2733	AMINES, FLAMMABLE, CORROSIVE, N.O.S. or POLYAMINES, FLAMMABLE, CORROSIVE, N.O.S.	3	FC	III	3 +8	274 544	5 L	E1	P001 IBC03 R001		MP19	T7	TP1 TP28
2734	AMINES, LIQUID, CORROSIVE, FLAMMABLE, N.O.S. or POLYAMINES, LIQUID, CORROSIVE, FLAMMABLE, N.O.S.	8	CF1	I	8 +3	274	0	E0	P001		MP8 MP17	T14	TP2 TP27
2734	AMINES, LIQUID, CORROSIVE, FLAMMABLE, N.O.S. or POLYAMINES, LIQUID, CORROSIVE, FLAMMABLE, N.O.S.	8	CF1	II	8 +3	274	1 L	E2	P001 IBC02		MP15	T11	TP2 TP27
2735	AMINES, LIQUID, CORROSIVE, N.O.S. or POLYAMINES, LIQUID, CORROSIVE, N.O.S.	8	C7	I	8	274	0	E0	P001		MP8 MP17	T14	TP2 TP27
2735	AMINES, LIQUID, CORROSIVE, N.O.S. or POLYAMINES, LIQUID, CORROSIVE, N.O.S.	8	C7	II	8	274	1 L	E2	P001 IBC02		MP15	T11	TP1 TP27
2735	AMINES, LIQUID, CORROSIVE, N.O.S. or POLYAMINES, LIQUID, CORROSIVE, N.O.S.	8	C7	III	8	274	5 L	E1	P001 IBC03 LP01 R001		MP19	T7	TP1 TP28
2738	N-BUTYLANILINE	6.1	T1	II	6.1		100 ml	E4	P001 IBC02		MP15	T7	TP2
2739	BUTYRIC ANHYDRIDE	8	C3	III	8		5 L	E1	P001 IBC03 LP01 R001		MP19	T4	TP1
2740	n-PROPYL CHLOROFORMATE	6.1	TFC	I	6.1 +3 +8		0	E0	P602		MP8 MP17	T20	TP2
2741	BARIUM HYPOCHLORITE with more than 22% available chlorine	5.1	OT2	II	5.1 +6.1		1 kg	E2	P002 IBC08	B4	MP2	T3	TP33
2742	CHLOROFORMATES, TOXIC, CORROSIVE, FLAMMABLE, N.O.S.	6.1	TFC	II	6.1 +3 +8	274 561	100 ml	E4	P001 IBC01		MP15		
2743	n-BUTYL CHLOROFORMATE	6.1	TFC	II	6.1 +3 +8		100 ml	E0	P001		MP15	T20	TP2
2744	CYCLOBUTYL CHLOROFORMATE	6.1	TFC	II	6.1 +3 +8		100 ml	E4	P001 IBC01		MP15	T7	TP2
2745	CHLOROMETHYL CHLOROFORMATE	6.1	TC1	II	6.1 +8		100 ml	E4	P001 IBC02		MP15	T7	TP2
2746	PHENYL CHLOROFORMATE	6.1	TC1	II	6.1 +8		100 ml	E4	P001 IBC02		MP15	T7	TP2
2747	tert-BUTYLCYCLOHEXYL CHLOROFORMATE	6.1	T1	III	6.1		5 L	E1	P001 IBC03 LP01 R001		MP19	T4	TP1
2748	2-ETHYLHEXYL CHLOROFORMATE	6.1	TC1	II	6.1 +8		100 ml	E4	P001 IBC02		MP15	T7	TP2
2749	TETRAMETHYLSILANE	3	F1	I	3		0	E0	P001		MP7 MP17	T14	TP2
2750	1,3-DICHLOROPROPANOL-2	6.1	T1	II	6.1		100 ml	E4	P001 IBC02		MP15	T7	TP2
2751	DIETHYLTHIO-PHOSPHORYL CHLORIDE	8	C3	II	8		1 L	E2	P001 IBC02		MP15	T7	TP2
2752	1,2-EPOXY-3-ETHOXYPROPANE	3	F1	III	3		5 L	E1	P001 IBC03 LP01 R001		MP19	T2	TP1
2753	N-ETHYLBENZYL-TOLUIDINES, LIQUID	6.1	T1	III	6.1		5 L	E1	P001 IBC03 LP01 R001		MP19	T7	TP1

ADR tank		Vehicle for tank carriage	Transport category (Tunnel restriction code)	Special provisions for carriage				Hazard identifi-cation No.	UN No.	Name and description
Tank code	Special provisions			Packages	Bulk	Loading, unloading and handling	Operation			
4.3	4.3.5, 6.8.4	9.1.1.2	1.1.3.6 (8.6)	7.2.4	7.3.3	7.5.11	8.5	5.3.2.3		3.1.2
(12)	(13)	(14)	(15)	(16)	(17)	(18)	(19)	(20)	(1)	(2)
L10CH	TU14 TE21	FL	1 (C/E)				S2 S20	338	2733	AMINES, FLAMMABLE, CORROSIVE, N.O.S. or POLYAMINES, FLAMMABLE, CORROSIVE, N.O.S.
L4BH		FL	2 (D/E)				S2 S20	338	2733	AMINES, FLAMMABLE, CORROSIVE, N.O.S. or POLYAMINES, FLAMMABLE, CORROSIVE, N.O.S.
L4BN		FL	3 (D/E)	V12			S2	38	2733	AMINES, FLAMMABLE, CORROSIVE, N.O.S. or POLYAMINES, FLAMMABLE, CORROSIVE, N.O.S.
L10BH		FL	1 (D/E)				S2 S14	883	2734	AMINES, LIQUID, CORROSIVE, FLAMMABLE, N.O.S. or POLYAMINES, LIQUID, CORROSIVE, FLAMMABLE, N.O.S.
L4BN		FL	2 (D/E)				S2	83	2734	AMINES, LIQUID, CORROSIVE, FLAMMABLE, N.O.S. or POLYAMINES, LIQUID, CORROSIVE, FLAMMABLE, N.O.S.
L10BH		AT	1 (E)				S20	88	2735	AMINES, LIQUID, CORROSIVE, N.O.S. or POLYAMINES, LIQUID, CORROSIVE, N.O.S.
L4BN		AT	2 (E)					80	2735	AMINES, LIQUID, CORROSIVE, N.O.S. or POLYAMINES, LIQUID, CORROSIVE, N.O.S.
L4BN		AT	3 (E)	V12				80	2735	AMINES, LIQUID, CORROSIVE, N.O.S. or POLYAMINES, LIQUID, CORROSIVE, N.O.S.
L4BH	TU15 TE19	AT	2 (D/E)			CV13 CV28	S9 S19	60	2738	N-BUTYLANILINE
L4BN		AT	3 (E)	V12				80	2739	BUTYRIC ANHYDRIDE
L10CH	TU14 TU15 TE19 TE21	FL	1 (C/D)			CV1 CV13 CV28	S2 S9 S14	668	2740	n-PROPYL CHLOROFORMATE
SGAN	TU3	AT	2 (E)	V11		CV24 CV28		56	2741	BARIUM HYPOCHLORITE with more than 22% available chlorine
L4BH	TU15 TE19	FL	2 (D/E)			CV13 CV28	S2 S9 S19	638	2742	CHLOROFORMATES, TOXIC, CORROSIVE, FLAMMABLE, N.O.S.
L4BH	TU15 TE19	FL	2 (D/E)			CV13 CV28	S2 S9 S19	638	2743	n-BUTYL CHLOROFORMATE
L4BH	TU15 TE19	FL	2 (D/E)			CV13 CV28	S2 S9 S19	638	2744	CYCLOBUTYL CHLOROFORMATE
L4BH	TU15 TE19	AT	2 (D/E)			CV13 CV28	S9 S19	68	2745	CHLOROMETHYL CHLOROFORMATE
L4BH	TU15 TE19	AT	2 (D/E)			CV13 CV28	S9 S19	68	2746	PHENYL CHLOROFORMATE
L4BH	TU15 TE19	AT	2 (E)	V12		CV13 CV28	S9	60	2747	tert-BUTYLCYCLOHEXYL CHLOROFORMATE
L4BH	TU15 TE19	AT	2 (D/E)			CV13 CV28	S9 S19	68	2748	2-ETHYLHEXYL CHLOROFORMATE
L4BN		FL	1 (D/E)				S2 S20	33	2749	TETRAMETHYLSILANE
L4BH	TU15 TE19	AT	2 (D/E)			CV13 CV28	S9 S19	60	2750	1,3-DICHLOROPROPANOL-2
L4BN		AT	2 (E)					80	2751	DIETHYLTHIO-PHOSPHORYL CHLORIDE
LGBF		FL	3 (D/E)	V12			S2	30	2752	1,2-EPOXY-3-ETHOXYPROPANE
L4BH	TU15 TE19	AT	2 (E)	V12		CV13 CV28	S9	60	2753	N-ETHYLBENZYL-TOLUIDINES, LIQUID

UN No.	Name and description	Class	Classifi-cation code	Packing group	Labels	Special provi-sions	Limited and excepted quantities		Packaging			Portable tanks and bulk containers	
									Packing instruc-tions	Special packing provisions	Mixed packing provisions	Instruc-tions	Special provisions
	3.1.2	2.2	2.2	2.1.1.3	5.2.2	3.3	3.4	3.5.1.2	4.1.4	4.1.4	4.1.10	4.2.5.2 7.3.2	4.2.5.3
(1)	(2)	(3a)	(3b)	(4)	(5)	(6)	(7a)	(7b)	(8)	(9a)	(9b)	(10)	(11)
2754	N-ETHYLTOLUIDINES	6.1	T1	II	6.1		100 ml	E4	P001 IBC02		MP15	T7	TP2
2757	CARBAMATE PESTICIDE, SOLID, TOXIC	6.1	T7	I	6.1	61 274 648	0	E5	P002 IBC07		MP18	T6	TP33
2757	CARBAMATE PESTICIDE, SOLID, TOXIC	6.1	T7	II	6.1	61 274 648	500 g	E4	P002 IBC08	B4	MP10	T3	TP33
2757	CARBAMATE PESTICIDE, SOLID, TOXIC	6.1	T7	III	6.1	61 274 648	5 kg	E1	P002 IBC08 LP02 R001	B3	MP10	T1	TP33
2758	CARBAMATE PESTICIDE, LIQUID, FLAMMABLE, TOXIC, flash-point less than 23 °C	3	FT2	I	3 +6.1	61 274	0	E0	P001		MP7 MP17	T14	TP2 TP27
2758	CARBAMATE PESTICIDE, LIQUID, FLAMMABLE, TOXIC, flash-point less than 23 °C	3	FT2	II	3 +6.1	61 274	1 L	E2	P001 IBC02 R001		MP19	T11	TP2 TP27
2759	ARSENICAL PESTICIDE, SOLID, TOXIC	6.1	T7	I	6.1	61 274 648	0	E5	P002 IBC07		MP18	T6	TP33
2759	ARSENICAL PESTICIDE, SOLID, TOXIC	6.1	T7	II	6.1	61 274 648	500 g	E4	P002 IBC08	B4	MP10	T3	TP33
2759	ARSENICAL PESTICIDE, SOLID, TOXIC	6.1	T7	III	6.1	61 274 648	5 kg	E1	P002 IBC08 LP02 R001	B3	MP10	T1	TP33
2760	ARSENICAL PESTICIDE, LIQUID, FLAMMABLE, TOXIC, flash-point less than 23 °C	3	FT2	I	3 +6.1	61 274	0	E0	P001		MP7 MP17	T14	TP2 TP27
2760	ARSENICAL PESTICIDE, LIQUID, FLAMMABLE, TOXIC, flash-point less than 23 °C	3	FT2	II	3 +6.1	61 274	1 L	E2	P001 IBC02 R001		MP19	T11	TP2 TP27
2761	ORGANOCHLORINE PESTICIDE, SOLID, TOXIC	6.1	T7	I	6.1	61 274 648	0	E5	P002 IBC07		MP18	T6	TP33
2761	ORGANOCHLORINE PESTICIDE, SOLID, TOXIC	6.1	T7	II	6.1	61 274 648	500 g	E4	P002 IBC08	B4	MP10	T3	TP33
2761	ORGANOCHLORINE PESTICIDE, SOLID, TOXIC	6.1	T7	III	6.1	61 274 648	5 kg	E1	P002 IBC08 LP02 R001	B3	MP10	T1	TP33
2762	ORGANOCHLORINE PESTICIDE, LIQUID, FLAMMABLE, TOXIC, flash-point less than 23 °C	3	FT2	I	3 +6.1	61 274	0	E0	P001		MP7 MP17	T14	TP2 TP27
2762	ORGANOCHLORINE PESTICIDE, LIQUID, FLAMMABLE, TOXIC, flash-point less than 23 °C	3	FT2	II	3 +6.1	61 274	1 L	E2	P001 IBC02 R001		MP19	T11	TP2 TP27
2763	TRIAZINE PESTICIDE, SOLID, TOXIC	6.1	T7	I	6.1	61 274 648	0	E5	P002 IBC07		MP18	T6	TP33
2763	TRIAZINE PESTICIDE, SOLID, TOXIC	6.1	T7	II	6.1	61 274 648	500 g	E4	P002 IBC08	B4	MP10	T3	TP33
2763	TRIAZINE PESTICIDE, SOLID, TOXIC	6.1	T7	III	6.1	61 274 648	5 kg	E1	P002 IBC08 R001	B3	MP10	T1	TP33
2764	TRIAZINE PESTICIDE, LIQUID, FLAMMABLE, TOXIC, flash-point less than 23 °C	3	FT2	I	3 +6.1	61 274	0	E0	P001		MP7 MP17	T14	TP2 TP27
2764	TRIAZINE PESTICIDE, LIQUID, FLAMMABLE, TOXIC, flash-point less than 23 °C	3	FT2	II	3 +6.1	61 274	1 L	E2	P001 IBC02 R001		MP19	T11	TP2 TP27
2771	THIOCARBAMATE PESTICIDE, SOLID, TOXIC	6.1	T7	I	6.1	61 274 648	0	E5	P002 IBC07		MP18	T6	TP33

	ADR tank		Vehicle for tank carriage	Transport category (Tunnel restriction code)	Special provisions for carriage				Hazard identifi-cation No.	UN No.	Name and description
Tank code	Special provisions				Packages	Bulk	Loading, unloading and handling	Operation			
4.3	4.3.5, 6.8.4	9.1.1.2	1.1.3.6 (8.6)	7.2.4	7.3.3	7.5.11	8.5	5.3.2.3		3.1.2
(12)	(13)	(14)	(15)	(16)	(17)	(18)	(19)	(20)	(1)	(2)
L4BH	TU15 TE19	AT	2 (D/E)			CV13 CV28	S9 S19	60	2754	N-ETHYLTOLUIDINES
S10AH L10CH	TU14 TU15 TE19 TE21	AT	1 (C/E)	V10		CV1 CV13 CV28	S9 S14	66	2757	CARBAMATE PESTICIDE, SOLID, TOXIC
SGAH L4BH	TU15 TE19	AT	2 (D/E)	V11		CV13 CV28	S9 S19	60	2757	CARBAMATE PESTICIDE, SOLID, TOXIC
SGAH L4BH	TU15 TE19	AT	2 (E)		VC1 VC2 AP7	CV13 CV28	S9	60	2757	CARBAMATE PESTICIDE, SOLID, TOXIC
L10CH	TU14 TU15 TE21	FL	1 (C/E)			CV13 CV28	S2 S22	336	2758	CARBAMATE PESTICIDE, LIQUID, FLAMMABLE, TOXIC, flash-point less than 23 °C
L4BH	TU15	FL	2 (D/E)			CV13 CV28	S2 S22	336	2758	CARBAMATE PESTICIDE, LIQUID, FLAMMABLE, TOXIC, flash-point less than 23 °C
S10AH L10CH	TU14 TU15 TE19 TE21	AT	1 (C/E)	V10		CV1 CV13 CV28	S9 S14	66	2759	ARSENICAL PESTICIDE, SOLID, TOXIC
SGAH L4BH	TU15 TE19	AT	2 (D/E)	V11		CV13 CV28	S9 S19	60	2759	ARSENICAL PESTICIDE, SOLID, TOXIC
SGAH L4BH	TU15 TE19	AT	2 (E)		VC1 VC2 AP7	CV13 CV28	S9	60	2759	ARSENICAL PESTICIDE, SOLID, TOXIC
L10CH	TU14 TU15 TE21	FL	1 (C/E)			CV13 CV28	S2 S22	336	2760	ARSENICAL PESTICIDE, LIQUID, FLAMMABLE, TOXIC, flash-point less than 23 °C
L4BH	TU15	FL	2 (D/E)			CV13 CV28	S2 S22	336	2760	ARSENICAL PESTICIDE, LIQUID, FLAMMABLE, TOXIC, flash-point less than 23 °C
S10AH L10CH	TU14 TU15 TE19 TE21	AT	1 (C/E)	V10		CV1 CV13 CV28	S9 S14	66	2761	ORGANOCHLORINE PESTICIDE, SOLID, TOXIC
SGAH L4BH	TU15 TE19	AT	2 (D/E)	V11		CV13 CV28	S9 S19	60	2761	ORGANOCHLORINE PESTICIDE, SOLID, TOXIC
SGAH L4BH	TU15 TE19	AT	2 (E)		VC1 VC2 AP7	CV13 CV28	S9	60	2761	ORGANOCHLORINE PESTICIDE, SOLID, TOXIC
L10CH	TU14 TU15 TE21	FL	1 (C/E)			CV13 CV28	S2 S22	336	2762	ORGANOCHLORINE PESTICIDE, LIQUID, FLAMMABLE, TOXIC, flash-point less than 23 °C
L4BH	TU15	FL	2 (D/E)			CV13 CV28	S2 S22	336	2762	ORGANOCHLORINE PESTICIDE, LIQUID, FLAMMABLE, TOXIC, flash-point less than 23 °C
S10AH L10CH	TU14 TU15 TE19 TE21	AT	1 (C/E)	V10		CV1 CV13 CV28	S9 S14	66	2763	TRIAZINE PESTICIDE, SOLID, TOXIC
SGAH L4BH	TU15 TE19	AT	2 (D/E)	V11		CV13 CV28	S9 S19	60	2763	TRIAZINE PESTICIDE, SOLID, TOXIC
SGAH L4BH	TU15 TE19	AT	2 (E)		VC1 VC2 AP7	CV13 CV28	S9	60	2763	TRIAZINE PESTICIDE, SOLID, TOXIC
L10CH	TU14 TU15 TE21	FL	1 (C/E)			CV13 CV28	S2 S22	336	2764	TRIAZINE PESTICIDE, LIQUID, FLAMMABLE, TOXIC, flash-point less than 23 °C
L4BH	TU15	FL	2 (D/E)			CV13 CV28	S2 S22	336	2764	TRIAZINE PESTICIDE, LIQUID, FLAMMABLE, TOXIC, flash-point less than 23 °C
S10AH L10CH	TU14 TU15 TE19 TE21	AT	1 (C/E)	V10		CV1 CV13 CV28	S9 S14	66	2771	THIOCARBAMATE PESTICIDE, SOLID, TOXIC

UN No.	Name and description	Class	Classifi-cation code	Packing group	Labels	Special provi-sions	Limited and excepted quantities		Packaging			Portable tanks and bulk containers	
									Packing instruc-tions	Special packing provisions	Mixed packing provisions	Instruc-tions	Special provisions
	3.1.2	2.2	2.2	2.1.1.3	5.2.2	3.3	3.4	3.5.1.2	4.1.4	4.1.4	4.1.10	4.2.5.2 7.3.2	4.2.5.3
(1)	(2)	(3a)	(3b)	(4)	(5)	(6)	(7a)	(7b)	(8)	(9a)	(9b)	(10)	(11)
2771	THIOCARBAMATE PESTICIDE, SOLID, TOXIC	6.1	T7	II	6.1	61 274 648	500 g	E4	P002 IBC08	B4	MP10	T3	TP33
2771	THIOCARBAMATE PESTICIDE, SOLID, TOXIC	6.1	T7	III	6.1	61 274 648	5 kg	E1	P002 IBC08 LP02 R001	B3	MP10	T1	TP33
2772	THIOCARBAMATE PESTICIDE, LIQUID, FLAMMABLE, TOXIC, flash-point less than 23 °C	3	FT2	I	3 +6.1	61 274	0	E0	P001		MP7 MP17	T14	TP2 TP27
2772	THIOCARBAMATE PESTICIDE, LIQUID, FLAMMABLE, TOXIC, flash-point less than 23 °C	3	FT2	II	3 +6.1	61 274	1 L	E2	P001 IBC02 R001		MP19	T11	TP2 TP27
2775	COPPER BASED PESTICIDE, SOLID, TOXIC	6.1	T7	I	6.1	61 274 648	0	E5	P002 IBC07		MP18	T6	TP33
2775	COPPER BASED PESTICIDE, SOLID, TOXIC	6.1	T7	II	6.1	61 274 648	500 g	E4	P002 IBC08	B4	MP10	T3	TP33
2775	COPPER BASED PESTICIDE, SOLID, TOXIC	6.1	T7	III	6.1	61 274 648	5 kg	E1	P002 IBC08 LP02 R001	B3	MP10	T1	TP33
2776	COPPER BASED PESTICIDE, LIQUID, FLAMMABLE, TOXIC, flash-point less than 23 °C	3	FT2	I	3 +6.1	61 274	0	E0	P001		MP7 MP17	T14	TP2 TP27
2776	COPPER BASED PESTICIDE, LIQUID, FLAMMABLE, TOXIC, flash-point less than 23 °C	3	FT2	II	3 +6.1	61 274	1 L	E2	P001 IBC02 R001		MP19	T11	TP2 TP27
2777	MERCURY BASED PESTICIDE, SOLID, TOXIC	6.1	T7	I	6.1	61 274 648	0	E5	P002 IBC07		MP18	T6	TP33
2777	MERCURY BASED PESTICIDE, SOLID, TOXIC	6.1	T7	II	6.1	61 274 648	500 g	E4	P002 IBC08	B4	MP10	T3	TP33
2777	MERCURY BASED PESTICIDE, SOLID, TOXIC	6.1	T7	III	6.1	61 274 648	5 kg	E1	P002 IBC08 LP02 R001	B3	MP10	T1	TP33
2778	MERCURY BASED PESTICIDE, LIQUID, FLAMMABLE, TOXIC, flash-point less than 23 °C	3	FT2	I	3 +6.1	61 274	0	E0	P001		MP7 MP17	T14	TP2 TP27
2778	MERCURY BASED PESTICIDE, LIQUID, FLAMMABLE, TOXIC, flash-point less than 23 °C	3	FT2	II	3 +6.1	61 274	1 L	E2	P001 IBC02 R001		MP19	T11	TP2 TP27
2779	SUBSTITUTED NITROPHENOL PESTICIDE, SOLID, TOXIC	6.1	T7	I	6.1	61 274 648	0	E5	P002 IBC07		MP18	T6	TP33
2779	SUBSTITUTED NITROPHENOL PESTICIDE, SOLID, TOXIC	6.1	T7	II	6.1	61 274 648	500 g	E4	P002 IBC08	B4	MP10	T3	TP33
2779	SUBSTITUTED NITROPHENOL PESTICIDE, SOLID, TOXIC	6.1	T7	III	6.1	61 274 648	5 kg	E1	P002 IBC08 LP02 R001	B3	MP10	T1	TP33
2780	SUBSTITUTED NITROPHENOL PESTICIDE, LIQUID, FLAMMABLE, TOXIC, flash-point less than 23 °C	3	FT2	I	3 +6.1	61 274	0	E0	P001		MP7 MP17	T14	TP2 TP27
2780	SUBSTITUTED NITROPHENOL PESTICIDE, LIQUID, FLAMMABLE, TOXIC, flash-point less than 23 °C	3	FT2	II	3 +6.1	61 274	1 L	E2	P001 IBC02 R001		MP19	T11	TP2 TP27
2781	BIPYRIDILIUM PESTICIDE, SOLID, TOXIC	6.1	T7	I	6.1	61 274 648	0	E5	P002 IBC07		MP18	T6	TP33
2781	BIPYRIDILIUM PESTICIDE, SOLID, TOXIC	6.1	T7	II	6.1	61 274 648	500 g	E4	P002 IBC08	B4	MP10	T3	TP33

ADR tank		Vehicle for tank carriage	Transport category (Tunnel restriction code)	Special provisions for carriage				Hazard identifi-cation No.	UN No.	Name and description
Tank code	Special provisions			Packages	Bulk	Loading, unloading and handling	Operation			
4.3	4.3.5, 6.8.4	9.1.1.2	1.1.3.6 (8.6)	7.2.4	7.3.3	7.5.11	8.5	5.3.2.3		3.1.2
(12)	(13)	(14)	(15)	(16)	(17)	(18)	(19)	(20)	(1)	(2)
SGAH L4BH	TU15 TE19	AT	2 (D/E)	V11		CV13 CV28	S9 S19	60	2771	THIOCARBAMATE PESTICIDE, SOLID, TOXIC
SGAH L4BH	TU15 TE19	AT	2 (E)		VC1 VC2 AP7	CV13 CV28	S9	60	2771	THIOCARBAMATE PESTICIDE, SOLID, TOXIC
L10CH	TU14 TU15 TE21	FL	1 (C/E)			CV13 CV28	S2 S22	336	2772	THIOCARBAMATE PESTICIDE, LIQUID, FLAMMABLE, TOXIC, flash-point less than 23 °C
L4BH	TU15	FL	2 (D/E)			CV13 CV28	S2 S22	336	2772	THIOCARBAMATE PESTICIDE, LIQUID, FLAMMABLE, TOXIC, flash-point less than 23 °C
S10AH L10CH	TU14 TU15 TE19 TE21	AT	1 (C/E)	V10		CV1 CV13 CV28	S9 S14	66	2775	COPPER BASED PESTICIDE, SOLID, TOXIC
SGAH L4BH	TU15 TE19	AT	2 (D/E)	V11		CV13 CV28	S9 S19	60	2775	COPPER BASED PESTICIDE, SOLID, TOXIC
SGAH L4BH	TU15 TE19	AT	2 (E)		VC1 VC2 AP7	CV13 CV28	S9	60	2775	COPPER BASED PESTICIDE, SOLID, TOXIC
L10CH	TU14 TU15 TE21	FL	1 (C/E)			CV13 CV28	S2 S22	336	2776	COPPER BASED PESTICIDE, LIQUID, FLAMMABLE, TOXIC, flash-point less than 23 °C
L4BH	TU15	FL	2 (D/E)			CV13 CV28	S2 S22	336	2776	COPPER BASED PESTICIDE, LIQUID, FLAMMABLE, TOXIC, flash-point less than 23 °C
S10AH L10CH	TU14 TU15 TE19 TE21	AT	1 (C/E)	V10		CV1 CV13 CV28	S9 S14	66	2777	MERCURY BASED PESTICIDE, SOLID, TOXIC
SGAH L4BH	TU15 TE19	AT	2 (D/E)	V11		CV13 CV28	S9 S19	60	2777	MERCURY BASED PESTICIDE, SOLID, TOXIC
SGAH L4BH	TU15 TE19	AT	2 (E)		VC1 VC2 AP7	CV13 CV28	S9	60	2777	MERCURY BASED PESTICIDE, SOLID, TOXIC
L10CH	TU14 TU15 TE21	FL	1 (C/E)			CV13 CV28	S2 S22	336	2778	MERCURY BASED PESTICIDE, LIQUID, FLAMMABLE, TOXIC, flash-point less than 23 °C
L4BH	TU15	FL	2 (D/E)			CV13 CV28	S2 S22	336	2778	MERCURY BASED PESTICIDE, LIQUID, FLAMMABLE, TOXIC, flash-point less than 23 °C
S10AH L10CH	TU14 TU15 TE19 TE21	AT	1 (C/E)	V10		CV1 CV13 CV28	S9 S14	66	2779	SUBSTITUTED NITROPHENOL PESTICIDE, SOLID, TOXIC
SGAH L4BH	TU15 TE19	AT	2 (D/E)	V11		CV13 CV28	S9 S19	60	2779	SUBSTITUTED NITROPHENOL PESTICIDE, SOLID, TOXIC
SGAH L4BH	TU15 TE19	AT	2 (E)		VC1 VC2 AP7	CV13 CV28	S9	60	2779	SUBSTITUTED NITROPHENOL PESTICIDE, SOLID, TOXIC
L10CH	TU14 TU15 TE21	FL	1 (C/E)			CV13 CV28	S2 S22	336	2780	SUBSTITUTED NITROPHENOL PESTICIDE, LIQUID, FLAMMABLE, TOXIC, flash-point less than 23 °C
L4BH	TU15	FL	2 (D/E)			CV13 CV28	S2 S22	336	2780	SUBSTITUTED NITROPHENOL PESTICIDE, LIQUID, FLAMMABLE, TOXIC, flash-point less than 23 °C
S10AH L10CH	TU14 TU15 TE19 TE21	AT	1 (C/E)	V10		CV1 CV13 CV28	S9 S14	66	2781	BIPYRIDILIUM PESTICIDE, SOLID, TOXIC
SGAH L4BH	TU15 TE19	AT	2 (D/E)	V11		CV13 CV28	S9 S19	60	2781	BIPYRIDILIUM PESTICIDE, SOLID, TOXIC

UN No.	Name and description	Class	Classifi-cation code	Packing group	Labels	Special provi-sions	Limited and excepted quantities		Packaging			Portable tanks and bulk containers	
									Packing instruc-tions	Special packing provisions	Mixed packing provisions	Instruc-tions	Special provisions
	3.1.2	2.2	2.2	2.1.1.3	5.2.2	3.3	3.4	3.5.1.2	4.1.4	4.1.4	4.1.10	4.2.5.2 7.3.2	4.2.5.3
(1)	(2)	(3a)	(3b)	(4)	(5)	(6)	(7a)	(7b)	(8)	(9a)	(9b)	(10)	(11)
2781	BIPYRIDILIUM PESTICIDE, SOLID, TOXIC	6.1	T7	III	6.1	61 274 648	5 kg	E1	P002 IBC08 LP02 R001	B3	MP10	T1	TP33
2782	BIPYRIDILIUM PESTICIDE, LIQUID, FLAMMABLE, TOXIC, flash-point less than 23 °C	3	FT2	I	3 +6.1	61 274	0	E0	P001		MP7 MP17	T14	TP2 TP27
2782	BIPYRIDILIUM PESTICIDE, LIQUID, FLAMMABLE, TOXIC, flash-point less than 23 °C	3	FT2	II	3 +6.1	61 274	1 L	E2	P001 IBC02 R001		MP19	T11	TP2 TP27
2783	ORGANOPHOSPHORUS PESTICIDE, SOLID, TOXIC	6.1	T7	I	6.1	61 274 648	0	E5	P002 IBC07		MP18	T6	TP33
2783	ORGANOPHOSPHORUS PESTICIDE, SOLID, TOXIC	6.1	T7	II	6.1	61 274 648	500 g	E4	P002 IBC08	B4	MP10	T3	TP33
2783	ORGANOPHOSPHORUS PESTICIDE, SOLID, TOXIC	6.1	T7	III	6.1	61 274 648	5 kg	E1	P002 IBC08 LP02 R001	B3	MP10	T1	TP33
2784	ORGANOPHOSPHORUS PESTICIDE, LIQUID, FLAMMABLE, TOXIC, flash-point less than 23 °C	3	FT2	I	3 +6.1	61 274	0	E0	P001		MP7 MP17	T14	TP2 TP27
2784	ORGANOPHOSPHORUS PESTICIDE, LIQUID, FLAMMABLE, TOXIC, flash-point less than 23 °C	3	FT2	II	3 +6.1	61 274	1 L	E2	P001 IBC02 R001		MP19	T11	TP2 TP27
2785	4-THIAPENTANAL	6.1	T1	III	6.1		5 L	E1	P001 IBC03 LP01 R001		MP19	T4	TP1
2786	ORGANOTIN PESTICIDE, SOLID, TOXIC	6.1	T7	I	6.1	61 274 648	0	E5	P002 IBC07		MP18	T6	TP33
2786	ORGANOTIN PESTICIDE, SOLID, TOXIC	6.1	T7	II	6.1	61 274 648	500 g	E4	P002 IBC08	B4	MP10	T3	TP33
2786	ORGANOTIN PESTICIDE, SOLID, TOXIC	6.1	T7	III	6.1	61 274 648	5 kg	E1	P002 IBC08 LP02 R001	B3	MP10	T1	TP33
2787	ORGANOTIN PESTICIDE, LIQUID, FLAMMABLE, TOXIC, flash-point less than 23 °C	3	FT2	I	3 +6.1	61 274	0	E0	P001		MP7 MP17	T14	TP2 TP27
2787	ORGANOTIN PESTICIDE, LIQUID, FLAMMABLE, TOXIC, flash-point less than 23 °C	3	FT2	II	3 +6.1	61 274	1 L	E2	P001 IBC02 R001		MP19	T11	TP2 TP27
2788	ORGANOTIN COMPOUND, LIQUID, N.O.S.	6.1	T3	I	6.1	43 274	0	E5	P001		MP8 MP17	T14	TP2 TP27
2788	ORGANOTIN COMPOUND, LIQUID, N.O.S.	6.1	T3	II	6.1	43 274	100 ml	E4	P001 IBC02		MP15	T11	TP2 TP27
2788	ORGANOTIN COMPOUND, LIQUID, N.O.S.	6.1	T3	III	6.1	43 274	5 L	E1	P001 IBC03 LP01 R001		MP19	T7	TP2 TP28
2789	ACETIC ACID, GLACIAL or ACETIC ACID SOLUTION, more than 80% acid, by mass	8	CF1	II	8 +3		1 L	E2	P001 IBC02		MP15	T7	TP2
2790	ACETIC ACID SOLUTION, not less than 50% but not more than 80% acid, by mass	8	C3	II	8		1 L	E2	P001 IBC02		MP15	T7	TP2
2790	ACETIC ACID SOLUTION, more than 10% and less than 50% acid, by mass	8	C3	III	8	597 647	5 L	E1	P001 IBC03 LP01 R001		MP19	T4	TP1
2793	FERROUS METAL BORINGS, SHAVINGS, TURNINGS or CUTTINGS in a form liable to self-heating	4.2	S4	III	4.2	592	0	E1	P003 IBC08 LP02 R001	PP20 B3 B6	MP14		

ADR tank		Vehicle for tank carriage	Transport category (Tunnel restriction code)	Special provisions for carriage				Hazard identification No.	UN No.	Name and description
Tank code	Special provisions			Packages	Bulk	Loading, unloading and handling	Operation			
4.3	4.3.5, 6.8.4	9.1.1.2	1.1.3.6 (8.6)	7.2.4	7.3.3	7.5.11	8.5	5.3.2.3		3.1.2
(12)	(13)	(14)	(15)	(16)	(17)	(18)	(19)	(20)	(1)	(2)
SGAH L4BH	TU15 TE19	AT	2 (E)		VC1 VC2 AP7	CV13 CV28	S9	60	2781	BIPYRIDILIUM PESTICIDE, SOLID, TOXIC
L10CH	TU14 TU15 TE21	FL	1 (C/E)			CV13 CV28	S2 S22	336	2782	BIPYRIDILIUM PESTICIDE, LIQUID, FLAMMABLE, TOXIC, flash-point less than 23 °C
L4BH	TU15	FL	2 (D/E)			CV13 CV28	S2 S22	336	2782	BIPYRIDILIUM PESTICIDE, LIQUID, FLAMMABLE, TOXIC, flash-point less than 23 °C
S10AH L10CH	TU14 TU15 TE19 TE21	AT	1 (C/E)	V10		CV1 CV13 CV28	S9 S14	66	2783	ORGANOPHOSPHORUS PESTICIDE, SOLID, TOXIC
SGAH L4BH	TU15 TE19	AT	2 (D/E)	V11		CV13 CV28	S9 S19	60	2783	ORGANOPHOSPHORUS PESTICIDE, SOLID, TOXIC
SGAH L4BH	TU15 TE19	AT	2 (E)		VC1 VC2 AP7	CV13 CV28	S9	60	2783	ORGANOPHOSPHORUS PESTICIDE, SOLID, TOXIC
L10CH	TU14 TU15 TE21	FL	1 (C/E)			CV13 CV28	S2 S22	336	2784	ORGANOPHOSPHORUS PESTICIDE, LIQUID, FLAMMABLE, TOXIC, flash-point less than 23 °C
L4BH	TU15	FL	2 (D/E)			CV13 CV28	S2 S22	336	2784	ORGANOPHOSPHORUS PESTICIDE, LIQUID, FLAMMABLE, TOXIC, flash-point less than 23 °C
L4BH	TU15 TE19	AT	2 (E)	V12		CV13 CV28	S9	60	2785	4-THIAPENTANAL
S10AH L10CH	TU14 TU15 TE19 TE21	AT	1 (C/E)	V10		CV1 CV13 CV28	S9 S14	66	2786	ORGANOTIN PESTICIDE, SOLID, TOXIC
SGAH L4BH	TU15 TE19	AT	2 (D/E)	V11		CV13 CV28	S9 S19	60	2786	ORGANOTIN PESTICIDE, SOLID, TOXIC
SGAH L4BH	TU15 TE19	AT	2 (E)		VC1 VC2 AP7	CV13 CV28	S9	60	2786	ORGANOTIN PESTICIDE, SOLID, TOXIC
L10CH	TU14 TU15 TE21	FL	1 (C/E)			CV13 CV28	S2 S22	336	2787	ORGANOTIN PESTICIDE, LIQUID, FLAMMABLE, TOXIC, flash-point less than 23 °C
L4BH	TU15	FL	2 (D/E)			CV13 CV28	S2 S22	336	2787	ORGANOTIN PESTICIDE, LIQUID, FLAMMABLE, TOXIC, flash-point less than 23 °C
L10CH	TU14 TU15 TE19 TE21	AT	1 (C/E)			CV1 CV13 CV28	S9 S14	66	2788	ORGANOTIN COMPOUND, LIQUID, N.O.S.
L4BH	TU15 TE19	AT	2 (D/E)			CV13 CV28	S9 S19	60	2788	ORGANOTIN COMPOUND, LIQUID, N.O.S.
L4BH	TU15 TE19	AT	2 (E)	V12		CV13 CV28	S9	60	2788	ORGANOTIN COMPOUND, LIQUID, N.O.S.
L4BN		FL	2 (D/E)				S2	83	2789	ACETIC ACID, GLACIAL or ACETIC ACID SOLUTION, more than 80% acid, by mass
L4BN		AT	2 (E)					80	2790	ACETIC ACID SOLUTION, not less than 50% but not more than 80% acid, by mass
L4BN		AT	3 (E)	V12				80	2790	ACETIC ACID SOLUTION, more than 10% and less than 50% acid, by mass
			3 (E)	V1	VC1 VC2 AP1			40	2793	FERROUS METAL BORINGS, SHAVINGS, TURNINGS or CUTTINGS in a form liable to self-heating

UN No.	Name and description	Class	Classification code	Packing group	Labels	Special provisions	Limited and excepted quantities		Packaging			Portable tanks and bulk containers	
									Packing instructions	Special packing provisions	Mixed packing provisions	Instructions	Special provisions
	3.1.2	2.2	2.2	2.1.1.3	5.2.2	3.3	3.4	3.5.1.2	4.1.4	4.1.4	4.1.10	4.2.5.2 7.3.2	4.2.5.3
(1)	(2)	(3a)	(3b)	(4)	(5)	(6)	(7a)	(7b)	(8)	(9a)	(9b)	(10)	(11)
2794	BATTERIES, WET, FILLED WITH ACID, electric storage	8	C11		8	295 598	1 L	E0	P801 P801a				
2795	BATTERIES, WET, FILLED WITH ALKALI, electric storage	8	C11		8	295 598	1 L	E0	P801 P801a				
2796	SULPHURIC ACID with not more than 51% acid or BATTERY FLUID, ACID	8	C1	II	8		1 L	E2	P001 IBC02		MP15	T8	TP2
2797	BATTERY FLUID, ALKALI	8	C5	II	8		1 L	E2	P001 IBC02		MP15	T7	TP2 TP28
2798	PHENYLPHOSPHORUS DICHLORIDE	8	C3	II	8		1 L	E0	P001 IBC02		MP15	T7	TP2
2799	PHENYLPHOSPHORUS THIODICHLORIDE	8	C3	II	8		1 L	E0	P001 IBC02		MP15	T7	TP2
2800	BATTERIES, WET, NON-SPILLABLE, electric storage	8	C11		8	238 295 598	1 L	E0	P003 P801a	PP16			
2801	DYE, LIQUID, CORROSIVE, N.O.S. or DYE INTERMEDIATE, LIQUID, CORROSIVE, N.O.S.	8	C9	I	8	274	0	E0	P001		MP8 MP17	T14	TP2 TP27
2801	DYE, LIQUID, CORROSIVE, N.O.S. or DYE INTERMEDIATE, LIQUID, CORROSIVE, N.O.S.	8	C9	II	8	274	1 L	E2	P001 IBC02		MP15	T11	TP2 TP27
2801	DYE, LIQUID, CORROSIVE, N.O.S. or DYE INTERMEDIATE, LIQUID, CORROSIVE, N.O.S.	8	C9	III	8	274	5 L	E1	P001 IBC03 LP01 R001		MP19	T7	TP1 TP28
2802	COPPER CHLORIDE	8	C2	III	8		5 kg	E1	P002 IBC08 LP02 R001	B3	MP10	T1	TP33
2803	GALLIUM	8	C10	III	8		5 kg	E0	P800	PP41	MP10	T1	TP33
2805	LITHIUM HYDRIDE, FUSED SOLID	4.3	W2	II	4.3		500 g	E2	P410 IBC04	PP40	MP14	T3	TP33
2806	LITHIUM NITRIDE	4.3	W2	I	4.3		0	E0	P403 IBC04		MP2		
2807	Magnetized material	9	M11				NOT SUBJECT TO ADR						
2809	MERCURY	8	CT1	III	8 +6.1	365	5 kg	E0	P800		MP15		
2810	TOXIC LIQUID, ORGANIC, N.O.S.	6.1	T1	I	6.1	274 315 614	0	E5	P001		MP8 MP17	T14	TP2 TP27
2810	TOXIC LIQUID, ORGANIC, N.O.S.	6.1	T1	II	6.1	274 614	100 ml	E4	P001 IBC02		MP15	T11	TP2 TP27
2810	TOXIC LIQUID, ORGANIC, N.O.S.	6.1	T1	III	6.1	274 614	5 L	E1	P001 IBC03 LP01 R001		MP19	T7	TP1 TP28
2811	TOXIC SOLID, ORGANIC, N.O.S.	6.1	T2	I	6.1	274 614	0	E5	P002 IBC07		MP18	T6	TP33
2811	TOXIC SOLID, ORGANIC, N.O.S.	6.1	T2	II	6.1	274 614	500 g	E4	P002 IBC08	B4	MP10	T3	TP33
2811	TOXIC SOLID, ORGANIC, N.O.S.	6.1	T2	III	6.1	274 614	5 kg	E1	P002 IBC08 LP02 R001	B3	MP10	T1	TP33
2812	Sodium aluminate, solid	8	C6				NOT SUBJECT TO ADR						
2813	WATER-REACTIVE SOLID, N.O.S.	4.3	W2	I	4.3	274	0	E0	P403 IBC99		MP2	T9	TP7 TP33
2813	WATER-REACTIVE SOLID, N.O.S.	4.3	W2	II	4.3	274	500 g	E2	P410 IBC07		MP14	T3	TP33
2813	WATER-REACTIVE SOLID, N.O.S.	4.3	W2	III	4.3	274	1 kg	E1	P410 IBC08 R001	B4	MP14	T1	TP33
2814	INFECTIOUS SUBSTANCE, AFFECTING HUMANS	6.2	I1		6.2	318	0	E0	P620		MP5		
2814	INFECTIOUS SUBSTANCE, AFFECTING HUMANS, in refrigerated liquid nitrogen	6.2	I1		6.2 +2.2	318	0	E0	P620		MP5		

ADR tank		Vehicle for tank carriage	Transport category (Tunnel restriction code)	Special provisions for carriage				Hazard identification No.	UN No.	Name and description
Tank code	Special provisions			Packages	Bulk	Loading, unloading and handling	Operation			
4.3	4.3.5, 6.8.4	9.1.1.2	1.1.3.6 (8.6)	7.2.4	7.3.3	7.5.11	8.5	5.3.2.3		3.1.2
(12)	(13)	(14)	(15)	(16)	(17)	(18)	(19)	(20)	(1)	(2)
			3 (E)		VC1 VC2 AP8			80	2794	BATTERIES, WET, FILLED WITH ACID, electric storage
			3 (E)		VC1 VC2 AP8			80	2795	BATTERIES, WET, FILLED WITH ALKALI, electric storage
L4BN		AT	2 (E)					80	2796	SULPHURIC ACID with not more than 51% acid or BATTERY FLUID, ACID
L4BN		AT	2 (E)					80	2797	BATTERY FLUID, ALKALI
L4BN		AT	2 (E)					80	2798	PHENYLPHOSPHORUS DICHLORIDE
L4BN		AT	2 (E)					80	2799	PHENYLPHOSPHORUS THIODICHLORIDE
			3 (E)		VC1 VC2 AP8			80	2800	BATTERIES, WET, NON-SPILLABLE, electric storage
L10BH		AT	1 (E)				S20	88	2801	DYE, LIQUID, CORROSIVE, N.O.S. or DYE INTERMEDIATE, LIQUID, CORROSIVE, N.O.S.
L4BN		AT	2 (E)					80	2801	DYE, LIQUID, CORROSIVE, N.O.S. or DYE INTERMEDIATE, LIQUID, CORROSIVE, N.O.S.
L4BN		AT	3 (E)	V12				80	2801	DYE, LIQUID, CORROSIVE, N.O.S. or DYE INTERMEDIATE, LIQUID, CORROSIVE, N.O.S.
SGAV		AT	3 (E)		VC1 VC2 AP7			80	2802	COPPER CHLORIDE
SGAV L4BN		AT	3 (E)		VC1 VC2 AP7			80	2803	GALLIUM
SGAN		AT	2 (D/E)	V1		CV23		423	2805	LITHIUM HYDRIDE, FUSED SOLID
			1 (E)	V1		CV23	S20		2806	LITHIUM NITRIDE
NOT SUBJECT TO ADR									2807	Magnetized material
L4BN		AT	3 (E)			CV13 CV28		86	2809	MERCURY
L10CH	TU14 TU15 TE19 TE21	AT	1 (C/E)			CV1 CV13 CV28	S9 S14	66	2810	TOXIC LIQUID, ORGANIC, N.O.S.
L4BH	TU15 TE19	AT	2 (D/E)			CV13 CV28	S9 S19	60	2810	TOXIC LIQUID, ORGANIC, N.O.S.
L4BH	TU15 TE19	AT	2 (E)	V12		CV13 CV28	S9	60	2810	TOXIC LIQUID, ORGANIC, N.O.S.
S10AH L10CH	TU15 TE19	AT	1 (C/E)	V10		CV1 CV13 CV28	S9 S14	66	2811	TOXIC SOLID, ORGANIC, N.O.S.
SGAH L4BH	TU15 TE19	AT	2 (D/E)	V11		CV13 CV28	S9 S19	60	2811	TOXIC SOLID, ORGANIC, N.O.S.
SGAH L4BH	TU15 TE19	AT	2 (E)		VC1 VC2 AP7	CV13 CV28	S9	60	2811	TOXIC SOLID, ORGANIC, N.O.S.
NOT SUBJECT TO ADR									2812	Sodium aluminate, solid
S10AN L10DH	TU4 TU14 TU22 TE21 TM2	AT	0 (B/E)	V1		CV23	S20	X423	2813	WATER-REACTIVE SOLID, N.O.S.
SGAN		AT	0 (D/E)	V1		CV23		423	2813	WATER-REACTIVE SOLID, N.O.S.
SGAN		AT	0 (E)	V1	VC1 VC2 AP3 AP4 AP5	CV23		423	2813	WATER-REACTIVE SOLID, N.O.S.
			0 (-)			CV13 CV25 CV26 CV28	S3 S9 S15		2814	INFECTIOUS SUBSTANCE, AFFECTING HUMANS
			0 (E)			CV13 CV25 CV26 CV28	S3 S9 S15		2814	INFECTIOUS SUBSTANCE, AFFECTING HUMANS, in refrigerated liquid nitrogen

UN No.	Name and description	Class	Classification code	Packing group	Labels	Special provisions	Limited and excepted quantities		Packaging			Portable tanks and bulk containers	
									Packing instructions	Special packing provisions	Mixed packing provisions	Instructions	Special provisions
	3.1.2	2.2	2.2	2.1.1.3	5.2.2	3.3	3.4	3.5.1.2	4.1.4	4.1.4	4.1.10	4.2.5.2 7.3.2	4.2.5.3
(1)	(2)	(3a)	(3b)	(4)	(5)	(6)	(7a)	(7b)	(8)	(9a)	(9b)	(10)	(11)
2814	INFECTIOUS SUBSTANCE, AFFECTING HUMANS (animal material only)	6.2	I1		6.2	318	0	E0	P620		MP5	BK1 BK2	
2815	N-AMINOETHYLPIPERAZINE	8	CT1	III	8 +6.1		5 L	E1	P001 IBC03 LP01 R001		MP19	T4	TP1
2817	AMMONIUM HYDROGENDIFLUORIDE SOLUTION	8	CT1	II	8 +6.1		1 L	E2	P001 IBC02		MP15	T8	TP2
2817	AMMONIUM HYDROGENDIFLUORIDE SOLUTION	8	CT1	III	8 +6.1		5 L	E1	P001 IBC03 R001		MP19	T4	TP1
2818	AMMONIUM POLYSULPHIDE SOLUTION	8	CT1	II	8 +6.1		1 L	E2	P001 IBC02		MP15	T7	TP2
2818	AMMONIUM POLYSULPHIDE SOLUTION	8	CT1	III	8 +6.1		5 L	E1	P001 IBC03 R001		MP19	T4	TP1
2819	AMYL ACID PHOSPHATE	8	C3	III	8		5 L	E1	P001 IBC03 LP01 R001		MP19	T4	TP1
2820	BUTYRIC ACID	8	C3	III	8		5 L	E1	P001 IBC03 LP01 R001		MP19	T4	TP1
2821	PHENOL SOLUTION	6.1	T1	II	6.1		100 ml	E4	P001 IBC02		MP15	T7	TP2
2821	PHENOL SOLUTION	6.1	T1	III	6.1		5 L	E1	P001 IBC03 LP01 R001		MP19	T4	TP1
2822	2-CHLOROPYRIDINE	6.1	T1	II	6.1		100 ml	E4	P001 IBC02		MP15	T7	TP2
2823	CROTONIC ACID, SOLID	8	C4	III	8		5 kg	E1	P002 IBC08 LP02 R001	B3	MP10	T1	TP33
2826	ETHYL CHLOROTHIOFORMATE	8	CF1	II	8 +3		0	E0	P001		MP15	T7	TP2
2829	CAPROIC ACID	8	C3	III	8		5 L	E1	P001 IBC03 LP01 R001		MP19	T4	TP1
2830	LITHIUM FERROSILICON	4.3	W2	II	4.3		500 g	E2	P410 IBC07		MP14	T3	TP33
2831	1,1,1-TRICHLOROETHANE	6.1	T1	III	6.1		5 L	E1	P001 IBC03 LP01 R001		MP19	T4	TP1
2834	PHOSPHOROUS ACID	8	C2	III	8		5 kg	E1	P002 IBC08 LP02 R001	B3	MP10	T1	TP33
2835	SODIUM ALUMINIUM HYDRIDE	4.3	W2	II	4.3		500 g	E0	P410 IBC04		MP14	T3	TP33
2837	BISULPHATES, AQUEOUS SOLUTION	8	C1	II	8		1 L	E2	P001 IBC02		MP15	T7	TP2
2837	BISULPHATES, AQUEOUS SOLUTION	8	C1	III	8		5 L	E1	P001 IBC03 LP01 R001		MP19	T4	TP1
2838	VINYL BUTYRATE, STABILIZED	3	F1	II	3	386	1 L	E2	P001 IBC02 R001		MP19	T4	TP1
2839	ALDOL	6.1	T1	II	6.1		100 ml	E4	P001 IBC02		MP15	T7	TP2
2840	BUTYRALDOXIME	3	F1	III	3		5 L	E1	P001 IBC03 LP01 R001		MP19	T2	TP1
2841	DI-n-AMYLAMINE	3	FT1	III	3 +6.1		5 L	E1	P001 IBC03 R001		MP19	T4	TP1

Tank code	Special provisions	Vehicle for tank carriage	Transport category (Tunnel restriction code)	Packages	Bulk	Loading, unloading and handling	Operation	Hazard identification No.	UN No.	Name and description
4.3	4.3.5, 6.8.4	9.1.1.2	1.1.3.6 (8.6)	7.2.4	7.3.3	7.5.11	8.5	5.3.2.3		3.1.2
(12)	(13)	(14)	(15)	(16)	(17)	(18)	(19)	(20)	(1)	(2)
			0 (E)			CV13 CV25 CV26 CV28	S3 S9 S15	606	2814	INFECTIOUS SUBSTANCE, AFFECTING HUMANS (animal material only)
L4BN		AT	3 (E)	V12				86	2815	N-AMINOETHYLPIPERAZINE
L4DH	TU14 TE21	AT	2 (E)			CV13 CV28		86	2817	AMMONIUM HYDROGENDIFLUORIDE SOLUTION
L4DH	TU14 TE21	AT	3 (E)	V12		CV13 CV28		86	2817	AMMONIUM HYDROGENDIFLUORIDE SOLUTION
L4BN		AT	2 (E)			CV13 CV28		86	2818	AMMONIUM POLYSULPHIDE SOLUTION
L4BN		AT	3 (E)	V12		CV13 CV28		86	2818	AMMONIUM POLYSULPHIDE SOLUTION
L4BN		AT	3 (E)	V12				80	2819	AMYL ACID PHOSPHATE
L4BN		AT	3 (E)	V12				80	2820	BUTYRIC ACID
L4BH	TU15 TE19	AT	2 (D/E)			CV13 CV28	S9 S19	60	2821	PHENOL SOLUTION
L4BH	TU15 TE19	AT	2 (E)	V12		CV13 CV28	S9	60	2821	PHENOL SOLUTION
L4BH	TU15 TE19	AT	2 (D/E)			CV13 CV28	S9 S19	60	2822	2-CHLOROPYRIDINE
SGAV L4BN		AT	3 (E)		VC1 VC2 AP7			80	2823	CROTONIC ACID, SOLID
L4BN		FL	2 (D/E)				S2	83	2826	ETHYL CHLOROTHIOFORMATE
L4BN		AT	3 (E)	V12				80	2829	CAPROIC ACID
SGAN		AT	2 (D/E)	V1		CV23		423	2830	LITHIUM FERROSILICON
L4BH	TU15 TE19	AT	2 (E)	V12		CV13 CV28	S9	60	2831	1,1,1-TRICHLOROETHANE
SGAV		AT	3 (E)		VC1 VC2 AP7			80	2834	PHOSPHOROUS ACID
SGAN		AT	2 (D/E)	V1		CV23		423	2835	SODIUM ALUMINIUM HYDRIDE
L4BN		AT	2 (E)					80	2837	BISULPHATES, AQUEOUS SOLUTION
L4BN		AT	3 (E)	V12				80	2837	BISULPHATES, AQUEOUS SOLUTION
LGBF		FL	2 (D/E)	V8			S2 S4 S20	339	2838	VINYL BUTYRATE, STABILIZED
L4BH	TU15 TE19	AT	2 (D/E)			CV13 CV28	S9 S19	60	2839	ALDOL
LGBF		FL	3 (D/E)	V12			S2	30	2840	BUTYRALDOXIME
L4BH	TU15	FL	3 (D/E)	V12		CV13 CV28	S2	36	2841	DI-n-AMYLAMINE

UN No.	Name and description	Class	Classification code	Packing group	Labels	Special provisions	Limited and excepted quantities		Packaging			Portable tanks and bulk containers	
									Packing instructions	Special packing provisions	Mixed packing provisions	Instructions	Special provisions
	3.1.2	2.2	2.2	2.1.1.3	5.2.2	3.3	3.4	3.5.1.2	4.1.4	4.1.4	4.1.10	4.2.5.2 7.3.2	4.2.5.3
(1)	(2)	(3a)	(3b)	(4)	(5)	(6)	(7a)	(7b)	(8)	(9a)	(9b)	(10)	(11)
2842	NITROETHANE	3	F1	III	3		5 L	E1	P001 IBC03 LP01 R001		MP19	T2	TP1
2844	CALCIUM MANGANESE SILICON	4.3	W2	III	4.3		1 kg	E1	P410 IBC08 R001	B4	MP14	T1	TP33
2845	PYROPHORIC LIQUID, ORGANIC, N.O.S.	4.2	S1	I	4.2	274	0	E0	P400		MP2	T22	TP2 TP7
2846	PYROPHORIC SOLID, ORGANIC, N.O.S.	4.2	S2	I	4.2	274	0	E0	P404		MP13		
2849	3-CHLOROPROPANOL-1	6.1	T1	III	6.1		5 L	E1	P001 IBC03 LP01 R001		MP19	T4	TP1
2850	PROPYLENE TETRAMER	3	F1	III	3		5 L	E1	P001 IBC03 LP01 R001		MP19	T2	TP1
2851	BORON TRIFLUORIDE DIHYDRATE	8	C1	II	8		1 L	E2	P001 IBC02		MP15	T7	TP2
2852	DIPICRYL SULPHIDE, WETTED with not less than 10% water, by mass	4.1	D	I	4.1	545	0	E0	P406	PP24	MP2		
2853	MAGNESIUM FLUOROSILICATE	6.1	T5	III	6.1		5 kg	E1	P002 IBC08 LP02 R001	B3	MP10	T1	TP33
2854	AMMONIUM FLUOROSILICATE	6.1	T5	III	6.1		5 kg	E1	P002 IBC08 LP02 R001	B3	MP10	T1	TP33
2855	ZINC FLUOROSILICATE	6.1	T5	III	6.1		5 kg	E1	P002 IBC08 LP02 R001	B3	MP10	T1	TP33
2856	FLUOROSILICATES, N.O.S.	6.1	T5	III	6.1	274	5 kg	E1	P002 IBC08 LP02 R001	B3	MP10	T1	TP33
2857	REFRIGERATING MACHINES containing non-flammable, non-toxic gases or ammonia solutions (UN 2672)	2	6A		2.2	119	0	E0	P003	PP32	MP9		
2858	ZIRCONIUM, DRY, coiled wire, finished metal sheets, strip (thinner than 254 microns but not thinner than 18 microns)	4.1	F3	III	4.1	546	5 kg	E1	P002 LP02 R001		MP11		
2859	AMMONIUM METAVANADATE	6.1	T5	II	6.1		500 g	E4	P002 IBC08	B4	MP10	T3	TP33
2861	AMMONIUM POLYVANADATE	6.1	T5	II	6.1		500 g	E4	P002 IBC08	B4	MP10	T3	TP33
2862	VANADIUM PENTOXIDE, non-fused form	6.1	T5	III	6.1	600	5 kg	E1	P002 IBC08 LP02 R001	B3	MP10	T1	TP33
2863	SODIUM AMMONIUM VANADATE	6.1	T5	II	6.1		500 g	E4	P002 IBC08	B4	MP10	T3	TP33
2864	POTASSIUM METAVANADATE	6.1	T5	II	6.1		500 g	E4	P002 IBC08	B4	MP10	T3	TP33
2865	HYDROXYLAMINE SULPHATE	8	C2	III	8		5 kg	E1	P002 IBC08 LP02 R001	B3	MP10	T1	TP33
2869	TITANIUM TRICHLORIDE MIXTURE	8	C2	II	8		1 kg	E2	P002 IBC08	B4	MP10	T3	TP33
2869	TITANIUM TRICHLORIDE MIXTURE	8	C2	III	8		5 kg	E1	P002 IBC08 LP02 R001	B3	MP10	T1	TP33
2870	ALUMINIUM BOROHYDRIDE	4.2	SW	I	4.2 +4.3		0	E0	P400		MP2	T21	TP7 TP33
2870	ALUMINIUM BOROHYDRIDE IN DEVICES	4.2	SW	I	4.2 +4.3		0	E0	P002	PP13	MP2		

Tank code (4.3)	Special provisions (4.3.5, 6.8.4)	Vehicle for tank carriage (9.1.1.2)	Transport category (Tunnel restriction code) (1.1.3.6) (8.6)	Packages (7.2.4)	Bulk (7.3.3)	Loading, unloading and handling (7.5.11)	Operation (8.5)	Hazard identification No. (5.3.2.3)	UN No. (1)	Name and description (3.1.2)
(12)	(13)	(14)	(15)	(16)	(17)	(18)	(19)	(20)	(1)	(2)
LGBF		FL	3 (D/E)	V12			S2	30	2842	NITROETHANE
SGAN		AT	3 (E)	V1	VC1 VC2 AP3 AP4 AP5	CV23		423	2844	CALCIUM MANGANESE SILICON
L21DH	TU14 TC1 TE21 TM1	AT	0 (B/E)	V1			S20	333	2845	PYROPHORIC LIQUID, ORGANIC, N.O.S.
			0 (E)	V1			S20		2846	PYROPHORIC SOLID, ORGANIC, N.O.S.
L4BH	TU15 TE19	AT	2 (E)	V12		CV13 CV28	S9	60	2849	3-CHLOROPROPANOL-1
LGBF		FL	3 (D/E)	V12			S2	30	2850	PROPYLENE TETRAMER
L4BN		AT	2 (E)					80	2851	BORON TRIFLUORIDE DIHYDRATE
			1 (B)				S14		2852	DIPICRYL SULPHIDE, WETTED with not less than 10% water, by mass
SGAH L4BH	TU15 TE19	AT	2 (E)		VC1 VC2 AP7	CV13 CV28	S9	60	2853	MAGNESIUM FLUOROSILICATE
SGAH L4BH	TU15 TE19	AT	2 (E)		VC1 VC2 AP7	CV13 CV28	S9	60	2854	AMMONIUM FLUOROSILICATE
SGAH L4BH	TU15 TE19	AT	2 (E)		VC1 VC2 AP7	CV13 CV28	S9	60	2855	ZINC FLUOROSILICATE
SGAH L4BH	TU15 TE19	AT	2 (E)		VC1 VC2 AP7	CV13 CV28	S9	60	2856	FLUOROSILICATES, N.O.S.
			3 (E)			CV9			2857	REFRIGERATING MACHINES containing non-flammable, non-toxic gases or ammonia solutions (UN 2672)
			3 (E)		VC1 VC2			40	2858	ZIRCONIUM, DRY, coiled wire, finished metal sheets, strip (thinner than 254 microns but not thinner than 18 microns)
SGAH	TU15 TE19	AT	2 (D/E)	V11		CV13 CV28	S9 S19	60	2859	AMMONIUM METAVANADATE
SGAH	TU15 TE19	AT	2 (D/E)	V11		CV13 CV28	S9 S19	60	2861	AMMONIUM POLYVANADATE
SGAH	TU15 TE19	AT	2 (E)		VC1 VC2 AP7	CV13 CV28	S9	60	2862	VANADIUM PENTOXIDE, non-fused form
SGAH	TU15 TE19	AT	2 (D/E)	V11		CV13 CV28	S9 S19	60	2863	SODIUM AMMONIUM VANADATE
SGAH	TU15 TE19	AT	2 (D/E)	V11		CV13 CV28	S9 S19	60	2864	POTASSIUM METAVANADATE
SGAV		AT	3 (E)		VC1 VC2 AP7			80	2865	HYDROXYLAMINE SULPHATE
SGAN		AT	2 (E)	V11				80	2869	TITANIUM TRICHLORIDE MIXTURE
SGAV		AT	3 (E)		VC1 VC2 AP7			80	2869	TITANIUM TRICHLORIDE MIXTURE
L21DH	TU14 TC1 TE21 TM1	AT	0 (B/E)	V1			S20	X333	2870	ALUMINIUM BOROHYDRIDE
			0 (E)	V1			S20		2870	ALUMINIUM BOROHYDRIDE IN DEVICES

UN No.	Name and description	Class	Classifi-cation code	Packing group	Labels	Special provi-sions	Limited and excepted quantities		Packaging			Portable tanks and bulk containers	
									Packing instruc-tions	Special packing provisions	Mixed packing provisions	Instruc-tions	Special provisions
	3.1.2	2.2	2.2	2.1.1.3	5.2.2	3.3	3.4	3.5.1.2	4.1.4	4.1.4	4.1.10	4.2.5.2 7.3.2	4.2.5.3
(1)	(2)	(3a)	(3b)	(4)	(5)	(6)	(7a)	(7b)	(8)	(9a)	(9b)	(10)	(11)
2871	ANTIMONY POWDER	6.1	T5	III	6.1		5 kg	E1	P002 IBC08 LP02 R001	B3	MP10	T1	TP33
2872	DIBROMOCHLORO-PROPANES	6.1	T1	II	6.1		100 ml	E4	P001 IBC02		MP15	T7	TP2
2872	DIBROMOCHLORO-PROPANES	6.1	T1	III	6.1		5 L	E1	P001 IBC03 LP01 R001		MP19	T4	TP1
2873	DIBUTYLAMINOETHANOL	6.1	T1	III	6.1		5 L	E1	P001 IBC03 LP01 R001		MP19	T4	TP1
2874	FURFURYL ALCOHOL	6.1	T1	III	6.1		5 L	E1	P001 IBC03 LP01 R001		MP19	T4	TP1
2875	HEXACHLOROPHENE	6.1	T2	III	6.1		5 kg	E1	P002 IBC08 LP02 R001	B3	MP10	T1	TP33
2876	RESORCINOL	6.1	T2	III	6.1		5 kg	E1	P002 IBC08 LP02 R001	B3	MP10	T1	TP33
2878	TITANIUM SPONGE GRANULES or TITANIUM SPONGE POWDERS	4.1	F3	III	4.1		5 kg	E1	P002 IBC08 LP02 R001	B3	MP11	T1	TP33
2879	SELENIUM OXYCHLORIDE	8	CT1	I	8 +6.1		0	E0	P001		MP8 MP17	T10	TP2
2880	CALCIUM HYPOCHLORITE, HYDRATED, or CALCIUM HYPOCHLORITE, HYDRATED MIXTURE, with not less than 5.5% but not more than 16% water	5.1	O2	II	5.1	314 322	1 kg	E2	P002 IBC08	B4 B13	MP10		
2880	CALCIUM HYPOCHLORITE, HYDRATED, or CALCIUM HYPOCHLORITE, HYDRATED MIXTURE, with not less than 5.5% but not more than 16% water	5.1	O2	III	5.1	314	5 kg	E1	P002 IBC08 R001	B4 B13	MP10		
2881	METAL CATALYST, DRY	4.2	S4	I	4.2	274	0	E0	P404		MP13	T21	TP7 TP33
2881	METAL CATALYST, DRY	4.2	S4	II	4.2	274	0	E0	P410 IBC06		MP14	T3	TP33
2881	METAL CATALYST, DRY	4.2	S4	III	4.2	274	0	E1	P002 IBC08 LP02 R001	B3	MP14	T1	TP33
2900	INFECTIOUS SUBSTANCE, AFFECTING ANIMALS only	6.2	I2		6.2	318	0	E0	P620		MP5		
2900	INFECTIOUS SUBSTANCE, AFFECTING ANIMALS only, in refrigerated liquid nitrogen	6.2	I2		6.2 +2.2	318	0	E0	P620		MP5		
2900	INFECTIOUS SUBSTANCE, AFFECTING ANIMALS only (animal material only)	6.2	I2		6.2	318	0	E0	P620		MP5	BK1 BK2	
2901	BROMINE CHLORIDE	2	2TOC		2.3 +5.1 +8		0	E0	P200		MP9	(M)	
2902	PESTICIDE, LIQUID, TOXIC, N.O.S.	6.1	T6	I	6.1	61 274 648	0	E5	P001		MP8 MP17	T14	TP2 TP27
2902	PESTICIDE, LIQUID, TOXIC, N.O.S.	6.1	T6	II	6.1	61 274 648	100 ml	E4	P001 IBC02		MP15	T11	TP2 TP27

ADR tank		Vehicle for tank carriage	Transport category (Tunnel restriction code)	Special provisions for carriage				Hazard identification No.	UN No.	Name and description
Tank code	Special provisions			Packages	Bulk	Loading, unloading and handling	Operation			
4.3	4.3.5, 6.8.4	9.1.1.2	1.1.3.6 (8.6)	7.2.4	7.3.3	7.5.11	8.5	5.3.2.3		3.1.2
(12)	(13)	(14)	(15)	(16)	(17)	(18)	(19)	(20)	(1)	(2)
SGAH L4BH	TU15 TE19	AT	2 (E)		VC1 VC2 AP7	CV13 CV28	S9	60	2871	ANTIMONY POWDER
L4BH	TU15 TE19	AT	2 (D/E)			CV13 CV28	S9 S19	60	2872	DIBROMOCHLORO-PROPANES
L4BH	TU15 TE19	AT	2 (E)	V12		CV13 CV28	S9	60	2872	DIBROMOCHLORO-PROPANES
L4BH	TU15 TE19	AT	2 (E)	V12		CV13 CV28	S9	60	2873	DIBUTYLAMINOETHANOL
L4BH	TU15 TE19	AT	2 (E)	V12		CV13 CV28	S9	60	2874	FURFURYL ALCOHOL
SGAH L4BH	TU15 TE19	AT	2 (E)		VC1 VC2 AP7	CV13 CV28	S9	60	2875	HEXACHLOROPHENE
SGAH L4BH	TU15 TE19	AT	2 (E)		VC1 VC2 AP7	CV13 CV28	S9	60	2876	RESORCINOL
SGAV		AT	3 (E)		VC1 VC2			40	2878	TITANIUM SPONGE GRANULES or TITANIUM SPONGE POWDERS
L10BH		AT	1 (C/D)			CV13 CV28	S14	X886	2879	SELENIUM OXYCHLORIDE
SGAN	TU3	AT	2 (E)	V11		CV24 CV35		50	2880	CALCIUM HYPOCHLORITE, HYDRATED, or CALCIUM HYPOCHLORITE, HYDRATED MIXTURE, with not less than 5.5% but not more than 16% water
SGAV	TU3	AT	3 (E)		VC1 VC2 AP6 AP7	CV24 CV35		50	2880	CALCIUM HYPOCHLORITE, HYDRATED, or CALCIUM HYPOCHLORITE, HYDRATED MIXTURE, with not less than 5.5% but not more than 16% water
		AT	0 (B/E)	V1			S20	43	2881	METAL CATALYST, DRY
SGAN		AT	2 (D/E)	V1				40	2881	METAL CATALYST, DRY
SGAN		AT	3 (E)	V1	VC1 VC2 AP1			40	2881	METAL CATALYST, DRY
			0 (-)			CV13 CV25 CV26 CV28	S3 S9 S15		2900	INFECTIOUS SUBSTANCE, AFFECTING ANIMALS only
			0 (E)			CV13 CV25 CV26 CV28	S3 S9 S15		2900	INFECTIOUS SUBSTANCE, AFFECTING ANIMALS only, in refrigerated liquid nitrogen
			0 (E)			CV13 CV25 CV26 CV28	S3 S9 S15	606	2900	INFECTIOUS SUBSTANCE, AFFECTING ANIMALS only (animal material only)
PxBH(M)	TA4 TT9	AT	1 (C/D)			CV9 CV10 CV36	S14	265	2901	BROMINE CHLORIDE
L10CH	TU14 TU15 TE19 TE21	AT	1 (C/E)			CV1 CV13 CV28	S9 S14	66	2902	PESTICIDE, LIQUID, TOXIC, N.O.S.
L4BH	TU15 TE19	AT	2 (D/E)			CV13 CV28	S9 S19	60	2902	PESTICIDE, LIQUID, TOXIC, N.O.S.

UN No.	Name and description	Class	Classifi-cation code	Packing group	Labels	Special provi-sions	Limited and excepted quantities		Packaging			Portable tanks and bulk containers	
									Packing instruc-tions	Special packing provisions	Mixed packing provisions	Instruc-tions	Special provisions
3.1.2		2.2	2.2	2.1.1.3	5.2.2	3.3	3.4	3.5.1.2	4.1.4	4.1.4	4.1.10	4.2.5.2 7.3.2	4.2.5.3
(1)	(2)	(3a)	(3b)	(4)	(5)	(6)	(7a)	(7b)	(8)	(9a)	(9b)	(10)	(11)
2902	PESTICIDE, LIQUID, TOXIC, N.O.S.	6.1	T6	III	6.1	61 274 648	5 L	E1	P001 IBC03 LP01 R001		MP19	T7	TP2 TP28
2903	PESTICIDE, LIQUID, TOXIC, FLAMMABLE, N.O.S., flash-point not less than 23 °C	6.1	TF2	I	6.1 +3	61 274	0	E5	P001		MP8 MP17	T14	TP2 TP27
2903	PESTICIDE, LIQUID, TOXIC, FLAMMABLE, N.O.S., flash-point not less than 23 °C	6.1	TF2	II	6.1 +3	61 274	100 ml	E4	P001 IBC02		MP15	T11	TP2 TP27
2903	PESTICIDE, LIQUID, TOXIC, FLAMMABLE, N.O.S., flash-point not less than 23 °C	6.1	TF2	III	6.1 +3	61 274	5 L	E1	P001 IBC03 R001		MP19	T7	TP2
2904	CHLOROPHENOLATES, LIQUID or PHENOLATES, LIQUID	8	C9	III	8		5 L	E1	P001 IBC03 LP01 R001		MP19		
2905	CHLOROPHENOLATES, SOLID or PHENOLATES, SOLID	8	C10	III	8		5 kg	E1	P002 IBC08 LP02 R001	B3	MP10	T1	TP33
2907	ISOSORBIDE DINITRATE MIXTURE with not less than 60% lactose, mannose, starch or calcium hydrogen phosphate	4.1	D	II	4.1	127	0	E0	P406 IBC06	PP26 PP80 B12	MP2		
2908	RADIOACTIVE MATERIAL, EXCEPTED PACKAGE - EMPTY PACKAGING	7				290	0	E0	See 1.7	See 4.1.9.1.3			
2909	RADIOACTIVE MATERIAL, EXCEPTED PACKAGE - ARTICLES MANUFACTURED FROM NATURAL URANIUM or DEPLETED URANIUM or NATURAL THORIUM	7				290	0	E0	See 1.7	See 4.1.9.1.3			
2910	RADIOACTIVE MATERIAL, EXCEPTED PACKAGE - LIMITED QUANTITY OF MATERIAL	7				290 368	0	E0	See 1.7	See 4.1.9.1.3			
2911	RADIOACTIVE MATERIAL, EXCEPTED PACKAGE - INSTRUMENTS or ARTICLES	7				290	0	E0	See 1.7	See 4.1.9.1.3			
2912	RADIOACTIVE MATERIAL, LOW SPECIFIC ACTIVITY (LSA-I), non fissile or fissile-excepted	7			7X	172 317 325	0	E0	See 2.2.7 and 4.1.9	See 4.1.9.1.3		T5 see 4.1.9.2.4	TP4
2913	RADIOACTIVE MATERIAL, SURFACE CONTAMINATED OBJECTS (SCO-I or SCO-II), non fissile or fissile-excepted	7			7X	172 317 336	0	E0	See 2.2.7 and 4.1.9	See 4.1.9.1.3		see 4.1.9.2.4	
2915	RADIOACTIVE MATERIAL, TYPE A PACKAGE, non-special form, non fissile or fissile-excepted	7			7X	172 317 325	0	E0	See 2.2.7 and 4.1.9	See 4.1.9.1.3			
2916	RADIOACTIVE MATERIAL, TYPE B(U) PACKAGE, non fissile or fissile excepted	7			7X	172 317 325 337	0	E0	See 2.2.7 and 4.1.9	See 4.1.9.1.3			
2917	RADIOACTIVE MATERIAL, TYPE B(M) PACKAGE, non fissile or fissile-excepted	7			7X	172 317 325 337	0	E0	See 2.2.7 and 4.1.9	See 4.1.9.1.3			
2919	RADIOACTIVE MATERIAL, TRANSPORTED UNDER SPECIAL ARRANGEMENT, non fissile or fissile-excepted	7			7X	172 317 325	0	E0	See 2.2.7 and 4.1.9	See 4.1.9.1.3			
2920	CORROSIVE LIQUID, FLAMMABLE, N.O.S.	8	CF1	I	8 +3	274	0	E0	P001		MP8 MP17	T14	TP2 TP27
2920	CORROSIVE LIQUID, FLAMMABLE, N.O.S.	8	CF1	II	8 +3	274	1 L	E2	P001 IBC02		MP15	T11	TP2 TP27
2921	CORROSIVE SOLID, FLAMMABLE, N.O.S.	8	CF2	I	8 +4.1	274	0	E0	P002 IBC05		MP18	T6	TP33
2921	CORROSIVE SOLID, FLAMMABLE, N.O.S.	8	CF2	II	8 +4.1	274	1 kg	E2	P002 IBC08	B4	MP10	T3	TP33
2922	CORROSIVE LIQUID, TOXIC, N.O.S.	8	CT1	I	8 +6.1	274	0	E0	P001		MP8 MP17	T14	TP2 TP27

ADR tank		Vehicle for tank carriage	Transport category (Tunnel restriction code)	Special provisions for carriage				Hazard identification No.	UN No.	Name and description
Tank code	Special provisions			Packages	Bulk	Loading, unloading and handling	Operation			
4.3	4.3.5, 6.8.4	9.1.1.2	1.1.3.6 (8.6)	7.2.4	7.3.3	7.5.11	8.5	5.3.2.3		3.1.2
(12)	(13)	(14)	(15)	(16)	(17)	(18)	(19)	(20)	(1)	(2)
L4BH	TU15 TE19	AT	2 (E)	V12		CV13 CV28	S9	60	2902	PESTICIDE, LIQUID, TOXIC, N.O.S.
L10CH	TU14 TU15 TE19 TE21	FL	1 (C/E)			CV1 CV13 CV28	S2 S9 S14	663	2903	PESTICIDE, LIQUID, TOXIC, FLAMMABLE, N.O.S., flash-point not less than 23 °C
L4BH	TU15 TE19	FL	2 (D/E)			CV13 CV28	S2 S9 S19	63	2903	PESTICIDE, LIQUID, TOXIC, FLAMMABLE, N.O.S., flash-point not less than 23 °C
L4BH	TU15 TE19	FL	2 (D/E)	V12		CV13 CV28	S2 S9	63	2903	PESTICIDE, LIQUID, TOXIC, FLAMMABLE, N.O.S., flash-point not less than 23 °C
L4BN		AT	3 (E)	V12				80	2904	CHLOROPHENOLATES, LIQUID or PHENOLATES, LIQUID
SGAV L4BN		AT	3 (E)		VC1 VC2 AP7			80	2905	CHLOROPHENOLATES, SOLID or PHENOLATES, SOLID
			2 (B)	V11			S14		2907	ISOSORBIDE DINITRATE MIXTURE with not less than 60% lactose, mannose, starch or calcium hydrogen phosphate
			4 (E)			CV33 (See 1.7.1.5.1)	S5 S21		2908	RADIOACTIVE MATERIAL, EXCEPTED PACKAGE - EMPTY PACKAGING
			4 (E)			CV33 (See 1.7.1.5.1)	S5 S21		2909	RADIOACTIVE MATERIAL, EXCEPTED PACKAGE - ARTICLES MANUFACTURED FROM NATURAL URANIUM or DEPLETED URANIUM or NATURAL THORIUM
			4 (E)			CV33 (See 1.7.1.5.1)	S5 S21		2910	RADIOACTIVE MATERIAL, EXCEPTED PACKAGE - LIMITED QUANTITY OF MATERIAL
			4 (E)			CV33 (See 1.7.1.5.1)	S5 S21		2911	RADIOACTIVE MATERIAL, EXCEPTED PACKAGE - INSTRUMENTS or ARTICLES
S2.65AN(+) L2.65CN(+)	TU36 TT7 TM7	AT	0 (E)		see 4.1.9.2.4	CV33	S6 S11 S21	70	2912	RADIOACTIVE MATERIAL, LOW SPECIFIC ACTIVITY (LSA-I), non fissile or fissile-excepted
			0 (E)		see 4.1.9.2.4	CV33	S6 S11 S21	70	2913	RADIOACTIVE MATERIAL, SURFACE CONTAMINATED OBJECTS (SCO-I or SCO-II), non fissile or fissile-excepted
			0 (E)			CV33	S6 S11 S12 S21	70	2915	RADIOACTIVE MATERIAL, TYPE A PACKAGE, non-special form, non fissile or fissile-excepted
			0 (E)			CV33	S6 S11 S21	70	2916	RADIOACTIVE MATERIAL, TYPE B(U) PACKAGE, non fissile or fissile excepted
			0 (E)			CV33	S6 S11 S21	70	2917	RADIOACTIVE MATERIAL, TYPE B(M) PACKAGE, non fissile or fissile-excepted
			0 (-)			CV33	S6 S11 S21	70	2919	RADIOACTIVE MATERIAL, TRANSPORTED UNDER SPECIAL ARRANGEMENT, non fissile or fissile-excepted
L10BH		FL	1 (D/E)				S2 S14	883	2920	CORROSIVE LIQUID, FLAMMABLE, N.O.S.
L4BN		FL	2 (D/E)				S2	83	2920	CORROSIVE LIQUID, FLAMMABLE, N.O.S.
S10AN L10BH		AT	1 (E)	V10			S14	884	2921	CORROSIVE SOLID, FLAMMABLE, N.O.S.
SGAN L4BN		AT	2 (E)	V11				84	2921	CORROSIVE SOLID, FLAMMABLE, N.O.S.
L10BH		AT	1 (C/D)			CV13 CV28	S14	886	2922	CORROSIVE LIQUID, TOXIC, N.O.S.

UN No.	Name and description	Class	Classification code	Packing group	Labels	Special provisions	Limited and excepted quantities		Packaging			Portable tanks and bulk containers	
									Packing instructions	Special packing provisions	Mixed packing provisions	Instructions	Special provisions
	3.1.2	2.2	2.2	2.1.1.3	5.2.2	3.3	3.4	3.5.1.2	4.1.4	4.1.4	4.1.10	4.2.5.2 7.3.2	4.2.5.3
(1)	(2)	(3a)	(3b)	(4)	(5)	(6)	(7a)	(7b)	(8)	(9a)	(9b)	(10)	(11)
2922	CORROSIVE LIQUID, TOXIC, N.O.S.	8	CT1	II	8 +6.1	274	1 L	E2	P001 IBC02		MP15	T7	TP2
2922	CORROSIVE LIQUID, TOXIC, N.O.S.	8	CT1	III	8 +6.1	274	5 L	E1	P001 IBC03 R001		MP19	T7	TP1 TP28
2923	CORROSIVE SOLID, TOXIC, N.O.S.	8	CT2	I	8 +6.1	274	0	E0	P002 IBC05		MP18	T6	TP33
2923	CORROSIVE SOLID, TOXIC, N.O.S.	8	CT2	II	8 +6.1	274	1 kg	E2	P002 IBC08	B4	MP10	T3	TP33
2923	CORROSIVE SOLID, TOXIC, N.O.S.	8	CT2	III	8 +6.1	274	5 kg	E1	P002 IBC08 R001	B3	MP10	T1	TP33
2924	FLAMMABLE LIQUID, CORROSIVE, N.O.S.	3	FC	I	3 +8	274	0	E0	P001		MP7 MP17	T14	TP2
2924	FLAMMABLE LIQUID, CORROSIVE, N.O.S.	3	FC	II	3 +8	274	1 L	E2	P001 IBC02		MP19	T11	TP2 TP27
2924	FLAMMABLE LIQUID, CORROSIVE, N.O.S.	3	FC	III	3 +8	274	5 L	E1	P001 IBC03 R001		MP19	T7	TP1 TP28
2925	FLAMMABLE SOLID, CORROSIVE, ORGANIC, N.O.S.	4.1	FC1	II	4.1 +8	274	1 kg	E2	P002 IBC06		MP10	T3	TP33
2925	FLAMMABLE SOLID, CORROSIVE, ORGANIC, N.O.S.	4.1	FC1	III	4.1 +8	274	5 kg	E1	P002 IBC06 R001		MP10	T1	TP33
2926	FLAMMABLE SOLID, TOXIC, ORGANIC, N.O.S.	4.1	FT1	II	4.1 +6.1	274	1 kg	E2	P002 IBC06		MP10	T3	TP33
2926	FLAMMABLE SOLID, TOXIC, ORGANIC, N.O.S.	4.1	FT1	III	4.1 +6.1	274	5 kg	E1	P002 IBC06 R001		MP10	T1	TP33
2927	TOXIC LIQUID, CORROSIVE, ORGANIC, N.O.S.	6.1	TC1	I	6.1 +8	274 315	0	E5	P001		MP8 MP17	T14	TP2 TP27
2927	TOXIC LIQUID, CORROSIVE, ORGANIC, N.O.S.	6.1	TC1	II	6.1 +8	274	100 ml	E4	P001 IBC02		MP15	T11	TP2 TP27
2928	TOXIC SOLID, CORROSIVE, ORGANIC, N.O.S.	6.1	TC2	I	6.1 +8	274	0	E5	P002 IBC05		MP18	T6	TP33
2928	TOXIC SOLID, CORROSIVE, ORGANIC, N.O.S.	6.1	TC2	II	6.1 +8	274	500 g	E4	P002 IBC06		MP10	T3	TP33
2929	TOXIC LIQUID, FLAMMABLE, ORGANIC, N.O.S.	6.1	TF1	I	6.1 +3	274 315	0	E5	P001		MP8 MP17	T14	TP2 TP27
2929	TOXIC LIQUID, FLAMMABLE, ORGANIC, N.O.S.	6.1	TF1	II	6.1 +3	274	100 ml	E4	P001 IBC02		MP15	T11	TP2 TP27
2930	TOXIC SOLID, FLAMMABLE, ORGANIC, N.O.S.	6.1	TF3	I	6.1 +4.1	274	0	E5	P002 IBC05		MP18	T6	TP33
2930	TOXIC SOLID, FLAMMABLE, ORGANIC, N.O.S.	6.1	TF3	II	6.1 +4.1	274	500 g	E4	P002 IBC08	B4	MP10	T3	TP33
2931	VANADYL SULPHATE	6.1	T5	II	6.1		500 g	E4	P002 IBC08	B4	MP10	T3	TP33
2933	METHYL 2-CHLOROPROPIONATE	3	F1	III	3		5 L	E1	P001 IBC03 LP01 R001		MP19	T2	TP1
2934	ISOPROPYL 2-CHLOROPROPIONATE	3	F1	III	3		5 L	E1	P001 IBC03 LP01 R001		MP19	T2	TP1
2935	ETHYL 2-CHLOROPROPIONATE	3	F1	III	3		5 L	E1	P001 IBC03 LP01 R001		MP19	T2	TP1
2936	THIOLACTIC ACID	6.1	T1	II	6.1		100 ml	E4	P001 IBC02		MP15	T7	TP2
2937	alpha-METHYLBENZYL ALCOHOL, LIQUID	6.1	T1	III	6.1		5 L	E1	P001 IBC03 LP01 R001		MP19	T4	TP1
2940	9-PHOSPHABICYCLO-NONANES (CYCLOOCTADIENE PHOSPHINES)	4.2	S2	II	4.2		0	E2	P410 IBC06		MP14	T3	TP33
2941	FLUOROANILINES	6.1	T1	III	6.1		5 L	E1	P001 IBC03 LP01 R001		MP19	T4	TP1

ADR tank		Vehicle for tank carriage	Transport category (Tunnel restriction code)	Special provisions for carriage				Hazard identification No.	UN No.	Name and description
Tank code	Special provisions			Packages	Bulk	Loading, unloading and handling	Operation			
4.3	4.3.5, 6.8.4	9.1.1.2	1.1.3.6 (8.6)	7.2.4	7.3.3	7.5.11	8.5	5.3.2.3		3.1.2
(12)	(13)	(14)	(15)	(16)	(17)	(18)	(19)	(20)	(1)	(2)
L4BN		AT	2 (E)			CV13 CV28		86	2922	CORROSIVE LIQUID, TOXIC, N.O.S.
L4BN		AT	3 (E)	V12		CV13 CV28		86	2922	CORROSIVE LIQUID, TOXIC, N.O.S.
S10AN L10BH		AT	1 (E)	V10		CV13 CV28	S14	886	2923	CORROSIVE SOLID, TOXIC, N.O.S.
SGAN L4BN		AT	2 (E)	V11		CV13 CV28		86	2923	CORROSIVE SOLID, TOXIC, N.O.S.
SGAV L4BN		AT	3 (E)		VC1 VC2 AP7	CV13 CV28		86	2923	CORROSIVE SOLID, TOXIC, N.O.S.
L10CH	TU14 TE21	FL	1 (C/E)				S2 S20	338	2924	FLAMMABLE LIQUID, CORROSIVE, N.O.S.
L4BH		FL	2 (D/E)				S2 S20	338	2924	FLAMMABLE LIQUID, CORROSIVE, N.O.S.
L4BN		FL	3 (D/E)	V12			S2	38	2924	FLAMMABLE LIQUID, CORROSIVE, N.O.S.
SGAN		AT	2 (E)	V11				48	2925	FLAMMABLE SOLID, CORROSIVE, ORGANIC, N.O.S.
SGAN		AT	3 (E)					48	2925	FLAMMABLE SOLID, CORROSIVE, ORGANIC, N.O.S.
SGAN		AT	2 (E)	V11		CV28		46	2926	FLAMMABLE SOLID, TOXIC, ORGANIC, N.O.S.
SGAN		AT	3 (E)			CV28		46	2926	FLAMMABLE SOLID, TOXIC, ORGANIC, N.O.S.
L10CH	TU14 TU15 TE19 TE21	AT	1 (C/E)			CV1 CV13 CV28	S9 S14	668	2927	TOXIC LIQUID, CORROSIVE, ORGANIC, N.O.S.
L4BH	TU15 TE19	AT	2 (D/E)			CV13 CV28	S9 S19	68	2927	TOXIC LIQUID, CORROSIVE, ORGANIC, N.O.S.
S10AH	TU14 TU15 TE19 TE21	AT	1 (C/E)	V10		CV1 CV13 CV28	S9 S14	668	2928	TOXIC SOLID, CORROSIVE, ORGANIC, N.O.S.
SGAH L4BH	TU15 TE19	AT	2 (D/E)	V11		CV13 CV28	S9 S19	68	2928	TOXIC SOLID, CORROSIVE, ORGANIC, N.O.S.
L10CH	TU14 TU15 TE19 TE21	FL	1 (C/D)			CV1 CV13 CV28	S2 S9 S14	663	2929	TOXIC LIQUID, FLAMMABLE, ORGANIC, N.O.S.
L4BH	TU15 TE19	FL	2 (D/E)			CV13 CV28	S2 S9 S19	63	2929	TOXIC LIQUID, FLAMMABLE, ORGANIC, N.O.S.
		AT	1 (C/E)	V10		CV1 CV13 CV28	S9 S14	664	2930	TOXIC SOLID, FLAMMABLE, ORGANIC, N.O.S.
SGAH L4BH	TU15 TE19	AT	2 (D/E)	V11		CV13 CV28	S9 S19	64	2930	TOXIC SOLID, FLAMMABLE, ORGANIC, N.O.S.
SGAH	TU15 TE19	AT	2 (D/E)	V11		CV13 CV28	S9 S19	60	2931	VANADYL SULPHATE
LGBF		FL	3 (D/E)	V12			S2	30	2933	METHYL 2-CHLOROPROPIONATE
LGBF		FL	3 (D/E)	V12			S2	30	2934	ISOPROPYL 2-CHLOROPROPIONATE
LGBF		FL	3 (D/E)	V12			S2	30	2935	ETHYL 2-CHLOROPROPIONATE
L4BH	TU15 TE19	AT	2 (D/E)			CV13 CV28	S9 S19	60	2936	THIOLACTIC ACID
L4BH	TU15 TE19	AT	2 (E)	V12		CV13 CV28	S9	60	2937	alpha-METHYLBENZYL ALCOHOL, LIQUID
SGAN		AT	2 (D/E)	V1				40	2940	9-PHOSPHABICYCLO-NONANES (CYCLOOCTADIENE PHOSPHINES)
L4BH	TU15 TE19	AT	2 (E)	V12		CV13 CV28	S9	60	2941	FLUOROANILINES

UN No.	Name and description	Class	Classification code	Packing group	Labels	Special provisions	Limited and excepted quantities		Packaging			Portable tanks and bulk containers	
									Packing instructions	Special packing provisions	Mixed packing provisions	Instructions	Special provisions
	3.1.2	2.2	2.2	2.1.1.3	5.2.2	3.3	3.4	3.5.1.2	4.1.4	4.1.4	4.1.10	4.2.5.2 7.3.2	4.2.5.3
(1)	(2)	(3a)	(3b)	(4)	(5)	(6)	(7a)	(7b)	(8)	(9a)	(9b)	(10)	(11)
2942	2-TRIFLUOROMETHYL-ANILINE	6.1	T1	III	6.1		5 L	E1	P001 IBC03 LP01 R001		MP19		
2943	TETRAHYDROFURFURYL-AMINE	3	F1	III	3		5 L	E1	P001 IBC03 LP01 R001		MP19	T2	TP1
2945	N-METHYLBUTYLAMINE	3	FC	II	3 +8		1 L	E2	P001 IBC02		MP19	T7	TP1
2946	2-AMINO-5-DIETHYLAMINOPENTANE	6.1	T1	III	6.1		5 L	E1	P001 IBC03 LP01 R001		MP19	T4	TP1
2947	ISOPROPYL CHLOROACETATE	3	F1	III	3		5 L	E1	P001 IBC03 LP01 R001		MP19	T2	TP1
2948	3-TRIFLUOROMETHYL-ANILINE	6.1	T1	II	6.1		100 ml	E4	P001 IBC02		MP15	T7	TP2
2949	SODIUM HYDROSULPHIDE, HYDRATED with not less than 25% water of crystallization	8	C6	II	8	523	1 kg	E2	P002 IBC08	B4	MP10	T7	TP2
2950	MAGNESIUM GRANULES, COATED, particle size not less than 149 microns	4.3	W2	III	4.3		1 kg	E1	P410 IBC08 R001	B4	MP14	T1 BK2	TP33
2956	5-tert-BUTYL-2,4,6-TRINITRO-m-XYLENE (MUSK XYLENE)	4.1	SR1	III	4.1	638	5 kg	E0	P409		MP2		
2965	BORON TRIFLUORIDE DIMETHYL ETHERATE	4.3	WFC	I	4.3 +3 +8		0	E0	P401		MP2	T10	TP2 TP7
2966	THIOGLYCOL	6.1	T1	II	6.1		100 ml	E4	P001 IBC02		MP15	T7	TP2
2967	SULPHAMIC ACID	8	C2	III	8		5 kg	E1	P002 IBC08 LP02 R001	B3	MP10	T1	TP33
2968	MANEB, STABILIZED or MANEB PREPARATION, STABILIZED against self-heating	4.3	W2	III	4.3	547	1 kg	E1	P002 IBC08 R001	B4	MP14	T1	TP33
2969	CASTOR BEANS or CASTOR MEAL or CASTOR POMACE or CASTOR FLAKE	9	M11	II	9	141	5 kg	E2	P002 IBC08	PP34 B4	MP10	T3 BK1 BK2	TP33
2977	RADIOACTIVE MATERIAL, URANIUM HEXAFLUORIDE, FISSILE	7			7X +7E +6.1 +8		0	E0	See 2.2.7 and 4.1.9	See 4.1.9.1.3			
2978	RADIOACTIVE MATERIAL, URANIUM HEXAFLUORIDE, non fissile or fissile-excepted	7			7X +6.1 +8	317	0	E0	See 2.2.7 and 4.1.9	See 4.1.9.1.3			
2983	ETHYLENE OXIDE AND PROPYLENE OXIDE MIXTURE, not more than 30% ethylene oxide	3	FT1	I	3 +6.1		0	E0	P001		MP7 MP17	T14	TP2 TP7
2984	HYDROGEN PEROXIDE, AQUEOUS SOLUTION with not less than 8% but less than 20% hydrogen peroxide (stabilized as necessary)	5.1	O1	III	5.1	65	5 L	E1	P504 IBC02 R001	PP10 B5	MP15	T4	TP1 TP6 TP24
2985	CHLOROSILANES, FLAMMABLE, CORROSIVE, N.O.S.	3	FC	II	3 +8	548	0	E0	P010		MP19	T14	TP2 TP7 TP27
2986	CHLOROSILANES, CORROSIVE, FLAMMABLE, N.O.S.	8	CF1	II	8 +3	548	0	E0	P010		MP15	T14	TP2 TP7 TP27
2987	CHLOROSILANES, CORROSIVE, N.O.S.	8	C3	II	8	548	0	E0	P010		MP15	T14	TP2 TP7 TP27
2988	CHLOROSILANES, WATER-REACTIVE, FLAMMABLE, CORROSIVE, N.O.S.	4.3	WFC	I	4.3 +3 +8	549	0	E0	P401	RR7	MP2	T14	TP2 TP7
2989	LEAD PHOSPHITE, DIBASIC	4.1	F3	II	4.1		1 kg	E2	P002 IBC08	B4	MP11	T3	TP33

ADR tank		Vehicle for tank carriage	Transport category (Tunnel restriction code)	Special provisions for carriage				Hazard identifi-cation No.	UN No.	Name and description
Tank code	Special provisions			Packages	Bulk	Loading, unloading and handling	Operation			
4.3	4.3.5, 6.8.4	9.1.1.2	1.1.3.6 (8.6)	7.2.4	7.3.3	7.5.11	8.5	5.3.2.3		3.1.2
(12)	(13)	(14)	(15)	(16)	(17)	(18)	(19)	(20)	(1)	(2)
L4BH	TU15 TE19	AT	2 (E)	V12		CV13 CV28	S9	60	2942	2-TRIFLUOROMETHYL-ANILINE
LGBF		FL	3 (D/E)	V12			S2	30	2943	TETRAHYDROFURFURYL-AMINE
L4BH		FL	2 (D/E)				S2 S20	338	2945	N-METHYLBUTYLAMINE
L4BH	TU15 TE19	AT	2 (E)	V12		CV13 CV28	S9	60	2946	2-AMINO-5-DIETHYLAMINOPENTANE
LGBF		FL	3 (D/E)	V12			S2	30	2947	ISOPROPYL CHLOROACETATE
L4BH	TU15 TE19	AT	2 (D/E)			CV13 CV28	S9 S19	60	2948	3-TRIFLUOROMETHYL-ANILINE
SGAN L4BN		AT	2 (E)	V11				80	2949	SODIUM HYDROSULPHIDE, HYDRATED with not less than 25% water of crystallization
SGAN		AT	3 (E)	V1	VC2 AP4 AP5	CV23		423	2950	MAGNESIUM GRANULES, COATED, particle size not less than 149 microns
			3 (D)			CV14	S24		2956	5-tert-BUTYL-2,4,6-TRINITRO-m-XYLENE (MUSK XYLENE)
L10DH	TU4 TU14 TU22 TE21 TM2	FL	0 (B/E)	V1		CV23	S2 S20	382	2965	BORON TRIFLUORIDE DIMETHYL ETHERATE
L4BH	TU15 TE19	AT	2 (D/E)			CV13 CV28	S9 S19	60	2966	THIOGLYCOL
SGAV		AT	3 (E)		VC1 VC2 AP7			80	2967	SULPHAMIC ACID
SGAN		AT	0 (E)	V1	VC1 VC2 AP3 AP4 AP5	CV23		423	2968	MANEB, STABILIZED or MANEB PREPARATION, STABILIZED against self-heating
SGAV		AT	2 (E)	V11	VC1 VC2			90	2969	CASTOR BEANS or CASTOR MEAL or CASTOR POMACE or CASTOR FLAKE
			0 (C)			CV33	S6 S11 S21	768	2977	RADIOACTIVE MATERIAL, URANIUM HEXAFLUORIDE, FISSILE
			0 (C)			CV33	S6 S11 S21	768	2978	RADIOACTIVE MATERIAL, URANIUM HEXAFLUORIDE, non fissile or fissile-excepted
L10CH	TU14 TU15 TE21	FL	1 (C/E)			CV13 CV28	S2 S22	336	2983	ETHYLENE OXIDE AND PROPYLENE OXIDE MIXTURE, not more than 30% ethylene oxide
LGBV	TU3 TC2 TE8 TE11 TT1	AT	3 (E)			CV24		50	2984	HYDROGEN PEROXIDE, AQUEOUS SOLUTION with not less than 8% but less than 20% hydrogen peroxide (stabilized as necessary)
L4BH		FL	2 (D/E)				S2 S20	X338	2985	CHLOROSILANES, FLAMMABLE, CORROSIVE, N.O.S.
L4BN		FL	2 (D/E)				S2	X83	2986	CHLOROSILANES, CORROSIVE, FLAMMABLE, N.O.S.
L4BN		AT	2 (E)					X80	2987	CHLOROSILANES, CORROSIVE, N.O.S.
L10DH	TU14 TU26 TE21 TM2 TM3	FL	0 (B/E)	V1		CV23	S2 S20	X338	2988	CHLOROSILANES, WATER-REACTIVE, FLAMMABLE, CORROSIVE, N.O.S.
SGAN		AT	2 (E)	V11				40	2989	LEAD PHOSPHITE, DIBASIC

UN No.	Name and description	Class	Classifi-cation code	Packing group	Labels	Special provi-sions	Limited and excepted quantities		Packaging			Portable tanks and bulk containers	
									Packing instruc-tions	Special packing provisions	Mixed packing provisions	Instruc-tions	Special provisions
	3.1.2	2.2	2.2	2.1.1.3	5.2.2	3.3	3.4	3.5.1.2	4.1.4	4.1.4	4.1.10	4.2.5.2 7.3.2	4.2.5.3
(1)	(2)	(3a)	(3b)	(4)	(5)	(6)	(7a)	(7b)	(8)	(9a)	(9b)	(10)	(11)
2989	LEAD PHOSPHITE, DIBASIC	4.1	F3	III	4.1		5 kg	E1	P002 IBC08 LP02 R001	B3	MP11	T1	TP33
2990	LIFE-SAVING APPLIANCES, SELF-INFLATING	9	M5		9	296 635	0	E0	P905				
2991	CARBAMATE PESTICIDE, LIQUID, TOXIC, FLAMMABLE, flash-point not less than 23 °C	6.1	TF2	I	6.1 +3	61 274	0	E5	P001		MP8 MP17	T14	TP2 TP27
2991	CARBAMATE PESTICIDE, LIQUID, TOXIC, FLAMMABLE, flash-point not less than 23 °C	6.1	TF2	II	6.1 +3	61 274	100 ml	E4	P001 IBC02		MP15	T11	TP2 TP27
2991	CARBAMATE PESTICIDE, LIQUID, TOXIC, FLAMMABLE, flash-point not less than 23 °C	6.1	TF2	III	6.1 +3	61 274	5 L	E1	P001 IBC03 R001		MP19	T7	TP2 TP28
2992	CARBAMATE PESTICIDE, LIQUID, TOXIC	6.1	T6	I	6.1	61 274 648	0	E5	P001		MP8 MP17	T14	TP2 TP27
2992	CARBAMATE PESTICIDE, LIQUID, TOXIC	6.1	T6	II	6.1	61 274 648	100 ml	E4	P001 IBC02		MP15	T11	TP2 TP27
2992	CARBAMATE PESTICIDE, LIQUID, TOXIC	6.1	T6	III	6.1	61 274 648	5 L	E1	P001 IBC03 LP01 R001		MP19	T7	TP2 TP28
2993	ARSENICAL PESTICIDE, LIQUID, TOXIC, FLAMMABLE, flash-point not less than 23 °C	6.1	TF2	I	6.1 +3	61 274	0	E5	P001		MP8 MP17	T14	TP2 TP27
2993	ARSENICAL PESTICIDE, LIQUID, TOXIC, FLAMMABLE, flash-point not less than 23 °C	6.1	TF2	II	6.1 +3	61 274	100 ml	E4	P001 IBC02		MP15	T11	TP2 TP27
2993	ARSENICAL PESTICIDE, LIQUID, TOXIC, FLAMMABLE, flash-point not less than 23 °C	6.1	TF2	III	6.1 +3	61 274	5 L	E1	P001 IBC03 R001		MP19	T7	TP2 TP28
2994	ARSENICAL PESTICIDE, LIQUID, TOXIC	6.1	T6	I	6.1	61 274 648	0	E5	P001		MP8 MP17	T14	TP2 TP27
2994	ARSENICAL PESTICIDE, LIQUID, TOXIC	6.1	T6	II	6.1	61 274 648	100 ml	E4	P001 IBC02		MP15	T11	TP2 TP27
2994	ARSENICAL PESTICIDE, LIQUID, TOXIC	6.1	T6	III	6.1	61 274 648	5 L	E1	P001 IBC03 LP01 R001		MP19	T7	TP2 TP28
2995	ORGANOCHLORINE PESTICIDE, LIQUID, TOXIC, FLAMMABLE, flash-point not less than 23 °C	6.1	TF2	I	6.1 +3	61 274	0	E5	P001		MP8 MP17	T14	TP2 TP27
2995	ORGANOCHLORINE PESTICIDE, LIQUID, TOXIC, FLAMMABLE, flash-point not less than 23 °C	6.1	TF2	II	6.1 +3	61 274	100 ml	E4	P001 IBC02		MP15	T11	TP2 TP27
2995	ORGANOCHLORINE PESTICIDE, LIQUID, TOXIC, FLAMMABLE, flash-point not less than 23 °C	6.1	TF2	III	6.1 +3	61 274	5 L	E1	P001 IBC03 R001		MP19	T7	TP2 TP28
2996	ORGANOCHLORINE PESTICIDE, LIQUID, TOXIC	6.1	T6	I	6.1	61 274 648	0	E5	P001		MP8 MP17	T14	TP2 TP27
2996	ORGANOCHLORINE PESTICIDE, LIQUID, TOXIC	6.1	T6	II	6.1	61 274 648	100 ml	E4	P001 IBC02		MP15	T11	TP2 TP27
2996	ORGANOCHLORINE PESTICIDE, LIQUID, TOXIC	6.1	T6	III	6.1	61 274 648	5 L	E1	P001 IBC03 LP01 R001		MP19	T7	TP2 TP28
2997	TRIAZINE PESTICIDE, LIQUID, TOXIC, FLAMMABLE, flash-point not less than 23 °C	6.1	TF2	I	6.1 +3	61 274	0	E5	P001		MP8 MP17	T14	TP2 TP27
2997	TRIAZINE PESTICIDE, LIQUID, TOXIC, FLAMMABLE, flash-point not less than 23 °C	6.1	TF2	II	6.1 +3	61 274	100 ml	E4	P001 IBC02		MP15	T11	TP2 TP27
2997	TRIAZINE PESTICIDE, LIQUID, TOXIC, FLAMMABLE, flash-point not less than 23 °C	6.1	TF2	III	6.1 +3	61 274	5 L	E1	P001 IBC03 R001		MP19	T7	TP2 TP28

ADR tank		Vehicle for tank carriage	Transport category (Tunnel restriction code)	Special provisions for carriage				Hazard identifi-cation No.	UN No.	Name and description
Tank code	Special provisions			Packages	Bulk	Loading, unloading and handling	Operation			
4.3	4.3.5, 6.8.4	9.1.1.2	1.1.3.6 (8.6)	7.2.4	7.3.3	7.5.11	8.5	5.3.2.3		3.1.2
(12)	(13)	(14)	(15)	(16)	(17)	(18)	(19)	(20)	(1)	(2)
SGAV		AT	3 (E)		VC1 VC2			40	2989	LEAD PHOSPHITE, DIBASIC
			3 (E)						2990	LIFE-SAVING APPLIANCES, SELF-INFLATING
L10CH	TU14 TU15 TE19 TE21	FL	1 (C/E)			CV1 CV13 CV28	S2 S9 S14	663	2991	CARBAMATE PESTICIDE, LIQUID, TOXIC, FLAMMABLE, flash-point not less than 23 °C
L4BH	TU15 TE19	FL	2 (D/E)			CV13 CV28	S2 S9 S19	63	2991	CARBAMATE PESTICIDE, LIQUID, TOXIC, FLAMMABLE, flash-point not less than 23 °C
L4BH	TU15 TE19	FL	2 (D/E)	V12		CV13 CV28	S2 S9	63	2991	CARBAMATE PESTICIDE, LIQUID, TOXIC, FLAMMABLE, flash-point not less than 23 °C
L10CH	TU14 TU15 TE19 TE21	AT	1 (C/E)			CV1 CV13 CV28	S9 S14	66	2992	CARBAMATE PESTICIDE, LIQUID, TOXIC
L4BH	TU15 TE19	AT	2 (D/E)			CV13 CV28	S9 S19	60	2992	CARBAMATE PESTICIDE, LIQUID, TOXIC
L4BH	TU15 TE19	AT	2 (E)	V12		CV13 CV28	S9	60	2992	CARBAMATE PESTICIDE, LIQUID, TOXIC
L10CH	TU14 TU15 TE19 TE21	FL	1 (C/E)			CV1 CV13 CV28	S2 S9 S14	663	2993	ARSENICAL PESTICIDE, LIQUID, TOXIC, FLAMMABLE, flash-point not less than 23 °C
L4BH	TU15 TE19	FL	2 (D/E)			CV13 CV28	S2 S9 S19	63	2993	ARSENICAL PESTICIDE, LIQUID, TOXIC, FLAMMABLE, flash-point not less than 23 °C
L4BH	TU15 TE19	FL	2 (D/E)	V12		CV13 CV28	S2 S9	63	2993	ARSENICAL PESTICIDE, LIQUID, TOXIC, FLAMMABLE, flash-point not less than 23 °C
L10CH	TU14 TU15 TE19 TE21	AT	1 (C/E)			CV1 CV13 CV28	S9 S14	66	2994	ARSENICAL PESTICIDE, LIQUID, TOXIC
L4BH	TU15 TE19	AT	2 (D/E)			CV13 CV28	S9 S19	60	2994	ARSENICAL PESTICIDE, LIQUID, TOXIC
L4BH	TU15 TE19	AT	2 (E)	V12		CV13 CV28	S9	60	2994	ARSENICAL PESTICIDE, LIQUID, TOXIC
L10CH	TU14 TU15 TE19 TE21	FL	1 (C/E)			CV1 CV13 CV28	S2 S9 S14	663	2995	ORGANOCHLORINE PESTICIDE, LIQUID, TOXIC, FLAMMABLE, flash-point not less than 23 °C
L4BH	TU15 TE19	FL	2 (D/E)			CV13 CV28	S2 S9 S19	63	2995	ORGANOCHLORINE PESTICIDE, LIQUID, TOXIC, FLAMMABLE, flash-point not less than 23 °C
L4BH	TU15 TE19	FL	2 (D/E)	V12		CV13 CV28	S2 S9	63	2995	ORGANOCHLORINE PESTICIDE, LIQUID, TOXIC, FLAMMABLE, flash-point not less than 23 °C
L10CH	TU14 TU15 TE19 TE21	AT	1 (C/E)			CV1 CV13 CV28	S9 S14	66	2996	ORGANOCHLORINE PESTICIDE, LIQUID, TOXIC
L4BH	TU15 TE19	AT	2 (D/E)			CV13 CV28	S9 S19	60	2996	ORGANOCHLORINE PESTICIDE, LIQUID, TOXIC
L4BH	TU15 TE19	AT	2 (E)	V12		CV13 CV28	S9	60	2996	ORGANOCHLORINE PESTICIDE, LIQUID, TOXIC
L10CH	TU14 TU15 TE19 TE21	FL	1 (C/E)			CV1 CV13 CV28	S2 S9 S14	663	2997	TRIAZINE PESTICIDE, LIQUID, TOXIC, FLAMMABLE, flash-point not less than 23 °C
L4BH	TU15 TE19	FL	2 (D/E)			CV13 CV28	S2 S9 S19	63	2997	TRIAZINE PESTICIDE, LIQUID, TOXIC, FLAMMABLE, flash-point not less than 23 °C
L4BH	TU15 TE19	FL	2 (D/E)	V12		CV13 CV28	S2 S9	63	2997	TRIAZINE PESTICIDE, LIQUID, TOXIC, FLAMMABLE, flash-point not less than 23 °C

UN No.	Name and description	Class	Classification code	Packing group	Labels	Special provisions	Limited and excepted quantities		Packaging			Portable tanks and bulk containers	
									Packing instructions	Special packing provisions	Mixed packing provisions	Instructions	Special provisions
	3.1.2	2.2	2.2	2.1.1.3	5.2.2	3.3	3.4	3.5.1.2	4.1.4	4.1.4	4.1.10	4.2.5.2 7.3.2	4.2.5.3
(1)	(2)	(3a)	(3b)	(4)	(5)	(6)	(7a)	(7b)	(8)	(9a)	(9b)	(10)	(11)
2998	TRIAZINE PESTICIDE, LIQUID, TOXIC	6.1	T6	I	6.1	61 274 648	0	E5	P001		MP8 MP17	T14	TP2 TP27
2998	TRIAZINE PESTICIDE, LIQUID, TOXIC	6.1	T6	II	6.1	61 274 648	100 ml	E4	P001 IBC02		MP15	T11	TP2 TP27
2998	TRIAZINE PESTICIDE, LIQUID, TOXIC	6.1	T6	III	6.1	61 274 648	5 L	E1	P001 IBC03 LP01 R001		MP19	T7	TP2 TP28
3005	THIOCARBAMATE PESTICIDE, LIQUID, TOXIC, FLAMMABLE, flash-point not less than 23 °C	6.1	TF2	I	6.1 +3	61 274	0	E5	P001		MP8 MP17	T14	TP2
3005	THIOCARBAMATE PESTICIDE, LIQUID, TOXIC, FLAMMABLE, flash-point not less than 23 °C	6.1	TF2	II	6.1 +3	61 274	100 ml	E4	P001 IBC02		MP15	T11	TP2 TP27
3005	THIOCARBAMATE PESTICIDE, LIQUID, TOXIC, FLAMMABLE, flash-point not less than 23 °C	6.1	TF2	III	6.1 +3	61 274	5 L	E1	P001 IBC03 R001		MP19	T7	TP2 TP28
3006	THIOCARBAMATE PESTICIDE, LIQUID, TOXIC	6.1	T6	I	6.1	61 274 648	0	E5	P001		MP8 MP17	T14	TP2
3006	THIOCARBAMATE PESTICIDE, LIQUID, TOXIC	6.1	T6	II	6.1	61 274 648	100 ml	E4	P001 IBC02		MP15	T11	TP2 TP27
3006	THIOCARBAMATE PESTICIDE, LIQUID, TOXIC	6.1	T6	III	6.1	61 274 648	5 L	E1	P001 IBC03 LP01 R001		MP19	T7	TP2 TP28
3009	COPPER BASED PESTICIDE, LIQUID, TOXIC, FLAMMABLE, flash-point not less than 23 °C	6.1	TF2	I	6.1 +3	61 274	0	E5	P001		MP8 MP17	T14	TP2 TP27
3009	COPPER BASED PESTICIDE, LIQUID, TOXIC, FLAMMABLE, flash-point not less than 23 °C	6.1	TF2	II	6.1 +3	61 274	100 ml	E4	P001 IBC02		MP15	T11	TP2 TP27
3009	COPPER BASED PESTICIDE, LIQUID, TOXIC, FLAMMABLE, flash-point not less than 23 °C	6.1	TF2	III	6.1 +3	61 274	5 L	E1	P001 IBC03 R001		MP19	T7	TP2 TP28
3010	COPPER BASED PESTICIDE, LIQUID, TOXIC	6.1	T6	I	6.1	61 274 648	0	E5	P001		MP8 MP17	T14	TP2 TP27
3010	COPPER BASED PESTICIDE, LIQUID, TOXIC	6.1	T6	II	6.1	61 274 648	100 ml	E4	P001 IBC02		MP15	T11	TP2 TP27
3010	COPPER BASED PESTICIDE, LIQUID, TOXIC	6.1	T6	III	6.1	61 274 648	5 L	E1	P001 IBC03 LP01 R001		MP19	T7	TP2 TP28
3011	MERCURY BASED PESTICIDE, LIQUID, TOXIC, FLAMMABLE, flash-point not less than 23 °C	6.1	TF2	I	6.1 +3	61 274	0	E5	P001		MP8 MP17	T14	TP2 TP27
3011	MERCURY BASED PESTICIDE, LIQUID, TOXIC, FLAMMABLE, flash-point not less than 23 °C	6.1	TF2	II	6.1 +3	61 274	100 ml	E4	P001 IBC02		MP15	T11	TP2 TP27
3011	MERCURY BASED PESTICIDE, LIQUID, TOXIC, FLAMMABLE, flash-point not less than 23 °C	6.1	TF2	III	6.1 +3	61 274	5 L	E1	P001 IBC03 R001		MP19	T7	TP2 TP28
3012	MERCURY BASED PESTICIDE, LIQUID, TOXIC	6.1	T6	I	6.1	61 274 648	0	E5	P001		MP8 MP17	T14	TP2 TP27
3012	MERCURY BASED PESTICIDE, LIQUID, TOXIC	6.1	T6	II	6.1	61 274 648	100 ml	E4	P001 IBC02		MP15	T11	TP2 TP27
3012	MERCURY BASED PESTICIDE, LIQUID, TOXIC	6.1	T6	III	6.1	61 274 648	5 L	E1	P001 IBC03 LP01 R001		MP19	T7	TP2 TP28
3013	SUBSTITUTED NITROPHENOL PESTICIDE, LIQUID, TOXIC, FLAMMABLE, flash-point not less than 23 °C	6.1	TF2	I	6.1 +3	61 274	0	E5	P001		MP8 MP17	T14	TP2 TP27
3013	SUBSTITUTED NITROPHENOL PESTICIDE, LIQUID, TOXIC, FLAMMABLE, flash-point not less than 23 °C	6.1	TF2	II	6.1 +3	61 274	100 ml	E4	P001 IBC02		MP15	T11	TP2 TP27

ADR tank		Vehicle for tank carriage	Transport category (Tunnel restriction code)	Special provisions for carriage				Hazard identification No.	UN No.	Name and description
Tank code	Special provisions			Packages	Bulk	Loading, unloading and handling	Operation			
4.3	4.3.5, 6.8.4	9.1.1.2	1.1.3.6 (8.6)	7.2.4	7.3.3	7.5.11	8.5	5.3.2.3		3.1.2
(12)	(13)	(14)	(15)	(16)	(17)	(18)	(19)	(20)	(1)	(2)
L10CH	TU14 TU15 TE19 TE21	AT	1 (C/E)			CV1 CV13 CV28	S9 S14	66	2998	TRIAZINE PESTICIDE, LIQUID, TOXIC
L4BH	TU15 TE19	AT	2 (D/E)			CV13 CV28	S9 S19	60	2998	TRIAZINE PESTICIDE, LIQUID, TOXIC
L4BH	TU15 TE19	AT	2 (E)	V12		CV13 CV28	S9	60	2998	TRIAZINE PESTICIDE, LIQUID, TOXIC
L10CH	TU14 TU15 TE19 TE21	FL	1 (C/E)			CV1 CV13 CV28	S2 S9 S14	663	3005	THIOCARBAMATE PESTICIDE, LIQUID, TOXIC, FLAMMABLE, flash-point not less than 23 °C
L4BH	TU15 TE19	FL	2 (D/E)			CV13 CV28	S2 S9 S19	63	3005	THIOCARBAMATE PESTICIDE, LIQUID, TOXIC, FLAMMABLE, flash-point not less than 23 °C
L4BH	TU15 TE19	FL	2 (D/E)	V12		CV13 CV28	S2 S9	63	3005	THIOCARBAMATE PESTICIDE, LIQUID, TOXIC, FLAMMABLE, flash-point not less than 23 °C
L10CH	TU14 TU15 TE19 TE21	AT	1 (C/E)			CV1 CV13 CV28	S9 S14	66	3006	THIOCARBAMATE PESTICIDE, LIQUID, TOXIC
L4BH	TU15 TE19	AT	2 (D/E)			CV13 CV28	S9 S19	60	3006	THIOCARBAMATE PESTICIDE, LIQUID, TOXIC
L4BH	TU15 TE19	AT	2 (E)	V12		CV13 CV28	S9	60	3006	THIOCARBAMATE PESTICIDE, LIQUID, TOXIC
L10CH	TU14 TU15 TE19 TE21	FL	1 (C/E)			CV1 CV13 CV28	S2 S9 S14	663	3009	COPPER BASED PESTICIDE, LIQUID, TOXIC, FLAMMABLE, flash-point not less than 23 °C
L4BH	TU15 TE19	FL	2 (D/E)			CV13 CV28	S2 S9 S19	63	3009	COPPER BASED PESTICIDE, LIQUID, TOXIC, FLAMMABLE, flash-point not less than 23 °C
L4BH	TU15 TE19	FL	2 (D/E)	V12		CV13 CV28	S2 S9	63	3009	COPPER BASED PESTICIDE, LIQUID, TOXIC, FLAMMABLE, flash-point not less than 23 °C
L10CH	TU14 TU15 TE19 TE21	AT	1 (C/E)			CV1 CV13 CV28	S9 S14	66	3010	COPPER BASED PESTICIDE, LIQUID, TOXIC
L4BH	TU15 TE19	AT	2 (D/E)			CV13 CV28	S9 S19	60	3010	COPPER BASED PESTICIDE, LIQUID, TOXIC
L4BH	TU15 TE19	AT	2 (E)	V12		CV13 CV28	S9	60	3010	COPPER BASED PESTICIDE, LIQUID, TOXIC
L10CH	TU14 TU15 TE19 TE21	FL	1 (C/E)			CV1 CV13 CV28	S2 S9 S14	663	3011	MERCURY BASED PESTICIDE, LIQUID, TOXIC, FLAMMABLE, flash-point not less than 23 °C
L4BH	TU15 TE19	FL	2 (D/E)			CV13 CV28	S2 S9 S19	63	3011	MERCURY BASED PESTICIDE, LIQUID, TOXIC, FLAMMABLE, flash-point not less than 23 °C
L4BH	TU15 TE19	FL	2 (D/E)	V12		CV13 CV28	S2 S9	63	3011	MERCURY BASED PESTICIDE, LIQUID, TOXIC, FLAMMABLE, flash-point not less than 23 °C
L10CH	TU14 TU15 TE19 TE21	AT	1 (C/E)			CV1 CV13 CV28	S9 S14	66	3012	MERCURY BASED PESTICIDE, LIQUID, TOXIC
L4BH	TU15 TE19	AT	2 (D/E)			CV13 CV28	S9 S19	60	3012	MERCURY BASED PESTICIDE, LIQUID, TOXIC
L4BH	TU15 TE19	AT	2 (E)	V12		CV13 CV28	S9	60	3012	MERCURY BASED PESTICIDE, LIQUID, TOXIC
L10CH	TU14 TU15 TE19 TE21	FL	1 (C/E)			CV1 CV13 CV28	S2 S9 S14	663	3013	SUBSTITUTED NITROPHENOL PESTICIDE, LIQUID, TOXIC, FLAMMABLE, flash-point not less than 23 °C
L4BH	TU15 TE19	FL	2 (D/E)			CV13 CV28	S2 S9 S19	63	3013	SUBSTITUTED NITROPHENOL PESTICIDE, LIQUID, TOXIC, FLAMMABLE, flash-point not less than 23 °C

UN No.	Name and description	Class	Classifi-cation code	Packing group	Labels	Special provi-sions	Limited and excepted quantities		Packaging			Portable tanks and bulk containers	
									Packing instruc-tions	Special packing provisions	Mixed packing provisions	Instruc-tions	Special provisions
3.1.2	3.1.2	2.2	2.2	2.1.1.3	5.2.2	3.3	3.4	3.5.1.2	4.1.4	4.1.4	4.1.10	4.2.5.2 7.3.2	4.2.5.3
(1)	(2)	(3a)	(3b)	(4)	(5)	(6)	(7a)	(7b)	(8)	(9a)	(9b)	(10)	(11)
3013	SUBSTITUTED NITROPHENOL PESTICIDE, LIQUID, TOXIC, FLAMMABLE, flash-point not less than 23 °C	6.1	TF2	III	6.1 +3	61 274	5 L	E1	P001 IBC03 R001		MP19	T7	TP2 TP28
3014	SUBSTITUTED NITROPHENOL PESTICIDE, LIQUID, TOXIC	6.1	T6	I	6.1	61 274 648	0	E5	P001		MP8 MP17	T14	TP2 TP27
3014	SUBSTITUTED NITROPHENOL PESTICIDE, LIQUID, TOXIC	6.1	T6	II	6.1	61 274 648	100 ml	E4	P001 IBC02		MP15	T11	TP2 TP27
3014	SUBSTITUTED NITROPHENOL PESTICIDE, LIQUID, TOXIC	6.1	T6	III	6.1	61 274 648	5 L	E1	P001 IBC03 LP01 R001		MP19	T7	TP2 TP28
3015	BIPYRIDILIUM PESTICIDE, LIQUID, TOXIC, FLAMMABLE, flash-point not less than 23 °C	6.1	TF2	I	6.1 +3	61 274	0	E5	P001		MP8 MP17	T14	TP2 TP27
3015	BIPYRIDILIUM PESTICIDE, LIQUID, TOXIC, FLAMMABLE, flash-point not less than 23 °C	6.1	TF2	II	6.1 +3	61 274	100 ml	E4	P001 IBC02		MP15	T11	TP2 TP27
3015	BIPYRIDILIUM PESTICIDE, LIQUID, TOXIC, FLAMMABLE, flash-point not less than 23 °C	6.1	TF2	III	6.1 +3	61 274	5 L	E1	P001 IBC03 R001		MP19	T7	TP2 TP28
3016	BIPYRIDILIUM PESTICIDE, LIQUID, TOXIC	6.1	T6	I	6.1	61 274 648	0	E5	P001		MP8 MP17	T14	TP2 TP27
3016	BIPYRIDILIUM PESTICIDE, LIQUID, TOXIC	6.1	T6	II	6.1	61 274 648	100 ml	E4	P001 IBC02		MP15	T11	TP2 TP27
3016	BIPYRIDILIUM PESTICIDE, LIQUID, TOXIC	6.1	T6	III	6.1	61 274 648	5 L	E1	P001 IBC03 LP01 R001		MP19	T7	TP2 TP28
3017	ORGANOPHOSPHORUS PESTICIDE, LIQUID, TOXIC, FLAMMABLE, flash-point not less than 23 °C	6.1	TF2	I	6.1 +3	61 274	0	E5	P001		MP8 MP17	T14	TP2 TP27
3017	ORGANOPHOSPHORUS PESTICIDE, LIQUID, TOXIC, FLAMMABLE, flash-point not less than 23 °C	6.1	TF2	II	6.1 +3	61 274	100 ml	E4	P001 IBC02		MP15	T11	TP2 TP27
3017	ORGANOPHOSPHORUS PESTICIDE, LIQUID, TOXIC, FLAMMABLE, flash-point not less than 23 °C	6.1	TF2	III	6.1 +3	61 274	5 L	E1	P001 IBC03 R001		MP19	T7	TP2 TP28
3018	ORGANOPHOSPHORUS PESTICIDE, LIQUID, TOXIC	6.1	T6	I	6.1	61 274 648	0	E5	P001		MP8 MP17	T14	TP2 TP27
3018	ORGANOPHOSPHORUS PESTICIDE, LIQUID, TOXIC	6.1	T6	II	6.1	61 274 648	100 ml	E4	P001 IBC02		MP15	T11	TP2 TP27
3018	ORGANOPHOSPHORUS PESTICIDE, LIQUID, TOXIC	6.1	T6	III	6.1	61 274 648	5 L	E1	P001 IBC03 LP01 R001		MP19	T7	TP2 TP28
3019	ORGANOTIN PESTICIDE, LIQUID, TOXIC, FLAMMABLE, flash-point not less than 23 °C	6.1	TF2	I	6.1 +3	61 274	0	E5	P001		MP8 MP17	T14	TP2 TP27
3019	ORGANOTIN PESTICIDE, LIQUID, TOXIC, FLAMMABLE, flash-point not less than 23 °C	6.1	TF2	II	6.1 +3	61 274	100 ml	E4	P001 IBC02		MP15	T11	TP2 TP27
3019	ORGANOTIN PESTICIDE, LIQUID, TOXIC, FLAMMABLE, flash-point not less than 23 °C	6.1	TF2	III	6.1 +3	61 274	5 L	E1	P001 IBC03 R001		MP19	T7	TP2 TP28
3020	ORGANOTIN PESTICIDE, LIQUID, TOXIC	6.1	T6	I	6.1	61 274 648	0	E5	P001		MP8 MP17	T14	TP2 TP27
3020	ORGANOTIN PESTICIDE, LIQUID, TOXIC	6.1	T6	II	6.1	61 274 648	100 ml	E4	P001 IBC02		MP15	T11	TP2 TP27
3020	ORGANOTIN PESTICIDE, LIQUID, TOXIC	6.1	T6	III	6.1	61 274 648	5 L	E1	P001 IBC03 LP01 R001		MP19	T7	TP2 TP28

ADR tank		Vehicle for tank carriage	Transport category (Tunnel restriction code)	Special provisions for carriage				Hazard identifi-cation No.	UN No.	Name and description
Tank code	Special provisions			Packages	Bulk	Loading, unloading and handling	Operation			
4.3	4.3.5, 6.8.4	9.1.1.2	1.1.3.6 (8.6)	7.2.4	7.3.3	7.5.11	8.5	5.3.2.3		3.1.2
(12)	(13)	(14)	(15)	(16)	(17)	(18)	(19)	(20)	(1)	(2)
L4BH	TU15 TE19	FL	2 (D/E)	V12		CV13 CV28	S2 S9	63	3013	SUBSTITUTED NITROPHENOL PESTICIDE, LIQUID, TOXIC, FLAMMABLE, flash-point not less than 23 °C
L10CH	TU14 TU15 TE19 TE21	AT	1 (C/E)			CV1 CV13 CV28	S9 S14	66	3014	SUBSTITUTED NITROPHENOL PESTICIDE, LIQUID, TOXIC
L4BH	TU15 TE19	AT	2 (D/E)			CV13 CV28	S9 S19	60	3014	SUBSTITUTED NITROPHENOL PESTICIDE, LIQUID, TOXIC
L4BH	TU15 TE19	AT	2 (E)	V12		CV13 CV28	S9	60	3014	SUBSTITUTED NITROPHENOL PESTICIDE, LIQUID, TOXIC
L10CH	TU14 TU15 TE19 TE21	FL	1 (C/E)			CV1 CV13 CV28	S2 S9 S14	663	3015	BIPYRIDILIUM PESTICIDE, LIQUID, TOXIC, FLAMMABLE, flash-point not less than 23 °C
L4BH	TU15 TE19	FL	2 (D/E)			CV13 CV28	S2 S9 S19	63	3015	BIPYRIDILIUM PESTICIDE, LIQUID, TOXIC, FLAMMABLE, flash-point not less than 23 °C
L4BH	TU15 TE19	FL	2 (D/E)	V12		CV13 CV28	S2 S9	63	3015	BIPYRIDILIUM PESTICIDE, LIQUID, TOXIC, FLAMMABLE, flash-point not less than 23 °C
L10CH	TU14 TU15 TE19 TE21	AT	1 (C/E)			CV1 CV13 CV28	S9 S14	66	3016	BIPYRIDILIUM PESTICIDE, LIQUID, TOXIC
L4BH	TU15 TE19	AT	2 (D/E)			CV13 CV28	S9 S19	60	3016	BIPYRIDILIUM PESTICIDE, LIQUID, TOXIC
L4BH	TU15 TE19	AT	2 (E)	V12		CV13 CV28	S9	60	3016	BIPYRIDILIUM PESTICIDE, LIQUID, TOXIC
L10CH	TU14 TU15 TE19 TE21	FL	1 (C/E)			CV1 CV13 CV28	S2 S9 S14	663	3017	ORGANOPHOSPHORUS PESTICIDE, LIQUID, TOXIC, FLAMMABLE, flash-point not less than 23 °C
L4BH	TU15 TE19	FL	2 (D/E)			CV13 CV28	S2 S9 S19	63	3017	ORGANOPHOSPHORUS PESTICIDE, LIQUID, TOXIC, FLAMMABLE, flash-point not less than 23 °C
L4BH	TU15 TE19	FL	2 (D/E)	V12		CV13 CV28	S2 S9	63	3017	ORGANOPHOSPHORUS PESTICIDE, LIQUID, TOXIC, FLAMMABLE, flash-point not less than 23 °C
L10CH	TU14 TU15 TE19 TE21	AT	1 (C/E)			CV1 CV13 CV28	S9 S14	66	3018	ORGANOPHOSPHORUS PESTICIDE, LIQUID, TOXIC
L4BH	TU15 TE19	AT	2 (D/E)			CV13 CV28	S9 S19	60	3018	ORGANOPHOSPHORUS PESTICIDE, LIQUID, TOXIC
L4BH	TU15 TE19	AT	2 (E)	V12		CV13 CV28	S9	60	3018	ORGANOPHOSPHORUS PESTICIDE, LIQUID, TOXIC
L10CH	TU14 TU15 TE19 TE21	FL	1 (C/E)			CV1 CV13 CV28	S2 S9 S14	663	3019	ORGANOTIN PESTICIDE, LIQUID, TOXIC, FLAMMABLE, flash-point not less than 23 °C
L4BH	TU15 TE19	FL	2 (D/E)			CV13 CV28	S2 S9 S19	63	3019	ORGANOTIN PESTICIDE, LIQUID, TOXIC, FLAMMABLE, flash-point not less than 23 °C
L4BH	TU15 TE19	FL	2 (D/E)	V12		CV13 CV28	S2 S9	63	3019	ORGANOTIN PESTICIDE, LIQUID, TOXIC, FLAMMABLE, flash-point not less than 23 °C
L10CH	TU14 TU15 TE19 TE21	AT	1 (C/E)			CV1 CV13 CV28	S9 S14	66	3020	ORGANOTIN PESTICIDE, LIQUID, TOXIC
L4BH	TU15 TE19	AT	2 (D/E)			CV13 CV28	S9 S19	60	3020	ORGANOTIN PESTICIDE, LIQUID, TOXIC
L4BH	TU15 TE19	AT	2 (E)	V12		CV13 CV28	S9	60	3020	ORGANOTIN PESTICIDE, LIQUID, TOXIC

UN No.	Name and description	Class	Classification code	Packing group	Labels	Special provisions	Limited and excepted quantities		Packaging			Portable tanks and bulk containers	
									Packing instructions	Special packing provisions	Mixed packing provisions	Instructions	Special provisions
	3.1.2	2.2	2.2	2.1.1.3	5.2.2	3.3	3.4	3.5.1.2	4.1.4	4.1.4	4.1.10	4.2.5.2 7.3.2	4.2.5.3
(1)	(2)	(3a)	(3b)	(4)	(5)	(6)	(7a)	(7b)	(8)	(9a)	(9b)	(10)	(11)
3021	PESTICIDE, LIQUID, FLAMMABLE, TOXIC, N.O.S., flash-point less than 23 °C	3	FT2	I	3 +6.1	61 274	0	E0	P001		MP7 MP17	T14	TP2 TP27
3021	PESTICIDE, LIQUID, FLAMMABLE, TOXIC, N.O.S., flash-point less than 23 °C	3	FT2	II	3 +6.1	61 274	1 L	E2	P001 IBC02 R001		MP19	T11	TP2 TP27
3022	1,2-BUTYLENE OXIDE, STABILIZED	3	F1	II	3	386	1 L	E2	P001 IBC02 R001		MP19	T4	TP1
3023	2-METHYL-2-HEPTANETHIOL	6.1	TF1	I	6.1 +3	354	0	E0	P602		MP8 MP17	T20	TP2
3024	COUMARIN DERIVATIVE PESTICIDE, LIQUID, FLAMMABLE, TOXIC, flash-point less than 23 °C	3	FT2	I	3 +6.1	61 274	0	E0	P001		MP7 MP17	T14	TP2 TP27
3024	COUMARIN DERIVATIVE PESTICIDE, LIQUID, FLAMMABLE, TOXIC, flash-point less than 23 °C	3	FT2	II	3 +6.1	61 274	1 L	E2	P001 IBC02 R001		MP19	T11	TP2 TP27
3025	COUMARIN DERIVATIVE PESTICIDE, LIQUID, TOXIC, FLAMMABLE, flash-point not less than 23 °C	6.1	TF2	I	6.1 +3	61 274	0	E5	P001		MP8 MP17	T14	TP2 TP27
3025	COUMARIN DERIVATIVE PESTICIDE, LIQUID, TOXIC, FLAMMABLE, flash-point not less than 23 °C	6.1	TF2	II	6.1 +3	61 274	100 ml	E4	P001 IBC02		MP15	T11	TP2 TP27
3025	COUMARIN DERIVATIVE PESTICIDE, LIQUID, TOXIC, FLAMMABLE, flash-point not less than 23 °C	6.1	TF2	III	6.1 +3	61 274	5 L	E1	P001 IBC03 R001		MP19	T7	TP1 TP28
3026	COUMARIN DERIVATIVE PESTICIDE, LIQUID, TOXIC	6.1	T6	I	6.1	61 274 648	0	E5	P001		MP8 MP17	T14	TP2 TP27
3026	COUMARIN DERIVATIVE PESTICIDE, LIQUID, TOXIC	6.1	T6	II	6.1	61 274 648	100 ml	E4	P001 IBC02		MP15	T11	TP2 TP27
3026	COUMARIN DERIVATIVE PESTICIDE, LIQUID, TOXIC	6.1	T6	III	6.1	61 274 648	5 L	E1	P001 IBC03 LP01 R001		MP19	T7	TP1 TP28
3027	COUMARIN DERIVATIVE PESTICIDE, SOLID, TOXIC	6.1	T7	I	6.1	61 274 648	0	E5	P002 IBC07		MP18	T6	TP33
3027	COUMARIN DERIVATIVE PESTICIDE, SOLID, TOXIC	6.1	T7	II	6.1	61 274 648	500 g	E4	P002 IBC08	B4	MP10	T3	TP33
3027	COUMARIN DERIVATIVE PESTICIDE, SOLID, TOXIC	6.1	T7	III	6.1	61 274 648	5 kg	E1	P002 IBC08 LP02 R001	B3	MP10	T1	TP33
3028	BATTERIES, DRY, CONTAINING POTASSIUM HYDROXIDE SOLID, electric storage	8	C11		8	295 304 598	2 kg	E0	P801 P801a				
3048	ALUMINIUM PHOSPHIDE PESTICIDE	6.1	T7	I	6.1	153 648	0	E0	P002 IBC07		MP18	T6	TP33
3054	CYCLOHEXYL MERCAPTAN	3	F1	III	3		5 L	E1	P001 IBC03 LP01 R001		MP19	T2	TP1
3055	2-(2-AMINOETHOXY) ETHANOL	8	C7	III	8		5 L	E1	P001 IBC03 LP01 R001		MP19	T4	TP1
3056	n-HEPTALDEHYDE	3	F1	III	3		5 L	E1	P001 IBC03 LP01 R001		MP19	T2	TP1
3057	TRIFLUOROACETYL CHLORIDE	2	2TC		2.3 +8		0	E0	P200		MP9	T50	TP21

ADR tank		Vehicle for tank carriage	Transport category (Tunnel restriction code)	Special provisions for carriage				Hazard identification No.	UN No.	Name and description
Tank code	Special provisions			Packages	Bulk	Loading, unloading and handling	Operation			
4.3	4.3.5, 6.8.4	9.1.1.2	1.1.3.6 (8.6)	7.2.4	7.3.3	7.5.11	8.5	5.3.2.3		3.1.2
(12)	(13)	(14)	(15)	(16)	(17)	(18)	(19)	(20)	(1)	(2)
L10CH	TU14 TU15 TE21	FL	1 (C/E)			CV13 CV28	S2 S22	336	3021	PESTICIDE, LIQUID, FLAMMABLE, TOXIC, N.O.S., flash-point less than 23 °C
L4BH	TU15	FL	2 (D/E)			CV13 CV28	S2 S22	336	3021	PESTICIDE, LIQUID, FLAMMABLE, TOXIC, N.O.S., flash-point less than 23 °C
LGBF		FL	2 (D/E)	V8			S2 S4 S20	339	3022	1,2-BUTYLENE OXIDE, STABILIZED
L10CH	TU14 TU15 TE19 TE21	FL	1 (C/D)			CV1 CV13 CV28	S2 S9 S14	663	3023	2-METHYL-2-HEPTANETHIOL
L10CH	TU14 TU15 TE21	FL	1 (C/E)			CV13 CV28	S2 S22	336	3024	COUMARIN DERIVATIVE PESTICIDE, LIQUID, FLAMMABLE, TOXIC, flash-point less than 23 °C
L4BH	TU15	FL	2 (D/E)			CV13 CV28	S2 S22	336	3024	COUMARIN DERIVATIVE PESTICIDE, LIQUID, FLAMMABLE, TOXIC, flash-point less than 23 °C
L10CH	TU14 TU15 TE19 TE21	FL	1 (C/E)			CV1 CV13 CV28	S2 S9 S14	663	3025	COUMARIN DERIVATIVE PESTICIDE, LIQUID, TOXIC, FLAMMABLE, flash-point not less than 23 °C
L4BH	TU15 TE19	FL	2 (D/E)			CV13 CV28	S2 S9 S19	63	3025	COUMARIN DERIVATIVE PESTICIDE, LIQUID, TOXIC, FLAMMABLE, flash-point not less than 23 °C
L4BH	TU15 TE19	FL	2 (D/E)	V12		CV13 CV28	S2 S9	63	3025	COUMARIN DERIVATIVE PESTICIDE, LIQUID, TOXIC, FLAMMABLE, flash-point not less than 23 °C
L10CH	TU14 TU15 TE19 TE21	AT	1 (C/E)			CV1 CV13 CV28	S9 S14	66	3026	COUMARIN DERIVATIVE PESTICIDE, LIQUID, TOXIC
L4BH	TU15 TE19	AT	2 (D/E)			CV13 CV28	S9 S19	60	3026	COUMARIN DERIVATIVE PESTICIDE, LIQUID, TOXIC
L4BH	TU15 TE19	AT	2 (E)	V12		CV13 CV28	S9	60	3026	COUMARIN DERIVATIVE PESTICIDE, LIQUID, TOXIC
S10AH L10CH	TU14 TU15 TE19 TE21	AT	1 (C/E)	V10		CV1 CV13 CV28	S9 S14	66	3027	COUMARIN DERIVATIVE PESTICIDE, SOLID, TOXIC
SGAH L4BH	TU15 TE19	AT	2 (D/E)	V11		CV13 CV28	S9 S19	60	3027	COUMARIN DERIVATIVE PESTICIDE, SOLID, TOXIC
SGAH L4BH	TU15 TE19	AT	2 (E)		VC1 VC2 AP7	CV13 CV28	S9	60	3027	COUMARIN DERIVATIVE PESTICIDE, SOLID, TOXIC
			3 (E)		VC1 VC2 AP8			80	3028	BATTERIES, DRY, CONTAINING POTASSIUM HYDROXIDE SOLID, electric storage
S10AH	TU15 TE19	AT	1 (C/E)	V10		CV1 CV13 CV28	S9 S14	642	3048	ALUMINIUM PHOSPHIDE PESTICIDE
LGBF		FL	3 (D/E)	V12			S2	30	3054	CYCLOHEXYL MERCAPTAN
L4BN		AT	3 (E)	V12				80	3055	2-(2-AMINOETHOXY) ETHANOL
LGBF		FL	3 (D/E)	V12			S2	30	3056	n-HEPTALDEHYDE
PxBH(M)	TA4 TT9	AT	1 (C/D)			CV9 CV10 CV36	S14	268	3057	TRIFLUOROACETYL CHLORIDE

UN No.	Name and description	Class	Classifi-cation code	Packing group	Labels	Special provi-sions	Limited and excepted quantities		Packaging			Portable tanks and bulk containers	
									Packing instruc-tions	Special packing provisions	Mixed packing provisions	Instruc-tions	Special provisions
	3.1.2	2.2	2.2	2.1.1.3	5.2.2	3.3	3.4	3.5.1.2	4.1.4	4.1.4	4.1.10	4.2.5.2 7.3.2	4.2.5.3
(1)	(2)	(3a)	(3b)	(4)	(5)	(6)	(7a)	(7b)	(8)	(9a)	(9b)	(10)	(11)
3064	NITROGLYCERIN, SOLUTION IN ALCOHOL with more than 1% but not more than 5% nitroglycerin	3	D	II	3	359	0	E0	P300		MP2		
3065	ALCOHOLIC BEVERAGES, with more than 70% alcohol by volume	3	F1	II	3		5 L	E2	P001 IBC02 R001	PP2	MP19	T4	TP1
3065	ALCOHOLIC BEVERAGES, with more than 24% but not more than 70% alcohol by volume	3	F1	III	3	144 145 247	5 L	E1	P001 IBC03 R001	PP2	MP19	T2	TP1
3066	PAINT (including paint, lacquer, enamel, stain, shellac, varnish, polish, liquid filler and liquid lacquer base) or PAINT RELATED MATERIAL (including paint thinning and reducing compound)	8	C9	II	8	163 367	1 L	E2	P001 IBC02		MP15	T7	TP2 TP28
3066	PAINT (including paint, lacquer, enamel, stain, shellac, varnish, polish, liquid filler and liquid lacquer base) or PAINT RELATED MATERIAL (including paint thinning and reducing compound)	8	C9	III	8	163 367	5 L	E1	P001 IBC03 R001		MP19	T4	TP1 TP29
3070	ETHYLENE OXIDE AND DICHLORODIFLUORO-METHANE MIXTURE with not more than 12.5% ethylene oxide	2	2A		2.2	662	120 ml	E1	P200		MP9	(M) T50	
3071	MERCAPTANS, LIQUID, TOXIC, FLAMMABLE, N.O.S. or MERCAPTAN MIXTURE, LIQUID, TOXIC, FLAMMABLE, N.O.S.	6.1	TF1	II	6.1 +3	274	100 ml	E4	P001 IBC02		MP15	T11	TP2 TP27
3072	LIFE-SAVING APPLIANCES NOT SELF-INFLATING containing dangerous goods as equipment	9	M5		9	296 635	0	E0	P905				
3073	VINYLPYRIDINES, STABILIZED	6.1	TFC	II	6.1 +3 +8	386	100 ml	E4	P001 IBC01		MP15	T7	TP2
3077	ENVIRONMENTALLY HAZARDOUS SUBSTANCE, SOLID, N.O.S.	9	M7	III	9	274 335 375 601	5 kg	E1	P002 IBC08 LP02 R001	PP12 B3	MP10	T1 BK1 BK2 BK3	TP33
3078	CERIUM, turnings or gritty powder	4.3	W2	II	4.3	550	500 g	E2	P410 IBC07		MP14	T3	TP33
3079	METHACRYLONITRILE, STABILIZED	6.1	TF1	I	6.1 +3	354 386	0	E0	P602		MP8 MP17	T20	TP2
3080	ISOCYANATES, TOXIC, FLAMMABLE, N.O.S. or ISOCYANATE SOLUTION, TOXIC, FLAMMABLE, N.O.S.	6.1	TF1	II	6.1 +3	274 551	100 ml	E4	P001 IBC02		MP15	T11	TP2 TP27
3082	ENVIRONMENTALLY HAZARDOUS SUBSTANCE, LIQUID, N.O.S.	9	M6	III	9	274 335 375 601	5 L	E1	P001 IBC03 LP01 R001	PP1	MP19	T4	TP1 TP29
3083	PERCHLORYL FLUORIDE	2	2TO		2.3 +5.1		0	E0	P200		MP9	(M)	
3084	CORROSIVE SOLID, OXIDIZING, N.O.S.	8	CO2	I	8 +5.1	274	0	E0	P002		MP18	T6	TP33
3084	CORROSIVE SOLID, OXIDIZING, N.O.S.	8	CO2	II	8 +5.1	274	1 kg	E2	P002 IBC06		MP10	T3	TP33
3085	OXIDIZING SOLID, CORROSIVE, N.O.S.	5.1	OC2	I	5.1 +8	274	0	E0	P503		MP2		
3085	OXIDIZING SOLID, CORROSIVE, N.O.S.	5.1	OC2	II	5.1 +8	274	1 kg	E2	P002 IBC06		MP2	T3	TP33
3085	OXIDIZING SOLID, CORROSIVE, N.O.S.	5.1	OC2	III	5.1 +8	274	5 kg	E1	P002 IBC08 R001	B3	MP2	T1	TP33
3086	TOXIC SOLID, OXIDIZING, N.O.S.	6.1	TO2	I	6.1 +5.1	274	0	E5	P002		MP18	T6	TP33
3086	TOXIC SOLID, OXIDIZING, N.O.S.	6.1	TO2	II	6.1 +5.1	274	500 g	E4	P002 IBC06		MP10	T3	TP33

ADR tank		Vehicle for tank carriage	Transport category (Tunnel restriction code)	Special provisions for carriage				Hazard identifi-cation No.	UN No.	Name and description
Tank code	Special provisions			Packages	Bulk	Loading, unloading and handling	Operation			
4.3	4.3.5, 6.8.4	9.1.1.2	1.1.3.6 (8.6)	7.2.4	7.3.3	7.5.11	8.5	5.3.2.3		3.1.2
(12)	(13)	(14)	(15)	(16)	(17)	(18)	(19)	(20)	(1)	(2)
			2 (B)				S2 S14		3064	NITROGLYCERIN, SOLUTION IN ALCOHOL with more than 1% but not more than 5% nitroglycerin
LGBF		FL	2 (D/E)				S2 S20	33	3065	ALCOHOLIC BEVERAGES, with more than 70% alcohol by volume
LGBF		FL	3 (D/E)	V12			S2	30	3065	ALCOHOLIC BEVERAGES, with more than 24% but not more than 70% alcohol by volume
L4BN		AT	2 (E)					80	3066	PAINT (including paint, lacquer, enamel, stain, shellac, varnish, polish, liquid filler and liquid lacquer base) or PAINT RELATED MATERIAL (including paint thinning and reducing compound)
L4BN		AT	3 (E)	V12				80	3066	PAINT (including paint, lacquer, enamel, stain, shellac, varnish, polish, liquid filler and liquid lacquer base) or PAINT RELATED MATERIAL (including paint thinning and reducing compound)
PxBN(M)	TA4 TT9	AT	3 (C/E)			CV9 CV10 CV36		20	3070	ETHYLENE OXIDE AND DICHLORODIFLUORO-METHANE MIXTURE with not more than 12.5% ethylene oxide
L4BH	TU15 TE19	FL	2 (D/E)			CV13 CV28	S2 S9 S19	63	3071	MERCAPTANS, LIQUID, TOXIC, FLAMMABLE, N.O.S. or MERCAPTAN MIXTURE, LIQUID, TOXIC, FLAMMABLE, N.O.S.
			3 (E)						3072	LIFE-SAVING APPLIANCES NOT SELF-INFLATING containing dangerous goods as equipment
L4BH	TU15 TE19	FL	2 (D/E)	V8		CV13 CV28	S2 S4 S9 S19	638	3073	VINYLPYRIDINES, STABILIZED
SGAV LGBV		AT	3 (-)	V13	VC1 VC2	CV13		90	3077	ENVIRONMENTALLY HAZARDOUS SUBSTANCE, SOLID, N.O.S.
SGAN		AT	2 (D/E)	V1		CV23		423	3078	CERIUM, turnings or gritty powder
L10CH	TU14 TU15 TE19 TE21	FL	1 (C/D)	V8		CV1 CV13 CV28	S2 S4 S9 S14	663	3079	METHACRYLONITRILE, STABILIZED
L4BH	TU15 TE19	FL	2 (D/E)			CV13 CV28	S2 S9 S19	63	3080	ISOCYANATES, TOXIC, FLAMMABLE, N.O.S. or ISOCYANATE SOLUTION, TOXIC, FLAMMABLE, N.O.S.
LGBV		AT	3 (-)	V12		CV13		90	3082	ENVIRONMENTALLY HAZARDOUS SUBSTANCE, LIQUID, N.O.S.
PxBH(M)	TA4 TT9	AT	1 (C/D)			CV9 CV10 CV36	S14	265	3083	PERCHLORYL FLUORIDE
S10AN L10BH		AT	1 (E)			CV24	S14	885	3084	CORROSIVE SOLID, OXIDIZING, N.O.S.
SGAN L4BN		AT	2 (E)	V11		CV24		85	3084	CORROSIVE SOLID, OXIDIZING, N.O.S.
			1 (E)			CV24	S20		3085	OXIDIZING SOLID, CORROSIVE, N.O.S.
SGAN	TU3	AT	2 (E)	V11		CV24		58	3085	OXIDIZING SOLID, CORROSIVE, N.O.S.
SGAN	TU3	AT	3 (E)			CV24		58	3085	OXIDIZING SOLID, CORROSIVE, N.O.S.
S10AH L10CH	TU14 TU15 TE19 TE21	AT	1 (C/E)			CV1 CV13 CV28	S9 S14	665	3086	TOXIC SOLID, OXIDIZING, N.O.S.
SGAH L4BH	TU15 TE19	AT	2 (D/E)	V11		CV13 CV28	S9 S19	65	3086	TOXIC SOLID, OXIDIZING, N.O.S.

UN No.	Name and description	Class	Classifi-cation code	Packing group	Labels	Special provi-sions	Limited and excepted quantities		Packaging			Portable tanks and bulk containers	
									Packing instruc-tions	Special packing provisions	Mixed packing provisions	Instruc-tions	Special provisions
3.1.2	3.1.2	2.2	2.2	2.1.1.3	5.2.2	3.3	3.4	3.5.1.2	4.1.4	4.1.4	4.1.10	4.2.5.2 7.3.2	4.2.5.3
(1)	(2)	(3a)	(3b)	(4)	(5)	(6)	(7a)	(7b)	(8)	(9a)	(9b)	(10)	(11)
3087	OXIDIZING SOLID, TOXIC, N.O.S.	5.1	OT2	I	5.1 +6.1	274	0	E0	P503		MP2		
3087	OXIDIZING SOLID, TOXIC, N.O.S.	5.1	OT2	II	5.1 +6.1	274	1 kg	E2	P002 IBC06		MP2	T3	TP33
3087	OXIDIZING SOLID, TOXIC, N.O.S.	5.1	OT2	III	5.1 +6.1	274	5 kg	E1	P002 IBC08 R001	B3	MP2	T1	TP33
3088	SELF-HEATING SOLID, ORGANIC, N.O.S.	4.2	S2	II	4.2	274	0	E2	P410 IBC06		MP14	T3	TP33
3088	SELF-HEATING SOLID, ORGANIC, N.O.S.	4.2	S2	III	4.2	274 665	0	E1	P002 IBC08 LP02 R001	B3	MP14	T1	TP33
3089	METAL POWDER, FLAMMABLE, N.O.S.	4.1	F3	II	4.1	552	1 kg	E2	P002 IBC08	B4	MP11	T3	TP33
3089	METAL POWDER, FLAMMABLE, N.O.S.	4.1	F3	III	4.1	552	5 kg	E1	P002 IBC08 R001	B4	MP11	T1	TP33
3090	LITHIUM METAL BATTERIES (including lithium alloy batteries)	9	M4		9A	188 230 310 376 377 636	0	E0	P903 P908 P909 P910 LP903 LP904				
3091	LITHIUM METAL BATTERIES CONTAINED IN EQUIPMENT or LITHIUM METAL BATTERIES PACKED WITH EQUIPMENT (including lithium alloy batteries)	9	M4		9A	188 230 310 360 376 377 636	0	E0	P903 P908 P909 P910 LP903 LP904				
3092	1-METHOXY-2-PROPANOL	3	F1	III	3		5 L	E1	P001 IBC03 LP01 R001		MP19	T2	TP1
3093	CORROSIVE LIQUID, OXIDIZING, N.O.S.	8	CO1	I	8 +5.1	274	0	E0	P001		MP8 MP17		
3093	CORROSIVE LIQUID, OXIDIZING, N.O.S.	8	CO1	II	8 +5.1	274	1 L	E2	P001 IBC02		MP15		
3094	CORROSIVE LIQUID, WATER-REACTIVE, N.O.S.	8	CW1	I	8 +4.3	274	0	E0	P001		MP8 MP17		
3094	CORROSIVE LIQUID, WATER-REACTIVE, N.O.S.	8	CW1	II	8 +4.3	274	1 L	E2	P001		MP15		
3095	CORROSIVE SOLID, SELF-HEATING, N.O.S.	8	CS2	I	8 +4.2	274	0	E0	P002		MP18	T6	TP33
3095	CORROSIVE SOLID, SELF-HEATING, N.O.S.	8	CS2	II	8 +4.2	274	1 kg	E2	P002 IBC06		MP10	T3	TP33
3096	CORROSIVE SOLID, WATER-REACTIVE, N.O.S.	8	CW2	I	8 +4.3	274	0	E0	P002		MP18	T6	TP33
3096	CORROSIVE SOLID, WATER-REACTIVE, N.O.S.	8	CW2	II	8 +4.3	274	1 kg	E2	P002 IBC06		MP10	T3	TP33
3097	FLAMMABLE SOLID, OXIDIZING, N.O.S.	4.1	FO	CARRIAGE PROHIBITED										
3098	OXIDIZING LIQUID, CORROSIVE, N.O.S.	5.1	OC1	I	5.1 +8	274	0	E0	P502		MP2		
3098	OXIDIZING LIQUID, CORROSIVE, N.O.S.	5.1	OC1	II	5.1 +8	274	1 L	E2	P504 IBC01		MP2		
3098	OXIDIZING LIQUID, CORROSIVE, N.O.S.	5.1	OC1	III	5.1 +8	274	5 L	E1	P504 IBC02 R001		MP2		
3099	OXIDIZING LIQUID, TOXIC, N.O.S.	5.1	OT1	I	5.1 +6.1	274	0	E0	P502		MP2		
3099	OXIDIZING LIQUID, TOXIC, N.O.S.	5.1	OT1	II	5.1 +6.1	274	1 L	E2	P504 IBC01		MP2		
3099	OXIDIZING LIQUID, TOXIC, N.O.S.	5.1	OT1	III	5.1 +6.1	274	5 L	E1	P504 IBC02 R001		MP2		
3100	OXIDIZING SOLID, SELF-HEATING, N.O.S.	5.1	OS	CARRIAGE PROHIBITED										
3101	ORGANIC PEROXIDE TYPE B, LIQUID	5.2	P1		5.2 +1	122 181 274	25 ml	E0	P520		MP4		

- 464 -

ADR tank		Vehicle for tank carriage	Transport category (Tunnel restriction code)	Special provisions for carriage				Hazard identification No.	UN No.	Name and description
Tank code	Special provisions			Packages	Bulk	Loading, unloading and handling	Operation			
4.3	4.3.5, 6.8.4	9.1.1.2	1.1.3.6 (8.6)	7.2.4	7.3.3	7.5.11	8.5	5.3.2.3		3.1.2
(12)	(13)	(14)	(15)	(16)	(17)	(18)	(19)	(20)	(1)	(2)
			1 (E)			CV24 CV28	S20		3087	OXIDIZING SOLID, TOXIC, N.O.S.
SGAN	TU3	AT	2 (E)	V11		CV24 CV28		56	3087	OXIDIZING SOLID, TOXIC, N.O.S.
SGAN	TU3	AT	3 (E)			CV24 CV28		56	3087	OXIDIZING SOLID, TOXIC, N.O.S.
SGAV		AT	2 (D/E)	V1				40	3088	SELF-HEATING SOLID, ORGANIC, N.O.S.
SGAV		AT	3 (E)	V1				40	3088	SELF-HEATING SOLID, ORGANIC, N.O.S.
SGAN		AT	2 (E)	V11				40	3089	METAL POWDER, FLAMMABLE, N.O.S.
SGAV		AT	3 (E)	V11	VC1 VC2			40	3089	METAL POWDER, FLAMMABLE, N.O.S.
			2 (E)						3090	LITHIUM METAL BATTERIES (including lithium alloy batteries)
			2 (E)						3091	LITHIUM METAL BATTERIES CONTAINED IN EQUIPMENT or LITHIUM METAL BATTERIES PACKED WITH EQUIPMENT (including lithium alloy batteries)
LGBF		FL	3 (D/E)	V12			S2	30	3092	1-METHOXY-2-PROPANOL
L10BH		AT	1 (E)			CV24	S14	885	3093	CORROSIVE LIQUID, OXIDIZING, N.O.S.
L4BN		AT	2 (E)			CV24		85	3093	CORROSIVE LIQUID, OXIDIZING, N.O.S.
L10BH		AT	1 (D/E)				S14	823	3094	CORROSIVE LIQUID, WATER-REACTIVE, N.O.S.
L4BN		AT	2 (E)					823	3094	CORROSIVE LIQUID, WATER-REACTIVE, N.O.S.
S10AN		AT	1 (E)				S14	884	3095	CORROSIVE SOLID, SELF-HEATING, N.O.S.
SGAN		AT	2 (E)	V11				84	3095	CORROSIVE SOLID, SELF-HEATING, N.O.S.
S10AN L10BH		AT	1 (E)				S14	842	3096	CORROSIVE SOLID, WATER-REACTIVE, N.O.S.
SGAN L4BN		AT	2 (E)	V11				842	3096	CORROSIVE SOLID, WATER-REACTIVE, N.O.S.
CARRIAGE PROHIBITED									3097	FLAMMABLE SOLID, OXIDIZING, N.O.S.
			1 (E)			CV24	S20		3098	OXIDIZING LIQUID, CORROSIVE, N.O.S.
			2 (E)			CV24			3098	OXIDIZING LIQUID, CORROSIVE, N.O.S.
			3 (E)			CV24			3098	OXIDIZING LIQUID, CORROSIVE, N.O.S.
			1 (E)			CV24 CV28	S20		3099	OXIDIZING LIQUID, TOXIC, N.O.S.
			2 (E)			CV24 CV28			3099	OXIDIZING LIQUID, TOXIC, N.O.S.
			3 (E)			CV24 CV28			3099	OXIDIZING LIQUID, TOXIC, N.O.S.
CARRIAGE PROHIBITED									3100	OXIDIZING SOLID, SELF-HEATING, N.O.S.
			1 (B)	V1 V5		CV15 CV20 CV22 CV24	S9 S17		3101	ORGANIC PEROXIDE TYPE B, LIQUID

UN No.	Name and description	Class	Classifi-cation code	Packing group	Labels	Special provi-sions	Limited and excepted quantities		Packaging			Portable tanks and bulk containers	
										Packing instruc-tions	Special packing provisions	Mixed packing provisions	Instruc-tions	Special provisions
	3.1.2	2.2	2.2	2.1.1.3	5.2.2	3.3	3.4	3.5.1.2	4.1.4	4.1.4	4.1.10	4.2.5.2 7.3.2	4.2.5.3
(1)	(2)	(3a)	(3b)	(4)	(5)	(6)	(7a)	(7b)	(8)	(9a)	(9b)	(10)	(11)
3102	ORGANIC PEROXIDE TYPE B, SOLID	5.2	P1		5.2 +1	122 181 274	100 g	E0	P520		MP4		
3103	ORGANIC PEROXIDE TYPE C, LIQUID	5.2	P1		5.2	122 274	25 ml	E0	P520		MP4		
3104	ORGANIC PEROXIDE TYPE C, SOLID	5.2	P1		5.2	122 274	100 g	E0	P520		MP4		
3105	ORGANIC PEROXIDE TYPE D, LIQUID	5.2	P1		5.2	122 274	125 ml	E0	P520		MP4		
3106	ORGANIC PEROXIDE TYPE D, SOLID	5.2	P1		5.2	122 274	500 g	E0	P520		MP4		
3107	ORGANIC PEROXIDE TYPE E, LIQUID	5.2	P1		5.2	122 274	125 ml	E0	P520		MP4		
3108	ORGANIC PEROXIDE TYPE E, SOLID	5.2	P1		5.2	122 274	500 g	E0	P520		MP4		
3109	ORGANIC PEROXIDE TYPE F, LIQUID	5.2	P1		5.2	122 274	125 ml	E0	P520 IBC520		MP4	T23	
3110	ORGANIC PEROXIDE TYPE F, SOLID	5.2	P1		5.2	122 274	500 g	E0	P520 IBC520		MP4	T23	TP33
3111	ORGANIC PEROXIDE TYPE B, LIQUID, TEMPERATURE CONTROLLED	5.2	P2		5.2 +1	122 181 274	0	E0	P520		MP4		
3112	ORGANIC PEROXIDE TYPE B, SOLID, TEMPERATURE CONTROLLED	5.2	P2		5.2 +1	122 181 274	0	E0	P520		MP4		
3113	ORGANIC PEROXIDE TYPE C, LIQUID, TEMPERATURE CONTROLLED	5.2	P2		5.2	122 274	0	E0	P520		MP4		
3114	ORGANIC PEROXIDE TYPE C, SOLID, TEMPERATURE CONTROLLED	5.2	P2		5.2	122 274	0	E0	P520		MP4		
3115	ORGANIC PEROXIDE TYPE D, LIQUID, TEMPERATURE CONTROLLED	5.2	P2		5.2	122 274	0	E0	P520		MP4		
3116	ORGANIC PEROXIDE TYPE D, SOLID, TEMPERATURE CONTROLLED	5.2	P2		5.2	122 274	0	E0	P520		MP4		
3117	ORGANIC PEROXIDE TYPE E, LIQUID, TEMPERATURE CONTROLLED	5.2	P2		5.2	122 274	0	E0	P520		MP4		
3118	ORGANIC PEROXIDE TYPE E, SOLID, TEMPERATURE CONTROLLED	5.2	P2		5.2	122 274	0	E0	P520		MP4		
3119	ORGANIC PEROXIDE TYPE F, LIQUID, TEMPERATURE CONTROLLED	5.2	P2		5.2	122 274	0	E0	P520 IBC520		MP4	T23	
3120	ORGANIC PEROXIDE TYPE F, SOLID, TEMPERATURE CONTROLLED	5.2	P2		5.2	122 274	0	E0	P520 IBC520		MP4	T23	TP33
3121	OXIDIZING SOLID, WATER-REACTIVE, N.O.S.	5.1	OW				CARRIAGE PROHIBITED						

ADR tank		Vehicle for tank carriage	Transport category (Tunnel restriction code)	Special provisions for carriage				Hazard identification No.	UN No.	Name and description
Tank code	Special provisions			Packages	Bulk	Loading, unloading and handling	Operation			
4.3	4.3.5, 6.8.4	9.1.1.2	1.1.3.6 (8.6)	7.2.4	7.3.3	7.5.11	8.5	5.3.2.3		3.1.2
(12)	(13)	(14)	(15)	(16)	(17)	(18)	(19)	(20)	(1)	(2)
			1 (B)	V1 V5		CV15 CV20 CV22 CV24	S9 S17		3102	ORGANIC PEROXIDE TYPE B, SOLID
			1 (D)	V1		CV15 CV20 CV22 CV24	S8 S18		3103	ORGANIC PEROXIDE TYPE C, LIQUID
			1 (D)	V1		CV15 CV20 CV22 CV24	S8 S18		3104	ORGANIC PEROXIDE TYPE C, SOLID
			2 (D)	V1		CV15 CV22 CV24	S19		3105	ORGANIC PEROXIDE TYPE D, LIQUID
			2 (D)	V1		CV15 CV22 CV24	S19		3106	ORGANIC PEROXIDE TYPE D, SOLID
			2 (D)	V1		CV15 CV22 CV24			3107	ORGANIC PEROXIDE TYPE E, LIQUID
			2 (D)	V1		CV15 CV22 CV24			3108	ORGANIC PEROXIDE TYPE E, SOLID
L4BN(+)	TU3 TU13 TU30 TE12 TA2 TM4	AT	2 (D)	V1		CV15 CV22 CV24		539	3109	ORGANIC PEROXIDE TYPE F, LIQUID
S4AN(+)	TU3 TU13 TU30 TE12 TA2 TM4	AT	2 (D)	V1		CV15 CV22 CV24		539	3110	ORGANIC PEROXIDE TYPE F, SOLID
			1 (B)	V8		CV15 CV20 CV21 CV22 CV24	S4 S9 S16		3111	ORGANIC PEROXIDE TYPE B, LIQUID, TEMPERATURE CONTROLLED
			1 (B)	V8		CV15 CV20 CV21 CV22 CV24	S4 S9 S16		3112	ORGANIC PEROXIDE TYPE B, SOLID, TEMPERATURE CONTROLLED
			1 (D)	V8		CV15 CV20 CV21 CV22 CV24	S4 S8 S17		3113	ORGANIC PEROXIDE TYPE C, LIQUID, TEMPERATURE CONTROLLED
			1 (D)	V8		CV15 CV20 CV21 CV22 CV24	S4 S8 S17		3114	ORGANIC PEROXIDE TYPE C, SOLID, TEMPERATURE CONTROLLED
			1 (D)	V8		CV15 CV21 CV22 CV24	S4 S18		3115	ORGANIC PEROXIDE TYPE D, LIQUID, TEMPERATURE CONTROLLED
			1 (D)	V8		CV15 CV21 CV22 CV24	S4 S18		3116	ORGANIC PEROXIDE TYPE D, SOLID, TEMPERATURE CONTROLLED
			1 (D)	V8		CV15 CV21 CV22 CV24	S4 S19		3117	ORGANIC PEROXIDE TYPE E, LIQUID, TEMPERATURE CONTROLLED
			1 (D)	V8		CV15 CV21 CV22 CV24	S4 S19		3118	ORGANIC PEROXIDE TYPE E, SOLID, TEMPERATURE CONTROLLED
L4BN(+)	TU3 TU13 TU30 TE12 TA2 TM4	AT	1 (D)	V8		CV15 CV21 CV22 CV24	S4	539	3119	ORGANIC PEROXIDE TYPE F, LIQUID, TEMPERATURE CONTROLLED
S4AN(+)	TU3 TU13 TU30 TE12 TA2 TM4	AT	1 (D)	V8		CV15 CV21 CV22 CV24	S4	539	3120	ORGANIC PEROXIDE TYPE F, SOLID, TEMPERATURE CONTROLLED
CARRIAGE PROHIBITED									3121	OXIDIZING SOLID, WATER-REACTIVE, N.O.S.

UN No.	Name and description	Class	Classification code	Packing group	Labels	Special provisions	Limited and excepted quantities		Packaging			Portable tanks and bulk containers	
									Packing instructions	Special packing provisions	Mixed packing provisions	Instructions	Special provisions
	3.1.2	2.2	2.2	2.1.1.3	5.2.2	3.3	3.4	3.5.1.2	4.1.4	4.1.4	4.1.10	4.2.5.2 7.3.2	4.2.5.3
(1)	(2)	(3a)	(3b)	(4)	(5)	(6)	(7a)	(7b)	(8)	(9a)	(9b)	(10)	(11)
3122	TOXIC LIQUID, OXIDIZING, N.O.S.	6.1	TO1	I	6.1 +5.1	274 315	0	E0	P001		MP8 MP17		
3122	TOXIC LIQUID, OXIDIZING, N.O.S.	6.1	TO1	II	6.1 +5.1	274	100 ml	E4	P001 IBC02		MP15		
3123	TOXIC LIQUID, WATER-REACTIVE, N.O.S.	6.1	TW1	I	6.1 +4.3	274 315	0	E0	P099		MP8 MP17		
3123	TOXIC LIQUID, WATER-REACTIVE, N.O.S.	6.1	TW1	II	6.1 +4.3	274	100 ml	E4	P001 IBC02		MP15		
3124	TOXIC SOLID, SELF-HEATING, N.O.S.	6.1	TS	I	6.1 +4.2	274	0	E5	P002		MP18	T6	TP33
3124	TOXIC SOLID, SELF-HEATING, N.O.S.	6.1	TS	II	6.1 +4.2	274	0	E4	P002 IBC06		MP10	T3	TP33
3125	TOXIC SOLID, WATER-REACTIVE, N.O.S.	6.1	TW2	I	6.1 +4.3	274	0	E5	P099		MP18	T6	TP33
3125	TOXIC SOLID, WATER-REACTIVE, N.O.S.	6.1	TW2	II	6.1 +4.3	274	500 g	E4	P002 IBC06		MP10	T3	TP33
3126	SELF-HEATING SOLID, CORROSIVE, ORGANIC, N.O.S.	4.2	SC2	II	4.2 +8	274	0	E2	P410 IBC05		MP14	T3	TP33
3126	SELF-HEATING SOLID, CORROSIVE, ORGANIC, N.O.S.	4.2	SC2	III	4.2 +8	274	0	E1	P002 IBC08 R001	B3	MP14	T1	TP33
3127	SELF-HEATING SOLID, OXIDIZING, N.O.S	4.2	SO					CARRIAGE PROHIBITED					
3128	SELF-HEATING SOLID, TOXIC, ORGANIC, N.O.S.	4.2	ST2	II	4.2 +6.1	274	0	E2	P410 IBC05		MP14	T3	TP33
3128	SELF-HEATING SOLID, TOXIC, ORGANIC, N.O.S.	4.2	ST2	III	4.2 +6.1	274	0	E1	P002 IBC08 R001	B3	MP14	T1	TP33
3129	WATER-REACTIVE LIQUID, CORROSIVE, N.O.S.	4.3	WC1	I	4.3 +8	274	0	E0	P402	RR7 RR8	MP2	T14	TP2 TP7
3129	WATER-REACTIVE LIQUID, CORROSIVE, N.O.S.	4.3	WC1	II	4.3 +8	274	500 ml	E0	P402 IBC01	RR7 RR8	MP15	T11	TP2 TP7
3129	WATER-REACTIVE LIQUID, CORROSIVE, N.O.S.	4.3	WC1	III	4.3 +8	274	1 L	E1	P001 IBC02 R001		MP15	T7	TP2 TP7
3130	WATER-REACTIVE LIQUID, TOXIC, N.O.S.	4.3	WT1	I	4.3 +6.1	274	0	E0	P402	RR4 RR8	MP2		
3130	WATER-REACTIVE LIQUID, TOXIC, N.O.S.	4.3	WT1	II	4.3 +6.1	274	500 ml	E0	P402 IBC01	RR4 RR8 BB1	MP15		
3130	WATER-REACTIVE LIQUID, TOXIC, N.O.S.	4.3	WT1	III	4.3 +6.1	274	1 L	E1	P001 IBC02 R001		MP15		
3131	WATER-REACTIVE SOLID, CORROSIVE, N.O.S.	4.3	WC2	I	4.3 +8	274	0	E0	P403		MP2	T9	TP7 TP33
3131	WATER-REACTIVE SOLID, CORROSIVE, N.O.S.	4.3	WC2	II	4.3 +8	274	500 g	E2	P410 IBC06		MP14	T3	TP33
3131	WATER-REACTIVE SOLID, CORROSIVE, N.O.S.	4.3	WC2	III	4.3 +8	274	1 kg	E1	P410 IBC08 R001	B4	MP14	T1	TP33
3132	WATER-REACTIVE SOLID, FLAMMABLE, N.O.S.	4.3	WF2	I	4.3 +4.1	274	0	E0	P403 IBC99		MP2		
3132	WATER-REACTIVE SOLID, FLAMMABLE, N.O.S.	4.3	WF2	II	4.3 +4.1	274	500 g	E2	P410 IBC04		MP14	T3	TP33
3132	WATER-REACTIVE SOLID, FLAMMABLE, N.O.S.	4.3	WF2	III	4.3 +4.1	274	1 kg	E1	P410 IBC06		MP14	T1	TP33
3133	WATER-REACTIVE SOLID, OXIDIZING, N.O.S.	4.3	WO					CARRIAGE PROHIBITED					
3134	WATER-REACTIVE SOLID, TOXIC, N.O.S.	4.3	WT2	I	4.3 +6.1	274	0	E0	P403		MP2		
3134	WATER-REACTIVE SOLID, TOXIC, N.O.S.	4.3	WT2	II	4.3 +6.1	274	500 g	E2	P410 IBC05		MP14	T3	TP33
3134	WATER-REACTIVE SOLID, TOXIC, N.O.S.	4.3	WT2	III	4.3 +6.1	274	1 kg	E1	P410 IBC08 R001	B4	MP14	T1	TP33
3135	WATER-REACTIVE SOLID, SELF-HEATING, N.O.S.	4.3	WS	I	4.3 +4.2	274	0	E0	P403		MP2		
3135	WATER-REACTIVE SOLID, SELF-HEATING, N.O.S.	4.3	WS	II	4.3 +4.2	274	0	E2	P410 IBC05		MP14	T3	TP33
3135	WATER-REACTIVE SOLID, SELF-HEATING, N.O.S.	4.3	WS	III	4.3 +4.2	274	0	E1	P410 IBC08	B4	MP14	T1	TP33

ADR tank		Vehicle for tank carriage	Transport category (Tunnel restriction code)	Special provisions for carriage				Hazard identifi-cation No.	UN No.	Name and description
Tank code	Special provisions			Packages	Bulk	Loading, unloading and handling	Operation			
4.3	4.3.5, 6.8.4	9.1.1.2	1.1.3.6 (8.6)	7.2.4	7.3.3	7.5.11	8.5	5.3.2.3		3.1.2
(12)	(13)	(14)	(15)	(16)	(17)	(18)	(19)	(20)	(1)	(2)
L10CH	TU14 TU15 TE19 TE21	AT	1 (C/E)			CV1 CV13 CV28	S9 S14	665	3122	TOXIC LIQUID, OXIDIZING, N.O.S.
L4BH	TU15 TE19	AT	2 (D/E)			CV13 CV28	S9 S19	65	3122	TOXIC LIQUID, OXIDIZING, N.O.S.
L10CH	TU14 TU15 TE19 TE21	AT	1 (C/E)			CV1 CV13 CV28	S9 S14	623	3123	TOXIC LIQUID, WATER-REACTIVE, N.O.S.
L4BH	TU15 TE19	AT	2 (D/E)			CV13 CV28	S9 S19	623	3123	TOXIC LIQUID, WATER-REACTIVE, N.O.S.
S10AH L10CH	TU14 TU15 TE19 TE21	AT	1 (C/E)			CV1 CV13 CV28	S9 S14	664	3124	TOXIC SOLID, SELF-HEATING, N.O.S.
SGAH L4BH	TU15 TE19	AT	2 (D/E)	V11		CV13 CV28	S9 S19	64	3124	TOXIC SOLID, SELF-HEATING, N.O.S.
S10AH L10CH	TU14 TU15 TE19 TE21	AT	1 (C/E)			CV1 CV13 CV28	S9 S14	642	3125	TOXIC SOLID, WATER-REACTIVE, N.O.S.
SGAH L4BH	TU15 TE19	AT	2 (D/E)	V11		CV13 CV28	S9 S19	642	3125	TOXIC SOLID, WATER-REACTIVE, N.O.S.
SGAN		AT	2 (D/E)	V1				48	3126	SELF-HEATING SOLID, CORROSIVE, ORGANIC, N.O.S.
SGAN		AT	3 (E)	V1				48	3126	SELF-HEATING SOLID, CORROSIVE, ORGANIC, N.O.S.
CARRIAGE PROHIBITED									3127	SELF-HEATING SOLID, OXIDIZING, N.O.S
SGAN		AT	2 (D/E)	V1		CV28		46	3128	SELF-HEATING SOLID, TOXIC, ORGANIC, N.O.S.
SGAN		AT	3 (E)	V1		CV28		46	3128	SELF-HEATING SOLID, TOXIC, ORGANIC, N.O.S.
L10DH	TU14 TE21 TM2	AT	0 (B/E)	V1		CV23	S20	X382	3129	WATER-REACTIVE LIQUID, CORROSIVE, N.O.S.
L4DH	TU14 TE21 TM2	AT	0 (D/E)	V1		CV23		382	3129	WATER-REACTIVE LIQUID, CORROSIVE, N.O.S.
L4DH	TU14 TE21 TM2	AT	0 (E)	V1		CV23		382	3129	WATER-REACTIVE LIQUID, CORROSIVE, N.O.S.
L10DH	TU14 TE21 TM2	AT	0 (B/E)	V1		CV23 CV28	S20	X362	3130	WATER-REACTIVE LIQUID, TOXIC, N.O.S.
L4DH	TU14 TE21 TM2	AT	0 (D/E)	V1		CV23 CV28		362	3130	WATER-REACTIVE LIQUID, TOXIC, N.O.S.
L4DH	TU14 TE21 TM2	AT	0 (E)	V1		CV23 CV28		362	3130	WATER-REACTIVE LIQUID, TOXIC, N.O.S.
S10AN L10DH	TU4 TU14 TU22 TE21 TM2	AT	0 (B/E)	V1		CV23	S20	X482	3131	WATER-REACTIVE SOLID, CORROSIVE, N.O.S.
SGAN		AT	0 (D/E)	V1		CV23		482	3131	WATER-REACTIVE SOLID, CORROSIVE, N.O.S.
SGAN		AT	0 (E)	V1		CV23		482	3131	WATER-REACTIVE SOLID, CORROSIVE, N.O.S.
			0 (E)	V1		CV23	S20		3132	WATER-REACTIVE SOLID, FLAMMABLE, N.O.S.
SGAN L4DH	TU14 TE21 TM2	AT	0 (D/E)	V1		CV23		423	3132	WATER-REACTIVE SOLID, FLAMMABLE, N.O.S.
SGAN L4DH	TU14 TE21 TM2	AT	0 (E)	V1		CV23		423	3132	WATER-REACTIVE SOLID, FLAMMABLE, N.O.S.
CARRIAGE PROHIBITED									3133	WATER-REACTIVE SOLID, OXIDIZING, N.O.S.
			0 (E)	V1		CV23 CV28	S20		3134	WATER-REACTIVE SOLID, TOXIC, N.O.S.
SGAN		AT	0 (D/E)	V1		CV23 CV28		462	3134	WATER-REACTIVE SOLID, TOXIC, N.O.S.
SGAN		AT	0 (E)	V1		CV23 CV28		462	3134	WATER-REACTIVE SOLID, TOXIC, N.O.S.
			1 (E)	V1		CV23	S20		3135	WATER-REACTIVE SOLID, SELF-HEATING, N.O.S.
SGAN L4DH	TU14 TE21 TM2	AT	2 (D/E)	V1		CV23		423	3135	WATER-REACTIVE SOLID, SELF-HEATING, N.O.S.
SGAN L4DH	TU14 TE21 TM2	AT	3 (E)	V1		CV23		423	3135	WATER-REACTIVE SOLID, SELF-HEATING, N.O.S.

UN No.	Name and description	Class	Classification code	Packing group	Labels	Special provisions	Limited and excepted quantities		Packaging			Portable tanks and bulk containers	
									Packing instructions	Special packing provisions	Mixed packing provisions	Instructions	Special provisions
	3.1.2	2.2	2.2	2.1.1.3	5.2.2	3.3	3.4	3.5.1.2	4.1.4	4.1.4	4.1.10	4.2.5.2 7.3.2	4.2.5.3
(1)	(2)	(3a)	(3b)	(4)	(5)	(6)	(7a)	(7b)	(8)	(9a)	(9b)	(10)	(11)
3136	TRIFLUOROMETHANE, REFRIGERATED LIQUID	2	3A		2.2	593	120 ml	E1	P203		MP9	T75	TP5
3137	OXIDIZING SOLID, FLAMMABLE, N.O.S.	5.1	OF	CARRIAGE PROHIBITED									
3138	ETHYLENE, ACETYLENE AND PROPYLENE MIXTURE, REFRIGERATED LIQUID containing at least 71.5% ethylene with not more than 22.5% acetylene and not more than 6% propylene	2	3F		2.1		0	E0	P203		MP9	T75	TP5
3139	OXIDIZING LIQUID, N.O.S.	5.1	O1	I	5.1	274	0	E0	P502		MP2		
3139	OXIDIZING LIQUID, N.O.S.	5.1	O1	II	5.1	274	1 L	E2	P504 IBC02		MP2		
3139	OXIDIZING LIQUID, N.O.S.	5.1	O1	III	5.1	274	5 L	E1	P504 IBC02 R001		MP2		
3140	ALKALOIDS, LIQUID, N.O.S. or ALKALOID SALTS, LIQUID, N.O.S.	6.1	T1	I	6.1	43 274	0	E5	P001		MP8 MP17		
3140	ALKALOIDS, LIQUID, N.O.S. or ALKALOID SALTS, LIQUID, N.O.S.	6.1	T1	II	6.1	43 274	100 ml	E4	P001 IBC02		MP15		
3140	ALKALOIDS, LIQUID, N.O.S. or ALKALOID SALTS, LIQUID, N.O.S.	6.1	T1	III	6.1	43 274	5 L	E1	P001 IBC03 LP01 R001		MP19		
3141	ANTIMONY COMPOUND, INORGANIC, LIQUID, N.O.S.	6.1	T4	III	6.1	45 274 512	5 L	E1	P001 IBC03 LP01 R001		MP19		
3142	DISINFECTANT, LIQUID, TOXIC, N.O.S.	6.1	T1	I	6.1	274	0	E5	P001		MP8 MP17		
3142	DISINFECTANT, LIQUID, TOXIC, N.O.S.	6.1	T1	II	6.1	274	100 ml	E4	P001 IBC02		MP15		
3142	DISINFECTANT, LIQUID, TOXIC, N.O.S.	6.1	T1	III	6.1	274	5 L	E1	P001 IBC03 LP01 R001		MP19		
3143	DYE, SOLID, TOXIC, N.O.S. or DYE INTERMEDIATE, SOLID, TOXIC, N.O.S.	6.1	T2	I	6.1	274	0	E5	P002 IBC07		MP18	T6	TP33
3143	DYE, SOLID, TOXIC, N.O.S. or DYE INTERMEDIATE, SOLID, TOXIC, N.O.S.	6.1	T2	II	6.1	274	500 g	E4	P002 IBC08	B4	MP10	T3	TP33
3143	DYE, SOLID, TOXIC, N.O.S. or DYE INTERMEDIATE, SOLID, TOXIC, N.O.S.	6.1	T2	III	6.1	274	5 kg	E1	P002 IBC08 LP02 R001	B3	MP10	T1	TP33
3144	NICOTINE COMPOUND, LIQUID, N.O.S. or NICOTINE PREPARATION, LIQUID, N.O.S.	6.1	T1	I	6.1	43 274	0	E5	P001		MP8 MP17		
3144	NICOTINE COMPOUND, LIQUID, N.O.S. or NICOTINE PREPARATION, LIQUID, N.O.S.	6.1	T1	II	6.1	43 274	100 ml	E4	P001 IBC02		MP15		
3144	NICOTINE COMPOUND, LIQUID, N.O.S. or NICOTINE PREPARATION, LIQUID, N.O.S.	6.1	T1	III	6.1	43 274	5 L	E1	P001 IBC03 LP01 R001		MP19		
3145	ALKYLPHENOLS, LIQUID, N.O.S. (including C2-C12 homologues)	8	C3	I	8		0	E0	P001		MP8 MP17	T14	TP2
3145	ALKYLPHENOLS, LIQUID, N.O.S. (including C2-C12 homologues)	8	C3	II	8		1 L	E2	P001 IBC02		MP15	T11	TP2 TP27
3145	ALKYLPHENOLS, LIQUID, N.O.S. (including C2-C12 homologues)	8	C3	III	8		5 L	E1	P001 IBC03 LP01 R001		MP19	T7	TP1 TP28
3146	ORGANOTIN COMPOUND, SOLID, N.O.S.	6.1	T3	I	6.1	43 274	0	E5	P002 IBC07		MP18	T6	TP33

ADR tank		Vehicle for tank carriage	Transport category (Tunnel restriction code)	Special provisions for carriage				Hazard identification No.	UN No.	Name and description
Tank code	Special provisions			Packages	Bulk	Loading, unloading and handling	Operation			
4.3	4.3.5, 6.8.4	9.1.1.2	1.1.3.6 (8.6)	7.2.4	7.3.3	7.5.11	8.5	5.3.2.3		3.1.2
(12)	(13)	(14)	(15)	(16)	(17)	(18)	(19)	(20)	(1)	(2)
RxBN	TU19 TA4 TT9	AT	3 (C/E)	V5		CV9 CV11 CV36	S20	22	3136	TRIFLUOROMETHANE, REFRIGERATED LIQUID
CARRIAGE PROHIBITED									3137	OXIDIZING SOLID, FLAMMABLE, N.O.S.
RxBN	TU18 TA4 TT9	FL	2 (B/D)	V5		CV9 CV11 CV36	S2 S17	223	3138	ETHYLENE, ACETYLENE AND PROPYLENE MIXTURE, REFRIGERATED LIQUID containing at least 71.5% ethylene with not more than 22.5% acetylene and not more than 6% propylene
			1 (E)			CV24	S20		3139	OXIDIZING LIQUID, N.O.S.
			2 (E)			CV24			3139	OXIDIZING LIQUID, N.O.S.
			3 (E)			CV24			3139	OXIDIZING LIQUID, N.O.S.
L10CH	TU14 TU15 TE19 TE21	AT	1 (C/E)			CV1 CV13 CV28	S9 S14	66	3140	ALKALOIDS, LIQUID, N.O.S. or ALKALOID SALTS, LIQUID, N.O.S.
L4BH	TU15 TE19	AT	2 (D/E)			CV13 CV28	S9 S19	60	3140	ALKALOIDS, LIQUID, N.O.S. or ALKALOID SALTS, LIQUID, N.O.S.
L4BH	TU15 TE19	AT	2 (E)	V12		CV13 CV28	S9	60	3140	ALKALOIDS, LIQUID, N.O.S. or ALKALOID SALTS, LIQUID, N.O.S.
L4BH	TU15 TE19	AT	2 (E)	V12		CV13 CV28	S9	60	3141	ANTIMONY COMPOUND, INORGANIC, LIQUID, N.O.S.
L10CH	TU14 TU15 TE19 TE21	AT	1 (C/E)			CV1 CV13 CV28	S9 S14	66	3142	DISINFECTANT, LIQUID, TOXIC, N.O.S.
L4BH	TU15 TE19	AT	2 (D/E)			CV13 CV28	S9 S19	60	3142	DISINFECTANT, LIQUID, TOXIC, N.O.S.
L4BH	TU15 TE19	AT	2 (E)	V12		CV13 CV28	S9	60	3142	DISINFECTANT, LIQUID, TOXIC, N.O.S.
S10AH L10CH	TU15 TE19	AT	1 (C/E)	V10		CV1 CV13 CV28	S9 S14	66	3143	DYE, SOLID, TOXIC, N.O.S. or DYE INTERMEDIATE, SOLID, TOXIC, N.O.S.
SGAH L4BH	TU15 TE19	AT	2 (D/E)	V11		CV13 CV28	S9 S19	60	3143	DYE, SOLID, TOXIC, N.O.S. or DYE INTERMEDIATE, SOLID, TOXIC, N.O.S.
SGAH L4BH	TU15 TE19	AT	2 (E)		VC1 VC2 AP7	CV13 CV28	S9	60	3143	DYE, SOLID, TOXIC, N.O.S. or DYE INTERMEDIATE, SOLID, TOXIC, N.O.S.
L10CH	TU14 TU15 TE19 TE21	AT	1 (C/E)			CV1 CV13 CV28	S9 S14	66	3144	NICOTINE COMPOUND, LIQUID, N.O.S. or NICOTINE PREPARATION, LIQUID, N.O.S.
L4BH	TU15 TE19	AT	2 (D/E)			CV13 CV28	S9 S19	60	3144	NICOTINE COMPOUND, LIQUID, N.O.S. or NICOTINE PREPARATION, LIQUID, N.O.S.
L4BH	TU15 TE19	AT	2 (E)	V12		CV13 CV28	S9	60	3144	NICOTINE COMPOUND, LIQUID, N.O.S. or NICOTINE PREPARATION, LIQUID, N.O.S.
L10BH		AT	1 (E)				S20	88	3145	ALKYLPHENOLS, LIQUID, N.O.S. (including C2-C12 homologues)
L4BN		AT	2 (E)					80	3145	ALKYLPHENOLS, LIQUID, N.O.S. (including C2-C12 homologues)
L4BN		AT	3 (E)	V12				80	3145	ALKYLPHENOLS, LIQUID, N.O.S. (including C2-C12 homologues)
S10AH L10CH	TU14 TU15 TE19 TE21	AT	1 (C/E)	V10		CV1 CV13 CV28	S9 S14	66	3146	ORGANOTIN COMPOUND, SOLID, N.O.S.

- 471 -

UN No.	Name and description	Class	Classifi-cation code	Packing group	Labels	Special provi-sions	Limited and excepted quantities		Packaging			Portable tanks and bulk containers	
										Special packing provisions	Mixed packing provisions	Instruc-tions	Special provisions
									Packing instruc-tions				
	3.1.2	2.2	2.2	2.1.1.3	5.2.2	3.3	3.4	3.5.1.2	4.1.4	4.1.4	4.1.10	4.2.5.2 7.3.2	4.2.5.3
(1)	(2)	(3a)	(3b)	(4)	(5)	(6)	(7a)	(7b)	(8)	(9a)	(9b)	(10)	(11)
3146	ORGANOTIN COMPOUND, SOLID, N.O.S.	6.1	T3	II	6.1	43 274	500 g	E4	P002 IBC08	B4	MP10	T3	TP33
3146	ORGANOTIN COMPOUND, SOLID, N.O.S.	6.1	T3	III	6.1	43 274	5 kg	E1	P002 IBC08 LP02 R001	B3	MP10	T1	TP33
3147	DYE, SOLID, CORROSIVE, N.O.S. or DYE INTERMEDIATE, SOLID, CORROSIVE, N.O.S.	8	C10	I	8	274	0	E0	P002 IBC07		MP18	T6	TP33
3147	DYE, SOLID, CORROSIVE, N.O.S. or DYE INTERMEDIATE, SOLID, CORROSIVE, N.O.S.	8	C10	II	8	274	1 kg	E2	P002 IBC08	B4	MP10	T3	TP33
3147	DYE, SOLID, CORROSIVE, N.O.S. or DYE INTERMEDIATE, SOLID, CORROSIVE, N.O.S.	8	C10	III	8	274	5 kg	E1	P002 IBC08 LP02 R001	B3	MP10	T1	TP33
3148	WATER-REACTIVE LIQUID, N.O.S.	4.3	W1	I	4.3	274	0	E0	P402	RR8	MP2	T13	TP2 TP7 TP38
3148	WATER-REACTIVE LIQUID, N.O.S.	4.3	W1	II	4.3	274	500 ml	E2	P402 IBC01	RR8	MP15	T7	TP2 TP7
3148	WATER-REACTIVE LIQUID, N.O.S.	4.3	W1	III	4.3	274	1 L	E1	P001 IBC02 R001		MP15	T7	TP2 TP7
3149	HYDROGEN PEROXIDE AND PEROXYACETIC ACID MIXTURE with acid(s), water and not more than 5% peroxyacetic acid, STABILIZED	5.1	OC1	II	5.1 +8	196 553	1 L	E2	P504 IBC02	PP10 B5	MP15	T7	TP2 TP6 TP24
3150	DEVICES, SMALL, HYDROCARBON GAS POWERED or HYDROCARBON GAS REFILLS FOR SMALL DEVICES with release device	2	6F		2.1		0	E0	P209		MP9		
3151	POLYHALOGENATED BIPHENYLS, LIQUID or HALOGENATED MONOMETHYLDIPHENYLMETHANES, LIQUID or POLYHALOGENATED TERPHENYLS, LIQUID	9	M2	II	9	203 305	1 L	E2	P906 IBC02		MP15		
3152	POLYHALOGENATED BIPHENYLS, SOLID or HALOGENATED MONOMETHYLDIPHENYLMETHANES, SOLID or POLYHALOGENATED TERPHENYLS, SOLID	9	M2	II	9	203 305	1 kg	E2	P906 IBC08	B4	MP10	T3	TP33
3153	PERFLUORO(METHYL VINYL ETHER)	2	2F		2.1	662	0	E0	P200		MP9	(M) T50	
3154	PERFLUORO(ETHYL VINYL ETHER)	2	2F		2.1	662	0	E0	P200		MP9	(M)	
3155	PENTACHLOROPHENOL	6.1	T2	II	6.1	43	500 g	E4	P002 IBC08	B4	MP10	T3	TP33
3156	COMPRESSED GAS, OXIDIZING, N.O.S.	2	1O		2.2 +5.1	274 655 662	0	E0	P200		MP9	(M)	
3157	LIQUEFIED GAS, OXIDIZING, N.O.S.	2	2O		2.2 +5.1	274 662	0	E0	P200		MP9	(M)	
3158	GAS, REFRIGERATED LIQUID, N.O.S.	2	3A		2.2	274 593	120 ml	E1	P203		MP9	T75	TP5
3159	1,1,1,2-TETRAFLUOROETHANE (REFRIGERANT GAS R 134a)	2	2A		2.2	662	120 ml	E1	P200		MP9	(M) T50	
3160	LIQUEFIED GAS, TOXIC, FLAMMABLE, N.O.S.	2	2TF		2.3 +2.1	274	0	E0	P200		MP9	(M)	
3161	LIQUEFIED GAS, FLAMMABLE, N.O.S.	2	2F		2.1	274 662	0	E0	P200		MP9	(M) T50	

ADR tank		Vehicle for tank carriage	Transport category (Tunnel restriction code)	Special provisions for carriage				Hazard identifi-cation No.	UN No.	Name and description
Tank code	Special provisions			Packages	Bulk	Loading, unloading and handling	Operation			
4.3	4.3.5, 6.8.4	9.1.1.2	1.1.3.6 (8.6)	7.2.4	7.3.3	7.5.11	8.5	5.3.2.3		3.1.2
(12)	(13)	(14)	(15)	(16)	(17)	(18)	(19)	(20)	(1)	(2)
SGAH L4BH	TU15 TE19	AT	2 (D/E)	V11		CV13 CV28	S9 S19	60	3146	ORGANOTIN COMPOUND, SOLID, N.O.S.
SGAH L4BH	TU15 TE19	AT	2 (E)		VC1 VC2 AP7	CV13 CV28	S9	60	3146	ORGANOTIN COMPOUND, SOLID, N.O.S.
S10AN L10BH		AT	1 (E)	V10			S20	88	3147	DYE, SOLID, CORROSIVE, N.O.S. or DYE INTERMEDIATE, SOLID, CORROSIVE, N.O.S.
SGAN L4BN		AT	2 (E)	V11				80	3147	DYE, SOLID, CORROSIVE, N.O.S. or DYE INTERMEDIATE, SOLID, CORROSIVE, N.O.S.
SGAV L4BN		AT	3 (E)		VC1 VC2 AP7			80	3147	DYE, SOLID, CORROSIVE, N.O.S. or DYE INTERMEDIATE, SOLID, CORROSIVE, N.O.S.
L10DH	TU14 TE21 TM2	AT	0 (B/E)	V1		CV23	S20	X323	3148	WATER-REACTIVE LIQUID, N.O.S.
L4DH	TU14 TE21 TM2	AT	0 (D/E)	V1		CV23		323	3148	WATER-REACTIVE LIQUID, N.O.S.
L4DH	TU14 TE21 TM2	AT	0 (E)	V1		CV23		323	3148	WATER-REACTIVE LIQUID, N.O.S.
L4BV(+)	TU3 TC2 TE8 TE11 TT1	AT	2 (E)			CV24		58	3149	HYDROGEN PEROXIDE AND PEROXYACETIC ACID MIXTURE with acid(s), water and not more than 5% peroxyacetic acid, STABILIZED
			2 (D)			CV9	S2		3150	DEVICES, SMALL, HYDROCARBON GAS POWERED or HYDROCARBON GAS REFILLS FOR SMALL DEVICES with release device
L4BH	TU15	AT	0 (D/E)		VC1 VC2 AP9	CV1 CV13 CV28	S19	90	3151	POLYHALOGENATED BIPHENYLS, LIQUID or HALOGENATED MONOMETHYLDIPHENYLMETHANES, LIQUID or POLYHALOGENATED TERPHENYLS, LIQUID
S4AH L4BH	TU15	AT	0 (D/E)	V11	VC1 VC2 AP9	CV1 CV13 CV28	S19	90	3152	POLYHALOGENATED BIPHENYLS, SOLID or HALOGENATED MONOMETHYLDIPHENYLMETHANES, SOLID or POLYHALOGENATED TERPHENYLS, SOLID
PxBN(M)	TA4 TT9	FL	2 (B/D)			CV9 CV10 CV36	S2 S20	23	3153	PERFLUORO(METHYL VINYL ETHER)
PxBN(M)	TA4 TT9	FL	2 (B/D)			CV9 CV10 CV36	S2 S20	23	3154	PERFLUORO(ETHYL VINYL ETHER)
SGAH	TU15 TE19	AT	2 (D/E)	V11		CV13 CV28	S9 S19	60	3155	PENTACHLOROPHENOL
CxBN(M)	TA4 TT9	AT	3 (E)			CV9 CV10 CV36		25	3156	COMPRESSED GAS, OXIDIZING, N.O.S.
PxBN(M)	TA4 TT9	AT	3 (C/E)			CV9 CV10 CV36		25	3157	LIQUEFIED GAS, OXIDIZING, N.O.S.
RxBN	TU19 TA4 TT9	AT	3 (C/E)	V5		CV9 CV11 CV36	S20	22	3158	GAS, REFRIGERATED LIQUID, N.O.S.
PxBN(M)	TA4 TT9	AT	3 (C/E)			CV9 CV10 CV36		20	3159	1,1,1,2-TETRAFLUOROETHANE (REFRIGERANT GAS R 134a)
PxBH(M)	TU6 TA4 TT9	FL	1 (B/D)			CV9 CV10 CV36	S2 S14	263	3160	LIQUEFIED GAS, TOXIC, FLAMMABLE, N.O.S.
PxBN(M)	TA4 TT9	FL	2 (B/D)			CV9 CV10 CV36	S2 S20	23	3161	LIQUEFIED GAS, FLAMMABLE, N.O.S.

UN No.	Name and description	Class	Classifi-cation code	Packing group	Labels	Special provi-sions	Limited and excepted quantities		Packaging			Portable tanks and bulk containers	
									Packing instruc-tions	Special packing provisions	Mixed packing provisions	Instruc-tions	Special provisions
3.1.2	3.1.2	2.2	2.2	2.1.1.3	5.2.2	3.3	3.4	3.5.1.2	4.1.4	4.1.4	4.1.10	4.2.5.2 7.3.2	4.2.5.3
(1)	(2)	(3a)	(3b)	(4)	(5)	(6)	(7a)	(7b)	(8)	(9a)	(9b)	(10)	(11)
3162	LIQUEFIED GAS, TOXIC, N.O.S.	2	2T		2.3	274	0	E0	P200		MP9	(M)	
3163	LIQUEFIED GAS, N.O.S.	2	2A		2.2	274 662	120 ml	E1	P200		MP9	(M) T50	
3164	ARTICLES, PRESSURIZED, PNEUMATIC or HYDRAULIC (containing non-flammable gas)	2	6A		2.2	283 371 594	120 ml	E0	P003		MP9		
3165	AIRCRAFT HYDRAULIC POWER UNIT FUEL TANK (containing a mixture of anhydrous hydrazine and methylhydrazine) (M86 fuel)	3	FTC	I	3 +6.1 +8		0	E0	P301		MP7		
3166	VEHICLE, FLAMMABLE GAS POWERED or VEHICLE, FLAMMABLE LIQUID POWERED or VEHICLE, FUEL CELL, FLAMMABLE GAS POWERED or VEHICLE, FUEL CELL, FLAMMABLE LIQUID POWERED	9	M11			312 385 666 667 669							
3167	GAS SAMPLE, NON-PRESSURIZED, FLAMMABLE, N.O.S., not refrigerated liquid	2	7F		2.1		0	E0	P201		MP9		
3168	GAS SAMPLE, NON-PRESSURIZED, TOXIC, FLAMMABLE, N.O.S., not refrigerated liquid	2	7TF		2.3 +2.1		0	E0	P201		MP9		
3169	GAS SAMPLE, NON-PRESSURIZED, TOXIC, N.O.S., not refrigerated liquid	2	7T		2.3		0	E0	P201		MP9		
3170	ALUMINIUM SMELTING BY-PRODUCTS or ALUMINIUM REMELTING BY-PRODUCTS	4.3	W2	II	4.3	244	500 g	E2	P410 IBC07		MP14	T3 BK1 BK2	TP33
3170	ALUMINIUM SMELTING BY-PRODUCTS or ALUMINIUM REMELTING BY-PRODUCTS	4.3	W2	III	4.3	244	1 kg	E1	P002 IBC08 R001	B4	MP14	T1 BK1 BK2	TP33
3171	BATTERY POWERED VEHICLE or BATTERY POWERED EQUIPMENT	9	M11			240 666 667 669							
3172	TOXINS, EXTRACTED FROM LIVING SOURCES, LIQUID, N.O.S.	6.1	T1	I	6.1	210 274	0	E5	P001		MP8 MP17		
3172	TOXINS, EXTRACTED FROM LIVING SOURCES, LIQUID, N.O.S.	6.1	T1	II	6.1	210 274	100 ml	E4	P001 IBC02		MP15		
3172	TOXINS, EXTRACTED FROM LIVING SOURCES, LIQUID, N.O.S.	6.1	T1	III	6.1	210 274	5 L	E1	P001 IBC03 LP01 R001		MP19		
3174	TITANIUM DISULPHIDE	4.2	S4	III	4.2		0	E1	P002 IBC08 LP02 R001	B3	MP14	T1	TP33
3175	SOLIDS or mixtures of solids (such as preparations and wastes) CONTAINING FLAMMABLE LIQUID, N.O.S. having a flash-point up to 60 °C	4.1	F1	II	4.1	216 274 601	1 kg	E2	P002 IBC06 R001	PP9	MP11	T3 BK1 BK2	TP33
3176	FLAMMABLE SOLID, ORGANIC, MOLTEN, N.O.S.	4.1	F2	II	4.1	274	0	E0				T3	TP3 TP26
3176	FLAMMABLE SOLID, ORGANIC, MOLTEN, N.O.S.	4.1	F2	III	4.1	274	0	E0				T1	TP3 TP26
3178	FLAMMABLE SOLID, INORGANIC, N.O.S.	4.1	F3	II	4.1	274	1 kg	E2	P002 IBC08	B4	MP11	T3	TP33
3178	FLAMMABLE SOLID, INORGANIC, N.O.S.	4.1	F3	III	4.1	274	5 kg	E1	P002 IBC08 LP02 R001	B3	MP11	T1	TP33
3179	FLAMMABLE SOLID, TOXIC, INORGANIC, N.O.S.	4.1	FT2	II	4.1 +6.1	274	1 kg	E2	P002 IBC06		MP10	T3	TP33

ADR tank		Vehicle for tank carriage	Transport category (Tunnel restriction code)	Special provisions for carriage				Hazard identification No.	UN No.	Name and description
Tank code	Special provisions			Packages	Bulk	Loading, unloading and handling	Operation			
4.3	4.3.5, 6.8.4	9.1.1.2	1.1.3.6 (8.6)	7.2.4	7.3.3	7.5.11	8.5	5.3.2.3		3.1.2
(12)	(13)	(14)	(15)	(16)	(17)	(18)	(19)	(20)	(1)	(2)
PxBH(M)	TU6 TA4 TT9	AT	1 (C/D)			CV9 CV10 CV36	S14	26	3162	LIQUEFIED GAS, TOXIC, N.O.S.
PxBN(M)	TA4 TT9	AT	3 (C/E)			CV9 CV10 CV36		20	3163	LIQUEFIED GAS, N.O.S.
			3 (E)			CV9			3164	ARTICLES, PRESSURIZED, PNEUMATIC or HYDRAULIC (containing non-flammable gas)
			1 (E)			CV13 CV28	S2 S19		3165	AIRCRAFT HYDRAULIC POWER UNIT FUEL TANK (containing a mixture of anhydrous hydrazine and methylhydrazine) (M86 fuel)
									3166	VEHICLE, FLAMMABLE GAS POWERED or VEHICLE, FLAMMABLE LIQUID POWERED or VEHICLE, FUEL CELL, FLAMMABLE GAS POWERED or VEHICLE, FUEL CELL, FLAMMABLE LIQUID POWERED
			2 (D)			CV9	S2		3167	GAS SAMPLE, NON-PRESSURIZED, FLAMMABLE, N.O.S., not refrigerated liquid
			1 (D)			CV9	S2		3168	GAS SAMPLE, NON-PRESSURIZED, TOXIC, FLAMMABLE, N.O.S., not refrigerated liquid
			1 (D)			CV9			3169	GAS SAMPLE, NON-PRESSURIZED, TOXIC, N.O.S., not refrigerated liquid
SGAN		AT	2 (D/E)	V1	VC1 VC2 AP2	CV23 CV37		423	3170	ALUMINIUM SMELTING BY-PRODUCTS or ALUMINIUM REMELTING BY-PRODUCTS
SGAN		AT	3 (E)	V1	VC1 VC2 AP2	CV23 CV37		423	3170	ALUMINIUM SMELTING BY-PRODUCTS or ALUMINIUM REMELTING BY-PRODUCTS
									3171	BATTERY POWERED VEHICLE or BATTERY POWERED EQUIPMENT
L10CH	TU14 TU15 TE19 TE21	AT	1 (C/E)			CV1 CV13 CV28	S9 S14	66	3172	TOXINS, EXTRACTED FROM LIVING SOURCES, LIQUID, N.O.S.
L4BH	TU15 TE19	AT	2 (D/E)			CV13 CV28	S9 S19	60	3172	TOXINS, EXTRACTED FROM LIVING SOURCES, LIQUID, N.O.S.
L4BH	TU15 TE19	AT	2 (E)	V12		CV13 CV28	S9	60	3172	TOXINS, EXTRACTED FROM LIVING SOURCES, LIQUID, N.O.S.
SGAN		AT	3 (E)	V1				40	3174	TITANIUM DISULPHIDE
		AT	2 (E)	V11	VC1 VC2 AP2			40	3175	SOLIDS or mixtures of solids (such as preparations and wastes) CONTAINING FLAMMABLE LIQUID, N.O.S. having a flash-point up to 60 °C
LGBV	TU27 TE4 TE6	AT	2 (E)					44	3176	FLAMMABLE SOLID, ORGANIC, MOLTEN, N.O.S.
LGBV	TU27 TE4 TE6	AT	3 (E)					44	3176	FLAMMABLE SOLID, ORGANIC, MOLTEN, N.O.S.
SGAN		AT	2 (E)	V11				40	3178	FLAMMABLE SOLID, INORGANIC, N.O.S.
SGAV		AT	3 (E)		VC1 VC2			40	3178	FLAMMABLE SOLID, INORGANIC, N.O.S.
SGAN		AT	2 (E)	V11		CV28		46	3179	FLAMMABLE SOLID, TOXIC, INORGANIC, N.O.S.

UN No.	Name and description	Class	Classification code	Packing group	Labels	Special provisions	Limited and excepted quantities		Packaging			Portable tanks and bulk containers	
									Packing instructions	Special packing provisions	Mixed packing provisions	Instructions	Special provisions
	3.1.2	2.2	2.2	2.1.1.3	5.2.2	3.3	3.4	3.5.1.2	4.1.4	4.1.4	4.1.10	4.2.5.2 7.3.2	4.2.5.3
(1)	(2)	(3a)	(3b)	(4)	(5)	(6)	(7a)	(7b)	(8)	(9a)	(9b)	(10)	(11)
3179	FLAMMABLE SOLID, TOXIC, INORGANIC, N.O.S.	4.1	FT2	III	4.1 +6.1	274	5 kg	E1	P002 IBC06 R001		MP10	T1	TP33
3180	FLAMMABLE SOLID, CORROSIVE, INORGANIC, N.O.S.	4.1	FC2	II	4.1 +8	274	1 kg	E2	P002 IBC06		MP10	T3	TP33
3180	FLAMMABLE SOLID, CORROSIVE, INORGANIC, N.O.S.	4.1	FC2	III	4.1 +8	274	5 kg	E1	P002 IBC06 R001		MP10	T1	TP33
3181	METAL SALTS OF ORGANIC COMPOUNDS, FLAMMABLE, N.O.S.	4.1	F3	II	4.1	274	1 kg	E2	P002 IBC08	B4	MP11	T3	TP33
3181	METAL SALTS OF ORGANIC COMPOUNDS, FLAMMABLE, N.O.S.	4.1	F3	III	4.1	274	5 kg	E1	P002 IBC08 LP02 R001	B3	MP11	T1	TP33
3182	METAL HYDRIDES, FLAMMABLE, N.O.S.	4.1	F3	II	4.1	274 554	1 kg	E2	P410 IBC04	PP40	MP11	T3	TP33
3182	METAL HYDRIDES, FLAMMABLE, N.O.S.	4.1	F3	III	4.1	274 554	5 kg	E1	P002 IBC04 R001		MP11	T1	TP33
3183	SELF-HEATING LIQUID, ORGANIC, N.O.S.	4.2	S1	II	4.2	274	0	E2	P001 IBC02		MP15		
3183	SELF-HEATING LIQUID, ORGANIC, N.O.S.	4.2	S1	III	4.2	274	0	E1	P001 IBC02 R001		MP15		
3184	SELF-HEATING LIQUID, TOXIC, ORGANIC, N.O.S.	4.2	ST1	II	4.2 +6.1	274	0	E2	P402 IBC02		MP15		
3184	SELF-HEATING LIQUID, TOXIC, ORGANIC, N.O.S.	4.2	ST1	III	4.2 +6.1	274	0	E1	P001 IBC02 R001		MP15		
3185	SELF-HEATING LIQUID, CORROSIVE, ORGANIC, N.O.S.	4.2	SC1	II	4.2 +8	274	0	E2	P402 IBC02		MP15		
3185	SELF-HEATING LIQUID, CORROSIVE, ORGANIC, N.O.S.	4.2	SC1	III	4.2 +8	274	0	E1	P001 IBC02 R001		MP15		
3186	SELF-HEATING LIQUID, INORGANIC, N.O.S.	4.2	S3	II	4.2	274	0	E2	P001 IBC02		MP15		
3186	SELF-HEATING LIQUID, INORGANIC, N.O.S.	4.2	S3	III	4.2	274	0	E1	P001 IBC02 R001		MP15		
3187	SELF-HEATING LIQUID, TOXIC, INORGANIC, N.O.S.	4.2	ST3	II	4.2 +6.1	274	0	E2	P402 IBC02		MP15		
3187	SELF-HEATING LIQUID, TOXIC, INORGANIC, N.O.S.	4.2	ST3	III	4.2 +6.1	274	0	E1	P001 IBC02 R001		MP15		
3188	SELF-HEATING LIQUID, CORROSIVE, INORGANIC, N.O.S.	4.2	SC3	II	4.2 +8	274	0	E2	P402 IBC02		MP15		
3188	SELF-HEATING LIQUID, CORROSIVE, INORGANIC, N.O.S.	4.2	SC3	III	4.2 +8	274	0	E1	P001 IBC02 R001		MP15		
3189	METAL POWDER, SELF-HEATING, N.O.S.	4.2	S4	II	4.2	274 555	0	E2	P410 IBC06		MP14	T3	TP33
3189	METAL POWDER, SELF-HEATING, N.O.S.	4.2	S4	III	4.2	274 555	0	E1	P002 IBC08 LP02 R001	B3	MP14	T1	TP33
3190	SELF-HEATING SOLID, INORGANIC, N.O.S.	4.2	S4	II	4.2	274	0	E2	P410 IBC06		MP14	T3	TP33
3190	SELF-HEATING SOLID, INORGANIC, N.O.S.	4.2	S4	III	4.2	274	0	E1	P002 IBC08 LP02 R001	B3	MP14	T1	TP33
3191	SELF-HEATING SOLID, TOXIC, INORGANIC, N.O.S.	4.2	ST4	II	4.2 +6.1	274	0	E2	P410 IBC05		MP14	T3	TP33
3191	SELF-HEATING SOLID, TOXIC, INORGANIC, N.O.S.	4.2	ST4	III	4.2 +6.1	274	0	E1	P002 IBC08 R001	B3	MP14	T1	TP33
3192	SELF-HEATING SOLID, CORROSIVE, INORGANIC, N.O.S.	4.2	SC4	II	4.2 +8	274	0	E2	P410 IBC05		MP14	T3	TP33
3192	SELF-HEATING SOLID, CORROSIVE, INORGANIC, N.O.S.	4.2	SC4	III	4.2 +8	274	0	E1	P002 IBC08 R001	B3	MP14	T1	TP33

ADR tank		Vehicle for tank carriage	Transport category (Tunnel restriction code)	Special provisions for carriage				Hazard identifi-cation No.	UN No.	Name and description
Tank code	Special provisions			Packages	Bulk	Loading, unloading and handling	Operation			
4.3	4.3.5, 6.8.4	9.1.1.2	1.1.3.6 (8.6)	7.2.4	7.3.3	7.5.11	8.5	5.3.2.3		3.1.2
(12)	(13)	(14)	(15)	(16)	(17)	(18)	(19)	(20)	(1)	(2)
SGAN		AT	3 (E)			CV28		46	3179	FLAMMABLE SOLID, TOXIC, INORGANIC, N.O.S.
SGAN		AT	2 (E)	V11				48	3180	FLAMMABLE SOLID, CORROSIVE, INORGANIC, N.O.S.
SGAN		AT	3 (E)					48	3180	FLAMMABLE SOLID, CORROSIVE, INORGANIC, N.O.S.
SGAN		AT	2 (E)	V11				40	3181	METAL SALTS OF ORGANIC COMPOUNDS, FLAMMABLE, N.O.S.
SGAV		AT	3 (E)		VC1 VC2			40	3181	METAL SALTS OF ORGANIC COMPOUNDS, FLAMMABLE, N.O.S.
SGAN		AT	2 (E)					40	3182	METAL HYDRIDES, FLAMMABLE, N.O.S.
SGAV		AT	3 (E)		VC1 VC2			40	3182	METAL HYDRIDES, FLAMMABLE, N.O.S.
L4DH	TU14 TE21	AT	2 (D/E)	V1				30	3183	SELF-HEATING LIQUID, ORGANIC, N.O.S.
L4DH	TU14 TE21	AT	3 (E)	V1				30	3183	SELF-HEATING LIQUID, ORGANIC, N.O.S.
L4DH	TU14 TE21	AT	2 (D/E)	V1		CV28		36	3184	SELF-HEATING LIQUID, TOXIC, ORGANIC, N.O.S.
L4DH	TU14 TE21	AT	3 (E)	V1		CV28		36	3184	SELF-HEATING LIQUID, TOXIC, ORGANIC, N.O.S.
L4DH	TU14 TE21	AT	2 (D/E)	V1				38	3185	SELF-HEATING LIQUID, CORROSIVE, ORGANIC, N.O.S.
L4DH	TU14 TE21	AT	3 (E)	V1				38	3185	SELF-HEATING LIQUID, CORROSIVE, ORGANIC, N.O.S.
L4DH	TU14 TE21	AT	2 (D/E)	V1				30	3186	SELF-HEATING LIQUID, INORGANIC, N.O.S.
L4DH	TU14 TE21	AT	3 (E)	V1				30	3186	SELF-HEATING LIQUID, INORGANIC, N.O.S.
L4DH	TU14 TE21	AT	2 (D/E)	V1		CV28		36	3187	SELF-HEATING LIQUID, TOXIC, INORGANIC, N.O.S.
L4DH	TU14 TE21	AT	3 (E)	V1		CV28		36	3187	SELF-HEATING LIQUID, TOXIC, INORGANIC, N.O.S.
L4DH	TU14 TE21	AT	2 (D/E)	V1				38	3188	SELF-HEATING LIQUID, CORROSIVE, INORGANIC, N.O.S.
L4DH	TU14 TE21	AT	3 (E)	V1				38	3188	SELF-HEATING LIQUID, CORROSIVE, INORGANIC, N.O.S.
SGAN		AT	2 (D/E)	V1				40	3189	METAL POWDER, SELF-HEATING, N.O.S.
SGAN		AT	3 (E)	V1	VC1 VC2 AP1			40	3189	METAL POWDER, SELF-HEATING, N.O.S.
SGAN		AT	2 (D/E)	V1				40	3190	SELF-HEATING SOLID, INORGANIC, N.O.S.
SGAN		AT	3 (E)	V1	VC1 VC2 AP1			40	3190	SELF-HEATING SOLID, INORGANIC, N.O.S.
SGAN		AT	2 (D/E)	V1		CV28		46	3191	SELF-HEATING SOLID, TOXIC, INORGANIC, N.O.S.
SGAN		AT	3 (E)	V1		CV28		46	3191	SELF-HEATING SOLID, TOXIC, INORGANIC, N.O.S.
SGAN		AT	2 (D/E)	V1				48	3192	SELF-HEATING SOLID, CORROSIVE, INORGANIC, N.O.S.
SGAN		AT	3 (E)	V1				48	3192	SELF-HEATING SOLID, CORROSIVE, INORGANIC, N.O.S.

UN No.	Name and description	Class	Classification code	Packing group	Labels	Special provisions	Limited quantities	Excepted quantities	Packing instructions	Special packing provisions	Mixed packing provisions	Portable tanks and bulk containers Instructions	Special provisions
	3.1.2	2.2	2.2	2.1.1.3	5.2.2	3.3	3.4	3.5.1.2	4.1.4	4.1.4	4.1.10	4.2.5.2 7.3.2	4.2.5.3
(1)	(2)	(3a)	(3b)	(4)	(5)	(6)	(7a)	(7b)	(8)	(9a)	(9b)	(10)	(11)
3194	PYROPHORIC LIQUID, INORGANIC, N.O.S.	4.2	S3	I	4.2	274	0	E0	P400		MP2		
3200	PYROPHORIC SOLID, INORGANIC, N.O.S.	4.2	S4	I	4.2	274	0	E0	P404		MP13	T21	TP7 TP33
3205	ALKALINE EARTH METAL ALCOHOLATES, N.O.S.	4.2	S4	II	4.2	183 274	0	E2	P410 IBC06		MP14	T3	TP33
3205	ALKALINE EARTH METAL ALCOHOLATES, N.O.S.	4.2	S4	III	4.2	183 274	0	E1	P002 IBC08 LP02 R001	B3	MP14	T1	TP33
3206	ALKALI METAL ALCOHOLATES, SELF-HEATING, CORROSIVE, N.O.S.	4.2	SC4	II	4.2 +8	182 274	0	E2	P410 IBC05		MP14	T3	TP33
3206	ALKALI METAL ALCOHOLATES, SELF-HEATING, CORROSIVE, N.O.S.	4.2	SC4	III	4.2 +8	182 274	0	E1	P002 IBC08 R001	B3	MP14	T1	TP33
3208	METALLIC SUBSTANCE, WATER-REACTIVE, N.O.S.	4.3	W2	I	4.3	274 557	0	E0	P403 IBC99		MP2		
3208	METALLIC SUBSTANCE, WATER-REACTIVE, N.O.S.	4.3	W2	II	4.3	274 557	500 g	E0	P410 IBC07		MP14	T3	TP33
3208	METALLIC SUBSTANCE, WATER-REACTIVE, N.O.S.	4.3	W2	III	4.3	274 557	1 kg	E1	P410 IBC08 R001	B4	MP14	T1	TP33
3209	METALLIC SUBSTANCE, WATER-REACTIVE, SELF-HEATING, N.O.S.	4.3	WS	I	4.3 +4.2	274 558	0	E0	P403		MP2		
3209	METALLIC SUBSTANCE, WATER-REACTIVE, SELF-HEATING, N.O.S.	4.3	WS	II	4.3 +4.2	274 558	0	E2	P410 IBC05		MP14	T3	TP33
3209	METALLIC SUBSTANCE, WATER-REACTIVE, SELF-HEATING, N.O.S.	4.3	WS	III	4.3 +4.2	274 558	0	E1	P410 IBC08 R001	B4	MP14	T1	TP33
3210	CHLORATES, INORGANIC, AQUEOUS SOLUTION, N.O.S.	5.1	O1	II	5.1	274 351	1 L	E2	P504 IBC02		MP2	T4	TP1
3210	CHLORATES, INORGANIC, AQUEOUS SOLUTION, N.O.S.	5.1	O1	III	5.1	274 351	5 L	E1	P504 IBC02 R001		MP2	T4	TP1
3211	PERCHLORATES, INORGANIC, AQUEOUS SOLUTION, N.O.S.	5.1	O1	II	5.1		1 L	E2	P504 IBC02		MP2	T4	TP1
3211	PERCHLORATES, INORGANIC, AQUEOUS SOLUTION, N.O.S.	5.1	O1	III	5.1		5 L	E1	P504 IBC02 R001		MP2	T4	TP1
3212	HYPOCHLORITES, INORGANIC, N.O.S.	5.1	O2	II	5.1	274 349	1 kg	E2	P002 IBC08	B4	MP10	T3	TP33
3213	BROMATES, INORGANIC, AQUEOUS SOLUTION, N.O.S.	5.1	O1	II	5.1	274 350	1 L	E2	P504 IBC02		MP2	T4	TP1
3213	BROMATES, INORGANIC, AQUEOUS SOLUTION, N.O.S.	5.1	O1	III	5.1	274 350	5 L	E1	P504 IBC02 R001		MP15	T4	TP1
3214	PERMANGANATES, INORGANIC, AQUEOUS SOLUTION, N.O.S.	5.1	O1	II	5.1	274 353	1 L	E2	P504 IBC02		MP2	T4	TP1
3215	PERSULPHATES, INORGANIC, N.O.S.	5.1	O2	III	5.1		5 kg	E1	P002 IBC08 LP02 R001	B3	MP10	T1	TP33
3216	PERSULPHATES, INORGANIC, AQUEOUS SOLUTION, N.O.S.	5.1	O1	III	5.1		5 L	E1	P504 IBC02 R001		MP15	T4	TP1 TP29
3218	NITRATES, INORGANIC, AQUEOUS SOLUTION, N.O.S.	5.1	O1	II	5.1	270 511	1 L	E2	P504 IBC02		MP15	T4	TP1
3218	NITRATES, INORGANIC, AQUEOUS SOLUTION, N.O.S.	5.1	O1	III	5.1	270 511	5 L	E1	P504 IBC02 R001		MP15	T4	TP1
3219	NITRITES, INORGANIC, AQUEOUS SOLUTION, N.O.S.	5.1	O1	II	5.1	103 274	1 L	E2	P504 IBC01		MP15	T4	TP1
3219	NITRITES, INORGANIC, AQUEOUS SOLUTION, N.O.S.	5.1	O1	III	5.1	103 274	5 L	E1	P504 IBC02 R001		MP15	T4	TP1
3220	PENTAFLUOROETHANE (REFRIGERANT GAS R 125)	2	2A		2.2	662	120 ml	E1	P200		MP9	(M) T50	
3221	SELF-REACTIVE LIQUID TYPE B	4.1	SR1		4.1 +1	181 194 274	25 ml	E0	P520	PP21	MP2		

ADR tank		Vehicle for tank carriage	Transport category (Tunnel restriction code)	Special provisions for carriage				Hazard identifi-cation No.	UN No.	Name and description
Tank code	Special provisions			Packages	Bulk	Loading, unloading and handling	Operation			
4.3	4.3.5, 6.8.4	9.1.1.2	1.1.3.6 (8.6)	7.2.4	7.3.3	7.5.11	8.5	5.3.2.3		3.1.2
(12)	(13)	(14)	(15)	(16)	(17)	(18)	(19)	(20)	(1)	(2)
L21DH	TU14 TC1 TE21 TM1	AT	0 (B/E)	V1			S20	333	3194	PYROPHORIC LIQUID, INORGANIC, N.O.S.
		AT	0 (B/E)	V1			S20	43	3200	PYROPHORIC SOLID, INORGANIC, N.O.S.
SGAN		AT	2 (D/E)	V1				40	3205	ALKALINE EARTH METAL ALCOHOLATES, N.O.S.
SGAN		AT	3 (E)	V1				40	3205	ALKALINE EARTH METAL ALCOHOLATES, N.O.S.
SGAN		AT	2 (D/E)	V1				48	3206	ALKALI METAL ALCOHOLATES, SELF-HEATING, CORROSIVE, N.O.S.
SGAN		AT	3 (E)	V1				48	3206	ALKALI METAL ALCOHOLATES, SELF-HEATING, CORROSIVE, N.O.S.
			1 (E)	V1		CV23	S20		3208	METALLIC SUBSTANCE, WATER-REACTIVE, N.O.S.
SGAN		AT	2 (D/E)	V1		CV23		423	3208	METALLIC SUBSTANCE, WATER-REACTIVE, N.O.S.
SGAN		AT	3 (E)	V1	VC1 VC2 AP3 AP4 AP5	CV23		423	3208	METALLIC SUBSTANCE, WATER-REACTIVE, N.O.S.
			1 (E)	V1		CV23	S20		3209	METALLIC SUBSTANCE, WATER-REACTIVE, SELF-HEATING, N.O.S.
SGAN		AT	2 (D/E)	V1		CV23		423	3209	METALLIC SUBSTANCE, WATER-REACTIVE, SELF-HEATING, N.O.S.
SGAN		AT	3 (E)	V1	VC1 VC2 AP3 AP4 AP5	CV23		423	3209	METALLIC SUBSTANCE, WATER-REACTIVE, SELF-HEATING, N.O.S.
L4BN	TU3	AT	2 (E)			CV24		50	3210	CHLORATES, INORGANIC, AQUEOUS SOLUTION, N.O.S.
LGBV	TU3	AT	3 (E)			CV24		50	3210	CHLORATES, INORGANIC, AQUEOUS SOLUTION, N.O.S.
L4BN	TU3	AT	2 (E)			CV24		50	3211	PERCHLORATES, INORGANIC, AQUEOUS SOLUTION, N.O.S.
LGBV	TU3	AT	3 (E)			CV24		50	3211	PERCHLORATES, INORGANIC, AQUEOUS SOLUTION, N.O.S.
SGAN	TU3	AT	2 (E)	V11		CV24		50	3212	HYPOCHLORITES, INORGANIC, N.O.S.
L4BN	TU3	AT	2 (E)			CV24		50	3213	BROMATES, INORGANIC, AQUEOUS SOLUTION, N.O.S.
LGBV	TU3	AT	3 (E)			CV24		50	3213	BROMATES, INORGANIC, AQUEOUS SOLUTION, N.O.S.
L4BN	TU3	AT	2 (E)			CV24		50	3214	PERMANGANATES, INORGANIC, AQUEOUS SOLUTION, N.O.S.
SGAV	TU3	AT	3 (E)		VC1 VC2 AP6 AP7	CV24		50	3215	PERSULPHATES, INORGANIC, N.O.S.
LGBV	TU3	AT	3 (E)			CV24		50	3216	PERSULPHATES, INORGANIC, AQUEOUS SOLUTION, N.O.S.
L4BN	TU3	AT	2 (E)			CV24		50	3218	NITRATES, INORGANIC, AQUEOUS SOLUTION, N.O.S.
LGBV	TU3	AT	3 (E)			CV24		50	3218	NITRATES, INORGANIC, AQUEOUS SOLUTION, N.O.S.
L4BN	TU3	AT	2 (E)			CV24		50	3219	NITRITES, INORGANIC, AQUEOUS SOLUTION, N.O.S.
LGBV	TU3	AT	3 (E)			CV24		50	3219	NITRITES, INORGANIC, AQUEOUS SOLUTION, N.O.S.
PxBN(M)	TA4 TT9	AT	3 (C/E)			CV9 CV10 CV36		20	3220	PENTAFLUOROETHANE (REFRIGERANT GAS R 125)
			1 (B)	V1		CV15 CV20 CV22	S9 S17		3221	SELF-REACTIVE LIQUID TYPE B

UN No.	Name and description	Class	Classifi-cation code	Packing group	Labels	Special provi-sions	Limited and excepted quantities		Packaging			Portable tanks and bulk containers	
									Packing instruc-tions	Special packing provisions	Mixed packing provisions	Instruc-tions	Special provisions
	3.1.2	2.2	2.2	2.1.1.3	5.2.2	3.3	3.4	3.5.1.2	4.1.4	4.1.4	4.1.10	4.2.5.2 7.3.2	4.2.5.3
(1)	(2)	(3a)	(3b)	(4)	(5)	(6)	(7a)	(7b)	(8)	(9a)	(9b)	(10)	(11)
3222	SELF-REACTIVE SOLID TYPE B	4.1	SR1		4.1 +1	181 194 274	100 g	E0	P520	PP21	MP2		
3223	SELF-REACTIVE LIQUID TYPE C	4.1	SR1		4.1	194 274	25 ml	E0	P520	PP21	MP2		
3224	SELF-REACTIVE SOLID TYPE C	4.1	SR1		4.1	194 274	100 g	E0	P520	PP21	MP2		
3225	SELF-REACTIVE LIQUID TYPE D	4.1	SR1		4.1	194 274	125 ml	E0	P520		MP2		
3226	SELF-REACTIVE SOLID TYPE D	4.1	SR1		4.1	194 274	500 g	E0	P520		MP2		
3227	SELF-REACTIVE LIQUID TYPE E	4.1	SR1		4.1	194 274	125 ml	E0	P520		MP2		
3228	SELF-REACTIVE SOLID TYPE E	4.1	SR1		4.1	194 274	500 g	E0	P520		MP2		
3229	SELF-REACTIVE LIQUID TYPE F	4.1	SR1		4.1	194 274	125 ml	E0	P520 IBC99		MP2	T23	
3230	SELF-REACTIVE SOLID TYPE F	4.1	SR1		4.1	194 274	500 g	E0	P520 IBC99		MP2	T23	
3231	SELF-REACTIVE LIQUID TYPE B, TEMPERATURE CONTROLLED	4.1	SR2		4.1 +1	181 194 274	0	E0	P520	PP21	MP2		
3232	SELF-REACTIVE SOLID TYPE B, TEMPERATURE CONTROLLED	4.1	SR2		4.1 +1	181 194 274	0	E0	P520	PP21	MP2		
3233	SELF-REACTIVE LIQUID TYPE C, TEMPERATURE CONTROLLED	4.1	SR2		4.1	194 274	0	E0	P520	PP21	MP2		
3234	SELF-REACTIVE SOLID TYPE C, TEMPERATURE CONTROLLED	4.1	SR2		4.1	194 274	0	E0	P520	PP21	MP2		
3235	SELF-REACTIVE LIQUID TYPE D, TEMPERATURE CONTROLLED	4.1	SR2		4.1	194 274	0	E0	P520		MP2		
3236	SELF-REACTIVE SOLID TYPE D, TEMPERATURE CONTROLLED	4.1	SR2		4.1	194 274	0	E0	P520		MP2		
3237	SELF-REACTIVE LIQUID TYPE E, TEMPERATURE CONTROLLED	4.1	SR2		4.1	194 274	0	E0	P520		MP2		
3238	SELF-REACTIVE SOLID TYPE E, TEMPERATURE CONTROLLED	4.1	SR2		4.1	194 274	0	E0	P520		MP2		
3239	SELF-REACTIVE LIQUID TYPE F, TEMPERATURE CONTROLLED	4.1	SR2		4.1	194 274	0	E0	P520		MP2	T23	
3240	SELF-REACTIVE SOLID TYPE F, TEMPERATURE CONTROLLED	4.1	SR2		4.1	194 274	0	E0	P520		MP2	T23	
3241	2-BROMO-2-NITROPROPANE-1,3-DIOL	4.1	SR1	III	4.1	638	5 kg	E1	P520 IBC08	PP22 B3	MP2		
3242	AZODICARBONAMIDE	4.1	SR1	II	4.1	215 638	1 kg	E0	P409		MP2	T3	TP33
3243	SOLIDS CONTAINING TOXIC LIQUID, N.O.S.	6.1	T9	II	6.1	217 274 601	500 g	E4	P002 IBC02	PP9	MP10	T3 BK1 BK2	TP33
3244	SOLIDS CONTAINING CORROSIVE LIQUID, N.O.S.	8	C10	II	8	218 274	1 kg	E2	P002 IBC05	PP9	MP10	T3 BK1 BK2	TP33
3245	GENETICALLY MODIFIED MICROORGANISMS or GENETICALLY MODIFIED ORGANISMS	9	M8		9	219 637	0	E0	P904 IBC08		MP6		
3245	GENETICALLY MODIFIED MICROORGANISMS or GENETICALLY MODIFIED ORGANISMS, in refrigerated liquid nitrogen	9	M8		9 +2.2	219 637	0	E0	P904 IBC08		MP6		

ADR tank		Vehicle for tank carriage	Transport category (Tunnel restriction code)	Special provisions for carriage				Hazard identification No.	UN No.	Name and description
Tank code	Special provisions			Packages	Bulk	Loading, unloading and handling	Operation			
4.3	4.3.5, 6.8.4	9.1.1.2	1.1.3.6 (8.6)	7.2.4	7.3.3	7.5.11	8.5	5.3.2.3		3.1.2
(12)	(13)	(14)	(15)	(16)	(17)	(18)	(19)	(20)	(1)	(2)
			1 (B)	V1		CV15 CV20 CV22	S9 S17		3222	SELF-REACTIVE SOLID TYPE B
			1 (D)	V1		CV15 CV20 CV22	S8 S18		3223	SELF-REACTIVE LIQUID TYPE C
			1 (D)	V1		CV15 CV20 CV22	S8 S18		3224	SELF-REACTIVE SOLID TYPE C
			2 (D)	V1		CV15 CV22	S19		3225	SELF-REACTIVE LIQUID TYPE D
			2 (D)	V1		CV15 CV22	S19		3226	SELF-REACTIVE SOLID TYPE D
			2 (D)	V1		CV15 CV22			3227	SELF-REACTIVE LIQUID TYPE E
			2 (D)	V1		CV15 CV22			3228	SELF-REACTIVE SOLID TYPE E
		AT	2 (D)	V1		CV15 CV22		40	3229	SELF-REACTIVE LIQUID TYPE F
		AT	2 (D)	V1		CV15 CV22		40	3230	SELF-REACTIVE SOLID TYPE F
			1 (B)	V8		CV15 CV20 CV21 CV22	S4 S9 S16		3231	SELF-REACTIVE LIQUID TYPE B, TEMPERATURE CONTROLLED
			1 (B)	V8		CV15 CV20 CV21 CV22	S4 S9 S16		3232	SELF-REACTIVE SOLID TYPE B, TEMPERATURE CONTROLLED
			1 (D)	V8		CV15 CV20 CV21 CV22	S4 S8 S17		3233	SELF-REACTIVE LIQUID TYPE C, TEMPERATURE CONTROLLED
			1 (D)	V8		CV15 CV20 CV21 CV22	S4 S8 S17		3234	SELF-REACTIVE SOLID TYPE C, TEMPERATURE CONTROLLED
			1 (D)	V8		CV15 CV21 CV22	S4 S18		3235	SELF-REACTIVE LIQUID TYPE D, TEMPERATURE CONTROLLED
			1 (D)	V8		CV15 CV21 CV22	S4 S18		3236	SELF-REACTIVE SOLID TYPE D, TEMPERATURE CONTROLLED
			1 (D)	V8		CV15 CV21 CV22	S4 S19		3237	SELF-REACTIVE LIQUID TYPE E, TEMPERATURE CONTROLLED
			1 (D)	V8		CV15 CV21 CV22	S4 S19		3238	SELF-REACTIVE SOLID TYPE E, TEMPERATURE CONTROLLED
		AT	1 (D)	V8		CV15 CV21 CV22	S4	40	3239	SELF-REACTIVE LIQUID TYPE F, TEMPERATURE CONTROLLED
		AT	1 (D)	V8		CV15 CV21 CV22	S4	40	3240	SELF-REACTIVE SOLID TYPE F, TEMPERATURE CONTROLLED
			3 (D)			CV14	S24		3241	2-BROMO-2-NITROPROPANE-1,3-DIOL
		AT	2 (D)			CV14	S24	40	3242	AZODICARBONAMIDE
SGAH	TU15 TE19	AT	2 (D/E)		VC1 VC2 AP7	CV13 CV28	S9 S19	60	3243	SOLIDS CONTAINING TOXIC LIQUID, N.O.S.
SGAV		AT	2 (E)		VC1 VC2 AP7			80	3244	SOLIDS CONTAINING CORROSIVE LIQUID, N.O.S.
			2 (E)			CV1 CV13 CV26 CV27 CV28	S17		3245	GENETICALLY MODIFIED MICROORGANISMS or GENETICALLY MODIFIED ORGANISMS
			2 (E)			CV1 CV13 CV26 CV27 CV28	S17		3245	GENETICALLY MODIFIED MICROORGANISMS or GENETICALLY MODIFIED ORGANISMS, in refrigerated liquid nitrogen

UN No.	Name and description	Class	Classification code	Packing group	Labels	Special provisions	Limited and excepted quantities		Packaging			Portable tanks and bulk containers	
									Packing instructions	Special packing provisions	Mixed packing provisions	Instructions	Special provisions
3.1.2	3.1.2	2.2	2.2	2.1.1.3	5.2.2	3.3	3.4	3.5.1.2	4.1.4	4.1.4	4.1.10	4.2.5.2 7.3.2	4.2.5.3
(1)	(2)	(3a)	(3b)	(4)	(5)	(6)	(7a)	(7b)	(8)	(9a)	(9b)	(10)	(11)
3246	METHANESULPHONYL CHLORIDE	6.1	TC1	I	6.1 +8	354	0	E0	P602		MP8 MP17	T20	TP2
3247	SODIUM PEROXOBORATE, ANHYDROUS	5.1	O2	II	5.1		1 kg	E2	P002 IBC08	B4	MP2	T3	TP33
3248	MEDICINE, LIQUID, FLAMMABLE, TOXIC, N.O.S.	3	FT1	II	3 +6.1	220 221 601	1 L	E2	P001		MP19		
3248	MEDICINE, LIQUID, FLAMMABLE, TOXIC, N.O.S.	3	FT1	III	3 +6.1	220 221 601	5 L	E1	P001 R001		MP19		
3249	MEDICINE, SOLID, TOXIC, N.O.S.	6.1	T2	II	6.1	221 601	500 g	E4	P002		MP10	T3	TP33
3249	MEDICINE, SOLID, TOXIC, N.O.S.	6.1	T2	III	6.1	221 601	5 kg	E1	P002 LP02 R001		MP10	T1	TP33
3250	CHLOROACETIC ACID, MOLTEN	6.1	TC1	II	6.1 +8		0	E0				T7	TP3 TP28
3251	ISOSORBIDE-5-MONONITRATE	4.1	SR1	III	4.1	226 638	5 kg	E0	P409		MP2		
3252	DIFLUOROMETHANE (REFRIGERANT GAS R 32)	2	2F		2.1	662	0	E0	P200		MP9	(M) T50	
3253	DISODIUM TRIOXOSILICATE	8	C6	III	8		5 kg	E1	P002 IBC08 LP02 R001	B3	MP10	T1	TP33
3254	TRIBUTYLPHOSPHANE	4.2	S1	I	4.2		0	E0	P400		MP2	T21	TP2 TP7
3255	tert-BUTYL HYPOCHLORITE	4.2	SC1			CARRIAGE PROHIBITED								
3256	ELEVATED TEMPERATURE LIQUID, FLAMMABLE, N.O.S. with flashpoint above 60° C, at or above its flashpoint and below 100° C	3	F2	III	3	274 560	0	E0	P099 IBC99		MP2	T3	TP3 TP29
3256	ELEVATED TEMPERATURE LIQUID, FLAMMABLE, N.O.S. with flashpoint above 60° C, at or above its flashpoint and at or above 100° C	3	F2	III	3	274 560	0	E0	P099 IBC99		MP2	T3	TP3 TP29
3257	ELEVATED TEMPERATURE LIQUID, N.O.S., at or above 100 °C and below its flash-point (including molten metals, molten salts, etc.), filled at a temperature higher than 190 °C	9	M9	III	9	274 643 668	0	E0	P099 IBC99			T3	TP3 TP29
3257	ELEVATED TEMPERATURE LIQUID, N.O.S., at or above 100 °C and below its flash-point (including molten metals, molten salts, etc.), filled at or below 190 °C	9	M9	III	9	274 643 668	0	E0	P099 IBC99			T3	TP3 TP29
3258	ELEVATED TEMPERATURE SOLID, N.O.S., at or above 240 °C	9	M10	III	9	274 643	0	E0	P099 IBC99				
3259	AMINES, SOLID, CORROSIVE, N.O.S. or POLYAMINES, SOLID, CORROSIVE, N.O.S.	8	C8	I	8	274	0	E0	P002 IBC07		MP18	T6	TP33
3259	AMINES, SOLID, CORROSIVE, N.O.S. or POLYAMINES, SOLID, CORROSIVE, N.O.S.	8	C8	II	8	274	1 kg	E2	P002 IBC08	B4	MP10	T3	TP33
3259	AMINES, SOLID, CORROSIVE, N.O.S. or POLYAMINES, SOLID, CORROSIVE, N.O.S.	8	C8	III	8	274	5 kg	E1	P002 IBC08 LP02 R001	B3	MP10	T1	TP33
3260	CORROSIVE SOLID, ACIDIC, INORGANIC, N.O.S.	8	C2	I	8	274	0	E0	P002 IBC07		MP18	T6	TP33
3260	CORROSIVE SOLID, ACIDIC, INORGANIC, N.O.S.	8	C2	II	8	274	1 kg	E2	P002 IBC08	B4	MP10	T3	TP33
3260	CORROSIVE SOLID, ACIDIC, INORGANIC, N.O.S.	8	C2	III	8	274	5 kg	E1	P002 IBC08 LP02 R001	B3	MP10	T1	TP33
3261	CORROSIVE SOLID, ACIDIC, ORGANIC, N.O.S.	8	C4	I	8	274	0	E0	P002 IBC07		MP18	T6	TP33

ADR tank		Vehicle for tank carriage	Transport category (Tunnel restriction code)	Special provisions for carriage				Hazard identifi-cation No.	UN No.	Name and description
Tank code	Special provisions			Packages	Bulk	Loading, unloading and handling	Operation			
4.3	4.3.5, 6.8.4	9.1.1.2	1.1.3.6 (8.6)	7.2.4	7.3.3	7.5.11	8.5	5.3.2.3		3.1.2
(12)	(13)	(14)	(15)	(16)	(17)	(18)	(19)	(20)	(1)	(2)
L10CH	TU14 TU15 TE19 TE21	AT	1 (C/D)			CV1 CV13 CV28	S9 S14	668	3246	METHANESULPHONYL CHLORIDE
SGAN	TU3	AT	2 (E)	V11		CV24		50	3247	SODIUM PEROXOBORATE, ANHYDROUS
L4BH	TU15	FL	2 (D/E)			CV13 CV28	S2 S19	336	3248	MEDICINE, LIQUID, FLAMMABLE, TOXIC, N.O.S.
L4BH	TU15	FL	3 (D/E)			CV13 CV28	S2	36	3248	MEDICINE, LIQUID, FLAMMABLE, TOXIC, N.O.S.
SGAH L4BH	TU15 TE19	AT	2 (D/E)			CV13 CV28	S9 S19	60	3249	MEDICINE, SOLID, TOXIC, N.O.S.
SGAH L4BH	TU15 TE19	AT	2 (E)		VC1 VC2 AP7	CV13 CV28	S9	60	3249	MEDICINE, SOLID, TOXIC, N.O.S.
L4BH	TU15 TC4 TE19	AT	0 (D/E)			CV13	S9 S19	68	3250	CHLOROACETIC ACID, MOLTEN
			3 (D)			CV14	S24		3251	ISOSORBIDE-5-MONONITRATE
PxBN(M)	TA4 TT9	FL	2 (B/D)			CV9 CV10 CV36	S2 S20	23	3252	DIFLUOROMETHANE (REFRIGERANT GAS R 32)
SGAV		AT	3 (E)		VC1 VC2 AP7			80	3253	DISODIUM TRIOXOSILICATE
		AT	0 (B/E)	V1			S20	333	3254	TRIBUTYLPHOSPHANE
CARRIAGE PROHIBITED									3255	tert-BUTYL HYPOCHLORITE
LGAV	TU35 TE24	FL	3 (D/E)				S2	30	3256	ELEVATED TEMPERATURE LIQUID, FLAMMABLE, N.O.S. with flashpoint above 60° C, at or above its flashpoint and below 100° C
LGAV	TU35 TE24	FL	3 (D/E)				S2	30	3256	ELEVATED TEMPERATURE LIQUID, FLAMMABLE, N.O.S. with flashpoint above 60° C, at or above its flashpoint and at or above 100° C
LGAV	TU35 TC7 TE6 TE14 TE18 TE24	AT	3 (D)		VC3			99	3257	ELEVATED TEMPERATURE LIQUID, N.O.S., at or above 100 °C and below its flash-point (including molten metals, molten salts, etc.), filled at a temperature higher than 190 °C
LGAV	TU35 TC7 TE6 TE14 TE24	AT	3 (D)		VC3			99	3257	ELEVATED TEMPERATURE LIQUID, N.O.S., at or above 100 °C and below its flash-point (including molten metals, molten salts, etc.), filled at or below 190 °C
			3 (D)		VC3			99	3258	ELEVATED TEMPERATURE SOLID, N.O.S., at or above 240 °C
S10AN L10BH		AT	1 (E)	V10			S20	88	3259	AMINES, SOLID, CORROSIVE, N.O.S. or POLYAMINES, SOLID, CORROSIVE, N.O.S.
SGAN L4BN		AT	2 (E)	V11				80	3259	AMINES, SOLID, CORROSIVE, N.O.S. or POLYAMINES, SOLID, CORROSIVE, N.O.S.
SGAV L4BN		AT	3 (E)		VC1 VC2 AP7			80	3259	AMINES, SOLID, CORROSIVE, N.O.S. or POLYAMINES, SOLID, CORROSIVE, N.O.S.
S10AN		AT	1 (E)	V10			S20	88	3260	CORROSIVE SOLID, ACIDIC, INORGANIC, N.O.S.
SGAN		AT	2 (E)	V11				80	3260	CORROSIVE SOLID, ACIDIC, INORGANIC, N.O.S.
SGAV		AT	3 (E)		VC1 VC2 AP7			80	3260	CORROSIVE SOLID, ACIDIC, INORGANIC, N.O.S.
S10AN L10BH		AT	1 (E)	V10			S20	88	3261	CORROSIVE SOLID, ACIDIC, ORGANIC, N.O.S.

UN No.	Name and description	Class	Classification code	Packing group	Labels	Special provisions	Limited and excepted quantities		Packaging			Portable tanks and bulk containers	
									Packing instructions	Special packing provisions	Mixed packing provisions	Instructions	Special provisions
3.1.2	3.1.2	2.2	2.2	2.1.1.3	5.2.2	3.3	3.4	3.5.1.2	4.1.4	4.1.4	4.1.10	4.2.5.2 7.3.2	4.2.5.3
(1)	(2)	(3a)	(3b)	(4)	(5)	(6)	(7a)	(7b)	(8)	(9a)	(9b)	(10)	(11)
3261	CORROSIVE SOLID, ACIDIC, ORGANIC, N.O.S.	8	C4	II	8	274	1 kg	E2	P002 IBC08	B4	MP10	T3	TP33
3261	CORROSIVE SOLID, ACIDIC, ORGANIC, N.O.S.	8	C4	III	8	274	5 kg	E1	P002 IBC08 LP02 R001	B3	MP10	T1	TP33
3262	CORROSIVE SOLID, BASIC, INORGANIC, N.O.S.	8	C6	I	8	274	0	E0	P002 IBC07		MP18	T6	TP33
3262	CORROSIVE SOLID, BASIC, INORGANIC, N.O.S.	8	C6	II	8	274	1 kg	E2	P002 IBC08	B4	MP10	T3	TP33
3262	CORROSIVE SOLID, BASIC, INORGANIC, N.O.S.	8	C6	III	8	274	5 kg	E1	P002 IBC08 LP02 R001	B3	MP10	T1	TP33
3263	CORROSIVE SOLID, BASIC, ORGANIC, N.O.S.	8	C8	I	8	274	0	E0	P002 IBC07		MP18	T6	TP33
3263	CORROSIVE SOLID, BASIC, ORGANIC, N.O.S.	8	C8	II	8	274	1 kg	E2	P002 IBC08	B4	MP10	T3	TP33
3263	CORROSIVE SOLID, BASIC, ORGANIC, N.O.S.	8	C8	III	8	274	5 kg	E1	P002 IBC08 LP02 R001	B3	MP10	T1	TP33
3264	CORROSIVE LIQUID, ACIDIC, INORGANIC, N.O.S.	8	C1	I	8	274	0	E0	P001		MP8 MP17	T14	TP2 TP27
3264	CORROSIVE LIQUID, ACIDIC, INORGANIC, N.O.S.	8	C1	II	8	274	1 L	E2	P001 IBC02		MP15	T11	TP2 TP27
3264	CORROSIVE LIQUID, ACIDIC, INORGANIC, N.O.S.	8	C1	III	8	274	5 L	E1	P001 IBC03 LP01 R001		MP19	T7	TP1 TP28
3265	CORROSIVE LIQUID, ACIDIC, ORGANIC, N.O.S.	8	C3	I	8	274	0	E0	P001		MP8 MP17	T14	TP2 TP27
3265	CORROSIVE LIQUID, ACIDIC, ORGANIC, N.O.S.	8	C3	II	8	274	1 L	E2	P001 IBC02		MP15	T11	TP2 TP27
3265	CORROSIVE LIQUID, ACIDIC, ORGANIC, N.O.S.	8	C3	III	8	274	5 L	E1	P001 IBC03 LP01 R001		MP19	T7	TP1 TP28
3266	CORROSIVE LIQUID, BASIC, INORGANIC, N.O.S.	8	C5	I	8	274	0	E0	P001		MP8 MP17	T14	TP2 TP27
3266	CORROSIVE LIQUID, BASIC, INORGANIC, N.O.S.	8	C5	II	8	274	1 L	E2	P001 IBC02		MP15	T11	TP2 TP27
3266	CORROSIVE LIQUID, BASIC, INORGANIC, N.O.S.	8	C5	III	8	274	5 L	E1	P001 IBC03 LP01 R001		MP19	T7	TP1 TP28
3267	CORROSIVE LIQUID, BASIC, ORGANIC, N.O.S.	8	C7	I	8	274	0	E0	P001		MP8 MP17	T14	TP2 TP27
3267	CORROSIVE LIQUID, BASIC, ORGANIC, N.O.S.	8	C7	II	8	274	1 L	E2	P001 IBC02		MP15	T11	TP2 TP27
3267	CORROSIVE LIQUID, BASIC, ORGANIC, N.O.S.	8	C7	III	8	274	5 L	E1	P001 IBC03 LP01 R001		MP19	T7	TP1 TP28
3268	SAFETY DEVICES, electrically initiated	9	M5		9	280 289	0	E0	P902 LP902				
3269	POLYESTER RESIN KIT, liquid base material	3	F3	II	3	236 340	5 L	E0	P302 R001				
3269	POLYESTER RESIN KIT, liquid base material	3	F3	III	3	236 340	5 L	E0	P302 R001				
3270	NITROCELLULOSE MEMBRANE FILTERS, with not more than 12.6% nitrogen, by dry mass	4.1	F1	II	4.1	237 286	1 kg	E2	P411		MP11		
3271	ETHERS, N.O.S.	3	F1	II	3	274	1 L	E2	P001 IBC02 R001		MP19	T7	TP1 TP8 TP28
3271	ETHERS, N.O.S.	3	F1	III	3	274	5 L	E1	P001 IBC03 LP01 R001		MP19	T4	TP1 TP29
3272	ESTERS, N.O.S.	3	F1	II	3	274 601	1 L	E2	P001 IBC02 R001		MP19	T7	TP1 TP8 TP28

ADR tank		Vehicle for tank carriage	Transport category (Tunnel restriction code)	Special provisions for carriage				Hazard identification No.	UN No.	Name and description
Tank code	Special provisions			Packages	Bulk	Loading, unloading and handling	Operation			
4.3	4.3.5, 6.8.4	9.1.1.2	1.1.3.6 (8.6)	7.2.4	7.3.3	7.5.11	8.5	5.3.2.3		3.1.2
(12)	(13)	(14)	(15)	(16)	(17)	(18)	(19)	(20)	(1)	(2)
SGAN L4BN		AT	2 (E)	V11				80	3261	CORROSIVE SOLID, ACIDIC, ORGANIC, N.O.S.
SGAV L4BN		AT	3 (E)		VC1 VC2 AP7			80	3261	CORROSIVE SOLID, ACIDIC, ORGANIC, N.O.S.
S10AN L10BH		AT	1 (E)	V10			S20	88	3262	CORROSIVE SOLID, BASIC, INORGANIC, N.O.S.
SGAN L4BN		AT	2 (E)	V11				80	3262	CORROSIVE SOLID, BASIC, INORGANIC, N.O.S.
SGAV L4BN		AT	3 (E)		VC1 VC2 AP7			80	3262	CORROSIVE SOLID, BASIC, INORGANIC, N.O.S.
S10AN L10BH		AT	1 (E)	V10			S20	88	3263	CORROSIVE SOLID, BASIC, ORGANIC, N.O.S.
SGAN L4BN		AT	2 (E)	V11				80	3263	CORROSIVE SOLID, BASIC, ORGANIC, N.O.S.
SGAV L4BN		AT	3 (E)		VC1 VC2 AP7			80	3263	CORROSIVE SOLID, BASIC, ORGANIC, N.O.S.
L10BH		AT	1 (E)				S20	88	3264	CORROSIVE LIQUID, ACIDIC, INORGANIC, N.O.S.
L4BN		AT	2 (E)					80	3264	CORROSIVE LIQUID, ACIDIC, INORGANIC, N.O.S.
L4BN		AT	3 (E)	V12				80	3264	CORROSIVE LIQUID, ACIDIC, INORGANIC, N.O.S.
L10BH		AT	1 (E)				S20	88	3265	CORROSIVE LIQUID, ACIDIC, ORGANIC, N.O.S.
L4BN		AT	2 (E)					80	3265	CORROSIVE LIQUID, ACIDIC, ORGANIC, N.O.S.
L4BN		AT	3 (E)	V12				80	3265	CORROSIVE LIQUID, ACIDIC, ORGANIC, N.O.S.
L10BH		AT	1 (E)				S20	88	3266	CORROSIVE LIQUID, BASIC, INORGANIC, N.O.S.
L4BN		AT	2 (E)					80	3266	CORROSIVE LIQUID, BASIC, INORGANIC, N.O.S.
L4BN		AT	3 (E)	V12				80	3266	CORROSIVE LIQUID, BASIC, INORGANIC, N.O.S.
L10BH		AT	1 (E)				S20	88	3267	CORROSIVE LIQUID, BASIC, ORGANIC, N.O.S.
L4BN		AT	2 (E)					80	3267	CORROSIVE LIQUID, BASIC, ORGANIC, N.O.S.
L4BN		AT	3 (E)	V12				80	3267	CORROSIVE LIQUID, BASIC, ORGANIC, N.O.S.
			4 (E)						3268	SAFETY DEVICES, electrically initiated
			2 (E)				S2 S20		3269	POLYESTER RESIN KIT, liquid base material
			3 (E)				S2		3269	POLYESTER RESIN KIT, liquid base material
			2 (E)						3270	NITROCELLULOSE MEMBRANE FILTERS, with not more than 12.6% nitrogen, by dry mass
LGBF		FL	2 (D/E)				S2 S20	33	3271	ETHERS, N.O.S.
LGBF		FL	3 (D/E)	V12			S2	30	3271	ETHERS, N.O.S.
LGBF		FL	2 (D/E)				S2 S20	33	3272	ESTERS, N.O.S.

UN No.	Name and description	Class	Classification code	Packing group	Labels	Special provisions	Limited and excepted quantities		Packaging			Portable tanks and bulk containers	
									Packing instructions	Special packing provisions	Mixed packing provisions	Instructions	Special provisions
3.1.2	3.1.2	2.2	2.2	2.1.1.3	5.2.2	3.3	3.4	3.5.1.2	4.1.4	4.1.4	4.1.10	4.2.5.2 7.3.2	4.2.5.3
(1)	(2)	(3a)	(3b)	(4)	(5)	(6)	(7a)	(7b)	(8)	(9a)	(9b)	(10)	(11)
3272	ESTERS, N.O.S.	3	F1	III	3	274 601	5 L	E1	P001 IBC03 LP01 R001		MP19	T4	TP1 TP29
3273	NITRILES, FLAMMABLE, TOXIC, N.O.S.	3	FT1	I	3 +6.1	274	0	E0	P001		MP7 MP17	T14	TP2 TP27
3273	NITRILES, FLAMMABLE, TOXIC, N.O.S.	3	FT1	II	3 +6.1	274	1 L	E2	P001 IBC02		MP19	T11	TP2 TP27
3274	ALCOHOLATES SOLUTION, N.O.S., in alcohol	3	FC	II	3 +8	274	1 L	E2	P001 IBC02		MP19		
3275	NITRILES, TOXIC, FLAMMABLE, N.O.S.	6.1	TF1	I	6.1 +3	274 315	0	E5	P001		MP8 MP17	T14	TP2 TP27
3275	NITRILES, TOXIC, FLAMMABLE, N.O.S.	6.1	TF1	II	6.1 +3	274	100 ml	E4	P001 IBC02		MP15	T11	TP2 TP27
3276	NITRILES, LIQUID, TOXIC, N.O.S.	6.1	T1	I	6.1	274 315	0	E5	P001		MP8 MP17	T14	TP2 TP27
3276	NITRILES, LIQUID, TOXIC, N.O.S.	6.1	T1	II	6.1	274	100 ml	E4	P001 IBC02		MP15	T11	TP2 TP27
3276	NITRILES, LIQUID, TOXIC, N.O.S.	6.1	T1	III	6.1	274	5 L	E1	P001 IBC03 LP01 R001		MP19	T7	TP1 TP28
3277	CHLOROFORMATES, TOXIC, CORROSIVE, N.O.S.	6.1	TC1	II	6.1 +8	274 561	100 ml	E4	P001 IBC02		MP15	T8	TP2 TP28
3278	ORGANOPHOSPHORUS COMPOUND, LIQUID, TOXIC, N.O.S.	6.1	T1	I	6.1	43 274 315	0	E5	P001		MP8 MP17	T14	TP2 TP27
3278	ORGANOPHOSPHORUS COMPOUND, LIQUID, TOXIC, N.O.S.	6.1	T1	II	6.1	43 274	100 ml	E4	P001 IBC02		MP15	T11	TP2 TP27
3278	ORGANOPHOSPHORUS COMPOUND, LIQUID, TOXIC, N.O.S.	6.1	T1	III	6.1	43 274	5 L	E1	P001 IBC03 LP01 R001		MP19	T7	TP1 TP28
3279	ORGANOPHOSPHORUS COMPOUND, TOXIC, FLAMMABLE, N.O.S.	6.1	TF1	I	6.1 +3	43 274 315	0	E5	P001		MP8 MP17	T14	TP2 TP27
3279	ORGANOPHOSPHORUS COMPOUND, TOXIC, FLAMMABLE, N.O.S.	6.1	TF1	II	6.1 +3	43 274	100 ml	E4	P001		MP15	T11	TP2 TP27
3280	ORGANOARSENIC COMPOUND, LIQUID, N.O.S.	6.1	T3	I	6.1	274 315	0	E5	P001		MP8 MP17	T14	TP2 TP27
3280	ORGANOARSENIC COMPOUND, LIQUID, N.O.S.	6.1	T3	II	6.1	274	100 ml	E4	P001 IBC02		MP15	T11	TP2 TP27
3280	ORGANOARSENIC COMPOUND, LIQUID, N.O.S.	6.1	T3	III	6.1	274	5 L	E1	P001 IBC03 LP01 R001		MP19	T7	TP1 TP28
3281	METAL CARBONYLS, LIQUID, N.O.S.	6.1	T3	I	6.1	274 315 562	0	E5	P601		MP8 MP17	T14	TP2 TP27
3281	METAL CARBONYLS, LIQUID, N.O.S.	6.1	T3	II	6.1	274 562	100 ml	E4	P001 IBC02		MP15	T11	TP2 TP27
3281	METAL CARBONYLS, LIQUID, N.O.S.	6.1	T3	III	6.1	274 562	5 L	E1	P001 IBC03 LP01 R001		MP19	T7	TP1 TP28
3282	ORGANOMETALLIC COMPOUND, LIQUID, TOXIC, N.O.S.	6.1	T3	I	6.1	274 562	0	E5	P001		MP8 MP17	T14	TP2 TP27
3282	ORGANOMETALLIC COMPOUND, LIQUID, TOXIC, N.O.S.	6.1	T3	II	6.1	274 562	100 ml	E4	P001 IBC02		MP15	T11	TP2 TP27
3282	ORGANOMETALLIC COMPOUND, LIQUID, TOXIC, N.O.S.	6.1	T3	III	6.1	274 562	5 L	E1	P001 IBC03 LP01 R001		MP19	T7	TP1 TP28
3283	SELENIUM COMPOUND, SOLID, N.O.S.	6.1	T5	I	6.1	274 563	0	E5	P002 IBC07		MP18	T6	TP33
3283	SELENIUM COMPOUND, SOLID, N.O.S.	6.1	T5	II	6.1	274 563	500 g	E4	P002 IBC08	B4	MP10	T3	TP33

ADR tank		Vehicle for tank carriage	Transport category (Tunnel restriction code)	Special provisions for carriage				Hazard identification No.	UN No.	Name and description
Tank code	Special provisions			Packages	Bulk	Loading, unloading and handling	Operation			
4.3	4.3.5, 6.8.4	9.1.1.2	1.1.3.6 (8.6)	7.2.4	7.3.3	7.5.11	8.5	5.3.2.3		3.1.2
(12)	(13)	(14)	(15)	(16)	(17)	(18)	(19)	(20)	(1)	(2)
LGBF		FL	3 (D/E)	V12			S2	30	3272	ESTERS, N.O.S.
L10CH	TU14 TU15 TE21	FL	1 (C/E)			CV13 CV28	S2 S22	336	3273	NITRILES, FLAMMABLE, TOXIC, N.O.S.
L4BH	TU15	FL	2 (D/E)			CV13 CV28	S2 S22	336	3273	NITRILES, FLAMMABLE, TOXIC, N.O.S.
L4BH		FL	2 (D/E)				S2 S20	338	3274	ALCOHOLATES SOLUTION, N.O.S., in alcohol
L10CH	TU14 TU15 TE19 TE21	FL	1 (C/D)			CV1 CV13 CV28	S2 S9 S14	663	3275	NITRILES, TOXIC, FLAMMABLE, N.O.S.
L4BH	TU15 TE19	FL	2 (D/E)			CV13 CV28	S2 S9 S19	63	3275	NITRILES, TOXIC, FLAMMABLE, N.O.S.
L10CH	TU14 TU15 TE19 TE21	AT	1 (C/E)			CV1 CV13 CV28	S9 S14	66	3276	NITRILES, LIQUID, TOXIC, N.O.S.
L4BH	TU15 TE19	AT	2 (D/E)			CV13 CV28	S9 S19	60	3276	NITRILES, LIQUID, TOXIC, N.O.S.
L4BH	TU15 TE19	AT	2 (E)	V12		CV13 CV28	S9	60	3276	NITRILES, LIQUID, TOXIC, N.O.S.
L4BH	TU15 TE19	AT	2 (D/E)			CV13 CV28	S9 S19	68	3277	CHLOROFORMATES, TOXIC, CORROSIVE, N.O.S.
L10CH	TU14 TU15 TE19 TE21	AT	1 (C/E)			CV1 CV13 CV28	S9 S14	66	3278	ORGANOPHOSPHORUS COMPOUND, LIQUID, TOXIC, N.O.S.
L4BH	TU15 TE19	AT	2 (D/E)			CV13 CV28	S9 S19	60	3278	ORGANOPHOSPHORUS COMPOUND, LIQUID, TOXIC, N.O.S.
L4BH	TU15 TE19	AT	2 (E)	V12		CV13 CV28	S9	60	3278	ORGANOPHOSPHORUS COMPOUND, LIQUID, TOXIC, N.O.S.
L10CH	TU14 TU15 TE19 TE21	FL	1 (C/D)			CV1 CV13 CV28	S2 S9 S14	663	3279	ORGANOPHOSPHORUS COMPOUND, TOXIC, FLAMMABLE, N.O.S.
L4BH	TU15 TE19	FL	2 (D/E)			CV13 CV28	S2 S9 S19	63	3279	ORGANOPHOSPHORUS COMPOUND, TOXIC, FLAMMABLE, N.O.S.
L10CH	TU14 TU15 TE19 TE21	AT	1 (C/E)			CV1 CV13 CV28	S9 S14	66	3280	ORGANOARSENIC COMPOUND, LIQUID, N.O.S.
L4BH	TU15 TE19	AT	2 (D/E)			CV13 CV28	S9 S19	60	3280	ORGANOARSENIC COMPOUND, LIQUID, N.O.S.
L4BH	TU15 TE19	AT	2 (E)	V12		CV13 CV28	S9	60	3280	ORGANOARSENIC COMPOUND, LIQUID, N.O.S.
L10CH	TU14 TU15 TE19 TE21	AT	1 (C/E)			CV1 CV13 CV28	S9 S14	66	3281	METAL CARBONYLS, LIQUID, N.O.S.
L4BH	TU15 TE19	AT	2 (D/E)			CV13 CV28	S9 S19	60	3281	METAL CARBONYLS, LIQUID, N.O.S.
L4BH	TU15 TE19	AT	2 (E)	V12		CV13 CV28	S9	60	3281	METAL CARBONYLS, LIQUID, N.O.S.
L10CH	TU14 TU15 TE19 TE21	AT	1 (C/E)			CV1 CV13 CV28	S9 S14	66	3282	ORGANOMETALLIC COMPOUND, LIQUID, TOXIC, N.O.S.
L4BH	TU15 TE19	AT	2 (D/E)			CV13 CV28	S9 S19	60	3282	ORGANOMETALLIC COMPOUND, LIQUID, TOXIC, N.O.S.
L4BH	TU15 TE19	AT	2 (E)	V12		CV13 CV28	S9	60	3282	ORGANOMETALLIC COMPOUND, LIQUID, TOXIC, N.O.S.
S10AH L10CH	TU14 TU15 TE19 TE21	AT	1 (C/E)	V10		CV1 CV13 CV28	S9 S14	66	3283	SELENIUM COMPOUND, SOLID, N.O.S.
SGAH L4BH	TU15 TE19	AT	2 (D/E)	V11		CV13 CV28	S9 S19	60	3283	SELENIUM COMPOUND, SOLID, N.O.S.

UN No.	Name and description	Class	Classification code	Packing group	Labels	Special provisions	Limited and excepted quantities		Packaging			Portable tanks and bulk containers	
									Packing instructions	Special packing provisions	Mixed packing provisions	Instructions	Special provisions
	3.1.2	2.2	2.2	2.1.1.3	5.2.2	3.3	3.4	3.5.1.2	4.1.4	4.1.4	4.1.10	4.2.5.2 7.3.2	4.2.5.3
(1)	(2)	(3a)	(3b)	(4)	(5)	(6)	(7a)	(7b)	(8)	(9a)	(9b)	(10)	(11)
3283	SELENIUM COMPOUND, SOLID, N.O.S.	6.1	T5	III	6.1	274 563	5 kg	E1	P002 IBC08 LP02 R001	B3	MP10	T1	TP33
3284	TELLURIUM COMPOUND, N.O.S.	6.1	T5	I	6.1	274	0	E5	P002 IBC07		MP18	T6	TP33
3284	TELLURIUM COMPOUND, N.O.S.	6.1	T5	II	6.1	274	500 g	E4	P002 IBC08	B4	MP10	T3	TP33
3284	TELLURIUM COMPOUND, N.O.S.	6.1	T5	III	6.1	274	5 kg	E1	P002 IBC08 LP02 R001	B3	MP10	T1	TP33
3285	VANADIUM COMPOUND, N.O.S.	6.1	T5	I	6.1	274 564	0	E5	P002 IBC07		MP18	T6	TP33
3285	VANADIUM COMPOUND, N.O.S.	6.1	T5	II	6.1	274 564	500 g	E4	P002 IBC08	B4	MP10	T3	TP33
3285	VANADIUM COMPOUND, N.O.S.	6.1	T5	III	6.1	274 564	5 kg	E1	P002 IBC08 LP02 R001	B3	MP10	T1	TP33
3286	FLAMMABLE LIQUID, TOXIC, CORROSIVE, N.O.S.	3	FTC	I	3 +6.1 +8	274	0	E0	P001		MP7 MP17	T14	TP2 TP27
3286	FLAMMABLE LIQUID, TOXIC, CORROSIVE, N.O.S.	3	FTC	II	3 +6.1 +8	274	1 L	E2	P001 IBC02		MP19	T11	TP2 TP27
3287	TOXIC LIQUID, INORGANIC, N.O.S.	6.1	T4	I	6.1	274 315	0	E5	P001		MP8 MP17	T14	TP2 TP27
3287	TOXIC LIQUID, INORGANIC, N.O.S.	6.1	T4	II	6.1	274	100 ml	E4	P001 IBC02		MP15	T11	TP2 TP27
3287	TOXIC LIQUID, INORGANIC, N.O.S.	6.1	T4	III	6.1	274	5 L	E1	P001 IBC03 LP01 R001		MP19	T7	TP1 TP28
3288	TOXIC SOLID, INORGANIC, N.O.S.	6.1	T5	I	6.1	274	0	E5	P002 IBC07		MP18	T6	TP33
3288	TOXIC SOLID, INORGANIC, N.O.S.	6.1	T5	II	6.1	274	500 g	E4	P002 IBC08	B4	MP10	T3	TP33
3288	TOXIC SOLID, INORGANIC, N.O.S.	6.1	T5	III	6.1	274	5 kg	E1	P002 IBC08 LP02 R001	B3	MP10	T1	TP33
3289	TOXIC LIQUID, CORROSIVE, INORGANIC, N.O.S.	6.1	TC3	I	6.1 +8	274 315	0	E5	P001		MP8 MP17	T14	TP2 TP27
3289	TOXIC LIQUID, CORROSIVE, INORGANIC, N.O.S.	6.1	TC3	II	6.1 +8	274	100 ml	E4	P001 IBC02		MP15	T11	TP2 TP27
3290	TOXIC SOLID, CORROSIVE, INORGANIC, N.O.S.	6.1	TC4	I	6.1 +8	274	0	E5	P002 IBC05		MP18	T6	TP33
3290	TOXIC SOLID, CORROSIVE, INORGANIC, N.O.S.	6.1	TC4	II	6.1 +8	274	500 g	E4	P002 IBC06		MP10	T3	TP33
3291	CLINICAL WASTE, UNSPECIFIED, N.O.S. or (BIO) MEDICAL WASTE, N.O.S. or REGULATED MEDICAL WASTE, N.O.S.	6.2	I3	II	6.2	565	0	E0	P621 IBC620 LP621		MP6	BK2	
3291	CLINICAL WASTE, UNSPECIFIED, N.O.S. or (BIO) MEDICAL WASTE, N.O.S. or REGULATED MEDICAL WASTE, N.O.S., in refrigerated liquid nitrogen	6.2	I3	II	6.2 +2.2	565	0	E0	P621 IBC620 LP621		MP6		
3292	BATTERIES, CONTAINING SODIUM, or CELLS, CONTAINING SODIUM	4.3	W3		4.3	239 295	0	E0	P408				
3293	HYDRAZINE, AQUEOUS SOLUTION with not more than 37% hydrazine, by mass	6.1	T4	III	6.1	566	5 L	E1	P001 IBC03 LP01 R001		MP19	T4	TP1

ADR tank		Vehicle for tank carriage	Transport category (Tunnel restriction code)	Special provisions for carriage				Hazard identification No.	UN No.	Name and description
Tank code	Special provisions			Packages	Bulk	Loading, unloading and handling	Operation			
4.3	4.3.5, 6.8.4	9.1.1.2	1.1.3.6 (8.6)	7.2.4	7.3.3	7.5.11	8.5	5.3.2.3		3.1.2
(12)	(13)	(14)	(15)	(16)	(17)	(18)	(19)	(20)	(1)	(2)
SGAH L4BH	TU15 TE19	AT	2 (E)		VC1 VC2 AP7	CV13 CV28	S9	60	3283	SELENIUM COMPOUND, SOLID, N.O.S.
S10AH L10CH	TU14 TU15 TE19 TE21	AT	1 (C/E)	V10		CV1 CV13 CV28	S9 S14	66	3284	TELLURIUM COMPOUND, N.O.S.
SGAH L4BH	TU15 TE19	AT	2 (D/E)	V11		CV13 CV28	S9 S19	60	3284	TELLURIUM COMPOUND, N.O.S.
SGAH L4BH	TU15 TE19	AT	2 (E)		VC1 VC2 AP7	CV13 CV28	S9	60	3284	TELLURIUM COMPOUND, N.O.S.
S10AH L10CH	TU14 TU15 TE19 TE21	AT	1 (C/E)	V10		CV1 CV13 CV28	S9 S14	66	3285	VANADIUM COMPOUND, N.O.S.
SGAH L4BH	TU15 TE19	AT	2 (D/E)	V11		CV13 CV28	S9 S19	60	3285	VANADIUM COMPOUND, N.O.S.
SGAH L4BH	TU15 TE19	AT	2 (E)		VC1 VC2 AP7	CV13 CV28	S9	60	3285	VANADIUM COMPOUND, N.O.S.
L10CH	TU14 TU15 TE21	FL	1 (C/E)			CV13 CV28	S2 S22	368	3286	FLAMMABLE LIQUID, TOXIC, CORROSIVE, N.O.S.
L4BH	TU15	FL	2 (D/E)			CV13 CV28	S2 S22	368	3286	FLAMMABLE LIQUID, TOXIC, CORROSIVE, N.O.S.
L10CH	TU14 TU15 TE19 TE21	AT	1 (C/E)			CV1 CV13 CV28	S9 S14	66	3287	TOXIC LIQUID, INORGANIC, N.O.S.
L4BH	TU15 TE19	AT	2 (D/E)			CV13 CV28	S9 S19	60	3287	TOXIC LIQUID, INORGANIC, N.O.S.
L4BH	TU15 TE19	AT	2 (E)	V12		CV13 CV28	S9	60	3287	TOXIC LIQUID, INORGANIC, N.O.S.
S10AH L10CH	TU14 TU15 TE19 TE21	AT	1 (C/E)	V10		CV1 CV13 CV28	S9 S14	66	3288	TOXIC SOLID, INORGANIC, N.O.S.
SGAH L4BH	TU15 TE19	AT	2 (D/E)	V11		CV13 CV28	S9 S19	60	3288	TOXIC SOLID, INORGANIC, N.O.S.
SGAH L4BH	TU15 TE19	AT	2 (E)		VC1 VC2 AP7	CV13 CV28	S9	60	3288	TOXIC SOLID, INORGANIC, N.O.S.
L10CH	TU14 TU15 TE19 TE21	AT	1 (C/E)			CV1 CV13 CV28	S9 S14	668	3289	TOXIC LIQUID, CORROSIVE, INORGANIC, N.O.S.
L4BH	TU15 TE19	AT	2 (D/E)			CV13 CV28	S9 S19	68	3289	TOXIC LIQUID, CORROSIVE, INORGANIC, N.O.S.
S10AH L10CH	TU15 TE19	AT	1 (C/E)	V10		CV1 CV13 CV28	S9 S14	668	3290	TOXIC SOLID, CORROSIVE, INORGANIC, N.O.S.
SGAH L4BH	TU15 TE19	AT	2 (D/E)	V11		CV13 CV28	S9 S19	68	3290	TOXIC SOLID, CORROSIVE, INORGANIC, N.O.S.
S4AH L4BH	TU15 TE19	AT	2 (-)	V1	VC3	CV13 CV25 CV28	S3	606	3291	CLINICAL WASTE, UNSPECIFIED, N.O.S. or (BIO) MEDICAL WASTE, N.O.S. or REGULATED MEDICAL WASTE, N.O.S.
			2 (-)	V1		CV13 CV25 CV28	S3		3291	CLINICAL WASTE, UNSPECIFIED, N.O.S. or (BIO) MEDICAL WASTE, N.O.S. or REGULATED MEDICAL WASTE, N.O.S., in refrigerated liquid nitrogen
			2 (E)	V1		CV23			3292	BATTERIES, CONTAINING SODIUM, or CELLS, CONTAINING SODIUM
L4BH	TU15 TE19	AT	2 (E)	V12		CV13 CV28	S9	60	3293	HYDRAZINE, AQUEOUS SOLUTION with not more than 37% hydrazine, by mass

UN No.	Name and description	Class	Classifi-cation code	Packing group	Labels	Special provi-sions	Limited and excepted quantities		Packaging			Portable tanks and bulk containers	
									Packing instruc-tions	Special packing provisions	Mixed packing provisions	Instruc-tions	Special provisions
	3.1.2	2.2	2.2	2.1.1.3	5.2.2	3.3	3.4	3.5.1.2	4.1.4	4.1.4	4.1.10	4.2.5.2 7.3.2	4.2.5.3
(1)	(2)	(3a)	(3b)	(4)	(5)	(6)	(7a)	(7b)	(8)	(9a)	(9b)	(10)	(11)
3294	HYDROGEN CYANIDE, SOLUTION IN ALCOHOL with not more than 45% hydrogen cyanide	6.1	TF1	I	6.1 +3	610	0	E0	P601		MP8 MP17	T14	TP2
3295	HYDROCARBONS, LIQUID, N.O.S.	3	F1	I	3		500 ml	E3	P001		MP7 MP17	T11	TP1 TP8 TP28
3295	HYDROCARBONS, LIQUID, N.O.S. (vapour pressure at 50 °C more than 110 kPa)	3	F1	II	3	640C	1 L	E2	P001		MP19	T7	TP1 TP8 TP28
3295	HYDROCARBONS, LIQUID, N.O.S. (vapour pressure at 50 °C not more than 110 kPa)	3	F1	II	3	640D	1 L	E2	P001 IBC02 R001		MP19	T7	TP1 TP8 TP28
3295	HYDROCARBONS, LIQUID, N.O.S.	3	F1	III	3		5 L	E1	P001 IBC03 LP01 R001		MP19	T4	TP1 TP29
3296	HEPTAFLUOROPROPANE (REFRIGERANT GAS R 227)	2	2A		2.2	662	120 ml	E1	P200		MP9	(M) T50	
3297	ETHYLENE OXIDE AND CHLOROTETRAFLUORO-ETHANE MIXTURE with not more than 8.8% ethylene oxide	2	2A		2.2	662	120 ml	E1	P200		MP9	(M) T50	
3298	ETHYLENE OXIDE AND PENTAFLUOROETHANE MIXTURE with not more than 7.9% ethylene oxide	2	2A		2.2	662	120 ml	E1	P200		MP9	(M) T50	
3299	ETHYLENE OXIDE AND TETRAFLUOROETHANE MIXTURE with not more than 5.6% ethylene oxide	2	2A		2.2	662	120 ml	E1	P200		MP9	(M) T50	
3300	ETHYLENE OXIDE AND CARBON DIOXIDE MIXTURE with more than 87% ethylene oxide	2	2TF		2.3 +2.1		0	E0	P200		MP9	(M)	
3301	CORROSIVE LIQUID, SELF-HEATING, N.O.S.	8	CS1	I	8 +4.2	274	0	E0	P001		MP8 MP17		
3301	CORROSIVE LIQUID, SELF-HEATING, N.O.S.	8	CS1	II	8 +4.2	274	0	E2	P001		MP15		
3302	2-DIMETHYLAMINOETHYL ACRYLATE	6.1	T1	II	6.1		100 ml	E4	P001 IBC02		MP15	T7	TP2
3303	COMPRESSED GAS, TOXIC, OXIDIZING, N.O.S.	2	1TO		2.3 +5.1	274	0	E0	P200		MP9	(M)	
3304	COMPRESSED GAS, TOXIC, CORROSIVE, N.O.S.	2	1TC		2.3 +8	274	0	E0	P200		MP9	(M)	
3305	COMPRESSED GAS, TOXIC, FLAMMABLE, CORROSIVE, N.O.S.	2	1TFC		2.3 +2.1 +8	274	0	E0	P200		MP9	(M)	
3306	COMPRESSED GAS, TOXIC, OXIDIZING, CORROSIVE, N.O.S.	2	1TOC		2.3 +5.1 +8	274	0	E0	P200		MP9	(M)	
3307	LIQUEFIED GAS, TOXIC, OXIDIZING, N.O.S.	2	2TO		2.3 +5.1	274	0	E0	P200		MP9	(M)	
3308	LIQUEFIED GAS, TOXIC, CORROSIVE, N.O.S.	2	2TC		2.3 +8	274	0	E0	P200		MP9	(M)	
3309	LIQUEFIED GAS, TOXIC, FLAMMABLE, CORROSIVE, N.O.S.	2	2TFC		2.3 +2.1 +8	274	0	E0	P200		MP9	(M)	
3310	LIQUEFIED GAS, TOXIC, OXIDIZING, CORROSIVE, N.O.S.	2	2TOC		2.3 +5.1 +8	274	0	E0	P200		MP9	(M)	
3311	GAS, REFRIGERATED LIQUID, OXIDIZING, N.O.S.	2	3O		2.2 +5.1	274	0	E0	P203		MP9	T75	TP5 TP22
3312	GAS, REFRIGERATED LIQUID, FLAMMABLE, N.O.S.	2	3F		2.1	274	0	E0	P203		MP9	T75	TP5
3313	ORGANIC PIGMENTS, SELF-HEATING	4.2	S2	II	4.2		0	E2	P002 IBC08	B4	MP14	T3	TP33

ADR tank		Vehicle for tank carriage	Transport category (Tunnel restriction code)	Special provisions for carriage				Hazard identifi-cation No.	UN No.	Name and description
Tank code	Special provisions			Packages	Bulk	Loading, unloading and handling	Operation			
4.3	4.3.5, 6.8.4	9.1.1.2	1.1.3.6 (8.6)	7.2.4	7.3.3	7.5.11	8.5	5.3.2.3		3.1.2
(12)	(13)	(14)	(15)	(16)	(17)	(18)	(19)	(20)	(1)	(2)
L15DH(+)	TU14 TU15 TE19 TE21	FL	0 (C/D)			CV1 CV13 CV28	S2 S9 S14	663	3294	HYDROGEN CYANIDE, SOLUTION IN ALCOHOL with not more than 45% hydrogen cyanide
L4BN		FL	1 (D/E)				S2 S20	33	3295	HYDROCARBONS, LIQUID, N.O.S.
L1.5BN		FL	2 (D/E)				S2 S20	33	3295	HYDROCARBONS, LIQUID, N.O.S. (vapour pressure at 50 °C more than 110 kPa)
LGBF		FL	2 (D/E)				S2 S20	33	3295	HYDROCARBONS, LIQUID, N.O.S. (vapour pressure at 50 °C not more than 110 kPa)
LGBF		FL	3 (D/E)	V12			S2	30	3295	HYDROCARBONS, LIQUID, N.O.S.
PxBN(M)	TA4 TT9	AT	3 (C/E)			CV9 CV10 CV36		20	3296	HEPTAFLUOROPROPANE (REFRIGERANT GAS R 227)
PxBN(M)	TA4 TT9	AT	3 (C/E)			CV9 CV10 CV36		20	3297	ETHYLENE OXIDE AND CHLOROTETRAFLUORO-ETHANE MIXTURE with not more than 8.8% ethylene oxide
PxBN(M)	TA4 TT9	AT	3 (C/E)			CV9 CV10 CV36		20	3298	ETHYLENE OXIDE AND PENTAFLUOROETHANE MIXTURE with not more than 7.9% ethylene oxide
PxBN(M)	TA4 TT9	AT	3 (C/E)			CV9 CV10 CV36		20	3299	ETHYLENE OXIDE AND TETRAFLUOROETHANE MIXTURE with not more than 5.6% ethylene oxide
PxBH(M)	TA4 TT9	FL	1 (B/D)			CV9 CV10 CV36	S2 S14	263	3300	ETHYLENE OXIDE AND CARBON DIOXIDE MIXTURE with more than 87% ethylene oxide
L10BH		AT	1 (E)				S14	884	3301	CORROSIVE LIQUID, SELF-HEATING, N.O.S.
L4BN		AT	2 (E)					84	3301	CORROSIVE LIQUID, SELF-HEATING, N.O.S.
L4BH	TU15 TE19	AT	2 (D/E)			CV13 CV28	S9 S19	60	3302	2-DIMETHYLAMINOETHYL ACRYLATE
CxBH(M)	TU6 TA4 TT9	AT	1 (C/D)			CV9 CV10 CV36	S14	265	3303	COMPRESSED GAS, TOXIC, OXIDIZING, N.O.S.
CxBH(M)	TU6 TA4 TT9	AT	1 (C/D)			CV9 CV10 CV36	S14	268	3304	COMPRESSED GAS, TOXIC, CORROSIVE, N.O.S.
CxBH(M)	TU6 TA4 TT9	FL	1 (B/D)			CV9 CV10 CV36	S2 S14	263	3305	COMPRESSED GAS, TOXIC, FLAMMABLE, CORROSIVE, N.O.S.
CxBH(M)	TU6 TA4 TT9	AT	1 (C/D)			CV9 CV10 CV36	S14	265	3306	COMPRESSED GAS, TOXIC, OXIDIZING, CORROSIVE, N.O.S.
PxBH(M)	TU6 TA4 TT9	AT	1 (C/D)			CV9 CV10 CV36	S14	265	3307	LIQUEFIED GAS, TOXIC, OXIDIZING, N.O.S.
PxBH(M)	TU6 TA4 TT9	AT	1 (C/D)			CV9 CV10 CV36	S14	268	3308	LIQUEFIED GAS, TOXIC, CORROSIVE, N.O.S.
PxBH(M)	TU6 TA4 TT9	FL	1 (B/D)			CV9 CV10 CV36	S2 S14	263	3309	LIQUEFIED GAS, TOXIC, FLAMMABLE, CORROSIVE, N.O.S.
PxBH(M)	TU6 TA4 TT9	AT	1 (C/D)			CV9 CV10 CV36	S14	265	3310	LIQUEFIED GAS, TOXIC, OXIDIZING, CORROSIVE, N.O.S.
RxBN	TU7 TU19 TA4 TT9	AT	3 (C/E)	V5		CV9 CV11 CV36	S20	225	3311	GAS, REFRIGERATED LIQUID, OXIDIZING, N.O.S.
RxBN	TU18 TA4 TT9	FL	2 (B/D)	V5		CV9 CV11 CV36	S2 S17	223	3312	GAS, REFRIGERATED LIQUID, FLAMMABLE, N.O.S.
SGAV		AT	2 (D/E)	V1				40	3313	ORGANIC PIGMENTS, SELF-HEATING

UN No.	Name and description	Class	Classification code	Packing group	Labels	Special provisions	Limited and excepted quantities		Packaging			Portable tanks and bulk containers	
									Packing instructions	Special packing provisions	Mixed packing provisions	Instructions	Special provisions
	3.1.2	2.2	2.2	2.1.1.3	5.2.2	3.3	3.4	3.5.1.2	4.1.4	4.1.4	4.1.10	4.2.5.2 7.3.2	4.2.5.3
(1)	(2)	(3a)	(3b)	(4)	(5)	(6)	(7a)	(7b)	(8)	(9a)	(9b)	(10)	(11)
3313	ORGANIC PIGMENTS, SELF-HEATING	4.2	S2	III	4.2		0	E1	P002 IBC08 LP02 R001	B3	MP14	T1	TP33
3314	PLASTICS MOULDING COMPOUND in dough, sheet or extruded rope form evolving flammable vapour	9	M3	III	None	207 633	5 kg	E1	P002 IBC08 R001	PP14 B3 B6	MP10		
3315	CHEMICAL SAMPLE, TOXIC	6.1	T8	I	6.1	250	0	E0	P099		MP8 MP17		
3316	CHEMICAL KIT or FIRST AID KIT	9	M11	II	9	251 340	See SP 251	See SP 340	P901				
3316	CHEMICAL KIT or FIRST AID KIT	9	M11	III	9	251 340	See SP 251	See SP 340	P901				
3317	2-AMINO-4,6-DINITROPHENOL, WETTED with not less than 20% water, by mass	4.1	D	I	4.1		0	E0	P406	PP26	MP2		
3318	AMMONIA SOLUTION, relative density less than 0.880 at 15 °C in water, with more than 50% ammonia	2	4TC		2.3 +8	23	0	E0	P200		MP9	(M) T50	
3319	NITROGLYCERIN MIXTURE, DESENSITIZED, SOLID, N.O.S. with more than 2% but not more than 10% nitroglycerin, by mass	4.1	D	II	4.1	272 274	0	E0	P099 IBC99		MP2		
3320	SODIUM BOROHYDRIDE AND SODIUM HYDROXIDE SOLUTION, with not more than 12% sodium borohydride and not more than 40% sodium hydroxide by mass	8	C5	II	8		1 L	E2	P001 IBC02		MP15	T7	TP2
3320	SODIUM BOROHYDRIDE AND SODIUM HYDROXIDE SOLUTION, with not more than 12% sodium borohydride and not more than 40% sodium hydroxide by mass	8	C5	III	8		5 L	E1	P001 IBC03 LP01 R001		MP19	T4	TP2
3321	RADIOACTIVE MATERIAL, LOW SPECIFIC ACTIVITY (LSA-II), non fissile or fissile-excepted	7			7X	172 317 325 336	0	E0	See 2.2.7 and 4.1.9	See 4.1.9.1.3		T5	TP4
3322	RADIOACTIVE MATERIAL, LOW SPECIFIC ACTIVITY (LSA-III), non fissile or fissile-excepted	7			7X	172 317 325 336	0	E0	See 2.2.7 and 4.1.9	See 4.1.9.1.3		T5	TP4
3323	RADIOACTIVE MATERIAL, TYPE C PACKAGE, non fissile or fissile-excepted	7			7X	172 317 325	0	E0	See 2.2.7 and 4.1.9	See 4.1.9.1.3			
3324	RADIOACTIVE MATERIAL, LOW SPECIFIC ACTIVITY (LSA-II), FISSILE	7			7X +7E	172 326 336	0	E0	See 2.2.7 and 4.1.9	See 4.1.9.1.3			
3325	RADIOACTIVE MATERIAL, LOW SPECIFIC ACTIVITY, (LSA-III), FISSILE	7			7X +7E	172 326 336	0	E0	See 2.2.7 and 4.1.9	See 4.1.9.1.3			
3326	RADIOACTIVE MATERIAL, SURFACE CONTAMINATED OBJECTS (SCO-I or SCO-II), FISSILE	7			7X +7E	172 336	0	E0	See 2.2.7 and 4.1.9	See 4.1.9.1.3			
3327	RADIOACTIVE MATERIAL, TYPE A PACKAGE, FISSILE, non-special form	7			7X +7E	172 326	0	E0	See 2.2.7 and 4.1.9	See 4.1.9.1.3			
3328	RADIOACTIVE MATERIAL, TYPE B(U) PACKAGE, FISSILE	7			7X +7E	172 326 337	0	E0	See 2.2.7 and 4.1.9	See 4.1.9.1.3			
3329	RADIOACTIVE MATERIAL, TYPE B(M) PACKAGE, FISSILE	7			7X +7E	172 326 337	0	E0	See 2.2.7 and 4.1.9	See 4.1.9.1.3			

ADR tank		Vehicle for tank carriage	Transport category (Tunnel restriction code)	Special provisions for carriage				Hazard identification No.	UN No.	Name and description
Tank code	Special provisions			Packages	Bulk	Loading, unloading and handling	Operation			
4.3	4.3.5, 6.8.4	9.1.1.2	1.1.3.6 (8.6)	7.2.4	7.3.3	7.5.11	8.5	5.3.2.3		3.1.2
(12)	(13)	(14)	(15)	(16)	(17)	(18)	(19)	(20)	(1)	(2)
SGAV		AT	3 (E)	V1				40	3313	ORGANIC PIGMENTS, SELF-HEATING
			3 (D/E)		VC1 VC2 AP2	CV36		90	3314	PLASTICS MOULDING COMPOUND in dough, sheet or extruded rope form evolving flammable vapour
			1 (E)			CV1 CV13 CV28	S9 S14		3315	CHEMICAL SAMPLE, TOXIC
			2 (E)						3316	CHEMICAL KIT or FIRST AID KIT
			3 (E)						3316	CHEMICAL KIT or FIRST AID KIT
			1 (B)				S14		3317	2-AMINO-4,6-DINITROPHENOL, WETTED with not less than 20% water, by mass
PxBH(M)	TA4 TT9	AT	1 (C/D)			CV9 CV10	S14	268	3318	AMMONIA SOLUTION, relative density less than 0.880 at 15 °C in water, with more than 50% ammonia
			2 (B)				S14		3319	NITROGLYCERIN MIXTURE, DESENSITIZED, SOLID, N.O.S. with more than 2% but not more than 10% nitroglycerin, by mass
L4BN		AT	2 (E)					80	3320	SODIUM BOROHYDRIDE AND SODIUM HYDROXIDE SOLUTION, with not more than 12% sodium borohydride and not more than 40% sodium hydroxide by mass
L4BN		AT	3 (E)	V12				80	3320	SODIUM BOROHYDRIDE AND SODIUM HYDROXIDE SOLUTION, with not more than 12% sodium borohydride and not more than 40% sodium hydroxide by mass
S2.65AN(+) L2.65CN(+)	TU36 TT7 TM7	AT	0 (E)			CV33	S6 S11 S21	70	3321	RADIOACTIVE MATERIAL, LOW SPECIFIC ACTIVITY (LSA-II), non fissile or fissile-excepted
S2.65AN(+) L2.65CN(+)	TU36 TT7 TM7	AT	0 (E)			CV33	S6 S11 S21	70	3322	RADIOACTIVE MATERIAL, LOW SPECIFIC ACTIVITY (LSA-III), non fissile or fissile-excepted
			0 (E)			CV33	S6 S11 S21	70	3323	RADIOACTIVE MATERIAL, TYPE C PACKAGE, non fissile or fissile-excepted
			0 (E)			CV33	S6 S11 S21	70	3324	RADIOACTIVE MATERIAL, LOW SPECIFIC ACTIVITY (LSA-II), FISSILE
			0 (E)			CV33	S6 S11 S21	70	3325	RADIOACTIVE MATERIAL, LOW SPECIFIC ACTIVITY, (LSA-III), FISSILE
			0 (E)			CV33	S6 S11 S21	70	3326	RADIOACTIVE MATERIAL, SURFACE CONTAMINATED OBJECTS (SCO-I or SCO-II), FISSILE
			0 (E)			CV33	S6 S11 S21	70	3327	RADIOACTIVE MATERIAL, TYPE A PACKAGE, FISSILE, non-special form
			0 (E)			CV33	S6 S11 S21	70	3328	RADIOACTIVE MATERIAL, TYPE B(U) PACKAGE, FISSILE
			0 (E)			CV33	S6 S11 S21	70	3329	RADIOACTIVE MATERIAL, TYPE B(M) PACKAGE, FISSILE

UN No.	Name and description	Class	Classifi-cation code	Packing group	Labels	Special provi-sions	Limited and excepted quantities		Packaging			Portable tanks and bulk containers	
									Packing instruc-tions	Special packing provisions	Mixed packing provisions	Instruc-tions	Special provisions
	3.1.2	2.2	2.2	2.1.1.3	5.2.2	3.3	3.4	3.5.1.2	4.1.4	4.1.4	4.1.10	4.2.5.2 7.3.2	4.2.5.3
(1)	(2)	(3a)	(3b)	(4)	(5)	(6)	(7a)	(7b)	(8)	(9a)	(9b)	(10)	(11)
3330	RADIOACTIVE MATERIAL, TYPE C PACKAGE, FISSILE	7			7X +7E	172 326	0	E0	See 2.2.7 and 4.1.9	See 4.1.9.1.3			
3331	RADIOACTIVE MATERIAL, TRANSPORTED UNDER SPECIAL ARRANGEMENT, FISSILE	7			7X +7E	172 326	0	E0	See 2.2.7 and 4.1.9	See 4.1.9.1.3			
3332	RADIOACTIVE MATERIAL, TYPE A PACKAGE, SPECIAL FORM, non fissile or fissile-excepted	7			7X	172 317	0	E0	See 2.2.7 and 4.1.9	See 4.1.9.1.3			
3333	RADIOACTIVE MATERIAL, TYPE A PACKAGE, SPECIAL FORM, FISSILE	7			7X +7E	172	0	E0	See 2.2.7 and 4.1.9	See 4.1.9.1.3			
3334	Aviation regulated liquid, n.o.s.	9	M11				NOT SUBJECT TO ADR						
3335	Aviation regulated solid, n.o.s.	9	M11				NOT SUBJECT TO ADR						
3336	MERCAPTANS, LIQUID, FLAMMABLE, N.O.S. or MERCAPTAN MIXTURE, LIQUID, FLAMMABLE, N.O.S.	3	F1	I	3	274	0	E0	P001		MP7 MP17	T11	TP2
3336	MERCAPTANS, LIQUID, FLAMMABLE, N.O.S. or MERCAPTAN MIXTURE, LIQUID, FLAMMABLE, N.O.S. (vapour pressure at 50 °C more than 110 kPa)	3	F1	II	3	274 640C	1 L	E2	P001		MP19	T7	TP1 TP8 TP28
3336	MERCAPTANS, LIQUID, FLAMMABLE, N.O.S. or MERCAPTAN MIXTURE, LIQUID, FLAMMABLE, N.O.S. (vapour pressure at 50 °C not more than 110 kPa)	3	F1	II	3	274 640D	1 L	E2	P001 IBC02 R001		MP19	T7	TP1 TP8 TP28
3336	MERCAPTANS, LIQUID, FLAMMABLE, N.O.S. or MERCAPTAN MIXTURE, LIQUID, FLAMMABLE, N.O.S.	3	F1	III	3	274	5 L	E1	P001 IBC03 LP01 R001		MP19	T4	TP1 TP29
3337	REFRIGERANT GAS R 404A (Pentafluoroethane, 1,1,1-trifluoroethane, and 1,1,1,2-tetrafluoroethane zeotropic mixture with approximately 44% pentafluoroethane and 52% 1,1,1-trifluoroethane)	2	2A		2.2	662	120 ml	E1	P200		MP9	(M) T50	
3338	REFRIGERANT GAS R 407A (Difluoromethane, pentafluoroethane, and 1,1,1,2-tetrafluoroethane zeotropic mixture with approximately 20% difluoromethane and 40% pentafluoroethane)	2	2A		2.2	662	120 ml	E1	P200		MP9	(M) T50	
3339	REFRIGERANT GAS R 407B (Difluoromethane, pentafluoroethane, and 1,1,1,2-tetrafluoroethane zeotropic mixture with approximately 10% difluoromethane and 70% pentafluoroethane)	2	2A		2.2	662	120 ml	E1	P200		MP9	(M) T50	
3340	REFRIGERANT GAS R 407C (Difluoromethane, pentafluoroethane, and 1,1,1,2-tetrafluoroethane zeotropic mixture with approximately 23%difluoromethane and 25% pentafluoroethane)	2	2A		2.2	662	120 ml	E1	P200		MP9	(M) T50	
3341	THIOUREA DIOXIDE	4.2	S2	II	4.2		0	E2	P002 IBC06		MP14	T3	TP33
3341	THIOUREA DIOXIDE	4.2	S2	III	4.2		0	E1	P002 IBC08 LP02 R001	B3	MP14	T1	TP33
3342	XANTHATES	4.2	S2	II	4.2		0	E2	P002 IBC06		MP14	T3	TP33

ADR tank		Vehicle for tank carriage	Transport category (Tunnel restriction code)	Special provisions for carriage				Hazard identification No.	UN No.	Name and description
Tank code	Special provisions			Packages	Bulk	Loading, unloading and handling	Operation			
4.3	4.3.5, 6.8.4	9.1.1.2	1.1.3.6 (8.6)	7.2.4	7.3.3	7.5.11	8.5	5.3.2.3		3.1.2
(12)	(13)	(14)	(15)	(16)	(17)	(18)	(19)	(20)	(1)	(2)
			0 (E)			CV33	S6 S11 S21	70	3330	RADIOACTIVE MATERIAL, TYPE C PACKAGE, FISSILE
			0 (-)			CV33	S6 S11 S21	70	3331	RADIOACTIVE MATERIAL, TRANSPORTED UNDER SPECIAL ARRANGEMENT, FISSILE
			0 (E)			CV33	S6 S11 S12 S21	70	3332	RADIOACTIVE MATERIAL, TYPE A PACKAGE, SPECIAL FORM, non fissile or fissile-excepted
			0 (E)			CV33	S6 S11 S21	70	3333	RADIOACTIVE MATERIAL, TYPE A PACKAGE, SPECIAL FORM, FISSILE
NOT SUBJECT TO ADR									3334	Aviation regulated liquid, n.o.s.
NOT SUBJECT TO ADR									3335	Aviation regulated solid, n.o.s.
L4BN		FL	1 (D/E)				S2 S20	33	3336	MERCAPTANS, LIQUID, FLAMMABLE, N.O.S. or MERCAPTAN MIXTURE, LIQUID, FLAMMABLE, N.O.S.
L1.5BN		FL	2 (D/E)				S2 S20	33	3336	MERCAPTANS, LIQUID, FLAMMABLE, N.O.S. or MERCAPTAN MIXTURE, LIQUID, FLAMMABLE, N.O.S. (vapour pressure at 50 °C more than 110 kPa)
LGBF		FL	2 (D/E)				S2 S20	33	3336	MERCAPTANS, LIQUID, FLAMMABLE, N.O.S. or MERCAPTAN MIXTURE, LIQUID, FLAMMABLE, N.O.S. (vapour pressure at 50 °C not more than 110 kPa)
LGBF		FL	3 (D/E)	V12			S2	30	3336	MERCAPTANS, LIQUID, FLAMMABLE, N.O.S. or MERCAPTAN MIXTURE, LIQUID, FLAMMABLE, N.O.S.
PxBN(M)	TA4 TT9	AT	3 (C/E)			CV9 CV10 CV36		20	3337	REFRIGERANT GAS R 404A (Pentafluoroethane, 1,1,1-trifluoroethane, and 1,1,1,2-tetrafluoroethane zeotropic mixture with approximately 44% pentafluoroethane and 52% 1,1,1-trifluoroethane)
PxBN(M)	TA4 TT9	AT	3 (C/E)			CV9 CV10 CV36		20	3338	REFRIGERANT GAS R 407A (Difluoromethane, pentafluoroethane, and 1,1,1,2-tetrafluoroethane zeotropic mixture with approximately 20% difluoromethane and 40% pentafluoroethane)
PxBN(M)	TA4 TT9	AT	3 (C/E)			CV9 CV10 CV36		20	3339	REFRIGERANT GAS R 407B (Difluoromethane, pentafluoroethane, and 1,1,1,2-tetrafluoroethane zeotropic mixture with approximately 10% difluoromethane and 70% pentafluoroethane)
PxBN(M)	TA4 TT9	AT	3 (C/E)			CV9 CV10 CV36		20	3340	REFRIGERANT GAS R 407C (Difluoromethane, pentafluoroethane, and 1,1,1,2-tetrafluoroethane zeotropic mixture with approximately 23%difluoromethane and 25% pentafluoroethane)
SGAV		AT	2 (D/E)	V1				40	3341	THIOUREA DIOXIDE
SGAV		AT	3 (E)	V1				40	3341	THIOUREA DIOXIDE
SGAV		AT	2 (D/E)	V1				40	3342	XANTHATES

UN No.	Name and description	Class	Classifi-cation code	Packing group	Labels	Special provisions	Limited and excepted quantities		Packaging			Portable tanks and bulk containers	
									Packing instruc-tions	Special packing provisions	Mixed packing provisions	Instruc-tions	Special provisions
	3.1.2	2.2	2.2	2.1.1.3	5.2.2	3.3	3.4	3.5.1.2	4.1.4	4.1.4	4.1.10	4.2.5.2 7.3.2	4.2.5.3
(1)	(2)	(3a)	(3b)	(4)	(5)	(6)	(7a)	(7b)	(8)	(9a)	(9b)	(10)	(11)
3342	XANTHATES	4.2	S2	III	4.2		0	E1	P002 IBC08 LP02 R001	B3	MP14	T1	TP33
3343	NITROGLYCERIN MIXTURE, DESENSITIZED, LIQUID, FLAMMABLE, N.O.S. with not more than 30% nitroglycerin, by mass	3	D		3	274 278	0	E0	P099		MP2		
3344	PENTAERYTHRITE TETRANITRATE (PENTAERYTHRITOL TETRANITRATE; PETN) MIXTURE, DESENSITIZED, SOLID, N.O.S. with more than 10% but not more than 20% PETN, by mass	4.1	D	II	4.1	272 274	0	E0	P099		MP2		
3345	PHENOXYACETIC ACID DERIVATIVE PESTICIDE, SOLID, TOXIC	6.1	T7	I	6.1	61 274 648	0	E5	P002 IBC07		MP18	T6	TP33
3345	PHENOXYACETIC ACID DERIVATIVE PESTICIDE, SOLID, TOXIC	6.1	T7	II	6.1	61 274 648	500 g	E4	P002 IBC08	B4	MP10	T3	TP33
3345	PHENOXYACETIC ACID DERIVATIVE PESTICIDE, SOLID, TOXIC	6.1	T7	III	6.1	61 274 648	5 kg	E1	P002 IBC08 LP02 R001	B3	MP10	T1	TP33
3346	PHENOXYACETIC ACID DERIVATIVE PESTICIDE, LIQUID, FLAMMABLE, TOXIC, flash-point less than 23 °C	3	FT2	I	3 +6.1	61 274	0	E0	P001		MP7 MP17	T14	TP2 TP27
3346	PHENOXYACETIC ACID DERIVATIVE PESTICIDE, LIQUID, FLAMMABLE, TOXIC, flash-point less than 23 °C	3	FT2	II	3 +6.1	61 274	1 L	E2	P001 IBC02 R001		MP19	T11	TP2 TP27
3347	PHENOXYACETIC ACID DERIVATIVE PESTICIDE, LIQUID, TOXIC, FLAMMABLE, flash-point not less than 23 °C	6.1	TF2	I	6.1 +3	61 274	0	E5	P001		MP8 MP17	T14	TP2 TP27
3347	PHENOXYACETIC ACID DERIVATIVE PESTICIDE, LIQUID, TOXIC, FLAMMABLE, flash-point not less than 23 °C	6.1	TF2	II	6.1 +3	61 274	100 ml	E4	P001 IBC02		MP15	T11	TP2 TP27
3347	PHENOXYACETIC ACID DERIVATIVE PESTICIDE, LIQUID, TOXIC, FLAMMABLE, flash-point not less than 23 °C	6.1	TF2	III	6.1 +3	61 274	5 L	E1	P001 IBC03 R001		MP19	T7	TP2 TP28
3348	PHENOXYACETIC ACID DERIVATIVE PESTICIDE, LIQUID, TOXIC	6.1	T6	I	6.1	61 274 648	0	E5	P001		MP8 MP17	T14	TP2 TP27
3348	PHENOXYACETIC ACID DERIVATIVE PESTICIDE, LIQUID, TOXIC	6.1	T6	II	6.1	61 274 648	100 ml	E4	P001 IBC02		MP15	T11	TP2 TP27
3348	PHENOXYACETIC ACID DERIVATIVE PESTICIDE, LIQUID, TOXIC	6.1	T6	III	6.1	61 274 648	5 L	E1	P001 IBC03 LP01 R001		MP19	T7	TP2 TP28
3349	PYRETHROID PESTICIDE, SOLID, TOXIC	6.1	T7	I	6.1	61 274 648	0	E5	P002 IBC07		MP18	T6	TP33
3349	PYRETHROID PESTICIDE, SOLID, TOXIC	6.1	T7	II	6.1	61 274 648	500 g	E4	P002 IBC08	B4	MP10	T3	TP33
3349	PYRETHROID PESTICIDE, SOLID, TOXIC	6.1	T7	III	6.1	61 274 648	5 kg	E1	P002 IBC08 LP02 R001	B3	MP10	T1	TP33
3350	PYRETHROID PESTICIDE, LIQUID, FLAMMABLE, TOXIC, flash-point less than 23 °C	3	FT2	I	3 +6.1	61 274	0	E0	P001		MP7 MP17	T14	TP2 TP27

ADR tank		Vehicle for tank carriage	Transport category (Tunnel restriction code)	Special provisions for carriage				Hazard identification No.	UN No.	Name and description
Tank code	Special provisions			Packages	Bulk	Loading, unloading and handling	Operation			
4.3	4.3.5, 6.8.4	9.1.1.2	1.1.3.6 (8.6)	7.2.4	7.3.3	7.5.11	8.5	5.3.2.3		3.1.2
(12)	(13)	(14)	(15)	(16)	(17)	(18)	(19)	(20)	(1)	(2)
SGAV		AT	3 (E)	V1				40	3342	XANTHATES
			0 (B)				S2 S14		3343	NITROGLYCERIN MIXTURE, DESENSITIZED, LIQUID, FLAMMABLE, N.O.S. with not more than 30% nitroglycerin, by mass
			2 (B)				S14		3344	PENTAERYTHRITE TETRANITRATE (PENTAERYTHRITOL TETRANITRATE; PETN) MIXTURE, DESENSITIZED, SOLID, N.O.S. with more than 10% but not more than 20% PETN, by mass
S10AH L10CH	TU14 TU15 TE19 TE21	AT	1 (C/E)	V10		CV1 CV13 CV28	S9 S14	66	3345	PHENOXYACETIC ACID DERIVATIVE PESTICIDE, SOLID, TOXIC
SGAH L4BH	TU15 TE19	AT	2 (D/E)	V11		CV13 CV28	S9 S19	60	3345	PHENOXYACETIC ACID DERIVATIVE PESTICIDE, SOLID, TOXIC
SGAH L4BH	TU15 TE19	AT	2 (E)		VC1 VC2 AP7	CV13 CV28	S9	60	3345	PHENOXYACETIC ACID DERIVATIVE PESTICIDE, SOLID, TOXIC
L10CH	TU14 TU15 TE21	FL	1 (C/E)			CV13 CV28	S2 S22	336	3346	PHENOXYACETIC ACID DERIVATIVE PESTICIDE, LIQUID, FLAMMABLE, TOXIC, flash-point less than 23 °C
L4BH	TU15	FL	2 (D/E)			CV13 CV28	S2 S22	336	3346	PHENOXYACETIC ACID DERIVATIVE PESTICIDE, LIQUID, FLAMMABLE, TOXIC, flash-point less than 23 °C
L10CH	TU14 TU15 TE19 TE21	FL	1 (C/E)			CV1 CV13 CV28	S2 S9 S14	663	3347	PHENOXYACETIC ACID DERIVATIVE PESTICIDE, LIQUID, TOXIC, FLAMMABLE, flash-point not less than 23 °C
L4BH	TU15 TE19	FL	2 (D/E)			CV13 CV28	S2 S9 S19	63	3347	PHENOXYACETIC ACID DERIVATIVE PESTICIDE, LIQUID, TOXIC, FLAMMABLE, flash-point not less than 23 °C
L4BH	TU15 TE19	FL	2 (D/E)	V12		CV13 CV28	S2 S9	63	3347	PHENOXYACETIC ACID DERIVATIVE PESTICIDE, LIQUID, TOXIC, FLAMMABLE, flash-point not less than 23 °C
L10CH	TU14 TU15 TE19 TE21	AT	1 (C/E)			CV1 CV13 CV28	S9 S14	66	3348	PHENOXYACETIC ACID DERIVATIVE PESTICIDE, LIQUID, TOXIC
L4BH	TU15 TE19	AT	2 (D/E)			CV13 CV28	S9 S19	60	3348	PHENOXYACETIC ACID DERIVATIVE PESTICIDE, LIQUID, TOXIC
L4BH	TU15 TE19	AT	2 (E)	V12		CV13 CV28	S9	60	3348	PHENOXYACETIC ACID DERIVATIVE PESTICIDE, LIQUID, TOXIC
S10AH L10CH	TU14 TU15 TE19 TE21	AT	1 (C/E)	V10		CV1 CV13 CV28	S9 S14	66	3349	PYRETHROID PESTICIDE, SOLID, TOXIC
SGAH L4BH	TU15 TE19	AT	2 (D/E)	V11		CV13 CV28	S9 S19	60	3349	PYRETHROID PESTICIDE, SOLID, TOXIC
SGAH L4BH	TU15 TE19	AT	2 (E)		VC1 VC2 AP7	CV13 CV28	S9	60	3349	PYRETHROID PESTICIDE, SOLID, TOXIC
L10CH	TU14 TU15 TE21	FL	1 (C/E)			CV13 CV28	S2 S22	336	3350	PYRETHROID PESTICIDE, LIQUID, FLAMMABLE, TOXIC, flash-point less than 23 °C

UN No.	Name and description	Class	Classifi-cation code	Packing group	Labels	Special provi-sions	Limited and excepted quantities		Packaging			Portable tanks and bulk containers	
									Packing instruc-tions	Special packing provisions	Mixed packing provisions	Instruc-tions	Special provisions
	3.1.2	2.2	2.2	2.1.1.3	5.2.2	3.3	3.4	3.5.1.2	4.1.4	4.1.4	4.1.10	4.2.5.2 7.3.2	4.2.5.3
(1)	(2)	(3a)	(3b)	(4)	(5)	(6)	(7a)	(7b)	(8)	(9a)	(9b)	(10)	(11)
3350	PYRETHROID PESTICIDE, LIQUID, FLAMMABLE, TOXIC, flash-point less than 23 °C	3	FT2	II	3 +6.1	61 274	1 L	E2	P001 IBC02 R001		MP19	T11	TP2 TP27
3351	PYRETHROID PESTICIDE, LIQUID, TOXIC, FLAMMABLE, flash-point not less than 23 °C	6.1	TF2	I	6.1 +3	61 274	0	E5	P001		MP8 MP17	T14	TP2 TP27
3351	PYRETHROID PESTICIDE, LIQUID, TOXIC, FLAMMABLE, flash-point not less than 23 °C	6.1	TF2	II	6.1 +3	61 274	100 ml	E4	P001 IBC02		MP15	T11	TP2 TP27
3351	PYRETHROID PESTICIDE, LIQUID, TOXIC, FLAMMABLE, flash-point not less than 23 °C	6.1	TF2	III	6.1 +3	61 274	5 L	E1	P001 IBC03 R001		MP19	T7	TP2 TP28
3352	PYRETHROID PESTICIDE, LIQUID, TOXIC	6.1	T6	I	6.1	61 274 648	0	E5	P001		MP8 MP17	T14	TP2 TP27
3352	PYRETHROID PESTICIDE, LIQUID, TOXIC	6.1	T6	II	6.1	61 274 648	100 ml	E4	P001 IBC02		MP15	T11	TP2 TP27
3352	PYRETHROID PESTICIDE, LIQUID, TOXIC	6.1	T6	III	6.1	61 274 648	5 L	E1	P001 IBC03 LP01 R001		MP19	T7	TP2 TP28
3354	INSECTICIDE GAS, FLAMMABLE, N.O.S.	2	2F		2.1	274 662	0	E0	P200		MP9	(M)	
3355	INSECTICIDE GAS, TOXIC, FLAMMABLE, N.O.S.	2	2TF		2.3 +2.1	274	0	E0	P200		MP9	(M)	
3356	OXYGEN GENERATOR, CHEMICAL	5.1	O3		5.1	284	0	E0	P500		MP2		
3357	NITROGLYCERIN MIXTURE, DESENSITIZED, LIQUID, N.O.S. with not more than 30% nitroglycerin, by mass	3	D	II	3	274 288	0	E0	P099		MP2		
3358	REFRIGERATING MACHINES containing flammable, non-toxic, liquefied gas	2	6F		2.1	291	0	E0	P003	PP32	MP9		
3359	FUMIGATED CARGO TRANSPORT UNIT	9	M11			302							
3360	Fibres, vegetable, dry	4.1	F1				NOT SUBJECT TO ADR							
3361	CHLOROSILANES, TOXIC, CORROSIVE, N.O.S.	6.1	TC1	II	6.1 +8	274	0	E0	P010		MP15	T14	TP2 TP7 TP27
3362	CHLOROSILANES, TOXIC, CORROSIVE, FLAMMABLE, N.O.S.	6.1	TFC	II	6.1 +3 +8	274	0	E0	P010		MP15	T14	TP2 TP7 TP27
3363	Dangerous goods in machinery or dangerous goods in apparatus	9	M11				NOT SUBJECT TO ADR [see also 1.1.3.1 (b)]							
3364	TRINITROPHENOL (PICRIC ACID), WETTED with not less than 10% water, by mass	4.1	D	I	4.1		0	E0	P406	PP24	MP2		
3365	TRINITROCHLOROBENZENE (PICRYL CHLORIDE), WETTED with not less than 10% water, by mass	4.1	D	I	4.1		0	E0	P406	PP24	MP2		
3366	TRINITROTOLUENE (TNT), WETTED with not less than 10% water, by mass	4.1	D	I	4.1		0	E0	P406	PP24	MP2		
3367	TRINITROBENZENE, WETTED with not less than 10% water, by mass	4.1	D	I	4.1		0	E0	P406	PP24	MP2		
3368	TRINITROBENZOIC ACID, WETTED with not less than 10% water, by mass	4.1	D	I	4.1		0	E0	P406	PP24	MP2		
3369	SODIUM DINITRO-o-CRESOLATE, WETTED with not less than 10% water, by mass	4.1	DT	I	4.1 +6.1		0	E0	P406	PP24	MP2		
3370	UREA NITRATE, WETTED with not less than 10% water, by mass	4.1	D	I	4.1		0	E0	P406	PP78	MP2		
3371	2–METHYLBUTANAL	3	F1	II	3		1 L	E2	P001 IBC02 R001		MP19	T4	TP1
3373	BIOLOGICAL SUBSTANCE, CATEGORY B	6.2	I4		6.2	319	0	E0	P650			T1	TP1

ADR tank		Vehicle for tank carriage	Transport category (Tunnel restriction code)	Special provisions for carriage				Hazard identification No.	UN No.	Name and description
Tank code	Special provisions			Packages	Bulk	Loading, unloading and handling	Operation			
4.3	4.3.5, 6.8.4	9.1.1.2	1.1.3.6 (8.6)	7.2.4	7.3.3	7.5.11	8.5	5.3.2.3		3.1.2
(12)	(13)	(14)	(15)	(16)	(17)	(18)	(19)	(20)	(1)	(2)
L4BH	TU15	FL	2 (D/E)			CV13 CV28	S2 S22	336	3350	PYRETHROID PESTICIDE, LIQUID, FLAMMABLE, TOXIC, flash-point less than 23 °C
L10CH	TU14 TU15 TE19 TE21	FL	1 (C/E)			CV1 CV13 CV28	S2 S9 S14	663	3351	PYRETHROID PESTICIDE, LIQUID, TOXIC, FLAMMABLE, flash-point not less than 23 °C
L4BH	TU15 TE19	FL	2 (D/E)			CV13 CV28	S2 S9 S19	63	3351	PYRETHROID PESTICIDE, LIQUID, TOXIC, FLAMMABLE, flash-point not less than 23 °C
L4BH	TU15 TE19	FL	2 (D/E)	V12		CV13 CV28	S2 S9	63	3351	PYRETHROID PESTICIDE, LIQUID, TOXIC, FLAMMABLE, flash-point not less than 23 °C
L10CH	TU14 TU15 TE19 TE21	AT	1 (C/E)			CV1 CV13 CV28	S9 S14	66	3352	PYRETHROID PESTICIDE, LIQUID, TOXIC
L4BH	TU15 TE19	AT	2 (D/E)			CV13 CV28	S9 S19	60	3352	PYRETHROID PESTICIDE, LIQUID, TOXIC
L4BH	TU15 TE19	AT	2 (E)	V12		CV13 CV28	S9	60	3352	PYRETHROID PESTICIDE, LIQUID, TOXIC
PxBN(M)	TA4 TT9	FL	2 (B/D)			CV9 CV10 CV36	S2 S20	23	3354	INSECTICIDE GAS, FLAMMABLE, N.O.S.
PxBH(M)	TU6 TA4 TT9	FL	1 (B/D)			CV9 CV10 CV36	S2 S14	263	3355	INSECTICIDE GAS, TOXIC, FLAMMABLE, N.O.S.
			2 (E)			CV24			3356	OXYGEN GENERATOR, CHEMICAL
			2 (B)				S2 S14		3357	NITROGLYCERIN MIXTURE, DESENSITIZED, LIQUID, N.O.S. with not more than 30% nitroglycerin, by mass
			2 (D)			CV9	S2		3358	REFRIGERATING MACHINES containing flammable, non-toxic, liquefied gas
			(-)						3359	FUMIGATED CARGO TRANSPORT UNIT
NOT SUBJECT TO ADR									3360	Fibres, vegetable, dry
L4BH	TU15 TE19	AT	2 (D/E)			CV13 CV28	S9 S19	68	3361	CHLOROSILANES, TOXIC, CORROSIVE, N.O.S.
L4BH	TU15 TE19	FL	2 (D/E)			CV13 CV28	S2 S9 S19	638	3362	CHLOROSILANES, TOXIC, CORROSIVE, FLAMMABLE, N.O.S.
NOT SUBJECT TO ADR [see also 1.1.3.1 (b)]									3363	Dangerous goods in machinery or dangerous goods in apparatus
			1 (B)				S14		3364	TRINITROPHENOL (PICRIC ACID), WETTED with not less than 10% water, by mass
			1 (B)				S14		3365	TRINITROCHLOROBENZENE (PICRYL CHLORIDE), WETTED with not less than 10% water, by mass
			1 (B)				S14		3366	TRINITROTOLUENE (TNT), WETTED with not less than 10% water, by mass
			1 (B)				S14		3367	TRINITROBENZENE, WETTED with not less than 10% water, by mass
			1 (B)				S14		3368	TRINITROBENZOIC ACID, WETTED with not less than 10% water, by mass
			1 (B)			CV13 CV28	S14		3369	SODIUM DINITRO-o-CRESOLATE, WETTED with not less than 10% water, by mass
			1 (B)				S14		3370	UREA NITRATE, WETTED with not less than 10% water, by mass
LGBF		FL	2 (D/E)				S2 S20	33	3371	2 –METHYLBUTANAL
L4BH	TU15 TU37 TE19	AT	(-)				S3	606	3373	BIOLOGICAL SUBSTANCE, CATEGORY B

UN No.	Name and description	Class	Classification code	Packing group	Labels	Special provisions	Limited and excepted quantities		Packaging			Portable tanks and bulk containers	
									Packing instructions	Special packing provisions	Mixed packing provisions	Instructions	Special provisions
	3.1.2	2.2	2.2	2.1.1.3	5.2.2	3.3	3.4	3.5.1.2	4.1.4	4.1.4	4.1.10	4.2.5.2 7.3.2	4.2.5.3
(1)	(2)	(3a)	(3b)	(4)	(5)	(6)	(7a)	(7b)	(8)	(9a)	(9b)	(10)	(11)
3373	BIOLOGICAL SUBSTANCE, CATEGORY B (animal material only)	6.2	I4		6.2	319	0	E0	P650			T1 BK1 BK2	TP1
3374	ACETYLENE, SOLVENT FREE	2	2F		2.1	662	0	E0	P200		MP9		
3375	AMMONIUM NITRATE EMULSION or SUSPENSION or GEL, intermediate for blasting explosives, liquid	5.1	O1	II	5.1	309	0	E2	P505 IBC02	B16	MP2	T1	TP1 TP9 TP17 TP32
3375	AMMONIUM NITRATE EMULSION or SUSPENSION or GEL, intermediate for blasting explosives, solid	5.1	O2	II	5.1	309	0	E2	P505 IBC02	B16	MP2	T1	TP1 TP9 TP17 TP32
3376	4-NITROPHENYL-HYDRAZINE, with not less than 30% water, by mass	4.1	D	I	4.1		0	E0	P406	PP26	MP2		
3377	SODIUM PERBORATE MONOHYDRATE	5.1	O2	III	5.1		5 kg	E1	P002 IBC08 LP02 R001	B3	MP10	T1 BK1 BK2 BK3	TP33
3378	SODIUM CARBONATE PEROXYHYDRATE	5.1	O2	II	5.1		1 kg	E2	P002 IBC08	B4	MP10	T3 BK1 BK2	TP33
3378	SODIUM CARBONATE PEROXYHYDRATE	5.1	O2	III	5.1		5 kg	E1	P002 IBC08 LP02 R001	B3	MP10	T1 BK1 BK2 BK3	TP33
3379	DESENSITIZED EXPLOSIVE, LIQUID, N.O.S.	3	D	I	3	274 311	0	E0	P099		MP2		
3380	DESENSITIZED EXPLOSIVE, SOLID, N.O.S.	4.1	D	I	4.1	274 311	0	E0	P099		MP2		
3381	TOXIC BY INHALATION LIQUID, N.O.S. with an LC_{50} lower than or equal to 200 ml/m^3 and saturated vapour concentration greater than or equal to 500 LC_{50}	6.1	T1 or T4	I	6.1	274	0	E0	P601		MP8 MP17	T22	TP2
3382	TOXIC BY INHALATION LIQUID, N.O.S. with an LC_{50} lower than or equal to 1000 ml/m^3 and saturated vapour concentration greater than or equal to 10 LC_{50}	6.1	T1 or T4	I	6.1	274	0	E0	P602		MP8 MP17	T20	TP2
3383	TOXIC BY INHALATION LIQUID, FLAMMABLE, N.O.S. with an LC_{50} lower than or equal to 200 ml/m^3 and saturated vapour concentration greater than or equal to 500 LC_{50}	6.1	TF1	I	6.1 +3	274	0	E0	P601		MP8 MP17	T22	TP2
3384	TOXIC BY INHALATION LIQUID, FLAMMABLE, N.O.S. with with an LC_{50} lower than or equal to 1000 ml/m^3 and saturated vapour concentration greater than or equal to 10 LC_{50}	6.1	TF1	I	6.1 +3	274	0	E0	P602		MP8 MP17	T20	TP2
3385	TOXIC BY INHALATION LIQUID, WATER-REACTIVE, N.O.S. with an LC_{50} lower than or equal to 200 ml/m^3 and saturated vapour concentration greater than or equal to 500 LC_{50}	6.1	TW1	I	6.1 +4.3	274	0	E0	P601		MP8 MP17	T22	TP2
3386	TOXIC BY INHALATION LIQUID, WATER-REACTIVE, N.O.S. with an LC_{50} lower than or equal to 1000 ml/m^3 and saturated vapour concentration greater than or equal to 10 LC_{50}	6.1	TW1	I	6.1 +4.3	274	0	E0	P602		MP8 MP17	T20	TP2
3387	TOXIC BY INHALATION LIQUID, OXIDIZING, N.O.S. with an LC_{50} lower than or equal to 200 ml/m^3 and saturated vapour concentration greater than or equal to 500 LC_{50}	6.1	TO1	I	6.1 +5.1	274	0	E0	P601		MP8 MP17	T22	TP2

ADR tank		Vehicle for tank carriage	Transport category (Tunnel restriction code)	Special provisions for carriage				Hazard identification No.	UN No.	Name and description
Tank code	Special provisions			Packages	Bulk	Loading, unloading and handling	Operation			
4.3	4.3.5, 6.8.4	9.1.1.2	1.1.3.6 (8.6)	7.2.4	7.3.3	7.5.11	8.5	5.3.2.3		3.1.2
(12)	(13)	(14)	(15)	(16)	(17)	(18)	(19)	(20)	(1)	(2)
L4BH	TU15 TU37 TE19	AT	– (-)				S3	606	3373	BIOLOGICAL SUBSTANCE, CATEGORY B (animal material only)
			2 (D)			CV9 CV10 CV36	S2 S20		3374	ACETYLENE, SOLVENT FREE
LGAV(+)	TU3 TU12 TU39 TE10 TE23 TA1 TA3	AT	2 (E)			CV24	S9 S23	50	3375	AMMONIUM NITRATE EMULSION or SUSPENSION or GEL, intermediate for blasting explosives, liquid
SGAV(+)	TU3 TU12 TU39 TE10 TE23 TA1 TA3	AT	2 (E)			CV24	S9 S23	50	3375	AMMONIUM NITRATE EMULSION or SUSPENSION or GEL, intermediate for blasting explosives, solid
			1 (B)	V1			S14		3376	4-NITROPHENYL-HYDRAZINE, with not less than 30% water, by mass
SGAV	TU3	AT	3 (E)		VC1 VC2 AP6 AP7	CV24		50	3377	SODIUM PERBORATE MONOHYDRATE
SGAV	TU3	AT	2 (E)	V11	VC1 VC2 AP6 AP7	CV24		50	3378	SODIUM CARBONATE PEROXYHYDRATE
SGAV	TU3	AT	3 (E)		VC1 VC2 AP6 AP7	CV24		50	3378	SODIUM CARBONATE PEROXYHYDRATE
			1 (B)				S2 S14		3379	DESENSITIZED EXPLOSIVE, LIQUID, N.O.S.
			1 (B)				S14		3380	DESENSITIZED EXPLOSIVE, SOLID, N.O.S.
L15CH	TU14 TU15 TE19 TE21	AT	1 (C/D)			CV1 CV13 CV28	S9 S14	66	3381	TOXIC BY INHALATION LIQUID, N.O.S. with an LC_{50} lower than or equal to 200 ml/m^3 and saturated vapour concentration greater than or equal to 500 LC_{50}
L10CH	TU14 TU15 TE19 TE21	AT	1 (C/D)			CV1 CV13 CV28	S9 S14	66	3382	TOXIC BY INHALATION LIQUID, N.O.S. with an LC_{50} lower than or equal to 1000 ml/m^3 and saturated vapour concentration greater than or equal to 10 LC_{50}
L15CH	TU14 TU15 TE19 TE21	FL	1 (C/D)			CV1 CV13 CV28	S2 S9 S14	663	3383	TOXIC BY INHALATION LIQUID, FLAMMABLE, N.O.S. with an LC_{50} lower than or equal to 200 ml/m^3 and saturated vapour concentration greater than or equal to 500 LC_{50}
L10CH	TU14 TU15 TE19 TE21	FL	1 (C/D)			CV1 CV13 CV28	S2 S9 S14	663	3384	TOXIC BY INHALATION LIQUID, FLAMMABLE, N.O.S. with with an LC_{50} lower than or equal to 1000 ml/m^3 and saturated vapour concentration greater than or equal to 10 LC_{50}
L15CH	TU14 TU15 TE19 TE21	AT	1 (C/D)			CV1 CV13 CV28	S9 S14	623	3385	TOXIC BY INHALATION LIQUID, WATER-REACTIVE, N.O.S. with an LC_{50} lower than or equal to 200 ml/m^3 and saturated vapour concentration greater than or equal to 500 LC_{50}
L10CH	TU14 TU15 TE19 TE21	AT	1 (C/D)			CV1 CV13 CV28	S9 S14	623	3386	TOXIC BY INHALATION LIQUID, WATER-REACTIVE, N.O.S. with an LC_{50} lower than or equal to 1000 ml/m^3 and saturated vapour concentration greater than or equal to 10 LC_{50}
L15CH	TU14 TU15 TE19 TE21	AT	1 (C/D)			CV1 CV13 CV28	S9 S14	665	3387	TOXIC BY INHALATION LIQUID, OXIDIZING, N.O.S. with an LC_{50} lower than or equal to 200 ml/m^3 and saturated vapour concentration greater than or equal to 500 LC_{50}

UN No.	Name and description	Class	Classification code	Packing group	Labels	Special provisions	Limited and excepted quantities		Packaging			Portable tanks and bulk containers	
									Packing instructions	Special packing provisions	Mixed packing provisions	Instructions	Special provisions
	3.1.2	2.2	2.2	2.1.1.3	5.2.2	3.3	3.4	3.5.1.2	4.1.4	4.1.4	4.1.10	4.2.5.2 7.3.2	4.2.5.3
(1)	(2)	(3a)	(3b)	(4)	(5)	(6)	(7a)	(7b)	(8)	(9a)	(9b)	(10)	(11)
3388	TOXIC BY INHALATION LIQUID, OXIDIZING, N.O.S. with an LC_{50} lower than or equal to 1000 ml/m^3 and saturated vapour concentration greater than or equal to 10 LC_{50}	6.1	TO1	I	6.1 +5.1	274	0	E0	P602		MP8 MP17	T20	TP2
3389	TOXIC BY INHALATION LIQUID, CORROSIVE, N.O.S. with an LC_{50} lower than or equal to 200 ml/m^3 and saturated vapour concentration greater than or equal to 500 LC_{50}	6.1	TC1 or TC3	I	6.1 +8	274	0	E0	P601		MP8 MP17	T22	TP2
3390	TOXIC BY INHALATION LIQUID, CORROSIVE, N.O.S. with an LC_{50} lower than or equal to 1000 ml/m^3 and saturated vapour concentration greater than or equal to 10 LC_{50}	6.1	TC1 or TC3	I	6.1 +8	274	0	E0	P602		MP8 MP17	T20	TP2
3391	ORGANOMETALLIC SUBSTANCE, SOLID, PYROPHORIC	4.2	S5	I	4.2	274	0	E0	P404	PP86	MP2	T21	TP7 TP33 TP36
3392	ORGANOMETALLIC SUBSTANCE, LIQUID, PYROPHORIC	4.2	S5	I	4.2	274	0	E0	P400	PP86	MP2	T21	TP2 TP7 TP36
3393	ORGANOMETALLIC SUBSTANCE, SOLID, PYROPHORIC, WATER-REACTIVE	4.2	SW	I	4.2 +4.3	274	0	E0	P404	PP86	MP2	T21	TP7 TP33 TP36 TP41
3394	ORGANOMETALLIC SUBSTANCE, LIQUID, PYROPHORIC, WATER-REACTIVE	4.2	SW	I	4.2 +4.3	274	0	E0	P400	PP86	MP2	T21	TP2 TP7 TP36 TP41
3395	ORGANOMETALLIC SUBSTANCE, SOLID, WATER-REACTIVE	4.3	W2	I	4.3	274	0	E0	P403		MP2	T9	TP7 TP33 TP36 TP41
3395	ORGANOMETALLIC SUBSTANCE, SOLID, WATER-REACTIVE	4.3	W2	II	4.3	274	500 g	E2	P410 IBC04		MP14	T3	TP33 TP36 TP41
3395	ORGANOMETALLIC SUBSTANCE, SOLID, WATER-REACTIVE	4.3	W2	III	4.3	274	1 kg	E1	P410 IBC06		MP14	T1	TP33 TP36 TP41
3396	ORGANOMETALLIC SUBSTANCE, SOLID, WATER-REACTIVE, FLAMMABLE	4.3	WF2	I	4.3 +4.1	274	0	E0	P403		MP2	T9	TP7 TP33 TP36 TP41
3396	ORGANOMETALLIC SUBSTANCE, SOLID, WATER-REACTIVE, FLAMMABLE	4.3	WF2	II	4.3 +4.1	274	500 g	E2	P410 IBC04		MP14	T3	TP33 TP36 TP41
3396	ORGANOMETALLIC SUBSTANCE, SOLID, WATER-REACTIVE, FLAMMABLE	4.3	WF2	III	4.3 +4.1	274	1 kg	E1	P410 IBC06		MP14	T1	TP33 TP36 TP41
3397	ORGANOMETALLIC SUBSTANCE, SOLID, WATER-REACTIVE, SELF-HEATING	4.3	WS	I	4.3 +4.2	274	0	E0	P403		MP2	T9	TP7 TP33 TP36 TP41
3397	ORGANOMETALLIC SUBSTANCE, SOLID, WATER-REACTIVE, SELF-HEATING	4.3	WS	II	4.3 +4.2	274	500 g	E2	P410 IBC04		MP14	T3	TP33 TP36 TP41
3397	ORGANOMETALLIC SUBSTANCE, SOLID, WATER-REACTIVE, SELF-HEATING	4.3	WS	III	4.3 +4.2	274	1 kg	E1	P410 IBC06		MP14	T1	TP33 TP36 TP41
3398	ORGANOMETALLIC SUBSTANCE, LIQUID, WATER-REACTIVE	4.3	W1	I	4.3	274	0	E0	P402		MP2	T13	TP2 TP7 TP36 TP41
3398	ORGANOMETALLIC SUBSTANCE, LIQUID, WATER-REACTIVE	4.3	W1	II	4.3	274	500 ml	E2	P001 IBC01		MP15	T7	TP2 TP7 TP36 TP41

ADR tank		Vehicle for tank carriage	Transport category (Tunnel restriction code)	Special provisions for carriage				Hazard identifi-cation No.	UN No.	Name and description
Tank code	Special provisions			Packages	Bulk	Loading, unloading and handling	Operation			
4.3	4.3.5, 6.8.4	9.1.1.2	1.1.3.6 (8.6)	7.2.4	7.3.3	7.5.11	8.5	5.3.2.3		3.1.2
(12)	(13)	(14)	(15)	(16)	(17)	(18)	(19)	(20)	(1)	(2)
L10CH	TU14 TU15 TE19 TE21	AT	1 (C/D)			CV1 CV13 CV28	S9 S14	665	3388	TOXIC BY INHALATION LIQUID, OXIDIZING, N.O.S. with an LC_{50} lower than or equal to 1000 ml/m^3 saturated vapour concentration greater than or equal to 10 LC_{50}
L15CH	TU14 TU15 TE19 TE21	AT	1 (C/D)			CV1 CV13 CV28	S9 S14	668	3389	TOXIC BY INHALATION LIQUID, CORROSIVE, N.O.S. with an LC_{50} lower than or equal to 200 ml/m^3 and saturated vapour concentration greater than or equal to 500 LC_{50}
L10CH	TU14 TU15 TE19 TE21	AT	1 (C/D)			CV1 CV13 CV28	S9 S14	668	3390	TOXIC BY INHALATION LIQUID, CORROSIVE, N.O.S. with an LC_{50} lower than or equal to 1000 ml/m^3 and saturated vapour concentration greater than or equal to 10 LC_{50}
L21DH	TU4 TU14 TU22 TC1 TE21 TM1	AT	0 (B/E)	V1			S20	43	3391	ORGANOMETALLIC SUBSTANCE, SOLID, PYROPHORIC
L21DH	TU4 TU14 TU22 TC1 TE21 TM1	AT	0 (B/E)	V1			S20	333	3392	ORGANOMETALLIC SUBSTANCE, LIQUID, PYROPHORIC
L21DH	TU4 TU14 TU22 TC1 TE21 TM1	AT	0 (B/E)	V1			S20	X432	3393	ORGANOMETALLIC SUBSTANCE, SOLID, PYROPHORIC, WATER-REACTIVE
L21DH	TU4 TU14 TU22 TC1 TE21 TM1	AT	0 (B/E)	V1			S20	X333	3394	ORGANOMETALLIC SUBSTANCE, LIQUID, PYROPHORIC, WATER-REACTIVE
S10AN L10DH	TU4 TU14 TU22 TE21 TM2	AT	1 (B/E)	V1		CV23	S20	X423	3395	ORGANOMETALLIC SUBSTANCE, SOLID, WATER-REACTIVE
SGAN L4DH	TU14 TE21 TM2	AT	2 (D/E)	V1		CV23		423	3395	ORGANOMETALLIC SUBSTANCE, SOLID, WATER-REACTIVE
SGAN L4DH	TU14 TE21 TM2	AT	3 (E)	V1		CV23		423	3395	ORGANOMETALLIC SUBSTANCE, SOLID, WATER-REACTIVE
S10AN L10DH	TU4 TU14 TU22 TE21 TM2	AT	0 (B/E)	V1		CV23	S20	X423	3396	ORGANOMETALLIC SUBSTANCE, SOLID, WATER-REACTIVE, FLAMMABLE
SGAN L4DH	TU14 TE21 TM2	AT	0 (D/E)	V1		CV23		423	3396	ORGANOMETALLIC SUBSTANCE, SOLID, WATER-REACTIVE, FLAMMABLE
SGAN L4DH	TU14 TE21 TM2	AT	0 (E)	V1		CV23		423	3396	ORGANOMETALLIC SUBSTANCE, SOLID, WATER-REACTIVE, FLAMMABLE
S10AN L10DH	TU14 TE21 TM2	AT	1 (B/E)	V1		CV23	S20	X423	3397	ORGANOMETALLIC SUBSTANCE, SOLID, WATER-REACTIVE, SELF-HEATING
SGAN L4DH		AT	2 (D/E)	V1		CV23		423	3397	ORGANOMETALLIC SUBSTANCE, SOLID, WATER-REACTIVE, SELF-HEATING
SGAN L4DH		AT	3 (E)	V1		CV23		423	3397	ORGANOMETALLIC SUBSTANCE, SOLID, WATER-REACTIVE, SELF-HEATING
L10DH	TU4 TU14 TU22 TE21 TM2	AT	0 (B/E)	V1		CV23	S20	X323	3398	ORGANOMETALLIC SUBSTANCE, LIQUID, WATER-REACTIVE
L4DH	TU14 TE21 TM2	AT	0 (D/E)	V1		CV23		323	3398	ORGANOMETALLIC SUBSTANCE, LIQUID, WATER-REACTIVE

UN No.	Name and description	Class	Classification code	Packing group	Labels	Special provisions	Limited and excepted quantities		Packaging			Portable tanks and bulk containers	
									Packing instructions	Special packing provisions	Mixed packing provisions	Instructions	Special provisions
	3.1.2	2.2	2.2	2.1.1.3	5.2.2	3.3	3.4	3.5.1.2	4.1.4	4.1.4	4.1.10	4.2.5.2 7.3.2	4.2.5.3
(1)	(2)	(3a)	(3b)	(4)	(5)	(6)	(7a)	(7b)	(8)	(9a)	(9b)	(10)	(11)
3398	ORGANOMETALLIC SUBSTANCE, LIQUID, WATER-REACTIVE	4.3	W1	III	4.3	274	1 L	E1	P001 IBC02		MP15	T7	TP2 TP7 TP36 TP41
3399	ORGANOMETALLIC SUBSTANCE, LIQUID, WATER-REACTIVE, FLAMMABLE	4.3	WF1	I	4.3 +3	274	0	E0	P402		MP2	T13	TP2 TP7 TP36 TP41
3399	ORGANOMETALLIC SUBSTANCE, LIQUID, WATER-REACTIVE, FLAMMABLE	4.3	WF1	II	4.3 +3	274	500 ml	E2	P001 IBC01		MP15	T7	TP2 TP7 TP36 TP41
3399	ORGANOMETALLIC SUBSTANCE, LIQUID, WATER-REACTIVE, FLAMMABLE	4.3	WF1	III	4.3 +3	274	1 L	E1	P001 IBC02 R001		MP15	T7	TP2 TP7 TP36 TP41
3400	ORGANOMETALLIC SUBSTANCE, SOLID, SELF-HEATING	4.2	S5	II	4.2	274	500 g	E2	P410 IBC06		MP14	T3	TP33 TP36
3400	ORGANOMETALLIC SUBSTANCE, SOLID, SELF-HEATING	4.2	S5	III	4.2	274	1 kg	E1	P002 IBC08		MP14	T1	TP33 TP36
3401	ALKALI METAL AMALGAM, SOLID	4.3	W2	I	4.3	182	0	E0	P403		MP2	T9	TP7 TP33
3402	ALKALINE EARTH METAL AMALGAM, SOLID	4.3	W2	I	4.3	183 506	0	E0	P403		MP2	T9	TP7 TP33
3403	POTASSIUM METAL ALLOYS, SOLID	4.3	W2	I	4.3		0	E0	P403		MP2	T9	TP7 TP33
3404	POTASSIUM SODIUM ALLOYS, SOLID	4.3	W2	I	4.3		0	E0	P403		MP2	T9	TP7 TP33
3405	BARIUM CHLORATE SOLUTION	5.1	OT1	II	5.1 +6.1		1 L	E2	P504 IBC02		MP2	T4	TP1
3405	BARIUM CHLORATE SOLUTION	5.1	OT1	III	5.1 +6.1		5 L	E1	P001 IBC02		MP2	T4	TP1
3406	BARIUM PERCHLORATE SOLUTION	5.1	OT1	II	5.1 +6.1		1 L	E2	P504 IBC02		MP2	T4	TP1
3406	BARIUM PERCHLORATE SOLUTION	5.1	OT1	III	5.1 +6.1		5 L	E1	P001 IBC02		MP2	T4	TP1
3407	CHLORATE AND MAGNESIUM CHLORIDE MIXTURE SOLUTION	5.1	O1	II	5.1		1 L	E2	P504 IBC02		MP2	T4	TP1
3407	CHLORATE AND MAGNESIUM CHLORIDE MIXTURE SOLUTION	5.1	O1	III	5.1		5 L	E1	P504 IBC02		MP2	T4	TP1
3408	LEAD PERCHLORATE SOLUTION	5.1	OT1	II	5.1 +6.1		1 L	E2	P504 IBC02		MP2	T4	TP1
3408	LEAD PERCHLORATE SOLUTION	5.1	OT1	III	5.1 +6.1		5 L	E1	P001 IBC02		MP2	T4	TP1
3409	CHLORONITROBENZENES, LIQUID	6.1	T1	II	6.1	279	100 ml	E4	P001 IBC02		MP15	T7	TP2
3410	4-CHLORO-o-TOLUIDINE HYDROCHLORIDE SOLUTION	6.1	T1	III	6.1		5 L	E1	P001 IBC03 R001		MP19	T4	TP1
3411	beta-NAPHTHYLAMINE SOLUTION	6.1	T1	II	6.1		100 ml	E4	P001 IBC02		MP15	T7	TP2
3411	beta-NAPHTHYLAMINE SOLUTION	6.1	T1	III	6.1		5 L	E1	P001 IBC02		MP19	T7	TP2
3412	FORMIC ACID with not less than 10% but not more than 85% acid by mass	8	C3	II	8		1 L	E2	P001 IBC02		MP15	T7	TP2
3412	FORMIC ACID with not less than 5% but less than 10% acid by mass	8	C3	III	8		5 L	E1	P001 IBC03 LP01 R001		MP19	T4	TP1
3413	POTASSIUM CYANIDE SOLUTION	6.1	T4	I	6.1		0	E5	P001		MP8 MP17	T14	TP2
3413	POTASSIUM CYANIDE SOLUTION	6.1	T4	II	6.1		100 ml	E4	P001 IBC02		MP15	T11	TP2 TP27
3413	POTASSIUM CYANIDE SOLUTION	6.1	T4	III	6.1		5 L	E1	P001 IBC03 LP01 R001		MP19	T7	TP2 TP28
3414	SODIUM CYANIDE SOLUTION	6.1	T4	I	6.1		0	E5	P001		MP8 MP17	T14	TP2

ADR tank		Vehicle for tank carriage	Transport category (Tunnel restriction code)	Special provisions for carriage				Hazard identification No.	UN No.	Name and description
Tank code	Special provisions			Packages	Bulk	Loading, unloading and handling	Operation			
4.3	4.3.5, 6.8.4	9.1.1.2	1.1.3.6 (8.6)	7.2.4	7.3.3	7.5.11	8.5	5.3.2.3		3.1.2
(12)	(13)	(14)	(15)	(16)	(17)	(18)	(19)	(20)	(1)	(2)
L4DH	TU14 TE21 TM2	AT	0 (E)	V1		CV23		323	3398	ORGANOMETALLIC SUBSTANCE, LIQUID, WATER-REACTIVE
L10DH	TU4 TU14 TU22 TE21 TM2	FL	0 (B/E)	V1		CV23	S2 S20	X323	3399	ORGANOMETALLIC SUBSTANCE, LIQUID, WATER-REACTIVE, FLAMMABLE
L4DH	TU4 TU14 TU22 TE21 TM2	FL	0 (D/E)	V1		CV23	S2	323	3399	ORGANOMETALLIC SUBSTANCE, LIQUID, WATER-REACTIVE, FLAMMABLE
L4DH	TU14 TE21 TM2	FL	0 (E)	V1		CV23	S2	323	3399	ORGANOMETALLIC SUBSTANCE, LIQUID, WATER-REACTIVE, FLAMMABLE
SGAN L4BN		AT	2 (D/E)	V1				40	3400	ORGANOMETALLIC SUBSTANCE, SOLID, SELF-HEATING
SGAN L4BN		AT	3 (E)	V1				40	3400	ORGANOMETALLIC SUBSTANCE, SOLID, SELF-HEATING
L10BN(+)	TU1 TE5 TT3 TM2	AT	1 (B/E)	V1		CV23	S20	X423	3401	ALKALI METAL AMALGAM, SOLID
L10BN(+)	TU1 TE5 TT3 TM2	AT	1 (B/E)	V1		CV23	S20	X423	3402	ALKALINE EARTH METAL AMALGAM, SOLID
L10BN(+)	TU1 TE5 TT3 TM2	AT	1 (B/E)	V1		CV23	S20	X423	3403	POTASSIUM METAL ALLOYS, SOLID
L10BN(+)	TU1 TE5 TT3 TM2	AT	1 (B/E)	V1		CV23	S20	X423	3404	POTASSIUM SODIUM ALLOYS, SOLID
L4BN	TU3	AT	2 (E)			CV24 CV28		56	3405	BARIUM CHLORATE SOLUTION
LGBV	TU3	AT	3 (E)			CV24 CV28		56	3405	BARIUM CHLORATE SOLUTION
L4BN	TU3	AT	2 (E)			CV24 CV28		56	3406	BARIUM PERCHLORATE SOLUTION
LGBV	TU3	AT	3 (E)			CV24 CV28		56	3406	BARIUM PERCHLORATE SOLUTION
L4BN	TU3	AT	2 (E)			CV24		50	3407	CHLORATE AND MAGNESIUM CHLORIDE MIXTURE SOLUTION
LGBV	TU3	AT	3 (E)			CV24		50	3407	CHLORATE AND MAGNESIUM CHLORIDE MIXTURE SOLUTION
L4BN	TU3	AT	2 (E)			CV24 CV28		56	3408	LEAD PERCHLORATE SOLUTION
LGBV	TU3	AT	3 (E)			CV24 CV28		56	3408	LEAD PERCHLORATE SOLUTION
L4BH	TU15 TE19	AT	2 (D/E)			CV13 CV28	S9 S19	60	3409	CHLORONITROBENZENES, LIQUID
L4BH	TU15 TE19	AT	2 (E)	V12		CV13 CV28	S9	60	3410	4-CHLORO-o-TOLUIDINE HYDROCHLORIDE SOLUTION
L4BH	TU15 TE19	AT	2 (D/E)			CV13 CV28	S9 S19	60	3411	beta-NAPHTHYLAMINE SOLUTION
L4BH	TU15 TE19	AT	2 (E)			CV13 CV28	S9	60	3411	beta-NAPHTHYLAMINE SOLUTION
L4BN		AT	2 (E)					80	3412	FORMIC ACID with not less than 10% but not more than 85% acid by mass
L4BN		AT	3 (E)	V12				80	3412	FORMIC ACID with not less than 5% but less than 10% acid by mass
L10CH	TU14 TU15 TE19 TE21	AT	1 (C/E)			CV1 CV13 CV28	S9 S14	66	3413	POTASSIUM CYANIDE SOLUTION
L4BH	TU15 TE19	AT	2 (D/E)			CV13 CV28	S9 S19	60	3413	POTASSIUM CYANIDE SOLUTION
L4BH	TU15 TE19	AT	2 (E)	V12		CV13 CV28	S9	60	3413	POTASSIUM CYANIDE SOLUTION
L10CH	TU14 TU15 TE19 TE21	AT	1 (C/E)			CV1 CV13 CV28	S9 S14	66	3414	SODIUM CYANIDE SOLUTION

UN No.	Name and description	Class	Classifi-cation code	Packing group	Labels	Special provi-sions	Limited and excepted quantities		Packaging			Portable tanks and bulk containers	
									Packing instruc-tions	Special packing provisions	Mixed packing provisions	Instruc-tions	Special provisions
	3.1.2	2.2	2.2	2.1.1.3	5.2.2	3.3	3.4	3.5.1.2	4.1.4	4.1.4	4.1.10	4.2.5.2 7.3.2	4.2.5.3
(1)	(2)	(3a)	(3b)	(4)	(5)	(6)	(7a)	(7b)	(8)	(9a)	(9b)	(10)	(11)
3414	SODIUM CYANIDE SOLUTION	6.1	T4	II	6.1		100 ml	E4	P001 IBC02		MP15	T11	TP2 TP27
3414	SODIUM CYANIDE SOLUTION	6.1	T4	III	6.1		5 L	E1	P001 IBC03 LP01 R001		MP19	T7	TP2 TP28
3415	SODIUM FLUORIDE SOLUTION	6.1	T4	III	6.1		5 L	E1	P001 IBC03 LP01 R001		MP19	T4	TP1
3416	CHLOROACETO-PHENONE, LIQUID	6.1	T1	II	6.1		0	E0	P001 IBC02		MP15	T7	TP2
3417	XYLYL BROMIDE, SOLID	6.1	T2	II	6.1		0	E4	P002 IBC08	B4	MP10	T3	TP33
3418	2,4-TOLUYLENEDIAMINE SOLUTION	6.1	T1	III	6.1		5 L	E1	P001 IBC03 LP01 R001		MP19	T4	TP1
3419	BORON TRIFLUORIDE ACETIC ACID COMPLEX, SOLID	8	C4	II	8		1 kg	E2	P002 IBC08	B4	MP10	T3	TP33
3420	BORON TRIFLUORIDE PROPIONIC ACID COMPLEX, SOLID	8	C4	II	8		1 kg	E2	P002 IBC08	B4	MP10	T3	TP33
3421	POTASSIUM HYDROGENDIFLUORIDE SOLUTION	8	CT1	II	8 +6.1		1 L	E2	P001 IBC02		MP15	T7	TP2
3421	POTASSIUM HYDROGENDIFLUORIDE SOLUTION	8	CT1	III	8 +6.1		5 L	E1	P001 IBC03 R001		MP19	T4	TP1
3422	POTASSIUM FLUORIDE SOLUTION	6.1	T4	III	6.1		5 L	E1	P001 IBC03 LP01 R001		MP19	T4	TP1
3423	TETRAMETHYL-AMMONIUM HYDROXIDE, SOLID	8	C8	II	8		1 kg	E2	P002 IBC08	B4	MP10	T3	TP33
3424	AMMONIUM DINITRO-o-CRESOLATE SOLUTION	6.1	T1	II	6.1		100 ml	E4	P001 IBC02		MP15	T7	TP2
3424	AMMONIUM DINITRO-o-CRESOLATE SOLUTION	6.1	T1	III	6.1		5 L	E1	P001 IBC02		MP19	T7	TP2
3425	BROMOACETIC ACID, SOLID	8	C4	II	8		1 kg	E2	P002 IBC08	B4	MP10	T3	TP33
3426	ACRYLAMIDE SOLUTION	6.1	T1	III	6.1		5 L	E1	P001 IBC03 LP01 R001		MP19	T4	TP1
3427	CHLOROBENZYL CHLORIDES, SOLID	6.1	T2	III	6.1		5 kg	E1	P002 IBC08 LP02 R001	B3	MP10	T1	TP33
3428	3-CHLORO-4-METHYLPHENYL ISOCYANATE, SOLID	6.1	T2	II	6.1		500 g	E4	P002 IBC08	B4	MP10	T3	TP33
3429	CHLOROTOLUIDINES, LIQUID	6.1	T1	III	6.1		5 L	E1	P001 IBC03 LP01 R001		MP19	T4	TP1
3430	XYLENOLS, LIQUID	6.1	T1	II	6.1		100 ml	E4	P001 IBC02		MP15	T7	TP2
3431	NITROBENZO-TRIFLUORIDES, SOLID	6.1	T2	II	6.1		500 g	E4	P002 IBC08	B4	MP10	T3	TP33
3432	POLYCHLORINATED BIPHENYLS, SOLID	9	M2	II	9	305	1 kg	E2	P906 IBC08	B4	MP10	T3	TP33
3434	NITROCRESOLS, LIQUID	6.1	T1	III	6.1		5 L	E1	P001 IBC03 LP01 R001		MP19	T4	TP1
3436	HEXAFLUOROACETONE HYDRATE, SOLID	6.1	T2	II	6.1		500 g	E4	P002 IBC08	B4	MP10	T3	TP33
3437	CHLOROCRESOLS, SOLID	6.1	T2	II	6.1		500 g	E4	P002 IBC08	B4	MP10	T3	TP33
3438	alpha-METHYLBENZYL ALCOHOL, SOLID	6.1	T2	III	6.1		5 kg	E1	P002 IBC08 LP02 R001	B3	MP10	T1	TP33

ADR tank		Vehicle for tank carriage	Transport category (Tunnel restriction code)	Special provisions for carriage				Hazard identifi-cation No.	UN No.	Name and description
Tank code	Special provisions			Packages	Bulk	Loading, unloading and handling	Operation			
4.3	4.3.5, 6.8.4	9.1.1.2	1.1.3.6 (8.6)	7.2.4	7.3.3	7.5.11	8.5	5.3.2.3		3.1.2
(12)	(13)	(14)	(15)	(16)	(17)	(18)	(19)	(20)	(1)	(2)
L4BH	TU15 TE19	AT	2 (D/E)			CV13 CV28	S9 S19	60	3414	SODIUM CYANIDE SOLUTION
L4BH	TU15 TE19	AT	2 (E)	V12		CV13 CV28	S9	60	3414	SODIUM CYANIDE SOLUTION
L4BH	TU15 TE19	AT	2 (E)	V12		CV13 CV28	S9	60	3415	SODIUM FLUORIDE SOLUTION
L4BH	TU15 TE19	AT	2 (D/E)			CV13 CV28	S9 S19	60	3416	CHLOROACETO-PHENONE, LIQUID
SGAH L4BH	TU15 TE19	AT	2 (D/E)	V11		CV13 CV28	S9 S19	60	3417	XYLYL BROMIDE, SOLID
L4BH	TU15 TE19	AT	2 (E)	V12		CV13 CV28	S9	60	3418	2,4-TOLUYLENEDIAMINE SOLUTION
SGAN L4BN		AT	2 (E)	V11				80	3419	BORON TRIFLUORIDE ACETIC ACID COMPLEX, SOLID
SGAN L4BN		AT	2 (E)	V11				80	3420	BORON TRIFLUORIDE PROPIONIC ACID COMPLEX, SOLID
L4DH	TU14 TE21	AT	2 (E)			CV13 CV28		86	3421	POTASSIUM HYDROGENDIFLUORIDE SOLUTION
L4DH	TU14 TE21	AT	3 (E)	V12		CV13 CV28		86	3421	POTASSIUM HYDROGENDIFLUORIDE SOLUTION
L4BH	TU15 TE19	AT	2 (E)	V12		CV13 CV28	S9	60	3422	POTASSIUM FLUORIDE SOLUTION
SGAN L4BN		AT	2 (E)	V11				80	3423	TETRAMETHYL-AMMONIUM HYDROXIDE, SOLID
L4BH	TU15 TE19	AT	2 (D/E)			CV13 CV28	S9 S19	60	3424	AMMONIUM DINITRO-o-CRESOLATE SOLUTION
L4BH	TU15 TE19	AT	2 (E)			CV13 CV28	S9	60	3424	AMMONIUM DINITRO-o-CRESOLATE SOLUTION
SGAN L4BN		AT	2 (E)	V11				80	3425	BROMOACETIC ACID, SOLID
L4BH	TU15 TE19	AT	2 (E)	V12		CV13 CV28	S9	60	3426	ACRYLAMIDE SOLUTION
SGAH L4BH	TU15 TE19	AT	2 (E)		VC1 VC2 AP7	CV13 CV28	S9	60	3427	CHLOROBENZYL CHLORIDES, SOLID
SGAH L4BH	TU15 TE19	AT	2 (D/E)	V11		CV13 CV28	S9 S19	60	3428	3-CHLORO-4-METHYLPHENYL ISOCYANATE, SOLID
L4BH	TU15 TE19	AT	2 (E)	V12		CV13 CV28	S9	60	3429	CHLOROTOLUIDINES, LIQUID
L4BH	TU15 TE19	AT	2 (D/E)			CV13 CV28	S9 S19	60	3430	XYLENOLS, LIQUID
SGAH L4BH	TU15 TE19	AT	2 (D/E)	V11		CV13 CV28	S9 S19	60	3431	NITROBENZO-TRIFLUORIDES, SOLID
S4AH L4BH	TU15	AT	0 (D/E)	V11	VC1 VC2 AP9	CV1 CV13 CV28	S19	90	3432	POLYCHLORINATED BIPHENYLS, SOLID
L4BH	TU15 TE19	AT	2 (E)	V12		CV13 CV28	S9	60	3434	NITROCRESOLS, LIQUID
SGAH L4BH	TU15 TE19	AT	2 (D/E)	V11		CV13 CV28	S9 S19	60	3436	HEXAFLUOROACETONE HYDRATE, SOLID
SGAH L4BH	TU15 TE19	AT	2 (D/E)	V11		CV13 CV28	S9 S19	60	3437	CHLOROCRESOLS, SOLID
SGAH L4BH	TU15 TE19	AT	2 (E)		VC1 VC2 AP7	CV13 CV28	S9	60	3438	alpha-METHYLBENZYL ALCOHOL, SOLID

UN No.	Name and description	Class	Classifi-cation code	Packing group	Labels	Special provi-sions	Limited and excepted quantities		Packaging			Portable tanks and bulk containers	
									Packing instruc-tions	Special packing provisions	Mixed packing provisions	Instruc-tions	Special provisions
3.1.2	3.1.2	2.2	2.2	2.1.1.3	5.2.2	3.3	3.4	3.5.1.2	4.1.4	4.1.4	4.1.10	4.2.5.2 7.3.2	4.2.5.3
(1)	(2)	(3a)	(3b)	(4)	(5)	(6)	(7a)	(7b)	(8)	(9a)	(9b)	(10)	(11)
3439	NITRILES, SOLID, TOXIC, N.O.S.	6.1	T2	I	6.1	274	0	E5	P002 IBC07		MP18	T6	TP33
3439	NITRILES, SOLID, TOXIC, N.O.S.	6.1	T2	II	6.1	274	500 g	E4	P002 IBC08	B4	MP10	T3	TP33
3439	NITRILES, SOLID, TOXIC, N.O.S.	6.1	T2	III	6.1	274	5 kg	E1	P002 IBC08 LP02 R001	B3	MP10	T1	TP33
3440	SELENIUM COMPOUND, LIQUID, N.O.S.	6.1	T4	I	6.1	274 563	0	E5	P001		MP8 MP17	T14	TP2 TP27
3440	SELENIUM COMPOUND, LIQUID, N.O.S.	6.1	T4	II	6.1	274 563	100 ml	E4	P001 IBC02		MP15	T11	TP2 TP27
3440	SELENIUM COMPOUND, LIQUID, N.O.S.	6.1	T4	III	6.1	274 563	5 L	E1	P001 IBC03 R001		MP19	T7	TP1 TP28
3441	CHLORODINITROBENZENES, SOLID	6.1	T2	II	6.1	279	500 g	E4	P002 IBC08	B4	MP10	T3	TP33
3442	DICHLOROANILINES, SOLID	6.1	T2	II	6.1	279	500 g	E4	P002 IBC08	B4	MP10	T3	TP33
3443	DINITROBENZENES, SOLID	6.1	T2	II	6.1		500 g	E4	P002 IBC08	B4	MP10	T3	TP33
3444	NICOTINE HYDROCHLORIDE, SOLID	6.1	T2	II	6.1	43	500 g	E4	P002 IBC08	B4	MP10	T3	TP33
3445	NICOTINE SULPHATE, SOLID	6.1	T2	II	6.1		500 g	E4	P002 IBC08	B4	MP10	T3	TP33
3446	NITROTOLUENES, SOLID	6.1	T2	II	6.1		500 g	E4	P002 IBC08	B4	MP10	T3	TP33
3447	NITROXYLENES, SOLID	6.1	T2	II	6.1		500 g	E4	P002 IBC08	B4	MP10	T3	TP33
3448	TEAR GAS SUBSTANCE, SOLID, N.O.S.	6.1	T2	I	6.1	274	0	E0	P002		MP18	T6	TP33
3448	TEAR GAS SUBSTANCE, SOLID, N.O.S.	6.1	T2	II	6.1	274	0	E0	P002 IBC08	B4	MP10	T3	TP33
3449	BROMOBENZYL CYANIDES, SOLID	6.1	T2	I	6.1	138	0	E5	P002		MP18	T6	TP33
3450	DIPHENYLCHLORO-ARSINE, SOLID	6.1	T3	I	6.1		0	E0	P002 IBC07		MP18	T6	TP33
3451	TOLUIDINES, SOLID	6.1	T2	II	6.1	279	500 g	E4	P002 IBC08	B4	MP10	T3	TP33
3452	XYLIDINES, SOLID	6.1	T2	II	6.1		500 g	E4	P002 IBC08	B4	MP10	T3	TP33
3453	PHOSPHORIC ACID, SOLID	8	C2	III	8		5 kg	E1	P002 IBC08 LP02 R001	B3	MP10	T1	TP33
3454	DINITROTOLUENES, SOLID	6.1	T2	II	6.1		500 g	E4	P002 IBC08	B4	MP10	T3	TP33
3455	CRESOLS, SOLID	6.1	TC2	II	6.1 +8		500 g	E4	P002 IBC08	B4	MP10	T3	TP33
3456	NITROSYLSULPHURIC ACID, SOLID	8	C2	II	8		1 kg	E2	P002 IBC08	B4	MP10	T3	TP33
3457	CHLORONITROTOLUENES, SOLID	6.1	T2	III	6.1		5 kg	E1	P002 IBC08 LP02 R001	B3	MP10	T1	TP33
3458	NITROANISOLES, SOLID	6.1	T2	III	6.1	279	5 kg	E1	P002 IBC08 LP02 R001	B3	MP10	T1	TP33
3459	NITROBROMOBENZENES, SOLID	6.1	T2	III	6.1		5 kg	E1	P002 IBC08 LP02 R001	B3	MP10	T1	TP33
3460	N-ETHYLBENZYL-TOLUIDINES, SOLID	6.1	T2	III	6.1		5 kg	E1	P002 IBC08 LP02 R001	B3	MP10	T1	TP33
3462	TOXINS, EXTRACTED FROM LIVING SOURCES, SOLID, N.O.S.	6.1	T2	I	6.1	210 274	0	E5	P002 IBC07		MP18	T6	TP33

ADR tank		Vehicle for tank carriage	Transport category (Tunnel restriction code)	Special provisions for carriage				Hazard identifi-cation No.	UN No.	Name and description
Tank code	Special provisions			Packages	Bulk	Loading, unloading and handling	Operation			
4.3	4.3.5, 6.8.4	9.1.1.2	1.1.3.6 (8.6)	7.2.4	7.3.3	7.5.11	8.5	5.3.2.3		3.1.2
(12)	(13)	(14)	(15)	(16)	(17)	(18)	(19)	(20)	(1)	(2)
S10AH L10CH	TU14 TU15 TE19 TE21	AT	1 (C/E)	V10		CV1 CV13 CV28	S9 S14	66	3439	NITRILES, SOLID, TOXIC, N.O.S.
SGAH L4BH	TU15 TE19	AT	2 (D/E)	V11		CV13 CV28	S9 S19	60	3439	NITRILES, SOLID, TOXIC, N.O.S.
SGAH L4BH	TU15 TE19	AT	2 (E)		VC1 VC2 AP7	CV13 CV28	S9	60	3439	NITRILES, SOLID, TOXIC, N.O.S.
L10CH	TU14 TU15 TE19 TE21	AT	1 (C/E)			CV1 CV13 CV28	S9 S14	66	3440	SELENIUM COMPOUND, LIQUID, N.O.S.
L4BH	TU15 TE19	AT	2 (D/E)			CV13 CV28	S9 S19	60	3440	SELENIUM COMPOUND, LIQUID, N.O.S.
L4BH	TU15 TE19	AT	2 (E)	V12		CV13 CV28	S9	60	3440	SELENIUM COMPOUND, LIQUID, N.O.S.
SGAH L4BH	TU15 TE19	AT	2 (D/E)	V11		CV13 CV28	S9 S19	60	3441	CHLORODINITROBENZENES, SOLID
SGAH L4BH	TU15 TE19	AT	2 (D/E)	V11		CV13 CV28	S9 S19	60	3442	DICHLOROANILINES, SOLID
SGAH L4BH	TU15 TE19	AT	2 (D/E)	V11		CV13 CV28	S9 S19	60	3443	DINITROBENZENES, SOLID
SGAH	TU15 TE19	AT	2 (D/E)	V11		CV13 CV28	S9 S19	60	3444	NICOTINE HYDROCHLORIDE, SOLID
SGAH	TU15 TE19	AT	2 (D/E)	V11		CV13 CV28	S9 S19	60	3445	NICOTINE SULPHATE, SOLID
SGAH L4BH	TU15 TE19	AT	2 (D/E)	V11		CV13 CV28	S9 S19	60	3446	NITROTOLUENES, SOLID
SGAH L4BH	TU15 TE19	AT	2 (D/E)	V11		CV13 CV28	S9 S19	60	3447	NITROXYLENES, SOLID
S10AH L10CH	TU14 TU15 TE19 TE21	AT	1 (C/E)			CV1 CV13 CV28	S9 S14	66	3448	TEAR GAS SUBSTANCE, SOLID, N.O.S.
SGAH L4BH	TU15 TE19	AT	2 (D/E)	V11		CV13 CV28	S9 S19	60	3448	TEAR GAS SUBSTANCE, SOLID, N.O.S.
S10AH L10CH	TU15 TE19	AT	1 (C/E)			CV1 CV13 CV28	S9 S14	66	3449	BROMOBENZYL CYANIDES, SOLID
S10AH L10CH	TU15 TE19	AT	1 (C/E)	V10		CV1 CV13 CV28	S9 S14	66	3450	DIPHENYLCHLORO-ARSINE, SOLID
SGAH L4BH	TU15 TE19	AT	2 (D/E)	V11		CV13 CV28	S9 S19	60	3451	TOLUIDINES, SOLID
SGAH L4BH	TU15 TE19	AT	2 (D/E)	V11		CV13 CV28	S9 S19	60	3452	XYLIDINES, SOLID
SGAV L4BN		AT	3 (E)		VC1 VC2 AP7			80	3453	PHOSPHORIC ACID, SOLID
SGAH L4BH	TU15 TE19	AT	2 (D/E)	V11		CV13 CV28	S9 S19	60	3454	DINITROTOLUENES, SOLID
SGAH L4BH	TU15 TE19	AT	2 (D/E)	V11		CV13 CV28	S9 S19	68	3455	CRESOLS, SOLID
SGAN L4BN		AT	2 (E)	V11				X80	3456	NITROSYLSULPHURIC ACID, SOLID
SGAH L4BH	TU15 TE19	AT	2 (E)		VC1 VC2 AP7	CV13 CV28	S9	60	3457	CHLORONITROTOLUENES, SOLID
SGAH L4BH	TU15 TE19	AT	2 (E)		VC1 VC2 AP7	CV13 CV28	S9	60	3458	NITROANISOLES, SOLID
SGAH L4BH	TU15 TE19	AT	2 (E)		VC1 VC2 AP7	CV13 CV28	S9	60	3459	NITROBROMOBENZENES, SOLID
SGAH L4BH	TU15 TE19	AT	2 (E)		VC1 VC2 AP7	CV13 CV28	S9	60	3460	N-ETHYLBENZYL-TOLUIDINES, SOLID
S10AH L10CH	TU15 TE19	AT	1 (C/E)	V10		CV1 CV13 CV28	S9 S14	66	3462	TOXINS, EXTRACTED FROM LIVING SOURCES, SOLID, N.O.S.

UN No.	Name and description	Class	Classification code	Packing group	Labels	Special provisions	Limited and excepted quantities		Packaging			Portable tanks and bulk containers	
									Packing instructions	Special packing provisions	Mixed packing provisions	Instructions	Special provisions
	3.1.2	2.2	2.2	2.1.1.3	5.2.2	3.3	3.4	3.5.1.2	4.1.4	4.1.4	4.1.10	4.2.5.2 7.3.2	4.2.5.3
(1)	(2)	(3a)	(3b)	(4)	(5)	(6)	(7a)	(7b)	(8)	(9a)	(9b)	(10)	(11)
3462	TOXINS, EXTRACTED FROM LIVING SOURCES, SOLID, N.O.S.	6.1	T2	II	6.1	210 274	500 g	E4	P002 IBC08	B4	MP10	T3	TP33
3462	TOXINS, EXTRACTED FROM LIVING SOURCES, SOLID, N.O.S.	6.1	T2	III	6.1	210 274	5 kg	E1	P002 IBC08 R001	B3	MP10	T1	TP33
3463	PROPIONIC ACID with not less than 90% acid by mass	8	CF1	II	8 +3		1 L	E2	P001 IBC02		MP15	T7	TP2
3464	ORGANOPHOSPHORUS COMPOUND, SOLID, TOXIC, N.O.S.	6.1	T2	I	6.1	43 274	0	E5	P002 IBC07		MP18	T6	TP33
3464	ORGANOPHOSPHORUS COMPOUND, SOLID, TOXIC, N.O.S.	6.1	T2	II	6.1	43 274	500 g	E4	P002 IBC08	B4	MP10	T3	TP33
3464	ORGANOPHOSPHORUS COMPOUND, SOLID, TOXIC, N.O.S.	6.1	T2	III	6.1	43 274	5 kg	E1	P002 IBC08 LP02 R001	B3	MP10	T1	TP33
3465	ORGANOARSENIC COMPOUND, SOLID, N.O.S.	6.1	T3	I	6.1	274	0	E5	P002 IBC07		MP18	T6	TP33
3465	ORGANOARSENIC COMPOUND, SOLID, N.O.S.	6.1	T3	II	6.1	274	500 g	E4	P002 IBC08	B4	MP10	T3	TP33
3465	ORGANOARSENIC COMPOUND, SOLID, N.O.S.	6.1	T3	III	6.1	274	5 kg	E1	P002 IBC08 LP02 R001	B3	MP10	T1	TP33
3466	METAL CARBONYLS, SOLID, N.O.S.	6.1	T3	I	6.1	274 562	0	E5	P002 IBC07		MP18	T6	TP33
3466	METAL CARBONYLS, SOLID, N.O.S.	6.1	T3	II	6.1	274 562	500 g	E4	P002 IBC08	B4	MP10	T3	TP33
3466	METAL CARBONYLS, SOLID, N.O.S.	6.1	T3	III	6.1	274 562	5 kg	E1	P002 IBC08 LP02 R001	B3	MP10	T1	TP33
3467	ORGANOMETALLIC COMPOUND, SOLID, TOXIC, N.O.S.	6.1	T3	I	6.1	274 562	0	E5	P002 IBC07		MP18	T6	TP33
3467	ORGANOMETALLIC COMPOUND, SOLID, TOXIC, N.O.S.	6.1	T3	II	6.1	274 562	500 g	E4	P002 IBC08	B4	MP10	T3	TP33
3467	ORGANOMETALLIC COMPOUND, SOLID, TOXIC, N.O.S.	6.1	T3	III	6.1	274 562	5 kg	E1	P002 IBC08 LP02 R001	B3	MP10	T1	TP33
3468	HYDROGEN IN A METAL HYDRIDE STORAGE SYSTEM or HYDROGEN IN A METAL HYDRIDE STORAGE SYSTEM CONTAINED IN EQUIPMENT or HYDROGEN IN A METAL HYDRIDE STORAGE SYSTEM PACKED WITH EQUIPMENT	2	1F		2.1	321 356	0	E0	P205		MP9		
3469	PAINT, FLAMMABLE, CORROSIVE (including paint, lacquer, enamel, stain, shellac, varnish, polish, liquid filler and liquid lacquer base) or PAINT RELATED MATERIAL, FLAMMABLE, CORROSIVE (including paint thinning and reducing compound)	3	FC	I	3 +8	163 367	0	E0	P001		MP7 MP17	T11	TP2 TP27
3469	PAINT, FLAMMABLE, CORROSIVE (including paint, lacquer, enamel, stain, shellac, varnish, polish, liquid filler and liquid lacquer base) or PAINT RELATED MATERIAL, FLAMMABLE, CORROSIVE (including paint thinning and reducing compound)	3	FC	II	3 +8	163 367	1 L	E2	P001 IBC02		MP19	T7	TP2 TP8 TP28

ADR tank		Vehicle for tank carriage	Transport category (Tunnel restriction code)	Special provisions for carriage				Hazard identification No.	UN No.	Name and description
Tank code	Special provisions			Packages	Bulk	Loading, unloading and handling	Operation			
4.3	4.3.5, 6.8.4	9.1.1.2	1.1.3.6 (8.6)	7.2.4	7.3.3	7.5.11	8.5	5.3.2.3		3.1.2
(12)	(13)	(14)	(15)	(16)	(17)	(18)	(19)	(20)	(1)	(2)
SGAH L4BH	TU15 TE19	AT	2 (D/E)	V11		CV13 CV28	S9 S19	60	3462	TOXINS, EXTRACTED FROM LIVING SOURCES, SOLID, N.O.S.
SGAH L4BH	TU15 TE19	AT	2 (E)		VC1 VC2 AP7	CV13 CV28	S9	60	3462	TOXINS, EXTRACTED FROM LIVING SOURCES, SOLID, N.O.S.
L4BN		FL	2 (D/E)				S2	83	3463	PROPIONIC ACID with not less than 90% acid by mass
S10AH L10CH	TU14 TU15 TE19 TE21	AT	1 (C/E)	V10		CV1 CV13 CV28	S9 S14	66	3464	ORGANOPHOSPHORUS COMPOUND, SOLID, TOXIC, N.O.S.
SGAH L4BH	TU15 TE19	AT	2 (D/E)	V11		CV13 CV28	S9 S19	60	3464	ORGANOPHOSPHORUS COMPOUND, SOLID, TOXIC, N.O.S.
SGAH L4BH	TU15 TE19	AT	2 (E)		VC1 VC2 AP7	CV13 CV28	S9	60	3464	ORGANOPHOSPHORUS COMPOUND, SOLID, TOXIC, N.O.S.
S10AH L10CH	TU14 TU15 TE19 TE21	AT	1 (C/E)	V10		CV1 CV13 CV28	S9 S14	66	3465	ORGANOARSENIC COMPOUND, SOLID, N.O.S.
SGAH L4BH	TU15 TE19	AT	2 (D/E)	V11		CV13 CV28	S9 S19	60	3465	ORGANOARSENIC COMPOUND, SOLID, N.O.S.
SGAH L4BH	TU15 TE19	AT	2 (E)		VC1 VC2 AP7	CV13 CV28	S9	60	3465	ORGANOARSENIC COMPOUND, SOLID, N.O.S.
S10AH L10CH	TU14 TU15 TE19 TE21	AT	1 (C/E)	V10		CV1 CV13 CV28	S9 S14	66	3466	METAL CARBONYLS, SOLID, N.O.S.
SGAH L4BH	TU15 TE19	AT	2 (D/E)	V11		CV13 CV28	S9 S19	60	3466	METAL CARBONYLS, SOLID, N.O.S.
SGAH L4BH	TU15 TE19	AT	2 (E)		VC1 VC2 AP7	CV13 CV28	S9	60	3466	METAL CARBONYLS, SOLID, N.O.S.
S10AH L10CH	TU14 TU15 TE19 TE21	AT	1 (C/E)	V10		CV1 CV13 CV28	S9 S14	66	3467	ORGANOMETALLIC COMPOUND, SOLID, TOXIC, N.O.S.
SGAH L4BH	TU15 TE19	AT	2 (D/E)	V11		CV13 CV28	S9 S19	60	3467	ORGANOMETALLIC COMPOUND, SOLID, TOXIC, N.O.S.
SGAH L4BH	TU15 TE19	AT	2 (E)		VC1 VC2 AP7	CV13 CV28	S9	60	3467	ORGANOMETALLIC COMPOUND, SOLID, TOXIC, N.O.S.
			2 (D)			CV9 CV10 CV36	S2 S20		3468	HYDROGEN IN A METAL HYDRIDE STORAGE SYSTEM or HYDROGEN IN A METAL HYDRIDE STORAGE SYSTEM CONTAINED IN EQUIPMENT or HYDROGEN IN A METAL HYDRIDE STORAGE SYSTEM PACKED WITH EQUIPMENT
L10CH	TU14 TE21	FL	1 (C/E)				S2 S20	338	3469	PAINT, FLAMMABLE, CORROSIVE (including paint, lacquer, enamel, stain, shellac, varnish, polish, liquid filler and liquid lacquer base) or PAINT RELATED MATERIAL, FLAMMABLE, CORROSIVE (including paint thinning and reducing compound)
L4BH		FL	2 (D/E)				S2 S20	338	3469	PAINT, FLAMMABLE, CORROSIVE (including paint, lacquer, enamel, stain, shellac, varnish, polish, liquid filler and liquid lacquer base) or PAINT RELATED MATERIAL, FLAMMABLE, CORROSIVE (including paint thinning and reducing compound)

UN No.	Name and description	Class	Classifi-cation code	Packing group	Labels	Special provi-sions	Limited and excepted quantities		Packaging			Portable tanks and bulk containers	
									Packing instruc-tions	Special packing provisions	Mixed packing provisions	Instruc-tions	Special provisions
	3.1.2	2.2	2.2	2.1.1.3	5.2.2	3.3	3.4	3.5.1.2	4.1.4	4.1.4	4.1.10	4.2.5.2 7.3.2	4.2.5.3
(1)	(2)	(3a)	(3b)	(4)	(5)	(6)	(7a)	(7b)	(8)	(9a)	(9b)	(10)	(11)
3469	PAINT, FLAMMABLE, CORROSIVE (including paint, lacquer, enamel, stain, shellac, varnish, polish, liquid filler and liquid lacquer base) or PAINT RELATED MATERIAL, FLAMMABLE, CORROSIVE (including paint thinning and reducing compound)	3	FC	III	3 +8	163 367	5 L	E1	P001 IBC03 R001		MP19	T4	TP1 TP29
3470	PAINT, CORROSIVE, FLAMMABLE (including paint, lacquer, enamel, stain, shellac, varnish, polish, liquid filler and liquid lacquer base) or PAINT RELATED MATERIAL, CORROSIVE, FLAMMABLE (including paint thinning and reducing compound)	8	CF1	II	8 +3	163 367	1 L	E2	P001 IBC02		MP15	T7	TP2 TP8 TP28
3471	HYDROGENDIFLUORIDES SOLUTION, N.O.S.	8	CT1	II	8 +6.1		1 L	E2	P001 IBC02		MP15	T7	TP2
3471	HYDROGENDIFLUORIDES SOLUTION, N.O.S.	8	CT1	III	8 +6.1		5 L	E1	P001 IBC03 R001		MP19	T4	TP1
3472	CROTONIC ACID, LIQUID	8	C3	III	8		5 L	E1	P001 IBC03 LP01 R001		MP19	T4	TP1
3473	FUEL CELL CARTRIDGES or FUEL CELL CARTRIDGES CONTAINED IN EQUIPMENT or FUEL CELL CARTRIDGES PACKED WITH EQUIPMENT containing flammable liquids	3	F3		3	328	1 L	E0	P004				
3474	1-HYDROXYBENZOTRIAZOLE MONOHYDRATE	4.1	D	I	4.1		0	E0	P406	PP48	MP2		
3475	ETHANOL AND GASOLINE MIXTURE or ETHANOL AND MOTOR SPIRIT MIXTURE or ETHANOL AND PETROL MIXTURE, with more than 10% ethanol	3	F1	II	3	333 664	1 L	E2	P001 IBC02		MP19	T4	TP1
3476	FUEL CELL CARTRIDGES or FUEL CELL CARTRIDGES CONTAINED IN EQUIPMENT or FUEL CELL CARTRIDGES PACKED WITH EQUIPMENT, containing water-reactive substances	4.3	W3		4.3	328 334	500 ml or 500 g	E0	P004				
3477	FUEL CELL CARTRIDGES or FUEL CELL CARTRIDGES CONTAINED IN EQUIPMENT or FUEL CELL CARTRIDGES PACKED WITH EQUIPMENT, containing corrosive substances	8	C11		8	328 334	1 L or 1 kg	E0	P004				
3478	FUEL CELL CARTRIDGES or FUEL CELL CARTRIDGES CONTAINED IN EQUIPMENT or FUEL CELL CARTRIDGES PACKED WITH EQUIPMENT, containing liquefied flammable gas	2	6F		2.1	328 338	120 ml	E0	P004				
3479	FUEL CELL CARTRIDGES or FUEL CELL CARTRIDGES CONTAINED IN EQUIPMENT or FUEL CELL CARTRIDGES PACKED WITH EQUIPMENT, containing hydrogen in metal hydride	2	6F		2.1	328 339	120 ml	E0	P004				
3480	LITHIUM ION BATTERIES (including lithium ion polymer batteries)	9	M4		9A	188 230 310 348 376 377 636	0	E0	P903 P908 P909 P910 LP903 LP904				

ADR tank		Vehicle for tank carriage	Transport category (Tunnel restriction code)	Special provisions for carriage				Hazard identification No.	UN No.	Name and description
Tank code	Special provisions			Packages	Bulk	Loading, unloading and handling	Operation			
4.3	4.3.5, 6.8.4	9.1.1.2	1.1.3.6 (8.6)	7.2.4	7.3.3	7.5.11	8.5	5.3.2.3		3.1.2
(12)	(13)	(14)	(15)	(16)	(17)	(18)	(19)	(20)	(1)	(2)
L4BN		FL	3 (D/E)	V12			S2	38	3469	PAINT, FLAMMABLE, CORROSIVE (including paint, lacquer, enamel, stain, shellac, varnish, polish, liquid filler and liquid lacquer base) or PAINT RELATED MATERIAL, FLAMMABLE, CORROSIVE (including paint thinning and reducing compound)
L4BN		FL	2 (D/E)				S2	83	3470	PAINT, CORROSIVE, FLAMMABLE (including paint, lacquer, enamel, stain, shellac, varnish, polish, liquid filler and liquid lacquer base) or PAINT RELATED MATERIAL, CORROSIVE, FLAMMABLE (including paint thinning and reducing compound)
L4DH	TU14 TE21	AT	2 (E)			CV13 CV28		86	3471	HYDROGENDIFLUORIDES SOLUTION, N.O.S.
L4DH	TU14 TE21	AT	3 (E)	V12		CV13 CV28		86	3471	HYDROGENDIFLUORIDES SOLUTION, N.O.S.
L4BN		AT	3 (E)	V12				80	3472	CROTONIC ACID, LIQUID
			3 (E)				S2		3473	FUEL CELL CARTRIDGES or FUEL CELL CARTRIDGES CONTAINED IN EQUIPMENT or FUEL CELL CARTRIDGES PACKED WITH EQUIPMENT containing flammable liquids
			1 (B)				S17		3474	1-HYDROXYBENZOTRIAZOLE MONOHYDRATE
LGBF		FL	2 (D/E)				S2 S20	33	3475	ETHANOL AND GASOLINE MIXTURE or ETHANOL AND MOTOR SPIRIT MIXTURE or ETHANOL AND PETROL MIXTURE, with more than 10% ethanol
			3 (E)	V1		CV23			3476	FUEL CELL CARTRIDGES or FUEL CELL CARTRIDGES CONTAINED IN EQUIPMENT or FUEL CELL CARTRIDGES PACKED WITH EQUIPMENT, containing water-reactive substances
			3 (E)						3477	FUEL CELL CARTRIDGES or FUEL CELL CARTRIDGES CONTAINED IN EQUIPMENT or FUEL CELL CARTRIDGES PACKED WITH EQUIPMENT, containing corrosive substances
			2 (D)			CV9 CV12	S2		3478	FUEL CELL CARTRIDGES or FUEL CELL CARTRIDGES CONTAINED IN EQUIPMENT or FUEL CELL CARTRIDGES PACKED WITH EQUIPMENT, containing liquefied flammable gas
			2 (D)			CV9 CV12	S2		3479	FUEL CELL CARTRIDGES or FUEL CELL CARTRIDGES CONTAINED IN EQUIPMENT or FUEL CELL CARTRIDGES PACKED WITH EQUIPMENT, containing hydrogen in metal hydride
			2 (E)						3480	LITHIUM ION BATTERIES (including lithium ion polymer batteries)

UN No.	Name and description	Class	Classi-fication code	Packing group	Labels	Special provi-sions	Limited and excepted quantities		Packaging			Portable tanks and bulk containers	
									Packing instruc-tions	Special packing provisions	Mixed packing provisions	Instruc-tions	Special provisions
	3.1.2	2.2	2.2	2.1.1.3	5.2.2	3.3	3.4	3.5.1.2	4.1.4	4.1.4	4.1.10	4.2.5.2 7.3.2	4.2.5.3
(1)	(2)	(3a)	(3b)	(4)	(5)	(6)	(7a)	(7b)	(8)	(9a)	(9b)	(10)	(11)
3481	LITHIUM ION BATTERIES CONTAINED IN EQUIPMENT or LITHIUM ION BATTERIES PACKED WITH EQUIPMENT (including lithium ion polymer batteries)	9	M4		9A	188 230 310 348 360 376 377 636	0	E0	P903 P908 P909 P910 LP903 LP904				
3482	ALKALI METAL DISPERSION, FLAMMABLE or ALKALINE EARTH METAL DISPERSION, FLAMMABLE	4.3	WF1	I	4.3 +3	182 183 506	0	E0	P402	RR8	MP2		
3483	MOTOR FUEL ANTI-KNOCK MIXTURE, FLAMMABLE	6.1	TF1	I	6.1 +3		0	E0	P602		MP8 MP17	T14	TP2
3484	HYDRAZINE AQUEOUS SOLUTION, FLAMMABLE with more than 37% hydrazine, by mass	8	CFT	I	8 +3 +6.1	530	0	E0	P001		MP8 MP17	T10	TP2
3485	CALCIUM HYPOCHLORITE, DRY, CORROSIVE or CALCIUM HYPOCHLORITE MIXTURE, DRY, CORROSIVE with more than 39% available chlorine (8.8% available oxygen)	5.1	OC2	II	5.1 +8	314	1 kg	E2	P002 IBC08	B4 B13	MP2		
3486	CALCIUM HYPOCHLORITE MIXTURE, DRY, CORROSIVE with more than 10% but not more than 39% available chlorine	5.1	OC2	III	5.1 +8	314	5 kg	E1	P002 IBC08 LP02 R001	B3 B13 L3	MP2		
3487	CALCIUM HYPOCHLORITE, HYDRATED, CORROSIVE or CALCIUM HYPOCHLORITE, HYDRATED MIXTURE, CORROSIVE with not less than 5.5% but not more than 16% water	5.1	OC2	II	5.1 +8	314 322	1 kg	E2	P002 IBC08	B4 B13	MP2		
3487	CALCIUM HYPOCHLORITE, HYDRATED, CORROSIVE or CALCIUM HYPOCHLORITE, HYDRATED MIXTURE, CORROSIVE with not less than 5.5% but not more than 16% water	5.1	OC2	III	5.1 +8	314	5 kg	E1	P002 IBC08 R001	B4 B13	MP2		
3488	TOXIC BY INHALATION LIQUID, FLAMMABLE, CORROSIVE, N.O.S. with an LC_{50} lower than or equal to 200 ml/m³ and saturated vapour concentration greater than or equal to 500 LC_{50}	6.1	TFC	I	6.1 +3 +8	274	0	E0	P601		MP8 MP17	T22	TP2
3489	TOXIC BY INHALATION LIQUID, FLAMMABLE, CORROSIVE, N.O.S. with an LC_{50} lower than or equal to 1000 ml/m³ and saturated vapour concentration greater than or equal to 10 LC_{50}	6.1	TFC	I	6.1 +3 +8	274	0	E0	P602		MP8 MP17	T20	TP2
3490	TOXIC BY INHALATION LIQUID, WATER-REACTIVE, FLAMMABLE, N.O.S. with an LC_{50} lower than or equal to 200 ml/m³ and saturated vapour concentration greater than or equal to 500 LC_{50}	6.1	TFW	I	6.1 +3 +4.3	274	0	E0	P601		MP8 MP17	T22	TP2
3491	TOXIC BY INHALATION LIQUID, WATER-REACTIVE, FLAMMABLE, N.O.S. with an LC_{50} lower than or equal to 1000 ml/m³ and saturated vapour concentration greater than or equal to 10 LC_{50}	6.1	TFW	I	6.1 +3 +4.3	274	0	E0	P602		MP8 MP17	T20	TP2
3494	PETROLEUM SOUR CRUDE OIL, FLAMMABLE, TOXIC	3	FT1	I	3 +6.1	343	0	E0	P001		MP7 MP17	T14	TP2
3494	PETROLEUM SOUR CRUDE OIL, FLAMMABLE, TOXIC	3	FT1	II	3 +6.1	343	1 L	E2	P001 IBC02		MP19	T7	TP2

ADR tank		Vehicle for tank carriage	Transport category (Tunnel restriction code)	Special provisions for carriage				Hazard identifi-cation No.	UN No.	Name and description
Tank code	Special provisions			Packages	Bulk	Loading, unloading and handling	Operation			
4.3	4.3.5, 6.8.4	9.1.1.2	1.1.3.6 (8.6)	7.2.4	7.3.3	7.5.11	8.5	5.3.2.3		3.1.2
(12)	(13)	(14)	(15)	(16)	(17)	(18)	(19)	(20)	(1)	(2)
			2 (E)						3481	LITHIUM ION BATTERIES CONTAINED IN EQUIPMENT or LITHIUM ION BATTERIES PACKED WITH EQUIPMENT (including lithium ion polymer batteries)
L10BN (+)	TU1 TE5 TT3 TM2	FL	1 (B/E)	V1		CV23	S2 S20	X323	3482	ALKALI METAL DISPERSION, FLAMMABLE or ALKALINE EARTH METAL DISPERSION, FLAMMABLE
L10CH	TU14 TU15 TE19 TE21 TT6	FL	1 (C/D)			CV1 CV13 CV28	S2 S9 S14	663	3483	MOTOR FUEL ANTI-KNOCK MIXTURE, FLAMMABLE
L10BH		FL	1 (C/D)			CV13 CV28	S2 S14	886	3484	HYDRAZINE AQUEOUS SOLUTION, FLAMMABLE with more than 37% hydrazine, by mass
SGAN	TU3	AT	2 (E)	V11		CV24 CV35		58	3485	CALCIUM HYPOCHLORITE, DRY, CORROSIVE or CALCIUM HYPOCHLORITE MIXTURE, DRY, CORROSIVE with more than 39% available chlorine (8.8% available oxygen)
SGAN	TU3	AT	3 (E)			CV24 CV35		58	3486	CALCIUM HYPOCHLORITE MIXTURE, DRY, CORROSIVE with more than 10% but not more than 39% available chlorine
SGAN	TU3	AT	2 (E)	V11		CV24 CV35		58	3487	CALCIUM HYPOCHLORITE, HYDRATED, CORROSIVE or CALCIUM HYPOCHLORITE, HYDRATED MIXTURE, CORROSIVE with not less than 5.5% but not more than 16% water
SGAN	TU3	AT	3 (E)			CV24 CV35		58	3487	CALCIUM HYPOCHLORITE, HYDRATED, CORROSIVE or CALCIUM HYPOCHLORITE, HYDRATED MIXTURE, CORROSIVE with not less than 5.5% but not more than 16% water
L15CH	TU14 TU15 TE19 TE21	FL	1 (C/D)			CV1 CV13 CV28	S2 S9 S14	663	3488	TOXIC BY INHALATION LIQUID, FLAMMABLE, CORROSIVE, N.O.S. with an LC_{50} lower than or equal to 200 ml/m³ and saturated vapour concentration greater than or equal to 500 LC_{50}
L10CH	TU14 TU15 TE19 TE21	FL	1 (C/D)			CV1 CV13 CV28	S2 S9 S14	663	3489	TOXIC BY INHALATION LIQUID, FLAMMABLE, CORROSIVE, N.O.S. with an LC_{50} lower than or equal to 1000 ml/m³ and saturated vapour concentration greater than or equal to 10 LC_{50}
L15CH	TU14 TU15 TE19 TE21	FL	1 (C/D)			CV1 CV13 CV28	S2 S9 S14	623	3490	TOXIC BY INHALATION LIQUID, WATER-REACTIVE, FLAMMABLE, N.O.S. with an LC_{50} lower than or equal to 200 ml/m³ and saturated vapour concentration greater than or equal to 500 LC_{50}
L10CH	TU14 TU15 TE19 TE21	FL	1 (C/D)			CV1 CV13 CV28	S2 S9 S14	623	3491	TOXIC BY INHALATION LIQUID, WATER-REACTIVE, FLAMMABLE, N.O.S. with an LC_{50} lower than or equal to 1000 ml/m³ and saturated vapour concentration greater than or equal to 10 LC_{50}
L10CH	TU14 TU15 TE21	FL	1 (C/E)			CV13 CV28	S2 S22	336	3494	PETROLEUM SOUR CRUDE OIL, FLAMMABLE, TOXIC
L4BH	TU15	FL	2 (D/E)			CV13 CV28	S2 S19	336	3494	PETROLEUM SOUR CRUDE OIL, FLAMMABLE, TOXIC

UN No.	Name and description	Class	Classification code	Packing group	Labels	Special provisions	Limited quantities	Excepted quantities	Packing instructions	Special packing provisions	Mixed packing provisions	Instructions	Special provisions
	3.1.2	2.2	2.2	2.1.1.3	5.2.2	3.3	3.4	3.5.1.2	4.1.4	4.1.4	4.1.10	4.2.5.2 7.3.2	4.2.5.3
(1)	(2)	(3a)	(3b)	(4)	(5)	(6)	(7a)	(7b)	(8)	(9a)	(9b)	(10)	(11)
3494	PETROLEUM SOUR CRUDE OIL, FLAMMABLE, TOXIC	3	FT1	III	3 +6.1	343	5 L	E1	P001 IBC03 R001		MP19	T4	TP1
3495	IODINE	8	CT2	III	8 +6.1	279	5 kg	E1	P002 IBC08 R001	B3	MP10	T1	TP33
3496	Batteries, nickel-metal hydride	9	M11				NOT SUBJECT TO ADR							
3497	KRILL MEAL	4.2	S2	II	4.2	300	0	E2	P410 IBC06		MP14	T3	TP33
3497	KRILL MEAL	4.2	S2	III	4.2	300	0	E1	P002 IBC08 LP02 R001	B3	MP14	T1	TP33
3498	IODINE MONOCHLORIDE, LIQUID	8	C1	II	8		1 L	E0	P001 IBC02		MP15	T7	TP2
3499	CAPACITOR, ELECTRIC DOUBLE LAYER (with an energy storage capacity greater than 0.3Wh)	9	M11		9	361	0	E0	P003				
3500	CHEMICAL UNDER PRESSURE, N.O.S.	2	8A		2.2	274 659	0	E0	P206		MP9	T50	TP4 TP40
3501	CHEMICAL UNDER PRESSURE, FLAMMABLE, N.O.S.	2	8F		2.1	274 659	0	E0	P206	PP89	MP9	T50	TP4 TP40
3502	CHEMICAL UNDER PRESSURE, TOXIC, N.O.S.	2	8T		2.2 +6.1	274 659	0	E0	P206	PP89	MP9	T50	TP4 TP40
3503	CHEMICAL UNDER PRESSURE, CORROSIVE, N.O.S.	2	8C		2.2 +8	274 659	0	E0	P206	PP89	MP9	T50	TP4 TP40
3504	CHEMICAL UNDER PRESSURE, FLAMMABLE, TOXIC, N.O.S.	2	8TF		2.1 +6.1	274 659	0	E0	P206	PP89	MP9	T50	TP4 TP40
3505	CHEMICAL UNDER PRESSURE, FLAMMABLE, CORROSIVE, N.O.S.	2	8FC		2.1 +8	274 659	0	E0	P206	PP89	MP9	T50	TP4 TP40
3506	MERCURY CONTAINED IN MANUFACTURED ARTICLES	8	CT3		8 +6.1	366	5 kg	E0	P003	PP90	MP15		
3507	URANIUM HEXAFLUORIDE, RADIOACTIVE MATERIAL, EXCEPTED PACKAGE, less than 0.1 kg per package, non-fissile or fissile-excepted	6.1		I	6.1 +8	317 369	0	E0	P603				
3508	CAPACITOR, ASYMMETRIC (with an energy storage capacity greater than 0.3Wh)	9	M11		9	372	0	E0	P003				
3509	PACKAGINGS, DISCARDED, EMPTY, UNCLEANED	9	M11		9	663	0	E0	P003 IBC08 LP02	RR9 BB3 LL1		BK2	
3510	ADSORBED GAS, FLAMMABLE, N.O.S.	2	9F		2.1	274	0	E0	P208		MP9		
3511	ADSORBED GAS, N.O.S.	2	9A		2.2	274	0	E0	P208		MP9		
3512	ADSORBED GAS, TOXIC, N.O.S.	2	9T		2.3	274	0	E0	P208		MP9		
3513	ADSORBED GAS, OXIDIZING, N.O.S.	2	9O		2.2 +5.1	274	0	E0	P208		MP9		

ADR tank		Vehicle for tank carriage	Transport category (Tunnel restriction code)	Special provisions for carriage				Hazard identifi-cation No.	UN No.	Name and description
Tank code	Special provisions			Packages	Bulk	Loading, unloading and handling	Operation			
4.3	4.3.5, 6.8.4	9.1.1.2	1.1.3.6 (8.6)	7.2.4	7.3.3	7.5.11	8.5	5.3.2.3		3.1.2
(12)	(13)	(14)	(15)	(16)	(17)	(18)	(19)	(20)	(1)	(2)
L4BH	TU15	FL	3 (D/E)	V12		CV13 CV28	S2	36	3494	PETROLEUM SOUR CRUDE OIL, FLAMMABLE, TOXIC
SGAV L4BN		AT	3 (E)		VC1 VC2 AP7	CV13 CV28		86	3495	IODINE
			NOT SUBJECT TO ADR						3496	Batteries, nickel-metal hydride
SGAN		AT	2 (D/E)	V1				40	3497	KRILL MEAL
SGAV		AT	3 (E)	V1	VC1 VC2 AP1			40	3497	KRILL MEAL
L4BN		AT	2 (E)					80	3498	IODINE MONOCHLORIDE, LIQUID
			4 (E)						3499	CAPACITOR, ELECTRIC DOUBLE LAYER (with an energy storage capacity greater than 0.3Wh)
		AT	3 (C/E)			CV9 CV10 CV12 CV36		20	3500	CHEMICAL UNDER PRESSURE, N.O.S.
		FL	2 (B/D)			CV9 CV10 CV12 CV36	S2	23	3501	CHEMICAL UNDER PRESSURE, FLAMMABLE, N.O.S.
		AT	1 (C/D)			CV9 CV10 CV12 CV28 CV36		26	3502	CHEMICAL UNDER PRESSURE, TOXIC, N.O.S.
		AT	1 (C/D)			CV9 CV10 CV12 CV36		28	3503	CHEMICAL UNDER PRESSURE, CORROSIVE, N.O.S.
		FL	1 (B/D)			CV9 CV10 CV12 CV28 CV36	S2	263	3504	CHEMICAL UNDER PRESSURE, FLAMMABLE, TOXIC, N.O.S.
		FL	1 (B/D)			CV9 CV10 CV12 CV36	S2	238	3505	CHEMICAL UNDER PRESSURE, FLAMMABLE, CORROSIVE, N.O.S.
			3 (E)			CV13 CV28			3506	MERCURY CONTAINED IN MANUFACTURED ARTICLES
			1 (D)		See SP 369		S21		3507	URANIUM HEXAFLUORIDE, RADIOACTIVE MATERIAL, EXCEPTED PACKAGE, less than 0.1 kg per package, non-fissile or fissile-excepted
			4 (E)						3508	CAPACITOR, ASYMMETRIC (with an energy storage capacity greater than 0.3Wh)
			4 (E)		VC2 AP10			90	3509	PACKAGINGS, DISCARDED, EMPTY, UNCLEANED
			2 (D)			CV9 CV10 CV36	S2		3510	ADSORBED GAS, FLAMMABLE, N.O.S.
			3 (E)			CV9 CV10 CV36			3511	ADSORBED GAS, N.O.S.
			1 (D)			CV9 CV10 CV36	S14		3512	ADSORBED GAS, TOXIC, N.O.S.
			3 (E)			CV9 CV10 CV36			3513	ADSORBED GAS, OXIDIZING, N.O.S.

UN No.	Name and description	Class	Classifi-cation code	Packing group	Labels	Special provi-sions	Limited and excepted quantities		Packaging			Portable tanks and bulk containers	
									Packing instruc-tions	Special packing provisions	Mixed packing provisions	Instruc-tions	Special provisions
3.1.2	3.1.2	2.2	2.2	2.1.1.3	5.2.2	3.3	3.4	3.5.1.2	4.1.4	4.1.4	4.1.10	4.2.5.2 7.3.2	4.2.5.3
(1)	(2)	(3a)	(3b)	(4)	(5)	(6)	(7a)	(7b)	(8)	(9a)	(9b)	(10)	(11)
3514	ADSORBED GAS, TOXIC, FLAMMABLE, N.O.S.	2	9TF		2.3 +2.1	274	0	E0	P208		MP9		
3515	ADSORBED GAS, TOXIC, OXIDIZING, N.O.S.	2	9TO		2.3 +5.1	274	0	E0	P208		MP9		
3516	ADSORBED GAS, TOXIC, CORROSIVE, N.O.S.	2	9TC		2.3 +8	274 379	0	E0	P208		MP9		
3517	ADSORBED GAS, TOXIC, FLAMMABLE, CORROSIVE, N.O.S.	2	9TFC		2.3 +2.1 +8	274	0	E0	P208		MP9		
3518	ADSORBED GAS, TOXIC, OXIDIZING, CORROSIVE, N.O.S.	2	9TOC		2.3 +5.1 +8	274	0	E0	P208		MP9		
3519	BORON TRIFLUORIDE, ADSORBED	2	9TC		2.3 +8		0	E0	P208		MP9		
3520	CHLORINE, ADSORBED	2	9TOC		2.3 +5.1 +8		0	E0	P208		MP9		
3521	SILICON TETRAFLUORIDE, ADSORBED	2	9TC		2.3 +8		0	E0	P208		MP9		
3522	ARSINE, ADSORBED	2	9TF		2.3 +2.1		0	E0	P208		MP9		
3523	GERMANE, ADSORBED	2	9TF		2.3 +2.1		0	E0	P208		MP9		
3524	PHOSPHORUS PENTAFLUORIDE, ADSORBED	2	9TC		2.3 +8		0	E0	P208		MP9		
3525	PHOSPHINE, ADSORBED	2	9TF		2.3 +2.1		0	E0	P208		MP9		
3526	HYDROGEN SELENIDE, ADSORBED	2	9TF		2.3 +2.1		0	E0	P208		MP9		
3527	POLYESTER RESIN KIT, solid base material	4.1	F4	II	4.1	236 340	5Kg	E0	P412				
3527	POLYESTER RESIN KIT, solid base material	4.1	F4	III	4.1	236 340	5Kg	E0	P412				
3528	ENGINE, INTERNAL COMBUSTION, FLAMMABLE LIQUID POWERED or ENGINE, FUEL CELL, FLAMMABLE LIQUID POWERED or MACHINERY, INTERNAL COMBUSTION, FLAMMABLE LIQUID POWERED or MACHINERY, FUEL CELL, FLAMMABLE LIQUID POWERED	3	F3		3	363 667 669	0	E0	P005				

ADR tank		Vehicle for tank carriage	Transport category (Tunnel restriction code)	Special provisions for carriage				Hazard identification No.	UN No.	Name and description
Tank code	Special provisions			Packages	Bulk	Loading, unloading and handling	Operation			
4.3	4.3.5, 6.8.4	9.1.1.2	1.1.3.6 (8.6)	7.2.4	7.3.3	7.5.11	8.5	5.3.2.3		3.1.2
(12)	(13)	(14)	(15)	(16)	(17)	(18)	(19)	(20)	(1)	(2)
			1 (D)			CV9 CV10 CV36	S2 S14		3514	ADSORBED GAS, TOXIC, FLAMMABLE, N.O.S.
			1 (D)			CV9 CV10 CV36	S14		3515	ADSORBED GAS, TOXIC, OXIDIZING, N.O.S.
			1 (D)			CV9 CV10 CV36	S14		3516	ADSORBED GAS, TOXIC, CORROSIVE, N.O.S.
			1 (D)			CV9 CV10 CV36	S2 S14		3517	ADSORBED GAS, TOXIC, FLAMMABLE, CORROSIVE, N.O.S.
			1 (D)			CV9 CV10 CV36	S14		3518	ADSORBED GAS, TOXIC, OXIDIZING, CORROSIVE, N.O.S.
			1 (D)			CV9 CV10 CV36	S14		3519	BORON TRIFLUORIDE, ADSORBED
			1 (D)			CV9 CV10 CV36	S14		3520	CHLORINE, ADSORBED
			1 (D)			CV9 CV10 CV36	S14		3521	SILICON TETRAFLUORIDE, ADSORBED
			1 (D)			CV9 CV10 CV36	S2 S14		3522	ARSINE, ADSORBED
			1 (D)			CV9 CV10 CV36	S2 S14		3523	GERMANE, ADSORBED
			1 (D)			CV9 CV10 CV36	S14		3524	PHOSPHORUS PENTAFLUORIDE, ADSORBED
			1 (D)			CV9 CV10 CV36	S2 S14		3525	PHOSPHINE, ADSORBED
			1 (D)			CV9 CV10 CV36	S2 S14		3526	HYDROGEN SELENIDE, ADSORBED
			2 (E)						3527	POLYESTER RESIN KIT, solid base material
			3 (E)						3527	POLYESTER RESIN KIT, solid base material
									3528	ENGINE, INTERNAL COMBUSTION, FLAMMABLE LIQUID POWERED or ENGINE, FUEL CELL, FLAMMABLE LIQUID POWERED or MACHINERY, INTERNAL COMBUSTION, FLAMMABLE LIQUID POWERED or MACHINERY, FUEL CELL, FLAMMABLE LIQUID POWERED

UN No.	Name and description	Class	Classifi-cation code	Packing group	Labels	Special provi-sions	Limited and excepted quantities		Packaging			Portable tanks and bulk containers	
									Packing instruc-tions	Special packing provisions	Mixed packing provisions	Instruc-tions	Special provisions
3.1.2	3.1.2	2.2	2.2	2.1.1.3	5.2.2	3.3	3.4	3.5.1.2	4.1.4	4.1.4	4.1.10	4.2.5.2 7.3.2	4.2.5.3
(1)	(2)	(3a)	(3b)	(4)	(5)	(6)	(7a)	(7b)	(8)	(9a)	(9b)	(10)	(11)
3529	ENGINE, INTERNAL COMBUSTION, FLAMMABLE GAS POWERED or ENGINE, FUEL CELL, FLAMMABLE GAS POWERED or MACHINERY, INTERNAL COMBUSTION, FLAMMABLE GAS POWERED or MACHINERY, FUEL CELL, FLAMMABLE GAS POWERED	2	6F		2.1	363 667 669	0	E0	P005				
3530	ENGINE, INTERNAL COMBUSTION or MACHINERY, INTERNAL COMBUSTION	9	M11		9	363 667 669	0	E0	P005				
3531	POLYMERIZING SUBSTANCE, SOLID, STABILIZED, N.O.S.	4.1	PM1	III	4.1	274 386	0	E0	P002 IBC07	PP92 B18		T7	TP4 TP6 TP33
3532	POLYMERIZING SUBSTANCE, LIQUID, STABILIZED, N.O.S.	4.1	PM1	III	4.1	274 386	0	E0	P001 IBC03	PP93 B19		T7	TP4 TP6
3533	POLYMERIZING SUBSTANCE, SOLID, TEMPERATURE CONTROLLED, N.O.S.	4.1	PM2	III	4.1	274 386	0	E0	P002 IBC07	PP92 B18		T7	TP4 TP6 TP33
3534	POLYMERIZING SUBSTANCE, LIQUID, TEMPERATURE CONTROLLED, N.O.S.	4.1	PM2	III	4.1	274 386	0	E0	P001 IBC03	PP93 B19		T7	TP4 TP6

ADR tank		Vehicle for tank carriage	Transport category (Tunnel restriction code)	Special provisions for carriage				Hazard identification No.	UN No.	Name and description
Tank code	Special provisions			Packages	Bulk	Loading, unloading and handling	Operation			
4.3	4.3.5, 6.8.4	9.1.1.2	1.1.3.6 (8.6)	7.2.4	7.3.3	7.5.11	8.5	5.3.2.3		3.1.2
(12)	(13)	(14)	(15)	(16)	(17)	(18)	(19)	(20)	(1)	(2)
									3529	ENGINE, INTERNAL COMBUSTION, FLAMMABLE GAS POWERED or ENGINE, FUEL CELL, FLAMMABLE GAS POWERED or MACHINERY, INTERNAL COMBUSTION, FLAMMABLE GAS POWERED or MACHINERY, FUEL CELL, FLAMMABLE GAS POWERED
									3530	ENGINE, INTERNAL COMBUSTION or MACHINERY, INTERNAL COMBUSTION
SGAN+	TU30 TE11	AT	2 (D)	V1		CV15 CV22		40	3531	POLYMERIZING SUBSTANCE, SOLID, STABILIZED, N.O.S.
L4BN+	TU30 TE11	AT	2 (D)	V1		CV15 CV22		40	3532	POLYMERIZING SUBSTANCE, LIQUID, STABILIZED, N.O.S.
SGAN+	TU30 TE11	AT	1 (D)	V8		CV15 CV21 CV22	S4	40	3533	POLYMERIZING SUBSTANCE, SOLID, TEMPERATURE CONTROLLED, N.O.S.
L4BN+	TU30 TE11	AT	1 (D)	V8		CV15 CV21 CV22	S4	40	3534	POLYMERIZING SUBSTANCE, LIQUID, TEMPERATURE CONTROLLED, N.O.S.

3.2.2   **Table B: Alphabetic index of substances and articles of ADR**

This index is an alphabetical list of the substances and articles which are listed in the UN numerical order in Table A of 3.2.1. It does not form an integral part of ADR. It has been submitted neither to the Working Party on the Transport of Dangerous Goods of the Inland Transport Committee for checking and approval nor to the Contracting Parties to ADR for formal acceptance. It has been prepared, with all necessary care by the Secretariat of the United Nations Economic Commission for Europe, in order to facilitate the consultation of Annexes A and B, but it cannot be relied upon as a substitute for the careful study and observance of the actual provisions of those annexes which, in case of conflict, are deemed to be authoritative.

NOTE 1: For the purpose of determining the alphabetical order the following information has been ignored, even when it forms part of the proper shipping name: numbers; Greek letters; the abbreviations "sec" and "tert"; and the letters "N" (nitrogen), "n" (normal), "o" (ortho), "m" (meta), "p" (para) and "N.O.S." (not otherwise specified).

NOTE 2: The name of a substance or article in block capital letters indicates a proper shipping name (see 3.1.2).

NOTE 3: The name of a substance or article in block capital letters followed by the word "see" indicates an alternative proper shipping name or part of a proper shipping name (except for PCBs) (see 3.1.2.1).

NOTE 4: An entry in lower case letters followed by the word "see" indicates that the entry is not a proper shipping name; it is a synonym.

NOTE 5: Where an entry is partly in block capital letters and partly in lower case letters, the latter part is considered not to be part of the proper shipping name (see 3.1.2.1).

NOTE 6: A proper shipping name may be used in the singular or plural, as appropriate, for the purposes of documentation and package marking (see 3.1.2.3).

NOTE 7: For the exact determination of a proper shipping name, see 3.1.2.

Name and description	UN No.	Class	Remarks	Name and description	UN No.	Class	Remarks
Accumulators, electric, see	2794	8		Activated carbon, see	1362	4.2	
	2795	8		Activated charcoal, see	1362	4.2	
	2800	8		ADHESIVES containing flammable liquid	1133	3	
	3028	8					
	3292	4.3					
ACETAL	1088	3		ADIPONITRILE	2205	6.1	
ACETALDEHYDE	1089	3		ADSORBED GAS, FLAMMABLE, N.O.S.	3510	2	
ACETALDEHYDE AMMONIA	1841	9					
ACETALDEHYDE OXIME	2332	3		ADSORBED GAS, N.O.S.	3511	2	
ACETIC ACID, GLACIAL	2789	8		ADSORBED GAS, OXIDIZING, N.O.S.	3513	2	
ACETIC ACID SOLUTION, more than 10% but not more than 80% acid, by mass	2790	8		ADSORBED GAS, TOXIC, CORROSIVE, N.O.S.	3516	2	
ACETIC ACID SOLUTION, more than 80% acid, by mass	2789	8		ADSORBED GAS, TOXIC, FLAMMABLE, CORROSIVE, N.O.S.	3517	2	
ACETIC ANHYDRIDE	1715	8		ADSORBED GAS, TOXIC, FLAMMABLE, N.O.S.	3514	2	
Acetoin, see	2621	3					
ACETONE	1090	3		ADSORBED GAS, TOXIC, N.O.S.	3512	2	
ACETONE CYANOHYDRIN, STABILIZED	1541	6.1		ADSORBED GAS, TOXIC, OXIDIZING, CORROSIVE, N.O.S.	3518	2	
ACETONE OILS	1091	3		ADSORBED GAS, TOXIC, OXIDIZING, N.O.S.	3515	2	
ACETONITRILE	1648	3					
ACETYL BROMIDE	1716	8		Aeroplane flares, see	0093	1	
ACETYL CHLORIDE	1717	3			0403	1	
ACETYLENE, DISSOLVED	1001	2			0404	1	
ACETYLENE, SOLVENT FREE	3374	2			0420	1	
Acetylene tetrabromide, see	2504	6.1			0421	1	
Acetylene tetrachloride, see	1702	6.1		AEROSOLS	1950	2	
ACETYL IODIDE	1898	8		AGENT, BLASTING, TYPE B	0331	1	
ACETYL METHYL CARBINOL	2621	3		AGENT, BLASTING, TYPE E	0332	1	
Acid butyl phosphate, see	1718	8		Air bag inflators, see	0503	1	
Acid mixture, hydrofluoric and sulphuric, see	1786	8			3268	9	
Acid mixture, nitrating acid, see	1796	8		Air bag modules, see	0503	1	
					3268	9	
Acid mixture, spent, nitrating acid, see	1826	8		AIR, COMPRESSED	1002	2	
Acraldehyde, inhibited, see	1092	6.1		Aircraft evacuation slides, see	2990	9	
ACRIDINE	2713	6.1		AIRCRAFT HYDRAULIC POWER UNIT FUEL TANK (containing a mixture of anhydrous hydrazine and methylhydrazine) (M86 fuel)	3165	3	
ACROLEIN DIMER, STABILIZED	2607	3					
ACROLEIN, STABILIZED	1092	6.1		Aircraft survival kits, see	2990	9	
ACRYLAMIDE, SOLID	2074	6.1		AIR, REFRIGERATED LIQUID	1003	2	
ACRYLAMIDE, SOLUTION	3426	6.1		ALCOHOLATES SOLUTION, N.O.S., in alcohol	3274	3	
ACRYLIC ACID, STABILIZED	2218	8		Alcohol, denatured, see	1986	3	
ACRYLONITRILE, STABILIZED	1093	3			1987	3	
Actinolite, see	2212	9		Alcohol, industrial, see	1986	3	
					1987	3	
				ALCOHOLS, N.O.S.	1987	3	

Name and description	UN No.	Class	Remarks	Name and description	UN No.	Class	Remarks
ALCOHOLS, FLAMMABLE, TOXIC, N.O.S.	1986	3		ALKYLSULPHONIC ACIDS, LIQUID with more than 5% free sulphuric acid	2584	8	
ALCOHOLIC BEVERAGES, with more than 24% but not more than 70% alcohol by volume	3065	3		ALKYLSULPHONIC ACIDS, LIQUID with not more than 5% free sulphuric acid	2586	8	
ALCOHOLIC BEVERAGES, with more than 70% alcohol by volume	3065	3		ALKYLSULPHONIC ACIDS, SOLID with more than 5% free sulphuric acid	2583	8	
Aldehyde, see	1989	3		ALKYLSULPHONIC ACIDS, SOLID with not more than 5% free sulphuric acid	2585	8	
ALDEHYDES, N.O.S.	1989	3					
ALDEHYDES, FLAMMABLE, TOXIC, N.O.S.	1988	3					
ALDOL	2839	6.1		ALKYLSULPHURIC ACIDS	2571	8	
ALKALI METAL ALCOHOLATES, SELF-HEATING, CORROSIVE, N.O.S.	3206	4.2		Allene, see	2200	2	
				ALLYL ACETATE	2333	3	
ALKALI METAL ALLOY, LIQUID, N.O.S.	1421	4.3		ALLYL ALCOHOL	1098	6.1	
ALKALI METAL AMALGAM, LIQUID	1389	4.3		ALLYLAMINE	2334	6.1	
				ALLYL BROMIDE	1099	3	
ALKALI METAL AMALGAM, SOLID	3401	4.3		ALLYL CHLORIDE	1100	3	
				Allyl chlorocarbonate, see	1722	6.1	
ALKALI METAL AMIDES	1390	4.3		ALLYL CHLOROFORMATE	1722	6.1	
ALKALI METAL DISPERSION	1391	4.3		ALLYL ETHYL ETHER	2335	3	
ALKALI METAL DISPERSION, FLAMMABLE	3482	4.3		ALLYL FORMATE	2336	3	
				ALLYL GLYCIDYL ETHER	2219	3	
Alkaline corrosive battery fluid, see	2797	8		ALLYL IODIDE	1723	3	
ALKALINE EARTH METAL ALCOHOLATES, N.O.S.	3205	4.2		ALLYL ISOTHIOCYANATE, STABILIZED	1545	6.1	
ALKALINE EARTH METAL ALLOY, N.O.S.	1393	4.3		ALLYLTRICHLOROSILANE, STABILIZED	1724	8	
ALKALINE EARTH METAL AMALGAM, LIQUID	1392	4.3		Aluminium alkyls, see	3394	4.2	
				Aluminium alkyl halides, liquid, see	3394	4.2	
ALKALINE EARTH METAL AMALGAM, SOLID	3402	4.3		Aluminium alkyl halides, solid, see	3393	4.2	
				Aluminium alkyl hydrides, see	3394	4.2	
ALKALINE EARTH METAL DISPERSION	1391	4.3		ALUMINIUM BOROHYDRIDE	2870	4.2	
ALKALINE EARTH METAL DISPERSION, FLAMMABLE	1391	4.3		ALUMINIUM BOROHYDRIDE IN DEVICES	2870	4.2	
ALKALOIDS, LIQUID, N.O.S.	3140	6.1		ALUMINIUM BROMIDE, ANHYDROUS	1725	8	
ALKALOIDS, SOLID, N.O.S.	1544	6.1		ALUMINIUM BROMIDE SOLUTION	2580	8	
ALKALOID SALTS, LIQUID, N.O.S.	3140	6.1		ALUMINIUM CARBIDE	1394	4.3	
ALKALOID SALTS, SOLID, N.O.S.	1544	6.1		ALUMINIUM CHLORIDE, ANHYDROUS	1726	8	
Alkyl aluminium halides, see	3394	4.2		ALUMINIUM CHLORIDE SOLUTION	2581	8	
ALKYLPHENOLS, LIQUID, N.O.S. (including C_2-C_{12} homologues)	3145	8					
ALKYLPHENOLS, SOLID, N.O.S. (including C_2-C_{12} homologues)	2430	8					

Name and description	UN No.	Class	Remarks	Name and description	UN No.	Class	Remarks
Aluminium dross, see	3170	4.3		AMMONIA, ANHYDROUS	1005	2	
ALUMINIUM FERROSILICON POWDER	1395	4.3		AMMONIA SOLUTION relative density between 0.880 and 0.957 at 15 °C in water, with more than 10% but not more than 35% ammonia	2672	8	
ALUMINIUM HYDRIDE	2463	4.3					
ALUMINIUM NITRATE	1438	5.1		AMMONIA SOLUTION, relative density less than 0.880 at 15 °C in water, with more than 35% but not more than 50% ammonia	2073	2	
ALUMINIUM PHOSPHIDE	1397	4.3					
ALUMINIUM PHOSPHIDE PESTICIDE	3048	6.1		AMMONIA SOLUTION, relative density less than 0.880 at 15 °C in water, with more than 50% ammonia	3318	2	
ALUMINIUM POWDER, COATED	1309	4.1					
ALUMINIUM POWDER, UNCOATED	1396	4.3					
ALUMINIUM REMELTING BY-PRODUCTS	3170	4.3		AMMONIUM ARSENATE	1546	6.1	
				Ammonium bichromate, see	1439	5.1	
ALUMINIUM RESINATE	2715	4.1		Ammonium bifluoride solid, see	1727	8	
ALUMINIUM SILICON POWDER, UNCOATED	1398	4.3		Ammonium bifluoride solution, see	2817	8	
				Ammonium bisulphate, see	2506	8	
ALUMINIUM SMELTING BY-PRODUCTS	3170	4.3		Ammonium bisulphite solution, see	2693	8	
Amatols, see	0082	1		AMMONIUM DICHROMATE	1439	5.1	
AMINES, FLAMMABLE, CORROSIVE, N.O.S.	2733	3		AMMONIUM DINITRO-o-CRESOLATE, SOLID	1843	6.1	
AMINES, LIQUID, CORROSIVE, N.O.S.	2735	8		AMMONIUM DINITRO-o-CRESOLATE, SOLUTION	3424	6.1	
AMINES, LIQUID, CORROSIVE, FLAMMABLE, N.O.S.	2734	8		AMMONIUM FLUORIDE	2505	6.1	
				AMMONIUM FLUORO-SILICATE	2854	6.1	
AMINES, SOLID, CORROSIVE, N.O.S.	3259	8		Ammonium hexafluorosilicate, see	2854	6.1	
Aminobenzene, see	1547	6.1		AMMONIUM HYDROGENDIFLUORIDE, SOLID	1727	8	
2-Aminobenzotrifluoruride, see	2942	6.1					
3-Aminobenzotrifluoruride, see	2948	6.1		AMMONIUM HYDROGENDIFLUORIDE SOLUTION	2817	8	
Aminobutane, see	1125	3					
2-AMINO-4-CHLOROPHENOL	2673	6.1		AMMONIUM HYDROGEN SULPHATE	2506	8	
2-AMINO-5-DIETHYL-AMINOPENTANE	2946	6.1					
2-AMINO-4,6-DINITROPHENOL, WETTED with not less than 20% water, by mass	3317	4.1		Ammonium hydrosulphide solution (treat as ammonium sulphide solution), see	2683	8	
				AMMONIUM METAVANADATE	2859	6.1	
2-(2-AMINOETHOXY) ETHANOL	3055	8		AMMONIUM NITRATE	0222	1	
N-AMINOETHYLPIPERAZINE	2815	8		AMMONIUM NITRATE with not more than 0.2% combustible substances, including any organic substance calculated as carbon, to the exclusion of any other added substance	1942	5.1	
1-Amino-2-nitrobenzene, see	1661	6.1					
1-Amino-3-nitrobenzene, see	1661	6.1					
1-Amino-4-nitrobenzene, see	1661	6.1					
AMINOPHENOLS (o-, m-, p-)	2512	6.1		AMMONIUM NITRATE EMULSION, intermediate for blasting explosives, liquid	3375	5.1	
AMINOPYRIDINES (o-, m-, p-)	2671	6.1					

Name and description	UN No.	Class	Remarks	Name and description	UN No.	Class	Remarks
AMMONIUM NITRATE EMULSION, intermediate for blasting explosives, solid	3375	5.1		Ammunition, blank, see	0014 0326 0327 0338 0413	1 1 1 1 1	
Ammonium nitrate explosive, see	0082 0331	1 1		Ammunition, fixed Ammunition, semi-fixed Ammunition, separate loading, see	0005 0006 0007 0321 0348 0412	1 1 1 1 1 1	
AMMONIUM NITRATE BASED FERTILIZER	2067	5.1					
Ammonium nitrate based fertilizer, uniform mixtures of the nitrogen/phosphate, nitrogen/potash or nitrogen/phosphate/potash type, containing not more than 70% ammonium nitrate and not more than 0.4% total combustible/organic material calculated as carbon or with not more than 45% ammonium nitrate and unrestricted combustible material	2071	9	Not subject to ADR	AMMUNITION, ILLUMINATING with or without burster, expelling charge or propelling charge	0171 0254 0297	1 1 1	
				AMMUNITION, INCENDIARY, liquid or gel, with burster, expelling charge or propelling charge	0247	1	
				AMMUNITION, INCENDIARY with or without burster, expelling charge or propelling charge	0009 0010 0300	1 1 1	
AMMONIUM NITRATE GEL, intermediate for blasting explosives, liquid	3375	5.1		Ammunition, incendiary (water-activated contrivances) with burster, expelling charge or propelling charge, see	0248 0249	1 1	
AMMONIUM NITRATE GEL, intermediate for blasting explosives, solid	3375	5.1					
AMMONIUM NITRATE, LIQUID hot concentrated solution, in a concentration of more than 80% but not more than 93%	2426	5.1		AMMUNITION, INCENDIARY, WHITE PHOSPHORUS with burster, expelling charge or propelling charge	0243 0244	1 1	
AMMONIUM NITRATE SUSPENSION, intermediate for blasting explosives, liquid	3375	5.1		Ammunition, industrial, see	0275 0276 0277 0278 0323 0381	1 1 1 1 1 1	
AMMONIUM NITRATE SUSPENSION, intermediate for blasting explosives, solid	3375	5.1					
AMMONIUM PERCHLORATE	0402 1442	1 5.1		Ammunition, lachrymatory, see	0018 0019 0301 2017	1 1 1 1	
Ammonium permanganate, see	1482	5.1		AMMUNITION, PRACTICE	0362 0488	1 1	
AMMONIUM PERSULPHATE	1444	5.1					
AMMONIUM PICRATE dry or wetted with less than 10% water, by mass	0004	1		AMMUNITION, PROOF	0363	1	
				AMMUNITION, SMOKE with or without burster, expelling charge or propelling charge	0015 0016 0303	1 1 1	
AMMONIUM PICRATE, WETTED with not less than 10% water, by mass	1310	4.1		Ammunition, smoke (water-activated contrivances), white phosphorus with burster, expelling charge or propelling charge, see	0248	1	
AMMONIUM POLYSULPHIDE SOLUTION	2818	8					
AMMONIUM POLYVANADATE	2861	6.1		Ammunition, smoke (water-activated contrivances), without white phosphorus or phosphides with burster, expelling charge or propelling charge, see	0249	1	
Ammonium silicofluoride, see	2854	6.1					
AMMONIUM SULPHIDE SOLUTION	2683	8					

Name and description	UN No.	Class	Remarks	Name and description	UN No.	Class	Remarks
AMMUNITION, SMOKE, WHITE PHOSPHORUS with burster, expelling charge or propelling charge	0245 0246	1 1		Aniline salt, see	1548	6.1	
				ANISIDINES	2431	6.1	
				ANISOLE	2222	3	
Ammunition, sporting, see	0012 0328 0339 0417	1 1 1 1		ANISOYL CHLORIDE	1729	8	
				Anthophyllite, see	2212	9	
				Antimonous chloride, see	1733	8	
AMMUNITION, TEAR-PRODUCING, NON-EXPLOSIVE without burster or expelling charge, non-fuzed	2017	6.1		ANTIMONY COMPOUND, INORGANIC, LIQUID, N.O.S.	3141	6.1	
				ANTIMONY COMPOUND, INORGANIC, SOLID, N.O.S.	1549	6.1	
AMMUNITION, TEAR-PRODUCING with burster, expelling charge or propelling charge	0018 0019 0301	1 1 1		Antimony hydride, see	2676	2	
				ANTIMONY LACTATE	1550	6.1	
				Antimony (III) lactate, see	1550	6.1	
AMMUNITION, TOXIC with burster, expelling charge or propelling charge	0020	1	Carriage prohibited	ANTIMONY PENTACHLORIDE, LIQUID	1730	8	
AMMUNITION, TOXIC with burster, expelling charge or propelling charge	0021	1	Carriage prohibited	ANTIMONY PENTACHLORIDE SOLUTION	1731	8	
				ANTIMONY PENTAFLUORIDE	1732	8	
Ammunition, toxic (water-activated contrivances) with burster, expelling charge or propelling charge, see	0248 0249	1 1		Antimony perchloride, liquid, see	1730	8	
				ANTIMONY POTASSIUM TARTRATE	1551	6.1	
AMMUNITION, TOXIC, NON-EXPLOSIVE without burster or expelling charge, non-fuzed	2016	6.1		ANTIMONY POWDER	2871	6.1	
				ANTIMONY TRICHLORIDE	1733	8	
Amosite, see	2212	9		A.n.t.u., see	1651	6.1	
Amphibole asbestos, see	2212	9		ARGON, COMPRESSED	1006	2	
AMYL ACETATES	1104	3		ARGON, REFRIGERATED LIQUID	1951	2	
AMYL ACID PHOSPHATE	2819	8		Arsenates, n.o.s., see	1556 1557	6.1 6.1	
Amyl aldehyde, see	2058	3					
AMYLAMINE	1106	3		ARSENIC	1558	6.1	
AMYL BUTYRATES	2620	3		ARSENIC ACID, LIQUID	1553	6.1	
AMYL CHLORIDE	1107	3		ARSENIC ACID, SOLID	1554	6.1	
n-AMYLENE, see	1108	3		ARSENICAL DUST	1562	6.1	
AMYL FORMATES	1109	3		Arsenical flue dust, see	1562	6.1	
AMYL MERCAPTAN	1111	3		ARSENICAL PESTICIDE, LIQUID, FLAMMABLE, TOXIC, flash-point less than 23 °C	2760	3	
n-AMYL METHYL KETONE	1110	3					
AMYL NITRATE	1112	3		ARSENICAL PESTICIDE, LIQUID, TOXIC	2994	6.1	
AMYL NITRITE	1113	3		ARSENICAL PESTICIDE, LIQUID, TOXIC, FLAMMABLE, flash-point not less than 23 °C	2993	6.1	
AMYLTRICHLOROSILANE	1728	8					
Anaesthetic ether, see	1155	3		ARSENICAL PESTICIDE, SOLID, TOXIC	2759	6.1	
ANILINE	1547	6.1					
Aniline chloride, see	1548	6.1		ARSENIC BROMIDE	1555	6.1	
ANILINE HYDROCHLORIDE	1548	6.1		Arsenic (III) bromide, see	1555	6.1	
Aniline oil, see	1547	6.1					

Name and description	UN No.	Class	Remarks	Name and description	UN No.	Class	Remarks
Arsenic chloride, see	1560	6.1		ARTICLES, PYROTECHNIC for technical purposes	0428 0429 0430 0431 0432	1 1 1 1 1	
ARSENIC COMPOUND, LIQUID, N.O.S., inorganic, including: Arsenates, n.o.s., Arsenites, n.o.s.; and Arsenic sulphides, n.o.s.	1556	6.1					
ARSENIC COMPOUND, SOLID, N.O.S., inorganic, including: Arsenates, n.o.s.; Arsenites, n.o.s.; and Arsenic sulphides, n.o.s.	1557	6.1		ARYLSULPHONIC ACIDS, LIQUID with more than 5% free sulphuric acid	2584	8	
				ARYLSULPHONIC ACIDS, LIQUID with not more than 5% free sulphuric acid	2586	8	
Arsenic (III) oxide, see	1561	6.1					
Arsenic (V) oxide, see	1559	6.1		ARYLSULPHONIC ACIDS, SOLID with more than 5% free sulphuric acid	2583	8	
ARSENIC PENTOXIDE	1559	6.1					
Arsenic sulphides, see	1556 1557	6.1 6.1		ARYLSULPHONIC ACIDS, SOLID with not more than 5% free sulphuric acid	2585	8	
ARSENIC TRICHLORIDE	1560	6.1		ASBESTOS, AMPHIBOLE	2212	9	
ARSENIC TRIOXIDE	1561	6.1		ASBESTOS, CHRYSOTILE	2590	9	
Arsenious chloride, see	1560	6.1		Asphalt, with a flash-point above 60 °C, at or above its flash-point, see	3256	3	
Arsenites, n.o.s., see	1556 1557	6.1 6.1					
Arsenous chloride, see	1560	6.1		Asphalt, at or above 100 °C and below its flash-point, see	3257	9	
ARSINE	2188	2		Aviation regulated liquid, n.o.s.	3334	9	Not subject to ADR
ARSINE, ADSORBED	3522	2					
ARTICLES, EEI, see	0486	1		Aviation regulated solid, n.o.s.	3335	9	Not subject to ADR
ARTICLES, EXPLOSIVE, EXTREMELY INSENSITIVE	0486	1					
ARTICLES, EXPLOSIVE, N.O.S.	0349 0350 0351 0352 0353 0354 0355 0356 0462 0463 0464 0465 0466 0467 0468 0469 0470 0471 0472	1 1 1 1 1 1 1 1 1 1 1 1 1 1 1 1 1 1 1		AZODICARBONAMIDE	3242	4.1	
				Bag charges, see	0242 0279 0414	1 1 1	
				Ballistite, see	0160 0161	1 1	
				Bangalore torpedoes, see	0136 0137 0138 0294	1 1 1 1	
				BARIUM	1400	4.3	
				BARIUM ALLOYS, PYROPHORIC	1854	4.2	
				BARIUM AZIDE, dry or wetted with less than 50% water, by mass	0224	1	
				BARIUM AZIDE, WETTED with not less than 50% water, by mass	1571	4.1	
ARTICLES, PRESSURIZED, HYDRAULIC (containing non-flammable gas)	3164	2		Barium binoxide, see	1449	5.1	
				BARIUM BROMATE	2719	5.1	
ARTICLES, PRESSURIZED, PNEUMATIC (containing non-flammable gas)	3164	2		BARIUM CHLORATE, SOLID	1445	5.1	
				BARIUM CHLORATE, SOLUTION	3405	5.1	
ARTICLES, PYROPHORIC	0380	1		BARIUM COMPOUND, N.O.S.	1564	6.1	
				BARIUM CYANIDE	1565	6.1	
				Barium dioxide, see	1449	5.1	

Name and description	UN No.	Class	Remarks	Name and description	UN No.	Class	Remarks
BARIUM HYPOCHLORITE with more than 22% available chlorine	2741	5.1		Benzyl chlorocarbonate, see	1739	8	
				BENZYL CHLOROFORMATE	1739	8	
BARIUM NITRATE	1446	5.1		Benzyl cyanide, see	2470	6.1	
BARIUM OXIDE	1884	6.1		BENZYLDIMETHYLAMINE	2619	8	
BARIUM PERCHLORATE, SOLID	1447	5.1		BENZYLIDENE CHLORIDE	1886	6.1	
BARIUM PERCHLORATE, SOLUTION	3406	5.1		BENZYL IODIDE	2653	6.1	
BARIUM PERMANGANATE	1448	5.1		BERYLLIUM COMPOUND, N.O.S.	1566	6.1	
BARIUM PEROXIDE	1449	5.1		BERYLLIUM NITRATE	2464	5.1	
Barium selenate, see	2630	6.1		BERYLLIUM POWDER	1567	6.1	
Barium selenite, see	2630	6.1		Bhusa	1327	4.1	Not subject to ADR
Barium superoxide, see	1449	5.1		BICYCLO[2.2.1]HEPTA-2,5-DIENE, STABILIZED	2251	3	
BATTERIES, CONTAINING SODIUM	3292	4.3		Bifluorides, n.o.s., see	1740	8	
BATTERIES, DRY, CONTAINING POTASSIUM HYDROXIDE SOLID, electric storage	3028	8		BIOLOGICAL SUBSTANCE, CATEGORY B	3373	6.2	
Batteries, nickel-metal hydride	3496	9	Not subject to ADR	(BIO) MEDICAL WASTE, N.O.S.	3291	6.2	
BATTERIES, WET, FILLED WITH ACID, electric storage	2794	8		BIPYRIDILIUM PESTICIDE, LIQUID, FLAMMABLE, TOXIC, flash-point less than 23 °C	2782	3	
BATTERIES, WET, FILLED WITH ALKALI, electric storage	2795	8		BIPYRIDILIUM PESTICIDE, LIQUID, TOXIC	3016	6.1	
BATTERIES, WET, NON-SPILLABLE, electric storage	2800	8		BIPYRIDILIUM PESTICIDE, LIQUID, TOXIC, FLAMMABLE, flash-point not less than 23 °C	3015	6.1	
BATTERY FLUID, ACID	2796	8		BIPYRIDILIUM PESTICIDE, SOLID, TOXIC	2781	6.1	
BATTERY FLUID, ALKALI	2797	8					
BATTERY POWERED EQUIPMENT	3171	9		BISULPHATES, AQUEOUS SOLUTION	2837	8	
BATTERY POWERED VEHICLE	3171	9		BISULPHITES, AQUEOUS SOLUTION, N.O.S.	2693	8	
BENZALDEHYDE	1990	9		Bitumen, with a flash-point above 60 °C, at or above its flash-point, see	3256	3	
BENZENE	1114	3					
BENZENESULPHONYL CHLORIDE	2225	8		Bitumen, at or above 100 °C and below its flash-point, see	3257	9	
Benzenethiol, see	2337	6.1		BLACK POWDER, COMPRESSED	0028	1	
BENZIDINE	1885	6.1		BLACK POWDER, granular or as a meal	0027	1	
Benzol, see	1114	3					
Benzolene, see	1268	3		BLACK POWDER, IN PELLETS	0028	1	
BENZONITRILE	2224	6.1		Blasting cap assemblies, see	0360	1	
BENZOQUINONE	2587	6.1			0361	1	
Benzosulphochloride, see	2225	8		Blasting caps, electric, see	0030	1	
BENZOTRICHLORIDE	2226	8			0255	1	
BENZOTRIFLUORIDE	2338	3			0456	1	
BENZOYL CHLORIDE	1736	8		Blasting caps, non electric, see	0029	1	
BENZYL BROMIDE	1737	6.1			0267	1	
					0455	1	
BENZYL CHLORIDE	1738	6.1		Bleaching powder, see	2208	5.1	

Name and description	UN No.	Class	Remarks	Name and description	UN No.	Class	Remarks
BOMBS with bursting charge	0033	1		BROMINE SOLUTION	1744	8	
	0034	1		BROMINE TRIFLUORIDE	1746	5.1	
	0035	1		BROMOACETIC ACID, SOLID	3425	8	
	0291	1		BROMOACETIC ACID, SOLUTION	1938	8	
Bombs, illuminating, see	0254	1					
BOMBS, PHOTO-FLASH	0037	1		BROMOACETONE	1569	6.1	
	0038	1		omega-Bromoacetone, see	2645	6.4	
	0039	1		BROMOACETYL BROMIDE	2513	8	
	0299	1		BROMOBENZENE	2514	3	
BOMBS, SMOKE, NON-EXPLOSIVE with corrosive liquid, without initiating device	2028	8		BROMOBENZYL CYANIDES, LIQUID	1694	6.1	
Bombs, target identification, see	0171	1		BROMOBENZYL CYANIDES, SOLID	3449	6.1	
	0254	1					
	0297	1		1-BROMOBUTANE	1126	3	
BOMBS WITH FLAMMABLE LIQUID with bursting charge	0399	1		2-BROMOBUTANE	2339	3	
	0400	1		BROMOCHLOROMETHANE	1887	6.1	
BOOSTERS WITH DETONATOR	0225	1		1-BROMO-3-CHLOROPROPANE	2688	6.1	
	0268	1		1-Bromo-2,3-epoxypropane, see	2558	6.1	
BOOSTERS without detonator	0042	1		Bromoethane, see	1891	6.1	
	0283	1		2-BROMOETHYL ETHYL ETHER	2340	3	
Borate and chlorate mixture, see	1458	5.1		BROMOFORM	2515	6.1	
BORNEOL	1312	4.1		Bromomethane, see	1062	2	
BORON TRIBROMIDE	2692	8		1-BROMO-3-METHYLBUTANE	2341	3	
BORON TRICHLORIDE	1741	2		BROMOMETHYLPROPANES	2342	3	
BORON TRIFLUORIDE ACETIC ACID COMPLEX, LIQUID	1742	8		2-BROMO-2-NITROPROPANE-1,3-DIOL	3241	4.1	
BORON TRIFLUORIDE ACETIC ACID COMPLEX, SOLID	3419	8		2-BROMOPENTANE	2343	3	
BORON TRIFLUORIDE	1008	2		BROMOPROPANES	2344	3	
BORON TRIFLUORIDE, ADSORBED	3519	2		3-BROMOPROPYNE	2345	3	
BORON TRIFLUORIDE DIETHYL ETHERATE	2604	8		BROMOTRIFLUOROETHYLENE	2419	2	
				BROMOTRIFLUOROMETHANE	1009	2	
BORON TRIFLUORIDE DIHYDRATE	2851	8		BRUCINE	1570	6.1	
BORON TRIFLUORIDE DIMETHYL ETHERATE	2965	4.3		BURSTERS, explosive	0043	1	
BORON TRIFLUORIDE PROPIONIC ACID COMPLEX, LIQUID	1743	8		BUTADIENES AND HYDROCARBON MIXTURE, STABILIZED, having a vapour pressure at 70 °C not exceeding 1.1 MPa (11 bar) and a density at 50 °C not lower than 0.525 kg/l	1010	2	
BORON TRIFLUORIDE PROPIONIC ACID COMPLEX, SOLID	3420	8					
BROMATES, INORGANIC, N.O.S.	1450	5.1		BUTADIENES, STABILIZED, (1,2-butadiene)	1010	2	
BROMATES, INORGANIC, AQUEOUS SOLUTION, N.O.S	3213	5.1		BUTADIENES, STABILIZED, (1,3-butadiene)	1010	2	
BROMINE	1744	8					
BROMINE CHLORIDE	2901	2		BUTANE	1011	2	
BROMINE PENTAFLUORIDE	1745	5.1		BUTANEDIONE	2346	3	

Name and description	UN No.	Class	Remarks	Name and description	UN No.	Class	Remarks
Butane-1-thiol, see	2347	3		n-BUTYL METHACRYLATE, STABILIZED	2227	3	
BUTANOLS	1120	3		BUTYL METHYL ETHER	2350	3	
1-Butanol, see	1120	3		BUTYL NITRITES	2351	3	
Butan-2-ol, see	1120	3		Butylphenols, liquid, see	3145	8	
Butanol, secondary, see	1120	3		Butylphenols, solid, see	2430	8	
Butanol, tertiary, see	1120	3		BUTYL PROPIONATES	1914	3	
Butanone, see	1193	3		p-tert-Butyltoluene, see	2667	6.1	
2-Butenal, see	1143	6.1		BUTYLTOLUENES	2667	6.1	
Butene, see	1012	2		BUTYLTRICHLOROSILANE	1747	8	
Bute-1-ene-3-one, see	1251	3		5-tert-BUTYL-2,4,6-TRINITRO-m-XYLENE	2956	4.1	
1,2-Buteneoxide, see	3022	3		BUTYL VINYL ETHER, STABILIZED	2352	3	
2-Buten-1-ol, see	2614	3		But-1-yne, see	2452	2	
BUTYL ACETATES	1123	3		1,4-BUTYNEDIOL	2716	6.1	
Butyl acetate, secondary, see	1123	3		2-Butyne-1,4-diol, see	2716	6.1	
BUTYL ACID PHOSPHATE	1718	8		BUTYRALDEHYDE	1129	3	
BUTYL ACRYLATES, STABILIZED	2348	3		BUTYRALDOXIME	2840	3	
Butyl alcohols, see	1120	3		BUTYRIC ACID	2820	8	
n-BUTYLAMINE	1125	3		BUTYRIC ANHYDRIDE	2739	8	
N-BUTYLANILINE	2738	6.1		Butyrone, see	2710	3	
sec-Butyl benzene, see	2709	3		BUTYRONITRILE	2411	3	
BUTYLBENZENES	2709	3		Butyroyl chloride, see	2353	3	
n-Butyl bromide, see	1126	3		BUTYRYL CHLORIDE	2353	3	
n-Butyl chloride, see	1127	3		Cable cutters, explosive, see	0070	1	
n-BUTYL CHLOROFORMATE	2743	6.1		CACODYLIC ACID	1572	6.1	
tert-BUTYLCYCLOHEXYL CHLOROFORMATE	2747	6.1		CADMIUM COMPOUND	2570	6.1	
BUTYLENES MIXTURE or 1-BUTYLENE or CIS-2-BUTYLENE or TRANS-2-BUTYLENE	1012	2		CAESIUM	1407	4.3	
				CAESIUM HYDROXIDE	2682	8	
1,2-BUTYLENE OXIDE, STABILIZED	3022	3		CAESIUM HYDROXIDE SOLUTION	2681	8	
Butyl ethers, see	1149	3		CAESIUM NITRATE	1451	5.1	
Butyl ethyl ether, see	1179	3		Caffeine, see	1544	6.1	
n-BUTYL FORMATE	1128	3		Cajeputene, see	2052	3	
tert-BUTYL HYPOCHLORITE	3255	4.2	Carriage prohibited	CALCIUM	1401	4.3	
N,n-BUTYLIMIDAZOLE	2690	6.1		CALCIUM ALLOYS, PYROPHORIC	1855	4.2	
N,n-Butyliminazole, see	2690	6.1		CALCIUM ARSENATE	1573	6.1	
n-BUTYL ISOCYANATE	2485	6.1		CALCIUM ARSENATE AND CALCIUM ARSENITE MIXTURE, SOLID	1574	6.1	
tert-BUTYL ISOCYANATE	2484	6.1					
Butyl lithium, see	3394	4.2		Calcium bisulphite solution, see	2693	8	
BUTYL MERCAPTAN	2347	3		CALCIUM CARBIDE	1402	4.3	

Name and description	UN No.	Class	Remarks
CALCIUM CHLORATE	1452	5.1	
CALCIUM CHLORATE, AQUEOUS SOLUTION	2429	5.1	
CALCIUM CHLORITE	1453	5.1	
CALCIUM CYANAMIDE with more than 0.1% calcium carbide	1403	4.3	
CALCIUM CYANIDE	1575	6.1	
CALCIUM DITHIONITE	1923	4.2	
CALCIUM HYDRIDE	1404	4.3	
CALCIUM HYDROSULPHITE, see	1923	4.2	
CALCIUM HYPOCHLORITE, DRY with more than 39% available chlorine (8.8% available oxygen)	1748	5.1	
CALCIUM HYPOCHLORITE, DRY, CORROSIVE with more than 39% available chlorine (8.8% available oxygen)	3485	5.1	
CALCIUM HYPOCHLORITE, HYDRATED with not less than 5.5% but not more than 16% water	2880	5.1	
CALCIUM HYPOCHLORITE, HYDRATED MIXTURE with not less than 5.5% but not more than 16% water	2880	5.1	
CALCIUM HYPOCHLORITE, HYDRATED, CORROSIVE with not less than 5.5% but not more than 16% water	3487	5.1	
CALCIUM HYPOCHLORITE, HYDRATED MIXTURE, CORROSIVE with not less than 5.5% but not more than 16% water	3487	5.1	
CALCIUM HYPOCHLORITE MIXTURE, DRY with more than 10% but not more than 39% available chlorine	2208	5.1	
CALCIUM HYPOCHLORITE MIXTURE, DRY with more than 39% available chlorine (8.8% available oxygen)	1748	5.1	
CALCIUM HYPOCHLORITE MIXTURE, DRY, CORROSIVE with more than 10% but not more than 39% available chlorine	3486	5.1	
CALCIUM HYPOCHLORITE MIXTURE, DRY, CORROSIVE with more than 39% available chlorine (8.8% available oxygen)	3485	5.1	
CALCIUM MANGANESE SILICON	2844	4.3	
CALCIUM NITRATE	1454	5.1	
Calcium oxide	1910	8	Not subject to ADR

Name and description	UN No.	Class	Remarks
CALCIUM PERCHLORATE	1455	5.1	
CALCIUM PERMANGANATE	1456	5.1	
CALCIUM PEROXIDE	1457	5.1	
CALCIUM PHOSPHIDE	1360	4.3	
CALCIUM, PYROPHORIC	1855	4.2	
CALCIUM RESINATE	1313	4.1	
CALCIUM RESINATE, FUSED	1314	4.1	
Calcium selenate, see	2630	6.1	
CALCIUM SILICIDE	1405	4.3	
Calcium silicon, see	1405	4.3	
Calcium superoxide, see	1457	5.1	
CAPACITOR, ASYMMETRIC, (with an energy storage capacity greater than 0.3Wh)	3508	9	
CAPACITOR, ELECTRIC DOUBLE LAYER (with an energy storage capacity greater than 0.3 Wh)	3499	9	
Camphanone, see	2717	4.1	
CAMPHOR OIL	1130	3	
CAMPHOR, synthetic	2717	4.1	
CAPROIC ACID	2829	8	
CARBAMATE PESTICIDE, LIQUID, FLAMMABLE, TOXIC, flash-point less than 23 °C	2758	3	
CARBAMATE PESTICIDE, LIQUID, TOXIC	2992	6.1	
CARBAMATE PESTICIDE, LIQUID, TOXIC, FLAMMABLE, flash-point not less than 23 °C	2991	6.1	
CARBAMATE PESTICIDE, SOLID, TOXIC	2757	6.1	
Carbolic acid, see	1671	6.1	
	2312	6.1	
	2821	6.1	
CARBON, animal or vegetable origin	1361	4.2	
CARBON, ACTIVATED	1362	4.2	
Carbon bisulphide, see	1131	3	
Carbon black (animal or vegetable origin), see	1361	4.2	
CARBON DIOXIDE	1013	2	
Carbon dioxide and ethylene oxide mixture, see	1041	2	
	1952	2	
	3300	2	
CARBON DIOXIDE, REFRIGERATED LIQUID	2187	2	

Name and description	UN No.	Class	Remarks	Name and description	UN No.	Class	Remarks
Carbon dioxide, solid	1845	9	Not subject to ADR - when used as a coolant, see 5.5.3	CARTRIDGES, SIGNAL	0054 0312 0405	1 1 1	
CARBON DISULPHIDE	1131	3		CARTRIDGES, SMALL ARMS	0012 0339 0417	1 1 1	
Carbonic anhydride, see	1013 1845 2187	2 9 2		CARTRIDGES, SMALL ARMS, BLANK	0014 0327 0338	1 1 1	
CARBON MONOXIDE, COMPRESSED	1016	2		Cartridges, starter, jet engine, see	0275 0276 0323 0381	1 1 1 1	
Carbon oxysulphide, see	2204	2.3					
CARBON TETRABROMIDE	2516	6.1		CASES, CARTRIDGE, EMPTY, WITH PRIMER	0055 0379	1 1	
CARBON TETRACHLORIDE	1846	6.1		CASES, COMBUSTIBLE, EMPTY, WITHOUT PRIMER	0446 0447	1 1	
Carbonyl chloride, see	1076	2		Casinghead gasoline, see	1203	3	
CARBONYL FLUORIDE	2417	2		CASTOR BEANS	2969	9	
CARBONYL SULPHIDE	2204	2		CASTOR FLAKE	2969	9	
Cartridge cases, empty, primed, see	0055 0379	1 1		CASTOR MEAL	2969	9	
Cartridges, actuating, for fire extinguisher or apparatus valve, see	0275 0276 0323 0381	1 1 1 1		CASTOR POMACE	2969	9	
				CAUSTIC ALKALI LIQUID, N.O.S.	1719	8	
Cartridges, explosive, see	0048	1		Caustic potash, see	1814	8	
CARTRIDGES, FLASH	0049 0050	1 1		Caustic soda, see	1824	8	
CARTRIDGES FOR TOOLS, BLANK	0014	1		Caustic soda liquor, see	1824	8	
				CELLS, CONTAINING SODIUM	3292	4.3	
CARTRIDGES FOR WEAPONS with bursting charge	0005 0006 0007 0321 0348 0412	1 1 1 1 1 1		CELLULOID in block, rods, rolls, sheets, tubes, etc., except scrap	2000	4.1	
				CELLULOID, SCRAP	2002	4.2	
				Cement, see	1133	3	
CARTRIDGES FOR WEAPONS, BLANK	0014 0326 0327 0338 0413	1 1 1 1 1		CERIUM, slabs, ingots or rods	1333	4.1	
				CERIUM, turnings or gritty powder	3078	4.3	
				Cer mishmetall, see	1323	4.1	
				Charcoal, activated, see	1362	4.1	
CARTRIDGES FOR WEAPONS, INERT PROJECTILE	0012 0328 0339 0417	1 1 1 1		Charcoal, non-activated, see	1361	4.2	
				CHARGES, BURSTING, PLASTICS BONDED	0457 0458 0459 0460	1 1 1 1	
Cartridges, illuminating, see	0171 0254 0297	1 1 1		CHARGES, DEMOLITION	0048	1	
CARTRIDGES, OIL WELL	0277 0278	1 1		CHARGES, DEPTH	0056	1	
CARTRIDGES, POWER DEVICE	0275 0276 0323 0381	1 1 1 1		Charges, expelling, explosive, for fire extinguishers, see	0275 0276 0323 0381	1 1 1 1	

Name and description	UN No.	Class	Remarks	Name and description	UN No.	Class	Remarks
CHARGES, EXPLOSIVE, COMMERCIAL without detonator	0442 0443 0444 0445	1 1 1 1		CHLORINE, ADSORBED	3520	2	
				CHLORINE PENTAFLUORIDE	2548	2	
				CHLORINE TRIFLUORIDE	1749	2	
CHARGES, PROPELLING	0271 0272 0415 0491	1 1 1 1		CHLORITES, INORGANIC, N.O.S.	1462	5.1	
				CHLORITE SOLUTION	1908	8	
				Chloroacetaldehyde, see	2232	6.1	
CHARGES, PROPELLING, FOR CANNON	0242 0279 0414	1 1 1		CHLOROACETIC ACID, MOLTEN	3250	6.1	
				CHLOROACETIC ACID, SOLID	1751	6.1	
CHARGES, SHAPED, FLEXIBLE, LINEAR	0237 0288	1 1		CHLOROACETIC ACID SOLUTION	1750	6.1	
CHARGES, SHAPED, without detonator	0059 0439 0440 0441	1 1 1 1		CHLOROACETONE, STABILIZED	1695	6.1	
				CHLOROACETONITRILE	2668	6.1	
				CHLOROACETOPHENONE, LIQUID	3416	6.1	
CHARGES, SUPPLEMENTARY, EXPLOSIVE	0060	1		CHLOROACETOPHENONE, SOLID	1697	6.1	
CHEMICAL KIT	3316	9		CHLOROACETYL CHLORIDE	1752	6.1	
CHEMICAL SAMPLE, TOXIC	3315	6.1		CHLOROANILINES, LIQUID	2019	6.1	
CHEMICAL UNDER PRESSURE, N.O.S.	3500	2		CHLOROANILINES, SOLID	2018	6.1	
				CHLOROANISIDINES	2233	6.1	
CHEMICAL UNDER PRESSURE, CORROSIVE, N.O.S.	3503	2		CHLOROBENZENE	1134	3	
CHEMICAL UNDER PRESSURE, FLAMMABLE, N.O.S.	3501	2		CHLOROBENZO-TRIFLUORIDES	2234	3	
				CHLOROBENZYL CHLORIDES, LIQUID	2235	6.1	
CHEMICAL UNDER PRESSURE, FLAMMABLE, CORROSIVE, N.O.S.	3505	2		CHLOROBENZYL CHLORIDES, SOLID	3427	6.1	
CHEMICAL UNDER PRESSURE, FLAMMABLE, TOXIC, N.O.S.	3504	2		1-Chloro-3-bromopropane, see	2688	6.1	
				1-Chlorobutane, see	1127	3	
CHEMICAL UNDER PRESSURE, TOXIC, N.O.S.	3502	2		2-Chlorobutane, see	1127	3	
Chile saltpetre, see	1498	5.1		CHLOROBUTANES	1127	3	
CHLORAL, ANHYDROUS, STABILIZED	2075	6.1		CHLOROCRESOLS, SOLUTION	2669	6.1	
				CHLOROCRESOLS, SOLID	3437	6.1	
CHLORATE AND BORATE MIXTURE	1458	5.1		CHLORODIFLUORO-BROMOMETHANE	1974	2	
CHLORATE AND MAGNESIUM CHLORIDE MIXTURE, SOLID	1459	5.1		1-CHLORO-1,1-DIFLUORO-ETHANE	2517	2	
CHLORATE AND MAGNESIUM CHLORIDE MIXTURE, SOLUTION	3407	5.1		CHLORODIFLUOROMETHANE	1018	2	
CHLORATES, INORGANIC, N.O.S.	1461	5.1		CHLORODIFLUOROMETHANE AND CHLORO-PENTAFLUOROETHANE MIXTURE with fixed boiling point, with approximately 49% chlorodifluoromethane	1973	2	
CHLORATES, INORGANIC, AQUEOUS SOLUTION, N.O.S.	3210	5.1					
CHLORIC ACID, AQUEOUS SOLUTION with not more than 10% chloric acid	2626	5.1		3-Chloro-1,2-dihydroxypropane, see	2689	6.1	
				Chlorodimethyl ether, see	1239	6.1	
CHLORINE	1017	2					

Name and description	UN No.	Class	Remarks	Name and description	UN No.	Class	Remarks
CHLORODINITROBENZENES, LIQUID	1577	6.1		CHLOROPICRIN AND METHYL BROMIDE MIXTURE, with more than 2% chloropicrin	1581	2	
CHLORODINITROBENZENES, SOLID	3441	6.1		CHLOROPICRIN AND METHYL CHLORIDE MIXTURE	1582	2	
2-CHLOROETHANAL	2232	6.1		CHLOROPICRIN MIXTURE, N.O.S.	1583	6.1	
Chloroethane, see	1037	2		CHLOROPLATINIC ACID, SOLID	2507	8	
Chloroethane nitrile, see	2668	6.1		CHLOROPRENE, STABILIZED	1991	3	
2-Chloroethanol, see	1135	6.1		1-CHLOROPROPANE	1278	3	
CHLOROFORM	1888	6.1		2-CHLOROPROPANE	2356	3	
CHLOROFORMATES, TOXIC, CORROSIVE, N.O.S.	3277	6.1		3-Chloro-propanediol-1,2, see	2689	6.1	
CHLOROFORMATES, TOXIC, CORROSIVE, FLAMMABLE, N.O.S.	2742	6.1		3-CHLOROPROPANOL-1	2849	6.1	
				2-CHLOROPROPENE	2456	3	
Chloromethane, see	1063	2		3-Chloropropene, see	1100	3	
1-Chloro-3-methylbutane, see	1107	3		3-Chloroprop-1-ene, see	1100	3	
2-Chloro-2-methylbutane, see	1107	3					
CHLOROMETHYL CHLOROFORMATE	2745	6.1		2-CHLOROPROPIONIC ACID	2511	8	
Chloromethyl cyanide, see	2668	6.1		2-CHLOROPYRIDINE	2822	6.1	
CHLOROMETHYL ETHYL ETHER	2354	3		CHLOROSILANES, CORROSIVE, N.O.S.	2987	8	
Chloromethyl methyl ether, see	1239	6.1		CHLOROSILANES, CORROSIVE, FLAMMABLE, N.O.S.	2986	8	
3-CHLORO-4-METHYLPHENYL ISOCYANATE, LIQUID	2236	6.1		CHLOROSILANES, FLAMMABLE, CORROSIVE, N.O.S.	2985	3	
3-CHLORO-4-METHYLPHENYL ISOCYANATE, SOLID	3428	6.1		CHLOROSILANES, TOXIC, CORROSIVE, N.O.S.	3361	6.1	
3-Chloro-2-methylprop-1-ene, see	2554	3		CHLOROSILANES, TOXIC, CORROSIVE, FLAMMABLE, N.O.S.	3362	6.1	
CHLORONITROANILINES	2237	6.1					
CHLORONITROBENZENES LIQUID	3409	6.1		CHLOROSILANES, WATER-REACTIVE, FLAMMABLE, CORROSIVE, N.O.S.	2988	4.3	
CHLORONITROBENZENES SOLID	1578	6.1					
CHLORONITROTOLUENES, LIQUID	2433	6.1		CHLOROSULPHONIC ACID (with or without sulphur trioxide)	1754	8	
CHLORONITROTOLUENES, SOLID	3457	6.1		1-CHLORO-1,2,2,2-TETRA-FLUOROETHANE	1021	2	
CHLOROPENTAFLUORO-ETHANE	1020	2		CHLOROTOLUENES	2238	3	
CHLOROPHENOLATES, LIQUID	2904	8		4-CHLORO-o-TOLUIDINE HYDROCHLORIDE, SOLID	1579	6.1	
CHLOROPHENOLATES, SOLID	2905	8		4-CHLORO-o-TOLUIDINE HYDROCHLORIDE, SOLUTION	3410	6.1	
CHLOROPHENOLS, LIQUID	2021	6.1		CHLOROTOLUIDINES LIQUID	3429	6.1	
CHLOROPHENOLS, SOLID	2020	6.1		CHLOROTOLUIDINES SOLID	3429	6.1	
CHLOROPHENYL-TRICHLOROSILANE	1753	8		1-CHLORO-2,2,2-TRIFLUORO-ETHANE	1983	2	
CHLOROPICRIN	1580	6.1		Chlorotrifluoroethylene, see	1082	2	

Name and description	UN No.	Class	Remarks	Name and description	UN No.	Class	Remarks
CHLOROTRIFLUOROMETHANE	1022	2		Collodion cottons, see	0340	1	
CHLOROTRIFLUOROMETHANE AND TRIFLUOROMETHANE AZEOTROPIC MIXTURE with approximately 60% chlorotrifluoromethane	2599	2			0341	1	
					0342	1	
					2059	3	
					2555	4.1	
					2556	4.1	
					2557	4.1	
Chromic acid, solid, see	1463	5.1		COMPONENTS, EXPLOSIVE TRAIN, N.O.S.	0382	1	
CHROMIC ACID SOLUTION	1755	8			0383	1	
Chromic anhydride, solid, see	1463	5.1			0384	1	
					0461	1	
CHROMIC FLUORIDE, SOLID	1756	8		Composition B, see	0118	1	
CHROMIC FLUORIDE SOLUTION	1757	8		COMPRESSED GAS, N.O.S.	1956	2	
Chromic nitrate, see	2720	5.1		COMPRESSED GAS, FLAMMABLE, N.O.S.	1954	2	
Chromium (VI) dichloride dioxide, see	1758	8		COMPRESSED GAS, OXIDIZING, N.O.S.	3156	2	
Chromium (III) fluoride, solid, see	1756	8		COMPRESSED GAS, TOXIC, N.O.S.	1955	2	
CHROMIUM NITRATE	2720	5.1		COMPRESSED GAS, TOXIC, CORROSIVE, N.O.S.	3304	2	
Chromium (III) nitrate, see	2720	5.1		COMPRESSED GAS, TOXIC, FLAMMABLE, N.O.S.	1953	2	
CHROMIUM OXYCHLORIDE	1758	8		COMPRESSED GAS, TOXIC, FLAMMABLE, CORROSIVE, N.O.S.	3305	2	
CHROMIUM TRIOXIDE, ANHYDROUS	1463	5.1					
CHROMOSULPHURIC ACID	2240	8		COMPRESSED GAS, TOXIC, OXIDIZING, N.O.S.	3303	2	
Chrysotile, see	2590	9					
Cinene, see	2052	3		COMPRESSED GAS, TOXIC, OXIDIZING, CORROSIVE, N.O.S.	3306	2	
Cinnamene, see	2055	3					
Cinnamol, see	2055	3		CONTRIVANCES, WATER-ACTIVATED with burster, expelling charge or propelling charge	0248	1	
CLINICAL WASTE, UNSPECIFIED, N.O.S.	3291	6.2			0249	1	
COAL GAS, COMPRESSED	1023	2					
COAL TAR DISTILLATES, FLAMMABLE	1136	3		COPPER ACETOARSENITE	1585	6.1	
Coal tar naphtha, see	1268	3		COPPER ARSENITE	1586	6.1	
Coal tar oil, see	1136	3		Copper (II) arsenite, see	1586	6.1	
COATING SOLUTION (includes surface treatments or coatings used for industrial or other purposes such as vehicle under coating, drum or barrel lining)	1139	3		COPPER BASED PESTICIDE, LIQUID, FLAMMABLE, TOXIC, flash-point less than 23 °C	2776	3	
				COPPER BASED PESTICIDE, LIQUID, TOXIC	3010	6.1	
COBALT NAPHTHENATES, POWDER	2001	4.1		COPPER BASED PESTICIDE, LIQUID, TOXIC, FLAMMABLE, flash-point not less than 23 °C	3009	6.1	
COBALT RESINATE, PRECIPITATED	1318	4.1		COPPER BASED PESTICIDE, SOLID, TOXIC	2775	6.1	
Cocculus, see	3172	6.1		COPPER CHLORATE	2721	5.1	
	3462	6.1		Copper (II) chlorate, see	2721	5.1	
				COPPER CHLORIDE	2802	8	
				COPPER CYANIDE	1587	6.1	

Name and description	UN No.	Class	Remarks	Name and description	UN No.	Class	Remarks
Copper selenate, see	2630	6.1		CORROSIVE SOLID, WATER-REACTIVE, N.O.S.	3096	8	
Copper selenite, see	2630	6.1		COTTON WASTE, OILY	1364	4.2	
COPRA	1363	4.2		COTTON, WET	1365	4.2	
CORD, DETONATING, flexible	0065	1		COUMARIN DERIVATIVE PESTICIDE, LIQUID, FLAMMABLE, TOXIC, flash-point less than 23 °C	3024	3	
	0289	1					
CORD, DETONATING, metal clad	0102	1					
	0290	1		COUMARIN DERIVATIVE PESTICIDE, LIQUID, TOXIC	3026	6.1	
CORD, DETONATING, MILD EFFECT, metal clad	0104	1		COUMARIN DERIVATIVE PESTICIDE, LIQUID, TOXIC, FLAMMABLE, flash-point not less than 23 °C	3025	6.1	
CORD, IGNITER	0066	1					
Cordite, see	0160	1					
	0161	1		COUMARIN DERIVATIVE PESTICIDE, SOLID, TOXIC	3027	6.1	
CORROSIVE LIQUID, N.O.S.	1760	8					
CORROSIVE LIQUID, ACIDIC, INORGANIC, N.O.S.	3264	8		Creosote, see	2810	6.1	
				Creosote salts, see	1334	4.1	
CORROSIVE LIQUID, ACIDIC, ORGANIC, N.O.S.	3265	8		CRESOLS, LIQUID	2076	6.1	
				CRESOLS, SOLID	3455	6.1	
CORROSIVE LIQUID, BASIC, INORGANIC, N.O.S.	3266	8		CRESYLIC ACID	2022	6.1	
CORROSIVE LIQUID, BASIC, ORGANIC, N.O.S.	3267	8		Crocidolite, see	2212	9	
CORROSIVE LIQUID, FLAMMABLE, N.O.S.	2920	8		CROTONALDEHYDE or CROTONALDEHYDE, STABILIZED	1143	6.1	
CORROSIVE LIQUID, OXIDIZING, N.O.S.	3093	8		CROTONIC ACID, LIQUID	3472	8	
CORROSIVE LIQUID, SELF-HEATING, N.O.S.	3301	8		CROTONIC ACID, SOLID	2823	8	
				Crotonic aldehyde / Crotonic aldehyde, stabilized, see	1143	6.1	
CORROSIVE LIQUID, TOXIC, N.O.S.	2922	8					
				CROTONYLENE	1144	3	
CORROSIVE LIQUID, WATER-REACTIVE, N.O.S.	3094	8		Crude naphtha, see	1268	3	
CORROSIVE SOLID, N.O.S.	1759	8		Cumene, see	1918	3	
CORROSIVE SOLID, ACIDIC, INORGANIC, N.O.S.	3260	8		Cupric chlorate, see	2721	5.1	
CORROSIVE SOLID, ACIDIC, ORGANIC, N.O.S.	3261	8		CUPRIETHYLENEDIAMINE SOLUTION	1761	8	
CORROSIVE SOLID, BASIC, INORGANIC, N.O.S.	3262	8		Cutback bitumen, with a flash-point not greater than 60 °C, see	1999	3	
CORROSIVE SOLID, BASIC, ORGANIC, N.O.S.	3263	8		Cutback bitumen, with a flash-point above 60 °C, at or above its flash-point, see	3256	3	
CORROSIVE SOLID, FLAMMABLE, N.O.S.	2921	8					
CORROSIVE SOLID, OXIDIZING, N.O.S.	3084	8		Cutback bitumen, at or above 100 °C and below its flash-point, see	3257	9	
CORROSIVE SOLID, SELF-HEATING, N.O.S.	3095	8		CUTTERS, CABLE, EXPLOSIVE	0070	1	
				CYANIDE SOLUTION, N.O.S.	1935	6.1	
CORROSIVE SOLID, TOXIC, N.O.S.	2923	8		CYANIDES, INORGANIC, SOLID, N.O.S.	1588	6.1	
				Cyanides, organic, flammable, toxic, n.o.s., see	3273	3	

Name and description	UN No.	Class	Remarks	Name and description	UN No.	Class	Remarks
Cyanides, organic, toxic, n.o.s., see	3276 3439	6.1		CYCLOOCTATETRAENE	2358	3	
				CYCLOPENTANE	1146	3	
Cyanides, organic, toxic, flammable, n.o.s., see	3275	6.1		CYCLOPENTANOL	2244	3	
				CYCLOPENTANONE	2245	3	
Cyanoacetonitrile, see	2647	6.1		CYCLOPENTENE	2246	3	
CYANOGEN	1026	2		CYCLOPROPANE	1027	2	
CYANOGEN BROMIDE	1889	6.1		CYCLOTETRAMETHYLENE-TETRANITRAMINE, DESENSITIZED	0484	1	
CYANOGEN CHLORIDE, STABILIZED	1589	2					
CYANURIC CHLORIDE	2670	8		CYCLOTETRAMETHYLENE-TETRANITRAMINE, WETTED with not less than 15% water, by mass	0226	1	
CYCLOBUTANE	2601	2					
CYCLOBUTYL CHLOROFORMATE	2744	6.1					
1,5,9-CYCLODODECATRIENE	2518	6.1		CYCLOTRIMETHYLENE-TRINITRAMINE AND CYCLOTETRAMETHYLENE-TETRANITRAMINE MIXTURE, DESENSITIZED with not less than 10% phlegmatiser by mass	0391	1	
CYCLOHEPTANE	2241	3					
CYCLOHEPTATRIENE	2603	3					
1,3,5-Cycloheptatriene, see	2603	3					
CYCLOHEPTENE	2242	3		CYCLOTRIMETHYLENE-TRINITRAMINE AND CYCLOTETRAMETHYLENE-TETRANITRAMINE MIXTURE, WETTED with not less than 15% water, by mass	0391	1	
1,4-Cyclohexadienedione, see	2587	6.1					
CYCLOHEXANE	1145	3					
Cyclehexanethiol, see	3054	3					
CYCLOHEXANONE	1915	3		CYCLOTRIMETHYLENE-TRINITRAMINE, DESENSITIZED	0483	1	
CYCLOHEXENE	2256	3					
CYCLOHEXENYLTRI-CHLOROSILANE	1762	8		CYCLOTRIMETHYLENE-TRINITRAMINE, WETTED with not less than 15% water, by mass	0072	1	
CYCLOHEXYL ACETATE	2243	3					
CYCLOHEXYLAMINE	2357	8		CYMENES	2046	3	
CYCLOHEXYL ISOCYANATE	2488	6.1		Cymol, see	2046	3	
CYCLOHEXYL MERCAPTAN	3054	3		Deanol, see	2051	8	
CYCLOHEXYLTRICHLORO-SILANE	1763	8		Dangerous goods in machinery or dangerous goods in apparatus	3363	9	Not subject to ADR [see also 1.1.3.1 (b)]
CYCLONITE AND CYCLOTETRAMETHYLENE-TETRANITRAMINE MIXTURE, WETTED with not less than 15% water, by mass or DESENSITIZED with not less than 10% phlegmatiser by mass, see	0391	1					
				DECABORANE	1868	4.1	
				DECAHYDRONAPHTHALENE	1147	3	
				Decalin, see	1147	3	
				n-DECANE	2247	3	
				DEFLAGRATING METAL SALTS OF AROMATIC NITRODERIVATIVES, N.O.S.	0132	1	
CYCLONITE, DESENSITIZED, see	0483	1		Depth charge, see	0056	1	
CYCLONITE, WETTED with not less than 15% water, by mass, see	0072	1		DESENSITIZED EXPLOSIVE, LIQUID, N.O.S.	3379	3	
CYCLOOCTADIENES	2520	3		DESENSITIZED EXPLOSIVE, SOLID, N.O.S.	3380	4.1	
CYCLOOCTADIENE PHOSPHINES, see	2940	4.2					

Name and description	UN No.	Class	Remarks	Name and description	UN No.	Class	Remarks
Detonating relays, see	0029	1		DICHLOROACETIC ACID	1764	8	
	0267	1		1,3-DICHLOROACETONE	2649	6.1	
	0360	1		DICHLOROACETYL CHLORIDE	1765	8	
	0361	1		DICHLOROANILINES, LIQUID	1590	6.1	
	0455	1		DICHLOROANILINES, SOLID	3442	6.1	
	0500	1		o-DICHLOROBENZENE	1591	6.1	
DETONATOR ASSEMBLIES, NON-ELECTRIC for blasting	0360	1		2,2'-DICHLORODIETHYL ETHER	1916	6.1	
	0361	1		DICHLORODIFLUORO-METHANE	1028	2	
	0500	1					
DETONATORS FOR AMMUNITION	0073	1		DICHLORODIFLUORO-METHANE AND DIFLUOROETHANE AZEOTROPIC MIXTURE with approximately 74% dichlorodifluoromethane	2602	2	
	0364	1					
	0365	1					
	0366	1					
DETONATORS, ELECTRIC for blasting	0030	1					
	0255	1		Dichlorodifluoromethane and ethylene oxide mixture, see	3070	2	
	0456	1					
DETONATORS, NON-ELECTRIC for blasting	0029	1		DICHLORODIMETHYL ETHER, SYMMETRICAL	2249	6.1	Carriage prohibited
	0267	1					
	0455	1		1,1-DICHLOROETHANE	2362	3	
DEUTERIUM, COMPRESSED	1957	2		1,2-Dichloroethane, see	1184	3	
DEVICES, SMALL, HYDROCARBON GAS POWERED with release device	3150	2		1,2-DICHLOROETHYLENE	1150	3	
				Di(2-chloroethyl) ether, see	1916	6.1	
DIACETONE ALCOHOL	1148	3		DICHLOROFLUOROMETHANE	1029	2	
DIALLYLAMINE	2359	3		alpha-Dichlorohydrin, see	2750	6.1	
DIALLYL ETHER	2360	3		DICHLOROISOCYANURIC ACID, DRY	2465	5.1	
4,4'-DIAMINODIPHENYL-METHANE	2651	6.1					
1,2-Diaminoethane, see	1604	8		DICHLOROISOCYANURIC ACID SALTS	2465	5.1	
Diaminopropylamine, see	2269	8		DICHLOROISOPROPYL ETHER	2490	6.1	
DI-n-AMYLAMINE	2841	3		DICHLOROMETHANE	1593	6.1	
DIAZODINITROPHENOL, WETTED with not less than 40% water, or mixture of alcohol and water, by mass	0074	1		1,1-DICHLORO-1-NITROETHANE	2650	6.1	
				DICHLOROPENTANES	1152	3	
Dibenzopyridine, see	2713	6.1		Dichlorophenol, see	2020	6.1	
DIBENZYLDICHLORO-SILANE	2434	8			2021	6.1	
DIBORANE	1911	2		DICHLOROPHENYL ISOCYANATES	2250	6.1	
1,2-DIBROMOBUTAN-3-ONE	2648	6.1		DICHLOROPHENYLTRI-CHLOROSILANE	1766	8	
DIBROMOCHLOROPROPANES	2872	6.1					
1,2-Dibromo-3-chloropropane, see	2872	6.1		1,2-DICHLOROPROPANE	1279	3	
DIBROMODIFLUOROMETHANE	1941	9		1,3-DICHLOROPROPANOL-2	2750	6.1	
DIBROMOMETHANE	2664	6.1		1,3-Dichloro-2-propanone, see	2649	6.1	
DI-n-BUTYLAMINE	2248	8		DICHLOROPROPENES	2047	3	
DIBUTYLAMINOETHANOL	2873	6.1		DICHLOROSILANE	2189	2	
2-Dibutylaminoethanol, see	2873	6.1		1,2-DICHLORO-1,1,2,2-TETRAFLUOROETHANE	1958	2	
N,N-Di-n-butylaminoethanol, see	2873	6.1					
DIBUTYL ETHERS	1149	3					

Name and description	UN No.	Class	Remarks	Name and description	UN No.	Class	Remarks
Dichloro-s-triazine-2,4,6-trione, see	2465	5.1		Difluorochloroethane, see	2517	2	
1,4-Dicyanobutane, see	2205	6.1		1,1-DIFLUOROETHANE	1030	2	
Dicycloheptadiene, see	2251	3		1,1-DIFLUOROETHYLENE	1959	2	
DICYCLOHEXYLAMINE	2565	8		DIFLUOROMETHANE	3252	2	
Dicyclohexylamine nitrite, see	2687	4.1		Difluoromethane, pentafluoroethane, and 1,1,1,2-tetrafluoroethane zeotropic mixture with approximately 10% difluoromethane and 70% pentafluoroethane, see	3339	2	
DICYCLOHEXYL-AMMONIUM NITRITE	2687	4.1					
DICYCLOPENTADIENE	2048	3					
1,2-DI-(DIMETHYLAMINO) ETHANE	2372	3		Difluoromethane, pentafluoroethane, and 1,1,1,2-tetrafluoroethane zeotropic mixture with approximately 20% difluoromethane and 40% pentafluoroethane, see	3338	2	
DIDYMIUM NITRATE	1465	5.1					
DIESEL FUEL	1202	3					
1,1-Diethoxyethane, see	1088	3					
1,2-Diethoxyethane, see	1153	3		Difluoromethane, pentafluoroethane, and 1,1,1,2-tetrafluoroethane zeotropic mixture with approximately 23%difluoromethane and 25% pentafluoroethane, see	3340	2	
DIETHOXYMETHANE	2373	3					
3,3-DIETHOXYPROPENE	2374	3					
DIETHYLAMINE	1154	3					
2-DIETHYLAMINOETHANOL	2686	8		DIFLUOROPHOSPHORIC ACID, ANHYDROUS	1768	8	
3-DIETHYLAMINO-PROPYLAMINE	2684	3		2,3-DIHYDROPYRAN	2376	3	
N,N-DIETHYLANILINE	2432	6.1		DIISOBUTYLAMINE	2361	3	
DIETHYLBENZENE	2049	3		DIISOBUTYLENE, ISOMERIC COMPOUNDS	2050	3	
Diethylcarbinol, see	1105	3		alpha-Diisobutylene, see	2050	3	
DIETHYL CARBONATE	2366	3		beta-Diisobutylene, see	2050	3	
DIETHYLDICHLOROSILANE	1767	8		DIISOBUTYL KETONE	1157	3	
Diethylenediamine, see	2579	8		DIISOOCTYL ACID PHOSPHATE	1902	8	
DIETHYLENEGLYCOL DINITRATE, DESENSITIZED with not less than 25% non-volatile, water-insoluble phlegmatizer, by mass	0075	1		DIISOPROPYLAMINE	1158	3	
				DIISOPROPYL ETHER	1159	3	
				DIKETENE, STABILIZED	2521	6.1	
				1,1-DIMETHOXYETHANE	2377	3	
DIETHYLENETRIAMINE	2079	8		1,2-DIMETHOXYETHANE	2252	3	
N,N-Diethylethanolamine, see	2686	3		Dimethoxystrychnine, see	1570	6.1	
DIETHYL ETHER	1155	3		DIMETHYLAMINE, ANHYDROUS	1032	2	
N,N-DIETHYLETHYLENE-DIAMINE	2685	8					
Di-(2-ethylhexyl) phosphoric acid, see	1902	8		DIMETHYLAMINE AQUEOUS SOLUTION	1160	3	
DIETHYL KETONE	1156	3		2-DIMETHYLAMINO-ACETONITRILE	2378	3	
DIETHYL SULPHATE	1594	6.1		2-DIMETHYLAMINOETHANOL	2051	8	
DIETHYL SULPHIDE	2375	3		2-DIMETHYLAMINOETHYL ACRYLATE	3302	6.1	
DIETHYLTHIOPHOSPHORYL CHLORIDE	2751	8		2-DIMETHYLAMINOETHYL METHACRYLATE	2522	6.1	
Diethylzinc, see	3394	4.2					
2,4-Difluoroaniline, see	2941	6.1		N,N-DIMETHYLANILINE	2253	6.1	

Name and description	UN No.	Class	Remarks	Name and description	UN No.	Class	Remarks
Dimethylarsenic acid, see	1572	6.1		DINITROPHENOLATES, alkali metals, dry or wetted with less than 15% water, by mass	0077	1	
N,N-Dimethylbenzylamine, see	2619	8					
2,3-DIMETHYLBUTANE	2457	3		DINITROPHENOLATES, WETTED with not less than 15% water, by mass	1321	4.1	
1,3-DIMETHYLBUTYLAMINE	2379	3					
DIMETHYLCARBAMOYL CHLORIDE	2262	8		DINITRORESORCINOL, dry or wetted with less than 15% water, by mass	0078	1	
DIMETHYL CARBONATE	1161	3					
DIMETHYLCYCLOHEXANES	2263	3		DINITRORESORCINOL, WETTED with not less than 15% water, by mass	1322	4.1	
N,N-DIMETHYLCYCLO-HEXYLAMINE	2264	8					
DIMETHYLDICHLOROSILANE	1162	3		DINITROSOBENZENE	0406	1	
DIMETHYLDIETHOXYSILANE	2380	3		Dinitrotoluene mixed with sodium chlorate, see	0083	1	
DIMETHYLDIOXANES	2707	3		DINITROTOLUENES, LIQUID	2038	6.1	
DIMETHYL DISULPHIDE	2381	3		DINITROTOLUENES, MOLTEN	1600	6.1	
Dimethylethanolamine, see	2051	8		DINITROTOLUENES, SOLID	3454	6.1	
DIMETHYL ETHER	1033	2		DIOXANE	1165	3	
N,N-DIMETHYLFORMAMIDE	2265	3		DIOXOLANE	1166	3	
DIMETHYLHYDRAZINE, SYMMETRICAL	2382	6.1		DIPENTENE	2052	3	
DIMETHYLHYDRAZINE, UNSYMMETRICAL	1163	6.1		DIPHENYLAMINE CHLOROARSINE	1698	6.1	
1,1-Dimethylhydrazine, see	1163	6.1		DIPHENYLCHLOROARSINE, LIQUID	1699	6.1	
N,N-Dimethyl-4-nitrosoaniline, see	1369	4.2		DIPHENYLCHLOROARSINE, SOLID	3450	6.1	
2,2-DIMETHYLPROPANE	2044	2					
DIMETHYL-N-PROPYLAMINE	2266	3		DIPHENYLDICHLOROSILANE	1769	8	
DIMETHYL SULPHATE	1595	6.1		DIPHENYLMETHYL BROMIDE	1770	8	
DIMETHYL SULPHIDE	1164	3		DIPICRYLAMINE, see	0079	1	
DIMETHYL THIOPHOSPHORYL CHLORIDE	2267	6.1		DIPICRYL SULPHIDE, dry or wetted with less than 10% water, by mass	0401	1	
Dimethylzinc, see	3394	4.2					
DINGU, see	0489	1		DIPICRYL SULPHIDE, WETTED with not less than 10% water, by mass	2852	4.1	
DINITROANILINES	1596	6.1					
DINITROBENZENES, LIQUID	1597	6.1		DIPROPYLAMINE	2383	3	
DINITROBENZENES, SOLID	3443	6.1		Dipropylene triamine, see	2269	8	
Dinitrochlorobenzene, see	1577 3441	6.1 6.1		DI-n-PROPYL ETHER	2384	3	
DINITRO-o-CRESOL	1598	6.1		DIPROPYL KETONE	2710	3	
DINITROGEN TETROXIDE	1067	2		DISINFECTANT, LIQUID, CORROSIVE, N.O.S.	1903	8	
DINITROGLYCOLURIL	0489	1		DISINFECTANT, LIQUID, TOXIC, N.O.S.	3142	6.1	
DINITROPHENOL, dry or wetted with less than 15% water, by mass	0076	1					
DINITROPHENOL SOLUTION	1599	6.1		DISINFECTANT, SOLID, TOXIC, N.O.S.	1601	6.1	
DINITROPHENOL, WETTED with not less than 15% water, by mass	1320	4.1		DISODIUM TRIOXOSILICATE	3253	8	
				DIVINYL ETHER, STABILIZED	1167	3	

Name and description	UN No.	Class	Remarks	Name and description	UN No.	Class	Remarks
DODECYLTRICHLOROSILANE	1771	8		Empty packaging, uncleaned			See 4.1.1.11, 5.1.3 and 5.4.1.1.6
Dry ice, see	1845	9	Not subject to ADR				
DYE INTERMEDIATE, LIQUID, CORROSIVE, N.O.S.	2801	8		Empty receptacle, uncleaned			See 5.1.3 and 5.4.1.1.6
DYE INTERMEDIATE, LIQUID, TOXIC, N.O.S.	1602	6.1		Empty tank, uncleaned			See 4.3.2.4, 5.1.3 and 5.4.1.1.6
DYE INTERMEDIATE, SOLID, CORROSIVE, N.O.S.	3147	8		Empty vehicle, uncleaned			See 5.1.3 and 5.4.1.1.6
DYE INTERMEDIATE, SOLID, TOXIC, N.O.S.	3143	6.1					
DYE, LIQUID, CORROSIVE, N.O.S.	2801	8		Enamel, see	1263 3066 3469 3470	3 8 3 8	
DYE, LIQUID, TOXIC, N.O.S.	1602	6.1					
DYE, SOLID, CORROSIVE, N.O.S.	3147	8		ENGINE, FUEL CELL, FLAMMABLE GAS POWERED	3529	2.1	
DYE, SOLID, TOXIC, N.O.S.	3143	6.1		ENGINE, FUEL CELL, FLAMMABLE LIQUID POWERED	3528	3	
Dynamite, see	0081	1					
Electric storage batteries, see	2794 2795 2800 3028	8 8 8 8		ENGINE, INTERNAL COMBUSTION	3530	9	
				ENGINE, INTERNAL COMBUSTION, FLAMMABLE GAS POWERED	3529	2	
Electrolyte (acid or alkaline) for batteries, see	2796 2797	8 8					
ELEVATED TEMPERATURE LIQUID, N.O.S., at or above 100 °C and below its flash-point (including molten metals, molten salts, etc.)	3257	9		ENGINE, INTERNAL COMBUSTION, FLAMMABLE LIQUID POWERED	3528	3	
				Engines, rocket, see	0250 0322	1 1	
ELEVATED TEMPERATURE LIQUID, FLAMMABLE, N.O.S. with flashpoint above 60 °C, at or above its flashpoint and at or above 100 °C	3256	3		ENVIRONMENTALLY HAZARDOUS SUBSTANCE, LIQUID, N.O.S.	3082	9	
				ENVIRONMENTALLY HAZARDOUS SUBSTANCE, SOLID, N.O.S.	3077	9	
ELEVATED TEMPERATURE LIQUID, FLAMMABLE, N.O.S. with flashpoint above 60 °C, at or above its flashpoint and below 100 °C	3256	3		EPIBROMOHYDRIN	2558	6.1	
				EPICHLOROHYDRIN	2023	6.1	
ELEVATED TEMPERATURE SOLID, N.O.S., at or above 240 °C	3258	9		1,2-Epoxybutane, stabilized, see	3022	3	
Empty battery-vehicle, uncleaned			See 4.3.2.4, 5.1.3 and 5.4.1.1.6	Epoxyethane, see	1040	2	
				1,2-EPOXY-3-ETHOXYPROPANE	2752	3	
Empty IBC, uncleaned			See 4.1.1.11, 5.1.3 and 5.4.1.1.6	2,3-Epoxy-1-propanal, see	2622	3	
				2,3-Epoxypropyl ethyl ether, see	2752	3	
				ESTERS, N.O.S.	3272	3	
Empty large packaging, uncleaned			See 4.1.1.11, 5.1.3 and 5.4.1.1.6	ETHANE	1035	2	
				ETHANE, REFRIGERATED LIQUID	1961	2	
Empty MEGC, uncleaned			See 4.3.2.4, 5.1.3 and 5.4.1.1.6	Ethanethiol, see	2363	3	
				ETHANOL	1170	3	

Name and description	UN No.	Class	Remarks	Name and description	UN No.	Class	Remarks
ETHANOL AND GASOLINE MIXTURE or ETHANOL AND MOTOR SPIRIT MIXTURE or ETHANOL AND PETROL MIXTURE, with more than 10% ethanol	3475	3		ETHYL CHLOROFORMATE	1182	6.1	
				ETHYL 2-CHLOROPROPIONATE	2935	3	
				Ethyl-alpha-chloropropionate, see	2935	3	
				ETHYL CHLOROTHIOFORMATE	2826	8	
ETHANOL SOLUTION	1170	3		ETHYL CROTONATE	1862	3	
ETHANOLAMINE	2491	8		ETHYLDICHLOROARSINE	1892	6.1	
ETHANOLAMINE SOLUTION	2491	8		ETHYLDICHLOROSILANE	1183	4.3	
Ether, see	1155	3		ETHYLENE, ACETYLENE AND PROPYLENE MIXTURE, REFRIGERATED LIQUID containing at least 71.5% ethylene with not more than 22.5% acetylene and not more than 6% propylene	3138	2	
ETHERS, N.O.S.	3271	3					
2-Ethoxyethanol, see	1171	3					
2-Ethoxyethyl acetate, see	1172	3					
Ethoxy propane-1, see	2615	3		ETHYLENE CHLOROHYDRIN	1135	6.1	
ETHYL ACETATE	1173	3		ETHYLENE	1962	2	
ETHYLACETYLENE, STABILIZED	2452	2		ETHYLENEDIAMINE	1604	8	
				ETHYLENE DIBROMIDE	1605	6.1	
ETHYL ACRYLATE, STABILIZED	1917	3		Ethylene dibromide and methyl bromide, liquid mixture, see	1647	6.1	
ETHYL ALCOHOL, see	1170	3		ETHYLENE DICHLORIDE	1184	3	
ETHYL ALCOHOL SOLUTION, see	1170	3		ETHYLENE GLYCOL DIETHYL ETHER	1153	3	
ETHYLAMINE	1036	2		ETHYLENE GLYCOL MONOETHYL ETHER	1171	3	
ETHYLAMINE, AQUEOUS SOLUTION with not less than 50% but not more than 70% ethylamine	2270	3		ETHYLENE GLYCOL MONOETHYL ETHER ACETATE	1172	3	
ETHYL AMYL KETONE	2271	3		ETHYLENE GLYCOL MONOMETHYL ETHER	1188	3	
N-ETHYLANILINE	2272	6.1					
2-ETHYLANILINE	2273	6.1		ETHYLENE GLYCOL MONOMETHYL ETHER ACETATE	1189	3	
ETHYLBENZENE	1175	3					
N-ETHYL-N-BENZYLANILINE	2274	6.1		ETHYLENEIMINE, STABILIZED	1185	6.1	
N-ETHYLBENZYLTOLUIDINES, LIQUID	2753	6.1		ETHYLENE OXIDE	1040	2	
N-ETHYLBENZYLTOLUIDINES, SOLID	3460	6.1		ETHYLENE OXIDE AND CARBON DIOXIDE MIXTURE with more than 87% ethylene oxide	3300	2	
ETHYL BORATE	1176	3		ETHYLENE OXIDE AND CARBON DIOXIDE MIXTURE with more than 9% but not more than 87% ethylene oxide	1041	2	
ETHYL BROMIDE	1891	6.1					
ETHYL BROMOACETATE	1603	6.1					
2-ETHYLBUTANOL	2275	3		ETHYLENE OXIDE AND CARBON DIOXIDE MIXTURE with not more than 9% ethylene oxide	1952	2	
2-ETHYLBUTYL ACETATE	1177	3					
ETHYL BUTYL ETHER	1179	3					
2-ETHYLBUTYRALDEHYDE	1178	3					
ETHYL BUTYRATE	1180	3		ETHYLENE OXIDE AND CHLOROTETRAFLUORO-ETHANE MIXTURE with not more than 8.8% ethylene oxide	3297	2	
ETHYL CHLORIDE	1037	2					
ETHYL CHLOROACETATE	1181	6.1					
Ethyl chlorocarbonate, see	1182	6.1					

Name and description	UN No.	Class	Remarks	Name and description	UN No.	Class	Remarks
ETHYLENE OXIDE AND DICHLORODIFLUORO-METHANE MIXTURE with not more than 12.5% ethylene oxide	3070	2		EXPLOSIVE, BLASTING, TYPE B	0082	1	
					0331	1	
				EXPLOSIVE, BLASTING, TYPE C	0083	1	
ETHYLENE OXIDE AND PENTAFLUOROETHANE MIXTURE with not more than 7.9% ethylene oxide	3298	2		EXPLOSIVE, BLASTING, TYPE D	0084	1	
				EXPLOSIVE, BLASTING, TYPE E	0241	1	
					0332	1	
ETHYLENE OXIDE AND PROPYLENE OXIDE MIXTURE, not more than 30% ethylene oxide	2983	3		Explosives, emulsion, see	0241	1	
					0332	1	
				Explosive, seismic, see	0081	1	
ETHYLENE OXIDE AND TETRAFLUOROETHANE MIXTURE with not more than 5.6% ethylene oxide	3299	2			0082	1	
					0083	1	
					0331	1	
				Explosive, slurry, see	0241	1	
ETHYLENE OXIDE WITH NITROGEN up to a total pressure of 1 MPa (10 bar) at 50 °C	1040	2			0332	1	
				Explosive, water gel, see	0241	1	
					0332	1	
ETHYLENE, REFRIGERATED LIQUID	1038	2		EXTRACTS, AROMATIC, LIQUID	1169	3	
ETHYL ETHER, see	1155	3		EXTRACTS, FLAVOURING, LIQUID	1197	3	
ETHYL FLUORIDE	2453	2		FABRICS, ANIMAL, N.O.S. with oil	1373	4.2	
ETHYL FORMATE	1190	3					
2-ETHYLHEXYLAMINE	2276	3		FABRICS IMPREGNATED WITH WEAKLY NITRATED NITROCELLULOSE, N.O.S.	1353	4.1	
2-ETHYLHEXYL CHLOROFORMATE	2748	6.1					
Ethylidene chloride, see	2362	3		FABRICS, SYNTHETIC, N.O.S. with oil	1373	4.2	
ETHYL ISOBUTYRATE	2385	3		FABRICS, VEGETABLE, N.O.S. with oil	1373	4.2	
ETHYL ISOCYANATE	2481	6.1					
ETHYL LACTATE	1192	3		FERRIC ARSENATE	1606	6.1	
ETHYL MERCAPTAN	2363	3		FERRIC ARSENITE	1607	6.1	
ETHYL METHACRYLATE, STABILIZED	2277	3		FERRIC CHLORIDE, ANHYDROUS	1773	8	
ETHYL METHYL ETHER	1039	2		FERRIC CHLORIDE SOLUTION	2582	8	
ETHYL METHYL KETONE	1193	3		FERRIC NITRATE	1466	5.1	
ETHYL NITRITE SOLUTION	1194	3		FERROCERIUM	1323	4.1	
ETHYL ORTHOFORMATE	2524	3		FERROSILICON with 30% or more but less than 90% silicon	1408	4.3	
ETHYL OXALATE	2525	6.1		FERROUS ARSENATE	1608	6.1	
ETHYLPHENYL-DICHLOROSILANE	2435	8		FERROUS METAL BORINGS in a form liable to self-heating	2793	4.2	
1-ETHYLPIPERIDINE	2386	3		FERROUS METAL CUTTINGS in a form liable to self-heating	2793	4.2	
ETHYL PROPIONATE	1195	3		FERROUS METAL SHAVINGS in a form liable to self-heating	2793	4.2	
ETHYL PROPYL ETHER	2615	3					
Ethyl silicate, see	1292	3		FERROUS METAL TURNINGS in a form liable to self-heating	2793	4.2	
Ethyl sulphate, see	1594	6.1					
N-ETHYLTOLUIDINES	2754	6.1		FERTILIZER AMMONIATING SOLUTION with free ammonia	1043	2	
ETHYLTRICHLOROSILANE	1196	3					
EXPLOSIVE, BLASTING, TYPE A	0081	1					

Name and description	UN No.	Class	Remarks	Name and description	UN No.	Class	Remarks
Fertilizer with ammonium nitrate, n.o.s., see	2067	5.1		FLAMMABLE LIQUID, TOXIC, N.O.S.	1992	3	
Fibres, animal, burnt wet or damp	1372	4.2	Not subject to ADR	FLAMMABLE LIQUID, TOXIC, CORROSIVE, N.O.S.	3286	3	
FIBRES, ANIMAL, N.O.S. with oil	1373	4.2		FLAMMABLE SOLID, CORROSIVE, INORGANIC, N.O.S.	3180	4.1	
FIBRES IMPREGNATED WITH WEAKLY NITRATED NITROCELLULOSE, N.O.S.	1353	4.1		FLAMMABLE SOLID, CORROSIVE, ORGANIC, N.O.S.	2925	4.1	
FIBRES, SYNTHETIC, N.O.S. with oil	1373	4.2		FLAMMABLE SOLID, INORGANIC, N.O.S.	3178	4.1	
Fibres, vegetable, burnt wet or damp	1372	4.2	Not subject to ADR	FLAMMABLE SOLID, ORGANIC, N.O.S.	1325	4.1	
Fibres, vegetable, dry	3360	4.1	Not subject to ADR	FLAMMABLE SOLID, ORGANIC, MOLTEN, N.O.S.	3176	4.1	
FIBRES, VEGETABLE, N.O.S. with oil	1373	4.2		FLAMMABLE SOLID, OXIDIZING, N.O.S.	3097	4.1	Carriage prohibited
Films, nitrocellulose base, from which gelatine has been removed; film scrap, see	2002	4.2		FLAMMABLE SOLID, TOXIC, INORGANIC, N.O.S.	3179	4.1	
FILMS, NITROCELLULOSE BASE, gelatin coated, except scrap	1324	4.1		FLAMMABLE SOLID, TOXIC, ORGANIC, N.O.S.	2926	4.1	
Filler, liquid, see	1263 3066 3469 3470	3 8 3 8		FLARES, AERIAL	0093 0403 0404 0420 0421	1 1 1 1 1	
FIRE EXTINGUISHER CHARGES, corrosive liquid	1774	8		Flares, aeroplane, see	0093 0403 0404 0420 0421	1 1 1 1 1	
Fire extinguisher charges, expelling, explosive, see	0275 0276 0323 0381	1 1 1 1					
FIRE EXTINGUISHERS with compressed or liquefied gas	1044	2		Flares, highway, Flares, distress, small, Flares, railway or highway, see	0191 0373	1 1	
FIRELIGHTERS, SOLID with flammable liquid	2623	4.1		FLARES, SURFACE	0092 0418 0419	1 1 1	
FIREWORKS	0333 0334 0335 0336 0337	1 1 1 1 1	See 2.2.1.1.7	Flares, water-activated, see	0248 0249	1 1	
FIRST AID KIT	3316	9		FLASH POWDER	0094 0305	1 1	
Fish meal, stabilized	2216	9	Not subject to ADR	Flue dusts, toxic, see	1562	6.1	
				Fluoric acid, see	1790	8	
FISH MEAL, UNSTABILIZED	1374	4.2		FLUORINE, COMPRESSED	1045	2	
Fish scrap, stabilized, see	2216	9	Not subject to ADR	FLUOROACETIC ACID	2642	6.1	
				FLUOROANILINES	2941	6.1	
FISH SCRAP, UNSTABILIZED, see	1374	4.2		2-Fluoroaniline, see	2941	6.1	
Flammable gas in lighters, see	1057	2		4-Fluoroaniline, see	2941	6.1	
FLAMMABLE LIQUID, N.O.S.	1993	3		o-Fluoroaniline, see	2941	6.1	
FLAMMABLE LIQUID, CORROSIVE, N.O.S.	2924	3		p-Fluoroaniline, see	2941	6.1	

Name and description	UN No.	Class	Remarks	Name and description	UN No.	Class	Remarks
FLUOROBENZENE	2387	3		FURALDEHYDES	1199	6.1	
FLUOROBORIC ACID	1775	8		FURAN	2389	3	
Fluoroethane, see	2453	2		FURFURYL ALCOHOL	2874	6.1	
Fluoroform, see	1984	2		FURFURYLAMINE	2526	3	
Fluoromethane, see	2454	2		Furyl carbinol, see	2874	6.1	
FLUOROPHOSPHORIC ACID, ANHYDROUS	1776	8		FUSE, DETONATING, metal clad	0102	1	
					0290	1	
FLUOROSILICATES, N.O.S.	2856	6.1		FUSE, DETONATING, MILD EFFECT, metal clad	0104	1	
FLUOROSILICIC ACID	1778	8		FUSE, IGNITER, tubular, metal clad	0103	1	
FLUOROSULPHONIC ACID	1777	8		FUSE, NON-DETONATING	0101	1	
FLUOROTOLUENES	2388	3		FUSEL OIL	1201	3	
FORMALDEHYDE SOLUTION with not less than 25% formaldehyde	2209	8		FUSE, SAFETY	0105	1	
FORMALDEHYDE SOLUTION, FLAMMABLE	1198	3		Fuze, combination, percussion or time, see	0106	1	
					0107	1	
					0257	1	
Formalin, see	1198	3			0316	1	
	2209	8			0317	1	
					0367	1	
Formamidine sulphinic acid, see	3341	4.2			0368	1	
FORMIC ACID with more than 85% acid by mass	1779	8		FUZES, DETONATING	0106	1	
					0107	1	
					0257	1	
FORMIC ACID with not more than 85% acid by mass	3412	8			0367	1	
Formic aldehyde, see	1198	3		FUZES, DETONATING with protective features	0408	1	
	2209	8			0409	1	
					0410	1	
2-Formyl-3,4-dihydro-2H-pyran, see	2607	3		FUZES, IGNITING	0316	1	
FRACTURING DEVICES, EXPLOSIVE without detonator, for oil wells	0099	1			0317	1	
					0368	1	
				GALLIUM	2803	8	
FUEL, AVIATION, TURBINE ENGINE	1863	3		GAS CARTRIDGES without a release device, non-refillable, see	2037	2	
FUEL CELL CARTRIDGES	3478	2		Gas drips, hydrocarbon, see	3295	3	
	3479	2		GAS OIL	1202	3	
	3473	3		GASOLINE	1203	3	
	3476	4.3		Gasoline and ethanol mixture, with more than 10% ethanol, see	3475	3	
	3477	8					
FUEL CELL CARTRIDGES CONTAINED IN EQUIPMENT	3478	2		Gasoline, casinghead, see	1203	3	
	3479	2					
	3473	3		GAS, REFRIGERATED LIQUID, N.O.S.	3158	2	
	3476	4.3					
	3477	8		GAS, REFRIGERATED LIQUID, FLAMMABLE, N.O.S.	3312	2	
FUEL CELL CARTRIDGES PACKED WITH EQUIPMENT	3478	2		GAS, REFRIGERATED LIQUID, OXIDIZING, N.O.S.	3311	2	
	3479	2					
	3473	3					
	3476	4.3		GAS SAMPLE, NON-PRESSURIZED, FLAMMABLE, N.O.S., not refrigerated liquid	3167	2	
	3477	8					
Fumaroyl dichloride, see	1780	3					
FUMARYL CHLORIDE	1780	8					
FUMIGATED CARGO TRANSPORT UNIT	3359	9					

Name and description	UN No.	Class	Remarks	Name and description	UN No.	Class	Remarks
GAS SAMPLE, NON-PRESSURIZED, TOXIC, N.O.S., not refrigerated liquid	3169	2		GUNPOWDER, granular or as a meal, see	0027	1	
				GUNPOWDER, IN PELLETS, see	0028	1	
GAS SAMPLE, NON-PRESSURIZED, TOXIC, FLAMMABLE, N.O.S., not refrigerated liquid	3168	2		Gutta percha solution, see	1287	3	
				HAFNIUM POWDER, DRY	2545	4.2	
Gelatin, blasting, see	0081	1		HAFNIUM POWDER, WETTED with not less than 25% water	1326	4.1	
Gelatin, dynamites, see	0081	1		HALOGENATED MONOMETHYLDIPHENYL-METHANES, LIQUID	3151	9	
GENETICALLY MODIFIED MICROORGANISMS	3245	9					
GENETICALLY MODIFIED ORGANISMS	3245	9		HALOGENATED MONOMETHYLDIPHENYL-METHANES, SOLID	3152	9	
GERMANE	2192	2		Hay	1327	4.1	Not subject to ADR
GERMANE, ADSORBED	3523	2					
Germanium hydride, see	2192	2		HEATING OIL, LIGHT	1202	3	
Glycer-1,3-dichlorohydrin, see	2750	6.1		Heavy hydrogen, see	1957	2	
GLYCEROL alpha-MONOCHLOROHYDRIN	2689	6.1		HELIUM, COMPRESSED	1046	2	
				HELIUM, REFRIGERATED LIQUID	1963	2	
Glyceryl trinitrate, see	0143	1		HEPTAFLUOROPROPANE	3296	2	
	0144	1		n-HEPTALDEHYDE	3056	3	
	1204	3		n-Heptanal, see	3056	3	
	3064	3		HEPTANES	1206	3	
GLYCIDALDEHYDE	2622	3		4-Heptanone, see	2710	3	
GRENADES, hand or rifle, with bursting charge	0284	1		n-HEPTENE	2278	3	
	0285	1		HEXACHLOROACETONE	2661	6.1	
	0292	1		HEXACHLOROBENZENE	2729	6.1	
	0293	1		HEXACHLOROBUTADIENE	2279	6.1	
Grenades, illuminating, see	0171	1		Hexachloro-1,3-butadiene, see	2279	6.1	
	0254	1		HEXACHLOROCYCLO-PENTADIENE	2646	6.1	
	0297	1					
GRENADES, PRACTICE, hand or rifle	0110	1		HEXACHLOROPHENE	2875	6.1	
	0318	1		Hexachloro-2-propanone, see	2661	6.1	
	0372	1		HEXADECYLTRICHLORO-SILANE	1781	8	
	0452	1					
Grenades, smoke, see	0015	1		HEXADIENES	2458	3	
	0016	1		HEXAETHYL TETRAPHOSPHATE	1611	6.1	
	0245	1					
	0246	1		HEXAETHYL TETRAPHOSPHATE AND COMPRESSED GAS MIXTURE	1612	2	
	0303	1					
GUANIDINE NITRATE	1467	5.1					
GUANYLNITROSAMINO-GUANYLIDENE HYDRAZINE, WETTED with not less than 30% water, by mass	0113	1		HEXAFLUOROACETONE	2420	2	
				HEXAFLUOROACETONE HYDRATE, LIQUID	2552	6.1	
GUANYLNITROSAMINO-GUANYLTETRAZENE, WETTED with not less than 30% water, or mixture of alcohol and water, by mass	0114	1		HEXAFLUOROACETONE HYDRATE, SOLID	3436	6.1	
GUNPOWDER, COMPRESSED, see	0028	1					

Name and description	UN No.	Class	Remarks	Name and description	UN No.	Class	Remarks
HEXAFLUOROETHANE	2193	2		HYDRAZINE AQUEOUS SOLUTION, with more than 37% hydrazine by mass	2030	8	
HEXAFLUOROPHOSPHORIC ACID	1782	8		HYDRAZINE, AQUEOUS SOLUTION with not more than 37% hydrazine, by mass	3293	6.1	
HEXAFLUOROPROPYLENE	1858	2					
Hexahydrocresol, see	2617	3		HYDRAZINE AQUEOUS SOLUTION, FLAMMABLE with more than 37% hydrazine, by mass	3484	8	
Hexahydromethyl phenol, see	2617	3					
HEXALDEHYDE	1207	3		Hydrides, metal, water-reactive, n.o.s., see	1409	4.3	
HEXAMETHYLENEDIAMINE, SOLID	2280	8					
				Hydriodic acid, anhydrous, see	2197	2	
HEXAMETHYLENEDIAMINE SOLUTION	1783	8		HYDRIODIC ACID	1787	8	
HEXAMETHYLENE DIISOCYANATE	2281	6.1		HYDROBROMIC ACID	1788	8	
				HYDROCARBON GAS MIXTURE, COMPRESSED, N.O.S.	1964	2	
HEXAMETHYLENEIMINE	2493	3		HYDROCARBON GAS MIXTURE, LIQUEFIED, N.O.S. such as mixtures A, A01, A02, A0, A1, B1, B2, B or C	1965	2	
HEXAMETHYLENETETRAMINE	1328	4.1					
Hexamine, see	1328	4.1					
HEXANES	1208	3					
HEXANITRODIPHENYLAMINE	0079	1		HYDROCARBON GAS REFILLS FOR SMALL DEVICES with release device	3150	2	
HEXANITROSTILBENE	0392	1					
Hexanoic acid, see	2829	8		HYDROCARBONS, LIQUID, N.O.S.	3295	3	
HEXANOLS	2282	3					
1-HEXENE	2370	3		HYDROCHLORIC ACID	1789	8	
HEXOGEN AND CYCLOTETRAMETHYLENE-TETRANITRAMINE MIXTURE, WETTED with not less than 15% water, by mass or DESENSITIZED with not less than 10% phlegmatiser by mass, see	0391	1		HYDROCYANIC ACID, AQUEOUS SOLUTION with not more than 20% hydrogen cyanide	1613	6.1	
				HYDROFLUORIC ACID with more than 60% but not more than 85% hydrogen fluoride	1790	8	
HEXOGEN, DESENSITIZED, see	0483	1		HYDROFLUORIC ACID with more than 85% hydrogen fluoride	1790	8	
HEXOGEN, WETTED with not less than 15% water, by mass, see	0072	1		HYDROFLUORIC ACID with not more than 60% hydrogen fluoride	1790	8	
HEXOLITE, dry or wetted with less than 15% water, by mass	0118	1		HYDROFLUORIC ACID AND SULPHURIC ACID MIXTURE	1786	8	
HEXOTOL, dry or wetted with less than 15% water, by mass, see	0118	1		Hydrofluoroboric acid, see	1775	8	
				Hydrofluorosilicic acid, see	1778	8	
HEXOTONAL	0393	1		HYDROGEN AND METHANE MIXTURE, COMPRESSED	2034	2	
HEXOTONAL, cast, see	0393	1					
HEXYL, see	0079	1		Hydrogen arsenide, see	2188	2	
HEXYLTRICHLOROSILANE	1784	8		HYDROGEN BROMIDE, ANHYDROUS	1048	2	
HMX, see	0391	1					
HMX, DESENSITIZED, see	0484	1		Hydrogen bromide solution, see	1788	8	
HMX, WETTED with not less than 15% water, by mass, see	0226	1		HYDROGEN CHLORIDE, ANHYDROUS	1050	2	
HYDRAZINE, ANHYDROUS	2029	8		HYDROGEN CHLORIDE, REFRIGERATED LIQUID	2186	2	Carriage prohibited

Name and description	UN No.	Class	Remarks	Name and description	UN No.	Class	Remarks
HYDROGEN, COMPRESSED	1049	2		HYDROGEN PEROXIDE, AQUEOUS SOLUTION, STABILIZED with more than 70% hydrogen peroxide	2015	5.1	
HYDROGEN CYANIDE, AQUEOUS SOLUTION with not more than 20% hydrogen cyanide, see	1613	6.1					
				HYDROGEN, REFRIGERATED LIQUID	1966	2	
HYDROGEN CYANIDE, SOLUTION IN ALCOHOL with not more than 45% hydrogen cyanide	3294	6.1		HYDROGEN SELENIDE, ADSORBED	3526	2	
				HYDROGEN SELENIDE, ANHYDROUS	2202	2	
HYDROGEN CYANIDE, STABILIZED containing less than 3% water	1051	6.1		Hydrogen silicide, see	2203	2	
				HYDROGEN SULPHIDE	1053	2	
HYDROGEN CYANIDE, STABILIZED, containing less than 3% water and absorbed in a porous inert material	1614	6.1		Hydroselenic acid, see	2202	2	
				Hydrosilicofluoric acid, see	1778	8	
				1-HYDROXYBENZOTRIAZOLE, ANHYDROUS, dry or wetted with less than 20% water, by mass	0508	1	
HYDROGENDIFLUORIDES, SOLID, N.O.S.	1740	8					
HYDROGENDIFLUORIDES SOLUTION, N.O.S.	3471	8		1-HYDROXYBENZOTRIAZOLE MONOHYDRATE	3474	4.1	
HYDROGEN FLUORIDE, ANHYDROUS	1052	8		3-Hydroxybutan-2-one, see	2621	3	
Hydrogen fluoride solution, see	1790	8		HYDROXYLAMINE SULPHATE	2865	8	
HYDROGEN IN A METAL HYDRIDE STORAGE SYSTEM	3468	2		1-Hydroxy-3-methyl-2-penten-4-yne, see	2705	8	
				3-Hydroxyphenol, see	2876	6.1	
HYDROGEN IN A METAL HYDRIDE STORAGE SYSTEM CONTAINED IN EQUIPMENT	3468	2		HYPOCHLORITES, INORGANIC, N.O.S.	3212	5.1	
				HYPOCHLORITE SOLUTION	1791	8	
HYDROGEN IN A METAL HYDRIDE STORAGE SYSTEM PACKED WITH EQUIPMENT	3468	2		IGNITERS	0121	1	
					0314	1	
					0315	1	
HYDROGEN IODIDE, ANHYDROUS	2197	2			0325	1	
					0454	1	
Hydrogen iodide solution, see	1787	8		3,3'-IMINODIPROPYLAMINE	2269	8	
HYDROGEN PEROXIDE AND PEROXYACETIC ACID MIXTURE with acid(s), water and not more than 5% peroxyacetic acid, STABILIZED	3149	5.1		Indiarubber, see	1287	3	
				INFECTIOUS SUBSTANCE, AFFECTING ANIMALS only	2900	6.2	
HYDROGEN PEROXIDE, AQUEOUS SOLUTION with not less than 8% but less than 20% hydrogen peroxide (stabilized as necessary)	2984	5.1		INFECTIOUS SUBSTANCE, AFFECTING HUMANS	2814	6.2	
				Ink, printer's, flammable, see	1210	3	
				INSECTICIDE GAS, N.O.S.	1968	2	
HYDROGEN PEROXIDE, AQUEOUS SOLUTION with not less than 20% but not more than 60% hydrogen peroxide (stabilized as necessary)	2014	5.1		INSECTICIDE GAS, FLAMMABLE, N.O.S.	3354	2	
				INSECTICIDE GAS, TOXIC, N.O.S.	1967	2	
				INSECTICIDE GAS, TOXIC, FLAMMABLE, N.O.S.	3355	2	
HYDROGEN PEROXIDE, AQUEOUS SOLUTION, STABILIZED with more than 60% hydrogen peroxide and not more than 70% hydrogen peroxide	2015	5.1		IODINE	3495	8	
				IODINE MONOCHLORIDE, LIQUID	3498	8	

Name and description	UN No.	Class	Remarks	Name and description	UN No.	Class	Remarks
IODINE MONOCHLORIDE, SOLID	1792	8		ISOCYANATES, TOXIC, N.O.S.	2206	6.1	
IODINE PENTAFLUORIDE	2495	5.1		ISOCYANATES, TOXIC, FLAMMABLE, N.O.S.	3080	6.1	
2-IODOBUTANE	2390	3		ISOCYANATE SOLUTION, FLAMMABLE, TOXIC, N.O.S.	2478	3	
Iodomethane, see	2644	6.1					
IODOMETHYLPROPANES	2391	3		ISOCYANATE SOLUTION, TOXIC, N.O.S.	2206	6.1	
IODOPROPANES	2392	3		ISOCYANATE SOLUTION, TOXIC, FLAMMABLE, N.O.S.	3080	6.1	
alpha-Iodotoluene, see	2653	6.1					
I.p.d.i., see	2290	6.1		ISOCYANATO-BENZOTRIFLUORIDES	2285	6.1	
Iron chloride, anhydrous, see	1773	8					
Iron (III) chloride, anhydrous, see	1773	8		3-Isocyanatomethyl-3,5,5-tri-methylcyclohexyl isocyanate, see	2290	6.1	
Iron chloride solution, see	2582	8		Isododecane, see	2286	3	
IRON OXIDE, SPENT obtained from coal gas purification	1376	4.2		ISOHEPTENE	2287	3	
				ISOHEXENE	2288	3	
IRON PENTACARBONYL	1994	6.1		Isooctane, see	1262	3	
Iron perchloride, anhydrous, see	1773	8		ISOOCTENE	1216	3	
Iron powder, pyrophoric, see	1383	4.2		Isopentane, see	1265	3	
Iron sesquichloride, anhydrous, see	1773	8		ISOPENTENES	2371	3	
IRON SPONGE, SPENT obtained from coal gas purification	1376	4.2		Isopentylamine, see	1106	3	
				Isopentyl nitrite, see	1113	3	
Iron swarf, see	2793	4.2		ISOPHORONEDIAMINE	2289	8	
ISOBUTANE	1969	2		ISOPHORONE DIISOCYANATE	2290	6.1	
ISOBUTANOL	1212	3		ISOPRENE, STABILIZED	1218	3	
Isobutene, see	1055	2		ISOPROPANOL	1219	3	
ISOBUTYL ACETATE	1213	3		ISOPROPENYL ACETATE	2403	3	
ISOBUTYL ACRYLATE, STABILIZED	2527	3		ISOPROPENYLBENZENE	2303	3	
ISOBUTYL ALCOHOL, see	1212	3		ISOPROPYL ACETATE	1220	3	
ISOBUTYL ALDEHYDE, see	2045	3		ISOPROPYL ACID PHOSPHATE	1793	8	
ISOBUTYLAMINE	1214	3		ISOPROPYL ALCOHOL, see	1219	3	
ISOBUTYLENE	1055	2		ISOPROPYLAMINE	1221	3	
ISOBUTYL FORMATE	2393	3		ISOPROPYLBENZENE	1918	3	
ISOBUTYL ISOBUTYRATE	2528	3		ISOPROPYL BUTYRATE	2405	3	
ISOBUTYL ISOCYANATE	2486	6.1		Isopropyl chloride, see	2356	3	
ISOBUTYL METHACRYLATE, STABILIZED	2283	3		ISOPROPYL CHLOROACETATE	2947	3	
				ISOPROPYL CHLOROFORMATE	2407	6.1	
ISOBUTYL PROPIONATE	2394	3		ISOPROPYL 2-CHLORO-PROPIONATE	2934	3	
ISOBUTYRALDEHYDE	2045	3					
ISOBUTYRIC ACID	2529	3		Isopropyl-alpha-chloropropionate, see	2934	3	
ISOBUTYRONITRILE	2284	3		Isopropyl ether, see	1159	3	
ISOBUTYRYL CHLORIDE	2395	3		Isopropylethylene, see	2561	3	
ISOCYANATES, FLAMMABLE, TOXIC, N.O.S.	2478	3		Isopropyl formate, see	1281	3	

Name and description	UN No.	Class	Remarks	Name and description	UN No.	Class	Remarks
ISOPROPYL ISOBUTYRATE	2406	3		Lead (II) cyanide	1620	6.1	
ISOPROPYL ISOCYANATE	2483	6.1		LEAD DIOXIDE	1872	5.1	
Isopropyl mercaptan, see	2402	3		LEAD NITRATE	1469	5.1	
ISOPROPYL NITRATE	1222	3		Lead (II) nitrate	1469	5.1	
ISOPROPYL PROPIONATE	2409	3		LEAD PERCHLORATE, SOLID	1470	5.1	
Isolpropyltoluene, see	2046	3		LEAD PERCHLORATE, SOLUTION	3408	5.1	
Isopropyltoluol, see	2046	3		Lead (II) perchlorate	1470	5.1	
ISOSORBIDE DINITRATE MIXTURE with not less than 60% lactose, mannose, starch or calcium hydrogen phosphate	2907	4.1			3408	5.1	
				Lead peroxide, see	1872	5.1	
ISOSORBIDE-5-MONONITRATE	3251	4.1		LEAD PHOSPHITE, DIBASIC	2989	4.1	
Isovaleraldehyde, see	2058	3		LEAD STYPHNATE, WETTED with not less than 20% water, or mixture of alcohol and water, by mass	0130	1	
JET PERFORATING GUNS, CHARGED, oil well, without detonator	0124	1					
	0494	1		LEAD SULPHATE with more than 3% free acid	1794	8	
Jet tappers, without detonator, see	0059	1		Lead tetraethyl, see	1649	6.1	
KEROSENE	1223	3		Lead tetramethyl, see	1649	6.1	
KETONES, LIQUID, N.O.S.	1224	3		LEAD TRINITRORESORCINATE, WETTED with not less than 20% water, or mixture of alcohol and water, by mass, see	0130	1	
KRILL MEAL	3497	4.2					
KRYPTON, COMPRESSED	1056	2					
KRYPTON, REFRIGERATED LIQUID	1970	2		LIFE-SAVING APPLIANCES NOT SELF-INFLATING containing dangerous goods as equipment	3072	9	
Lacquer, see	1263	3					
	3066	8		LIFE-SAVING APPLIANCES, SELF-INFLATING	2990	9	
	3469	3					
	3470	8		LIGHTER REFILLS containing flammable gas	1057	2	
Lacquer base, liquid, see	1263	3					
	3066	8		LIGHTERS containing flammable gas	1057	2	
	3469	3					
	3470	8		LIGHTERS, FUSE	0131	1	
Lacquer base or lacquer chips, nitrocellulose, dry, see	2557	4.1		Limonene, inactive, see	2052	3	
				LIQUEFIED GAS, N.O.S.	3163	2	
Lacquer base or lacquer chips, plastic, wet with alcohol or solvent, see	1263	3		LIQUEFIED GAS, FLAMMABLE, N.O.S.	3161	2	
	2059	3					
	2555	4.1		LIQUEFIED GASES, non-flammable, charged with nitrogen, carbon dioxide or air	1058	2	
	2556	4.1					
LEAD ACETATE	1616	6.1					
Lead (II) acetate, see	1616	6.1		LIQUEFIED GAS, OXIDIZING, N.O.S.	3157	2	
LEAD ARSENATES	1617	6.1					
LEAD ARSENITES	1618	6.1		LIQUEFIED GAS, TOXIC, N.O.S.	3162	2	
LEAD AZIDE, WETTED with not less than 20% water, or mixture of alcohol and water, by mass	0129	1		LIQUEFIED GAS, TOXIC, CORROSIVE, N.O.S.	3308	2	
				LIQUEFIED GAS, TOXIC, FLAMMABLE, N.O.S.	3160	2	
Lead chloride, solid, see	2291	6.1					
LEAD COMPOUND, SOLUBLE, N.O.S.	2291	6.1					
LEAD CYANIDE	1620	6.1					

Name and description	UN No.	Class	Remarks	Name and description	UN No.	Class	Remarks
LIQUEFIED GAS, TOXIC, FLAMMABLE, CORROSIVE, N.O.S.	3309	2		LITHIUM METAL BATTERIES PACKED WITH EQUIPMENT (including lithium alloy batteries)	3091	9	
LIQUEFIED GAS, TOXIC, OXIDIZING, N.O.S.	3307	2		LITHIUM NITRATE	2722	5.1	
				LITHIUM NITRIDE	2806	4.3	
LIQUEFIED GAS, TOXIC, OXIDIZING, CORROSIVE, N.O.S.	3310	2		LITHIUM PEROXIDE	1472	5.1	
Liquefied petroleum gas, see	1075	2		Lithium silicide, see	1417	4.3	
Liquid filler, see	1263	3		LITHIUM SILICON	1417	4.3	
	3066	8		L.n.g., see	1972	2	
	3469	3		LONDON PURPLE	1621	6.1	
	3470	8		L.p.g., see	1075	2	
Liquid lacquer base, see	1263	3		Lye, see	1823	8	
	3066	8		Lythene, see	1268	3	
	3469	3		MACHINERY, FUEL CELL, FLAMMABLE GAS POWERED	3529	2	
	3470	8		MACHINERY, FUEL CELL, FLAMMABLE LIQUID POWERED	3528	3	
LITHIUM	1415	4.3					
Lithium alkyls, liquid, see	3394	4.2					
Lithium alkyls, solid, see	3393	4.2		MACHINERY, INTERNAL COMBUSTION,	3530	9	
LITHIUM ALUMINIUM HYDRIDE	1410	4.3					
LITHIUM ALUMINIUM HYDRIDE, ETHEREAL	1411	4.3		MACHINERY, INTERNAL COMBUSTION, FLAMMABLE GAS POWERED	3529	2	
LITHIUM BOROHYDRIDE	1413	4.3					
LITHIUM FERROSILICON	2830	4.3		MACHINERY, INTERNAL COMBUSTION, FLAMMABLE LIQUID POWERED	2528	3	
LITHIUM HYDRIDE	1414	4.3					
LITHIUM HYDRIDE, FUSED SOLID	2805	4.3		MAGNESIUM in pellets, turnings or ribbons	1869	4.1	
LITHIUM HYDROXIDE	2680	8		Magnesium alkyls, see	3394	4.2	
LITHIUM HYDROXIDE SOLUTION	2679	8		MAGNESIUM ALLOYS with more than 50% magnesium in pellets, turnings or ribbons	1869	4.1	
LITHIUM HYPOCHLORITE, DRY	1471	5.1					
LITHIUM HYPOCHLORITE MIXTURE	1471	5.1		MAGNESIUM ALLOYS POWDER	1418	4.3	
Lithium in cartouches, see	1415	4.3		MAGNESIUM ALUMINIUM PHOSPHIDE	1419	4.3	
LITHIUM ION BATTERIES (including lithium ion polymer batteries)	3480	9		MAGNESIUM ARSENATE	1622	6.1	
				Magnesium bisulphite solution, see	2693	8	
LITHIUM ION BATTERIES CONTAINED IN EQUIPMENT (including lithium ion polymer batteries)	3481	9		MAGNESIUM BROMATE	1473	5.1	
				MAGNESIUM CHLORATE	2723	5.1	
LITHIUM ION BATTERIES PACKED WITH EQUIPMENT (including lithium ion polymer batteries)	3481	9		Magnesium chloride and chlorate mixture, see	1459	5.1	
					3407	5.1	
				MAGNESIUM DIAMIDE	2004	4.2	
				Magnesium diphenyl, see	3393	4.2	
LITHIUM METAL BATTERIES (including lithium alloy batteries)	3090	9		MAGNESIUM FLUORO-SILICATE	2853	6.1	
LITHIUM METAL BATTERIES CONTAINED IN EQUIPMENT (including lithium alloy batteries)	3091	9					

Name and description	UN No.	Class	Remarks	Name and description	UN No.	Class	Remarks
MAGNESIUM GRANULES, COATED, particle size not less than 149 microns	2950	4.3		MEDICINE, LIQUID, TOXIC, N.O.S.	1851	6.1	
				MEDICINE, SOLID, TOXIC, N.O.S.	3249	6.1	
MAGNESIUM HYDRIDE	2010	4.3		p-Mentha-1,8-diene, see	2052	8	
MAGNESIUM NITRATE	1474	5.1		MERCAPTANS, LIQUID, FLAMMABLE, N.O.S.	3336	3	
MAGNESIUM PERCHLORATE	1475	5.1		MERCAPTANS, LIQUID, FLAMMABLE, TOXIC, N.O.S.	1228	3	
MAGNESIUM PEROXIDE	1476	5.1					
MAGNESIUM PHOSPHIDE	2011	4.3		MERCAPTANS, LIQUID, TOXIC, FLAMMABLE, N.O.S.	3071	6.1	
MAGNESIUM POWDER	1418	4.3					
Magnesium scrap, see	1869	4.1		MERCAPTAN MIXTURE, LIQUID, FLAMMABLE, N.O.S.	3336	3	
MAGNESIUM SILICIDE	2624	4.3					
Magnesium silicofluoride, see	2853	6.1		MERCAPTAN MIXTURE, LIQUID, FLAMMABLE, TOXIC, N.O.S.	1228	3	
Magnetized material	2807	9	Not subject to ADR				
MALEIC ANHYDRIDE	2215	8		MERCAPTAN MIXTURE, LIQUID, TOXIC, FLAMMABLE, N.O.S.	3071	6.1	
MALEIC ANHYDRIDE, MOLTEN	2215	8					
Malonic dinitrile, see	2647	6.1		2-Mercaptoethanol, see	2966	6.1	
Malonodinitrile, see	2647	6.1		2-Mercaptopropionic acid, see	2936	6.1	
MALONONITRILE	2647	6.1		5-MERCAPTOTETRAZOL-1-ACETIC ACID	0448	1	
MANEB	2210	4.2					
MANEB PREPARATION with not less than 60% maneb	2210	4.2		MERCURIC ARSENATE	1623	6.1	
				MERCURIC CHLORIDE	1624	6.1	
MANEB PREPARATION, STABILIZED against self-heating	2968	4.3		MERCURIC NITRATE	1625	6.1	
				MERCURIC POTASSIUM CYANIDE	1626	6.1	
MANEB, STABILIZED against self-heating	2968	4.3					
				Mercuric sulphate, see	1645	6.1	
Manganese ethylene-di-dithiocarbamate, see	2210	4.2		Mercurol, see	1639	6.1	
				Mercurous bisulphate, see	1645	6.1	
Manganese ethylene-1,2-dithiocarbamate, see	2210	4.2		Mercurous chloride, see	2025	6.1	
				MERCUROUS NITRATE	1627	6.1	
MANGANESE NITRATE	2724	5.1		Mercurous sulphate, see	1645	6.1	
Manganese (II) nitrate, see	2724	5.1		MERCURY	2809	8	
MANGANESE RESINATE	1330	4.1		MERCURY ACETATE	1629	6.1	
Manganous nitrate, see	2724	5.1		MERCURY AMMONIUM CHLORIDE	1630	6.1	
MANNITOL HEXANITRATE, WETTED with not less than 40% water, or mixture of alcohol and water, by mass	0133	1					
				MERCURY BASED PESTICIDE, LIQUID, FLAMMABLE, TOXIC, flash-point less than 23 °C	2778	3	
MATCHES, FUSEE	2254	4.1					
MATCHES, SAFETY (book, card or strike on box)	1944	4.1		MERCURY BASED PESTICIDE, LIQUID, TOXIC	3012	6.1	
				MERCURY BASED PESTICIDE, LIQUID, TOXIC, FLAMMABLE, flash-point not less than 23 °C	3011	6.1	
MATCHES, "STRIKE ANYWHERE"	1331	4.1					
MATCHES, WAX "VESTA"	1945	4.1					
MEDICAL WASTE, N.O.S.	3291	6.2		MERCURY BASED PESTICIDE, SOLID, TOXIC	2777	6.1	
MEDICINE, LIQUID, FLAMMABLE, TOXIC, N.O.S.	3248	3		MERCURY BENZOATE	1631	6.1	

Name and description	UN No.	Class	Remarks	Name and description	UN No.	Class	Remarks
Mercury bichloride, see	1624	6.1		METALLIC SUBSTANCE, WATER-REACTIVE, N.O.S.	3208	4.3	
MERCURY BROMIDES	1634	6.1		METALLIC SUBSTANCE, WATER-REACTIVE, SELF-HEATING, N.O.S.	3209	4.3	
MERCURY COMPOUND, LIQUID, N.O.S.	2024	6.1					
MERCURY COMPOUND, SOLID, N.O.S.	2025	6.1		METAL POWDER, FLAMMABLE, N.O.S.	3089	4.1	
MERCURY CONTAINED IN MANUFACTURED ARTICLES	3506	8		METAL POWDER, SELF-HEATING, N.O.S.	3189	4.2	
MERCURY CYANIDE	1636	6.1		METAL SALTS OF ORGANIC COMPOUNDS, FLAMMABLE, N.O.S.	3181	4.1	
MERCURY FULMINATE, WETTED with not less than 20% water, or mixture of alcohol and water, by mass	0135	1		METHACRYLALDEHYDE, STABILIZED	2396	3	
MERCURY GLUCONATE	1637	6.1		METHACRYLIC ACID, STABILIZED	2531	8	
MERCURY IODIDE	1638	6.1		METHACRYLONITRILE, STABILIZED	3079	6.1	
MERCURY NUCLEATE	1639	6.1					
MERCURY OLEATE	1640	6.1		METHALLYL ALCOHOL	2614	3	
MERCURY OXIDE	1641	6.1		Methanal, see	1198	3	
MERCURY OXYCYANIDE, DESENSITIZED	1642	6.1			2209	8	
MERCURY POTASSIUM IODIDE	1643	6.1		Methane and hydrogen mixture, see	2034	2	
MERCURY SALICYLATE	1644	6.1		METHANE, COMPRESSED	1971	2	
MERCURY SULPHATE	1645	6.1		METHANE, REFRIGERATED LIQUID	1972	2	
MERCURY THIOCYANATE	1646	6.1		METHANESULPHONYL CHLORIDE	3246	6.1	
Mesitylene, see	2325	3					
MESITYL OXIDE	1229	3		METHANOL	1230	3	
Metal alkyl halides, water-reactive, n.o.s. / Metal aryl halides, water-reactive, n.o.s., see	3394	4.2		2-Methoxyethyl acetate, see	1189	3	
Metal alkyl hydrides, water-reactive, n.o.s. / Metal aryl hydrides, water-reactive, n.o.s., see	3394	4.2		METHOXYMETHYL ISOCYANATE	2605	6.1	
				4-METHOXY-4-METHYLPENTAN-2-ONE	2293	3	
Metal alkyls, water-reactive, n.o.s. / Metal aryls, water-reactive, n.o.s., see	3393	4.2		1-Methoxy-2-nitrobenzene, see	2730	6.1	
					3458	6.1	
METAL CARBONYLS, LIQUID, N.O.S.	3281	6.1		1-Methoxy-3-nitrobenzene, see	2730	6.1	
					3458	6.1	
METAL CARBONYLS, SOLID, N.O.S.	3466	6.1		1-Methoxy-4-nitrobenzene, see	2730	6.1	
					3458	6.1	
METAL CATALYST, DRY	2881	4.2		1-METHOXY-2-PROPANOL	3092	3	
METAL CATALYST, WETTED with a visible excess of liquid	1378	4.2		METHYL ACETATE	1231	3	
METALDEHYDE	1332	4.1		METHYLACETYLENE AND PROPADIENE MIXTURE, STABILIZED such as mixture P1 or mixture P2	1060	2	
METAL HYDRIDES, FLAMMABLE, N.O.S.	3182	4.1					
METAL HYDRIDES, WATER-REACTIVE, N.O.S.	1409	4.3		beta-Methyl acrolein, see	1143	6.1	
				METHYL ACRYLATE, STABILIZED	1919	3	
				METHYLAL	1234	3	

Name and description	UN No.	Class	Remarks	Name and description	UN No.	Class	Remarks
Methyl alcohol, see	1230	3		Methyl alpha-chloropropionate, see	2933	3	
Methyl allyl alcohol, see	2614	3		METHYLCHLOROSILANE	2534	2	
METHYLALLYL CHLORIDE	2554	3		Methyl cyanide, see	1648	3	
METHYLAMINE, ANHYDROUS	1061	2		METHYLCYCLOHEXANE	2296	3	
METHYLAMINE, AQUEOUS SOLUTION	1235	3		METHYLCYCLOHEXANOLS, flammable	2617	3	
METHYLAMYL ACETATE	1233	3		METHYLCYCLOHEXANONE	2297	3	
Methyl amyl alcohol, see	2053	3		METHYLCYCLOPENTANE	2298	3	
Methyl amyl ketone, see	1110	3		METHYL DICHLOROACETATE	2299	6.1	
N-METHYLANILINE	2294	6.1		METHYLDICHLOROSILANE	1242	4.3	
Methylated spirit, see	1986	3		Methylene bromide, see	2664	6.1	
	1987	3		Methylene chloride, see	1593	6.1	
alpha-METHYLBENZYL ALCOHOL, LIQUID	2937	6.1		Methylene chloride and methyl chloride mixture, see	1912	2	
alpha-METHYLBENZYL ALCOHOL, SOLID	3438	6.1		Methylene cyanide, see	2647	6.1	
				p,p'-Methylene dianiline, see	2651	6.1	
METHYL BROMIDE with not more than 2% chloropicrin	1062	2		Methylene dibromide, see	2664	6.1	
Methyl bromide and chloropicrin mixture, with more than 2% chloropicrin, see	1581	2		2,2'-Methylene-di-(3,4,6-trichlorophenol), see	2875	6.1	
				Methyl ethyl ether, see	1039	2	
METHYL BROMIDE AND ETHYLENE DIBROMIDE MIXTURE, LIQUID	1647	6.1		METHYL ETHYL KETONE, see	1193	3	
				2-METHYL-5-ETHYLPYRIDINE	2300	6.1	
METHYL BROMOACETATE	2643	6.1		METHYL FLUORIDE	2454	2	
2-METHYLBUTANAL	3371	3		METHYL FORMATE	1243	3	
3-METHYLBUTAN-2-ONE	2397	3		2-METHYLFURAN	2301	3	
2-METHYL-1-BUTENE	2459	3		Methyl glycol, see	1188	3	
2-METHYL-2-BUTENE	2460	3		Methyl glycol acetate, see	1189	3	
3-METHYL-1-BUTENE	2561	3		2-METHYL-2-HEPTANETHIOL	3023	6.1	
N-METHYLBUTYLAMINE	2945	3		5-METHYLHEXAN-2-ONE	2302	3	
METHYL tert-BUTYL ETHER	2398	3		METHYLHYDRAZINE	1244	6.1	
METHYL BUTYRATE	1237	3		METHYL IODIDE	2644	6.1	
METHYL CHLORIDE	1063	2		METHYL ISOBUTYL CARBINOL	2053	3	
Methyl chloride and chloropicrin mixture, see	1582	2		METHYL ISOBUTYL KETONE	1245	3	
				METHYL ISOCYANATE	2480	6.1	
METHYL CHLORIDE AND METHYLENE CHLORIDE MIXTURE	1912	2		METHYL ISOPROPENYL KETONE, STABILIZED	1246	3	
METHYL CHLOROACETATE	2295	6.1		METHYL ISOTHIOCYANATE	2477	6.1	
Methyl chlorocarbonate, see	1238	6.1		METHYL ISOVALERATE	2400	3	
Methyl chloroform, see	2831	6.1		METHYL MAGNESIUM BROMIDE IN ETHYL ETHER	1928	4.3	
METHYL CHLOROFORMATE	1238	6.1					
METHYL CHLOROMETHYL ETHER	1239	6.1		METHYL MERCAPTAN	1064	2	
METHYL 2-CHLORO-PROPIONATE	2933	3		Methyl mercapto-propionaldehyde, see	2785	6.1	

Name and description	UN No.	Class	Remarks	Name and description	UN No.	Class	Remarks
METHYL METHACRYLATE MONOMER, STABILIZED	1247	3		Missiles, guided, see	0180	1	
					0181	1	
4-METHYLMORPHOLINE	2535	3			0182	1	
					0183	1	
N-METHYLMORPHOLINE, see	2535	3			0295	1	
					0397	1	
METHYL NITRITE	2455	2	Carriage prohibited		0398	1	
					0436	1	
METHYL ORTHOSILICATE	2606	6.1			0437	1	
					0438	1	
METHYLPENTADIENE	2461	3		Mixtures A, A01, A02, A0, A1, B1, B2, B or C, see	1965	2	
Methylpentanes, see	1208	3					
2-METHYLPENTAN-2-OL	2560	3		Mixture F1, mixture F2 or mixture F3, see	1078	2	
4-Methylpentan-2-ol, see	2053	3		MIXTURES OF 1,3-BUTADIENE AND HYDROCARBONS, STABILIZED, having a vapour pressure at 70 °C not exceeding 1.1 MPa (11 bar) and a density at 50 °C not lower than 0.525 kg/l	1010	2	
3-Methyl-2-penten-4ynol, see	2705	8					
METHYLPHENYL-DICHLOROSILANE	2437	8					
2-Methyl-2-phenylpropane, see	2709	3					
1-METHYLPIPERIDINE	2399	3		Mixture P1 or mixture P2, see	1060	2	
METHYL PROPIONATE	1248	3		MOLYBDENUM PENTACHLORIDE	2508	8	
Methylpropylbenzene, see	2046	3		Monochloroacetic acid, see	1750	6.1	
METHYL PROPYL ETHER	2612	3			1751	6.1	
METHYL PROPYL KETONE	1249	3		Monochlorobenzene, see	1134	3	
Methyl pyridines, see	2313	3		Monochlorodifluoromethane, see	1018	2	
Methylstyrene, inhibited, see	2618	3		Monochlorodifluoromethane and monochloropentafluoroethane mixture, see	1973	2	
alpha-Methylstyrene, see	2303	3					
Methyl sulphate, see	1595	6.1		Monochlorodifluoromonobromo-methane, see	1974	2	
Methyl sulphide, see	1164	3					
METHYLTETRAHYDROFURAN	2536	3		Monochloropentafluoroethane and monochlorodifluoromethane mixture, see	1973	2	
METHYL TRICHLOROACETATE	2533	6.1					
METHYLTRICHLOROSILANE	1250	3		Monoethylamine, see	1036	2	
alpha-METHYLVALERAL-DEHYDE	2367	3		MONONITROTOLUIDINES, see	2660	6.1	
Methyl vinyl benzene, inhibited, see	2618	3		Monopropylamine, see	1277	3	
METHYL VINYL KETONE, STABILIZED	1251	6.1		MORPHOLINE	2054	8	
M.i.b.c., see	2053	3		MOTOR FUEL ANTI-KNOCK MIXTURE	1649	6.1	
MINES with bursting charge	0136	1		MOTOR FUEL ANTI-KNOCK MIXTURE, FLAMMABLE	3483	6.1	
	0137	1					
	0138	1		MOTOR SPIRIT	1203	3	
	0294	1		Motor spirit and ethanol mixture, with more than 10% ethanol, see	3475	3	
Mirbane oil, see	1662	6.1					
				Muriatic acid, see	1789	8	
				MUSK XYLENE, see	2956	4.1	
				Mysorite, see	2212	9	
				Naphta, see	1268	3	
				Naphta, petroleum, see	1268	3	

Name and description	UN No.	Class	Remarks	Name and description	UN No.	Class	Remarks
Naphta, solvent, see	1268	3		NICOTINE PREPARATION, SOLID, N.O.S.	1655	6.1	
NAPHTHALENE, CRUDE	1334	4.1		NICOTINE SALICYLATE	1657	6.1	
NAPHTHALENE, MOLTEN	2304	4.1		NICOTINE SULPHATE, SOLID	3445	6.1	
NAPHTHALENE, REFINED	1334	4.1		NICOTINE SULPHATE, SOLUTION	1658	6.1	
alpha-NAPHTHYLAMINE	2077	6.1		NICOTINE TARTRATE	1659	6.1	
beta-NAPHTHYLAMINE, SOLID	1650	6.1		NITRATES, INORGANIC, N.O.S.	1477	5.1	
beta-NAPHTHYLAMINE, SOLUTION	3411	6.1		NITRATES, INORGANIC, AQUEOUS SOLUTION, N.O.S.	3218	5.1	
NAPHTHYLTHIOUREA	1651	6.1		NITRATING ACID MIXTURE with more than 50% nitric acid	1796	8	
1-Naphthylthiourea, see	1651	6.1		NITRATING ACID MIXTURE with not more than 50% nitric acid	1796	8	
NAPHTHYLUREA	1652	6.1		NITRATING ACID MIXTURE, SPENT, with more than 50% nitric acid	1826	8	
NATURAL GAS, COMPRESSED with high methane content	1971	2		NITRATING ACID MIXTURE, SPENT, with not more than 50% nitric acid	1826	8	
NATURAL GAS, REFRIGERATED LIQUID with high methane content	1972	2		NITRIC ACID, other than red fuming, with at least 65% but not more than 70% nitric acid	2031	8	
Natural gasoline, see	1203	3					
Neohexane, see	1208	3		NITRIC ACID, other than red fuming, with less than 65% nitric acid	2031	8	
NEON, COMPRESSED	1065	2					
NEON, REFRIGERATED LIQUID	1913	2		NITRIC ACID, other than red fuming, with more than 70% nitric acid	2031	8	
Neothyl, see	2612	3					
NICKEL CARBONYL	1259	6.1		NITRIC ACID, RED FUMING	2032	8	
NICKEL CYANIDE	1653	6.1		NITRIC OXIDE, COMPRESSED	1660	2	
Nickel (II) cyanide, see	1653	6.1		NITRIC OXIDE AND DINITROGEN TETROXIDE MIXTURE	1975	2	
NICKEL NITRATE	2725	5.1					
Nickel (II) nitrate, see	2725	5.1					
NICKEL NITRITE	2726	5.1		NITRIC OXIDE AND NITROGEN DIOXIDE MIXTURE, see	1975	2	
Nickel (II) nitrite, see	2726	5.1					
Nickelous nitrate, see	2725	5.1		NITRILES, FLAMMABLE, TOXIC, N.O.S.	3273	3	
Nickelous nitrite, see	2726	5.1					
Nickel tetracarbonyl, see	1259	6.1		NITRILES, LIQUID, TOXIC, N.O.S.	3276	6.1	
NICOTINE	1654	6.1					
NICOTINE COMPOUND, LIQUID, N.O.S	3144	6.1		NITRILES, SOLID, TOXIC, N.O.S.	3439	6.1	
NICOTINE COMPOUND, SOLID, N.O.S.	1655	6.1		NITRILES, TOXIC, FLAMMABLE, N.O.S.	3275	6.1	
NICOTINE HYDROCHLORIDE, LIQUID	1656	6.1		NITRITES, INORGANIC, N.O.S.	2627	5.1	
NICOTINE HYDROCHLORIDE, SOLID	3444	6.1		NITRITES, INORGANIC, AQUEOUS SOLUTION, N.O.S.	3219	5.1	
NICOTINE HYDROCHLORIDE, SOLUTION	1656	6.1		NITROANILINES (o-, m-, p-)	1661	6.1	
NICOTINE PREPARATION, LIQUID, N.O.S.	3144	6.1		NITROANISOLES, LIQUID	2730	6.1	
				NITROANISOLES, SOLID	3458	6.1	

Name and description	UN No.	Class	Remarks	Name and description	UN No.	Class	Remarks
NITROBENZENE	1662	6.1		NITROCELLULOSE WITH WATER (not less than 25% water, by mass)	2555	4.1	
Nitrobenzene bromide, see	2732	6.1					
NITROBENZENESULPHONIC ACID	2305	8		Nitrochlorobenzenes, see	1578 3409	6.1	
Nitrobenzol, see	1662	6.1		3-NITRO-4-CHLOROBENZO-TRIFLUORIDE	2307	6.1	
5-NITROBENZOTRIAZOL	0385	1					
NITROBENZOTRIFLUORIDES, LIQUID	2306	6.1		NITROCRESOLS, LIQUID	3434	6.1	
				NITROCRESOLS, SOLID	2446	6.1	
NITROBENZOTRIFLUORIDES, SOLID	3431	6.1		NITROETHANE	2842	3	
NITROBROMOBENZENES, LIQUID	2732	6.1		NITROGEN, COMPRESSED	1066	2	
				NITROGEN DIOXIDE, see	1067	2	
NITROBROMOBENZENES, SOLID	3459	6.1		NITROGEN, REFRIGERATED LIQUID	1977	2	
NITROCELLULOSE, dry or wetted with less than 25% water (or alcohol), by mass	0340	1		NITROGEN TRIFLUORIDE	2451	2	
				NITROGEN TRIOXIDE	2421	2	Carriage prohibited
NITROCELLULOSE, unmodified or plasticized with less than 18% plasticizing substance, by mass	0341	1		NITROGLYCERIN, DESENSITIZED with not less than 40% non-volatile water-insoluble phlegmatizer, by mass	0143	1	
NITROCELLULOSE MEMBRANE FILTERS, with not more than 12.6% nitrogen, by dry mass	3270	4.1		NITROGLYCERIN MIXTURE, DESENSITIZED, LIQUID, N.O.S. with not more than 30% nitroglycerin, by mass	3357	3	
NITROCELLULOSE, with not more than 12.6% nitrogen, by dry mass, MIXTURE WITH PLASTICIZER, WITH PIGMENT	2557	4.1		NITROGLYCERIN MIXTURE, DESENSITIZED, LIQUID, FLAMMABLE, N.O.S. with not more than 30% nitroglycerin, by mass	3343	3	
NITROCELLULOSE, with not more than 12.6% nitrogen, by dry mass, MIXTURE WITH PLASTICIZER, WITHOUT PIGMENT	2557	4.1					
NITROCELLULOSE, with not more than 12.6% nitrogen, by dry mass, MIXTURE WITHOUT PLASTICIZER, WITH PIGMENT	2557	4.1		NITROGLYCERIN MIXTURE, DESENSITIZED, SOLID, N.O.S. with more than 2% but not more than 10% nitroglycerin, by mass	3319	4.1	
NITROCELLULOSE, with not more than 12.6% nitrogen, by dry mass, MIXTURE WITHOUT PLASTICIZER, WITHOUT PIGMENT	2557	4.1		NITROGLYCERIN, SOLUTION IN ALCOHOL with more than 1% but not more than 5% nitroglycerin	3064	3	
				NITROGLYCERIN SOLUTION IN ALCOHOL with more than 1% but not more than 10% nitroglycerin	0144	1	
NITROCELLULOSE, PLASTICIZED with not less than 18% plasticizing substance, by mass	0343	1		NITROGLYCERIN SOLUTION IN ALCOHOL with not more than 1% nitroglycerin	1204	3	
NITROCELLULOSE SOLUTION, FLAMMABLE with not more than 12.6% nitrogen, by dry mass, and not more than 55% nitrocellulose	2059	3		NITROGUANIDINE, dry or wetted with less than 20% water, by mass	0282	1	
				NITROGUANIDINE, WETTED with not less than 20% water, by mass	1336	4.1	
NITROCELLULOSE, WETTED with not less than 25% alcohol, by mass	0342	1					
				NITROHYDROCHLORIC ACID	1798	8	Carriage prohibited
NITROCELLULOSE WITH ALCOHOL (not less than 25% alcohol, by mass, and not more than 12.6% nitrogen, by dry mass)	2556	4.1		NITROMANNITE, WETTED, see	0133	1	
				NITROMETHANE	1261	3	

Name and description	UN No.	Class	Remarks	Name and description	UN No.	Class	Remarks
Nitromuriatic acid, see	1798	8		OCTOL, dry or wetted with less than 15% water, by mass, see	0266	1	
NITRONAPHTHALENE	2538	4.1		OCTOLITE, dry or wetted with less than 15% water, by mass	0266	1	
NITROPHENOLS (o-, m-, p-)	1663	6.1					
4-NITROPHENYL-HYDRAZINE, with not less than 30% water, by mass	3376	4.1		OCTONAL	0496	1	
				OCTYL ALDEHYDES	1191	3	
NITROPROPANES	2608	3		tert-Octyl mercaptan, see	3023	6.1	
p-NITROSODIMETHYLANILINE	1369	4.2		OCTYLTRICHLOROSILANE	1801	8	
NITROSTARCH, dry or wetted with less than 20% water, by mass	0146	1		Oenanthol, see	3056	3	
				OIL GAS, COMPRESSED	1071	2	
NITROSTARCH, WETTED with not less than 20% water, by mass	1337	4.1		Oleum, see	1831	8	
NITROSYL CHLORIDE	1069	2		ORGANIC PEROXIDE TYPE B, LIQUID	3101	5.2	
NITROSYLSULPHURIC ACID, LIQUID	2308	8		ORGANIC PEROXIDE TYPE B, LIQUID, TEMPERATURE CONTROLLED	3111	5.2	
NITROSYLSULPHURIC ACID, SOLID	3456	8		ORGANIC PEROXIDE TYPE B, SOLID	3102	5.2	
NITROTOLUENES, LIQUID	1664	6.1					
NITROTOLUENES, SOLID	3446	6.1		ORGANIC PEROXIDE TYPE B, SOLID, TEMPERATURE CONTROLLED	3112	5.2	
NITROTOLUIDINES	2660	6.1					
NITROTRIAZOLONE	0490	1		ORGANIC PEROXIDE TYPE C, LIQUID	3103	5.2	
NITRO UREA	0147	1		ORGANIC PEROXIDE TYPE C, LIQUID, TEMPERATURE CONTROLLED	3113	5.2	
NITROUS OXIDE	1070	2					
NITROUS OXIDE, REFRIGERATED LIQUID	2201	2		ORGANIC PEROXIDE TYPE C, SOLID	3104	5.2	
NITROXYLENES, LIQUID	1665	6.1		ORGANIC PEROXIDE TYPE C, SOLID, TEMPERATURE CONTROLLED	3114	5.2	
NITROXYLENES, SOLID	3447	6.1					
Non-activated carbon, see	1361	4.2					
Non-activated charcoal, see	1361	4.2		ORGANIC PEROXIDE TYPE D, LIQUID	3105	5.2	
NONANES	1920	3		ORGANIC PEROXIDE TYPE D, LIQUID, TEMPERATURE CONTROLLED	3115	5.2	
NONYLTRICHLOROSILANE	1799	8					
2,5-NORBORNADIENE, STABILIZED, see	2251	3		ORGANIC PEROXIDE TYPE D, SOLID	3106	5.2	
Normal propyl alcohol, see	1274	3		ORGANIC PEROXIDE TYPE D, SOLID, TEMPERATURE CONTROLLED	3116	5.2	
NTO, see	0490	1					
OCTADECYLTRICHLORO-SILANE	1800	8		ORGANIC PEROXIDE TYPE E, LIQUID	3107	5.2	
OCTADIENE	2309	3		ORGANIC PEROXIDE TYPE E, LIQUID, TEMPERATURE CONTROLLED	3117	5.2	
OCTAFLUOROBUT-2-ENE	2422	2					
OCTAFLUOROCYCLOBUTANE	1976	2		ORGANIC PEROXIDE TYPE E, SOLID	3108	5.2	
OCTAFLUOROPROPANE	2424	2					
OCTANES	1262	3		ORGANIC PEROXIDE TYPE E, SOLID, TEMPERATURE CONTROLLED	3118	5.2	
OCTOGEN, see	0226	1					
	0391	1					
	0484	1					

Name and description	UN No.	Class	Remarks	Name and description	UN No.	Class	Remarks
ORGANIC PEROXIDE TYPE F, LIQUID	3109	5.2		ORGANOMETALLIC SUBSTANCE, LIQUID, PYROPHORIC, WATER-REACTIVE	3394	4.2	
ORGANIC PEROXIDE TYPE F, LIQUID, TEMPERATURE CONTROLLED	3119	5.2		ORGANOMETALLIC SUBSTANCE, SOLID, PYROPHORIC, WATER-REACTIVE	3393	4.2	
ORGANIC PEROXIDE TYPE F, SOLID	3110	5.2		ORGANOMETALLIC SUBSTANCE, LIQUID, WATER-REACTIVE	3398	4.3	
ORGANIC PEROXIDE TYPE F, SOLID, TEMPERATURE CONTROLLED	3120	5.2		ORGANOMETALLIC SUBSTANCE, SOLID, WATER-REACTIVE	3395	4.3	
Organic peroxides, see 2.2.52.4 for an alphabetic list of currently assigned organic peroxides and see	3101 to 3120	5.2		ORGANOMETALLIC SUBSTANCE, LIQUID, WATER-REACTIVE, FLAMMABLE	3399	4.3	
ORGANIC PIGMENTS, SELF-HEATING	3313	4.2		ORGANOMETALLIC SUBSTANCE, SOLID, WATER-REACTIVE, FLAMMABLE	3396	4.3	
ORGANOARSENIC COMPOUND, LIQUID, N.O.S.	3280	6.1		ORGANOMETALLIC SUBSTANCE, SOLID, WATER-REACTIVE, SELF-HEATING	3397	4.3	
ORGANOARSENIC COMPOUND, SOLID, N.O.S.	3465	6.1		ORGANOPHOSPHORUS COMPOUND, LIQUID, TOXIC, N.O.S.	3278	6.1	
ORGANOCHLORINE PESTICIDE, LIQUID, FLAMMABLE, TOXIC, flash-point less than 23 °C	2762	3		ORGANOPHOSPHORUS COMPOUND, SOLID, TOXIC, N.O.S.	3464	6.1	
ORGANOCHLORINE PESTICIDE, LIQUID, TOXIC	2996	6.1		ORGANOPHOSPHORUS COMPOUND, TOXIC, FLAMMABLE, N.O.S.	3279	6.1	
ORGANOCHLORINE PESTICIDE, LIQUID, TOXIC, FLAMMABLE, flash-point not less than 23 °C	2995	6.1		ORGANOPHOSPHORUS PESTICIDE, LIQUID, FLAMMABLE, TOXIC, flash-point less than 23 °C	2784	3	
ORGANOCHLORINE PESTICIDE, SOLID, TOXIC	2761	6.1		ORGANOPHOSPHORUS PESTICIDE, LIQUID, TOXIC	3018	6.1	
ORGANOMETALLIC COMPOUND, LIQUID, TOXIC, N.O.S.	3282	6.1		ORGANOPHOSPHORUS PESTICIDE, LIQUID, TOXIC, FLAMMABLE, flash-point not less than 23 °C	3017	6.1	
ORGANOMETALLIC COMPOUND, SOLID, TOXIC, N.O.S.	3467	6.1		ORGANOPHOSPHORUS PESTICIDE, SOLID, TOXIC	2783	6.1	
Organometallic compound, solid, water-reactive, flammable, n.o.s., see	3396	4.3		ORGANOTIN COMPOUND, LIQUID, N.O.S.	2788	6.1	
Organometallic compound or Organometallic compound solution or Organometallic compound dispersion, water-reactive, flammable, n.o.s., see	3399	4.3		ORGANOTIN COMPOUND, SOLID, N.O.S.	3146	6.1	
ORGANOMETALLIC SUBSTANCE, LIQUID, PYROPHORIC	3392	4.2		ORGANOTIN PESTICIDE, LIQUID, FLAMMABLE, TOXIC, flash-point less than 23 °C	2787	3	
ORGANOMETALLIC SUBSTANCE, SOLID, PYROPHORIC	3391	4.2		ORGANOTIN PESTICIDE, LIQUID, TOXIC	3020	6.1	
ORGANOMETALLIC SUBSTANCE, SOLID, SELF-HEATING	3400	4.2					

Name and description	UN No.	Class	Remarks	Name and description	UN No.	Class	Remarks
ORGANOTIN PESTICIDE, LIQUID, TOXIC, FLAMMABLE, flash-point not less than 23 °C	3019	6.1		PARAFORMALDEHYDE	2213	4.1	
				PARALDEHYDE	1264	3	
ORGANOTIN PESTICIDE, SOLID, TOXIC	2786	6.1		PCBs, see	2315	9	
					3432	9	
Orthophospohoric acid, see	1805	8		PENTABORANE	1380	4.2	
OSMIUM TETROXIDE	2471	6.1		PENTACHLOROETHANE	1669	6.1	
OXIDIZING LIQUID, N.O.S.	3139	5.1		PENTACHLOROPHENOL	3155	6.1	
OXIDIZING LIQUID, CORROSIVE, N.O.S.	3098	5.1		PENTAERYTHRITE TETRANITRATE with not less than 7% wax, by mass	0411	1	
OXIDIZING LIQUID, TOXIC, N.O.S.	3099	5.1		PENTAERYTHRITE TETRANITRATE, DESENSITIZED with not less than 15% phlegmatizer, by mass	0150	1	
OXIDIZING SOLID, N.O.S.	1479	5.1					
OXIDIZING SOLID, CORROSIVE, N.O.S.	3085	5.1					
OXIDIZING SOLID, FLAMMABLE, N.O.S.	3137	5.1	Carriage prohibited	PENTAERYTHRITE TETRANITRATE MIXTURE, DESENSITIZED, SOLID, N.O.S. with more than 10% but not more than 20% PETN, by mass	3344	4.1	
OXIDIZING SOLID, SELF-HEATING, N.O.S.	3100	5.1	Carriage prohibited				
OXIDIZING SOLID, TOXIC, N.O.S.	3087	5.1		PENTAERYTHRITE TETRANITRATE, WETTED with not less than 25% water, by mass	0150	1	
OXIDIZING SOLID, WATER-REACTIVE, N.O.S.	3121	5.1	Carriage prohibited	PENTAERYTHRITOL TETRANITRATE, see	0150	1	
					0411	1	
Oxirane, see	1040	2			3344	4.1	
OXYGEN, COMPRESSED	1072	2		PENTAFLUOROETHANE	3220	2	
OXYGEN DIFLUORIDE, COMPRESSED	2190	2		Pentafluoroethane, 1,1,1-trifluoroethane, and 1,1,1,2-tetrafluoroethane zeotropic mixture with approximately 44% pentafluoroethane and 52% 1,1,1-trifluoroethane, see	3337	2	
OXYGEN GENERATOR, CHEMICAL	3356	5.1					
OXYGEN, REFRIGERATED LIQUID	1073	2					
1-Oxy-4-nitrobenzene, see	1663	6.1		PENTAMETHYLHEPTANE	2286	3	
PACKAGINGS, DISCARDED, EMPTY, UNCLEANED	3509	9		Pentanal, see	2058	3	
PAINT (including paint, lacquer, enamel, stain, shellac, varnish, polish, liquid filler and liquid lacquer base)	1263	3		PENTANE-2,4-DIONE	2310	3	
	3066	8		PENTANES, liquid	1265	3	
	3469	3		n-Pentane, see	1265	3	
	3470	8		PENTANOLS	1105	3	
PAINT RELATED MATERIAL (including paint thinning and reducing compound)	1263	3		3-Pentanol, see	1105	3	
	3066	8		1-PENTENE	1108	3	
	3469	3		1-PENTOL	2705	8	
	3470	8		PENTOLITE, dry or wetted with less than 15% water, by mass	0151	1	
Paint thinning and reducing compound, see	1263	3					
	3066	8		Pentyl nitrite, see	1113	3	
	3469	3		PERCHLORATES, INORGANIC, N.O.S.	1481	5.1	
	3470	8					
PAPER, UNSATURATED OIL TREATED, incompletely dried (including carbon paper)	1379	4.2		PERCHLORATES, INORGANIC, AQUEOUS SOLUTION, N.O.S.	3211	5.1	
Paraffin, see	1223	3					

Name and description	UN No.	Class	Remarks	Name and description	UN No.	Class	Remarks
PERCHLORIC ACID with more than 50% but not more than 72% acid, by mass	1873	5.1		PETROLEUM DISTILLATES, N.O.S.	1268	3	
PERCHLORIC ACID with not more than 50% acid, by mass	1802	8		Petroleum ether, see	1268	3	
Perchlorobenzene, see	2729	6.1		PETROLEUM GASES, LIQUEFIED	1075	2	
Perchlorocyclopentadiene, see	2646	6.1		Petroleum naphtha, see	1268	3	
Perchloroethylene, see	1897	6.1		Petroleum oil, see	1268	3	
PERCHLOROMETHYL MERCAPTAN	1670	6.1		PETROLEUM PRODUCTS, N.O.S.	1268	3	
PERCHLORYL FLUORIDE	3083	2		Petroleum raffinate, see	1268	3	
Perfluoroacetylchloride, see	3057	2		PETROLEUM SOUR CRUDE OIL, FLAMMABLE, TOXIC	3494	3	
PERFLUORO (ETHYL VINYL ETHER)	3154	2		Petroleum spirit, see	1268	3	
PERFLUORO (METHYL VINYL ETHER)	3153	2		PHENACYL BROMIDE	2645	6.1	
				PHENETIDINES	2311	6.1	
Perfluoropropane, see	2424	2		PHENOLATES, LIQUID	2904	8	
PERFUMERY PRODUCTS with flammable solvents	1266	3		PHENOLATES, SOLID	2905	8	
PERMANGANATES, INORGANIC, N.O.S.	1482	5.1		PHENOL, MOLTEN	2312	6.1	
				PHENOL, SOLID	1671	6.1	
PERMANGANATES, INORGANIC, AQUEOUS SOLUTION, N.O.S.	3214	5.1		PHENOL SOLUTION	2821	6.1	
				PHENOLSULPHONIC ACID, LIQUID	1803	8	
PEROXIDES, INORGANIC, N.O.S.	1483	5.1		PHENOXYACETIC ACID DERIVATIVE PESTICIDE, LIQUID, FLAMMABLE, TOXIC, flash-point less than 23 °C	3346	3	
PERSULPHATES, INORGANIC, N.O.S.	3215	5.1					
PERSULPHATES, INORGANIC, AQUEOUS SOLUTION, N.O.S.	3216	5.1		PHENOXYACETIC ACID DERIVATIVE PESTICIDE, LIQUID, TOXIC	3348	6.1	
PESTICIDE, LIQUID, FLAMMABLE, TOXIC, N.O.S., flash-point less than 23 °C	3021	3		PHENOXYACETIC ACID DERIVATIVE PESTICIDE, LIQUID, TOXIC, FLAMMABLE, flash-point not less than 23 °C	3347	6.1	
PESTICIDE, LIQUID, TOXIC, N.O.S.	2902	6.1					
PESTICIDE, LIQUID, TOXIC, FLAMMABLE, N.O.S., flash-point not less than 23 °C	2903	6.1		PHENOXYACETIC ACID DERIVATIVE PESTICIDE, SOLID, TOXIC	3345	6.1	
PESTICIDE, SOLID, TOXIC, N.O.S.	2588	6.1		PHENYLACETONITRILE, LIQUID	2470	6.1	
Pesticide, toxic, under compressed gas, n.o.s, see	1950	2		PHENYLACETYL CHLORIDE	2577	8	
				Phenylamine, see	1547	6.1	
PETN, see	0150	1		1-Phenylbutane, see	2709	3	
	0411	1		2-Phenylbutane, see	2709	3	
	3344	4.1		PHENYLCARBYLAMINE CHLORIDE	1672	6.1	
PETN/TNT, see	0151	1		PHENYL CHLOROFORMATE	2746	6.1	
PETROL	1203	3		Phenyl cyanide, see	2224	6.1	
Petrol and ethanol mixture, with more than 10% ethanol, see	3475	3		PHENYLENEDIAMINES (o-, m-, p-)	1673	6.1	
				Phenylethylene, see	2055	3	
PETROLEUM CRUDE OIL	1267	3		PHENYLHYDRAZINE	2572	6.1	

Name and description	UN No.	Class	Remarks	Name and description	UN No.	Class	Remarks
PHENYL ISOCYANATE	2487	6.1		Phosphorus (V) sulphide, free from yellow and white phosphorus, see	1340	4.3	
Phenylisocyanodichloride, see	1672	6.1		Phosphorus sulphochloride, see	1837	8	
PHENYL MERCAPTAN	2337	6.1		PHOSPHORUS TRIBROMIDE	1808	8	
PHENYLMERCURIC ACETATE	1674	6.1		PHOSPHORUS TRICHLORIDE	1809	6.1	
PHENYLMERCURIC COMPOUND, N.O.S.	2026	6.1		PHOSPHORUS TRIOXIDE	2578	8	
PHENYLMERCURIC HYDROXIDE	1894	6.1		PHOSPHORUS TRISULPHIDE, free from yellow and white phosphorus	1343	4.1	
PHENYLMERCURIC NITRATE	1895	6.1		PHOSPHORUS, WHITE, DRY	1381	4.2	
PHENYLPHOSPHORUS DICHLORIDE	2798	8		PHOSPHORUS, WHITE IN SOLUTION	1381	4.2	
PHENYLPHOSPHORUS THIODICHLORIDE	2799	8		PHOSPHORUS, WHITE, MOLTEN	2447	4.2	
2-Phenylpropene, see	2303	3		PHOSPHORUS, WHITE, UNDER WATER	1381	4.2	
PHENYLTRICHLOROSILANE	1804	8		PHOSPHORUS, YELLOW, DRY	1381	4.2	
PHOSGENE	1076	2		PHOSPHORUS, YELLOW, IN SOLUTION	1381	4.2	
9-PHOSPHABICYCLO-NONANES	2940	4.2		PHOSPHORUS, YELLOW, UNDER WATER	1381	4.2	
PHOSPHINE	2199	2		Phosphoryl chloride, see	1810	6.1	
PHOSPHINE, ADSORBED	3525	2		PHTHALIC ANHYDRIDE with more than 0.05% of maleic anhydride	2214	8	
Phosphoretted hydrogen, see	2199	2		PICOLINES	2313	3	
PHOSPHORIC ACID, SOLUTION	1805	8		PICRAMIDE, see	0153	1	
PHOSPHORIC ACID, SOLID	3453	8		PICRIC ACID, WETTED, see	1344	4.1	
Phosphoric acid, anhydrous, see	1807	8			3364	4.1	
PHOSPHOROUS ACID	2834	8		PICRITE, see	0282	1	
PHOSPHORUS, AMORPHOUS	1338	4.1		PICRITE, WETTED, see	1336	4.1	
Phosphorus bromide, see	1808	8		Picrotoxin, see	3172	6.1	
Phosphorus chloride, see	1809	6.1			3462	6.1	
PHOSPHORUS HEPTASULPHIDE, free from yellow and white phosphorus	1339	4.1		PICRYL CHLORIDE, see	0155	1	
PHOSPHORUS OXYBROMIDE	1939	8		PICRYL CHLORIDE, WETTED, see	3365	4.1	
PHOSPHORUS OXYBROMIDE, MOLTEN	2576	8		alpha-PINENE	2368	3	
PHOSPHORUS OXYCHLORIDE	1810	6.1		PINE OIL	1272	3	
PHOSPHORUS PENTABROMIDE	2691	8		PIPERAZINE	2579	8	
PHOSPHORUS PENTACHLORIDE	1806	8		PIPERIDINE	2401	8	
PHOSPHORUS PENTAFLUORIDE	2198	2		Pivaloyl chloride, see	2438	6.1	
PHOSPHORUS PENTAFLUORIDE, ADSORBED	3524	2		Plastic explosives, see	0084	1	
PHOSPHORUS PENTASULPHIDE, free from yellow and white phosphorus	1340	4.3		PLASTICS MOULDING COMPOUND in dough, sheet or extruded rope form evolving flammable vapour	3314	9	
PHOSPHORUS PENTOXIDE	1807	8		PLASTICS, NITROCELLULOSE-BASED, SELF-HEATING, N.O.S.	2006	4.2	
PHOSPHORUS SESQUISULPHIDE, free from yellow and white phosphorus	1341	4.1					

Name and description	UN No.	Class	Remarks	Name and description	UN No.	Class	Remarks
Polish, see	1263	3		POTASSIUM BROMATE	1484	5.1	
	3066	8		POTASSIUM CHLORATE	1485	5.1	
	3469	3		POTASSIUM CHLORATE, AQUEOUS SOLUTION	2427	5.1	
	3470	8					
POLYAMINES, FLAMMABLE, CORROSIVE, N.O.S.	2733	3		Potassium chlorate mixed with mineral oil, see	0083	1	
POLYAMINES, LIQUID, CORROSIVE, N.O.S.	2735	8		POTASSIUM CUPROCYANIDE	1679	6.1	
POLYAMINES, LIQUID, CORROSIVE, FLAMMABLE, N.O.S.	2734	8		POTASSIUM CYANIDE, SOLID	1680	6.1	
				POTASSIUM CYANIDE, SOLUTION	3413	6.1	
POLYAMINES, SOLID, CORROSIVE, N.O.S.	3259	8		Potassium dicyanocuprate (I), see	1679	6.1	
POLYCHLORINATED BIPHENYLS, LIQUID	2315	9		POTASSIUM DITHIONITE	1929	4.2	
POLYCHLORINATED BIPHENYLS, SOLID	3432	9		POTASSIUM FLUORIDE, SOLID	1812	6.1	
				POTASSIUM FLUORIDE, SOLUTION	3422	6.1	
POLYESTER RESIN KIT, liquid base material	3269	3		POTASSIUM FLUOROACETATE	2628	6.1	
POLYESTER RESIN KIT, solid base material	3527	4.1		POTASSIUM FLUOROSILICATE	2655	6.1	
				Potassium hexafluorosilicate, see	2655	6.1	
POLYHALOGENATED BIPHENYLS, LIQUID	3151	9		Potassium hydrate, see	1814	8	
POLYHALOGENATED BIPHENYLS, SOLID	3152	9		POTASSIUM HYDROGENDIFLUORIDE, SOLID	1811	8	
POLYHALOGENATED TERPHENYLS, LIQUID	3151	9		POTASSIUM HYDROGENDIFLUORIDE, SOLUTION	3421	8	
POLYHALOGENATED TERPHENYLS, SOLID	3152	9		POTASSIUM HYDROGEN SULPHATE	2509	8	
POLYMERIC BEADS, EXPANDABLE, evolving flammable vapour	2211	9		POTASSIUM HYDROSULPHITE, see	1929	4.2	
POLYMERIZING SUBSTANCE, LIQUID, STABILIZED, N.O.S.	3532	4.1		Potassium hydroxide, liquid, see	1814	8	
POLYMERIZING SUBSTANCE, LIQUID, TEMPERATURE CONTROLLED, N.O.S.	3534	4.1		POTASSIUM HYDROXIDE, SOLID	1813	8	
				POTASSIUM HYDROXIDE SOLUTION	1814	8	
POLYMERIZING SUBSTANCE, SOLID, STABILIZED, N.O.S.	3531	4.1		POTASSIUM METAL ALLOYS, LIQUID	1420	4.3	
POLYMERIZING SUBSTANCE, SOLID, TEMPERATURE CONTROLLED, N.O.S	3533	4.1		POTASSIUM METAL ALLOYS, SOLID	3403	4.3	
				POTASSIUM METAVANADATE	2864	6.1	
Polystyrene beads, expandable, see	2211	9		POTASSIUM MONOXIDE	2033	8	
POTASSIUM	2257	4.3		POTASSIUM NITRATE	1486	5.1	
POTASSIUM ARSENATE	1677	6.1		Potassium nitrate and sodium nitrate mixture, see	1499	5.1	
POTASSIUM ARSENITE	1678	6.1		POTASSIUM NITRATE AND SODIUM NITRITE MIXTURE	1487	5.1	
Potassium bifluoride, see	1811	8					
Potassium bisulphate, see	2509	8		POTASSIUM NITRITE	1488	5.1	
Potassium bisulphite solution, see	2693	8		POTASSIUM PERCHLORATE	1489	5.1	
POTASSIUM BOROHYDRIDE	1870	4.3					

Name and description	UN No.	Class	Remarks	Name and description	UN No.	Class	Remarks
POTASSIUM PERMANGANATE	1490	5.1		PROJECTILES, inert with tracer	0345	1	
POTASSIUM PEROXIDE	1491	5.1			0424	1	
POTASSIUM PERSULPHATE	1492	5.1			0425	1	
POTASSIUM PHOSPHIDE	2012	4.3		PROJECTILES with burster or expelling charge	0346	1	
Potassium selenate, see	2630	6.1			0347	1	
Potassium selenite, see	2630	6.1			0426	1	
Potassium silicofluoride, see	2655	6.1			0427	1	
POTASSIUM SODIUM ALLOYS, LIQUID	1422	4.3			0434	1	
					0435	1	
POTASSIUM SODIUM ALLOYS, SOLID	3404	4.3		PROJECTILES with bursting charge	0167	1	
					0168	1	
POTASSIUM SULPHIDE with less than 30% water of crystallization	1382	4.2			0169	1	
					0324	1	
POTASSIUM SULPHIDE, ANHYDROUS	1382	4.2			0344	1	
				PROPADIENE, STABILIZED	2200	2	
POTASSIUM SULPHIDE, HYDRATED with not less than 30% water of crystallization	1847	8		Propadiene and methyl acetylene mixture, stabilized, see	1060	2	
				PROPANE	1978	2	
POTASSIUM SUPEROXIDE	2466	5.1		PROPANETHIOLS	2402	3	
Potassium tetracyanomercurate (II), see	1626	6.1		n-PROPANOL	1274	3	
POWDER CAKE, WETTED with not less than 17% alcohol, by mass	0433	1		PROPELLANT, LIQUID	0495	1	
					0497	1	
POWDER CAKE, WETTED with not less than 25% water, by mass	0159	1		PROPELLANT, SOLID	0498	1	
					0499	1	
POWDER PASTE, see	0159	1			0501	1	
	0433	1		Propellant with a single base, Propellant with a double base, Propellant with a triple base, see	0160	1	
POWDER, SMOKELESS	0160	1			0161	1	
	0161	1		Propene, see	1077	2	
	0509	1		PROPIONALDEHYDE	1275	3	
Power devices, explosive, see	0275	1		PROPIONIC ACID with not less than 10% and less than 90% acid by mass	1848	8	
	0276	1					
	0323	1					
	0381	1		PROPIONIC ACID with not less than 90% acid by mass	3463	8	
PRIMERS, CAP TYPE	0044	1					
	0377	1		PROPIONIC ANHYDRIDE	2496	8	
	0378	1		PROPIONITRILE	2404	3	
Primers, small arms, see	0044	1		PROPIONYL CHLORIDE	1815	3	
PRIMERS, TUBULAR	0319	1		n-PROPYL ACETATE	1276	3	
	0320	1		PROPYL ALCOHOL, NORMAL, see	1274	3	
	0376	1					
PRINTING INK, flammable or PRINTING INK RELATED MATERIAL (including printing ink thinning or reducing compound), flammable	1210	3		PROPYLAMINE	1277	3	
				n-PROPYLBENZENE	2364	3	
				Propyl chloride, see	1278	3	
				n-PROPYL CHLOROFORMATE	2740	6.1	
Projectiles, illuminating, see	0171	1		PROPYLENE	1077	2	
	0254	1		PROPYLENE CHLOROHYDRIN	2611	6.1	
	0297	1		1,2-PROPYLENEDIAMINE	2258	8	
				Propylene dichloride, see	1279	3	

Name and description	UN No.	Class	Remarks	Name and description	UN No.	Class	Remarks
PROPYLENEIMINE, STABILIZED	1921	3		RADIOACTIVE MATERIAL, EXCEPTED PACKAGE - EMPTY PACKAGING	2908	7	
PROPYLENE OXIDE	1280	3					
PROPYLENE TETRAMER	2850	3		RADIOACTIVE MATERIAL, EXCEPTED PACKAGE - INSTRUMENTS or ARTICLES	2911	7	
Propylene trimer, see	2057	3					
PROPYL FORMATES	1281	3		RADIOACTIVE MATERIAL, EXCEPTED PACKAGE - LIMITED QUANTITY OF MATERIAL	2910	7	
n-PROPYL ISOCYANATE	2482	6.1					
Propyl mercaptan, see	2402	3					
n-PROPYL NITRATE	1865	3		RADIOACTIVE MATERIAL, LOW SPECIFIC ACTIVITY (LSA-I), non fissile or fissile-excepted	2912	7	
PROPYLTRICHLOROSILANE	1816	8					
Pyrazine hexahydride, see	2579	8		RADIOACTIVE MATERIAL, LOW SPECIFIC ACTIVITY (LSA-II), FISSILE	3324	7	
PYRETHROID PESTICIDE, LIQUID, FLAMMABLE, TOXIC, flash-point less than 23 °C	3350	3					
				RADIOACTIVE MATERIAL, LOW SPECIFIC ACTIVITY (LSA-II), non fissile or fissile-excepted	3321	7	
PYRETHROID PESTICIDE, LIQUID, TOXIC	3352	6.1					
PYRETHROID PESTICIDE, LIQUID, TOXIC, FLAMMABLE, flash-point not less than 23 °C	3351	6.1		RADIOACTIVE MATERIAL, LOW SPECIFIC ACTIVITY, (LSA-III), FISSILE	3325	7	
PYRETHROID PESTICIDE, SOLID, TOXIC	3349	6.1		RADIOACTIVE MATERIAL, LOW SPECIFIC ACTIVITY (LSA-III), non fissile or fissile-excepted	3322	7	
PYRIDINE	1282	3					
PYROPHORIC ALLOY, N.O.S.	1383	4.2		RADIOACTIVE MATERIAL, SURFACE CONTAMINATED OBJECTS (SCO-I or SCO-II), FISSILE	3326	7	
Pyrophoric organometallic compound, water-reactive, n.o.s., liquid, see	3394	4.2					
Pyrophoric organometallic compound, water-reactive, n.o.s., solid, see	3393	4.2		RADIOACTIVE MATERIAL, SURFACE CONTAMINATED OBJECTS (SCO-I or SCO-II), non fissile or fissile-excepted	2913	7	
PYROPHORIC LIQUID, INORGANIC, N.O.S.	3194	4.2		RADIOACTIVE MATERIAL, TRANSPORTED UNDER SPECIAL ARRANGEMENT, FISSILE	3331	7	
PYROPHORIC LIQUID, ORGANIC, N.O.S.	2845	4.2					
PYROPHORIC METAL, N.O.S.	1383	4.2		RADIOACTIVE MATERIAL, TRANSPORTED UNDER SPECIAL ARRANGEMENT, non fissile or fissile-excepted	2919	7	
PYROPHORIC SOLID, INORGANIC, N.O.S.	3200	4.2					
PYROPHORIC SOLID, ORGANIC, N.O.S.	2846	4.2		RADIOACTIVE MATERIAL, TYPE A PACKAGE, FISSILE, non-special form	3327	7	
PYROSULPHURYL CHLORIDE	1817	8					
Pyroxylin solution, see	2059	3		RADIOACTIVE MATERIAL, TYPE A PACKAGE, non-special form, non fissile or fissile-excepted	2915	7	
PYRROLIDINE	1922	3					
QUINOLINE	2656	6.1		RADIOACTIVE MATERIAL, TYPE A PACKAGE, SPECIAL FORM, FISSILE	3333	7	
Quinone, see	2587	6.1					
RADIOACTIVE MATERIAL, EXCEPTED PACKAGE - ARTICLES MANUFACTURED FROM NATURAL URANIUM or DEPLETED URANIUM or NATURAL THORIUM	2909	7		RADIOACTIVE MATERIAL, TYPE A PACKAGE, SPECIAL FORM, non fissile or fissile-excepted	3332	7	

Name and description	UN No.	Class	Remarks	Name and description	UN No.	Class	Remarks
RADIOACTIVE MATERIAL, TYPE B(M) PACKAGE, FISSILE	3329	7		REFRIGERANT GAS R 133a, see	1983	2	
				REFRIGERANT GAS R 134a, see	3159	2	
RADIOACTIVE MATERIAL, TYPE B(M) PACKAGE, non fissile or fissile-excepted	2917	7		REFRIGERANT GAS R 142b, see	2517	2	
				REFRIGERANT GAS R 143a, see	2035	2	
RADIOACTIVE MATERIAL, TYPE B(U) PACKAGE, FISSILE	3328	7		REFRIGERANT GASR 152a, see	1030	2	
				REFRIGERANT GAS R 161, see	2453	2	
RADIOACTIVE MATERIAL, TYPE B(U) PACKAGE, non fissile or fissile-excepted	2916	7		REFRIGERANT GAS R 218, see	2424	2	
				REFRIGERANT GAS R 227, see	3296	2	
RADIOACTIVE MATERIAL, TYPE C PACKAGE, FISSILE	3330	7		REFRIGERANT GAS R 404A	3337	2	
				REFRIGERANT GAS R 407A	3338	2	
RADIOACTIVE MATERIAL, TYPE C PACKAGE, non fissile or fissile-excepted	3323	7		REFRIGERANT GAS R 407B	3339	2	
				REFRIGERANT GAS R 407C	3340	2	
RADIOACTIVE MATERIAL, URANIUM HEXAFLUORIDE, FISSILE	2977	7		REFRIGERANT GAS R 500, see	2602	2	
				REFRIGERANT GAS R 502, see	1973	2	
RADIOACTIVE MATERIAL, URANIUM HEXAFLUORIDE, non fissile or fissile-excepted	2978	7		REFRIGERANT GAS R 503, see	2599	2	
				REFRIGERANT GAS R 1132a, see	1959	2	
				REFRIGERANT GAS R 1216, see	1858	2	
Rags, oily	1856	4.2	Not subject to ADR	REFRIGERANT GAS R 1318, see	2422	2	
RDX, see	0072	1		REFRIGERANT GAS RC 318, see	1976	2	
	0391	1		REFRIGERATING MACHINES containing flammable, non-toxic, liquefied gas	3358	2	
	0483	1					
RECEPTACLES, SMALL, CONTAINING GAS without a release device, non-refillable	2037	2		REFRIGERATING MACHINES containing non-flammable, non-toxic, gases or ammonia solutions (UN 2672)	2857	2	
Red phosphorus, see	1338	4.1					
REFRIGERANT GAS, N.O.S., such as mixture F1, mixture F2 or mixture P2	1078	2		REGULATED MEDICAL WASTE, N.O.S.	3291	6.2	
REFRIGERANT GAS R 12, see	1028	2		RELEASE DEVICES, EXPLOSIVE	0173	1	
REFRIGERANT GAS R 12B1, see	1974	2		RESIN SOLUTION, flammable	1866	3	
REFRIGERANT GAS R 13, see	1022	2		Resorcin, see	2876	6.1	
REFRIGERANT GAS R 13B1, see	1009	2		RESORCINOL	2876	6.1	
REFRIGERANT GAS R 14, see	1982	2		RIVETS, EXPLOSIVE	0174	1	
REFRIGERANT GAS R 21, see	1029	2		Road oil, with a flash-point not greater than 60 °C, see	1999	3	
REFRIGERANT GAS R 22, see	1018	2					
REFRIGERANT GAS R 23, see	1984	2		Road oil, with a flash-point above 60 °C, at or above its flash-point, see	3256	3	
REFRIGERANT GAS R 32, see	3252	2					
REFRIGERANT GAS R 40, see	1063	2		Road oil, at or above 100 °C and below its flash-point, see	3257	9	
REFRIGERANT GAS R 41, see	2454	2		ROCKET MOTORS	0186	1	
REFRIGERANT GAS R 114, see	1958	2			0280	1	
					0281	1	
REFRIGERANT GAS R 115, see	1020	2			0510	1	
REFRIGERANT GAS R 116, see	2193	2		ROCKET MOTORS, LIQUID FUELLED	0395	1	
REFRIGERANT GAS R 124, see	1021	2			0396	1	
REFRIGERANT GAS R 125, see	3220	2					

Name and description	UN No.	Class	Remarks	Name and description	UN No.	Class	Remarks
ROCKET MOTORS WITH HYPERGOLIC LIQUIDS with or without expelling charge	0250 0322	1 1		SELENIUM COMPOUND, SOLID, N.O.S.	3283	6.1	
ROCKETS with bursting charge	0180 0181 0182 0295	1 1 1 1		SELENIUM DISULPHIDE	2657	6.1	
				SELENIUM HEXAFLUORIDE	2194	2	
				SELENIUM OXYCHLORIDE	2879	8	
ROCKETS with expelling charge	0436 0437 0438	1 1 1		SELF-HEATING LIQUID, CORROSIVE, INORGANIC, N.O.S.	3188	4.2	
ROCKETS with inert head	0183 0502	1 1		SELF-HEATING LIQUID, CORROSIVE, ORGANIC, N.O.S.	3185	4.2	
ROCKETS, LINE-THROWING	0238 0240 0453	1 1 1		SELF-HEATING LIQUID, INORGANIC, N.O.S.	3186	4.2	
ROCKETS, LIQUID FUELLED with bursting charge	0397 0398	1 1		SELF-HEATING LIQUID, ORGANIC, N.O.S.	3183	4.2	
ROSIN OIL	1286	3		SELF-HEATING LIQUID, TOXIC, INORGANIC, N.O.S.	3187	4.2	
RUBBER SCRAP, powdered or granulated	1345	4.1		SELF-HEATING LIQUID, TOXIC, ORGANIC, N.O.S.	3184	4.2	
RUBBER SHODDY, powdered or granulated	1345	4.1		SELF-HEATING SOLID, CORROSIVE, INORGANIC, N.O.S.	3192	4.2	
RUBBER SOLUTION	1287	3		SELF-HEATING SOLID, CORROSIVE, ORGANIC, N.O.S.	3126	4.2	
RUBIDIUM	1423	4.3		SELF-HEATING SOLID, INORGANIC, N.O.S.	3190	4.2	
RUBIDIUM HYDROXIDE	2678	8		SELF-HEATING SOLID, ORGANIC, N.O.S.	3088	4.2	
RUBIDIUM HYDROXIDE SOLUTION	2677	8		SELF-HEATING SOLID, OXIDIZING, N.O.S	3127	4.2	Carriage prohibited
Rubidium nitrate, see	1477	5.1					
SAFETY DEVICES, electrically initiated	3268	9		SELF-HEATING SOLID, TOXIC, INORGANIC, N.O.S.	3191	4.2	
SAFETY DEVICES, PYROTECHNIC	0503	1		SELF-HEATING SOLID, TOXIC, ORGANIC, N.O.S.	3128	4.2	
Saltpetre, see	1486	5.1		SELF-REACTIVE LIQUID TYPE B	3221	4.1	
SAMPLES, EXPLOSIVE, other than initiating explosive	0190	1		SELF-REACTIVE LIQUID TYPE B, TEMPERATURE CONTROLLED	3231	4.1	
Sand acid, see	1778	8		SELF-REACTIVE LIQUID TYPE C	3223	4.1	
Seat-belt pretensioners, see	0503 3268	1 9		SELF-REACTIVE LIQUID TYPE C, TEMPERATURE CONTROLLED	3233	4.1	
SEED CAKE with more than 1.5% oil and not more than 11% moisture	1386	4.2		SELF-REACTIVE LIQUID TYPE D	3225	4.1	
SEED CAKE with not more than 1.5% oil and not more than 11% moisture	2217	4.2		SELF-REACTIVE LIQUID TYPE D, TEMPERATURE CONTROLLED	3235	4.1	
Seed expellers, see	1386 2217	4.2 4.2		SELF-REACTIVE LIQUID TYPE E	3227	4.1	
SELENATES	2630	6.1		SELF-REACTIVE LIQUID TYPE E, TEMPERATURE CONTROLLED	3237	4.1	
SELENIC ACID	1905	8					
SELENITES	2630	6.1					
SELENIUM COMPOUND, LIQUID, N.O.S.	3440	6.1		SELF-REACTIVE LIQUID TYPE F	3229	4.1	

Name and description	UN No.	Class	Remarks	Name and description	UN No.	Class	Remarks
SELF-REACTIVE LIQUID TYPE F, TEMPERATURE CONTROLLED	3239	4.1		SILICON TETRAFLUORIDE	1859	2	
SELF-REACTIVE SOLID TYPE B	3222	4.1		SILICON TETRAFLUORIDE, ADSORBED	3521	2	
SELF-REACTIVE SOLID TYPE B, TEMPERATURE CONTROLLED	3232	4.1		SILVER ARSENITE	1683	6.1	
SELF-REACTIVE SOLID TYPE C	3224	4.1		SILVER CYANIDE	1684	6.1	
SELF-REACTIVE SOLID TYPE C, TEMPERATURE CONTROLLED	3234	4.1		SILVER NITRATE	1493	5.1	
SELF-REACTIVE SOLID TYPE D	3226	4.1		SILVER PICRATE, WETTED with not less than 30% water, by mass	1347	4.1	
SELF-REACTIVE SOLID TYPE D, TEMPERATURE CONTROLLED	3236	4.1		SLUDGE ACID	1906	8	
SELF-REACTIVE SOLID TYPE E	3228	4.1		SODA LIME with more than 4% sodium hydroxide	1907	8	
SELF-REACTIVE SOLID TYPE E, TEMPERATURE CONTROLLED	3238	4.1		SODIUM	1428	4.3	
SELF-REACTIVE SOLID TYPE F	3230	4.1		Sodium aluminate, solid	2812	8	Not subject to ADR
SELF-REACTIVE SOLID TYPE F, TEMPERATURE CONTROLLED	3240	4.1		SODIUM ALUMINATE SOLUTION	1819	8	
SHALE OIL	1288	3		SODIUM ALUMINIUM HYDRIDE	2835	4.3	
Shaped charges, see	0059	1		SODIUM AMMONIUM VANADATE	2863	6.1	
	0439	1					
	0440	1		SODIUM ARSANILATE	2473	6.1	
	0441	1		SODIUM ARSENATE	1685	6.1	
Shellac, see	1263	3		SODIUM ARSENITE, AQUEOUS SOLUTION	1686	6.1	
	3066	8					
	3469	3		SODIUM ARSENITE, SOLID	2027	6.1	
	3470	8		SODIUM AZIDE	1687	6.1	
SIGNAL DEVICES, HAND	0191	1		Sodium bifluoride, see	2439	8	
	0373	1		Sodium binoxide, see	1504	5.1	
SIGNALS, DISTRESS, ship	0194	1		Sodium bisulphite solution, see	2693	8	
	0195	1		SODIUM BOROHYDRIDE	1426	4.3	
	0505	1		SODIUM BOROHYDRIDE AND SODIUM HYDROXIDE SOLUTION, with not more than 12% sodium borohydride and not more than 40% sodium hydroxide by mass	3320	8	
	0506	1					
Signals, distress, ship, water-activated, see	0249	1					
SIGNALS, RAILWAY TRACK, EXPLOSIVE	0192	1					
	0193	1					
	0492	1					
	0493	1					
SIGNALS, SMOKE	0196	1		SODIUM BROMATE	1494	5.1	
	0197	1		SODIUM CACODYLATE	1688	6.1	
	0313	1		SODIUM CARBONATE PEROXYHYDRATE	3378	5.1	
	0487	1					
	0507	1		SODIUM CHLORATE	1495	5.1	
SILANE	2203	2		SODIUM CHLORATE, AQUEOUS SOLUTION	2428	5.1	
Silicofluoric acid, see	1778	8					
Silicofluorides, n.o.s., see	2856	6.1		Sodium chlorate mixed with dinitrotoluene, see	0083	1	
Silicon chloride, see	1818	8					
SILICON POWDER, AMORPHOUS	1346	4.1		SODIUM CHLORITE	1496	5.1	
				SODIUM CHLOROACETATE	2659	6.1	
SILICON TETRACHLORIDE	1818	8					

Name and description	UN No.	Class	Remarks	Name and description	UN No.	Class	Remarks
SODIUM CUPROCYANIDE, SOLID	2316	6.1		SODIUM NITRATE	1498	5.1	
SODIUM CUPROCYANIDE SOLUTION	2317	6.1		SODIUM NITRATE AND POTASSIUM NITRATE MIXTURE	1499	5.1	
SODIUM CYANIDE, SOLID	1689	6.1		SODIUM NITRITE	1500	5.1	
SODIUM CYANIDE, SOLUTION	3414	6.1		Sodium nitrite and potassium nitrate mixture, see	1487	5.1	
Sodium dicyanocuprate (I), solid, see	2316	6.1		SODIUM PENTACHLORO-HENATE	2567	6.1	
Sodium dicyanocuprate (I) solution, see	2317	6.1		SODIUM PERBORATE MONOHYDRATE	3377	5.1	
Sodium dimethylarsenate, see	1688	6.1		SODIUM PERCHLORATE	1502	5.1	
SODIUM DINITRO-o-CRESOLATE, dry or wetted with less than 15% water, by mass	0234	1		SODIUM PERMANGANATE	1503	5.1	
SODIUM DINITRO-o-CRESOLATE, WETTED with not less than 10% water, by mass	3369	4.1		SODIUM PEROXIDE	1504	5.1	
				SODIUM PEROXOBORATE, ANHYDROUS	3247	5.1	
SODIUM DINITRO-o-CRESOLATE, WETTED with not less than 15% water, by mass	1348	4.1		SODIUM PERSULPHATE	1505	5.1	
				SODIUM PHOSPHIDE	1432	4.3	
Sodium dioxide, see	1504	5.1		SODIUM PICRAMATE, dry or wetted with less than 20% water, by mass	0235	1	
SODIUM DITHIONITE	1384	4.2					
SODIUM FLUORIDE, SOLID	1690	6.1		SODIUM PICRAMATE, WETTED with not less than 20% water, by mass	1349	4.1	
SODIUM FLUORIDE, SOLUTION	3415	6.1					
SODIUM FLUOROACETATE	2629	6.1		Sodium potassium alloys, liquid, see	1422	4.3	
SODIUM FLUOROSILICATE	2674	6.1		Sodium selenate, see	2630	6.1	
Sodium hexafluorosilicate, see	2674	6.1		Sodium selenite, see	2630	6.1	
Sodium hydrate, see	1824	8		Sodium silicofluoride, see	2674	6.1	
SODIUM HYDRIDE	1427	4.3		SODIUM SULPHIDE, ANHYDROUS	1385	4.2	
Sodium hydrogen 4-amino-phenylarsenate, see	2473	6.1		SODIUM SULPHIDE with less than 30% water of crystallization	1385	4.2	
SODIUM HYDROGEN-DIFLUORIDE	2439	8		SODIUM SULPHIDE, HYDRATED with not less than 30% water	1849	8	
SODIUM HYDROSULPHIDE with less than 25% water of crystallization	2318	4.2					
				SODIUM SUPEROXIDE	2547	5.1	
SODIUM HYDROSULPHIDE, HYDRATED with not less than 25% water of crystallization	2949	8		SOLIDS CONTAINING CORROSIVE LIQUID, N.O.S.	3244	8	
				SOLIDS or mixtures of solids (such as preparations and wastes) CONTAINING FLAMMABLE LIQUID, N.O.S. having a flash-point up to 60 °C	3175	4.1	
SODIUM HYDROSULPHITE, see	1384	4.2					
SODIUM HYDROXIDE, SOLID	1823	8					
SODIUM HYDROXIDE SOLUTION	1824	8					
Sodium metasilicate pentahydrate, see	3253	8		SOLIDS CONTAINING TOXIC LIQUID, N.O.S.	3243	6.1	
				Solvents, flammable, n.o.s., see	1993	3	
SODIUM METHYLATE	1431	4.2		Solvents, flammable, toxic, n.o.s., see	1992	3	
SODIUM METHYLATE SOLUTION in alcohol	1289	3					
SODIUM MONOXIDE	1825	8					

Name and description	UN No.	Class	Remarks	Name and description	UN No.	Class	Remarks
SOUNDING DEVICES, EXPLOSIVE	0204 0296 0374 0375	1 1 1 1		Substances liable to spontaneous combustion, n.o.s., see	2845 2846 3194 3200	4.2 4.2 4.2 4.2	
Squibs, see	0325 0454	1 1		SUBSTITUTED NITROPHENOL PESTICIDE, LIQUID, FLAMMABLE, TOXIC, flash-point less than 23 °C	2780	3	
Stain, see	1263 3066 3469 3470	3 8 3 8		SUBSTITUTED NITROPHENOL PESTICIDE, LIQUID, TOXIC	3014	6.1	
STANNIC CHLORIDE, ANHYDROUS	1827	8		SUBSTITUTED NITROPHENOL PESTICIDE, LIQUID, TOXIC, FLAMMABLE, flash-point not less than 23 °C	3013	6.1	
STANNIC CHLORIDE PENTAHYDRATE	2440	8					
STANNIC PHOSPHIDES	1433	4.3		SUBSTITUTED NITROPHENOL PESTICIDE, SOLID, TOXIC	2779	6.1	
Steel swarf, see	2793	4.2		SULPHAMIC ACID	2967	8	
STIBINE	2676	2		SULPHUR	1350	4.1	
Straw	1327	4.1	Not subject to ADR	SULPHUR CHLORIDES	1828	8	
Strontium alloys, pyrophoric, see	1383	4.2		Sulphur dichloride, see	1828	8	
STRONTIUM ARSENITE	1691	6.1		SULPHUR DIOXIDE	1079	2	
STRONTIUM CHLORATE	1506	5.1		Sulphuretted hydrogen, see	1053	2	
Strontium dioxide, see	1509	5.1		SULPHUR HEXAFLUORIDE	1080	2	
STRONTIUM NITRATE	1507	5.1		SULPHURIC ACID with more than 51% acid	1830	8	
STRONTIUM PERCHLORATE	1508	5.1		SULPHURIC ACID with not more than 51% acid	2796	8	
STRONTIUM PEROXIDE	1509	5.1					
STRONTIUM PHOSPHIDE	2013	4.3		SULPHURIC ACID, FUMING	1831	8	
STRYCHNINE	1692	6.1		SULPHURIC ACID, SPENT	1832	8	
STRYCHNINE SALTS	1692	6.1		Sulphuric and hydrofluoric acid mixture, see	1786	8	
STYPHNIC ACID, see	0219 0394	1 1		SULPHUR, MOLTEN	2448	4.1	
STYRENE MONOMER, STABILIZED	2055	3		Sulphur monochloride, see	1828	8	
SUBSTANCES, EVI, N.O.S., see	0482	1		SULPHUROUS ACID	1833	8	
SUBSTANCES, EXPLOSIVE, N.O.S.	0357 0358 0359 0473 0474 0475 0476 0477 0478 0479 0480 0481 0485	1 1 1 1 1 1 1 1 1 1 1 1 1		SULPHUR TETRAFLUORIDE	2418	2	
				SULPHUR TRIOXIDE, STABILIZED	1829	8	
				SULPHURYL CHLORIDE	1834	6.1	
				SULPHURYL FLUORIDE	2191	2	
				Table Tennis Balls, see	2000	4.1	
				Talcum with tremolite and/or actinolite, see	2212	9	
				TARS, LIQUID, including road oils, and cutback bitumens, with a flash-point not greater than 60 °C	1999	3	
SUBSTANCES, EXPLOSIVE, VERY INSENSITIVE, N.O.S.	0482	1					
				Tars, liquid, with a flash-point above 60 °C, at or above its flash-point, see	3256	3	

Name and description	UN No.	Class	Remarks	Name and description	UN No.	Class	Remarks
Tars, liquid, at or above 100 °C and below its flash-point, see	3257	9		TETRAMETHYLSILANE	2749	3	
				TETRANITROANILINE	0207	1	
Tartar emetic, see	1551	6.1		TETRANITROMETHANE	1510	6.1	
TEAR GAS CANDLES	1700	6.1		TETRAPROPYL ORTHOTITANATE	2413	3	
TEAR GAS SUBSTANCE, LIQUID, N.O.S.	1693	6.1		TETRAZENE, WETTED with not less than 30% water, or mixture of alcohol and water, by mass, see	0114	1	
TEAR GAS SUBSTANCE, SOLID, N.O.S.	3448	6.1					
TELLURIUM COMPOUND, N.O.S.	3284	6.1		TETRAZOL-1-ACETIC ACID	0407	1	
TELLURIUM HEXAFLUORIDE	2195	2		1H-TETRAZOLE	0504	1	
TERPENE HYDROCARBONS, N.O.S.	2319	3		TETRYL, see	0208	1	
				Textile waste, wet	1857	4.2	Not subject to ADR
TERPINOLENE	2541	3		THALLIUM CHLORATE	2573	5.1	
TETRABROMOETHANE	2504	6.1		Thallium (I) chlorate, see	2573	5.1	
1,1,2,2-TETRACHLOROETHANE	1702	6.1		THALLIUM COMPOUND, N.O.S.	1707	6.1	
TETRACHLOROETHYLENE	1897	6.1		THALLIUM NITRATE	2727	6.1	
TETRAETHYL DITHIO-PYROPHOSPHATE	1704	6.1		Thallium (I) nitrate, see	2727	6.1	
TETRAETHYLENEPENTAMINE	2320	8		Thallous chlorate, see	2573	5.1	
Tetraethyl lead, see	1649	6.1		4-THIAPENTANAL	2785	6.1	
TETRAETHYL SILICATE	1292	3		Thia-4-pentanal, see	2785	6.1	
Tetraethyoxysilane, see	1292	3		THIOACETIC ACID	2436	3	
Tetrafluorodichloroethane, see	1958	2		THIOCARBAMATE PESTICIDE, LIQUID, FLAMMABLE, TOXIC, flash-point less than 23 °C	2772	3	
1,1,1,2-TETRAFLUOROETHANE	3159	2					
TETRAFLUOROETHYLENE, STABILIZED	1081	2		THIOCARBAMATE PESTICIDE, LIQUID, TOXIC	3006	6.1	
TETRAFLUOROMETHANE	1982	2		THIOCARBAMATE PESTICIDE, LIQUID, TOXIC, FLAMMABLE, flash-point not less than 23 °C	3005	6.1	
1,2,3,6-TETRAHYDRO-BENZALDEHYDE	2498	3					
TETRAHYDROFURAN	2056	3		THIOCARBAMATE PESTICIDE, SOLID, TOXIC	2771	6.1	
TETRAHYDRO-FURFURYLAMINE	2943	3		THIOGLYCOL	2966	6.1	
Tetrahydro-1,4-oxazine, see	2054	3		THIOGLYCOLIC ACID	1940	8	
TETRAHYDROPHTHALIC ANHYDRIDES with more than 0.05% of maleic anhydride	2698	8		THIOLACTIC ACID	2936	6.1	
				THIONYL CHLORIDE	1836	8	
1,2,3,6-TETRAHYDROPYRIDINE	2410	3		THIOPHENE	2414	3	
TETRAHYDROTHIOPHENE	2412	3		Thiophenol, see	2337	6.1	
Tetramethoxysilane, see	2606	6.1		THIOPHOSGENE	2474	6.1	
TETRAMETHYLAMMONIUM HYDROXIDE SOLID	3423	8		THIOPHOSPHORYL CHLORIDE	1837	8	
				THIOUREA DIOXIDE	3341	4.2	
TETRAMETHYLAMMONIUM HYDROXIDE SOLUTION	1835	8		Tin (IV) chloride, anhydrous, see	1827	8	
				Tin (IV) chloride pentahydrate, see	2440	8	
Tetramethylene, see	2601	2		TINCTURES, MEDICINAL	1293	3	
Tetramethylene cyanide, see	2205	6.1		Tin tetrachloride, see	1827	8	
Tetramethyl lead, see	1649	6.1					

Name and description	UN No.	Class	Remarks	Name and description	UN No.	Class	Remarks
TITANIUM DISULPHIDE	3174	4.2		TOXIC BY INHALATION LIQUID, N.O.S. with an LC_{50} lower than or equal to 1000 ml/m^3 and saturated vapour concentration greater than or equal to 10 LC_{50}	3382	6.1	
TITANIUM HYDRIDE	1871	4.1					
TITANIUM POWDER, DRY	2546	4.2					
TITANIUM POWDER, WETTED with not less than 25% water	1352	4.1		TOXIC BY INHALATION LIQUID, CORROSIVE, N.O.S. with an LC_{50} lower than or equal to 200 ml/m^3 and saturated vapour concentration greater than or equal to 500 LC_{50}	3389	6.1	
TITANIUM SPONGE GRANULES	2878	4.1					
TITANIUM SPONGE POWDERS	2878	4.1					
TITANIUM TETRACHLORIDE	1838	6.1		TOXIC BY INHALATION LIQUID, CORROSIVE, N.O.S. with an LC_{50} lower than or equal to 1000 ml/m^3 and saturated vapour concentration greater than or equal to 10 LC_{50}	3390	6.1	
TITANIUM TRICHLORIDE MIXTURE	2869	8					
TITANIUM TRICHLORIDE MIXTURE, PYROPHORIC	2441	4.2					
TITANIUM TRICHLORIDE, PYROPHORIC	2441	4.2		TOXIC BY INHALATION LIQUID, FLAMMABLE, N.O.S. with an LC_{50} lower than or equal to 200 ml/m^3 and saturated vapour concentration greater than or equal to 500 LC_{50}	3383	6.1	
TNT, see	0209	1					
	0388	1					
	0389	1					
TNT mixed with aluminium, see	0390	1		TOXIC BY INHALATION LIQUID, FLAMMABLE, N.O.S. with an LC_{50} lower than or equal to 1000 ml/m^3 and saturated vapour concentration greater than or equal to 10 LC_{50}	3384	6.1	
TNT, WETTED with not less than 30% water, by mass, see	1356	4.1					
TNT, WETTED with not less than 10% water, by mass, see	3366	4.1					
Toe puffs, nitrocellulose base, see	1353	4.1		TOXIC BY INHALATION LIQUID, FLAMMABLE, CORROSIVE, N.O.S. with an LC_{50} lower than or equal to 200 ml/m³ and saturated vapour concentration greater than or equal to 500 LC_{50}	3488	6.1	
TOLUENE	1294	3					
TOLUENE DIISOCYANATE	2078	6.1					
TOLUIDINES, LIQUID	1708	6.1					
TOLUIDINES, SOLID	3451	6.1		TOXIC BY INHALATION LIQUID, FLAMMABLE, CORROSIVE, N.O.S. with an LC_{50} lower than or equal to 1000 ml/m³ and saturated vapour concentration greater than or equal to 10 LC_{50}	3489	6.1	
Toluol, see	1294	3					
2,4-TOLUYLENEDIAMINE, SOLID	1709	6.1					
2,4-TOLUYLENEDIAMINE, SOLUTION	3418	6.1					
Toluylene diisocyanate, see	2078	6.1		TOXIC BY INHALATION LIQUID, OXIDIZING, N.O.S. with an LC_{50} lower than or equal to 200 ml/m^3 and saturated vapour concentration greater than or equal to 500 LC_{50}	3387	6.1	
Tolylene diisocyanate, see	2078	6.1					
Tolylethylene, inhibited, see	2618	3					
TORPEDOES with bursting charge	0329	1		TOXIC BY INHALATION LIQUID, OXIDIZING, N.O.S. with an LC_{50} lower than or equal to 1000 ml/m^3 and saturated vapour concentration greater than or equal to 10 LC_{50}	3388	6.1	
	0330	1					
	0451	1					
TORPEDOES, LIQUID FUELLED with inert head	0450	1					
TORPEDOES, LIQUID FUELLED with or without bursting charge	0449	1		TOXIC BY INHALATION LIQUID, WATER-REACTIVE, N.O.S. with an LC_{50} lower than or equal to 200 ml/m^3 and saturated vapour concentration greater than or equal to 500 LC_{50}	3385	6.1	
TOXIC BY INHALATION LIQUID, N.O.S. with an LC_{50} lower than or equal to 200 ml/m^3 and saturated vapour concentration greater than or equal to 500 LC_{50}	3381	6.1					

Name and description	UN No.	Class	Remarks	Name and description	UN No.	Class	Remarks
TOXIC BY INHALATION LIQUID, WATER-REACTIVE, N.O.S. with an LC_{50} lower than or equal to 1000 ml/m^3 and saturated vapour concentration greater than or equal to 10 LC_{50}	3386	6.1		TRACERS FOR AMMUNITION	0212 0306	1 1	
				Tremolite, see	2212	9	
				TRIALLYLAMINE	2610	3	
				TRIALLYL BORATE	2609	6.1	
TOXIC BY INHALATION LIQUID, WATER-REACTIVE, FLAMMABLE, N.O.S. with an LC_{50} lower than or equal to 200 ml/m^3 and saturated vapour concentration greater than or equal to 500 LC_{50}	3490	6.1		TRIAZINE PESTICIDE, LIQUID, FLAMMABLE, TOXIC, flash-point less than 23 °C	2764	3	
				TRIAZINE PESTICIDE, LIQUID, TOXIC	2998	6.1	
				TRIAZINE PESTICIDE, LIQUID, TOXIC, FLAMMABLE, flash-point not less than 23 °C	2997	6.1	
TOXIC BY INHALATION LIQUID, WATER-REACTIVE, FLAMMABLE, N.O.S. with an LC_{50} lower than or equal to 1000 ml/m^3 and saturated vapour concentration greater than or equal to 10 LC_{50}	3491	6.1		TRIAZINE PESTICIDE, SOLID, TOXIC	2763	6.1	
				Tribromoborane, see	2692	8	
				TRIBUTYLAMINE	2542	6.1	
				TRIBUTYLPHOSPHANE	3254	4.2	
TOXIC LIQUID, CORROSIVE, INORGANIC, N.O.S.	3289	6.1		Trichloroacetaldehyde, see	2075	6.1	
TOXIC LIQUID, CORROSIVE, ORGANIC, N.O.S.	2927	6.1		TRICHLOROACETIC ACID	1839	8	
TOXIC LIQUID, FLAMMABLE, ORGANIC, N.O.S.	2929	6.1		TRICHLOROACETIC ACID SOLUTION	2564	8	
TOXIC LIQUID, INORGANIC, N.O.S.	3287	6.1		Trichlororaceticaldehyde, see	2075	6.1	
TOXIC LIQUID, ORGANIC, N.O.S.	2810	6.1		TRICHLOROACETYL CHLORIDE	2442	8	
TOXIC LIQUID, OXIDIZING, N.O.S.	3122	6.1		TRICHLOROBENZENES, LIQUID	2321	6.1	
				TRICHLOROBUTENE	2322	6.1	
TOXIC LIQUID, WATER-REACTIVE, N.O.S.	3123	6.1		1,1,1-TRICHLOROETHANE	2831	6.1	
				TRICHLOROETHYLENE	1710	6.1	
TOXIC SOLID, CORROSIVE, INORGANIC, N.O.S.	3290	6.1		TRICHLOROISOCYANURIC ACID, DRY	2468	5.1	
TOXIC SOLID, CORROSIVE, ORGANIC, N.O.S.	2928	6.1		Trichloronitromethane, see	1580	6.1	
TOXIC SOLID, FLAMMABLE, ORGANIC, N.O.S.	2930	6.1		TRICHLOROSILANE	1295	4.3	
TOXIC SOLID, INORGANIC, N.O.S.	3288	6.1		1,3,5-Trichloro-s-triazine-2,4,6-trione, see	2468	5.1	
TOXIC SOLID, ORGANIC, N.O.S.	2811	6.1		2,4,6-Trichloro-1,3,5- triazine, see	2670	8	
TOXIC SOLID, OXIDIZING, N.O.S.	3086	6.1		TRICRESYL PHOSPHATE with more than 3% ortho isomer	2574	6.1	
TOXIC SOLID, SELF-HEATING, N.O.S.	3124	6.1		TRIETHYLAMINE	1296	3	
TOXIC SOLID, WATER-REACTIVE, N.O.S.	3125	6.1		Triethyl borate, see	1176	3	
				TRIETHYLENETETRAMINE	2259	8	
TOXINS, EXTRACTED FROM LIVING SOURCES, LIQUID, N.O.S.	3172	6.1		Triethyl orthoformate, see	2524	3	
				TRIETHYL PHOSPHITE	2323	3	
				TRIFLUOROACETIC ACID	2699	8	
TOXINS, EXTRACTED FROM LIVING SOURCES, SOLID, N.O.S.	3462	6.1		TRIFLUOROACETYL CHLORIDE	3057	2	
				Trifluorobromomethane, see	1009	2	

Name and description	UN No.	Class	Remarks	Name and description	UN No.	Class	Remarks
Trifluorochloroethane, see	1983	2		TRINITROBENZOIC ACID, WETTED with not less than 10% water, by mass	3368	4.1	
TRIFLUOROCHLORO-ETHYLENE, STABILIZED, REFRIGERANT GAS R 1113	1082	2		TRINITROBENZOIC ACID, WETTED with not less than 30% water, by mass	1355	4.1	
Trifluorochloromethane, see	1022	2		TRINITROCHLOROBENZENE	0155	1	
1,1,1-TRIFLUOROETHANE	2035	2		TRINITROCHLOROBENZENE WETTED with not less than 10% water, by mass	3365	4.1	
TRIFLUOROMETHANE	1984	2					
TRIFLUOROMETHANE, REFRIGERATED LIQUID	3136	2		TRINITRO-m-CRESOL	0216	1	
2-TRIFLUOROMETHYLANILINE	2942	6.1		TRINITROFLUORENONE	0387	1	
3-TRIFLUOROMETHYLANILINE	2948	6.1		TRINITRONAPHTHALENE	0217	1	
TRIISOBUTYLENE	2324	3		TRINITROPHENETOLE	0218	1	
TRIISOPROPYL BORATE	2616	3		TRINITROPHENOL, dry or wetted with less than 30% water, by mass	0154	1	
TRIMETHYLACETYL CHLORIDE	2438	6.1					
TRIMETHYLAMINE, ANHYDROUS	1083	2		TRINITROPHENOL (PICRIC ACID), WETTED with not less than 30% water, by mass	1344	4.1	
TRIMETHYLAMINE, AQUEOUS SOLUTION, not more than 50% trimethylamine, by mass	1297	3		TRINITROPHENOL WETTED with not less than 10% water, by mass	3364	4.1	
1,3,5-TRIMETHYLBENZENE	2325	3		TRINITROPHENYL-METHYLNITRAMINE	0208	1	
TRIMETHYL BORATE	2416	3		TRINITRORESORCINOL, dry or wetted with less than 20% water, or mixture of alcohol and water, by mass	0219	1	
TRIMETHYLCHLOROSILANE	1298	3					
TRIMETHYLCYCLOHEXYL-AMINE	2326	8					
Trimethylene chlorobromide, see	2688	6.1		TRINITRORESORCINOL, WETTED with not less than 20% water, or mixture of alcohol and water, by mass	0394	1	
TRIMETHYLHEXA-METHYLENEDIAMINES	2327	8					
TRIMETHYLHEXAMETHYLENE DIISOCYANATE	2328	6.1		TRINITROTOLUENE (TNT), dry or wetted with less than 30% water, by mass	0209	1	
2,4,4-Trimethylpentene-1, see	2050	3					
2,4,4-Trimethylpentene-2, see	2050	3		TRINITROTOLUENE AND HEXANITROSTILBENE MIXTURE	0388	1	
TRIMETHYL PHOSPHITE	2329	3					
TRINITROANILINE	0153	1		TRINITROTOLUENE MIXTURE CONTAINING TRINITROBENZENE AND HEXANITROSTILBENE	0389	1	
TRINITROANISOLE	0213	1					
TRINITROBENZENE, dry or wetted with less than 30% water, by mass	0214	1					
TRINITROBENZENE, WETTED with not less than 10% water, by mass	3367	4.1		TRINITROTOLUENE AND TRINITROBENZENE MIXTURE	0388	1	
TRINITROBENZENE, WETTED with not less than 30% water, by mass	1354	4.1		TRINITROTOLUENE, WETTED with not less than 10% water, by mass	3366	4.1	
TRINITROBENZENE-SULPHONIC ACID	0386	1		TRINITROTOLUENE, WETTED with not less than 30% water, by mass	1356	4.1	
TRINITROBENZOIC ACID, dry or wetted with less than 30% water, by mass	0215	1		TRIPROPYLAMINE	2260	3	
				TRIPROPYLENE	2057	3	

Name and description	UN No.	Class	Remarks	Name and description	UN No.	Class	Remarks
TRIS-(1-AZIRIDINYL) PHOSPHINE OXIDE SOLUTION	2501	6.1		VINYL ACETATE, STABILIZED	1301	3	
				Vinylbenzene, see	2055	3	
TRITONAL	0390	1		VINYL BROMIDE, STABILIZED	1085	2	
Tropilidene, see	2603	3		VINYL BUTYRATE, STABILIZED	2838	3	
TUNGSTEN HEXAFLUORIDE	2196	2		VINYL CHLORIDE, STABILIZED	1086	2	
TURPENTINE	1299	3		VINYL CHLOROACETATE	2589	6.1	
TURPENTINE SUBSTITUTE	1300	3		VINYL ETHYL ETHER, STABILIZED	1302	3	
UNDECANE	2330	3					
URANIUM HEXAFLUORIDE, RADIOACTIVE MATERIAL, EXCEPTED PACKAGE, less than 0.1 kg per package, non-fissile or fissile-excepted	3507	6.1		VINYL FLUORIDE, STABILIZED	1860	2	
				VINYLIDENE CHLORIDE, STABILIZED	1303	3	
				VINYL ISOBUTYL ETHER, STABILIZED	1304	3	
UREA HYDROGEN PEROXIDE	1511	5.1		VINYL METHYL ETHER, STABILIZED	1087	2	
UREA NITRATE, dry or wetted with less than 20% water, by mass	0220	1		VINYLPYRIDINES, STABILIZED	3073	6.1	
UREA NITRATE, WETTED with not less than 10% water, by mass	3370	4.1		VINYLTOLUENES, STABILIZED	2618	3	
UREA NITRATE, WETTED with not less than 20% water, by mass	1357	4.1		VINYLTRICHLOROSILANE	1305	3	
Valeral, see	2058	3		Warheads for guided missiles, see	0286 0287 0369 0370 0371	1 1 1 1 1	
VALERALDEHYDE	2058	3					
n-Valeraldehyde, see	2058	3					
Valeric aldehyde, see	2058	3		WARHEADS, ROCKET with burster or expelling charge	0370 0371	1 1	
VALERYL CHLORIDE	2502	8					
VANADIUM COMPOUND, N.O.S.	3285	6.1		WARHEADS, ROCKET with bursting charge	0286 0287 0369	1 1 1	
Vanadium (IV) oxide sulphate, see	2931	6.1					
Vanadium oxysulphate, see	2931	6.1					
VANADIUM OXYTRICHLORIDE	2443	8		WARHEADS, TORPEDO with bursting charge	0221	1	
VANADIUM PENTOXIDE, non-fused form	2862	6.1		WATER-REACTIVE LIQUID, N.O.S.	3148	4.3	
VANADIUM TETRACHLORIDE	2444	8		WATER-REACTIVE LIQUID, CORROSIVE, N.O.S.	3129	4.3	
VANADIUM TRICHLORIDE	2475	8		WATER-REACTIVE LIQUID, TOXIC, N.O.S.	3130	4.3	
VANADYL SULPHATE	2931	6.1		WATER-REACTIVE SOLID, N.O.S.	2813	4.3	
Varnish, see	1263 3066 3469 3470	3 8 3 8		WATER-REACTIVE SOLID, CORROSIVE, N.O.S.	3131	4.3	
				WATER-REACTIVE SOLID, FLAMMABLE, N.O.S.	3132	4.3	
VEHICLE, FLAMMABLE GAS POWERED	3166	9		WATER-REACTIVE SOLID, OXIDIZING, N.O.S.	3133	4.3	Carriage prohibited
VEHICLE, FLAMMABLE LIQUID POWERED	3166	9		WATER-REACTIVE SOLID, SELF-HEATING, N.O.S.	3135	4.3	
VEHICLE, FUEL CELL, FLAMMABLE GAS POWERED	3166	9		WATER-REACTIVE SOLID, TOXIC, N.O.S.	3134	4.3	
VEHICLE, FUEL CELL, FLAMMABLE LIQUID POWERED	3166	9					
Villiaumite, see	1690	6.1		White arsenic, see	1561	6.1	

Name and description	UN No.	Class	Remarks	Name and description	UN No.	Class	Remarks
White spirit, see	1300	3		Zinc hexafluorosilicate, see	2855	6.1	
WOOD PRESERVATIVES, LIQUID	1306	3		ZINC HYDROSULPHITE, see	1931	9	
				ZINC NITRATE	1514	5.1	
Wool waste, wet	1387	4.2	Not subject to ADR	ZINC PERMANGANATE	1515	5.1	
				ZINC PEROXIDE	1516	5.1	
XANTHATES	3342	4.2		ZINC PHOSPHIDE	1714	4.3	
XENON	2036	2		ZINC POWDER	1436	4.3	
XENON, REFRIGERATED LIQUID	2591	2		ZINC RESINATE	2714	4.1	
				Zinc selenate, see	2630	4.1	
XYLENES	1307	3		Zinc selenite, see	2630	4.1	
XYLENOLS, LIQUID	3430	6.1		Zinc silicofluoride, see	2855	6.1	
XYLENOLS, SOLID	2261	6.1		ZIRCONIUM, DRY, coiled wire, finished metal sheets, strip (thinner than 254 microns but not thinner than 18 microns)	2858	4.1	
XYLIDINES, LIQUID	1711	6.1					
XYLIDINES, SOLID	3452	6.1					
Xylols, see	1307	3		ZIRCONIUM, DRY, finished sheets, strip or coiled wire	2009	4.2	
XYLYL BROMIDE, LIQUID	1701	6.1					
XYLYL BROMIDE, SOLID	3417	6.1		ZIRCONIUM HYDRIDE	1437	4.1	
ZINC AMMONIUM NITRITE	1512	5.1		ZIRCONIUM NITRATE	2728	5.1	
ZINC ARSENATE	1712	6.1		ZIRCONIUM PICRAMATE, dry or wetted with less than 20% water, by mass	0236	1	
ZINC ARSENATE AND ZINC ARSENITE MIXTURE	1712	6.1					
ZINC ARSENITE	1712	6.1		ZIRCONIUM PICRAMATE, WETTED with not less than 20% water, by mass	1517	4.1	
ZINC ASHES	1435	4.3					
Zinc bisulphite solution, see	2693	8		ZIRCONIUM POWDER, DRY	2008	4.2	
ZINC BROMATE	2469	5.1		ZIRCONIUM POWDER, WETTED with not less than 25% water	1358	4.1	
ZINC CHLORATE	1513	5.1					
ZINC CHLORIDE, ANHYDROUS	2331	8		ZIRCONIUM SCRAP	1932	4.2	
ZINC CHLORIDE SOLUTION	1840	8		ZIRCONIUM SUSPENDED IN A FLAMMABLE LIQUID	1308	3	
ZINC CYANIDE	1713	6.1					
ZINC DITHIONITE	1931	9		ZIRCONIUM TETRACHLORIDE	2503	8	
ZINC DUST	1436	4.3					
ZINC FLUOROSILICATE	2855	6.1					

CHAPTER 3.3

SPECIAL PROVISIONS APPLICABLE TO CERTAIN ARTICLES OR SUBSTANCES

3.3.1   When Column (6) of Table A of Chapter 3.2 indicates that a special provision is relevant to a substance or article, the meaning and requirements of that special provision are as set forth below. Where a special provision includes a requirement for package marking, the provisions of 5.2.1.2 (a) and (b) shall be met. If the required mark is in the form of specific wording indicated in quotation marks, such as "Damaged Lithium Batteries", the size of the mark shall be at least 12 mm, unless otherwise indicated in the special provision or elsewhere in ADR.

16   Samples of new or existing explosive substances or articles may be carried as directed by the competent authorities (see 2.2.1.1.3) for purposes including: testing, classification, research and development, quality control, or as a commercial sample. Explosive samples which are not wetted or desensitized shall be limited to 10 kg in small packages as specified by the competent authorities. Explosive samples which are wetted or desensitized shall be limited to 25 kg.

23   Even though this substance has a flammability hazard, it only exhibits such hazard under extreme fire conditions in confined areas.

32   This substance is not subject to the requirements of ADR when in any other form.

37   This substance is not subject to the requirements of ADR when coated.

38   This substance is not subject to the requirements of ADR when it contains not more than 0.1% calcium carbide.

39   This substance is not subject to the requirements of ADR when it contains less than 30% or not less than 90% silicon.

43   When offered for carriage as pesticides, these substances shall be carried under the relevant pesticide entry and in accordance with the relevant pesticide provisions (see 2.2.61.1.10 to 2.2.61.1.11.2).

45   Antimony sulphides and oxides which contain not more than 0.5% of arsenic calculated on the total mass are not subject to the requirements of ADR.

47   Ferricyanides and ferrocyanides are not subject to the requirements of ADR.

48   The carriage of this substance, when it contains more than 20% hydrocyanic acid, is prohibited.

59   These substances are not subject to the requirements of ADR when they contain not more than 50% magnesium.

60   If the concentration is more than 72%, the carriage of this substance is prohibited.

61   The technical name which shall supplement the proper shipping name shall be the ISO common name (see also ISO 1750:1981 "*Pesticides and other agrochemicals - common names*", as amended), other name listed in the WHO "*Recommended Classification of Pesticides by Hazard and Guidelines to Classification*" or the name of the active substance (see also 3.1.2.8.1 and 3.1.2.8.1.1).

62   This substance is not subject to the requirements of ADR when it contains not more than 4% sodium hydroxide.

65   Hydrogen peroxide aqueous solutions with less than 8% hydrogen peroxide are not subject to the requirements of ADR.

66   Cinnabar is not subject to the requirements of ADR.

103   The carriage of ammonium nitrites and mixtures of an inorganic nitrite with an ammonium salt is prohibited.

105   Nitrocellulose meeting the descriptions of UN No. 2556 or UN No. 2557 may be classified in Class 4.1.

113   The carriage of chemically unstable mixtures is prohibited.

119   Refrigerating machines include machines or other appliances which have been designed for the specific purpose of keeping food or other items at a low temperature in an internal compartment, and air conditioning units. Refrigerating machines and refrigerating machine components are not subject to the provisions of ADR if they contain less than 12 kg of gas in Class 2, group A or O according to 2.2.2.1.3, or if they contain less than 12 litres ammonia solution (UN No. 2672).

122   The subsidiary risks, control and emergency temperatures if any, and the UN number (generic entry) for each of the currently assigned organic peroxide formulations are given in 2.2.52.4, 4.1.4.2 packing instruction IBC520 and 4.2.5.2.6 portable tank instruction T23.

123   *(Reserved)*

127   Other inert material or inert material mixture may be used, provided this inert material has identical phlegmatizing properties.

131   The phlegmatized substance shall be significantly less sensitive than dry PETN.

135   The dihydrated sodium salt of dichloroisocyanuric acid does not meet the criteria for inclusion in Class 5.1 and is not subject to ADR unless meeting the criteria for inclusion in another Class.

138   p-Bromobenzyl cyanide is not subject to the requirements of ADR.

141   Products which have undergone sufficient heat treatment so that they present no hazard during carriage are not subject to the requirements of ADR.

142   Solvent extracted soya bean meal containing not more than 1.5% oil and 11% moisture, which is substantially free of flammable solvent, is not subject to the requirements of ADR.

144   An aqueous solution containing not more than 24% alcohol by volume is not subject to the requirements of ADR.

145   Alcoholic beverages of packing group III, when carried in receptacles of 250 litres or less, are not subject to the requirements of ADR.

152   The classification of this substance will vary with particle size and packaging, but borderlines have not been experimentally determined. Appropriate classifications shall be made in accordance with 2.2.1.

153   This entry applies only if it is demonstrated, on the basis of tests, that the substances when in contact with water are not combustible nor show a tendency to auto-ignition and that the mixture of gases evolved is not flammable.

162   *(Deleted)*

163   A substance mentioned by name in Table A of Chapter 3.2 shall not be carried under this entry. Substances carried under this entry may contain 20% or less nitrocellulose provided the nitrocellulose contains not more than 12.6% nitrogen (by dry mass).

168   Asbestos which is immersed or fixed in a natural or artificial binder (such as cement, plastics, asphalt, resins or mineral ore) in such a way that no escape of hazardous quantities of respirable asbestos fibres can occur during carriage is not subject to the requirements of ADR. Manufactured articles containing asbestos and not meeting this provision are nevertheless not subject to the requirements of ADR when packed so that no escape of hazardous quantities of respirable asbestos fibres can occur during carriage.

169   Phthalic anhydride in the solid state and tetrahydrophthalic anhydrides, with not more than 0.05% maleic anhydride, are not subject to the requirements of ADR. Phthalic anhydride molten at a temperature above its flash-point, with not more than 0.05% maleic anhydride, shall be classified under UN No. 3256.

172   Where a radioactive material has (a) subsidiary risk(s):

(a)   The substance shall be allocated to packing group I, II or III, if appropriate, by application of the packing group criteria provided in Part 2 corresponding to the nature of the predominant subsidiary risk;

(b)   Packages shall be labelled with subsidiary risk labels corresponding to each subsidiary risk exhibited by the material; corresponding placards shall be affixed to cargo transport units in accordance with the relevant provisions of 5.3.1;

(c)   For the purposes of documentation and package marking, the proper shipping name shall be supplemented with the name of the constituents which most predominantly contribute to this (these) subsidiary risk(s) and which shall be enclosed in parenthesis;

(d)   The dangerous goods transport document shall indicate the label model number(s) corresponding to each subsidiary risk in parenthesis after the Class number "7" and, where assigned the packing group as required by 5.4.1.1.1 (d).

For packing, see also 4.1.9.1.5.

177   Barium sulphate is not subject to the requirements of ADR.

178   This designation shall be used only when no other appropriate designation exists in Table A of Chapter 3.2, and only with the approval of the competent authority of the country of origin (see 2.2.1.1.3).

181   Packages containing this type of substance shall bear a label conforming to model No. 1 (see 5.2.2.2.2) unless the competent authority of the country of origin has permitted this label to be dispensed with for the specific packaging employed because test data have proved that the substance in this packaging does not exhibit explosive behaviour (see 5.2.2.1.9).

182   The group of alkali metals includes lithium, sodium, potassium, rubidium and caesium.

183   The group of alkaline earth metals includes magnesium, calcium, strontium and barium.

186   In determining the ammonium nitrate content, all nitrate ions for which a molecular equivalent of ammonium ions is present in the mixture shall be calculated as ammonium nitrate.

188   Cells and batteries offered for carriage are not subject to other provisions of ADR if they meet the following:

(a)   For a lithium metal or lithium alloy cell, the lithium content is not more than 1 g, and for a lithium ion cell, the Watt-hour rating is not more than 20 Wh;

(b)   For a lithium metal or lithium alloy battery the aggregate lithium content is not more than 2 g, and for a lithium ion battery, the Watt-hour rating is not more than 100 Wh. Lithium ion batteries subject to this provision shall be marked with the Watt-hour rating on the outside case, except those manufactured before 1 January 2009;

(c)   Each cell or battery meets the provisions of 2.2.9.1.7 (a) and (e);

(d)   Cells and batteries, except when installed in equipment, shall be packed in inner packagings that completely enclose the cell or battery. Cells and batteries shall be protected so as to prevent short circuits. This includes protection against contact with conductive materials within the same packaging that could lead to a short circuit. The inner packagings shall be packed in strong outer packagings which conform to the provisions of 4.1.1.1, 4.1.1.2 and 4.1.1.5;

(e)   Cells and batteries when installed in equipment shall be protected from damage and short circuit, and the equipment shall be equipped with an effective means of preventing accidental activation. This requirement does not apply to devices which are intentionally active in carriage (radio frequency identification (RFID) transmitters, watches, sensors, etc.) and which are not capable of generating a dangerous evolution of heat. When batteries are installed in equipment, the equipment shall be packed in strong outer packagings constructed of suitable material of adequate strength and design in relation to the packaging's capacity and its intended use unless the battery is afforded equivalent protection by the equipment in which it is contained;

(f)   Each package shall be marked with the appropriate lithium battery mark, as illustrated in 5.2.1.9;

This requirement does not apply to:

(i)   Packages containing only button cell batteries installed in equipment (including circuit boards); and

(ii)   Packages containing no more than four cells or two batteries installed in equipment, where there are not more than two packages in the consignment;

(g)   Except when batteries are installed in equipment, each package shall be capable of withstanding a 1.2 m drop test in any orientation without damage to cells or batteries contained therein, without shifting of the contents so as to allow battery to battery (or cell to cell) contact and without release of contents; and

(h)   Except when batteries are installed in or packed with equipment, packages shall not exceed 30 kg gross mass.

As used above and elsewhere in ADR, "lithium content" means the mass of lithium in the anode of a lithium metal or lithium alloy cell.

Separate entries exist for lithium metal batteries and lithium ion batteries to facilitate the carriage of these batteries for specific modes of carriage and to enable the application of different emergency response actions.

A single cell battery as defined in Part III, sub-section 38.3.2.3 of the *Manual of Tests and Criteria* is considered a "cell" and shall be carried according to the requirements for "cells" for the purpose of this special provision.

190   Aerosol dispensers shall be provided with protection against inadvertent discharge. Aerosols with a capacity not exceeding 50 ml containing only non-toxic constituents are not subject to the requirements of ADR.

191   Receptacles, small, with a capacity not exceeding 50 ml, containing only non-toxic constituents are not subject to the requirements of ADR.

194   The control and emergency temperatures, if any, and the UN number (generic entry) for each of the currently assigned self-reactive substances are given in 2.2.41.4.

196   Formulations which in laboratory testing neither detonate in the cavitated state nor deflagrate, which show no effect when heated under confinement and which exhibit no explosive power may be carried under this entry. The formulation must also be thermally stable (i.e. the SADT is 60 °C or higher for a 50 kg package). Formulations not meeting these criteria shall be carried under the provisions of Class 5.2, (see 2.2.52.4).

198   Nitrocellulose solutions containing not more than 20% nitrocellulose may be carried as paint, perfumery products or printing ink, as applicable (see UN Nos. 1210, 1263, 1266, 3066, 3469 and 3470).

199   Lead compounds which, when mixed in a ratio of 1:1000 with 0.07M hydrochloric acid and stirred for one hour at a temperature of 23 °C ± 2 °C, exhibit a solubility of 5% or less (see ISO 3711:1990 *"Lead chromate pigments and lead chromate -molybdate pigments – Specifications and methods of test"*) are considered insoluble and are not subject to the requirements of ADR unless they meet the criteria for inclusion in another class.

201 Lighters and lighter refills shall comply with the provisions of the country in which they were filled. They shall be provided with protection against inadvertent discharge. The liquid portion of the gas shall not exceed 85% of the capacity of the receptacle at 15 °C. The receptacles, including the closures, shall be capable of withstanding an internal pressure of twice the pressure of the liquefied petroleum gas at 55 °C. The valve mechanisms and ignition devices shall be securely sealed, taped or otherwise fastened or designed to prevent operation or leakage of the contents during carriage. Lighters shall not contain more than 10 g of liquefied petroleum gas. Lighter refills shall not contain more than 65 g of liquefied petroleum gas.

NOTE: For waste lighters collected separately see Chapter 3.3, special provision 654.

203 This entry shall not be used for polychlorinated biphenyls, liquid, UN No. 2315 and polychlorinated biphenyls, solid, UN No.3432.

204 *(Deleted)*

205 This entry shall not be used for UN No. 3155 PENTACHLOROPHENOL.

207 Plastics moulding compounds may be made from polystyrene, poly(methyl methacrylate) or other polymeric material.

208 The commercial grade of calcium nitrate fertilizer, when consisting mainly of a double salt (calcium nitrate and ammonium nitrate) containing not more than 10% ammonium nitrate and at least 12% water of crystallization, is not subject to the requirements of ADR.

210 Toxins from plant, animal or bacterial sources which contain infectious substances, or toxins that are contained in infectious substances, shall be classified in Class 6.2.

215 This entry only applies to the technically pure substance or to formulations derived from it having an SADT higher than 75 °C and therefore does not apply to formulations which are self-reactive substances (for self-reactive substances, see 2.2.41.4). Homogeneous mixtures containing not more than 35% by mass of azodicarbonamide and at least 65% of inert substance are not subject to the requirements of ADR unless criteria of other classes are met.

216 Mixtures of solids which are not subject to the requirements of ADR and flammable liquids may be carried under this entry without first applying the classification criteria of Class 4.1, provided there is no free liquid visible at the time the substance is loaded or at the time the packaging or cargo transport unit is closed. Sealed packets and articles containing less than 10 ml of a packing group II or III flammable liquid absorbed into a solid material are not subject to ADR provided there is no free liquid in the packet or article.

217 Mixtures of solids which are not subject to the requirements of ADR and toxic liquids may be carried under this entry without first applying the classification criteria of Class 6.1, provided there is no free liquid visible at the time the substance is loaded or at the time the packaging or cargo transport unit is closed. This entry shall not be used for solids containing a packing group I liquid.

218 Mixtures of solids which are not subject to the requirements of ADR and corrosive liquids may be carried under this entry without first applying the classification criteria of Class 8, provided there is no free liquid visible at the time the substance is loaded or at the time the packaging or cargo transport unit is closed.

219 Genetically modified microorganisms (GMMOs) and genetically modified organisms (GMOs) packed and marked in accordance with packing instruction P904 of 4.1.4.1 are not subject to any other requirements of ADR.

If GMMOs or GMOs meet the criteria for inclusion in Class 6.1 or 6.2 (see 2.2.61.1 and 2.2.62.1) the requirements in ADR for the carriage of toxic substances or infectious substances apply.

220 Only the technical name of the flammable liquid component of this solution or mixture shall be shown in parentheses immediately following the proper shipping name.

221 Substances included under this entry shall not be of packing group I.

224   Unless it can be demonstrated by testing that the sensitivity of the substance in its frozen state is no greater than in its liquid state, the substance shall remain liquid during normal transport conditions. It shall not freeze at temperatures above -15 °C.

225   Fire extinguishers under this entry may include installed actuating cartridges (cartridges, power device of classification code 1.4C or 1.4S), without changing the classification of Class 2, group A or O according to 2.2.2.1.3 provided the total quantity of deflagrating (propellant) explosives does not exceed 3.2 g per extinguishing unit. Fire extinguishers shall be manufactured, tested, approved and labelled according to the provisions applied in the country of manufacture.

        NOTE: *"Provisions applied in the country of manufacture" means the provisions applicable in the country of manufacture or those applicable in the country of use.*

        Fire extinguishers under this entry include:

        (a)   portable fire extinguishers for manual handling and operation;

        (b)   fire extinguishers for installation in aircraft;

        (c)   fire extinguishers mounted on wheels for manual handling;

        (d)   fire extinguishing equipment or machinery mounted on wheels or wheeled platforms or units carried similar to (small) trailers, and

        (e)   fire extinguishers composed of a non-rollable pressure drum and equipment, and handled e.g. by fork lift or crane when loaded or unloaded.

        NOTE: *Pressure receptacles which contain gases for use in the above-mentioned fire extinguishers or for use in stationary fire-fighting installations shall meet the requirements of Chapter 6.2 and all requirements applicable to the relevant dangerous goods when these pressure receptacles are carried separately.*

226   Formulations of this substance containing not less than 30% non-volatile, non-flammable phlegmatizer are not subject to the requirements of ADR.

227   When phlegmatized with water and inorganic inert material the content of urea nitrate may not exceed 75% by mass and the mixture shall not be capable of being detonated by the Series 1, type (a), test in the *Manual of Tests and Criteria*, Part 1.

228   Mixtures not meeting the criteria for flammable gases (see 2.2.2.1.5) shall be carried under UN No. 3163.

230   Lithium cells and batteries may be carried under this entry if they meet the provisions of 2.2.9.1.7.

235   This entry applies to articles which contain Class 1 explosive substances and which may also contain dangerous goods of other classes. These articles are used to enhance safety in vehicles, vessels or aircraft – e.g. air bag inflators, air bag modules, seat-belt pretensioners, and pyromechanical devices.

236   Polyester resin kits consist of two components: a base material (either Class 3 or Class 4.1, packing group II or III) and an activator (organic peroxide). The organic peroxide shall be type D, E, or F, not requiring temperature control. The packing group shall be II or III, according to the criteria of either Class 3 or Class 4.1, as appropriate, applied to the base material. The quantity limit shown in column (7a) of Table A of Chapter 3.2 applies to the base material.

237   The membrane filters, including paper separators, coating or backing materials, etc., that are present in carriage, shall not be liable to propagate a detonation as tested by one of the tests described in the *Manual of Tests and Criteria*, Part I, Test series 1 (a).

        In addition the competent authority may determine, on the basis of the results of suitable burning rate tests taking account of the standard tests in the *Manual of Tests and Criteria*, Part III, sub-section 33.2.1, that nitrocellulose membrane filters in the form in which they are to be carried are not subject to the requirements applicable to flammable solids in Class 4.1.

238 (a) Batteries can be considered as non-spillable provided that they are capable of withstanding the vibration and pressure differential tests given below, without leakage of battery fluid.

Vibration test: The battery is rigidly clamped to the platform of a vibration machine and a simple harmonic motion having an amplitude of 0.8 mm (1.6 mm maximum total excursion) is applied. The frequency is varied at the rate of 1 Hz/min between the limits of 10 Hz and 55 Hz. The entire range of frequencies and return is traversed in 95 ± 5 minutes for each mounting position (direction of vibration) of the battery. The battery is tested in three mutually perpendicular positions (to include testing with fill openings and vents, if any, in an inverted position) for equal time periods.

Pressure differential test: Following the vibration test, the battery is stored for six hours at 24 °C ± 4 °C while subjected to a pressure differential of at least 88 kPa. The battery is tested in three mutually perpendicular positions (to include testing with fill openings and vents, if any, in an inverted position) for at least six hours in each position.

(b) Non-spillable batteries are not subject to the requirements of ADR if, at a temperature of 55 °C, the electrolyte will not flow from a ruptured or cracked case and there is no free liquid to flow and if, as packaged for carriage, the terminals are protected from short circuit.

239 Batteries or cells shall not contain dangerous substances other than sodium, sulphur or sodium compounds (e.g. sodium polysulphides and sodium tetrachloroaluminate). Batteries or cells shall not be offered for carriage at a temperature such that liquid elemental sodium is present in the battery or cell unless approved and under the conditions established by the competent authority of the country of origin. If the country of origin is not a Contracting Party to ADR, the approval and conditions of carriage shall be recognized by the competent authority of the first country Contracting Party to ADR reached by the consignment.

Cells shall consist of hermetically sealed metal casings which fully enclose the dangerous substances and which are so constructed and closed as to prevent the release of the dangerous substances under normal conditions of carriage.

Batteries shall consist of cells secured within and fully enclosed by a metal casing so constructed and closed as to prevent the release of the dangerous substances under normal conditions of carriage.

240 This entry only applies to vehicles powered by wet batteries, sodium batteries, lithium metal batteries or lithium ion batteries and equipment powered by wet batteries or sodium batteries carried with these batteries installed. Lithium batteries shall meet the requirements of 2.2.9.1.7, except as otherwise provided for in special provision 667.

For the purpose of this special provision, vehicles are self-propelled apparatus designed to carry one or more persons or goods. Examples of such vehicles are electrically-powered cars, motorcycles, scooters, three- and four-wheeled vehicles or motorcycles, trucks, locomotives, bicycles (pedal cycles with an electric motor) and other vehicles of this type (e.g. self-balancing vehicles or vehicles not equipped with at least one seating position), wheel chairs, lawn tractors, self-propelled farming and construction equipment, boats and aircraft. This includes vehicles carried in a packaging. In this case some parts of the vehicle may be detached from its frame to fit into the packaging.

Examples of equipment are lawnmowers, cleaning machines or model boats and model aircraft. Equipment powered by lithium metal batteries or lithium ion batteries shall be assigned to the entries UN 3091 LITHIUM METAL BATTERIES CONTAINED IN EQUIPMENT or UN 3091 LITHIUM METAL BATTERIES PACKED WITH EQUIPMENT or UN 3481 LITHIUM ION BATTERIES CONTAINED IN EQUIPMENT or UN 3481 LITHIUM ION BATTERIES PACKED WITH EQUIPMENT, as appropriate. Hybrid electric vehicles powered by both an internal combustion engine and wet batteries, sodium batteries, lithium metal batteries or lithium ion batteries, carried with the battery(ies) installed shall be assigned to the entries UN 3166 VEHICLE, FLAMMABLE GAS POWERED or UN 3166 VEHICLE, FLAMMABLE LIQUID POWERED, as appropriate. Vehicles which contain a fuel cell shall be assigned to the entries UN 3166 VEHICLE, FUEL CELL, FLAMMABLE

GAS POWERED or UN 3166 VEHICLE, FUEL CELL, FLAMMABLE LIQUID POWERED, as appropriate.

Vehicles may contain other dangerous goods than batteries (e.g. fire extinguishers, compressed gas accumulators or safety devices) required for their functioning or safe operation without being subject to any additional requirements for these other dangerous goods, unless otherwise specified in ADR.

241   The formulation shall be prepared so that it remains homogeneous and does not separate during carriage. Formulations with low nitrocellulose contents and not showing dangerous properties when tested for their liability to detonate, deflagrate or explode when heated under defined confinement by tests of Test series 1 (a), 2 (b) and 2 (c) respectively in the *Manual of Tests and Criteria*, Part I and not being a flammable solid when tested in accordance with test No. 1 in the *Manual of Tests and Criteria*, Part III, sub-section 33.2.1.4 (chips, if necessary, crushed and sieved to a particle size of less than 1.25 mm) are not subject to the requirements of ADR.

242   Sulphur is not subject to the requirements of ADR when it has been formed to a specific shape (e.g. prills, granules, pellets, pastilles or flakes).

243   Gasoline, motor spirit and petrol for use in spark-ignition engines (e.g. in automobiles, stationary engines and other engines) shall be assigned to this entry regardless of variations in volatility.

244   This entry includes e.g. aluminium dross, aluminium skimmings, spent cathodes, spent potliner, and aluminium salt slags.

247   Alcoholic beverages containing more than 24% alcohol but not more than 70% by volume, when carried as part of the manufacturing process, may be carried in wooden barrels with a capacity of more than 250 litres and not more than 500 litres meeting the general requirements of 4.1.1, as appropriate, on the following conditions:

(a)   The wooden barrels shall be checked and tightened before filling;

(b)   Sufficient ullage (not less than 3%) shall be left to allow for the expansion of the liquid;

(c)   The wooden barrels shall be carried with the bungholes pointing upwards;

(d)   The wooden barrels shall be carried in containers meeting the requirements of the CSC. Each wooden barrel shall be secured in custom-made cradles and be wedged by appropriate means to prevent it from being displaced in any way during carriage.

249   Ferrocerium, stabilized against corrosion, with a minimum iron content of 10% is not subject to the requirements of ADR.

250   This entry may only be used for samples of chemicals taken for analysis in connection with the implementation of the Convention on the Prohibition of the Development, Production, Stockpiling and Use of Chemical Weapons and on their Destruction. The carriage of substances under this entry shall be in accordance with the chain of custody and security procedures specified by the Organisation for the Prohibition of Chemical Weapons.

The chemical sample may only be carried providing prior approval has been granted by the competent authority or the Director General of the Organisation for the Prohibition of Chemical Weapons and providing the sample complies with the following provisions:

(a)   It shall be packed according to packing instruction 623 in the ICAO Technical Instructions (see S-3-8 of the Supplement); and

(b)   During carriage, a copy of the document of approval for transport, showing the quantity limitations and the packing provisions shall be attached to the transport document.

251 The entry CHEMICAL KIT or FIRST AID KIT is intended to apply to boxes, cases etc. containing small quantities of various dangerous goods which are used for example for medical, analytical or testing or repair purposes. Such kits may not contain dangerous goods for which the quantity "0"has been indicated in Column (7a) of Table A of Chapter 3.2.

Components shall not react dangerously (see "dangerous reaction" in 1.2.1). The total quantity of dangerous goods in any one kit shall not exceed either 1 *l* or 1 kg. The packing group assigned to the kit as a whole shall be the most stringent packing group assigned to any individual substance in the kit.

Where the kit contains only dangerous goods to which no packing group is assigned, no packing group need be indicated on the dangerous goods transport document.

Kits which are carried on board vehicles for first-aid or operating purposes are not subject to the requirements of ADR.

Chemical kits and first aid kits containing dangerous goods in inner packagings which do not exceed the quantity limits for limited quantities applicable to individual substances as specified in Column (7a) of Table A of Chapter 3.2 may be carried in accordance with Chapter 3.4.

252 Provided the ammonium nitrate remains in solution under all conditions of carriage, aqueous solutions of ammonium nitrate, with not more than 0.2% combustible material, in a concentration not exceeding 80%, are not subject to the requirements of ADR.

266 This substance, when containing less alcohol, water or phlegmatizer than specified, shall not be carried unless specifically authorized by the competent authority (see 2.2.1.1).

267 Any explosives, blasting, type C containing chlorates shall be segregated from explosives containing ammonium nitrate or other ammonium salts.

270 Aqueous solutions of Class 5.1 inorganic solid nitrate substances are considered as not meeting the criteria of Class 5.1 if the concentration of the substances in solution at the minimum temperature encountered during carriage is not greater than 80% of the saturation limit.

271 Lactose or glucose or similar materials, may be used as a phlegmatizer provided that the substance contains not less than 90%, by mass, of phlegmatizer. The competent authority may authorize these mixtures to be classified in Class 4.1 on the basis of a test Series 6(c) of Section 16 of Part I of the *Manual of Tests and Criteria* on at least three packages as prepared for carriage. Mixtures containing at least 98%, by mass, of phlegmatizer are not subject to the requirements of ADR. Packages containing mixtures with not less than 90%, by mass, of phlegmatizer need not bear a label conforming to model No. 6.1.

272 This substance shall not be carried under the provisions of Class 4.1 unless specifically authorized by the competent authority (see UN No. 0143 or UN No. 0150 as appropriate).

273 Maneb and maneb preparations stabilized against self-heating need not be classified in Class 4.2 when it can be demonstrated by testing that a cubic volume of 1 m^3 of substance does not self-ignite and that the temperature at the centre of the sample does not exceed 200 °C, when the sample is maintained at a temperature of not less than 75 °C $\pm$ 2 °C for a period of 24 hours.

274 The provisions of 3.1.2.8 apply.

278 These substances shall not be classified and carried unless authorized by the competent authority on the basis of results from Series 2 tests and a Series 6(c) test of Part I of the *Manual of Tests and Criteria* on packages as prepared for carriage (see 2.2.1.1). The competent authority shall assign the packing group on the basis of 2.2.3 criteria and the package type used for the Series 6(c) test.

279 The substance is assigned to this classification or packing group based on human experience rather than the strict application of classification criteria set out in ADR.

280 This entry applies to safety devices for vehicles, vessels or aircraft, e.g. air bag inflators, air bag modules, seat-belt pretensioners, and pyromechanical devices, which contain dangerous goods of Class 1 or of other classes, when carried as component parts and if these articles as

presented for carriage have been tested in accordance with Test Series 6(c) of Part 1 of the Manual of Tests and Criteria, with no explosion of the device, no fragmentation of device casing or pressure receptacle, and no projection hazard nor thermal effect which would significantly hinder fire-fighting or emergency response efforts in the immediate vicinity. This entry does not apply to life saving appliances described in special provision 296 (UN Nos. 2990 and 3072).

282   *(Deleted)*

283   Articles, containing gas, intended to function as shock absorbers, including impact energy-absorbing devices, or pneumatic springs are not subject to the requirements of ADR provided:

(a)   Each article has a gas space capacity not exceeding 1.6 litres and a charge pressure not exceeding 280 bar where the product of the capacity (litres) and charge pressure (bars) does not exceed 80 (i.e. 0.5 litres gas space and 160 bar charge pressure, 1 litre gas space and 80 bar charge pressure, 1.6 litres gas space and 50 bar charge pressure, 0.28 litres gas space and 280 bar charge pressure);

(b)   Each article has a minimum burst pressure of 4 times the charge pressure at 20 °C for products not exceeding 0.5 litres gas space capacity and 5 times charge pressure for products greater than 0.5 litres gas space capacity;

(c)   Each article is manufactured from material which will not fragment upon rupture;

(d)   Each article is manufactured in accordance with a quality assurance standard acceptable to the competent authority; and

(e)   The design type has been subjected to a fire test demonstrating that the article relieves its pressure by means of a fire degradable seal or other pressure relief device, such that the article will not fragment and that the article does not rocket.

See also 1.1.3.2 (d) for equipment used for the operation of the vehicle.

284   An oxygen generator, chemical, containing oxidizing substances shall meet the following conditions:

(a)   The generator when containing an explosive actuating device shall only be carried under this entry when excluded from Class 1 in accordance with the NOTE under paragraph 2.2.1.1.1 (b);

(b)   The generator, without its packaging, shall be capable of withstanding a 1.8 m drop test onto a rigid, non-resilient, flat and horizontal surface, in the position most likely to cause damage, without loss of its contents and without actuation;

(c)   When a generator is equipped with an actuating device, it shall have at least two positive means of preventing unintentional actuation.

286   Nitrocellulose membrane filters covered by this entry, each with a mass not exceeding 0.5 g, are not subject to the requirements of ADR when contained individually in an article or a sealed packet.

288   These substances shall not be classified and carried unless authorized by the competent authority on the basis of results from Series 2 tests and a Series 6(c) test of Part I of the *Manual of tests and Criteria* on packages as prepared for carriage (see 2.2.1.1).

289   Safety devices, electrically initiated and safety devices, pyrotechnic installed in vehicles, wagons, vessels or aircraft or in completed components such as steering columns, door panels, seats, etc. are not subject to ADR.

290   When this radioactive material meets the definitions and criteria of other classes as defined in Part 2, it shall be classified in accordance with the following:

(a)   Where the substance meets the criteria for dangerous goods in excepted quantities as set out in Chapter 3.5, the packagings shall be in accordance with 3.5.2 and meet the testing requirements of 3.5.3. All other requirements applicable to radioactive material, excepted packages as set out in 1.7.1.5 shall apply without reference to the other class;

(b)   Where the quantity exceeds the limits specified in 3.5.1.2 the substance shall be classified in accordance with the predominant subsidiary risk. The transport document shall describe the substance with the UN number and proper shipping name applicable to the other class supplemented with the name applicable to the radioactive excepted package according to Column (2) of Table A of Chapter 3.2, and the substance shall be carried in accordance with the provisions applicable to that UN number. An example of the information shown on the transport document is:

"UN 1993, Flammable liquid, n.o.s. (ethanol and toluene mixture), Radioactive material, excepted package – limited quantity of material, 3, PG II".

In addition, the requirements of 2.2.7.2.4.1 shall apply;

(c)   The provisions of Chapter 3.4 for the carriage of dangerous goods packed in limited quantities shall not apply to substances classified in accordance with sub-paragraph (b);

(d)   When the substance meets a special provision that exempts this substance from all dangerous goods provisions of the other classes it shall be classified in accordance with the applicable UN number of Class 7 and all requirements specified in 1.7.1.5 shall apply.

291   Flammable liquefied gases shall be contained within refrigerating machine components. These components shall be designed and tested to at least three times the working pressure of the machinery. The refrigerating machines shall be designed and constructed to contain the liquefied gas and preclude the risk of bursting or cracking of the pressure retaining components during normal conditions of carriage. Refrigerating machines and refrigerating-machine components are not subject to the requirements of ADR if they contain less than 12 kg of gas.

292   *(Deleted)*

293   The following definitions apply to matches:

(a)   Fusee matches are matches the heads of which are prepared with a friction-sensitive igniter composition and a pyrotechnic composition which burns with little or no flame, but with intense heat;

(b)   Safety matches are matches which are combined with or attached to the box, book or card that can be ignited by friction only on a prepared surface;

(c)   Strike anywhere matches are matches that can be ignited by friction on a solid surface;

(d)   Wax Vesta matches are matches that can be ignited by friction either on a prepared surface or on a solid surface.

295   Batteries need not be individually marked and labelled if the pallet bears the appropriate mark and label.

296   These entries apply to life-saving appliances such as life rafts, personal flotation devices and self-inflating slides. UN No. 2990 applies to self-inflating appliances and UN No. 3072 applies to life-saving appliances that are not self-inflating. Life-saving appliances may contain:

(a)   Signal devices (Class 1) which may include smoke and illumination signal flares packed in packagings that prevent them from being inadvertently activated;

(b)   For UN No. 2990 only, cartridges, power device of Division 1.4, compatibility group S, may be contained for purposes of the self-inflating mechanism and provided that the quantity of explosives per appliance does not exceed 3.2 g;

(c)   Class 2 compressed or liquefied gases, group A or O, according to 2.2.2.1.3;

(d)   Electric storage batteries (Class 8) and lithium batteries (Class 9);

(e)   First aid kits or repair kits containing small quantities of dangerous goods (e.g.: substances of Class 3, 4.1, 5.2, 8 or 9); or

(f)   "Strike anywhere" matches packed in packagings that prevent them from being inadvertently activated.

Life-saving appliances packed in strong rigid outer packagings with a total maximum gross mass of 40 kg, containing no dangerous goods other than compressed or liquefied gases of Class 2, group A or group O, in receptacles with a capacity not exceeding 120 ml, installed solely for the purpose of the activation of the appliance, are not subject to the requirements of ADR.

298   *(Deleted)*

300   Fish meal, fish scrap and krill meal shall not be loaded if the temperature at the time of loading exceeds 35 °C or 5 °C above the ambient temperature whichever is higher.

302   Fumigated cargo transport units containing no other dangerous goods are only subject to the provisions of 5.5.2.

303   Receptacles shall be assigned to the classification code of the gas or mixture of gases contained therein determined in accordance with the provisions of section 2.2.2.

304   This entry may only be used for the transport of non-activated batteries which contain dry potassium hydroxide and which are intended to be activated prior to use by addition of an appropriate amount of water to the individual cells.

305   These substances are not subject to the requirements of ADR when in concentrations of not more than 50 mg/kg.

306   This entry may only be used for substances that are too insensitive for acceptance into Class 1 when tested in accordance with Test Series 2 (see *Manual of Tests and Criteria*, Part I).

307   This entry may only be used for uniform mixtures containing ammonium nitrate as the main ingredient within the following composition limits:

(a)   Not less than 90% ammonium nitrate with not more than 0.2% total combustible/organic material calculated as carbon and with added matter, if any, which is inorganic and inert towards ammonium nitrate; or

(b)   Less than 90% but more than 70% ammonium nitrate with other inorganic materials or more than 80% but less than 90% ammonium nitrate mixed with calcium carbonate and/or dolomite and/or mineral calcium sulphate and not more than 0.4% total combustible/organic material calculated as carbon; or

(c)   Nitrogen type ammonium nitrate based fertilizers containing mixtures of ammonium nitrate and ammonium sulphate with more than 45% but less than 70% ammonium nitrate and not more than 0.4% total combustible/organic material calculated as carbon such that the sum of the percentage compositions of ammonium nitrate and ammonium sulphate exceeds 70%.

309   This entry applies to non sensitized emulsions, suspensions and gels consisting primarily of a mixture of ammonium nitrate and fuel, intended to produce a Type E blasting explosive only after further processing prior to use.

The mixture for emulsions typically has the following composition: 60-85% ammonium nitrate, 5-30% water, 2-8% fuel, 0.5-4% emulsifier agent, 0-10% soluble flame suppressants, and trace additives. Other inorganic nitrate salts may replace part of the ammonium nitrate.

The mixture for suspensions and gels typically has the following composition: 60-85% ammonium nitrate, 0-5% sodium or potassium perchlorate, 0-17% hexamine nitrate or monomethylamine nitrate, 5-30% water, 2-15% fuel, 0.5-4% thickening agent, 0-10% soluble flame suppressants, and trace additives. Other inorganic nitrate salts may replace part of the ammonium nitrate.

Substances shall satisfactorily pass Tests 8 (a), (b) and (c) of Test Series 8 of the *Manual of Tests and Criteria*, Part I, Section 18 and be approved by the competent authority.

310 The testing requirements in the Manual of Tests and Criteria, part III sub-section 38.3 do not apply to production runs, consisting of not more than 100 cells and batteries, or to pre-production prototypes of cells and batteries when these prototypes are carried for testing when packaged in accordance with packing instruction P910 of 4.1.4.1

The transport document shall include the following statement: "Carriage in accordance with special provision 310".

Damaged or defective cells, batteries, or cells and batteries contained in equipment shall be carried in accordance with special provision 376 and packaged in accordance with packing instructions P908 of 4.1.4.1 or LP904 of 4.1.4.3, as applicable.

Cells, batteries or cells and batteries contained in equipment carried for disposal or recycling may be packaged in accordance with special provision 377 and packing instruction P909 of 4.1.4.1.

311 Substances shall not be carried under this entry unless approved by the competent authority on the basis of the results of appropriate tests according to Part I of the *Manual of Tests and Criteria*. Packaging shall ensure that the percentage of diluent does not fall below that stated in the competent authority approval, at any time during carriage.

312 Vehicles powered by a fuel cell engine shall be assigned to UN No. 3166 VEHICLE, FUEL CELL, FLAMMABLE GAS POWERED or UN No. 3166 VEHICLE, FUEL CELL, FLAMMABLE LIQUID POWERED, as appropriate. These entries include hybrid electric vehicles powered by both a fuel cell and an internal combustion engine with wet batteries, sodium batteries, lithium metal batteries or lithium ion batteries, carried with the battery(ies) installed.

Other vehicles which contain an internal combustion engine shall be assigned to the entries UN 3166 VEHICLE, FLAMMABLE GAS POWERED or UN 3166 VEHICLE, FLAMMABLE LIQUID POWERED, as appropriate. These entries include hybrid electric vehicles powered by both an internal combustion engine and wet batteries, sodium batteries, lithium metal batteries or lithium ion batteries, carried with the battery(ies) installed.

Lithium batteries shall meet the requirements of 2.2.9.1.7, except as otherwise provided for in special provision 667.

313 *(Deleted)*

314 (a) These substances are liable to exothermic decomposition at elevated temperatures. Decomposition can be initiated by heat or by impurities (e.g. powdered metals (iron, manganese, cobalt, magnesium) and their compounds);

(b) During the course of carriage, these substances shall be shaded from direct sunlight and all sources of heat and be placed in adequately ventilated areas.

315 This entry shall not be used for Class 6.1 substances which meet the inhalation toxicity criteria for packing group I described in 2.2.61.1.8.

316 This entry applies only to calcium hypochlorite, dry, when carried in non friable tablet form.

317 "Fissile-excepted" applies only to those fissile material and packages containing fissile material which are excepted in accordance with 2.2.7.2.3.5.

318 For the purposes of documentation, the proper shipping name shall be supplemented with the technical name (see 3.1.2.8). When the infectious substances to be carried are unknown, but suspected of meeting the criteria for inclusion in Category A and assignment to UN No. 2814 or 2900, the words "suspected Category A infectious substance" shall be shown, in parentheses, following the proper shipping name on the transport document.

319 Substances packed and packages which are marked in accordance with packing instruction P650 are not subject to any other requirements of ADR.

320 *(Deleted)*

321 These storage systems shall always be considered as containing hydrogen.

322 When carried in non-friable tablet form, these goods are assigned to packing group III.

323 *(Reserved)*

324 This substance needs to be stabilized when in concentrations of not more than 99%.

325 In the case of non-fissile or fissile excepted uranium hexafluoride, the material shall be classified under UN No. 2978.

326 In the case of fissile uranium hexafluoride, the material shall be classified under UN No. 2977.

327 Waste aerosols consigned in accordance with 5.4.1.1.3 may be carried under this entry for the purposes of reprocessing or disposal. They need not be protected against movement and inadvertent discharge provided that measures to prevent dangerous build up of pressure and dangerous atmospheres are addressed. Waste aerosols, other than those leaking or severely deformed, shall be packed in accordance with packing instruction P207 and special provision PP87, or packing instruction LP200 and special packing provision L2. Leaking or severely deformed aerosols shall be carried in salvage packagings provided appropriate measures are taken to ensure there is no dangerous build up of pressure.

NOTE: For maritime carriage, waste aerosols shall not be carried in closed containers.

328 This entry applies to fuel cell cartridges including when contained in equipment or packed with equipment. Fuel cell cartridges installed in or integral to a fuel cell system are regarded as contained in equipment. Fuel cell cartridge means an article that stores fuel for discharge into the fuel cell through (a) valve(s) that control(s) the discharge of fuel into the fuel cell. Fuel cell cartridges, including when contained in equipment, shall be designed and constructed to prevent fuel leakage under normal conditions of carriage.

Fuel cell cartridge design types using liquids as fuels shall pass an internal pressure test at a pressure of 100 kPa (gauge) without leakage.

Except for fuel cell cartridges containing hydrogen in metal hydride which shall be in compliance with special provision 339, each fuel cell cartridge design type shall be shown to pass a 1.2 meter drop test onto an unyielding surface in the orientation most likely to result in failure of the containment system with no loss of contents.

When lithium metal or lithium ion batteries are contained in the fuel cell system, the consignment shall be consigned under this entry and under the appropriate entries for UN 3091 LITHIUM METAL BATTERIES CONTAINED IN EQUIPMENT or UN 3481 LITHIUM ION BATTERIES CONTAINED IN EQUIPMENT.

329 *(Reserved)*

330 *(Deleted)*

331 *(Reserved)*

332 Magnesium nitrate hexahydrate is not subject to the requirements of ADR.

333 Ethanol and gasoline, motor spirit or petrol mixtures for use in spark-ignition engines (e.g. in automobiles, stationary engines and other engines) shall be assigned to this entry regardless of variations in volatility.

334 A fuel cell cartridge may contain an activator provided it is fitted with two independent means of preventing unintended mixing with the fuel during carriage.

335 Mixtures of solids which are not subject to the requirements of ADR and environmentally hazardous liquids or solids shall be classified as UN 3077 and may be carried under this entry provided there is no free liquid visible at the time the substance is loaded or at the time the packaging or cargo transport unit is closed. Each cargo transport unit shall be leakproof when used for carriage in bulk. If free liquid is visible at the time the mixture is loaded or at the time the packaging or cargo transport unit is closed, the mixture shall be classified as UN 3082.

Sealed packets and articles containing less than 10 ml of an environmentally hazardous liquid, absorbed into a solid material but with no free liquid in the packet or article, or containing less than 10 g of an environmentally hazardous solid, are not subject to the requirements of ADR.

336   A single package of non-combustible solid LSA-II or LSA-III material, if carried by air, shall not contain an activity greater than 3 000 A_2.

337   Type B(U) and Type B(M) packages, if carried by air, shall not contain activities greater than the following:

(a)   For low dispersible radioactive material: as authorized for the package design as specified in the certificate of approval;

(b)   For special form radioactive material: 3 000 A_1 or 100 000 A_2, whichever is the lower; or

(c)   For all other radioactive material: 3 000 A_2.

338   Each fuel cell cartridge carried under this entry and designed to contain a liquefied flammable gas shall:

(a)   Be capable of withstanding, without leakage or bursting, a pressure of at least two times the equilibrium pressure of the contents at 55 °C;

(b)   Not contain more than 200 ml liquefied flammable gas, the vapour pressure of which shall not exceed 1 000 kPa at 55 °C; and

(c)   Pass the hot water bath test prescribed in 6.2.6.3.1.

339   Fuel cell cartridges containing hydrogen in a metal hydride carried under this entry shall have a water capacity less than or equal to 120 ml.

The pressure in the fuel cell cartridge shall not exceed 5 MPa at 55 °C. The design type shall withstand, without leaking or bursting, a pressure of twice the design pressure of the cartridge at 55 °C or 200 kPa more than the design pressure of the cartridge at 55 °C, whichever is greater. The pressure at which this test is conducted is referred to in the drop test and the hydrogen cycling test as the "minimum shell burst pressure".

Fuel cell cartridges shall be filled in accordance with procedures provided by the manufacturer. The manufacturer shall provide the following information with each fuel cell cartridge:

(a)   Inspection procedures to be carried out before initial filling and before refilling of the fuel cell cartridge;

(b)   Safety precautions and potential hazards to be aware of;

(c)   Method for determining when the rated capacity has been achieved;

(d)   Minimum and maximum pressure range;

(e)   Minimum and maximum temperature range; and

(f)   Any other requirements to be met for initial filling and refilling including the type of equipment to be used for initial filling and refilling.

The fuel cell cartridges shall be designed and constructed to prevent fuel leakage under normal conditions of carriage. Each cartridge design type, including cartridges integral to a fuel cell, shall be subjected to and shall pass the following tests:

Drop test

A 1.8 metre drop test onto an unyielding surface in four different orientations:

(a)   Vertically, on the end containing the shut-off valve assembly;

(b)   Vertically, on the end opposite to the shut-off valve assembly;

(c)   Horizontally, onto a steel apex with a diameter of 38 mm, with the steel apex in the upward position; and

(d)   At a 45° angle on the end containing the shut-off valve assembly.

There shall be no leakage, determined by using a soap bubble solution or other equivalent means on all possible leak locations, when the cartridge is charged to its rated charging pressure. The fuel cell cartridge shall then be hydrostatically pressurized to destruction. The recorded burst pressure shall exceed 85% of the minimum shell burst pressure.

Fire test

A fuel cell cartridge filled to rated capacity with hydrogen shall be subjected to a fire engulfment test. The cartridge design, which may include a vent feature integral to it, is deemed to have passed the fire test if:

(a)   The internal pressure vents to zero gauge pressure without rupture of the cartridge; or

(b)   The cartridge withstands the fire for a minimum of 20 minutes without rupture.

Hydrogen cycling test

This test is intended to ensure that a fuel cell cartridge design stress limits are not exceeded during use.

The fuel cell cartridge shall be cycled from not more than 5% rated hydrogen capacity to not less than 95% rated hydrogen capacity and back to not more than 5% rated hydrogen capacity. The rated charging pressure shall be used for charging and temperatures shall be held within the operating temperature range. The cycling shall be continued for at least 100 cycles.

Following the cycling test, the fuel cell cartridge shall be charged and the water volume displaced by the cartridge shall be measured. The cartridge design is deemed to have passed the hydrogen cycling test if the water volume displaced by the cycled cartridge does not exceed the water volume displaced by an uncycled cartridge charged to 95% rated capacity and pressurized to 75% of its minimum shell burst pressure.

Production leak test

Each fuel cell cartridge shall be tested for leaks at 15 °C ± 5 °C, while pressurized to its rated charging pressure. There shall be no leakage, determined by using a soap bubble solution or other equivalent means on all possible leak locations.

Each fuel cell cartridge shall be permanently marked with the following information:

(a)   The rated charging pressure in MPa;

(b)   The manufacturer's serial number of the fuel cell cartridges or unique identification number; and

(c)   The date of expiry based on the maximum service life (year in four digits; month in two digits).

340   Chemical kits, first aid kits and polyester resin kits containing dangerous substances in inner packagings which do not exceed the quantity limits for excepted quantities applicable to individual substances as specified in column (7b) of Table A of Chapter 3.2, may be carried in accordance with Chapter 3.5. Class 5.2 substances, although not individually authorized as excepted quantities in column (7b) of Table A of Chapter 3.2, are authorized in such kits and are assigned Code E2 (see 3.5.1.2).

341   *(Reserved)*

342   Glass inner receptacles (such as ampoules or capsules) intended only for use in sterilization devices, when containing less than 30 ml of ethylene oxide per inner packaging with not more than 300 ml per outer packaging, may be carried in accordance with the provisions in

Chapter 3.5, irrespective of the indication of "E0" in column (7b) of Table A of Chapter 3.2 provided that:

(a) After filling, each glass inner receptacle has been determined to be leak-tight by placing the glass inner receptacle in a hot water bath at a temperature, and for a period of time, sufficient to ensure that an internal pressure equal to the vapour pressure of ethylene oxide at 55 °C is achieved. Any glass inner receptacle showing evidence of leakage, distortion or other defect under this test shall not be carried under the terms of this special provision;

(b) In addition to the packaging required by 3.5.2, each glass inner receptacle is placed in a sealed plastics bag compatible with ethylene oxide and capable of containing the contents in the event of breakage or leakage of the glass inner receptacle; and

(c) Each glass inner receptacle is protected by a means of preventing puncture of the plastics bag (e.g. sleeves or cushioning) in the event of damage to the packaging (e.g. by crushing).

343 This entry applies to crude oil containing hydrogen sulphide in sufficient concentration that vapours evolved from the crude oil can present an inhalation hazard. The packing group assigned shall be determined by the flammability hazard and inhalation hazard, in accordance with the degree of danger presented.

344 The provisions of 6.2.6 shall be met.

345 This gas contained in open cryogenic receptacles with a maximum capacity of 1 litre constructed with glass double walls having the space between the inner and outer wall evacuated (vacuum insulated) is not subject to ADR provided each receptacle is carried in an outer packaging with suitable cushioning or absorbent materials to protect it from impact damage.

346 Open cryogenic receptacles conforming to the requirements of packing instruction P203 of 4.1.4.1 and containing no dangerous goods except for UN No. 1977 nitrogen, refrigerated liquid, which is fully absorbed in a porous material are not subject to any other requirements of ADR.

347 This entry shall only be used if the results of Test series 6 (d) of Part I of the Manual of Tests and Criteria have demonstrated that any hazardous effects arising from functioning are confined within the package.

348 Batteries manufactured after 31 December 2011 shall be marked with the Watt-hour rating on the outside case.

349 Mixtures of a hypochlorite with an ammonium salt are not to be accepted for carriage. UN No. 1791 hypochlorite solution is a substance of Class 8.

350 Ammonium bromate and its aqueous solutions and mixtures of a bromate with an ammonium salt are not to be accepted for carriage.

351 Ammonium chlorate and its aqueous solutions and mixtures of a chlorate with an ammonium salt are not to be accepted for carriage.

352 Ammonium chlorite and its aqueous solutions and mixtures of a chlorite with an ammonium salt are not to be accepted for carriage.

353 Ammonium permanganate and its aqueous solutions and mixtures of a permanganate with an ammonium salt are not to be accepted for carriage.

354 This substance is toxic by inhalation.

355 Oxygen cylinders for emergency use carried under this entry may include installed actuating cartridges (cartridges, power device of Division 1.4, Compatibility Group C or S), without changing the classification in Class 2 provided the total quantity of deflagrating (propellant) explosives does not exceed 3.2 g per oxygen cylinder. The cylinders with the installed

actuating cartridges as prepared for carriage shall have an effective means of preventing inadvertent activation.

356 Metal hydride storage systems intended to be installed in vehicles, wagons, vessels or aircraft shall be approved by the competent authority of the country of manufacture[1] before acceptance for carriage. The transport document shall include an indication that the package was approved by the competent authority of the country of manufacture[1] or a copy of the competent authority of the country of manufacture[1] approval shall accompany each consignment.

357 Petroleum crude oil containing hydrogen sulphide in sufficient concentration that vapours evolved from the crude oil can present an inhalation hazard shall be consigned under the entry UN 3494 PETROLEUM SOUR CRUDE OIL, FLAMMABLE, TOXIC.

358 Nitroglycerin solution in alcohol with more than 1% but not more than 5% nitroglycerin may be classified in Class 3 and assigned to UN No. 3064 provided all the requirements of packing instruction P300 of 4.1.4.1 are complied with.

359 Nitroglycerin solution in alcohol with more than 1% but not more than 5% nitroglycerin shall be classified in Class 1 and assigned to UN No. 0144 if not all the requirements of packing instruction P300 of 4.1.4.1 are complied with.

360 Vehicles only powered by lithium metal batteries or lithium ion batteries shall be classified under the entry UN 3171 battery-powered vehicle.

361 This entry applies to electric double layer capacitors with an energy storage capacity greater than 0.3 Wh. Capacitors with an energy storage capacity of 0.3 Wh or less are not subject to ADR. Energy storage capacity means the energy held by a capacitor, as calculated using the nominal voltage and capacitance. All capacitors to which this entry applies, including capacitors containing an electrolyte that does not meet the classification criteria of any class of dangerous goods, shall meet the following conditions:

    (a) Capacitors not installed in equipment shall be carried in an uncharged state. Capacitors installed in equipment shall be carried either in an uncharged state or protected against short circuit;

    (b) Each capacitor shall be protected against a potential short circuit hazard in carriage as follows:

        (i) When a capacitor's energy storage capacity is less than or equal to 10Wh or when the energy storage capacity of each capacitor in a module is less than or equal to 10 Wh , the capacitor or module shall be protected against short circuit or be fitted with a metal strap connecting the terminals; and

        (ii) When the energy storage capacity of a capacitor or a capacitor in a module is more than 10 Wh, the capacitor or module shall be fitted with a metal strap connecting the terminals;

    (c) Capacitors containing dangerous goods shall be designed to withstand a 95 kPa pressure differential;

    (d) Capacitors shall be designed and constructed to safely relieve pressure that may build up in use, through a vent or a weak point in the capacitor casing. Any liquid which is released upon venting shall be contained by the packaging or by the equipment in which a capacitor is installed; and

    (e) Capacitors shall be marked with the energy storage capacity in Wh.

Capacitors containing an electrolyte not meeting the classification criteria of any class of dangerous goods, including when installed in equipment, are not subject to other provisions of ADR.

---

[1] *If the country of manufacture is not a Contracting Party to ADR, the approval shall be recognized by the competent authority of a Contracting Party to ADR.*

Capacitors containing an electrolyte meeting the classification criteria of any class of dangerous goods, with an energy storage capacity of 10 Wh or less are not subject to other provisions of ADR when they are capable of withstanding a 1.2 metre drop test unpackaged on an unyielding surface without loss of contents.

Capacitors containing an electrolyte meeting the classification criteria of any class of dangerous goods that are not installed in equipment and with an energy storage capacity of more than 10 Wh are subject to ADR.

Capacitors installed in equipment and containing an electrolyte meeting the classification criteria of any class of dangerous goods, are not subject to other provisions of ADR provided the equipment is packaged in a strong outer packaging constructed of suitable material, and of adequate strength and design in relation to the packaging's intended use and in such a manner as to prevent accidental functioning of capacitors during carriage. Large robust equipment containing capacitors may be offered for carriage unpackaged or on pallets when capacitors are afforded equivalent protection by the equipment in which they are contained.

NOTE: Capacitors which by design maintain a terminal voltage (e.g. asymmetrical capacitors) do not belong to this entry.

362   *(Reserved)*

363 (a)   This entry applies to engines or machinery, powered by fuels classified as dangerous goods via internal combustion systems or fuel cells (e.g. combustion engines, generators, compressors, turbines, heating units, etc.), in quantities above those specified in column (7a) of Table A of Chapter 3.2, except vehicle equipment assigned to UN No. 3166 referred to in special provision 666.

   NOTE: This entry does not apply to equipment referred to in 1.1.3.2 (a), (d) and (e), 1.1.3.3 and 1.1.3.7.

(b)   Engines or machinery which are empty of liquid or gaseous fuels and which do not contain other dangerous goods, are not subject to ADR.

   NOTE 1:   An engine or machinery is considered to be empty of liquid fuel when the liquid fuel tank has been drained and the engine or machinery cannot be operated due to a lack of fuel. Engine or machinery components such as fuel lines, fuel filters and injectors do not need to be cleaned, drained or purged to be considered empty of liquid fuels. In addition, the liquid fuel tank does not need to be cleaned or purged.

   NOTE 2:   An engine or machinery is considered to be empty of gaseous fuels when the gaseous fuel tanks are empty of liquid (for liquefied gases), the pressure in the tanks does not exceed 2 bar and the fuel shut-off or isolation valve is closed and secured.

(c)   Engines and machinery containing fuels meeting the classification criteria of Class 3, shall be assigned to the entries UN No. 3528 ENGINE, INTERNAL COMBUSTION, FLAMMABLE LIQUID POWERED or UN No. 3528 ENGINE, FUEL CELL, FLAMMABLE LIQUID POWERED or UN No. 3528 MACHINERY, INTERNAL COMBUSTION, FLAMMABLE LIQUID POWERED or UN No. 3528 MACHINERY, FUEL CELL, FLAMMABLE LIQUID POWERED, as appropriate.

(d)   Engines and machinery containing fuels meeting the classification criteria of flammable gases of Class 2, shall be assigned to the entries UN No. 3529 ENGINE, INTERNAL COMBUSTION, FLAMMABLE GAS POWERED or UN No. 3529 ENGINE, FUEL CELL, FLAMMABLE GAS POWERED or UN No. 3529 MACHINERY, INTERNAL COMBUSTION, FLAMMABLE GAS POWERED or UN No. 3529 MACHINERY, FUEL CELL, FLAMMABLE GAS POWERED, as appropriate.

   Engines and machinery powered by both a flammable gas and a flammable liquid shall be assigned to the appropriate UN No. 3529 entry.

(e)   Engines and machinery containing liquid fuels meeting the classification criteria of 2.2.9.1.10 for environmentally hazardous substances and not meeting the classification criteria of any other class shall be assigned to the entries UN No. 3530 ENGINE,

INTERNAL COMBUSTION or UN No. 3530 MACHINERY, INTERNAL COMBUSTION, as appropriate.

(f)   Engines or machinery may contain other dangerous goods than fuels (e.g. batteries, fire extinguishers, compressed gas accumulators or safety devices) required for their functioning or safe operation without being subject to any additional requirements for these other dangerous goods, unless otherwise specified in ADR. However, lithium batteries shall meet the requirements of 2.2.9.1.7, except as provided for in special provision 667.

(g)   The engines or machinery are not subject to any other requirements of ADR if the following requirements are met:

(i)   The engine or machinery, including the means of containment containing dangerous goods, shall be in compliance with the construction requirements specified by the competent authority of the country of manufacture[2];

(ii)   Any valves or openings (e.g. venting devices) shall be closed during carriage;

(iii)   The engines or machinery shall be oriented to prevent inadvertent leakage of dangerous goods and secured by means capable of restraining the engines or machinery to prevent any movement during carriage which would change the orientation or cause them to be damaged;

(iv)   For UN No. 3528 and UN No. 3530:

Where the engine or machinery contains more than 60 *l* of liquid fuel and has a capacity of more than 450 *l* but not more than 3 000 *l*, it shall be labelled on two opposite sides in accordance with 5.2.2.

Where the engine or machinery contains more than 60 *l* of liquid fuel and has a capacity of more than 3 000 *l*, it shall be placarded on two opposite sides. Placards shall correspond to the labels required in Column (5) of Table A of Chapter 3.2 and shall conform to the specifications given in 5.3.1.7. Placards shall be displayed on a background of contrasting colour, or shall have either a dotted or solid outer boundary line.

(v)   For UN No. 3529:

Where the fuel tank of the engine or machinery has a water capacity of more than 450 *l* but not more than 1 000 *l*, it shall be labelled on two opposite sides in accordance with 5.2.2.

Where the fuel tank of the engine or machinery has a water capacity of more than 1 000 *l*, it shall be placarded on two opposite sides. Placards shall correspond to the labels required in Column (5) of Table A of Chapter 3.2 and shall conform to the specifications given in 5.3.1.7. Placards shall be displayed on a background of contrasting colour, or shall have either a dotted or solid outer boundary line.

(vi)   A transport document in accordance with 5.4.1 is required only when the engine or machinery contains more than 1 000 *l* of liquid fuels, for UN 3528 and UN 3530, or the fuel tank has a water capacity of more than 1 000 *l*, for UN 3529.

This transport document shall contain the following additional statement "Transport in accordance with special provision 363".

364   This article may only be carried under the provisions of Chapter 3.4 if, as presented for carriage, the package is capable of passing the test in accordance with Test Series 6(d) of Part I of the Manual of Tests and Criteria as determined by the competent authority.

---

[2]   *For example, compliance with the relevant provisions of Directive 2006/42/EC of the European Parliament and of the Council of 17 May 2006 on machinery, and amending Directive 95/16/EC (Official Journal of the European Union No. L 157 of 9 June 2006, pp. 0024-0086).*

365   For manufactured instruments and articles containing mercury, see UN No. 3506.

366   Manufactured instruments and articles containing not more than 1 kg of mercury are not subject to ADR.

367   For the purposes of documentation:

The proper shipping name "Paint related material" may be used for consignments of packages containing "Paint" and "Paint related material" in the same package;

The proper shipping name "Paint related material, corrosive, flammable" may be used for consignments of packages containing "Paint, corrosive, flammable" and "Paint related material, corrosive, flammable" in the same package;

The proper shipping name "Paint related material, flammable, corrosive" may be used for consignments of packages containing "Paint, flammable, corrosive" and "Paint related material, flammable, corrosive" in the same package; and

The proper shipping name "Printing ink related material" may be used for consignments of packages containing "Printing ink" and "Printing ink related material" in the same package.

368   In the case of non-fissile or fissile-excepted uranium hexafluoride, the material shall be classified under UN No. 3507 or UN No. 2978.

369   In accordance with 2.1.3.5.3 (a), this radioactive material in an excepted package possessing toxic and corrosive properties is classified in Class 6.1 with radioactivity and corrosivity subsidiary risks.

Uranium hexafluoride may be classified under this entry only if the conditions of 2.2.7.2.4.1.2, 2.2.7.2.4.1.5, 2.2.7.2.4.5.2 and, for fissile-excepted material, of 2.2.7.2.3.5 are met.

In addition to the provisions applicable to the carriage of Class 6.1 substances with a corrosivity subsidiary risk, the provisions of 5.1.3.2, 5.1.5.2.2, 5.1.5.4.1 (b), 7.5.11 CV33 (3.1), (5.1) to (5.4) and (6) shall apply.

No Class 7 label is required to be displayed.

370   This entry applies to:

-     ammonium nitrate with more than 0.2% combustible substances, including any organic substance calculated as carbon, to the exclusion of any added substance; and

-     ammonium nitrate with not more than 0.2% combustible substances, including any organic substance calculated as carbon, to the exclusion of any added substance, that gives a positive result when tested in accordance with Test Series 2 (see Manual of Tests and Criteria, Part I). See also UN No. 1942.

371 (1)   This entry also applies to articles, containing a small pressure receptacle with a release device. Such articles shall comply with the following requirements:

(a)   The water capacity of the pressure receptacle shall not exceed 0.5 litres and the working pressure shall not exceed 25 bar at 15 °C;

(b)   The minimum burst pressure of the pressure receptacle shall be at least four times the pressure of the gas at 15 °C;

(c)   Each article shall be manufactured in such a way that unintentional firing or release is avoided under normal conditions of handling, packing, carriage and use. This may be fulfilled by an additional locking device linked to the activator;

(d)   Each article shall be manufactured in such a way as to prevent hazardous projections of the pressure receptacle or parts of the pressure receptacle;

(e)   Each pressure receptacle shall be manufactured from material which will not fragment upon rupture;

(f)   The design type of the article shall be subjected to a fire test. For this test, the provisions of paragraphs 16.6.1.2 except letter g, 16.6.1.3.1 to 16.6.1.3.6, 16.6.1.3.7 (b) and 16.6.1.3.8 of the Manual of Tests and Criteria shall be applied. It shall be demonstrated that the article relieves its pressure by means of a fire degradable seal or other pressure relief device, in such a way that the pressure receptacle will not fragment and that the article or fragments of the article do not rocket more than 10 metres;

(g)   The design type of the article shall be subjected to the following test. A stimulating mechanism shall be used to initiate one article in the middle of the packaging. There shall be no hazardous effects outside the package such as disruption of the package, metal fragments or a receptacle which passes through the packaging.

(2)   The manufacturer shall produce technical documentation of the design type, manufacture as well as the tests and their results. The manufacturer shall apply procedures to ensure that articles produced in series are made of good quality, conform to the design type and are able to meet the requirements in (1). The manufacturer shall provide such information to the competent authority on request.

372   This entry applies to asymmetric capacitors with an energy storage capacity greater than 0.3 Wh. Capacitors with an energy storage capacity of 0.3 Wh or less are not subject to ADR.

Energy storage capacity means the energy stored in a capacitor, as calculated according to the following equation,

$$Wh = 1/2C_N(U_R^2-U_L^2) \times (1/3600),$$

using the nominal capacitance (C_N), rated voltage (U_R) and rated lower limit voltage (U_L).

All asymmetric capacitors to which this entry applies shall meet the following conditions:

(a)   Capacitors or modules shall be protected against short circuit;

(b)   Capacitors shall be designed and constructed to safely relieve pressure that may build up in use, through a vent or a weak point in the capacitor casing. Any liquid which is released upon venting shall be contained by packaging or by equipment in which a capacitor is installed;

(c)   Capacitors shall be marked with the energy storage capacity in Wh; and

(d)   Capacitors containing an electrolyte meeting the classification criteria of any class of dangerous goods shall be designed to withstand a 95 kPa pressure differential;

Capacitors containing an electrolyte not meeting the classification criteria of any class of dangerous goods, including when configured in a module or when installed in equipment are not subject to other provisions of ADR.

Capacitors containing an electrolyte meeting the classification criteria of any class of dangerous goods, with an energy storage capacity of 20 Wh or less, including when configured in a module, are not subject to other provisions of ADR when the capacitors are capable of withstanding a 1.2 metre drop test unpackaged on an unyielding surface without loss of contents.

Capacitors containing an electrolyte meeting the classification criteria of any class of dangerous goods that are not installed in equipment and with an energy storage capacity of more than 20 Wh are subject to ADR.

Capacitors installed in equipment and containing an electrolyte meeting the classification criteria of any class of dangerous goods, are not subject to other provisions of ADR provided that the equipment is packaged in a strong outer packaging constructed of suitable material, and of adequate strength and design, in relation to the packaging's intended use and in such a manner as to prevent accidental functioning of capacitors during carriage. Large robust equipment containing capacitors may be offered for carriage unpackaged or on pallets when capacitors are afforded equivalent protection by the equipment in which they are contained.

NOTE: Notwithstanding the provisions of this special provision, nickel-carbon asymmetric capacitors containing Class 8 alkaline electrolytes shall be carried as UN 2795 BATTERIES, WET, FILLED WITH ALKALI, electric storage.

373   Neutron radiation detectors containing non-pressurized boron trifluoride gas may be carried under this entry provided that the following conditions are met:

(a)   Each radiation detector shall meet the following conditions.

(i)   The pressure in each detector shall not exceed 105 kPa absolute at 20 °C;

(ii)   The amount of gas shall not exceed 13 g per detector;

(iii)   Each detector shall be manufactured under a registered quality assurance programme;

NOTE: ISO 9001 may be used for this purpose.

(iv)   Each neutron radiation detector shall be of welded metal construction with brazed metal to ceramic feed through assemblies. These detectors shall have a minimum burst pressure of 1800 kPa as demonstrated by design type qualification testing; and

(v)   Each detector shall be tested to a 1×10^{-10} cm^3/s leaktightness standard before filling.

(b)   Radiation detectors carried as individual components shall be carried as follows:

(i)   Detectors shall be packed in a sealed intermediate plastics liner with sufficient absorbent or adsorbent material to absorb or adsorb the entire gas contents;

(ii)   They shall be packed in strong outer packaging. The completed package shall be capable of withstanding a 1.8 m drop test without leakage of gas contents from detectors;

(iii)   The total amount of gas from all detectors per outer packaging shall not exceed 52 g.

(c)   Completed neutron radiation detection systems containing detectors meeting the conditions of paragraph (a) shall be carried as follows:

(i)   The detectors shall be contained in a strong sealed outer casing;

(ii)   The casing shall contain sufficient absorbent or adsorbent material to absorb or adsorb the entire gas contents;

(iii)   The completed systems shall be packed in strong outer packagings capable of withstanding a 1.8 m drop test without leakage unless a system's outer casing affords equivalent protection.

Packing instruction P200 of 4.1.4.1 is not applicable.

The transport document shall include the following statement "Transport in accordance with special provision 373".

Neutron radiation detectors containing not more than 1 g of boron trifluoride, including those with solder glass joints, are not subject to ADR provided they meet the requirements in paragraph (a) and are packed in accordance with paragraph (b). Radiation detection systems containing such detectors are not subject to ADR provided they are packed in accordance with paragraph (c).

374   *(Reserved)*

375   These substances when carried in single or combination packagings containing a net quantity per single or inner packaging of 5 l or less for liquids or having a net mass per single or inner

packaging of 5 kg or less for solids, are not subject to any other provisions of ADR provided the packagings meet the general provisions of 4.1.1.1, 4.1.1.2 and 4.1.1.4 to 4.1.1.8.

376   Lithium ion cells or batteries and lithium metal cells or batteries identified as being damaged or defective such that they do not conform to the type tested according to the applicable provisions of the Manual of Tests and Criteria shall comply with the requirements of this special provision.

For the purposes of this special provision, these may include, but are not limited to:

- Cells or batteries identified as being defective for safety reasons;

- Cells or batteries that have leaked or vented;

- Cells or batteries that cannot be diagnosed prior to carriage; or

- Cells or batteries that have sustained physical or mechanical damage.

NOTE: In assessing a battery as damaged or defective, the type of battery and its previous use and misuse shall be taken into account.

Cells and batteries shall be carried according to the provisions applicable to UN No. 3090, UN No. 3091, UN No. 3480 and No. UN 3481, except special provision 230 and as otherwise stated in this special provision.

Packages shall be marked "DAMAGED/DEFECTIVE LITHIUM-ION BATTERIES" or "DAMAGED/DEFECTIVE LITHIUM METAL BATTERIES", as applicable.

Cells and batteries shall be packed in accordance with packing instructions P908 of 4.1.4.1 or LP904 of 4.1.4.3, as applicable.

Cells and batteries liable to rapidly disassemble, dangerously react, produce a flame or a dangerous evolution of heat or a dangerous emission of toxic, corrosive or flammable gases or vapours under normal conditions of carriage shall not be carried except under conditions approved by the competent authority of any ADR Contracting Party who may also recognize an approval granted by the competent authority of a country which is not an ADR Contracting Party provided that this approval has been granted in accordance with the procedures applicable according to RID, ADR, ADN, the IMDG Code or the ICAO Technical Instructions. In this case the cells and batteries are assigned to transport category 0.

377   Lithium ion and lithium metal cells and batteries and equipment containing such cells and batteries carried for disposal or recycling, either packed together with or packed without non-lithium batteries, may be packaged in accordance with packing instruction P909 of 4.1.4.1.

These cells and batteries are not subject to the requirements of 2.2.9.1.7 (a) to (e).

Packages shall be marked "LITHIUM BATTERIES FOR DISPOSAL" or "LITHIUM BATTERIES FOR RECYCLING".

Identified damaged or defective batteries shall be carried in accordance with special provision 376 and packaged in accordance with packing instruction P908 of 4.1.4.1 or packing instruction LP904 of 4.1.4.3, as applicable.

378   Radiation detectors containing this gas in non-refillable pressure receptacles not meeting the requirements of Chapter 6.2 and packing instruction P200 of 4.1.4.1 may be carried under this entry provided:

(a)   The working pressure in each receptacle does not exceed 50 bar;

(b)   The receptacle capacity does not exceed 12 litres;

(c)   Each receptacle has a minimum burst pressure of at least 3 times the working pressure when a relief device is fitted and at least 4 times the working pressure when no relief device is fitted;

(d)   Each receptacle is manufactured from material which will not fragment upon rupture;

(e)   Each detector is manufactured under a registered quality assurance programme;

   NOTE: *ISO 9001 may be used for this purpose.*

(f)   Detectors are carried in strong outer packagings. The complete package shall be capable of withstanding a 1.2 metre drop test without breakage of the detector or rupture of the outer packaging. Equipment that includes a detector shall be packed in strong outer packaging unless the detector is afforded equivalent protection by the equipment in which it is contained; and

(g)   The transport document includes the following statement "Transport in accordance with special provision 378".

Radiation detectors, including detectors in radiation detection systems, are not subject to any other requirements of ADR if the detectors meet the requirements in (a) to (f) above and the capacity of detector receptacles does not exceed 50 ml.

379   Anhydrous ammonia adsorbed or absorbed on a solid contained in ammonia dispensing systems or receptacles intended to form part of such systems are not subject to the other provisions of ADR if the following conditions are observed:

(a)   The adsorption or absorption presents the following properties:

   (i)   The pressure at a temperature of 20 °C in the receptacle is less than 0.6 bar;

   (ii)   The pressure at a temperature of 35 °C in the receptacle is less than 1 bar;

   (iii)   The pressure at a temperature of 85 °C in the receptacle is less than 12 bar.

(b)   The adsorbent or absorbent material shall not have dangerous properties listed in classes 1 to 8;

(c)   The maximum contents of a receptacle shall be 10 kg of ammonia; and

(d)   Receptacles containing adsorbed or absorbed ammonia shall meet the following conditions:

   (i)   Receptacles shall be made of a material compatible with ammonia as specified in ISO 11114-1:2012;

   (ii)   Receptacles and their means of closure shall be hermetically sealed and able to contain the generated ammonia;

   (iii)   Each receptacle shall be able to withstand the pressure generated at 85 °C with a volumetric expansion no greater than 0.1%;

   (iv)   Each receptacle shall be fitted with a device that allows for gas evacuation once pressure exceeds 15 bar without violent rupture, explosion or projection; and

   (v)   Each receptacle shall be able to withstand a pressure of 20 bar without leakage when the pressure relief device is deactivated.

When carried in an ammonia dispenser, the receptacles shall be connected to the dispenser in such a way that the assembly is guaranteed to have the same strength as a single receptacle.

The properties of mechanical strength mentioned in this special provision shall be tested using a prototype of a receptacle and/or dispenser filled to nominal capacity, by increasing the temperature until the specified pressures are reached.

The test results shall be documented, shall be traceable and shall be communicated to the relevant authorities upon request.

380 and 381   *(Reserved)*

382   Polymeric beads may be made from polystyrene, poly (methyl methacrylate) or other polymeric material. When it can be demonstrated that no flammable vapour, resulting in a flammable atmosphere, is evolved according to test U1 (Test method for substances liable to evolve flammable vapours) of Part III, sub-section 38.4.4 of the Manual of Tests and Criteria, polymeric beads, expandable need not be classified under this UN number. This test should only be performed when de-classification of a substance is considered.

383   Table tennis balls manufactured from celluloid are not subject to ADR where the net mass of each table tennis ball does not exceed 3.0 g and the total net mass of table tennis balls does not exceed 500 g per package.

384   *(Reserved)*

385   This entry applies to vehicles powered by flammable liquid or gas internal combustion engines or fuel cells.

Hybrid electric vehicles powered by both, an internal combustion engine and wet batteries, sodium batteries, lithium metal batteries or lithium ion batteries, carried with the batteries installed shall be assigned to this entry. Vehicles powered by wet batteries, sodium batteries, lithium metal batteries or lithium ion batteries, carried with the batteries installed, shall be assigned to the entry UN No. 3171 BATTERYPOWERED VEHICLE (see special provision 240).

For the purpose of this special provision, vehicles are self-propelled apparatus designed to carry one or more persons or goods. Examples of such vehicles are cars, motorcycles, trucks, locomotives, scooters, three- and four-wheeled vehicles or motorcycles, lawn tractors, self-propelled farming and construction equipment, boats and aircraft.

Dangerous goods such as batteries, air bags, fire extinguishers, compressed gas accumulators, safety devices and other integral components of the vehicle that are necessary for the operation of the vehicle or for the safety of its operator or passengers, shall be securely installed in the vehicle and are not otherwise subject to ADR. However, lithium batteries shall meet the requirements of 2.2.9.1.7, except as otherwise provided for in special provision 667.

386   When substances are stabilized by temperature control, the provisions of 2.2.41.1.17, special provision V8 of Chapter 7.2, special provision S4 of Chapter 8.5 and the requirements of Chapter 9.6 apply. When chemical stabilization is employed, the person offering the packaging, IBC or tank for carriage shall ensure that the level of stabilization is sufficient to prevent the substance in the packaging, IBC or tank from dangerous polymerization at a bulk mean temperature of 50 °C, or, in the case of a portable tank, 45 °C. Where chemical stabilization becomes ineffective at lower temperatures within the anticipated duration of carriage, temperature control is required. In making this determination factors to be taken into consideration include, but are not limited to, the capacity and geometry of the packaging, IBC or tank and the effect of any insulation present, the temperature of the substance when offered for carriage, the duration of the journey and the ambient temperature conditions typically encountered in the journey (considering also the season of year), the effectiveness and other properties of the stabilizer employed, applicable operational controls imposed by regulation (e.g. requirements to protect from sources of heat, including other cargo carried at a temperature above ambient) and any other relevant factors.

387-499   *(Reserved)*

500   *(Deleted)*

501   For naphthalene, molten, see UN No. 2304.

502   UN No. 2006 plastics, nitrocellulose-based, self-heating, n.o.s., and 2002 celluloid scrap are substances of Class 4.2.

503   For phosphorus, white, molten, see UN No. 2447.

504   UN No. 1847 potassium sulphide, hydrated with not less than 30% water of crystallization, UN No. 1849 sodium sulphide, hydrated with not less than 30% water of crystallization and UN No. 2949 sodium hydrosulphide hydrated with not less than 25% water of crystallization are substances of Class 8.

505   UN No. 2004 magnesium diamide is a substance of Class 4.2.

506   Alkaline earth metals and alkaline earth metal alloys in pyrophoric form are substances of Class 4.2.

       UN No. 1869 magnesium or magnesium alloys containing more than 50% magnesium as pellets, turnings or ribbons, are substances of Class 4.1.

507   UN No. 3048 aluminium phosphide pesticides, with additives inhibiting the emission of toxic flammable gases are substances of Class 6.1.

508   UN No. 1871 titanium hydride and UN No. 1437 zirconium hydride are substances of Class 4.1. UN No. 2870 aluminium borohydride is a substance of Class 4.2.

509   UN No. 1908 chlorite solution is a substance of Class 8.

510   UN No. 1755 chromic acid solution is a substance of Class 8.

511   UN No. 1625 mercuric nitrate, UN No. 1627 mercurous nitrate and UN No. 2727 thallium nitrate are substances of Class 6.1. Thorium nitrate, solid, uranyl nitrate hexahydrate solution and uranyl nitrate, solid are substances of Class 7.

512   UN No. 1730 antimony pentachloride, liquid, UN No. 1731 antimony pentachloride solution, UN No. 1732 antimony pentafluoride and UN No. 1733 antimony trichloride are substances of Class 8.

513   UN No. 0224 barium azide, dry or wetted with less than 50% water, by mass, is a substance of Class 1. UN No. 1571 barium azide, wetted with not less than 50% water, by mass, is a substance of Class 4.1. UN No. 1854 barium alloys, pyrophoric, are substances of Class 4.2. UN No. 1445 barium chlorate, solid, UN No. 1446 barium nitrate, UN No. 1447 barium perchlorate, solid, UN No. 1448 barium permanganate, UN No. 1449 barium peroxide, UN No. 2719 barium bromate, UN No. 2741 barium hypochlorite with more than 22% available chlorine, UN No. 3405 barium chlorate, solution and UN No. 3406 barium perchlorate, solution, are substances of Class 5.1. UN No. 1565 barium cyanide and UN No. 1884 barium oxide are substances of Class 6.1.

514   UN No. 2464 beryllium nitrate is a substance of Class 5.1.

515   UN No. 1581 chloropicrin and methyl bromide mixture and UN No. 1582 chloropicrin and methyl chloride mixture are substances of Class 2.

516   UN No. 1912 methyl chloride and methylene chloride mixture is a substance of Class 2.

517   UN No. 1690 sodium fluoride, solid, UN No. 1812 potassium fluoride, solid, UN No. 2505 ammonium fluoride, UN No. 2674 sodium fluorosilicate, UN No. 2856 fluorosilicates, n.o.s., UN No. 3415 sodium fluoride, solution and UN No. 3422 potassium fluoride, solution, are substances of Class 6.1.

518   UN No. 1463 chromium trioxide, anhydrous (chromic acid, solid) is a substance of Class 5.1.

519   UN No. 1048 hydrogen bromide, anhydrous, is a substance of Class 2.

520   UN No. 1050 hydrogen chloride, anhydrous, is a substance of Class 2.

521   Solid chlorites and hypochlorites are substances of Class 5.1.

522   UN No. 1873 perchloric acid aqueous solution with more than 50% but not more than 72% pure acid, by mass are substances of Class 5.1. Perchloric acid solutions containing more than 72% pure acid, by mass, or mixtures of perchloric acid with any liquid other than water, are not to be accepted for carriage.

523   UN No. 1382 anhydrous potassium sulphide and UN No. 1385 anhydrous sodium sulphide and their hydrates with less than 30% water of crystallization, and UN No. 2318 sodium hydrosulphide with less than 25% water of crystallization are substances of Class 4.2.

524 UN No. 2858 finished zirconium products of a thickness of 18 µm or more are substances of Class 4.1.

525 Solutions of inorganic cyanides with a total cyanide ion content of more than 30% shall be classified in packing group I, solutions with a total cyanide ion content of more than 3% and not more than 30% in packing group II and solutions with a cyanide ion content of more than 0.3% and not more than 3% in packing group III.

526 UN No. 2000 celluloid is assigned to Class 4.1.

528 UN No. 1353 fibres or fabrics impregnated with weakly nitrated cellulose, non-self heating are substances of Class 4.1.

529 UN No. 0135 mercury fulminate, wetted with not less than 20% water, or mixture of alcohol and water, by mass, is a substance of Class 1. Mercurous chloride (calomel) is a substance of Class 6.1 (UN No. 2025).

530 UN No. 3293 hydrazine, aqueous solution with not more than 37% hydrazine, by mass, is a substance of Class 6.1.

531 Mixtures having a flash-point below 23 °C and containing more than 55% nitrocellulose, whatever its nitrogen content or containing not more than 55% nitrocellulose with a nitrogen content above 12.6% (by dry mass), are substances of Class 1 (see UN Nos. 0340 or 0342) or of Class 4.1 (UN Nos. 2555, 2556 or 2557).

532 UN No. 2672 ammonia solution containing not less than 10% but not more than 35% ammonia is a substance of Class 8.

533 UN No. 1198 formaldehyde solutions, flammable are substances of Class 3. Formaldehyde solutions, non-flammable, with less than 25% formaldehyde are not subject to the requirements of ADR.

534 While in some climatic conditions, petrol (gasoline) may have a vapour pressure at 50 °C of more than 110 kPa (1.10 bar) but not more than 150 kPa (1.50 bar) it is to continue to be considered as a substance having a vapour pressure at 50 °C of not more than 110 kPa (1.10 bar).

535 UN No. 1469 lead nitrate, UN No. 1470 lead perchlorate, solid and UN No. 3408 lead perchlorate, solution, are substances of Class 5.1.

536 For naphthalene, solid, see UN No. 1334.

537 UN No. 2869 titanium trichloride mixture, not pyrophoric, is a substance of Class 8.

538 For sulphur (in the solid state), see UN No. 1350.

539 Solutions of isocyanates having a flash-point of not less than 23 °C are substances of Class 6.1.

540 UN No. 1326 hafnium powder, wetted, UN No. 1352 titanium powder, wetted or UN No. 1358 zirconium powder, wetted, with not less than 25% water, are substances of Class 4.1.

541 Nitrocellulose mixtures with a water content, alcohol content or plasticizer content lower than the stated limits are substances of Class 1.

542 Talc containing tremolite and/or actinolite is covered by this entry.

543 UN No. 1005 ammonia, anhydrous, UN No. 3318 ammonia solution with more than 50% ammonia and UN No. 2073 ammonia solution, with more than 35% but not more than 50% ammonia, are substances of Class 2. Ammonia solutions with not more than 10% ammonia are not subject to the requirements of ADR.

544 UN No. 1032 dimethylamine, anhydrous, UN No. 1036 ethylamine, UN No. 1061 methylamine, anhydrous and UN No. 1083 trimethylamine, anhydrous, are substances of Class 2.

545   UN No. 0401 dipicryl sulphide, wetted with less than 10% water by mass is a substance of Class 1.

546   UN No. 2009 zirconium, dry, finished sheets, strip or coiled wire, in thicknesses of less than 18 μm, is a substance of Class 4.2. Zirconium, dry, finished sheets, strip or coiled wire, in thicknesses of 254 μm or more, is not subject to the requirements of ADR.

547   UN No. 2210 maneb or UN No. 2210 maneb preparations in self-heating form are substances of Class 4.2.

548   Chlorosilanes which, in contact with water, emit flammable gases, are substances of Class 4.3.

549   Chlorosilanes having a flash-point of less than 23 °C and which, in contact with water, do not emit flammable gases are substances of Class 3. Chlorosilanes having a flash-point equal to or greater than 23 °C and which, in contact with water, do not emit flammable gases are substances of Class 8.

550   UN No. 1333 cerium in slabs, rods or ingots is a substance of Class 4.1.

551   Solutions of these isocyanates having a flash-point below 23 °C are substances of Class 3.

552   Metals and metal alloys in powdered or other flammable form, liable to spontaneous combustion, are substances of Class 4.2. Metals and metal alloys in powdered or other flammable form which, in contact with water, emit flammable gases are substances of Class 4.3.

553   This mixture of hydrogen peroxide and peroxyacetic acid shall, in laboratory testing (see *Manual of Tests and Criteria*, Part II, section 20), neither detonate in the cavitated state nor deflagrate at all and shall show no effect when heated under confinement nor any explosive power. The formulation shall be thermally stable (self-accelerating decomposition temperature 60 °C or higher for a 50 kg package), and a liquid compatible with peroxyacetic acid shall be used for desensitization. Formulations not meeting these criteria are to be regarded as substances of Class 5.2 (see *Manual of Tests and Criteria*, Part II, paragraph 20.4.3(g)).

554   Metal hydrides which, in contact with water, emit flammable gases are substances of Class 4.3. UN No. 2870 aluminium borohydride or UN No. 2870 aluminium borohydride in devices is a substance of Class 4.2.

555   Dust and powder of metals in non-spontaneously combustible form, non-toxic which nevertheless, in contact with water, emit flammable gases, are substances of Class 4.3.

556   Organometallic compounds and their solutions which ignite spontaneously are substances of Class 4.2. Flammable solutions with organometallic compounds in concentrations which, in contact with water, neither emit flammable gases in dangerous quantities nor ignite spontaneously are substances of Class 3.

557   Dust and powder of metals in pyrophoric form are substances of Class 4.2.

558   Metals and metal alloys in pyrophoric form are substances of Class 4.2. Metals and metal alloys which, in contact with water, do not emit flammable gases and are not pyrophoric or self-heating, but which are easily ignited, are substances of Class 4.1.

559   *(Deleted)*

560   An elevated temperature liquid, n.o.s. at or above 100 °C (including molten metals and molten salts) and, for a substance having a flashpoint, at a temperature below its flashpoint, is a substance of Class 9 (UN No. 3257).

561   Chloroformates having predominantly corrosive properties are substances of Class 8.

562   Spontaneously combustible organometallic compounds are substances of Class 4.2. Water-reactive organometallic compounds, flammable, are substances of Class 4.3.

563   UN No. 1905 selenic acid is a substance of Class 8.

564　UN No. 2443 vanadium oxytrichloride, UN No. 2444 vanadium tetrachloride and UN No. 2475 vanadium trichloride are substances of Class 8.

565　Unspecified wastes resulting from medical/veterinary treatment of humans/animals or from biological research, and which are unlikely to contain substances of Class 6.2 shall be assigned to this entry. Decontaminated clinical wastes or wastes resulting from biological research which previously contained infectious substances are not subject to the requirements of Class 6.2.

566　UN No. 2030 hydrazine aqueous solution, with more than 37% hydrazine, by mass, is a substance of Class 8.

567　*(Deleted)*

568　Barium azide with a water content lower than the stated limit is a substance of Class 1, UN No. 0224.

569-579　*(Reserved)*

580　*(Deleted)*

581　This entry covers mixtures of propadiene with 1 to 4% methylacetylene as well as the following mixtures:

Mixture	Content, % by volume			Permitted technical name for purposes of 5.4.1.1
	Methylacetylene and propadiene, not more than	Propane and propylene, not more than	C4-saturated hydrocarbons, not less than	
P1	63	24	14	"Mixture P1"
P2	48	50	5	"Mixture P2"

582　This entry covers, <u>inter alia</u>, mixtures of gases indicated by the letter R ..., with the following properties:

Mixture	Maximum vapour pressure at 70 °C (MPa)	Minimum density at 50 °C (kg/l)	Permitted technical name for purposes of 5.4.1.1
F1	1.3	1.30	"Mixture F1"
F2	1.9	1.21	"Mixture F2"
F3	3.0	1.09	"Mixture F3"

NOTE 1*: Trichlorofluoromethane (refrigerant R 11), 1,1,2-trichloro-1,2,2-trifluoroethane (refrigerant R 113), 1,1,1-trichloro-2,2,2-trifluoroethane (refrigerant R 113a), 1-chloro-1,2,2-trifluoroethane (refrigerant R 133) and 1-chloro-1,1,2-trifluoroethane (refrigerant R 133b) are not substances of Class 2. They may, however, enter into the composition of mixtures F1 to F3.*

NOTE 2*: The reference densities correspond to the densities of dichlorofluoromethane (1.30 kg/l), dichlorodifluoromethane (1.21 kg/l) and chlorodifluoromethane (1.09 kg/l).*

583   This entry covers, <u>inter alia</u>, mixtures of gases with the following properties:

Mixture	Maximun vapour pressure at 70 °C (MPa)	Minimun density at 50 °C (kg/l)	Permitted technical namea for purposes of 5.4.1.1
A	1.1	0.525	"Mixture A" or "Butane"
A01	1.6	0.516	"Mixture A01" or "Butane"
A02	1.6	0.505	"Mixture A02" or "Butane"
A0	1.6	0.495	"Mixture A0" or "Butane"
A1	2.1	0.485	"Mixture A1"
B1	2.6	0.474	"Mixture B1"
B2	2.6	0.463	"Mixture B2"
B	2.6	0.450	"Mixture B"
C	3.1	0.440	"Mixture C" or "Propane"

a   *For carriage in tanks, the trade names "Butane" or "Propane" may be used only as a complement*

584   This gas is not subject to the requirements of ADR when:

-    It contains not more than 0.5% air in the gaseous state;

-    It is contained in metal capsules (sodors, sparklets) free from defects which may impair their strength;

-    The leakproofness of the closure of the capsule is ensured;

-    A capsule contains not more than 25 g of this gas;

-    A capsule contains not more than 0.75 g of this gas per cm^3 of capacity.

585   *(Deleted)*

586   Hafnium, titanium and zirconium powders shall contain a visible excess of water. Hafnium, titanium and zirconium powders, wetted, mechanically produced, of a particle size of 53 μm and over, or chemically produced, of a particle size of 840 μm and over, are not subject to the requirements of ADR.

587   Barium stearate and barium titanate are not subject to the requirements of ADR.

588   Solid hydrated forms of aluminium bromide and aluminium chloride are not subject to the requirements of ADR.

589   *(Deleted)*

590   Ferric chloride hexahydrate is not subject to the requirements of ADR.

591   Lead sulphate with not more than 3% free acid is not subject to the requirements of ADR.

592   Uncleaned empty packagings (including empty IBCs and large packagings), empty tank-vehicles, empty demountable tanks, empty portable tanks, empty tank-containers and empty small containers which have contained this substance are not subject to the requirements of ADR.

593   This gas, intended for the cooling of e.g. medical or biological specimens, if contained in double wall receptacles which comply with the provisions of packing instruction P203, paragraph (6) for open cryogenic receptacles of 4.1.4.1 is not subject to the requirements of ADR except as specified in 5.5.3.

594   The following articles, manufactured and filled according to the provisions applied in the country of manufacture, are not subject to the requirements of ADR:

(a)   UN No. 1044 fire extinguishers provided with protection against inadvertent discharge, when:

–   they are packaged in a strong outer packaging; or

–   they are large fire extinguishers which meet the requirements of special packing provision PP91 of packing instruction P003 in 4.1.4.1;

(b)   UN No. 3164 articles, pressurized pneumatic or hydraulic, designed to withstand stresses greater than the internal gas pressure by virtue of transmission of force, intrinsic strength or construction, when they are packaged in a strong outer packaging.

NOTE: *"Provisions applied in the country of manufacture" means the provisions applicable in the country of manufacture or those applicable in the country of use.*

596   Cadmium pigments, such as cadmium sulphides, cadmium sulphoselenides and cadmium salts of higher fatty acids (e.g. cadmium stearate), are not subject to the requirements of ADR.

597   Acetic acid solutions with not more than 10% pure acid by mass, are not subject to the requirements of ADR.

598   The following are not subject to the requirements of ADR:

(a)   New storage batteries when:

-   they are secured in such a way that they cannot slip, fall or be damaged;

-   they are provided with carrying devices, unless they are suitably stacked, e.g. on pallets;

-   there are no dangerous traces of alkalis or acids on the outside;

-   they are protected against short circuits;

(b)   Used storage batteries when:

-   their cases are undamaged;

-   they are secured in such a way that they cannot leak, slip, fall or be damaged, e.g. by stacking on pallets;

-   there are no dangerous traces of alkalis or acids on the outside of the articles;

-   they are protected against short circuits.

"Used storage batteries" means storage batteries carried for recycling at the end of their normal service life.

599   *(Deleted)*

600   Vanadium pentoxide, fused and solidified, is not subject to the requirements of ADR.

601   Pharmaceutical products (medicines) ready for use, which are substances manufactured and packaged for retail sale or distribution for personal or household consumption are not subject to the requirements of ADR.

602   Phosphorus sulphides which are not free from yellow and white phosphorus are not to be accepted for carriage.

603   Anhydrous hydrogen cyanide not meeting the description for UN No. 1051 or UN No. 1614 is not to be accepted for carriage. Hydrogen cyanide (hydrocyanic acid) containing less than 3% water is stable, if the pH-value is 2.5 ± 0.5 and the liquid is clear and colourless.

604-606   *(Deleted)*

607   Mixtures of potassium nitrate and sodium nitrite with an ammonium salt are not to be accepted for carriage.

608   *(Deleted)*

609 Tetranitromethane not free from combustible impurities is not to be accepted for carriage.

610 The carriage of this substance, when it contains more than 45% hydrogen cyanide is prohibited.

611 Ammonium nitrate containing more than 0.2% combustible substances (including any organic substance calculated as carbon) is not to be accepted for carriage unless it is a constituent of a substance or article of Class 1.

612 *(Reserved)*

613 Chloric acid solution containing more than 10% chloric acid and mixtures of chloric acid with any liquid other than water is not to be accepted for carriage.

614 2,3,7,8-tetrachlorodibenzo-p-dioxin (TCDD) in concentrations considered highly toxic according to the criteria in 2.2.61.1 is not to be accepted for carriage.

615 *(Reserved)*

616 Substances containing more than 40% liquid nitric esters shall satisfy the exudation test specified in 2.3.1.

617 In addition to the type of explosive, the commercial name of the particular explosive shall be marked on the package.

618 In receptacles containing 1,2-butadiene, the oxygen concentration in the gaseous phase shall not exceed 50 ml/m^3.

619-622 *(Reserved)*

623 UN No. 1829 sulphur trioxide shall be inhibited. Sulphur trioxide, 99.95% pure or above, may be carried without inhibitor in tanks provided that its temperature is maintained at or above 32.5 °C. For the carriage of this substance without inhibitor in tanks at a minimum temperature of 32.5 °C, the specification "**Transport under minimum temperature of the product of 32.5 °C**" shall appear in the transport document.

625 Packages containing these articles shall be clearly marked as follows: "**UN 1950 AEROSOLS**".

626-627 *(Reserved)*

632 Considered to be spontaneously flammable (pyrophoric).

633 Packages and small containers containing this substance shall bear the following mark: "**Keep away from any source of ignition**". This mark shall be in an official language of the forwarding country, and also, if that language is not English, French or German, in English, French or German, unless any agreements concluded between the countries concerned in the transport operation provide otherwise.

634 *(Deleted)*

635 Packages containing these articles need not bear a label conforming to model No. 9 unless the article is fully enclosed by packaging, crates or other means that prevent the ready identification of the article.

636 (a) Cells contained in equipment shall not be capable of being discharged during carriage to the extent that the open circuit voltage falls below 2 volts or two thirds of the voltage of the undischarged cell, whichever is the lower.

(b) Up to the intermediate processing facility:

– lithium cells and batteries with a gross mass of not more than 500 g each or lithium ion cells with a Watt-hour rating of not more than 20 Wh, lithium ion batteries with a Watt-hour rating of not more than 100 Wh, lithium metal cells with a lithium content of not more than 1 g and lithium metal batteries with an aggregate lithium content of not more than 2 g, not contained in equipment, collected and handed over for carriage for sorting, disposal or recycling; as well as

– lithium cells and batteries contained in equipment from private households collected and handed over for carriage for depollution, dismantling, recycling or disposal;

NOTE: "Equipment from private households" means equipment which comes from private households and equipment which comes from commercial, industrial, institutional and other sources which, because of its nature and quantity, is similar to that from private households. Equipment likely to be used by both private households and users other than private households shall in any event be considered to be equipment from private households.

are not subject to the other provisions of ADR including special provision 376 and paragraph 2.2.9.1.7, if they meet the following conditions:

(i) The provisions of packing instruction P909 of 4.1.4.1 apply except for the additional requirements 1 and 2;

(ii) A quality assurance system is in place to ensure that the total amount of lithium cells or batteries per transport unit does not exceed 333 kg;

NOTE: The total quantity of lithium cells and batteries in the mix may be assessed by means of a statistical method included in the quality assurance system. A copy of the quality assurance records shall be made available to the competent authority upon request.

(iii) Packages are marked "LITHIUM BATTERIES FOR DISPOSAL" or "LITHIUM BATTERIES FOR RECYCLING" as appropriate.

If equipment containing lithium cells or batteries is carried unpackaged or on pallets in accordance with packing instruction P 909 (3) of 4.1.4.1, this mark may alternatively be affixed to the external surface of the vehicles or containers.

637 Genetically modified microorganisms and genetically modified organisms are those which are not dangerous for humans and animals, but which could alter animals, plants, microbiological substances and ecosystems in such a way as cannot occur naturally. Genetically modified microorganisms and genetically modified organisms are not subject to the requirements of ADR when authorized for use by the competent authorities of the countries of origin, transit and destination[3].

Live vertebrate or invertebrate animals shall not be used to carry these substances classified under this UN number unless the substance can be carried in no other way.

For the carriage of easily perishable substances under this UN number appropriate information shall be given, e.g.: **"Cool at +2 °/+4 °C"** or **"Carry in frozen state"** or **"Do not freeze"**.

638 Substances related to self-reactive substances (see 2.2.41.1.19).

639 See 2.2.2.3, classification code 2F, UN No. 1965, Note 2.

640 The physical and technical characteristics mentioned in column (2) of Table A of Chapter 3.2 determine different tank codes for the carriage of substances of the same packing group in ADR tanks.

---

[3] *See in particular Part C of Directive 2001/18/EC of the European Parliament and of the Council on the deliberate release into the environment of genetically modified organisms and repealing Council Directive 90/220/EEC (Official Journal of the European Communities, No. L 106, of 17 April 2001, pp. 8-14), which sets out the authorization procedures for the European Community.*

In order to identify these physical and technical characteristics of the product carried in the tank, the following shall be added, to the particulars required in the transport document, only in case of carriage in ADR tanks:

"Special provision 640X" where "X" is the applicable capital letter appearing after the reference to special provision 640 in column (6) of Table A of Chapter 3.2.

These particulars may, however, be dispensed with in the case of carriage in the type of tank which, for substances of a specific packing group of a specific UN number, meets at least the most stringent requirements.

642   Except as authorized under 1.1.4.2, this entry of the UN Model Regulations shall not be used for the carriage of fertilizer ammoniating solutions with free ammonia.

643   Stone or aggregate asphalt mixture is not subject to the requirements for Class 9.

644   This substance is admitted for carriage provided that:

- The pH is between 5 and 7 measured in an aqueous solution of 10% of the substance carried;

- The solution does not contain more than 0.2% combustible material or chlorine compounds in quantities such that the chlorine level exceeds 0.02%.

645   The classification code as mentioned in Column (3b) of Table A of Chapter 3.2 shall be used only with the approval of the competent authority of a Contracting Party to ADR prior to carriage. The approval shall be given in writing as a classification approval certificate (see 5.4.1.2.1 (g)) and shall be provided with a unique reference. When assignment to a division is made in accordance with the procedure in 2.2.1.1.7.2, the competent authority may require the default classification to be verified on the basis of test data derived from Test Series 6 of the Manual of Tests and Criteria, Part I, Section 16.

646   Carbon made by steam activation process is not subject to the requirements of ADR.

647   The carriage of vinegar and acetic acid food grade with not more than 25% pure acid by mass is subject only to the following requirements:

(a)   Packagings, including IBCs and large packagings, and tanks shall be manufactured from stainless steel or plastic material which is permanently resistant to corrosion of vinegar/acetic acid food grade;

(b)   Packagings, including IBCs and large packagings, and tanks shall be subjected to a visual inspection by the owner at least once a year. The results of the inspections shall be recorded and the records kept for at least one year. Damaged packagings, including IBCs and large packagings, and tanks shall not be filled;

(c)   Packagings, including IBCs and large packagings, and tanks shall be filled in a way that no product is spilled or adheres to the outer surface;

(d)   Seals and closures shall be resistant to vinegar/acetic acid food grade. Packagings, including IBCs and large packagings, and tanks shall be hermetically sealed by the packer or the filler so that under normal conditions of carriage there will be no leakage;

(e)   Combination packagings with inner packaging made of glass or plastic (see packing instruction P001 in 4.1.4.1) which fulfil the general packing requirements of 4.1.1.1, 4.1.1.2, 4.1.1.4, 4.1.1.5, 4.1.1.6, 4.1.1.7 and 4.1.1.8 may be used;

The other provisions of ADR do not apply.

648   Articles impregnated with this pesticide, such as fibreboard plates, paper strips, cotton-wool balls, sheets of plastics material, in hermetically closed wrappings, are not subject to the provisions of ADR.

649   *(Deleted)*

650　Waste consisting of packaging residues, solidified residues and liquid residues of paint may be carried under the conditions of packing group II. In addition to the provisions of UN No. 1263 packing group II, the waste may also be packed and carried as follows:

(a)　The waste may be packed in accordance with packing instruction P002 of 4.1.4.1 or to packing instruction IBC06 of 4.1.4.2;

(b)　The waste may be packed in flexible IBCs of types 13H3, 13H4 and 13H5 in overpacks with complete walls;

(c)　Testing of packagings and IBCs indicated under (a) or (b) may be carried out in accordance with the requirements of Chapters 6.1 or 6.5, as appropriate, in relation to solids, at the packing group II performance level.

　　　The tests shall be carried out on packagings and IBCs, filled with a representative sample of the waste, as prepared for carriage;

(d)　Carriage in bulk in sheeted vehicles, closed containers or sheeted large containers, all with complete walls is allowed. The body of vehicles or containers shall be leakproof or rendered leakproof, for example by means of a suitable and sufficiently stout inner lining;

(e)　If the waste is carried under the conditions of this special provision, the goods shall be declared in accordance with 5.4.1.1.3 in the transport document, as follows:
"UN 1263 WASTE PAINT, 3, II, (D/E)", or
"UN 1263 WASTE PAINT, 3, PG II, (D/E)".

651　Special provision V2 (1) does not apply if the net explosive mass per transport unit does not exceed 4 000 kg, provided that the net explosive mass per vehicle does not exceed 3 000 kg.

652　Austenitic stainless steel, ferritic and austenitic steel (Duplex steel) and welded titanium receptacles which do not meet the requirements of Chapter 6.2 but have been constructed and approved in accordance with national aviation provisions for use as hot air balloon or hot air airship fuel receptacles, brought into service (date of initial inspection) before 1 July 2004, may be carried by road provided they meet the following conditions:

(a)　The general provisions of 6.2.1 shall be complied with;

(b)　The design and construction of the receptacles shall have been approved for aviation use by a national air transport authority;

(c)　As an exemption from 6.2.3.1.2, the calculation pressure shall be derived from a reduced maximum ambient temperature of +40 °C; in this case:

(i)　as an exemption from 6.2.5.1, cylinders may be manufactured from rolled and annealed commercially pure titanium with the minimum requirements of $R_m > 450$ MPa, $\varepsilon_A > 20\%$ (ε_A = elongation after fracture);

(ii)　austenitic stainless steel and ferritic and austenitic steel (Duplex steel) cylinders may be used with a stress level up to 85% of the minimum guaranteed yield strength (Re) at a calculation pressure derived from a reduced maximum ambient temperature of +40 °C;

(iii)　the receptacles shall be equipped with a pressure relief device having a nominal set pressure of 26 bar; the test pressure of these receptacles shall be not less than 30 bar;

(d)　When the exemptions from (c) are not applied, the receptacles shall be designed for a reference temperature of 65 °C and shall be equipped with pressure relief devices with a nominal set pressure specified by the competent authority of the country of use;

(e)　The main body of the receptacles shall be covered by an outer, water-resistant protective layer at least 25 mm thick made from structural cellular foam or similar material;

(f)   During carriage, the receptacle shall be firmly secured in a crate or an additional safety device;

(g)   The receptacles shall be marked with a clear, visible label stating that the receptacles are for use only in hot air balloons and hot air airships;

(h)   The duration of service (from the date of initial inspection) shall not exceed 25 years.

653   The carriage of this gas in cylinders having a test pressure capacity product of maximum 15.2 MPa.litre (152 bar.litre) is not subject to the other provisions of ADR if the following conditions are met:

-     The provisions for construction and testing of cylinders are observed;

-     The cylinders are contained in outer packagings which at least meet the requirements of Part 4 for combination packagings. The general provisions of packing of 4.1.1.1, 4.1.1.2 and 4.1.1.5 to 4.1.1.7 shall be observed;

-     The cylinders are not packed together with other dangerous goods;

-     The total gross mass of a package does not exceed 30 kg; and

-     Each package is clearly and durably marked with "UN 1006" for argon compressed, "UN 1013" for carbon dioxide, "UN 1046" for helium compressed or "UN 1066" for nitrogen compressed. This mark is displayed within a diamond-shaped area surrounded by a line that measures at least 100 mm by 100 mm.

654   Waste lighters collected separately and consigned in accordance with 5.4.1.1.3 may be carried under this entry for the purposes of disposal. They need not be protected against inadvertent discharge provided that measures are taken to prevent the dangerous build up of pressure and dangerous atmospheres.

Waste lighters, other than those leaking or severely deformed, shall be packed in accordance with packing instruction P003. In addition the following provisions shall apply:

-     Only rigid packagings of a maximum capacity of 60 litres shall be used;

-     The packagings shall be filled with water or any other appropriate protection material to avoid any ignition;

-     Under normal conditions of carriage all ignition devices of the lighters shall fully be covered by the protection material;

-     The packagings shall be adequately vented to prevent the creation of flammable atmosphere and the build up of pressure;

-     The packages shall only be carried in ventilated or open vehicles or containers.

Leaking or severely deformed lighters shall be carried in salvage packagings, provided appropriate measures are taken to ensure there is no dangerous build up of pressure.

NOTE: Special provision 201 and special packing provisions PP84 and RR5 of packing instruction P002 in 4.1.4.1 do not apply to waste lighters.

655 Cylinders and their closures designed, constructed, approved and marked in accordance with Directive 97/23/EC[4] or Directive 2014/68 EU[5] and used for breathing apparatus may be carried without conforming to Chapter 6.2, provided that they are subject to inspections and tests specified in 6.2.1.6.1 and the interval between tests specified in packing instruction P200 in 4.1.4.1 is not exceeded. The pressure used for the hydraulic pressure test is the pressure marked on the cylinder in accordance with Directive 97/23/EC[4] or Directive 2014/68 EU[5].

656 *(Deleted)*

657 This entry shall be used for the technically pure substance only; for mixtures of LPG components, see UN No. 1965 or see UN No. 1075 in conjunction with NOTE 2 in 2.2.2.3.

658 UN No. 1057 LIGHTERS complying with standard EN ISO 9994:2006 + A1:2008 "Lighters – Safety Specification" and UN No. 1057 LIGHTER REFILLS, may be carried subject only to the provisions of 3.4.1 (a) to (h), 3.4.2 (except for the total gross mass of 30 kg), 3.4.3 (except for the total gross mass of 20 kg), 3.4.11 and 3.4.12, provided the following conditions are met:

   (a)   The total gross mass of each package is not more than 10 kg;

   (b)   Not more than 100 kg gross mass of such packages is carried in a vehicle or large container; and

   (c)   Each outer packaging is clearly and durably marked with "UN 1057 LIGHTERS" or "UN 1057 LIGHTER REFILLS", as appropriate.

659 Substances to which PP86 or TP7 are assigned in Column (9a) and Column (11) of Table A in Chapter 3.2 and therefore require air to be eliminated from the vapour space, shall not be used for carriage under this UN number but shall be carried under their respective UN numbers as listed in Table A of Chapter 3.2.

NOTE: See also 2.2.2.1.7.

660 For the carriage of fuel gas containment systems designed to be fitted in motor vehicles containing this gas the provisions of sub-section 4.1.4.1, Chapter 5.2, Chapter 5.4 and Chapter 6.2 of ADR need not be applied, provided the following conditions are met:

---

[4]   *Directive 97/23/EC of the European Parliament and of the Council of 29 May 1997 on the approximation of the laws of the Member States concerning pressure equipment (PED) (Official Journal of the European Communities No. L 181 of 9 July 1997, p. 1 - 55).*
[5]   *Directive 2014/68/EU of the European Parliament and of the Council of 15 May 2014 on the harmonisation of the laws of the Member States relating to the making available on the market of pressure equipment (PED) (Official Journal of the European Union No. L 189 of 27 June 2014, p. 164 - 259).*

(a) The fuel gas containment systems shall meet the requirements of ECE Regulation No. 67 Revision 2[6], ECE Regulation No. 110 Revision 1[7] or ECE Regulation No. 115[8] or Regulation (EC) No. 79/2009[9] in combination with Regulation (EU) No. 406/2010[10], as applicable.

(b) The fuel gas containment systems shall be leakproof and shall not exhibit any signs of external damage which may affect their safety.

NOTE 1: Criteria may be found in standard ISO 11623:2002 Transportable gas cylinders – Periodic inspection and testing of composite gas cylinders (or ISO DIS 19078 Gas cylinders – Inspection of the cylinder installation, and requalification of high pressure cylinders for the on-board storage of natural gas as a fuel for automotive vehicles).

NOTE 2: If the fuel gas containment systems are not leakproof or overfilled or if they exhibit damage that could affect their safety, they shall only be carried in salvage pressure receptacles in conformity with ADR.

(c) If the fuel gas containment system is equipped with two valves or more integrated in line, two valves shall be so closed as to be gastight under normal conditions of carriage. If only one valve exists or only one valve works properly all openings with the exception of the opening of the pressure relief device shall be so closed as to be gastight under normal conditions of carriage.

(d) Fuel gas containment systems shall be carried in such a way as to prevent obstruction of the pressure relief device or any damage to the valves and any other pressurised part of the fuel gas containment systems and unintentional release of the gas under normal conditions of carriage. The fuel gas containment system shall be secured so as to prevent slipping, rolling or vertical movement.

(e) Fuel gas containment systems shall satisfy the provisions of 4.1.6.8 (a), (b), (c), (d) or (e).

(f) The marking and labelling provisions of Chapter 5.2 shall be met, unless fuel gas containment systems are consigned in a handling device. If so, the marks and danger labels shall be affixed to the handling device.

(g) Documentation

Every consignment that is carried in accordance with this special provision shall be accompanied by a transport document, containing at least the following information:

(i) The UN number of the gas contained in the fuel gas containment systems, preceded by the letters "UN";

(ii) The proper shipping name of the gas;

---

[6] *ECE Regulation No. 67 (Uniform provisions concerning the approval of: I. Approval of specific equipment of vehicles of category M and N using liquefied petroleum gases in their propulsion system. II. Approval of vehicles of category M and N fitted with specific equipment for the use of liquefied petroleum gases in their propulsion system with regard to the installation of such equipment.).*

[7] *ECE Regulation No. 110 (Uniform provisions concerning the approval of: I. Specific components of motor vehicles using compressed natural gas (CNG) and/or liquefied natural gas (LNG) in their propulsion system; II. Vehicles with regard to the installation of specific components of an approved type for the use of compressed natural gas (CNG) and/or liquefied natural gas (LNG) in their propulsion system.)*

[8] *ECE Regulation No. 115 (Uniform provisions concerning the approval of: I. Specific LPG (liquefied petroleum gases) retrofit systems to be installed in motor vehicles for the use of LPG in their propulsion system; II. Specific CNG (compressed natural gas) retrofit systems to be installed in motor vehicles for the use of CNG in their propulsion system).*

[9] *Regulation (EC) No 79/2009 of the European Parliament and of the Council of 14 January 2009 on type-approval of hydrogen-powered motor vehicles, and amending Directive 2007/46/EC.*

[10] *Commission Regulation (EU) No 406/2010 of 26 April 2010 implementing Regulation (EC) No 79/2009 of the European Parliament and of the Council on type-approval of hydrogen-powered motor vehicles.*

(iii)   The label model number;

(iv)   The number of fuel gas containment systems;

(v)   In the case of liquefied gases the net mass in kg of the gas of each fuel gas containment system and in the case of compressed gases the water capacity in litres of each fuel gas containment system followed by the nominal working pressure;

(vi)   The names and the addresses of the consignor and the consignee.

(i) to (v) shall appear according to one of the following examples:

Example 1: UN 1971 natural gas, compressed, 2.1, 1 fuel gas containment system of 50 l in total, 200 bar.

Example 2: UN 1965 hydrocarbon gas mixture, liquefied, n.o.s., 2.1, 3 fuel gas containment systems, each of 15 kg net mass of gas.

NOTE: *All other provisions of ADR shall be applied.*

661   *(Deleted)*

662   Cylinders not conforming to the provisions of Chapter 6.2 which are used exclusively on board a ship or aircraft, may be carried for the purpose of filling or inspection and subsequent return, provided the cylinders are designed and constructed in accordance with a standard recognized by the competent authority of the country of approval and all the other relevant requirements of ADR are met including:

(a)   The cylinders shall be carried with valve protection in conformity with 4.1.6.8;

(b)   The cylinders shall be marked and labelled in conformity with 5.2.1 and 5.2.2; and

(c)   All the relevant filling requirements of packing instruction P200 of 4.1.4.1 shall be complied with.

The transport document shall include the following statement: "Carriage in accordance with special provision 662".

663   This entry may only be used for packagings, large packagings or IBCs, or parts thereof, which have contained dangerous goods which are carried for disposal, recycling or recovery of their material, other than reconditioning, repair, routine maintenance, remanufacturing or reuse, and which have been emptied to the extent that only residues of dangerous goods adhering to the packaging parts are present when they are handed over for carriage.

Scope:

Residues present in the packagings, discarded, empty, uncleaned shall only be of dangerous goods of classes 3, 4.1, 5.1, 6.1, 8 or 9. In addition, they shall not be:

-    Substances assigned to packing group I or that have "0" assigned in Column (7a) of Table A of Chapter 3.2; nor

-    Substances classified as desensitized explosive substances of Class 3 or Class 4.1; nor

-    Substances classified as self-reactive substances of Class 4.1; nor

-    Radioactive material; nor

-    Asbestos (UN 2212 and UN 2590), polychlorinated biphenyls (UN 2315 and UN 3432) and polyhalogenated biphenyls, halogenated monomethyldiphenylmethanes or polyhalogenated terphenyls (UN 3151 and UN 3152).

General provisions:

Packagings, discarded, empty, uncleaned with residues presenting a risk or a subsidiary risk of Class 5.1 shall not be packed together with other packagings, discarded, empty, uncleaned, or loaded together with other packagings, discarded, empty, uncleaned in the same container, vehicle or bulk container.

Documented sorting procedures shall be implemented on the loading site to ensure compliance with the provisions applicable to this entry.

NOTE: All the other provisions of ADR apply.

664   When substances under this entry are carried in fixed tanks (tank-vehicles) or demountable tanks, these tanks may be equipped with additive devices.

Additive devices:

-       are part of the service equipment for dispensing additives of UN 1202, UN 1993 packing group III, UN 3082 or non-dangerous substances during discharge of the tank;

-       consist of elements such as connecting pipes and hoses, closing devices, pumps and dosing devices which are permanently connected to the discharge device of the tank's service equipment;

-       include means of containment which are an integral part of the shell, or permanently fixed to the exterior of the tank or tank-vehicle.

Alternatively, additive devices may have connectors for connecting packagings. In this latter case, the packaging itself is not considered part of the additive device.

The following requirements shall apply depending on the configuration:

(a)   Construction of the means of containment:

(i)   As an integral part of the shell (e.g. a tank compartment), they shall meet the relevant provisions of Chapter 6.8.

(ii)   When permanently fixed to the exterior of the tank or to the tank-vehicle, they are not subject to the construction provisions of ADR provided they comply with the following provisions:

They shall be made of a metallic material and comply with the following minimum wall thickness requirements:

Material	Minimum wall thickness^a
Austenitic stainless steels	2.5 mm
Other steels	3 mm
Aluminium alloys	4 mm
Pure aluminium of 99.80%	6 mm

^{a} *For means of containment made with double walls, the aggregate thickness of the outer metal wall and the inner metal wall shall correspond to the wall thickness prescribed.*

Welding shall be carried out in accordance with the first paragraph of 6.8.2.1.23, except that other suitable methods may be applied to confirm the quality of the welding.

(iii)   Packagings which are connectable to the additive device shall be metal packagings and meet the relevant construction requirements of Chapter 6.1, as applicable for the additive concerned.

(b) Tank approval

For tanks equipped or intended to be equipped with additive devices, where the additive device is not included in the original type approval of the tank, the provisions of 6.8.2.3.4 shall apply.

(c) Use of means of containment and additive devices

(i) In case of (a) (i) above, no additional requirements.

(ii) In case of (a) (ii) above, the total capacity of the means of containment shall not exceed 400 litres per vehicle.

(iii) In case of (a) (iii) above, 7.5.7.5 and 8.3.3 shall not apply. The packagings may only be connected to the additive device during discharge of the tank. During carriage, the closures and connectors shall be closed so as to be leaktight.

(d) Testing for additive devices

The provisions of 6.8.2.4 shall apply to the additive device. However, in case of (a) (ii) above, at the time of the initial, intermediate or periodic inspection of the tank, the means of containment of the additive device shall only be subject to an external visual inspection and a leakproofness test. The leakproofness test shall be carried out at a test pressure of at least 0.2 bar.

NOTE: For the packagings described in (a) (iii) above, the relevant provisions of ADR shall apply.

(e) Transport document

Only the information required in accordance with 5.4.1.1.1 (a) to (d) needs to be added to the transport document for the additive concerned. In this case, the remark "additive device" shall be added to the transport document.

(f) Training of drivers

Drivers who have been trained in accordance with 8.2.1 for carriage of this substance in tanks need no additional training for the carriage of the additives.

(g) Placarding or marking

Placarding or marking of the fixed tank (tank-vehicle) or demountable tank for the carriage of substances under this entry in accordance with Chapter 5.3 is not affected by the presence of an additive device or the additives contained therein.

665 Unground hard coal, coke and anthracite, meeting the classification criteria of Class 4.2, packing group III, are not subject to the requirements of ADR.

666 Vehicles assigned to UN No. 3166 or UN No. 3171 and battery powered equipment assigned to UN 3171 in conformity with special provisions 240, 312 and 385, as well as any dangerous goods they contain that are necessary for their operation or the operation of their equipment, when carried as a load, are not subject to any other provisions of ADR, provided the following conditions are met:

(a) For liquid fuels, any valves between the engine or equipment and the fuel tank shall be closed during carriage unless it is essential for the equipment to remain operational. Where appropriate, the vehicles shall be loaded upright and secured against falling;

(b) For gaseous fuels, the valve between the gas tank and engine shall be closed and the electric contact open unless it is essential for the equipment to remain operational;

(c) Metal hydride storage systems shall be approved by the competent authority of the country of manufacture. If the country of manufacture is not a contracting party to ADR the approval shall be recognized by the competent authority of a contracting party to ADR;

(d) The provisions of (a) and (b) do not apply to vehicles which are empty of liquid or gaseous fuels,

NOTE 1: *A vehicle is considered to be empty of liquid fuel when the liquid fuel tank has been drained and the vehicle cannot be operated due to a lack of fuel. Vehicle components such as fuel lines, fuel filters and injectors do not need to be cleaned, drained or purged to be considered empty of liquid fuels. In addition, the liquid fuel tank does not need to be cleaned or purged.*

NOTE 2: *A vehicle is considered to be empty of gaseous fuels when the gaseous fuel tanks are empty of liquid (for liquefied gases), the pressure in the tanks does not exceed 2 bar and the fuel shut-off or isolation valve is closed and secured.*

667 (a) The requirements of 2.2.9.1.7 (a) do not apply when pre-production prototype lithium cells or batteries or lithium cells or batteries of a small production run, consisting of not more than 100 cells or batteries, are installed in the vehicle, engine or machinery;

(b) The requirements of 2.2.9.1.7 do not apply to lithium cells or batteries in damaged or defective vehicles, engine or machinery. In such cases the following conditions shall be met:

(i) If the damage or defect has no significant impact on the safety of the cell or battery, damaged and defective vehicles, engines or machinery, may be carried under the conditions defined in special provisions 363 or 666, as appropriate;

(ii) If the damage or defect has a significant impact on the safety of the cell or battery, the lithium cell or battery shall be removed and carried according to special provision 376.

However if it is not possible to safely remove the cell or battery or it is not possible to verify the status of the cell or battery, the vehicle, engine or machinery may be towed or carried as specified in (i).

668 Elevated temperature substances for the purpose of applying road markings are not subject to the requirements of ADR, provided that the following conditions are met:

(a) They do not fulfil the criteria of any class other than Class 9;

(b) The temperature of the outer surface of the boiler does not exceed 70 °C;

(c) The boiler is closed in such a way that any loss of product is prevented during carriage;

(d) The maximum capacity of the boiler is limited to 3 000 *l*.

669 A trailer fitted with equipment, powered by a liquid or gaseous fuel or an electric energy storage and production system, that is intended for use during carriage operated by this trailer as a part of a transport unit, shall be assigned to UN numbers 3166 or 3171 and be subject to the same conditions as specified for these UN numbers, when carried as a load on a vehicle, provided that the total capacity of the tanks containing liquid fuel does not exceed 500 litres.

CHAPTER 3.4

DANGEROUS GOODS PACKED IN LIMITED QUANTITIES

3.4.1   This Chapter provides the provisions applicable to the carriage of dangerous goods of certain classes packed in limited quantities. The applicable quantity limit for the inner packaging or article is specified for each substance in Column (7a) of Table A of Chapter 3.2. In addition, the quantity "0" has been indicated in this column for each entry not permitted to be carried in accordance with this Chapter.

Limited quantities of dangerous goods packed in such limited quantities, meeting the provisions of this Chapter are not subject to any other provisions of ADR except the relevant provisions of:

(a)   Part 1, Chapters 1.1, 1.2, 1.3, 1.4, 1.5, 1.6, 1.8, 1.9;

(b)   Part 2;

(c)   Part 3, Chapters 3.1, 3.2, 3.3 (except special provisions 61, 178, 181, 220, 274, 625, 633 and 650 (e));

(d)   Part 4, paragraphs 4.1.1.1, 4.1.1.2, 4.1.1.4 to 4.1.1.8;

(e)   Part 5, 5.1.2.1(a) (i) and (b), 5.1.2.2, 5.1.2.3, 5.2.1.10, 5.4.2;

(f)   Part 6, construction requirements of 6.1.4 and paragraphs 6.2.5.1 and 6.2.6.1 to 6.2.6.3;

(g)   Part 7, Chapter 7.1 and 7.2.1, 7.2.2, 7.5.1 (except 7.5.1.4), 7.5.2.4, 7.5.7, 7.5.8 and 7.5.9;

(h)   8.6.3.3 and 8.6.4.

3.4.2   Dangerous goods shall be packed only in inner packagings placed in suitable outer packagings. Intermediate packagings may be used. In addition, for articles of Division 1.4, Compatibility Group S, the provisions of section 4.1.5 shall be fully complied with. The use of inner packagings is not necessary for the carriage of articles such as aerosols or "receptacles, small, containing gas". The total gross mass of the package shall not exceed 30 kg.

3.4.3   Except for articles of Division 1.4, Compatibility Group S, shrink-wrapped or stretch-wrapped trays meeting the conditions of 4.1.1.1, 4.1.1.2 and 4.1.1.4 to 4.1.1.8 are acceptable as outer packagings for articles or inner packagings containing dangerous goods carried in accordance with this Chapter. Inner packagings that are liable to break or be easily punctured, such as those made of glass, porcelain, stoneware or certain plastics, shall be placed in suitable intermediate packagings meeting the provisions of 4.1.1.1, 4.1.1.2 and 4.1.1.4 to 4.1.1.8, and be so designed that they meet the construction requirements of 6.1.4. The total gross mass of the package shall not exceed 20 kg.

3.4.4   Liquid goods of Class 8, packing group II in glass, porcelain or stoneware inner packagings shall be enclosed in a compatible and rigid intermediate packaging.

3.4.5 and 3.4.6   *(Reserved)*

3.4.7   **Marking of packages containing limited quantities**

3.4.7.1   Except for air transport, packages containing dangerous goods in limited quantities shall bear the mark shown in Figure 3.4.7.1:

Figure 3.4.7.1

Mark for packages containing limited quantities

The mark shall be readily visible, legible and able to withstand open weather exposure without a substantial reduction in effectiveness.

The mark shall be in the form of a square set at an angle of 45° (diamond-shaped). The top and bottom portions and the surrounding line shall be black. The centre area shall be white or a suitable contrasting background. The minimum dimensions shall be 100 mm x 100 mm and the minimum width of the line forming the diamond shall be 2 mm. Where dimensions are not specified, all features shall be in approximate proportion to those shown.

3.4.7.2   If the size of the package so requires, the minimum outer dimensions shown in Figure 3.4.7.1 may be reduced to be not less than 50 mm x 50 mm provided the mark remains clearly visible. The minimum width of the line forming the diamond may be reduced to a minimum of 1 mm.

3.4.8   Marking of packages containing limited quantities conforming to Part 3, Chapter 4 of the ICAO Technical Instructions

3.4.8.1   Packages containing dangerous goods packed in conformity with the provisions of Part 3, Chapter 4 of the ICAO Technical Instructions may bear the mark shown in Figure 3.4.8.1 to certify conformity with these provisions:

Figure 3.4.8.1

Mark for packages containing limited quantities conforming to Part 3, Chapter 4
of the ICAO Technical Instructions

The mark shall be readily visible, legible and able to withstand open weather exposure without a substantial reduction in effectiveness.

The mark shall be in the form of a square set at an angle of 45° (diamond-shaped). The top and bottom portions and the surrounding line shall be black. The centre area shall be white or a suitable contrasting background. The minimum dimensions shall be 100 mm x 100 mm and the minimum width of the line forming the diamond shall be 2 mm. The symbol "Y" shall be placed in the centre of the mark and shall be clearly visible. Where dimensions are not specified, all features shall be in approximate proportion to those shown.

3.4.8.2   If the size of the package so requires, the minimum outer dimensions shown in Figure 3.4.8.1 may be reduced to be not less than 50 mm x 50 mm provided the mark remains clearly visible. The minimum width of the line forming the diamond may be reduced to a minimum of 1 mm. The symbol "Y" shall remain in approximate proportion to that shown in Figure 3.4.8.1.

3.4.9   Packages containing dangerous goods bearing the mark shown in 3.4.8 with or without the additional labels and marks for air transport shall be deemed to meet the provisions of section 3.4.1 as appropriate and of sections 3.4.2 to 3.4.4 and need not bear the mark shown in 3.4.7.

3.4.10   Packages containing dangerous goods in limited quantities bearing the mark shown in 3.4.7 and conforming with the provisions of the ICAO Technical Instructions, including all necessary marks and labels specified in Parts 5 and 6, shall be deemed to meet the provisions of section 3.4.1 as appropriate and of sections 3.4.2 to 3.4.4.

3.4.11   **Use of overpacks**

For an overpack containing dangerous goods packed in limited quantities, the following applies:

Unless the marks representative of all dangerous goods in an overpack are visible, the overpack shall be:

–   marked with the word "OVERPACK". The lettering of the "OVERPACK" mark shall be at least 12 mm high. The mark shall be in an official language of the country of origin and also, if that language is not English, French or German, in English, French or German, unless agreements, if any, concluded between the countries concerned in the transport operation provide otherwise; and

–   marked with the marks required by this Chapter.

Except for air transport, the other provisions of 5.1.2.1 apply only if other dangerous goods which are not packed in limited quantities are contained in the overpack and only in relation to these other dangerous goods.

3.4.12   In advance of carriage, consignors of dangerous goods packed in limited quantities shall inform the carrier in a traceable form of the total gross mass of such goods to be consigned.

3.4.13   (a)   Transport units with a maximum mass exceeding 12 tonnes carrying dangerous goods packed in limited quantities shall be marked in accordance with 3.4.15 at the front and at the rear except when the transport unit contains other dangerous goods for which orange-coloured plate marking in accordance with 5.3.2 is required. In this latter case, the transport unit may display the required orange-coloured plate marking only, or both the orange-coloured plate marking in accordance with 5.3.2 and the marks in accordance with 3.4.15.

(b)   Containers carrying dangerous goods packed in limited quantities, on transport units with a maximum mass exceeding 12 tonnes, shall be marked in accordance with 3.4.15 on all four sides except when the container contains other dangerous goods for which placarding in accordance with 5.3.1 is required. In this latter case, the container may display the required placards only, or both the placards in accordance with 5.3.1 and the marks in accordance with 3.4.15.

The carrying transport unit need not be marked, except when the marks affixed to the containers are not visible from outside this carrying transport unit. In this latter case, the same marks shall be affixed at the front and at the rear of the transport unit.

3.4.14   The marks specified in 3.4.13 may be dispensed with, if the total gross mass of the packages containing dangerous goods packed in limited quantities carried does not exceed 8 tonnes per transport unit.

3.4.15   The marks specified in 3.4.13 shall be the same as the one required in 3.4.7, except that their minimum dimensions shall be 250 mm × 250 mm. These marks shall be removed or covered if no dangerous goods in limited quantities are carried.

CHAPTER 3.5

DANGEROUS GOODS
PACKED IN EXCEPTED QUANTITIES

3.5.1   **Excepted quantities**

3.5.1.1   Excepted quantities of dangerous goods of certain classes, other than articles, meeting the provisions of this Chapter are not subject to any other provisions of ADR except for:

(a)   The training requirements in Chapter 1.3;

(b)   The classification procedures and packing group criteria in Part 2;

(c)   The packaging requirements of 4.1.1.1, 4.1.1.2, 4.1.1.4 and 4.1.1.6.

NOTE: In the case of radioactive material, the requirements for radioactive material in excepted packages in 1.7.1.5 apply.

3.5.1.2   Dangerous goods which may be carried as excepted quantities in accordance with the provisions of this Chapter are shown in column (7b) of Table A of Chapter 3.2 list by means of an alphanumeric code as follows:

Code	Maximum net quantity per inner packaging (in grams for solids and ml for liquids and gases)	Maximum net quantity per outer packaging (in grams for solids and ml for liquids and gases, or sum of grams and ml in the case of mixed packing)
E0	Not permitted as Excepted Quantity	
E1	30	1000
E2	30	500
E3	30	300
E4	1	500
E5	1	300

For gases, the volume indicated for inner packagings refers to the water capacity of the inner receptacle and the volume indicated for outer packagings refers to the combined water capacity of all inner packagings within a single outer packaging.

3.5.1.3   Where dangerous goods in excepted quantities for which different codes are assigned are packaged together the total quantity per outer packaging shall be limited to that corresponding to the most restrictive code.

3.5.1.4   Excepted quantities of dangerous goods assigned to codes E1, E2, E4 and E5 with a maximum net quantity of dangerous goods per inner packaging limited to 1 ml for liquids and gases and 1 g for solids and a maximum net quantity of dangerous goods per outer packaging which does not exceed 100 g for solids or 100 ml for liquids and gases are only subject to:

(a)   The provisions of 3.5.2, except that an intermediate packaging is not required if the inner packagings are securely packed in an outer packaging with cushioning material in such a way that, under normal conditions of carriage, they cannot break, be punctured, or leak their contents; and for liquids, the outer packaging contains sufficient absorbent material to absorb the entire contents of the inner packagings; and

(b)   The provisions of 3.5.3.

3.5.2   **Packagings**

Packagings used for the carriage of dangerous goods in excepted quantities shall be in compliance with the following:

(a)   There shall be an inner packaging and each inner packaging shall be constructed of plastic (with a minimum thickness of 0.2 mm when used for liquids), or of glass, porcelain, stoneware, earthenware or metal (see also 4.1.1.2) and the closure of each inner packaging shall be held securely in place with wire, tape or other positive means; any receptacle having a neck with moulded screw threads shall have a leak proof threaded type cap. The closure shall be resistant to the contents;

(b)   Each inner packaging shall be securely packed in an intermediate packaging with cushioning material in such a way that, under normal conditions of carriage, they cannot break, be punctured or leak their contents. For liquid dangerous goods, the intermediate or outer packaging shall contain sufficient absorbent material to absorb the entire contents of the inner packagings. When placed in the intermediate packaging, the absorbent material may be the cushioning material. Dangerous goods shall not react dangerously with cushioning, absorbent material and packaging material or reduce the integrity or function of the materials. Regardless of its orientation, the package shall completely contain the contents in case of breakage or leakage;

(c)   The intermediate packaging shall be securely packed in a strong, rigid outer packaging (wooden, fibreboard or other equally strong material);

(d)   Each package type shall be in compliance with the provisions in 3.5.3;

(e)   Each package shall be of such a size that there is adequate space to apply all necessary marks; and

(f)   Overpacks may be used and may also contain packages of dangerous goods or goods not subject to the requirements of ADR.

3.5.3   **Tests for packages**

3.5.3.1   The complete package as prepared for carriage, with inner packagings filled to not less than 95% of their capacity for solids or 98% for liquids, shall be capable of withstanding, as demonstrated by testing which is appropriately documented, without breakage or leakage of any inner packaging and without significant reduction in effectiveness:

(a)   Drops onto a rigid, non-resilient flat and horizontal surface from a height of 1.8 m:

(i)   Where the sample is in the shape of a box, it shall be dropped in each of the following orientations:

-    flat on the base;

-    flat on the top;

-    flat on the longest side;

-    flat on the shortest side;

-    on a corner;

(ii)   Where the sample is in the shape of a drum, it shall be dropped in each of the following orientations:

-    diagonally on the top chime, with the centre of gravity directly above the point of impact;

-    diagonally on the base chime;

-    flat on the side;

NOTE: Each of the above drops may be performed on different but identical packages.

(b)   A force applied to the top surface for a duration of 24 hours, equivalent to the total weight of identical packages if stacked to a height of 3 m (including the sample).

3.5.3.2   For the purposes of testing, the substances to be carried in the packaging may be replaced by other substances except where this would invalidate the results of the tests. For solids, when another substance is used, it must have the same physical characteristics (mass, grain size, etc.) as the substance to be carried. In the drop tests for liquids, when another substance is used, its relative density (specific gravity) and viscosity should be similar to those of the substance to be carried.

3.5.4   Marking of packages

3.5.4.1   Packages containing excepted quantities of dangerous goods prepared in accordance with this Chapter shall be durably and legibly marked with the mark shown in 3.5.4.2. The first or only label number indicated in column (5) of Table A of Chapter 3.2 for each of the dangerous goods contained in the package shall be shown in the mark. Where the name of the consignor or consignee is not shown elsewhere on the package this information shall be included within the mark.

3.5.4.2   *Excepted quantities mark*

Figure 3.5.4.2

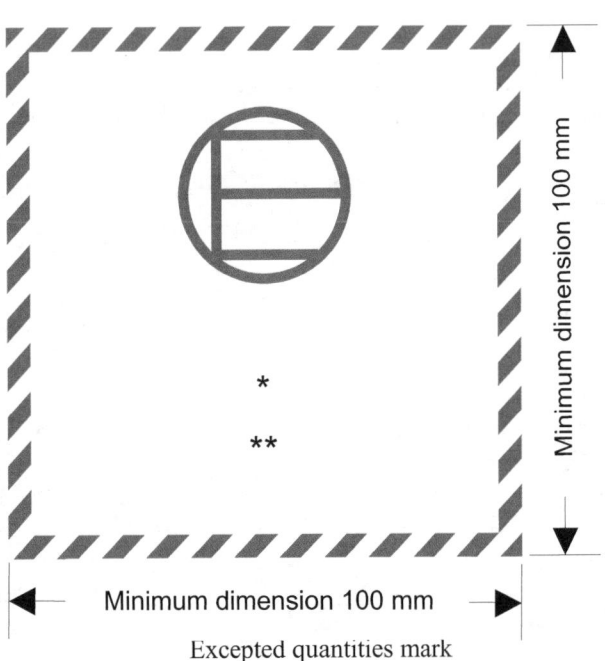

Excepted quantities mark

*   The first or only label number indicated in column (5) of Table A of Chapter 3.2 shall be shown in this location.

**   The name of the consignor or of the consignee shall be shown in this location if not shown elsewhere on the package.

The mark shall be in the form of a square. The hatching and symbol shall be of the same colour, black or red, on white or suitable contrasting background. The minimum dimensions shall be 100 mm x 100 mm. Where dimensions are not specified, all features shall be in approximate proportion to those shown.

3.5.4.3   *Use of overpacks*

For an overpack containing dangerous goods packed in excepted quantities, the following applies:

Unless the marks representative of all dangerous goods in an overpack are visible, the overpack shall be:

–   marked with the word "OVERPACK". The lettering of the "OVERPACK" mark shall be at least 12 mm high. The mark shall be in an official language of the country of origin and also, if that language is not English, French or German, in English, French or German, unless agreements, if any, concluded between the countries concerned in the transport operation provide otherwise; and

–   marked with the marks required by this Chapter.

The other provisions of 5.1.2.1 apply only if other dangerous goods which are not packed in excepted quantities are contained in the overpack and only in relation to these other dangerous goods.

3.5.5   **Maximum number of packages in any vehicle or container**

The number of packages in any vehicle or container shall not exceed 1 000.

3.5.6   **Documentation**

If a document or documents (such as a bill of lading, air waybill or CMR/CIM consignment note) accompanies(y) dangerous goods in excepted quantities, at least one of these documents shall include the statement "Dangerous Goods in Excepted Quantities" and indicate the number of packages.